Jane's
UNDERWATER WARFARE SYSTEMS

Edited by Anthony J Watts

Eleventh Edition
1999-2000

ISBN 0 7106 1920 0
"Jane's" is a registered trade mark

Copyright © 1999 by Jane's Information Group Limited, Sentinel House, 163 Brighton Road, Coulsdon, Surrey CR5 2YH, UK

In the USA and its dependencies
Jane's Information Group Inc, 1340 Braddock Place, Suite 300, Alexandria, Virginia 22314-1651, USA

DPA
DIRECTORY & DATABASE
PUBLISHERS ASSOCIATION
M E M B E R

British Library Cataloguing-in-Publication Data.
A catalogue record for this book is available from the British Library.

Printed and bound in Great Britain by Biddles Ltd, Guildford and King's Lynn

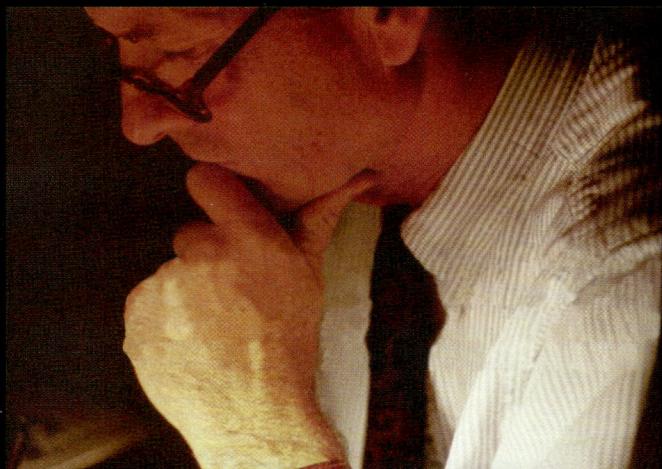

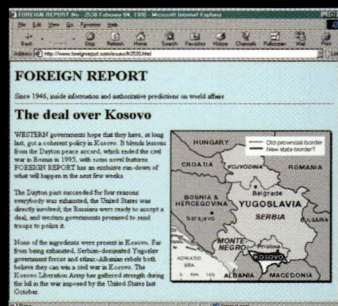

Contents

CONTENTS

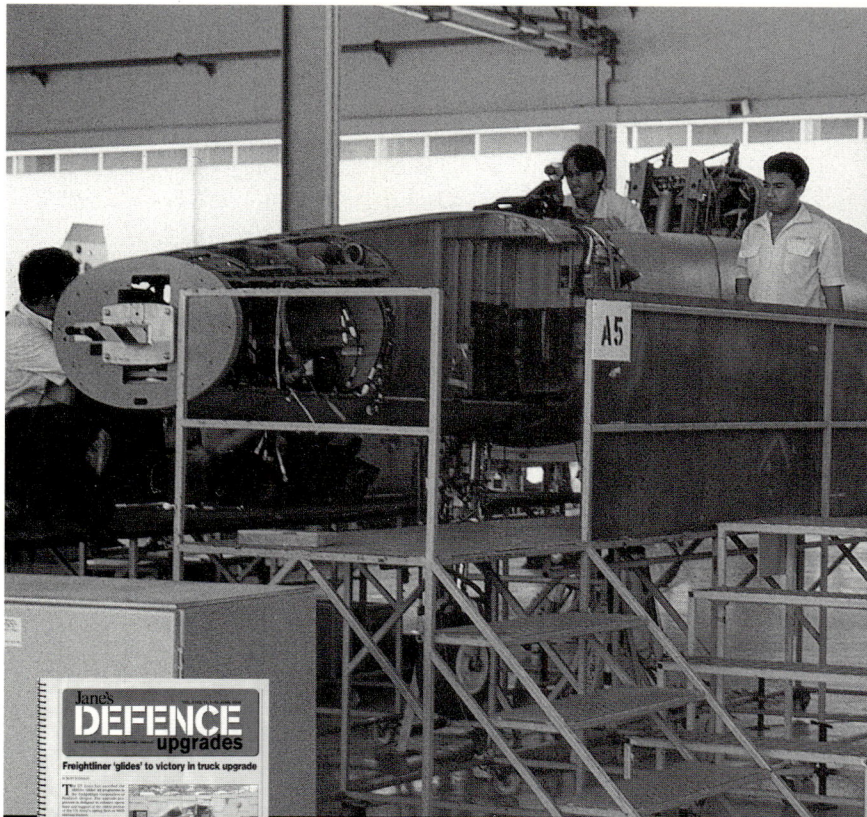

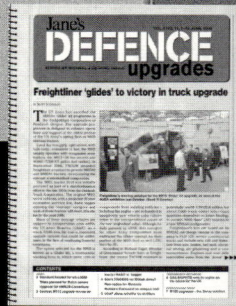

[5]

How to use this book

The book is divided into six main sections covering Submarines, Anti-submarine warfare, Underwater weapons, Mine countermeasures vessels, Mine warfare and Associated underwater warfare systems.

These main divisions are further divided into subsections covering various related systems associated with that specific aspect of underwater warfare, for example, command and control and weapon systems, sonar systems, torpedoes, mines and so on. The section on Submarines covers submarine forces, aspects of submarine design and design descriptions of a number of submarines. The section on Mine countermeasures vessels covers mine countermeasures forces and tables of construction and mine countermeasures vessel designs. The section on Associated underwater warfare systems covers various disciplines which, although not directly related to the other main sections, nevertheless have an important bearing on the conduct of underwater warfare. These primarily concern equipments used in the measuring of environmental conditions which determine the quality of sound propagation through water. Other subsections cover such important areas as training systems, associated navigation equipments, such as echo-sounders, radars and so on. Equipment and systems are listed under the country of origin which is entered alphabetically. Each entry gives details of the type of equipment, a technical description, its operational status and, in many cases, a history of the

development of the equipment and its specifications, as well as a final note listing the main contractor.

Type As a quick ready-reference this specifies, within each subsection, the type and role of that particular equipment.

Development Where applicable this heading covers the history of the development of a specified item of equipment up to the point where it entered service and, in some instances, details of any major development after entry into service.

Description Under this heading a detailed technical description is given of each system, insofar as military and commercial confidentiality allows. Where equipments integrate with other equipment or systems, this is noted.

Specification Under this heading the main technical parameters of systems are listed, including dimensions and weight (water displacement in the case of ship specifications) operating parameters (where available) and, in the case of weapons, destructive charge and so on.

Operational status This lists the current status of the system - that is, under development, on order, undergoing trials and so on.

To help users of this title evaluate the published data, Jane's Information Group has divided entries into three categories:-

● *VERIFIED* The Editor has made a detailed examination of the entry's content, checking its relevance and accuracy for publication in the new edition to the best of his ability.

● *UPDATED* During the verification process, significant changes to content have b een made to reflect the latest position known to Jane's at the time of publication.

● *NEW ENTRY* Information on new equipment and/or systems appearing for the first time in the title.

New pictures this year are dated 1999. Some illustrations are followed by a seven digit number for ease of reference by the Jane's image library.

This edition contains a total of 1,188 entries of which 514 have been updated and 76 are new. 87 new illustrations (including both drawings and photographs) have been added. The number of entries removed from the last edition is 76.

Glossary of acronyms and abbreviations

AA	active adjunct (sonar)
AC	alternating current
ACINT	acoustic intelligence
ACTD	American coastal tactical defense (Magic Lantern system) (USA)
ADCAP	advanced capabilities (USA)
ADP	automatic data processing
AF	audio frequency
AFC	automatic frequency control
AGC	automatic gain control
Ah	ampere hour
AI	artificial intelligence
AIO	action information organisation
AIP	air independent propulsion
ALCS	airborne launch control system
ALFS	airborne low-frequency sonar
ALISS	advanced lightweight influence sweep system (USA)
AM	amplitude modulation
AMCM	advanced mine countermeasures (USA)
ARPA	automatic radar plotting aid
ASROC	anti-submarine rocket (USA)
ASV	anti-surface vessel
ASW	anti-submarine warfare
ATE	automatic test equipment
ATM	asynchronous transfer mode
AUV	autonomous underwater vehicle
bhp	brake horse power
BIBS	built-in breathing set
BITE	built-in test equipment
CAC	computer aided classification
CAD	computer aided detection
CASS	command active sonobuoy system
CCD	charge-coupled device
CCIS	command and control information system
CCTV	closed-circuit television system
CEP	circular error probable
CFAR	constant false alarm rate
CIC	combat information centre
CIU	control interface unit
COTS	commercial off the shelf
cp	controllable pitch (propeller)
CPA	closest point of approach
CPU	central processing unit
CRT	cathode ray tube
CSDT	control for submarinedischarge torpedo
CSLT	control for surface-launched torpedoes
CSU	control selection unit
CW	continuous wave
CZ	convergence zone (sonar)
DARPA	Defence Advanced Research Projects Agency (USA)
dB	decibels
DC	direct current
DCN	Direction des Constructions Navales (France)
DDS	dry deck shelter (for submarines)
DED	docking and essential defects
Demon	demodulated noise (sonar processing)
DF	direction-finding
DGPS	digital global positioning system
DHS	data handling system
DICASS	directional command active sonobuoy system
DIFAR	direction low-frequency analysis and recording
DMT	deep mobile target
DoD	Department of Defense (USA)
DPM	digital plotter map
DRA	Defence Research Agency (UK)
DRTS	detecting, ranging and tracking system
DSP	digital speech processor
ECAN	Etablissement des Constructions et Armes Navales (France)
ECCM	electronic counter-countermeasures
ECDIS	electronic chart display system
ECG	electrocardiogram
ECM	electronic countermeasures
ECR	embedded computer resources
EDM	engineering development model
EEG	electroencephalogram
EEZ	economic exclusion zone
EFS	electronic function select
EHF	extremely high frequency
ELF	extremely low frequency
ELFE	extremely low frequency electromagnetic
ELINT	electronic intelligence
EOD	explosive ordnance disposal
EPROM	erasable programmable read-only memory
ESM	electronic support measures
EW	electronic warfare
FCS	fire-control system
FFT	fast Fourier transform
FLIR	forward-looking infra-red
flops	floating point operations per second
FOM	figure-of-merit
fp	fixed pitch (propeller)
FRAS	free-rocket anti-submarine (Russia)
FRP	fibre-reinforced plastic
FSD	full-scale development
FSK	frequency shift keying
g	grams
GMDSS	global maritime distress system for ships
GPS	global positioning system (formerly NAVSTAR)
GRP	glass-reinforced plastic
GT	gas turbine
HBO	hyperbasic oxygen
HE	high explosive
HF	high frequency
hp	horse power
HP	high pressure
HTP	high test peroxide
ICBM	intercontinental ballistic missile
IF	intermediate frequency
IFCS	integrated fire-control system
IFF	identification, friend or foe
IFM	instantaneous frequency measurement
IHO	International Hydrographic Organisation
II	image intensifier
IMO	International Maritime Organisation
INS	inertial navigation system
IR	infra-red
IRBM	intermediate range ballistic missile
ITT	invitation to tender
IUSS	integrated undersea surveillance system (USA)
kWb	kilowatt brake
kWh	kilowatt hour
LAMPS	light airborne multipurpose system (USA)
LAN	local area network
LANTIRN	low-altitude navigation and targeting infra-red night system (USA)
LAPADS	lightweight acoustic processing and display system (UK)
LBL	long baseline
LCD	liquid crystal display
LED	light emitting diode/display
LFA	low frequency active
LLTV	low-light television
LLLTV	low-light level television
LOFAR	low-frequency analysis and recording
LOS	line of sight
LOX	liquid oxygen
LRU	line replacement unit
mA	milli amps
MAD	magnetic anomaly detector
Mbar	millibar
MCAC	multimission craft, air cushion (USA)
MCDV	maritime coastal defence vessel
MCM	mine countermeasures
MCMV	mine countermeasures vessel
Megaflops	million floating point operations per second
MF	medium frequency
MFC	multifunction console
MIRV	multiple independently targetable re-entry vehicle
M1	millilitre
mmho	milli ohms (conductivity)
MNS	mine neutralisation system
MoD	Ministry of Defence (UK)
MoU	Memorandum of Understanding
MPA	maritime patrol aircraft
MRV	multiple re-entry vehicle
MSBS	submarine-launched ballistic missile (mer-sol balistique stratégique) (France)
MSC	coastal minesweeper
MSK	minimum shift keying
MSO	ocean-going minesweeper
msw	metres seawater
MTBF	mean time between failures
MTI	moving target indication
mV	milli volts
NAVSTAR	navigation satellite timing and ranging
nm	nanometre
NOAA	National Oceanic and Atmospheric Administration (USA)
ns	nano seconds
NSF	National Science Foundation (USA)
NSR	naval staff requirement (UK)
NSWG	Naval Surface Warfare Group (USA)
oa	overall
ONI	Office of Naval Intelligence
ONR	Office of Naval Research (USA)
ORBAT	order of battle
OTPI	on top position indicator
OTT	out to tender
PCB	printed circuit board
PCU	power control unit
PD	project definition
PPI	plan position indicator
PRF	pulse repetition frequency
psi	pounds per square inch
PSP	programmable signal processor
PVDF	polyvinyl difluoride
RAM	radar absorbent material
RAM	random access memory
RCS	radar cross-section
RDSS	rapid deployable surveillance system (USA)
RDT&E	research, development, test and evaluation
RF	radio frequency
RfP	request for proposal
RFS	radio frequency surveillance
RLG	ring laser gyro
ROV	remotely operated vehicle
rpm	revolutions per minute
rpm	rounds per minute
RPV	remotely piloted vehicle
RWR	radar warning receiver
Rx	receive
SACU	stand-alone digital communications units
SAV	surface auxiliary vessel
SAW	surface acoustic wave
SCS	submarine control system
SDV	swimmer delivery vehicle
SEPADS	sonar environmental prediction and display system (UK)
SEWACO	sensor weapon control and command (Netherlands)
SFOC	specific oil fuel consumption
SHF	super high frequency
shp	shaft horse power
SINS	ship's inertial navigation system
SLBM	submarine-launched ballistic missile
SLCM	submarine-launched cruise missile
SLMM	sea-launched mobile mine (USA)
SNLE	nuclear-powered ballistic missile submarine (sous-marin nucléaire lanceur d'engins balistique) (France)
SOSUS	sound surveillance system (USA)
SPAT	self-propelled acoustic target (UK)
SRDT	single rotation directional transmission
SSB	conventional powered ballistic missile submarine
SSB	single sideband
SSBN	nuclear-powered ballistic missile submarine
SSE	submerged signal ejector
SSK	submarine (non-nuclear)
SSKP	single-shot kill probability
SSN	nuclear-powered attack submarine
STC	sequential turbocharging
SURTASS	surveillance towed array sensor system (USA)
SUT	surface and underwater target (UK)
SVGA	super video graphics array
t/m³	tonnes per cubic metre
TACTAS	tactical towed array sonar
TAS	towed array sonar
TMA	target motion analysis
TRDT	triple rotation directional transmission
TUP	transfer under pressure
TWS	track-while-scan (radar)
TWT	travelling-wave tube
Tx	transmit
UAV	unmanned aerial vehicle
UCSS	undersea coastal surveillance system (USA)
UHF	ultra high frequency
USB	upper sideband
UV	ultra-violet
VCS	versatile console system
VDS	variable depth sonar
VGA	video graphics array
VHF	very high frequency
VHSIC	very high-speed integrated circuit
VLA	vertical launch ASROC
VLAD	vertical line array difar
VLF	very low frequency
VLSI	very large-scale integration (electronic circuits)
Wh	watt hour
XBT	expendable bathythermograph
XSV	expendable sound velocity system
YIG	yttrium indium garnet
μPa	micro Pascal
μs	micro seconds

New entries in this edition

Entries deleted from this edition

(Page references are from the 1998-99 edition)

ADMINISTRATION

Director, Defence Business Unit:	*Alan Condron*
Publisher:	*Karen Heffer*
Managing Editor:	*Simon Michell*
Database Manager/Data Administrator:	*Ruth Simmance*
Team Leader:	*Steve McCarthy*

EDITORIAL OFFICE

Jane's Information Group Limited, Sentinel House, 163 Brighton Road,
Coulsdon, Surrey CR5 2YH, UK
Tel: (+44 181) 700 37 00 Fax: (+44 181) 700 37 88
e-mail: juws@janes.co.uk

SALES OFFICE

Jane's Information Group Limited, Sentinel House, 163 Brighton Road,
Coulsdon, Surrey CR5 2YH, UK
Tel: (+44 181) 700 37 00 Fax: (+44 181) 763 10 06
e-mail: info@janes.co.uk

Send enquiries to: *Jo Moon – Senior Sales and Customer Service Manager*
Florina Shons (Continental Europe/Middle East/Africa)
Julie Reeder (UK & Scandinavia)

Send USA enquiries to: *Robert Loughman – Vice-President Product Sales*
Jane's Information Group Inc, 1340 Braddock Place, Suite 300, Alexandria,
Virginia 22314-1651, USA
Tel: (+1 703) 683 37 00 Fax: (+1 703) 836 00 29 Telex: 6819193
Tel: (+1 800) 824 07 68 Fax: (+1 800) 836 02 97

Send Asia enquiries to: *David Fisher*
Jane's Information Group Asia, 60 Albert Street, #15-01 Albert Complex,
Singapore 189969
Tel: (+65) 336 62 80 Fax: (+65) 336 99 21

Send Australia/New Zealand enquiries to: *Pauline Roberts*
Jane's Information Group, PO Box 3502, Rozelle,
New South Wales 2039, Australia
Tel: (+61 2) 85 87 79 00 Fax: (+61 2) 85 87 79 01

ADVERTISEMENT SALES OFFICES

Australia: *Richard West* (UK Head Office)

Austria: *Richard West* (UK Head Office)

Benelux: *Karen Lewis, Stephen Judge* (UK Head Office)

Channel Islands: *Karen Lewis, Stephen Judge* (UK Head Office)

Eastern Europe: *Richard West* (UK Head Office)

Egypt: *Karen Lewis, Stephen Judge* (UK Head Office)

France: *Patrice Février*
BP 418, 35 avenue MacMahon,
F-75824 Paris Cedex 17
Tel: (+33 1) 45 72 33 11 Fax: (+33 1) 45 72 17 95
e-mail: patrice.fevrier@wanadoo.fr

Germany: *Richard West* (UK Head Office)

Greece: *Karen Lewis, Stephen Judge* (UK Head Office)

India: *Richard West* (UK Head Office)

Ireland: *Karen Lewis, Stephen Judge* (UK Head Office)

Israel: *Oreet Ben-Yaacov*
Oreet International Media, 15 Kinneret Street,
IL-51201 Bene Berak
Tel: (+972 3) 570 65 27 Fax: (+972 3) 570 65 26
e-mail: oreetimc@netvision.net.il

Italy and Switzerland: Ediconsult Internazionale Srl
Piazza Fontane Marose 3, I-16123 Genoa, Italy
Tel: (+39 010) 58 36 84 Fax: (+39 010) 56 65 78
e-mail: ediconsult@iol.it

Korea, South: *Young Seoh Chinn*
JES Media International, 6th Floor Donghye Building,
47-16 Myungil-Dong, Kangdong-Gu, Seoul 134 070
Tel: (+82 2) 481 34 11 Fax: (+82 2) 481 34 14
e-mail: jesmedia@unitel.co.kr

Pakistan: *Richard West* (UK Head Office)

Russian Federation and Associated States (CIS): *Simon Kay*
33 St John's Street, Crowthorne, Berkshire RG45 7NQ, UK
Tel: (+44 1344) 77 71 23 Fax: (+44 1344) 77 58 85
e-mail: crowkay@msn.com

Saudi Arabia: *Karen Lewis, Stephen Judge* (UK Head Office)

Scandinavia: *Gillian Thompson*
The Falsten Partnership, 11 Chardmore Road, Stamford Hill,
London N16 6JA, UK
Tel: (+44 181) 806 23 01 Fax: (+ 44 181) 806 81 37
e-mail: falsten@dial.pipex.com

Singapore: *Richard West* (UK Head Office)

South Africa: *Karen Lewis, Stephen Judge* (UK Head Office)

Spain: *Michael Andrade*
Via Exclusivas SL, Viriato 69SC, 28010Madrid
Tel: (+34 91) 448 76 22 Fax: (+34 91) 446 01 98
e-mail: via@varenga.com

Thailand: *Richard West* (UK Head Office)

Turkey: *Jay Morton* (UK Head Office)

United Arab Emirates: *Karen Lewis, Stephen Judge* (UK Head Office)

UK – South, West, North and Middle England, Wales and Scotland:
Richard West, Jay Morton

UK – South East, East and London: *Karen Lewis, Stephen Judge*

UK Head Office
Advertisement Sales Manager: *Karen Lewis*
Jane's Information Group Limited, Sentinel House, 163 Brighton Road,
Coulsdon, Surrey CR5 2YH, UK
Tel: (+44 181) 700 38 55 Fax: (+44 181) 700 38 59
e-mail: karen.lewis@janes.co.uk

Advertisement Sales Manager: *Richard West*
Jane's Information Group Limited, Sentinel House, 163 Brighton Road,
Coulsdon, Surrey CR5 2YH, UK
Tel: (+44 181) 700 37 39 Fax: (+44 181) 700 37 44
e-mail: richard.west@janes.co.uk

Advertisement Sales Executive: *Stephen Judge*
Jane's Information Group Limited, Sentinel House, 163 Brighton Road,
Coulsdon, Surrey CR5 2YH, UK
Tel: (+44 181) 700 38 53 Fax: (+44 181) 700 38 59
e-mail: stephen.judge@janes.co.uk

Advertisement Sales Executive: *Jay Morton*
Jane's Information Group Limited, Sentinel House, 163 Brighton Road,
Coulsdon, Surrey CR5 2YH, UK
Tel: (+44 181) 700 39 61 Fax: (+44 181) 700 37 44
e-mail: jay.morton@janes.co.uk

USA and Canada: *Ronald R. Lichtinger, III,* Advertising Sales Director
Jane's Information Group Inc, 1340 Braddock Place, Suite 300,
Alexandria, Virginia 22314-1651, USA
Tel: (+1 703) 683 37 00 Fax: (+1 703) 836 55 37
e-mail: lichtinger@janes.com

North Eastern USA and East Canada: *Harry Carter*
Jane's Information Group Inc, 1340 Braddock Place, Suite 300,
Alexandria, Virginia 22314-1651, USA
Tel: (+1 703) 683 37 00 Fax: (+1 703) 836 55 37
e-mail: carter@janes.com

South Eastern USA: *Kristin D. Schulze*
5370 Eastbay Drive, Suite 104,
Clearwater, Florida 33764, USA
Tel: (+1 727) 524 77 41 Fax: (+1 727) 524 75 62
e-mail: kristin@intnet.net

Western USA and West Canada: *Richard L. Ayer*
127 Avenida del Mar, Suite 2A,
San Clemente, California 92672, USA
Tel: (+1 949) 366 84 55 Fax: (+1 949) 366 92 89
e-mail: ayercomm@earthlink.net

Administration:
USA and Canada: *Maureen Nute*
Jane's Information Group Inc, 1340 Braddock Place, Suite 300,
Alexandria, Virginia 22314-1651, USA
Tel: (+1 703) 683 37 00 Fax: (+1 703) 836 00 29
e-mail: nute@janes.com

UK and Rest of World: *Joni Beeden*
Jane's Information Group Limited, Sentinel House, 163 Brighton Road,
Coulsdon, Surrey CR5 2YH, UK
Tel: (+44 181) 700 37 42 Fax: (+44 181) 700 38 59
e-mail: joni.beeden@janes.co.uk

Users' Charter

This publication is brought to you by Jane's Information Group, a global company with more than 100 years of innovation and an unrivalled reputation for impartiality, accuracy and authority.

Our collection and output of information and images is not dictated by any political or commercial affiliation. Our reportage is undertaken without fear of, or favour from, any government, alliance, state or corporation.

We publish information that is collected overtly from unclassified sources, although much could be regarded as extremely sensitive or not publicly accessible.

Our validation and analysis aims to eradicate misinformation or disinformation as well as factual errors; our objective is always to produce the most accurate and authoritative data.

In the event of any significant inaccuracies, we undertake to draw these to the readers' attention to preserve the highly valued relationship of trust and credibility with our customers worldwide.

If you believe that these policies have been breached by this title, you are invited to contact the Editor.

A copy of Jane's Information Group's Code of Conduct for its editorial teams is available from the publisher.

INVESTOR IN PEOPLE

Jane's

Sea
family of titles

Jane's Amphibious Warfare Capabilities
Evaluate the organisational structure and capabilities of amphibious forces around the globe. Comprehensive information on equipment and logistics, with country by country overviews including recent operations and funding, providing you with both the range and depth of coverage you need.

Jane's Fighting Ships
The ultimate reference source of the world's navies. Provides current information on warships, auxiliaries and paramilitary vessels in service and under construction. Covers complete and up-to-date technical details for each entry from dimensions and main machinery to speed and crew size.

Jane's Major Warships
Provides intelligence on the construction, equipment and fighting performance of every warship class from systems and sensors to dimensions and structure. Includes hundreds of photographs and line drawings plus lists of pennants and full contact information for contractors worldwide.

Jane's Marine Propulsion
Detailed information on marine engines currently in production for naval or commercial use. Includes engine, transmission, propellers and drive systems, along with main specifications and development history.

Jane's Merchant Ships
Allows easy recognition and identification of cargo and passenger carrying vessels around the world via use of the revised Talbot-Booth system. Textual details are provided with over 8,000 accompanying illustrations.

Jane's Naval Construction and Retrofit Markets
Helps you to identify market opportunities and trends by detailing naval constructions, retrofitting and major refitting programmes currently being developed, or planned for the future.

Jane's Naval Weapons Systems
Covers new and upgraded weaponry to provide you with a full picture of the current naval weapon technology market. Descriptions range from naval guns and missiles to torpedoes and anti-submarine weapons.

Jane's Underwater Technology
In-depth information on the latest sub-sea equipment, allowing evaluation of underwater hardware, systems, technologies and applications. Over 600 photographs and drawings provide full visual reference in a clear and concise format.

Jane's Underwater Warfare Systems
An indispensable guide to the technologies and systems required to equip modern submarines. Details the current status of equipment, review of potential for refitting and modernisation and evaluation of fleet strengths.

Jane's Naval/Maritime Special Reports
Trends in Maritime Violence
Patrol Craft Markets
Naval Mine Warfare Markets

Other Jane's titles

Air
Jane's Aero-Engines
Jane's Aircraft Component Manufacturers
Jane's Aircraft Upgrades
Jane's Air-Launched Weapons
Jane's All the World's Aircraft
Jane's Avionics
Jane's Helicopter Markets and Systems
Jane's Space Directory
Jane's Unmanned Aerial Vehicles and Targets
Jane's World Air Forces

Land
Jane's Ammunition Handbook
Jane's Armour and Artillery
Jane's Armour and Artillery Upgrades
Jane's Infantry Weapons
Jane's Land-Based Air Defence
Jane's Military Vehicles and Logistics
Jane's Mines and Mine Clearance
Jane's NBC Defence Systems
Jane's Police and Security Equipment
Jane's World Armies

Systems
Jane's Airborne Electronic Mission Systems
Jane's C^{4}I Systems
Jane's Electro-Optic Systems
Jane's Military Communications
Jane's Radar and Electronic Warfare Systems
Jane's Simulation and Training Systems
Jane's Strategic Weapon Systems

Industry
Jane's International ABC Aerospace Directory
Jane's International Defence Directory
Jane's World Defence Industry

Transport
Jane's Air Traffic Control
Jane's Airports and Handling Agents
Jane's Airports, Equipment and Services
Jane's High-Speed Marine Transportation
Jane's Road Traffic Management
Jane's Urban Transport Systems
Jane's World Airlines
Jane's World Railways

Geopolitical
Jane's Chem-Bio Handbook
Jane's Chemical-Biological Defence Guidebook
Jane's Counter Terrorism
Jane's Intelligence Watch Report
Jane's Sentinel Security Assessments
Jane's Terrorism Watch Report
Jane's World Insurgency and Terrorism

Magazines
Jane's Defence Weekly
Jane's Defence Industry
Jane's Defence Upgrades
Jane's Foreign Report
Jane's Intelligence Review
Jane's International Defense Review
Jane's Missiles and Rockets
Jane's Airport Review

Oksoy *undergoing shock trials* (Kvaerner)

Foreword

'**Whoever controls the Indian Ocean dominates Asia. The ocean is the key to the seven seas. In the 20th century, the destiny of the world will be decided on its waters**'. As we approach the millenium it is perhaps worth remembering those words of the renowned American naval strategist Alfred Thayer Mahan.

Already this century, the outcome of two world wars has been virtually decided at sea. Both during the First and Second World Wars, submarine campaigns undertaken by Germany almost forced Allied nations to the negotiating table. But after a long struggle both submarine campaignes were defeated. These submarine campaigns were further backed up by mining campaigns that considerably interfered with the passage of shipping, both mercantile and military. These campaigns were conducted by both sides and in Asia in the Second World War, the United States of America virtually strangled Japan with a mining and submarine campaign. Even within the last decade, the underwater threat has had a major impact on naval operations in the Gulf region.

Although Mahan was writing before these events came to pass, perhaps his prophetic statement might now be embellished with the words 'Indeed it will be *beneath* those waters that the world's destiny will be decided'.

Underwater warfare will continue to heavily influence maritime strategy, particularly at present, with the breakdown of the Soviet Communist block and the ending of the Cold War and the proliferation of submarine forces and mine warfare capability world-wide. Whereas during the Cold War, the West and NATO could clearly define the main threat (primarily the Soviet underwater threat) now collective security forces like NATO, together with individual nations, face a very obtuse threat that may suddenly appear from any direction. It is now essential to be ready to meet all eventualities.

In the face of a very nebulous threat therefore, what currently dominates maritime philosophy?

Undoubtedly one of the major aspects is the rapidity with which technology is developing and advancing, particularly with the computer. With this comes the fact that the military is beginning to move from a resource-based age to a knowledge-based information age. With this idea now taking hold, so it will become clear that war will move from the current notion of asset-based platform centric warfare to the concept of collaborative network centric warfare. That this will be so is being dictated by cost-conscious defence agencies around the world endeavouring to meet their political masters demands for massive cuts in defence expenditure, with resulting major reductions in platform assets in what is erroneously referred to as the peacetime environment or age. Hence, when it becomes necessary to mobilise forces in a time of rising tension of outright hostilities, those forces are now frequently multinational in character. The days when a nation could mobilise its own forces to meet a major threat have gone. Today it is necessary both politically and from sheer necessity regarding available assets to join with other nations in assembling and presenting a cohesive, all-embracing counter to any major threat.

Quite what the concept of collaborative network centric warfare means no one has yet fully defined. In some quarters, it is considered that such a concept relates solely to the notion of information exchange, while to others it is viewed as a concept of operations. However, it is not just a question of whether collaborative network centric warfare relates to information exchange or operations.

Events during the last decade have shown that it is no longer feasible or practical, in most instances, for nations to carry out operations in isolation. National force levels alone dictate that this is no longer a practical solution to dealing with a sudden emergency. Multinational forces are therefore now convened to deal with emergencies. Time will not allow mobilisation of additional available industrial resources to provide extra assets to meet a contingency. It will be a question of mobilising all available underwater warfare assets from the outset, particularly mine warfare resources. This will be necessary in the concept of collaborative network centric warfare if underwater warfare in the littoral zone is to be effectively conducted (Royal Netherlands Navy)

1999/0017915

For collaborative network centric warfare to be effective, it requires that all forms of communication, data and instruction can be easily understood by all members of a multinational force. This problem will have to be addressed at the highest level if aims and objectives are to be effectively implemented and optimum use made of available assets (Newport News Shipbuilding) **1999**/0017916

Of much greater concern is how such a concept will affect different forms of warfare to which it is applied. This means that not only different forms of warfare but also the whole concept of warfare itself may have to be redefined. This will require detailed study into the way in which war is to be conducted in the future and ultimately the type of platforms that are to be used to conduct warfare. In fact, the entire notion of collaborative network centric warfare is so complex and fraught with so many uncertainties that it seems that only a few people are willing to try to unravel the Gordian knot.

Of particular relevance to this book is the question of how the concept will affect underwater warfare.

Collaborative network centric warfare is currently in the very early stages of evolution and the overall implications for warfare are not yet fully understood. That the concept will continue to evolve as military planners devise new strategic, tactical and operational concepts based on both new and emerging technologies and changing trends in national and international relations is a fact. At its most basic level, operations, collaborative network centric warfare is defined as a linking of sensors and decision makers to exchange by communication and information systems coherent tactical information. This will enable accurate assessment of the tactical situation by all involved parties and the selection of the appropriate response by those parties. This will be achieved by both vertical and horizontal transfer of sensor information.

Currently the concept already operates at individual asset level and, somewhat, at national force level. Where it is currently not operating and not yet developed to any major degree is at international level. That this is now an urgent priority has been highlighted by the continuing operations against Iraq and multinational operations in the former Yugoslavia.

The main advantage perceived in the concept of collaborative network centric warfare is to eradicate the possibility of confusion over main objectives that arise when international forces operate together. In addition, it will offer the capability within multinational forces to overcome shortfalls experienced by nations within the force. For example, some countries will only have at their disposal lower levels of technology with which to face state-of-the-art systems possessed by a hostile force. In underwater warfare capability, they may lack specific assets such as minehunters, active towed array sonars, submarines and so on. It is in addressing the shortfalls that collaborative network centric warfare may be able to make a major contribution by calling on the assets and technology available to larger, more powerful navies and co-ordinating their integration into the multi-national force.

What collaborative network centric warfare urgently requires, and

what is lacking at present, is flexibility within and between international forces operating together. This demands an organisational structure that will allow full horizontal and vertical interactive communications between forces and units. This, in turn, demands decentralised and co-operative command and control to allow rapid decision making and response to any situation. At present, to introduce such a concept is fraught with problems and will require boldness both on the part of field commanders and command structures at national and international level. To what degree are these commands willing to co-operate and agree?

Information management and communications
In the field of underwater warfare, the failure to implement collaborative network centric warfare could have dire consequences in any future conflict. For example, to what degree are international ASW forces willing to engage in the *full* exchange of data on a current tactical situation in order to enable effective prosecution of an ASW operation? Furthermore, even if they are willing to engage in collaborative network centric warfare, do they have the capability to distribute and access the wealth of information and data available? With the level of information currently available in underwater operations, the stage is rapidly being reached where many units could be reaching information overload, which will then lead to even greater risks and uncertainty in the 'fog' of war. If this potential information overload is then embedded into a collaborative network centric concept, the crisis point could be reached where the network suddenly jams through the exchange and sharing of these vast quantities of information. There is, therefore, a parallel need to address the problem of information management and in a collaborative network centric concept, the way in which this could best be achieved, without overloading individual units, forces and commanders with too much information.

Communications and management of communications, therefore, are central to the concept of collaborative network centric warfare. The ability to pass data and information between units, forces and nations will depend wholly on the communications networks in place and the way in which they are managed and how the data and information is handled. The same requirement holds true for forces engaged in underwater warfare.

In fact it will probably be in underwater warfare that the whole concept of collaborative network centric warfare and its problems will probably first be tested. That this will be so is evidenced by the fact that current maritime philosophy has focussed its attention on expeditionary warfare in the littoral scenario. Any nation wishing to defend itself against the possibility of expeditionary warfare in the

A problem of growing proportions that must be addressed by the very closest co-operation between industry and the military is information overload. The command in the field must not be presented with so many alternatives and so much data that the human brain ceases to function effectively. Decisions at sea have literally to be taken within a few seconds. There is no time to sit and analyse a whole set of information gathered from every available source. It is a question of providing the right information, at the right time, in the right format so that the right snap decision can be made. For example in a complex air/surface/subsurface scenario, the presence of an approaching hostile torpedo must be highlighted immediately for effective countermeasures to be undertaken (BAeSEMA) **1999**/0017917

littoral zone will concentrate its efforts on developing its underwater warfare capability. This is understandable, as underwater warfare is the most cost-effective way of defending territorial rights. This means that the cheapest and most effective form of underwater warfare, the mine and the midget submarine, will continue to proliferate. Any future multinational force wishing to engage in expeditionary warfare in the littoral zone will first have to neutralise the enemy's capability to conduct underwater warfare. This primarily will be the hostile force's ability to engage in mine warfare and the conduct of submarine operations in the littoral zone.

Collaborative network centric warfare will form a vital element in these operations. This will require that all available ASW and MCM assets be put at the disposal of the expeditionary force commander. For them to be fully effective, they must be completely integrated into the Expeditionary Force under the direction of the Force commander. Without adequate and secure communication links in both vertical and horizontal dimensions right from the political level down to individual unit level, the force commander will be unable to deploy these forces effectively in the conduct of ASW and MCM operations.

The key to successful collaborative network warfare is, as noted above, communication. That is, communication in all its forms, including secure and uncoded communication, satellite communication, language communication and so on. Communication is at the heart of underwater warfare, especially submarine warfare. Here, the need for adequate communication with all friendly forces and shore commands is seen as vital to the successful conduct of underwater operations - both from the submarine's point of view and from that of ASW forces engaged in hunting for submarines.

The most pressing need is for effective underwater communications, for which industry is striving to develop systems that will prove effective at lower and lower frequencies, down in the region of the Extremely Low Frequency (ELF) waveband. ELF communications, however, do suffer from certain limitations as noted in the introduction to the Communications section later in this

book. An alternative communication system using a satellite laser has been investigated and airborne laser communications may offer a radical solution to the problem of maintaining secure and continuous communications with a submarine and with other surface ASW forces.

The proliferating underwater threat

While the command of forces and technology continues to exercise military planners, the developing strategic situation is currently dominating political minds and ultimately how defence budgets should be allocated. The western powers have now established that Russia is no longer the world's major threat. Although, in spite of heavy cuts and no money, the Russian navy and in particular the submarine force, could still pose a potential threat that could not be ignored in a time of tension. The threat today is now perceived as the maverick totalitarian regimes. Some of these have been able, since the ending of the Cold War, to acquire quantities of highly sophisticated weaponry. This has occurred as the scramble to earn desperately needed foreign currency and the need to bolster bankrupt industries has forced former communist countries to sell much of their military equipment. This is equipment they can no longer afford to maintain or sustain, or which is no longer considered relevant to their needs in the new world order.

With the perceived need among the democratic nations of the world to be able to counter the newly armed totalitarian regimes has come the realisation that the way of warfighting developed during the Cold War is no longer relevant. Consequently, new strategies are being worked out using new technologies to combat these perceived threats. Expeditionary warfare and warfare in the littoral zone are the new buzzwords which have entered the military vocabulary. Implicit in both these concepts is the fact that concepts of underwater warfare developed since the 1950s must be re-examined. No longer is it just a matter of providing for deep open ocean warfare. The current need, and one which has been recognised by many nations, is to provide adequate resources equipped with suitable technology to conduct underwater warfare within the littoral zone and right up to the surf zone to provide the optimum safety for the movement of any amphibious forces which may be needed to go ashore. Although the need has been recognised, it is not always possible in times of financial restraints, to provide the necessary funding within the tight timescales required, to provide those necessary resources, including the high technology now demanded by such warfare. This includes technology such as active towed array sonars capable of operating within the littoral zone and mine countermeasures sonars capable of penetrating the seabed to detect buried mines and so on. Endemic in this period of financial restraints has been the slowing down in many major naval powers of the replacement of obsolescent platforms and equipment. These are not necessarily restraints that have affected the totalitarian regimes to anywhere near the same degree. Thus, while the UK, USA and Russia, to name but three major navies, are drastically reducing their underwater forces and delaying the acquisition of new units and equipment, other smaller navies are steadily building up their submarine and mine warfare forces. Neither is the major powers upgrading their surface ASW capabilities to match the increasing submarine threat worldwide at the rate which prudence might require.

To return to Mahan's quote at the start of this foreword, it will be in the Indian Ocean and in the Asian region that the greatest danger lies. In these areas, underwater forces are developing at an enormous rate and, should any conflict occur there in the future, nations who may wish to prosecute ASW and MCM operations there may be hard pressed to do so in the future. In spite of the severe economic crisis that affected much of the Asian region early in 1998, there remains a determination to build up naval power in the region, particularly underwater assets. Politically, the region continues to exhibit instability and numerous potential flashpoints remain which could erupt at any time. Of growing concern is the development of the PLA Navy. Although technologically far behind many other nations in the region, the PLA Navy is making great strides to build up its naval power and, in particular, its underwater capability. It is a truism in underwater warfare that the only submarine that can be safely ignored is one that has been sunk. No matter what the state of the technology, if manned by a dedicated crew, every submarine is an extremely lethal weapon. In this respect, China stands astride the Asian region, deploying not only ocean-going diesel electric submarines but also nuclear powered attack and strategic missile submarines (only one SSBN is currently operational although a new design is currently under study).

With regard to the Indian Ocean, both Pakistan and India are busily developing their underwater capability. Both nations are acquiring new submarines and the potential for conflict in this region is never far away. To this must now be added the maverick state of Iran which also has three submarines operational. This too is another potential threat that cannot be ignored. Of possibly even greater significance to littoral warfare is the increasing proliferation of midget submarines, particularly in the two regions mentioned above. These small craft, which are extremely difficult to detect unless prior intelligence has warned of their imminence, could prove to be a major threat to any planned amphibious operations. Also to be considered is their potential for covert operations.

Industry at a crossroads

Industry is now hard put to meet the emerging requirements of politicians, naval planners and military strategists. The current decade has seen enormous inroads made into industry's capability to supply what the military needs. This has largely been due to the ending of the Cold War and the cry for the Peace Dividend. That Dividend may about to be called in. To meet the dramatic downturn in defence spending over the last nine years, industry has radically slimmed down its operations. Numerous companies involved in defence manufacturing now no longer exist, or have merged with other, larger companies to form mega defence manufacturing consortia. Within this rationalisation, and there was indeed a need for some rationalisation, the capacity of industry to meet a sudden demand for increased production of material such as MCMVs, towed array sonars, torpedoes and so on may no longer be available. The day of the national defence industry may be ending. The future may well see continental industries evolving to meet the demands of the military. Already the European defence industry is moving towards such a goal with major mergers not just between national companies such as British Aerospace and the defence arm of GEC-Marconi, but possibly also between British Aerospace and German and French

Sophisticated technology is becoming available on an increasing scale to even small navies. The submarine and mine threat is such, that continuous effort must be devoted to developing new technologies to combat the threat. Hence the need for a new generation of low frequency active towed array sonars to detect submarines and a parametric sonar to detect buried mines (L3 Communications, Ocean Systems)

1999/0017918

partners. The resulting combine may even outstrip the giant American defence companies such as Lockheed Martin that has subsumed a number of small American defence companies over the last decade.

While all these moves make for a slimmer, more effective and commercially more viable operation, will they be able to provide what may be needed by the military in the future? Already research and development has been severely affected by the downturn in military budgets. Defence budgets can no longer afford the same level of financial support for R&D that they formerly undertook and industry itself, in the constrained financial world of the 1990s, has been forced to severely prune its own R&D efforts. Undoubtedly R&D will continue to be implemented, certainly in the more advanced nations of the world, but at a much slower pace than hitherto. What this will mean in the long term is that nations with a lower level of technology will gradually catch up on the technologically more advanced nations to the point where they are almost level.

In terms of underwater warfare and the concept of collaborative network centric warfare, this could be beneficial or counterproductive. Certainly, for multinational forces involving a wide range of national assets, much of the underwater resources may be very similar in capability. On the other hand, they may find themselves facing a hostile threat that is equal, and possibly superior in some areas. In the light of the present status of the defence industries of Europe and America, will they be able to provide the necessary capacity and capability to meet any such short-term developments? Even in the medium term, will industry be able to keep up with the demands of the military for even more sophisticated technology with which to fight any future conflict? These are questions that must be addressed, not just by industry or the military, but also by the politicians.

Conclusion

So, what will be the future as we enter the new millennium? It will certainly be a time of increasing instability around the world. This will demand increasing awareness and an ability to make rapid decisions as to the type of reaction required to meet any new situation. In this context, the concept of collaborative network centric warfare must loom large and in particular within the political establishment. For peace to be maintained, this will require a willingness and openness on the part of all who seek to work together for peace. This will demand an even greater acceptance of collaboration within the secretive world of underwater warfare, together with a willingness to maintain the democratic nations' capability to develop new technologies to combat the underwater threat.

Acknowledgements

An editor always seeks ways and means of improving the content of a reference work, the mainstay of which is the amount of new information and illustrations that can be included. Not only is it necessary to expand and develop new coverage but it is also necessary to ensure that existing entries are as up to date as possible. Any reference work is only as accurate as the day each entry goes to press and inevitably in such a work as this, some entries will be out of date when they appear in print. In spite of the fact that changes to equipment do not occur overnight, systems are continually undergoing development and hence change, which it is the duty of the editor to record. New orders are also being placed which we would like to record in this title.

The work on revising entries continues throughout the year.

To achieve this, the editor relies both on the suppliers and the users of equipment to provide new information and illustrations. This is supplemented by information gleaned from other published works; news and information supplied by various contributors around the world and the editor's own extensive records and photo library.

To assist the editor in updating the yearbook, a copy of the current entry is sent to the appropriate contractor each year with a request for amendments. It is essential that these are returned promptly to ensure that the yearbook is accurate and up to date. It is always difficult and sometimes impossible to incorporate material that is received late. Because of the volume of change now being incorporated into the title, major contributors are asked to forward any new information as soon as it becomes available. This will help to spread the workload and ease the burden of carrying out major revisions on a vast number of entries at the last minute.

The editor always welcomes information and photographs from anyone with an interest in underwater warfare. Material can be sent in a variety of ways to the editor or the publishers at the following addresses:

Anthony J Watts
Editor Jane's Underwater Warfare Systems
Jane's Information Group
Sentinal House
163 Brighton Road
Coulsdon
Surrey, CR5 2NH
UK
Fax: (+44 1306) 63 12 26
Email: anthony.watts@virgin.net

Compiling such a work as this has to be a team effort, and it is a pleasure to acknowledge the help and support from other editors and contributors and the team at Jane's. Apart from primary sources, many of Jane's publications prove invaluable in updating and confirming information; in particular *Jane's Fighting Ships* and *Jane's Naval Weapon Systems* and their editors Captain Richard Sharpe and Ted Hooton. Jane's magazines, including *Jane's Navy International*, *Jane's Defence Weekly* and *International Defence Review* are also invaluable sources of reference. Other major sources include the Japanese magazine *Ships of the World*, the German *Marine Forum*, the Italian *Rivista Marittima*, and the United States *Naval Institute Proceedings*.

Finally, it would be impossible effectively to compile this work without the help and support of the excellent team at Jane's. Under Simon Michelle, the in-house editorial team, led by Steve McCarthy and with Elizabeth Glendinning, Carol Offer and Tracy Johnson, performs a major task and help to sort out many problems. In revising, it is easy to miss spelling errors and quirks of phraseology that creep in, not to mention understandable typographical errors in the mass of data and specifications peculiar to the underwater engineer. It is an encouragement to know that another opinion is always available and that other checks are being made. Standardising the large number of entries and simplifying and explaining complex technical terms and technology in a work such as this is a huge task, which is greatly eased with such a valuable team.

Thanks are also due to the staff at Biddles Ltd who are responsible for the finished product.

Anthony J Watts
Newdigate
February 1999

Farncomb, the second 'Collins' class boat for the Royal Australian Navy. This programme is now progressing towards the half way stage (Courtesy of Royal Australian Navy)

1999/0024653

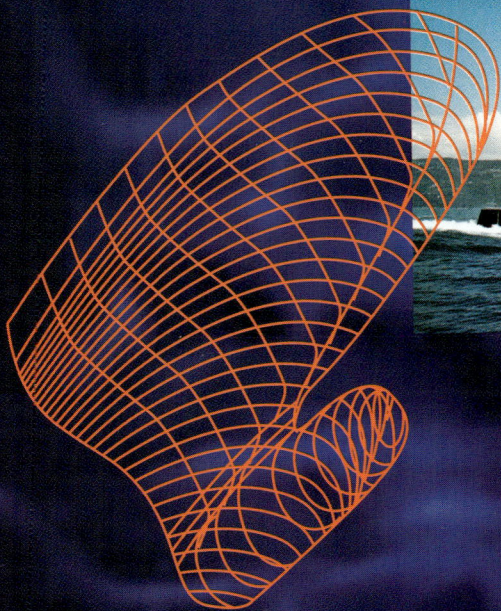

FINCANTIERI INSTINCT OF THE SEA

Advanced technology and great experience, accumulated in more than 200 years of activity and the building of hundreds of ships for the Italian and foreign navies: these are the main features of the naval vessels built by Fincantieri, an integrated Group, in the forefront the world over, able to satisfy every requirement of those concerned with defence and security.

FINCANTIERI

NAVAL vessels
Naval Vessel Business Unit:
Via Cipro, 11
16129 Genova (Italy)
Tel. +39 010 59951
Fax +39 010 5995379
Telex 270168 FINCGE I
www.fincantieri.com

SUBMARINES

THE SUBMARINE

Any country with a littoral, and even more so island states are almost entirely dependent on sea lines of communication for their continued existence. The protection of those sea-lines against underwater attack in particular is vital to ensure the safe carriage of imports and exports (food and materials and, during hostilities, reinforcements to support the war effort), as was evidenced during the First and Second World Wars. It is also vital to a continental power such as the United States since, if it were to become engaged in hostilities, it would need to support its forces wherever they were engaged. The enormous movements of troops and supplies prior to and during Operation Desert Storm, the war with Iraq in 1991, highlighted that such would be the case.

The modern submarine is one of the most powerful weapons that a navy can count in its order of battle. It provides considerable fighting potential, and ought to form a vital and integral element of any homogenous fleet. In recent years the underwater speed of even diesel-electric submarines has increased enormously. Modern submarines are also extremely quiet, particularly when running at slow speeds, and the latest types use acoustic cladding to further reduce their signature. The high underwater speed, extended cruising range, powerful weapons, and acoustic discretion enable the submarine to take maximum advantage of the underwater environment. As such, therefore, the submarine is, and will remain, a viable and extremely potent weapon system.

Types of submarine

Four main categories of submarine are currently identified. Of these, the most powerful is undoubtedly the strategic submarine (SSBN) deploying long- and medium-range strategic nuclear weapons. At present only the major powers (China, France, the Russian Federation and Associated States (RFAS), the UK, and the USA) have such lethal weapons at their disposal. However, as some smaller powers begin to acquire nuclear-powered submarines, then it is possible that at some point during the next century, they too, may acquire the ability to deploy submarine-launched strategic missiles, unless they are willing to abide by the strategic arms limitation treaties.

The second category is the nuclear-powered attack submarine (SSN). This too, is a very powerful weapon, capable of launching not only conventional torpedoes, but also long-range anti-ship missiles, and long-range, stand-off, conventionally armed land attack missiles. However, the primary role of the SSN is Anti-Submarine Warfare (ASW), for which it is eminently suited with its high speed, powerful torpedoes, and long-range capability. For the SSN this entails tracking, at extended ranges, other SSNs using its highly sensitive integrated sonar suite with powerful towed and flank array sonars.

But the SSN does not always operate as a lone unit. It also has the ability to co-operate with other forces, both surface and subsurface and is suitable for use with integrated task groups, providing an advanced protective screen against hostile submarines and surface forces.

The SSN has become a key player in strike warfare, as has been proven in the Arabian Gulf region where it showed itself well suited to the task of theatre missions, deploying long-range land attack missiles such as the US Tomahawk. In this guise, and with its extended range anti-ship missiles and long-range torpedoes, the SSN offers a covert force multiplier to the surface task group, and has been shown to be fully capable of dominating the battlespace. This has now become one of the SSNs primary missions made possible by the ability to download target data from a satellite, enabling it to conduct over-the-horizon engagements, hopefully well out of range of hostile surface forces.

A new role which is now increasing in importance and to which SSNs are being assigned is that of clandestine operations. For this a number of British and US submarines are being fitted with Dry Deck Shelters for the carriage of Swimmer Delivery Vehicles (SDVs). In this role the SSN will be used to support expeditionary warfare operations.

Of the major powers, Russia still possesses a submarine force of some considerable size, which continues to receive high priority in the defence budget in spite of massive cutbacks in expenditure. However, many of the former Soviet Union's submarines are now being set aside for scrap. In an effort to draw in foreign currency reserves and assure its position in the market place, Russia is now engaged in a much more intensive marketing exercise worldwide, and is achieving success in exporting its 'Kilo' design.

Except for the navies of China, France, Russia, the UK and the USA, the SSN has been considered to be well beyond the financial and operating capabilities of most navies. However, some navies are now actively pursuing the acquisition/construction of SSNs. India and Brazil, for example, are developing suitable nuclear propulsion plants for submarines, and developing the ability to build such vessels, and examples might well begin to enter service around the year 2010.

There are, however, some tasks which fall quite outside the scope of SSNs, and to which are really unsuited. Such tasks can really only be undertaken with any real possibility of achievement by SSKs (non-nuclear submarines).

Because of its smaller size and greater stealth, the SSK is much more suited to relatively shallow water operations than the SSN, and is now considered to be a primary platform for operations in the littoral zone (up to a distance of 100 miles from the shoreline).

The standard role of the SSK is to counter hostile surface forces. In this, the boat either acts alone or in conjunction with other SSKs or friendly surface and shore-based forces. Other tasks to which the SSK is highly suited include the conduct of economic warfare, a task for which it proved to be more than ideally suited in two World Wars. Furthermore, it is an ideal platform from which to carry out clandestine operations, although these roles are now being increasingly undertaken by SSNs as well.

Two other roles for which the SSK is eminently suitable include minelaying and surveillance, although, on occasions, SSNs are also tasked with these missions. Minelaying is a task that is not relished by submariners, for while engaged in such operations the submarine is precluded, by the nature and requirements of the task and the lack of space for other weapons, from carrying out any other mission, except perhaps that of surveillance. The SSK is most effective in the minelaying role when used to prevent or hinder

The SSN is a very powerful weapon when armed with long-range cruise missiles. 'Los Angeles' class submarine Greeneville *is seen here on trials in May 1995* (Newport News Shipbuilding) **1998**/0007204

Large numbers of SSBNs in the American and RFAS fleets are being decommissioned and under the strategic arms limitations treaties, will not be replaced. The British fleet of SSBNs is being replaced by the new 'Vanguard' class, the lead ship of which is illustrated (Photograph via Anchor Consultancy Photo Library) **1996**

Proliferation of submarine forces is likely to continue in the forseeable future, with units equipped with increasingly sophisticated technology (Courtesy Kockums)   ***1999**/0024655*

The type of submarine which is in most widespread service throughout the world is the SSK. In this category the most popular type, and one which will remain operational for some years to come, is the German Type 209. The Colombian Type 209/1200 Tayrona *is illustrated* (Courtesy HDW)   ***1998**/0007205*

the passage of high-value units or for laying coastal minefields in the face of hostile forces for defensive or offensive purposes.

Likewise, in a surveillance or reconnaissance role, until the mission is accomplished, the SSK is precluded from undertaking other roles. In some countries, where the threat of amphibious assault is considered very high, the SSK is used as an advanced line of defence to counter any such hostile operation.

Other tactical roles for submarines include convoy protection. For fairly obvious reasons this is also a role that is not greatly relished by submariners, for it is fraught with the potential danger of mistaken identity by friendly forces. Even so, the submarine with its high speed and extended sonar range can provide an important dimension to convoy protection.

The fourth category of submarine is the midget. This is a highly specialised craft designed to undertake very specific clandestine missions. Being very small, these craft are able to penetrate very close inshore virtually undetected. They are, therefore, ideally suited to laying mines in vulnerable points such as the entrances to anchorages and major ports, to attack invasion forces with mines or torpedoes, and to infiltrate specialised personnel such as saboteurs, spies, or advance reconnaissance parties prior to an amphibious assault. Twice in recent years North Korean small displacement submarines have made headlines when captured off the South Korean coast. In each instance the submarine was carrying a number of North Korean infiltrators, all of whom were killed before they could carry out their mission.

Midget submarines are fitted with the minimum of equipment, sufficient merely to carry out their assigned tasks. They have limited range capabilities, but these are adequate enough to allow them to operate unaided by other surface forces. Weapons stowage is very limited, but they can transport mines and carry a limited amount of military cargo such as explosives as well as

personnel. The crew is small in number and expected to perform a variety of functions commensurate with the task in hand. Stealth is of the utmost importance to the conduct of midget submarine operations.

Whatever the tactical role and strategic requirement for the submarine, there is no doubt that, being a covert weapon, it can exercise an influence out of all proportion to its size. Never was such potential more clearly displayed than in the South Atlantic War of 1982. Although the Argentine SSKs failed to achieve any real success, their presence in the area of conflict forced the Royal Navy to expend a major ASW effort on the protection of the surface warship groups and the merchant ships sailing with the assault force. Likewise the presence of UK submarines (both SSNs and SSKs) exerted a paralysing effect on the Argentine Navy, particularly after the sinking of the *Belgrano*.

Fleet strengths

The accompanying tables list those countries currently operating submarines and the state of the fleets. At the time of publication a total of 45 countries boasted a submarine inventory of 599 boats (including known midget submarines in service). Five countries had in service 182 nuclear powered submarines while 43 countries possesed 417 SSKs.

Following the acceptance of the strategic nuclear weapons treaties, both the Russian and US navies have been progressively decommissioning large numbers of older SSBNs. Once construction of the latest generation of SSBNs in the American, British and French fleets has been concluded, little construction in this category of vessel will be undertaken until at least the second quarter or middle of the next century.

In all the navies currently operating SSNs, emphasis is now centred on developing the next generation of boats, for which design studies are now well under way. However, the early generation of SSNs is now in considerable need of modernisation (particularly with

regard to command systems and sensors), or replacement by new construction. This is at present being undertaken in the Russian, British and American SSN fleets.

The USA, because of her geographical and strategic position, is firmly wedded to the doctrine of the all-SSN fleet. However, the ending of the Cold War has led to a major reappraisal of the USN's submarine requirement. The SSN-21 'Seawolf' programme will now end at just three boats. For the future, the USN is studying plans for a less sophisticated SSN.

The UK has ordered the first three boats of the 'Astute' class (previously referred to as the Batch II 'Trafalgar' design). Work is also well in hand for a major upgrade to be undertaken on the earlier 'Trafalgar' class SSNs and some of the 'Swiftsure' class, known as the 'S & T' Update. The ending of the Cold War has led to a reappraisal of the need for SSKs and the four boats of the 'Upholder' class are to be leased to Canada.

France has modified her position with respect to a mixed SSN/SSK fleet. The SSKs are being withdrawn from service and will not be replaced by SSKs. Development of the SSN fleet will, however, continue, with a new generation of boats now in the early definition stage. However, the French have decided not to continue construction of the 'Amethyste' class, but the existing boats and the 'Rubis' class will be modernised to the 'Amethyste' standard.

In Russia, construction of SSNs will continue at a very much reduced pace.

Most other major Western powers either cannot afford the SSN, or do not see a role for it. European NATO navies are therefore either in the middle of, or just about to commence, modernising existing SSK fleets, or are embarking on new construction.

It is of note that in smaller navies many SSKs have reached, or passed, the midway point in their lives. To continue to remain effective they require major updating of their electronic systems and weapons. Other boats have long since reached the end of their lives and need replacing. Some navies are seeking to introduce SSNs into their order of battle with a view to projecting a much more powerful underwater capability: the most notable of these are India and Brazil.

In the Far East there is a noticeable move towards recognising the role of the SSK. Those navies currently operating SSKs are giving careful consideration to the need to modernise their existing fleet, or to seek replacement construction, while others are seeking to develop a submarine force from scratch. Australia is replacing her existing SSK fleet with a new fleet built to a Swedish design. South Korea is now a major submarine operator in the region, while Singapore has embarked on the development of a submarine force.

In the Middle and Near East those navies currently operating SSKs are giving careful consideration as to whether to modernise existing units or to replace them with new construction. A new entrant into the submarine field is Iran with three operational SSKs, and developments there are being watched with considerable interest.

Two major factors have to be borne in mind when seeking to develop a submarine force. First, the minimum number of boats needed to ensure an effective and operationally available force at all times is generally considered to be at least three and preferably four or five. The reasons are fairly straightforward. At any one time, one boat may be expected to be in dock undergoing maintenance or refit, one will be working up having completed a refit, and the remainder will be operational. Of these at least one would be on patrol, one going out on patrol and one returning from patrol.

The second factor concerns manpower. The submarine is a technologically advanced system which, because of the environment in which it operates, cannot afford to carry inexperienced crew members who, through lack of training and proper maintenance, by a simple mistake could cause the loss of the boat and her crew. Adequate and continuing training must, of necessity, be a major factor in developing a submarine force.

It can be seen, therefore, that up until the turn of the century there will be considerable potential, primarily for updating existing submarines, and to a more limited degree in the present stringent financial climate, for building replacement boats.

Modernisation

Although submarine forces worldwide are no longer

expanding at the rate they were before the ending of the Cold War they are, nevertheless, carrying out extensive modernisation of their fleets. These modernisation programmes will extend the life of existing boats to almost 30 years, making them highly effective units in a modern naval scenario. New forces are being organised and those with obsolete boats are seeking to acquire replacements.

Altogether there is a very considerable market for both modernisation and new build. The most important aspect to be considered must be the weapon control, sensor and command functions. To be capable of carrying out effective missions the modern submarine must be equipped with a fully integrated combat suite comprising sensors, AIO and fire-control system. With budgets tight, however, some areas may have to be left unmodernised. If this is the case then priority has to be given to updating the command and sonar systems. It would also be advisable, wherever possible, to update the optical and EW capabilities of a boat. Major technological advances have been made in all these areas and current models are far superior to those of even 10 years ago. In addition, to maintain the stealth capabilities of the platform, considerable attention needs to be paid today to noise reducing techniques – especially in the light of developments with low-frequency active towed array sonars.

As far as propulsion systems in SSKs are concerned, the diesels, as long as they are in reasonable condition, are probably worth retaining, for they are very hard wearing and have a long life expectancy. Likewise the electric motors could be retained. With the batteries, however, a different situation exists. There have been considerable advances in recent years in battery design. When modernising an older boat, consideration should be given to replacing old batteries with new designs capable of charging at a high rate in a short time. In addition the need to extend the battery compartment to give the boat a greater submerged endurance should be considered.

One development that is now attracting increasing consideration is the closed-cycle Air Independent Engine (AIP). Already, Germany and Sweden have developed such systems and versions have undergone trials at sea. Sweden now has AIP systems operational at sea, and Germany has ordered a new class of submarine that will be fitted with an AIP system. The Netherlands and France have systems under development and Pakistan has ordered three new 'Agosta 90B' type boats from France, the last of which will be fitted during build with the Mesma AIP system, while the two earlier boats will be retrofitted with the system later.

UPDATED

TABLE I

Submarine construction 1990-1998

Name (Pt No)	Builder	Ordered	Laid down	Launched	Commissioned	Remarks
Australia						
SSK CLASS - 'Collins'						
Collins (S 73)	ASC		14 Feb 1990	28 Aug 1993	30 July 1996	1st of 6
Farncomb (S 74)	ASC		1 Mar 1991	15 Dec 1995	31 Jan 1998	2nd of 6
Waller (S 75)	ASC		19 Mar 1992	14 Mar 1997	1999	3rd of 6
Dechaineux (S 76)	ASC		4 Mar 1993	12 Mar 1998	Aug 1999	4th of 6
Sheean (S 77)	ASC		17 Feb 1994	Mar 1999	Jun 2000	5th of 6
Rannkin (S 78)	ASC		12 May 1995	Mar 2000	Jun 2001	6th of 6
Brazil						
SSN CLASS						
Riachuelo	Rio de Janeiro DY					Planned
SSK CLASS - Type 209/1400						
Tamoio (S 31)	Rio de Janeiro DY		15 July 1986	18 Nov 1993	12 Dec 1994	2nd of 4
Timbira (S 32)	Rio de Janeiro DY		15 Sep 1987	5 Jan 1996	16 Dec 1996	3rd of 4
Tapajo (S 33)	Rio de Janeiro DY		6 Mar 1996	Jun 1998	1999	4th of 4
SSK CLASS - Improved 'Tupi' SNAC-1						
Tikuna (S 34)	Rio de Janeiro DY	1994	1999	2002	2003	1st of 2
Chile						
SSK CLASS- Scorpene						
One unnamed	DCNI/Bazan	Dec 1997	1998		2003/4	1st of 2
One unnamed	DCNI/Bazan	Dec 1997			2003/4	2nd of 2
China						
SSBN CLASS - Project 094						
Unnamed		1997/98				1st of 4
SSN CLASS - Project 093						
Unnamed	Bohai SY		1994	2000	2001	
Unnamed					2004	
SSN CLASS - 'Han'						
No 405	Huludao SY		1997	8 Apr 1990	Dec 1990	5th of 5
SSK CLASS - 'Kilo'						
Unknown Type 877 (364)	Admiralty Yd	Mid 1993			Feb 1995	1st of 4
Unknown Type 877 (365)	Admiralty Yd	Mid 1993			Nov 1995	2nd of 4
Unnamed Type 636 (366)	Admiralty Yd	Mid 1993			Sep 1997	3rd of 4
Unnamed Type 636 (367)	Admiralty Yd	Mid 1993		17 Jun 1998	Oct 1998	4th of 4
SSK CLASS - 'Song' Project 039						
Unnamed (320)	Wuhan SY			May 1994	Aug 1998	
Unnamed (321)	Wuhan SY		1995	1998	1998	
2 unnamed	Wuhan SY			Mid 1997		Building
SSK CLASS - 'Ming'						
2 unnamed				1998	1998	
4 unnamed						Building
France						
SSBN CLASS -'Le Triomphant'						
Le Triomphant (S 616)	Cherbourg	10 Mar 1986		16 July 1993	24 Oct 1996	1st of 4
Temeraire (S 617)	Cherbourg	18 Oct 1989	1991	8 Aug 1997	Apr 1999	2nd of 4
Le Vigilant (S 618)	Cherbourg	27 May 1993		Mar 2002	Dec 2003	3rd of 4
Unnamed (S.619)	Cherbourg			Nov 2005	July 2007	4th of 4
SSN CLASS - 'Rubis'						
Amethyste (S 605)	Cherbourg		11 Oct 1994	14 May 1988	3 Mar 1992	5th of 6
Perle (S 606)	Cherbourg		27 Mar 1987	22 Sep 1990	7 July 1993	6th of 6
Germany						
SSK CLASS - Type 212						
U 31	GSC	7 July 1994	1 July1998	Oct 2001	Sep 2003	1st of 4
U 32	GSC	7 July 1994	11 July 2000	Nov 2003	2005	2nd of 4
U 33	GSC	7 July 1994	30 Apr 2001	Aug 2004	2006	3rd of 4
U 34	GSC	7 July 1994	Dec 2001	May 2005	2006	4th of 4
Greece						
SSK CLASS - Type 209/1200						
2 unnamed						Funds allocated 1997

Name (Pt No)	Builder	Ordered	Laid down	Launched	Commissioned	Remarks
India						
SSN CLASS - ATV (Advanced Technology Vessel)						
Unnamed	Vishakapatnam	2001/2	2006/7			1st of 5 planned
SSK CLASS - 'Kilo' Project 877EKM						
Sindhukiri (S 61)	Admiralty Yd				4 Mar 1990	7th of 12
Sindhuvijay (S 62)	Admiralty Yd				8 Mar 1991	8th of 12
Sindhurakshak (S 63)	Admiralty Yd	Jan 1997			24 Dec 1997	9th of 12
Two unnamed Type 636	India					11/12th of 12
SSK CLASS - Type 209/1500						
Shalki (S 46)	Mazagon Dock	5 June 1984		30 Sep 1989	7 Feb 1992	3rd of 6
Shankul (S 47)	Mazagon Dock	3 Sep 1989		21 Mar 1992	28 May 1994	4th of 6
SSK CLASS - Project 75 - possibly Type 209						
2 unnamed	Mazagon Dock					1/2nd of 2 proposed
Iran						
SSK CLASS - 'Kilo' Project 877EKM						
Tareq (901)	Admiralty Yd		1988	1991	21 Nov 1992	1st of 3
Noor (902)	Admiralty Yd		1989	1992	6 June 1993	2nd of 3
Yunes (903)	Admiralty Yd		1990	1993	25 Nov 1996	3rd of 3
Israel						
SSK CLASS - 'Dolphin'						
Dolphin	GSC	April 1991	15 Feb 1992	15 Apr 1996	Mar 1998	1st of 3
Leviathan	GSC	April 1991	Aug 1992	27 May 1997	Late1999	2nd of 3
Tekuma	GSC	July 1994	July 1994	9 July 1998	1999	3rd of 3
Italy						
SSK CLASS - Improved 'Sauro'						
Primo Longobardo (S 524)	Fincantieri		19 Dec 1991	20 June 1992	20 May 1994	3rd of 4
Gianfranco Gazzana Priaroggia (S 525)	Fincantieri		12 Nov 1992	26 June 1993	12 Apr 1995	4th of 4
SSK CLASS - Type 212A						
One unnamed	Fincantieri	1997	July 1998	Oct/Nov 2002	Nov 2004	1st of 2
One unnamed	Fincantieri	1997	Dec 2000	Apr 2004	Feb 2006	2nd of 2
Japan						
SSK CLASS - 'Harushio'						
Harushio (583)	Mitsubishi		21 Apr 1987	26 July 1989	30 Nov 1990	1st of 7
Natsushio (584)	Kawasaki		8 Apr 1988	20 Mar 1990	20 Mar 1991	2nd of 7
Hayashio (585)	Mitsubishi		9 Dec 1988	17 Jan 1991	25 Mar 1992	3rd of 7
Arashio (586)	Kawasaki		8 Jan 1990	17 Mar 1992	17 Mar 1993	4th of 7
Wakashio (587)	Mitsubishi		12 Dec 1990	22 Jan 1993	1 Mar 1994	5th of 7
Fuyushio (588)	Kawasaki		12 Dec 1991	16 Feb 1994	7 Mar 1995	6th of 7
Asashio (589)	Mitsubishi		24 Dec 1992	12 July 1995	12 Mar 1997	7th of 7
SSK CLASS - 'Oyashio'						
Oyashio (590)	Kawasaki		26 Jan 1994	15 Oct 1996	16 Mar 1998	1st
Michishio (591)	Mitsubishi		16 Feb 1995	18 Sep 1997	Mar 1999	2nd
Job No 8107 (592)	Kawasaki		6 Mar 1996	Nov 1998	Mar 2000	3rd
(593)	Mitsubishi		26 Mar 1997	Sep 1999	Mar 2001	4th
(594)	Kawasaki		Mar 1998	Nov 2000	Mar 2002	5th
Netherlands						
SSK CLASS - 'Walrus'						
Walrus (S 801)	RDM		11 Oct 1979	26 Oct 1985	25 Mar 1992	1st of 4
Zeeleeuw (S 802)	RDM		24 Sep 1981	20 June 1987	25 Apr 1990	2nd of 4
Dolfijn (S 808)	RDM		12 June 1986	25 Apr 1990	29 Jan 1993	3rd of 4
Bruinvis (S 810)	RDM		14 Apr 1988	25 Apr 1992	5 July 1994	4th of 4
Norway						
SSK CLASS - 'Ula'						
Utsira (S 301)	Thyssen		15 June 1990	21 Nov 1991	30 Apr 1992	2nd of 6
Utstein (S 302)	Thyssen		6 Dec 1989	26 Apr 1991	14 Nov 1991	3rd of 6
Utvaer (S 303)	Thyssen		8 Dec 1988	20 Apr 1990	8 Nov 1990	4th of 6
Uthaug (S 304)	Thyssen		15 June 1989	18 Oct 1990	7 May 1991	5th of 6
Uredd (S 305)	Thyssen		23 June 1988	22 Sep 1989	3 May 1990	6th of 6
Pakistan						
SSK CLASS - 'Agosta 90B'						
Unnamed (S 137)	Cherbourg		1997	Aug 1998	Apr 1999	3rd of 5
Unnamed (S 138)	Cherbourg/Karachi	Sep 1992	1998	2001	Feb 2002	4th of 5
Unnamed (S 139)	Karachi		1999	2002	Oct 2002	5th of 5
Portugal						
SSK CLASS						
3/4 Unnamed		1999		2004		1/3rd of 3. RfP issued
RFAS						
SSBN CLASS - 'Bory' Project 995						
Yuri Dolgoruky	Severodvinsk		2 Nov 1996	2001	2003	1st of 2
SSBN CLASS - 'Delta IV' Project 667BDRM						
K117	Severodvinsk		Sep 1987	Sep 1988	Mar 1990	5th of 7
K18	Severodvinsk		Sep 1988	Nov 1989	Sep 1991	6th of 7
K407	Severodvinsk		Nov 1989	Jan 1991	20 Feb 1992	7th of 7
SSGN CLASS - 'Oscar II' Project 949A						
Smolensk (K 410)	Severodvinsk			Dec 1989	Dec 1990	5th of 12
Pekov (K 442)	Severodvinsk			Jan 1990	Jan 1991	6th of 12
Kasatka (K 456)	Severodvinsk			Dec 1991	Nov 1992	7th of 12
Orel (K 266)	Severodvinsk			Jan 1992	Jan 1993	8th of 12
Omsk (K 186)	Severodvinsk			May 1993	Dec 1993	9th of 12
Kursk (K 141)	Severodvinsk			May 1994	Oct 1994	10th of 12
Tomsk (K 512)	Severodvinsk			18 Jul 1995	May 1997	11th of 12
Belgorod (K 530)	Severodvinsk			1998	1999	12th of 12

Name (Pt No)	Builder	Ordered	Laid down	Launched	Commissioned	Remarks
SSN CLASS - 'Akula I/II' Project 971/ 971M						
Pantera (K 317)	Severodvinsk			May 1990	Oct 1990	7th of 10
Narwhal (K 331)	Komsomolsk			June 1990	Sep 1990	8th of 10
Wolf (K 461)	Severodvinsk			June 1991	Dec 1991	9th of 10
Morz (K 419)	Komsomolsk			May 1992	Aug 1992	10th of 10
Leopard (K 328)	Severodvinsk			June 1992	Oct 1992	11th of 10
Tigr (K 157)	Severodvinsk			June 1993	Oct 1993	12th of 10
Drakon (K 267)	Komsomolsk			Sep 1994	29 July 1995	13th of 10
Vepr (II)	Severodvinsk			Dec 1994	July 1995	14th of 10
Gepard (II)	Komsomolsk			1997	1997	15th of 10
Bison (II)	Severodvinsk			1997	1998	16th of 10
SSN CLASS - 'Sierra II' Project 945A						
Pskov (K 534)	Nizhny Novgorod			July 1988	Oct 1980	1st of 2
Nizny Novgorod (K 336)	Nizhny Novgorod			July 1992	12 Aug 1993	2nd of 2
SSN CLASS - Severodvinsk Project 885						
Severodvinsk	Severodvinsk		21 Dec 1993	2000	2001	1st of 7
Unnamed	Severodvinsk		Oct 1996	2001	2002	2nd of 7
Unnamed	Severodvinsk		Jan 1998	2003	2004	3rd of 7
SSK CLASS - 'Lada' Project 677 (export variant Amur)						
St Petersburg	Admiralty Yd		26 Dec1997		2001	1st of 3
Unnamed	Admiralty Yd		Dec 1997		2002	Export 2nd of 3
South Korea						
SSK CLASS - Type 209/1200						
Chang Bogo (061)	HDW		1989	18 June 1992	2 June 1993	1st of 9
Yi Chon (062)	Daewoo		1990	14 Oct 1992	30 Apr 1994	2nd of 9
Choi Muson (063)	Daewoo		1991	25 Aug 1993	27 Feb 1995	3rd of 9
Pakui (065)	Daewoo		1992	20 May 1994	3 Feb 1996	4th of 9
Lee Jongmu (066)	Daewoo		1993	17 Apr 1995	29 Aug 1996	5th of 9
Jeongun (067)	Daewoo		1994	7 May 1996	Mar 1998	6th of 9
Unnamed (068)	Daewoo		1995	Feb 1998	Dec 1998	7th of 9
Unnamed (069)	Daewoo		1996	Jun 1998	Aug 1999	8th of 9
Unnamed (071)	Daewoo		1997	Mar 1999	Apr 2000	9th of 9
Sweden						
SSK CLASS - 'Vastergotland'						
Ostergotland (Gd)	Kockums		1986	9 Dec 1988	12 Jan 1990	4th of 4
SSK CLASS - 'Gotland'						
Gotland (Gld)	Kockums	28 Mar 1990	20 Nov 1992	2 Feb 1994	July 1996	1st of 3
Uppland	Kockums	28 Mar 1990	14 Jan 1994	9 Feb 1996	Sep 1997	2nd of 3
Halland (Hnd)	Kockums	28 Mar 1990	21 Oct 1994	27 Sep 1996	Oct 1997	3rd of 3
Turkey						
SSK CLASS - 'Atilay' Type 209/1200						
Dolunay (S 352)	Golcuk DY		9 Mar 1981	22 July 1988	29 June 1990	6th of 6
SSK CLASS - 'Preveze' Type 209/1400						
Preveze (S 353)	Golcuk DY		12 Sep 1989	22 Oct 1993	28 July 1994	1st of 8
Sakarya (S 354)	Golcuk DY		1 Feb 1990	28 July 1994	12 Dec 1995	2nd of 8
18 Mart (S 355)	Golcuk DY		28 July 1994	25 Aug 1997	Aug 1998	3rd of 8
Anafartalar (S 356)	Golcuk DY		1 Aug 1995	Aug 1998	2000	4th of 8
4 Unnamed	Golcuk DY	1998				6/8th of 8
6 Unnamed						Projected
UK						
SSBN CLASS - 'Vanguard'						
Vanguard (S 28)	VSEL	30 Apr 1986	3 Sep 1986	5 Mar 1992	14 Aug 1993	1st of 4
Victorious (S 29)	VSEL	6 Oct 1987	3 Dec 1987	29 Sep 1993	7 Jan 1995	2nd of 4
Vigilant (S 30)	VSEL	13 Nov 1990	16 Feb 1991	14 Oct 1995	2 Nov 1996	3rd of 4
Vengeance (S 31)	VSEL	7 July 1992	1 Feb 1993	Aug 1998	Jul 1999	4th of 4
SSN CLASS - 'Trafalgar'						
Talent (S 92)	VSEL	10 Sep 1994	13 May 1986	15 Apr 1988	12 May 1990	6th of 7
Triumph (S 93)	VSEL	3 Jan 1986	2 Feb 1987	16 Feb 1991	12 Oct 1991	7th of 7
SSN CLASS - 'Astute'						
Astute	VSEL	17 Mar 1997				1st of 5
Ambush	VSEL	17 Mar 1997				2nd of 5
Artful	VSEL	17 Mar 1997				3rd of 5
2 unnamed	VSEL					4/5th of 5
USA						
SSBN CLASS - 'Ohio'						
West Virginia (SSBN 736)	General Dynamics			14 Oct 1989	20 Oct 1990	11th of 18
Kentucky (SSBN 737)	General Dynamics			11 Aug 1990	13 July 1991	12th of 18
Maryland (SSBN 738)	General Dynamics			10 Aug 1991	13 June 1992	13th of 18
Nebraska (SSBN 734)	General Dynamics			15 Aug 1992	10 July 1993	14th of 18
Rhode Island (SSBN 740)	General Dynamics			17 July 1993	9 July 1994	15th of 18
Maine (SSBN 741)	General Dynamics		1990	16 July 1994	29 July 1995	16th of 18
Wyoming (SSBN 742)	General Dynamics		1991	15 July 1995	13 July 1996	17th of 18
Louisiana (SSBN 743)	General Dynamics		1992	27 July 1996	6 Sep 1997	18th of 18
SSN CLASS - 'Los Angeles'						
Albany (SSN 753)	Newport News SB		22 Apr 1985	13 June 1987	7 Apr 1990	
Miami (SSN 755)	General Dynamics		24 Oct 1986	12 Nov 1988	9 June 1990	
Scranton (SSN 756)	Newport News SB		29 June 1986	3 July 1989	26 Jan 1991	
Alexandria (SSN 757)	General Dynamics		19 June 1987	23 June 1990	29 June 1991	
Asheville (SSN 758)	Newport News SB		1 Jan 1987	28 Oct 1989	28 Sep 1991	
Jefferson City (SSN 759)	Newport News SB		21 Sep 1987	24 Mar 1990	30 Jan 1992	
Annapolis (SSN 760)	General Dynamics		15 June 1988	18 May 1991	10 Apr 1992	
Springfield (SSN 761)	General Dynamics		29 Jan 1990	4 Jan 1992	9 Jan 1993	
Columbus (SSN 762)	General Dynamics		9 Jan 1991	1 Aug 1992	31 July 1993	
Santa Fe (SSN 763)	General Dynamics		9 July 1991	12 Dec 1992	8 Jan 1994	
Boise (SSN 764)	Newport News SB		28 Aug 1988	20 Oct 1990	7 Nov 1992	

Name (Pt No)	Builder	Ordered	Laid down	Launched	Commissioned	Remarks
Montpelier (SSN 765)	Newport News SB		19 May 1989	6 Apr 1991	13 Mar 1993	
Charlotte (SSN 766)	Newport News SB		17 Aug 1990	3 Oct 1992	16 Sep 1994	
Hampton (SSN 767)	Newport News SB		2 Mar 1990	28 Sep 1991	6 Nov 1993	
Hartford (SSN 768)	General Dynamics		27 Apr 1992	4 Dec 1993	10 Dec 1994	
Toledo (SSN 769)	Newport News SB		6 May 1991	28 Aug 1993	28 Feb 1995	
Tucson (SSN 770)	Newport News SB		15 Aug 1991	15 Mar 1994	9 Sep 1995	
Columbia (SSN 771)	General Dynamics		24 Apr 1993	24 Sep 1994	9 Oct 1995	
Greeneville (SSN 772)	Newport News SB		28 Feb 1992	17 Sep 1994	16 Feb 1996	
Cheyenne (SSN 773)	Newport News SB		6 July 1992	3 Apr 1995	13 Sep 1996	
SSN CLASS - 'Seawolf'						
Seawolf (SSN 21)	General Dynamics	9 Jan 1989	25 Oct 1989	24 June 1995	19 Jul 1997	1st of 3
Connecticut (SSN 22)	General Dynamics	3 May 1991	14 Sep 1992	15 Sep 1997	Dec 1998	2nd of 3
Jimmy Carter	General Dynamics	July 1996	5 Dec 1995	June 2000	Dec 2001	3rd of 3
NSSN CLASS-'Virginia'						
Virginia (SSN 774)	General Dynamics	1998	30 Sep	2003	2004	1st of 4
3 unnamed						2nd/4th of 4 proposed

NOTES
Shipyards
Admiralty (Sudomekh) Yd - Admiralty Yard, St Petersburg, Russia
ASC - Australian Submarine Corporation, Adelaide, Australia
Bazan - Bazan Shipyard, Cartagena, Spain
Cherbourg - Cherbourg Dockyard, France
Daewoo - Daewoo, Okpo, South Korea
DCN - DCN, Cherbourg Naval Dockyard, France
Fincantieri - Fincantieri, Monfalcone, Italy
General Dynamics - General Dynamics, Groton, USA
Golcuk - Golcuk Dockyard, Kocaeli, Turkey
GSC - German Submarine Consortium, Germany (comprising HDW and Thyssen)
HDW - Howaldtswerke, Kiel, Germany

Huludao - Huludao Shipyard, China
Karachi - Karachi Dockyard, Pakistan
Kawasaki - Kawasaki, Kobe, Japan
Kockums - Kockums, Malmo, Sweden
Komsomolsk - Komsomolsk, Russia
Mazagon - Mazagon Dock, Bombay, India
Mitsubishi - Mitsubishi, Kobe, Japan
Newport News SB - Newport News Shipbuilding, USA
Nizhny Novgorod - Gorky, Russian Federation
RDM - Rotterdamsche Droogdok Maatschappij, Rotterdam, Netherlands
Rio de Janeiro - Arsenal de Marinha, Rio de Janeiro Dockyard, Brazil
Severodvinsk - Severodvinsk 492 Shipyard, Russia
Sudomekh - Sudomekh Shipyard, Russian Federation
Thyssen - Thyssen Werften GmbH, Emden, Germany

VSEL - Vickers Shipbuilding & Engineering, Barrow, UK
Wuhan - Wuhan Shipyard, China

Under current construction practice laying down dates for submarines are not often quoted, the date usually quoted being that for the cutting of the first steel. A similar practice relates to the launching date. Most submarines are now rolled out of the berth in a ceremony, and floating out dates are not normally listed.

GLOSSARY
ITT - Invitation to tender
PD - Project Definition
RfP - Request for Proposal

UPDATED

TABLE II

Operational submarines 1999

Navy	No	Type	Country of build	Class	Entered service	Remarks
Albania	4	SSK	RFAS	'Whiskey V'	1960-61	Obsolete. May be non-operational
Algeria	2	SSK	RFAS	'Kilo' 877E	1987-88	Refitted 1990s
Argentina	2	SSK	Germany	TR 1700	1984	
	1	SSK	Germany	Type 209/1200	1974	Modernised with new engines, electronics and weapons
Australia	4	SSK	Australia	'Collins'	1995-99	Two more building
Brazil	3	SSK	Germany	Type 209/1400	1989-1994	One more building
	1	SSK	UK	'Oberon'	1977	New hull sonars fitted 1996-97
Bulgaria	2	SSK	RFAS	'Romeo'	1958-61	Obsolete
Canada	3	SSK	UK	'Oberon'	1965-68	Modernised
Chile	2	SSK	Germany	Type 209/1300	1984	Modernised early 1990s
	2	SSK	UK	'Oberon'	1976	New sonar fitted 1992
China	1	SSBN	China	'Xia'	1987	Completing major refit
	1	SSGN	China	'Golf'	1966	Refitted 1995 to deploy JL-2 missile
	5	SSN	China	'Han'	1970-90	
	2	SSK	China	'Song'	1996-97	Two more building
	2	SSK	RFAS	'Kilo' 877 EKM	1995-97	
	2	SSK	RFAS	'Kilo' 636	1995-97	More may be acquired
	39	SSK	RFAS/China	'Romeo'	1962-87	Early vessels obsolete. Many in reserve
	15	SSK	China	'Ming'	1975-97	Construction continuing
Colombia	2	SSK	Germany	Type 209/1200	1975	Refitted 1990-91
	2	Midget	Italy	Cosmos	1972	Rebuilt early 1980s
Croatia	1	Midget	Yugoslavia	'Una'	1985	One more building
Cuba	1	SSK	RFAS	'Foxtrot'	1979-84	Obsolete
Denmark	2	SSK	Denmark	Narhvalen	1970	Modernised 1993-98. New sonar and optronics mast
	3	SSK	Germany	Type 207	1964-65	Modernised 1992-93. New sonar
Ecuador	2	SSK	Germany	Type 209-1300	1977-78	Require major refit
Egypt	4	SSK	China	'Romeo'	1982-84	Modernised with new sonar, FCS and weapons
France	1	SSBN	France	'Le Triomphant'	1996	Two more building
	3	SSBN	France	'L'Inflexible'	1976-85	
	6	SSN	France	'Rubis'	1983-93	
	2	SSK	France	'Agosta'	1977-78	Modernised
	1	SSK	France	'Daphne'	1969	Obsolete
Germany	12	SSK	Germany	Type 206A	1973-75	Modernised with new sonar, AIO and periscopes
	2	SSK	Germany	Type 205	1968-69	Obsolete
Greece	4	SSK	Germany	Type 209/1100	1971-72	Being modernised
	4	SSK	Germany	Type 209/1200	1979-80	Modernised
India	9	SSK	RFAS	'Kilo' 877 EM	1986-91	
	4	SSK	Germany	Type 209/1500	1986-94	
	4	SSK	RFAS	'Foxtrot'	1970-74	Obsolete

Navy	No	Type	Country of build	Class	Entered service	Remarks
Indonesia	2	SSK	Germany	Type 209/1300	1981	
	2	SSK	Germany	Type 206		Transferred 1997
Iran	3	SSK	RFAS	'Kilo' 877 EKM	1992-96	
	3	Midget		Various	1988-	May be some local construction
Israel	3	SSK	UK	Type 540	1977	Modernised
	2	SSK	Germany	'Dolphin'	1997	One more completing
Italy	4	SSK	Italy	Improved 'Sauro'	1988-95	
	4	SSK	Italy	'Sauro'	1979-82	Modernised early 1990s
Japan	2	SSK	Japan	'Oyashio'	1998-99	Construction continuing
	7	SSK	Japan	'Harushio'	1990-97	
	10	SSK	Japan	'Yuushio'	1980-89	New sonars fitted
Korea North	22	SSK	N. Korea	'Romeo'	1973-75	
	20	SSK	N. Korea	'Sang-O'	1991-99	Construction continuing
	45	Midget	N. Korea	'Yugo'	1960-69	
Korea South	8	SSK	S. Korea	Type 209/1200	1993-99	
	3	Midget	S. Korea	KSS-1	1983	
	8	Midget	Italy	'Dolphin'		
Libya	4	SSK	RFAS	'Foxtrot'	1976-83	Obsolete
Netherlands	4	SSK	Holland	'Walrus'	1992-94	
Norway	6	SSK	Germany	Type 207	1964-67	Modernised early 1990s
	6	SSK	Germany	'Ula' (Type 210)	1990-92	
Pakistan	1	SSK	France	'Agosta 90B'	1999	Two more building
	2	SSK	France	'Agosta'	1979-80	
	4	SSK	France	'Daphne'	1969-70	Obsolescent
	3	Midget	Pakistan	MG		
Peru	6	SSK	Germany	Type 209/1200	1975-83	Modernised with new FCS late 1980s
	2	SSK	USA	'Marlin' Type	1954	Obsolete
Poland	1	SSK	RFAS	'Kilo' 877E	1986	
	2	SSK	RFAS	'Foxtrot'	1987-88	
Portugal	3	SSK	France	'Daphne'	1967-69	Obsolete. New radar installed 1993-94
Romania	1	SSK	RFAS	'Kilo' 877E	1986	
RFAS	6	SSBN	RFAS	'Typhoon'	1981-89	Being modernised to deploy SS-N-28
	7	SSBN	RFAS	'Delta IV'	1985-92	
	9	SSBN	RFAS	'Delta III'	1976-82	Paying, off
	4	SSBN	RFAS	'Delta I'	1972-74	Scrapping
	12	SSGN	RFAS	'Oscar II'	1986-98	Construction continuing
	12	SSN	RFAS	'Akula I/II'	1986-97	Early boats paying off
	1	SSN	RFAS	'Sierra I'	1984-87	
	2	SSN	RFAS	'Sierra II'	1990-93	
	11	SSN	RFAS	'Victor III'	1978-92	Scrapping
	1	SSN	RFAS	'Yankee Notch'		Converted 1983 from SSBN
	15	SSK	RFAS	'Kilo'	1982-93	Construction continuing for export only
	6	SSK	RFAS	'Tango'	1972-82	Obsolescent
	2	SSK	RFAS	'Foxtrot'	1971	Obsolete
	2	Midget	RFAS	'Pyranja'	1988-91	
Singapore	4	SSK	Sweden	'Sjoormen'	1969	Acquired 1997
South Africa	3	SSK	France	'Daphne'	1970-71	Modernised 1988 onwards to extend life to 2005.
Spain	4	SSK	Spain	'Agosta'	1983-86	Being modernised
	4	SSK	Spain	'Daphne'	1973-75	Refitting to deploy F-17 torpedo
Sweden	3	SSK	Sweden	'Gotland'	1996-97	AIP installed
	4	SSK	Sweden	'Vastergotland'	1987-90	
	3	SSK	Sweden	'Nacken'	1980-81	Modernised. One fitted with AIP
Taiwan	2	SSK	Netherlands	'Hai Lung'	1987-88	
	2	SSK	USA	'Guppy II'	1945-46	Obsolete
Turkey	3	SSK	Turkey	Type 209/1400	1994-95	
	6	SSK	Germany/Turkey	Type 209/1200	1976-90	Early boats need updating
	2	SSK	USA	'Tang' Type	1952	Obsolete
	5	SSK	USA	'Guppy IIA/III'	1944-45	Obsolete
UK	4	SSBN	UK	'Vanguard'	1993-96	One more building
	7	SSN	UK	'Trafalgar'	1983-91	To be modernised with new AIO and sensors
	5	SSN	UK	'Swiftsure'	1974-81	To be modernised with new AIO and sensors
USA	18	SSBN	USA	'Ohio'	1981-97	
	2	SSN	USA	'Seawolf'	1997	One more building
	53	SSN	USA	'Los Angeles'	1976-96	Early boats being modernised
	8	SSN	USA	'Sturgeon'	1970-75	Being decommissioned
	2	SSN	USA	'Benjamin Franklin'		
Venezuela	2	SSK	Germany	Type 209/1300	1976-77	Modernised 1992-93 with new AIO
Yugoslavia	2	SSK	Yugoslavia	'Sava'	1978-81	
	2	SSK	Yugoslavia	'Heroj'	1968-70	
	35	Midget	Yugoslavia	'Una'	1985-89	

NOTE: The RFAS also has a number of auxiliary submarines not listed in this table as they carry no weapons.

UPDATED

TABLE III

Submarine age 1998

Age (yrs)	No	Class	Navies	Age (yrs)	No	Class	Navies
0-5 (1994-98)				**16-20 (1979-83)**			
	13	Type 209	Brazil (2), India (1), South Korea (7), Turkey (3)		11	Type 209	Greece (4), Indonesia (2), Peru (4), Turkey (1)
	4	'Collins'	Australia 1 Agosta 90B Pakistan		4	'Agosta'	Pakistan (2), Spain (2)
	2	'Dolphin'	Israel		3	'Foxtrot'	Libya (3*)
	3	'Gotland'	Sweden		1	'Ming'	China
	3	'Harushio'	Japan		3	'Nacken'	Sweden
	10	'Kilo'	China (4), India (1), Iran (3), RFAS (2)		22	'Romeo'	China (15), Egypt (2*), North Korea (5*)
	11	'Ming'	China		2	'Sauro'	Italy
	2	'Oyashio'	Japan		6	'Tango'	RFAS
	17	'Sang-O'	North Korea		4	'Yuushio'	Japan
	2	'Song'	China		1	'Sava'	Yugoslavia
	1	'Walrus'	Netherlands	TOTAL	57		
TOTAL	68				8	'DELTA III'	RFAS
	4	'AKULA I/II'	RFAS		1	'HAN'	China
	8	'LOS ANGELES'	USA		1	'L'INFLEXIBLE'	France
	4	'OHIO'	USA		14	'LOS ANGELES'	USA
	3	'OSCAR II'	RFAS		3	'OHIO'	USA
	2	'SEAWOLF'	USA		1	'RUBIS'	France
	2	'LE TRIOMPHANT'	France		2	'SWIFTSURE'	UK
	3	'VANGUARD'	UK		2	'TYPHOON'	RFAS
TOTAL	26				2	'VICTOR III'	RFAS
6-10 (1989-93)				TOTAL	34		
	4	Type 209	Brazil (1), India (1), South Kores (1), Turkey (1)	**21-25 (1974-78)**			
	4	'Harushio'	Japan		12	206A	Germany
	13	'Kilo'	India (4), RFAS (9)		12	Type 209	Argentina (1), Colombia (2), Ecuador (2), Peru (2), Turkey (3), Venezuela (2)
	3	'Ming'	China		2	'Agosta'	France
	2	'Romeo'	North Korea (2)		2	'Daphne'	Spain
	3	'Sang-O'	North Korea		1	'Foxtrot'	Libya (1*)
	4	Improved 'Sauro'	Italy		3	'Gal' Type 540	Israel
	6	'Ula'	Norway		3	'Oberon'	Brazil (1), Chile (2)
	2	'Vastergotland'	Sweden		2	'Sauro'	Italy
	3	'Walrus'	Netherlands		1	'Sava'	Yugoslavia
	1	'Yuushio'	Japan		17	'Romeo'	China (9), North Korea (8*)
TOTAL	45			TOTAL	55		
	7	'AKULA I/II'	RFAS		2	'DELTA I'	RFAS
	4	'DELTA IV'	RFAS		1	'DELTA III'	RFAS
	1	'HAN'	China		1	'HAN'	China
	15	'LOS ANGELES'	USA		2	'L'INFLEXIBLE'	France
	5	'OHIO'	USA		2	'LOS ANGELES'	USA
	6	'OSCAR II'	RFAS		3	'STURGEON'	USA
	2	'RUBIS'	France		3	'SWIFTSURE'	UK
	2	'SIERRA II'	RFAS	TOTAL	14		
	3	'TRAFALGAR'	UK	**Over 25 (pre-1974)**			
	1	'TYPHOON'	RFAS		2	Type 205	Germany
	3	'VICTOR III'	RFAS		2	Type 206	Indonesia
TOTAL	49				9	Type 207	Denmark (3), Norway (6)
11-15 (1984-88)					4	Type 209	Greece
	2	TR 1700	Argentina		2	'Abtao'	Peru
	5	Type 209	Chile (2), India (2), Turkey (1)		13	'Daphne'	France (1), Pakistan (4), Portugal (3), South Africa (3), Spain (2)
	2	'Agosta'	Spain		6	'Foxtrot'	India (4*), RFAS (2)
	3	'Foxtrot'	Cuba (1*), Poland (1*)		7	Ex-US 'Guppy'	Taiwan (2), Turkey (5)
	2	'Hai Lung'	Taiwan		2	'Heroj'	Yugoslavia
	12	'Kilo'	Algeria (2), India (4), Poland (1), Romania (1*) RFAS (4)		1	'Ming'	China
					2	'Narhvalen'	Denmark
	22	'Romeo'	China (15), Egypt (2*), North Korea (5)		3	'Oberon'	Canada (3)
					4	'Romeo'	Bulgaria (2), North Korea (2*)
	2	'Vastergotland'	Sweden		4	'Sjoormen'	Singapore
	5	'Yuushio'	Japan		2	'Tang'	Turkey
TOTAL	55				4	'Whiskey'	Albania (4*)
	1	'AKULA I/II'	RFAS	TOTAL	67		
	3	'DELTA IV'	RFAS		2	'B FRANKLIN'	USA
	2	'HAN'	China		2	'DELTA I'	RFAS
	14	'LOS ANGELES'	USA		1	'GOLF'	China
	6	'OHIO'	USA		5	'STURGEON'	USA
	3	'OSCAR II'	RFAS		1	'Yankee Notch'	RFAS
	3	'RUBIS'	France	TOTAL	11		
	1	'SIERRA I'	RFAS				
	3	'TRAFALGAR'	UK				
	3	'TYPHOON'	RFAS				
	6	'VICTOR III'	RFAS				
	1	'XIA'	China				
TOTAL	47						

* – Indicates approximate date transferred

NOTE: Submarines listed in CAPITALS are nuclear powered. The age of many Chinese and Russian boats cannot be determined with any certainty and ages are approximate. Some ex-Russian submarines such as the 'Foxtrot' and so on, may be older than indicated, the age given being that in service with the navy concerned.

UPDATED

SUBMARINE FORCES

ALBANIA
In mid-1996 the three 'Whiskey V' class boats were all based alongside with the fourth boat being used for harbour training. These boats are now considered obsolete and probably incapable of diving.

ALGERIA
The two Type 677E 'Kilo' class SSKs underwent refits in Russia between 1993 and 1996.

ARGENTINA
Since deciding to expand her submarine arm with TR 1700 German-designed boats, four of which were to be built locally in addition to two built in Germany severe financial problems have forced Argentina to cease construction of the submarines. The third and fourth units were in various stages of construction and work on the fifth had begun. It has been reported that the third and fourth boats may be scrapped while material assembled for the fifth and sixth units may be used to maintain the first two boats. Type 209 boat *Salta* has completed her mid-life modernisation, but work on the *San Luis* was suspended and the boat is no longer operational. Upgrading included installing new machinery and weapons and renewing electrical installations.

AUSTRALIA
Construction of the Swedish-designed Type 471 'Collins' class is now well under way and four boats have been commissioned. The navy has begun to establish plans for the enhancement of its submarine capability. To be undertaken in phases the enhancements will include: the capability to operate Special Forces by the 'Collins'; a mid-life update for the 'Collins' and the possibility of acquiring a new submarine class to enter service between 2010-12. Proposals to upgrade the 'Collins' combat system have already been approved, while the possibility of adding an AIP system will be given detailed consideration and for this purpose Stirling 4V-275R AIP engines have been acquired for shore trials.

BRAZIL
The fourth and last Type 209/1400 'Tupi' class has been launched at the Arsenal de Marinha do Rio de Janeiro. Effort is now being concentrated on the improved 'Tupi' (S-NAC-1) boats, one of which are on order. This design will form a bridge between the 'Tupi' class and the succeeding nuclear (S-NAC-2) design. Work on this design is progressing and considerable financial resources have been poured into the project, although it has now been announced that construction will not start until 2010. A prototype nuclear propulsion plant (IPEN-MB-1) has been built for installation in a hull of about 2,800 tonnes. The two 'Oberon' class boats have been upgraded with the STN ATLAS Elektronik CSU 90 sonar system.

BULGARIA
With the ending of the Cold War, the future of the submarine force with two obsolete 'Romeo's' must be in doubt since a projected order for two 'Kilo' class submarines was subsequently cancelled.

CANADA
The Canadian government has finally announced that, subject to final negotiations, it will lease (with an option to purchase) the four British 'Upholder' class SSKs for a period of eight years. VSEL will reactivate the boats and provide Canada with logistic and technical support together with training. The boats will be adapted to meet Canadian requirements which include modification to fire Mk 48 Mod 4 torpedoes; operate the Canadian new towed array sonar; installing a new communications suite and transferring the Lockheed Martin Librascope fire-control system from the 'Oberons'. The first boat is expected to become operational about the end of 2000. Long-term plans may include retrofitting an AIP fuel-cell system that is currently under development by Ballard Power Systems.

Farncomb, the second 'Collins' class boat for the Royal Australian Navy. This programme is now progressing towards the half way stage (Courtesy of Royal Australian Navy) **1999**/0024653

CHILE
In December 1997 the navy ordered from DCN of France and Empresa Nacional Bazan of Spain two 'Scorpene' class submarines to replace the 'Oberons'. The boats will be delivered to the Chilean Navy early next century. The pressure hulls will be built by DCN at Cherbourg with the forward sections being assembled and fitted out at Cherbourg and the aft section being assembled and fitted out at Cartagena. Final outfitting of the lead boat, for which the first steel was cut during 1998, will be carried out by Cherbourg and of the second at Cartagena. The two 'Oberon' boats have been refitted with the STN ATLAS Elektronik CSU 90 system that will be transferred to the Type 209s when the 'Oberons' pay off.

CHINA
The PLA Navy is expanding its strategic nuclear capability and a new SSBN design, Type 094, is being developed to deploy a new longer-range strategic missile, the CSS-NX-4. It is anticipated that construction of the first of class would commence in 1997 with an in-service date of approximately 2005 onwards. In addition to the new SSBN, the PLA Navy is also working on a new SSN design, Type 093. Construction of this design, which is to succeed the 'Han' class, has a higher priority than the Type 094, and commenced in late 1994 with completion due in 2001. Construction of SSKs continues with the Type 039 'Song' class, the first of which was completed in 1995. This boat is now undergoing a second series of intense trials, including submerged launches of a new anti-ship cruise missile, the YJ 8-2, preparatory to entering service. A second boat has been completed and a further two 'Song' class are under construction. Construction of the 'Ming' class SSKs is continuing. Three 'Kilos' are operational and a fourth was delivered in the autumn of 1998. Further acquisition of 'Kilos' will possibly depend on entry into service of numbers of the new 'Song' class that is in series construction. Some 62 Chinese-built Soviet-derived 'Romeo' class boats remain operational or in reserve and China has supplied a number of these boats to foreign navies. Electronics and warfighting capabilities are limited, especially where Chinese technology has been used. The US Office of Naval Intelligence (ONI) has stated that China 'will pose the most complex submarine challenge outside Russia as a result of its commitment to increased training, the steadily expanding scope and complexity of its exercises, and an active acquisition programme targeted at modern technology'. Another ONI report *World Submarine Challenge* notes that 'Submarines will have an increasing role in China's deterrence strategy'.

COLOMBIA
Plans exist for the acquisition of two more Type 209 boats. So far, funding problems have prevented implementation of any further acquisitions. The navy also operates two midget submarines.

DENMARK
The two 'Narhvalen' class boats are being modernised with SAGEM non-penetrating optronic sensor masts, new periscopes, Racal Sea Lion ESM and improved radar and sonar. The three 'Kobben' boats were fitted with new sonar in 1992-93. All five boats will be withdrawn from service between 2001-2006. A government committee examining the restructuring of the Danish Navy has recommended that they be replaced with four boats fitted with AIP systems and capable of operating outside Scandinavian and arctic waters. To meet a future requirement the navy has commenced PD studies for a new class.

ECUADOR
The two Type 209/1300 boats were scheduled to undergo a major refit, their second, in 1992-93, but work has not yet begun.

EGYPT
In a government to government deal Egypt will obtain funding under the FMF scheme for the acquisition and refit of two ex-Netherlands 'Zwaardvis' class submarines, with options to subsequently purchase two new-build ROM Moray 1400 boats. The 'Zwaardvis' class boats will be modernised by ingalls in the USA; refitting the Lockheed Martin SUBICS 900 command and control system. Updating of the four Chinese 'Romeo's' has been completed and the boats are now capable of firing Sub-Harpoon missiles and Mk 37 torpedoes. Other new equipment installed includes Lockheed Martin active sonar, STN ATLAS Elektronik CSU 83 passive sonar, a new fire-control system, and ARGOSystems Phoenix AR-700 ESM system.

FRANCE
The SSBN programme has been reduced from six 'Le Triomphant' class boats to four, all of which will eventually deploy the new M51 missile that is planned to enter service in 2010. The order for the fourth unit has been delayed from 2000 until 2003. A programme exists to define the requirements for a new generation of SSBN. Work has commenced on the design of a new SSN (the SMAF - Sous-Marin d'Attaque Future) of about 6,000 tonnes that will be fitted with a vertical launch anti-ship missile. Only one 'Agosta' class boat will remain operational after 1999.

U 16, a Type 206A submarine of the German Navy (Courtesy HDW)   **1998**/0007206

GERMANY

The 12-boat Type 212 programme has been reduced to four boats. Further boats will not be funded until after 2005. Steel for the first boat was cut in July 1998. The first of class is scheduled to enter service in 2003. The type 206A boats will remain in service longer than originally planned. Type 205 and unmodernised Type 206 boats are to be decommissioned and the latter are being sold off.

GREECE

Modernisation of the four German-built 'Glavkos' Type 209/1100 boats is now almost finished under Project Neptune. The update includes STN ATLAS Elektronik flank array sonar, a Kanaris fire-control system, new ARGOSystems ESM and the capability to fire Sub-Harpoon. The class will remain effective for another 20 years. There is a requirement for a new class of submarines and funds have been made available to acquire two more SSKs.

INDIA

Current projections indicate that India will have nine submarines in service in 2010 and only three in 2015 unless more submarines are acquired either from overseas purchases or from indigenous construction. Work on the design of an SSN (known as the ATV - Advanced Technology Vessel) based on the Russian 'Charlie I' class, is well under way with the keel due to be laid in 2001/2 and launched in 2006/7. The design will be based on a Russian SSN, and fitted with a Russian/Indian-designed nuclear reactor which will shortly undergo shore trials. The SSN will deploy the Sagarika cruise missile currently under development by the Defence Research and Development Organisation. Five boats are planned. The Navy also plans to build, in Mazagon Docks Ltd, two more submarines of a type yet to be decided. The Defence Research and Development Organisation is also developing a submarine-launched ballistic missile (Dhanush), whiich is still in the conceptual design stage.

INDONESIA

Although funding had been approved for the purchase of five Type 206 submarines from Germany, the financial crisis in the region that severely affected Indonesia has led to the cancellation of the purchase. The boats had already been formally handed over and renamed towards the end of September 1997. Subsequently, only one boat, the *Nagabanda*, has been accepted into service, the Navy postponing the takeover of the remaining five Type 206 boats.

IRAN

Iran has taken delivery of three Russian-built 'Kilo' class SSKs. The Iranians complained of problems with the 'Kilos' and persistent trouble has been experienced with the batteries. Exercises involving the 'Kilo' class boats conducted in the Persian Gulf in the spring of 1998 indicate that the problems have now been overcome. During the exercises live ammunition was fired at mock targets.

ISRAEL

First of class *Dolphin* has been undergoing further sea trials, while the second of the three 'Dolphin' class submarines, *Leviathan*, has begun initial sea trials. An

option on a third boat was taken up in 1994. The 'Gal' class is also being modernised with new sensors and a fire-control system.

ITALY

Following a technological and project MoU signed on 22 April 1996 between Italy and Germany, the Italian Navy ordered two new boats to be built to the German Type 212A design and fitted with an AIP system. The first boat may be laid down towards the end of 1998. The improved 'Sauro' class submarines *Primo Longobardo* and *Gianfranco Gassana Priaroggia* are scheduled to be modernised with the German STN ATLAS Elektronik ISUS 90 integrated combat system. The old submarine *Toti* paid off for the last time on 16 August, 1997 having been in service for 30 years. She has gone into reserve for disarming. The Italian Navy is also studying the possibility of defining a requirement to replace these boats with a small design of around 700 tonnes.

JAPAN

Construction of the latest 'Oyashio' class of four boats will continue for a number of years to maintain a fleet strength of between 12 and 14 boats.

KOREA SOUTH

Seven Type 209/1200 boats have been completed and are in service. Plans are being developed for a follow-on class Type 1400 possibly incorporating auxiliary closed-cycle AIP propulsion and capable of launching Sub Harpoon. The design will aim to match the performance of the latest Japanese submarines. About nine midget submarines are also operational.

KOREA NORTH

Building of 'Romeo' class boats has ceased at 22 units with construction being switched to the 250 tonne 'Sang-O' class, two of which have now been seized off South Korea. The navy also has about 45 midget submarines as well as 20 of the 'Sang-O class', construction of which is continuing.

MALAYSIA

Plans to acquire four SSKs sometime during the 1990s have been shelved in favour of the OPV programme. However, the navy still has a requirement for a submarine force of two new boats and two refurbished boats. Personnel are undergoing training in Australia, India and Pakistan.

NETHERLANDS

The 'Moray' project developed by RDM continues to develop using limited funding provided by the government with the proviso that companies involved in developing AIP technology are involved in the design. This will enable the Dutch to retain essential submarine expertise in design and construction and allow RDM to continue seeking markets in the export field. The submarine *Zeehond* was transferred to RDM in 1990 for use as a trials vessel for AIP related to the development of the 'Moray' design. The two 'Zwaardvis' class SSKs have been put up for sale and may be purchased by Egypt. Although over 20 years old the boats have been very well maintained.

NORWAY

Work is progressing through the early design stages to define the requirements for a new replacement SSK, but there is some uncertainty over the question of funding. Norway is collaborating with Denmark and Sweden on the Viking Project. A study has been prepared to compare the costs of two alternative solutions: a Norwegian proposal, which envisages a total of 10 submarines, and a Danish/Swedish solution, which proposes six submarines.

PAKISTAN

Work has begun in France on the construction of the first of three new 'Agosta 90B' class boats, the first boat being floated out of dock to complete fitting out in the summer of 1998. Trials were due to commence around the end of 1998 and the beginning of 1999. The second boat will be assembled at Karachi under a technology transfer arrangement, while the third will be built entirely in Pakistan, being fitted with a MESMA AIP system. The two earlier boats will then be subsequently retrofitted with the AIP system. These boats are fitted to fire anti-ship missiles. Four 'Daphne' class submarines and three midgets are also operational.

PORTUGAL

Replacement of the three 'Daphnes' is now considered to be an urgent priority, and an RfP for 3/4 boats was issued with five proposals due on 3 June 1998. The designs being offered are 'Scorpene' (DCNI and Bazan), Type 209/1400 (GSC), 'Sauro' derivative (Fincantieri), 'Gotland' derivative (Kockums), and 'Moray' type (RDM). The Portuguese require that the boats be capable of operating in Atlantic waters and must be able to deploy Sub Harpoon. An option for an AIP system is also specified. A shortlist of contenders was due to be announced by the autumn of 1998 with an order being placed in 1999. The first boat is scheduled to enter service in 2004.

RUSSIA

The strength of the strategic missile fleet will be severely pruned, leaving a force of about 25 boats of

The Pakistan Navy submarine Hashmat (pictured in 1990) will shortly be joined by the first of three 'Agosta 90B' type boats (Anchor Consultancy Photograph Library)   **1999**/0024652

Large numbers of Russian 'Foxtrot' (left) and 'Whiskey' (right) class submarines have been scrapped since the ending of the Cold War (H&L van Ginderen) **1996**

Construction of Russian 'Kilo' class SSKs for domestic use has ceased, but the design is now being exported in increasing numbers (H&L van Ginderen) **1996**

the 'Delta IV' (Project 667 BDRM) and 'Typhoon' (Project 941) classes in service. However, a new ballistic missile, the SS-N-2B, is under development and a new class of SSBN, the 'Bory' (Project 955) is being developed to deploy the new missile. The six 'Typhoon' class boats are being slowly modernised to deploy the SS-N-2B. Construction of 'Oscar II' (Project 949A) class missile submarines continues. Construction of Improved 'Akula' class (Project 971) SSNs in the White Sea yards also continues. A new-generation SSN, the 'Severodvinsk' (Project 885), designed to succeed the 'Akula' class was laid down at Severodvinsk in December 1993. Building of the new class is being severely hampered by funding problems. The submarines are larger than the 'Akula' enabling them to incorporate new stealth measures. All non-nuclear boats are rapidly being withdrawn from service and by the end of the decade only some 'Kilo' class boats, together with the new 'Lada' (Project 1450) will remain operational. The first sections of the new Project 1450 submarine (St Petersburg) have already been laid down at the Admiralty (Sudomekh) yard at St Petersburg. The first sections of a second boat of the 'Amur' (Project 1650, the export variant of the 'Lada') have also been laid down for an overseas customer. It is estimated that at the start of the next millennium the RFAS ORBAT will include 19 SSBNs, 14 SSGNs, 45 SSNs and 42 SSKs.

SINGAPORE
Under a Joint Technology Development Fund agreed between Sweden and Singapore. The Singapore Minister for Defence announced on 23 September 1995 that the navy was to acquire the submarine Sjobjornen for training. The boat initially renamed Riken and sebsequently Challenger, was delivered at the end of 1997 and refurbishment completed in February 1998. The government has also acquired the four other boats of the Swedish 'Sjoormen' class. Three of the boats will be put into service with the fourth being used for spares.

SOUTH AFRICA
Plans exist to acquire a force of three submarines to replace the 'Daphnes' and feasibility studies have been undertaken. An order has been placed with the German Submarine Consortium (GSC) for three Type 209 boats. The 'Daphne' class has been modernised to extend its operational life to around 2005. Included in the update are two multifunction sonar operator consoles, providing a new long-range passive sonar display, a new triple transducer intercept sonar, new triple transducer array passive ranging sonar and new sonar classifier system, with library: two multifunction interchangeable displays for time-bearing and tactical plots, new electronic charting navigation sub-system, new ESM sensor head and console, an E/O search periscope upgrade with split image range finding and low light and day video recording: attack periscope upgrade. Much of the work has been undertaken by UEC.

SPAIN
Plans are in hand to build four new boats to replace the

obsolete 'Daphne' class. Construction is due to begin at the end of the decade and the design may be based on the 'Scorpene' being developed in collaboration with France. The 'Agosta' class boats are undergoing a major upgrading with new ESM, towed array sonar, periscopes with IR imaging and improved fire-control system. The fire-control system of the 'Daphnes' is being modernised to enable them to launch F17 torpedoes.

SWEDEN
Under the Resolution on Defence of 1996 the Swedish submarine fleet will be reduced from 12 to 9 units over the next 10 years. Project definition studies on a new design being developed under the 'Viking' Project in collaboration with Norway and Denmark, and dubbed Submarine 2000, are due to complete in mid-1999. It is anticipated that orders for this new class will not be placed until early next century with an in-service date beyond 2005. The aim is to produce a modularised design so that each of the participating members can adapt it to meet their own requirements. The three boats of the 'Nacken' class have recently been modernised with new sonars and fire-control systems, enabling them to remain in service until around 2008. Under a long-term plan, studies are in hand to develop a submarine launched variant of the RBS-15 missile which might begin to enter service aboard the 'Vastergotland' class from 2005 onwards.

TAIWAN
Taiwan remains determined to acquire more SSKs and up to 12 boats are projected. The navy has now stated that it will build new boats locally. This would probably have to involve some form of technology transfer as no national shipbuilders have a submarine capability. The navy has said that the design will be based on the 'Hai Lung' design already in service.

THAILAND
Although previous plans to acquire submarines have been cancelled, there is still a requirement for three to four submarines. Seven solutions were proposed to meet the previous plans: DCN International, HDW (Type 209), Kockums ('Gotland'), ROM ('Moray'), and Russia ('Kilo').

TURKEY
Three 'Preveze' class boats are in service and one more is under construction. Four of the six boats of the 'Atilay' class (Type 209/1200) are also to be modernised up to 'Preveze' standard. Germany and Turkey have signed an agreement for the construction of four more 'Preveze' class boats, with construction due to commence in 1999. In addition six more submarines are projected to replace the obsolete ex-American boats.

UK
The publication of the Labour government's Strategic Defence Review reveals that the submarine force will be reduced from 12 to 10 SSNs, the Splendid and Spartan being decommissioned early. All remaining SSNs will be adapted to fire the Tomahawk cruise missile. Construction of the 'Vanguard' class SSBN is continuing. Seven 'Trafalgar' class boats are now operational and the first three boats of the succeeding 'Astute' class have been ordered. These will incorporate the more modern PWR2 reactor (the same as in the 'Vanguard' class), the SMCS command system, Type 2076 integrated sonar system, optronics

All seven UK 'Trafalgar' class SSNs are now operational. Turbulent is illustrated. Plans are now in hand for the Batch 2 design (Courtesy of VSEL) **1996**

sensors and weapon system improvements. The 'Trafalgar' class is being modernised with SMCS, Type 2076 sonar and the capability to fire Tomahawk cruise missiles. 'Swiftsure' class boats will also be upgraded to the same degree under the S & T Update. *Splendid* deployed the first full warload of Tomahawk missiles for her class in 1998, sailing from Rosyth on 1 May 1998. Early SSNs are being retired and stored pending a decision on how to dispose of their nuclear reactors and prefeasibility studies are in hand for a replacement boat for the 'Trafalgar' class. Referred to as the Future Attack Submarine (FASM), the boats would begin to enter service from about 2015.

UKRAINE

It was anticipated that the Ukraine would receive three obsolete ex-Russian Navy 'Foxtrots' in 1997, but to date this has not been confirmed.

USA

The ending of the Cold War has led to a major reappraisal of the USNs submarine requirement. Under the FY97 defence budget, funds have been approved to complete SSN-23, the third 'Seawolf'. A new follow-on class of SSN referred to as the 'New Attack Submarine' (NSSN) class is under design and authorisation has been given to begin construction on the first two units. Construction of the lead ship is due to begin in 1998 and the second in 1999. The design will incorporate some 75 per cent of the 'Seawolf' capability. A revolutionary approach to the design will offer a variety of functions on a common hull, propulsion and command system. A replaceable module will offer additional cruise missiles, ballistic missiles, or the ability to transport up to 200 Special Forces troops. The 'Los Angeles' class SSNs are gradually being upgraded with the Raytheon Combat Control System Mk 2 (CCS Mk 2) which is being integrated with the AN/BQQ-5E and TB-29 sonars and as it is fitted, the AN/BQG-5 Wide Aperture Array (WAA). Of the 18 'Ohio' class SSBNs, four of the earlier boats will be converted to a tactical role, while the four remaining boats deploying the Trident 1 strategic missile will be upgraded to deploy the Trident 11 (D-5) missile.

VENEZUELA

The two Type 209 boats have very recently completed a half-life refit with new engines, fire control, sonar and attack periscopes being fitted. The hull has been slightly lengthened to accommodate the new equipment.

UPDATED

ASPECTS OF SUBMARINE DESIGN

15

ASPECTS OF SUBMARINE DESIGN

The design of a submarine can generally be considered to comprise three main areas: the propulsion system; combat information area; and the weapons handling area - all encompassed within the hull.

Integrated into these areas are crew accommodation, and messing facilities, ballast tanks, fuel compartments and so on. The arrangement of these areas falls into a distinct pattern: the forward end houses the weapons systems and ballast/trim tanks; the central, slightly forward portion accommodates the combat information area; while the propulsion system forms the central to after part, together with other ballast/trim tanks. The arrangement in strategic nuclear submarines differs slightly with the strategic missile stowage usually sited forward of the machinery section and abaft the combat information area. The torpedo weapons handling area remains in the forward end which also encompasses the combat information area and crew accommodation.

Operational requirements
Size has always been a major factor in submarine design. On the whole submariners prefer a boat to be as small as possible, in spite of the fact that this results in extremely cramped accommodation. A small boat presents much less of a target to hostile forces, as well as being more manoeuvrable, than a large one. However, size is to a very large degree dictated by the space requirements of propulsion, combat information and sensors and weapons systems. The submarine's role also affects its size. Diesel-electric submarines (SSKs) which are intended for long-range patrols require large fuel reserves, large weapons stowage space and larger crews to provide a full watch-keeping system.

Such limitations do not affect nuclear submarines (SSNs) to the same degree, for they are generally much larger than SSKs, and there is adequate space for accommodation and even recreation to ensure acceptable crew habitability. SSKs designed for short patrols and coastal duties, on the other hand, can be much smaller. Large fuel reserves are not required, a weapon reload capability may not be essential or could be much smaller, and a much smaller crew can be accommodated as a simplified watch-keeping system can be adopted. All this is based on the fact that the coastal boat will frequently return to its base or mobile mother ship for replenishment of disposable stores, allowing the crew the chance to relax ashore. They may not even be expected to sleep on board, merely embarking to undertake the mission.

Currently the trend is towards larger SSKs. This has, to some extent, been dictated by the increasing sophistication of weapons and combat systems and the perceived need for increased weapons stores. Of greater significance, however, has been the demand by a number of navies for SSKs with ocean-going patrol capabilities, rather than boats designed more for coastal patrol missions. On the other hand, modern technology has enabled systems to be considerably reduced in size and weight, together with savings in manpower resulting from the greater use of automation. However, at the forefront of all considerations must remain the need to ensure the safety of the crew and the boat. The full impact of Air Independent Propulsion (AIP) systems on SSK design has yet to be assimilated. This may well have an effect on the future size of SSKs, as the systems become more reliable and possibly, in the future, completely replace diesel engines. In addition the development of AIP will offer the SSK increased deep submergence endurance, a capability which it currently lacks, and in which the SSN has proved to be superior.

Hull design
The two most significant features of a submarine's design are its ability to withstand pressure and shock. These two factors determine to what depths the boat can operate with safety and, more importantly, what is the safety factor or depth below the normal operating depth to which it can sink before the hull implodes or is crushed. This can be some considerable depth below the normal operating depth (usually considered to be in the region of 1.5 to 2 times normal operating depth), and is relied on in emergency to evade hostile attack.

In recent years improvements in calculating various stress factors and new construction techniques have removed many areas of uncertainty relating to the behaviour of bodies underwater, and there is now a tendency to accept the lower safety factor. The shape of a submarine also affects its safety factor. For example, in long submarines, changes in trim may lead to the bow or stern being at a much greater depth than the rest of the submarine and so the safety factor should also take into account the overall length of the submarine. Underwater explosions can have a peculiar effect on submarine hulls, creating sudden and wide-ranging increases and decreases in pressure. The hull must be built, therefore, so that it is able to flex itself to absorb these violent pressure changes without fracturing and without weld joints cracking and so on.

To avoid being damaged, equipment must be carried on special shock absorbing mountings and individual components must exhibit a very low failure rate. To further ensure safe operation, equipment must incorporate a high degree of redundancy. Hence, wherever possible, vital systems are duplicated.

The shock and pressure factors depend not only on the shape of the hull, but also on the material used in its construction. It is now common practice to use high-tensile steel for hull construction and highly specialised welding techniques are used to ensure that joints maintain the same strength as the rest of the hull. To ensure that welding has been correctly carried out, it is necessary to ultrasonically check all pressure hull welds and preferably to X-ray them as well. While steel is always used for the pressure hull and free-flooding end section, GRP or aluminium is invariably used for the casings and the sail.

As well as having the ability to withstand sudden changes in pressure and violent shockwaves, the submarine must also be able to manoeuvre adeptly at both high and low speeds. To achieve this a perfect hydrodynamic shape is required. Because the modern submarine is not expected to have to transit for any distance on the surface, the familiar 'teardrop' design has been developed to exploit to the full the submerged performance and handling characteristics.

Manoeuvrability is further enhanced by the cruciform arrangement of rudders and hydroplanes at the stern. However, some designs favour an X-form arrangement for the control planes. To achieve optimum manoeuvrability demands a low length/beam ratio in the region of 6 to 8. This enables a higher speed at a given displacement/power ratio to be achieved; decreases the surface area, so reducing the sonar target strength for a given displacement; and allows greater flexibility in the use of available space.

The ability of a submarine to carry out high-speed manoeuvres in both horizontal and vertical planes is governed not just by its maximum speed, but by overall performance levels in all areas. To carry out a crash

The modern submarine is fitted with a wide range of masts to carry a variety of antennas relating to various functions. Reading from right to left these are: radar, attack periscope and communications mast alongside each other, EW mast, communications mast, search periscope and snorkel induction masts (Courtesy Racal Radar Defence Systems) **1998**/0007207

dive at high speed, a manoeuvre possibly required to avoid a torpedo attack, demands complete freedom of movement in the vertical plane, which in turn requires a greater permissible diving depth.

In addition, by achieving greater diving depth a submarine is enabled to make greater use of thermal layers that hinder hostile sonar operations. Conversely, by using a deep-lying sound channel the submarine can use its sonars to detect hostile forces at much greater ranges. The ability to operate at greater depths also offers the submarine greater protection against attack as it is more difficult to detect and weapons may not be so effective at great depths.

Minimum target signature

Apart from the need to ensure that the submarine can withstand pressure and shock, the third major feature to be considered by the submarine designer is stealth. In order to avoid being detected and to enhance its own ability to detect targets, the submarine must be as stealthy as possible in all respects. Signature management is therefore now a primary consideration in submarine design, and stringent measures are necessary in order to minimise any unwanted signatures. Not only must the submarine be as acoustically silent as possible but all non-acoustic signatures must also be reduced to a practicable minimum. This is necessary for Anti-Submarine Warfare (ASW) forces will use every means at their disposal in the fight against the submarine, and this includes a variety of non-acoustic methods of detection.

Included among the latter are the following features:
- IR detection of heat emitted by the submarine and also detection of any change in surface temperature caused by the passage of a submarine and its resultant wake, and the detection of the disturbed nature of the surface itself. Conventional propulsion systems emit hot exhaust gases and hot cooling water whilst snorkelling. These can be reduced by feeding hot cooling water into a ballast tank and then feeding it out through the wake of the propeller, or cooling the gases by water injection and expelling the mix just beneath the surface. However, both methods still leave a trace of warm water in the wake of the boat surrounded by the cooler sea water that remains for some time after the boat has passed and which can be detected.
- Radar to detect any part of the submarine which pierces the surface (any masts and so on) and which may create a radar return and also to detect surface disturbance created by a submarine operating near the surface. Radar cross section on masts is reduced by the use of special coatings. Careful and judicious use of masts, however, is one of the surest ways of reducing radar cross section.
- Chemical detection to sense the presence of any exhaust fumes from a snorkel, or any other fumes from waste or changes in the chemical composition of the sea in the vicinity where a submarine has passed.
- Magnetic Anomaly Detector (MAD) to detect any changes in the earth's magnetic field caused by the presence or passage of a submarine. This can be reduced by the use of non-magnetic steel in construction and by the installation of a degaussing system.
- Changes in the electrical potential in the vicinity of a submarine that is creating an ELF signature from its machinery and hull.
- Finally, the means to detect changes in bioluminescence in the water caused by the passage of a submarine.

To overcome the capability of the next generation of ASW sonars (low-frequency active down to 1 kHz) every effort is being devoted to reducing low-frequency narrowband noise radiation. This frequency band allows sensors to detect targets at extremely long ranges, especially if they are nuclear-powered boats or are SSKs recharging batteries using a snorkel. To overcome this problem the machinery producing discrete frequencies or tonals has to be acoustically treated or alternative equipments selected.

However, a submarine cannot be completely 'silent' for even when 'shut down' some machinery must still remain in operation. Minimal power must be applied to the propellers, or the boat will gradually sink and other

systems too must remain operative, for example, hydroplanes and rudders, the hydraulic system, trim and compensating tanks and air conditioning plant. These systems produce what is termed 'pulse' noises (or transients) which can be detected using low-frequency sonar arrays with special processing and memory comparison libraries.

The bow area is one of the most significant areas of hull design. Not only does it affect the pressure and sound signatures, but it is also a significant design factor that can affect the boat's ability to acoustically detect. The foremost bow section houses the sonar, and to achieve maximum detection capability its aperture must be as wide as possible. Furthermore, this area of the hull has to accommodate the weapons discharge and handling equipment and trim tanks. To incorporate all this equipment into a highly sensitive hydrodynamic and acoustic area poses considerable constraints on the designer.

Likewise, the after body of the submarine housing the shafting and propeller has to be very carefully designed to reduce cavitation and self-generated noise to an absolute minimum.

The overall aim of the designer, therefore, is to obtain the best possible flow of water from the bow, along the hull, to the propeller and to ensure maximum efficiency of operation and optimum manoeuvrability.

In the United States new technologies are being investigated for the next generation of SSNs. At the Naval Undersea Warfare Center experiments have shown that the effects of the boundary layer between water and the hull can be manipulated. Because this effect follows a distinct pattern, scientists would be able to characterise and manipulate the water/hull interface to provide significant benefits in the form of noise reduction and reducing drag and so improve both fuel efficiency, hydrodynamic manoeuvrability and stealth. Control can be exercised through the addition of a special external skin consisting of a large number of magnets and electrodes arranged orthogonally and energised to create electromagnetic forces which can control the turbulence. This technology, known as Electro-Magnetic Turbulence signature Control (EMTC), has shown that drag can be reduced by as much as 45 per cent.

By siting all propulsion and machinery aft, the designer achieves a measurable improvement in noise reduction and considerably improves the boat's ability to use its acoustic sensors. The maximum amount of noise reduction from the propulsion system is achieved by connecting the electric motor directly to the shaft. The noise factor is further reduced by the almost universal adoption of the highly skewed seven-bladed propeller revolving at very low revolutions per minute.

To reduce the target echo strength and noise radiated by a submarine, hulls and spaces where noise is created are treated with tiling and cladding. In the latest generation of submarines this acoustic cladding is also being fitted internally as well as externally. Two types of stealth control are used – anechoic tiles to reduce the target echo strength (that is the backscatter return from hostile active sonars) and decoupling materials that create an impedance mismatch between the boat's hull and the surrounding water. This reduces self-radiated noise and any energy transmitted as noise from the hull to the near-field region of the flank array sonar, which degrades the sonar's performance and reduces the range at which the submarine can detect a target.

This treatment is frequently composed of compressible layers of visco-elastic and other materials which undergo changes in thickness as the submarine changes its depth in the water. As a result, the acoustic performance of the treatment may vary. This compression can, at great depths (between about 300 and 500 m) substantially degrade the performance of baffles in specific areas on the hull and lead to an overall degradation in the platform's stealth with regard to radiated noise and flank array sonar self-noise.

Such is the importance attached to achieving minimum sonar target strength that it is now essential for submarines also to be equipped with some form of noise monitoring and diagnostic system, and these are being increasingly retrofitted to operational submarines, as well as being incorporated in new designs.

To achieve a satisfactory measure of signature control with regard to noise requires a careful balance

in deciding where to place the two types of treatment. In some cases, for example around the machinery compartment and weapon discharge areas, it will be necessary to apply both types of material in a hybrid arrangement.

Also under investigation in the US by the Defense Advanced Research Projects Agency (DARPA), the Office of Naval Research (ONR) and the Space and Naval Warfare Systems Command (SPARWAR) are a new generation of submarine antennas equipped with multifunction conformal arrays. These will also improve stealth and offer increased performance. These new arrays are needed to provide increased data rate submarine communications in the SHF band (phased array antennas incorporated in a conformal or folding configuration). They will also offer improved intelligence data and millimetric wave imaging to provide the command team with an increasing availability of intelligence relating to the strategic and tactical scenario. Furthermore conformal sonar arrays would drastically reduce the space currently required by the current generation of large, heavy spherical bow arrays. This would enable the bow section of the submarine to be completely redesigned further improving hydrodynamic and stealth capability as well as improving acoustic sensor capability.

It is also being claimed in some sectors in the US that to improve still further on the stealth capabilities of 'Seawolf' and NSSN, the US Navy will have to give consideration to electric propulsion. This would be in the form of permanent electric magnet motors and advanced high-power electronics.

To overcome the problems briefly outlined here it is necessary for the submarine to be designed as a single unit - platform, machinery and sonar sensor suite.

Self-defence

An area that is now receiving increasing attention in the submarine community is self-defence. In the face of increasingly sophisticated anti-submarine weapons launched from the air, surface and subsurface, submarines are having to employ increasingly sophisticated methods to evade attack. Self-defence against both torpedoes and helicopters is of primary concern to submariners and systems that provide detection and countermeasures will become increasingly important. Already systems designed to attack helicopters, while a submarine remains submerged, are under development.

Habitability and crew safety

Because of the nature of the submarine, its role, and the design constraints imposed by the medium in which it operates, crew accommodation has always been very cramped. However, accommodation should be to the highest possible standards commensurate with the boat's safety and mission profile in order to ensure that the crew remains fit for action both physically and psychologically.

Undoubtedly SSNs offer much higher standards of habitability than SSKs, but conditions in both types of submarine have been considerably improved in recent years. One of the major benefits has been the consolidation of modern technology to improve life support systems.

Boats should be fitted with efficient air conditioning systems and air purification equipment, including oxygen generating plant and extractors to remove contaminants from the air. Future systems may apply biotechnology with plant cells absorbing carbon dioxide and releasing oxygen.

In larger submarines it is considered better for officers and crew to have separate messing and washing facilities, and cabins and bunks should be sited as far away as possible from noise and working areas so that the crew can get the maximum amount of rest during off-duty periods.

Safety in a submarine is of prime importance and in general safety standards applied to submarines are considerably better than those of surface ships. The one safety factor most vital to the submariner is that of adequate escape facilities. In the past crews have had to rely on the standard depth-limited free ascent method which, in a modern submarine, may not be at all practicable to use. In addition, there are considerable risks attached to this method of escape. It places a severe physical strain on the body that, in the case of an injured man, may well prove fatal.

U214

Proven technology in advance of our age

Today fuel cell (FC) systems are by common consent admitted to be the ideal solution for air-independent propulsion for conventional submarines. The new submarine class 214 which has been developed by HDW is a fully integrated FC submarine. Employment of the noiseless fuel cell propulsion system increases the submarine's submerged cruising range considerably and leads to a reduction of her indiscretion rate, thus making the FC submarine virtually impossible to detect even more so than any other conventional submarine. The submarine class 214 is also available as a conventional diesel-electric submarine which may later be upgraded with a fuel cell plant.

SIEMENS

The Future has already begun:
New Technologies for Tomorrow´s Submarines

Saving energy is important, especially on board submarines. Because reducing energy consumption increases the operating range... while cutting thermal losses. And acting with the submarine of the future in mind, Siemens develops and supplies advanced new technologies for power generation and propulsion systems.

"Direct electrical energy" is the principle behind the fuel cell, which is compact, quiet, and dissipates minimal heat. Fuel cells convert hydrogen and oxygen directly into electrical energy, leaving a residue of pure water. We've already field-tested shipboard fuel cell systems of up to 100 kW. And development work on an appropriate fuel cell module with a solid electrolyte (PEM) has already been successfully completed.

In drive technology, each kilowatt-hour saved is just as important as low-noise operation. Our answer to these challenges is the PERMASYN® motor, a propeller drive some 40% smaller and lighter than conventional DC motors.

The PERMASYN motor and fuel cell together on board submarine class 212 form the optimal combination for a considerably extended submerged endurance along with reduced signatures.

To find out more about our new technologies call or write:

Siemens AG
Industrial Projects and
Technical Services Group
Marine Engineering
P. O. Box 32 40
91050 Erlangen/Germany
Tel. +49/91 31/7-2 71 79
Fax +49/91 31/7-2 32 53
marine.engineering@erl9.
siemens.de
www.siemens.de

Industrial Projects
and Technical Services

your success is our goal

159U450597A4

For escape from depths below 150 m it is now becoming standard practice for submarines to be fitted with special attachments to enable a Deep Submersible Rescue Vehicle (DSRV) to clamp itself on to the escape hatch. This allows the crew to exit from the submarine at great depths and in the dry.

To further improve safety, all except the smallest boats are now fitted with a pressure tight bulkhead that divides the boat into two pressurised compartments. Thus, if one compartment is damaged and flooded, at least the remainder of the crew in the other compartment are enabled to survive and have a reasonable chance of escape using either free ascent or the DSRV.

UPDATED

SUBMARINE DESIGNS

CROATIA

Velebit

Type
Midget submarine.

Description
A number of naval units were captured from Yugoslavia during the civil war and put into service by the newly independent state of Croatia. Included among these was a 'Una' class midget submarine, one of six built during the mid-1980s.

The single hull boats are built of micro-alloy steel with a glass fibre sail. The hull is divided into three compartments with two watertight bulkheads. The forward compartment provides accommodation for the crew of four and six combat divers, as well as command and control for all the boat's systems. The centre compartment houses the lock in/lock out chamber with the main hatch cover and the after compartment houses the propulsion system.

The hull of *Velebit* has been lengthened by 2.3 m to incorporate the 105 kW diesel generator that further extends the boat's capabilities and range. This enables the batteries to be charged at sea, rather than in harbour as with the five other boats in the series which still remain in service with the Yugoslav Navy. The generator powers two newly developed permanent magnet DC electric motors with a combined output of about 80 kW controlled by an electronic converter. These drive a specially designed single 7-bladed skewed low noise propeller giving the boat a maximum speed of 7 kt submerged. The new propulsion system is claimed to develop a much lower noise signature than the previous system.

Platform control is exercised through an X-plane configuration at the stern which is controlled by autopilot and two bow-mounted dive planes. The craft is fitted with a lock in/lock out chamber for divers and can carry four heavy ground mines or four SDVs with limpet mines.

The boat is manned by a crew of six and can carry up to eight combat divers. Space is provided for a demolition team with four berths and storage for their equipment. Other payloads can include four R-1 SDVs and as an option two torpedoes. Alternatively the boat can carry four large demolition charges or 12 light limpet mines which can be stored in the divers lockout. Other options include four heavyweight or six lightweight seabed mines.

A small STN ATLAS Elektronik high-frequency passive/active cylindrical search sonar is fitted. Recent reports indicate that the boat has been fitted with a snorkel induction mast and exhaust. It is reported that the Croatians have developed plans to build further similar types of craft that may be armed with torpedo tubes and four SDVs, although to date there is no indication of new construction taking place.

Operational status
One unit in service.

Specifications
Displacement: n/k
Length: 21.9 m
Beam: 2.7 m
Draught: n/k

Contractor
BrodoSplit, Split.

A Croatian 'Una' class midget submarine　　　　*1997*　　　　**UPDATED**

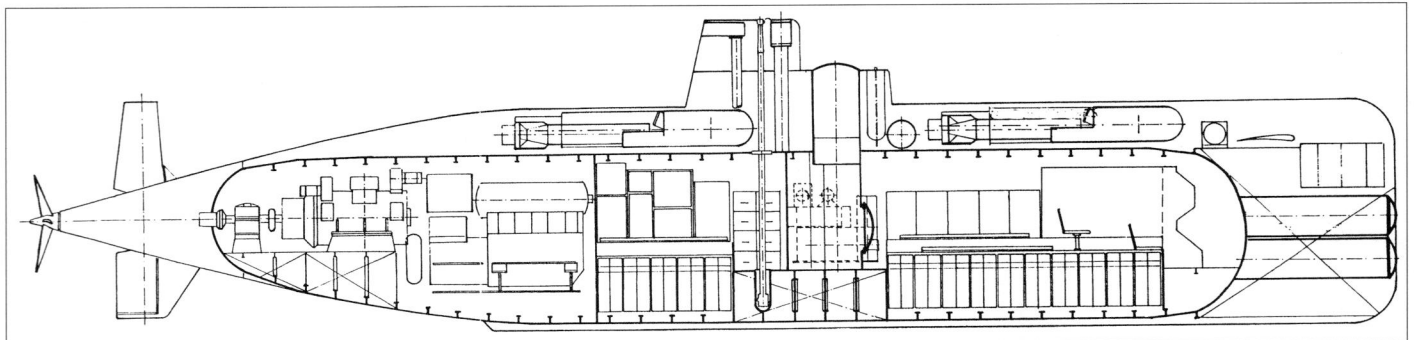

General arrangement layout of Yugoslav Navy 'Una' class midget　　　　*1998*/0007208

FRANCE

'Agosta' class

Type
Diesel electric patrol submarine.

Description
The 'Agosta' features a circular outer hull of 6.8 m maximum, which completely surrounds the pressure hull. At the bow the cross-section of the outer hull narrows to an oval shape, ensuring the minimum underwater hydrodynamic resistance. Considerable efforts have been made to reduce self-generated noise resulting in a clean, streamlined casing and noise damping of all equipment. An array of 36 hydrophones is fitted all round the hull to measure the radiated noise level and provide immediate identification of areas of self-generated noise.

The boats are armed with four bow torpedo tubes fitted with a pneumatic ram discharge system. The tubes allow weapon discharge, irrespective of the speed of the submarine, down to its maximum diving

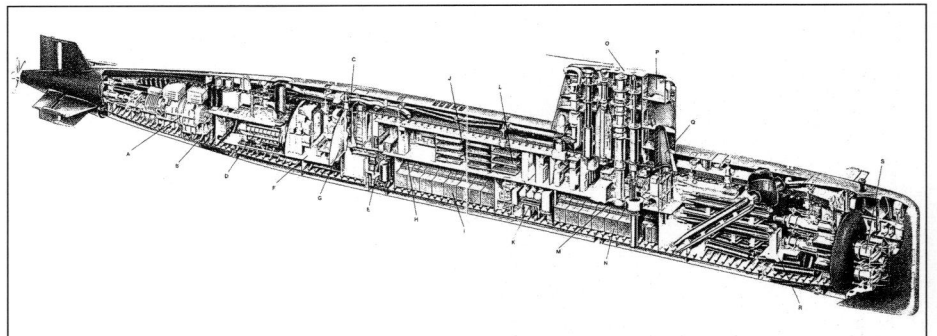

General arrangement of the 'Agosta' class submarine (Courtesy Anchor Consultancy Photo Library)
1999/0024656

depth. The design also incorporates a rapid reload system. Units of the French Navy have been modified to fire the Aerospatiale submarine-launched SM 39 Exocet anti-ship missile. The Pakistan units were modified to fire the Sub-Harpoon missile, but the USA has banned sales of this equipment to Pakistan.

An 'Agosta' class submarine (Courtesy Marine Nationale)   **1998**/0007209

Fire control is exercised through a single, centralised computer using the DLA 2A system. This features an automatic navigational plotting system. Sensors include an 8 kHz active search/attack DUUA 2D sonar, passive ranging and passive towed array sonars. Arur and Arud intercept and warning systems are also fitted, except in the Spanish boats which are fitted with the British Manta system.

The propulsion system comprises two SEMT-Pielstick 16 PA4 V 185 VG diesels developing 2.65 MW driving two Jeumont Schneider 1.7 MW alternators. A single water-cooled 3.4 MW electric motor with double armature directly drives the propeller. There is no intermediate clutch in order to reduce noise level to a minimum. In addition, a 23 kW electric motor is provided for cruise speeds. The two independent banks of batteries of 320 elements use an electrolyte agitation system with water circuit cooling to extend service life and improve efficiency. Two 850 kW generators on flexible mountings provide hotel services, the alternating current being provided by means of five converter sets with two solid-state inverters.

The space between the external and pressure hull accommodates four pairs of ballast tanks. On diving, the air in each pair of tanks is released by a vent valve operated from the central control position. The expulsion of water ballast, likewise operated from this position is effected by means of compressed air. The two bow tanks are also provided with a special blowing system which, in the event of excessive dip, makes possible the rapid expulsion of the water ballast from the bow, even at great depth.

A single operator facing the bows exercises submarine control. The joystick control operates simultaneously the steering, rudder and depth rudder in 'coupled' control. Opposite the operator are various indicators and a synthesiser presenting a symbolic image on the screen enabling the pilot to move the control intuitively.

The steering and diving system functions in either:
Automatic control (steering and diving)
Manual computer assisted control
Manual and emergency control.

Crew accommodation includes a bunk for every crew member each with ventilation and light. The boats are fully air conditioned for operation in the tropics and are equipped with oxygen generating units and independent CO_2 absorption units. Air is conditioned by two installations generating 27,000 frigories/hour, which are used to cool the vessel. During diving, air is regenerated through the chemical burning and absorption of CO_2 granules based on soda-rich lime.

The 'Agosta 90B', building for Pakistan is an improved variant of the basic 'Agosta' class already in service with the Pakistan Navy. The boats are fitted with a UDS SUBTICS 6-console integrated combat system with an associated sonar suite comprising bow, ranging and intercept arrays, an active array and a clip-on towed array. The ESM system is the DR-3000U. A SAGEM Minicin Mk 3 inertial navigation unit will aid navigation. Propulsion is provided by two 850 kW diesel engines or a single 2,200 kW electric propulsion motor. Control of the boat will be carried out with the SAGEM SS Mk 1 control station to control heading and submersion either manually or through the autopilot. The last boat in the series will be fitted with the MESMA AIP system (which will extend the hull length by 9 m). This system will also be retrofitted to the first two boats in the series. These units will also have much improved acoustic quieting and a fully integrated sonar suite including flank, intercept and towed arrays. The hulls are being constructed of HLES 80 steel (equivalent to HY100) that should allow the boats to reach diving depths of 350 m.

Operational status
The 'Agosta' design is operational in service with the French, Pakistan and Spanish navies, and an updated variant, the 'Agosta 90B', has been ordered by the Pakistan Navy to replace its ageing 'Daphne' class. 'Agosta' class units of the French Navy will have paid off by the turn of the century. Spanish Navy units are being modernised with new weapon fire-control system, towed array sonars, new ESM and periscopes with IR imaging capability.

Specifications
Displacement: 1,230/1,510 t (Spanish boats - 1,490/1,740 t, Pakistan boats - 1,510/1,760 t)
Length: 67.6 m
Beam: 6.8 m
Draught: 5.4 m

Contractor
DCN International, Cherbourg.

UPDATED

GERMANY

Type 209

Type
Diesel electric patrol submarine.

Description
German SSK design continues to dominate the world market, having been principally developed by one company, Ingenieur Kontor Lübeck (IKL) and with boats built by HDW. The principal design is the Type 209 which is offered and built in a number of versions ranging from 970 tonnes standard displacement up to 1,500 tonnes. Although the design was originally developed some years ago. HDW, together with other German companies has now developed a series of modifications which, when retrofitted to existing boats, will enable them to extend their operational life to around 30 years.

The single-hull design incorporates two ballast tanks and fore and aft trim tanks. The 1500 series features a central bulkhead with an IKL-designed integrated escape sphere which can accommodate up to 40 of the crew. The sphere is supplied with 8 hours supply of oxygen and can withstand pressures down to the maximum diving depth of the boat.

The boats have been fitted with a variety of combat systems, while acoustic sensors are usually of the STN ATLAS Elektronik type. Sonars generally comprise a cylindrical bow-mounted passive array and an active array at the fore edge of the sail, together with three passive ranging arrays mounted along the port and starboard sides of the hull.

EW systems are generally either of French or American origin, although some units, notably the

A Type 209 submarine (Courtesy HDW)   **1998**/0007210

Turkish boats, have been fitted with British ESM systems. Eight swimout discharge torpedo tubes are fitted with accurately sited launch rails whose position can be adjusted during construction to accommodate different types of weapon. Up to 14 torpedoes are carried. A recent development allows boats to carry out submerged launch of anti-ship missiles such as Sub-Harpoon.

Machinery almost invariably comprises four MTU diesels driving four alternators. Submerged power is provided through a single Siemens electric motor developing around 3.7 MW and driving a single slow revolution propeller. Management is exercised through a remote machinery control system. The very high capacity GRP lead-acid cells incorporate a battery cooling system. All boats are fitted with snort masts. Submerged speed is around 21 to 22 kt with a speed of 11 kt maintained while snorting.

The Type 209s built for the Chilean Navy feature escape hatches in the engine room and the forward torpedo room, the latter being equipped to take a DSRV. An innovation in the design is a hatch aft of the sail providing a machinery exit route. This incorporates a pressure reactive hatch seal whose watertight integrity increases with an increase in external pressure, thus improving the safety factor in the event of underwater explosion. The sail has been lengthened and longer masts fitted in the Chilean boats to cope with the exceptional wave heights experienced in that part of the world.

Operational status

Variants in service with the navies of Argentina, Brazil, Chile, Colombia, Ecuador, Greece, India, Indonesia, South Korea, Peru, Turkey and Venezuela.

Specifications

Type	1100	1200	1300	1400	1500
Displacement	1,100/1,210 t	1,248/1,440 t[1]	1,260/1,390 t[2]	1,260/1,440 t[3]	1,660/1,850 t
Length	54.4 m	55.9-56 m	59.5-61.2 m	61-62 m	64.4 m
Beam	6.2 m	6.2-6.3 m	6.2-6.3 m	6.2 m	6.5 m
Draught	5.5 m	5.5 m	5.4-5.5 m	5.5 m	6.0 m

[1] Varies between 980-1,280/1,185-1,400
[2] Varies between 1,260-1,285/1,390-1,600
[3] Varies between 1,260-1,454/1,440-1,586

Contractor

HDW, Kiel.

UPDATED

TR 1700 series

Type

Diesel electric patrol submarine.

Description

The TR 1700 developed by Thyssen Werften features a high underwater sprint speed (in excess of 25 kt) but with average transit speed, high submerged endurance and low indiscretion, and high survival capability combined with considerable offensive capacity. The electronics in the Argentine boats are a mixture of American, Dutch, French and German systems. The boats are fitted with six swimout discharge torpedo tubes, and the Dutch Sinbads fire-control system can control three torpedoes simultaneously and track up to five targets. The total number of torpedoes carried is 22.

Because of its great structural strength, the TR 1700 can dive to below depths of 300 m with safety. A pressuretight bulkhead adds considerably to survival capability and crew safety is futher enhanced by the provision of facilities to handle a Deep Submersible Rescue Vehicle (DSRV). The TR 1700 is designed to operate with a small crew (about 30), and a very high standard of accommodation has been provided to enable long duration patrols (70 days) to be undertaken with some degree of comfort.

The propulsion arrangement comprises four MTU 16V 652 81 diesels driving four 970 kW generators to charge the eight 120 cell batteries. These power the 6.6 MW Siemens electric motor, to give a submerged speed of around 25 kt. Cruising range is 12,000 n miles at 8 kt surfaced, and the diving depth is in excess of 300 m.

Operational status

Two units in service with the Argentine Navy. Four other boats in various stages of construction will probably now be scrapped and used for spares.

Specifications
Displacement 2,116/2,264 t
Length: 66 m
Beam: 7.3 m
Draught: 6.5 m

Contractor
Thyssen Werften GmbH, Emden.

The TR 1700 SSK Santa Cruz *of the Argentine Navy seen here on contractor's trials* (Thyssen Werften)
1996

VERIFIED

Type 201 - 'Ula' class

Type

Diesel electric patrol submarine.

Description

Design work on the Norwegian 'Ula' class submarines began under a Memorandum of Understanding (MoU) signed between Germany and Norway in March 1979. Under the terms of the MoU Norway was to purchase STN ATLAS Elektronik (then Atlas Elektronik) sonars, Carl Zeiss periscopes and STN ATLAS torpedoes. In return, Germany was to purchase the Kongsberg Gruppen AS (formerly NFT) MSI-90U command system for the German Type 211 submarines (subsequently cancelled and replaced by the Type 212 design). The MSI-90U underwent extensive acceptance trials in 1989-90, including firing of the new DM2A3 torpedo with which the Norwegian submarines are armed. The combat system is designed to engage multiple targets in quick succession. Sensors feeding the combat system include the STN ATLAS DBQS-21DN (CSU-83) sonar, Carl Zeiss type SERO 14 and SERO 15 attack and search periscopes mounted in non-penetrating masts. The boats are armed with 14 × DM2A3

The Norwegian SSK Ula *of the Type 201 'Ula' class* (Thyssen Werften)   *1996*

torpedoes discharged from eight torpedo tubes. Propulsion is provided by two MTU 16V 396 SB83 diesels and two NEBB electric generators. The main electric propulsion motor is a 4.5 MW Siemens model with power supplied by Anker batteries. The navigation radar is the Kelvin Hughes Type 1007.

An updated export version of the 'Ula' design is the Thyssen TR 100 model. This is a medium-sized ocean-going submarine capable of operating both in the attack and submarine hunter-killer roles in a high-threat ASW environment. Particular attention has been paid to

minimising acoustic signature to enhance the capability of the platform's acoustic sensors and to reduce its own acoustic target level. This is achieved by use of sound absorbing materials and sound insulation, together with very stringent noise emission characteristics on all equipment. Indiscretion has been minimised by use of high battery charging capacity leading to reduced snorting time and the IR signature is minimised by diffusion of exhaust gases before emission to the sea. Crew safety is enhanced by the use of a hatch designed for DSRV docking.

Operational status

Six boats in service with the Norwegian Navy.

Specifications
Displacement: 1,040/1,150 t
Length: 59 m
Beam: 5.4 m
Draught: 4.6 m

Contractor
Thyssen Werften GmbH, Emden.

VERIFIED

Type 212

Type
Hybrid diesel-electric/AIP (fuel cell) submarines.

Description
The Type 212 forms part of Germany's new fleet programme for the next century. The first batch of four units has been authorised and will enter service beginning early within the next decade. Construction of the first boat started on 1 July 1998 at HDW.

The Type 212 is about three times the size of the Type 206. This enables it to develop a much larger radius of action and increased endurance, which will be considerably enhanced with the use of a hybrid diesel/fuel cell AIP system based on the liquid electrolyte system already successfully tested in the submarine U1. Power is being considerably increased over the system tested in U1, by replacing the liquid electrolyte with a solid polymer technology. Siemens KWU in Erlangen has developed the polymer electrolyte fuel cell (PEM-FC) AIP system. The H_2 metal hydride storage cylinders as well as the O_2 tanks for the fuel cell system will be mounted outside the pressure hull and inside the outer casing, allowing the size of the pressure hull to be reduced. With the extended range and endurance, the boats will be suitable for operations in the whole operating area of the German fleet with emphasis on European waters and the North Atlantic.

In addition to the fuel cell battery, the boats will be fitted with a modular sodium sulphide high-energy battery system driving a Siemens permanent magnet excited electric motor of the new Permasyn type, which is much smaller in size and weight than present motors. A single MTU 16V 396-diesel generator is mounted in a new acoustically treated motor room. The motive power will drive a 7-bladed skewed propeller.

Considerable effort has been devoted to incorporating the latest stealth technology. The sophisticated hydrodynamic design has resulted in a carefully shaped hull and sail with no straight lines giving rise to a very low target echo strength. Sensors will be fitted for the detection and analysis of self-generated noise. The use of amagnetic steel in the construction gives a very low magnetic signature and heat emission will be considerably reduced resulting in a much-reduced IR wake signature. The boats will be equipped with an integrated ship control and monitoring system, steering being exercised through a one-man console with two engineering consoles manned by a single crewmember sited alongside.

The first batch of boats will be fitted with STN ATLAS Elektronik DBQS 40 integrated sonar system. This will include a cylindrical medium frequency bow array operating in the 0.3 to 12 kHz band, a passive flank array (FAS-3), a passive ranging sonar (PRS), an intercept sonar and a low frequency passive towed array sonar winch system and the navigation and obstacle avoidance sonar. Optronic sensors will be the Zeiss SERO 15 (with laser rangefinder) and SERO 14 (with GPS antenna). The EW capability will be provided by a Daimler Chryster Aerospace FL 1800U system. In addition the boats will be fitted with own-noise monitoring equipment and the TAU 2000 torpedo countermeasures system. Control of the sensors is exercised through a Kongsberg MSI-9Ou command system. Navigation equipment includes a Kelvin Hughes Type 1007 radar while the integrated communications suite will comprise HF, VHF, UHF and VLF, Inmarsat-C, UHF SATCOM and GMDSS. Other systems include inertial platform, attitude and heading reference equipment, EM-log, echo sounder and GPS. The control room will be mounted on a frame elastically connected to the hull, while the electronics will hang from a frame above.

Torpedoes will be discharged through six bow torpedo tubes using an HDW water ram discharge system for launching the DM2 A4 heavyweight torpedo. In addition, the boats will be provided with fittings for carrying a minelaying belt.

Forward, the hull features a two-deck arrangement that results in considerably improved habitability standards.

Boats for Italy will be of the same configuration (apart from the weapons load which will be Italian torpedoes) as per the MoU signed between the German and Italian governments in April 1996. The Italian contract includes an option on a further two boats.

Operational status
On order for the German and Italian navies. Steel for U.31 was cut at HDW on 1 July 1998.

Specifications
Displacement: 1,450/1,800 t
Length: 56 m
Beam: 7 m
Draught: 5.8 m

Contractors
HDW, Kiel.
Thyssen Nordseewerke, Emden.
Fincantieri, Monfalcone (for Italy).

UPDATED

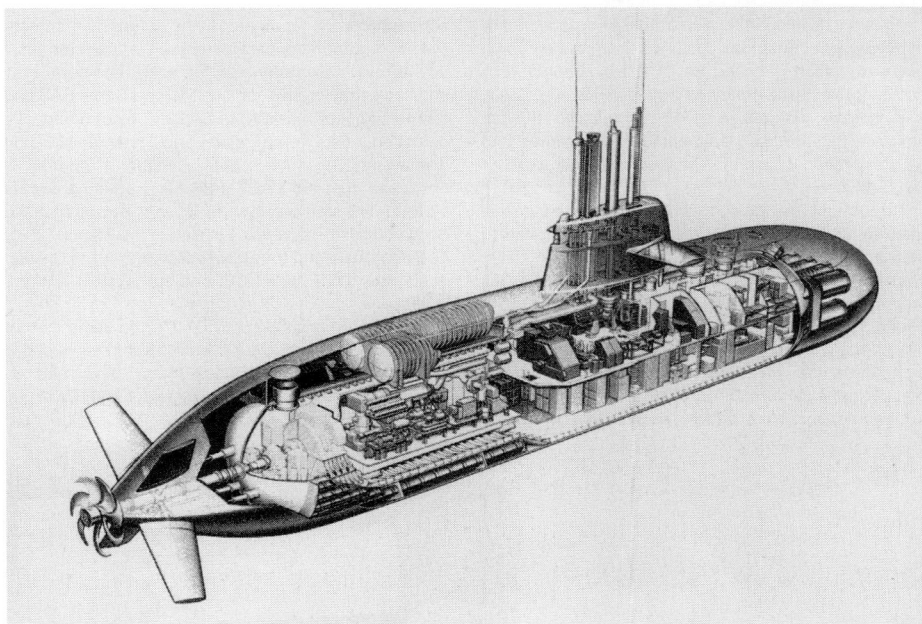

General arrangement layout of the Type 212 (Courtesy HDW)   **1999**/0045142

'Dolphin' class

Type
Diesel electric patrol submarine.

Description
The design developed for the Israel Navy, the 1,550 tonne Dolphin features considerable stealth technology, with emphasis being laid on its interdiction capabilities. For this the boats will be fitted with a wet and dry facility for special forces operations and an extensive payload of torpedoes and Sub-Harpoon missiles for the 4 ×533 mm and 2 ×650 mm bow torpedo tubes. However, conflicting reports are circulating concerning the actual weapon fit, and a photograph of the second unit, Leviathan, out of the

Dolphin (Courtesy Michael Nitz)   **1998**/0007211

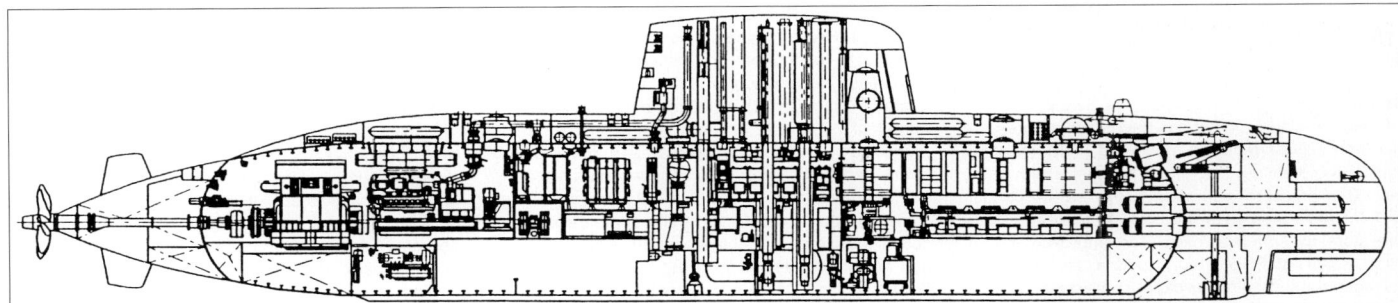

Dolphin   **1998**/0007212

water shows 10 weapons openings in two horizontal rows of five in the bow. The size of the openings is difficult to determine, but it is speculated that the four outer tubes may be for SDVs and the remainder for the mix of cruise missiles and torpedoes.

Sensors will include STN ATLAS Elektronik CSU-90 sonar which will be integrated with PRS 3 passive ranging and FAS 3 flank array sonars, all integrated with the STN ATLAS ISUS 90-1 weapon control system. The ESM system is the Timnex 4CH(V)2. Periscopes comprise the Kollmorgen Model 90 attack and a Kollmorgan search type.

Propulsion will comprise three 3.12 MW (total) MTU

diesel generators (16V 396 SE 84) and a single Siemens 2.85 MW electric motor driving a single skewed propeller. Control surfaces comprise an X-form rudder with hydroplanes on the forward casing, control being exercised through a GEC-Marconi control platform. The AC current is rectified to feed two 216-cell banks of Hagen batteries. Submerged speed will be about 20 kt, 12 kt while snorkelling.

Operational status
On order for the Israel Navy. *Dolphin* is due to be delivered in early 1999 and *Leviathan* around the end of 1999.

Specifications
Displacement: 1,640/1,800 t
Length: 57.3 m
Beam: 6.8 m
Draught: 6.2 m

Contractor
HDW, Kiel.

UPDATED

INTERNATIONAL

'Scorpene'

Type
Diesel electric patrol submarine.

Description
In a 50/50 partnership, DCN of France and Bazan of Spain have developed for export the 'Scorpene' family of advanced submarines designed around a common hull form which is based on DCN's experience with nuclear-powered submarines both SSBNs and SSNs. The design features a hydrodynamically efficient albacore hullform with fin-mounted hydroplanes and X-configuration tailplane. The hull is constructed of HLES 80 steel.

Throughout the design, the major emphasis has been to achieve maximum stealth capabilities, and to this end the design features very low acoustic, magnetic, electromagnetic and IR signatures. All equipment is mounted on flexible couplings and in addition, the single deck is also carried on flexible mountings. The single hull is divided into three compartments with two decks. The forward compartment houses 6 ×533 mm positive discharge torpedo tubes and a battery section; the mid-ships

compartment contains accommodation, control room, header tanks and auxiliary systems; and the aft compartment houses the machinery and a second battery.

Certain equipment is common to all the family such as the SAGEM integrated navigation system which is developed from the Minicin system; the external communications equipment; the integrated combat system incorporates passive/active sonars (cylindrical bow array, active array, ranging array, intercept array and flank array) from Thomson Marconi Sonar SAS and optronic sensor masts from SAGEM; electromagnetic detection, data processing and tactical situation display and weapon control. Platform control is exercised from the SAGEM SS Mk 1 steering control system. The combat system is very powerful being derived from the systems installed in the French SSBN *Le Triomphant* and the SSN *Amethyste*. In the Chilean boats the UDS SUBTICS combat system will be installed.

The electric propulsion system is the same as that used in the French nuclear-powered submarines and is a 2,800 kW AC synchronous motor developed by Jeumont-Schneider, powered by two 1,100 kW diesel generators and driving a single shaft. A DC motor could

replace the aft part of the submarine. Diesel power is provided by SEMT-Pielstick turbocharged diesels. Chile has specified four MTU diesels for her version of the 'Scorpene' with a Jeumont Schneider permanent magnet electric motor. The battery configuration is based around a 360-cell bank.

A major feature of the design is the provision for an AIP propulsion unit based on the French MESMA system which is being developed by Bertin. To allow for easy installation of the AIP system the design features a cofferdam that is sited immediately forward of the machinery compartment and the MESMA AIP system would be inserted at this point. The boats under construction for Chile will not be fitted with the AIP system.

Even without the MESMA AIP the boat has an extensive submerged range, and at 20 kt the submerged endurance is in excess of two hours.

The boats are to be built in sections at different sites and brought together at the shipyard for final assembly.

The design allows for the 18 weapons carried to be launched from six torpedo tubes using a pneumatic ram, but two turbine pumps could be used to expel weapons or a swimout system adopted. Strachan & Henshaw in the UK have carried out a study into the feasibility of integrating an air turbine pump torpedo discharge system into the design.

Diving depth is in excess of 300 m. The crew totals about 30 to 32 and considerable emphasis has been placed on the crew environment, including a very high standard of comfort and accommodation to ensure they remain fit throughout the mission.

Specifications
Displacement: 1,450/1,590 t
Length: 63.5 m
Beam: 6.2 m
Draught: n/k

Operational status
Two units are on order for the Chilean Navy.

Contractors
DCN International, Paris, France.
Bazan, Spain.

Model of the Scorpene DCNI   ***1999**/0024654*   *UPDATED*

ITALY

'Sauro' class

Type
Diesel electric patrol submarine.

Description
The 'Sauro' class submarines have a single pressure hull with ballast tanks at bow and stern mounted outside the pressure hull, and a buoyancy tank in the sail. The pressure hull is constructed in circular sections of HY80 high-tensile steel and closed at bow and stern by spherical caps. The improved 'Sauro' class boats (the last four) incorporate a central bulkhead for emergency escape. The last two boats in the series are fitted with anechoic tiles.

The propulsion system comprises three Fincantieri Diesel Engine Division A 210 16 NM (SM in the Improved 'Sauro') diesels rigidly connected to three Marelli DC generators for battery charging or for feeding the electric motor or both. The electric motor is a Marelli double armature DC motor with a continuous output of 2.4 MW at 225 rpm (for short periods the motor can develop 3.14 MW at 244 rpm), directly connected to the shaft. The batteries are sited in two compartments each of 148 cells. All diesel generators, auxiliaries and propulsion motor compartments are sound insulated and all systems are elastic mounted and flexibly connected in order to reduce structure-borne noise. Machinery control is extensively automated. The snort mast features a specially

designed wave contour head with a very low radar profile. The *Fecia di Cossato* will be used to test an AIP system.

Six bow swimout torpedo tubes for guided or unguided weapons are fitted. Torpedo tubes are loaded through a hydraulically powered loading hatch in the torpedo room and weapons are fired hydropneumatically. Studies have been undertaken to adapt the B512 swimout torpedo discharge tube to launch encapsulated sub-launched anti-ship missiles. The last two boats in the improved series have been lengthened to accommodate missiles and normal weapon payload is 12 ×Type184 torpedoes.

The improved 'Sauros' feature a completely new control room equipped with Alenia multifunction digital consoles with double or single displays and a fully integrated sonar system for one-man operation combining surveillance, detection, tracking, sonar intercept, passive ranging, spectral analysis and classification. Sonars include the Alenia IPD 70/S and MD 100S. This is integrated with the SMA SACTIS weapon control system, also using multifunction consoles.

The concept of the new control room is aimed at improving standardisation, maximising automation, and developing maximum integration of systems leading to greatly reduced manning requirements. Other modifications include siting all unmanned parts of systems outside the control room, and installing a semi-automatic microprocessor-controlled plotting table.

The machinery in the improved 'Sauros' is the Fincantieri Diesel engine Division A 210 16SM driving three alternators developing 2.16 MW. The electric motor is the same as in the earlier boats. General characteristics and hull configuration remain the same as in the earlier boats, but internal spaces have been redesigned to allow better location of equipment. The hull is of HY 80 steel and features a central bulkhead for emergency escape purposes.

The last two boats in the Improved series are to be modernised with the STN ATLAS Elektronik ISUS 90 combat system, modified to handle the upgraded A184 Mod 3 to be supplied to the Italian Navy by Whitehead Alenia Sistemi Subacquei (WASS).

Operational status
In service with the Italian Navy. Modernisation contract due to be signed with Fincantieri in mid-1998 had not been concluded at time of going to press.

Specifications
'Sauro'
Displacement: 1,456/1,631 t
Length: 63.9 m
Beam: 6.8 m
Draught: 5.7 m
Improved 'Sauro'
Displacement 1,476-1,653/1,662-1,862 t
Length: 64.4-66.4 m
Beam: 6.8 m
Draught: 5.6 m

Contractor
Fincantieri, Monfalcone.

UPDATED

MG 120/ER

Type
Shallow Water Attack Submarine (SWATS).

Description
This MG 120/ER SWATS is designed to carry out both offensive and defensive operations in shallow waters where larger submarines are unable to fully exploit their potential capabilities.

Because of its small size and low target echo strength, the submarine is extremely difficult to detect. The mission profile of the SWATS means that it operates in an environment characterised by strong surface and bottom reverberations where hostile active sonars are severely degraded in their capability to detect the small target. The submarine is also able to crawl along the seabed further enhancing its stealth capability.

The SWATS was initially conceived as a special craft for covert operations, a mission which it still retains. However, the design has been further developed featuring significant standoff capabilities and comprehensive attack and passive equipment which make it suitable for a wide range of tactical missions.

In peacetime the submarine can be used to train both submarine crews and commanding officers and also to train forces in ASW procedures and tactics.

Two of the outstanding characteristics of the MG 120/ER SWATS is its range, which exceeds 2,000 n miles surfaced, and its low indiscretion rate. The submarine is also fitted with an AIP system called the Underwater Auxiliary Propulsion Engine (UAPE) comprising a closed circuit diesel fuelled by liquid oxygen. This confers a submerged radius of 400 n miles.

Armament configurations can be varied according to the mission in hand and may comprise: up to four wire-guided torpedoes; two chariots; 16 mini-torpedoes; four remote-controlled Chariots; up to 12 ground influence (magnetic/acoustic) mines; 20 limpet mines; or commando gear containers.

The boat is manned by a crew of six and can carry up to 15 combat divers.

Operational status
In service with several armed forces.

Contractor
Cosmos SpA, Livorno.

VERIFIED

Launch of the third Improved 'Sauro' class SSK Primo Longobardo (Fincantieri)   *1996*

An SWAT MG 120/ER submarine running on the surface   *1998*/0007213

JAPAN

'Harushio' designs

Type
Diesel electric patrol submarine.

Description
The 'Harushio' design, the latest to be developed by Japan, was first approved in the 1986 estimates. The design is a natural evolution from the preceding 'Yuushio' class from which it differs in having slightly increased dimensions. Improved stealth technology has been incorporated into the design including a greater degree of noise control, mainly affecting propeller cavitation and machinery silencing, as well as the fitting of anechoic tiles. Other new features include a towed array sonar and improved communications system. The last boat of the class has a slightly larger displacement and a small reduction in crew numbers resulting from greater system automation.

Much of the equipment is of indigenous design, based on American systems. The sonar suite is manufactured by the Oki Corporation using technology supplied by the Hughes Corporation. The boats deploy the American Sub-Harpoon missile together with

Japanese designed and manufactured Type 89 wire-guided torpedoes. The propulsion system consists of two Kawasaki 12V25/25S diesels developing 4.1 MW driving two Kawasaki alternators developing 3.7 MW. A single Fuji electric motor develops a maximum of 5.3 MW to drive a single shaft.

The improved 'Harushio' design ('Oyashio' class) feature a higher degree of machinery automation than the 'Harushio', with a further reduction in manning requirement. Length has been increased by about 1 m. The submarines are being equipped with a large flank array sonar, which is said to be the reason for the increase in displacement over the 'Harushio' class. The double hull is treated with anechoic tiles. Two diesels provide a total output of 5,520 bhp while the two electric motors have a combined output of 7,750 hp.

Under the terms of an agreement concluded with Kockums in Sweden, Kawasaki Heavy Industries has carried out experiments with an 800 hp Stirling AIP engine. Kawasaki has also investigated the fuel cell method of energy production and, on a commercial basis, a superconducting propulsion system. As yet, however, no decision has been taken to install a new system of propulsion in the improved 'Harushio' class, and it seems unlikely that the early boats of this class will be fitted with such a system. However, the next-generation submarines will undoubtedly be equipped with some form of AIP system.

Operational status
In service with the Japanese Maritime Self-Defense Force.

Specifications
Displacement: 2,450/2,750 t
Length: 77 m
Beam: 10 m
Draught: 7.7 m

Contractor
Kawasaki, Kobe.
Mitsubishi, Kobe.

UPDATED

NETHERLANDS

'Walrus' class

Type
Diesel electric patrol submarine.

Description
The 'Walrus' class SSK, built for the Royal Netherlands Navy by RDM and designed in close co-operation with the naval architects of NEVESBU, features improved manoeuvrability and higher speed compared to the previous 'Zwaardvis' class. Diving depth has also been increased by about 50 per cent, resulting from the use of MAREL high-yield tensile steel for hull construction. The centre section of the hull is of the single-body type while the ends are of double-hull design.

The design features a high degree of automation and monitoring which results in a considerable reduction in manpower requirement. The advanced Integrated Platform Monitoring and Control System embodies the complete integration of platform data information and display, with the possibility of operating and supervising platform systems from one or several positions. The control room also houses the computer-based steering control system. The boats are fitted with the Signaal SEWACO VIII command system, the Thomson Marconi Type 2026 low frequency towed array sonar, and Thomson Marconi TSM 2272 Eledone medium/low frequency passive sonar and the DUUX-5 passive ranging sonar. The weapons system comprises four torpedo tubes for launching American NT.37C/E or Mk 48 Mod 4 torpedoes and Sub-Harpoon missiles. A total of 20 weapons is carried, or alternatively 40 mines may be carried in lieu of the torpedoes. The first two units are powered by SEMT-Pielstick diesels, and the second two units by Brons-Werkspoor diesels driving a seven-bladed propeller. The 5.1 MW electric motor is by Holec.

Operational status
Four boats in service with the Royal Netherlands Navy.

Specifications
Displacement: 2,465/2,800 t
Length: 67.7 m
Beam: 8.4 m
Draught: 7.0 m

Contractor
RDM, Rotterdam.

UPDATED

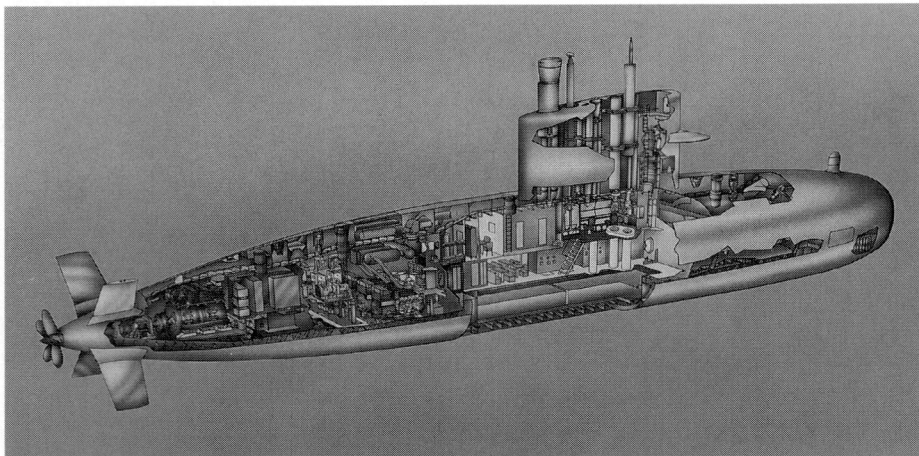
General arrangement layout of the Netherlands 'Walrus' class (RDM)   *1996*

Moray

Type
Diesel electric patrol submarine.

Description
RDM has also developed the 'Moray' design. The 'Moray' is a derivative of the 'Walrus' design, and offers a wide range of customised systems and degrees of system integration to accommodate specific requirements for endurance and weapons capabilities according to customer requirements. Three basic designs displaying different displacements and combat capabilities are on offer. The single-hull design is built of high-tensile steel, of comparatively low weight but resistant to high pressure. The design operating depth of all versions is in the region of 300 m.

The Type 1800 (1,907 tonnes submerged) features provision for a clipped-on towed array sonar and the capability to deploy sub-launched anti-ship missiles.

The 1800 also features the capability to incorporate an AIP system which comprises two 380 kW diesel generator sets with CCD exhaust gas treatment and water management, LOX storage and auxiliaries. These are in addition to the standard three 980 kW diesel generators. This will increase the submerged endurance to 477 hours at 6 kt speed.

Propulsion is provided by three diesel generators giving an indiscretion rate of 13 per cent at a speed in advance of 6 kt. The large fuel reserves provide for a range of 11,000 n miles at 6 kt and endurance is rated at 65 days. Battery capacity provides for a submerged endurance of 40 hours at 6 kt. The powerful main electric motor allows for 1 hour submerged speed at 20 kt. Operational diving depth is 300 m. A Signaal integrated combat system is proposed and the weapon payload of 20 torpedoes is discharged through six bow torpedo tubes with turbine ejection pumps. The crew totals 38.

Specifications
Basic version:
Displacement: 1,451/1,907 t
Length: 66.5 m
Beam: 6.4 m
Draught: n/k
AIP variant:
Displacement: 1,626/2,233 t
Length: 75.9 m
Beam: 6.4 m
Draught: n/k

Contractor
RDM, Rotterdam.

UPDATED

RUSSIAN FEDERATION AND ASSOCIATED STATES (CIS)

'Kilo' class

Type
Diesel electric patrol submarine.

Description
The 'Kilo' class design dates back to the 1970s and the first vessel, for the then Soviet Navy, was launched in 1979. Since then it has undergone continual improvements. The 'Kilo' has been developed from the previous 'Foxtrot' and 'Tango' designs, but shows an improved hull form. However, it is still fairly basic compared to its modern Western counterparts. The basic variant is the Project 877; the Project 877K has an improved fire-control system; while the Project 877M is fitted to fire wire-guided torpedoes from two tubes. The latest variant is the Project 636, which is available for export. This model is 1.2 m longer than previous variants and features improved stealth technology with a redesigned propulsion system which is claimed to generate half the noise of its earlier variants. Capability has been improved with an automated combat information system that can track up to six targets simultaneously and provide simultaneous fire-control data on two targets. The forward hydroplanes are mounted on the hull just forward of the fin. The pressure hull is divided into six compartments separated by pressure bulkheads and has a reserve of buoyancy of 32 per cent at normal load and is heavily compartmented, the boat remaining buoyant with any compartment flooded. Normal diving depth is 240 m and maximum depth is 350 m.

Electronic equipment is all of Russian manufacture, the sonar suite comprising the hull-mounted, low-/medium-frequency passive search and attack Shark Teeth (MGK-400EM), and the hull-mounted, high-frequency active search/attack sonar Mouse Roar. The Shark Teeth, although primarily a passive search and attack sonar, also has some active capability. The Indian boats are additionally fitted with the low-/medium-frequency, passive search Whale series. The electronic warfare suite consists of either the Brick Group, Stop Light or Squid Head radar warning system and Quad Loop D/F. For navigation the ships are fitted with the I-band Snoop Tray radar.

Standard armament comprises six torpedo tubes firing a mixture of TEST-71ME/TEST 71/96 wire-guided ASW active/passive homing and Type 53-65 ASV passive wake homing torpedoes with a total of 18 weapons being carried. At least two torpedo tubes are equipped to fire wire-guided anti-submarine weapons. As with most submarines, mines can be carried in lieu of the torpedoes, up to a total of 24. In addition, some vessels are fitted to carry an SA-N-5/8/10 shoulder-held SAM launcher with 6-8 missiles. The containerised portable missile launcher is carried in a well between the snort and communications masts.

Kilo at St Petersburg (H M Steele)   *1997*

Propulsion is provided by two 4-2AA-42M diesels (in the export variants and the latest variant the Type 636 in service with the Russian Navy) developing 2.68 MW powering two generators. The single shaft is driven by a single electric motor developing 4.34 MW and powering a slow turning seven-bladed propeller. In addition, two small MT-168 auxiliary motors developing 150 kW are fitted and a low powered electric motor of 95 kW for economic running and slow speed operations (6 kt) in ultra quiet mode. Two 120 cell storage batteries are accommodated in the first and third compartments. Battery capacity is 9,700 kW h. Fuel reserves total 51.6 tonnes normal and 172 tonnes maximum. The Iranians experienced difficulties with their batteries which, having been designed for operation in cold water regions, suffered from overheating in the warm waters of the Gulf. The Indians are said to have suffered similar problems with their 'Kilos' and from their experience, have assisted the Iranians to overcome their difficulties with modifications to the battery cooling system. The Indians are also said to be considering changing the diesels in their boats.

Operational status
In service with the Russian Navy. The first country to receive an export 'Kilo' was India and since then the design has been exported to Algeria, China, Iran, Poland and Romania. The latest export variant of the 'Kilo' is the Type 877EKM which is being delivered to Iran and China. This variant features improved weapons system management and fire control, together with a basic combat information suite. Accommodation has been improved compared to earlier boats in the series. Export variants are distinguished by the letter E after the type or project designator. Export construction is now said to be running at about three boats a year. India has recently been offered the Type 636.

Specifications
Displacement: 2,356/3,076 t
Length: 72.6 m (73.8 m in some E and EKM variants and some boats in Russian Navy)
Beam: 9.9 m
Draught: 6.6 m

Contractor
Admiralty Yard, St Petersburg.
Sudomekh, St Petersburg.
Krasnoje, Gorki.
Gorki, Nizhny Novgorod.
Komsomolsk, Komsomolsk.

UPDATED

'Lada' class (Project 677) 'Amur' class (Project 1450)

Type
Patrol submarine.

Description
The latest design to be developed by the Central Design Bureau Marine Engineering (Rubin) is the 'Lada (Project 677)/'Amur' (Project 1450) export variant. The design has been developed for introduction as a gradual replacement for the 'Kilo' class.

This fourth generation SSK incorporates all new design technology and weapons. Compared to the 'Kilo', the 'Amur' has reduced the noise factor still further by a considerable factor. This has been achieved by stringent control of acoustic emission from equipment, double elastic mounting of all equipment and the use of acoustic tiles. Weapons payload includes the latest types of Russian torpedo, anti-submarine rocket torpedo, cruise missile and mines. Torpedoes can be fired both singly and in salvoes of up to six weapons. An automatic torpedo loading system permits prompt reloading of torpedo tubes for subsequent firing within less than 20 seconds. Sensors include extremely sensitive passive transducer

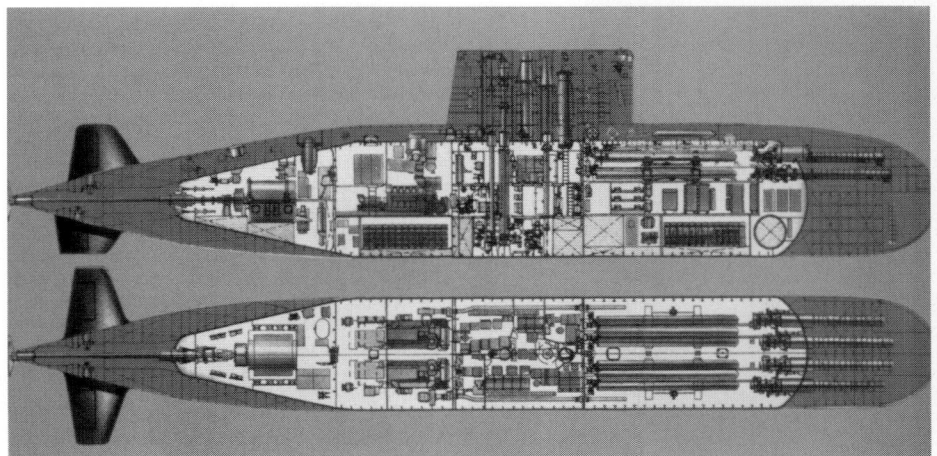

General arrangement drawing of the 'Amur' class (Anchor Consultancy)   *1999*/0024688

arrays that are much more effective as a result of the boat's very low noise signature. In addition to the hull-mounted sonars the boat can deploy a towed array that is streamed through the sail. The usual array of ESM, radar and periscopes is fitted and the commander's periscope features, in addition to the normal viewing system, a night vision facility and a laser range-finder.

Propulsion comprises a full electric propulsion system consisting of two diesel generators and a single electric motor developing 2,000 kW together with storage battery, driving a single shaft. The electric motor and diesels can be used simultaneously to achieve maximum propulsion speed. Power supply is in AC configuration. Provision has been made, in the design, for fitting an AIP system should a customer require this.

All machinery control and submarine attitude, together with damage control and general boat system management is controlled either from a central console in the main control room or other control stations around the boat.

The design is fully tropicalised for operation in any part of the world.

Operational status
Two boats are under construction, one, the *St Petersburg*, a 'Lada' for the Russian Navy, and the second an 'Amur' export version for an unnamed overseas customer.

Contractor
Admiralty Yard, St Petersburg.

NEW ENTRY

'Pyranja' Project 865

Type
Midget submarine.

Description
A product of the Malachite Marine Engineering Bureau, the 'Pyranja' 250 ton diesel-electric powered midget submarine features a corrosion resistant hull.

A single 160 kW diesel generator provides power to a 60 kW electric motor to drive the single shaft. The design also provides for the insertion of an AIP unit, should this be required, which would increase length by about 15.2 m.

Sensors provided include a basic sonar system together with periscopes with radar and ESM. In the larger AIP design a full active/passive sonar suite would be installed.

The design provides for the carriage of up to six divers who can enter and leave the boat by a lock-in/lock/out chamber. Weapons such as mines and torpedoes are discharged through two torpedo tubes fitted outside the pressure hull and angled outwards. Two unpressurised outboard compartments provide for the carriage of special diver vehicles or equipment for the divers.

Operational status
Two units are in service with the Russian Navy.

Specifications
Displacement: 221/257 t
Length: 29.2 m
Beam: 5.0 m
Draught: 3.9 m

Contractor
Admiralty Yard, St Petersburg.

NEW ENTRY

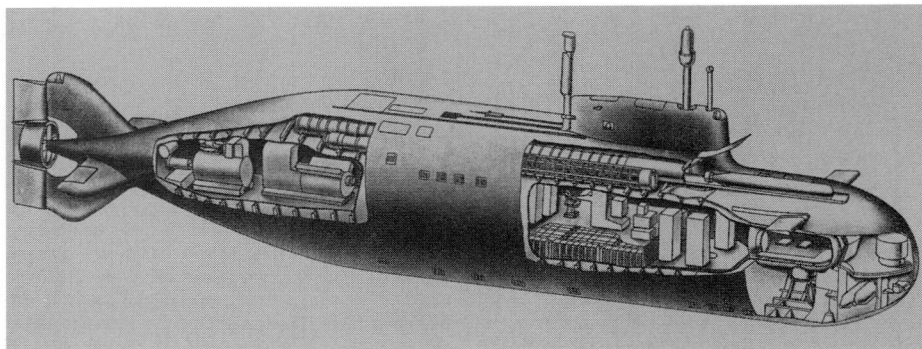

General arrangement of the Pyranja
1999/0024697

SWEDEN

'Vastergotland' class

Type
Diesel electric patrol submarine.

Description
The four 'Vastergotland' A-17 boats built by Kockums for the Swedish Navy have been designed to carry out a variety of roles to provide defence against invasion. Among the missions envisaged for the boats are attack, mining, surveillance, anti-submarine and interjection of special forces.

The two-deck single-hull design features two watertight compartments divided by a centre watertight bulkhead. The pressure hull is of circular cross-section ending in truncated cones. Fore and aft the hull extends into ballast tanks. The forward compartment houses accommodation, stores, communications room and control room on the upper level, with weapons handling and nine torpedo tubes (six conventional 533 mm with 12 torpedoes and three 400 mm tubes with six weapons), forward battery section and auxiliary machinery on the lower deck. Immediately aft of the watertight bulkhead is a cylindrical tank section with a passage aft and an escape lock which connects to a rescue vehicle or bell. The after section houses the electrical control centre with the aft battery section, diesel generators and propulsion motor on the deck below. Manoeuvrability is exercised through an X-configuration rudder/after hydroplane design.

The boat carries a comprehensive range of sensors including periscopes, Terma radar, the American ARGOSystems AR-700-S5 ESM, and a wide range of STN ATLAS Elektronik sonars, including the CSU 83. Sonar arrays are mounted on and under the casing. The boats are equipped with an Ericsson IPS-17 weapons control system developed from the NIBS system fitted in the 'Nacken' class.

The resiliently mounted propulsion motor is a Jeumont-Schneider water-cooled, single-armature 1.32 MW DC motor. Lead-acid storage batteries developed by Tudor provide electrical power, charging being accomplished with Jeumont-Schneider alternators with built-in rectifiers. The alternators are powered by two Hedemora V12A/15 turbocharged diesels. All equipment is resiliently mounted on steel springs or rubber mounts, and the complete control room is placed on rubber mounts permitting a simple mounting of units directly on to the platform.

Operational status
Four boats in service with the Swedish Navy.

Specifications
Displacement: 1,070/1,143 t
Length: 48.5 m
Beam: 6.1 m
Draught: 5.6 m

Contractor
Kockums, Malmo.

UPDATED

The Ostergotland, *fourth unit of the 'Vastergotland' class* (H&L van Ginderen)   *1996*

Bow view of 'Vastergotland' showing sail-mounted hydroplanes and intercept sonar array housing (Courtesy of Kockums)   *1999*/0024690

'Gotland' class

Type
Diesel electric with AIP patrol submarine.

Description
Kockums has just completed building three Type A-19 'Gotland' class 1,200 tonne boats to replace the 'Sjoormen' class boats built in the 1960s. The main roles envisaged for the 'Gotland' boats are attack, surveillance, minelaying and ASW. Externally the single-hull 'Gotland' is very similar to the Type A-17, but slightly larger with improved performance characteristics in certain areas, and with increased stealth capability afforded by anechoic tiles which are fitted to the hull. All equipment is resiliently mounted and the control room, accommodation and propulsion systems and all equipment are carried on rubber mountings to isolate them from the hull. The hull is divided into two watertight compartments separated by a tank section, which incorporates a one-man escape chamber accessible from both compartments. The section is fitted with an escape trunk to which can be mated a DSRV or a rescue bell. The forward compartment is divided into two decks, the lower deck housing the four 533 mm and two 400 mm bow torpedo tubes and associated weapon handling gear and reload racks. Beneath this compartment is the battery space and auxiliary machinery. The upper deck houses the control room, accommodation and other equipment rooms associated with the sensors and communications. The lower deck in the aft compartment houses the diesel generator sets and more battery space, while much of the auxiliary machinery is mounted on the upper deck. This aft section also contains the AIP machinery and electric motor.

In the aft section are installed two Stirling V4-275R AIP systems developed by Kockums using LOX and diesel in a helium environment. The engines together develop about 150 kW under continuous run conditions. This enables the boats to carry out extended periods of submerged patrol, independent of snorkelling and to transit at slow to moderate speeds. Experience with the trials installation in the submarine *Nacken* has shown that it is possible to operate submarines for up to 90 per cent of sailing time using the AIP, the remainder using the diesels for propulsion. Space is available for the addition of two more Stirling engines at a future date should it be proved necessary that a high AIP transit speed is required. Behind the Stirling engines is the electrical centre for controlling the batteries, and the boat's electrical requirements. The standard propulsion system comprises two MTU 16V 396 turbocharged diesels driving two Jeumont-Schneider alternators, with a Jeumont-Schneider electric motor. Battery power is provided by Varta lead-acid batteries. All machinery is controlled by two persons, a machinist and a motorman.

The aft control surfaces are arranged in an X-configuration, being individually and independently controlled. The forward control surfaces are mounted on a common shaft on the sail. Steering control is supervised and exercised through a Van Rietschoten & Houwens system single console system.

Sensor capability is also improved compared to earlier classes, and comprises an STN ATLAS Elektronik CSU-90 integrated sonar system incorporating hull-mounted, cylindrical bow, LF flank, and intercept arrays and passive search and attack, and a Racal Radar Defence Systems (formerly Thorn EMI Electronics/MEL) MANTA ESM system. Only one periscope is mounted, a US Kollmorgen unit. The second two boats in the series will be fitted with the Pilkington CK038 system. The periscope is the only mast that penetrates the hull. The Sesub 940 combat system is a derivative of the CelsiusTech 9LV Mk 3 based on a LAN configuration and using a series of intelligent nodes which provide access for various subsystems into the databus. The navigation radar is supplied by Terma.

The 'Gotland' is designed to deploy the new Torpedo 2000, although pending entry into service of this weapon the boats will be able to deploy the Type 613.

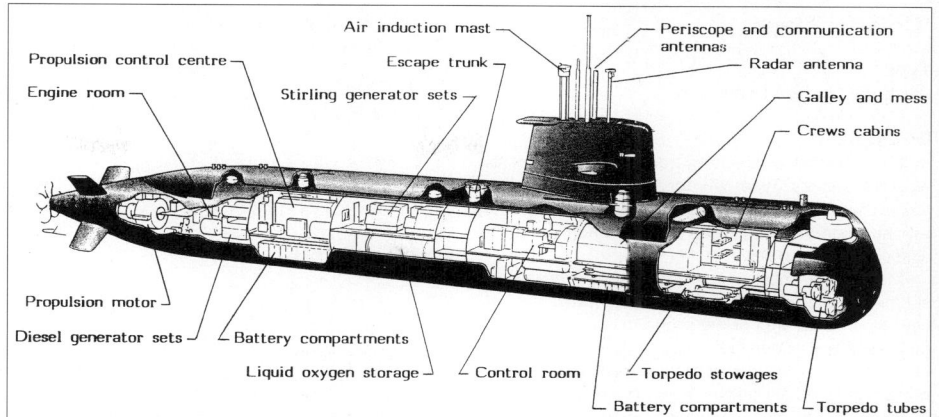
General arrangement layout of the 'Gotland' class (Kockums)   *1998*/0007214

Gotland on the synchrolift ready for launch (Kockums)   *1996*

Gotland showing array of masts   *1999*/0024689

With its dual-torpedo capability, the 'Gotland' class can deploy the 400 mm 43X2 torpedo from either the 400 mm tubes or the 533 mm tubes, the latter being able to house two 43X2 weapons in lieu of one Torpedo 2000. The tubes may also be used to lay mines, although Kockums has developed a mine pannier which can be strapped on to the side of the hull in a matter of hours. These panniers are capable of carrying up to 25 mines each.

Operational status
Three boats in service with the Swedish Navy.

Specifications
Displacement: 1,240/1,490 t
Length: 60 m
Beam: 6.1 m
Draught: 5.6 m

Contractor
Kockums, Malmo

UPDATED

'Collins' class

Type

Diesel electric patrol submarine.

Description

In June 1987 Kockums was awarded the contract for six new submarines Type 471 for the Royal Australian Navy. These large SSKs are designed for ocean patrol with an extended submerged (using snorkel) radius of action of 9,000 n miles at 10 kt.

One of the most important design areas in a submarine is noise and in particular the prevention of cavitation in the area of the propeller and around the hull. SSPA of Sweden has carried out detailed studies on the design to ensure that it features an extremely low noise signature. To ensure noise is maintained at a minimum level all equipment is double elastic mounted, which also provides for excellent shock absorption. In addition, to minimise detection by hostile sonars, the hull is coated with Australian designed and manufactured anechoic tiles. The hull is constructed of micro-alloy OX812 steel manufactured under licence in Australia.

The propulsion system comprises three Hedemora VB210 18-cylinder 4-stroke turbocharged diesel generators with a combined output of 4.425 MW. The generators and electric motors are being supplied by Jeumont-Schneider of France and ASC. A single Jeumont-Schneider 5.4 MW water-cooled, double-armature DC shunt electric motor is powered by 400 Varta-designed lead-acid battery cells driving a single shaft fitted with a 4.2 m skew-back 7-bladed propeller. Maximum submerged speed is in excess of 20 kt. In addition, a McTaggart Scott DM 43006 retractable hydraulic motor is fitted for emergency propulsion. Later boats may be fitted with a Stirling AIP system, but no definite decision has been made by the Australian government on the acquisition of such a system for the 'Collins' class.

The single hull is built of 690 MPa high tensile micro-alloy steel, with two continuous decks.

Armament comprises six bow torpedo tubes firing Mk 48 torpedoes or Sub-Harpoon missiles. A total of 23 weapons can be carried by each submarine. The torpedo tubes are arranged into three banks to port and three to starboard. Each bank of tubes is capable of independent operation in all respects because all discharge, control, and operating equipment is duplicated. The discharge system, designed by Strachan & Henshaw (part of the Weir Group), uses the Air Turbine Pump (ATP) as its prime mover for the weapons. Strachan & Henshaw has also designed and built the Submarine Signal and Decoy Ejector (SSDE) for the 'Collins' class, two of which are fitted. The SSDEs are capable of discharging a wide range of pyrotechnics, decoys, countermeasures and communications buoys over the full speed and depth range of the submarine. Other stores which can be ejected from the SSDE are expendable communication buoys, intelligence gathering probes, emergency signalling devices, flares and grenades. The SSDE is integrated within the submarine's Combat System.

Fire control is exercised through a Loral Librascope weapons control system, with data derived from Thomson Marconi Scylla integrated sonar system (developed from the Eledone sonar system) and the

The leadship of the 'Collins' class. The long tube at the stern is for dispensing the towed array (Australian Submarine Corporation)
1996

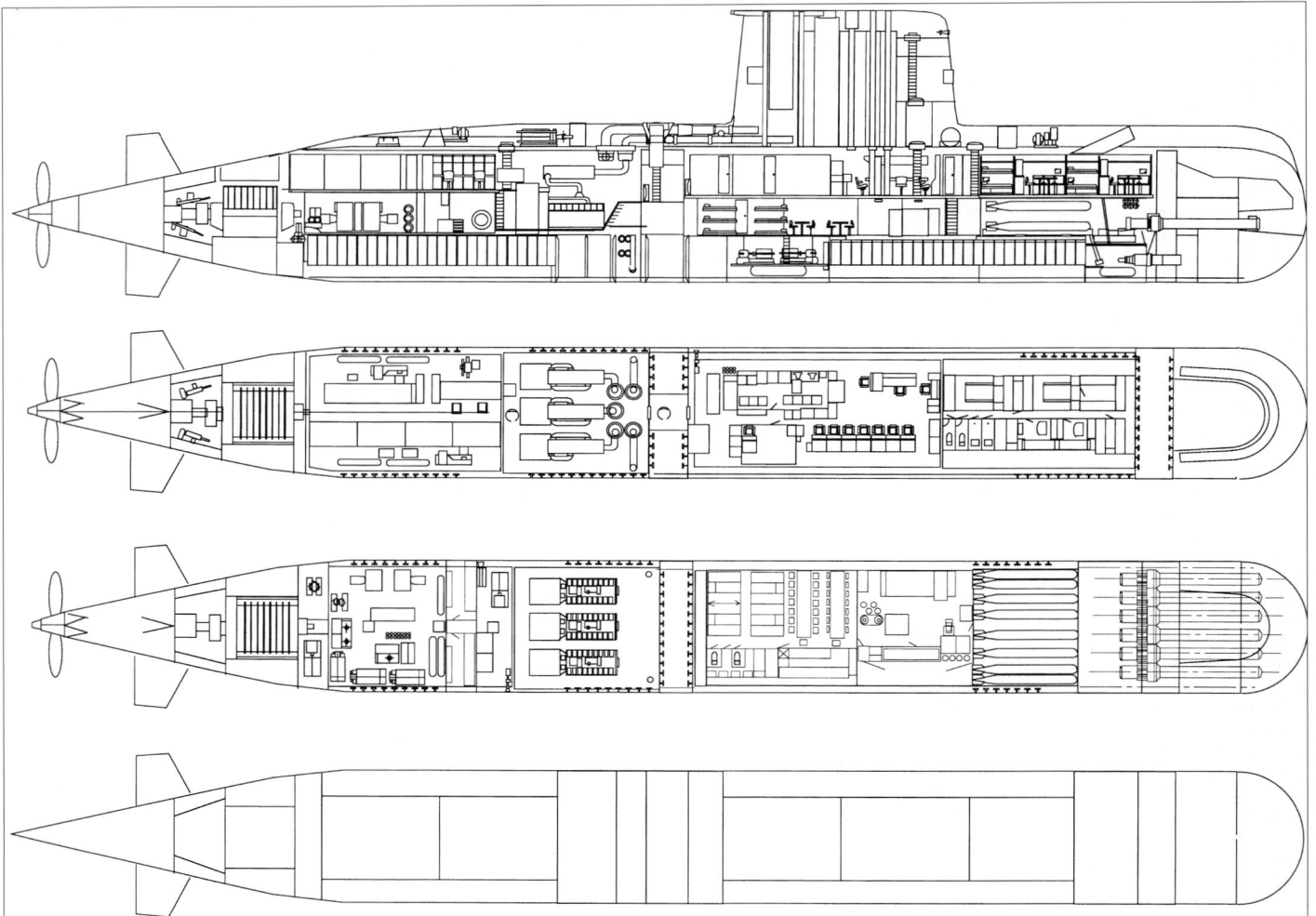

General arrangement layout of the 'Collins' class (Courtesy RAN)
1996

Australian GEC-Marconi Systems Pty Ltd Kariwara retractable thin-line towed array for which McTaggart Scott of the UK is providing the towed array handling equipment. Later boats in the class may be fitted with the Thomson Marconi Narama retractable thin-line passive towed array. Other sonar sensors comprise the TSM 2253 LF passive flank array, TSM 2225 passive ranging array, intercept array and active and passive mine avoidance arrays.

Rockwell Ship Systems Australia (now Boeing Australia) is responsible for the Combat System design, assembly, test, integration and support activities; while Computer Sciences of Australia is responsible for the design, development and integration of the system software.

The non-hull penetrating masts are being supplied by Riva Calzoni of Italy. The first search (CK 43) and attack (CH 93) periscopes will be built in the UK by Pilkington

Optronics, the remaining sets being manufactured in Australia. US ARGOSystems and British Aerospace Australia, together with Watkins Johnson of the USA and the Australian company Stanlite, are responsible for the development and manufacture of the EW system which will include the ARGOSystems AR 740 radar warning ESM system. The CH 93 periscope is to be replaced by a Pilkington CM010 optronic non-hull penetrating periscope.

The Integrated Ship Control, Management and Monitoring System (ISCMMS) is designed by Saab Instruments of Sweden and is based upon its SCC-200 steering control system. The system uses special and general purpose processors linked by a dual-redundant Enternet databus. Manoeuvrability is exercised through four aft control surfaces arranged in an X-configuration, each individually actuated. Fin-mounted hydroplanes confer accurate depth keeping.

UK GEC-Marconi Radar & Control Systems, in conjunction with GEC-Marconi Australia, is supplying the degaussing system.

Operational status
Entering service with the Royal Australian Navy.

Specifications
Displacement: 3,050/3,350 t
Length: 77.8 m
Beam: 7.8 m
Draught: 6.8 m

Contractor
Kockums, Malmo (design authority).
Australian Submarine Corporation, Port Adelaide (builder).

UPDATED

Submarine 2000

Type
Diesel-electric patrol submarine.

Description
Kockums, under the direction of the Swedish FMV (the Swedish Defence Materiel Administration), is also involved in prefeasibility studies on the Submarine 2000, a new design for the next century to follow on

from the 'Gotland' class. The outline of the submarine will be significantly different from current designs, with considerable emphasis being placed on stealth technology and manoeuvrability. The advanced hydrodynamic design will probably exhibit minimum sail area, much of which will be faired well down into the hull casing; a feature made possible by new mast technology and optronic sensors, all of which will be of the non-hull penetrating type. The command and control system will probably feature extensive

knowledge-based system architecture. The design may also be of the single watertight compartment type. New propulsion technology will also probably be incorporated and may include a pumpjet propulsion as well as AIP.

Contractor
Kockums, Malmo.

VERIFIED

UNITED KINGDOM

'Upholder' class

Type
Diesel-electric submarine.

Development
The decision to prepare an outline requirement for 'Upholder' class was taken in the mid-1970s. Two of the most important factors to be considered in the new design were the need to optimise noise reduction and the need to reduce snorting time required to recharge the batteries. By reducing these factors to an absolute minimum, the possibility of the boat being detected would be considerably reduced and its discretion improved.

The original Staff Target for the new design outlined a displacement of 2,250 tonnes submerged, later increased to 2,400 tonnes, which was endorsed in late 1977.

Discussions between MoD and the shipbuilder (VSEL) resulted in agreed characteristics that were the basis of the Naval Staff Requirement (NSR) submission that was finally approved in 1980. Among the specific characteristics laid down for the design were a displacement of 2,400 tonnes (hence the Type 2400 designation), a maximum diving depth well in excess of 200 metres, a long patrol radius with a time on patrol of 28 days and a 49 day stores endurance. Maximum underwater speed was to be over 20 kt with a maximum snort speed of over 10 kt, and surface speed of over 12 kt. Taking into account the normal underwater cruising speed together with the snort speed, and the frequency and period of the snort operation, the designers calculated that the boat could achieve an average speed of advance of 8 kt with a very low indiscretion rate. The detailed design and full contract definition were completed between 1981 and 1982.

To carry out its intended role, it was deemed necessary that the new SSK should embark the most advanced combat system available. The combat system covered not only the boat's combat system and weapons (which included torpedoes, missiles and mines both current and proposed), but also its navigation, sensors and comprehensive worldwide communication suite. Design flexibility was also to be extended to other areas such as the propulsion system. Also emphasised was the requirement to keep the crew size to less than 50. For the Royal Navy the significant consideration was a complement large enough to

achieve continuous and effective operation of the combat systems in extreme North Atlantic conditions. On the other hand it had to be low enough to be comfortably accommodated in the design. These criteria necessitated the extensive use of remote-control facilities and automation.

Description
To achieve the optimum underwater performance criteria laid down in the NSR a single-skinned, tear drop shaped hull with a beam: length ratio similar to that of the latest classes of SSN was adopted. As well as providing excellent manoeuvrability characteristics this also enabled a two-deck arrangement to be incorporated. Ballast tanks are sited fore and aft.

One of the most striking aspects of the 'Upholder' design is its excellent signal-noise ratio that makes it ideal for use as a sensor platform. Operating conditions have been optimised by concentrating on the stealth aspect and carefully selecting appropriate operating frequencies for the equipment. Stealth has been optimised by mounting elastomeric acoustic tiles on the outer hull, technology transferred from the nuclear submarine programme.

The pressure hull is constructed of HY-80 high tensile steel in three main watertight compartments.

Internal high tensile steel frames stiffen the circular hull, the after section being built of truncated conical sections conforming to the optimum hydrodynamic shape of the vessel. The fore and aft sections of the pressure hull terminate in dome bulkheads. The after watertight compartment contains all the machinery (diesels, air conditioning plant, electric motor and so on). Outside the pressure hull are Nos 3 and 4 main ballast tanks; there is space in one of the tanks for a secondary drive motor if required. The extreme aft end of the structure is free flooding and contains the hydroplane and rudder bearings, shaft, and linkages that are of cruciform arrangement.

The propeller shaft runs the full length of the aft external structure, passing through the aft dome bulkhead and main ballast tank bulkheads. Beneath the machinery deck in the after compartment are fuel, lubricating oil and compensating tanks; all the fuel and lubricating oil is carried within the pressure hull.

The after compartment is provided with an escape tower sited in the middle of the compartment, which also serves as an access. There is also an easily removed shipping opening, large enough to permit

quick removal and shipping of a complete diesel engine, which gives unrestricted access during maintenance periods.

The amidships watertight compartment houses the control room which includes communications, ESM, and COs quarters on the upper deck level with accommodation and auxiliary machinery spaces below. Underneath these two decks are oil fuel, fresh water and compensating tanks and No 2 battery compartment.

Two hatches allow access to the amidships compartment, one aft of the fin providing general access and battery loading, with the other sited inside the GRP-clad fin exiting to the navigation position and the five-man lock-out chamber which is used by divers when the boat is submerged.

The forward watertight compartment houses the main weapon stowage and handling and discharge equipment on the upper deck to full magazine standards, with ratings accommodation and extensive sonar processing equipment on the lower deck. Beneath is sited No 1 battery compartment with oil fuel tanks built into the wings.

The compartment is provided with three hatches, one of which is set at an angle for weapon loading. Embarkation rails are permanently fitted together with powered movement for handling the weapons. The other two hatches in the forward compartment are a one-man escape tower and a general access hatch.

The torpedo tubes themselves penetrate through the forward dome bulkhead of the pressure hull into the free flooding bow space; beneath is the large cylindrical sonar array that is protected by a GRP fairing. Also housed in this bow section is the water transfer equipment for the positive air turbine (pressurised water) discharge system of the torpedo tubes. Weapons' handling is fully automated, both to speed up loading and to reduce manpower requirements.

The forward free-flooding external structure also houses Nos 1 and 2 main ballast tanks. No 2 main ballast tank incorporates a recess housing the hydroplanes, which retract within the lines of the hull.

Both the escape towers are fitted with docking seats built into the flat top of the casing capable of taking a deep submergence rescue vehicle.

The single, seven-bladed, fixed pitch propeller is directly driven by a single air-cooled twin armature DC electric motor with shunt fields fed by two DC/DC static

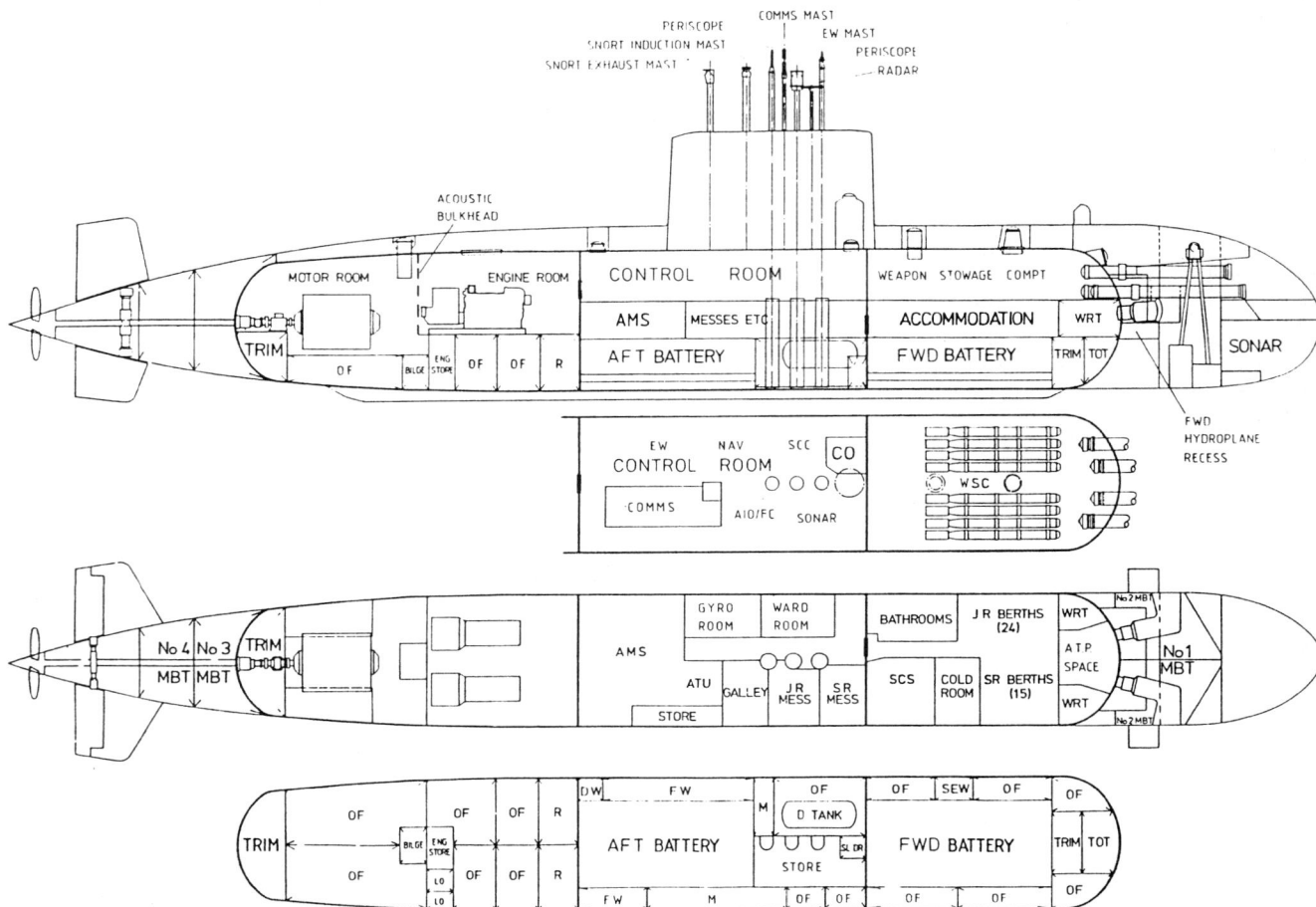

General arrangement layout of 'Upholder'

1999/0024691

converters and driving the shaft via a thrust block. A third converter is available on standby for use in emergency. The motor is separated from the diesel generators by a thermal and acoustic bulkhead. The motor operates in an oil-free environment using an open-loop air cooling system; water heat exchangers cooling the hot air as it leaves the motor.

The motor, which avoids compensating windings in order to minimise noise and simplify maintenance, can be connected either in series or in parallel, as can the two battery sections. In conjunction with shunt field control this gives a continuous speed range of approximately 20 to 100 per cent. Below 20 per cent control is achieved by means of armature voltage and field control. Selection of motor/battery combinations is achieved through motor-driven camshaft-operated contractors, with circuit breakers for fault protection. The system can be operated in a number of ways ranging from fully automatic, closed-loop control of shaft speed, through reducing automatic modes covering failure of system components to direct manual control of camshaft and field armature regulators.

The high battery capacity required is divided between two battery compartments, each of which contains 240 lead-acid cells developed by Chloride Power Storage. The internal design of the cell uses modern lead-acid technology.

The batteries provide power for the ship's electrical system, which operates at battery voltage (340-720V) and at three other voltages through converters: -
a) 440V 60Hz 3-phase, transformed and reduced to 115V
b) 115V 400Hz 3-phase and single phase
c) 24v dc

Battery charging is accomplished in three stages. Firstly at constant current or power when the battery is recharged to the point where it starts to gas. It is then charged at constant voltage to top up the battery and the current gradually falls to a low level. Charging can then be completed at a constant low level of current to ensure that all the cells are charged to the same level. Each charge cycle can be automatic.

High power is achieved with two new high-speed ALSTOM, the lightweight VALENTA RPA 200SZ diesels driving GEC 1.4MW ac generators with built-in rectifiers. These are multiphase machines using

brushless exciters and closed air cooling with dedicated seawater heat exchanger. The 4-stroke 16 cylinder diesels are of the direct injection mechanically supercharged and inter-cooled V-type. They are directly connected to the very short, single bearing AC generators, each with the rectifier mounted on top and overhanging the bearing. The generator is a self-excited double three-phase alternator, the two three-phase systems being displaced by 30° and through bridge rectifiers, paralleled on the DC side.

A single gear-driven centrifugal compressor drives the supercharger at the free end of the engine through an intermediate fluid coupling. Using a mechanically driven supercharger allows well over twice the engine power to be produced compared with normally aspirated engines in submarine conditions. It also leads to good cylinder scavenging under all service conditions.

The diesel exhausts include inboard silencers and independent surface and snorkel outlets. A cooling facility is provided to reduce the IR signature of the exhaust gases. All engines and generators are carried on resilient mountings to achieve the required noise and shock attenuation levels.

Remote control for most, and surveillance of all, ship and machinery control systems is centralised at two positions. The ship control console, sited in the control room, incorporates as well as machinery control surveillance, a new one-man ship control system. The centralised control and surveillance systems comprise three major units:
a) Surveillance panel, including VDUs
b) Electrical, diesels and battery panel
c) Ship control panel

Automatic (closed-loop) systems are available for depth, course, speed, and battery charging. Other systems (for example trimming) make use of semi-automatic control in which the operator decides when a response is required.

In the event of failure, comprehensive local instrumentation and manual controls are provided to enable safety and integrity to be maintained.

The advantages gained using these automated systems lead to a reduction in watchkeeping effort and ensure a high degree of efficiency and safety.

The main armament comprises six positive discharge 533 mm bow torpedo tubes that are capable

of handling the latest weapons. Up to a maximum of 12 full size reload weapons in varying combinations are carried on shock-protected mountings. The weapon handling equipment can handle weapons of varying lengths and diameters, and can be operated in fully automatic, power assisted, or manual mode, depending on user requirements. The weapon handling and discharge facilities (designed by Strachan and Henshaw) are similar to those developed for the 'Trafalgar' class SSN, but adapted to the 'Upholder' design.

The positive discharge system gives compatibility between operation of torpedoes and Sub Harpoon. A swim out system would have necessitated a separate launch system for the missile, as well as for the discharge of mines.

Discharge is microprocessor controlled, using fibre optic sensors to feed back data on the state of the tube and the weapon inside. This provides the combat system with rapid, up-to-date information on the weapon state. By using this system full control of the weapon, including actual discharge, can be exercised from the control room-based command system rather than from the tubes themselves. A sequence of hydraulic and mechanical interlocks ensures safety of operation at all times. This again results in considerable saving in manpower requirements.

The 'Upholder' class is fitted with the DCC command system that features a dual data bus system that enables updates and alterations to be achieved quickly and effectively. Secondary data handling facilities (outfit DCG) provide oceanographic calculations, additional target analysis and a data link terminal.

A number of sonar arrays are fitted, including a bow cylindrical array Type 2040 (French Thomson-Marconi Sonar SAS Argonaute) consisting of three layers of transducers set in 48 staves, providing both intercept and range data. Data gathered by the sonar is digitally processed covering adaptive, active, medium range passive and intercept modes simultaneously. Using adaptive processing simultaneously over LF/HF bands in passive mode, good target discrimination over long ranges is obtained, even under adverse conditions. An important feature of the sonar bow array is its size (3m high and 3m in diameter), the height of the array being very important for frequency operation and so on. The vertical aperture can also be varied.

In addition to the bow sonar the 'Upholder' class also carries a Type 2046 VLF passive search sonar, Type 2007 passive LF ranging sonar with three flank arrays on either side of the hull and a Passive/Active Ranging and Intercept System (PARIS).

Simultaneous passive tracking of up to 12 targets can be achieved with digital databus transfer of information to the multidisplay AIO.

For surface search the Type 2400 is fitted with Pilkington CK035 search periscope with a Racal passive ESM receiving antenna on top of the periscope; and/or an active vertically polarised antenna for signals in the communications bands. The attack periscope is the CH085 fitted with LLTV and thermal imaging giving it complete day/night capability.

The communications suite for the 'Upholder' provides comprehensive coverage for transmission and reception across the military HF and UHF and the International VHF band. In addition reception facilities will cover VLF/LF and MF bands. Equipment includes an HF transmitter (with spare); two MF/HF receivers; two VLF receivers; a UHF transceiver (with spare); VHF transceiver for International Marine band. Also carried are two teleprinters, and three remote voice user positions are sited around the boat. The MF/HF section incorporates a facility for full remote frequency channel selection, the first time this capability has been fitted in a submarine.

Modernisation
As part of the deal with Canada, the 'Upholders' will be upgraded to enable them to meet Canadian requirements. Part of the plan is to reuse some of the systems already installed in Canada's 'Oberon' class boats, such as the combat system, in order to provide compatibility with the existing heavyweight torpedoes in Maritime Command's inventory. Also included in the upgrade will be a new communications suite to provide standardisation with Canadian communications equipment.

Before handover the 'Upholder' boats will have new batteries installed and their machinery reconditioned.

In Canada the boats will have their weapon handling and discharge system modified to allow them to deploy the Mk 48 Mod 4 heavyweight torpedo. This will include replacing Outfit DCC currently fitted in the 'Upholder' boats with the Lockheed Martin Librascope SFCS Mk1 Mod C fire-control system currently installed in the 'Oberon' boats. The existing sonar suite will be retained with the addition of a retrofitted Canadian Subtass tail with a Hermes Electronics acoustic array which is currently installed in the 'Oberon' boats. An ESM system suitable for Maritime Command still remains to be selected.

Operational status
Entered service with the Royal Navy in the early 1990s, laid up 1994-95 under defence review cuts and subsequently put up for sale. Leased to the Canadian Maritime Forces 1998 with option to purchase.

Specifications
Displacement 2,167/2,455 t
Length: 70.3 m
Beam: 7.6 m
Draught: 5.5 m

Contractor
Cammell Laird, Birkenhead (VSEL).

NEW ENTRY

UNITED STATES OF AMERICA

'Los Angeles' class

Type
Nuclear attack submarine.

Development
Work on designing the 'Los Angeles' class SSN (SSN-688) began in 1968. Since then ship design has undergone a revolution and the technology of the 1960s and 1970s has been replaced by the computer and computer-aided design and manufacturing technology, which has done much to assist in the upgrading of the class throughout its life. Modular construction technology has also been introduced during the construction of the class. This involved prefabricating eight large cylindrical sections which were extensively outfitted prior to assembly as the main hull. The design was configured to meet the requirement for SSNs to operate with the carrier battle groups and to counter the Russian 'Victor' class SSNs. To accomplish this an extra 5 kt speed compared to the preceding 'Sturgeon' class was requested, which resulted in an increase in displacement of some 2,300 tonnes. The other major aspect of the design related to the combat system in which a fundamental change was made, when digital computer technology was introduced to replace analogue technology. This enabled the boats to handle more complex tactical situations than previously.

The design has been progressively upgraded since the first boat put to sea in 1976 and major combat capabilities have been incorporated since 1985, beginning with USS *Providence* (SSN-719) and subsequent units. Among additions incorporated into the design are 12 vertical launch tubes mounted externally for deploying Tomahawk cruise missiles and the AN/BSY-1 advanced combat system and Mk 117 fire-control system which was installed in 1986 in the USS *San Juan* (SSN-751) and subsequent units. Other upgrades installed include the addition of under ice navigation equipment and minelaying capabilities. These changes have resulted in three distinct subclasses. The modifications have left virtually no volume or weight margins for additional equipment, which has led to the need for a follow-on design. The first six units in the class have now been decommissioned.

Description
The 'Los Angeles' class was designed to be built of HY-100 steel, but were actually constructed of HY-80. From *San Juan* (SSN-751) on the sail-mounted planes have been repositioned near the bow and made retractable for under ice navigation, while the sail has been hardened to cope with the stresses imposed as the submarine surfaces through the Arctic ice.

'Los Angeles' class submarine Tucson *on sea trials June 1995* (Courtesy Newport News Shipbuilding)
1998/0007215

As originally designed the class was armed with four torpedo tubes for launching Mk 48 torpedoes (subsequently replaced by Mk 48 ADCAP), the Subroc missile and Sub-Harpoon. It was intended that as Subroc became obsolescent during the 1980s it would be replaced by a new missile, the Sea Lance. However, the latter was cancelled and Subroc was withdrawn from service in 1990. The first units to be modified to fire the Tomahawk missile from a torpedo tube was the *La Jolla* (SSN-701), with the *Atlanta* (SSN-7122) being the first to operationally deploy the missile. The normal payload comprised eight Tomahawk missiles, but for specific operations this number was increased, for example during Operation Desert Storm. The introduction of the Tomahawk missile meant that the analogue Mk 113 fire-control system could no longer be used, and so the Mk 117 digital fire-control system was introduced on later boats and retrofitted to earlier vessels in the class. From USS *Providence* (SSN-719) onwards all boats have been fitted with 12 vertical launch tubes for the Tomahawk missile. These are sited just behind the bow sonar and in front of the forward dome shaped bulkhead. With special launch tubes allocated for the Tomahawk, the boats have been able to redress the reduction in numbers of torpedoes carried when the missiles were fired from the torpedo tubes, as well as increasing the standard number of missiles carried from eight to 12. This does not preclude the boats from carrying additional Tomahawk missiles for launching from the torpedo tubes should it be deemed necessary.

The original sonar sensor outfit comprised the large BQQ-5 spherical bow array (since upgraded through various modifications to the latest BQQ-5E) and the BQR-15 clip-on towed array. The towed array is stored in a large channel on the starboard side of the casing. The ice navigation sonar BQS-15 was subsequently installed. Beginning in 1998 it is intended to install the Near-term Mine Reconnaissance System (NMRS) and the accompanying AQS-14 side scan sonar which will be deployed from a torpedo tube. Six of the class will also be fitted to deploy the ASDS.

The class is fitted with the S6G pressurised water reactor (26 MW). The last boat in the series, *Cheyenne*, is fitted with the prototype propulsion plant for the 'Seawolf' class. Noise reduction improvements have also been incorporated in many of the boats, including anechoic tiles which have been applied to the outer hull.

As the last 'Sturgeon' class SSN are decommissioned, the 'Los Angeles' class are being fitted with the Dry Dock Shelter (DDS) carriers, to take on the role previously performed by the 'Sturgeon' boats. The DDS houses a Swimmer Delivery Vehicle (SDV), or inflatable boats or a platoon of SEALS.

Contractors
General Dynamics, Groton, Connecticut.
Newport News Shipbuilding, Newport News, Virginia.

UPDATED

'Seawolf' class

Type
Nuclear attack submarine.

Development
Work on defining requirements for a new nuclear-powered attack submarine design began in 1982, and a programme for the new design, subsequently known as the 'Seawolf' (SSN-21), was formally initiated in 1983. This would require a construction programme of some three to four boats a year. It was planned to start work on building the lead ship of the new class in 1989 with delivery due at end of 1994 or beginning of 1995.

The aim of the project was to define a multimission platform whose primary task would be to provide a counter to Soviet SSNs deployed in protecting their SSBNs and the Soviet landmass, or engaged in attacking US or allied shipping, naval battle groups or land targets.

To ensure continuity of capability over a period of time the design had to exhibit significant growth potential so that major improvements could be incorporated which would enable the boats to counter more sophisticated threats which might develop in the future. Hence the design had to be physically evolutionary and revolutionary in capability. This led to the programme team (operating under the name Group Tango) to define six goals (*not* requirements) for potential performance parameters which were to be the subject of detailed study and development for incorporation in the design and which might be achieved during the boat's service life between 1995 and 2035. These six goals or ideal parameters were identified as being: quietness, speed, depth, weapons load, launch tubes, and ability to operate under ice in the Arctic.

Originally the US Navy intended to acquire a force of 30 of the new submarines to supplement the 'Los Angeles' class (SSN-688). Following the ordering of the lead ship in 1989 under FY89 Budget, it was planned that two boats would be acquired in FY91 and then an average of three a year would be ordered until 2000, when the target goal of 30 boats would be reached. Then, in 1990, with the lead ship already under construction, the programme ran into a major funding problem when the Senate Armed Services Committee voted to delete all funds for the construction of 'Seawolf' boats under the FY91 and future programmes. This vote was rigorously opposed by the House Armed Services Committee and on 13 August 1990 the then Secretary of Defense, Dick Cheney, approved funding for one more 'Seawolf' (SSN-22) under the FY91 Budget.

A major controversy surrounded the programme with a number of objections being voiced. Among the objections raised were the cost of the programme, the level of technology incorporated in the design, the impact of the cost on the numbers planned and the industrial base which would supply all the equipment, and the capability of the design. At this point some observers noted that the design would only meet two of the goals set out in the original plan, namely quieting and the ability to operate under ice in the Arctic.

In June 1991 the US Navy discovered serious cracks in the welding in the hull of SSN-21 which would require replacement of all welds completed up to that date. This resulted in a major delay in the construction of the lead ship.

Finally, in late January 1992, it was announced that after the third 'Seawolf' submarine (SSN-23), to be built under the FY92 Budget, no more submarines would be authorised for construction for a period of seven years.

Although authorised under the FY92 Budget, approval for partial funding of US$700 million for SSN-23 was not finally given until the FY96 Budget. Also under the FY96 Budget, funds were allocated for the completion of 'Seawolf' (SSN-21) and 'Connecticut' (SSN-22).

The decision to discontinue construction of the highly controversial and very expensive 'Seawolf' was in the main influenced by a very much constrained defence budget, technical problems and to a degree the perceived diminished threat from the Russian Fleet.

Description
The hull is constructed of HY-100 steel enabling the boat to dive to much greater depths (in the region of 600 m) than the 'Los Angeles' SSN-688 class. Although many of the improvements which have been embodied in the later units of the 'Los Angeles' class have been incorporated in the design, the 'Seawolf' shows a significant improvement over the SSN-688 design in a number of areas, achieving a greater diving depth, higher speed, and exhibiting a much lower overall acoustic signature (it was planned that the 'Seawolf' should be 15 decibels quieter) aided by the full use of acoustic cladding, and with 25 per cent greater weapons payload (50 weapons compared to the 37 of the SSN-688). The significant reduction in acoustic signature has resulted in the 'Seawolf' being able to maintain speeds in excess of 20 kt while still being able to effectively operate a passive sonar (this is referred to as the maximum acoustic speed). The only other submarines capable of achieving such maximum acoustic speeds are the Russian 'Akula' and presumably 'Severodvinsk' classes.

Among the new features are a smaller length beam ratio which improves manoeuvrability. This has resulted in a teardrop hull that is nearly 30 per cent larger than the SSN-688. Manoeuvrability is further aided by a fin leading into the lower part of the faired sail, bow planes which can be withdrawn into the boat for under ice operations, and six stern fins.

The increased volume of the 'Seawolf' has allowed for the incorporation of modern sound-shielding materials and enhanced radiation safety features. The propulsion system comprises a newly designed pressurised water reactor, the S6W developed by General Electric. The system has been designed to meet stringent noise and weight reduction targets which have been achieved by the development and application of new, high-strength, lightweight materials in the major reactor components, piping and foundations which have also reduced the possibility of corrosion. Developing 52,000 hp (38.8 MW) to give a maximum quoted speed of 35 kt, the reactor drives two turbines feeding a single shaft pumpjet. Secondary propulsion is provided by a Westinghouse motor.

The US Navy examined various alternative designs for the stern arrangement and eventually selected a British-designed pumpjet propulsor, similar to those fitted in the Royal Navy's 'Trafalgar' class.

The heart of the 'Seawolf' design is the AN/BSY-2 fully distributed combat system which incorporates extensive use of modular software and local processors to provide targeting, weapon and mine setting, launch processing and after launch control, over-the-horizon targeting and combat system management. This is the first fully distributed system to enter service with the US Navy and is based around UYK-44 computers with a fibre optic LAN databus in the combat centre.

Sensors integrated by the BSY-2 include: the BQG-5 wide aperture hull-mounted passive flank array providing long-range detection and localisation; the TB-16 passive array (part of the BQQ-5 sonar system) which is stored in a sheath along the boat's hull; the TB-23 thin-line array (part of the upgraded BQQ-5D) which is reeled into the ship's main ballast tanks; a large, bow-mounted, spherical MF active long-range panoramic sonar with listening mode; and the short-range, active, high resolution, HF BQS-24 sail-mounted sonar capable of detecting small targets including mines (MIDAS - Mine and Ice Detection and Avoidance System).

Armament comprises eight 660 mm torpedo tubes mounted in the hull abaft the large, bow-mounted sonar array. These large tubes (outside dimensions 762 mm) are designed to enable them to accommodate various combinations of 50 weapons including cruise missiles, various ASW weapons including mines and at some time in the future UUVs. The weapon mix includes: TOMAHAWK (both land attack and anti-ship versions) cruise missiles; HARPOON anti-ship missiles; and ADCAP Mk 48 torpedoes. Alternatively up to 100 mines, including the CAPTOR, can be carried in lieu of the HARPOON and torpedoes.

Other systems fitted include a Kollmorgen Type 18 periscope and the AN/WQC-6 underwater communications system.

The 'Seawolf' is manned by a crew of 133, including 12 officers.

The delay in starting work on the third boat in the series has enabled it to be reconfigured for the operation of Special Operations Forces (SEALs). This has meant reconfiguring the torpedo room to provide a much more habitable area for the SEALs. Also incorporated in the design is a lock in/lock out chamber capable of accommodating an 8-man SEAL team. The boat will also be able to host the Advanced Swimmer Delivery System (ASDS), in essence a mini-submarine. It is said that these features can be easily retrofitted into SSN-21 and SSN-22 if required.

Operational status
The lead ship 'Seawolf' is now operational. Although scheduled for completion in 1995, construction was delayed when defective welds were discovered in the hull in July 1991. The problem was traced to unduly high carbon content in the wire used to weld the HY-100 steel. This led to brittle and cracked welds throughout some 16 per cent of the hull that had been completed at that time. In addition, concern over the titanium torpedo hatches led to a change to HY-100 steel in the construction of eight breech doors. Sea trials in late 1996 showed that the boat exceeded expectations for speed, stealth, and sensor performance. A further delay then resulted before the boat could be commissioned in early 1997, when some contour panels over the wide aperture sonar array became detached from the hull and had to be refitted.

Specifications
Displacement: 7,460/9,137 t
Length: 107.6 (oa) m
Beam: 12.9 m diameter
Draught: 10.97 m

Contractor
General Dynamics Electric Boat Division, Groton, Connecticut.

UPDATED

NSSN

Type
Nuclear attack submarine.

Development
Following the controversy surrounding the 'Seawolf' class, the US Navy embarked on a study to examine a new attack submarine design as a low-cost complement to the 'Seawolf'.

Referred to originally as the 'Centurion' and now known as the NSSN, this design too is proving to be highly controversial. There was no denying the need to develop a new attack submarine to replace the 'Los Angeles' class, the question was 'what should the new submarine be like?' Under the FY96 National Defense Authorization Act, US$704.5 million was specifically allocated for long-lead and advance procurement items for a prototype submarine to be ordered under FY98 and US$100 million for long-lead and advance construction costs for an SSN to be authorised in FY99 and built by Newport News. Plans also envisaged funding being provided for the construction of four attack submarines to be authorised and funded in FY98-01. It was envisaged that this might eventually result in a class of 29 boats. A further US$100 million was to be made available to ARPA for the development and demonstration of advanced technologies including electric drive, hydrodynamic quieting and ship control automation. Under the FY96 budget, Congress provided additional funding for advanced procurement of materials for the first NSSN and gave authorisation for the acquisition of four NSSN. This was to provide contracts for General Dynamics Electric Boat Division for one NSSN each in FY98 and FY2000 and to Newport News Shipbuilding for one boat each in FY99 and FY01. The Bill also stipulated that new technologies were to be incorporated into each boat each year as they became available and their value proven.

The US DoD has agreed that NSSN is the most affordable solution to meet the replacement requirement and to preserve the US Navy's critical undersea warfare capabilities.

With the ending of the Cold War, one of the US Navy's redefined primary mission requirements is the ability to covertly insert military force ashore in a hostile environment. Potential missions for the NSSN missions have been evaluated as: collection of critical intelligence using advanced electronic sensors; management of the underwater battle space using advanced platform-mounted sonars and special offboard systems; the detection of minefields in the path of the surface battle or amphibious group; monitoring of other threats and targets to ensure that campaign objectives are achieved; the support of a wide range of covert special warfare missions including: SAR in hostile territory, gathering of intelligence, sabotage and diversionary attacks; other clandestine missions.

Hence the decision to implement from the outset the ability to covertly deploy Special Forces in the design.

However a US General Accounting Report published in the spring of 1998 casts some doubt on the potential effectiveness of the design. The report noted that some of the NSSN system requirements had been reduced. Of primary concern in the light of recent advances in open ocean, ASW capabilities are EW and acoustic intercept. Funding cuts had led to minimal loads of capability. Other areas of concern noted in the report were the choice of propulsor which still remained to be decided, the external communications outfit and the TB 29 towed array threat detection sonar which the navy considered too expensive.

Description

To meet the criteria for the incorporation of new technologies, and to meet the demand for a very wide range of mission possibilities, the whole design concept of the NSSN is centred on modularity. As such, therefore, the design and manufacturing processes will have to be sufficiently flexible to adapt to future mission requirements which may not at present be fully anticipated.

The design will feature three main areas of modularity: construction — the boats will be assembled from self-contained subsystems which themselves will be assembled and tested prior to integration with the main hull; technological — subsystems will be assembled using open architecture components offering an adaptable configuration which will facilitate maintenance and the incorporation of new technology throughout service life of the boat; operational — the baseline design will emphasise the capability to reconfigure the boat to undertake a number of different missions, either through construction of different variants of the basic design or through post-construction upgrades.

To achieve the required modularity the boats will be assembled using a number of fully self-contained modules, incorporating key internal systems which can be tested in the factory prior to insertion in the main hull. The main modules are: pressure hull, non-pressure hull, command and control system, weapons handling, machinery room, auxiliary machinery room, accommodation area, and sail. It is this last modularity

area which forms the really revolutionary aspect of the NSSN. Using this concept boats can be assembled incorporating mission-specific hull modules either during construction or at a later date during refit or overhaul. In this way a boat can be reconfigured for different tasks during its lifetime without causing it to undergo a major rebuild.

The design will incorporate a new modular isolated deck structure to house a number of structurally integrated enclosures to protect computerised combat systems and accommodate commercial equipment without need for MILSPECs. This will enable extensive use of open architecture COTS electronic systems, particularly in the command, control, communications and intelligence (C³I) area to be fitted. As a result the NSSNs combat system will develop more than 20× the signal processing capacity and 50× the data processing capacity of 'Seawolf'. Anticipating the development of even more powerful computers and new weapons systems in the future, the NSSN will be equipped with a new asynchronous databus. As this will be built into the boat's modular units, it will be much easier to change the databus during the boat's lifetime. The fully distributed, open architecture databus combat system will enable UUV/UAV's to be incorporated into the boat for littoral warfare, and also enable the deployment of the army's new tactical missile system, to deliver smart anti-tank weapons in direct support of forces ashore.

The weapons payload, although less extensive than that of the 'Seawolf', will, nevertheless, be very impressive incorporating a mix of land-attack weapons in 12 vertical launch cells and 26 other torpedo tube-launched weapons including: Mk 48 ADCAP torpedoes, anti-ship cruise missiles, mines and for certain missions Unmanned Underwater Vehicles (UUVs) launched from the four torpedo tubes which will be adapted from those fitted in the 'Seawolf'. Other weapons under evaluation for possible use include a submerged launch version of the US Army's enhanced fibre optic-guided (E-FOGM) anti-helicopter and anti-ship missile and Unmanned Aerial Vehicles (UAVs) to provide real-time monitoring of the near and overland battle space. For self-defence the boats will carry a wide range of countermeasures including an internal reloadable ejector and 14 external non-reloadable ejectors.

To provide for the deployment of SEALs and Special Operations Forces of other services and their equipment for extended periods in forward areas, the design will feature a dry deck shelter and the Advanced Swimmer Delivery System (ASDS). A special lock in/lock out chamber will accommodate a SEAL team of eight or nine members. The torpedo room will also be reconfigurable to provide for improved habitability for Special Forces during transit to the operational area.

To provide the necessary capability for the new boat

to undertake the wide range of missions envisaged, it will be equipped with an advanced photonics mast fitted with improved night vision, enhanced image recognition and a laser rangefinder in place of the standard periscope. Signals from the photonic and advanced ASTECS ESM sensors will be relayed to the Combat Control System Mk 2 and AN/BSY-2 command system via a fibre optic link. All masts will be of the non-hull penetrating type which will lead to a complete redesign of the sail configuration compared to existing boats, which are all fitted with hull penetrating masts. Sonar sensors will include a large spherical active/passive bow passive/active array sonar, the AN/BQG-5 lightweight wide aperture array, a towed array and a large flank array, as well as passive ranging sonar, intercept sonar and HF sail- and chin-mounted active sonars for mine detection and ice navigation. It is envisaged that the NSSN will be able to deploy the advanced fibre optic Acoustic Deployable System (ADS) that can detect and locate quiet submarines and mining activity in shallow water regions over a long period of time.

Provision is also being made to fit an advanced submarine communications incorporating a mast carrying SHF frequency antenna for satellite links for secure, high volume, high data rate with instantaneous transmission of images to other ships and submarines. The communications suite will also enable the boat to communicate with offboard underwater, surface, airborne and space-based assets, giving it the ability to operate with other platforms at far greater ranges than is currently possible. This is vital if the boats are to be properly integrated with the entire battle force for both strategic and tactical missions in both shallow and deep waters. With the focus which the US Navy now places on operations in littoral waters and particularly for operations in potentially mined areas, the design will incorporate an electromagnetic silencing system and advanced mine detection and avoidance systems.

The NSSN will be powered by an S9G PWR reactor adapted from the 'Seawolf' power plant. This will be a much more compact system than previous plants, and is the first specifically designed not to need refuelling during the entire 30-year service life of the boat.

Specifications
Displacement: 7,700 t submerged
Length: 115 (oa) m
Beam: 10.5 m
Draught: 9.3 m

Contractor
General Dynamics Electric Boat Division, Groton, Connecticut.
Newport News Shipbuilding, Newport News, Virginia.

UPDATED

TABLE IV

Submarine designs - specifications and electronics

Class	Navy	Displacement	Dimensions	Weapon Control	Sonars	Radar	EW
Type 205	Germany	419/450	43.9 × 4.6 × 4.3	Mk 8	SRS M1H	Calypso II	ESM
Type 206	Germany	450/498	48.6 × 4.6 × 4.5	Mk 8	DUUX 2, 410 A4	Calypso II	DR 2000U
Type 206A	Germany	450/498	48.6 × 4.6 × 4.5	CSU 83	DUUX 2, DBQS-21D	Calypso II	DR 2000U + Sarie 2
Type 207 (Tumleren)	Denmark	459/524	47.4 × 4.6 × 4.3	Terma	PSU 83	Terma	Sea Lion
Type 207 (Kobben)	Norway	459[1]/524	47.4 × 4.6 × 4.3	MSI-90(U)	CSU 83	Type 1007	ArgoSystems
Type 209/1100-1200 (Glavkos)	Greece	1,100/1,210	54.4 × 6.2 × 5.5 (Type 1200 55.9 m)	Kanaris	CSU 83, CSU 3-4	Calypso II	AR-700-S5
Type 209/1200 (Salta)	Argentina	1,248/1,440	55.9 × 6.3 × 5.5	M8	CSU 3, DUUX 2C, DUUG 1D	Calypso II	DR 2000
Type 209/1200 (Pijao)	Colombia	1,180/1,285	55.9 × 6.3 × 5.4	M8/24	CSU 3-2, PRS 3-4	Calypso II	
Type 209/1200 (Chang Bogo)	South Korea	1,100/1,285	56 × 6.2 × 5.5	ISUS 83	CSU 83		ArgoSystems
Type 209/1200 (Casma)	Peru	1,185/1,290	56 × 6.2 × 5.5	Sepa Mk 3[2]	CSU 3, DUUX 2C3	Calypso	ESM
Type 209/1200 (Atilay)	Turkey	980/1,185	61.2 × 6.2 × 5.5	Sinbads[4]	CSU 3	S33B	DR 2000
Type 209/1300 (Thomson)	Chile	1,260/1,390	59.5 × 6.2 × 5.5		CSU 3	Calypso II	DR 2000U
Type 209/1300 (Shyri)	Ecuador	1,285/1,390	59.5 × 6.3 × 5.4	M8 Mod 24	CSU 3, DUUX 2	Calypso	DR 2000U
Type 209/1300 (Cakra)	Indonesia	1,285/1,390	59.5 × 6.2 × 5.4	Sinbads	CSU 3-2, PRS-3/4	Calypso	DR 2000U

Class	Navy	Displacement	Dimensions	Weapon Control	Sonars	Radar	EW
Type 209/1300 (Cabalo)	Venezuela	1,285/1,600	61.2 × 6.2 × 5.5	STN ATLAS	CSU 3-32, DUUX 2	Scanter Mil	DR 2000U
Type 209/1400 (Tupi)	Brazil	1,260/1,440	61 × 6.2 × 5.5	KAFS	CSU 83/1	Calypso III	DR-4000
Type 209/1400 (Preveze)	Turkey	1,454/1,586	62 × 6.2 × 5.5	ISUS 83-2	CSU 83, TAS-3	S 63B	Porpoise
Type 209/1500 (Shishumar)	India	1,660/1,850	64.4 × 6.5 × 6	Mk 1	CSU 83, DUUX-5	Calypso	Phoenix II AR 700 or Sea Sentry
Type 212	Germany	1,450/1,800	56 × 7 × 5.8	MSI-90U	DBQS-40, FAS-3, TAS-3, FMS 52, MOA 3070	Type 1007	FL 1800U
Abtao	Peru	825[1]/1,400	74.1 × 6.7 × 4.3		Eledone 1102/5	SS-2A	ESM
Agosta	France	1,230/1,510	67.6 × 6.8 × 5.4	DLA 2A	DSUV 22, DUUA 2D, DUUA 1D, DUUX 2, DSUV 62A	DRUA 33	Arur, Arud
Agosta (Hashmat)	Pakistan	1,490/1,740	67.6 × 6.8 × 5.4	Subics Mk 2	DSUV 2H, DUUA 2A/2B, DUUX 2A, DUUA 1D	DRUA 33	Arud or DR-3000U
Agosta (Galerna)	Spain	1,490/1,740	67.6 × 6.8 × 5.4	DLA 2A	DSUV 22, DUUA 2A/2B, DUUX 2A/5, DSUV-62	DRUA 33C	Manta
Agosta 90B	Pakistan	1,510/1,760	67.57 × 6.8 × nk	SUBTICS	n/k	n/k	DR3000U
Amur	RFAS	1,450/2,100	58 × 7.2 × 4.5		MF & LF	I-band	ESM
Collins	Australia	3,051/3,353	77.5 × 7.8 × 7	Rockwell	Scylla, Kariwara	Type 1007	AR 740
Daphne	France	860/1,038	57.8 × 6.8 × 4.6	DLT D3	DSUV 2, DUUA 2, DUUX 2	Calypso	
Daphne (Hangor)	Pakistan	869/1,043	57.8 × 6.8 × 4.6		DSUV 1	DRUA 31	Arud
Daphne (Albacora)	Portugal	869/1,043	57.8 × 6.8 × 5.2	DLT D3	DSUV 2, DUUA 2	KH 1007	Arur
Daphne	South Africa	869/1,043	57.8 × 6.8 × 4.6	DCSC-2	DUUA 2, UEC	Calypso II	Arud
Delfin (Daphne)	Spain	869/1,043	57.8 × 6.8 × 4.6	DLA 2A	DSUV 22, DUUA 2A, DUUX 2A, DSUV 62[5]	DRUA 33C	Manta E
Dolphin	Israel	1,640/1,800	57.3 × 6.8 × 6.2	ISUS 90-1	CSU-90, PRS-3, FAS-3	Elta	Timnex 4CH(V)2
Dolphin**	South Korea	70/83	25 × 2.1	–	HF	–	–
Foxtrot (Type 641)	Cuba	1,950/2,475	91.3 × 7.5 × 6		Herkules/Feniks	Snoop Tray	Stop Light
Foxtrot (Type 641)	India	1,952/2,475	91.3 × 7.5 × 6		Herkules/Feniks	Snoop Tray	Stop Light
Foxtrot (Type 641)	Libya	1,950/2,475	91.3 × 7.5 × 6		Herkules/Feniks	Snoop Tray	Stop Light
Foxtrot (Type 641)	Poland	1,952/2,475	91.3 × 7.5 × 6		Herkules	Snoop Tray	Stop Light
Foxtrot (Type 641)	RFAS	1,952/2,475	91.3 × 7.5 × 6		Pike Jaw	Snoop Tray or Snoop Plate	Stop Light
Gotland (A 19)	Sweden	1,240/1,490	60 × 6 × 5.6	IPS-19	CSU-90	Terma	Manta S
Gal (Vickers Type 540)	Israel	420/600	45 × 4.7 × 3.7	STN ATLAS	STN ATLAS, EDO 1110	Plessey	Timnex 4CH(V)1
Guppy II	Taiwan	1,870[1]/2,420	93.7 × 8.3 × 5.5		BQR 2B, BQS 4C, DUUG 1B	SS 2	WLR-1/3
Guppy IIA	Turkey	1,848/2,440	93.2 × 8.2 × 5.2	Mk 106	BQR 2B, BQS 4, BQG 3	SS 2A	
Guppy III	Turkey	1,975[1]/2,450	99.5 × 8.2 × 5.2		BQR 2B, BQG 4	SS 2A	
Hai Lung	Taiwan	2,376/2,660	66.9 × 8.4 × 6.7	Sinbads M	SIASS-Z	ZW 06	AR 700SF, Timnex 4CH(V)2
Harushio	Japan	2,450[1]/2,750	77 × 10 × 7.7		ZQQ 5B, ZQR 1	ZPS 6	ZLR 3-6
Heroj	Yugoslavia	615/705	50.4 × 4.7 × 4.5		Eledone	Snoop Group	Stop Light
Inrepido**	Colombia	58/70	23 × 4	–	–	–	–
Kilo (Project 877E)	Algeria	2,325/3,076	73.8 × 9.9 × 6.6		Shark Teeth, Shark Fin, Mouse Roar	Snoop Tray	Brick Group
Kilo (Project 877EKM/636)	China	2,325/3,076	73.8 × 9.9 × 6.6		Shark Teeth, Mouse Roar	Snoop Tray	Squid Head/Brick Pulp
Kilo (Project 877EM)	India	2,325/3,076	72.6 × 9.9 × 6.6		Shark Teeth, Shark Fin, Mouse Roar	Snoop Tray	Stop Light
Kilo (Project 877EKM)	Iran	2,356/3,076	73.8 × 9.9 × 6.6		Shark Teeth, Mouse Roar	Snoop Tray	Squid Head
Kilo (Project 877E)	Poland	2,325/3,076	73.8 × 9.9 × 6.6		Shark Teeth, Whale	Snoop Tray	Brick Group
Kilo (Project 877E)	Romania	2,325/3,076	72.6 × 9.9 × 6.6		Shark Teeth, Shark Fin, Mouse Roar	Snoop Tray	Brick Group
Kilo (Project 877/877K/877M)	RFAS	2,325/3,076	72.6 × 9.9 × 6.6		Shark Teeth, Shark Fin, Mouse Roar	Snoop Tray	Squid Head or Brick Pulp
LOS ANGELES	USA	6,082/6,927	110.3 × 10 × 10	CCS Mk2[6], BSY-1, Mk 117	BQQ-5D/E, BQG-5D[7], BQR-23/25, BQS-15, MIDAS	BPS 15	BRD-7, WLR-8(V)2, WLR-10
MG 110**	Pakistan	n/k/118	27.8 × 5.6	–	–	–	–
Ming (Type 035)[8]	China	1,584/2,113	76 × 7.6 × 5.1		Pike Jaw, DUUX 5	Snoop Tray	
Moray	Netherlands						
NSSN	USA	n/k/7,700	115 × 10.5 × 9.3	CCSM	TB16, TB29	BPS 16	WLQ-4(U)
Nacken (A 14)	Sweden	1,015/1,085	49.5[9] × 5.7 × 5.5	IPS-17	Thomson Sintra	Terma	AR-700-S5
Narhvalen	Denmark	420/450	44.3 × 4.6 × 4.2	M8	PSU 83	Terma	Sea Lion
Oberon	Australia	2,030/2,410	90 × 8.1 × 5.5	SFCS Mk 1	CSU 3-41, Type 2007, BQQ 4	Type 1006	Mavis ODU
Oberon (Humaita)	Brazil	2,030/2,410	90 × 8.1 × 5.5	DCH	CSU 90, Type 2007	Type 1006	UA 4
Oberon	Canada	2,030/2,410	90 × 8.1 × 5.5	Librascope	Triton[10], Type 2007, BQG 501[11]	Type 1006	Guardian Star
Oberon	Chile	2,030/2,410	90 × 8.1 × 5.5	STN ATLAS	Type 2007, CSU 90	Type 1006	DRU 2000U
Oyashio	Japan	2,700/3,000	81.7 × 9 × 9		ZQQ 5B, ZQR 1	ZPS 6	ZLR 7
Pyranja*	RFAS	218/253	29.2 × 5 × 3.9	–	HF	–	–
Romeo	Bulgaria	1,475/1,830	76.6 × 6.7 × 4.9			Snoop Plate	Stop Light
Modified Romeo ES5G	China	1,650/2,100	76.6 × 6.7 × 5.2		Herkules or Pike Jaw	Snoop Plate[12]	
Romeo (Type 033)	China	1,475/1,830	76.6 × 6.7 × 5.2	–	Herkules or Tamir 5, DUUX 5	Snoop Plate[13]	
Romeo (Type 031)	North Korea	1,475/1,830	76.6 × 6.7 × 5.2		Tamir 5L, Feniks	Snoop Plate	
Romeo	Egypt	1,475/1,830	76.6 × 6.7 × 4.9	Mk 2	CSU 83, Loral		AR-700
Sang-O	North Korea	256/277	35.5 × 3.8 × 3.7	–	n/k	I-band	
Santa Cruz (TR 1700)	Argentina	2,116/2,264	66 × 7.3 × 6.5	Sinbads	CSU 3/4, DUUX 5	Calypso IV	Sea Sentry III
Improved Sauro	Italy	1,476/1,662	64.4[14] × 6.8 × 5.6	BSN 716(V)2[15]	IPD 70/S, MD 100S	BPS 704	BLD-727
Sauro (Type 1081)	Italy	1,456/1,631	63.9 × 6.8 × 5.7	BSN 716(V)1[15]	IPD 70/S, MD 100	BPS 704	BLD 727
Sava	Yugoslavia	830/960	55.7 × 7.2 × 5.1		PRS-3	Snoop Group	Stop Light
Scorpene	Chile	1,450/1,590	63.5 × 6.2 × nk	SUBTICS		Type 1007	AR-900
Sea Horse[a]	Taiwan	n/k/52	14.5 × 2.3	–	n/k	–	–
SEAWOLF	USA	7,460/9,137	107.5 × 13 × 11	BSY-2	BQQ-5D, TB16, TB23, BQS-24	BPS 16	WLQ-4(U)

Class	Navy	Displacement	Dimensions	Weapon Control	Sonars	Radar	EW
Sjoormen (A 12)	Singapore	1,130/1,210	51 × 6.1 × 5.8	IPS-12	Hydra	Terma	
Sjoormen (A 12)	Sweden	1,130/1,210	51 × 6.1 × 5.8	IPS-12	Hydra	Terma	
Song	China	1,700/2,250	75 × 8.4 × 5.3	n/k	MF & LF	I-bard	Type 921-A
Spiggen II*	Sweden	n/k/14	11 × 1.7 × 1.4	–	–	–	–
Tang	Turkey	2,100/2,700	87.4 × 8.3 × 5.8	Mk 106	BQR 2B, BQS 4, BQG 4	BPS 12	WLR-1
Tango (Som) (Type 641B)	RFAS	3,000/3,800	91 × 9.1 × 7.2		Shark Teeth/Shark Fin, Mouse Roar	Snoop Tray	Squid Head or Brick Group
Tolgorae*	South Korea	150/175	n/k	–	HF	–	–
Improved Tupi[16]	Brazil	1,850/2,425	67 × 8 × 5.5	KAFS	CSU 83/1	Calypso III	DR-4000
Ula (Type P 6071)	Norway	1,040/1,150	59 × 5.4 × 4.6	MSI-90(U)	CSU 83	Type 1007	Sealion
Una*	Yugoslavia	76/88	18.8 × 2.7 × 2.5		HF	–	–
Upholder	Canada	2,168/2,455	70.3 × 7.6 × 5.5	SFCS Mk1	Type 2006 Type 2007 Type 2040	Type 1007	
Vastergotland (A 17)	Sweden	1,070/1,143	48.5 × 6.1 × 5.6	IPS-17	CSU 83	Terma	AR-700-S5
Velebit	Croatia	88/99	21.09 × 2.7 × 2.4	–	STN	–	–
Walrus	Netherlands	2,465/2,800	67.7 × 8.4 × 7	SEWACO VIII	TSM 2272, Type 2026, DUUX 5	ZW 07	ARGOS 700
Whiskey V	Albania	1,080/1,350	76 × 6.5 × 4.9		Tamir	Snoop Plate	–
Yugo [Ψ]	North Korea	90/110	20 × 3.1 × 4.6	–	–	–	–
Yuushio	Japan	2,200[1]/2,450	76 × 9.9 × 7.4		ZQQ 5, ZQR 1	ZPS 6	ZLR 3-6
Not known[β]	Iran	76/90	19 × 2.8	–	–	–	–

* Midget submarine
** Midget submarine. Italian COSMOS type
[Ψ] Midget submarine. Yugoslav design
[α] Midget submarine. German design
[β] Midget submarine. North Korean design based on Italian design
Names in capitals are nuclear boats described in text
[1] Standard

[2] Or Sinbad M8/24
[3] Or PRS 3
[4] M8 in S347/348
[5] Not in S74
[6] SSN 688/750
[7] SSN 710, 751/773
[8] Were Type ES5C/D. Modified ES5E design
[9] Nacken 55.5 m

[10] Type 2051
[11] Micropuffs
[12] Also Snoop Tray for missile control
[13] Or Snoop Tray
[14] Last two boats 66.4 m
[15] Sactis
[16] SNAC-1

UPDATED

TABLE V

Submarine designs - weapons

Class	Navy	Torpedo tubes	Torpedoes	Missiles	Mines[1]
Type 205	Germany	8/533	8*	–	
Type 206/206A	Germany	8/533	8	–	16 (+ 24[2])
Type 207 (Tumleren)	Denmark	8/533	8	–	
Type 207 (Kobben)	Norway	8/533	8	–	
Type 209/1100~1200 (Glavkos)	Greece	8/533	14	Sub-Harpoon[3]	
Type 209/1200 (Salta)	Argentina	8/533	14	–	
Type 209/1200 (Pijao)	Colombia	8/533	14	–	
Type 209/1200 (Chang Bogo)	South Korea	8/533	14	–	28
Type 209/1200 (Casma)	Peru	8/533	14	–	
Type 209/1200 (Atilay)	Turkey	8/533	14		
Type 209/1300 (Thomson)	Chile	8/533	14	–	–
Type 209/1300 (Shyri)	Ecuador	8/533	14	–	
Type 209/1300 (Cakra)	Indonesia	8/533	14	–	
Type 209/1300 (Cabalo)	Venezuela	8/533	14	–	20
Type 209/1400 (Tupi)	Brazil	8/533	16	–	–
Type 209/1400 (Preveze)	Turkey	8/533	14[4]	Sub-Harpoon	–
Type 209/1500 (Shishumar)	India	8/533	14		
Type 212	Germany	6/533	24		24[2]
Abtao	Peru	6/533[5]	n/k	–	
Agosta	France	4/533	20[4]	SM 39 Exocet	36
Agosta (Hashmat)	Pakistan	4/533	20	Sub-Harpoon	Stonefish
Agosta (Galerna)	Spain	4/533	20	–	19[6]
Agosta 90B	Pakistan	4/533	16	SM39 Exocet	n/k
Amur	RFAS	6/533	16	SS-N-21	n/k
Collins	Australia	6/533	22[4]	Sub-Harpoon	44
Daphne	France	12/550	12	–	12
Daphne (Hangor)	Pakistan	12/550	12	Sub-Harpoon	Stonefish
Daphne (Albacora)	Portugal	12/550	12	–	12
Daphne	South Africa	12/550	12	–	12
Daphne (Delfin)	Spain	12/550	12	–	12
Dolphin	Israel	6/650, 4/533	14	Sub-Harpoon	n/k
Dolphin**	South Korea	2/533	2	–	n/k
Foxtrot (Type 641)	Cuba	10/533	22	–	44
Foxtrot (Type 641)	India	10/533	22	–	44
Foxtrot (Type 641)	Libya	10/533	22	–	44
Foxtrot (Type 641)	Poland	10/533	22	–	32
Foxtrot (Type 641)	RFAS	10/533	22	–	44
Gal (Vickers Type 540)	Israel	8/533	10	Sub-Harpoon	

Class	Navy	Torpedo tubes	Torpedoes	Missiles	Mines[1]
Gotland (A 19)	Sweden	4/533	12	–	12 (+48[2])
		2/400	6		
Guppy II	Taiwan	10/533		–	
Guppy IIA	Turkey	10/533	21	–	40
Guppy III	Turkey	10/533	21	–	40
Hai Lung	Taiwan	6/533	20		
Harushio	Japan	6/533	20	Sub-Harpoon	
Heroj	Yugoslavia	4/533	6	–	12
Intrepido**	Colombia	–	–	–	n/k
Kilo (Project 877E)	Algeria	6/533	18	6-8 SA-N-5/8?	24
Kilo (Project 877EKM/636)	China	6/533	18	SA-N-8	24
Kilo (Project 877EM)	India	6/533	18	SA-N-8	24
Kilo (Project 877EKM)	Iran	6/533	18	SA-N-10 SAM?	24
Kilo (Project 877E)	Poland	6/533	18	–	24
Kilo (Project 877E)	Romania	6/533	18	6-8 × SA-N-5/8	24
Kilo (Project 877/877K/877M)	RFAS	6/533	18	6-8 × SA-N-5/8	24
LOS ANGELES	USA	4/533	26	Tomahawk[7] Sub-Harpoon	n/k
MG 110**	Pakistan	2 × 533	2	–	8 limpet
Ming (Type 035)[8]	China	8/533	16	–	32
Moray	–	6 × 533	20		
NSSN	USA	4/533	28	Tomahawk Sub-Harpoon	
Nacken (A 14)	Sweden	6/533	8		(48[2])
		2/400	4		
Narhvalen	Denmark	8/533	–	–	
Oberon	Australia	6/533	20[4]	Sub-Harpoon	
Oberon (Humaita)	Brazil	8/533	24	–	–
Oberon	Canada	6/533	20	–	–
Oberon	Chile	8/533	22	–	–
Oyashio	Japan	6/544	20	Sub-Harpoon	
Pyranja*	RFAS	2/533	2	–	2
Romeo	Bulgaria	8/533	14	–	28
Modified Romeo ES5G	China	8/533	22[9]	YJ-1 Eagle Strike	28
Romeo (Type 033)	China	8/533	14	–	28
Romeo	Egypt	8/533	14	Sub-Harpoon	28
Romeo (Type 031)	North Korea	8/533	14	–	28
Sang-O	North Korea	2 or 4/533	2/4	–	16
Santa Cruz (TR 1700)	Argentina	6/533	22	–	34
Improved Sauro	Italy	6/533	12	–	
Sauro (Type 1081)	Italy	6/533	12	–	
Sava	Yugoslavia	6/533	10	–	20
Scorpene	Chile	6/533	18	–	30
Sea Horse[a]	Taiwan	–	–	–	n/k
SEAWOLF	USA	8/533	52	Tomahawk Sub-Harpoon	100
Sjoormen (A 12)	Singapore	4/533	10		n/k
		2/400	4		
Sjoormen (A 12)	Sweden	4/533	10		n/k
		2/400	4		
Song	China	6/533	n/k	–	n/k
Spiggen II*	Sweden	–	–	–	–
Tang	Turkey	8/533	21	–	n/k
Tango (Type 641B)	RFAS	6/533	24		
Tolgorae*	South Korea	2 × 406	2	–	n/k
Improved Tupi[10]	Brazil	8/533	16	–	32
Ula (Type P 6071)	Norway	8/533	14	–	
Una*	Yugoslavia	–	–	–	4
Upholder	Canada	6/533	12	–	n/k
Vastergotland (A 17)	Sweden	6/533	12	–	48[2]
Velebit	Croatia	2/533	2	–	4
		3/400	6		
Walrus	Netherlands	4/533	20	Sub-Harpoon	40
Whiskey V	Albania	6/533	12	–	24
Yugo[Ψ]		2/n/k	2	–	
Yuushio	Japan	6/533	20	Sub-Harpoon	
Not known[β]	Iran	–	–	–	–

* Midget submarine
** Midget submarine. Italian COSMOS type
[Ψ] Midget submarine. Yugoslav design
[α] Midget submarine. German design
[β] Midget submarine. North Korean design based on Italian design
Names in capitals are nuclear boats described in text

[1] In lieu of torpedoes
[2] Carried in containers either side of hull or in external girdle
[3] Can only be launched from 4 tubes
[4] Including missiles
[5] Also 1 × 127 mm gun
[6] With torpedo load of 9 weapons

[7] 12 VLS tubes in SSN 719/773
[8] Were Type ES5C/D. Modified ES5E design
[9] Including 6 missiles
[10] SNAC-1
n/k: not known

UPDATED

TABLE VI

Submarine designs - machinery

Class	Navy	Diesels[1]	EM[1]	Shafts	Speed[2]	Range[3]	Depth[4]	Endurance[5]	Crew
Type 205	Germany	2/882 kW	1/1.32	1	10/17/?		159		22
Type 206/206A	Germany	2/882 kW	1/1.32	1	10/17/?	4,500/5			22
Type 207 (Kobben)	Norway	2/880 kW	1/1.32	1	12/18/?	5,000/8[6]	200		18
Type 207 (Tumleren)	Denmark	2/880 kW	1/1.25	1	12/18/?	5,000/8[6]	200		24
Type 209/1100~1200 (Glavkos)	Greece	4/1.76	1/3.38	1	11/21.5/n/k		250	50	31
Type 209/1200 (Salta)	Argentina	4/1.76	1/3.36	1	10/22/11	6,000/8//400/4	250		31
Type 209/1200 (Pijao)	Colombia	4/1.76	1/3.38	1	11/22/n/k	8,000/8//400/4	250		34
Type 209/1200 (Chang Bogo)	South Korea	4/2.8	1/3.38	1	11/22/11	7,500/8//n/k	250		33
Type 209/1200 (Casma)	Peru	4/1.76	1/3.38	1	11/21.5/11	n/k//240/8	250	50	35
Type 209/1200 (Atilay)	Turkey	4/1.76	1/3.38	1	11/22/n/k	7,500/8//n/k	250	50	38
Type 209/1300 (Cabalo)	Venezuela	4/1.76	1/3.38	1	10/22/n/k	7,500/10//n/k	250	50	33
Type 209/1300 (Thomson)	Chile	4/1.76	1/3.38	1	11/21.5/n/k	8,200/8[6]//400/4			32
Type 209/1300 (Cakra)	Indonesia	4/1.76	1/3.38	1	11/21.5/n/k	8,200/8	240	50	34
Type 209/1300	Ecuador	4/1.76	1/3.38	1	11/21.5/11				33
Type 209/1400 (Tupi)	Brazil	4/1.76	1/3.36	1	11/21.5/11	8,200/8//400/4	250		30
Type 209/1400 (Preveze)	Turkey	4/2.8	1/3.38	1	15/21.5/15	8,200/8//400/4	280	50	30
Type 209/1500 (Shishumar)	India	4/1.76	1/3.38	1	11/22/n/k	13,000/10//8,000/8[6]	260		40
Type 212	Germany	1/3.12 1-AIP/300kW	1/2.85	1	12/20/n/k	8,000/8//420/8		30	27
Abtao	Peru	2/1.8	2/n/k	2	16/10/n/k	5,000/10//n/k			40
Agosta	France	2/2.65	1/3.4	1	12/20/n/k	8,500/9[6]//350/3.5	320		54
Agosta (Hashmat)	Pakistan	2/2.65	1/3.4	1	12/20/n/k	8,500/9[6]//350/3.5	300		54
Agosta (Galerna)	Spain	2/2.7	1/3.4	1	12/20/n/k	8,500/9[6]//350/3.5	300	45	54
Agosta 90B	Pakistan	n/k	1/2,200 kW	1	n/k/>20/n/k	n/k	350	68	36
Amur	RFAS	2/n/k	1/n/k	1	11/17 /n/k	n/k//300	250	30	34
Collins	Australia	3/4.42	1/5.4	1	10/20/10	11,500/10//n/k		70	42
Daphne	France	2/1.8	2/1.9	2	13.5/16/n/k	10,000/7//n/k	300		53
Daphne (Hangor)	Pakistan	2/1.8	2/1.9	2	13/15.5/n/k	4,500/5//n/k	300		45
Daphne (Albacora)	Portugal	2/1.8	2/1.9	2	13.5/16/n/k	2,710/12//2,130/10[6]	300		55
Daphne (Delfin)	Spain	2/1.8	2/1.9	2	13.2/15.5/?	2,710/12//4,300/7.5[6]	300		47
Daphne	South Africa	2/1.8	2/1.9	2	13.5/16/?	2,700/12//4,500/5[6]	300		47
Dolphin	Israel	3/3.12	1/2.85	1	?/20/11	8,000/8//420/8	350	30	30
Dolphin**	South Korea	1/n/k	1/n/k	1	9/6	n/k	n/k	n/k	6
Foxtrot (Type 641)	Cuba	3/4.4	3/3.97	3	16/15/9	20,000/8//380/2	250		75
Foxtrot (Type 641)	India	3/4.4	3/3.97[7]	3	16/15/?	20,000/8//380/2	250		75
Foxtrot (Type 641)	Libya	3/4.4	3/3.97	3	16/15	20,000/8//380/2	250		75
Foxtrot (Type 641)	Poland	3/4.4	3/3.97	3	16/15/9	20,000/8//380/2	250		75
Foxtrot (Type 641)	RFAS	3/4.4	3/3.97	3	16/15/9	20,000/8//380/2	250		75
Gal (Vickers Type 540)	Israel	2/882 kW	1/1.32	1	11/17/n/k	n/k			22
Gotland (A 19)	Sweden	2/n/k[8]	1/n/k	1	11/20	n/k			28
Guppy II	Taiwan	3/3.3	2/4	2	18/15/n/k	8,000/12//n/k			75
Guppy IIA	Turkey	3/3.4	2/3.6	2	17/15/n/k	12,000/10//n/k			85
Guppy III	Turkey	4/4.41	2/4.2	2	17.5/15/n/k	10,000/10//n/k			86
Hai Lung	Taiwan	3/3	1/3.74	1	12/20/n/k	10,000/9//n/k			67
Harushio	Japan	2/4.1	1/5.3	1	12/25	n/k	350		74
Heroj	Yugoslavia	2/880 kW	1/1.15	1	10/15/?	n/k//4,100/10[6]	150		31
Intrepido**	Colombia	1/n/k	1/221kW	1	11/6	1,200/n/k//60/n/k	n/k		4
Kilo (Project 877E)	Algeria	2/2.68	1/4.34[9]	1	10/17/9	6,000/7[6]//400/3	350	45	52
Kilo (Project 877EKM)	China	2/2.68	1/4.34[9]	1	10/17/8-10	6,000/7[6]//400/3	350	45	52
Kilo (Project 877EM)	India	2/2.68	1/4.34[9]	1	10/17/9	6,000/7[6]//400/3	350	45	52
Kilo (Project 877EKM)	Iran	2/2.68	1/4.05[9]	1	10/17/8-10	6,000/7[6]//400/3	350	45	53
Kilo (Project 877E)	Poland	2/2.68	1/4.34[9]	1	10/17/9	6,000/7[6]//400/3	350	45	52
Kilo (Project 877E)	Romania	2/2.68	1/4.34[9]	1	10/20/9	6,000/7[6]//400/3	350	45	52
Kilo (Project 877/877K/877M)	RFAS	2/2.68	1/4.34[9]	1	10/17/9	6,000/7[6]//400/3	350	45	52
LOS ANGELES	USA	1/26[9]	1/242kW[v]	1	-/32/-	Unlimited	450		133
MG 110**	Pakistan	1/n/k	1/n/k	1	n/k/7	1,200/n/k//60/n/k	150		6
Ming (Type 035)[10]	China	2/3.82	n/k	2	15/18/10	8,000/8[6]//330/4	300		57
Moray	–	3/n/k	n/k	2	n/k/20/n/k	11,000/6//n/k	300	65	38
NSSN	USA	1/17.9[9]	1/n/k[v]	1	n/k/28/-	Unlimited	n/k		100+
Nacken (A 14)	Sweden	1/1.27[8]	1/1.32	1	12/20		300		27
Narhvalen	Denmark	2/1.62	1/882 kW	1	12/17/n/k				24
Oberon	Australia	2/2.74	2/4.48	2	12/17/11	9,000/12			64
Oberon (Humaita)	Brazil	2/2.74	2/4.48	2	12/17/10	9,000/12			70
Oberon	Canada	2/2.74	2/4.48	2	12/17/10	9,000/12	200		65
Oberon	Chile	2/2.74	2/4.48	2	12/17/10				65
Oyashio	Japan	2/4.1	2/5.7	1	12/20				69
Pyranja	RFAS	1/160 kW	1/60 kW	1	6.5/6.5	1,450/4	200	10	4
Romeo	Bulgaria	2/2.94	2/1.98	2	16/13/n/k	9,000/9		50	54
Modified Romeo ES5G	China	2/2.94	2/1.98	2	15/13/10				54
Romeo (Type 033)	China	2/2.94	2/1.98	2	15.2/13/10	9,000/9	300	–	54
Romeo	Egypt	2/2.94	2/1.98	2	16/13/n/k	9,000/9			54
Romeo (Type 031)	North Korea	2/2.94	2/1.98	2	15/13/n/k	9,000/9//n/k			54
Sang-O	North Korea	1/n/k	1/n/k	1	7.6/8.8/7.2	2,700/7//n/k	180		19
Santa Cruz (TR 1700)	Argentina	4/4.94	1/6.6	1	15/25/15	12,000/8//460/6	270	70	29
Improved Sauro	Italy	3/27	1/2.3	1	11/19/12	11,000/11//250/4	300	45	50
Sauro (Type 1081)	Italy	3/2.46	1/2.36	1	11/19/12	11,000/11//250/4	300	35	49
Sava	Yugoslavia	2/1.18	1/1.5	1	10/16/n/k		300		35
Scorpene	Chile	4/2,500 kW	n/k	1	n/k/20/n/k	6,400/n/k//n/k	>300	50	32
Sea Horse[a]	Taiwan	1/n/k	1/80 kW	1	n/k/5	400/n/k//35/n/k	n/k		4

Class	Navy	Diesels[1]	EM[1]	Shafts	Speed[2]	Range[3]	Depth[4]	Endurance[5]	Crew
SEAWOLF	USA	1/38.8[◊]	1/n/k[v]	1	n/k/35/ –	Unlimited	c610		133
Sjoormen (A 12)	Singapore	2/1.62	1/1.1	1	12/20/n/k		150	21	23
Sjoormen (A 12)	Sweden	2/1.62	1/1.1	1	12/20/n/k		150	21	23
Song	China	3/n/k	1/n/k	1	15/22/n/k	n/k	n/k		60
Spiggen II*	Sweden	1/n/k		1	n/k/5		100	14	6
Tang	Turkey	3/3.4	2/4.2	2	16/16/n/k	7,600/15//n/k			87
Tango (Type 641B)	RFAS	3/4.6	3/3.8	3	13/16/n/k	14,000/7[6]/500/3	250		62
Tolgorae*	South Korea	1/n/k	1/n/k	1	9/6	n/k			6
Toti (Type 1075)	Italy	2/1.62	1/1.62	1	14/15	3,000/5//n/k	180		26
Improved Tupi[11]	Brazil	4/1.76	1/3.38	1	11/22/11	11,000/8//400/4	300	60	39
Ula (Type P 6071)	Norway	2/1.98	1/4.41	1	11/23/n/k	5,000/8//n/k	250		18-20
Una*	Yugoslavia	–	2/36 kW	1	6/7	n/k//200/4	120		4
Upholder	Canada	2/2.7	1/4	1	–	n/k/8,000/8	>200	49	47
Vastergotland (A 17)	Sweden	2/1.62	1/1.32	1	11/20		300		28
Velebit	Croatia	1/105 kW	2/80 kW	1	6/7	n/k//200/4	120	6	4
Walrus	Netherlands	3/4.63	1/5.1	1	12/20/n/k	10,000/9[6]	300		52
Whiskey V	Albania	2/2.94	2/1.98	2	18/14/7	8,500/10//n/k	150	–	54
Yugo[Ψ]	North Korea	2/236 kW	n/k	1	12/8	550/10//50/4			4
Yuushio	Japan	2/5	2/5.3	1	12/20/n/k		275		75
Not known[β]	Iran	2/236 kW	n/k	1	12/8	1,200/6/n/k	c100		3

* Midget submarine
** Midget submarine. Italian COSMOS type
[Ψ] Midget submarine. Yugoslav design
[α] Midget submarine. German design
[β] Midget submarine. North Korean design based on Italian design
Names in capitals are nuclear boats described in text
[◊] PWR S6G power plant in Los Angeles. Seawolf PWR S6W, NSSN PWR S9G nuclear systems
[v] Auxiliary or secondary motor

[1] Number/MW output (total output listed)
[2] Speed: surfaced/dived/snorting in kt
[3] Range: n miles at kt surfaced/submerged
[4] Maximum depth in m under normal conditions. Crushing depth is considerably greater. Operating depth gradually reduces as a boat ages to the extent that boats beyond 30 years are usually restricted to operating at periscope depth
[5] Days
[6] Under snorkel and 7,500 @kt in the Project 636

[7] Plus 1 auxiliary motor 103 kW
[8] Also 2 Stirling AIP 150 kW
[9] Also 2 auxiliary motors 150 kW, and 1 economic speed motor 95 kW
[10] Were Type ES5C/D. Modified ES5E design
[11] SNAC-1
n/k: not known

UPDATED

MACHINERY

Machinery

One of the most significant developments since World War II, apart from nuclear propulsion for submarines, has been the adoption of the standard high-speed diesel for SSKs. Because it takes considerable time, effort and expense to develop a diesel, the main diesel manufacturers have tended to develop and adapt from existing ranges, engines suitable for submarine propulsion. However, while the diesel will continue to be the prime mover in a non-nuclear submarine, air independent propulsion (AIP) systems are now a reality and by the turn of the century will be at sea in a number of boats of different navies.

Since weight and space problems are of paramount importance to the submarine designer, the naturally aspired diesel engines used during and after World War II have now been replaced by higher performance supercharged diesels. Diesel engines can be supercharged in one of three ways: by exhaust gas-driven compressors (turbochargers); by mechanically driven compressors (superchargers); or by a combination of both.

Continuing efforts are being made to improve diesel engine performance, bearing in mind the problems of greater exhaust back pressures and lower induction pressures developed when snorkelling. It is claimed that the mechanical supercharger is less sensitive to back pressure and develops a more consistent performance under varying conditions than the turbocharger. The latter, on the other hand, is claimed to have improved power:weight and power:volume ratios, as well as improved fuel consumption.

However, there are problems associated with supercharging a diesel while snorkelling. These include:

- a considerable drop in air intake pressure due to air pressure losses (between 0.10 and 0.15 bar) along the suction pipe inside the snort mast;
- high exhaust counter-pressure (currently >0.6 bar) due to the underwater level position of the gas discharge and the pressure losses inside pipes and the discharge valve;
- starting and running for a few seconds against the closed exhaust circuit leading to an exhaust counter-pressure increasing to about 1.5 bar until the exhaust discharge valve opens;
- and high air inlet temperature resulting from the fact that combustion air is used to renew the air in the submarine.

Furthermore, transient conditions induced by the sea swell period or result from the submarine's pitching, lead to a correlative increase in exhaust counter-pressure. This results from the increase of immersion and a decrease in inlet pressure after the air suction valve is closed.

From the first three conditions it is obvious that the engine has to act initially as a pump in order to counterbalance the loss in air intake pressure and the increase in gas outlet pressure. This places an additional workload on the engine compared to open air operating engines.

As a result a submarine's diesel cannot be supercharged in the same way as a surface vessel's diesel. This is because it is necessary to provide special timing on the engine valves with nearly complete suppression of the valve overlap period during which both inlet and exhaust valves are opened to prevent a counter-sweeping of the combustion chamber. Because of this, air cooling of the engine during the scavenging phase is reduced, leading to a higher temperature of the combustion chamber components in a submarine diesel than in a conventional diesel. This results in a power derating for thermal stresses.

Up to now most submarine engines have been supercharged by mechanically driven compressors. The advantages are:

- a positive system for maintaining air flow through the engine, which is largely unaffected by the snorting conditions, with a consistent performance over the whole output range;
- easy adjustment of the air flow;
- good load acceptance under all conditions;
- reduction of engine fouling;
- and tolerance to salt water.

However, since air compression work is taken on the engine crankshaft, there has to be a trade-off that leads to rather high specific fuel consumption in the region of 285 to 300 g/kWh at full load. This increases still more on part loads due to the fact that at constant engine speed there is constant compression work.

As a result of this limitation some manufacturers have adopted the exhaust gas driven turbocharger. Using this system it is possible to operate the diesel under snorting conditions by opening the exhaust valves a little earlier than on a naturally aspired engine. This results in sufficient power being available on the turbine shaft of the turbocharger that is necessary to carry out compression. This results in improved specific fuel consumption figures in the range of 250 g/kWh.

However, three main problems are encountered with turbocharged engines:

(1) Firstly the boost pressure obtained by the compressor is very much affected by the difference between the absolute air inlet pressure and the absolute exhaust back pressure. Furthermore, when these two parameters suddenly change due to increased wave motion (the snort valve closes and immersion increases) the boost pressure is drastically reduced requiring a quick unloading of the engine to prevent overheating. Thus the turbocharger must be 'over matched' to overcome this problem which in turn results in surging as conditions improve. Then a waste gate is necessary to discharge the excess air that occurs in calm conditions.

(2) When engine power is reduced (at the end of the battery charging) the boost pressure drastically falls off and becomes insufficient to prevent the counter-scavenging of the engine by exhaust gases leading to rapid fouling of the engine.

(3) Inside the turbocharger the boost pressure is normally used as a seal to prevent the passage of exhaust gases towards the airside. As the pressure of the gases is higher than the air pressure, a separate compressed air supply is required to supply this seal.

The propulsion motor of the SSK is the low rpm DC electric motor connected directly to the propeller shaft and using two or four armatures. Speed control is achieved by grouping the batteries feeding the motor either in series or in parallel, variation between being achieved by altering the strength of the field. Developments in recent years have concentrated on increasing the power output and improving the cooling system using either direct water cooling or combined air/salt water-cooling. Work is now in progress on developing motors based on permanent magnets or a superconducting arrangement. The use of permanent magnets could lead to considerable reduction in weight and saving in space.

The other main machinery area in the SSK is the battery compartment.

The introduction of the silver-zinc battery has resulted in a considerable increase in the energy storage capacity. Experiments have been undertaken in the United States of America with a lithium thionyl chloride battery, but the expense of such batteries precludes their consideration at the present time.

Batteries are extremely heavy and are placed in the bottom of the boat where they help to provide essential stability, accounting for about a quarter of the total displacement. Batteries have improved considerably in recent years, particularly in the areas of efficiency and service life. To provide control, batteries are divided into a number of sections (usually four) which are coupled in series or parallel.

The batteries are recharged by generators driven by V-form 4-stroke diesels that usually develop three-phase current which is rectified, although AC/DC units with static controls are also common. These AC generators have proved much more reliable with higher speed and higher output, resulting in a shorter recharging time than the older DC generators.

Continuing efforts to improve the discretion rate of conventional submarines and extend their range has spawned a whole new field of expertise relating to AIP systems. Diesel electric submarines are limited in submerged endurance by their battery capacity. Battery charging with the diesels is the most critical operational situation for submarines on patrol. The signatures of a submarine while battery charging are rapidly becoming more and more detectable, be it the radiated diesel noise, fumes and heat, or the emerging snorkel head.

A low-noise AIP is of vital interest, therefore, in being able to extend the submerged endurance, some sources quoting a doubling of range at a reasonable underwater cruising speed. Currently the most popular forms of AIP systems are fuel cells, the closed-cycle diesel engine, the Stirling engine and the closed-cycle gas turbine. Some systems have been proven at sea, others are still under development.

The fuel cell uses an alloy as a catalyst with a mixture of liquid oxygen and hydrogen resulting in cold combustion to produce a direct current and water through chemical reaction. An alternative to the fuel cell is the closed-cycle engine such as the Stirling which uses cryogenic oxygen and various types of fuel. The disadvantage of such a system is the exhaust gases that have to be extracted and removed.

NEW ENTRY

FRANCE

12 PA4 Series

Type
Supercharged diesel.

Description
Currently six different types of submarine from six different countries are fitted with SEMT-Pielstick PA diesels. The PA diesel installed in submarines differs from the standard engine in having a high exhaust back pressure in the exhaust manifold.

There are two standard ranges of submarine diesel: the PA4 185 SM and the PA4 200 SM, both series being V-form diesels with 12 and 16 cylinders for the PA4 185 SM, and 8 cylinders for the PA4 200 SM and 12 cylinders for the PA4 200 SMDS. The engines cover the range 500 to 1,380 mechanical kW. The two series are very similar using the same welded tunnel type of frame with V-form at 90°, same single camshaft and monobloc injection pumps, same bearing design and diameter, same variable geometry combustion system and same compressor drive. They differ in that the bore of the 200 series has been increased from 185mm to 200mm (stroke remaining the same at 210mm), increased diameter of crankshaft pins, an oblique cut type connecting rod design which enables the piston to be dismantled with its connecting rod through the cylinder liner, and a stronger, stiffer cylinder head design with improved air and gas circulation induction with new valves. The 12 PA4 200 SM uses a motor-driven compressor for supercharging to deliver 1,030kW at 1,300rpm at its coupling flange.

The 12 PA4 200 SMDS is supercharged with a compound system using both turbocharger and mechanically driven centrifugal compressor working in series on the combustion air feeding line. The combination comprises a conventional turbo-compressor in the first stage and a mechanically driven blower for the high-pressure stage. The work of the blower consists in re-establishing a pressure level at engine admission equivalent to the exhaust counter-pressure, and then to increase the power by using a

A 12 PA4 200 SM diesel as fitted in the 'Walrus' class of the Royal Netherlands Navy (Courtesy SEMT-Pielstick)
1999/0024692

turbo-compressor. In this configuration the engine delivers 1,325kW at 1,300rpm at the coupling flange.

In PA4 200 SMDS each of the two compressors is fitted with a pair of air intercoolers so that the supercharged air can recover a specific gravity consistent with a good filling of the combustion chambers after its compression.

The mechanically driven compressor level delivers the compressed air flow at a pressure suitable to build up the barrier against the passage of gases inside the turbocharger.

For surface operation a butterfly valve opens a bypass to discharge a part of the excess air coming out of the mechanically driven compressor to bring it back to the turbocharger air inlet, thus avoiding any surging of the compressor.

The two compressors in series can deliver a supercharging air pressure as high as that in surface engines. Consequently engine derating is limited to its minimum figure. Furthermore the engine is fairly insensitive to variations in exhaust back pressure and loss of air suction pressure due to the action of the mechanically driven compressor.

To obtain maximum efficiency the mechanically driven blower must be adjusted so that its compression ratio can restore the surface conditions, as seen, that is, from the turbocharger side. For instance, if the absolute air inlet pressure is 900 mbar and the absolute exhaust pressure is 1,620 mbar, the compression ratio must be: 1,620/900=1.8.

The power:weight ratio is improved by 20% as the engine weight is only increased by about 10% compared to a mechanically supercharged engine, while the power is increased by 30%.

The specific fuel consumption (SFOC) is improved by so much that the mechanically driven compressor work is reduced. For a normal compression ratio the SFOC figures are between 250 to 275 g/kWh.

The greatest advantage of this engine, however, is that it is insensitive to any deterioration in snorting conditions.

Tests have demonstrated that no derating is required up to 1,800 mbar at exhaust and 850 mbar at inlet with a 50° inlet temperature.

Operational status
The 12 PA4 V185 engine is in service aboard the 'Daphne' class boats of France, Pakistan, Portugal, South Africa and Spain, while the 16 cylinder variant is operational aboard the 'Agosta' class of France, Pakistan and Spain and is being delivered for the new 'Agosta 90B' boats building for Pakistan. The 12 PA 4 200 SM engine is in service in the Netherlands 'Walrus' class boats, while the 8 cylinder version is operational aboard all the SSNs of the French Navy.

Contractor
SEMT-Pielstick, St Denis.

NEW ENTRY

MESMA

Type
AIP closed Rankine cycle engine.

Description
The MESMA (Module d'Energie Sous Marine) closed, cycle engine is being developed by DCN (prime contractor, secondary loop and systems integration) in partnership with Bertin & Cie (combustion chamber and primary loop), l'Air Liquide (LOX storage and cryogenic pumps), Empresa Nacional Bazan (ethanol storage and integration), Termodyne (turbine) and Technicatome (safety aspects). The system features a heat generation loop integrated with a Rankine cycle steam turbine.

In the MESMA system, thermal energy is produced by the combustion of a gaseous mixture of liquid oxygen and ethanol. The liquid oxygen is stored cryogenically at low temperature (-185° C) and low pressure (that is in liquid form) before being pumped into a vaporiser which turns it into gaseous form before combustion. In the combustion chamber the gaseous oxygen is mixed with ethanol to produce a thermal output of 700° C with a thermal energy of about 300 Whe/kg at around 60 bar. This thermal energy is used to heat a secondary circuit. The gases are passed through a heat exchanger (producing steam at 500° C and 18 bar pressure and a steam generator. The secondary circuit comprises a Rankine-cycle system (steam generator, turbine generator and condenser) which is used to drive a high-speed turbo alternator-rectifier to supply electrical energy for the batteries, the propulsion plant and ancillaries. The oxygen and ethanol flow rates are monitored so that power production can be varied according to the submarine's requirements. Operating at about 95 per cent efficiency the steam turbine has an output ranging between 150 and 600 kW. Having passed through the steam generator and heater the hot gases are expelled directly outside the submarine at 60 bar pressure, which allows discharge at any diving depth without the use of an auxiliary pump. Part of these gases are bled off at the exit of the steam generator and returned to the combustion chamber to regulate its temperature.

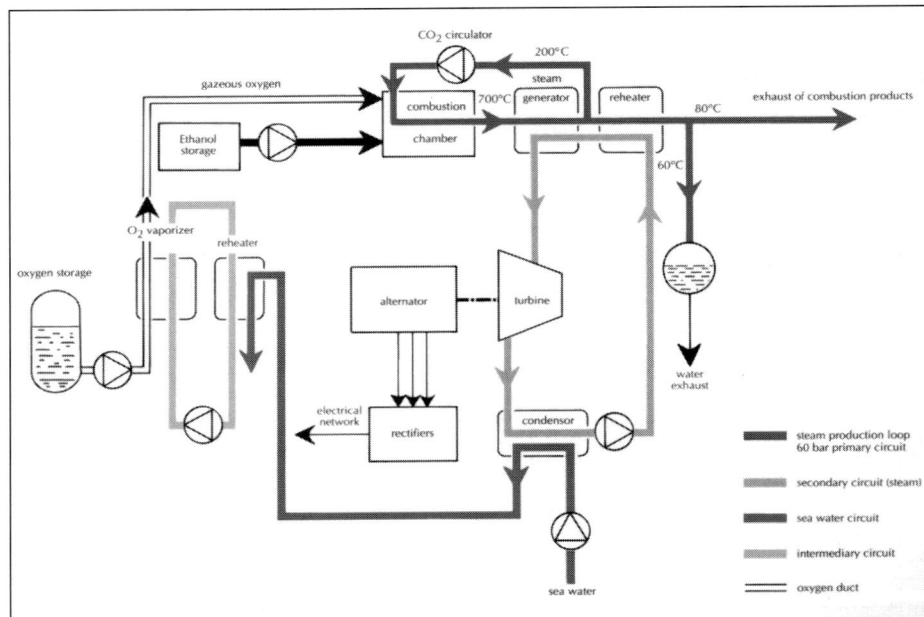

Schematic layout of the MESMA AIP principle
1999/0024694

The liquid oxygen and ethanol are stored separately in a dedicated section of the boat. The system has a high discretion value resulting from the silent and continuous combustion, the action of the high-speed turbine generator, and the constant-pressure waste disposal configuration.

With an output of around 200 kW, the MESMA AIP allows a submerged submarine to increase its submerged endurance by a factor of around five at a speed of 4 kt.

An 8 to 10 m hull section module will house the entire MESMA closed-cycle system including storage units for the ethanol and liquid oxygen, steam module and turbine.

Operational status
Combustion tests on a 30-tonne core steam land-based prototype module of the MESMA AIP have been carried out by DCN. Full prototype testing of the MESMA system commenced in May 1998 with the first production system due to be tested in the latter part of 1999. The first production module will then be shipped to Pakistan for installation in the third Agosta 90B in 2000. A further two systems will then be installed in the previous two 'Agosta 90B' boats.

Contractor
DCN Indret, Nantes.
Bertin, Plaisir.

Electric motor

Type
Electric motor.

Description
The Jeumont-Schneider motor is of the totally enclosed pressurised type with a circuit water cooling system. The motor comprises one or two armatures coupled directly to the single shaft. A feature of the design is the ability to mount an auxiliary motor on the propeller shaft for emergency use or for extremely silent running at very slow speed.

Various methods are used to power the motor to efficiently provide a range of speeds. For low speeds the motor is powered by one or two batteries connected in parallel through a DC/DC converter, adjusting the field current and armature voltage. In the medium range, speed is adjusted by altering the field current, power again being supplied by batteries connected in parallel. For high speed, the batteries are connected in series again, adjustments of the field current providing variations in shaft speed. Speeds between the three fixed ranges are obtained by connecting the two halves of the motor in series or in parallel.

While batteries are being charged through the diesel generators the low and medium speed ranges are available. This is achieved by connecting one or two generators in parallel to supply the motor and having the batteries also connected in parallel with the generators. For diesel/electric propulsion when surfaced, either or both generators are used to power the motor directly, the batteries being disconnected. In this case armature voltage is adjusted by controlling generator voltage via the generator field.

Operational status
In widespread service around the world.

Contractors
Jeumont-Schneider, Puteux.

NEW ENTRY

GERMANY

396 Series

Type
Turbocharged diesel for submarines.

Description
In the 396 Series MTU has developed a modular concept of turbocharging, enabling various combinations of turbine/compressor units to be matched to the specific needs of individual applications. The 396 Series is available in 6, 8, 12 and 16 cylinder versions offering power rating up to 1,060 kW under snorkel conditions which meets the power demand of medium-sized SSKs, and can be delivered either for twin- or triple-engine installations.

These 4-cycle exhaust gas turbocharged diesels use the familiar 'V' configuration with direct injection and external charge air-seawater cooling. Other features include direct liquid cooling of exhaust pipes and turbines resulting in a maximum surface temperature of less than 100°C without the need for additional insulation and providing full sealing against exhaust gas leakages. Emergency air shut-off flaps are integrated into the charge-air manifolds to provide protection against the engine running at excessive speed. In the event of excessive intake air vacuum the engine shuts down automatically.

The diesel is started against closed exhaust flaps with an exhaust back pressure of 2.4 bar absolute, and operates under snorkel conditions at 0.86 bar absolute intake-air depression, and 1.8 bar absolute back pressure. During operation, exhaust flow prevents seawater from entering via the exhaust outlet

The composite pistons are oil-cooled with steel crown and light-metal skirt and valve rotators are fitted on each of the two inlet and two exhaust valves on each cylinder. Valve control is exercised through two camshafts located high in the engine 'V'. Exhaust gases pass through a water-cooled silencer and water-cooled exhaust ducting for ejection, usually under the surface of the water. High-performance sleeve bearings and high-pressure fuel injection is used for improved combustion, the multi-plunger injection pumps being located in the 'V'. The crankcase and bell housing are of nodular cast iron, with a light-metal oil pan, the crankshaft running in sleeve bearings with connecting rods grouped in pairs.

The turbochargers incorporate simple, radial turbine and radial compressor wheels that are integral cast pieces. They are ess susceptible to blade vibrations and consequently permit higher circumferential speeds. Gas tightness between the turbine and compressor is achieved by a piston ring sealing.

Operational status
MTU 396 Series diesels are in service with Type 209 submarines of South Korea and Turkey, the Swedish 'Gotland' class, the Norwegian 'Ula' class, the Israeli 'Dolphin' class, and are being installed in the Type 212 boats for the German and Italian navies.

Contractor
MTU, Friedrichshafen.

NEW ENTRY

PEM AIP System

Type
Proton Exchange Membrane (PEM) fuel cell.

Development
Since the early 1980s a consortium comprising Ferrostaal AG, IKL and HDW has funded research and development into a fuel cell AIP propulsion system for submarines. Exhaustive development tests with a land-based system were completed in 1986 and the German Navy Type 205 submarine *U.1* was refitted at HDW's yard with a prototype propulsion system installed in a specially inserted hull section. Work on the refit began in March 1987 and the boat began sea trials in November 1988. Full-scale development of the fuel cell by Siemens was completed during the early part of 1995. The first cell module was delivered to HDW in August 1998 for integration into a shore-based total propulsion system for the Type 212 submarine.

Description
In the PEM fuel cell propulsion system a chemical reaction occurs to change the fuel (oxygen and hydrogen) directly into DC electric power with waste heat and water as a by-product. This method of energy conversion gives the submarine important operational advantages in that no noise is radiated and very low heat is generated due to the high efficiency level (up to 60 to 70 per cent). Compared to conventional power plants, fuel cells are more efficient and produce less pollution. In addition they offer low operating temperature (80°C), low voltage degradation, long service life, favourable load, temperature and overload behaviour, and do not require a liquid corrosive electrolyte.

The fuel cell uses an alloy as a catalyst with a mixture of liquid oxygen and hydrogen which is stored in tanks sited between the pressure and outer hulls. The oxygen is stored cryogenically in liquid form in specially insulated double-walled tanks while the hydrogen is stored in chemically bound form in low temperature metal hydrides. The chemical reaction between the hydrogen and oxygen at low-temperature in the Siemens fuel cell modules produces a direct current. Because of the high efficiency of the fuel cell the quantity of oxygen required for conversion is only about half that required in combustion engines.

The heart of the fuel cell is the membrane electrode unit that carries out the energy conversion. The membrane electrode unit comprises the polymer electrolyte, gas diffusion electrodes with platinum catalyst and carbon sheets on each side. In the chemical reaction the protons move from the anode to the cathode where liquid water is formed as a by-product. The theoretical voltage of the fuel cell is 1.48 V (referred to as the upper calorific value of hydrogen). At no load conditions slightly more than 1 V per cell is available. The cooling unit transfers reactants to the electrodes, removes heat from the electrodes, removes the unwanted water produced in the cell, and seals the various media from one another.

The fuel cell module consists of the fuel cell stack with about 70 electrochemical cells, a humidifier, a separator for the removal of product water, and the electromechanical and electronic control elements. The fuel cell module is installed in a pressurised container and operated in a protective gas atmosphere of 3.5 bar. A power source may comprise several fuel cell modules that can be connected electrically either in series or in parallel as required. Cooling water absorbs heat given off during energy conversion, part of which is used as dehydration energy for the hydrogen and part as evaporation energy for the very cold oxygen.

The plant is controlled from two switchboards, one for the module electronics and the other for the plant itself and the safety system. The overall system is operated from the central submarine control console. In an emergency the fuel cell can be operated directly from the switchboard.

The German Navy has decided that its new Type 212 class submarines will be fitted with the fuel cell technology as an integral part of a hybrid propulsion system. This decision was based on a successful development phase of 10 years carried out by HDW/ Ferrostaal/IKL and supported by the German MoD/ BWB. In the Type 212 the fuel cell will power a Siemens Permasyn electric propulsion motor to drive a seven-bladed, low-speed, single-skewed propeller. The fuel cell modules will deliver an output range of between 30 kW and 50 kW, with a nominal output of 34 kW.

Operational status
On order for installation in the Type 212 boats for the German and Italian navies.

Contractors
HDW, Kiel.
Siemens, Erlangen.

NEW ENTRY

Permasyn

Type
Permanent magnet field electric motor.

Description
Siemens has developed the Permasyn permanent magnetic field static converter-fed inverter-controlled synchronous propulsion motor for the German Type 212 submarines. The motor incorporates a permanent excited rotor and a polyphase stator winding with winding phases separately supplied by converters. These are pulse-controlled inverters with gate turn-off thyristors driven by a buffered DC power supply. The inverters control the armature voltage by means of DC voltage pulses and supply rotor position dependent AC to the stator windings. The pulse-controlled inverter functions both as voltage controller and as commutator.

In this way the motor handles changing speeds in both forward and reverse directions with ease and offers stepless speed variation over the whole forward and astern range. The output range is up to 5 MW at a rated speed of 120 rpm.

The rotor consists of a bell-shaped magnetic yoke that is in the form of a hollow cylinder with high-energy capacity permanent magnets of samarium and cobalt mounted on the outer circumference and modular power electronics mounted inside the rotating cylinder. With only low losses from the rotor complex noisy air cooling apparatus is not required and so direct fresh water-cooling can be employed.

The stator consists of a solid, closed-yoke ring and is multiphase wound, each phase being electrically separated. Moving parts of the stator and the

electronics are cooled via cooling water ducts and cooling plates.

Built-in switching elements allow grouping both in number and circuit configuration. The magnitude and curve shaping of phase currents is achieved using pulse-controlled inverters to achieve minimum noise levels.

The main advantage of the Permasyn motor compared to a conventional DC commutator motor developing the same torque, is that it is much smaller (weight and volume saving of some 60 per cent). It can therefore be run at significantly lower speeds (thus improving submerged endurance), while a much larger diameter propeller can be fitted. This results in overall

lower noise emission and better propeller efficiency allowing revolutions to be reduced from 200 to 120 rpm at the same output.

Operational status
Scheduled for delivery and installation in the Type 212 submarines.

Contractor
Siemens, Erlangen.

NEW ENTRY

Double-Decker

Type
Submarine battery.

Description
Varta employs two different design principles in its batteries that are dependent on the requirements for the battery. In addition to cells of normal design with long, continuous full height plates, Double-Decker cells have been developed in which the negative and positive plates are only half as tall as in normal cells. This configuration allows the cell design characteristics to be varied within limits so that it can be adapted to meet specific demands with regard to low- or high-current capacity.

In the Double-Decker design two plate elements connected in parallel are arranged one above the other in one cell container. Lead insulated copper conductors join the two sets of cell connectors, which leads to greatly reduced internal resistance of the cell. The Double-Decker cell employs low-antimony alloy technology to produce low inner resistance and hence good discharge performance, especially at medium- and high-power discharges. This results in more efficient charging leading to reduced snorting time.

Each plate element forms a unit made up of positive and negative plate groups, and their separators. The positive plate is of tubular design. The negative plate is a grid plate where the grid is both carrier of the active material and current conductor. The active material encloses the electrically conductive lead rods. The separator is a partitioning device mounted between the positive and negative plates.

The pole bridge is the physical connection between the plates and cell poles. Its main carrier consists of a several millimetre thick lead-coated copper tube, serving as a heat exchanger for connection to an external cooling system.

An important requirement for the function of submarine batteries is to achieve as uniform a distribution of acid density and heat within the cell as possible. As a result the acid must be circulated in the cell, especially during charging. In the Varta Double-Decker this is achieved using the 'air lift' principle.

Using an insulation-measuring sensor the insulation value between the measuring sensor and cell poles as well as between the measuring sensor and boat's hull can be determined. Openings in the cell cover accommodate acid level indicators, service plugs and vent plugs. The vent plug prevents explosion and flooding and covers against drying out of the acidic cell gas. An inspection glass is fitted to indicate the electrolyte flow during the acid circulation

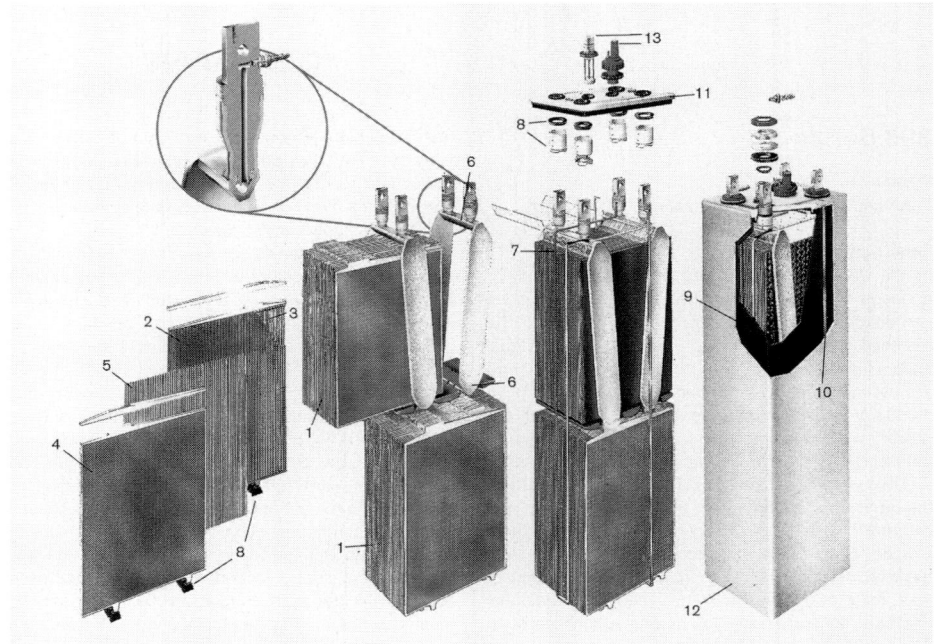

General arrangement of Double-Decker cell. (1)Plate element (2)Positive plate (3)Positive grid (4)Negative plate (5)Separator (6)Pole bridges (7)Acid circulation system in the cell (8)Shock absorption elements (9)Rubber liner (10)Insulation measuring sensor (11)Cell cover (12)Cell container (13)Service fittings **1999**/0024699

Cell protection is provided by spring elements, which support the plates to insulate them against shock, while the cell cover is of glass fibre reinforced material, which is extremely impact resistant, but at the same time highly flexible. The cell container is made of flame-resistant GRP resin, resistant to shocks and diluted sulphuric acid and prevents separate standing cells from bulging.

A computer developed jointly by Siemens and Varta automatically monitors the state of the cell. The computer collates, monitors, calculates, evaluates and registers data on battery and all voltage, energy charge and discharge, the state of the charge, acid density, acid temperature and energy remaining. More than 1,100 measurement points are collated in a 15 second cycle to present the operator with data such as ampère hours, charging factor, number of cycles and capacity available in the battery for differing speed requirements. Data is available automatically or on call-up on video display or high-speed printout.

The Double-Decker Plus battery differs from the standard Double-Decker design in that a copper conductor has been introduced to optimise the negative plate. This results in higher capacity during discharge, higher current capability in charging and a marked reduction in the heat produced in the cell.

Operational status
Double-Decker batteries have been installed in submarines of Argentina, Australia, Brazil, Colombia, Ecuador, Germany, Greece, Indonesia, Netherlands, Peru, Poland, Taiwan and Venezuela.

Contractor
VHB Industriebatterien GmbH, Hagen.

NEW ENTRY

CSM Cell

Type
Submarine battery.

Description
Hagen has developed the CSM (Copper Stretch Metal) cell, in which the whole negative grid is made of expanded copper with a thin coating of lead, resulting in ×15 higher conductivity. The CSM is used to hold the active mass of fine crystalline porous lead to produce the best possible conductor, internal electrical resistance being reduced by about 60 to 70 per cent. It is possible to use copper for the negative grid because

it is not electrochemically corroded, being in a voltage range of immunity. By reducing grid resistance to the point where voltage drop in the cell is lowered, current, distribution becomes more uniform. The spines of the positive plate are electronically coated with low antimony lead to reduce the creation of hydrogen and achieve optimum charge during the first state of charging. The lead spines are surrounded by active material held by acid-resistant tubes made of glass-web. The top frames are composed of aluminium inlay to provide improved conductivity.

The electrolyte is diluted sulphuric acid using the highest possible concentration (1.3 g/m). Above this

level corrosion is accelerated and performance would decrease.

Cells are housed in slim-walled high-performance GRP containers; the plate packages being entirely sealed by a rubber lining. Cell armatures provide easy access for control and maintenance.

Temperature differences in and between the battery rooms adversely affect the internal resistances of the cells. A single cell voltage can vary with 70 mV/10° change of temperature. Voltage differences in serial connections may add up in the partial batteries and subsequently affect deviating currents between these parallel-switched partial batteries. To avoid this

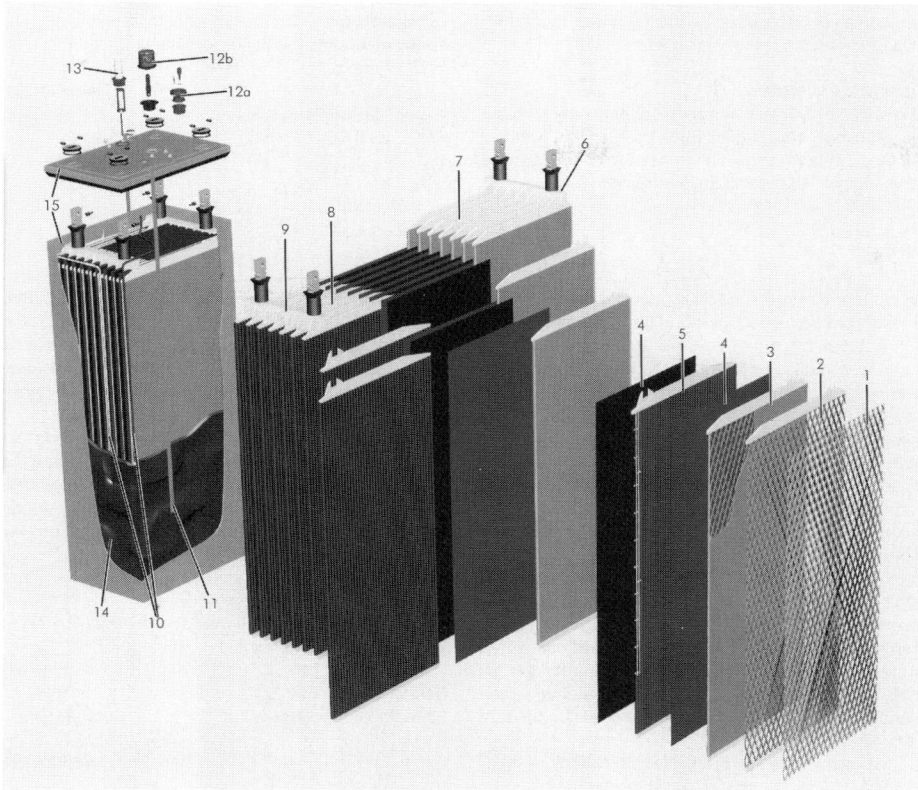

undesirable effect, cooling water is lead through pole-bridge channels of each cell.

To prevent the electrolyte from developing an increased density in the lower volume of very tall cells an auxiliary piping system is fitted to feed compressed air to the cells to agitate the electrolyte and ensure even density throughout the cell.

Operational status
In service with the navies of Chile, India, Israel, Netherlands, Norway and South Korea.

Contractor
Hagen, Soest.

NEW ENTRY

General arrangement of the CSM cell. (1)Negative grid of CSM cell (2)Lead-coated negative grid (3) Single negative plate (4)Separator (5)Single positive tubular plate (6)Negative pole bridge (7)Negative plate set (8)Positive plate set (9)Positive pole bridge (10)Acid agitation pipes (11)Lead strip for insulation measurement (12a)Service plug (vent) (12b)Backflash arrestor (alternative vent) (13)Acid level indicator (14)Rubber protection bag (15) Glass fibre container and cover
1999/0024700

SWEDEN

V4-275

Type
Stirling AIP engine.

Development
Kockums began testing a prototype Stirling engine in 1983 and carried out tests on a floating platform in 1985. Two Mk 1 engines were then exhaustively trailed aboard the Swedish submarine *Nacken*. An 8m long section containing two V4-275 SUB Mk 1 Stirling engines, two large Liquid Oxygen (LOX) tanks, and shock mounts and sound isolation equipment was inserted in the submarine. In the configuration adopted by the Swedish Navy the Stirling AIP plant is intended as an auxiliary low-power, long-endurance system to complement (but not replace) the conventional diesel/electric plant. The main function of the AIP source is to power the boat (with minimum rediated noise) during low-speed surveillance operations, allowing the main battery to remain fully charged and available for engagements demanding full speed.

Description
In the Stirling AIP system heat energy provided by high-pressure diesel oil/liquid oxygen combustion is converted by the engine into mechanical power through a thermodynamic process, and then converted into DC electric power by a generator. The efficiency of the engine is around 38 per cent. The V4-275R engine uses one part diesel fuel to four parts LOX, the two fuels being combined and burned in a circular combustion unit.

The heat so generated is continuously transferred to the working gas (helium) contained in a closed-circuit heat exchanger inside a large bell-shaped chamber. A regenerator reclaims the heat energy, storing much of the heat contained in the gas after expansion and feeding it back into the cycle as the gas reverses direction. The working gas expands above each piston when heated and compresses below it when cooled, driving the piston up and down to rotate the crankshaft and provide mechanical output. The pistons in a double-action Stirling system have a dual function, each piston operating simultaneously in two cycles so

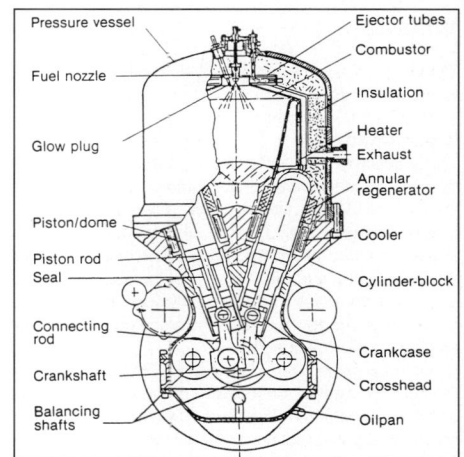

Block schematic of the Stirling engine
1999/0024696

that the hot (upper) surface of a piston works within the same cycle of the cold (lower) surface of the adjacent piston.

Engine exhaust is dissolved in the surrounding seawater without creating bubbles by using an absorber that mixes cooling water with the exhaust gases. The exhaust-cooling unit is integrated into the engine module and cools the gas from 800°C to around 25°C. Combustion pressure is maintained at 20 to 30 bar, independent of depth down to 300 m. This being higher than the surrounding seawater pressure, allows the exhaust products (which are soluble in saltwater) to be discharged overboard without using a compressor. This results in a noiseless and traceless system. The engine can be adapted to operate at depths beyond 300 m (down to 600 m), but only if an exhaust gas compressor were added, which would increase noise and fuel consumption. The pressure-proof combustion chamber is housed in a spherical casing for reasons of weight and strength.

Stirling generators are always run in parallel with the boat's batteries which simplifies power control and provides the submarine with complete independence in the event of a failure of the Stirling system.

The V4-275 is a 4-cylinder engine in a 'V' configuration, delivering a maximum output of some 75 kW with 110 kW of heat at 2,000 rpm and at a water

A V4-275 Mk 1 Stirling engine (Kockums)
1999/0024695

temperature of 50°C. Fuel consumption is 250 g/kWh and oxygen of 950 g/kWh. The engine weighs 750 kg, has a swept volume of 275 cm³ and measures 0.8 × 0.8 × 1.4 m. The complete system, including compensating water tanks and all auxiliary systems are fitted in an autonomous hull section. Noise levels compared to diesel engines of similar size are some 15 to 25 dB lower at higher frequencies and not less than 8 to 10 dB lower at shaft frequencies. Submerged endurance is determined primarily by the amount of LOX stored in the cryogenic tanks.

Operational status
The V4-275R Mk 2 is in service aboard the Swedish navy 'Gotland' class boats. Each boat is fitted with two engines, although originally intended to carry three (the third engine was deleted for cost reasons).

Contractor
Kockums Submarine Systems, Malmo.

NEW ENTRY

SCC-200 and ISCMMS

Type
Submarine steering control and ship management systems.

Description
The SAAB SCC-200 steering control system, already in service with the Swedish Navy, is now operational aboard the Norwegian 'Ula' class SSKs and an upgraded variant equips the Australian 'Collins' class SSKs. The one-man control system is a dual-channel system providing course and depth control, trim and compensation water control and speed control of the main motor. Control is carried out automatically or semi-automatically and the operator is kept informed of the full state of the boat on the display.

The much modified variant of the SCC-200 developed for the Australian Navy 'Collins' class is referred to as the ISCMMS (Integrated Ship Control Management and Monitoring System). The system configuration comprises control consoles, general control consoles, manoeuvring control console, local control console, substations and the databus system.

Two types of control console are used - a general purpose console and a limited purpose console. The general control console consists of two full-colour VDU screens, a keyboard, hardwired controls and a general computer unit. The VDUs are provided with function keys integrated into the lower part of the bazels, adjacent to the screens. The keys are used to adapt the control possibilities of the system to the actual status of the system at any time. The manoeuvring console is a general control console configured for the task of manoeuvring. It is provided with autopilot software and reinforced redundancy. The local control consoles have limited functionality and consist of a monochrome display and command keyboard. They are connected to host computers. These consoles are intended for local control of the host nodes and all connected platform equipment. The databus is a dual-redundant Ethernet system.

The system uses distributed architecture with functionality being provided at substation level.

The display features a tiled window layout that ensures all displayed information is visible to the operator at all times. The separation of commands and detailed data (such as alert delays and object names) from the mimic diagrams leads to less cluttered displays and less operator overload.

The system is capable of displaying a large number of pages or mimic diagrams. Each page provides the operator with a clear and precise view of the platform system under scrutiny using computer-generated graphic colour mimic diagrams. Objects can be quickly selected using the cursor symbol.

Operational status
ISCMMS is now operational aboard the Australian 'Collins' class submarines.

Contractor
Saab Dynamics, Jonkoping.

NEW ENTRY

UNITED KINGDOM

Valenta

Type
Submarine diesel.

Description
The lightweight Valenta RPA 200SZ 4-stroke 16-cylinder mechanically driven, direct supercharged, intercooled V-type diesel was developed for the 'Upholder'-class SSK. The use of mechanically driven superchargers, insensitive to exhaust back pressure enables well over twice the engine power to be produced compared with normally aspirated engines and gives good cylinder scavenging under all submarine service conditions.

A single gear-driven centrifugal compressor drives the supercharger at the free end of the engine through an intermediate fluid coupling. This supplies air to the engine at a boost pressure ratio of approximately 2.4:1 through a seawater-cooled intercooler. The gear drive to the supercharger also drives the seawater pump that is mounted on the engine.

The supercharger comprises two main units each of moderate size and weight and easily handles within the confines of a submarine hull. This configuration maintains interchangeability with the standard Valenta engine.

The drive increases the 1,350 crankshaft rpm to 24,000 rpm in two stages. The primary speed increase is by epicyclic gears that drive into a fluid coupling before the secondary spur gear stage. The fluid coupling isolates the secondary stage gears from torsional vibration in the low-speed train arising from crankshaft vibrations and effectively isolates the high-speed train from these vibrations. It also allows the very high-speed supercharger rotor to overrun in the event of the engine being suddenly shut down. The engine output is 1,518 kW at 1,350 rpm.

Valenta RPA 200SZ as supplied to the 'Upholder' class (ALSTOM)   **1999**/0024686

One notable design feature of the Valenta is the arrangement used to extract any salt water that may have entered via the exhaust manifolds and open valves when submerged. Before starting, the engine is rotated at low rpm while air servo cylinders mounted on the valve covers depress the engine inlet valves enabling any sea water in the engine and exhaust system to be purged through drain valves in the induction manifold. The exhausts incorporate inboard silencers and independent surface-use and snort outlets. A cooling facility is provided to reduce the IR signature of the exhaust gases.

Operational status
The Valenta equips the 'Upholder' class submarines that have now been leased to the Canadian government.

Contractor
ALSTOM Engines Ltd, Colchester.

NEW ENTRY

Batteries

Type
Lead acid submarine batteries.

Description
Hawker Chloride Industrial Batteries manufactures both flat plate and tubular plate lead-acid batteries for diesel/electric and nuclear-powered submarines. The batteries can be recharged without requiring rest periods on the surface, which greatly improves the boats' direction.

The flat pasted positive plate cells comprise a lead alloy lattice grid with a paste of lead oxide (the active material) and sulphuric acid pressed into the matrices.

Shedding on the positive grid is overcome by using a felted glass wool separator placed against the faces of the positive plates (which effectively retain the lead dioxide active material) as well as the usual microporous polyvinyl chloride separator fitted against the faces of the negative plates. The oxidation-resistant and highly porous glass wool is capable of compression during the assembly of the cell, and provides good retention for the active material when placed next to the positive grid. These allow free circulation of the electrolyte to the active material and ensure a mechanical barrier against the possibility of short circuits.

Flat plate (pasted plate) batteries are currently used in all Royal Navy SSNs. Various designs are available with capacities ranging from 5,150 to 11,750 Ah per cell.

The tubular positive plate cell of 8,850 Ah capacity is a comb-shaped one-piece lead alloy casting consisting of a horizontal bar supporting a series of vertical spines. Each spine is enclosed in a porous acid-resistant glass/polyester fabric tube filled with the active material (lead oxide powder). The negative plate is a flat pasted grid. An internal water cooling system is fitted which limits the temperature to a consistent level below 35°C and regulates the temperature in all the cells to offer enhanced performance. A simple air pump is also fitted to produce rotational agitation of the acid in the cell from the bottom to top. This ensures effective circulation of the electrolyte to ensure that all the acid is used. The water has to be exceptionally pure to prevent the conduction of current between the cells.

This cell design uses an expanded copper negative plate that improves performance at high-discharge rates and enables the cell to maintain its capacity at lower rates. The battery is available with either stud- or blade-type terminals. The cell can also be supplied with conventional cast negative grids.

The plates in both flat plate and tubular cells are suspended which gives enhanced resistance to shock and vibration and reduces the risk of short circuits

Cutaway drawing of the new BLC submarine cell (Hawker Chloride Industrial Batteries)
1999/0024687

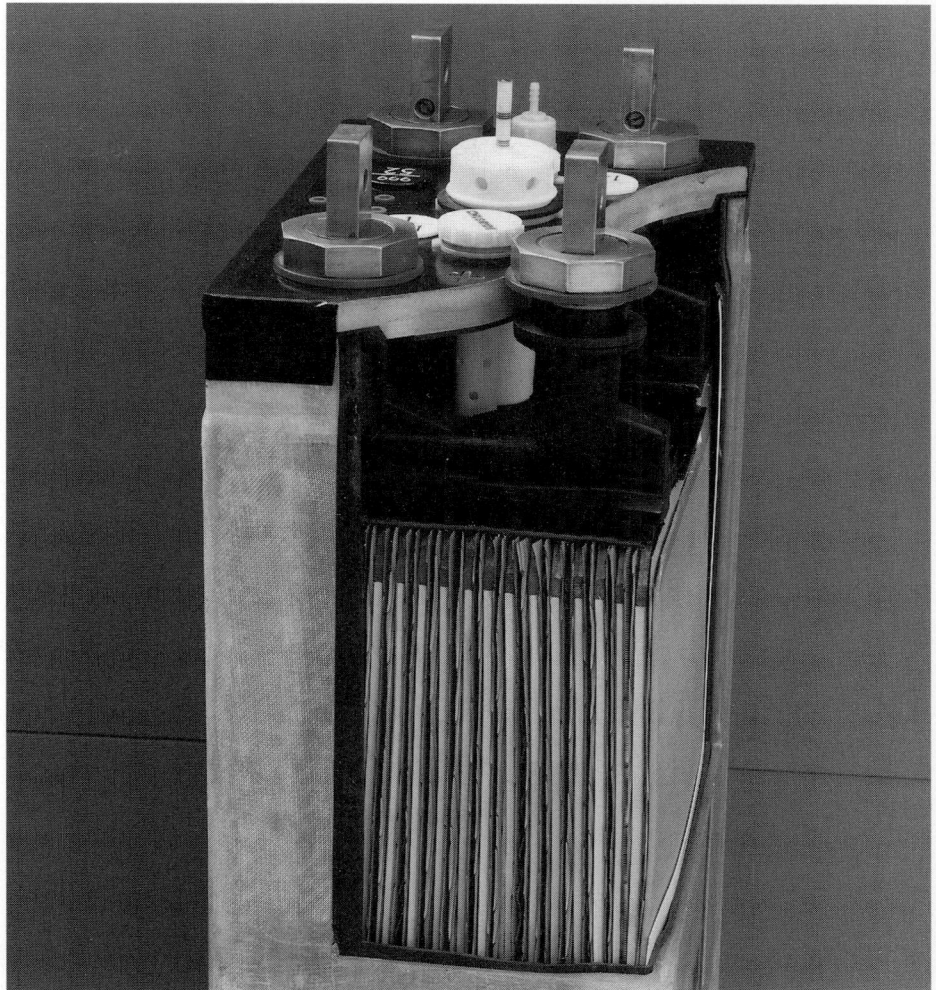

Cutaway view of the tubular plate submarine cell (Hawker Chloride Industrial Batteries)   **1999**/0024685

caused by sediment. To provide the necessary support for the plates, the body of the battery is made of tough GRP. This also provides a reliable enclosure for the electrolyte and greater resistance to mechanical damage.

The production of hydrogen in cells occurs as a result of a chemico-electrical reaction. This problem is overcome by the use of the single casting that allows the use of low-antimony lead alloys that reduce the evolution of hydrogen under open-circuit conditions.

In its latest development of the lead-acid battery, Hawker Chloride claims a 22 per cent improvement in performance at the high-discharge rates (necessary for high speeds) and a 6 to 8 per cent improvement at the low-discharge range. This new cell (BLC - Bottom Lug Connection) offers increased speed or range. This is achieved by the introduction of a high-conductivity electrical connection from the bottom of the plate in a cell to the main power take-off lug at the top. In this way it is possible to make use of the under-used stored energy in the lower part of the cell, especially under operating conditions which impose a heavy demand on the cells. Normally this stored energy is not used until the later stages of discharge, by which time the cell voltage has begun to fall so rapidly that it is no longer able to deliver sufficient power. In a further development a new thin-walled container construction (a glass fibre reinforced plastic material into which the resin is injected under pressure) allows the volume of acid in the cell to be increased and leads to improved performance at low-discharge rates. For a mid-sized cell this results in an increase of around 8 per cent of the total volume (or an additional 7 litres of electrolyte).

Operational status

Batteries are installed in 'Upholder' class submarines and a large number of other submarines around the world.

Contractor

Hawker Chloride Industrial Batteries, Manchester.

NEW ENTRY

ANTI-SUBMARINE WARFARE

Command and control and weapon control systems
Submarine & surface ship ASW combat information systems
Weapon control systems

Sonar systems
Surface ship sonar systems
Submarine sonar systems
Airborne dipping sonars
Sonobuoys
Airborne acoustic processing and receiving systems
Static detection systems

Countermeasures
Acoustic decoys
Electronic warfare

Underwater communications
Submarine communications systems
Underwater telephones
Communications buoys

Electro-optical sensors

Magnetic anomaly detection systems

COMMAND AND CONTROL AND WEAPON CONTROL SYSTEMS

Considerable advances have been made in combat information systems for submarines resulting primarily from major developments in the field of electronics. With the new generation of combat information systems using the latest developments in electronics, weapons can be deployed much more effectively. Improvements in sensors, particularly sonar systems, also provide the modern combat system with a much greater volume of more accurate data from which targets can be detected, tracked and localised, which further aids the effective deployment of weapons.

The sum total of these developments has had a considerable effect on the design and capability of combat information systems. The constant reduction in size and weight of equipments has had a two-fold effect on the submarine. First, it has enabled the smaller displacement submarine to be equipped with a fairly sophisticated combat system, not always previously possible because of the weight and space requirements demanded by such systems. Second, it has enabled larger ocean-going submarines to be fitted with very sophisticated combat information systems, which considerably enhance their capability.

The advent of passive ranging sonars (broadband, wideband and narrowband), intercept sonars, passive cylindrical bow hydrophone arrays, flank array sonars, active sonars, low-frequency sonars and towed array sonars, together with other vastly improved sensors such as EW, optical and optronic systems, all provide the combat suite with a huge volume of data which is used to provide the command team with all the necessary information from which to make its decisions.

Another area of considerable importance which affects the overall mission capability of the submarine is its navigation systems. Many submarines now fit inertial, satellite, GPS and other navigation systems, as well as the usual gyrocompass and log. All these assist in providing a highly accurate position-finding capability - essential if modern weapons are to be directed accurately onto their target.

This huge bank of data must be assimilated and processed at high speed to generate an accurate picture of the complete underwater and surface environment. It requires programmable high-capacity computers which have the ability to retrieve and handle vast amounts of data on a number of targets simultaneously, from a wide variety of sources, process that data and distribute it to a number of different destinations.

The best and most effective way of handling the data in this way is undoubtedly the databus. The ability to have data passing round a system has enabled the multipurpose console to be developed which provides enormous flexibility and redundancy, each console being able to undertake a variety of different functions at the command of the operator, depending on the priority and nature of the requirement. Current command systems are therefore characterised by organising the various elements around a LAN (Local Area Network) interconnecting processors distributed around the system. Many systems rely on proprietary computers developed by the contractor, but this situation is now changing as commercial-off-the-shelf (COTS) technology becomes more acceptable and in many cases self-imposed due to finanical considerations - a case of not reinventing the wheel.

This enormous potential and flexibility has enabled designers to develop and introduce the concept of the totally integrated combat system for the submarine. This incorporates in one total package sensors, combat suite and weapon control. To treat these three prime areas as a single system is of considerable value as it overcomes many of the problems formerly experienced in interfacing different systems.

With the advent of the fully integrated combat suite and the extensive use of automation, designers are now integrating all command and control and weapon control functions within the confines of the submarine control room. This brings enormous advantages to the submarine commander who is now fully aware at all times of all that is happening to his boat, and is in complete control of the entire environment relating to the submarine.

Other developments related to masts and sensors, particularly optronic sensor systems and non-hull penetrating masts, will enable the number of masts to be reduced, leading possibly to a smaller sail and improved hydrodynamic performance. The operations room will be free of all mast encumbrances, which will enable designers to develop an entirely new concept for the layout of the control room and the position and relationship of the various members of the command team. The use of fibre optic links between mast and control room may also enable the sail to be resited in a better position to enhance hydrodynamic performance. Finally, future systems will incorporate the latest developments in Artificial Intelligence (AI) and expert systems architecture which will greatly assist in reducing the time and possibly increase accuracy in carrying out classification and identification of targets.

As a result of these developments, a complete redesign of the submarine's control room is a possible consideration for the next generation of submarine. Such a concept will, however, only be possible with a high degree of automation and with systems which are completely reliable.

For the future the increasing ranges of sensors, growing ranges of weapons and the changing role and endurance of submarines are all areas that will have to be addressed within the context of an integrated combat system. Flowing from this will be the need to address the presentation of data to the command team, for new technologies will not only offer a much larger volume of processed data, but it will be handled at a much higher rate than hitherto.

UPDATED

SUBMARINE & SURFACE SHIP ASW COMBAT INFORMATION SYSTEMS

AUSTRALIA

Collins Combat System

Type
Submarine combat system.

Development
Prime contractor for the combat system of the 'Collins' class submarines of the Royal Australian Navy is Boeing Australia (formerly Rockwell Systems Australia). Boeing is responsible for overall programme and systems management, combat system installation and integration, as well as the supply of the internal and external communications suite. Other members of the team include the Autonetics & Missile Systems Division of Boeing North America with responsibility for system supervisory units, databus hardware and communications software, Lockheed Martin Librascope who is supplying the weapons data converters, multifunction consoles and command plot, Thomson Marconi Sonar SAS providing the sonar arrays and signal processing and CSA who have responsibility for supplying combat system software and setting up the shore facilities.

Description
The combat system architecture is based on a fully distributed CWCS using seven multifunction consoles and a command plot table, interconnected and integrated to weapons and sensors via a high-capacity ESSDMS (Expanded Service Shipboard Data Multiplex System) fibre optic databus. The primary areas of capability are surveillance, track prosecution and support. The system can give positions and projected positions on up to 25 targets. Applications software is written in Ada and executed on Motorola 68040 processors. Sonar sensors integrated by the combat system include the Thomson Marconi Sonar Scylla sonar suite which incorporates: a bow-mounted cylindrical passive array; active and passive mine-avoidance arrays; TSM 2253 low-frequency PVDF passive flank planar arrays; an intercept array; and TSM 2225 ranging arrays. A towed array sonar will be added at a later date and the American TB23 and Thomson Marconi Sonar Narama and Kariwara systems are all being considered. Other allied systems integrated by the command system include: an ARGOSystems AR 740 ESM, Pilkington Optronics CKO43 search periscope with LLTV and thermal imager and a CHO93 attack periscope also with LLTV, Link 11, a Kelvin Hughes Type 1007 radar, an L3 Communications ELAC Nautik echo-sounder, an EDO Model 3040 Doppler navigation system, two Litton SINS inertial navigation systems and Rockwell-Collins GPS.

Sensor inputs are integrated within the combat system at data level, enabling any operator to display and correlate information simultaneously from up to eight independent sensors. Any one of the seven multifunction consoles can be configured to display sonar, radar, tactical picture, fire control or navigation data.

Operational status
A fully operational combat system became functional aboard HMAS *Collins* in 1998. A second and final contracted phase of fully functional software should be delivered to *Collins* by mid-1999. A 12 month period of trials will follow before *Collins* is accepted into operational service in mid-2000.

Contractor
Boeing Australia, Boeing North America, Autonetics & Missile Systems Division.
Lockheed Martin Librascope Corporation, Glendale, California.
Thomson Marconi Sonar SAS, Sophia Antipolis.

UPDATED

DENMARK

Sub-TDS

Type
Tactical data system for submarines.

Description
The Terma Tactical Data System for submarines (Sub-TDS) is designed to meet the requirements of combat information systems for small submarines.

Functions performed include: picture compilation, situation assessment, tactical manoeuvre calculations and weapons deployment. It provides a standard tactical datalink to exchange track and ESM information as well as plain text messages, utilising existing onboard radio equipment, but may also be supplied with NATO Link 11 or other customer specified datalink.

The Sub-TDS supports the Terma Torpedo Fire-Control System (TFCS), which has its own processor cabinet and torpedo control panels. The system provides control of up to four wire guided torpedoes simultaneously. It handles readiness status reports from the tubes and torpedoes, and controls run-up and firing as well as course and depth guidance when under way. The TFCS also supports fire-control solutions for straight running torpedoes.

Sub-TDS consists of the following basic modules which may be added in any number to suit the specific submarine type:
(1) the Graphic Display (GD), which is a 20 in high-resolution raster scan colour monitor, portrait oriented, the lower part being utilised for information readout. The GD can display three different types of picture:
  (a) General Operational Plot (GOP), which compiles and displays a complete situation picture with maps, tracks, ESM data, and an extended selection of tactical patterns and tools. The entire system relates to 'own ship's' navigational position and readout positions in latitude/longitude, colour grid, or true or relative bearing/range as desired. The GOP has a wide scale of editing tools, and may off-centre and zoom to view any local geographical position
  (b) Contact Evaluation Plot (CEP), which plots contacts as a function of time and allows the system to produce a situation picture and an attack evaluation based on inputs from passive sensors
  (c) Attack Plot (ATP), which extracts information from the two above mentioned plots, and produces the local attack situation; it also includes all the necessary vector diagrams, forecasts and recommendations for torpedo attack on selected targets
(2) the System Terminal forms the man/machine interface with a keyboard containing a 'QWERTY' keypad and 60 dedicated function keys, and a high-precision rollerball
(3) Data Processing Unit (DPU), which serves as the main TDS data co-ordination unit and holds all interfaces to the internal TDS modules as well as external interfaces like navigational fix systems, ESM, communications, sonars, hydrophones, periscopes, radar and depth gauge
(4) Torpedo Guidance Unit (TGU), which is a processor that handles all calculations for launch, guidance and torpedo position determination. The unit interfaces torpedoes and tubes to the entire combat system, and also handles the interface of the Torpedo Fire-Control Panel, Tube Order Panel, and the Torpedo Distribution Unit, which are peripheral equipments to the TGU
(5) peripheral equipment includes tape cassette or disk recorder, printer and colour plotter. The online functions are the recording of all events within the system and the loading of maps for display. Offline the system may produce playback to the screens for lessons learned, and printed records of various selections and formats, such as track records or narratives
(6) The modules are housed in water-cooled cabinets, shaped to fit into the narrow space of the hull.

Operational status
In service with the Royal Danish Navy 'Kobben' class submarines.

Contractor
Terma Elektronik AS, Lystrup.

VERIFIED

FRANCE

SUBTICS

Type
Submarine tactical integrated combat system.

Development
Based on 20 years' experience in the development of sonar subsystems (DMUX80, DSUV22, DMUX20, TSM 2233, TSM 2933) as well as command and weapon control subsystems (SYTAC, TITAC, LAT-NG), DCN and Thomson Sintra ASM (now Thomson Marconi Sonar SAS) have developed an integrated approach for both data analysis and system manning.

Both companies have made extensive use of commercial-off-the-shelf (COTS) technologies and worldwide recognised hardware and software standards (such as workstations, power PC, UNIX, Ethernet) to further improve performance and simultaneously reduce production and life cycle costs of their equipments.

Using this expertise, the two companies have together developed an efficient and competitive range of products named SUBTICS (SUBmarine Tactical Integrated Combat Systems) which is now being offered for both refit and new building submarine programmes.

To achieve this objective and using the accumulated experience of both companies, DCN and Thomson Sintra ASM (now Thomson Marconi Sonar SAS) have established UDS International, a joint company in charge of manufacture and marketing of SUBTICS in France and on the export market.

Description
The main subsystems of SUBTICS are:
(a) a comprehensive set of sensors which include: bow cylindrical or conformal arrays; towed arrays; and flank or distributed arrays that multiply the detection capacity of the submarine against silent targets
(b) all types of heavyweight wire-guided torpedoes and anti-surface submarine-launched missiles
(c) an integrated architecture which links both sensors and weapons to the system's communication and data handling core through the databusses of the system network.

The core of SUBTICS is an open and modular architecture based on standard data processors (TMS 320 C30 processors on a speed ring network),

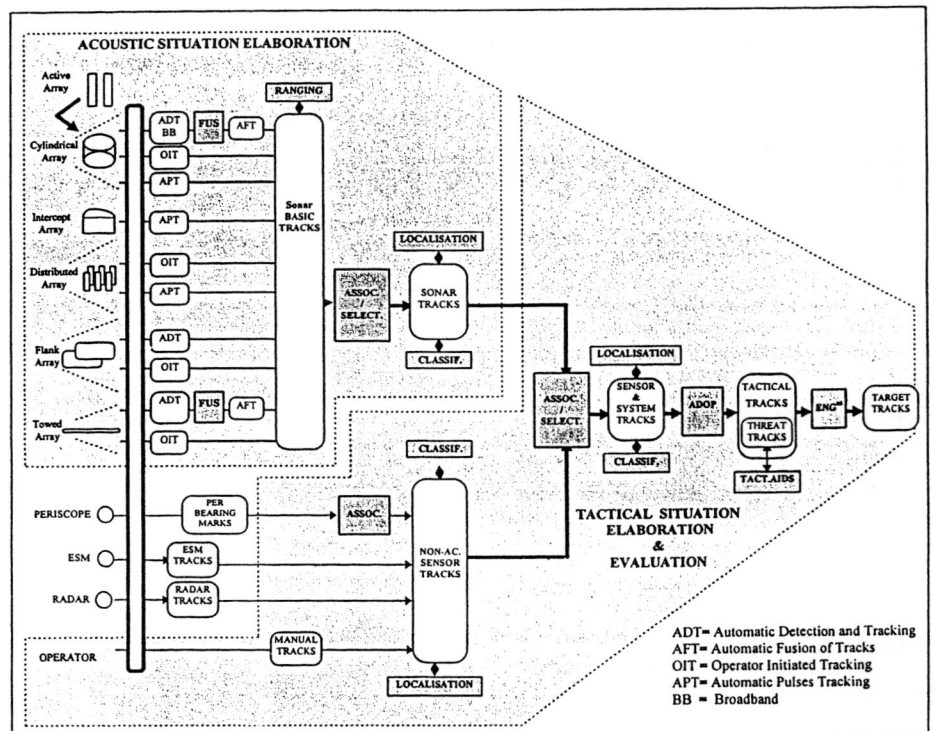

Track management data flow of SUBTICS

1998/0007216

redundant databusses and six multifunction common consoles equipped with two high-definition 19 in colour monitors.

The dual redundant Ethernet databus offers reliable and continuous communications in the event of damage, the multifunction common consoles allow manning and configuration flexibility, and together they provide function and data operational availability.

The choice of an open structure, together with the standard and sizeable processing and display resources, allows existing or new operational components to be easily integrated with minimum life cycle costs.

The basic components of the sonar subsystem include a set of acoustic arrays and appropriate detection (or transmission), tracking, and analysis and localisation processing modules, together with a set of common processing modules for contact motion analysis, classification, identification and track management using broadband, narrowband, demon and pulse (passive interception and ranging) processing channels.

The basic set of acoustic arrays includes a cylindrical or conformal bow array, an active array, a distributed array, an intercept array, a towed array, a flank array, and optionally an obstacle avoidance array.

The contact motion analysis module includes automatic and interactive processing functions. The

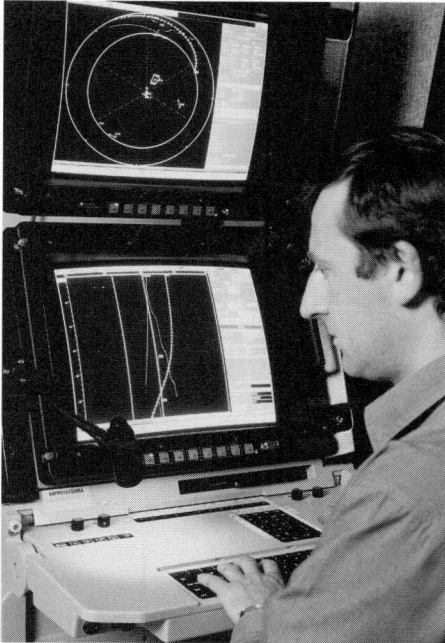

Typical display on a SUBTICS console (Courtesy UDS International)   *1999*/0024684

SUBTICS consoles on test (Courtesy UDS International)   *1999*/0024683

classification and identification module includes audio and spectrum analysis processing functions plus interactive hypothesis generation and verification functions. The track management module includes interactive association and fusion processing functions plus sorting of best representative acoustic track functions.

The basic components of the command and weapon control subsystem include a set of situation elaboration modules, a set of command and decision modules, and a set of engagement and launching modules.

The set of situation elaboration modules includes acoustic and non-acoustic sensors track association and fusion functions, interactive target motion analysis functions, and track management functions able to sort out tracks of particular tactical interest among a set of 100 recorded tracks.

The set of command and decision modules includes threat evaluation of localised and classified tracks, plus attack and escape manoeuvre planning tools.

Engagement and launching modules give the submarine capacity to launch simultaneously torpedoes in wire-guided mode or in fire-and-forget mode and anti-surface missiles.

Operational performances are primarily related to SUBTICS' sensors and weapons capabilities, which multiply the detection and action area of the submarine. Performance is further enhanced as a result of the fully integrated architecture of SUBTICS. This means that as a result of full integration every element, either data processing, computers or multifunction common consoles, are connected to redundant databusses.

As a result the system's processing functions and the operators are offered unique accessibility to all information, together with a unique availability of information due to reconfiguration capabilities and redundancies.

SUBTICS full integration also means that every operational function is coherently associated in a

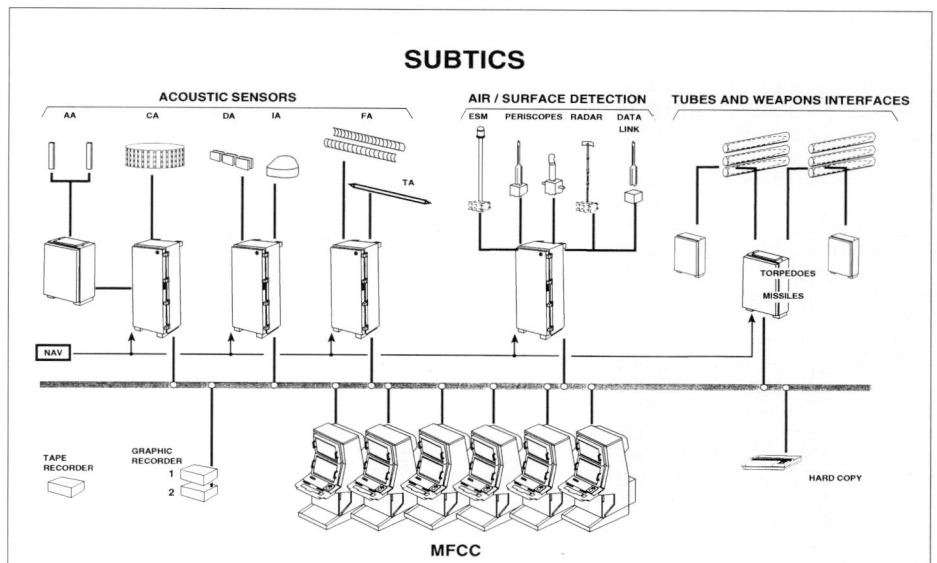

Typical SUBTICS block schematic   *1998*/0019007

continuous track building process along which tracks are continuously summarised and displayed together with their associated localisation and classification attributes. As a result, clear and comprehensive tactical situation displays are presented allowing the command to take fast, accurate and correct decisions.

With adaptability and evolutionary requirements in mind, SUBTICS can be fitted on any type of submarine both for new construction as well as for modernisation programmes. The system responds to the present and future operational needs.

Operational status

Three systems were ordered in early 1995 for Pakistan's new 'Agosta' boats and are currently being assembled. The system has also been ordered in 1997 for the Chilean 'Scorpene' submarines. SUBTICS is currently in series production.

Contractor

UDS International, Sophia Antipolis.

UPDATED

SYVA

Type

Tactical ASW system for frigates.

Description

SYVA is a tactical ASW system for surface vessels integrating a suite of sensors (passive and active sonars and so on) and incorporating new-generation hardware and a specially developed software suite. The version scheduled to be installed as part of the modernisation of the French Navy 'Tourville' class frigates features the SLASM VLF active sonar.

Initially developed for SLASM, SYVA is able to accept a variety of new and different interfaces to integrate different generations of sonars such as DSBV 61A, DSBV 62C and UTSS 1A currently operational aboard French Navy warships.

The system is easily tailored to the requirements of any naval force and can be fitted or retrofitted to all types of surface combatants. It provides digital interfaces with: navigation systems; active and passive sonars; environmental data acquisition systems; and command support systems.

As an ASW tactical decision making aid, SYVA helps duty officers by providing a comprehensive overview of the ASW tactical situation using:
- tactical decision making toolkit
- target motion analysis using passive sonar detection (bearings and frequencies),
- acoustic environment analysis.

The target motion analysis tool allows: ranging measurements to be calculated using bearing and frequency rate measurements; interactive TMA which is based either on a maximum likelihood estimation process or Kalman filtering using, when available,

operator inputs such as target course, range, speed and fundamental frequency; and operator TMA using simultaneous bearing/frequency deviations.

The acoustic environment analysis tool manages environmental data when this is available or statistical data such as sound velocity profiles, sea states, bottom depth and type and ambient noise. Sonar parameters and self-generated noise are also accessed to calculate the effectiveness of own ship sonar within the environment. From this data, loss and range predictions are calculated.

An online database provides operational data on threats.

The tactical situation displays own ship tracks and target tracks as detected by the ship's sonars (DSBV 61A, DSBV 62C, UTSS 1A or SLASM) which are then evaluated by the TMA function. The system also displays and plots tracks provided by the command support system or surface/air combat system (SENIT).

SYVA also provides an interactive, configurable control system and multitarget onboard simulation.

SYVA is based on an open architecture and distributed software configuration which allows different implementations to be designed to suit various operational requirements and associated with suitable equipment packaging (COTS, ruggedised, or full military specification) and can be configured from single workstation to complete system.

The SYVA software suite is based on industry software standards, including: Unix; C-language; X11/Motif; RISC graphic workstations; and Ethernet LAN.

All SYVA, are designed around hardware featuring: commercial-off-the-shelf components; component repackaging to ruggedised requirements; full compliance with military specifications.

Operational status
The system is currently in service aboard seven French ASW frigates (two retrofitted with SLASM) and two

shore-based training centres. The system is fully operational and the success achieved during sea trials of the first SYVA package led to a decision to retrofit all French ASW frigates irrespective of their sonar equipment configuration.

Contractors
DCN International, Paris.
Matra Cap Systemes, Six-Fours-Les Plages.

VERIFIED

GERMANY

ISUS

Type
Integrated combat system.

Description
ISUS is an integrated submarine combat system designed to acquire, process, analyse and display data from all sensors fitted in the boat. This interactive process leads from the display of raw sensor information to an easy-to-read display of the combat situation, providing the command team with optimum assistance in the decision making process.

The system integrates a wide variety of acoustic arrays including: cylindrical or conformal; flank; towed; passive ranging; cylindrical transducer; and intercept, as well as own noise hydrophones and accelerometers; Doppler and EM log; underwater telephone; radar; ESM; optic and optronic sensors and navigation sensors.

The latest variant, ISUS 90, is an open architecture system based on a modular Ethernet databus linking DEC compatible computers with high-speed RISC signal processors, VME backplanes, Ada software and data transfer from sensors via a fibre optic link. The system being supplied to the 'Dolphin' class submarines is believed to incorporate eight multifunction consoles.

The system performs the following functions on data either sequentially by one operator or in a task sharing concept simultaneously by several operators: broadband detection over 360° with high bearing accuracy; sonar pulse detection over 360° in a very broad frequency band; detection of low-frequency spectral lines over a long range with high bearing accuracy; transient detection and storage; accurate range measurement in two 120° sectors; monitoring of own hydrodynamic and machinery noise; location of targets by single pulse active sonar; navigation; integration of periscope and radar displays into the multifunction consoles; automatic and interactive tracking of targets (the target track is controlled by one main sensor and supported by auxiliary sensors); audio broadband and narrowband analysis (the analysis leads to computer-aided classification suggestion); threat analysis; launching suggestion for weapons; integration of the torpedo sonar information from wire-guided torpedoes; launching of underwater missiles; control of wire-guided torpedoes and display of the target/weapon situation.

Operational status
Variants of the ISUS system are in service aboard the German 206A class, the Type 209 submarines of the South Korean (ISUS 83) and Turkish (ISUS 83-2) navies, and is being fitted in the Israeli Navy 'Dolphin' class

Consoles of the ISUS 90 integrated combat system   1996

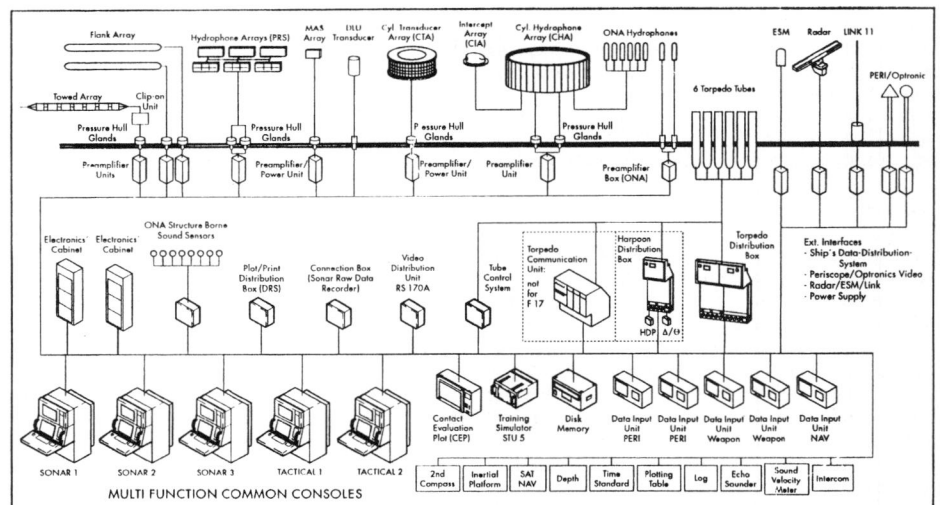

Typical ISUS 90 configuration   1996

(ISUS 90) submarines. The ISUS 90-1 system is also being installed in the Type 212 boats. A variant is also operational in the Venezuelan Type 209 boats.

Contractor
STN ATLAS Elektronik GmbH, Bremen.

UPDATED

INTERNATIONAL

Isaacs

Type
Integrated submarine sonar and weapon control system.

Description
The advanced operational capabilities of submarines demand the evaluation of track data using an

integrated multisensor command system. Such data includes information gathered by acoustic sensors, optical sensors, navigation systems and electromagnetic sensors. Such data must be in a flexible format so that digital, analogue and graphic interfaces can be used with control and monitoring of the majority of sensors being performed by multifunction consoles.

The Isaacs system, being developed by Terma

Elektronik A/S of Denmark and L3 Communications ELAC Nautik of Germany, is designed in a modular way to achieve this requirement and can be configured with all or some of the following elements:
(a) passive sonar with cylindrical or conformal (long line) array
(b) passive sonar with flank array
(c) passive ranging sonar
(d) intercept sonar

(e) active sonar
(f) obstacle avoidance and navigation sonar
(g) echo-sounder/environmental data collector
(h) noise monitoring system
(i) underwater communication
(j) sonar beacon equipment
(k) radar electronics (excluding mast and antenna)
(l) weapon control consoles
(m) torpedo tube interface.

This complex and highly flexible system renders additional equipment consoles, such as those for radar and periscopes, superfluous. By using, wherever possible, existing modules and units, Isaacs qualifies as a low-risk system.

The passive sonar array consists of a steel cylinder with double baffles with 96 or 128 stave assemblies, each one carrying four Type KE 4 hydrophones. This also serves as the antenna for the active sonar. Other configurations are available instead of the cylindrical

array, for example a conformal array with integrated preamplifier, or conformal long line array. The intercept sonar array consists of four Type KE 6 omnidirectional hydrophones with integrated preamplifier. The two flank arrays consist of Type KE 5 low-frequency hydrophones, while Type KE 4 hydrophones provide measurement of cavitation and self-generated noise. Both the flank array and noise measurement hydrophones are mounted in a decoupled installation to overcome structure-borne noise. The passive ranging sonar consists of six arrays of compact design with baffles.

The central control and monitoring consoles provide simultaneous control of torpedoes independently aimed at different targets, handling of wire guided and torpedo homing parameters, full control and monitoring capability, including general operational plot, passive contact evaluation plot, attack plot, attack simulations, datalink, full range of sensor interfaces,

data recording and playback. Target motion analysis on more than 10 tracks can be performed simultaneously.

Specifications
Operating frequencies
Passive sonar: 0.3-12 kHz
Passive ranging sonar: 1-10 kHz
Intercept sonar: 10-100 kHz
Obstacle avoidance sonar: about 195 kHz
Active sonar: 10 kHz
Passive sonar flank array: 0.1-1,000 Hz

Contractors
L3 Communications ELAC Nautik GmbH, Kiel, Germany.
Terma Elektronik A/S, Lystrup, Denmark.

VERIFIED

ITALY

MM/BSN-716

Type
Submarine action information system.

Development
SACTIS is the abbreviated title given to the submarine action information system developed by SMA for Italian Navy 'Sauro' class submarines in collaboration with Datamat. Two versions are produced, known as SACTIS 1 and 2, or by the Italian Navy designations MM/BSN-716(V)1 and MM/BSN-716(V)2, respectively. The two versions are intended for fitting in all the 'Sauro' class boats.

Description
In both versions the principal functions include:
(a) automatic data acquisition from all sensors (for example, hydrophone, sonar, passive ranging system, search and attack periscopes, radar, ESM, navigation system, depth-sounder)
(b) manual data input
(c) real-time computation of ship's position and target position (from bearing data only)
(d) data display (raw and processed) with five different presentations: unfiltered situation; tactical situation; time/bearing display; acoustic propagation display; tactical operation tabular evaluation
(e) calculation and display of data for typical manoeuvres such as screen or barrage penetration, evasion, target interception or collision course, approach and divergence routes, Closest Point of Approach (CPA)
(f) data and event recording
(g) playback for training and analysis.

The hardware comprises: two or three display groups which have vertical CRT displays and keyboards for data input and system management, with

Three-console SACTIS system

separate alphanumeric 'tote' displays above the situation display CRTs; a central processor unit based on a Rolm MSE 14 digital computer; disk memories; a printer; a specially designed interface unit for input/output between SACTIS and the various sensors. The displays are arranged side-by-side to form an operating console and up to 30 separate targets can be displayed, from which individual track histories can be selected for examination and 10 of them presented as filtered targets. Main sensors connected to the system include the BPS-704 radar, IPD-70S sonar system and the Thetis electronic warfare system.

The later SACTIS 2 version includes provision for connection with the submarine's A184 torpedo fire-control system, inclusion of the ELT/810 sonar

prediction system and connection to a Link 11 receiver. There is also extended computer capacity and a third operator position at the SACTIS console.

Operational status
Italian 'Sauro' class submarines are fitted with the MM/BSN-716 system.

Contractors
Officine Galileo SpA (main), Florence.
Datamat, Ingegneria dei Sistemi SpA, Rome (software).

VERIFIED

FCS Mk 3

Type
Submarine command and control systems.

Description
The FIAT Componenti e Impianti per l'Energia e l'Industria SpA – UTI SEPA part of Gilardini, the lead company of the FIAT Group's industrial component sector offers naval fire-control systems for various applications. Such digital systems are mostly based on the use of SEPA militarised mini- and microprocessors of the company's ULP Series. For example, a developed version of the torpedo and weapons control subsystem for SICS is available separately as the FCS Mk 3 for submarine command and control, and is installed in the later 'Sauro' class submarines.

The FCS Mk 3 is a fully integrated submarine command and control system having as its main functions:
(a) acoustic sensor performance prediction
(b) prediction of counter-detection range

(c) target motion analysis
(d) threat evaluation
(e) target designation for attack
(f) weapons control (torpedoes and missiles)
(g) countermeasures control.

Sensor inputs to the system include radar, navigation and attitude information, periscope and acoustic data from the integrated active/passive sonar system.

Target and 'own ship's' data from these sources are used to calculate and display target positions and target vectors, impact point predictions, tactical situation, a launch and guidance display, and wire guidance signals. The local control and switching box connects the signals from the FCS to the selected torpedoes, provides launching tube control, and feeds electric power to torpedoes in the tubes. Within the torpedo, the electronic control unit carries out wire signal interfacing, torpedo steering in accordance with FCS commands, computation of target data from the torpedo acoustic head, and torpedo homing.

Inputs to the system include: active and passive sonar data, navigational information, radar and

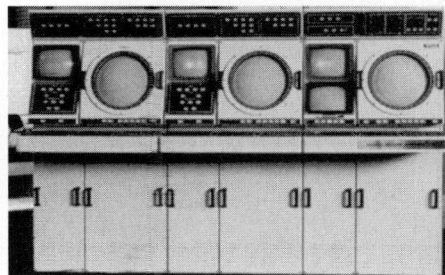

Consoles for the SEPA FCS Mk 3

ESM/EW sensors, and periscope. Weapons include A184 torpedo tubes, which are connected via a local control and switching box.

Operational status
In service aboard Italian 'Sauro' class submarines and Type 209 boats of the Peruvian Navy.

Contractor
FIAT Componenti e Impianti per l'Energia e l'Industria
SpA – Unita Tecnologico Industriale SEPA, Turin.

UPDATED

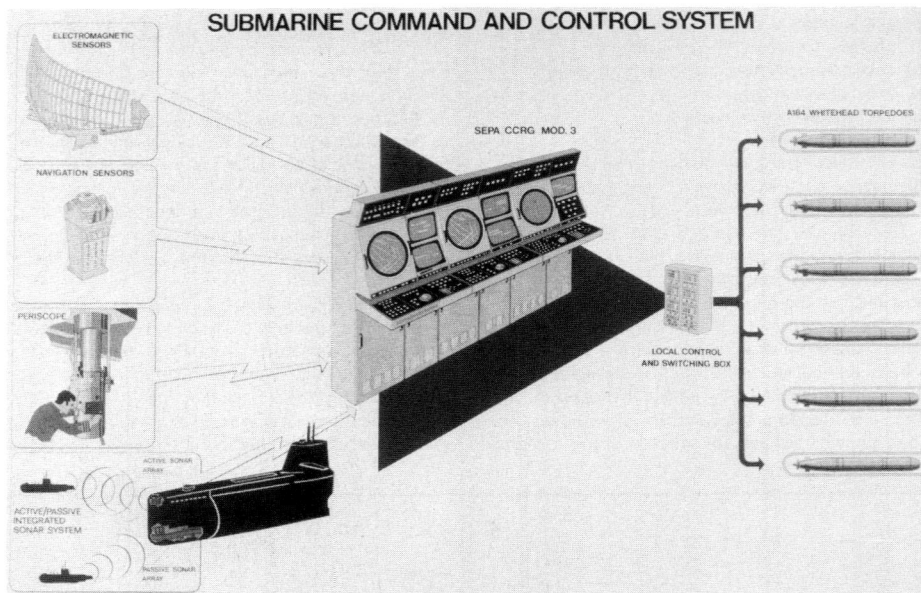

Schematic diagram of FCS Mk 3 for submarines

NETHERLANDS

SEWACO VIII

Type
Submarine data handling and weapon control system.

Description
SEWACO is an automated data handling and weapon control system designed for the 'Walrus' class submarines of the Royal Netherlands Navy. The system forms the link between the submarine's sensors and its weapons and consists of seven identical Display and Computer Consoles (DaCCs). The built-in computer is of the SMR-MU type; the 16 in (406 mm) Plan View Display (PVD) provides a high-load, synthetic picture together with compressed radar video or sonar video, and the control panel is a multipurpose unit. The use of identical DaCCs offers maximum flexibility.

A Central Control Unit (CCU) is used to regulate the mutual data transfer between the DaCCs and the sensors and weapons. For this function two SMR-MU computers are provided, one active and the other a hot standby machine. The large amount of data from the sonars is handled by two extra SMR-MUs, also housed in the CCU. Three types of sonar are fitted (towed array, flank array and circular array), and other sensors and data sources include a noise analyser, ESM facilities, radar, periscopes and position-finding equipment.

The weapon control system can control a mixed load of weapons for subsurface and surface engagements, and the ship's launching system consists of two mutually independent sections, each of which is controlled from a Launching System Control Panel (LSCP). The interface with the weapons (modern subsurface missile) is formed by two identical distribution cabinets. All hardware necessary for the integration of these weapons is included in the distribution cabinet, avoiding the necessity for additional equipment.

The main functions of SEWACO are:
(a) sonar display and control. This makes it possible to omit the original sonar displays and controls
(b) contact evaluation. This entails displaying information obtained from all sensors in a time/bearing format
(c) classification, where ESM, ASM and noise information is compared with cassette tape stored libraries to provide a rapid classification of the target
(d) contact motion analysis. Modern tracking filters are used for the automatic determination of target movements
(e) tactical plot and general plot functions, giving an up-to-date survey of the tactical and/or navigational situation or an historical situation survey
(f) weapon control, whereby the weapon systems are provided with the requisite aiming data
(g) simulation and test to verify the system's operability and to aid in operator training for the functions (a) to (f).
Additional functions can be inserted easily because the system is highly software oriented.

Operational status
Four systems have been delivered and equip the boats of the 'Walrus' class of the Royal Netherlands Navy. A more basic system has been supplied for shore-based training.

Contractor
Hollandse Signaalapparaten BV, Hengelo.

VERIFIED

SINBADS

Type
Submarine integrated battle and data system.

Description
SINBADS is a compact data handling and weapon control system which supersedes the Signaal M8 Series of torpedo FCSs for submarines. The computer used is the Signaal SMR-MU, a fourth-generation general purpose machine, and the complete system combines the weapon control and data handling functions.

The data handling function covers sensor display and sensor selection for track initiation. The raw sensor data of all sensors can be displayed. Each individual sensor is indicated with a unique label. The development of SINBADS is based on the application of modern passive sonars, capable of automatic tracking of targets. The tracking algorithm is based on modern filtering techniques and the tracking system basically includes as its main mode of operation a 'bearing-only' analysis which, however, also accepts other inputs of target information. The complete tracking system functions as an interactive system in which the command can intervene in the tracking process. For this purpose there are four display formats available.

SINBADS can handle five targets and three torpedoes simultaneously. These can be guided, unguided or a mixture thereof. The system also includes data recording for on- and offline use, weapon simulation for training, online failure monitoring, and an emergency mode for which hand controls are provided to enable torpedoes to be set, fired and controlled in case of a computer or display failure.

Operational status
In service with the navies of: Argentina (two TR 1700 type); Indonesia (two 'Cakra' class); Peru (two 'Casma' class); Turkey (six 'Atilay' class) and Taiwan (two 'Hai Lung' class).

Contractor
Hollandse Signaalapparaten BV, Hengelo.

VERIFIED

NORWAY

MSI-90U

Type
Submarine basic command and weapons control system.

Development
The MSI-90U has been designed to comply with the requirements of the Royal Norwegian Navy and the German Navy for their next generation of submarines, the Norwegian 'Ula' class and German 'U212' class.

The basic design philosophy has been to achieve high, all-round capabilities, system flexibility and availability, with limited support costs through a structured design to the functional parts of the system. The basic design highlights the importance of redundancy and independent operation to the different subsystems.

The MSI-90U has undergone an extensive technical and operational test programme as part of the Royal Norwegian Navy's test of the 'Ula' class.

Description

The MSI-90U is a software-based command and weapons control system which uses distributed processing, a high-capacity serial data transmission system (LAN), and multifunction operator consoles to achieve a high degree of capability, flexibility and availability. The main features of the system are:

(a) distributed data processing using the 32-bit KS-900 general purpose computer designed for real-time processing, based on the commonly used Motorola 68020/68040 range of microprocessors. The modular KS-900 is designed for programming in Ada, C and Pascal

(b) a high-capacity Local Area Network (LAN) is used for data communication between subsystems of MSI-90U, between MSI-90U and sensors, and sensor to sensor communication

(c) multifunction operator consoles allowing every operator to have access to all information in the system as well as permitting the number of operators to be adjusted to suit the current tactical situation. This means that one operator can operate the complete system with one console in, for example, a patrol situation. In the standard MSI-90U configuration four identical multifunction operator consoles are used, which can be configured for different console (work) modes. The number of consoles can be adapted to customer requirements

(d) built-in redundancy allowing subsystems to operate separately in graceful degradation fallback mode.

The LAN, named BUDOS, is based on Ethernet, and can interface sensors/subsystems via Ethernet, RS-422 or NATO STANAG 4156. Via BUDOS, any subsystem can communicate with any other connected subsystem. Connection of new subsystems can easily be made by means of spare interfaces on the existing BUDOS multiplexers, or by adding more multiplexers.

Four multifunction operator consoles (each with its own console KS-900 computer), three main KS-900 computers and two or three KS-900 weapon computers (depending on weapon type and customer requirements) ensures adequate redundancy.

Information is presented on two high-resolution colour raster scan displays. The man/machine interface is via a programmable entry panel (a plasma display with touch-sensitive overlay matrix), trackerball with associated control buttons and a standard alphanumeric keyboard. The MMI concept of the system, including layout of display pictures, has been subject to extensive MMI studies and trials in co-operation with the Royal Norwegian Navy and the German Navy.

The system is designed to carry out sensor integration covering target motion analysis, classification and identification, and weapons assignment and control. A number of supplementary facilities is also provided, such as: tactical evaluation and navigation, threat evaluation, engagement analysis, preprogrammed movements, sound trajectory calculations and presentations, predicted sonar ranges for own ship and hostile ships, presentation of geographical fixed points and areas, data recording, and simulation for training purposes and so on.

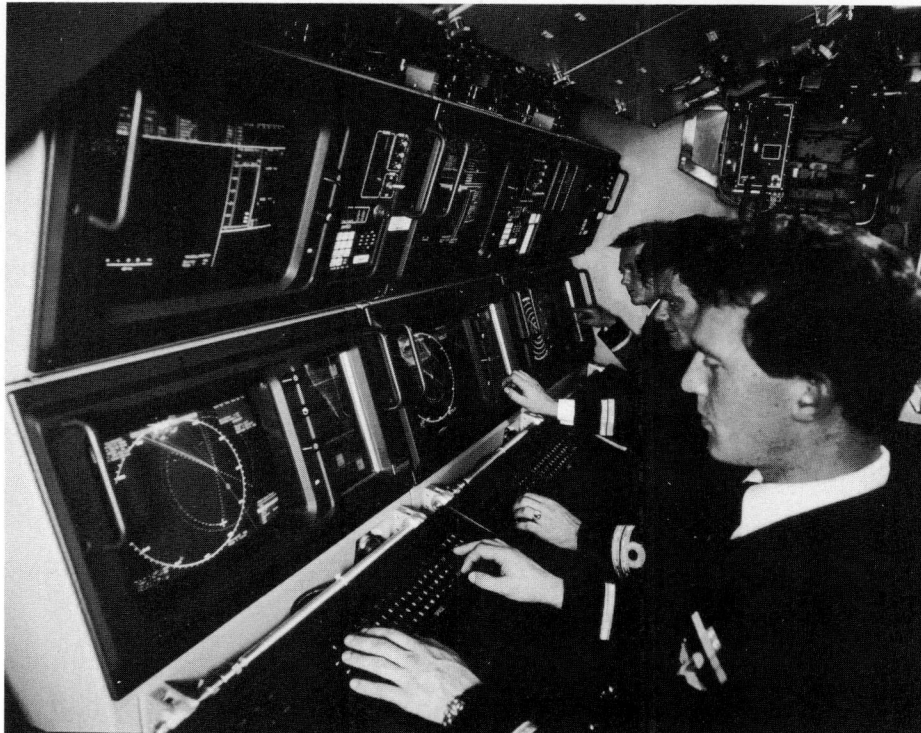

MSI-90U in operation aboard the Norwegian Navy Ula

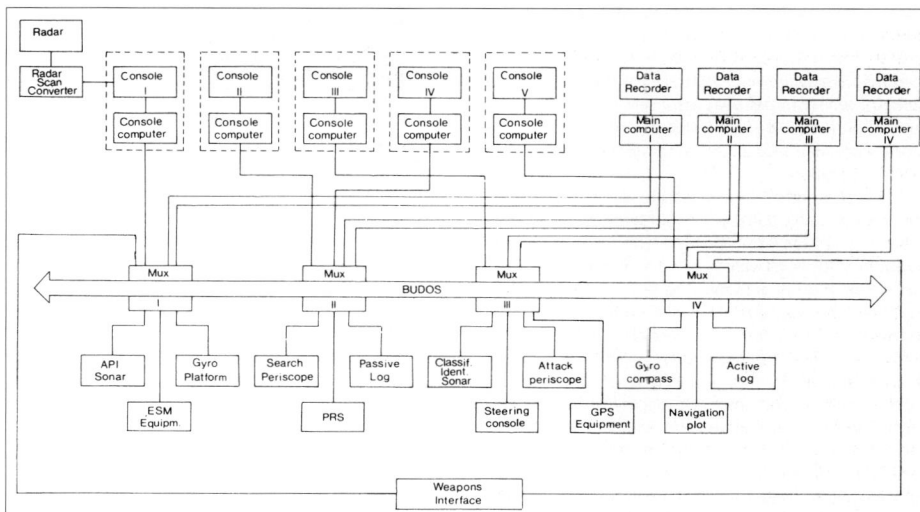

Block diagram of a typical MSI-90U configuration

The system is capable of handling more than 25 targets simultaneously and controlling up to eight torpedoes in different salvo combinations simultaneously.

Operational status

The MSI-90U is now operational on board all six Norwegian 'Ula' class submarines. The MSI-90U is currently undergoing a joint German/Norwegian upgrading programme. Series deliveries of the MSI-90U for the German 'U 212' class, as well as upgrading of the systems on board the Norwegian 'Ula'

class, are under contract and are being undertaken until 2005. The system has also been contracted for the Italian Navy U 212I submarines. Series deliveries will take place between 2002 and 2004.

Contractor

Kongsberg Defence & Aerospace, AS, Kongsberg.

UPDATED

SOUTH AFRICA

Project Nickels

Type

Submarine command and surveillance system.

Description

Altech Defence Systems, has indigenously developed a new submarine command and surveillance system under the Nickels Project. The system has been designed as part of the modernisation programme for the three 'Daphne' class submarines.

The fully integrated suite features software-driven systems and maximum use of COTS items. Processing is carried out using Multibus II Intel 486 microprocessors with Multibus II TMS 340 (AFGIS and TIGA variants) as graphics engines. The system is based on a conventional star architecture configuration using point-to-point serial and parallel interfaces to link the various subsystems and equipment to the command system.

The system comprises two multifunction consoles with large colour screen displays, plasma panels and

touch keys mounted either side of the existing MF sonar consoles. The consoles present time/bearing and tactical plot displays (including own boat position in relation to target plots) and the system carries out automatic computer-aided target motion analysis. It also performs calculation of torpedo fire solutions and allocation of targets, ray path tracing and passage planning. Data from the new long-range passive sonar, a new triple-transducer intercept sonar, a new triple-transducer array passive ranging sonar and a new classification system are all integrated by the command system, together with data from ESM, optical sensors and navigation equipment and the torpedo fire-control equipment.

Operational status
The prototype has been carrying out user evaluation and operational trials aboard the submarine *Emily Hobhouse* since 1994. An industrialised version is scheduled to be installed towards the end of 1998.

Contractor
Altech Defence Systems.

UPDATED

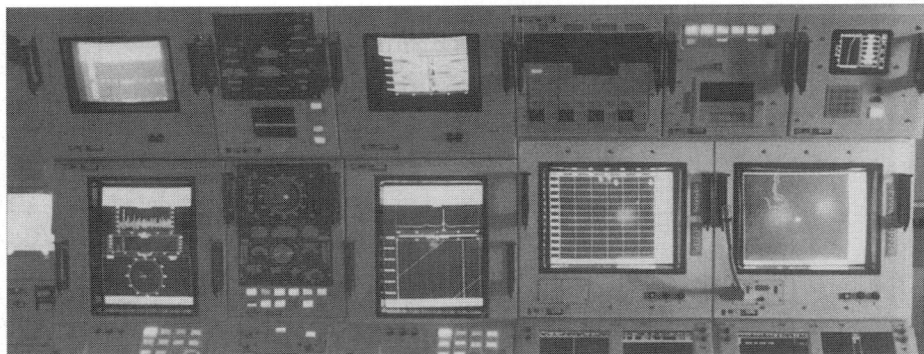

UEC Projects' Submarine Command and Surveillance System (SCAS) for the 'Daphne' update　　**1997**

SWEDEN

NEDPS (NIBS)

Type
Submarine action information and fire-control system.

Description
The NEDPS (Näcken Electronic Data Processing System (in Swedish, NIBS)), was developed for use in the 'Näcken' class submarines of the Swedish Navy.

The fully integrated system is designed to acquire, process and display information for tactical evaluation forming a basis for decisions regarding selected targets and torpedo guidance. The system includes complete hardware and software for controlling wire-guided homing torpedoes. Also included is computer control and monitoring of ship's functions, such as propulsion, steering, depth keeping, trim and storage battery condition.

The system has a dual-computer configuration, the information being gathered through an extensive data collecting system from the different surveillance and weapon systems, as well as from the navigation system. The information is presented to the operators on two separate displays, each of which is equipped with input facilities in the form of special keyboards and trackerballs. Each operator console incorporates three weapon panels in the upper part of the console, together with an alphanumeric display which presents supplementary information such as datalink messages and ESM data. After target acquisition the tracking is carried out automatically and relevant fire-control data are calculated continuously. Torpedoes are normally controlled fully automatically after firing with graphic and alphanumeric presentation of all relevant information. Manual override of the tactical and fire-control displays is available at all times. The operational plot provides an extensive range of supplementary information, including: own submarine operational area; own submarine movements together with navigation reference points and times; operational areas for other submarines; minefields; coastal radar stations with radar parameters. The plot also displays up to 50 internal labelled targets, 10 target channels for TMA; 240 external (via datalink) targets with target numbers and identities; and 32 ESM lines.

The tactical plot is used for the passive sonar TMA calculations on up to 10 targets simultaneously. On the 'Näcken' class this plot is manned by the executive officer who supervises the fire-control operator as well as the sonar, radar and ESM operators. He also handles the periscope observations made by the commander.

Operational status
The NEDPS Mk 2, which uses software developed from the earlier Mk 1 version, with a single multifunction console, is in service with the four 'Västergötland' class submarines, and has been retrofitted to the 'Näcken' class submarines as part of their upgrade programme. One system has also been installed in a training simulator at Berga Naval College. The 'Sjöormen' class boats being transferred to Singapore also carry the NEDPS system.

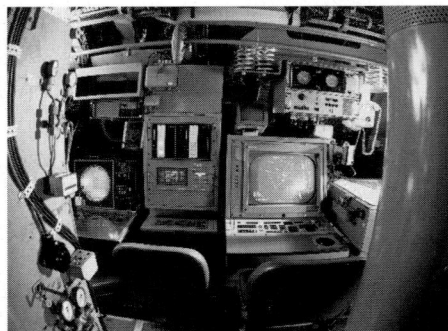

The Mk 2 action information/fire-control system fitted as part of the mid-life update on the 'Sjöormen' class submarines

The Mk 2 action information/fire-control system consoles as fitted in the 'Västergötland' class submarines

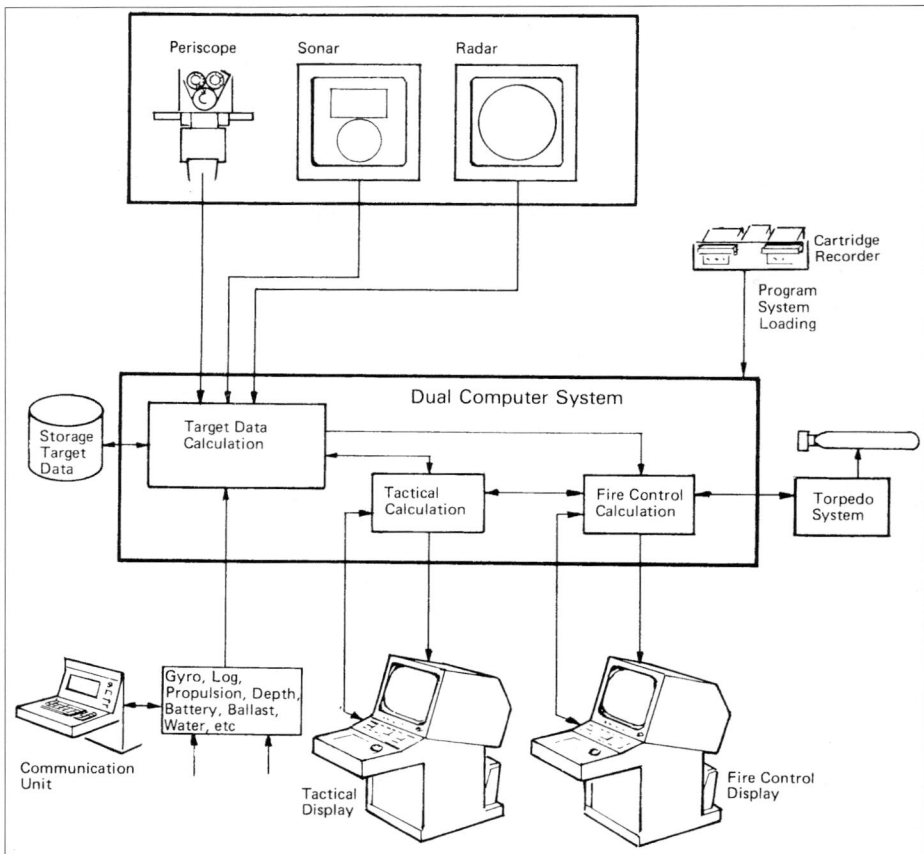

NEDPS Mk 2

Contractor
CelsiusTech Naval Systems AB, Järfälla.

UPDATED

Sesub 940A

Type
Submarine combat system.

Description
The combat system in the 'Gotland' class is a derivative of the CelsiusTech 9SCS Mk 3 based on a distributed configuration using duplicated LAN and a series of intelligent nodes which provide access for various subsystems into the databus. The system is designated Sesub 940A by the Swedish Navy. Sesub 940A uses Ada software which is based extensively on software already developed for the 9LV Mk 3 operational in the 'Anzac' class frigates of Australia and New Zealand and other vessels of the Danish, Finnish and Swedish navies. The suite is based around three Terma-built Type IID multifunction consoles providing command and control, communications and weapons control functions. The system is connected on a dual Ethernet copper wire LAN.

The system features multisensor Target Motion Analysis (TMA) and simultaneous control of all torpedoes. TMA is calculated based on passive sonar bearings only, with optional supplementary target data from other sensors such as periscope and radar.

The system offers full support for simultaneous control of all torpedoes of any type that the submarine can launch.

Operational status
In service aboard the Swedish Navy's new 'Gotland' class submarines.

Contractor
CelsiusTech Naval Systems AB, Järfälla.

UPDATED

The combat system aboard a 'Gotland' class submarine (CelsiusTech Naval Systems)   *1998*/0007217

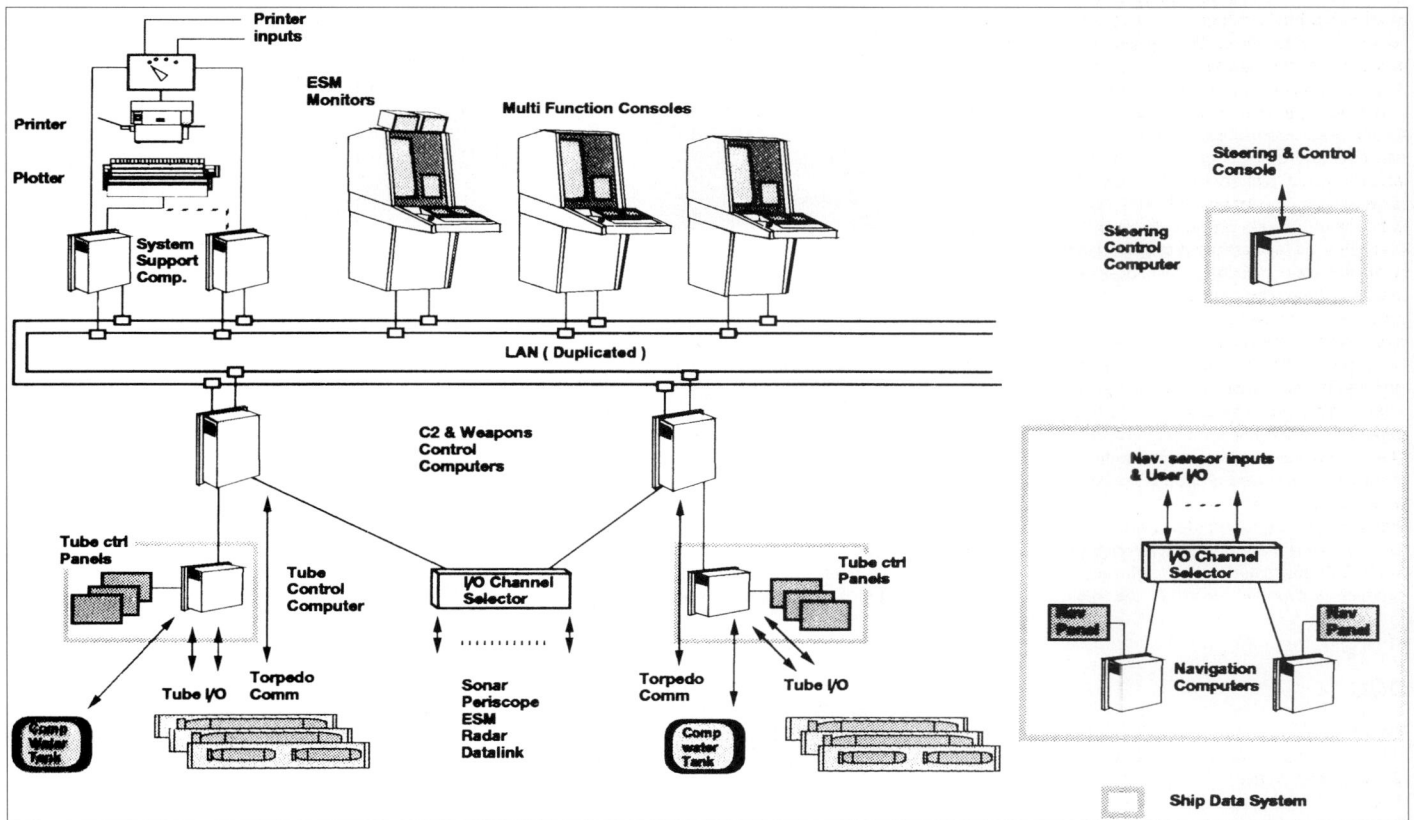

Block schematic of the 9SCS Mk 3 (Sesub 940A) submarine combat system   *1998*/0007218

UNITED KINGDOM

DCB

Type
Submarine tactical data handling and integrated fire-control system.

Development
To reduce the manpower requirements of attack teams, the fitting of digital computer systems in the Royal Navy's nuclear submarine fleet began in the early 1970s with the combat information System DCA. System DCB was developed from System DCA in the late 1970s when a second digital computer, for fire control, was integrated into the system. Development of DCB has continued with regular software updates to incorporate new sensors, weapons and tactics. System DCB is in service in all existing nuclear-powered Royal Navy submarines.

Description
System DCB is a well-proven tactical data handling system with integrated fire-control facilities. The system comprises a two-position combat information console and a two-position fire-control console installed in the control room; weapon interface and battery monitoring equipments installed in the weapon compartment; and the combat information and fire-control computers (two F2420 computers) and associated electronics installed in the computer room. All console positions are provided with two CRT displays, one pictorial and the other for tabular tote information. The main operator interaction with the system is through light pen selections on the displays, supplemented by dedicated switches.

The submarine's sonars, periscopes, ESM and navigation equipments are linked to the combat information computer by S³ (Serial Signalling System) interfaces, and synchro and resolver inputs which are converted in computer peripheral units and input as digital data. The computer processes the received sensor data into tracks. Control of track association can be exercised by the operator. Additional information such as classification and intercept data can also be input by the operator in coded digital form from the keyboards. Target motion analysis, using bearing-only techniques, is carried out for all contacts. The tracking solutions and the sensor data received are compiled into a tactical picture that is shown in plan, time/bearing or time/frequency form on the combat information console pictorial displays. To assist in the evaluation of the tactical picture a variety of tactical calculations can be carried out at operator request. The results of these calculations, together with amplifying data on contacts, can be presented on the tote displays. Contact data are also output to the Automatic Contact Evaluation Plotter (ACEP). Digital data on selected targets are passed to the fire-control computer where fire-control solutions for either wire-guided torpedoes or anti-ship air flight missiles are calculated. The solutions are shown in plan form on the pictorial displays at the fire-control console, while the weapon settings are shown in tabular form on the tote displays. Weapon settings, tube orders and weapon orders are passed to the weapon interface equipment as digital words on duplicated serial highways. The weapon interface equipment decodes the digital data and

System DCB console

applies power, switch signals, discharge gear drive pulses and guidance commands to the weapons and tubes as necessary.

After firing, data feedback received from the wire-guided torpedoes is passed to the fire-control computer on duplicated serial highways. The fire-control computer decodes the feedback data and generates steer commands to the weapon if required, as well as updating the fire-control information on the displays.

The duplication of the serial highways for weapon data and the cross-linking between the weapon compartment units provides a choice of fallback modes to maintain a weapon firing capability.

Operational status
In service with Royal Navy 'Trafalgar' and 'Swiftsure' class submarines. Being replaced by SMCS under mid-life refit programme.

Contractors
GEC-Marconi Radar and Defence Systems,Command and Information Systems Division, Camberley.
Ultra Electronics. Loudwater.

VERIFIED

DCC

Type
Submarine tactical data handling and integrated fire-control system.

Development
System DCC is the tactical data handling and fire-control system installed in the 'Upholder' class SSKs for the Royal Navy. The system is derived from System DCB which is in service with Royal Navy nuclear submarines.

Description
System DCC comprises a three-position console (with electronics in the lower compartments) in the control room, supported by computer equipment in the computer room and weapon interface equipment in the

forward part of the submarine. The main system computer (FM 1600E) is redundant, a second computer being provided to run the system should a fault occur in the primary computer, thus ensuring that the facilities of DCC are maintained.

Each operator position provides two circular 12 in cursive display units, one for pictorial displays and the other for tabular tote data, a light pen and a fire-control order and status unit keyboard. The main human computer interface with the system is provided by the light pens, but certain selection and switching facilities for fire control are provided on the console keyboards. Each operator position provides full combat information and fire-control facilities, allowing increased flexibility in the manning of the system, such that a single operator can control the complete system. There is close integration between the sensors and System DCC with sensor data being passed to DCC via

a digital databus. The sensors passing tactical and environmental data to DCC include: long-range passive sonar, medium-range passive/active sonar with cylindrical array located in the bow, sonar intercept, passive ranging sonar using three aligned arrays on each side of the hull, underwater communications, bathythermograph, and sound speed-measuring system. The periscope masts are provided with facilities for fitting a variety of electronic and electro-optical devices. A Type 1006/7 surface surveillance and navigation radar and omnidirectional and directional warning antennas for the ESM are carried on a separate mast or incorporated in the search periscope.

The sensor data are collated, processed and presented in plan, time/bearing or time/frequency form on the pictorial displays of the console, with supplementary or amplifying data being shown on the tote displays. Contact data are also output to the Automatic Contact Evaluation Plotter (ACEP). The processing, which includes target motion analysis and tactical calculations, is provided to support the command in the assessment of the tactical situation and the decision making processes.

Using advanced algorithms, DCC calculates the ranges on all known targets and carries out target motion analysis on up to 35 contacts simultaneously. The system is such that the Type 2400 is capable of engaging both surface and sub-surface targets simultaneously and engaging multitarget threats.

The AIO retains a fully automated time/bearing display function, but with a facility for manual injection if required. On being provided with data on sonar contacts, the computer works out probable position data (a calculated position derived from passive sonar information) and presents this as a ppi display on the tactical display. To improve the accuracy of this position (reducing the ellipse of uncertainty), further, more accurate data from other systems (including systems external to the boat) can be injected into the

display either automatically or manually using a light pen. The system is so designed that the operator can carry out manual operations using the display, while the computer continues to work out the real situation completely automatically.

Coverage of both target and weapon envelopes is provided together with relevant tactical information on these, that is, the firing envelope necessary to engage the target, and weapons position relative to the target are calculated. In the fully automatic mode DCC can guide up to two wire-guided torpedoes to separate target areas within the weapon envelope, when the weapon takes over autonomous control of the engagement, and a salvo firing of up to four Sub Harpoon missiles. However, a manual override is available if required.

The weapon fit for the 'Upholder' class includes dual-purpose, wire-guided torpedoes (anti-submarine and anti-ship), anti-ship missiles, and mines. The fire-control facilities for these weapons are integrated into the system and include target designation, weapon and tube preparation, weapon launch and, where

appropriate, post-launch guidance. Weapon settings, tube orders and weapon orders are passed to the weapon interface equipments which apply power and switch signals to the tubes and weapons during the preparation and launch phases. After firing, the data feedback from the wire-guided torpedoes is passed to the central computer which generates steer commands as required, and also updates the fire-control information on the displays.

Operational status

DCC is installed in the four 'Upholder' class SSKs which are being leased to Canada. The system is due to be replaced by the SFCS NG1.

Contractors

GEC-Marconi Radar and Defence Systems, Command and Information Systems Division, Camberley.
Ultra Electronics, Loudwater.

UPDATED

KAFS

Type

Submarine combat information and fire-control system.

Development

KAFS is the submarine system variant of the Ferranti Modular Combat System range developed for the export market in the mid-1980s. The development of KAFS reflects Ferranti's experience in submarine combat information and fire-control systems, gained from the development of Systems DCA, DCB, DCC and DCH for the Royal Navy.

Description

KAFS is suitable for conventional submarines of all sizes. The system uses up-to-date processing techniques and software algorithms to provide a wide range of operational facilities including sensor data handling, target motion analysis, picture compilation, fire control, tactical calculations and data recording. A comprehensive onboard training facility enabling training in both combat information and fire-control procedures to be carried out in harbour or at sea is also included. The system hardware and software has been specifically designed in a compact, modular manner to provide a range of facilities which can be adapted or expanded as necessary to meet individual customer requirements.

The equipment configuration is a console in the control room linked by a MIL-STD-1553B databus to the weapon control equipment in the torpedo room. The torpedo room equipment comprises a local control panel which provides common services together with two weapon interface units and two tube switching units. The weapon interface and tube switching units are provided one each side to port and starboard, so that in the event of a failure in the equipment of one side the other remains operational. Data from the sonar, search and attack periscopes, ESM, radar and navigation sensors are also passed to the system via the databus.

The control room console provides two identical operator positions mounted above two electrical sections which house the central computer and all the associated electronics. Full combat information and fire-control facilities are provided at each console operator position and either operator can control the whole system, thereby allowing one position to be shut down in patrol state. Each console position has two CRT displays, two command data panels, and a weapon control and system status panel. The two display areas provide the operator with a label plan, time/bearing or vertical section display and a tote display providing pages of data and selections to enable him to interact with the system using a light pen. Track data, intercept alarms and tube and weapon status are displayed on the combat information and fire-control command data panels. Tube and weapon preparation orders are passed to the torpedo room using the weapon control and system status panel. Weapon guidance is fully automatic, although KAFS also has facilities for manual guidance post-launch and system control, as well as status indications.

The weapon control equipment in the torpedo room, which is microprocessor-based, provides the interfaces

KAFS console

to the tubes and weapons. Tube and weapon status is displayed on the local control panel which also provides weapon discharge and control facilities in fallback mode, independent of the central processor and facilities for weapon battery heating together with monitoring to detect hazardous conditions. The weapon interface and tube switching units apply switch settings and power supplies to the tube and weapons during the preparation and launch phases. After launch, guidance commands are also routed to the weapons via these units.

Operational status

In 1984, the Brazilian Navy placed an order for two systems and a shore training system followed by an order for two further systems in 1987; the systems are

fitted in HDW Type 209 submarines. Three systems are now fully operational including successful torpedo firings and the shore system has been installed and accepted in Brazil. The last two systems were built under subcontract by SFB Informatica in Rio de Janeiro and have been delivered to the Marinha do Brasil. Two more systems are being installed in the Improved 'Tupi' class boats.

Contractors

GEC-Marconi Radar and Defence Systems, Command and Information Systems Division, Camberley.
SFB Informatica, Rio de Janeiro.

VERIFIED

Manta

Type
Submarine tactical data handling system.

Development
Manta is a tactical data processing system which has been designed, developed and built to support the command system in selected UK submarines. The system is installed and operational as System DCG(R), a two-user variant of Manta. The system exploits commercially available computing units to achieve performance, portability and reliability within a low lifetime cost and short procurement lead times.

Manta replicates the full functionality of its precursor, System DCC, including its interface to the DCB AIO computer. The system adopts COTS hardware to provide high performance and operability allied with future upgrade potential.

Description
Manta can operate independently although it is linked to the main submarine command (action information) computer from which it derives data describing the submarine's own movement and the perceived environment. Manta subjects these data to rigorous and intensive analysis to produce a more accurate picture of the tactical situation. While the system was developed originally for UK submarines, connections to any automated data source including: command systems, sensors, communications equipment and weapons, may readily be achieved using standard interfaces. Alternatively, the manual data entry capability enables Manta to be used independently.

The system is based entirely on COTS hardware repackaged to meet environmental and electromagnetic compatibility requirements, allied with substantial reuse of existing software. It provides a distributed open systems processing environment, based on networked SUN Sparc workstations running UNIX. The network is easily reconfigurable to provide inbuilt redundancy and to deliver high system availability and reliability.

Each user is able to select from various applications within Manta's open environment, including the original DCD functionality. The initial configuration of the system includes a number of new utilities including

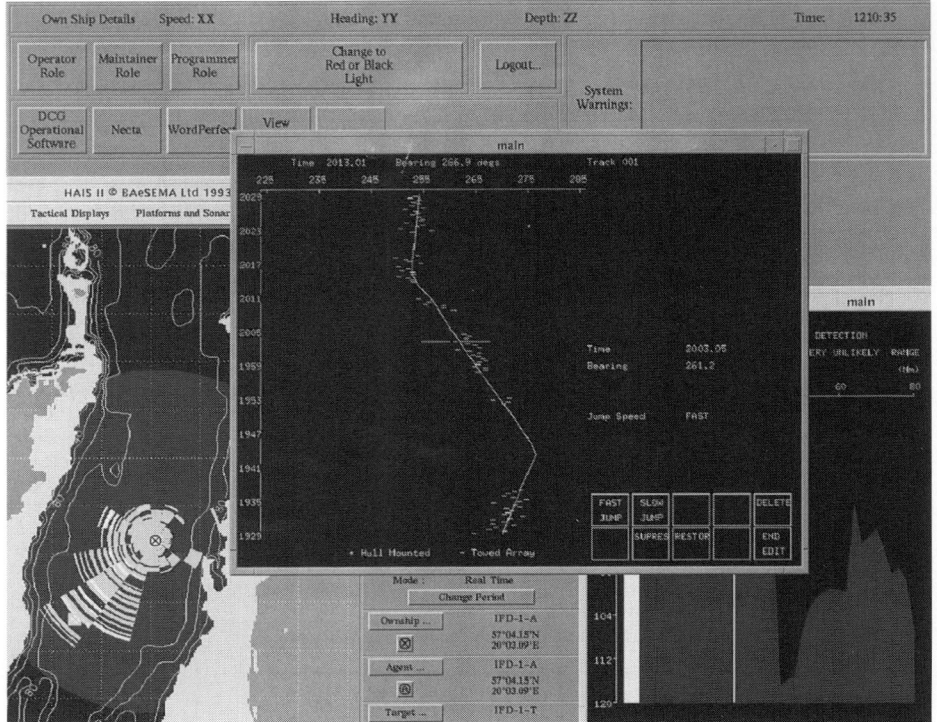

DCG(R) Tactical display   *1996*

advanced environmental and oceanographic analysis and production facilities. The design also includes substantial expansion capability, permitting future inclusion of new control room facilities either as integrated functions with access to tactical picture data, or as stand-alone functions running within the overall environment.

This gives the system great potential for supporting the implementation and trials at sea of prototype software and technology demonstrators. Commonality with the next-generation SubMarine Command System (SMCS Releases 6 and 7) is another key feature of Manta's design, which offers a cost-effective migration path for software ultimately destined for inclusion in SMCS.

The software design includes a substantial suite of existing and proven tactical applications and incorporates a high-performance object-oriented track database. This is combined with the use of Windows, point and click user interaction, high screen resolution and colour displays to provide a powerful yet easy-to-use operating environment.

Operational status
In service with the Royal Navy.

Contractor
BAeSEMA, Dorchester.

VERIFIED

SMCS

Type
Submarine command system.

Development
Advances in both submarine design and sensor technology have led to significant increases in the quantity and diversity of tactical data available to the combat team of a modern submarine. To manage the increased data processing, SMCS (SubMarine Command System), based on BAeSEMA's SUCCESSOR architecture, was selected by the Royal Navy as the new command system for fitting or retrofitting to the 'Vanguard', 'Trafalgar' and 'Swiftsure' class submarines. It is currently fitted in HMS *Vanguard*.

Description
SMCS is a high-capacity open architecture distributed processing command system based on a dual-redundant fibre optic Local Area Network (LAN). It is designed to meet the needs of submarine command systems well into the next century and takes full advantage of state-of-the-art commercially available technology and the most up-to-date design techniques. SMCS interfaces to the submarine's sensors and tactical weapons, providing the control of the weapon discharge systems and functions to support the full range of combat system management tasks. The functions include: Track Motion Analysis (TMA); tactical picture compilation; oceanographic data analysis; weapon management; weapon command tactical aids; onboard training; data recording and replay.

The capacity of SMCS for handling and storing data is orders of magnitude greater than that of any current system. Track processing has been specially developed, taking full advantage of many years of

SMCS console   *1998*/0007219

experience in this area. Many simultaneous TMA solutions can be handled. An important consideration in the functional design has been to automate as much of the routine of track management as possible, leaving operators free to concentrate on the more critical aspects.

Operator interface is by high-resolution colour graphics displays at a number of MultiFunction Consoles (MFCs), with interactive plasma panels and 'pucks' (similar to a mouse). Secondary workstations are also available where space is particularly limited.

SMCS technology is based on recognised and widely used commercial components, ruggedised to suit the submarine environment. The use of commercial standards ensures widespread and long-term availability and support for the key components, as well as ease of upgrade. The main processor is the Intel 80386/486, with the possibility of upgrading as this family of processors is enhanced, while fast processing is performed by Inmos transputers, which can also be upgraded. The system architecture allows this substantial processing power to be easily increased, if necessary, to accommodate future requirements, and there is sufficient spare physical capacity to do so.

SMCS software is written in Ada, while Occam is used for the transputers. This approach yields a highly modular system, which facilitates enhancements.

Some 200 processors with 2 Gbyte of storage are used to handle the data.

Fundamental to the SMCS architecture is the high bandwidth, dual-redundant, fibre optic Local Area Network (LAN) based on the IEEE 802.5 token passing protocol, with built-in self-healing. To the LAN are connected the processing nodes and the MFCs, which also contain processing capability.

An important feature of SMCS is its inherent resilience to failure. All processing nodes are duplicated, with one in use and the other on standby and the MFCs are identical. The standby nodes are automatically switched into use in the event of a failure being detected in the other.

A number of key features of SMCS include:
(1) a high degree of flexibility and expandability
(2) a substantial increase in usable processing power
(3) improved operability, to ease the growing burden on the crew in the modern submarine environment
(4) improved reliability to provide a 'non-stop' system
(5) the capability to match envisaged and likely improvements in weapons and sensor technology.

The flexibility of the basic system architecture has been demonstrated, not only by the different submarine fits being produced, but also by the adoption of SUCCESSOR-based systems for the Royal Navy's surface ships, where the detailed operational requirements are very different.

Versions of SMCS can be readily produced to suit

almost any type of submarine, to form the key central element of the combat system. A second production order for a further nine SubMarine Command Systems was placed by the UK MoD in 1994.

BAeSEMA was also awarded two further development contracts with a value of over £50 million. The first was for an additional SMCS software release to interface to improved sonars that are part of the initial phase of the 'Swiftsure' and 'Trafalgar' class ('S & T') update programme. The second contract was to cover the major phase of the 'S & T' update programme and will enable SMCS to respond to the introduction of the new 2076 sonar suite and marks a significant evolution of the BAeSEMA SUCCESSOR technology. New SMCS displays will be developed using high-performance graphics packages that conform to the X-Windows standard and run under the general purpose UNIX operating system.

Operational status
In service with the Royal Navy. Export variants are also available. The second production order for nine systems was placed with BAeSEMA in 1994, and is being installed in Swiftsure', 'Trafalgar' and 'Vanguard' class submarines.

Contractor
BAeSEMA, New Malden.

UPDATED

UNITED STATES OF AMERICA

SUBICS-900

Type
Submarine combat control system.

Development
The SUBICS-900 (SUBmarine Integrated Combat System) is a variant of the equipment developed by Lockheed Martin Librascope for the Royal Australian Navy's Type 471 'Collins' class submarines. The system is designed to integrate sonars/sensors designed and developed by Alliant-ELAC, which are currently operational on diesel-electric submarines.

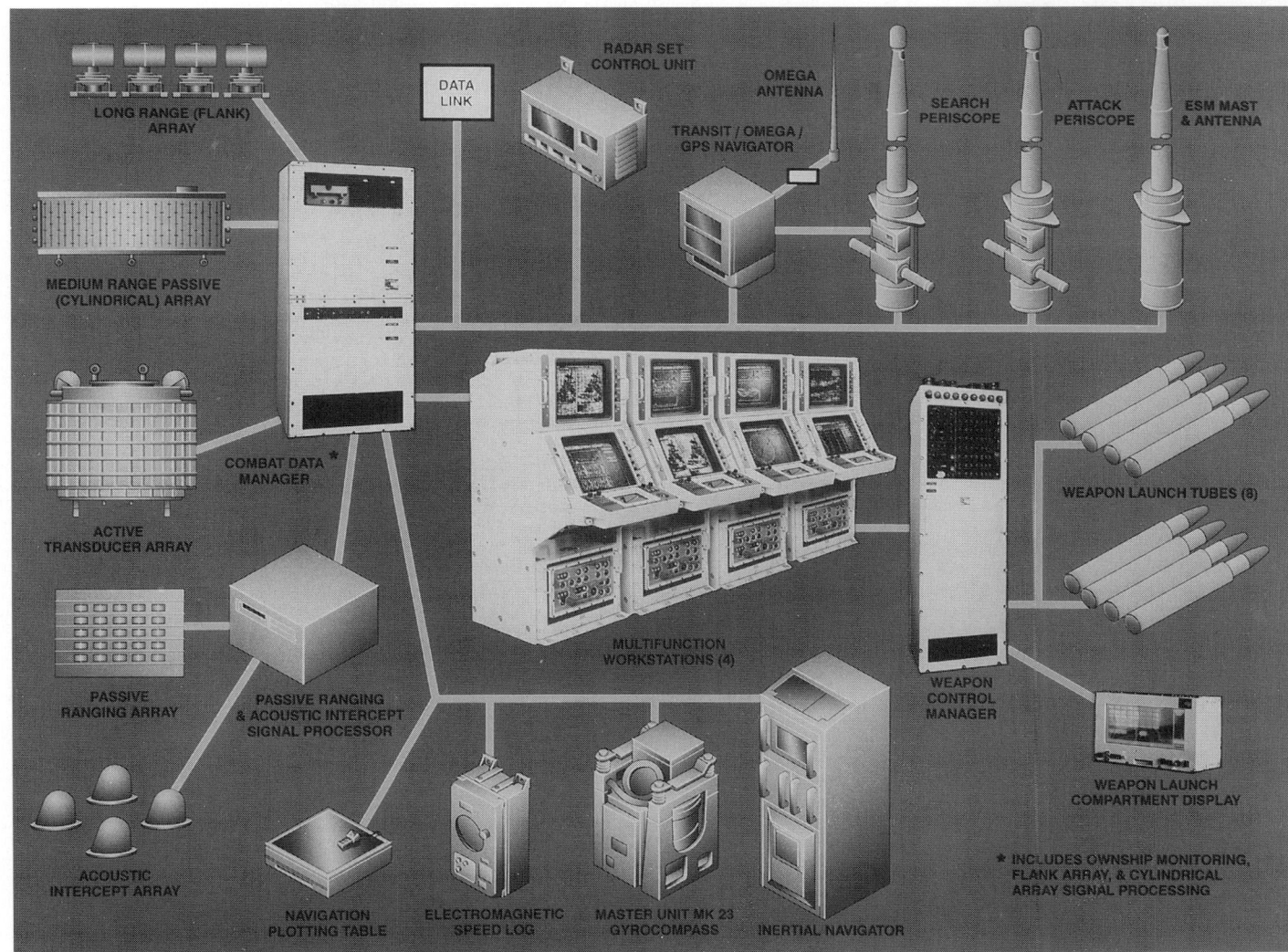

Block schematic of the submarine integrated combat system SUBICS-900

Description

SUBICS-900 is an advanced, totally integrated combat system which meets multimission requirements for modern diesel-electric submarines. The tactical functions provided for are: tactical evaluation and planning; integrated surveillance and threat prosecution; and combat navigation. Using these functions the combat system performs: threat identification and enables tactical evaluation and planning to be performed; evaluates possible responses; gathers data and processes this to provide contact information in the form of a tactical display; performs torpedo/missile control functions; and displays for the command the geographical situation with navigation functions and alerts the command to approaching hazards.

To carry out these functions the combat suite integrates acoustic (flank array, cylindrical array, passive ranging, intercept, and active ranging sonars and cavitation and self-noise monitoring equipment), electromagnetic (ESM, surface search radar and communication datalink), and optronic (search and attack periscopes) sensors with the capability to track 68 targets simultaneously. Threat prosecution is carried out against four targets simultaneously using three wire-guided torpedoes and a salvo of underwater-launched missiles. Optical sensors provide search/attack TV images showing up to eight tracks. Via datalink 120 contact reports can be maintained, together with up to eight tracks. The combat navigation capability provides the display of navigation charts, own ship position fixing, dead reckoning, recommended course and speed manoeuvres based on the tactical situation, and automatic alerts to natural and manmade hazards. It integrates Transit/Omega/GPS navigator, inertial navigator, gyrocompass, electromagnetic log, echo-sounder and navigation plotting table.

The combat system is centred around a Combat Data Manager; four multifunction consoles; and a Weapon Control Manager. These elements are used to: integrate sensor data; convert it into contact information; identify threats; and control weapons as required. The combat/weapon control equipment comprises an open system architecture Combat Data Manager, VME-based multifunction consoles, a Weapon Control Manager and an Emergency Weapon Control Display.

COMBAT DATA MANAGER   MULTIFUNCTION WORKSTATIONS   WEAPON CONTROL MANAGER

Combat control system of the SUBICS-900   *1996*

The Combat Data Manager combines data from the navigation sensors, acoustic sensors, electromagnetic sensors and optical sensors and processes it into contact information. The unit also provides audiotape recording and distribution of acoustic and electromagnetic audio and periscope video to the multifunction consoles.

The multifunction consoles provide the operators with interactive displays for combat navigation, tactical planning, integrated surveillance, threat prosecution, performance monitoring/fault localisation, onboard training and logging/retrieval. Information is displayed on two 19 in diagonal, $1,280 \times 1,024$ picture element resolution, 60 Hz refresh rate, touch interactive colour monitors.

The Weapon Control Manager provides automatic threat prosecution to aid operators in the control of torpedoes and missiles. Operators can set torpedoes and missiles before launch, control their launch, and monitor and guide torpedoes after launch. The launch tube interface includes switching circuitry for interfacing with either torpedoes or missiles in each of eight launch tubes. The processing computes trajectories for three torpedo attacks and one missile

salvo of one to four missiles simultaneously. Post-launch control includes telemetry from the torpedoes and automatic and interactive guidance commands to the torpedoes.

Operational status

The SUBICS-900 system is Lockheed Martin Librascope's advanced next-generation totally integrated submarine combat system designed and developed for international export. This system is currently in the development phase. However, the major elements of the combat control system and sonar are currently deployed and operational on modern diesel-electric submarines.

SUBICS-900 is a variant of the system designed and developed for the Royal Australian Navy's Type 471 'Collins' class submarines.

Contractor

Lockheed Martin Librascope Corporation, Glendale, California.

VERIFIED

Combat Control System Mark 2 (CCS Mk 2)

Type

Submarine integrated weapon control system.

Development

The CCS Mk 2 is a major US Navy upgrade that reduces to four the many submarine combat control system variants currently in the US fleet.

In late 1988, Raytheon won the competition to develop and manage the CCS Mk 2 upgrade. The first system became operational in 1995. As prime contractor for much of the currently deployed CCS Mk 1 system, Raytheon used its experience to retain approximately 80 per cent of the CCS Mk 1 functionality in the CCS Mk 2 system.

The CCS Mk 2 provides commonality across combat systems on the 'Los Angeles' class attack submarines and 'Ohio' class ballistic missile submarines. The system integrates data from advanced submarine sensors and controls weapon targeting and launch. The system incorporates significant potential for growth to handle additional submarine mission requirements. This growth potential was realised through the incorporation of a distributed network and open system architecture subsystems in the Block 1 upgrade, which is operational.

The CCS Mk 2 programme also implements a common in-service support programme for both the 'Los Angeles' and 'Ohio' class platforms. Through the adoption of common systems, CCS Mk 2 has significantly reduced support costs.

Description

The upgrade programme is a major step toward the US Navy goal of evolving fire-control equipment systems into comprehensive combat control systems. The system consolidates the functions of sonar, target motion analysis, tactical displays, weapon order control and submarine navigation and communications.

The improvement provides common equipment and software in the 'Ohio' class and 'Los Angeles' class submarines, including those with vertical launch capability. The upgrade will also be installed aboard new 'Los Angeles' class submarines equipped with the AN/BSY-1 sonar system in 1996.

With the installation of CCS Mk 2, the many different combat systems currently in the fleet have been reduced to four variants, which achieve nearly complete hardware and software commonality. In replacing the AN/UYK-7 standard computers used in older combat control systems with the AN/UYK-43 model, Raytheon provided increased processing performance with reserve capacity for future enhancements. In addition, the CCS Mk 1 OTH processor was replaced by the AN/UYK-44 and a parallel processor. The parallel processor provides a significant increase in responsiveness for the track/correlation function.

The CCS Mk 2 also includes an upgrade of all cursive displays to modern raster displays. Based on commercial technology from Silicon Graphics Inc, the display system includes a sophisticated graphics engine and a high-speed display processor. With a local processing capability of eight million instructions per second (Mips), the display processor decreases

dependency on the central processing unit by more than 50 per cent, providing significant processing reserve for future fire-control system enhancements.

The design of the tactical system has been paralleled by the development of a common logistics and in-service support programme. This support programme has realised efficiencies and reduced the overall cost of the support programme. Due to the commonality of the combat systems and the common support programme, the training programmes for the submarine crew have been combined and streamlined.

Operational status

Raytheon Company won a US$405 million competitive contract for the system upgrade in September 1988 from the US Navy's Naval Sea Systems Command. Additional functional improvements and other contract changes have increased the contract's value to over US$500 million. As prime contractor, Raytheon leads the programme's management and technical effort, which includes a major software upgrade. The company's Marlborough, Massachusetts facility is providing the raster displays.

Contractor

Raytheon Company, Electronic Systems, Portsmouth, Rhode Island.

VERIFIED

AN/BSY-1(V)

Type
Submarine combat system.

Development
The BSY-1(V) totally integrated combat system has been developed as a result of restructuring the SUBACS programme. As part of the redesigned SUBACS, the fibre optic databus distribution system was replaced by the traditional copper cable together with already developed hardware in order to accomplish data distribution. The system is designed to equip the 'Los Angeles' class attack submarines.

Description
The system is the first submarine combat system in the US Navy to integrate navigation, sonar and weapons system data for improved target detection, classification, localisation, combat control and weapons launch. The BSY-1(V) integrates the medium- to low-frequency bow-mounted Submarine Active Detection System (SADS) sonar used for the detection and fire-control solutions on hostile subsurface to surface targets. This operates over 360° in long-range search, and provides a passive listening mode. The system also integrates the high-frequency, active Mine and Ice Detection Avoidance System (MIDAS) mounted in the sail and which is used for close-range detection of mines and polar navigation. Finally the system integrates two towed passive arrays: the TB-16 and the lightweight, long thin-line TB-23 towed array, the latter providing long-range submarine target detection. The TB-23 can be reeled into the submarine's ballast tanks.

The integrated combat system will carry out target motion analysis and calculate a fire-control solution for Mk 48 Adcap torpedoes, Harpoon and Tomahawk cruise missiles.

The system is designed to improve data processing and management capabilities using new and more capable computers, new data displays and additional software and increased automation in areas such as surveillance, detection and tracking of targets. This will enable operators to perform multiple tasks and handle multiple targets simultaneously. The aim is to reduce the response time between initial detection and launching of the weapon.

The distributed processing architecture system comprises 117 units, including 64 general purpose and 35 specialised processors, digital beamformers, signal conditioners, displays and disk storage devices. Some 3.6 million lines of tactical software code are used in this system.

Operational status
The first unit designated to receive the BSY-1(V) system was the 'Los Angeles' class submarine *San Juan*. The system was installed during the summer of 1987. Coincident with the first system going to sea, the US Navy commissioned a shore trainer. Since then, 21 submarines have received the system and all 21 of the acoustic subsystems have been upgraded with the Navy's new standard AN/UYK-43 computer (which will replace the old AN/UYK-7). The latest version of the system, referred to as 'ECI-010 Level', which incorporates the latest version of the Tomahawk fire control and acoustic improvements, completed certification at the contractor's facility in September 1993. As of August 1995, 19 submarines have been upgraded to the ECI-010 level. Through FY90 contracts, a total of 25 AN/BSY-1 systems has been awarded, including two for non-shipboard use (training and so on).

As of summer 1995, Lockheed Martin Federal Systems had delivered 24 of the 25 AN/BSY-1 sets, including a Maintenance Trainer. In addition, two Team Trainers and four Weapons Launch Systems Operator Trainers have been delivered. Boats currently fitted with AN/BSY-1 are *San Juan, Pasadena, Albany, Topeka, Miami, Scranton, Alexandria, Asheville, Jefferson City, Annapolis, Boise, Springfield, Columbus, Montpelier, Santa Fe, Hampton, Charlotte, Hartford, Toledo, Tucson* and *Columbia* (in order of completion).

Contractors
Lockheed Martin Federal Systems, Manassas, Virginia (production).
Lockheed Martin, Syracuse, New York (towed array).
Raytheon Company, Electronic Systems, Portsmouth, Rhode Island (hull array).
EdgeTech, Rockville, Maryland (system engineering).

VERIFIED

AN/BSY-2

Type
Advanced submarine combat system.

Development
The BSY-2 combat system was instituted after the SUBACS programme was cancelled in FY86. It was developed for the new SSN-21 'Seawolf' nuclear-powered attack submarine of the US Navy. The development of BSY-2 is considered to be one of the largest computer software efforts ever undertaken for a submarine. Considerable effort was devoted to developing and integrating a massive amount of software involving some 5.5 million lines of code, the majority of which is written in Ada. Responsibility for developing the software was shared among seven development organisations under the direction of the prime contractor. Four of these form part of the prime contractor and three are subcontractors.

Description
The BSY-2 is an integrated sensor and fire-control system combining active sonar, passive flank array sonar, passive towed array sonar and combat control system with an advanced computer system which is designed to detect, classify, track and launch weapons at enemy subsurface, surface and land targets. The boat's sensors include a wide aperture array, long thin-line towed array (TB 29), a hemispherical transmit array, a standard towed array (TB-16), a Mine and Ice Detection Avoidance Sonar (MIDAS), and large spherical array. Other hardware includes a tactical situation plotter, transmit group, a large vertical display screen and a large horizontal display screen, multifunction combat system display consoles, weapon launch system, multi-array conditioner.

The BSY-2 also integrates the BQG-5 wide aperture array (installed on the submarine USS *Augusta* and USS *Cheyenne*) offering significant performance improvement over existing sensors.

The system has been developed to counter the submarine threat of the 21st century, and as such is being designed to enable the submarine to: detect targets in a much shorter time than is currently possible; allow operators to perform multiple tasks and handle multiple targets simultaneously; and to greatly reduce the time between detecting a threat and the launch of weapons.

The new UYS-2 acoustic processor, which has several times the capacity of the UYS-1 is used, and the BSY-2's distributed architecture is specifically

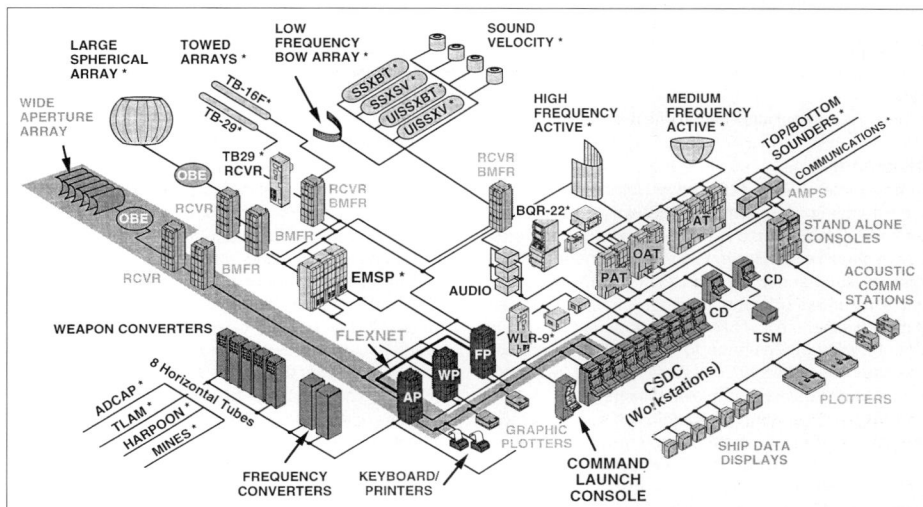

Block schematic of the AN/BSY-2 combat system   *1998*/0007220

designed to cope with the increased processing requirements of the acoustic array suite. The system comprises 92 Motorola 68030 32-bit processors which carry out the various functions within the combat suite. The design specification requires the processors to be grouped in four task areas: acoustics, command and control, weapons and display. Redundancy is provided in various ways. In the weapon cluster a designated back-up spare will be available for each of the four processors. Thus, if any processor fails it will be automatically replaced by its spare, providing 100 per cent redundancy. The remaining three clusters will not, however, achieve the same level of redundancy. For these, various back-up support options will be available should one or more processors in a cluster fail. At the first level of redundancy six spare processors will be available for electronic substitution in the event of failure. As a second line of defence, other processors in the affected cluster can be available for use as spares, if they are not being used for their own function at the time. Finally, if all available processors are in use and failure occurs, then the system will suffer gradual degradation with some loss in operational capability. If operational capability is lost a prioritised list of functions will be implemented, assuring that minimum defensive functions (self-protect functions) are maintained at all times. These will afford the submarine protection against hostile threats, but will not provide it with the full capability required to carry out an attack mission. Finally, should a cluster begin to lose operational capability, then spare processors carried on board can be manually inserted into the system, but this takes time which in certain threat situations may not be available.

Operational status
Contract for full-scale development and production was awarded in March 1988. Total development and procurement costs for four BSY-2 and two BQG-5 systems was US$1.7 billion. Integration and testing of the system took place between 1989 and 1994, the first system was due to be delivered in November 1993. However, this first system did not possess full capabilities as additional software, needed to increase performance to full operational capability, did not undergo testing and integration until April 1993, for delivery in June 1994. The system is being installed on the 'Seawolf' class and has had two block upgrades before commissioning in July 1997.

Contractor
Lockheed Martin, Syracuse, New York.

VERIFIED

NSSN Combat System

Type
Submarine combat system.

Development
In April 1996, Lockheed Martin Federal Systems received an initial US$160 million contract from the US Naval Sea Systems Command to develop the command, control, communication and intelligence system for the US Navy's New Attack Submarine Type NSSN. Development is being implemented in three increments to allow for additional technology insertion without impacting on the overall test and integration schedule.

Development and subsystem level testing for sonar and architecture will be completed at Lockheed Martin Federal Systems' Manassas site, while combat control development and subsystem testing will be completed by Raytheon Electronics Systems at Portsmouth, Rhode Island. Other members of the team led by Lockheed Martin include Northrop Grumman and Digital Systems Resources.

Description
The full combat suite solution will encompass sonar, combat control and architecture subsystems, plus the integration of all additional combat suite electronics on board the NSSN platform. These include ESM, radar, external and internal communications, submarine defensive warfare systems, navigation, total ship monitoring, periscope/imaging, navigation sensor system interface, tactical support devices and special purpose subsystems.

The design will make maximum use of COTS and adopt an open system architecture which will facilitate future technology integration and performance improvements. COTS equipment will account for 78 per cent of all system hardware and 76 per cent of software code.

The architectural approach provides a loosely coupled environment for each subsystem. Processed data and information is shared via a hierarchy of fibre optic networks implemented using asynchronous transfer mode communications, while sonar sensor data is distributed to processing hardware via fibre optic fibre channel standard technology. Standardised interfaces at the configuration item (drawer level) allow for technology upgrades without major system-level impact.

The open system architecture will be built around the AN/UYQ-70(V) (Q70) computing and display infrastructure, TAC-X computer technology and asynchronous transfer mode networking technology. The combat system will also be fully compatible with the Joint Maritime Command Information System (to allow NSSNs to interoperate with other ships, submarines, aircraft, ground units and command activities) and the Advanced Tomahawk Weapon Control System.

The Q-70 Colour Common Display Console will form the common workstation for all combat departments including command and control, sonar, photonics, ESM, external communications and defined special-purpose applications. A command workstation configuration provides a deck-mounted stand-alone capability, while vertical and horizontal large screen displays will provide a group viewing and planning facility. All Q-70 versions specified for NSSN will use a 100 MHz Hewlett-Packard single-board 6U VME processor, and run on HP-UX or HP-RT operating systems.

To match NSSN's modular construction philosophy, the combat system electronics will be mounted in a command and control system module employing a modular integrated deck structure to provide a benign environment for COTS.

Operational status
Inter-subsystem integration will initially be performed at Lockheed Martin's Manassas facility, moving to the command and control system module off-hull assembly and test Site at General Dynamics' Electric Boat in Groton. The command and control system module is to be assembled off-hull, with individual subsystems installed in structurally integrated enclosures before being end-loaded into the module's prefabricated hull section.

Contractors
Lockheed Martin Federal Systems, Manassas, Virginia.
Raytheon Electronics Systems, Portsmouth, Rhode Island.
Northrop Grumman, Baltimore, Maryland.

VERIFIED

PISCES

Type
Submarine combat information and fire-control system.

Development
Paramax International Submarine Combat information and Engagement System (PISCES) was developed by Lockheed Martin Tactical Defense Systems (formerly, Unisys Government Systems Group) to enable today's modern submarines to exploit their advantage of stealth by allowing for the rapid collection and analysis of data of either tactical or strategic importance. This, in turn, allows the submarine to accomplish its primary purposes of intelligence gathering and weapons employment against potential adversaries.

PISCES is the third-generation submarine combat information and fire-control system developed by Unisys. It is the culmination of 30 years of experience in the development of submarine combat and fire-control systems. PISCES was produced using the Unisys design philosophy of open systems and open architecture design. To support this philosophy, PISCES incorporates the latest in proven leading-edge technologies such as VME bus structure, Motorola 680X0 processors, RISC processors, UNIX, Vx works and C.

Description
PISCES tasks, in support of the overall submarine mission, involve processing data gathered by own ship tracking sensors (sonar, radar, ESM, CCTV and periscopes), navigation system sensors (speed log, gyrocompass, GPS and Inertial Navigation System (INS)) and communications systems (datalink, RDF and IFF). This data is then recorded, analysed and manipulated by PISCES operators to assess potential adversary capabilities and co-ordinate manoeuvring of the submarine and deployment of its weapons.

PISCES consists of two Modular Multifunction Consoles (MMCs), Firing Distribution Unit (FDU) cabinet and the Remote Firing Control Panel (RFCP). The main PISCES control station is located typically in the submarine attack centre and consists of two identical MMCs. The MMCs are the primary operator stations for the collection, manipulation and recording of data from the combat system sensors. In addition, the MMCs are the primary stations for weapon selection, preset orders generated, testing, launching and guidance functions. The FDU, which is typically located in the torpedo room, provides data conversion and launch for the applicable submarine's weapons. The RFCP, also located in the torpedo room, provides the man/machine interface with the FDU and the torpedo tubes for tube load status, local weapon presetting, launching and guidance control, as well as for system offline diagnostics.

PISCES is capable of managing and tracking in excess of 200 tracks from multiple sources using automatic and manual Target Motion Analysis (TMA) techniques, while the weapon launch and control portion of PISCES provides simultaneous control of multiple weapon types in virtually any mix (including salvos of the same weapon) for use against a single or multiple threats. Weapons that PISCES supports include: NT-37 torpedo versions C to F, V56 ASUW, VA53 ASUW, V53 ASW, SUT, SST-4, Mk-37 Mods 0-3 torpedoes and the Federated Harpoon Missile. In addition to track management and weapon capabilities, PISCES provides a data extraction and playback function for recording of mission-critical data for later replay and analysis, system training function to

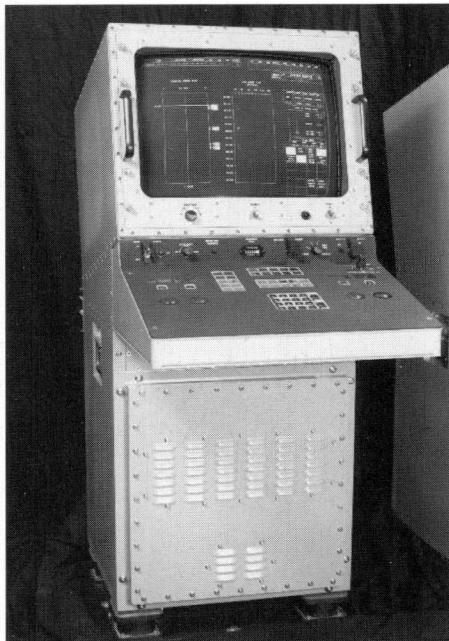

The MMC for PISCES   *1994*

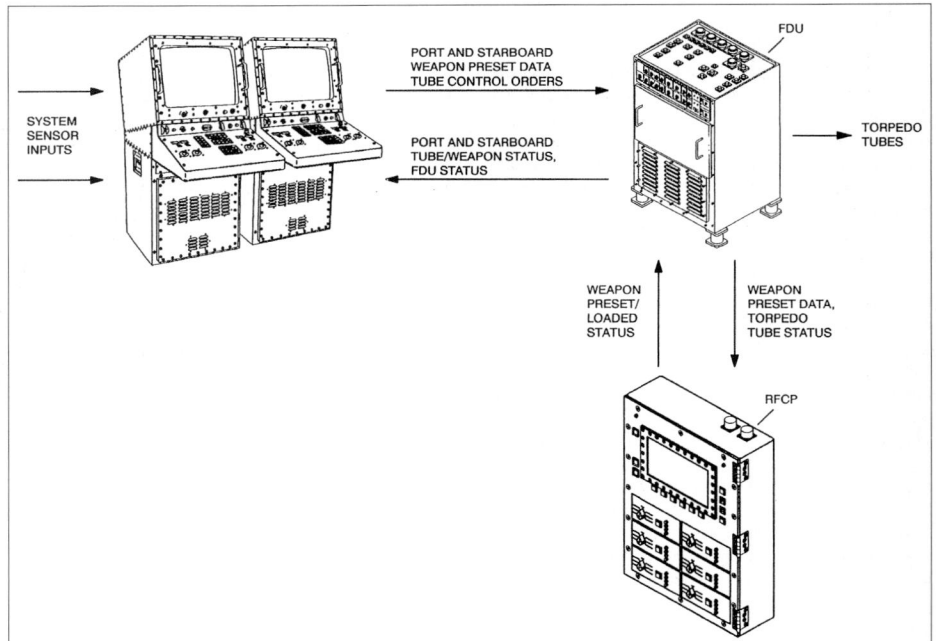

Block schematic of PISCES   *1994*

maintain system operator proficiency, piloting and prediction functions to aid in the manoeuvring and navigation of the ship, environmental/oceanography functions to aid in determining ship operating environment and an optional overlay/cartographic function.

PISCES operator console
Two Modular Multifunction Consoles (MMCs) provide the primary control of the PISCES. Both MMCs are identical, each maintaining identical but independent system databases through the use of an intercomputer channel and redundant communication paths to the remainder of the combat system. In the event of a casualty, one MMC will meet all PISCES system requirements. Each MMC is capable of displaying multicolour radar video, CCTV data and high-speed graphics such as tracks and other tactical symbology. It also displays line segments, status reports, system alerts and other alphanumeric data. The MMC consists of a monitor assembly, a processor unit and an Operator Entry Panel (OEP).

Monitor assembly: The monitor assembly presents tactical video, graphic and alphanumeric data on a militarised, high-resolution, 19 in, 1,024 × 1,280 pixel, raster scan, colour Cathode Ray Tube (CRT). The monitor assembly also provides Finger On Glass (FOG) touchscreen capability interfacing with the graphics/display processors, which are components of the MMC processor unit.

Processor unit: The processor unit is a self-contained, self-cooled unit capable of operating independently of the other units of the MMC. The processor unit houses the graphics processor, display generator, applications processor and external I/O interface, each of which is described in the following paragraphs.

The PISCES applications processor provides the processing necessary for program loading, scheduling,

data management, I/O control, target tracking, weapons calculation, weapons control, own ship navigation, environmental data processing and online maintenance.

The graphics/display processor provides the processing necessary for display request management, display initialisation and termination, as well as periodic display management and operator action management processing.

The external I/O interface provides the PISCES interfaces to the combat system's sensor subsystems. The external I/O interface unit is controlled by the applications processor. External I/O interfaces are provided on Euro-Card VME bus plug-in modules in a section of the VME chassis reserved for application unique functions. Interfaces supported include:
Synchro
RS-232C
RS-422
MIL-STD-1397 A to E
MIL-STD-188
MIL-STD-1553
SAFENET
IEEE 802.3.

Operator Entry Panel (OEP)
The OEP is the primary device for operator control of PISCES. The OEP provides momentary action push-buttons for target selection, mode and weapon control, a numeric keypad for numeric data entry and four trackballs, for entering variable/encoder data. The OEP can be tailored to meet customer requirements.

Firing and Distribution Unit (FDU)
The FDU provides the interface between the weapon/tubes and the remainder of PISCES. The FDU provides data conversion and control of weapon data flow for all ship weapons. Weapon orders are transmitted from the

FDU to the weapon, and read-back data returns from the weapon to the FDU. The FDU is divided into two identically functional sides (port and starboard), each communicating with one of the PISCES MMCs using non-redundant cables. In addition, each functional side of the FDU can communicate and control up to four weapons using identical but redundant interface electronics and cabling channels. Each side of the FDU communicates with the Remote Firing Control Panel (RFCP).

Remote Firing Control Panel (RFCP)
The RFCP provides the man/machine interface with the FDU and the torpedo tubes. It also provides PISCES with an emergency firing capability when the interfaces with the MMCs have failed or at command discretion. The RFCP provides the operator with the capability of weapon/tube selection and system monitoring during normal modes of operation. In the emergency mode, the RFCP provides the operator with the capability to preset, fire and guide the system torpedoes and Harpoon missiles. In addition, the RFCP provides the FDU diagnostic interface.

Operational status
Four systems were ordered, produced and accepted by the Egyptian Navy for the Romeo-C Submarine Modernisation Programme. All four systems have been successfully installed and integrated with the other combat system equipment.

Contractor
Lockheed Martin, Tactical Defense Systems, Great Neck, New York.

VERIFIED

SOAS (Submarine Operational Automation System)

Type
Submarine command support software programme.

Description
The demand to process and present the growing volume of constantly changing data provided by current and next-generation advanced sensors, covering both the tactical and ocean environment as well as the mass of data on own ship systems, and the increasingly lower acoustic signatures generated by modern submarines, is constantly increasing. To meet this demand the US Defense Advanced Research Projects Agency (DARPA) is studying software requirements for the next generation of submarine command systems.

Referred to as SOAS (Submarine Operational Automation System), the initial phase of the programme involves the development of a series of prototype software programs. These will form the key element in the next generation of software architecture for use in submarine command systems designed to provide highly integrated information and data management to assist the commander in tactical and ship control decision making. Using advanced computer hardware designs based on a series of parallel computers capable of carrying out millions of calculations simultaneously, and incorporating neural networks and artificial intelligence-based programs,

SOAS will integrate signature and vulnerability management, ship monitoring and control, situation assessment, tactical planning, and the ship-crew interface.

Unlike the development of the BSY-1(V) system which was designed to meet a specific set of operational requirements, SOAS adopts a 'system extension' approach in which operational requirements are added as development of the system proceeds.

To meet this requirement General Electric was awarded the contract to develop software prototypes in which programs are developed to satisfy an increasingly demanding tactical scenario. The first prototype, P1, completed in December 1990, covered the requirement for a submarine to carry out surveillance of hostile forces and reconnaissance in an area of low hostile activity. A land-based test system was established at General Electric's facility at Moorestown, where elements of the software undergo tests to prove the concept and where new concepts are evaluated and successful prototype designs are then integrated with the established architecture.

Subsequent developments envisaged a scenario where the number of contacts is significantly increased and incorporated stealth signature management programs being developed at the Applied Physics Laboratory of Johns Hopkins University.

At the end of FY93, DARPA declared a proof of principle completion on the SOAS programme. Subsequently, Lockheed Martin Laboratories, Camden has adapted technologies developed under SOAS into

multiple programmes. Examples include: knowledge-based correlation to Rotorcraft Pilot's Associate (Army); knowledge-based correlation and scene understanding to Electronic Warfare Advisor (DARPA); scene understanding and event monitoring to AEIWS Studies (Navy); contract management risk reduction to AN/BSY-2 (Navy); Case-Based Response Planner and Advanced Ship Self-Defense Combat Systems (Navy); primary man/machine interface, test simulation and primary communication substrate to Submarine Defensive Warfare System (Navy).

Operational status
Phase 1 contracts for development plans and demonstration of capabilities were awarded to General Electric, Lockheed-Georgia, McDonnell Douglas and Martin Marietta in January 1989. In February 1990, General Electric won the Phase 2 development contract, the design phase for which was completed in May 1991. Software coding was completed during the summer of 1991 and full software integration was scheduled for October/November 1991. Demonstration of the integrated P1 and P2 software architectures was conducted in December 1991 and the programme was completed in 1995.

Contractor
Lockheed Martin, Camden, New Jersey.

VERIFIED

AN/SQQ-89

Type
Surface ship Anti-Submarine Warfare (ASW) combat system.

Development
The US Navy recognised in the early 1970s that, despite improvements in the performance of ASW fire-control systems such as the Mk 116, the effectiveness of ASW warships in the face of quieter and higher-performance submarines was limited. Towed array sonar systems were being developed with more sophisticated signal processors which made detection at longer ranges possible, while the development of the LAMPS (Light Airborne MultiPurpose System) Mk III

helicopter improved the capability for localisation and engagements of targets beyond the horizon.

It was clear that the best solution would be to create an integrated ASW combat system which would incorporate the Mk 116 as well as data from all the shipborne and airborne sonars. Research and development of the system, later designated AN/SQQ-89 Surface Ship ASW Combat System, began in 1976. The initial system technical evaluation trials were conducted in 1982 on board the 'Spruance' class destroyer *Moosbrugger* (DD 980). The first production system, AN/SQQ-89(V)2 was installed in the 'Oliver Hazard Perry' class frigate (FFG 7) USS *Curts* (FFG 38) in 1985.

The SQQ-89 was also selected for the later 'Ticonderoga' class cruisers (CG 47) and the first

SQQ-89(V)3 was installed in the USS *San Jacinto* (CG 56) which was commissioned in January 1988. The first SQQ-89(V)1 was installed in the 'Arleigh Burke' class destroyer *Arleigh Burke* (DDG 51) in July 1991. Every 'Arleigh Burke' class destroyer after DDG 51 (DDG 52 through to DDG 78) will have the SQQ-89(V)6.

Two companies have supplied the AN/SQQ-89 ASW combat system to the US Navy. In 1990 and 1992, Northrop Grumman won competitive contracts for the production, logistics support and design agent activities. In 1991 and 1996, Lockheed Martin won competitive contracts for the same activities.

Description
The SQQ-89 is the US Navy's premier surface ship ASW combat system designed to detect, locate, track and

engage submarine targets. It transmits and/or receives acoustic signals using a variety of sensors to provide target classification, as well as performing and controlling Target Motion Analysis (TMA) and controlling the setting of 'own ship' ASW weapons. In addition, it provides multisensor track correlation, track management control and forwards track data to the ship's combat direction system or command and decision system.

The system is an integrated combat system which consists of two or three sonar sensors which interface either with the Mk 116 fire-control system or a Weapons Alternate Processor (WAP). This integration provides maximum operational effectiveness while requiring the minimum manpower to operate the equipment. The SQQ-89 permits the sharing of display consoles when multiple sensors are in use, with the system using between two and six consoles. One or two of these consoles are usually for the Mk 116 system and the remainder are for the sonar systems, with OL-190 acoustic data converters interfacing the consoles with the computers. The AN/SQQ-89(V)1, 3, 5 and 8 have a total of five display consoles, while the (V)4, 6 and 7 have six. There is no Mk 116 in the SQQ-89(V)2 system aboard the 'Oliver Hazard Perry' class ships. Two display consoles are used for the acoustic sensors in the FFG 7s.

The sensors integrated in the SQQ-89 are the AN/SQS-53B/CD hull-mounted sonar, the AN/SQR-19 TACtical Towed Array Sonar (TACTAS), the AN/SQQ-28 LAMPS Mk III Sonar Signal Processing System (SSPS), all supported by the AN/UYQ-25 Sonar *In situ* Mode Assessment System (SIMAS). The fire-control equipment is the Mk 116, except for the FFG 7 class which uses the WAP. In addition to the operational equipment, the SQQ-89 includes an OnBoard Trainer (OBT), the AN/SQQ-89(V)T. The OBT provides realistic training and front-end stimulation of SQQ-89 ASW and AN/SLQ-32(V) electronic warfare sensors. It features a trainer console, LAMPS helicopter navigation simulator, signal generator/processor and various peripheral systems.

The AN/SQR-19 is a passive, long-range omnidirectional computer-aided system for detecting, classifying and tracking surface and submarine targets. The computer-aided detection features provide operator alerts and optimised search and classification displays allowing operators to handle high-density, multiple contact situations. The SQR-19 is a digital system providing high reliability and low maintenance. The OA-9056/SQR-19 towed array group consists of an 82 mm diameter array which is towed on a 1 n mile (1.7 km) cable to depths of 365 m. The two other major subsystems are the Ship-based Electronic Subsystem (SES) and the OK-410/SQR-19 Handling and Stowage Group (H&SG). Early versions of the SQR-19 use the AN/UYK-20 data processing system, but systems produced from 1992 onwards can use the AN/UYK-44-based Signal Data Processing Unit (SDPU).

The SQQ-28 SSPS was originally designed as the shipboard processor for the LAMPS Mk III weapon

Class	Variant	Hull Numbers
CG Ticonderoga (CG 47)	(V)2	CG 54-55
	(V)3	CG 56-64
	A(V)3	CG 65
	(V)6	CG 68-73
	(V)7	CG 66-67
DD Spruance (DD 963)	(V)1	DD 965, 980, 992
	(V)3	DD 971, 981
	(V)5	DD 963, 964, 966, 967, 968, 970, 973, 975, 991
	(V)6	DD 987
	A(V)6	978, 982, 984, 985, 986, 988, 989, 990, 993, 994, 995, 996
	(V)8	DD 969, 972, 974, 976, 977, 979, 983, 997
DDG Arleigh Burke (DDG 51)	(V)4	DDG 51
	(V)6	DDG 52-78
	(V)10	DDG 79-88
FFG Oliver Hazard Perry (FFG 7)	(V)2	FFG 7-15, 24, 27-61
	(V)9	FFG 16-23, 25, 26

system. It processes raw data from sonobuoys linked from the SH-60B Seahawk ASW helicopter to the ship via the AN/ARQ-44 radio link and received by the AN/SRQ-4. The SH-60 LAMPS Mk III helicopter and SQQ-28 provide a long-range sensor capability along with a long-range offboard weapons delivery system. The SQQ-28 processes all active and passive sonobuoy data. A secure two-way datalink between the helicopter and ship contains acoustic or radar and ESM data along with helicopter command and control communications. LAMPS Mk I acoustic data is received by the AN/SKR-4B(V) receiver. Modifications to the SKR-4 interface allow LAMPS Mk I and LAMPS Mk III interoperability. LAMPS Mk I/III interoperability kits are installed in 'Ticonderoga' class cruisers beginning with the USS *San Jacinto* (CG 56) and in all of the 'Arleigh Burke' class destroyers (DDG 51). When SQQ-28 is installed in ships with the SQR-19, it shares the UYK-44-based Signal Data Processing Unit (SDPU). In older, stand-alone configurations, it uses the AN/UYK-20 DPS.

The AN/UYK-25 and associated SIMAS provide a computer-based capability for monitoring the environmental sensors for providing sonar *in situ* performance predictions and mode selection to provide the most effective SQQ-89 system performance in various environmental conditions. The system receives environmental data (water temperature versus depth, windspeed and so on) and computes the expected acoustic propagation conditions. Based on these predictions, SIMAS provides sensor setting recommendations to optimise sensor performance. The UYQ-25 is hosted in a commercial-off-the-shelf (COTS) processor and display system. The UYQ-25A(V)2 is associated with all SQQ-89 variants with the exception of 'Ticonderoga' class cruisers (CG 47 to CG 54). The updated system is the AN/UYQ-25B(V)1.

The Mk 116 ASW Command control System

(ASWCS) is a computerised system that integrates acoustic sensor data and non-acoustic data to perform multiple contact management, contact correlation, and contact localisation through the use of automated TMA techniques. The Mk 116 directly interfaces with shipboard acoustic sensors (SQS-53C, SQR-19 and SQQ-28), non-acoustic sensors (radar, ESM) and receives offboard sensor data via datalink and the ship's combat direction system. The Mk 116 provides ASW weapon control functions for the Vertical Launch ASROC (VLA) weapon system and shipboard torpedo tubes and generates fire-control solutions for targets of interest.

The OBT provides integrated shipboard training in port or at sea, generating target scenarios and directly stimulating the acoustic sensors to provide operational ASW team training.

The AN/SQQ-89 surface ship ASW combat system is a fully integrated combat system. The individual subsystems (SQS-53C, SQR-19 and so on) are, however, capable of stand-alone operation. Each subsystem can be installed alone or in various combinations, depending on each set of ship requirements.

Operational status
Over 135 SQQ-89 systems have been ordered and the system is still in production. A new variant, the AN/SQQ-89(V)10, will be introduced beginning with the last 'Arleigh Burke' class destroyer procured in FY94, DDG 79. This variant will incorporate the Enhanced Modular Signal Processor (EMSP), but will not include the AN/SQR-19 TACTAS.

Contractor
Northrop Grumman Corporation, Electric Sensors and Systems Division, Annapolis, Maryland.

UPDATED

ISCWS

Type
Integrated sonar, command and weapon system.

Description
ISCWS is based around a distributed architecture providing high availability of all system capabilities which will allow it to meet changing requirements well into the 21st century. The distributed architecture features dual redundant high-speed databus, redundant processors, multifunction operator consoles and a standardised set of modules.

Logistics and maintenance are simplified by the use of the same hydrophone and preamplifier in all passive arrays. The ISCWS also integrates all the submarine's other sensors and functions including ESM, radar, navigation, communication and weapons to result in a completely integrated submarine combat system.

The system features lightweight fibre optic cabling for interconnections which reduces weight and increases reliability. In addition, a common system processor cabinet is used for all signal and data processing which improves Mean Time Between Failure (MTBF). Three types of computer resources are used in the system. The digital processing unit uses

both system and array processors, while workstation processors are used in the operator consoles.

There is considerable scope for enhanced capabilities which would be available through future upgrades.

Contractor
Raytheon Electronic Systems, Naval Systems, Portsmouth, Rhode Island.

VERIFIED

TABLE VII

Operational combat & weapon control systems

System	Manufacturer	Country	Installed on	Navy	Units
SUB-TDS	Terma	Denmark	Tumleren Type 207	Denmark	3
DLA 2A		France	Agosta	France	3
DLT D3	Thomson Marconi Sonar SAS	France	Daphne	France	1
DLT D3	Thomson Marconi Sonar SAS	France	Daphne (Hangor)	Pakistan	4
DLT D3	Thomson Marconi Sonar SAS	France	Daphne (Albacora)	Portugal	3
DLT D3	Thomson Marconi Sonar SAS	France	Daphne (Delfin)	Spain	4
Subtics	Thomson Marconi Sonar SAS	France	Agosta (Hashmat), Scorpene	Pakistan, Chile	5‡
FCS Mk3	SEPA	Italy	Sauro, Type 209	Italy, Peru	8***
TFCS	STN ATLAS Elektronik	Germany	Oberon	Chile	2
TFCS	STN ATLAS Elektronik	Germany	Gal (Vickers Type 540)	Israel	3
CSU 83	STN ATLAS Elektronik	Germany	Type 206A, Type 209/1300 (Cabalo)	Germany, Venezuela	14
Isus 83	STN ATLAS Elektronik	Germany	Type 209/1200 (Chang Bogo)	Korea, South	6
Isus 83-2	STN ATLAS Elektronik	Germany	Type 209/1400 (Preveze)	Turkey	2
Isus 90-1	STN ATLAS Elektronik	Germany	Dolphin, Type 212	Germany, Israel	2 + 3‡
FCS Mk 3	Sepa	Italy	Sauro, Type 209	Italy, Peru	8
Mk 3	Sepa	Italy	Type 209/1200 (Casma)	Peru	4
Sactis (V2)	SMA	Italy	Improved Sauro	Italy	4
Sactis (V1)	SMA	Italy	Sauro Type 1081	Italy	4
M8	Signaal	Netherlands	Narhvalen	Denmark	2
M8	Signaal	Netherlands	Type 209/1200 (Salta)	Argentina	1
M8	Signaal	Netherlands	Type 206	Germany	2
M8	Signaal	Netherlands	Type 205	Germany	2
M8**	Signaal	Netherlands	Type 209/1200 (Atilay)	Turkey	2
M8	Signaal	Netherlands	Type 206	Indonesia	4
M8/24	Signaal	Netherlands	Type 209/1200 (Pijao)	Colombia	2
M8/24	Signaal	Netherlands	Type 209/1300 (Shirya)	Ecuador	2
M8/24	Signaal	Netherlands	Type 209/1200 (Casma)	Peru	2
M8	Signaal	Netherlands	Type 209/1100 (Glavkos)	Greece	4
M8	Signaal	Netherlands	Type 209/1300 (Thomson)	Chile	2
Sewaco VIII	Signaal	Netherlands	Walrus	Netherlands	4
Sinbads	Signaal	Netherlands	Type 209/1200 (Casma)	Peru	2
Sinbads	Signaal	Netherlands	Type 209/1300 (Cakra)	Indonesia	2
Sinbads	Signaal	Netherlands	Santa Cruz TR 1700	Argentina	2
Sinbads	Signaal	Netherlands	Type 209/1200 (Atilay)	Turkey	4
Sinbads M	Signaal	Netherlands	Hai Lung	Taiwan	2
MSI-90(U)	Kongsberg	Norway	Type 212	Germany	4‡
MSI-90(U)	Kongsberg	Norway	Ula Type P 6071	Norway	6
MSI-90(U)	Kongsberg	Norway	Type 207 (Kobben)	Norway	6
IPS-12 (Nibs)	Ericsson	Sweden	Sjoormen A 12	Singapore	1
IPS-12 (Nibs)	Ericsson	Sweden	Sjoormen A 12	Sweden	2
IPS-17 (Nibs)	Ericsson	Sweden	Nacken A 14	Sweden	3
IPS-17 (Nibs)	Ericsson	Sweden	Vastergotland A 17	Sweden	4
IPS-19 (Sesub)	CelsiusTech	Sweden	Gotland A 19	Sweden	3
DCB	GEC-Marconi S3I	UK	Swiftsure	UK	5
DCB	GEC-Marconi S3I	UK	Trafalgar	UK	7
DCC	GEC-Marconi S3I	UK	Upholder	UK	4
DCH	Ferranti	UK	Oberon (Humaita)	Brazil	3
KAFS	Ferranti	UK	Type 209/1400 (Tupi)	Brazil	3
	Rockwell	USA	Collins	Australia	1
TFCS	Loral Librascope	USA	Oberon	Canada	3
Mk 106		USA	Guppy IIA	Turkey	5
Mk 106		USA	Tang	Turkey	2
CCS Mk 2	Raytheon	USA	Los Angeles	USA	33
SFCS Mk 1	Lockheed Martin Librascope	USA	Type 209/1500 (Shishumar)	India	4
SFCS Mk 1	Lockheed Martin Librascope	USA	Oberon	Australia	2
Kanaris§§	Lockheed Martin	USA	Type 209/1200 (Glavkos)	Greece	4
AN/BSY-1(V)	Lockheed Martin	USA	Los Angeles	USA	23
AN/BSY-2	Lockheed Martin	USA	Seawolf	USA	1
Pisces	Lockheed Martin	USA	Romeo	Egypt	4

TFCS Torpedo Fire-Control System   §§ After modernisation
** S 347-348   *** Not in Casma and Antofagasta of Peru
‡ Building

UPDATED

WEAPON CONTROL SYSTEMS

AUSTRALIA

Launcher Control System (LCS)

Type

Weapon launch control system.

Description

Designed, developed and manufactured by British Aerospace Australia, under contract to Strachan & Henshaw of the UK, the Launcher Control System (LCS) provides weapon launch control for the Royal Australian Navy's 'Collins' class submarines.

British Aerospace Australia has designed and developed both hardware and software for the LCS, to monitor depth, speed, air and hydraulic pressures, launcher door positions and provide executive control over the firing sequence. The LCS interfaces directly into the integrated ship's control monitoring and management system, providing details of launch system status, events, faults and weapons discharge. In consultation with the boat's combat system, the LCS assists in the selection of the most efficient weapon discharge characteristics.

Each submarine is fitted with two separate, independent LCS systems, each operating a bank of three torpedo tubes. Each LCS consists of a central launch controller, three local tube control panels and a local system operating panel.

The LCS uses a complete VME-based system developed from the high-level specification. Detailed electronic hardware design utilising 68000 processor technology and fail-safe analogue techniques was undertaken from the system design.

The full suite of mission-critical operating software was fully designed and developed using the Ada programming language in accordance with DoD-STD-2167A.

As part of the contract, British Aerospace Australia has also developed an extensive ILS analysis covering FMECA, LSA/LSAR, RAM, MTBF and spares provisioning. Full system level EMIC testing and validation has been performed to ensure compatibility with other ships' systems.

British Aerospace Australia has also developed the associated simulators and set-to-work equipment for the functional testing of the system and for integration of the LCS into the land-based submarine Combat and Ship's Management System test facility.

Operational status

Production of six ship sets, together with spares, commenced in 1992 with final delivery in September 1994. The first two systems are now in operational service in 'Collins' class submarines.

Contractor

British Aerospace Australia, Technology Park, South Australia.

UPDATED

FRANCE

TSM 2072

Type

Submarine Tactical Data Handling System (TDHS) and Weapon Control System (WCS).

Description

The TSM 2072 results from the combination of the DLT-D3 (see below) and the M8 and SINBADS produced by Hollandse Signaalapparaten BV (part of the Thomson Marconi Sonar SAS organisation). The system can be integrated into any existing or newly developed combat system.

The TSM 2072 is capable of launching a wide range of weapons (torpedoes and missiles) of various origins (for example, France, Germany, UK and the US).

The system functions include:
(a) MultiSensor Situation Elaboration (MSSE): providing management of tracks supplied by acoustic and non-acoustic (optical, optronic, ESM, radar) sensors, determination of the tactical situation using interactive multisensor localisation, identification and designation procedures;
(b) Situation Threat Assessment and Planning (STAP): providing evaluation of the tactical situation and the production of navigation plans (route and tactics) and engagement plans (selection of weapons and optimisation of launch and guidance procedures);
(c) control of weapons interfaces and weapon guidance.

The system uses 68020 and 68040 microprocessors communicating across a standard VME-type bus and uses structured software developed in C and Ada. It includes multifunction workstations with one or two high-definition 19 in colour monitors (COLIBRI or MOC console).

Contractor

Thomson Marconi Sonar SAS, Sophia Antipolis.

VERIFIED

DLT-D3

Type

Submarine torpedo fire-control system.

Description

The DLT-D3 torpedo fire-control system is used in French Navy submarines. All types of torpedo employed by the French Navy can be launched, including wire-guided. The system may also be expanded for anti-surface missile applications.

Target data are fed to the system from onboard sensors, which comprise fore and aft sonars, acoustic rangefinder, and attack and surveillance periscopes.

The system employs a general purpose digital computer, associated with a CRT display terminal.

The following functions are performed by the system:
(a) updating of the tactical situation from the data delivered by the various sensor systems fitted and the navigation equipment
(b) assistance in calculating target components by the use of special recorded programmes
(c) weapon control; computation of the firing path, remote setting of torpedoes, and firing sequence control. The system is designed for launching any type of torpedo
(d) maintaining a chronological record
(e) maintenance assistance using test programmes.

The DLT-D3 operating programmes permit tracking of eight targets, simultaneous guidance of two wire-guided torpedoes and preparation of a third for launching. Each of the three displays of the terminal is dedicated to the presentation of the following data, in accordance with the programme implemented: tactical situation, firing path, and alphanumeric display of parameters (tote) and decoding of the designations and functions of the two common keyboards.

DLT-D3 submarine weapon fire-control system

Communication between the operator(s) and the computer is through the keyboards. Emergency launch of torpedoes is possible from either bow or stern station.

The system is designed to be manned by one or two operators at the operations centre, and one operator at the bow station with possibly another at the stern station. The equipment arrangement, typically, is one CIMSA 15M125 digital computer (QTD), a monitoring and control console (VIC), an azimuth relay (RZ), a true-bearing diagram (GZ) in the operations centre, a tube selection panel (FAT), and tube servicing station (PST) in the bow and/or aft torpedo tube compartments.

Operational status

The system is in service aboard the 'Daphne' class submarines of Portugal.

Contractor

Thomson Marconi Sonar SAS, Sophia Antipolis.

UPDATED

INTERNATIONAL

Kanaris

Type

Submarine fire-control system.

Development

This system has been developed by Lockheed Martin Tactical Defense Systems in collaboration with the Hellenic Navy's Research and Development Centre (GETEN) for the first four units of the Greek Navy's 'Glavkos' class (Type 209) submarines.

Description

Each system consists of a two-cabinet firing distribution unit and the main control unit with two consoles. The latter are the Lockheed Martin (formerly Unisys) Tactical Modular Displays (TMDs) which feature embedded AN/UYK-44 computers permitting the system to control four weapons simultaneously.

Operational status

Five systems are operational: four for shipborne use and the fifth as a land-based training unit.

Contractors

Lockheed Martin, Tactical Defense Systems, Great Neck, New York.
GETEN (Hellenic Navy Research and Development, Centre).

VERIFIED

Weapon control – shipborne

Type

Shipborne torpedo control systems.

Description

These systems allow the remote selection, presetting and launch of lightweight torpedoes such as Stingray, A 244/S, A 244/S Mod 1, Mk 44, Mk 46, IMPACT-MU90 and others. The systems can be supplied in three different configurations:
Configuration 1 – the system is manually controlled by the operator by means of a remote-control panel installed in the ship's operations room;

Configuration 2 – the remote-control panel is interfaced with the combat information system. In this case the optimum presetting values which vary with the tactical situation are determined and updated by the combat information computer which automatically sends them to the selected torpedo through the control panel;
Configuration 3 – is intended for light vessels not fitted with a combat information system. The remote-control panel is directly interfaced with the sonar: it determines and updates the target's course and speed and, on the basis of these data, computes the optimum presetting parameters which are sent to the selected torpedo. A display is incorporated in the control panel for

representing the tactical situation and the recommended attacking manoeuvre.

Operational status

In production. In service with the Italian Navy and 13 other navies.

Contractors

Eurotorp, Sophia Antipolis, France.
DEN, St-Tropez, France.
Whitehead Alenia Sistemi Subacquei, Genoa, Italy (production).

UPDATED

Weapon control – airborne

Type

Airborne torpedo control systems.

Description

These miniaturised systems are designed to allow remote selection, presetting and launch of the A 244/S Mod 1 torpedo and/or MU90 Impact and other existing types of lightweight torpedo, from any type of helicopter or fixed-wing aircraft. The systems are normally supplied in one of two configurations:
Configuration 1 – for helicopters fitted with dunking

sonar. The torpedo control can be carried out either in hovering or in forward flight. In the former case, data from the dunking sonar is processed by the computer unit of the system to determine the target course and speed, and the computer then determines the optimum torpedo presetting values which are sent automatically to the selected torpedo. The hit probability is also displayed on the control panel to assist in decision making;
Configuration 2 – for fixed-wing aircraft and helicopters not fitted with dunking sonar. This configuration is intended for torpedo launch in forward flight rather than both the hovering and forward flight conditions.

Operational status

In production and in service in many types of ASW aircraft and helicopters.

Contractors

Eurotorp, Sophia Antipolis, France.
DCN, St-Tropez, France.
Whitehead Alenia Sistemi Subacquei, Italy (production).

UPDATED

APS series control systems

Type

Torpedo control systems.

Description

These miniaturised systems are designed to allow remote selection, presetting and launch of the A 244/S and/or MU90 Impact torpedo and other existing types of lightweight torpedo, from any type of helicopter or fixed-wing aircraft. The systems are normally supplied in one of two configurations:
Configuration 1 – for helicopters fitted with dunking sonar. The torpedo control can be carried out either in the hover or in forward flight. In the former case, data from the dunking sonar are processed by the control panel computer to determine the target course and speed, and the computer then determines the optimum

torpedo presetting values which are sent automatically to the selected torpedo. The hit probability is also calculated and displayed on the control panel to assist in decision making;
Configuration 2 – for fixed-wing aircraft and helicopters not fitted with dunking sonar. This configuration is intended for torpedo launch in forward flight rather than both the hover and forward flight conditions.

Operational status

In production.

Contractors

Eurotorp, Sophia Antipolis, France.
DCN, St-Tropez, France.
Whitehead Alenia Sistemi Subacquei, Genoa, Italy (production).

UPDATED

Panel of APS 102 airborne torpedo control system for A 244/S torpedoes

NETHERLANDS

M8

Type

Submarine torpedo fire-control system.

Development

The M8 series originated in a Royal Netherlands Navy contract awarded in 1955 for the development of a torpedo FCS for the 'Walrus' class submarines. German interest in the system gave additional impetus to subsequent development and led to the successful development of systems for use in both submarines and surface ships. Two types of torpedo were involved: the German Seal and Seeschlange, for use against

surface vessels and submarines respectively. This successful collaboration with the German industry continued and M8 series systems are standard fitting on all submarines produced by Howaldtswerke Deutsche Werft in Kiel. The ultimate development is the SINBADS system, but other related developments are the M9 systems in the German 'Köln' and 'Thetis' class vessels, and the M11 systems for two new Argentine fast patrol boats.

Description

The M8 is a digital computer-based fire-control system for use in submarines for the direction of torpedoes against either surface shipping or submerged targets. It

was produced in several versions with designations ranging from the M8/0 of the mid-1950s prototype to the latest. M8 has been succeeded by SINBADS, Gipsy and Submarine SEWACO. The basic system comprises a torpedo display control and computer console, a sound path display unit, amplifier and supply unit, distribution box, local control panel(s), and gyro angle setting units. Complete system weight is about 900 kg. The system may be operated by one man, or two men if the submarine is operating with consorts.

The system will accept target data inputs from a range of sensors which includes radar, sonar, passive sound detection systems, periscope observation and consort reports. Ship's own navigational data is also fed into the M8 computer. The display, which has range scale settings for 20, 10 and 5 km, presents the positions of all contacts from all sensors simultaneously. One or more sensors may be connected to the computer for torpedo engagement, and up to three targets may be attacked simultaneously. The computer is programmed to provide firing data for wire-guided, programmed, conventional, and other types of torpedo, and performs automatic calculation of target position, course and speed. The CRT display can give true motion, relative motion or off-centred presentation of the tactical situation.

Operational status
No longer in production. The following list is believed to record accurately the known installations:

Argentina	Type 209 (1)
Colombia	Type 209 (2)
Denmark	Narhvalen (2)
Ecuador	Type 209 (2)
Germany	Type 205 (2)
	Type 206 (2)
Indonesia	Type 206 (4)
Peru	Type 209 (2)
Turkey	Type 209 (2)

Contractor
Hollandse Signaalapparaten BV, Hengelo.

VERIFIED

UNITED KINGDOM

Torpedo launch controller

Type
Electronic launch controller for submarine discharge systems.

Description
The launch controller is a microprocessor-based system which monitors and controls the torpedo firing sequences from loading to discharge. It also integrates the discharge instrumentation system with comprehensive self-diagnostic and fault tolerant interfaces to the torpedo tube. This system allows remote control and monitoring of the torpedo tubes from the command centre without operator assistance, as well as fault recognition.

Operational status
In service 'Vanguard' class submarines (UK).

Contractor
Ultra Electronics, Loudwater.

UPDATED

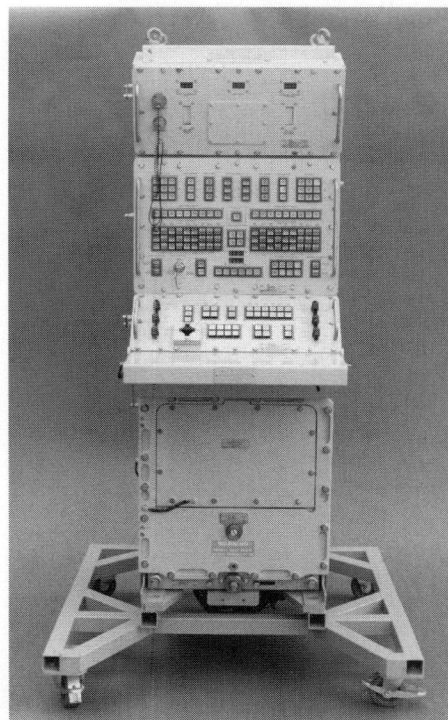

Torpedo launch controller installed in Strachan & Henshaw weapon control and discharge cabinet

DCM

Type
Weapons interface equipment.

Description
The purpose of Outfit DCM is to enable the preparation, firing and control of the submarine's tactical weapons. It can provide monitoring and data feedback in normal and reversionary modes through the interlinking of the command/fire-control system of both the weapons discharge system and the weapons themselves. The system accepts tactical data and firing and control commands from the command/fire-control system via data highways and feeds them, as appropriate, to the weapon discharge system and the weapons themselves. It also feeds tactical data back from the weapons for analysis and action by the command system. The system monitors and reports the weapons fitted, their condition and readiness state.

Full reliability is assured by the use of parallel paths, dual-redundant elements and interconnections and the duplication of critical subsystems.

The system is currently being modified to enable it to interface to the Tomahawk missile and its associated weapon control. The system was originally included within the DCB and DCC command systems and has been designed as an equipment in its own right for use with the SMCS in the 'Vanguard' class and post major refits in designated 'Swiftsure' and 'Trafalgar' class boats.

Operational status
In service in all Royal Navy submarines with minor variants dependant on the class and command system installed. Proposed for the UK 'Astute' class.

Contractor
Ultra Electronics Command and Control Systems, Loudwater.

VERIFIED

UNITED STATES OF AMERICA

Fire-Control System Mk 113

Type
Submarine weapon control system.

Description
The Fire-Control System Mk 113 Mods 6, 8 and 10 are used in SSN attack class submarines of the US Fleet. Target bearings from the BQQ-5 sonar suite are prefiltered, and together with own ship motion data, are processed in the Mk 130 digital computer, the first electronic digital computer to be used aboard attack class submarines. With input of own ship state vector information, the target range, course and speed can be computed from passive bearings only. The operator interacts with the Target Motion Analysis (TMA) function at the analyser console Mk 51.

The FCS Mk 113 consists of the following equipment:
(a) the attack director Mk 75 receives own ship and target information to compute torpedo ballistics and wire guidance controls, and performs position-keeping on both target and torpedo
(b) the attack control console Mk 50 is the system firing panel, tactical display, and torpedo room status panel.

The addition of two torpedo control panels Mk 66, two Tone Signal Generators (TSG) Mk 47, together with major modifications to existing Mod 2 equipment, resulted in FCS Mk 113 Mod 6 and Mod 8 to accommodate the Mk 48 torpedo. Torpedo Control Console (TCC) Mk 66 (which represented the first application of MSI/LSI circuit technology on submarines) acts as a preset panel while TSG Mk 47 generates the signals for transmission of data to the torpedo. The Fire-Control System Mk 113 Mod 6 is a field-modified Mod 2 while the Mod 8 comes directly from the factory with modifications built in.

Operational status
The FCS Mk 113 has been operational since 1962. It is the weapon control system deployed aboard the US Navy nuclear attack class and, with the addition of an Analyser Console Mk 78 (FCS Mk 113 Mod 9), SSBN fleet ballistic missile submarines. In the late 1960s additional equipment required for the Mk 48 torpedo was developed. The attack class submarines have been upgraded with the FCS Mk 117 and Combat Control System (CCS) Mk 1 equipments. Only six FCS Mk 113 Mod 10 systems remain operational on the earliest 'Los Angeles' class SSNs.

The Fire-Control System Mk 113 Mod 9 is the heart of the defensive weapon system of the two 'Benjamin Franklin' class submarines. The system controls preparation, status, launch and guidance of the Mk 48 heavyweight torpedo.

Librascope's system refurbishment contract made provisions for an embedded AN/UYK-44 computer and a new state-of-the-art CRT electronics section for the Analyzer Console Mk 78, which computes TMA data. These modifications will extend the useful life of the Mk 113 Mod 9 system through the 1990s.

Contractor
Lockheed Martin Librascope Corporation, Glendale, California.

VERIFIED

Attack control console of Mk 113 fire-control system

Trident FCS

Type
Submarine-launched strategic missile fire-control system.

Development
General Dynamics Defense Systems (formerly General Electric, then Martin Marietta, then Lockheed Martin) is a key supplier for the Trident strategic weapons system.

Description
The fire-control technology serves the primary purpose of preparing the missile guidance system for flight and controlling the missile launch sequence.

Support equipment consists of: guidance system test equipment for shore-based activities and nuclear submarine strategic weapons facilities; guidance system containers and handling equipment; and equipment to support fire-control and guidance operations at all maintenance levels. In addition, Defense Systems provides the equipment which controls the strategic weapons Trident training facilities.

Operational status
The system is in service aboard the 'Ohio' class SSBNs of the US Navy.

Contractor
General Dynamics Defense Systems, Pittsfield, Massachusetts.

UPDATED

Mk 116 ASW control system

Type
ASW fire-control system ASWCS.

Development
The Mk 116 was developed as an underwater fire-control system in the early 1980s. The concept resulted in a significant approach to performing fire-control functions in a separate computer interfaced with the Combat Direction System (CDS)/Command and Decision System (C&DS) computer. Previous Mk 116 configurations were resident in the CDS/C&DS computer processor. The ASWCS modifications to previous versions of the fire control came in anticipation of further improvements in submarine quieting technologies. These modifications, incorporated in the AN/SQQ-89 surface ship ASW combat system, are known as Mods 5, 6 or 7. Classes equipped with the Mk 116 Mods 5, 6 or 7 include 'Ticonderoga' class cruisers, and the 'Arleigh Burke' and 'Spruance' classes of destroyers. The Mk 116 ASWCS is not incorporated in the 'Oliver Hazard Perry' class frigates, SQQ-89 variants (V)2 and (V)9. Fire-control functions are performed by the weapons alternate processor in those variants.

Description
The Mk 116 ASWCS is a mainframe computer system which is designed to provide battle planning, threat evaluation, tactical data processing, contact management, target engagement processing, and weapon fire control. It is organised into two subsystems: the Computer Processing Subsystem (CPS); and the Weapon Control and Setting Subsystem (WCSS). Data is received from a number of external sensors and distributed around the system by an integral switchboard or data converter. The system

Class/Type	Hull numbers	Mk 116
'Virginia' CGN 38	CGN 41	Mod 1
'Kidd'	DDG 993-996	Mod 2
'Ticonderoga'	CG 47-53	Mod 4
AN/SQQ-89 ship classes		
'Ticonderoga'	CG 54-64	Mod 5
	CG 65-73	Mod 7
'Arleigh Burke'	DDG 51-78	Mod 7
'Spruance'	963, 964, 966-973, 975, 977-978, 980-989, 991, 992, 997	Mod 7
	965	Mod 8
	974, 976, 979, 990	Mod 9

directly interfaces with shipboard acoustic sensors (typically AN/SQS-53, AN/SQR-19 and AN/SQQ-28), radar, ESM, and receives offboard sensor data via datalink.

The CPS provides displays of tactical data, computes and stores track data, calculates weapon control firing solutions for selected targets and generates weapon recommendations and launcher orders. Most systems are based upon a three bay AN/UYK-7 computer and AN/UYA-4 family OJ-194 display console. The UYK-7 has a memory of 256,000 32-bit words and an optimum processing power of 667,000 operations. The newest systems have two AN/UYK-43B computers and, from the AN/UYQ-21 family, the OJ-452 display console. The UYK-43B has two central processing units each with a memory of 2,560,000 words and an optimum processing power of 3,002,000 operations. The software in both systems is in the US Navy's own CMS-2 language and Mod 7 has 215,000 lines of code.

Before the Mod 6 version, the WCSS is basically the same in all the Mk 116 systems. It consists of the Mk 329 weapon control panel which selects the fire-control mode, the sensor/weapon pairing, launcher, and torpedo presetting. It is located close to, and operates

in conjunction with, the OJ-194 display system except in the Mod 5 and later versions which use the OJ-452 consoles.

Other WCSS equipment includes the unmanned Mk 330 missile and Mk 331 torpedo setting panels for selecting, presetting and launching ASROC/torpedo attacks. The Mk 330 is not used in the Mk 116 Mod 4 systems installed in the 'Ticonderoga' class cruisers CG 52-55. There are two Mk 332 weapon status and approval panels to provide command level display of ASROC weapon selection and launcher rail status. Also, on the bridge is the Mk 333 bridge display panel.

The external sensors all include an electromagnetic log and the Type F wind indicator. The other sensors and the data distribution units are as shown below.

Operational status
A total of 75 Mk 116 ASWCS had been produced by January 1994 and it remains in production for incorporation into the AN/SQQ-89 as shown above.

Contractor
Northrop Grumman Corporation, Electronic Sensors and Systems Division, Annapolis, Maryland.

Mod	Sonar SQS-53	Switchboard	Data converter	Gyros	Bathytherm	Fathometer
1	A	Mk 70	–	Mk 19	–	–
2	A	Mk 70	–	Mk 19	–	–
3	C	Mk 34	–	Type 11	AN/BQH-7A-SSQ-61	AN/UQN-4
4	A	–	Mk 18	Type 11	AN/BQH-7A-SSQ-61	AN/UQN-4
5	A	Mk 59	–	Type 11	AN/BQH-7A-SSQ-61	AN/UQN-4
6	B	Mk 59	–	Type 11	AN/BQH-7A-SSQ-61	AN/UQN-4
7	C	–	OL-190	Type 11	AN/BQH-7A-SSQ-61	AN/UQN-4
8	B	Mk 59	–	Type 11	AN/BQH-7A-SSQ-61	AN/UQN-4
9	A	Mk 59	–	Type 11	AN/BQH-7A-SSQ-61	AN/UQN-4

UPDATED

SONAR SYSTEMS

This section covers all types of sonar used for ASW and deployed by surface ship, submarine and helicopter, as well as sonobuoys, airborne acoustic processing systems and static sonar systems. Shipborne sonars used for naval hydrographic purposes, training systems and simulators are described in the Hydrographic survey systems section.

In carrying out ASW, sonars are assigned to the following tasks: detection long-/medium-range passive; classification – medium-range active/passive using high-speed, highly accurate signal processing, Doppler correlation and digital analysis; localisation – shorter range active/passive for tracking and final computations for controlling/directing weapon deployment.

Despite the great variety of available systems, employing the latest technology in sound detection and data processing, the detection of a hostile submarine is still an extremely difficult and slow process and ties up large numbers and forms of ASW assets.

The large static systems moored on the ocean bed are operated mainly by the United States and Russia, while a number of other countries are installing less sophisticated systems. Static systems tend to be used in the most strategically important and sensitive areas and in particular at choke points, that is those areas which are limited in width by land masses, but which serve as entry and exit lanes for submarines and in particular the strategic missile submarines. Attack submarines whose role it is to detect hostile submarines and neutralise them also patrol these areas. Should a hostile submarine escape this first line of defence, it must then be detected and attacked by a second line of defence, such as surface ship and aircraft using towed array, hull-mounted or variable depth sonar, dipping sonar, sonobuoys, magnetic anomally detectors and wake detection systems, torpedoes and depth charges.

Although great strides have been made in the design of transducers and other sensing devices, the major developments in conventional sonars today are those related to the processing and display systems and their integration into a combat system. The advent of digital computers, microprocessors and even more sophisticated electronic components, has meant that large amounts of information can be extracted from relatively weak signals. This is then processed at high speed and presented in a carefully arranged display from which rapid assessments and decisions can be drawn concerning the next stage in the operation.

Both hull-mounted and variable depth sonars are used for passive and active detection while towed arrays are employed for longer-range passive surveillance. Active towed arrays are now beginning to be deployed in an effort to overcome the increasing use of stealth technology in submarine design that is making it very difficult to carry out initial detection of a target. The towed array has many advantages over the hull-mounted system in that it is deployed a considerable distance behind the ship and is thus not affected by any noise emanating from the towing vessel. Towed arrays also achieve very long detection ranges by operating at very low frequencies where propagation losses are lower, enabling the low-frequency sound emanating from propeller cavitation and machinery of a hostile boat to be detected. The towed array does, however, suffer from a number of disadvantages, including: being unable to determine the range of a contact; ambiguity in bearing; directional uncertainty because of sideways movement of the array and the towing cable; flexing of the hydrophone array; and a number of other physical factors. These factors have been the subject of intensive research and development in recent years and are now being largely overcome. In addition, a large winch and handling gear is required to deploy the array. This is not a major problem on surface ships (except from a weight point of view), but is impractical in all but the larger classes of submarines.

In some cases the array has to be clamped on after the submarine leaves harbour and removed just before re-entering, making the boat extremely vulnerable when this operation is being carried out. Some submarines overcome this problem by stowing the array on the side of the hull. However, despite its disadvantages the towed array is of immense value in long-range detection and a considerable amount of development effort is currently being devoted to the design of thin line towed arrays to provide a much lighter and more manoeuvrable system.

The primary airborne detection system is the sonobuoy which is produced in its tens of thousands for deployment from fixed-wing aircraft and helicopters. Since they are expendable, sonobuoys have to be relatively cheap to manufacture but reliable in operation. Large numbers of these devices are in current service and include both passive and active buoys, directional and non-directional, large size and small size, all of which transmit information back to the aircraft for processing and display.

Another underwater threat that is of increasing importance is the free swimmer and small submersible. Harbours and offshore platforms are particularly vulnerable to this form of attack and a number of static sonar systems have been developed to detect such intruders. The sensors of these systems are usually bottom or cable moored and connected to data processors and displays in a shore-based centre. In many cases an integrated system using a variety of radar, acoustics, electro-optical and/or TV-based sensors is employed. To counter this threat, an increasing number of small static and portable detection systems are being developed.

Detection problems

Passive sonars rely for detection on flow noise and cavitation caused by movement in the water of hull and propellers which creates noise over a wide frequency band; on regular low-frequency sound created by machinery; and on hostile active sonars.

Active sonar is bound by the same physical laws which affect passive sonars, except that the effect is doubled, for the sound has to travel in two directions – out and back – before signal processing can be undertaken. Active pulses also lose energy because of the increased distance they have to travel and by absorption in the hull of the target. Active sonars therefore suffer high propagation losses and their effective range is not as great as a passive sonar.

The range problem can be overcome by transmitting a low-frequency pulse (typically 1 kHz with a wavelength of approximately 1.6 m), but it is difficult to focus the beam accurately down a single point bearing. The spread of the pulse at low frequency also reduces accuracy by which range can be determined. Range can be improved by increasing transmitting power, but the relation is not linear. Advantages are that low-frequency pulses can penetrate anechoic coatings. It is also more difficult to detect low-frequency pulses at great distances (typically in the order of hundreds of miles).

Data handling

On the data processing side there has been a significant increase in automated detection, classification, tracking and so on and the presentation of data in synthetic form. Processing power is available in abundance to cope with the demands that will be made on it in the future. The problem will be knowing what to do with the vast amount of information gathered in order to process it in the right way, to present the command with the right type of picture best suited to particular functions and from which accurate assessments can be made as to future tactics.

In the future, command systems must be fully integrated, which includes the sonar plot. This integration can be achieved in one of two ways:

(a) by passing fully processed data on all contacts to the command system for evaluation, a task which will require the application of artificial intelligence
(b) by relegating some functions of the command system back to the sonar processing system so that the command system receives heavily processed track data via a distributed system.

Factors affecting detection

Sonars are all affected by the same environmental factors; for example, temperature, salinity, pressure and seasonal variations in these factors as well as the geographical area in which operations are to be conducted. In addition, efficiency depends on whether one is operating in passive or active mode. All ships create self-generated noise from their movement through the water (cavitation) and from their machinery, as noted above. Own ship machinery-generated noise is more of a problem when the sonar is operating in passive mode, while cavitation tends to dissipate the energy of the transmitted pulse. In active mode, effectiveness can be curtailed further by temperature variations which may restrict range capability.

These problems can be overcome in two ways, either singly or combined, depending on the requirements. One is to increase the transmitting power. This requires increased volume and weight capability in the ship together with the necessary generating power and a larger transducer array. The second solution is to mount the transducer as far away from noise sources as possible; for example, it would have to be a towed, variable depth or bow-mounted array. Bow arrays also suffer disadvantages, for it is not unknown for ships to be manoeuvred forgetting the extra projection at the bow, resulting in damage to the protective cover and even the array itself. To further improve the efficiency of the sonar it is necessary to ensure that the ship be as quiet as possible. This can be achieved by providing a self-measuring noise diagnostic system which will enable the signal processing to cancel out self-generated noise frequencies.

Solutions to temperature variations and self-generated noise in passive mode are overcome by using variable depth sonars and the latter by towed arrays. All sounds possess their own very distinctive characteristics or 'footprints', for example frequency, amplitude and so on. Classification of targets can therefore be greatly aided if received echoes can be matched against a known library of sounds.

Taking advantage of the various properties of the medium through which the sound passes in order to extend and improve the capabilities of the sonar, requires various ancillary systems which can measure environmental factors and provide propagation predictions. Using this equipment, an operator can decide how best to use the sonar to obtain more accurate and positive results. For example, under the right conditions, range can be extended by use of bottom bounce, but it suffers in that the beam cannot be refocused after reflection and hence scattering and divergency occur.

Alternatively, the convergence zone mode of operation can be used which does not require so much power, but which may not always give such an extended range. The mode of operation is selected according to conditions and the tactical situation, but smaller sonar sets may not have the capability to operate in the bottom bounce mode.

Sonars usually operate on a variety of frequencies. For example, some older types of sonar work on a short range of about 2,750 m using two narrow beams with phase shift covering the 17 to 30 kHz band. Medium-range sets use a wider beam and operate at a lower frequency to overcome the problem of attenuation and so increase range. The latest very low frequency active towed array sonar operate on a frequency below 3 kHz. A number of research systems have been built over the last 15 years and the concept is now beginning to enter operational service.

Modern sonars use electronic scanning to provide a panoramic picture (except for the small area round the stern which would be blanked by propeller noise). These are usually low-frequency sets with very high power to achieve long-range detection. Torpedo detection is now an essential and vitally important function of any sonar, for without it a ship is virtually defenceless against torpedo attack. Torpedo alert function can be provided by a dedicated intercept sonar or integrated into a passive sonar array.

Probably one of the most difficult areas for sonar operations in the future will be the littoral zone where depths generally are in the order of 100 fathoms (600 ft/183 m) or less. Here the effect of surface waves, high density of surface traffic, a disproportionate effect of bottom absorption, scattering and reflection – depending on the composition of the seabed – all combine to create enormous problems for the sonar designer and operator. In this scenario, only active sonars offer any certain possibility of detecting the latest ultra stealthy SSK and these, generally, can only reliably offer detection ranges in the order of 5 n miles.

This problem is now beginning to be addressed with the development of special LF active sonars designed to take advantage of the shallow water channel. However, there is always a trade-off. While the

detection probability factor can be increased, the classification factor is diminished.

UPDATED

SURFACE SHIP SONAR SYSTEMS

AUSTRALIA

Mulloka sonar

Type
Hull-mounted scanning sonar system.

Description
Mulloka is a lightweight, relatively high-frequency active hull-mounted scanning sonar for installation in Royal Australian Navy (RAN) anti-submarine escorts and designed for operation in coastal waters around Australia. The solid-state forced air-cooled equipment is of high reliability and has been designed using a standardised packaging approach to aid maintainability and keep production simple.

The equipment includes a dedicated online computer which is used to control transmission parameters, perform automatic signal processing functions and conduct checkout procedures. All signal interfaces have been grouped at a single location, so that modifications necessary to interface Mulloka with other ship equipment configurations are simplified.

The sonar beam-forming process involves the development of a large cylindrical transducer array consisting of 96 staves with 25 transducers per stave. Each driven in differing combinations and phases

under computer control to form the required sonar beams electronically. The hydrophone elements of the array are driven by 480 power amplifiers during the transmit period, the length of which is a function of the pulse length of the transmitted signal. The transmit period is selectable over the range of approximately 1 to 4 seconds. The transmitter beams are formed sequentially during this period and cover the ocean for 360° in azimuth. The formation of the beam transmit pattern is random moving between transmissions to the next in order to prevent the submarine commander having any indication of detection. Immediately following the transmission, the sonar reverts to the listening mode for a period determined by the range scale selected by the operator. Range is about 9,000 m.

Operational status
In 1979, a prototype Mulloka sonar was formally handed over to the RAN after undergoing trials in HMAS *Yarra* since 1975. A production programme to build systems to equip six RAN 'River' class destroyer escorts was then completed. One 'River' class frigate, *Torrens* remains in service, but is scheduled to be decommissioned in 1998.

In April 1980, it was announced by the Australian Ministry of Defence that an initial contract had been awarded to Honeywell Inc in America to provide a transducer array for the Mulloka sonar. Two Australian companies, GEC-Marconi Systems Pty Ltd and Dunlop Industrial and Aviation, were also participating. A contract for the electronics of the system was awarded to THORN EMI (Australia) Ltd in 1979.

In 1985, contracts were awarded for two systems to be fitted to two 'FFG 7' class frigates being built in Australia (*Melbourne* and *Newcastle*). As a result of the transferred technology, Plessey Australia was contracted to establish a production facility and to manufacture two transducer arrays and spares which were delivered over the period 1987 to 1989. In mid-1987, THORN EMI (Australia) Ltd was awarded a A$6 million contract to upgrade the system electronics, including a new display system.

Contractor
THORN EMI (Australia) Ltd

VERIFIED

Narama and Kariwara towed thin-line arrays

Type
Thin-line towed arrays for submarine and surface ship operation.

Description
Following the joint venture of the former Thomson Sintra Pacific and GEC-Marconi Systems companies, Thomson Marconi Sonar now supply two thin-line winchable towed arrays for military application. These rugged and robust thin-line towed arrays offer substantial savings in space and weight when compared to the traditional liquid filled arrays. These technologies are particularly well suited for winching and offer submarines a unique capability of a towed array that can be remotely winched and fully stowed on

board. Both technologies offer excellent performance over a wide range of frequencies and tow speeds, and can be configured to meet specific customer requirements.

Narama is based on enhanced liquid filled array technology and can be supplied in a range of diameters from 40 to 65 mm. A revolutionary design concept gives the array a very robust nature and provides for reliable winching and long life. The simple and uncomplicated structure of Narama provides excellent acoustic performance. The array can be fitted with digital telemetry for large hydrophone channel populations, or with hard-wired preamplifiers for a very simple low-cost streamer. The small diameter allows a much smaller and lighter handling system to be used and confers on the system a very low self-noise, due to the improved signal-to-noise ratio and high-operating speed. The array incorporates advanced vibration

isolation modules. It can be operated at speeds up to 15 kt without degradation to acoustic signals. Available array designs include short lengths for tactical use and longer surveillance arrays which use digital telemetry. The arrays supplied for the 'Oberon' class submarines are of a self-streaming configuration, but the system is also compatible with the new 'Collins' class submarines.

Kariwara uses a solid filled array technology and has been tailored for 40 mm diameter. The hybrid solid/liquid filled construction of the array provides an extremely robust construction while maintaining excellent acoustic performance. The construction uses a combination of tough buoyancy-controlled polymers assembled in modular functional sections which are joined by titanium electromechanical connectors. Not only is this solid construction easily winched and highly resistant to damage, but it provides graceful degradation by containing and localising the effects of damage. Kariwara is supplied with a Digital Telemetry System.

Operational status
Two Narama systems were delivered to the RAN for its 'Oberon' class submarines in 1997 and SAES of Spain has also purchased the Narama as part of an R&D Project for the Spanish Navy. Both orders relate to 40 mm diameter versions.

Loading a Narama thin-line array on to an 'Oberon' class submarine   *1998*/0007240

The winch assembly used with the Kariwara thin-line array in 'Collins' class submarines
1998/0007221

Kariwara is currently in service with the RAN on both 'Collins' and 'Oberon' class submarines and has been trialled by an RAN frigate.

Specifications
Diameter: 25-50 mm
Length: up to 5 km

Channels: up to 1,000
Operating speed: up to 15 kt, max >20 kt
Survival speed: 30 kt
Axial stiffness: 1% strain at 15 kt
Min bend radius: 0.5 m (50 mm diameter array)
Tensile load: exceeds 100 kn (>10 t for 50 mm diameter)

Max crush load: >50 kn (5 t)/m

Contractor
Thomson Marconi Sonar Pty Ltd, Rydalmere, New South Wales.

VERIFIED

Petrel/TSM 5424

Type
Hull-mounted active volumetric acoustic processing sonar.

Development
The Petrel 5424 three-dimensional (3-D) sonar design is the result of a three year development undertaken in Australia by Thomson Marconi Sonar Pty in conjunction with the Royal Australian Navy (RAN) and the Australian Department of Defence Industrial Development programme. This development grew out of the RAN's operational requirement to optimise mine avoidance in coastal waters, by use of a high-frequency, high-resolution system. RAN analysis found traditional lower frequency systems were less effective in Australian coastal waters. The provision of a capability to provide real-time 3-D mapping of the seabed was also seen as important, given the large areas of shallow and unsurveyed waters within Australia's area of direct military interest.

Description
The Petrel has been designed to provide an instantaneous, high-fidelity resolution of range, bearing and elevation, within a 60° azimuth by 12° elevation sector ahead of the vessel. The sonar creates this volume by use of patented volumetric acoustic processing technology. This capability allows the separation of sea surface, seabed and water column returns, and allows for the reduction of reverberation and other artefacts by isolation of these into separate matrix areas. The latter capability is important as it means that Petrel is insensitive to changing water

conditions and hence achieves good performance in warm coastal waters. Detection probability is rated as 95 per cent 2 sigma while extremely low false alarm rates are germane at 1 in 50 hours. The data is processed to display a real-time spatial three-dimensional data set of seabed data ahead of the vessel and also icon prompted representation of point contacts such as mines.

The system is mechanically stabilised and can be operated at speeds up to 12 kt, in up to Sea State 4.

The system currently operates using X-Windows but will be fitted with Windows NT.

The sonar is designed for navigational safety and mine avoidance in both surface and submarine military applications. The primary display and operator interface is on the bridge and it is designed to minimise time demand on the seaman by providing automatic monitoring of features of a depth less than an operator selected limit by providing continuous search for point contacts with icon prompting and alarm for rapid operator recognition. Minimum detection range is 650 m and maximum is 1,000 m. Petrel is designed for the following roles:

1) **Mine and natural obstacle avoidance;** Petrel detects surface, moored and seabed mines and will provide high resolution real-time spatial mapping for safe navigation in poorly charted or coral waters.
2) **Shallow water hydrographic feature detection;** Petrel detects seabed features as small as 1 m³ for subsequent examination by multibeam echo-sounding systems. This detection is undertaken ahead of the vessel for optimised survey vessel safety and mission efficiency.
3) **Amphibious warfare;** Petrel provides real-time intuitive data on landing beach gradient and

amphibious entry/egress routes. Petrel is fully retractable and can be operated by beaching craft.
4) **Warship defence from swimmer attack at anchor;** Petrel will scan 180° in 6 seconds and will detect the diver's lung cavity and diving apparatus. Attacking divers can be tracked in 3-D and the availability of depth data permits optimised anti-diver charge fuse timing.
5) **Shallow water anti-submarine warfare;** Petrel will detect bottomed small submarines including bottom crawling clandestine craft in shallow water.

Operational status
Petrel is commercially available with an 18 month delivery time in both the surface ship (warship and survey ship) and submarine variants. The distinction in variants is mechanical only, reflecting the need for mechanical roll stabilisation in small to medium surface warships, which operate in a highly dynamic environment. The submarine variant does not have mechanical stabilisation, with the transducer array fixed behind an acoustic aperture. A commercial variant comprising the same system fixed to the bow of VLCC and other larger vessels is being considered.

Petrel is to be fitted to the six 'Adelaide' class FFG 7 frigates and the 'Anzac' class frigates. The seven survey launches began fitting Petrel towards the end of 1998.

Contractor
Thomson Marconi Sonar Pty Ltd, Rydalmere, New South Wales.

UPDATED

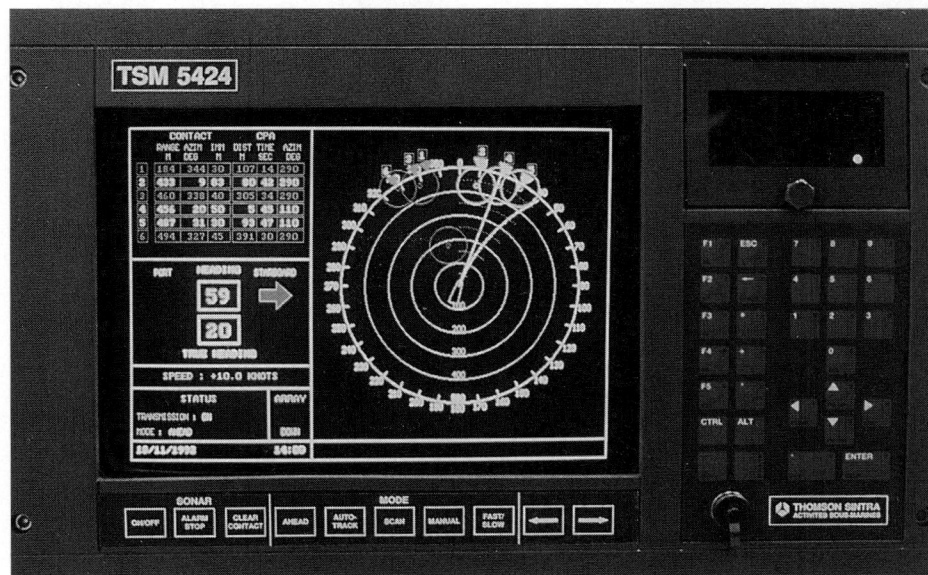

The Petrel helmsman indicator displays a safe course to avoid mines when transiting a minefield **1996**

The wet end of the Petrel TSM 5424 **1998**/0007222

CANADA

AN/SQQ-504 towing condition monitor

Type
Monitoring system for VDS sonar systems.

Description
The AN/SQQ-504 towing condition monitor aids ship's personnel in the successful launch, tow and recovery of the Variable Depth Sonar (VDS) body. With the

introduction of this type of equipment, the command and sonar operations are provided with means to optimise operations and minimise risk to VDS equipment, due to the variability of sea conditions, ship's speed and motion and so on.

The monitor provides VDS towcable strain information (actual strain when the body is under tow, or predicted strain when the body has not been launched or is to be towed at a different cable length) to the operator and ship's command. The system also

provides information by which the VDS operator can judge the performance of the VDS hoist system and enables personnel to monitor the degrading effect of ship's motion on the VDS sonar detection capability.

The system consists of a group of sensors, a processor unit, two recording units, three identical remote display units and a VDS operator's control and display unit. These items perform the functions of data acquisition, signal conditioning, data processing, display and recording. Sensors include boom and deck accelerometers, a cable strain gauge, a boom switch, a cable length synchro and ship's speed data.

Operational status
The AN/SQQ-504 is fitted to Canadian Navy destroyers.

Contractor
Northrop Grumman – Canada Ltd, Burlington, Ontario.

VERIFIED

AN/SQQ-504 towing condition monitor, showing primary indicator (left) and remote indicator (right)

21 HS-WB series

Type
Bistatic sonars.

Description
The 21 HS-WB bistatic sonar series are compact, high-performance systems with a multimode active and wideband passive capability, designed to meet the needs of a wide variety of platforms and operational roles. ASW missions include search, detection and tracking of submarines and torpedoes in both deep and shallow water environments. The sonars make use of operator interactive COTS-based computer-aided techniques, for enhanced signal processing and automatic detection and tracking. Active capability supports multiple search strategies, using CW, FM coded pulse and shallow water waveforms. The wideband passive capability supports the detection and warning of torpedoes.

The 21 HS-WB bistatic sonars are manned by a single operator using a console with two high-definition colour, menu-driven X-Windows displays that can be user customised.

Operational status
The system is ready for full scale production.

Contractor
Northrop Grumman – Canada Ltd, Burlington, Ontario.

VERIFIED

AN/SQR-501 CANTASS

Type
Surface ship towed array sonar system.

Description
The AN/SQR-501 is a passive, critical angle towed array which uses a Computing Devices Canada display and processing subsystem, the Lockheed Martin AN/SQR-19 towed array and Indal Technologies OK-410 handling gear. The bearing-stabilised sonar display subsystem uses two, four or six high-definition TV monitors for this application. The acoustic data processing system uses the Computing Devices AN/UYS-501 signal processor, modular interfacing and data management electronics. The acoustic data processing subsystem also provides for the storage and playback of raw acoustic data by means of a Sony DR 1000L high-density digital recorder.

The system provides frequency and bearing analysis of acoustic emissions from long ranges and is consistent in both shallow water and beyond the second convergence zone. Data is presented to the operator on programmable CRT displays in a variety of alternative formats. Audio data is presented as a further aid to detection and classification. Processing algorithms were developed by the Canadian Defence Research Establishment Atlantic and implemented by Computing Devices Canada Ltd.

Processing and data manipulation is carried out entirely in software using the Ada programming language. The modular design offers a growth capability and adaptability to meet changing requirements. Developed for a surface ship application, the design of the interfaces, display and processing subsystems allows for adaptability to a variety of sensors and platforms including submarines.

Operational status
A contract with the Canadian Navy for 15 systems was signed in August 1990 for the new Canadian 'Halifax' class frigates and the 'Annapolis' class destroyers. All systems have now been installed and are in full operational use.

Contractor
Computing Devices Canada Ltd, Ottawa, Ontario.

VERIFIED

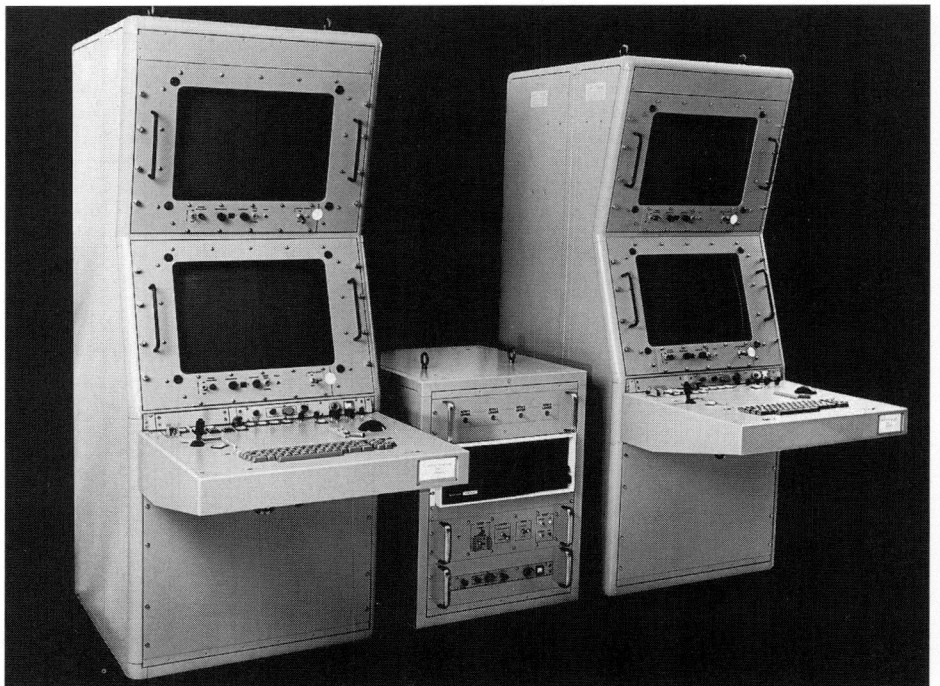

Display consoles of the AN/SQR-501 CANTASS　　1997

Electronic cabinets of the AN/SQR-501 CANTASS
1997

AN/SQS-510 sonar system

Type
Medium-frequency active sonar system.

Description
The AN/SQS-510 is an adaptable, medium-frequency, hull-mounted or variable depth, active sonar system, which was created as a major redesign of the AN/SQS-505. It integrates the existing transmitter group, transducer and hull outfit or towed body equipment with a new programmable digital receiver, processing and display groups and modern digital signal and display processing algorithms to create a modern high-performance sonar system. It provides simultaneous active and passive digital signal processing, display processing and Computer-Aided Detection (CAD) for rapid detection and localisation of active and broadband passive contacts. In addition to normal blue water operations, this sonar is optimised for performance in three specific scenarios; shallow water submarine detection, mine avoidance and torpedo detection. The sonar system is controlled from a dual-screen operator's console. The high-resolution digital colour monitors provide synthesised displays of processed active and passive acoustic data. The operator is alerted to active and passive contacts by sophisticated CAD and tracking algorithms. Passive alerting is provided in both active and listen states.

Operational status
The AN/SQS-510 is installed on the Portuguese 'Vasco da Gama' and 'Joao Belo' class frigates and on the Canadian destroyers *Nipigon* (as a hull-mounted sonar) and *Terra Nova* (as a VDS). Currently, Computing Devices Canada is under contract to provide AN/SQS-510 to the Canadian 'Halifax' class frigates

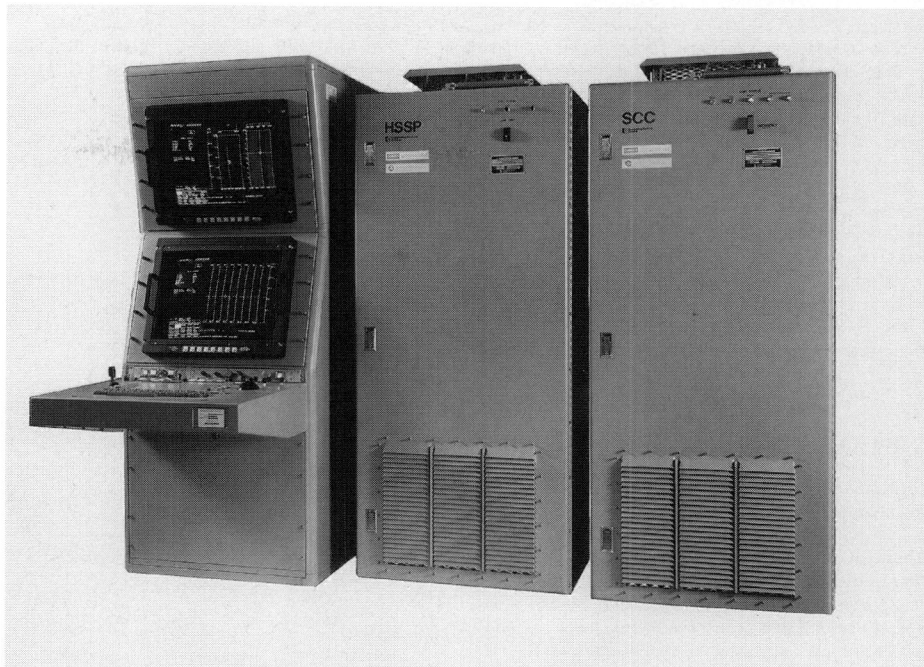

Display console and electronics cabinets of the AN/SQS-510 sonar   1997

(Canadian Patrol Frigate Programme) in a hull-mounted mode and the 'Iroquois' class DDH-280 destroyers in both hull-mounted and VDS. In January 1997, Computing Devices Canada was awarded a contract by the Belgian Navy to supply the AN/SQS-510 for the upgrading of the 'Wielingen' class frigates.

Contractor
Computing Devices Canada Ltd, Ottawa, Ontario.

VERIFIED

Hydra MSS

Type
Integrated ASW/MCM sonar suite.

Description
The Hydra MSS (Multi Sonar System) integrated sonar suite is on order for the Swedish Navy to carry out both ASW and MCM surveillance for surface vessels.

The total integrated system comprises towed array sonar, variable depth sonar, hull-mounted sonar and sonars for ROVs and sonobuoys.

The system will use the latest technology including digital CD-ROM.

Operational status
On order to equip the four 'Visby' class corvettes and a

complementary system ordered for the 'Goteborg' class corvettes.

Contractor
Computing Devices Canada Ltd, Ottawa, Ontario.

NEW ENTRY

CTS-24 ASW OMNI sonar

Type
Active hull-mounted sonar.

Description
The CTS-24 is a medium-frequency sonar designed for navies operating in shallow as well as deep water. The narrow vertical beam, 6°, maintains excellent performance in shallow water, while the 24 kHz operating frequency provides extended detection ranges in blue water.

Applications for the CTS-24 are for shallow and deep water ASW, mine detection/avoidance (surface, moored and bottom mines), MCM support roles (ROV/AUV tracking) and general surveillance.

The main features of the sonar are:
(a) automatic detection

(b) automatic tracking of up to three targets, target range, bearing, depth, speed and heading are automatically updated
(c) electronic stabilisation of receive beams for ship's roll and pitch
(d) electronic tilt for target interrogation in the water column, as well as ocean bottom
(e) high-resolution (0.25 m) zoom window (250 m wide) can be set through the full operator selectable range scale
(f) source level: 223 dB/μPa/m
(g) 13.5 kW omni transmit power.
Various modes of operation are available providing:
Omni-active: simultaneous audio and video presentation of CW or FM modes. Three target tracker channels provide automatic tracking of targets. Own ship track is also displayed.

Omni passive: omni video presentation with steerable audio sector for classification.
Surveillance: provides a 12° sector azimuth steerable operation for minimising effects of reverberation in shallow water. Video and audio presentation.
Maintaining close contact: provides a broad 30° depressed beam for maintaining contact on close in targets without the need for operator interaction.
High-resolution: provides a 250 m window with a 0.25 m range resolution, adjustable to any position of the selected operating range.

Contractor
C-Tech Ltd, Cornwall, Ontario.

VERIFIED

CTS-36/39 OMNI sonar

Type
Hull-mounted sonar.

Description
The CTS-36/39 Omni sonar is a multipurpose hull-mounted ship system for use in surveillance, ASW and MCM support roles in coastal waters and has been optimised for shallow water operation.

The scanning and multibeam sonar uses narrow horizontal and vertical receiving beams (6°), high-speed scanning and digital processing techniques to provide instantaneous detection, classification and target data through a full 360° field of view. Sector transmission is operator selectable in 24° sector increments for operation in areas such as harbours or

narrow channels. Transmit power and pulse length are also operator selectable to provide optimum setting for any area of operation. Simultaneous video and audio information is available in both active and passive modes, with the audio sector being steerable in passive. Automatic detection and target tracking are provided in active mode using selectable CW or FM transmission. Three independent receive channels (each 60 x 6° at −3dB points) can each track automatically one, two or three targets. Target range, depth, course, speed and heading are instantaneously displayed and updated for each target. Own ship's track is also displayed.

The CTS-36/39 also provides:
(a) Detection of various types of targets such as divers, mines, submarines and so on
(b) Detection of targets in the water column as well as

bottom targets such as wrecks and rocks on the seabed using electronic tilt
(c) Detection and location of large targets at long ranges
(d) Electronic roll and pitch stabilisation of receive beams
(e) Interface to ship's own command/communications and control (C² or C³) systems.

In surveillance mode the operator can select the transmit sector width and position, using a 12° receive beam stepped through the operator selected sector. Maintaining contact on close in targets is aided through the use of a broad receive beam at a fixed, depressed angle of 30°. The electronic tilt feature enables extended detection range to counter the effects of ray refraction in thermocline conditions, as well as providing enhanced capability in detecting and identifying seabed targets in shallow areas exhibiting high reverberation. Automatic detection and target tracking are provided with a resolution of 1m in range and 0.75° in bearing. Target range, depth, bearing, speed and heading are constantly updated and displayed. Target cursor and markers provide a history track on the display.

The MMI comprises touch screen panel and trackball with a 20 in high-resolution display monitor. This provides PPI, B-scan, passive and surveillance sector displays for operator selection.

The system can function as a stand-alone unit or be integrated into a full combat system.

The transducer is contained in a self-contained retractable dome that is stored in the retracted position inside a well tube that forms the interface with the ship's hull. The hoist unit is located inside the ship compartment where the well tube is welded to the hull.

Operational status
The CTS-36/39 is operational with the Danish Navy 'Thetis' class and 'Flyvefisken' class ships. A total of 19 ship sets has been delivered to date.

Specifications
Total weight: 650 kg
Operating frequency: dual-frequency 36 kHz/39 kHz operator selectable; CW mode; FM chirp.
Ranges: 250 m, 500 m, 1,000 m, 2,000 m, 4,000 m operator selectable

Electronic tilt: from +8 to −24°
Electronic stabilisation: ± 25° roll, ± 10° pitch
Receiving beamwidth: 6° in horizontal and vertical
Transmitter: 270 channels; power output/channel 50 W max, 13.5 kW omni total
Source level: 223 dB/μ Pa/m
Source power: 10 kW

Contractor
C-Tech Ltd, Cornwall, Ontario.

UPDATED

CANTASS PAS

Type
Post-mission analysis system.

Description
CANTASS PAS (Canadian Towed Array Sonar System Post-Analysis System) is an advanced post-mission analysis system that processes, stores, analyses and classifies sonar data collected by towed arrays. Data is assimilated and processed from tape in three times real time. The system randomly accesses up to 60 hours of processed sonar data. The operator workstations display data on a high-resolution monitor, where analysts tag interesting time ranges of data for future access. Processed data is displayed on up to six operator workstations at up to eight times real time. The displays are Windows-based and graphically oriented. These features help analysts distinguish events from background noise. Display formats available to the operators include: search summary; single-beam/ bearing; triiple-beam/bearing; broadband acoustic; amplitude line integration; and full-screen user-configurable display format. Several of the formats have subformats which allow the operator to view many displays of the same format type simultaneously. The signal processing system takes raw acoustic data beam-formed in 43 beams and applies advanced special analysis to produce narrowband, broadband,

CANTASS PAS workstation and electronics cabinets   *1998*/0007223

amplitude line integration and demodulated noise (Demon).

Operational status
In service with the Canadian Navy at ADAC Halifax and Vancouver.

Contractors
Array Systems Computing Inc, North York, Ontario.
Litton Systems Canada Ltd.

VERIFIED

Indal Technologies sonar handling equipment

Type
Sonar towed bodies, handling equipment and sonar domes.

Description
Indal Technologies Inc specialises in the provision of sonar towed bodies, handling equipment and sonar domes, primarily for underwater ASW and MCM applications.

For VDS, there is a range of systems that includes the hydrodynamic towed body housing the sonar transducer, an electromechanical towcable equipped with a low drag, rigid Flexnose fairing, a launch/ recovery system for the towed body and a tow winch which includes a no 'fleet-angle' level wind system and passive and active motion compensation systems. In addition, there is a mechanical device known as Sidewinder, which provides electrical and optical (multifibre) continuity for power and signal without the need for slip rings. A large inventory of VDS handling systems is available ranging from LFA (Low Frequency Active) sonar transmitters to mine identification high-frequency sonar.

The company is supplying the Second Acoustic System (SAS) handling system; a very large handling system used for deploying, towing and recovering acoustic systems from a US Navy T-AGOS 23 SURTASS ship.

The OK-410 (V) Handling and Stowage Group (H&SG) deploys, streams and recovers a tactical towed line array on the Canadian Navy's CANTASS programme.

The Subtow handling and storage group has been designed and produced for the ESISS (Experimental Submarine Integrated Sonar System) towed array handling system installed in a flooded compartment on a Canadian Navy 'Oberon' class submarine.

Indal Technologies Inc designs and manufactures hull-mounted sonar domes, both for the Canadian SQS-505 and the Raytheon SQS-56/DE1160 sonar systems. The domes typically include the C-5 hull outfit which raises and lowers the sonar transducer out of the dome and into the ship's hull for repair or protection.

Operational status
VDS handling and towing systems: More than 35 systems in operational service with Brazil, Canada, Finland, India, Italy, Norway, Singapore and Sweden.
OK-410 (V) H&SG: 'Halifax' class and 'Annapolis' class.
SAS handling system – In production. T-AGOS 23.
Sonar domes: various models in operational service with Canada, Greece, Morocco, Portugal, Saudi Arabia, Spain, Turkey and the US.
Subtow – In operation.
Nixie torpedo decoy handling system – In operational service with the Canadian Navy.

Contractor
Indal Technologies Inc, Mississauga, Ontario.

UPDATED

Model 15-750

Type
Variable depth sonar handling and towing system.

Description
The model 15-750 provides a launch, variable depth towing and recovery capability for an Indal Technologies towed body, weighing approximately 4,000 kg, fitted with a sonar transducer and environmental and heading sensor package.

The system comprises an electromechanical cable fitted with low drag Flexnose fairings, a hydraulically driven single-layer winch, a cable tension stabiliser/ spooler assembly, an overboard handling assembly and a towed body. During launch and recovery of the towed body, the winch and overboard handling unit operations are automatically integrated.

The winch consists of a single, large, chain-driven drum mounted on a static shaft. The chain is driven from a sprocket mounted on the output shaft of a planetary gearbox. Input to the gearbox is from a fixed displacement motor powered through a multidisc, fail-safe brake. The starboard side of the drum is recessed to allow ready access to the Sidewinder mechanism mounted on the shaft inside the drum. This provides a continuous electrical connection between the towcable and the deck-mounted cabling to the sonar processing equipment.

The cable tension stabiliser consists of a sheave mounted at the end of a swinging arm which is supported by a hydropneumatic spring system. This spring system comprises a quantity of nitrogen gas at a precharged pressure, contained in an extendable bladder within an accumulator. By injecting hydraulic oil into the accumulator, the volume of the bladder is adjusted, thus changing the effective pressure and spring rate of the gas. Pressure exerted by the gas spring is coupled through a hydraulic circuit into two cylinders which position and restrain the swinging arm sheave, thus resisting tension in the cable.

The system is controlled by a single operator from a control station and is suitable for either 'open deck' or 'between deck' installation. The control console is designed for location adjacent the winch from where the operator has a clear view of the ocean, enabling him to visually monitor as well as control deployment and recovery of the towed body.

Operational status

The Model 15-750 system is in service with the Indian and Italian navies.

Specifications

Tow speeds: Launch and recovery 8-14 kt
Operation 6-20 kt
Survival 30 kt
Cable length: 229 m
Sea state operation: Launch/recovery – 4

Contractor

Indal Technologies Inc, Mississauga, Ontario.

UPDATED

FINLAND

Sonac PTA

Type

Passive towed array system.

Description

The Sonac PTA passive towed array sonar has been optimised for use in archipelagos and other areas of shallow water where environmental conditions make submarine detection difficult. The simple, lightweight hardware is suitable for installation in small inshore patrol vessels.

The sonar features 24 hydrophone channels feeding digital information to an onboard processor for display and threat analysis. Data transmission is through fibre optic cables mounted in the towcable. Automatic electronic cancellation of propeller and machinery noise allows a short towing cable without loss of detection sensitivity. In addition, the array itself is only 78 m long and vibration isolation modules in the array and fairing of the towcable prevent interference due to turbulence. The configuration allows the maximum manoeuvrability during search operations in restricted waters. Nominal towing speed is 3 to 12 kt.

Array depth is regulated by an active depressor which is controlled by the operator. As well as providing precise depth control in shallow waters, the depressor also allows the array to penetrate thermal layers.

The omnidirectional hydrophones each incorporate an integral electronic unit which filters, amplifies and digitises the sound signals before transmission to the onboard processor.

Onboard processing equipment is housed in a single cabinet which also incorporates the operator's control panel and two VDUs. Dedicated processors perform signal analysis on the sonar returns while separate general purpose computers handle the operator interface and navigation calculations, also providing BITE functions.

In broadband mode, surveillance signals from all directions are presented together with direction and signal power indicated for each source covering the preceding 30 minutes. In narrowband mode the spectral estimates of each signal are displayed. The compact processor unit makes installation in a small patrol vessel simple. Optional peripheral equipment includes a no-break power supply and also a data recorder to store incoming information from the array in unprocessed form for post-operation analysis.

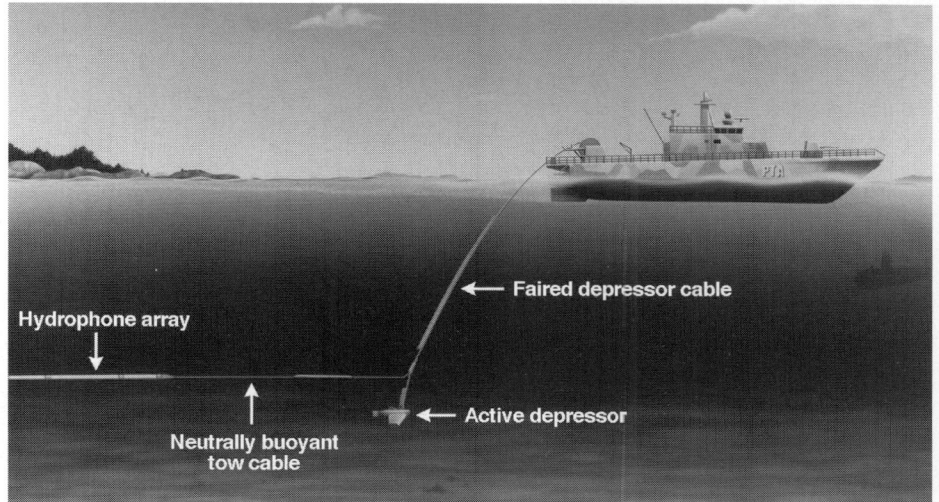

Impression of Sonac PTA deployed from small coastal vessel   *1995*

Sonac PTA provides the operator with easy access to all essential information. The six main displays offer:
(1) broadband surveillance: sound intensity versus bearing with waterfall history
(2) narrowband surveillance: sound spectra versus bearing with waterfall history
(3) more detailed information of targets being tracked are analysed and displayed
(4) audio information from a selected direction is generated for headphones and/or for classification purposes
(5) alarm and replay of unexpected events, transients and so on is performed
(6) navigation information with tracked target information is presented.

The system features a menu-based interface which guides the operator in all phases of the operation. Access to the system is via a trackerball and a few special keys. Operation of the peripherals, such as active depressor, raw data recorder and so on are included in the user interface.

Signals are digitised in the towed array. In order to optimise the operation in different environments, the parameters of the array (amplification, own ship noise cancellation) are remotely controlled from the

processing unit within a large dynamic range. The data cannot be affected by ship-generated EMI disturbances because they are digitally transmitted to the processing unit via a fibre optic link. The signals are interpolated, filtered, beam-formed and analysed in the compact processing unit located in the presentation cabinet. The results are presented on two-colour video monitors by means of clear waterfall history displays. Open architecture is used both in software and hardware which guarantees easy interfacing with other systems as well as with upgrading.

Operational status

In service with the Finnish Navy on the 'Helsinki', 'Rauma' and 'Ruissalo' classes since 1993.

Specifications

Array length: 78 m
Towcable length: 600 m
Depressor cable length: 120 m
Operational speed: 3-12 kt
Operating depth: 100 m (max)

Contractor

Patria Finavitec, Tampere.

VERIFIED

Sonar winch of the Sonac PTA   *1996*

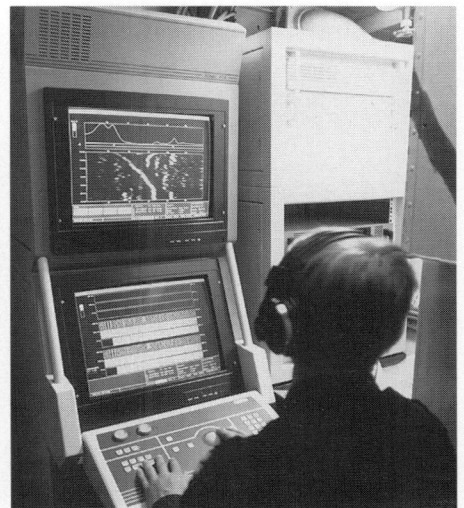

The operator console of the Sonac PTA   *1996*

FRANCE

DSBV 61B sonar

Type

Surface ship towed array sonar.

Description

The DSBV 61B is a very low frequency passive towed array sonar system consisting of a linear array, a towing cable which is also used for data transmission, a winch for handling and storage of array and cable and the necessary signal processing and data extraction electronics. The latter is made up of five cabinets plus two display consoles associated with two TOTES. The design allows two selectable towing modes: critical angle towing or towing behind the DUBV 43C VDS body.

The display consoles feature three-colour display of target information and can display, on request, the various sonar memory pages. A built-in fault display allows a quick reconfiguration and/or rapid maintenance of the system.

Operational status

Operational on the F70 type ASW corvettes D644-646. A derivative known as Anaconda is described briefly under a separate entry.

Contractor

Thomson Marconi Sonar SAS, Sophia Antipolis.

VERIFIED

DSBV 61B linear towed array sonar (serial equipment) on board the Primauguet *F-70 type ASW destroyer*

DSBV 62C sonar

Type

Surface ship very low frequency passive towed array.

Description

This stand-alone sonar, which can be installed on board surface ships or submarines (DSUVL62C see under France in Submarine sonar systems section), is the latest VLF passive towed array sonar in operation in the French Navy. The DSBV 62C uses the same analogue array as in the submarine version, plus an additional section for digitisation of signals and transmission via the heavy cable, which is used in a critical angle towing configuration.

The DSBV 62C is the highest performance passive linear array in operation in the French Navy. It incorporates both broadband and narrowband processing, with ADT, panoramic threat verniers and narrowband track fusion. Lofar and Demon are also available to the operator. The basic system has the capacity to process signals from various array architectures.

Operational status

Delivered and at sea on 'Tourville' class frigates. The export version is the TSM 2932.

Contractor

Thomson Marconi Sonar SAS, Sophia Antipolis.

VERIFIED

DUBA 25 sonar

Type

Hull-mounted active panoramic sonar system.

Description

The DUBA 25 medium-frequency attack sonar fitted in the French 'Aviso' class ships is an active sonar with a panoramic transducer array and assembly with sonar dome supplied by the French Navy. A hull-mounted sonar, the DUBA 25 provides 'Aviso' class ships with all-round surveillance, acquisition and attack facilities for ASW operations. The power employed is sufficient for operating ranges of several kilometres. The transducer array consists of 36 staves arranged to form a cylinder 115 cm in diameter and provided with roll stabilisation.

Operational status

Operational since 1975 on board 16 Type A69 frigates and the destroyer *Cassard* (DUBA 25A). No longer in production. More detailed information is given in earlier editions of *Jane's Naval Weapon Systems*.

Contractor

Thomson Marconi Sonar SAS, Sophia Antipolis.

UPDATED

DUBV 23D sonar

Type

Surface ship medium-frequency search sonar.

Description

The DUBV 23D is a bow-mounted, medium-frequency, panoramic sonar for anti-submarine operations. The 48-column transducer array is housed in a bulb at the fore part of the ship, the bulb being of streamlined design to reduce parasitic noises to permit listening at high speeds. The panoramic sonar is intended for both search and attack roles.

In addition to the transducer array, the equipment includes the transmitter/receiver unit, a computer section for the processing of data being fed to weapons and control and display consoles at the anti-submarine attack station.

The DUBV 23D is of identical design to the towed sonar DUBV 43B/C and in French vessels the two sonars are used together for ASW.

Operating modes provide for: panoramic surveillance, sector surveillance, step surveillance, passive surveillance at sonar frequency, panoramic attack transmission, or 'searchlight' attack transmission. In addition to the system's own display devices, the DUBV 23D provides for target data outputs to other ships' systems and repeater PPIs.

Operational status

The DUBV 23D is in service in the 'Suffren' and F 67 and F 70 (DD 640-643) classes of the French Navy. The sonar is also installed on the Chinese 'Luhu' class frigates. The system is no longer in production, but systems in service are being upgraded through the SLASM programme (see separate entry).

DUBV 23D sonar control console

Specifications
Transmitter
Frequencies: four operating frequencies in the neighbourhood of 5 kHz, of which two are operational
Power: 96 kW (2 × 48 kW)
Type: CW, FM (linear frequency modulation with non-coherent data processing at reception)

Duration: 4, 30, 150 or 700 ms
Scatter echo: with or without rejection
Doppler effect correction: on all 48 channels
Cadence: adjustable step by step from approximately 1,370-43,875 m
Receiver: panoramic, directional, passive listening in sonar band

Contractor
Thomson Marconi Sonar SAS, Sophia Antipolis.

VERIFIED

DUBV 43B/C sonar

Type
Variable depth sonar system.

Description
The DUBV 43B/C medium-frequency (5 kHz) Variable Depth Sonar (VDS) comprises a streamlined towed body containing the sonar transducer array. It is equipped with stabilisers providing for control in roll, pitch and depth. The submerged weight of the fish is around 8 tonnes. The range of towing speeds is 4 to 30 kt and detection ranges of up to 25 km are quoted. The VDS can be towed at depths between 10 and 200 m at speeds up to 24 kt (depths of 700 m in the C variant). The length of tow cable is 600 m.

Operational status
The DUBV 43B is in service in the 'Suffren' class and Type F 67 and F 70 (DD 640-642). The DUBV 43C is in service with the remaining Type F 70 frigates of the French Navy. It is also installed on the Chinese 'Luhu' class.
    Two updates have been added to the system:
(a) UTCS 1B which includes improvement of the signal processor
(b) a URDT torpedo alarm receiver.

Contractors
Direction des Constructions Navales, Paris.
Thomson Marconi Sonar SAS, Sophia Antipolis.
Safare Crouzet, Nice.

UPDATED

DUBV 43C VDS towed transducer and towing gear aboard the French corvette Georges Leygues

Anaconda/TSM 2980 TACTAS system

Type
Towed array tactical sonar.

Description
Thomson Sintra Activités Sous-Marines has been contracted to supply an export version of a tactical towed array sonar derived from the DSBV 61B, currently in service with the French Navy, with an advanced technology sonar receiver built around the Mustang processor. This latter equipment was developed by Thomson Sintra Activités Sous-Marines to increase the performance of all sonars, particularly in automatic detection, narrowband analysis, automatic tracking, adaptive beam-forming and self-noise cancellation.

Operational status
Eight systems have been delivered to the Royal Netherlands Navy 'Karel Doorman' class frigates.

Contractor
Thomson Marconi Sonar SAS, Sophia Antipolis.

VERIFIED

Diodon sonar/TSM 2630

Type
Hull-mounted and variable depth sonar systems.

Description
Diodon is a panoramic medium-frequency sonar for small- and medium-sized surface ships. It offers high-level performance in signal and data processing, target classification, automatic detection and tracking and man/machine interface.
    The active sonar is as efficient in shallow waters as in deep and is very easy to use. Control is exercised through a one-man console using a dual-screen

Diodon sonar transducer array   1998/0007224   *Diodon sonar system, showing (left to right) display console, receiver and power supply*

high-resolution colour display. The array can be fitted within a fixed dome or supplied as a Variable Depth Sonar (VDS); in case of a combined system, the same electronics suite can operate both the hull-mounted and VDS arrays. Embedded facilities include passive listening, sound ray tracer, an acoustic propagation prediction for the day and a recording system that can be used for debriefing and training.

In the VDS version, the Diodon is fitted with a Swan 2000 towing system. Diodon can be combined with other active or passive TSM sonars.

Operational status
In service with a number of navies including: Argentina, Ecuador, Portugal and Saudi Arabia.

Specifications
Acoustic array: 24 identical staves; weight 450 kg
Frequency: 11, 12, or 13 kHz (selectable)
Transmitter: omnidirectional transmission in CW and FM modes
Receiver: 36 preformed beams; advanced processing with replica correlator in FM and spectrum analysis (FFT) in CW

Video processing: data processing in FM and CW modes: automatic detection and tracking; simultaneous multitrack extraction (up to 64); target analysis (apparent length, Doppler measurement). The sonar is easy to link to the own ship combat system
Display: high-resolution flicker-free colour TV monitors
Total weight: 1,500 kg (additional 8 t for the winch and handling system in the VDS version)

Contractor
Thomson Marconi Sonar SAS, Sophia Antipolis.
UPDATED

Spherion B sonar/TSM 2633

Type
Bow-mounted sonar systems.

Description
Spherion is a long-range, medium-frequency active sonar for ASW surface combatants. It comprises a spherical bow-mounted array with an electronic processing system that allows for real-time computation in transmission and reception modes; three-dimensional stabilisation to compensate for pitch, roll and yaw in the parent vessel; and a beam-tilting capability to counter adverse sound velocity profiles and strong reverberation. Spherion also includes a torpedo alert function.

The sonar can be combined with the medium-frequency Diodon VDS, sharing the same electronic cabinets and console.

Data processing in FM and CW modes provides automatic detection and tracking and simultaneous multitrack extraction on up to 64 targets; target analysis is carried out using apparent length and Doppler measurement. The sonar is easy to link to the own ship combat system

Operational status
In production and operational at sea. Spherion has been chosen by the Royal Norwegian Navy to equip the 'Oslo' class frigates and also by the Australian and New Zealand navies to equip the new 'Anzac' class frigates. Three other navies (India, Malaysia and Taiwan) have also selected the Spherion system. A total of 25 systems has been ordered.

Specifications
Acoustic array: Spherical with 160 individual transducers
Frequency: 7 kHz (approx)
Transmitter: omnidirectional and sectoral transmissions in CW and FM modes, various modes including TRDT
Receiver: 36 preformed beams stabilised in azimuth and elevation; advanced processing with replica correlator in FM and spectrum analysis (FFT) in CW
Display: high-resolution, flicker-free colour TV monitors
Total weight: 4,000 kg

Contractor
Thomson Marconi Sonar SAS, Sophia Antipolis.
UPDATED

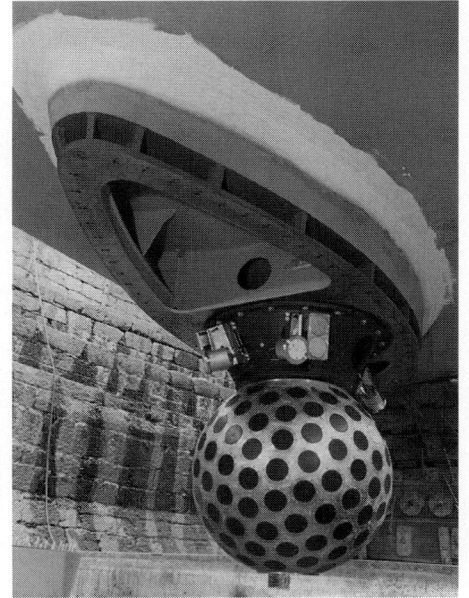

Bow-mounted array of Spherion sonar   1996

Lamproie sonar/TSM 2932

Type
Passive towed array sonar system.

Description
The Lamproie linear towed array is the commercial variant of the DSBV 62 sonar (see above). The array is used for passive detection, classification and position fixing of underwater targets using a very low frequency. Panoramic broadband and narrowband analysis of the detected signals is performed continuously and displayed on two high-resolution colour TV screens.

The 74 m long, 80 mm diameter array is designed to be towed by either a submarine or surface ship behind

a 300 m long cable. When used by a surface ship the array can be towed either behind a long, heavy cable (critical angle tow), or behind a composite system. In the latter case, the array is towed by a long, neutrally buoyant cable connected to a depressor, which itself is then connected to the towing vessel through a short, heavy cable.

Lamproie is able to track several targets concurrently, either in broad or narrowband, with a very fine bearing accuracy. Tracking can be automatic or manual. An automatic rejection system for narrowband countermeasures protects trackers from sonar jamming. Narrowband analysis can be performed concurrently on several beams, steered either under automatic tracking or manually. Lofar analysis includes

refined vernier capabilities. Demon analysis demodulates signals for cavitation detection. A high-fidelity audio channel improves target identification. Automatic torpedo warning is also provided. The sonar array is handled by an Angler winch.

Operational status
Systems delivered to two countries.

Contractor
Thomson Marconi Sonar SAS, Sophia Antipolis.
UPDATED

Salmon sonar/TSM 2640

Type
Hull-mounted and variable depth sonar systems.

Description
The Salmon sonar is a compact, medium-range, high-frequency active sonar designed for light patrol vessels conducting ASW operations mainly in shallow waters. It offers high-detection probability in severe noise environment and includes sophisticated signal and data processing, automatic detection and tracking functions and operator aids.

The basic version consists of an acoustic array of 24 staves housed in a streamlined towfish body, a towing winch, a transmitter and a receiver unit and an operator console equipped with a multicolour display. Embedded facilities include passive listening, sound ray tracer, integrated onboard simulator and an acoustic propagation prediction for the day and a recording system.

A Salmon VDS sonar fitted on the Danish frigate Thetis. The illustration shows the vessel before having her stern modified to a transom configuration

Data processing in FM and CW modes provides automatic detection and tracking and simultaneous multitrack extraction on up to 64 targets; target analysis is carried out using apparent length and Doppler measurement. The sonar is easy to link to the own ship combat system

A hull-mounted version is also available.

Operational status
In production. Two systems have been installed on the 'Stockholm' class and four of the improved variant on the new 'Goteborg' class corvettes of the Royal Swedish Navy. Salmon systems are also being supplied to the Royal Danish Navy for fitting on board the Stanflex 2000 'Thetis' class fishery patrol vessels and the Stanflex 300 'Flyvefisken' class patrol vessels.

Specifications
Acoustic array: 24 identical staves of 7 transducers; weight 260 kg
Frequency: 19 kHz (approx)
Transmitter: omnidirectional transmission in CW or FM modes
Receiver: 36 preformed beams; advanced processing coherent in FM with pulse compression and spectrum analysis (FFT) in CW
Display: two high-resolution flicker-free colour TV monitors
Total weight: 7,630 kg

Contractor
Thomson Marconi Sonar SAS, Sophia Antipolis.

UPDATED

SLASM TSM 2670

Type
High-performance integrated sonar system.

Description
SLASM (also referred to by the French Navy as ATBF) is a high-performance integrated sonar system for large ASW surface ships (frigates or destroyers), comprising the following main subsystems or functions:
(a) a very low frequency, deep diving, active towed array sonar
(b) an active low frequency (5 kHz) hull-mounted sonar
(c) an active medium-frequency variable depth sonar
(d) a passive very low frequency towed array sonar
(e) a torpedo alert system, using the VDS or towed array
(f) sonobuoy processing.
The system consists of:
(a) a large, deep diving, towed fish housing the very low frequency transmitter array and the low-frequency transmitter/receiver arrays
(b) a long linear array, to be towed clipped on to the towfish, or towed independently
(c) a hull-mounted low-frequency transmitter/receiver array
(d) a large towing and handling system, a by-product of the existing DUBV 43 system.
The towfish houses the 1 kHz transmitter and 5 kHz transceiver. Behind the towfish is towed the VLF receiver and a torpedo warning receiver. The transducers in the towfish are mounted on two vertical panels covering 120 to 150° on either side.

Signal and data processing are performed through three identical cabinets, each being able to process any function of the system; two more cabinets house the recording and simulator system, the post-processing and the system controller. Detection is performed out to the first and second convergence zones.

The man/machine interface consists of nine multifunction consoles and two dedicated monitors.

The array is towed out to 600 m astern of the vessel down to depths of 200 m. A very low frequency tactical passive array (DSBV 62C) can be towed 200 m astern of the fish, both arrays forming part of the new SLASM which can operate in bistatic mode.

Towed body for the SLASM system (Courtesy DCN Lorient)   *1999*/0024702

Total weight of the system is 12 tonnes. The ATBF-2 under development is much lighter (2 to 4 tonnes), omnidirectional array operating at between 1 to 1.5 kHz.

Claimed performances are a towing speed slightly in excess of 15 kt and a range in excess of 30 km.

Operational status
The prototype was trialled on the frigate *Cdt Riviere* in 1998.

The SLASM has been ordered by the French Navy to upgrade the F 67 ASW frigates. The first operational unit entered service aboard the French frigate *Tourville* in mid-1995. The sonar has undergone trials in the Mediterranean and Atlantic.

A second version is foreseen, probably through a collaborative venture with some other European navies; this second batch is to be installed in the French Navy aboard the F 70 ASW frigates, by the end of the century.

Contractor
Thomson Marconi Sonar SAS, Sophia Antipolis, (industrial prime contractor).

UPDATED

Gudgeon sonar/TSM 2362

Type
Lightweight panoramic active surveillance sonar.

Description
The Gudgeon TSM 2362 is a lightweight panoramic sonar designed for use in shallow water conditions on small surface ships of 150 tonnes and above. The single-operator system performs surveillance, tracking, classification aid and ASW operations.

Sound ray tracing and maintenance test functions are integrated. The small acoustic unit can be mounted in a fixed or retractable dome, or can be carried in a towed fish.

The basic configuration comprises a dome, cylindrical acoustic array, power supply/transmitter cabinet, receiver cabinet, display console and, in the VDS version, a Swan 1000 type towing system.

The acoustic array, an adaptation of the HS12 dipping sonar array, comprises a cylinder consisting of 12 staves each of six transducers. The transmitter operates in CW or FM in omnidirectional or sectoral modes around the 12 kHz frequency band.

The receiver produces 18 preformed beams with bearing stabilisation. It carries out coherent FM processing with pulse compression and CW spectral analysis. Post-processing includes normalisation, echo sorting, correlation in time and space and automatic target location and tracking. It also features directive audio for echo classification. The sonar uses processing algorithms developed for ATAS and Spherion.

The console features a dual, multicolour, non-flicker high-definition TV monitor, data being presented on a PPI display with multiping or extraction video and a tactile touch-sensitive keyboard. The console also features a double zoom target analysis function measuring apparent length plus Doppler measurement. Automatic tracking on up to 64 targets is achieved with range scales of 1, 2, 4, 8, 12 and 16 km.

The sonar also features integrated sound ray tracing and prediction of detection range.

Onboard recording and training functions are also incorporated.

Hull-mounted and VDS versions of Gudgeon are easily connected and, depending on the associated circuitry, the two sonars can operate simultaneously or alternately.

Operational status
As of March 1993, seven hull-mounted sonars had been sold to Singapore to equip the first six units of the 'Fearless' class and one shore-based unit for training.

Contractor
Thomson Marconi Sonar SAS, Sophia Antipolis.

UPDATED

Albatros

Type
Stand-alone torpedo alert system.

Description
The Albatros torpedo alert system is an advanced modular sonar system specially designed for torpedo detection and classification.

It includes a permanent, panoramic and completely automatic torpedo surveillance function which delivers pre-alert information. This warns the operator of a threat, allowing him to start and control the final classification phase performed in the direction given by the prealert.

The prealert function is achieved using a low-level panoramic beam selector, Sycat and a medium-level classification stage which sifts actual torpedo signals from false alarms.

The Sycat neural algorithm has been designed to provide a quick assessment of torpedo/non-torpedo classification, with special attention being paid to ships which radiate high-frequency spectral lines.

The basic system comprises:
(a) a short special towed array
(b) a small-size Angler-type towing and handling system
(c) a processing cabinet
(d) a display.

The main characteristics of the system are:
(a) large azimuth coverage
(b) permanent surveillance
(c) automatic prealert
(d) bearing warning without port/starboard ambiguity
(e) final classification and decision under operator control.

Operational status
Development has been completed.

Contractor
Thomson Marconi Sonar SAS, Sophia Antipolis.

UPDATED

Muross (Al Spherat)

Type
Sonar suite to equip frigates with ASW and MCM functions.

Description
The Muross (MUltiROle Sonar System) sonar suite comprises a number of acoustic sensors including:
(a) Spherion B/TSM 2633 active hull-mounted sonar
(b) ATAS V1, 2, or 3/TSM 2681, 2682 or 2683 active towed array sonar
(c) Petrel/TSM 5424 mine avoidance and safe route computation system (under development)
(d) Satar torpedo alert system.

Muross is being developed to provide frigates with the complete set of functions necessary to face modern threats. This suite 'federates' existing sensors, detection and classification means through the use of an upgraded version of the Thomson Marconi Sonar SAS standard Colibri console.

Operational status
Concept proposal.

Contractor
Thomson Marconi Sonar SAS, Sophia Antipolis.

VERIFIED

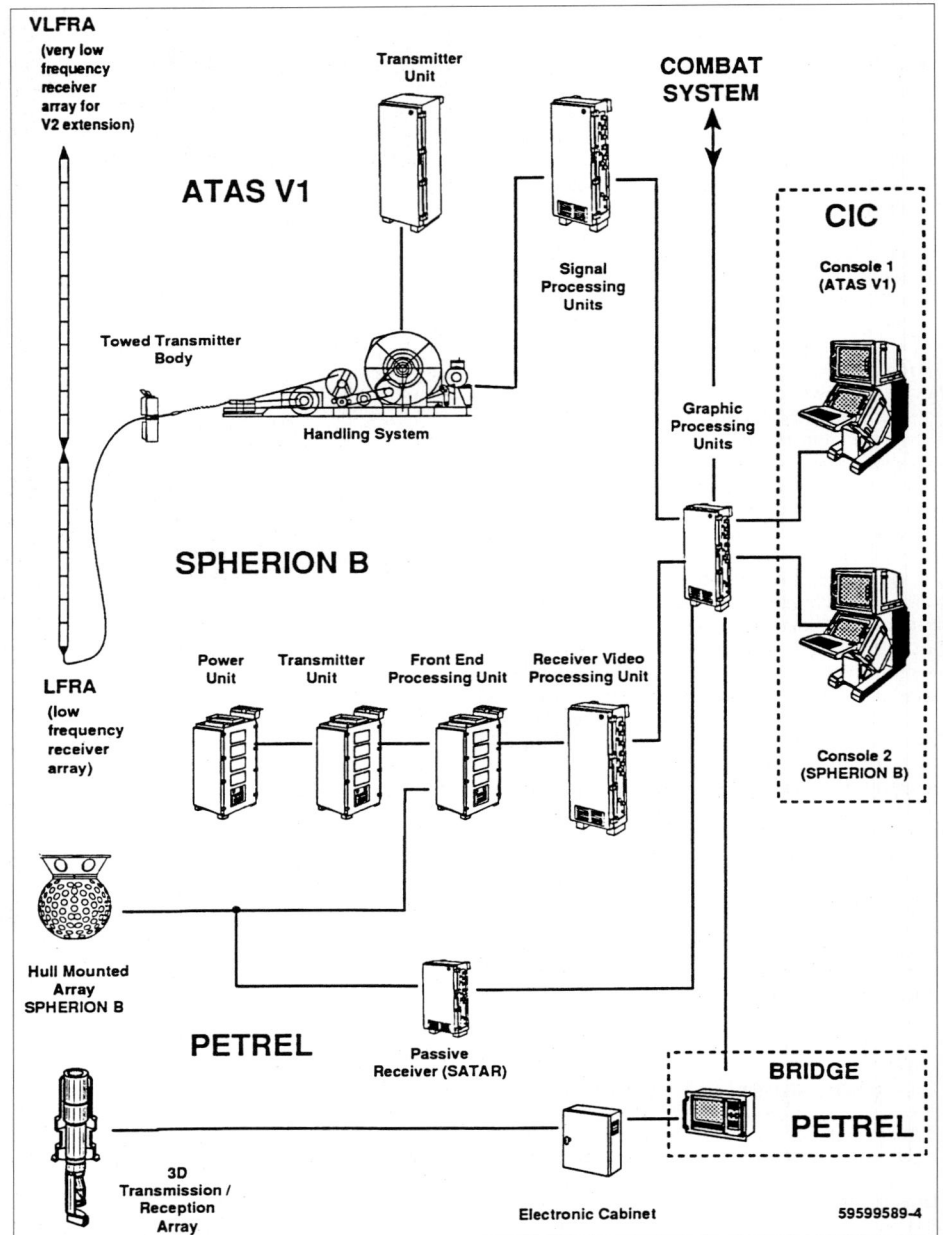

Block schematic of the Muross sonar suite
1996

CAPTAS 20/TSM 2651/TSM 2652

Type
Combined Active/Passive Towed Array Sonar (CAPTAS).

Description
CAPTAS 20 is a low-frequency active/passive sonar for surface ships. Two configurations are available which feature different functions:
(a) Version V1: TSM 2651, ASW active sonar (frequency range 1.4 to 2.2 kHz) concurrently operating with a passive sonar for self-defence against torpedoes.
(b) Version V2: TSM 2652, same functions as version V1, plus a tactical very low frequency passive towed array sonar (frequency range 10 to 2,000 Hz).

The CAPTAS 20 is one of the basic sensors of MUROSS (MUltiROle Sonar System, see separate description), which 'federates' several sonars (hull-mounted and variable depth sonars) controlled by the ASWTA (Anti-Submarine Warfare Tactical Assistance).

The sonar has been specially designed to cover deep water as well as shallow water applications. The handling system and cable length make it possible to deploy the arrays at depths well below most of the surface layers (30 to 1,000 m operating depths). The towed body shows a remarkable stability at any selected depth which enables its use in very shallow coastal waters (around 50 m).

The CAPTAS 20 V1 uses a special linear passive array for receiving echoes from active array transmissions and for detecting radiated noise of torpedoes. The use of trielement hydrophone units

permits target bearings to be measured in real time with no left-right ambiguity, from 30 to 150°, to port and starboard. This major capability has been sea-proven on the ATAS VI. This feature is critical under torpedo attack when detection and response time is of the utmost importance or when detecting at long range in the convergence zone, when the number of perceived echoes is often low and not compatible with a ship's manoeuvres.

CAPTAS V2 has a similar functionality to V1, but with the capability of detecting submarines passively with the lower frequency receive towed array which has greater gain at the lower frequencies.

The transmitting body houses wideband ceramic transducers in a free flooding ring configuration which provides a high source level for detection under low noise conditions and a wide FM bandwidth (1 loctave wide with 3 dB source level) for operation in conditions of strong reverberation. The unit operates in either radial or cavity mode, or both modes can operate simultaneously.

The winch is of new design and moves back on rails to lower the body into the water which leaves only the cable to be paid out.

The towing cable set comprises a cable fitted with fairings whose very low drag makes it possible for the transmit array to prove highly stable when towed at a great depth and a neutrally buoyant horizontal cable to position the receive linear array far behind the ship, which strongly reduces the influence of ship radiated noise.

The twin ceramic ring version weighs 900 kg.

Operational status
Under development and a prototype has undergone sea trials. The system has been sold to Saudi Arabia.

Contractor
Thomson Marconi Sonar SAS, Sophia Antipolis.

UPDATED

Towed body of CAPTAS 20

1998/0007225

MFS Horizon

Type
Common new-generation hull-mounted sonar for the Horizon programme.

Description
The MFS Horizon hull-mounted, medium-frequency sonar is being developed by Thomson Marconi Sonar SAS. It is a compact, modular system which derives maximum benefit from the use of COTS technology and is intended to support all surface ship ASW operations in both deep and shallow waters.

The multichannel sonar has two configurations, designed to provide appropriate levels of performance:
(a) the S version, featuring a spherical array
(b) the C version, featuring a large cylindrical array.

It represents a comprehensive synthesis of features and capabilities derived from Sonar 2050 (the Royal Navy's principal surface ship sonar) and Spherion B (25 systems in service in a number of navies around the world).

MFS Horizon utilises the latest technologies making it possible to obtain a significant reduction in size of the inboard equipment and overall costs of support.

Both versions feature a wide range of operational functions which match the various missions/situations encountered at sea. These include:
(1) anti-submarine warfare in active and passive modes
(2) self-defence against moored mines (active mode) and torpedoes (passive and active modes)
(3) navigation aid in unknown areas
(4) intelligence collection on surface ship signatures using over-the-horizon intercept
(5) acoustic communication via underwater telephone.

Both versions feature fully stabilised transmit/receive beam-forming in all modes together with vertically steerable beams to take account of the environmental conditions. The sonar comprises one amplifier per transducer. The various active transmissions are carried out in Omni, Sector and TRDT modes.

The receiver processes two active channels simultaneously, providing significant flexibility in the use of various transmit codes allowing scope for optimal matching to the threat type and environment.

The sonar is controlled by a single operator at a twin-display X-Windows console.

Operational status
Under development.

Contractor
Thomson Marconi Sonar SAS, Sophia Antipolis.

VERIFIED

U/RDT-1A passive sonar

Type
Wideband passive torpedo detecting sonar.

Description
The U/RDT-1A is a preformed beam passive sonar, designed specifically for the detection of the noise level radiated by torpedoes. The level and frequency spectrum is determined by an operator with a colour video screen for panoramic surveillance, a monochrome classification video screen and an audio system for received-noise listening. An automatic alarm alerts the operator when a new source is detected. The system normally uses the array of the active sonar (VDS on hull-mounted sonar), but can use an independent passive array. The system detects torpedo-generated noise at ranges up to 10 km. It can be adapted to most existing types of array:
(a) 24, 36, 48 (and so on) staves
(b) 1 to 10 kHz bandwidth

Operational status
In service with the French Navy.

Contractor
Safare Crouzet, Nice.

VERIFIED

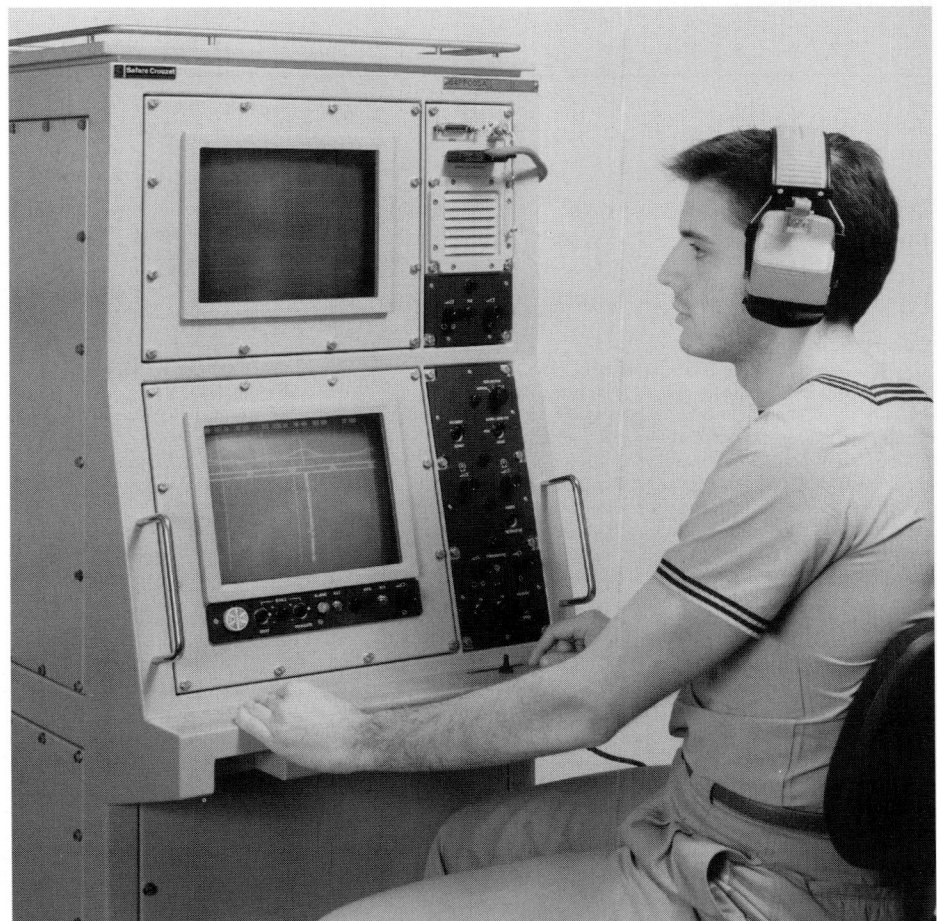

The operator console of the U/RDT-1A sonar

Satar

Type
Shipboard automatic torpedo alert receiver.

Description
Satar, an enhanced version of the URDT-1A torpedo alert system currently in service with the French Navy, adds automatic torpedo alert capabilities to existing hull-mounted or variable depth sonar systems. It integrates recent developments in automatic alert technology and forms an essential part of the anti-torpedo defence system for surface vessels.

The system provides two main functions:
(1) fully automatic, permanent, panoramic surveillance, operating simultaneously with active sonar mode to provide automatic early warning of torpedo attack and
(2) classification, operating in either manual or automatic mode. In manual mode the system operator has control of audio, Lofar, Demon and the results of automatic torpedo early warning

detection. The automatic mode draws on the results obtained by automatic torpedo early warning detection.

Torpedo alert can be initiated in one of two ways:
(1) by the operator, in manual mode or;
(2) automatically. In automatic mode, Satar uses the neural network processor Sycat.

The system comprises two units – a signal preprocessing cabinet located in the hold which processes signals from the sonar array; and a signal and graphic processing unit and display console in the CIC. The latter performs signal processing, data processing and display processing functions. The system connects to the existing array via a junction box.

In fully automatic mode the display presents:
(a) panoramic broadband waterfall in azimuth/time co-ordinates with panoramic instantaneous energy level curve. The waterfall shows both raw data and synthetic track data
(b) a second waterfall corresponding to the Lofar and Demon of the tracks currently in preclassification and classification processing with associated

classification marks. In manual mode, the second waterfall displays the Lofar and Demon of the beam in classification and steered by the operator.

Satar is designed for use with various types of Thomson Marconi Sonar SAS sonar (for example Spherion, Diodon, Gudgeon and so on) and can also be used with other types of hull-mounted or variable depth sonar (depending on the array characteristics).

Operational status
Development completed. The Satar is a functional part of existing sonars. It is proposed as a component of the Muross (MUltiROle Sonar System), a complete system intended to equip frigates.

Contractors
Thomson Marconi Sonar SAS, Sophia Antipolis.
Safare Crouzet, Nice.

VERIFIED

Swan

Type
Towing and handling systems.

Description
Swan towing and handling systems constitute a family of equipments adapted to various sizes of VDS. Swan systems include:
(a) towed bodies, adapted to sonar array sizes
(b) towed faired cables
(c) winch assemblies (winches, poles, hydraulic power units)
(d) winch operator consoles.

Main characteristics of the Swan family are:
(a) active stabilisation of the towing system, which compensates for the ship's stern movements. In conjunction with a highly stable towed body, it enables the VDS to operate in rough seas
(b) high survival towing speed
(c) safe and easy operations, with secured launching and recovery sequences (only one man is required for operations)
(d) compact size.

To date, the Swan family consists of three models: Swan 200, Swan 1000 and Swan 2000, respectively adapted to Gudgeon, Salmon and Diodon sonars. Derivatives are possible for other sonar types.

Three versions have been ordered or delivered to date, in order to meet own ship characteristics:
(a) a turntable version, to be installed upon the deck
(b) a sliding or railed version, to be installed below the helicopter deck
(c) a containerised version.

Operational status
Operational at sea.

Contractor
Thomson Marconi Sonar SAS, Sophia Antipolis.

VERIFIED

Angler

Type
Towing and handling systems.

Description
Angler towing systems constitute a family of equipments designed to be adapted to various types of linear towed arrays.

Main characteristics are:
(a) compact and lightweight design
(b) system operated by a single operator most of the time
(c) easy and fast onboard installation
(d) a slip ring allows permanent sonar operation during handling.

The Angler family comprises different models adapted to the various towed sonars (cables, arrays),

one of them being the TSM Lamproie (TSM 2932) passive towed array sonar.

Operational status
Operational at sea.

Contractor
Thomson Marconi Sonar SAS, Sophia Antipolis.

VERIFIED

GERMANY

DSQS-21 (AS0 83-86) series sonars

Type
Hull-mounted and variable depth sonar systems.

Description
The DSQS-21 series sonars (also known as the AS083 to 86 series) are designed for operation in surface ships as part of an anti-submarine weapon system. For operations against submarines below the thermal layer, the sonar can be supplemented with a towing system providing a Variable Depth Sonar (VDS) capability.

Computer-aided detection techniques are used for classification and tracking and the information is presented on colour CRT displays to permit Doppler coding and the discrimination of data on the integrated

Specifications

Sonar		Diameter of transducer arrays		No of preformed beams		Class of ship
		HMS	VDS	HMS	VDS	
DSQS-21	B/BZ	1.8 m		64		Corvettes, frigates
	BZ/VDS	1.8 m	1 m	64	32	Corvettes, frigates
DSQS-21	C/CZ	1 m		32		Corvettes, frigates
	CZ/VDS	1 m	1 m	32	32	Corvettes, frigates

displays. A 'Z' in the designation indicates variants equipped with electronic stabilisation to minimise the effects of ship's motion.

The VDS is designed to alternate with the hull-mounted sonar and the displays of the two sonars are integrated. Fitted with automatic depth control and heave compensation for stability of the 'towfish', the VDS can be handled automatically or manually.

Operational status
In service but no longer in production. Fitted to destroyers and frigates of the German Navy. More than 60 systems have been delivered to 15 countries.

Contractor
STN ATLAS Elektronik GmbH, Bremen.

UPDATED

ASO 90 series

Type
Hull-mounted and variable depth sonar systems.

Development
The ASO 90 series (comprising DSQS-23 and DSQS-24) has been developed from experience with the DSQS-21 systems which were built in various versions.

Description
The ASO 90 systems feature electronically stabilised cylindrical transmit and receive arrays with a large number of transducers. Arrays are either hull-mounted or variable depth types (ACTAS, TAS or VDS) and are variously designed for low-, medium- and high-frequency applications, according to ship type. Variable depth type sonars, either in a combination with hull-mounted or as stand-alone systems, may be used in the event of adverse sound propagation conditions.

Advanced detection techniques for parallel, active and passive mode operation over 360° azimuth combined with spatial and temporal signal processing using the latest technology, enable the detection of submarine echoes even in difficult waters. The usual techniques of Doppler filter bank (continuous wave) and pulse compression (frequency modulation) are supplemented by special pre- and post-processing of the signal data for full-colour presentation on a multifunction operator's console which incorporates a window display surface and open system architecture, typical for all ASO 90 systems. Incorporating automatic fault detection and location functions, complete stand-alone or integrated assemblies can be configured for interface to other ships' systems including onboard or ship-based simulators, as well as recording/replay units.

The ASO 90 is designed for computer-aided detection, localisation and classification over an entire 360° azimuth. The modular design of the systems enables them to provide, as required: underwater telephone; automatic target tracking; classification; ray trace; intercept; torpedo warning; mine avoidance; vertical stabilisation; simulation; interface to own ship's fire-control system; interface to ACTAS, TAS or VDS systems. ASO 90 can be extended to a fully integrated ASW system.

Specifications

Sonar system	HMS array	VDS array	Frequency	Recommended ship size
ASO 95	2.5 m	1.2 m	Low	Destroyers, frigates
ASO 94	1.0/1.2 m	1.0/0.6 m	Medium	Frigates, corvettes
ASO 93	0.6 m	0.6 m	Medium	Corvettes, off-shore patrol vessels
ASO 94 C	1.0 m	-	Medium	Off-shore
ASO 93 C	0.6 m	-	Medium	Patrol vessels

Operational status

Current range of sonars in service with the German Navy as DSQS-23 and other customers. Contracts received for DSQS-24 type sonars for the new German Type 124 ('Sachsen' class) frigates and Netherlands 'De Zeven Provincien' class frigates. Four DSQS 24C systems have been ordered for the Netherlands frigates with delivery due in 2000.

Contractor

STN ATLAS Elektronik GmbH, Bremen.

UPDATED

ASA 92 series

Type

Active towed array sonar system.

Description

The ASA 92 Activated Towed Array Sonar (ACTAS) is a compact, lightweight system intended to support anti-submarine operations and surface ship surveillance.

Among the main features of the new system are:

(a) a long-range detection capability using low-frequency transmission (the pilot system is understood to have achieved detection ranges at around 45 km)

(b) an ability to operate in thermal layers at various depths by employing submersible antennas for transmitting and receiving

(c) the detection of quiet submarines using active transmission

(d) the ability to use the passive mode simultaneously with the active mode for the detection of noisy targets such as torpedoes, surface ships and noisy submarines

(e) shallow and deep water operation using an optimised array and handling system configuration.

The ASA 92 ACTAS can be operated fully in parallel, in both active and passive mode.

In active mode the sonar operates in a low-frequency range with omnidirectional transmission including instantaneous right/left discrimination achieved through the twin line receiving array arrangement. The active array transmits at around 2 kHz (with a 400 Hz bandwidth) with an output power in excess of 200 dB. The passive receive array covers the 50 Hz to 2.5 kHz frequencies with a selectable bandwidth of 1,300 Hz. Both CW and FM pulses are available for the detection and classification of fast-moving targets with high Doppler, as well as slow-moving and stationary targets with low or zero Doppler.

Targets are displayed in true motion, ship centre PPI display and simultaneously in an echogram, that is a rectangular bearing/range display. Additional display modes are provided for classification.

The passive mode allows full panoramic surveillance to be carried out. Broadband and narrowband processing can be performed in a frequency range typical for towed array sonars. Displays include bearing/time records and – for target related noise analysis (Demon, Lofar) – frequency/time records.

A torpedo warning function is included.

Targets detected in active or passive mode can be assigned to automatic target trackers.

Other system features include an audio channel, recording/replay of sonar data and sonar performance prediction.

The body can be towed at depths between 20 and 350 m.

The ASA 92 ACTAS towing system is a compact design of approximately 10 tonnes total weight and is designed to be mounted below a helicopter flight deck, as well as on the deck aft. To facilitate paying out the receive array and cable, the tail of the towed body separates. Once the receive array is deployed the tail is bolted back on to the main body. The ASA 92 ACTAS system can be integrated with any ASO 90 type of sonar.

Operational status

The ACTAS system underwent sea trials off Norway between September 1996 and February 1997 and in Chile in July 1998.

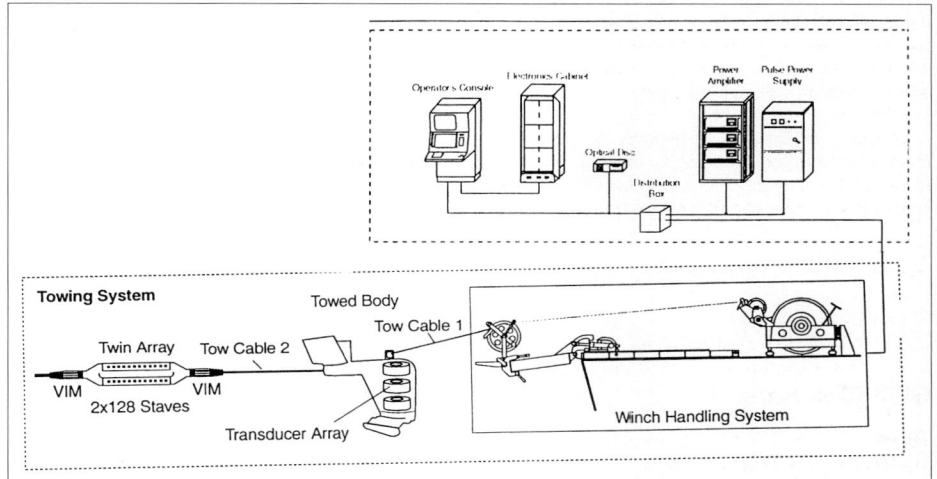

ASA 92 ACTAS general arrangement   *1998*/0007228

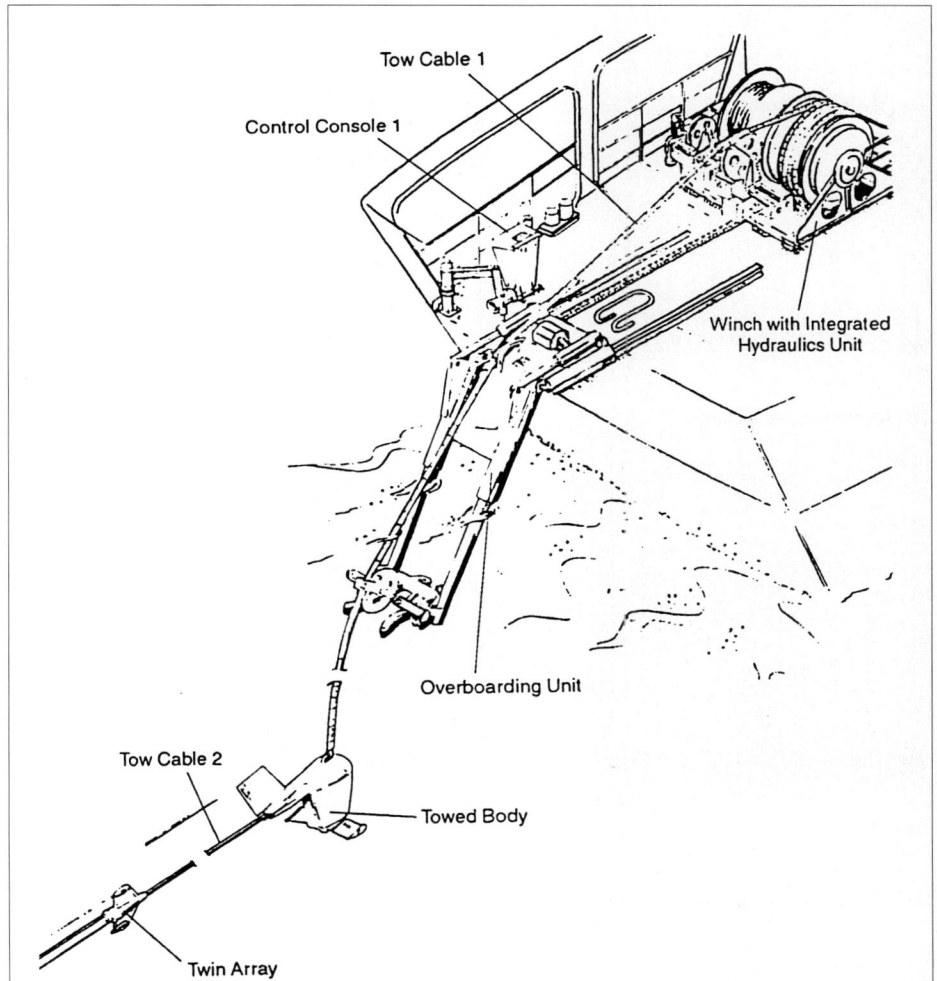

ASA 92 ACTAS towing arrangement   *1998*/0007229

Contractor

STN ATLAS Elektronik GmbH, Bremen.

UPDATED

TAS 90 series

Type

Passive towed array system.

Description

The TAS 90 series is designed for installation on surface ships and operates down to very low frequencies.

The sonar uses the latest spatial and temporal signal processing techniques and can detect weak signals from submarines, torpedoes and surface ships even in difficult acoustic conditions. Data extraction and presentation is on full-colour, raster display and provides a permanent overview of the tactical situation. Detailed analysis of detected targets is carried out using Lofar, Demon and SPECTRUM techniques. Detection, tracking and analysis operations by the operator are facilitated by special features such as surveillance channels, independent broadband/ narrowband facilities, frequency/time records with long time history and magnifier and transient detection. The TAS 90 series is fully compatible with the ASO 90 active sonars either as a stand-alone system or an integrated version.

Various combinations in lengths of cable and streamer stored on an easy to operate winch are provided. The streamer has a long acoustic aperture, with a large number of preformed beams.

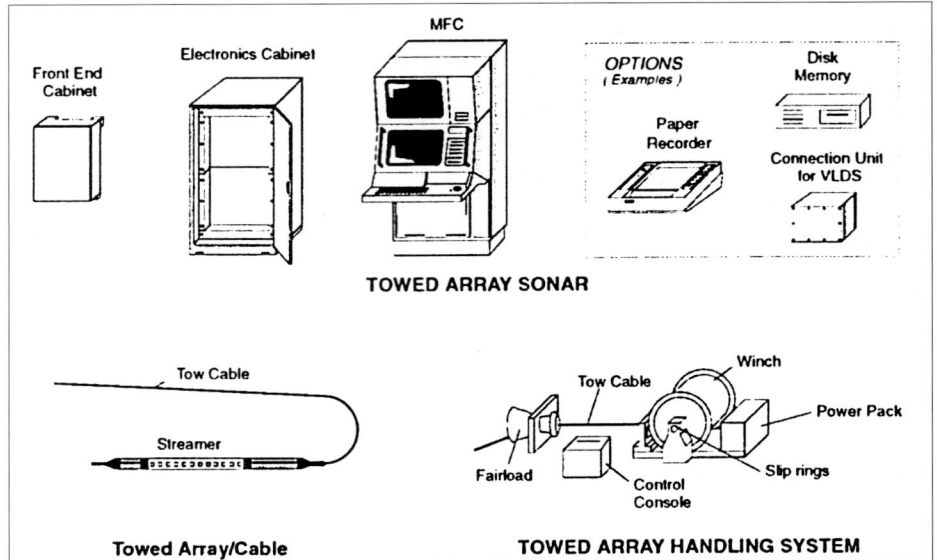

Passive TAS system arrangement   *1998*/0007230

Contractor

STN ATLAS Elektronik GmbH, Bremen.

UPDATED

INTERNATIONAL

GETAS sonars

Type

Towed array passive sonar system.

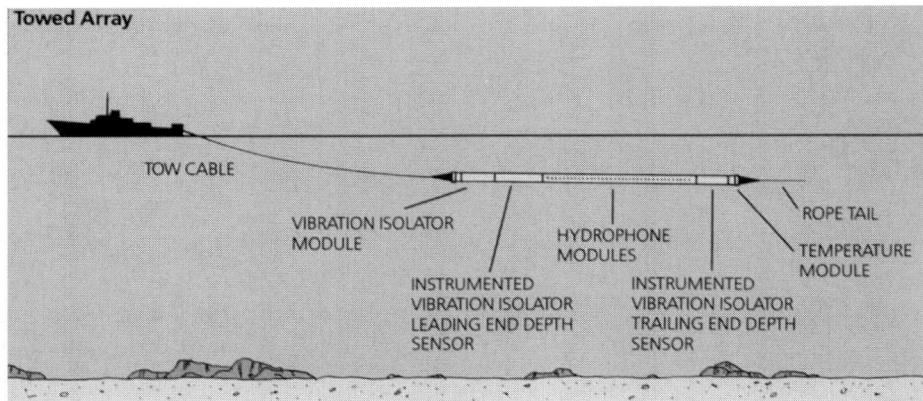

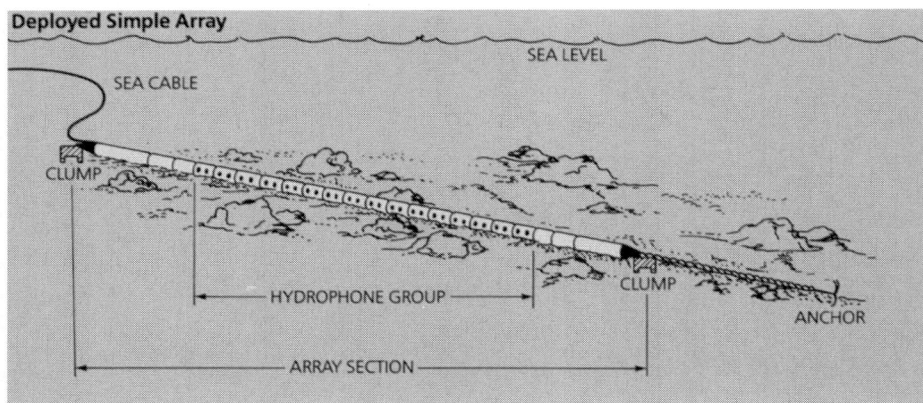

GETAS diagrams

Description

The GETAS series of passive sonars has been developed and proven against quiet submarine and surface targets in both a towed array and bottom-laid version. The systems provide a fully integrated comprehensive solution to passive sonar detection, tracking and classification. They are designed for single-operator manning.

Thomson Marconi Sonar Ltd provides the 'dry end' processing and display equipment and Geophysical Company of Norway (GECO A/S) provides the 'wet end' components.

The onboard or shore-based equipment comprises a signal conditioning unit, broad and narrowband data processors, touch-sensitive control panels and waterfall displays housed in a single cabinet. A number of features has been incorporated including BITE, automatic track followers, remote hard-copy and tape recording outputs. In addition, real-time clock and date, array heading, depth and tension and seawater temperature and salinity are available on the display.

There are two variants of the shipborne 'wet end', each consisting of a winch and cable with acoustic vibrator, isolator and instrumentation array sections. The all-up weight of the standard system is 8,550 kg and of the small ship version 2,155 kg.

A submarine version is also available. The bottom-laid version is similar to the standard but modified for abrasion resistance and negative buoyancy. All systems are easily deployed and recovered.

Specifications

Acoustic section
Overall length: 100 m
Operational speed: 8-12 kt
Operating depth: 200 m
Survival depth: 1,200 m
Towcable
Length: 1,000 m
Processors
Broadband: display mode waterfall; X-axis bearing, Y-axis time
Narrowband: waterfall display

Contractors

Thomson Marconi Sonar Ltd, Templecombe, UK.
GECO A/S, Kjorbokollen, Norway.

VERIFIED

ISRAEL

CORIS-TAS

Type
Passive towed array sonar for surface vessels.

Description
The CORIS-TAS LF, critical angle towed array sonar is based on a towed array fitted with wide-aperture hydrophones receiving LF acoustic signals radiated from surface vessels and submarines. Three main functions are performed: detection, tracking and classification.

It provides full azimuth coverage and performance capabilities over the 10 to 1,600 Hz frequency band, divided into four sub-bands for maximum array efficiency, with no degradation in the lower sub-band. Broadband detection is carried out with automatic normalisation and jammer rejection and narrowband detection with 0.3 kHz frequency resolution in the lower sub-band. Over-the-horizon detection is achieved by combining LF acoustic detection with advanced signal processing. Broadband, narrowband and Demon tracking are provided and accurate classification is achieved through the relevant spectral displays.

Contractor
Rafael, Haifa.

VERIFIED

ITALY

DE 1167 sonar system

Type
Hull-mounted or variable depth sonar.

Description
The DE 1167 sonar is manufactured under a licence agreement with Raytheon (see separate entry in the USA section). The system configuration includes 7.5 kHz hull-mounted 36 stave transducer arrays, a transducer array dome, four cabinets and a single operator console.

The hull-mounted DE 1167 is an active/passive preformed beam, omni and directional transmission sonar which uses three independent 600 Hz wide FM bands centred at 7.5 kHz and a spatial Polarity Coincidence Correlation (PCC) receiver. The passive mode, which is selected automatically when transmissions are halted, is primarily useful for torpedo detection. A half-frequency simultaneous PCC receiver (bandwidth 3 to 8 kHz) is also part of the system. Optional items include a performance prediction subsystem, an auxiliary display, a remote display and a training/test target remote-control unit. Signal reception and beam-forming are accomplished by broadband analogue circuitry followed by clipper amplifiers for perfect data normalisation. Detection processing, display processing system control/timing and waveform generation are carried out digitally.

Two microprocessors perform display ping history, cursor ground stabilisation and target motion estimation functions for torpedo direction. The modular air-cooled 12 kW transmitter uses highly efficient class A/D power transistor techniques.

Several operational features are unique to a sonar of this size and range:
(a) the display processing incorporates a ping history mode through which the sonar data obtained in three of the previous ping cycles may be retained on the viewing surface, allowing the operator to readily differentiate between randomly spaced noise events and geographically consistent valid acoustic reflectors
(b) the clipped PCC processing permits accurate thresholding of all signals, such that the false alarm rate, or number of random noise indications on the screen remains relatively low and constant over all variations in background noise and reverberation levels, further facilitating contact detection
(c) ground stabilised cursors and target motion analysis permit rapid determination of contact motion over the bottom, an excellent clue to the nature of the contact.

These three operational features, combined with extremely accurate tracking displays, a built-in fault detection/localisation subsystem, performance verification software and test/training provides ease of maintenance, operation and support.

An improved version with three simultaneous linear receivers offers improved shallow water operation and obstacle avoidance.

Operational status
Eight systems are in operational service with the Italian Navy 'Minerva' class corvettes, with one system at the training centre. Four systems are under contract for the Italian Navy.

Specifications
Centre frequency: 7.5 kHz
Receiver type: spatial polarity correlator between 36 pairs of half beams
Beam characteristics: 36 sets of right and left half beams for active and passive detection. Selectable 15° horizontal ×12.5° vertical sum beam for audio listening. 1.25° bearing interpolation for display. Selectable 24° horizontal ×21° vertical passive sum beam for audio listening. 2.5° bearing interpolation for display
Active display: 300 range cells, 288 bearing cells; single and multiple echo history; 4 intensity levels
Passive display: electronic bearing time recorder with medium time averaging. LTA/STA with passive receiver
Track displays: sector scan indicator (1,000 yd × 10°) and target Doppler indicator (1,000 yd × ± 60 kt)
Target data: range 8 yd (6.1 m) resolution. Bearing 1.25° resolution plus active search display 0.1° and 3.3 yd on SSI display. ± 60 kt of Doppler at 1 kt steps on the TDI
Data format: standard: NTDS slow (digital)
Power requirements: active 20 kVA (pulse) at 15% (max) duty cycle 400 V 60 Hz 3-phase
Weight: hull-mounted 7.5 kHz transducer 1,600 kg; GRP dome 750 kg

Contractor
Whitehead Alenia Sistemi Subacquei SpA, Genoa.

VERIFIED

DMSS-2000 sonar systems family

Type
Hull-mounted or variable depth sonar.

Description
The DMSS-2000 sonar systems family is an integrated hull-mounted and variable depth active and passive, search, detection, classification, localisation, tracking and small object avoidance sonar. It provides active direct path sonar capabilities in shallow water and long-range direct path and convergence zone sonar capabilities in deep water environments, as well as passive torpedo detection. In addition, the system incorporates a fully integrated fault detection/fault localisation capability.

The active detection function includes active search, Computer-Aided Detection (CAD) and tracking, sonar target management and contact evaluation capabilities. Ping history and stabilisation in true bearing and in ground are available for displays and controls. The active search function is available both in sector and omnidirectional sectors in FM or CW transmission modes. There are three simultaneous receivers: replica correlator, segmented replica correlator and energy detector.

The passive detection function provides broadband, narrowband and Demon processing for detection and as support to the torpedo detection function. Performance prediction provides the operator with an interactive capability to predict range/depth regions.

The system is capable of full integration within the ship's combat system through NTDS or an Ethernet link and provides a link with navigation sensors (gyrocompass, GPS, log) and underwater telephone equipment.

Both hull-mounted and variable depth configurations can be provided in various frequency combinations: hull-mounted – 3,750 and 7,500 Hz: variable depth sonar – 7,500 and 12,000 Hz.

The system configuration includes the array, transducer dome and a single-operator console both for control and display purposes. Depending on the array configuration, frequency and power, the number of the main cabinets varies from four to 12.

The DMSS-2000 sonar systems family is a joint venture between Whitehead Alenia Sistemi Subacquei and Raytheon in the USA.

Operational status
Prototypes have already been evaluated at sea and the Italian Navy has awarded Whitehead Alenia a contract to supply a system which is currently in production.

Contractors
Whitehead Alenia Sistemi Subacquei SpA, Genoa.
Raytheon Company, Electronic Systems, Portsmouth, USA.

VERIFIED

ISO 100

Type
Wideband intercept sonar.

Description
The ISO is a wideband, high-probability and reliable bearing precision sonar interceptor. The system is based on the use of a multimode type sensor and provides frequency and pulse length measurement across the complete waveband together with accurate bearing data.

Two different configurations of the system are possible:
(1) completely integrated with the main sonar;
(2) an autonomous system with numerical display.

The system also provides an online automatic test function.

Alenia Elsag is also developing a configuration based on the use of a double sensor. In this system it is anticipated that bearing accuracy and detection range will be improved by a factor of 1.5.

Operational status
The ISO 100 system has undergone sea trials during which it has proved its capability, even at high-platform speed.

Specifications
Coverage: 360° (horizontal), 40° (vertical)
Bearing accuracy: better than 5° across the full operating band
Frequency measuring tolerance: 4%
Pulse length tolerance: 2 ms

Contractor
Whitehead Alenia Sistemi Subacquei SpA, Genoa.

VERIFIED

JAPAN

OQS series

Type
Hull-mounted sonar systems.

Description
Most early Japanese ASW ships were fitted with various types of US sonar systems. Based on these US sonars, the Japanese have developed a series of indigenous sonars. These sonars use state-of-the-art digital processing and display technology developed in Japan. The OQS-3 is a licence-built Japanese version of the US SQS-23 sonar. The low-frequency, active, search and attack sonar uses bottom bounce detection. It equips the 'Haruna' and 'Minegumo' class

destroyers. Built by NEC, the OQS-4 (I), which is based on the OQS-101, is a low-frequency, hull-mounted active search and attack sonar which equips the 'Hatakaze' class destroyers. It is a solid-state system with improved signal processing and electronics. The OQS-4A(II) is a medium-frequency sonar which equips the 'Asagiri' class destroyers.

The OQS-101, built by NEC and which equips the 'Shirane' class destroyers, is a low-frequency sonar equivalent to the US SQS-53 system and is mounted in a bow dome. The OQS-102, also manufactured by NEC, is a low-frequency sonar equivalent to the US SQS-53C system and equips the 'Kongo' class AEGIS destroyers. The OQS-5, also manufactured by NEC, is a low-frequency sonar that equips the 'Murasame' class

destroyers. The OQR-1 is a passive towed array very low frequency sonar (TACTASS) which equips the 'Asagiri' and 'Murasame' classes and which is being retrofitted to the 'Hatsuyuki' class.

Operational status
Sonars are operational on Japanese destroyers and frigates.

Contractor
NEC Corporation, Tokyo.

UPDATED

NETHERLANDS

PHS-32 sonar

Type
Hull-mounted search and attack sonar system.

Description
The PHS-32 is a medium-range high-performance search and attack sonar in which the newest technological developments for signal processing and operation are employed and aided by the use of a general purpose computer. The computer yields a compact, lightweight sonar for corvettes from 200 tonnes up to frigate size ships.

The signal processing facilities provided include fast Fourier transformation processing of all preformed beam receiving channels. All data are presented on a single TV-type display, while operation has been much simplified by the use of light pen control. Automatic tracking of up to four targets is provided. These features, combined with a high accuracy, make this sonar very useful in an attack.

A circular transducer permits all-round coverage in various modes of transmission such as: Omni, TRDT, MCC (wide vertical beam), LISTEN (passive with time/ bearing recorder presentation). An audio beam is also available. The system can be delivered in a fixed, retractable or VDS-dome outfit.

The features of the PHS-32 include: single-operator control; display data processing for continuous presentation and memory mode (ping history);

four-target automatic tracking; built-in energy storage for lower peak power demands on ships' mains; integrated online and offline test systems.

Automatic co-ordination of transmitting pulse and vertical beamwidth in rough weather reduces the effect of roll and pitch.

Operational status
In operational service on board ships from fast patrol boats to frigates of Indonesia and the 'Ulsan', 'Po Hang' and 'Dong Hae' classes of South Korea.

Specifications
Frequencies: 3
Pulse lengths: 12.5, 25, 50, 100 ms and 400 ms long pulse (CW or FM)
Detection range: > 10,000 yd
Vertical beamwidth: 12 or 20°
Bearing accuracy: 1° RMS
Range accuracy: ± 0.5-2%
Notch filtering: selectable rejection bandwidth
Own Doppler correction: on all 60 channels
Weights
Electronic cabinets: (sonar console, duplexer and amplifier cabinet, transmitter cabinet) 1,041 kg
Retractable hull-outfit including transducers: 7,900 kg
Fixed hull-outfit including transducer: 2,500 kg
Variable depth system: 5,000-8,000 kg

Lightweight modular transducer of PHS-32 sonar

Contractor
Hollandse Signaalapparaten BV, Hengelo.

UPDATED

PHS-36 sonar

Type
Hull-mounted search and attack sonar systems.

Description
The PHS-36 series of sonar systems is modular in design and intended for worldwide operation on any type of surface ship. The PHS-36 which will be fitted in the M-frigates of the Royal Netherlands Navy is stated to be the first system in the world which is fully integrated into the SEWACO (Sensor Weapon Control and Command) system.

The PHS-36 provides information for the data handling functions of active sonar, passive sonar, acoustic support and track management and other functions such as equipment monitoring. The sound

velocity processing and recording system also provides data for acoustic support.

A colour-coded presentation on a high-resolution display gives a graphic picture of sonar information.

The dual-processing system performs FM and CW processing simultaneously to cope with both noise-limited and reverberation-limited conditions, as well as with a variety of submarine running conditions as provided for ASW scenarios. The main functions performed by the sonar processor are auto-detection, auto-track initiation and auto-track processing.

The flexibility of the modular PHS-36 sonar processor allows easy add on of a towed line array with or without active adjunct or other types of ASW sensor.

Operational status
In operational service with the Royal Netherlands Navy 'Karel Doorman' class frigates.

Specifications
Transmission
Frequencies: two frequencies around 7 kHz
Pulse type: CW, FM, FM + CW (a sequential transmission per interval)
Pulse length: 75-1,200 ms depending on range-scale/ pulse type selections
Reception
Preformed beams: 32
Bandwidth passive mode: Band I low; Band II high

Contractor
Hollandse Signaalapparaten BV, Hengelo.

VERIFIED

NORWAY

SS105 scanning sonar

Type
Hull-mounted active sonar system.

Description
This sonar is intended to fulfil the sonar requirements of

modern ocean-going coastguard vessels. It is a 360° scanning sonar with 48 preformed receiving beams with 11° beamwidth and 'split-beam' processing in each beam. The working frequency is 14 kHz. The sonar system consists of an operator's console, transmitter unit, receiver unit, hydraulic power unit and hull unit.

The transmitter consists of 48 switching type amplifiers, 600 W each. Total output power is approximately 15 kW. Omni, Single RDT and Triple RDT are available transmission modes. The main display has a CRT PPI, 280 mm in diameter, with scale ranges from 2 to 16 km. Markers provided are target cursor, stern cursor, transmitting and receiving sectors.

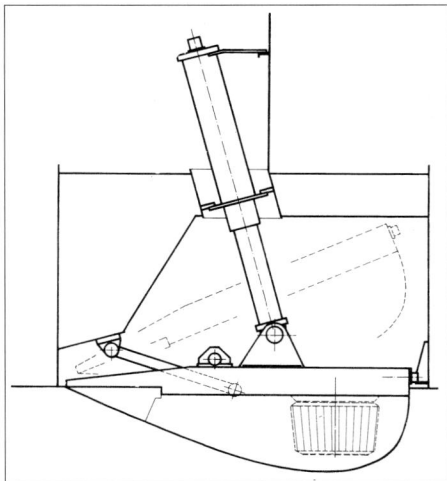

SS105 hull unit

An LED display shows target range, relative/true target bearing and ship's speed. Target data are transmitted digitally to other systems on board as required.

The 48 stave transducer is installed in a streamlined retractable sonar dome, which will take ship's speed up to 25 kt. A fixed dome arrangement is also possible.

Operational status
The SS105 is fitted to coastguard vessels of the Royal Norwegian Navy, the Estonian Border Guard and the Finnish Frontier Guard service. This sonar is no longer in production.

Specifications
Transmitter
Type: Class S amplifier
Number of channels: 48
Max power output per channel: 600 W
Pulse lengths: 10, 30 and 60 ms
Frequency: 14 kHz
Transducer
Number of staves: 48 (circular-mounted)
Active face per stave: 225 cm^2
Beamwidth vertical plane: 12 ± 1°
Resonance frequency: 14 kHz
Tilt: 6° (mechanical)
Transmitting modes and performance
Omni
Beamwidth vertical plane: 12 ± 1°

Beamwidth horizontal plane: 360°
Max output level: 219 dB/μPa/m
Directivity index: 9.5 dB
SRDT
Beamwidth vertical plane: 12 ± 1°, 1 beam
Beamwidth horizontal plane: 8.5 ± 1°, scanning a sector variable frcm 10 to 115°
Directivity index: 25 dB
Max output level: 230 dB/μPa/m
TRDT
Beamwidth vertical plane: 12 ± 1°
Beamwidth horizontal plane: 8.5 ± 1°, 3 beams each scanning a sector of 120°
Max output level: 230 dB/μPa/m
Directivity index: 25 dB
Receiving performance
Number of simultaneous beams: 48
Bandwidths: 400 and 800 Hz
Beamwidth vertical plane: 12 ± 1°
Beamwidth horizontal plane: 11 ± 1°
Directivity index: 26 dB

Contractor
Simrad AS, Horten.

UPDATED

SS 245 series

Type
Series of multibeam, hull-mounted sonars.

Description
The SS 245 series is a family of active high-resolution multibeam sonars designed for naval vessels. The relatively low operating frequency of 24 kHz and high-source level ensures long-range detection. The transmitting modes are omnidirectional, or directive with a single trainable beam. In all modes, both transmission and reception beams are tiltable from +10 to −20° in 1° steps and also stabilised against roll and pitch by the supplied vertical reference units. In addition, the system contains an MCC-mode (Maintenance of Close Contact) with a broad vertical beam in both transmission and reception. Signal processing for CW and FM as well as beam-forming is carried out by a high-speed digital processor using the full dynamic range of the signals. The echoes are presented in 64 beams, interpolated to 128 beams on the display unit.

The various sonars of the series comprise the following units: Control and Display Unit (CDU), Transceiver Unit (TRU), Vertical Reference Unit (VRU) and a selection of different Hull Units (HLU):
SS 245 – hoistable transducer without dome
SS 246 – hoistable transducer with fixed GRP dome
SS 247 – fixed transducer and GRP dome
SS 248 – hoistable transducer and GRP dome
SS 249 – hoistable transducer with inflatable rubber dome.

The SS 245, SS 246 and SS 249 hull units are equipped with a motor-driven screw to hoist and lower the transducer in and out of its operative position. The SS 249 uses a water pump to inflate and deflate the rubber dome. The SS 248 uses a hydraulic system for raising and lowering the transducer and dome. The weight of the hull unit is approximately 150 to 350 kg depending on configuration.

All command and control functions are performed on the control and display unit which incorporates a 20 in raster scan high-resolution colour display, an operator panel and a rack assembly with the display and operator processing circuits.

The MMI uses dedicated control buttons on the CDU keyboard which offer direct access to primary sonar

Control/display unit of the SS 245 sonar system

functions such as range, gain, tracking and so on. Three targets can be tracked simultaneously. Secondary functions are menu-controlled. The menu is displayed alongside the sonar picture and commands are selected via a joystick-controlled cursor on the display.

Two main display modes are available: relative or true motion. Zoom and aspect scan facilities are available in each mode as well as Doppler analysis.

Other options available include passive Demon analysis, ray trace based on a temperature probe input and video recording of the screen.

Operational status
In operation on a number of vessels.

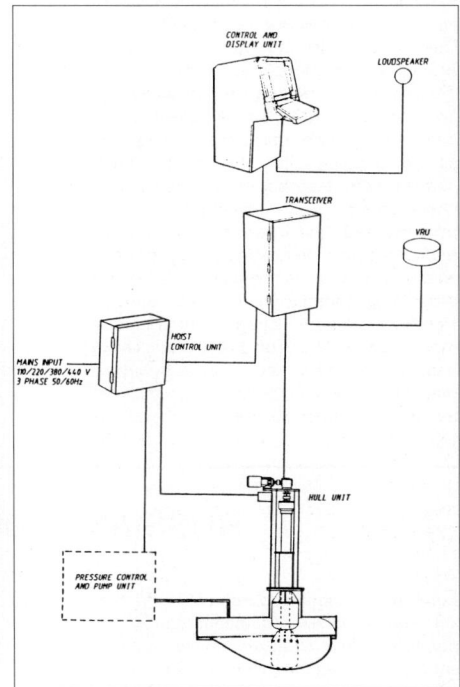

General arrangement of the SS 249 sonar **1995**

Specifications
Frequency: 24 kHz
Beam-forming: 64 digitally formed beams
Beamwidth transmission/source level: single beam – 12° (horizontal): 15° (vertical)/220 dB. Omni – 360° (horizontal); 15° (vertical)/213 dB
Pulse length: CW – 1.25-200 ms (range dependent) FM – (subpulse) 1.25-80 ms (range dependent)
Number of subpulses: 2-8 (selectable)
Range scale: 500-12,000 m

Contractor
Simrad AS, Horten.

VERIFIED

SS 575 sonar series

Type
Medium-range hull-mounted active sonar systems.

Description
The SS 575 series of active high-resolution sonars is designed for medium-range detection and tracking of submarine targets. Detection is achieved using medium-frequency (57 kHz) and high source level

combined with CW or FM signal processing. The transducer elements are mounted in a semi-spherical array offering unlimited possibilities for stabilised beam configurations under and around the vessel.

The transmitting modes are omnidirectional or directive with a single trainable beam. In all modes, both transmission and reception beams are tiltable from +1 to −90° in 1° steps and also stabilised against roll and pitch by the supplied vertical reference unit. The sonar images are presented either in relative or

true motion on the full screen and with the option of a split screen operation with vertical display. Within the different modes the operator can select zoom, aspect scan or Doppler analysis submodes. Signal processing and beam-forming are performed by a high-speed digital processor using the full dynamic range of the signals. The display is a 20 in high-resolution colour monitor.

Three types of hull units are available, each with its own designation; SS 575 with a standard hoisting length of 770 mm and the SS 576 with an extended maximum hoisting length of 2,250 or 3,100 mm.

The sonar comprises the following units: Control and Display Unit (CDU); Transceiver Unit (TRU); Hull unit with transducer (HLU); Hoist Control Unit (HCU); Transducer Matching Unit (TMU) and Vertical Reference Unit (VRU).

The man/machine interface uses dedicated control buttons on the CDU keyboard which offer direct access to primary sonar functions such as range, gain, tracking and so on. Secondary functions are menu-controlled.

The menu is displayed alongside the sonar picture and commands are selected via a joystick-controlled cursor on the display.

Operational status
In operation on several vessels.

Specifications
Frequency: 57 kHz
Beam-forming: 64 digitally formed beams
Beamwidth transmission: single beam – 11°

(horizontal); 11° (vertical). Omni – 360° omni; 11° (vertical)
Pulse length: CW – 0.6-160 ms (range dependent). FM (Subpulses) – 0.6-5 ms (range and mode dependent)
Number of subpulses: 2-8 operator selectable
Range scale: 125-6,000 m

Contractor
Simrad AS, Horten.

VERIFIED

ST 240 Toadfish

Type
Dipping and variable depth sonar.

Description
The ST 240 Toadfish is a lightweight, active, single frequency, high-resolution sonar system designed for ASW and for detection of other submerged targets. The transducer array is located in a submersible body, which may either be towed or dipped.

The system comprises a Control and Display Unit (CDU), Power and Beamformer Unit (PBU), towed body (Toadfish) and a hydraulic crane system and winch system. The ST 240 is an omni and sector transmitting system with multibeam reception covering 360°. Operating modes include omni, sector B-seam echogram and Doppler and passive with Demon. Tracking is implemented in active modes. The Toadfish is equipped with a CID sensor transferring temperature and salinity to present a real-time ray trace on the screen during launching. Transmission modes are CW and FM, with the possibility for Doppler analysis In CW. The power and beamformer unit receives preprocessed digital data from the towed body and performs signal processing (detection, normalisation and ping-to-ping correlation) with a fast digital processing system using the full dynamic range of the signals. Tracking is achieved in the omni and sector modes and presented on a special display window. The omni, sector and passive modes display windows for bearing, time and temperature. The unit interfaces with the single-operator control and display unit in which

presentation is performed on a raster scan, high-resolution 20 in colour monitor. Interfaces to log and gyrocompass are provided, as well as temperature input from an external probe for display of the present ray pattern without changing the towed body depth. The Toadfish also has its own compass to give the true fish bearing during towing or dipping. The ST 240 can be integrated with other sonars in the Simrad range using a common dual display, single-operator control and display unit.

A similar system called ST 570 with an operational frequency of 57 kHz is also available.

Operational status
In operation on several vessels including the Finnish Navy 'Rauma' class.

Specifications
Frequency: 24 kHz
Beam-forming: 64 digitally formed beams
Beamwidth transmission:
Single beam: 8° (horizontal); 12° (vertical)
Sector: 30°, 60°, 120° (horizontal omni); 12° (vertical)
Pulse length: CW – 1.3-200 ms (range dependent) FM (subpulse) – 1.3-83.2 (range dependent)
Number of subpulses: 2-8 (selectable)
Range scale: 500-12,000 m
Operating depth: 100 m

Contractor
Simrad AS, Horten.

VERIFIED

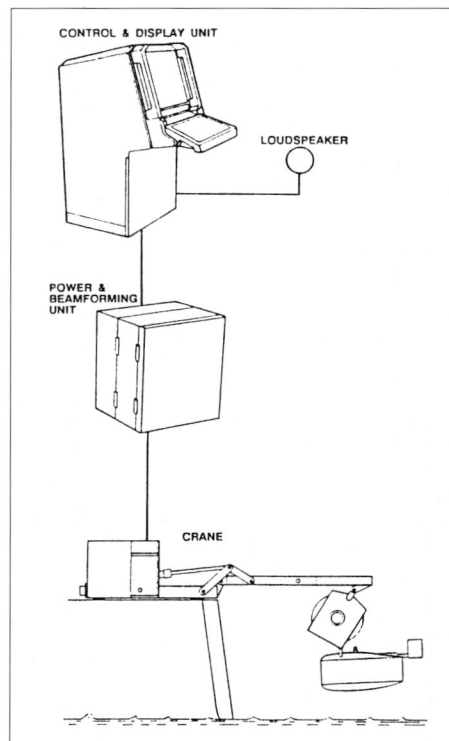

General layout of ST 240 Toadfish   **1995**

RUSSIAN FEDERATION AND ASSOCIATED STATES (CIS)

Because of security constraints and the generally inaccessible location of sonar equipment, it is not possible to treat RFAS underwater equipment in the same manner as has been adopted for that of other nations. Instead, the following notes summarise what little has been gathered. While this inevitably falls well short of the ideal treatment, it is hoped that readers will find the ensuing paragraphs of some help. The table in this entry gives a list of the types of sonar carried by various RFAS surface ship classes.

Underwater acoustic experiments by what was formerly the Soviet Union, initially in relation to communications applications, are widely agreed to date back to the years immediately before the First World War, but the earliest references to anything resembling what is now known as sonar occur in the 1930s when research into hydrophones for submarines is mentioned. At this general time, the possibilities of thermal detection devices for both aircraft and surface shipping were pursued. In the same period, the then Soviet Union is credited with the production of passive seabed acoustic detector equipment. Early Soviet records claim that at the start of the Second World War, the naval forces had a variety of sonar equipment available for shore and ship installations.

Most, if not all, of those reported were apparently passive devices, those for submarine fitting consisting of an elliptical array made of 8, 12 or 16 hydrophone elements. There were also passive sonar sets for surface ships, one of these being named *Tamir* which began sea trials in 1940. This equipment is stated by the former Soviet Union to have become the standard sonar employed in the anti-submarine campaign of that time. Neither the then Soviet Union's former allies, nor her enemies, appear to have been unduly impressed by the results achieved, according to historians, official

Winch assembly and towed body of ST 240

and unofficial. By the post-Second World War period, the nation had gained access to sonar technology originating in America, Britain and Germany, either by gift or as booty.

Since then, there has been steady and impressive growth of RFAS interest in all aspects of submarine warfare, and sonar equipment has been accorded a priority within these activities at least as great as that given to it by the Western navies. The advent of ASW helicopter and aircraft carriers, such as the *Moskva*, *Leningrad* and *Kiev*, in advance of comparable ships being commissioned in Western navies, might imply a higher priority.

The appearance of these ships was preceded by smaller vessels designed for ASW operations and special purpose aircraft and helicopters for naval duties, the latter being deployed in both land-based and embarked formations. These developments occurred in the mid-1950s to mid-1960s period.

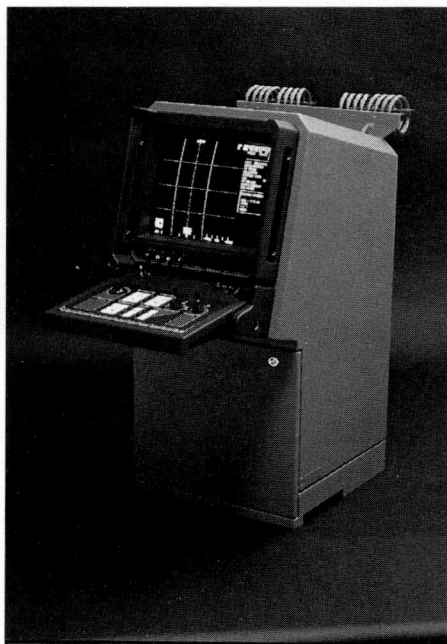

Single-operator display console of ST 240

Among the classes of vessel which the Russians rate as having a primary ASW function, all are fitted with hull-mounted sonar suites and variable depth sonars. These classes include the 'Kara' class cruisers, 'Kashin' and 'Udaloy' class destroyers, 'Neustrashimy', 'Gepard' and 'Krivak' class frigates and among smaller classes

This is probably the most detailed unclassified view of a modern four-unit group of a low-frequency hull search and probable medium-frequency VDS sonar. The ASW fire-control computer and switchboard are at the far end of the control space. This arrangement could be typical of the sonar layout found on 'Krivak' class destroyers (Jane's Intelligence Review)

dipping sonar, sonobuoys and MAD.

One sonar on which some information is available is the Bull Horn (NATO codename) system carried by the 'Sovremenny' class destroyers. This system probably operates in the 20-30 kHz frequency band and comprises a unit of four pull-out chassis, with a small (77 mm) CRT behind an orange filter screen. The lower three racks feature simple on/off toggle switches and push-buttons, only two multiple selector switches being visible. The only outputs seem to be to the small CRT and a meter which implies a simplistic sonar technology from the 1960s.

Much more recently, a certain amount of information has become available on acoustic and non-acoustic ASW systems. The RFAS appears to be making a considerable effort to develop non-acoustic sensors, including surface signature detection systems which sense the disturbance on the ocean surface caused by a passing submarine.

Wake detection systems that sense the turbulent wakes, internal wave wake, or contaminant (radioactive or chemical) wake of submerged submarines are also believed to be under development, as are magnetic or electric field detection systems that pick up the submarine or the subsurface disturbances caused by its passage. These wake and disturbance detection systems must operate at high altitude to offer effective surveillance and must therefore be either airborne or spaceborne.

In the traditional acoustic sensing field, the RFAS is understood to be continuing to deploy 'Cluster Lance' planar acoustic arrays in the Pacific waters near the mainland where broad area surveillance is required. It is believed that they may also be deploying static barrier arrays at entry and exit points in the Barents, Greenland and Kara seas. Russia is also believed to be evaluating a surveillance towed array, similar to the US SURTASS system.

The table of surface ship sonar types represents the latest information available. Information on submarine sonars is given in the Submarine sonar systems section of this book.

The Analysis section contains a list of current sonar systems under their NATO designations.

of vessels the 'Grisha' and 'Parchim' class frigates. In addition, many smaller patrol vessels carry sonar and ASW armaments and most larger warships are fitted with some form of sonar purely for self-defence purposes, or as back up to support vessels with a primary ASW function. Most of the vessels with a primary ASW role also carry Ka 25 (Hormone) or Ka 27 (Helix) helicopters with an ASW capability including

Sonar*	Type	Frequency band	Mode	Function	Class	Number	Type
Bull Horn	Hull-mounted	LF/MF	Active	Search/attack	Kuznetsov	1	Carrier
					Slava	3	Cruiser
					Sovremenny	16α	Destroyer
					Parchim	12	Frigate
Bull Nose	Hull-mounted	LF/MF	Active	Search/attack	Kara	2	Cruiser
					Grisha	1	Frigate (Lithuania)
					Grisha	47	Frigate
					Grisha	5	Frigate (Ukraine)
					Kashin	1	Frigate (Poland)
					Kashin	2	Frigate
					Koni	2	Frigate (Cuba)
					Kotor	2	Frigate (Yugoslavia)
					Krivak	24	Frigate
					Krivak	1	Frigate (Ukraine)
Herkules	Hull-mounted	MF	Active	Search/attack	Kynda	1	Cruiser
					Petya	4	Frigate (India)
					Petya	2	Frigate (Syria)
Horse Jaw	Hull-mounted	LF/MF	Active	Search/attack	Kiev	1	Carrier
					Kuznetsov	1	Carrier
					Kirov	4	Cruiser
					Udaloy I/II	8	Destroyer
					Gepard	1	Frigate
Steer Hide	Hull-mounted	LF/MF	Active	Search/attack	Slava	3	Cruiser
					Krivak 1	?	Frigate (after modernisation)
Whale Tongue	Hull-mounted	MF	Active	Search/attack	Sovremenny	16 α	Destroyer
					Neustrashimy	1β	Frigate
Elk Tail	VDS	HF	Active	Search	Grisha	47	Frigate
					Grisha	47	Frigate
					Grisha	5δ	Frigate (Ukraine)
					Parchim	1	Frigate (Indonesia – in some units only)
Foal Tail	VDS	HF	Active	Search/attack	Babochka	1	Patrol
					Mukha	2	Patrol
					Pauk	35	Patrol
					Stenka	50	Patrol (in some)
					Tarantul	47	Patrol
					Tarantul	4	Patrol (Poland)
					Turya	13	Patrol
Horse Tail	VDS	MF	Active	Search	Kiev	1	Carrier
					Kirov	4	Cruiser
					Udaloy II	1	Destroyer

Sonar*	Type	Frequency band	Mode	Function	Class	Number	Type
Lamb Tail	VDS	HF	Active	Search	Parchim	12	Frigate
Mare Tail	VDS	MF	Active	Search	Kara	2	Cruiser
					Kashin	2	Destroyers
					Kashin	5	Frigate (India)
					Kashin	1	Frigate (Poland)
					Krivak	24	Frigate (in some)
Mouse Tail	VDS	MF	Active	Search	Udaloy I	7	Destroyer
					Ivan Rogov	1	Amphibious assault ship
					Barnbuk	1	Survey
					Smolny	3	Training ship
Ox Tail	VDS		Active	Search	Neustrashimy	1β	Frigate
					Gepard	1δ	Frigate
Ox Yoke	Hull-mounted	MF	Active	Search/attack	Neustrashimy	1β	Frigate
Rat Tail	VDS	HF	Active	Attack	Muravey	14	Patrol
					Pauk	1	Patrol (Bulgaria)
					Pauk	1	Patrol (Cuba)
					Pauk	4	Patrol (India)
					Svetlyak	27	Patrol

* NATO designation
α Plus one building for Russia and 2 building for China
δ Plus one building
β Plus two building

UPDATED

UNITED KINGDOM

Type 2016 sonar

Type
Hull-mounted fleet escort active and passive search and attack sonar with bow- or keel-mounted variants.

Development
Design was initiated in the late 1960s and an experimental equipment underwent tests in HMS *Matapan*. A development contract was awarded in 1973 and the first prototype was completed in 1978 with fleet weapons acceptance in 1983. A bow-mounted variant, incorporating electronic beam stabilisation, was developed in 1983.

Description
The Royal Navy's Type 2016 fleet escort sonar is a replacement for the existing RN sonars Types 177 and 184. It employs computer-aided techniques which are operator interactive, have automatic detection and tracking, enhanced signal processing and displays complemented by an information storage facility.

The Type 2016 is a hull-mounted panoramic surveillance and attack equipment with facilities for classification and multiple target tracking. Interference between nearby vessels using the Type 2016 sonar is largely eliminated by the use of a new type of broadband transducer. The sonar display console is designed for manning by a single operator under normal cruise conditions, this crew member being able to initiate the preparatory actions necessary for urgent action.

Digital data processing facilities are based on use of Ferranti computer equipment, with other subcontractors including Marconi Radar.

The Type 2016 system comprises the following main elements: four active receiver cabinets, one passive and control cabinet and one T/R switch and beam-former cabinet, all housed in Marconi MC70 type cabinets. The computer suite consists of three D.811A cabinets and a separate cabinet for computer spares. The two transmitters are housed in two standard RN/AUWE cabinets. The solid-state electronics are based on a modular system approach using medium-scale integrated devices, printed circuit back planes with wire wrap connections and standard line-drive receivers for all inter-cabinet signal connections.

Extensive use is made of hybrid circuit techniques to reduce volume.

The Type 2016 array system, which is roll stabilised, is fitted in a ribless monocoque glass-reinforced plastic sonar dome within a fixed hull outfit. Depending on the ship fit, the array is mounted either in a keel or bow dome; array stabilisation is either hydraulic or by electronic beam-steering.

The display console contains three main displays and a top-mounted Versatile Console System (VCS). The right-hand display is dedicated to passive and auxiliary data, the passive data are displayed with two different integration times, enabling both torpedoes and more distant targets to be tracked. The auxiliary data is used for data logging and system information including the display of such items as course, speed, weather, sea bottom conditions and bathymetric information. System monitoring information is also provided here and on a maintainer's teleprinter or intelligent terminal, which enables faults to be pinpointed down to board level. The other two displays can be used to view surveillance, classification or target history information and are fully interchangeable. Special keyboards are provided at both positions for interaction with the automatic tracking process, display selection and data transmission to the command/combat information system.

Operational status
In service with the Brazilian Navy and in the Royal Navy in Type 22 frigates, 'Invincible' class aircraft carriers and Type 42 destroyers. Production was completed in 1987. Being replaced by sonar 2050 in Royal Navy.

Contractor
Thomson Marconi Sonar Ltd, Templecombe.

UPDATED

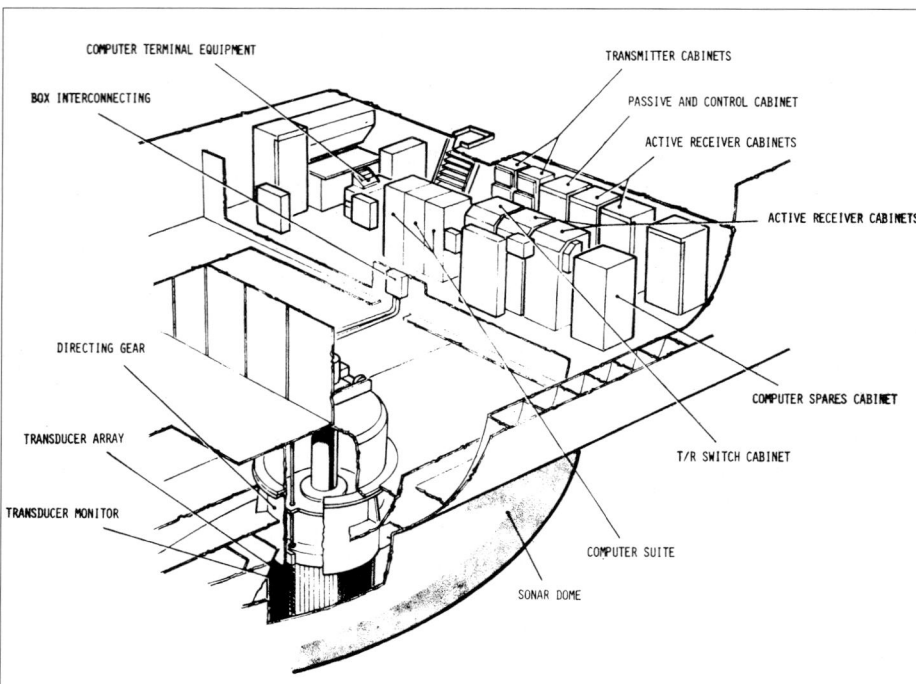

Arrangement of Type 2016 fleet escort sonar equipment and hull outfit compartment

Console of Type 2016 sonar aboard an RN frigate. Three displays (left to right) are for classification, surveillance and passive presentation

Type 2031Z sonar

Type
Towed array sonar system.

Description
The Type 2031Z is a towed array sonar system fitted to ASW frigates of the Royal Navy. Using the innovative 'Curtis' architecture developed at the UK Admiralty Research Establishment in Portland, Dorset, the 2031Z electronic signal processing suite has, because of its compactness, a considerably lower installed cost than other systems, yet provides a multi-octave broad and narrowband analysis facility capable of tracking the most elusive threats.

The array itself, is towed behind the ship and has a very low self-noise characteristic and is considerably longer than that used by any other self-contained sonar. It provides good bearing accuracy, even down to the very low frequency end of the acoustic spectrum. The combination of an advanced signal processing architecture and a high-performance towed array ensures that the system is a dual-capability sensor, able to fulfil both the tactical and surveillance role. A 2031Z equipped frigate can, therefore, act as an ASW picket or close escort.

The modular nature of the 2031Z, both as regards the digital signal processing electronics and the towed array itself, makes it possible to configure other types of sonar systems from the same range of modules. As an example, a variant of the 2031Z signal processing, suitable for submarine applications, is also available and has been delivered to the Royal Navy.

Operational status
The Type 2031Z is in series production for all Royal Navy Type 22 (Batches 2 and 3) and Type 23 frigates (first 10 ships). A total of 24 systems have been built or are on order.

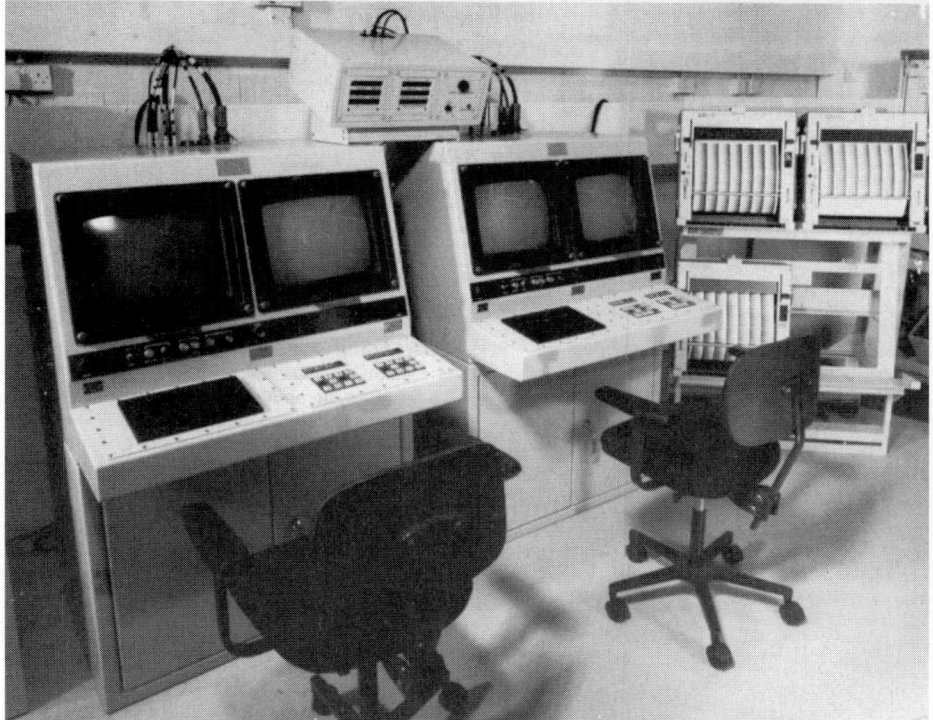

Sonar 2031Z units, including dual-display consoles, non-acoustic data displays and hard-copy recorders

Contractors
Thomson Marconi Sonar Ltd, Templecombe.
Ultra Electronics, Ocean Systems (overall system design authority and manufacturer of the inboard electronics suite).

Basys Marine, Barnstaple (towed array).
Clarke Chapman Limited, Gateshead (winching equipment).

UPDATED

Type 2050 sonar

Type
Hull-mounted active search sonar system.

Description
The Type 2050 is the medium-range, medium frequency hull-mounted attack sonar for the Royal Navy and is being fitted to the Type 23 and retrofitted to Type 42 and Type 22 frigates. It is a successor to the Type 2016 and is compatible with both bow and keel variants of the Type 2016 array. The equipment has been developed from the Thomson Marconi Sonar Ltd FMS series (see separate entry) and includes digital signal processing and distributed data processing using Digital Signal Processing (DSP) cards and multiple Argus M700 processors.

The provision of extensive data processing facilities and improved man/machine interfaces allows one operator to control the complete system. The equipment will interface to the combat system data highway feeding data to the CACS action-information system. The suite consists of five processing cabinets and the system uses the same display consoles as will be used in the Type 2054 equipped Trident submarines. These are air-cooled and in monochrome, but colour is available in the system for possible export applications.

The UK Mod has planned to update Type 2050 to incorporate an adjunct wideband array and new passive processing technology to provide a torpedo alert capability.

Under the update a Wideband Interstitial Passive Array (WIPA) developed by Northrop Grumman of Canada will be mounted around the existing array. The new array comprises 32 staves of six wideband passive hydrophones.

Inboard electronics will be modified with a new processing card (Passive Identification and Notification of Torpedoes – PAINT) developed by Thomson Marconi Sonar Ltd. The new card will handle digital

Type 2050 hull-mounted sonar system

data from the WIPA and process it to provide automatic torpedo alert.

Operational status
The UK MoD has ordered 34 Type 2050 sonar systems for the Royal Navy and these are being fitted in Type 23 frigates and retrofitted into Type 22 frigates and some Type 42 destroyers. An export version, the FMS 21, is described in the FMS series entry. Four Type 2050 sonars are in service with the Brazilian Navy onboard ex-Royal Navy Type 22 frigates sold to Brazil.

Contractor
Thomson Marconi Sonar Ltd, Templecombe.

UPDATED

ATAS V1/V2/V3 – TSM 2681/2682/2683

Type
Active/passive towed array sonar.

Description
ATAS is an active (FM) towed array sonar incorporating the best features of active variable depth sonars and passive towed arrays. The system provides both active detection of targets and simultaneous torpedo warning capability (ATAS V1). Provision is also made to permit the addition of a conventional, low-frequency passive towed array (ATAS V2) and integration with other hull-mounted sonars and command systems. It is possible to operate simultaneously in the active and the VLF

passive modes (V3) by adding one cabinet, one display and some additional electronics (one rack) to the V2 version.

The ATAS system comprises a high-power, low-frequency transmitter, operating in three frequencies around the 3 kHz band, which can be deployed at up to 900 m behind the towing platform, together with an in-line receiver array towed a further 300 m behind the transmitter; all of which may be towed at depths suited to prevailing oceanographic conditions. The port/starboard ambiguity normally associated with a towed array is automatically and instantaneously resolved within the system without the need for ship manoeuvre. ATAS can be operated at depths of 235 m and speeds up to 16 kt.

The transmitter has a high-power acoustic sound source consisting of a vertical stave of 10 in low-frequency flextensional transducers. Transmissions in CW or FM are omnidirectional in the horizontal plane with a vertical beamwidth of 25°. The receiver array is a 40 m long flexible tube housing 32 sets of hydrophones and associated electronics to condition and digitise hydrophone signals. The receiver gives full azimuth cover and in the broadside direction, a bearing resolution of up to 0.5°. The receiver uses 96 preformed beams with bearing stabilisation. ATAS can hold information on as many as 200 potential targets and can highlight the 24 most likely ones.

The processing equipment uses current technology to provide high-processing capability within a single compact unit.

The compact nature of ATAS permits it to be deployed from non-specialised ships of 200 tonnes and above, including ships of opportunity or ships taken up from trade, with a minimum of modification to the platform. Hull penetration is eliminated and a space of only 1.8 m by 5.6 m is required on the stern of the ship for the self-contained winch and handling equipment which weighs 4.7 tonnes (excluding in-water equipment). Other shipborne equipment comprises two electronics cabinets and a display unit

ATAS sonar showing the deployment mechanism and the transmitter

incorporating dual high-resolution colour monitors. The display console can be installed remotely from all other equipment; it displays active and passive search, localisation and classification information in addition to performance of the day and performance monitoring data. All system equipment can be supplied in containerised form.

Operational status

Development has been completed and ATAS is now in full-scale production for two major customers. Taiwan

is acquiring ATAS for the later units of the 'Cheng Kung' class frigates and the new French-built 'La Fayette' class. Pakistan is acquiring ATAS for its six ex-Royal Navy Type 21 frigates and Oman may also purchase ATAS. Eight systems are now under production.

Contractor
BAeSEMA, Bristol.

UPDATED

Bearing Ambiguity Resolution Sonar (BARS)

Type
Torpedo detection towed array sonar.

Description
The Bearing Ambiguity Resolution Sonar's (BARS) technology provides a unique proven solution to the ambiguity problem inherent in conventional towed array sonar systems. The passive, high-frequency BARS sonar is optimised for torpedo detection and provides the instantaneous bearing resolution which is essential for rapid response, without the need to resort to platform manoeuvres. Its shipborne signal processor is compact and VME compliant. BARS technology has

been proven in many sea trials around the world and is currently in production as part of the ATAS system. Bearing ambiguity is overcome by using a trielement hydrophone configuration which allows processing in two dimensions. The horizontal aperture allows for the alteration of the ambiguous array response.

BARS is suitable for installation on both ships and submarines and may be configured as a stand-alone system or as an adjunct to existing sensors. As a stand-alone system, BARS will provide threat detection without the need for extensive modifications, while as an adjunct sensor BARS may be retrofitted to enhance existing capabilities.

The system is designed to be compatible with the AN/SLQ-25A Nixie towed acoustic countermeasure.

Operational status
BARS technology is in service in the ATAS sonar. BARS has now been selected by the US Navy for a Foreign Comparative Testing programme allied to the UK/US SSTD project. This programme will demonstrate the benefits that can be achieved by incorporating BARS within a torpedo defence system when exercised against representative threats.

Contractor
BAeSEMA Ltd, Bristol.

UPDATED

Type 2087

Type
Low-frequency active and passive towed array sonar.

Development
Type 2087 will be a towed, LF active and passive sonar with high-power, long-range detection capabilities. Type 2087 will be a combined active, low-frequency (down to 500 Hz) transmitter with a wide aperture passive receiver (down to 100 Hz). Two teams have been invited to bid into the parallel Project Definition studies which will last 18 months before moving on to the Full Development and Initial Production Contract.

One of the teams formed to bid for the sonar Type

2087 programme for the Royal Navy is led by Babcock Defence Systems of Weymouth. The other international members of this team include Northrop Grumman Corporation (USA), Indal Technology (Canada), STN ATLAS Elektronik (Germany) and Lockheed Martin of the USA who joined the team in mid-1998. The other UK members are Data Sciences, Cogent, BMT, Reliability Consultants Ltd, PMES and BASYS Marine.

The second team led by Thomson Marconi Sonar Ltd (formerly GEC-Marconi Sonar System and Ferranti Thomson Sonar Systems) (responsible for the inboard design), includes Thomson Marconi Sonar SAS (responsible for the wet end), GEC-Marine Yarrow Shipbuilders (shipfitting), Orincon Corporation (support for data processing, performance prediction

and environmental aspects) and Strachan & Henshaw (responsible for the array handling).

Operational status
Two teams have been awarded a contract to prepare a Project Definition study as the first stage of a broader competition to secure the contract for design, development and manufacture which will be awarded in mid-1998.

Contractor
Thomson Marconi Sonar Ltd, Templecombe.
Babcock Defence Systems, Weymouth.

UPDATED

PMS 26 and PMS 27 sonars

Type
Hull-mounted active sonar systems.

Description
Hull-mounted ship sonars designed for single-operator control in both the surveillance and attack roles. The PMS 26 and PMS 27 are also used as 'dunking' sonars

in certain helicopters under the designation Type 195.

The PMS 26 10 kHz sonar is a self-contained system for ships and patrol craft down to 150 tonnes. It provides full 360° coverage in four steps of 90° and may be manually controlled to cover a particular sector, or set to carry out automatic search procedure. It incorporates a 'maintenance of close contact' facility for tracking close or deep targets. The single-operator controls the sonar through a special console. He is

provided with three sources of sonar information: audio, visual Doppler and visual sector.

The Doppler facility provides increased initial detection range and improved classification capabilities compared with conventional small ship sonars. The PMS 26 transducer array is mounted within a hull outfit with a glass-reinforced plastic dome.

The PMS 27 differs from the PMS 26 only in its associated hull outfit. The PMS 27 transducer array is mounted in a Royal Navy hull outfit 19 or similar, which makes the equipment suitable for installation in small escorts down to about 650 tonnes displacement. The PMS 27 can be used as a surveillance sonar in association with a separate fire-control sonar within the same hull outfit.

This enables the ship to continue surveillance for new threats while engaging a target already detected.

Operational status

In operational service but no longer in production. A total of nine PMS 26 systems remains in service on board four Danish frigates ('Niels Juel' class and 'Beskytteren'), the Irish OPV *Eithne*, a Nigerian Mk 9 corvette, Turkish patrol craft and two Venezuelan coastguard vessels.

Contractor

Thomson Marconi Sonar Ltd, Templecombe.

UPDATED

MS 56 – ASW sonar

Type

Small ship hull-mounted or variable depth sonar.

Description

The MS 56 is a lightweight, compact shipborne sonar, one of the Series 5 range of modular sonar systems. The MS 56 provides all the facilities required for ASW operations by small ships and patrol craft. It is available with fixed, retractable or variable depth arrays for installation on vessels of up to 300 tonnes displacement and with maximum speeds of up to 30 kt.

The system includes a high-resolution colour display which provides the single operator with facilities for search, detection and classification. The man/machine interface, provided by pull-down menus, is designed for ease of use and reduced training requirements.

The lightweight panoramic array uses transmit and receive hydrophones for maximum efficiency. Incorporating the latest technology, all the transmit and receive electronics are contained in two lightweight air-cooled shelves.

Operational status

Currently being evaluated by several navies.

Specifications

Modes: CW, FM, passive, aural
Range scales: 625-20,000 yd
Pulse lengths: 5-1,000 ms
Weights: 220 kg (console); 50 kg (transmitter rack); 385 kg (hull outfit and array)

Contractor

Thomson Marconi Sonar Ltd, Templecombe.

VERIFIED

MS 56 lightweight panoramic array

COMTASS

Type

Compact towed array sonar system.

Description

COMTASS is a compact towed array system belonging to the Thomson Marconi Sonar Ltd Series 5 family of modular sonar systems. It requires the minimum of structural changes for installation and can be fitted to a vessel in less than half a day. The high-performance array has been designed to withstand high-speed towing. The processing and electronics are contained in two 19 in units, each with power supplies and cooling fans and flexibility to interface to any command system. The total modularity of the display and processor gives scope for additional processing if required.

The standard console, which requires only one operator, allows for the simultaneous display of broadband and narrowband passive data for the long-range surveillance of surface and submarine targets. It is also possible to integrate active sonars into the COMTASS console, with flexible control of the displayed data.

The system provides passive detection and tracking of targets on a multifunction operator's console with detection ranges given as 30+ miles, with an extensive over-the-horizon ships' surveillance capability. The console consists of two high-resolution, colour, refreshed raster-type displays in a variety of formats, according to the customer's specification.

The handling system, consisting of winch and control panel, is provided by Strachan & Henshaw. The system is designed for minimum maintenance and semi-automatic deployment. To provide further flexibility it can be palletised or containerised for easy removal and transfer between vessels and for quick fitment to reserve or auxiliary vessels. The critical angle tow lends itself to ease of handling, deployment and retrieval.

The sonar can integrate with hull-mounted sonars.

Operational status

COMTASS has been evaluated by a number of navies and has been fully validated by the UK defence authorities. Now in service with an Asian navy.

Contractors

Thomson Marconi Sonar Ltd, Templecombe.
Strachan & Henshaw, Bristol (handling system).

UPDATED

COMTASS towed array system fitted on a trials ship

COMTASS lightweight array and handling system

Series 5

Type
Family of modular sonar systems.

Description
In terms of private venture, the experience gained in MoD and other programmes has been applied to generate a new family of modular sonar systems known as Series 5. These systems consist of a number of mutually compatible modules covering beamformers, signal and data processing, man/machine interfaces and data fusion.

Combinations of these elements provide cost-effective optimum solutions for all surface ship and submarine applications. COMTASS and GETAS are examples. Other combinations cover requirements from the equivalent of a full Royal Navy fit to a modest single octave trainer. They can all be tailored to suit individual technical specifications and budgets.

Research into fibre optic technology has been progressing for some time and technology demonstrator towed array systems are now on trials at sea.

Contractor
Thomson Marconi Sonar Ltd, Templecombe.

VERIFIED

FMS series sonar systems

Type
Family of passive and active sonar systems.

Description
The FMS series is an integrated range of low-cost, compact, high-performance systems, developed as a private venture to meet the needs of a wide variety of platforms and operational roles and is based on the new generation of sonar designs selected by the Royal Navy. The series has been designed to optimise performance, simplify installation and in-service support and reduce through-life costs.

The systems within the series consist of:
FMS 12 – passive narrowband (2 octave)
FMS 13 – passive narrowband (3 octave)
FMS 15 – passive narrowband (5 octave)
FMS 20 – active (4 ft hull array with VDS option)
FMS 21 – active (6 ft hull array with VDS option)
FMS 30 – passive broadband (single shelf – 3 ft array)
FMS 31 – passive broadband
FMS 52 – navigation and obstacle avoidance sonar.

All FMS systems employ the latest advanced distributed digital signal and data processing techniques to provide significant cost, space and performance improvements over the majority of current equipments. These techniques are subject to continuing development to further improve these factors.

Signal processing, based on advanced 'Curtis' techniques, is achieved using programmable digital signal processing modules which can be easily configured to meet the optimum performance requirements of specific array outfits and for easy through-life enhancement.

The data processing equipment is based on multiple M700 processors, arranged in distributed arrays. The use of large numbers of these interconnected in the data processing system gives very high levels of computer assistance and automatic operation. Features such as automatic detection, tracking and classification of multiple targets are provided. Standard programmable modules for both signal processing and data processing functions enable significant reduction in spares holding, simplify maintenance and reduce overall cost.

The ergonomically designed operator consoles allow single operator control of the complete system. Each contains two 1,000 line high-resolution raster displays and an operator's desk. The man/machine interface includes a trackerball, nudge keys and a touch-sensitive plasma panel. From the plasma panel the operator can select various displays, focus on selected beams, scroll around beams, set verniers, annotate features on the display and select a variety of cursors. The trackerball and nudge keys are used for fast and accurate positioning of the cursor to provide specific displays of frequency, bearing, time and harmonic ratio information.

Each sonar in the FMS series is configured by selecting from a range of standard modules. This allows active or passive sonars for ships and submarines, large or small, to be proposed to suit customers' specific requirements. The modules are the same as, or closely related to, equipment in or ordered for RN service.

The processing equipment can be housed in air- or water-cooled equipment practice cabinets to meet customers' individual requirements.

Contractor
Thomson Marconi Sonar Ltd, Templecombe.

VERIFIED

FMS 12/ FMS 13/ FMS 15

Type
Hull mounted or towed array sonar.

Description
The FMS 12, 13 and 15 are high-resolution narrowband surveillance sonars designed to derive signal inputs from either towed or hull-mounted line arrays. Each system consists of a single electronics cabinet which weighs 300 kg and a single-operator's console weighing 350 kg.

The FMS 12 provides frequency cover over two octaves, FMS 13 over three octaves and FMS 15 over five. Features common to both systems include: 360° bearing cover, 32 beam resolution and multiple beam surveillance display. They also provide high-resolution vernier analysis classification, with sonar aural and audio communications interface. Both systems have built-in test equipment, an air-cooled operator console and electronics cabinet.

Operational status
An enhanced version of FMS 12 was included in Project Triton – the Royal Navy's Type 2051 sonar update programme for the 'Oberon' class submarines. FMS 12 also provides the basis for the RN Type 2046 passive sonar system, 26 of which have been ordered for updating the current and next generation of nuclear-powered attack submarines.

The Royal New Zealand Navy ordered an FMS 15/2 system in March 1989 which was supplied in January 1990.

Contractor
Thomson Marconi Sonar Ltd, Templecombe.

UPDATED

FMS 20

Type
Hull mounted active sonar.

Description
The FMS 20 is an active sonar system which combines the range resolution of an FM sweep with the Doppler discrimination of a CW pulse. It also has a passive capability for a built-in torpedo alarm facility. In its most comprehensive form, FMS 20 consists of four electronic cabinets, each with a weight of 765 kg and a two display operator console weighing 340 kg. The associated transducer arrays may be hull-mounted (4 ft diameter) with or without an associated variable depth array.

The equipment provides a 32 beam resolution, electronic array motion compensation, 360° bearing cover, variable range scales and a variable centre frequency of operation. The display formats available to the operator as standard features include active FM and CW information, classification of targets, track and history totes, environmental and ray tracing and monitoring of the equipment's status.

Using modular technology, the number of electronic cabinets required to drive the FMS 20 system can be reduced to two, albeit with some reduction in overall capability.

Operational status
FMS 20 is currently being considered by several NATO and other navies for inclusion in both new ship and update programmes.

Contractor
Thomson Marconi Sonar Ltd, Templecombe.

VERIFIED

FMS 21

Type
Active sonar.

Description
This system is derived from and is the export variant of, the Royal Navy's new Type 2050 sonar. It is an active sonar system which, like FMS 20, combines the range resolution of an FM sweep with the Doppler discrimination of a CW pulse. Associated transducer arrays may be hull-mounted (6 ft diameter), with or without an associated variable depth array and provide a 64 beam resolution.

The four cabinets which contain the electronics for the FMS 21 system and its two-display operator console (providing single-operator control) weigh 765 kg and 340 kg respectively. Equipped with a variety of display formats as standard features, the operator has active FM and CW information, classification of targets, track and history totes at his fingertips, as well as environmental, ray tracing and system status information. FMS 21 also has a passive capability with an automatic torpedo warning alarm.

Operational status
17 systems of Type 2050 have been ordered by the RN.

Contractor
Thomson Marconi Sonar Ltd, Templecombe.

UPDATED

FMS 52

Type
Obstacle avoidance sonar.

Description
The FMS 52 is a high-frequency, active navigation and obstacle avoidance sonar which minimises the chances of collision with all types of subsurface obstacles, including moored mines. The 64 element transducer array can be either hull-mounted (for surface vessels), fin- or hull-mounted (for submarines), or have fixed installation for harbour surveillance. The use of electronic beam-steering in preference to mechanical methods increases the accuracy and data update of the system.

Other features of the FMS 52 include:
(a) steerable transmitter beams, designed to ensure that transmission patterns can be optimised for searches at different depths and in varying water conditions
(b) bearing resolution of 3°
(c) expansibility – up to four arrays giving full 360° coverage
(d) capability of unmanned operation.

Contractor
Thomson Marconi Sonar Ltd, Templecombe.

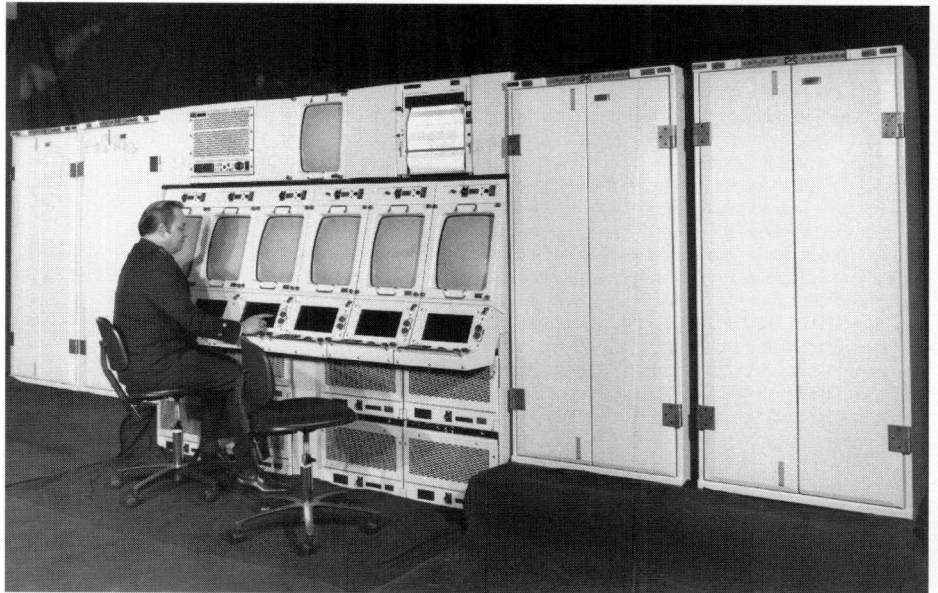

A purpose-built combination of the FMS 15 and FMS 21 sonar systems

VERIFIED

Type 184M/P sonar

Type
Active and passive hull-mounted search sonar.

Description
The Type 184M was one of the Royal Navy's primary surface ship anti-submarine search and attack sonars.

It is a 360° scanning sonar incorporating both active and passive modes of operation; it provides range, bearing and target Doppler data to the fire-control computer. The Type 184M was designed in association with the Admiralty Underwater Warfare Establishment (AUWE) and is fitted in many RN surface ships. The equipment has also been supplied to a number of other navies.

Few technical details have been cleared for publication, but the equipment provides a dual-frequency transmission and three receiver systems. The latter comprise:
(a) an all-round search and tracking PPI system with eight receiving channels
(b) a 4 beam sector search Doppler system with B-scan display
(c) a continuous torpedo warning system with its own display.
A circular 32 stave transducer array is employed.

Solid-state modernisation kits (184P) to improve the performance and reliability of this sonar are now in production for the Royal Navy. This programme is being implemented in three phases. The first phase introduced a 16 channel PPI system and replaced all electromechanical selectors and timers with solid-state electronics. The second phase offers improved detection capability against moving targets. The 31 digital filters in each of the four Doppler beams measure up to ± 40 kt of target speed with a resolution of better than 3 kt. The third phase provides separate Doppler, PPI and hydrophone effect geographically stabilised memory colour displays, with special attention paid to man/machine interfaces.

The PPI display offers improved detection through noise discrimination; 4 beam sector search Doppler display provides moving target detection capability; and the hydrophone effect display in the passive mode presents the output of the hydrophone effect processor on the periphery of a circular plot with maximum noise signal directed towards the centre.

Operational status
No longer in production. No longer in service with the Royal Navy. Ships fitted with the system have been sold to Chile ('County' and 'Leander' classes), Ecuador ('Leander' class) and Pakistan ('Leander' class).

Contractor
Graseby Dynamics Ltd, Watford.

Type 184 sonar control console

UPDATED

G 750 sonar

Type
Active and passive hull-mounted search sonar.

Description
The G 750 is a search and attack all-round sonar for ships of frigate size and above. It is a modernised and improved version of the Type 184M, providing accurate fire-control data from two independent automatic tracking systems, while simultaneously maintaining all-round surveillance, independent Doppler search and classification and continuous torpedo warning.

Three separate processing and display systems are incorporated for:
(a) all-round search and tracking (PPI)
(b) Doppler
(c) passive search.
Digital readout displays are used extensively and visual displays are supplemented by audio in all three systems. Nine cabinets are required for the complete electronics. Three are display consoles; three contain transmitter and control equipment, two house receivers and one is used for monitoring purposes. For installation in smaller ships or submarines, single passive or PPI display systems with smaller transducers are produced. In the standard G 750 system, an improved version of the Type 184M cylindrical 32 stave transducer is used to form a large number of beams. Range and bearing measurement is by in-beam scanning, obviating the need for separate expanded displays.

Target range and bearing are measured by either of two methods:
(a) by means of a manually positioned marker technique, in which the circular markers are placed over the target echo, it being the marker position that is measured; or
(b) by echo gating, where the echo range and bearing

are measured directly. This form of tracking is normally on acquisition of a possible target and is available on either computer-assisted or non-computer-assisted basis. Marker boxes representing the gated area replace the circles. In both methods the targets are individually identified by dotted or full markers.

Provisions are made for interfacing the G 750 sonar with computer-based ASW and tactical information systems to provide a source of range, bearing and target Doppler data. Operational capabilities of the equipment include active detection of submarines throughout 360° and the ability to track two targets simultaneously. Similar detection coverage and tracking capabilities are available in respect of torpedo targets, with suitable outputs to torpedo warning and avoidance systems. There is also an active search capability over a 45° arc with facilities for determining

target Doppler. Four 11° beams forming a 45° 'searchlight' are providing a 360° search facility capable of measuring target Doppler up to ± 40 kt. The system is designed for operation in adverse reverberation conditions, with ripple or omnidirection dual-frequency modulated and Doppler CW transmission modes. The system can be applied as an HF or LF variant. Targets can be tracked to a maximum range of 20 km.

Operational status
No longer in production. Six systems delivered to the Indian Navy for 'Godavan' and 'Nilgiri' classes.

Contractor
Graseby Dynamics Ltd, Watford.

UPDATED

Type G 750 sonar control console

CHIP (Concurrent High-speed Integrated Processor)

Type
Fast time acoustic analyser.

Description
The CHIP analysis equipment has been developed using the Thomson Marconi Sonar Ltd Series 5 technology, for the detailed spectral (Demon and broadband) analysis of surface, subsurface or airborne tape recorded acoustic data. It incorporates automatic scaling of all processing parameters so that non-real-time processing can be undertaken without the need

for manual correction. This enables tapes to be played at 0.5, 1, 2, 4 and 8 times the recording speed, while maintaining the correct frequency/time relationships and presentation on the display. It has 16 input channels and eight processing channels, each of which is independently configurable to a comprehensive set of processing parameters. A large input storage facility is provided to capture data from tape for immediate replay and reprocessing.

Control is exercised with an easy-to-use menu system and a high-definition colour VDU to control processing which is carried out by an array of digital signal processors.

An operator can set the equipment to undertake an

initial exploratory analysis unattended, returning to assess the resultant data by viewing time histories stored in the display memory or inspection of the continuous hard-copy printouts.

Operational status
In service with a NATO navy.

Contractor
Thomson Marconi Sonar Ltd, Templecombe.

VERIFIED

UNITED STATES OF AMERICA

AN/SQR-17A signal processor

Type
Integrated shipboard submarine detection, classification, signal processing, display and recording system.

Description
The AN/SQR-17A is a totally integrated submarine detection and classification system for shipboard and

land-based applications. The (V)2 system configuration consists of the eight channel AN/SQR-17A sonar signal processor and display unit, the 28 track RD-420B tape recording system, the AN/ARR-75 sonobuoy receiver interface unit and the antenna and microphone access unit.

The AN/SQR-17A can process signals detected by Difar, Dicass, VLAD, Lofar, active and BT sonobuoys and displays raw acoustic in either A- or B-scan format. Data is then enhanced, incorporating alphanumeric

designations and graphics with human factor considerations, to aid personnel in target analysis and tactical co-ordination.

The system has a 19 in acoustic data display and two 9 in auxiliary video-monitor displays for menu prompting, for selection of the appropriate system operating mode and for monitoring the status of multiple acoustic sensors.

The high-density solid-state mass-memory display provides full multitarget detection, tracking and classification for ASW encounters. The display uses real-time, high-resolution, TV raster formats and is designed modularly with electrographic hard-copy capabilities in standard-gram format. The system can interface with towed arrays, hull-mounted sonars, helicopter data downlink and onboard sonobuoy receivers.

The AN/SQR-17A has gone through a series of upgrades and improvements through the (V)1, (V)2 and (V)3 versions.

Operational status
Over 100 units have been sold to the US Navy. The company has also sold units to friendly navies and continues to provide installation and refurbishment services. Upgraded systems have been supplied to the US Navy for the Mobile Inshore Underwater Warfare (MIUW) programme.

Contractor
DRS Electronic Systems Inc, a subsidiary of DRS Technologies Inc, Gaithersburg, Maryland.

VERIFIED

(Left to right) AN/ARR-75 sonobuoy receiver/interface unit, AN/SQR-17A sonar signal processing system, RD-420B recorder/reproducer

AN/SQR-18A(V) TACTAS sonar

Type
Towed array passive and active receive sonar system.

Description
The AN/SQR-18A(V) Tactical Towed Array Sonar (TACTAS) has been designed to enhance the air/sea warfare capabilities of the US Navy surface ship fleet. It provides long-range passive detection and classification of submarine threats. The system is a high-technology, single-operator equipment, providing full azimuth coverage in both narrowband and broadband search modes.

The 800 ft long array consists of an acoustic section, a vibration isolation module and a rope drogue. In the short tow version, it is towed behind the Variable Depth Sonar (VDS) body of the AN/SQS-35V (see separate entry). Preamplifiers are in the array, while signal processing and other electronics are on board the ship.

The original AN/SQR-18A has now been replaced by the AN/SQR-18A(V)1 version, which with a new low noise array provides increased detection range and allows effective operation at high own ship's speed. An adaptive processor provides interference cancellation,

AN/SQR-18A console/display

making the system virtually immune to tow ship noise and allows operation near other noisy ships. An improved tracker provides accurate bearing data for effective target motion analysis solutions, even on weak targets. The most significant improvement made to the basic system has been the installation of an operator auto alert capability which provides target detection for operator selected frequencies. The system is capable of providing acoustic data to an external recorder and playing back recorded data.

The AN/SQR-18(V)2 critical angle towing version has its own towing and handling capability permitting it to be used on ships without a VDS. This variant has recently been completely modernised and utilises an Advanced Modular Signal Processor (AMSP), a programmable system using interactive software design. The new processor provides additional processing capability used in conjunction with an improved operator automatic alert system which can be optimised for each individual threat and alert level.

In the US Navy, the AN/SQR-18A(V)1 and (V)2 systems are integrated with the AN/SQR-17 sonobuoy processor and AN/SQS-26 hull-mounted sonar. This combination provides an effective combat suite for the LAMPS Mk-1 ASW suite. The AN/SQR-18A(V)1 or (V)2 are capable of effective operation as stand-alone systems or can be integrated with other combat information systems.

The AN/SQR-18A(V)1 system electronics consist of three major and five small units, while the (V)2 consists of two major and six small units.

A signal conditioner provides gain, equalisation, interfacing, auto-ranging gain control and readies the hydrophone signal for digitisation. This is the only analogue part of the AN/SQR-18A(V).

An embedded trainer is included which provides realistic targets in dynamic interactive scenarios with high-fidelity classification clues that allow onboard training of both the operator and the target localisation team.

Digital signal processing techniques are used in both variants to present narrowband and broadband low-frequency information. The system may be operated in both narrow and broadband simultaneously. Tracking is provided by an automatically stabilised tracker beam. The automatic target following circuits keep the beam nulled on a target through own ship and target manoeuvres. All processed data is stored in a modular electronic solid-state memory. Narrowband outputs are presented on the CRT screens in Lofar format. The AN/SQR-18A(V)1 provides two independent CRT screens, while the (V)2 provides an additional CRT to present situation summary information. Controls are provided to measure the frequency and amplitude of tonals and harmonic ratio of target frequencies. Results of

measurements, bearings of beams displayed and tag numbers are displayed in alphanumeric format.

Development of the AN/SQR-18A(V)3 has been completed. The fully digital system uses the latest modular digital signal processing technology and standard computer subassemblies, combined with a standard console and colour displays. This architecture provides a system that can be readily reprogrammed and upgraded. The AN/SQR-18A(V)3 uses digitised hydrophone data from an array and critical angle towcable identical to the AN/SQR-19. Greatly increased performance over earlier variants is achieved.

The AN/SQR-18A(V)3 has also been reprogrammed, under contract to the US Navy, to operate as a low-frequency active receiver using the towed array as a sensor. Coherent processing of selected multiple active waveform types is easily accommodated. Digital processing, target trackers and display features are provided as aids to minimise false alarms and simplify operation. The system can be fully integrated into shipboard combat systems using standard digital bus formats. Data is fed to a tactical display which integrates reduced false alarm acoustic data with the ship's combat information. A PC-based passive/active test set and trainer is included with multiple target features and active echo types. Up to four dynamic targets can be inserted simultaneously to enhance operator and combat system team training.

Operational status
In use with the US Navy and with the Royal Netherlands and Japanese navies. Forty AN/SQR-18A systems have been delivered to the US Navy and have been updated to an AN/SQR-18A(V)1 standard. Seven AN/SQR-18A (V)2 systems have been delivered with these being designated for the US Navy FF 1052 frigates not equipped with VDS, as well as some FFG-7 class frigates.

Now in service with various countries that have leased/purchased units of the FF1052 'Knox' class frigates including: Egypt, Taiwan, Thailand and FFG-7 'Oliver Hazard Perry' class frigates: Bahrain, Egypt, Turkey (currently building) and the Taiwanese 'Cheng Kung' class.

The AN/SQR-18A(V)3 active receiver has completed initial sea trials on two US Navy FFG-7 class frigates.

Contractors
EDO Corporation, Combat Systems Division, College Point, New York.
Lockheed Martin, Syracuse, New York (towed array and handling and storage subsystem for the (V)2).

UPDATED

AN/SQR-19 TACTAS sonar

Type
Passive towed array sonar system.

Description
The AN/SQR-19 is a passive towed sensor array designed to give surface vessels long-range detection and tracking of both submarines and surface ships.

It includes: a towed sensor array with improved ranging ability and reduced self-noise; a handling system which improves system performance by maintaining array position and depth; and a sophisticated signal processing and automated tracking system providing more sonar information with reduced operator workload. The 'wet end' consists of a linear array and towcable, plus the associated handling and stowage equipment. Coaxial conductors within the towcable carry the flow of telemetered acoustic signals, plus heading, depth and temperature data from the array to the ship and also handles the commands to the array. Located below decks are the signal and data processing equipments, plus the display and control

consoles. The displays are shared with the AN/SQS-53 hull-mounted sonar and the LAMPS III ASW helicopter, so that the ship is provided with a single integrated ASW system (the AN/SQQ-89).

The AN/SQR-19 provides omnidirectional long-range passive detection and classification of submarine threats at 'tactically significant' own ship speeds in seas up to state 4, using an array towed on a 1,700 m cable to provide tow depths down to 365 m.

The array features a multiple convergence zone capability. The 82 mm diameter array is 242 m long and incorporates 16 modules. Low flow noise maximises detection opportunities which are conditioned by environmental features and known or anticipated threat operating depths. Array acoustic data is processed into 43 beams and provides narrowband, broadband and demodulated noise (DEMON) signals.

The system uses computer aided detection, visual alerts, optimum search display formats, special classification display formats and complex data processing.

The TACTAS shares data processing equipment with the AN/SQQ-28 sonar signal processing subsystem

and sonar displays with the AN/SQS-53C and AN/SQQ-28 to provide a single, integrated ASW combat system.

Operational status
The first operational system was installed on board the USS *Moosbrugger* early in 1982 and has completed both technical and operational evaluation. The system has been installed on about 98 USN platforms including: CG 47 'Ticonderoga' class cruisers; DDG-51 Flight I and II 'Arleigh Burke', DDG 993 'Kidd' and DD 963 'Spruance' class destroyers; and FFG-7 'Oliver Hazard Perry' class frigates. In addition, there have been Foreign Military Sales (FMS) of the SQR-19 and AN/SQQ-28 to Spain and sales of 'wet end' elements of the SQR-19 to Canada where it is fitted in the CANTASS towed array.

Contractors
Northrop Grumman Corporation, Electronic Sensors and Systems Division, Sykesville, Maryland.
Lockheed Martin, Syracuse, New York.

UPDATED

AN/SQS-35, AN/SQS-36, AN/SQS-38 sonars

Type
Hull-mounted and variable depth sonar systems.

Description
These systems have been designed to detect submarines at medium ranges in both deep and shallow waters. The AN/SQS-35 and the AN/SQS-38

sonars have been in service with the US Navy since the mid-1960s.

The US systems are improved miniaturised solid-state versions of previously developed vacuum-tube equipments manufactured for the US, Italian, Norwegian and Japanese navies.

Some versions combine both variable depth capability and hull sonar capability, selectable by the sonar operator at the control console.

Weapons associated with the system are the Mk 44 torpedo and ASROC (USA).

Operational status
Now in service with various countries that have leased/purchased units of the FF 1052 'Knox' class frigates including: Greece, Taiwan, Thailand and Turkey.

Contractor
EDO Corporation, Combat Systems Division, College Point, New York.

UPDATED

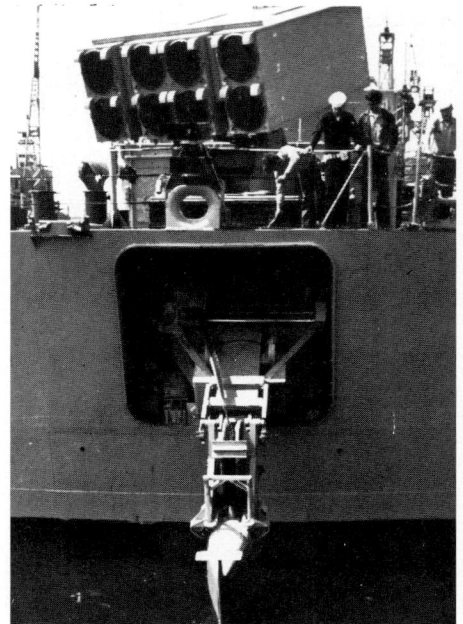

AN/SQS-35(V) variable depth sonar seen aboard escort ship USS Francis Hammond

AN/SQS-53/26 sonars

Type
Hull-mounted passive and active sonar system.

Description
Claimed to be the most advanced surface ship ASW sonar in the US Navy inventory, the AN/SQS-53 is a hull-mounted, active and passive high-power (190 kW), long-range system evolved from the AN/SQS-26CX. Functions of the system include: the detection, tracking and classification of underwater targets; underwater communications; countermeasures against acoustic underwater weapons; and certain oceanographic recording uses. Target data obtained by the sonar are transmitted to the ship's Mk 116 digital underwater fire-control system.

The AN/SQS-53 can detect, classify and track multiple targets and is the first USN surface ship sonar designed to interface directly with a vessel's digital computers. The system has a cylindrical array of 576 transducer elements housed in a large bulb dome below the waterline of the ship's bow. Operating frequency is 3 kHz. There are 37 cabinets of signal processing, transmitting and display equipment. Passive and active operating modes are possible.

There are three active modes:
(a) surface duct
(b) bottom bounce
(c) convergence zone.

The surface duct mode depends upon sound energy being transmitted essentially in the horizontal plane. Due to the high level of noise introduced into the return signals near the surface, this mode is useful only for relatively short distances. Nevertheless, this method is conventional for many surface ship sonars and the high-transmission power of the AN/SQS-53 is stated to provide longer-range capability in the surface duct mode than previous sonars.

In the bottom bounce mode the sound energy is directed obliquely toward the seabed. The energy is reflected upward from the ocean floor toward the surface at considerable distances from the ship. Submarine echoes are received via a similar return path. This method is useful in waters of more than certain minimum depths and where the seabed has the requisite favourable characteristics.

Convergence zone mode operation takes advantage of the characteristics of very deep water. The sound energy is refracted downward due to the temperature and pressure conditions near the surface, but, as depth increases, these physical effects change and the sound path alters direction to cause the energy to return to the surface in a coarsely focused convergence zone. This zone can form at great distances from the ship and provides the longest range of coverage for the sonar when the water conditions are favourable for this mode of operation.

Passive detection gives the bearings of targets based on their own noise generation, rather than by echo location. It has proved a valuable method on the currently deployed AN/SQS-26 equipped ships, especially at low speed. The 'Spruance' class destroyers and her successors have improved passive detection capabilities at higher speeds due to improved noise suppression measures. The passive mode can be operated simultaneously with the active modes.

Operational status
The SQS-53 is installed in all ships of the DD 963 'Spruance' class destroyers and in all 27 ships in the CG 47 'Ticonderoga' class cruisers (either in A, B, or C variants).

The latest version of the system is the AN/SQS-53C Battle Group Sonar. It is a much more advanced system using 22 fewer cabinets with a consequent reduction in cabling and associated support requirements. The new version uses seven AN/UYK-44(V) computers linked together in a multiple embedded configuration. The first SQS-53C was fitted in USS *Stump* in August 1986 for evaluation purposes. Testing was completed in mid-1989. The 53C is being fitted aboard all the ships in the DDG 51 'Arleigh Burke' class destroyers, as well as some of the units of the 'Spruance' class destroyers and 'Ticonderoga' class cruisers.

Bow housing of AN/SQS-26 sonar

Bow dome cutaway showing partial view of AN/SQS-53 cylindrical transducer

The SQS-53C incorporates a number of improvements, including a better performance in the active mode, simultaneous operation in active and passive mode, multiple target automatic tracking and improved displays. The system's ability to provide small object avoidance capability inshore has also been enhanced by the Kingfisher system (see below). The

enhancement was successfully used by a number of ships during Operation Desert Storm. In 1992, Northrop Grumman was awarded the contract for all production and system design agent services for the SQS-53C as part of its SQQ-89 integrated surface ship ASW combat system.

Contractors

Northrop Grumman Corporation, Electronic Sensors and Systems Division, Maryland, Sykesville (Type 53C).

Hughes Aircraft, Naval and Maritime Systems, Mukilteo, Washington.

UPDATED

AN/SQS-56/DE1160 Sonar

Type

Hull-mounted active and passive sonar system.

Description

The AN/SQS-56 is a modern hull-mounted sonar developed as a company funded product by Raytheon Electronic Systems for the US Navy's 'Oliver Hazard Perry' class frigates. The US Navy has provided the AN/SQS-56, via the Foreign Military Sales (FMS) programme, to Saudi Arabia for its PCG ships, to Australia for its 'FFG-7' class frigates and to Turkey for its MEKO 200 programme. Versions commercially exported with the designation DE1160 are operational on the Italian Navy's 'Lupo' class frigates and the Spanish, Moroccan and Egyptian navies' 'Descubierta' class frigates and are being considered for installation in new construction ASW ships of several other navies. The DE1160, when configured with 36 kW transmitters, is identical to the AN/SQS-56. Outfitted with a VDS array and handling subsystem, it becomes Raytheon's DE1164 sonar (see separate entry) and is installed in the Italian Navy's 'Maestrale' and 'Alpino' class ships.

The DE1160, when equipped with a larger, low-frequency transducer array and three additional transmitter cabinets, is designated DE1160LF and is capable of convergence zone performance. The DE1160LF was delivered to the Italian Navy for the helicopter carrier *Giuseppe Garibaldi* early in 1984. This system has also been ordered by the Spanish Navy and will upgrade its 'Baleares' class frigates.

A VDS version of the DE1160LF will combine the convergence zone performance of the *Garibaldi* sonar with the environmental adaptability of the DE1164 under the denomination DE1164LF, a sonar system for major ASW combatants. This system is installed on the Italian Navy's 'Animoso' class ships.

The AN/SQS-56 sonar features digital implementation, system control by a built-in mini-computer and an advanced display system. Digital implementation allows packaging of the complete multifunction active and passive sonar in five medium-size electronics cabinets and one operator's console. Computer-controlled functions provide a system which is extremely flexible and easy to operate. The computer is also used to provide automated fault detection and localisation and a built-in training capability. The human-engineered display ensures proper interpretation by operators, even by those with relatively low levels of training.

The sonar is an active/passive, preformed beam, digital sonar providing panoramic echo ranging and panoramic (DIMUS) passive surveillance. All signal processing, except transducer received signal amplification and linear transducer transmit drive, is accomplished in digital hardware, most of which is implemented using US Navy SEMP (Standard Electronic Module Programme) components in compact water-cooled cabinets small enough to allow installation through standard size hatches. All visual data are presented on flicker-free, digitally refreshed television type raster scan CRT displays. Complete symbol and alphanumeric facilities are included. System timing, control and interface communication are accomplished by a general purpose mini-computer which is a component of the basic system. Both 400 Hz synchro and MIL-STD-1397 Type A or C, Cat. II digital interfaces are available in the basic sonar system. Except for the 400 Hz synchro reference power, the

A US Navy sonar operator using the AN/SQS-56 dual-sonar display

entire sonar operates from 440 to 480 V, three-phase 60 Hz ship's power.

The basic system includes the transducer array, transducer junction box, five electronics cabinets (array interface, transmitter(s), receiver and controller), operator console, sonar dome and control unit. Options in production include a loudspeaker/intercom, a water-cooling unit, a remote display and a performance prediction subsystem.

A single operator can search, track, classify and designate multiple targets from the active system while simultaneously maintaining anti-torpedo surveillance on the passive display. Computer-assisted system control permits the operator to concentrate on the sonar data being displayed rather than on the system control.

The improved version of the system features a new digital signal processor and colour display integrated with the existing sonar transmitter and array face. New algorithms combine the outputs of multiple receivers providing a low false alarm rate colour display which has proved to be effective in shallow waters. The sonar also features a small object avoidance capability demonstrated in Operation Desert Storm and a torpedo detection function. The system is available in 3.75, 7.5 and 12 kHz transmit frequencies and can be provided with optional activated towed transmit array and receive arrays. Other options available include sonobuoy

processing, TMA and integration of non-acoustic sensors.

Operational status

Approval for service use was issued early in 1980 by the US Navy following successful completion of final operational test and evaluation. By 1997, systems ordered for the AN/SQS-56/DE1160 totalled more than 100 and included 53 systems and 10 trainers for the US Navy; four systems for Australia (AN/SQS-56); seven systems for Saudi Arabia (AN/SQS-56); six systems for Turkey (AN/SQS-56); six systems for Italy (four DE1160, one DE1160LF and one trainer); 18 systems for Spain (six DE1160, five DE1160LF, six AN/SQS-56SP and one trainer for 'Oliver Hazard Perry' class ships); one DE1160 for Morocco; and four DE1160 HM/VDS for Greece (for the MEKO frigates).

In January 1990, Raytheon announced a contract to supply AN/SQS-56 hull-mounted sonar systems to Taiwan for installation on the new 'Cheng Kung' class frigates. Six systems have been delivered and a seventh is in production.

Contractor

Raytheon Company, Electronic Systems, Portsmouth, Rhode Island.

VERIFIED

DE1160 (I) sonar systems

Type

Hull-mounted active and passive sonar system.

Description

The DE1160 (I) is a derivative of the successful AN/

SQS-56 and DE1160 Series sonars produced by Raytheon. The system utilises the same active transducers and transmitters of the DE1160 sonar and incorporates an open architecture processing and display system using colour display processing. All the systems developed under the DE1160 and DE1167 Series (3.75 kHz, 7.5 kHz and 12 kHz systems) can be

upgraded to this newer technology. In addition to the hull-mounted systems, this system can also be implemented in a VDS mode. The DE1160 (I) system features automatic computer-aided detection and tracking, multiple simultaneous receivers for enhanced shallow water performance, a small object avoidance detection function which was demonstrated in

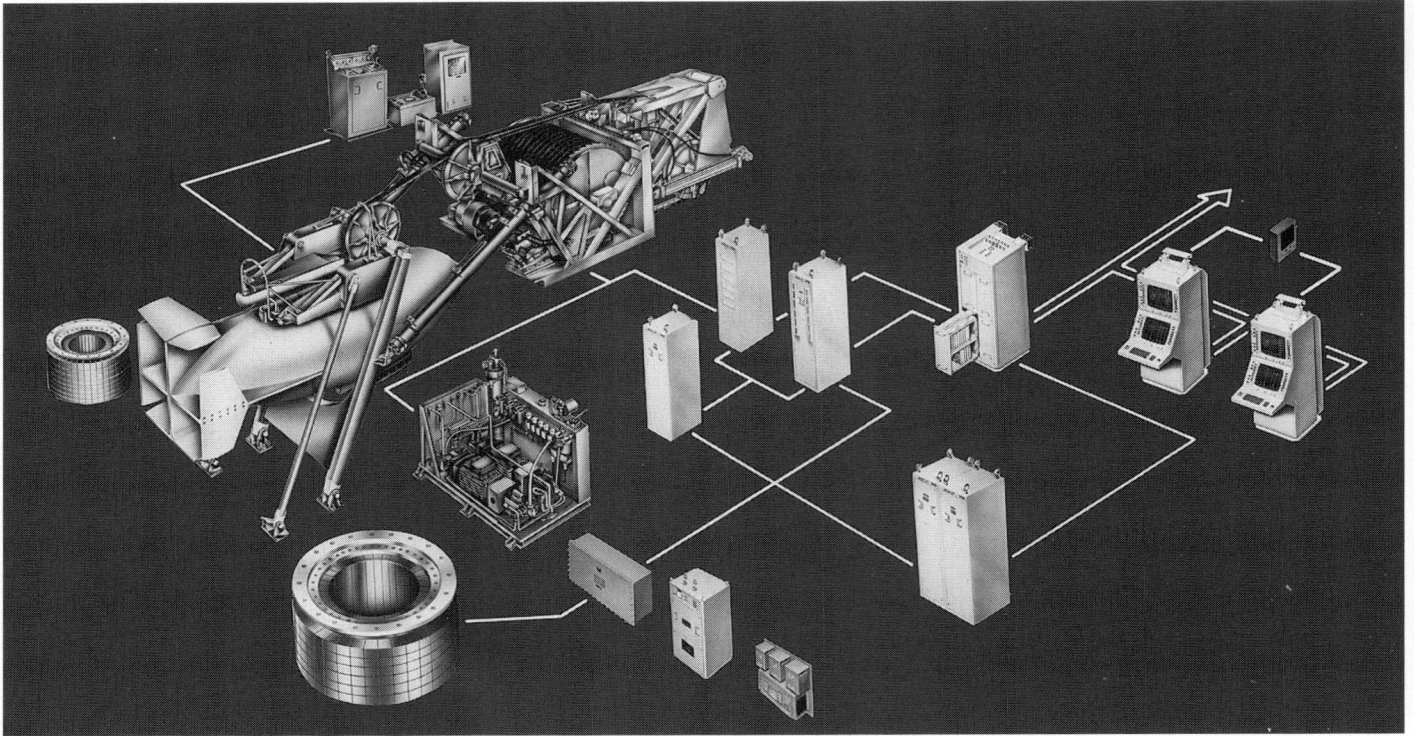

Schematic layout of the DE1160 (I) sonar system   *1996*

Operation Desert Storm and colour displays. Geographic stabilised PPI displays and B-scan displays are used for detection and enhanced A-scan and range-crossrange windows are used as classification aids. A rapid replay feature is provided to allow the operator to evaluate target motion and discriminate against false alarms. The system has the ability to replay up to 20 pings from memory for classification analysis.

The design of the system is implemented in ruggedised COTS hardware which can be either water-cooled or air-cooled. The Digital Processor Unit (DPU) is composed of a number of separate, dedicated processors connected with the redundant system-wide pair of Ethernet buses. Each processor within the DPU contains a multipurpose module set in a VME baseplate. Modules are designed in the 6U Eurocard format and are based on industry standards – many are available as third party off-the-shelf modules.

All software is developed to DOD-STD-2167A requirement and a Raytheon standardised development methodology.

Operational status
A prototype system was tested aboard the Italian frigate *Grecale* in the summer of 1994.

Contractor
Raytheon Company, Electronic Systems, Portsmouth, Rhode Island.

VERIFIED

DE1164 sonar

Description
The DE1164 sonar consists of the Raytheon DE1160 hull-mounted sonar augmented by a fully integrated variable depth sonar subsystem. All sonar functions of the DE1164 are identical to those of the DE1160. However, the DE1164 provides transmission and reception via various combinations of the hull-mounted and/or the towed Variable Depth Sonar (VDS) transducer arrays. Addition of the VDS subsystem improves overall sonar operational flexibility and allows the VDS transducer array to operate at acoustically favourable depths and in a much quieter environment.

In addition to the components of the full DE1160 hull-mounted system, the DE1164 includes one extra cabinet of electronics, a VDS towed body with associated cable and the electrohydraulic mechanism associated with launching, towing and retrieving the VDS body. The VDS handling equipment provides for one-man operation for launching and retrieving and unattended no-power towing. For reliability, two independent hydraulic power supplies are provided, either of which may support the entire operation; an emergency retrieval system is also available as an option. VDS body weight, cable size, length and careful attention to drag provide a VDS depth capability greater than 200 m at 20 kt of ship's speed.

Both the hull-mounted and VDS arrays use the common set of DE1160 transmitting, receiving and display electronics. Selection of the particular combination of transmit/receive array functions is ordered by the operator via the sonar console input keyboard; the system computer then sets up the required sequence. Alphanumeric symbols on the display inform the operator about which particular array is in use during any specific ping-cycle.

During normal operation, the power requirements of the DE1164 are identical to those of the DE1160C. A maximum of 74 kW additional power is required from the 440 V, three-phase, 60 Hz power mains during VDS retrieval or launching. The hull-mounted sonar may be

VDS configuration of DE1164 sonar on the Libeccio *of the Italian Navy*

operated as a DE1160 when the VDS is stowed or being launched/recovered.

Operational status
A total of 10 DE1164 systems have been delivered to the Italian Navy and are in operational service on 'Maestrale' and 'Alpino' class ships. Further deliveries include a VDS trainer and two DE1164LF systems which are in service aboard the Italian Navy 'De La Penne' class guided missile destroyers. The primary electronics system for this latest DE1164LF award were

built in Italy under co-production and licence agreements. The mechanical hoist was built by Fincantieri Naval Shipbuilding Division in Genoa and part of the transmitter and transducers by Elsag and its subsidiaries.

Contractor
Raytheon Company, Electronic Systems, Portsmouth, Rhode Island.

VERIFIED

DE1167 sonar

Type
Hull-mounted and variable depth sonar systems.

Description
Raytheon's DE1167 family of sonars implements the proven features of the AN/SQS-56/DE1160 systems using advanced microprocessor architecture and state-of-the-art display and transmitter technology. The DE1167 series is based on large size modules and air-cooled cabinets, which permit the production of smaller, simpler, cheaper sonar systems and facilitate in-country manufacturing/repair participation where required. The DE1167 family is designed to satisfy the requirements of most ASW platforms and missions. Configurations include hull-mounted, VDS and integrated HM/VDS systems, featuring a 12 kHz VDS and either 12 or 7.5 kHz hull-mounted 36 stave transducer arrays. Configurations using 48 stave arrays or a frequency lower than 7.5 kHz are possible.

Like the Raytheon DE1160B/C and DE1164 sonars, the DE1167 features primarily digital electronics and an advanced control and display system. The standard inboard electronics consist of three cabinets and a single operator console. Outboard units consist of a transducer array and 2.74 m long dome for the hull-mounted 12 kHz installation (4.17 m long dome for 7.5 kHz) and/or the VDS winch, overboarding assembly, control station, hydraulic power supply, faired cable and towed body for the 12 kHz variable depth subsystem.

The basic DE1167 HM is an active/passive, preformed beam, omni and directional transmission sonar which uses three non-interfering 600 Hz wide FM transmission bands, centred at 12 or 7.5 kHz and has a spatial Polarity Coincidence Correlation (PCC) receiver. The passive mode, which is selected automatically when transmissions are stopped, is primarily useful for torpedo detection. Optional items in production include: a performance prediction subsystem, an auxiliary half-frequency passive receiver, an auxiliary display, a remote display and a training/test target remote-control unit. Signal reception and beam-forming are accomplished by broadband analogue circuitry followed by clipper amplifiers for perfect data normalisation. Detection processing, display processing, system control/timing and waveform generation are done digitally. Two microprocessors perform display ping history, cursor ground stabilisation and target-motion estimation functions for torpedo direction. The modular air-cooled 12 kW transmitter uses highly efficient class A/D power transistor techniques.

Several operational features are unique to a sonar of this size and range. First, the display processing incorporates a ping history mode through which the sonar data obtained in three of the previous ping cycles may be retained on the viewing surface, allowing the operator to differentiate readily between randomly spaced noise samples and geographically consistent, valid acoustic reflectors. Secondly, the clipped PCC processing permits accurate thresholding of all signals such that the false alarm rate, or number of random noise indications on the screen, remains relatively low and constant over all variations in background noise

Operator's display/control console of Raytheon DE1167 sonar

and reverberation levels, further facilitating contact detection. Thirdly, ground-stabilised cursors and target motion analysis permit rapid determination of contact motion over the bottom, an excellent clue as to the nature of the contact. These three operational features, combined with extremely accurate tracking displays, a built-in fault detection/localisation subsystem, performance verification software and test/training, result in a high-performance system which is easy to operate, maintain and support.

Operational status
As of November 1988, 41 systems were under contract, including 26 DE1167 for the Korean Navy, two DE1167LF/VDS systems for Spanish-built corvettes for the Egyptian Navy, nine DE1167LF systems for the Italian Navy and four DE1167LF systems for Japan. ELSAG has implemented a licence agreement with Raytheon to manufacture DE1167 sonars for the Italian Navy corvette programme.

Specifications
Centre frequency: 12 kHz (HM and VDS), 7.5 kHz (LF and HM)
Source level: TRDT 227 dB (HM), omni 217 dB
Pulse type: 600, 2,000 Hz FM sweep; 100, 200, 50, 6 ms pulse lengths
Receiver type: Spatial Polarity Coincidence Correlator (PCC) between 36 pairs of half-beams
Beam characteristics: 36 sets of right and left half-

beams for active and passive detection. Selectable 10°H × 13°V scm beam for audio listening. 1.25° bearing interpolation for fine search display
Active display: 300 range cells, 288 bearing cells; single and multiple echo history; 4 intensity levels; flicker-free, bit-image memory technique
Passive display: Electronic Bearing Time Recorder (EBTR) with med um time averaging. DIMUS-type LTA/STA with optional passive receiver
Track displays: Sector scan indicator (1,000 yd × 10°) and target Doppler indicator (1,000 yd × ± 60 kt)
Target data: range: 8 yd (6.1 m) resolution. Bearing 1.25° resolution plus active search display – 0.1° and 3.3 yd on SSI display. ± 60 kt of Doppler at 1 kt steps on the TDI
Data format: standard: NTDS ANEW (digital). Optional: NTDS slow, Fast Serial D/S synchro converters
Power requirements: passive 800 W. Active 20 kVA (pulsed) at 10% (max) duty cycle 440 V, 60 Hz, 3-phase, VDS launch/retrieve 75 kVA (max) 440 V, 60 Hz, 3-phase
Weights: Hull-mounted 12 kHz; 1,500 kg (nominal); VDS 10,000 kg (nominal) with 200 m cable length

Contractor
Raytheon Company, Electronic Systems, Portsmouth, Rhode Island.
Note: For further information see Italy.

VERIFIED

DE1191 sonar modernisation

Description
The DE1191 sonar upgrades and improves the performance of older AN/SSQ-23 and AN/SQS-23 sonars which are being used by navies worldwide aboard numerous FRAM II destroyer class ships.

The DE1191 features advanced inboard electronics which, when coupled with the existing AN/SQS-23 dome and array, comprise a newly configured, lightweight system with the capability to outperform most current modern surface ship sonars. The AN/SQS-23 is a sonar of 1950s design. The system has the potential for excellent performance, with a large low-frequency transducer array situated in a quiet location and on a relatively quiet platform.

Replacement of the AN/SQS-23, inboard electronics with DE1191 hardware, a solid-state transmitter and a modern receiver/display subsystem, eliminates weight totalling 12 tons and reduces onboard space requirements by 450 ft³.

This massive equipment reduction greatly enhances

long-term reliability, maintainability and logistic support for both transmitter and receiver/display functions. Furthermore the DE1191's computer-assisted receiver/display, a slight modification of Raytheon's DE1167 receiver/display, has notably superior detection range and operability features, resulting in an impressive 14 dB improvement in sonar performance.

DE1191 inboard electronics are solid-state, ensuring a longer and more economical life cycle as compared to the vacuum tubes currently used by AN/SQS-23 systems. Improved equipment and the reduction in hardware minimise the need for large and costly spare part inventories, both on board and at depot. In direct contrast to the AN/SQS-23, where spare inventories are soon to be completely phased out by the US Navy, the DE1191 equipment is common to several other Raytheon sonars and availability is guaranteed for 20-30 years.

Modernisation can be achieved incrementally in two major stages: the Solid-State Transmitter or SST, which is always installed first and later the receiver/display – or all at once, as time and funding permit. The

DE1191's streamlined configuration will reduce both time and expense necessary to conduct maintenance and operator training.

The modern integrated circuits of the DE1191, proven system techniques and high-volume production have all contributed to low-cost equipment that is economical to support. It is available to the international market directly through Raytheon or, via foreign military sales arrangements with the US government.

Operational status
In production. n February 1990, Raytheon reached agreement with the Greek Navy for production of three DE1191 systems for use aboard former US Navy 'Charles F Adams' class destroyers. These will use the existing AN/SQS-23 dome and array, but will employ new lightweight electronics.

Contractor
Raytheon Company, Electronic Systems, Portsmouth, Rhode Island.

VERIFIED

Solid-State Transmitter (SST)

Type
Sonar transmitter subsystem.

Description
The Raytheon SST provides sonar systems with the benefits of modern, solid-state technology, offering greatly improved reliability and maintainability, plus very significant space savings relative to vacuum tube transmitters. This transmitter subsystem is made available for system modernisation or for new system applications.

Configured to meet a specific system's requirements, the SST consists of multiple 1 kW modules with associated power supplies. As configured for the AN/SQQ-23 and AN/SQS-23 sonars upgrade, the subsystem consists of two cabinets of 24 modules each and a performance monitor/system interface cabinet. It replaces more than 30 units of the previous transmitter, including all the energy storage motor generators.

Operational status
Production includes transmitters for seven navies. Similar transmitters are used on the AN/BQS-13, AN/SQS-56, DE1160 series and the AN/SQQ/SQS-23 sonar systems. An air-cooled version of the SST has been delivered in a programme to update the AN/BQS-4 submarine sonar.

Specifications
Standard 1 kW module (1 per channel)
Power output: 1 kW
Operating frequency: ± 1.5 kHz (nominal)
Distortion: 6% max
Load: nominal ± 100%
Duty cycle: 15%
Linearity: ± 1 dB from 0—12 dB
Gain adjustment: 3 dB
Protection: short-circuit and over-temperature
Mean time to repair: 3 min
Weight and size: 1,651 kg in 1.2 × 1.8 m of deck space
SST as manufactured for AN/SQQ/SQS-23
Power requirements: 115 V AC, 60 Hz 1-ph, 0.37 kVA, 0.92 PF; 440 V AC, 60 Hz, 3-ph, 128 kW (Max 0.9 PF)
Modes of operation
(a) transmit: same as any associated system such as omni, sector, FM or CW
(b) performance monitoring: automatic and manual
(c) self-test: receiving system not required.

Contractor
Raytheon Company, Electronic Systems, Portsmouth, Rhode Island.

VERIFIED

Raytheon standard sonar transmitter cabinet

Model 610 sonar

Type
Hull-mounted active sonar system.

Development
The Model 610 was developed by the EDO Corporation as a private venture starting in 1965. The first prototype was completed in 1966 and the first production model completed its sea trials in 1969. The Model 610 has been continuously improved and the 610E model is of solid-state construction.

Modernisation packages are now offered which are based on advances made in the EDO Model 780 Series and the US Navy AN/SQR-18A(V)3 active receiver. These incorporate items such as operator selectable three-frequency operation, raster displays, digital scan switches, performance prediction and new console switches.

The latest improvement package, which is identified as the Model 610E-Mod 1, incorporates modern COTS digital signal processing and colour display technology. The Mod 1 variant upgrade replaces the receiver/processor/display portion of the system and reduces the system manning requirement to one operator. This improvement upgrade is being used to upgrade the Brazilian 'Niteroi' class 610 sonars which started in 1998.

Description
Designed for the long-range detection of submarines in deep and shallow water, the EDO Model 610 36-stave, scanning sonar has two active consoles, enabling it to perform a search-while-track function. Facilities offered include a search capability in three 120° sectors, passive correlation and reverberation processing. The transmitter and receiver beams are preformed. Output is available for a fire-control system. All mode changes and range scale changes are controlled by console push-buttons and displays include a Doppler display on each of the active consoles and a passive sonar bearing time recorder display.

Operational status
The basic system is no longer in production. Model 610 systems are in operational service with: the Royal Netherlands Navy (on the 'Tromp' class as the CWE 610), the ex-Netherlands 'Van Speijk' class frigates of the Indonesian Navy (as the CWE 610), the Italian 'Audace', the Peruvian cruiser 'Almirante Gray' and 'Lupo' class frigates, the Venezuelan 'Lupo' class frigates (in a modified version as the SQS-29) and on the Brazilian 'Niteroi' class frigates.

Contractor
EDO Corporation, Combat Systems Division, College Point, New York.

UPDATED

Model 700 series sonars

Type
Hull-mounted and variable depth sonar system.

Development
Like the Model 610, the Model 700 series was developed by the EDO Corporation as a private venture for sale on the international market. Modern upgrades are available using the latest in digital signal processing and display hardware and software techniques. These upgrades have a high degree of commonality with the EDO Model 610E-Mod 1 sonar modernisation and have been given the designation of EDO Model 983 and Model 993.

Description
The model 700E is a medium-range hull-mounted sonar. The Model 700/702 hull-mounted VDS uses common 700E electronics with a hull-mounted transducer and a lightweight VDS hoist. Selection of hull-mounted or VDS operation is by push-button on the operator's console. The Model 700/701 is a VDS which provides a capability to detect deep targets when bathythermal conditions are unfavourable for hull-mounted sonars.

The basic equipment has a 254 mm panoramic CRT display and a Doppler display. All mode and range scale changes are made by push-button controls on the operator's console.

Operational status
The basic system is no longer in production. Model 700 sonars are in operational service with the Brazilian Navy (on the ASW frigates of the 'Niteroi' class).

Contractor
EDO Corporation, Combat Systems Division, College Point, New York.

VERIFIED

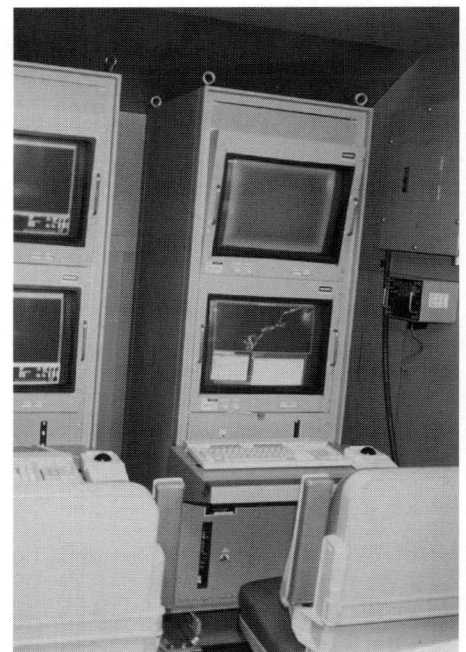
Typical colour standard console used on new system configurations of the EDO Model 610/700/780 Series
1995

Model 780 series sonar

Type
Hull-mounted and variable depth sonar systems.

Development
The Model 780 series has been developed by the EDO Corporation for sale on the international market. New variants of the 780 sonar family is being offered which is designed and fabricated using latest technology COTS Digital Signal Processing (DSP) and colour display workstation hardware and software. Active sonar signal processing and operator interfaces are based on

techniques developed for the US Navy's AN/SQR-18A (V)3 active receiver.

Description

The EDO Model 780 series sonar is a family of high-performance computer-based sonars designed to maximise the ASW capability of ships ranging from high-speed patrol craft to ASW frigates. Configured as a modular system, the Model 780 series can be assembled to match ship size constraints and required ASW capability for hull-mounted or variable depth sonar operation. The following models are available:

780: 13 kHz Variable Depth Sonar (VDS)
786: 13 kHz hull-mounted
795: 5 kHz hull-mounted
796: 7 kHz hull-mounted
7860: combined 13 kHz hull-mounted with 13 kHz VDS
7867: combined 13 kHz hull-mounted with 7 kHz VDS
7950: combined 5 kHz hull-mounted with 13 kHz VDS
7960: combined 7 kHz hull-mounted with 13 kHz VDS
7967: combined 7 kHz hull-mounted with 7 kHz VDS.

The system features simultaneous active and passive search, tracking and classification. The operator can conduct 360° active/passive search operations while automatically tracking and analysing targets of interest. Display formats include: panoramic B-scan detection; panoramic B-scan tracking and identification; expanded range versus speed target Doppler indicator; expanded range versus bearing sector scan indicator; 360° passive bearing versus time search; expanded passive bearing versus time bearing deviation indicator; passive analysis target noise indicator; and acoustic performance ray path plots. Alerts of hostile sonar emissions detected by the integral acoustic intercept receiver are prominently displayed on the CRTs. A variety of transmission modes allows the operator to select the appropriate transmission mode for the tactical situation.

At sea tests of the new linear FM coherent signal processor have resulted in exceptional long-range detections under difficult high-reverberation, shallow water conditions.

The Model 780 series' electronics are flexibly designed. The latest Model 796 Mod 1 sonar has been adapted to use a customer furnished standard console in place of the original control console. This system incorporates the latest improvements, among which are automatic multiple target tracking, multicolour display, increased target capacity and flexible software-based display formats.

The VDS versions have exceptionally lightweight handling systems which permit installation on ships as small as 250 tons. The Model 787 features a new design compact 7 kHz transducer which has the same weight and dimensions as the 13 kHz Model 780. This provides for greater detection ranges without the weight and space penalties associated with conventional 7 kHz VDS systems. The Model 780/787 VDS tow body is very similar to the AN/SQS-35 tow body and is therefore capable of towing the AN/SQR-18A(V)1 tactical towed array system.

The combined hull-mounted/VDS systems are also single-operator systems. The combined electronics allow the one operator to simultaneously operate both

780 series VDS

Model 780 sonar system equipment comprising sonar control console, data storage and control, transmitter/power supply and sonar receiver

the hull-mounted and VDS sonars including VDS depth control.

Operational status

Hull-mounted and VDS versions are currently deployed operationally. The Model 796 Mod 1 has completed initial sea trials and is in service aboard the 'Victory' class corvettes of the Singapore Navy. The sonar has been fitted to some of the Israeli 'Reshef' class vessels and the Model 796 to the three 'Eilat' class.

Contractor

EDO Corporation, Combat Systems Division, College Point, New York.

UPDATED

Model 997 series sonar

Type

Hull-mounted active sonar system.

Development

The Model 997 series has been developed for sale on the international market. The systems being offered use the latest COTS/OSA hardware and software. These modern sonars are new technology upgrades of the Model 610E sonars and was previously designated the Model 610E Mod 1.

Other hull sonar variants which share the identical system architecture are:

Model 993   13 kHz centre frequency
Model 993 Mod 1   13 kHz compact series
Model 995   5 kHz centre frequency
Note: All variants are provided with operator selectable three-frequency band operation.

Description

The Model 997 is a nominal 7 kHz, 36 stave hull sonar which is suitable for corvette and larger displacement ships. Three separate frequency bands are selectable from the single-operator control console. A panoramic and various directional transmission modes are standard. The hardware consists of one console unit, one receiver/processor unit, a small signal conditioner interface box and either two or three transmitter units. The console is a standard US navy type workstation and is provided with two high resolution colour monitors. All variants can be provided with one or two flat panel type colour displays to further reduce the system 'footprint' on small hulls. Major features are coherent signal processing of both FM and CW waveforms, either separately or simultaneously and post-processing to reduce false alarms. Broadband passive detection operates concurrently and utilises its own frequency band below the active bands. The use of standard digital interfacing standards simplifies the integration of the sonar with the ship's combat system.

Displays include 12 ping history active search and expanded zoom formats for each preformed beam, target classification aides, passive broadband bearing/time history and a tactical (geographic) format. Display overlays are provided for performance prediction, bottom topography and land mass outlines. The tactical display is formatted to easily interface to other onboard or offboard sensors.

Operational status

A contract has been placed for six systems with an option for an additional system. The systems are in production and installation of the first system will take place in 1998.

Contractor

EDO Corporation, Combat Systems Division, Chesapeake, Virginia.

VERIFIED

Model 980 series sonar

Type
Active towed array sonar system.

Development
The Model 980 (ALOFTS) Series has been developed by the EDO Corporation for sale on the international market. The systems being offered use the latest COTS/OSA hardware and software techniques. These modern sonars are new technology low-frequency upgrades of the AN/SQS-35(V)/SQR-ISA(v) sonars. The low-frequency concept demonstration model was developed with EDO funding and tested aboard a US Navy frigate in 1983. The normal 13 kHz transducer was replaced with a low-frequency flextensional transmitting array. Further development was funded by the US Navy and provided as a triple array tow from the VDS towed body. This system was tested by the US Navy in 1989.

Two variants, which share the identical system architecture, are offered as follows:

Model 980-1	1.1 kHz nominal centre frequency
Model 980-2	2.2 kHz nominal centre frequency

Note: All variants are provided with operator selectable two frequency band operation. Other centre frequencies can be provided.

Description
The Model 980 uses the SQA-13 VDS body and SQR-18 style towed array combination which has been towed at speeds up to 30 kt on the US 'Knox' class frigates. The system is suitable for corvette and larger size ships.

The Model 980 is controlled from a single console by one operator who selects two separate frequency bands from the console. Port-starboard bearing ambiguity is resolved through the use of directional transmissions. The highly directional transmissions are produced at 219 dB source level. The hardware consists of one console unit, one receiver/processor unit, a small signal conditioner interface box and either one or two transmitter units. The console is a standard US Navy type workstation and is provided with two high resolution colour monitors. Alternatively, the system can be provided with one or two flat panel type colour displays to further reduce the system footprint on small hulls. Modern COTS digital signal processors and computers are standardised for all variants. The system features coherent signal processing of both FM and CW waveforms, separate or simultaneously and post-processing to reduce false alarms. Broadband and/or narrowband passive detection operates concurrently and utilises its own frequency band separate from the active bands. The use of standard digital interfacing standards simplifies the integration of the sonar with the ship's combat system.

Displays include 12 ping history active search and expanded zoom formats for each preformed beam, target classification aids, passive broadband bearing/time history and tactical (geographic) format. Display overlays are provided for performance prediction, bottom topography and land mass outlines. The tactical display is formatted to easily interface to other onboard or offboard sensors using standard symbology.

Optional features include passive broadband search and track, passive classification using DEMON and a training controller.

Contractor
EDO Corporation, Combat Systems Division, Chesapeake, Virginia.

UPDATED

5951 submarine classification sonar system

Type
Side scan sonar system for detection of bottomed or hovering submarines.

Description
Klein side scan sonar systems provide the ASW operator with the capability to classify bottomed or hovering submarines. The sonar provides high-resolution images of the target area permitting high-probability classification of targets that standard low-frequency ASW sonars do not have the resolution to resolve.

The 5951 side scan sonar system provides the operator with sufficiently high-resolution images of the target area to resolve bottomed or hovering submarine targets from geological or biological targets without difficulty.

The system consists of a lightweight VDS transducer which is towed in the vicinity of the target to be classified. Sound energy is projected outward from the transducer in a very narrow horizontal beam pattern, 1° or less depending on frequency of operation, to provide extremely high along-track resolution with a resulting high-definition image of the target. The vertical beam pattern is large, typically 40° to provide the maximum insonification of a large volume of the sea.

The 5951 sonar system is lightweight (weighing less than 100 kg (225 lb) complete, including the VDS transducer), operates on 110/220 V AC or 24 V DC and consumes less than 200 W of power, permitting operation from various platforms as large as destroyer type vessels down to as small as 10 m patrol craft. Due to the low weight and power requirements, installation of the sonar system has minimal impact on the stability of the vessel or the power generation system.

Various operating frequencies are available, including 50 kHz for long-range, medium-resolution operations, 100 kHz for high-resolution operations, 500 kHz for very high-resolution and simultaneous 100/500 kHz operation for multispectral imaging for resolution of the more difficult targets.

Record of a US 'Guppy I' class submarine showing stern quarter aspect (stern to the left of photograph) from a Klein 500 kHz side scan sonar. Note the conning station on the forward (right) edge of the conning tower and possible antenna bracket on rear of conning tower. The picture was gathered at a range of 50 m (164 ft) and depth of 35 m (115 ft)

In addition to the ASW applications, the system can provide a secondary capability when used for the detection of mine and mine-like targets.

Operational status
All versions of the sonar are currently in production and are in service with navies worldwide for anti-submarine warfare, mine countermeasures and oceanographic related applications.

Contractor
Klein Associates Incorporated, Salem, New Hampshire.

VERIFIED

UQQ-2 SURTASS

Type
Mobile towed array surveillance system.

Development
SURTASS (Surface Towed Array Sonar Segment) commenced full-scale development in 1976. Testing was carried out at sea in 1977. Technical evaluation of SURTASS was completed in March 1978 and operational evaluation took place in May and June 1978. The first full tests of the system, including the satellite link and data processing, took place in July 1979 and the system was declared ready for operational evaluation in March 1980. Introduction to US fleet service began in late 1984.

Description
SURTASS is the successor system to the AN/BQR-15 Towed Array Surveillance System (TASS) already in operation with the US fleet. This mobile surveillance system complements fixed networks by providing the essential flexibility to respond to changes in Soviet submarine deployment patterns and by extending coverage to remote ocean areas not monitored by fixed systems. It can also serve as an emergency reserve facility if fixed networks are disabled. There are two shore-based data processing centres (one on the east coast of the USA and one on the west coast) to which acoustic target information gathered by the towed arrays is relayed by satellite. After processing, target information is transmitted back to operational ships at sea. Several ships are planned for each of the two processing centres, operating in waters to the east and west of the North American continent respectively.

Whereas the AN/BQR-15 is towed by submarines, slow surface ships will tow long SURTASS arrays back and forth over designated patrol lines. Funds have been set aside and a special type of platform to operate SURTASS, designated T-AGOS, has been built. More detailed information on these ships is given in *Jane's Fighting Ships*.

The basic portion of SURTASS is a 2,500 m long hydrophone array at the end of a cable 1,800 m long. Five AN/UYK-20 data processors and an AN/UYS-1 signal processor handle data processing on board the T-AGOS ships before the data is relayed via the satellite. The sonar element also includes the winch, handling equipment and associated electronics. The other three major elements include: the communication/navigation system for acoustic data transmission, command and control and vessel navigation; the T-AGOS ships themselves and the shore processing stations. It is intended that the T-AGOS ships will operate worldwide in 90 day missions.

A number of improvements has been introduced into the product line and, during 1984, a product improvement programme was implemented. This applied to the AN/UYS-1 processor and other classified units and, during 1985-86, specifications and designs for classification improvements for the AN/UYS-2 enhanced modular signal processor were developed.

Details concerning the Low-frequency active Transmit Subsystem (LTS) can be found in the section on Transducers.

The latest development of the system, the SURTASS Twin line and LFA is a much shorter twin-line array that can be used for both active and VLF passive operations. The system has been tested both in very shallow water (50 m) and against high clutter. Trials have been undertaken in real operations in the Adriatic.

The twin lines are maintained at a constant distance apart by the active towing system.

Much of the electronics used in this system are based on open architecture and COTS components.

Operational status
A total of 14 T-AGOS ocean surveillance ships is now operational with the US Navy. In mid-1990, Hughes was awarded a US$24 million contract for block upgrades to the basic SURTASS equipment. In 1986, the US Navy called for a design revision in the T-AGOS programme and the 'Victorious' class ships were completed to a Small Waterplane Area Twin Hull (SWATH) design. In mid-1990, Hughes was awarded a US$45 million contract to provide additional capabilities to additional SURTASS systems on the SWATH ships. This is now valued at about US$80 million. The programme is scheduled to complete in 1995.

A separate block upgrade to the existing platforms is under way. It is valued at US$27 million and was

One of four 'Victorious' class ocean surveillance ships used to deploy the UQQ-2 SURTASS towed array system. The winch and array handling control position can be seen behind the tripod mast **1998**/0007231

scheduled to complete in 1995. A concurrent production contract to support the upgrade is valued at an additional US$24 million.

Hughes also supplies the O & M technician crews under a contract valued at US$33 million, with funding approved in increments.

Plans to acquire five 'Impeccable' class T-AGOS-23 type ships have been abandoned following the collapse of the Warsaw Pact, while work on the first of class, *Impeccable,* has been stopped following difficulties at the shipyard.

The Japanese Maritime Self-Defence Force has built two 'Hibiki' class SWATH vessels, which are similar to the US T-AGOS type vessels. These ships are fitted with a complete UQQ-2 SURTASS package supplied by Hughes.

The SURTASS Twin-line and LFA began fleet operations during 1998.

Contractor
Hughes Aircraft Company, Fullerton, California.

UPDATED

SURTASS RDA

Type
Passive towed array.

Development
In December 1987, a contract was awarded for the design and development of a Reduced Diameter Array (RDA) for the SURTASS system. The aim was to provide a surface towed array with improved acoustic performance through reduced self-noise, improved dynamic range and rapid circuit recovery time to accommodate operation in the presence of strong transients. This was accomplished when the first engineering development model was delivered in May 1992 and successfully passed First Article testing.

Description
The SURTASS RDA system includes a reduced diameter towed array with multichannel digital telemetry, fibre optic towcable with fairing, deck cables, winch junction box, plug storage box and telemetry receiver unit. RDA is a centre-nested array providing longer apertures for improved frequency response. An improved coupler that connects each array module greatly reduces maintenance time at sea and increases reliability. The length of the array is 2,378 m and there are repeaters in every third module to regenerate the signal. These repeater modules, as well as others, also measure non-acoustic data such as heading, depth, temperature and current. Tuned vibration isolation modules provide rejection of towcable vibration to the array. There are fewer module types due to a universal

spares concept. The telemetry receiver unit provides power to the array, performs PKL/FL functions and conditions the signal for transmission to the beam-forming processor.

Operational status
Four RDA systems have been delivered to the US Navy and are in operational service.

Contractor
Lucent Technologies Advanced Technology Systems, Greensboro, North Carolina.

VERIFIED

TAS 2019

Type
Surface ship towed array sonar.

Description
The passive towed array TAS 2019, a variant of the US Navy's AN/SQR-19 sonar, provides long-range detection and classification of submarines, over-the-horizon surveillance of surface combatants and other surface shipping and automatic torpedo alert. The hydrophones and electronics that comprise the system are packaged in a manner designed to minimise flow noise to maintain sonar performance at tactically significant 'own ship' speeds.

The system features four nested acoustic apertures covering seven octaves with a spectral bandwidth necessary to detect acoustic signals from current and projected submarine and torpedo threats, as well as surface combatants and other surface vessels.

The sonar incorporates a sophisticated signal processing design providing increased sonar information with reduced operator workload. It features CAD (Computer Aided Detection), automatic threat alerts and automatic target tracking which allows operators to work at high-performance levels, even in coastal waters where a high volume of traffic is anticipated. The sonar includes the US Navy-supported towed array group and hydraulically driven winch assembly used by the AN/SQR-19. To minimise size,

weight and cost, the TAS 2019 uses state-of-the art signal processing which is implemented using modern, high-performance COTS computers in a single cabinet. Standard AN/SQR-19 display formats are presented on a full-colour display console workstation. With the expandable, open architecture of its commercial computers, the TAS 2019 can be easily integrated into a ship's combat suite.

Contractor
Lockheed Martin, Ocean, Radar and Sensor System, Syracuse, New York.

VERIFIED

LFATS

Type
Low-Frequency Active Towed Sonar (LFATS).

Description
LFATS uses technology derived from the HELRAS helicopter dipping sonar and comprises a towed body integrating separate transmit and receive arrays operating at 1.38 kHz, a compact integrated winch and handling system and inboard processing subsystem. The inboard electronics comprise the transmit power amplifier providing a very high source level and sonar processing units which are designed around an open architecture configuration to allow future technology and performance improvements to be embodied using COTS components. The transmit array consists of 16 compact projector elements contained in the body of the towed vehicle. An innovative array extension and retraction mechanism enables the system to develop high transmit directivity. Return signals are received by up to four small diameter liquid filled linear towed arrays. Convergence zone capabilities have been proved and trials have shown the system to have an extended detection range capability in excess of 30 miles. Two variants are available, a heavier unit which incorporates a high-powered winch to tow the system at 15 kt for installation in frigates and a lighter unit for installation on smaller vessels down to 100 tonnes. The array and handling system weighs 3.3 tonnes.

The low frequency results in a very long wavelength which should not be affected by any internal acoustic treatment in the submarine and is able to penetrate any currently known acoustic treatment.

The sonar is capable of very high resolution of very low doppler.

The sonar is able to operate at depths between 15 and 300 m.

Operational status
Contractor trials have been undertaken and a demonstration unit has begun to carry out in-water demonstrations.

Specifications
Source level: 222 dB/1μPa.
Frequency: 1.38 kHz.
Operating speed: 5-15 kt
Survival speed: 30 kt
Operating depth: 15-300 m (15-100 m @15 kt)
Weight: 3,270 kg (array and handling system)

Contractor
L3 Communications, Ocean Systems, Sylmar, California.

LFATS undergoing sea trials. Note the linear towed arrays extending from the fairing on top of the body.
(Photograph courtesy of L3 Communications)   *1998*/0007235   **UPDATED**

21HS series

Type
Commercial COTS sonar.

Description
The 21st Century Hull Sonar (21HS) family is a multimode hull, bow or VDS active and passive sonar systems. The sonar is designed for installation on surface ships of varying sizes and hull configurations. It provides long-range search, detection and tracking using direct path, surface duct and variable depression modes. The 21HS provides active and passive receive processing at three distinct centre frequencies: 3.5, 7 and 13 kHz determined by the customer's sonar requirements and constrained by the specific ship's available power and weight capability.

The 21HS active capability provides multiple search strategies using CW, FM and SW (shallow water) waveforms. Passive capability provides detection of broadband, DEMON and narrowband radiated noise. The system utilises colour, X-Windows displays that can be user customised. The 21HS offers a very small footprint for a high-power sonar.

Operational status
Trained sonar operators were used to evaluate the 21HS MMI under operational conditions during tests conducted aboard a US Navy DDG 51 destroyer in June 1996. All of the sea test objectives were met or exceeded.

Northrop Grumman has been awarded a contract by Ingalls Shipbuilding, a division of Litton Industries Inc, to refit two Venezuelan 'Mariscal Sucre' class ('Lupo' type frigates) with the 21HS. The systems are being delivered during 1999. Venezuela plans to equip four more frigates in 2000.

Specifications
21HS-3
Frequency: 3.5 kHz
Cabinets: 7
Footprint: 5.6 m²
Electrical weight: 6,000 kg
Array weight: 11,000 kg
21HS-7
Frequency: 7 kHz
Cabinets: 5
Footprint: 5 m²
Electrical weight: 4,000 kg

Array weight: 2,700 kg
21HS-13
Frequency: 13 kHz
Cabinets: 2
Footprint: 1.2 m²
Electrical weight: 455 kg
Array weight: 227 kg

Contractor
Northrop Grumman Corporation, Annapolis, Maryland.

                    UPDATED

AMT 21 Sonar series

Type
Hull-mounted active and passive sonar system.
Mine Avoidance and Torpedo Defence, with Optional Activated Towed Array.

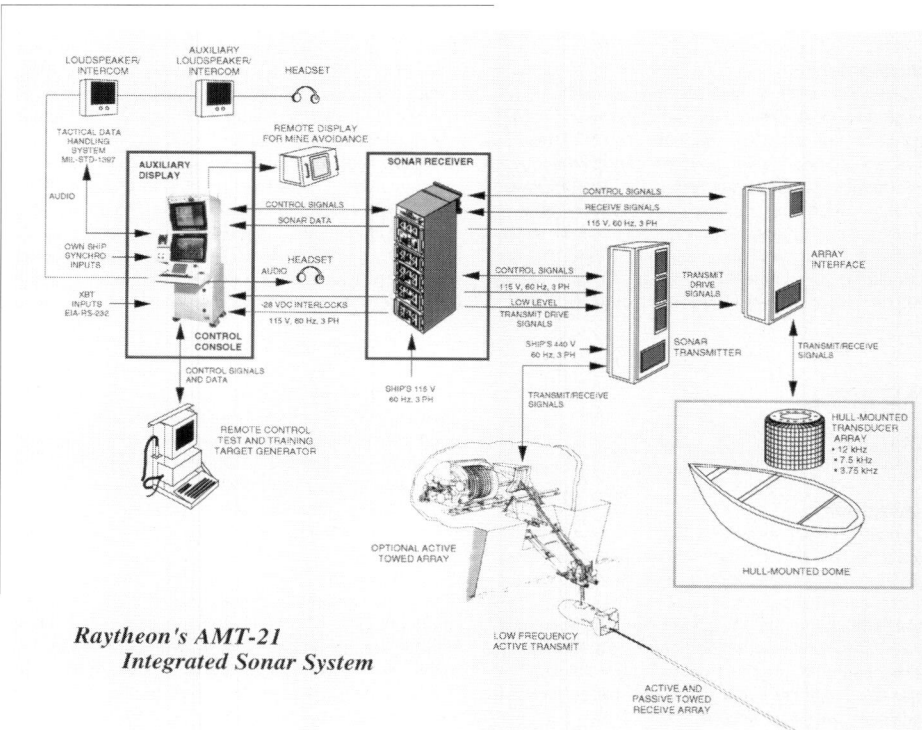

Raytheon's AMT-21 Integrated Sonar System

Schematic arrangement of the AMT-21 integrated sonar system   *1999*/0024701

Description

The AMT 21 is a state-of-the-art modern sonar system that can be implemented utilising a hull transducer or a towed active transducer or both. The sonar is an integrated ASW, torpedo defence and mine avoidance underwater system.

The sonar is available in frequencies centred at 3.75 Khz, 7.5 Khz and 12 Khz and uses proven transducer and transmitter technology. It features an open architecture processing and display system using colour display processing. Other features include automatic computer-aided detection and tracking, multiple simultaneous receivers for enhanced shallow water performance, a small object avoidance detection function (which was demonstrated in Operation Desert Storm) and torpedo defence algorithms.

Geographic stabilised PP displays and B-scan displays are used for detection and enhanced A-scan and range-crossrange windows that display target Doppler information which are used as classification aids. A rapid replay feature is provided to allow the operator to evaluate target motion and discriminate against false alarms. The system has the ability to replay up to 20 pings from memory for classification analysis.

The design of the system is implemented in ruggedised COTS hardware that can be either water or air-cooled. The Digital Processor Unit (DPU) comprises a number of separate, dedicated processors connected with the redundant system-wide pair of Ethernet buses. Each processor within the DPU contains a multipurpose module set in a VME baseplate. Modules are designed in the 6U Eurocard format and are based on industry standards – many are available as third party off-the-shelf modules.

All software is developed to DOD-STD-2167A requirement and a Raytheon standardised methodology.

Operational status

The AMT 21 3.75 Khz Sonar is currently being installed in the Spanish Navy F100 frigate.

Contractor

Raytheon Company, Electronic Systems, Portsmouth, Rhode Island.

NEW ENTRY

AN/SLX-1 (MSTRAP)

Type

System to provide automatic alertment of incoming torpedo threats.

Description

Designed for installation aboard surface ships and submarines, the MultiSensor Torpedo Recognition and Alertment Processor (MSTRAP) COTS-based system provides warning of an incoming torpedo threat on an easy-to-read display. This multifunction system provides information, signal processing and the controls necessary to detect, classify and localise threat torpedoes. It also offers command and control functions, TMA, threat evaluation and tactical advice. Displays with automatic visual and audio alarm of torpedo threats allow sufficient time to make tactical decisions and effective deployment of countermeasures and appropriate ship evasive manoeuvres. Data is received simultaneously from either an existing ASW hull sonar, towed array or sonobuoys, or the AN/SLQ-25B towed array sensor module. MSTRAP's open system architecture enables easy installation, hardware/software upgrades and future improvements to be implemented.

Operational status

The first unit of MSTRAP was delivered only 14 months after receipt of contract. MSTRAP has been operationally tested and proven at sea. Fourteen units have been delivered with six more in production. Options can be exercised to take the total production to 60 units. MSTRAP forms the Block I upgrade to the AN/SQQ-89(V) 6/10 system and will be augmented with the LEAD (Launched Expendable Acoustic Device) torpedo countermeasure.

Specifications

Power: 115 VAC, 60 Hz, single phase
Size: 660 lb, 72 in high, 25 in wide, 22 in deep
Note: The AN/SLX-1 (MSTRAP) cabinet has significant excess space available. This can be used for other electronics or the cabinet can be reduced in size to fit specific needs.

Contractor

Northrop Grumman Corporation, Annapolis, Maryland.

UPDATED

AN/SLX-1 MSTRAP hardware comprises the processing enclosure, workstation display and interconnecting LAN equipment.
1998/0007232

Kingfisher

Type

Small obstacle avoidance sonar modification.

Development

Kingfisher has been developed using COTS technology to modify hull-mounted sonars to provide command teams with surveillance data to enable them to safely navigate around hazardous small objects such as mines. The initial model of Kingfisher was an analogue hardware unit developed for the 'Oliver H Perry' (FFG 7) class AN/SQS-56 sonar in response to the mine threat highlighted when the *Samuel B Roberts* (FFG 58) struck a mine in the Persian Gulf in April 1988. The first prototype kit was successfully tested aboard the USS *Halyburton* (FFG 40) just four months after the *Samuel B Roberts* incident.

Subsequently, six copies of an enhanced second-generation AN/SQS-56 prototype were built and tested aboard 'Oliver H Perry' class frigates in the Persian Gulf.

Description

Initial Kingfisher modification kits provided 30° of coverage in three 10° beams and modified the sonar's transmit and receive characteristics to enable own ship detection of small objects. The adaptation featured a short pulse with a high repetition rate and multiping history. An additional feature of the initial modification kit used a unique colour display format and GPS navigation that allowed the sonar system to distinguish between stationary and non-stationary objects in shallow water and high reverberation.

The initial kits were followed by a further modification (referred to as SWAK – Shallow Water Active Kit) which featured additional display features and an expanded search sector covering 90°. Subsequent modifications allowed the SWAK workstation to be plugged directly into the ship's sonar system rather than having to remove parts of existing hardware to install Kingfisher.

Further modifications were then introduced to enable Kingfisher to be used with the SQS-53A and SQS-53B sonars.

The latest development is a modification to allow Kingfisher to be installed with the SQS-53C sonar integrated within the AN/SQQ-89 system. This allows all US Navy surface ships with a primary ASW mission to be Kingfisher capable. This purely software modification features a short pulse length and a wide bandwidth within the 'variable depression' mode of the SQS-53C. This modification provides 120° coverage using three overlapping 45° sectors, each of nine beams. Kingfisher allows for short range scale (2.5, 5, 10 and 20,000 yards) selections which provide rapid display updates for close-in detection.

The latest upgrades to the SQS-53A/B and SQS-56 sonars feature a laserplot electronic navigation system. Defense Mapping Agency CD-ROM charts are fed via desk-top computer to the WRN-6 satellite navigation receiver which allows charts to be presented on colour displays on sonar consoles and on the bridge. This allows contacts/objects to be immediately localised in a position relative to own ship as well as in latitude and longitude co-ordinates.

Contractor

Raytheon Company, Electronic Systems, Portsmouth, Rhode Island.

VERIFIED

AN/SQQ-28

Type

Sonar signal processing system.

Description

The AN/SQQ-28 sonar signal processing subsystem processes sonobuoy acoustic data down-linked to the ship from the SH-60B LAMPS Mk III ASW helicopter. The system provides the shipboard processing and contact management required to accomplish detection of a target and to prosecute parts of the ASW mission.

The system processes data from passive and active sonobuoys including DIFAR, VLAD and DICASS buoys. Up to eight omnidirectional or four directional/active buoys can be processed simultaneously. The processing subsystem incorporates an acoustic target tracker that automates the target tracking process. This automated tracker averages around 10 minutes to determine the target course. Multiple display formats and automated detection aids enhance operator performance.

Operational status

In service with the US Navy.

Contractor

Northrop Grumman Corporation, Electronic Sensors and Systems Division, Sykesville, Maryland.

VERIFIED

TABLE VIII

Surface ship sonar equipments

System	Frequency	Type	Manufacturer	Country	Installed on	Navy	Units
Model 610	MF	HM	EDO	USA	Audace	Italy	2
(CWE 610)					Almirante Grau	Peru	1
(AN/SQS-29)					Lupo	Peru	4
					Lupo	Venezuela	6[00]
					Niteroi	Brazil	6
					Tromp	Holland	2
					Van Speijk	Indonesia	6
Model 700	MF	HM	EDO	USA	Niteroi	Brazil	2
Model 780	MF	HM	EDO	USA	Reshef	Israel	6
Model 796	MF	M	EDO	USA	Eilat	Israel	3
Type 162	HF	HM	Kelvin Hughes	UK	County	Chile	4
					Godavari	India	5
					Hercules	Argentina	2
					Leander	Chile	4
					Leander	Ecuador	2
					Leander	New Zealand	2
					Leander	Pakistan	2
					Nilgiri	India	5
					Tribal	Indonesia	3
					Type 21	Pakistan	6
					Type 42	UK	12
Type 170	HF	HM	Graseby	UK	Alvand	Iran	3
					Leander	Chile	4
					Nilgiri	India	5
					Leander	Pakistan	2
					Rahmat	Malaysia	1
					Salisbury	Bangladesh	1
					Tribal	Indonesia	3
Type 174	MF	HM	Graseby	UK	Alvand	Iran	3
					Rahmat	Malaysia	1
					Salisbury	Bangladesh	1
Type 177	MF	HM	Graseby	UK	Leander	New Zealand	1
					Tribal	Indonesia	3
Type 184	MF	HM	Graseby	UK	County	Chile	4
					Leander	Chile	4
					Leander	Ecuador	2
					Leander	Pakistan	2
					Type 21	Pakistan	6
					Type 42	Argentina	2
					Viraat	India	1
Type 2016	MF	HM	Thomson Marconi Sonar	UK	Invincible	UK	3
					Type 22	UK	10*
					Type 22	Brazil	4
					Type 42	UK	12*
Type 2031	LF	TAS	Thomson Marconi Sonar	UK	Type 22	UK	10
					Type 23	UK	11
Type 2050		HM	Thomson Marconi Sonar	UK	Type 22	UK	n/k
					Type 23	UK	13
					Type 42	UK	n/k
1BV	HF	HM	L-3 Communications ELAC	Germany	Thetis	Greece	5
AN/SQR-18	LF	TAS	EDO	USA	Cheng Kung	Taiwan	6
					FFG 7	Bahrain	1
					FFG 7	Turkey	3
					Knox	Egypt	2
					Knox	Taiwan	9
					Knox	Thailand	2
					Takatsuki	Japan	2
			EDO/NEC	USA/Japan	Shirane	Japan	2
AN/SQR-19	VLF	TAS	Northrop Grumman	USA	Arleigh Burke	USA	23
					FFG 7	USA	42
					FFG 7	Spain	6
					Kidd	USA	4
					Spruance	USA	31
					Ticonderoga	USA	27
AN/SQR-501	LF	TAS	Computing Devices	Canada	Annapolis	Canada	2
					Halifax	Canada	12
AN/SQS-17	HF	HM	EDO	USA	Bayandor	Iran	2
					Ctd Riviere	Uruguay	3
					Parvin	Iran	3
					Tacoma	Thailand	2
AN/SQS-23	MF	M	Sangamo/Raytheon	USA	Carpenter	Turkey	2
					Gearing	South Korea	5
					Perth	Australia	3
					Vittorio Veneto	Italy	1
AN/SQS-26	MF	HM	EDO	USA	Bronstein	Mexico	2
					California	USA	1
					Garcia	Brazil	4
					Knox	Egypt	2
					Knox	Greece	3
					Knox	Taiwan	9
					Knox	Thailand	2
					Knox	Turkey	8

System	Frequency	Type	Manufacturer	Country	Installed on	Navy	Units
AN/SQS-29	HF	HM	Sangamo	USA	Berk	Turkey	2
					Gearing	South Korea	2
AN/SQS-35	MF	VDS	EDO	USA	Baleares	Spain	5
					Knox	Greece	3
			EDO/Nec	USA/Japan	Shirane	Japan	2
				Japan	Yamagumo	Japan	2
AN/SQS-36	LF/MF	HM	Nec	Japan	Ishikari	Japan	1
		VDS	EDO	USA	Minegumo	Japan	1
		HM	Nec	Japan	Yubari	Japan	2
AN/SQS-39V	MF/HF	HM	EDO	USA	Claud Jones	Indonesia	1
AN/SQS-42V	MF/HF	HM	EDO	USA	Claud Jones	Indonesia	1
AN/SQS-43	MF/HF	HM			Allen M Sumner	Iran	1
AN/SQS-44	MF/HF	HM			Allen M Sumner	Iran	1
AN/SQS-45V	MF/HF	HM	EDO	USA	Claud Jones	Indonesia	1
AN/SQS-53	MF	HM	Various	USA	FFG 7	Turkey	3
					Kidd	USA	4
					Spruance	USA	31
					Ticonderoga	USA	27
					Virginia	USA	1
AN/SQS-56 (DE 1160)	MF	HM	Raytheon	USA	Badr	Saudi Arabia	3
					Baleares	Spain	5
					Barbaros	Turkey	3
					Cheng Kung	Taiwan	7
					Descubierta	Egypt	2
					Descubierta	Morocco	1
					Descubierta	Spain	6
					FFG 7	Australia	4
					FFG 7	Bahrain	1
					FFG 7	Egypt	3
					FFG 7	Spain	6
					FFG 7	USA	42
					Hydra	Greece	2
					Garibaldi	Italy	1
					Lupo	Italy	4
					Yavuz	Turkey	3
AN/SQS-501	HF	HM		Canada	Annapolis	Canada	2
					Restigouche	Canada	2
AN/SQS-505	MF	HM	Westinghouse	US/Canada	Annapolis	Canada	1
					Clemenceau	France	2
					Nilgiri	India	5
					Kortenaer	Greece	5
					Kortenaer	UAE	2
					Restigouche	Canada	1
					Wielingen	Belgium	3
AN/SQS-509	MF	HM	Westinghouse	USA	Jacob Heemskerck	Holland	2
					Kortenaer	Holland	4
AN/SQS-510	MF	HM	Computing Devices	Canada	Annapolis	Canada	1
		HM			Halifax	Canada	12
		VDS/HM			Iroquois	Canada	4
		HM			Cdt Joao Belo	Portugal	4
		VDS/HM			Restigouche	Canada	1
		HM			Vasco da Gama	Portugal	3
ASO 4	MF	HM	STN ATLAS Elektronik	Germany	Almirante Padilla	Colombia	4
					Espora	Argentina	4
ASO 84-41	MF	HM	STN ATLAS Elektronik	Germany	Assad	Malaysia	4
APSOH			Bharat	India	Godavari	India	4
					Nilgiri	India	1
ATAS	LF	TAS	BAeSEMA	UK	Cheng Kung	Taiwan	6
					La Fayette	Taiwan	3
					Type 21	Pakistan	6
Bull Horn	MF/LF	HM		RFAS	Kuznetsov	RFAS	1
					Parchim	RFAS	12
					Slava	RFAS	3
					Sovremenny	China	2
					Sovremenny	RFAS	16
Bull Nose	LF/MF/HF	HM		RFAS	Grisha III	Lithuania	2
					Grisha	RFAS	67
					Grisha	Ukraine	6
					Kara	RFAS	2
					Kashin	Poland	1
					Kotor	Yugoslavia	2
					Krivak	RFAS	24
					Krivak	Ukraine	1
CTS 36/39	HF	HM	C-Tech	Canada	Thetis	Denmark	4
					Flyvefisken	Denmark	n/k
CWE 10N	MF	HM			De Ruyter	Peru	1
COMTASS					n/k	Far East	n/k
DE 1164	MF	HM	Raytheon	USA	Alpino	Italy	2
	LF/MF	HM/VDS			De La Penne	Italy	2
	MF/HF	HM/VDS			Maestrale	Italy	8
DE 1167	MF	VDS	Raytheon	USA	Descubierta	Egypt	2
		HM	Raytheon/Alenia	US/Italy	Minerva	Italy	8
DE 1191	MF	HM	Raytheon		Charles F Adams	Greece	4
DSBV 61B	VLF	TAS	Thomson Marconi SAS	France	Georges Leygues	France	3
DSBV 62C	VLF	TAS	Thomson Marconi SAS	France	Tourville	France	3

System	Frequency	Type	Manufacturer	Country	Installed on	Navy	Units
DSQS 21	MF	HM	STN ATLAS	Germany	Almirante Brown	Argentina	4
					Bremen	Germany	8
					Gearing	Taiwan	2
					Inhauma	Brazil	4
					Khamronsin	Thailand	2
					Kasturi	Malaysia	2
					Lutjens	Germany	3
					Makut Rajakumar	Thailand	1
					Rattanakosin	Thailand	2
					Tapi	Thailand	2
DUBA 3A	HF	HM	Thomson Marconi SAS	France	Cdt Joao Belo	Portugal	4
					Cdt Riviere	Uruguay	3
DUBA 25	MF	HM	Thomson Marconi SAS	France	Type A69	France	16
					Cassard	France	1
DUBV 23	MF	HM	Thomson Marconi SAS	France	Suffren	France	2
					Tourville	France	3
					Georges Leygues	France	4
					Luhu	China	2
DUBV 24	MF	HM	Thomson Marconi SAS	France	Cassard	France	1
					Georges Leygues	France	3
					Jeanne d'Arc	France	1
DUBV 25	MF	HM	Thomson Marconi SAS	France	Cassard	France	1
DUBV 43B/C	MF	VDS	Thomson Marconi SAS	France	Suffren	France	2
					Tourville	France	1
					Georges Leygues	France	7
					Luhu	China	2
Echo Type 5	HF	HM		China	Daxin	China	1
	MF				Jiangwei	China	4
					Jianghu I/II	China	28
					Jianghu I	Egypt	1
					Osman	Bangladesh	1
Elk Trail	HF	VDS		RFAS	Grisha	RFAS	67
					Grisha	Ukraine	6
					Parchim I	Indonesia	n/k
Foal Tail	HF	VDS		RFAS	Babochka	RFAS	1
					Mukha	RFAS	2
					Mukha	Ukraine	1
					Pauk	RFAS	32
					Stenka	RFAS	n/k
					Tarantul	RFAS	47
					Turya	RFAS	13
G 750	MF	HM	Graseby	UK	Godavari	India	1
					Leander	India	4
					Leander	New Zealand	2
Herkules	MF/HF	HM		RFAS	Kaszub	Poland	1
					Koni	Algeria	3
					Koni	Bulgaria	1
					Koni	Cuba	2
					Koni	Libya	2
					Koni	Yugoslavia	2
					Kynda	RFAS	1
					Parchim I	Indonesia	16
					Petya II	India	4
					Petya	Syria	2
Horse Jaw	LF/MF	HM		RFAS	Kiev	RFAS	1
					Kirov	RFAS	4
					Kuznetsov	RFAS	1
					Udaloy I/II	RFAS	8
Horse Tail	MF	VDS		RFAS	Kiev	RFAS	1
					Kirov	RFAS	4
					Udaloy II	RFAS	1
Lamb Tail	HF	VDS		RFAS	Parchim	RFAS	12
MG 329M	HF	Dipping		RFAS	Kaszub	Poland	1
Mare Tail	MF	VDS		RFAS	Kara	RFAS	2
					Kashin II	India	5
					Kashin	RFAS	2
					Krivak	RFAS	21
Mouse Tail	MF	VDS		RFAS	Ivan Rogov	RFAS	1
					Udaloy I	RFAS	7
Mulloka	MF	HM	Thorn	Australia	River	Australia	1
					FFG 7	Australia	2
OQR 1	VLF	TAS			Asagiri	Japan	8
					Hatsuyuki	Japan	n/k
					Murasame	Japan	2
OQR 2 (SQR-19)	TAS	VLF	Oki	Japan	Kongo	Japan	4
OQS 3	MF	HM	Sangamo/Mitsubishi	USA/Japan	Haruna	Japan	2
			Nec	Japan	Minegumo	Japan	1
OQS 3A	MF	HM	Hitachi	Japan	Chikugo	Japan	8
	LF		Nec		Tachikaze	Japan	3
	MF		Nec		Yamagumo	Japan	3
OQS 4/A (SQS-23)	LF	HM	Mitsubishi	Japan	Asagiri	Japan	8
	MF		Nec		Hatakaze	Japan	2
	LF		Nec		Hatsuyuki	Japan	12
OQS 5	LF	HM	Mitsubishi	Japan	Murasame	Japan	2
OQS 8	MF	HM	Hitachi	Japan	Abukuma	Japan	6
OQS 101	LF	HM	Nec	Japan	Shirane	Japan	2

System	Frequency	Type	Manufacturer	Country	Installed on	Navy	Units
OQS 102 (SQS-53)	MF	HM	Nec	Japan	Kongo	Japan	3
Ox Tail	MF	VDS		RFAS	Gepard	RFAS	1
					Neustrashimy	RFAS	2
Ox Yoke	MF	HM		RFAS	Gepard	RFAS	1
					Neustrashimy	RFAS	2
PHS 32	MF	HM	Signaal	Holland	Dewantara	Indonesia	1
					Dong Hae	South Korea	4
					Fatahillah	Indonesia	3
					Po Hang	South Korea	20
					Ulsan	South Korea	9
PHS 36	MF	HM	Signaal	Holland	Karel Doorman	Holland	8
PMS 26	MF	HM	Thomson Marconi Sonar	UK	Alm. Clemente	Venezuela	2
					Beskytteren	Denmark	1
					Eithne	Ireland	1
					Erinomi	Nigeria	1
					Girne	Turkey	1
					Niels Juel	Denmark	3
					Turk	Turkey	12
Pegas	HF	HM		RFAS	Luda	China	15
RIZ	HF	HM			Kralj	Croatia	1
Rat Tail	HF	VDS		RFAS	Muravey	RFAS	14
					Pauk II	Cuba	1
					Pauk II	India	4
					Svetlyak	RFAS	27
SA 950	HF	HM	Simrad AS	Norway	Alta	Norway	5
					Goteborg	Sweden	4
					Hugin	Sweden	8
					Stockholm	Sweden	2
SJD 5A	MF	HM		China	Chao Phraya	Thailand	4
SJD 7	MF	HM		China	Naresuan	Thailand	2
SQS-3D	HF	HM	Simrad AS	Norway	Mirna	Croatia	4
					Mirna	Yugoslavia	6
SS 105	MF	HM	Simrad AS	Norway	Hauk	Norway	14
					Nordkapp	Norway	3
					Silma	Estonia	1
					Tursas	Finland	2
					Valpas	Finland	1
SS 242	HF	HM	Simrad AS	Norway	Lokki	Finland	4
SS 304	HF	HM/VDS	Simrad AS	Norway	Helsinki	Finland	4
ST 570	HF	VDS	Simrad AS	Norway	Hugin	Sweden	8
ST 240 (Toadfish)		VDS	Simrad AS		Rauma	Finland	5
Sonac PTA			Patria Finavitec	Finland	Ruissalo	Finland	2
					Helsinki	Finland	4
					Rauma	Finland	5
Stag Horn	HF			RFAS	Najin	North Korea	2
					Sariwon	North Korea	3
					Soho	North Korea	1
					Tral	North Korea	2
Steer Hide	LF/MF	HM		RFAS	Krivak	RFAS	3
					Slava	RFAS	3
TSM 2064		VDS	Thomson Marconi SAS	France	Victory	Singapore	6
TSM 2362 (Gudgeon)	MF	HM	Thomson Marconi SAS	France	Fearless	Singapore	6
TSM 2633 (Spherion)	MF	HM	Thomson Marconi SAS	France	ANZAC	Australia	8
					ANZAC	New Zealand	1
					La Fayette	Taiwan	3
					Lekiu	Malaysia	2
					Oslo	Norway	4
TSM 2635 (Diodon)	MF	HM	Thomson Marconi SAS	France	Baptista Andrade	Portugal	4
					Drummond	Argentina	3
					Esmeraldas	Ecuador	6
		HM/VDS			Madina	Saudi Arabia	4
TSM 2643 (Salmon)	MF	VDS	Thomson Marconi SAS	France	Flyvefisken	Denmark	n/k
					Goteborg	Sweden	4
					Stockholm	Sweden	2
		HM/VDS			Thetis	Denmark	4
TSM 2670 (SLASM/ATBF 1/2)	VLF	TAS	Thomson Marconi SAS	France	Tourville	France	2
TSM 2980 (Anaconda)	LF	TAS	Thomson Marconi SAS	France	Karel Doorman	Holland	8
Tamir	HF			RFAS	Luda	China	15
Terne III	HF	HM	Kongsberg Simrad	Norway	Oslo	Norway	4
UQQ-2	LF	TAS	Hughes Aircraft	USA	Hibiki	Japan	2
					Stalwart	USA	6
					Victorious	USA	4
Vycheda	MF			India	Kashin II	India	5
					Petya	Vietnam	5
Whale Tongue	MF	HM		RFAS	Neustrashimy	RFAS	2
					Sovermenny	China	2
					Sovremenny	RFAS	16

Note: HM – Hull-Mounted Sonar, TAS – Towed Array Sonar, VDS – Variable Depth Sonar, VLF – Very Low Frequency (around and below 1 kHz), LF – Low Frequency (below 3 kHz), MF – Medium Frequency (3-14 kHz), HF – High Frequency (above 14 kHz)

[00] To be replaced by 21 HS
* Being replaced by Type 2050

UPDATED

SUBMARINE SONAR SYSTEMS

FRANCE

TSM 2233

Type
Submarine passive and active sonar system.

Description
The TSM 2233 is a modular, integrated sonar system designed to fit any size of submarine and to fulfil any operational requirement.

The system includes any or all of the following acoustic sensors: linear towed array, flank arrays, bow array (cylindrical or conformal), intercept array, distributed arrays, active array, obstacle avoidance array.

The TSM 2233 sonar system provides the following set of functions:

Passive detection
Broadband, narrowband and transient processing are concurrently performed so as to match with the noise characteristics of any vessel. For each operational function, sonar data from the various arrays and processing are gathered and associated to reduce the operator workload while maintaining maximum sensitivity.

Automatic anti-jamming
The anti-jamming feature automatically rejects narrowband jammers, thereby hardening tracks and enhancing Contact Motion Analysis (CMA) performance (range, accuracy and convergence delay).

Passive adaptive processing
The adaptive processing function is based on an optimal array processing theory which minimises the effect of jamming by strong signals when listening to low-level signals. It is a particularly useful facility in discriminating between two targets in a common limited sector.

Automatic detection and tracking
By automatically initiating tracks, this feature allows the operator to avoid repetitive tasks so that he can deal with more complex situations.

Interception
This function allows interception of all active sonar pulses from low-frequency surface ship sonars to high-frequency torpedo acoustic heads. Very early warning of hostile sonars with a low false alarm rate is provided. Interception warnings are integrated on the passive listening scope. Accurate parameters of pulses are presented on digital readouts.

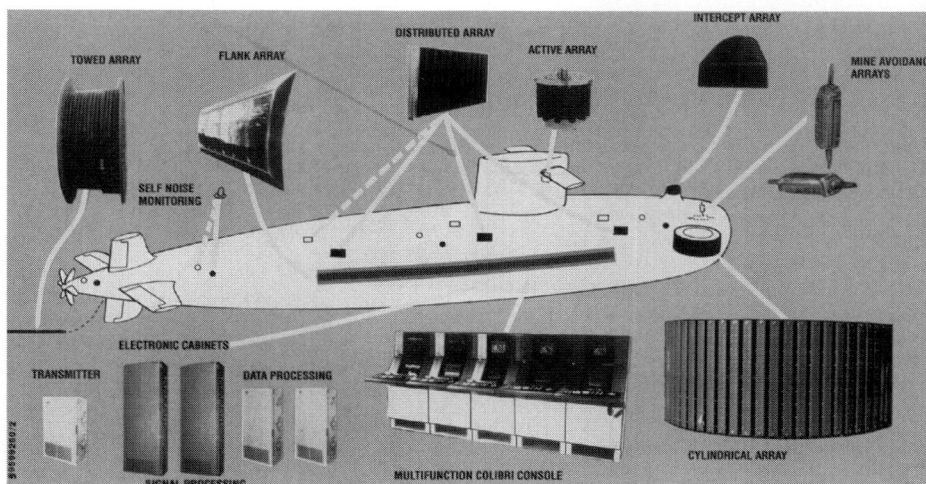

TSM 2233 submarine sonar suite   *1995*

Active capability
Although initial bearings are provided by the passive detection function, the operator can select the active mode to determine range accurately. In this mode, reception is performed by the passive listening array, and is associated with high-detection sensitivity.

Classification
The spectrum analysis facility (Lofar and Demon) complements the audio function by providing the operator with specific data in target classification, which is computer-aided through an interactive database.

Hostile weapon alarm
A specific algorithm sorts out the contacts taking into account various target parameters and warns the operator in the event of a hostile weapon.

Contact Motion Analysis (CMA)
Contact localisation and motion analysis of contacts are performed automatically on all tracks.

Key features of the system are: modular signal processing which makes it possible to adapt the configuration to suit new vessel designs and modernisation (possibly keeping existing arrays); reduced operator workload due to the high level of automation of the detection, tracking and localisation functions; and the flexibility and ease of adaptability to

submarine size and mission requirements resulting from the coherent set of complementary sensors.

The system uses TMS 320, C30 and 68040 microprocessors for signal processing in MIMD-type host systems communicating internally across high-speed ring networks and externally on standard VME-type buses.

Programs use structured software developed in C and Ada. Multifunction workstations are used with one or two high-definition 19 in colour monitors (Colibri or MOC consoles).

Operational status
TSM 2233 is the latest generation of the well-known Eledone family, some versions of which are in service in the French Navy (DSUV 22 for the 'Agosta' class), the Royal Navy (2040 Argonaute on *Upholder*), the Royal Netherlands Navy ('Walrus' class TSM 2272) and the Spanish Navy ('Agosta' and 'Daphne' classes).

Variants of the TSM 2233 are installed on the Royal Australian Navy's 'Collins' class *(Scylla)* and French Navy 'Rubis' class (DMUX 20).

Contractor
Thomson Marcon Sonar SAS, Sophia Antipolis.

UPDATED

TSM 2225

Type
Panoramic surveillance and direct passive ranging sonar.

Description
The TSM 2225, which is a module of the TSM 2233 sonar system, consists of a set of six arrays (three on each side of the submarine), one cabinet and one

console. The sonar performs panoramic surveillance and pulse interception, target and automatic pulse tracking and ranging, and target localisation.

Simultaneous tracking on up to eight noise sources and eight sonar transmissions is carried out using Lofar, Demon and pulse mode analysis.

TMS 320, C30 and 68040 microprocessors are used for signal processing as in the TSM 2233 (see previous entry) and data is handled in the same way as in the TSM 2233.

Operational status
The TSM 2225 is the latest version of the DUUX 5 and is installed on the Australian 'Collins' class, Argentina 'Santa Cruz' class and Pakistan Navy 'Agosta' class submarines as well as on Chinese 'Han' and 'Ming' class submarines.

Contractor
Thomson Marconi Sonar SAS, Sophia Antipolis.

UPDATED

TSM 2253

Type
Submarine planar flank array sonar.

Description
The TSM 2253 is a low-frequency passive sonar using two planar arrays installed on the submarine's flanks. Each array consists of up to 64 planar hydrophones (0.5 m wide, 1 m in height, 0.1 m thick) composed of polyvinyldifluoride (PVDF) films moulded in

polyurethane. Compared with linear configured ceramic hydrophones, the planar array shows an excellent resistance to shocks, offers directivity in elevation, achieves noise cancellation due to the surface effect and involves very low installation overheads.

The sonar performs panoramic surveillance through simultaneous processing of broadband and narrowband and automatically initiates tracking on up to 64 targets. In addition, the operator can initiate up to eight tracks. Signal analysis is carried out using Lofar,

Installation of planar flank array on a submarine
1996

Demon and audio with computer-aided classification. Localisation is through specific CMA. The sonar can interface with any weapons control system.

The processing system is the same as that in the TSM 2233 (see above).

Operational status
A preceding generation (TSM 2285) of the TSM 2253 is installed on the Norwegian Navy 'Ula' class. Planar flank array sonars are installed on French Navy nuclear submarines and equip the Australian 'Collins' class.

Contractor
Thomson Marconi Sonar SAS, Sophia Antipolis.

VERIFIED

TSM 2933

Type
Submarine linear towed array sonar.

Description
The TSM 2933 passive linear towed array sonar is produced, either as a module of the TSM 2233 sonar system or, in a stand-alone configuration with its own processing, operating and interfacing facilities.

The TSM 2933 performs panoramic surveillance through simultaneous processing of broadband and narrowband in several fixed or adjustable frequency bands, and automatically initiates tracking on up to 64 targets. In addition, the operator can initiate up to eight tracks. Signal analysis is carried out using Lofar, Demon and audio with computer-aided classification. Localisation is through specific TMA. The sonar can interface with any weapons control system.

The linear array is 100 m long and 0.08 m in diameter and is towed behind a 400 m long neutrally buoyant cable. The TSM 2933 can also be fitted with a thin-line array (Narama).

The processing system is the same as in the TSM 2233 (see above).

Operational status
The TSM 2933 is currently in service in the French Navy (as the DSUV 62 version) and in some foreign navies.

TSM 2933 linear towed array during recovery on a 1,200 tonne submarine

The TSM 2933 version designed for surface ships differs from the submarine version only in towing subassembly configuration and is in service in several navies.

Contractor
Thomson Marconi Sonar SAS, Sophia Antipolis.

VERIFIED

DUUX-5 (Fenelon) sonar

Type
Panoramic surveillance and direct passive ranging sonar.

Description
Fenelon is the name given to a panoramic sonar capable of automatic and simultaneous tracking of three targets. The equipment incorporates a passive acoustic rangefinder for measuring the range of three targets within 120° sectors, and a panoramic sonar interceptor measuring the true bearing of all sonar transmissions received within the 2 to 15 kHz band.

The DUUX-5 equipment was developed by Alcatel as a successor to the DUUX-2 series, of which at least 120 equipments were built and supplied to 14 navies for fitting in eight types of submarine. The Fenelon equipment enables range and bearing information on targets to be obtained by the submarine without the need for any transmissions and with minimum delay. Speed of operation permits target course and speed to be computed rapidly, also allowing any changes in either speed or course to be detected without delay. Four targets can be tracked simultaneously (three on radiated self-noise, one on sonar pulses), and there is a

continuous panoramic bearing display over 360°. Range information is provided over arcs of 120° on each side of the submarine. There are facilities for transmission of target data automatically to the ship's weapon control system and plotting table.

Performance characteristics include high accuracy and discrimination, immunity against sonar pulse interference, simplified calibration, and integrated test facilities. Under normal conditions, results in the middle sector are 0.3° for bearing accuracy, 2° discrimination accuracy between two targets and 5 per cent of range on radiated noise for a target at a distance of 10 km. Two types of hydrophones are available. The hydrophonic unit is composed of two bases, with three hydrophones each on the starboard and port sides of the submarine. All other technical details remain classified.

Operational status
Has replaced the DUUX-2 passive acoustic rangefinder. At sea on 'L'Inflexible', 'Agosta', TR 1700 and Type 209/1500 class submarines.

Contractor
Thomson Marconi Sonar SAS, Sophia Antipolis.
1960

UPDATED

Fenelon: DUUX-5 control and display console

TSM 5421B

Type
Submarine mine and obstacle avoidance sonar.

Description
The TSM 5421B carries out panoramic search by scanning a 30° sector to detect moored mines and

obstacles such as piers and close-range platforms. The system consists of a wet end (diameter 0.5 m, height 0.5 m) and a receiver/display cabinet (0.5 m × 0.6 m × 0.6 m).

Operational status
In service at sea.

Contractor
Thomson Marconi Sonar SAS, Sophia Antipolis.

VERIFIED

TSM 5423

Type
Submarine mine and obstacle avoidance sonar.

Description
The sonar carries out detection and classification of any type of mine, as well as the detection of underwater

obstacles. Detection is carried out in a 60° bearing and 30° elevation sector ahead of the submarine. The sonar comprises two arrays (0.6 m × 0.15 m × 0.15 m) and a receiver/display cabinet.

Operational status
In service at sea.

Contractor
Thomson Marconi Sonar SAS, Sophia Antipolis.

VERIFIED

Velox M7

Type
Sonar intercept receiver.

Description
The Velox M7 sonar intercept receiver is designed to intercept signals in the frequency range 2.5 to 100 kHz. Detected signals are automatically measured to provide direction, level, frequency and pulse length and to identify recurring identical pulses. The system is also capable of monitoring the noise radiated by ships.

The receiver is designed for integration into a federated combat system, or can be used in a stand-alone configuration.

The receiver uses multiband processing based on a constant false alarm rate operation and provides a permanent digital display of the parameters of the last two intercepts; synthetic analogue display of the threat direction; extensive audio facilities, including direct, compressed or heterodyned and anti-saturation features.

Velox M7 comprises a small receiving array and an electronics cabinet or rack drawer. A second identical array can be installed in another location for manual selection to improve the horizontal coverage.

As an option, the frequency range can be extended to 0.5 kHz and/or up to 250 kHz.

Operational status
In service with the French Navy.

Contractor
Safare Crouzet, Nice.

VERIFIED

Display panel of the Velox M7 intercept sonar receiver

GERMANY

CSU 83 sonar system

Type
Passive/active sonar system.

Description
The CSU 83 (DBQS-21) submarine sonar is intended for boats of 400 tonnes and upwards. It is the primary sensor of the submarine fire-control and command system of the German Navy. It comprises a passive bow array operating in the 0.3 to 12 kHz band, a passive ranging array and a sonar intercept array. In addition, two sensors to detect the submarine's own noises are included.

The packaging of the electronics is based on a modular design and provides easy access to all components for servicing, as well as low size and weight. The basic features of the CSU 83 are:
(a) high probability of detection over 360°
(b) computer-aided processing and evaluation
(c) discrimination of data on integrated displays
(d) simple operation by two operators via interactive, colour-coded control/displays
(e) sequential digital signal processing
(f) automatic fault detection and localisation.

Operational status
The CSU 83 has been installed in 12 Type 206A SSKs of the German Navy and is in service in other navies worldwide (see table IX).

Contractor
STN ATLAS Elektronik GmbH, Bremen.

VERIFIED

PSU 83 sonar

Type
Passive surveillance and detection sonar.

Description
The PSU 83 is designed for new submarines of small and medium sizes, and can also be retrofitted to older submarines to improve their detection facilities. High-technology preprocessing units with newly designed preamplifiers and filter units ensure sensitive and undisturbed reception.

The very good detection performance is achieved in two independent detection channels: one for surveillance tasks with short integration times, and the other for long-range detection using longer integration times. Both detection results are displayed simultaneously to provide a comparison facility. The sonar is equipped with eight independent automatic target tracking channels.

For redundancy purposes and to improve performance, the preprocessing unit and the beamformer are split into two independent electronics units. Additionally, a redundant beamformer and a separate direct signal path to the operator's console are provided. A built-in target data analyser and a bearing prediction facility improve the operational value of the sonar.

All the signal processing electronics of the PSU 83 are combined in one operator's console and two preprocessing units. The operator's console consists of software-driven buttons, trackerball for accurate cursor movement, high-resolution raster scan colour display and a video output for slave display. For extended functions the operator's console is available with two displays.

The PSU 83 can be extended with a number of options, including:

(a) intercept sonar
(b) flank array sonar including Lofar detection, tracking and analysis
(c) Demon detection, tracking and analysis
(d) mass storage unit for target data and display recording
(e) sonar graphic recorder
(f) onboard training simulator
(g) own noise analysis.

When combined with a fire-control system, the PSU 83 forms a complete submarine warfare system.

Operational status
In service in submarines of the Royal Danish and other navies.

Contractor
STN ATLAS Elektronik GmbH, Bremen.

UPDATED

CSU 90 sonar system

Type
Attack, intercept, passive ranging, mine avoidance, active and flank array sonar.

Description
The basic version of the CSU 90 system comprises three main sonars; an attack panoramic passive sonar, an intercept passive sonar and a low-frequency flank array sonar. The attack sonar has a 2.8 m cylindrical hydrophone array and full azimuth coverage, a Demon detection facility uses the array of the attack sonar, and the flank array is used for long-range detection. Any target detected by one of these sonars initiates an automatic process of data collection.

CSU 90 is equipped with eight independent compound tracks. Each provides the warfare system with a complete set of tactical data. Tracks are automatically incorporated in a compound track and each sensor can initiate a compound track. The total set of compound tracks comprises:
(a) eight ATT channels for the attack sonar
(b) eight ATT channels for the Demon path, each with eight ALTs
(c) the Target Motion Analyser (TMA) which controls eight compound tracks and generates a sonar tactical display
(d) TMA controls providing further manual tracks
(e) flank array sonar with eight ATT channels and eight MTs with eight ALTs each.

Optional extensions include a passive ranging sonar, an own noise analyser, acoustic passive classification, automatic warning channel, target motion analyser, a disk memory unit, plotting devices and simulation facilities. The disk memory unit allows all displays and sonar data to be recorded for subsequent analysis and for training. It also handles the data, for comparison purposes, of the sonar information processor and the classification file for the intercept sonar.

Operational status
In service aboard the Israeli submarines *Dolphin* and *Leviathan* and Swedish 'Gotland' class, Greek Navy 'Glavkos' class, Chilean 'Oberon' class and has been selected for the German and Italian U212 class submarines.

Contractor
STN ATLAS Elektronik GmbH, Bremen.

UPDATED

FAS 3-1 sonar

Type
Stand-alone flank array sonar.

Description
The FAS 3-1 flank array sonar is designed to perform long-range early detection and classification of enemy surface ships and submarines by picking up their low-frequency noise emissions and carrying out classification on the basis of evaluated Lofar/Demon/Spectrum data. For directional long-range detection, the FAS 3-1 has two flank arrays (linear arrays mounted at the sides of the pressure hull). Because of the direct coupling to the submarine, the flank arrays are specially designed for maximum decoupling from unwanted own noise.

The system consists of the two flank arrays, electronics cabinet, operator's console and recording systems.

FAS 3-1 covers the frequency band from 10 to 2.5 kHz. The arrays are 20 to 48 m long, depending on the length of the submarine. Bearing accuracy is 1° on a port/starboard sector of 90° between relative bearings port/starboard 45 to 135°. This can be extended with reduced performance to port/starboard 10 to 170°.

Operational status
In service.

Contractor
STN ATLAS Elektronik GmbH, Bremen.

UPDATED

TAS towed array sonar

Type
Very low frequency towed array sonar system.

Description
STN ATLAS Elektronik has developed and manufactured a towed array sonar that fulfils all the requirements of submarine sonar for long-range detection, target classification and target motion analysis.

The main features of the sonar are:
(a) early detection of targets
(b) independence of bathythermal conditions, in which the towed array is operated in the optimum layer
(c) detection of the acoustic spectral lines of quiet submarines
(d) detection of transients
(e) detection of low-frequency noise by cavitating surface vessels
(f) target motion analysis computed by algorithms specially adapted to the towed array sonar.

To perform the above tasks the TAS can be fitted with different 'wet ends' of various diameters, and task-oriented electronics. Signals received are fed to a north-stabilised beamformer which creates 192 different beams over the azimuth. Separate paths are used for signal enhancement for the detection display and audio channel. Up to eight targets can be tracked automatically and a separate signal path is used for frequency analysis, using Lofar techniques. Automatic multiline trackers can be set on characteristic frequency lines of the target's signature. Together, the automatic target tracker and multiline tracker can follow a target even at long ranges or under extremely severe conditions such as target crossing or convoy acoustic environment. The built-in multipoint divider, cursor and other support measures assist the operator in defining and classifying the type of engines, gearbox ratio and other information. This is achieved using Lofar transient noise detection, audio analysis/classification and bearing/time record.

Operational status
In service.

Contractor
STN ATLAS Elektronik GmbH, Bremen.

UPDATED

CSU 3-4 sonar

Type
Active and passive search and attack sonar.

Description
The CSU 3-4 is a conventional submarine sonar designed for a single operator. It is a fast-scanning medium-range active sonar which can be used as a long-range passive device, an intercept sonar and as an underwater telephone. In its active mode, the sonar operates using a single beam. The equipment has a cylindrical transducer array with electronic beam-steering. The sonar is equipped with automatic interference suppression against acoustic countermeasures.

The CSU 3-4 is able to track four targets simultaneously. The basic functions of the CSU 3-4 are described below:
Passive mode
(a) simultaneous azimuth coverage over 360°
(b) panorama presentation of target information on a CRT display with true or ship-relative bearing
(c) bearing/time recorder for long-term plotting of all contacts
(d) high-bearing accuracy and angular resolution
(e) automatic target tracking of up to four targets simultaneously.
Active mode
(a) single ping operation
(b) presentation of a 30° sector (waterfall) on the CRT, with continuous memory-refreshed display of target echoes
(c) reverberation threshold for high-detection performance
(d) magnified CRT display
(e) cylindrical transducer array with electronic beam-steering.
Intercept mode
(a) presentation of up to eight external pulses at the centre of the CRT
(b) additionally, a numerical indication of bearing, frequency, signal level and pulse duration of one of the eight pulses is provided
(c) determination of target elevation angle for estimation of distance and optimum sound ray path
(d) presentation of the received pulses on the audio channel.

The system is capable of evaluating sonar data in the following ways:
(a) analysis of noise spectra in various frequency bands
(b) automatic line tracking
(c) automatic manoeuvre detection
(d) high resolution of target bearings
(e) automatic target tracking of targets set by the sonar, including target crossing situations
(f) measurement of range, bearing and Doppler in the active mode.

Using these features, the system can determine certain target parameters such as speed, propeller shaft revolution rate and number of blades or a characteristic low-frequency spectrum. This data can be stored in an optional database for later recall for comparison with current target data.

Operational status
In service.

Contractor
STN ATLAS Elektronik GmbH, Bremen.

UPDATED

PRS 3-15 sonar

Type
Stand-alone passive ranging sonar.

Description
The PRS 3-15 is a submarine passive ranging sonar designed to be operated in conjunction with the CSU, or as an independent system. It can be used for passive detection of targets and for passive rangefinding of submarines. The 15 stave array with a high-directivity index provides a good signal-to-noise ratio, and suppression of adjacent targets that are not of interest. Six individual arrays are mounted along the hull, three on the starboard side and three on the port side.

The PRS 3-15 performs the following functions:
(a) panoramic detection with high-array gain
(b) measurement of target bearing
(c) automatic tracking of target bearing on up to four targets simultaneously
(d) range measurement on up to four targets
(e) calculation of target course and speed without the need for manoeuvres by the host submarine
(f) narrowband audio channel via PRS array or CSU cylindrical array
(g) automatic calculation of a confidence factor for the range data.

The signals received by the arrays are correlated with each other, the sampling values of the correlation function are integrated according to range and target dynamics, and are then analysed by the processor in order to estimate the time delays of the signals. From the time delay measurements the range is calculated. In addition, the variance of the time delay estimations is determined and converted to a confidence factor of the range value.

Operational status
In service aboard some Peruvian 'Casma' class Type 209/1200 submarines.

Contractor
STN ATLAS Elektronik GmbH, Bremen.

UPDATED

Lopas

Type
Submarine passive sonar.

Description
The Lopas (Low-cost passive sonar) has been developed for the upgrading of older types of submarines with a low-cost state-of-the-art system to improve their capabilities considerably. Special attention has been paid to achieve a minimum of mechanical and electrical disturbance on board the submarine, and most existing interfaces can be retained.

The sonar provides simultaneous reception from 96 preformed beams, and up to eight targets can be simultaneously and automatically tracked with concurrent spectral analysis (FFT), together with display and evaluation of the signals of all preformed beams.

The results of signal processing and analysis of 15 minutes of history are continuously stored and can be called off and displayed at any time. Additionally, interesting signals and data can be continuously stored by means of an integrated cassette recorder.

Other advantages include increased system flexibility (due to software adaptation), BITE, and failure localisation which increases availability of the system. The control and display console allows definition of useful standard operating modes and assists the operator by offering these modes for selection via the menu.

The design of the system allows the existing hydrophone array of the passive sonar to be retained. New preamplifiers incorporating pc-boards are fitted inside the pressure hull, and are designed as charge amplifiers.

The hydrophone signals are divided into groups, combined and transmitted to a multiplex component. The beamformer is of modular design and composed of several submodules. The production of directional signals is based on fast calculation of linear-combined scanned signal values which are time-shifted. By choosing adequate stored coefficients of the linear combination, the parameters of the directivity patterns can be influenced. The signals of the 96 preformed beams are digitally sampled at the output of the beamformer.

Further processing is focused on the task to extract the parameters of significant targets from the flood of raw data.

Signal processing embraces a digital energy detector which calculates the current energy of the signal of each preformed beam. Display of the results with differing integration times allows recognition of fast alterations of time parameters and better detection due to the different frequencies of environment and own ship noises. Correlation detection can be displayed alternatively or additionally. All preformed beams are submitted to an FFT analysis with a resolution of 100 Hz. Simultaneous display of all or selected spectra is possible. Detectability of lines in the

target spectrum is considerably increased by means of the reduced bandwidth (20 dB). Comparison of the stored lines of the self-noise level against the received spectrum lines can be used to eliminate the self-noise interference. An increased FFT frequency resolution (0.25 Hz) enables the determination of basic and harmonic frequencies of line families and thence classification characteristics. This extraction of target data is supported by the electronics of the signal processor in interactive communication with the operator.

Up to eight targets can be allocated to Automatic Target Tracking (ATT) channels and an ATT algorithm defines expected values for future measurements.

All control and display functions are managed by a central processor. The high-resolution 14 in colour monitor presents both graphically the results of signal processing and alphanumerically the parameters of signals and status. The system is controlled by a keyboard and trackerball. Fixed programmed operating and display modes can be selected. The sound velocity profile can be displayed and input of sound velocity parameters can be accomplished manually or via interface directly from a sound velocity sensor, if installed. Displays include: polar diagram; bearing/amplitude; bearing/amplitude history; bearing/frequency history; and FFT bearing/frequency presentation of all 96 preformed beams; and FFT analysis of a single channel (for example ATT or cursor).

Operational status
Installed on Type 206 submarines purchased by Indonesia.

Console with bearing/time recorder for Lopas
1995

Contractor
L-3 Communications ELAC Nautik GmbH, Kiel.

UPDATED

ITALY

IPD 70/S integrated sonar

Type
Passive and active surveillance and attack sonar.

Description
The IPD 70/S integrated sonar suite is an improved version of the IPD 70 currently installed on board 'Sauro' class submarines. The system features a passive sonar, and an active sonar for search and attack, and various optional equipment such as:
(a) IN-100A high-precision LF interceptor
(b) ISO-100 panoramic HF interceptor
(c) TS-100 long-range directional/omnidirectional underwater telephone
(d) MD-100 passive rangefinding sonar (see separate entry).

A digital console and processor with two high-resolution raster scan monitors with image memories are used to control operations, display processed information and interface the command and control subsystem.

The IPD 70/S sensor configuration includes:
(a) a large passive, bow-mounted conformal array for long-range, high-accuracy surveillance, detection and tracking
(b) an integrated cylindrical array to perform the functions of passive sensing, passive interception, active sonar and underwater telephone
(c) the MD-100 linear array equipment for passive rangefinding
(d) a multimode sensor for panoramic broadband interception.

The passive sonar component provides for detection and tracking of targets (up to four per subsystem) over

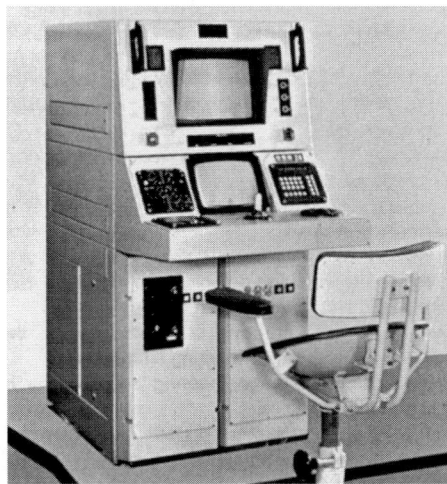

Control console for IPD 70 integrated sonar

a wide frequency coverage (200 Hz to 7.5 kHz), divided in adjacent bands. Two separate beam-formers and a high-resolution compensator allow long detection range with high bearing accuracy.

In the active mode, panoramic search in sectors can be selected.

The IN-100A LF interceptor provides for detection of active sonar emissions at ranges of several tens of kilometres, and measurement of bearing, pulse length, frequency and repetition rate to classify enemy sonars.

The ISO-100 panoramic HF interceptor has the capability to detect sonar emissions in the medium- and high-frequency ranges, evaluating bearing, frequency, amplitude, length and repetition rates of intercepted pulses.

The TS-100 underwater telephone is used for long-range directional or omnidirectional communications. Operating frequency (15 kHz) and modulation techniques comply with NATO standards.

The receiving subsystem includes the CM10 set, a figure-of-merit evaluation equipment. Based on an estimation of the most significant acoustic parameters which affect the performance of both the active and passive systems, the CM10 provides synthetic data displayed at the operator's request.

The display console, based on two high-performance ESA-24 digital processors, features an improved design man/machine interface, with two high-resolution raster scan displays with memories. These provide presentation of easy-to-read graphic and alphanumeric data, and a variety of ergonomic controls (keyboards, trackerballs, and so on) for single-operator monitoring and management of all system components.

An extensive built-in self-test and diagnostic feature is provided.

Operational status
In operational service with the Italian Navy 'Sauro' class submarines.

Contractor
Whitehead Alenia Sistemi Subacquei SpA, Genoa.

UPDATED

MD-100 Mk 1 rangefinder set

Type
Passive rangefinding sonar system.

Description
The MD-100 Mk 1 is an upgraded version of the MD-100. It is designed for installation on small- or medium-size submarines and can be supplied either as a stand-alone device or as an optional component of the integrated sonar set IPD70/S. In the latter case, all

necessary space allocation, as well as command and display facilities, are shared on a common basis with the host system.

The operational capability of the equipment includes:
(a) passive panoramic, high-resolution surveillance
(b) automatic passive tracking of targets in range and bearing.

The MD-100 Mk 1 sensor configuration consists of an independent set of acoustic sensors arranged as two

linear arrays, each composed of three high-directivity streamlined transducer arrays.

The signals collected by the transducers, after appropriate conditioning, are fed to the main unit for all the relevant stages of signal processing. Handling of signals entails the evaluation of correlation and coherence relationships existing among the various couples of waveforms involved, so as to obtain bearing and distance estimates of a noise source with high accuracy.

The MD-100 features simultaneous tracking in bearing and range of up to four targets.

The display unit, in both the stand-alone and in the integrated versions, is based on a high-resolution raster scan type screen, which allows constant monitoring of trajectories and all relevant data (both in graphic and alphanumeric form) pertaining to the targets being tracked. In addition, continuous indication is given of special parameters, such as reliability coefficients of processed data, tracking losses and so on, which could be valuable in deciding the operator's course of action.

Operational status

The MD-100 system is installed on all 'Sauro' class submarines of the Italian Navy.

Contractor

Whitehead Alenia Sistemi Subacquei SpA, Genoa.

UPDATED

SARA system

Type

Spectral analysis and classification system.

Description

SARA is a spectral analysis and classification equipment developed for Italian Navy submarines. The system can be supplied as an add on to an integrated sonar set. The operational capabilities of the system are to perform narrowband analysis of acoustic signatures to identify and classify targets.

The processor, after initial conditioning, performs automatic and/or manual detection and extraction of spectral lines, via Lofar and Demon analysis, and automatic comparison, based on inference rules, of signatures stored in appropriate databases which are also generated during the process.

A high-resolution colour display and associated processor is used to present the output of the analysis (Lofargrams, Demon, vernier) in operator selectable formats. These can then be stored on magnetic tape, if required.

The total system configuration is composed of an onboard equipment and a land-based system for further processing, archiving and updating of the database.

Operational status

In development for the Italian Navy.

Contractor

Whitehead Alenia Sistemi Subacquei SpA, Genoa.

VERIFIED

D-100 WOLVES

Type

Submarine retrofit and upgrade sonar family.

Description

The D-100 sonar family is based on a modular architecture which consists of three, operationally proven units: a pressure-resistant front end conditioner; signal and data processor; and display and control unit. These units make use of high integration technologies, which offer high processing capability and BITE. Each unit is also designed for automatic reconfiguration in case of failure.

The main operational functions are: true time delay interpolation beam-forming; fixed beams equispaced or independent of the geometry of the hydrophone array; independently steerable adaptive beams; audio beam with spectral analysis and Demon; multiband operation; constant false alarm rate; Track-While-Scan (TWS); Target Motion Analysis (TMA); sonar data fusion; and interactive acoustic classification.

The unit features 16-bit Sigma delta analogue-to-digital conversion and provides an optical link between the signal conditioner and signal processor. The operator console incorporates two independent colour monitors driven by a video graphic processor offering 256 colours on 1,280 × 1,024 pixels.

Contractor

Whitehead Alenia Sistemi Subacquei SpA, Genoa.

UPDATED

JAPAN

ZQQ series

Type

Various submarine type sonars.

Description

Most early Japanese ASW submarines were fitted with various types of US sonar systems. Based on these US sonars, the Japanese have developed a series of indigenous sonars. These sonars use state-of-the-art digital processing and display technology as developed in Japan. The ZQQ 1 passive sonar manufactured by Hughes/Oki was comparable to the US BQQ-2. The ZQQ 2 built by Hughes/Oki was a cylindrical bow array which equipped the 'Uzushio' class, now out of service. The passive/active search and attack sonar is a medium-/low-frequency system which can also operate in an acoustic intercept mode. The ZQQ 5 built by Hughes/Oki is a medium-/low-frequency, bow-mounted spherical, passive/active search and attack sonar which equips the 'Yuushio' class. The latest sonar is the Hughes/Oki ZQQ 5B medium-/low-frequency, active/passive search and attack sonar comprising spherical bow and flank arrays. The sonar also incorporates a clip-on towed linear passive intercept array. The sonar is equipping the submarines of the 'Harushio' class and is also being retrofitted to the 'Yuushio' class.

The ZQR 1 is a very low frequency, passive, towed array sonar similar to the US AN/BQR-15.

Operational status

Sonars are operational on Japanese submarines: ZQQ-5 on 'Yuushio' class and ZQQ-5B on 'Harushio' and 'Oyashio' classes. The ZQR 1 TAS equips the 'Harushio', 'Yuushio' and 'Oyashio' classes.

UPDATED

NETHERLANDS

SIASS-1 and -2 sonars

Type

Passive and active attack and surveillance sonar.

Description

The Submarine Integrated Attack and Surveillance Sonar (SIASS), developed by Signaal, is available in two versions: SIASS-1 and SIASS-2. The former is a stand-alone system with its own sonar consoles interfacing with a submarine data handling and fire-control system. SIASS-2 is an integrated part of the Signaal Submarine Sensor Weapon and Command System (SEWACO).

Both SIASS-1 and SIASS-2 perform and integrate the following tasks:

(a) low (long-range) and medium (medium-range) surveillance of noise radiating targets through the use of broadband and multichannel narrowband signal processing for surveillance, analysis and tracking (Lofargrams, ZOOM-FFT, Demon, ALI displays)
(b) passive range detection
(c) automatic target tracking for a multitude of targets, including automatic line tracking
(d) contact motion analysis (position determination on bearings-only information)
(e) active sonar

(f) LF, MF and HF intercept
(g) classification
(h) noise level monitoring and cavitation indication.

The sonar suite is divided into three groups of equipment, each with its own processing cabinet:

(a) a cylindrical passive array performing the functions of narrowband and broadband operation in the medium-frequency band, low-frequency acoustic intercept, and active operation by means of an associated active transmit array
(b) port and starboard flank arrays to carry out narrowband and broadband passive operations in the low-frequency band
(c) port and starboard passive ranging arrays, together with a high-frequency intercept sonar.

Each sonar can operate autonomously, in combination with one or more display consoles, if other parts of the system are unavailable. This allows, for example, the passive ranger to operate as a back-up for the cylindrical array sonar. A flank array sonar which performs low-frequency coverage and full primary sonar operations is available.

SIASS can store 35 tracks and performs automatic contact motion analysis on each sonar target being tracked, up to the maximum that can be handled by the system. The broadband subsystems of the cylindrical array and flank array sets, each have four operating channels with eight preselected frequencies. The high-frequency and low-frequency acoustic-intercept sonars each have 16 channels, and the active sonar has four; in these cases tracking is based on the use of associated algorithms. The electronic classification for each sonar can hold information on up to 300 specific platforms.

The cylindrical array sonar acts as the primary attack sensor. A combination of broadband and narrowband processing is used to extend detection ranges, to enhance tracking, and to provide quick and accurate classification of contacts. Simultaneous panoramic coverage is achieved by the use of preformed beams.

The flank array sonar is a low-frequency passive equipment that can act as the primary attack sonar within its frequency range. The port and starboard line arrays, used in conjunction with real-time delay beam-forming, provide coverage of large bearing sectors on each side of the submarine. The processing chain is similar to that of the cylindrical array sonar and can be matched to interface with a towed array if required.

Each of the preformed beams produced by the cylindrical array sonar and the 64 beams produced from the flank array undergo online spectral analysis in parallel, a complex fast Fourier transform being used to give reliable detection and tracking of noise emitted by a contact. The frequency-analysed preformed beam signals from the cylindrical array processor are used as

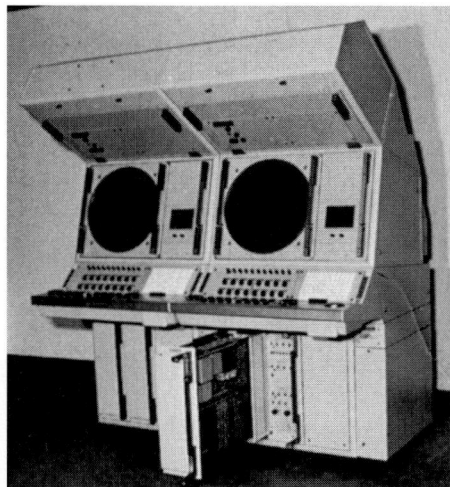

SIASS display and computer consoles

a low-frequency intercept sonar that can associate a large number of different transmissions simultaneously, and incorporate extensive algorithms to reduce false alarms.

The cylindrical array and flank array sonars also have enhanced classification channels for high-resolution vernier analysis, and to enable frequencies below 100 Hz to be measured. They can also provide multichannel automatic tracking.

The active sonar can be operated in directional single ping, and omnidirectional single ping or multiping modes. Return echoes are detected and processed by the cylindrical array sonar, and when operating in the multiping mode, the active sonar can automatically track multiple targets.

The passive ranging sonar operates in the lower part of the frequency band and is normally slaved to the cylindrical array. In this slaved operating mode, range is automatically calculated once a broadband contact that is being tracked on the cylindrical array has entered the ranging sector. The passive ranger's signal processing uses a method of frequency-domain correlation to measure wavefront curvatures and thus provide range information.

The high-frequency intercept sonar provides coverage up to 100 kHz and employs a fully digital compressor using fast Fourier transform techniques to detect, correlate and classify intercepts of torpedoes or sonars, and to warn the operator automatically.

Control is exercised from multipurpose consoles which may be integrated parts of a total SEWACO system. These consoles are multifunction operator workstations equipped with a high-resolution raster scan colour display. As an optional extension, signal processing for a towed array can be offered.

Operational status

In operational service on the Taiwanese 'Hai Lung' class.

Contractor

Hollandse Signaalapparaten BV, Hengelo.

VERIFIED

RUSSIAN FEDERATION AND ASSOCIATED STATES (CIS)

Submarine sonar systems

Early sonars appeared during the Second World War. By 1958, two sonar systems appeared that became the standard on all following classes of diesel electric submarines for nearly two decades. These were the Feniks and Herkules, and a 'top-hat' looking underwater telephone, codenamed Fez. The wraparound Feniks HF bow array required at least a crude compensator switch to steer listening beams. The medium-sized Herkules dome on deck is a combination active/passive system, and is very similar to that fitted in surface ships.

Four submarine classes had a larger size topside sonar dome and, since these classes all carried new 16 in ASW torpedo tubes aft, this dome was probably related to the fire-control computers for these tubes. Since passive fire-control solutions require triangulation, it is quite likely that the vertical fin aft on many later submarine classes may include sonar hydrophones.

The 'November' class SSNs were the first with a rakishly streamlined bow array, although this was probably of a medium-frequency type. The first low-frequency sonar appeared in the early 1960s on the 'Echo II' and 'Juliet' classes, and by 1967 streamlined bow shapes with sharply sloped low-frequency bow array windows had appeared.

The 'Victor III', 'Sierra' and 'Akula' classes have a teardrop shaped pod on top of the aft rudder fin. Most references state that it houses either a towed array

Sonar	Type	Frequency band	Mode	Function	Class	Number	Type
Feniks	Hull-mounted		Passive	Search	Foxtrot	4	SSK (Libya)
					Romeo	22	SSK (North Korea)
Herkules	Hull-mounted	MF/HF	Active	Search & attack	Romeo	39*	SSK (China)
Mouse Roar	Hull-mounted	HF	Active	Attack	Akula I/II	12	SSN
					Delta I	4	SSBN
					Delta III	9	SSBN
					Kilo	15	SSK
					Kilo	9	SSK (India)
					Kilo	3	SSK (Iran)
					Kilo	2	SSK (Algeria)
					Oscar II	12	SSGN
					Sierra I/II	3	SSN
					Tango	6	SSK
					Typhoon	6	SSBN
					Victor III	11	SSN
					Yankee Notch	1	SSN
Pelamida	TAS	VLF	Passive	Search	Delta III	9	SSBN
					Oscar II	12	SSGN
					Typhoon	6	SSBN
Pike Jaw	Hull-mounted	HF	Passive/active	Search & attack	Foxtrot	2	SSK
Shark Fin	Hull-mounted	MF	Passive/active	Search & attack	Kilo	15	SSK
					Kilo	9	SSK (India)
					Tango	6	SSK
Shark Gill	Hull-mounted	LF/MF	Passive/active	Search & attack	Akula I/II	12	SSN
					Typhoon	6	SSBN
					Delta IV	7	SSBN
					Oscar II	12	SSGN
					Sierra I/II	3	SSN
					Victor III	11	SSN
Shark Hide	Flank	LF	Passive	Search	Delta III	9	SSBN
					Delta IV	7	SSBN
Shark Rib	Flank	LF	Passive	Search	Oscar II	12	SSGN
					Sierra I/II	3	SSN
					Typhoon	6	SSBN
					Victor III	11	SSN
Shark Teeth	Hull-mounted	LF/MF	Passive/active	Search & attack	Delta I	4	SSBN
					Delta III	9	SSBN
					Kilo	15	SSK
					Kilo	9	SSK (India)
					Kilo	3	SSK (Iran)
					Tango	6	SSK
					Yankee Notch	1	SSN
Skat 3	TAS	VLF	Passive	Search	Sierra I/II	3	SSN
					Victor III	11	SSN
Tamir	Hull-mounted	HF	Passive	Search & attack	Romeo	39	SSK (China)
					Whiskey V	4	SSK (Albania)
Whale Series	Hull-mounted	MF	Passive	Search	Kilo	1	SSK (Poland)

* In some units

Bow sonar on 'Kilo' class submarine

'Foxtrot' class submarine bow sonars

'Sierra' class SSN bow with missing plates behind bow sonar belt

indicating a high level of signal processing power and beam-steering, or perhaps a wire communications antenna. The later 'Oscar' class also has a vertical aft fin that has towed array capability.

Very few details of submarine sonar systems have emerged and the accompanying list gives only the types of sonar that are understood to be fitted to various current classes of submarines. All systems are passive/attack sonars, the principal difference between classes being the frequency. No information is available on the use of towed arrays by RFAS submarines.

The 'Kilo' class is fitted with the MGK-400 EM digital sonar system. This is a passive/active sonar with UWT and telegraph capabilities over both long and short ranges. Echo ranging is provided over a ±30° sector of the target's relative bearing.

UPDATED

Large sonar arrays feature prominently on bows of RFAS submarines

UNITED KINGDOM

Type 2007 sonar

Type
Passive search and detection sonar.

Description
The Type 2007 is a conventional submarine sonar which uses steerable beams, enabling the submarine to maintain a straight course while the beams are steered to search the waters for detection and tracking of hostile submarines. The Type 2007 search beams listen at a frequency of between 1 and approximately 3 kHz and are used for initial long-range target detection and for gaining a bearing resolution. The sonar listens for propeller cavitation, nuclear cooling pumps and turbine reduction gears for initial target detection.

The Type 2007 is a more modern version of the original Type 186 passive conformal hydrophone system. This latter equipment consisted of a series of hydrophones evenly spaced along both sides of the hull.

Operational status
The Type 2007 is installed in 'Oberon' class submarines operated by Australia, Brazil, Canada and Chile. It has also been fitted to the Royal Navy's 'Trafalgar' class submarines. No longer in production. Still in operational use. Graseby now only manufactures transducers for the system.

Contractor
Graseby Dynamics Ltd, Watford.

UPDATED

Type 2019

Type
Submarine intercept sonar.

Description
The Type 2019 is the standard intercept sonar mounted on Royal Navy submarines.

Under a contract awarded to PMES in 1993, 12 Type 2019 sets have been upgraded with interactive plasma panel displays, new faster processors and improved recording facilities. These will offer the operator a greatly enhanced operability with the option of automatic transfer of target data to the boat's combat system highway. These improvements to the MMI are important details relating to the human factors programme. Ergonomically designed consoles and easy-to-read colour displays minimise operator fatigue and enhance operator performance.

Cogent have been awarded a contract to provide a replacement array for the 2019 referred to as Hull Outfit 51R. Housed in a free-flooding carbon fibre dome, the array comprises individually replaceable HF transducers mounted beneath a compact replaceable LF array unit. The broadband passive array

New free-flooding carbon fibre dome for Type 2019 (Courtesy Cogent Defence Systems)
1999/0024708

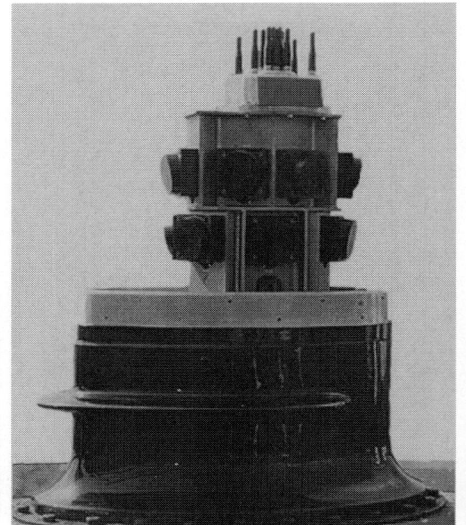

Array configuration for replacement Type 2019 array (Courtesy Cogent Defence Systems)
1999/0024707

is a multi-octave system providing bearing and azimuth coverage through 360°. The improved system will be installed in boats as they come alongside dock. All current systems are due to be upgraded by 1999.

W & J Tod Ltd of Weymouth have been awarded a contract to design and build two sonar domes for the update array system currently being developed by Cogent and which is undergoing trials.

Contractors
PMES Ltd, Rugeley.
Thomson Marconi Sonar SAS, Sophia Antipolis.
Cogent Defence Systems, Newport.

UPDATED

Type 2020 sonar

Type
Passive and active bow-mounted sonar.

Description
The Type 2020 is fitted in some Royal Navy hunter-killer nuclear submarines, it provides both passive and active detection capabilities from its bow-mounted array. Although considerable emphasis has been placed on passive detection, the equipment incorporates a highly effective computer-aided active capability.

Passive detection is subdivided into low-frequency and high-frequency bands with information presented on dual-speed multipen recorders. The system is intended primarily for operation in a computer-aided mode but can be automatically reconfigured to a fallback mode of operation, with target information being passed to the action information organiser in both modes. Facilities for expansion are included in the design to allow interfacing with future sonar systems. Solid-state switching is used for the transmit/receive function and electronic beam-steering techniques are employed. The design provides high-detection sonar performance with a full 360° coverage within a frequency range reported to be from 2 to 16 kHz for passive operation. The active transmitter uses an automatic beam-steering unit with a manual override capability. It transmits in the low-/medium-frequency bands, reportedly between 3 and 13 kHz.

A computer-controlled automatic monitoring subsystem is incorporated in the equipment, with a central monitoring unit driving local monitoring units within each equipment cabinet. Diagnosis of computer

Display/console of Type 2020 submarine sonar

faults and programme loading is accelerated by the use of a floppy disk.

Operational status
The first production model was accepted by the Royal Navy in October 1982 and production deliveries are now complete. In some 'Trafalgar' class submarines Type 2020 is being replaced by Type 2074 (see below).

Contractor
Thomson Marconi Sonar Ltd, Templecombe.

VERIFIED

Type 2026 sonar

Type
Passive towed array sonar.

Type 2026 submarine sonar

Description
The Type 2026 sonar is a submarine passive towed array sonar and consists of a multi-octave towed array, a towcable and an inboard processor and display system. The system provides surveillance and classification on data from both towed and flank arrays. It contains tracking, combat information system interface, data recording, Demon, aural and hard-copy facilities. The signal processor may also be used to provide a signal analysis and display function for data from other sonars.

The sonar may be operated by one or two operators and is controlled using a touchscreen, joystick and switches. It includes BITE and may be reconfigured in the event of a fault to minimise its effects.

Data is displayed on three acoustic data displays, with comprehensive cursor and annotation facilities. Other facilities for driving remote displays (such as in the submarine's control room) or external video recorders are provided. The signal processor is cooled

by an onboard chilled water supply, with an optional heat exchanger. The operator's console may be fitted in the sound room, with the signal processor in a remote cabinet space.

The Type 2026 sonar was developed in conjunction with the Admiralty Research Establishment by GEC Avionics and Plessey Naval Systems, both now part of Thomson Marconi Sonar Ltd. GEC Avionics manufactured the signal processor and display system, Cogent, the towcable and Plessey, the array and terminal unit.

Operational status
Systems are operational on a number of Royal Navy SSNs and a variant has been supplied to the Royal Netherlands Navy and equips the 'Walrus' class.

Contractor
Thomson Marconi Sonar Ltd, Templecombe.

UPDATED

Type 2032 processor

Type
Electronics updating package.

Description
The Type 2032 is an electronics processing package designed to be integrated with the bow-mounted active

array to give a limited narrowband capability. As well as certain electronic processing equipment, the Type 2032 add-on package includes a number of special transducer elements, a beamformer cabinet and control facilities. It is believed to be complementary to the existing Type 2020 array and is understood to be intended for retrofit and new-build applications in the Royal Navy's 'Trafalgar' and 'Swiftsure' class SSNs.

Operational status
In service aboard Royal Navy SSNs.

Contractor
Thomson Marconi Sonar Ltd, Templecombe.

VERIFIED

Type 2046 submarine sonar

Type
Passive hull-mounted sonar.

Description
The Type 2046 sonar is an integrated, passive system designed for the Royal Navy nuclear-powered attack submarines, and is also fitted into the Type 2400 'Upholder' class of conventional submarines.

Developed from advanced 'Curtis' digital signal processing technology, and including FMS 12 display processing with Ferranti M700 Argus processors, the Type 2046 provides long-range target detection, classification and tracking information.

The system provides 360° bearing cover, multiple beam resolution and broadband surveillance with a high-resolution narrowband frequency expansion

facility. The operator interface consists of four-colour displays in a two-operator console suite with touch-sensitive plasma panels.

The Type 2046 is built to a modular design to make maximum use of the latest technology, to save space and cost, and to assist through-life enhancement of the system. It is also equipped with fully integrated built-in test equipment for ease of maintenance.

Operational status
A total of 27 systems has been ordered for Royal Navy submarines together with a follow-on order for a further two systems for training purposes. All systems have now been delivered. Forms part of the 2054 sonar suite on 'Vanguard' class SSBNs, and equips some 'Trafalgar' class SSNs.

Contractor
Thomson Marconi Sonar Ltd, Templecombe.

UPDATED

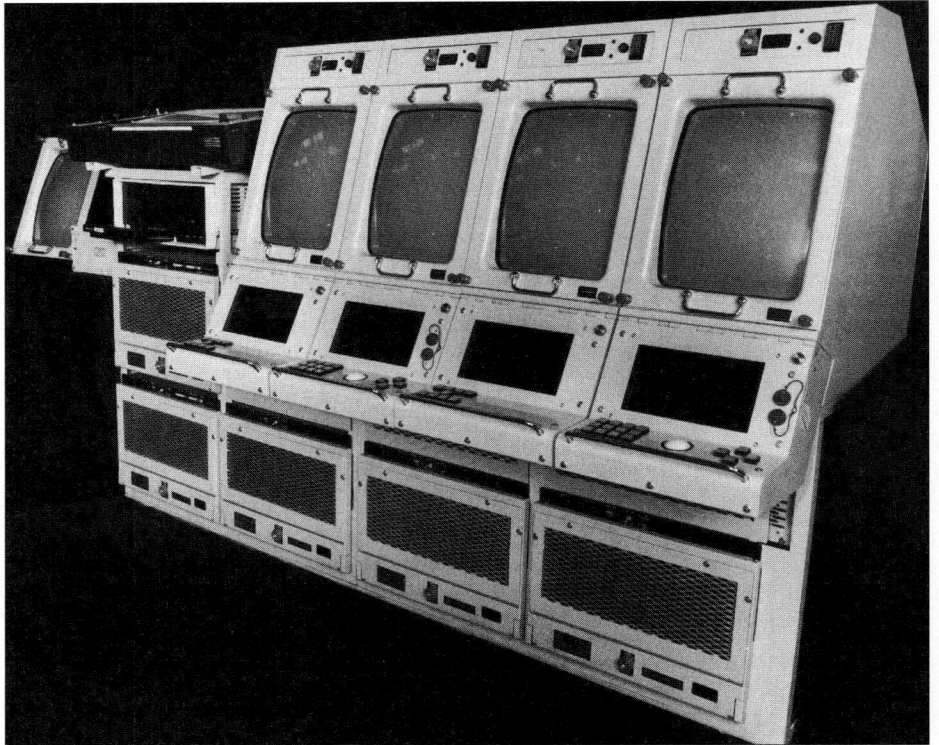

Operators' interface console of Type 2046 sonar

Type 2051 sonar

Type
Passive and active sonar system.

Description
The Type 2051 sonar system was provided for the update programme of the Royal Navy 'Oberon' class submarines. The Triton fit was designed to bring the 'Oberon' class submarines up to date with the latest advances in sonar technology, including computer-assisted detection and classification.

The main elements of the system are a compact narrowband sonar which is linked to the submarine's existing flank array or towed array to provide all-round surveillance and classification, broadband passive sonars for both bow and flank arrays, an active sonar with variable power and sector capabilities, and an intercept sonar. A newly designed bow sonar dome, designed to improve the hydrodynamic performance of the submarine and reduce water flow noise, has been fitted.

Operational status
In service with the Royal Canadian Navy.

Contractor
Thomson Marconi Sonar Ltd, Templecombe (prime contractor).

VERIFIED

Type 2054 sonar

Type
Passive/active intercept and towed array sonar suite.

Description
Sonar 2054 is the major sonar for the new Royal Navy 'Vanguard' class Trident ballistic missile carrying submarines. The complete equipment, which is the largest system in the Royal Navy, consists of a suite incorporating passive, active, intercept and towed array sonars.

The former GEC-Marconi, Sonar Systems Division (now Thomson Marconi Sonar Ltd), the prime contractor and system design authority, supplies: the bow-mounted array with the necessary beam-forming and signal processing equipment for narrowband, broadband and active; the passive ranging equipment; intercept equipment; sonar controller's equipment; sonar maintainer equipment; and the towed array. Major subcontractors are: the former Ferranti-Thomson Sonar Systems (also now Thomson Marconi Sonar Ltd) for the consoles, displays and data processing for broadband, narrowband and active, and Cogent for the towed array handling system.

The processing in the Type 2054 is almost entirely digital, the arrays being sampled by the preprocessor and immediately digitised before being passed to the beam-forming/signal processing. The system provides surveillance and classification on data from all arrays, which include bearing, range, target motion and torpedo detection capabilities. The total system comprises 18 cabinets and seven consoles (two broadband, two narrowband, one passive ranging, one intercept and one sonar controller, with the option of either a broadband or narrowband being used for active). The cabinets are interconnected by a 1553 standard data highway, which also provides the interface to the outside world.

Operational status
A total of six systems was ordered, two being used as shore-based systems. All have now been delivered.

Contractor
Thomson Marconi Sonar Ltd, Templecombe.

VERIFIED

Type 2074 sonar system

Type
Passive and active hull-mounted sonar system.

Description
The Type 2074 passive and active, low-frequency bow sonar is being fitted to eight of the Royal Navy's nuclear-powered hunter-killer submarines. It replaces the Types 2001 and 2020 in these vessels. The sonar makes full use of the advantages of Thomson Marconi Sonar Ltd Series 5 family of modular sonars and is incorporated into a well established equipment practice, also developed by this company and used in all major new Royal Navy sonar development programmes.

The Type 2074 provides improved sonar performance and is significantly smaller (by a ratio of 5:1) than the sonar it replaces.

Due to the modular flexible nature of its design, the sonar can be readily configured to suit a customer's particular requirements and offers stretch potential for future enhancements. The total system consists of five cabinets and five multifunction consoles.

Operational status
In production for the Royal Navy. The modular nature of the design is used as the basis for the company's current export activities. The 2074 LRE is the main active and passive bow sonar in some 'Swiftsure' and 'Trafalgar' class submarines where it is gradually replacing the Type 2020.

Contractor
Thomson Marconi Sonar Ltd, Templecombe.

UPDATED

Type 2076 sonar

Type
Integrated passive/active sonar.

Description
The Type 2076 is an integrated passive/active search and attack sonar suite designed as a mid-life update for the Royal Navy 'Trafalgar' class submarines.

The integrated sonar suite incorporates a comprehensive set of new arrays and processing, feeding the command and control system, Installation of the system will occur during normal refit periods without interruption to the normal operational cycle.

Operational status
The initial contract is for four systems with options for a further two, plus additional options for a further four or six for the 'Astute' class.

Delivery of a shore-based trial integration system was concluded by the end of 1995.

Type 2076 is replacing Types 2074 and 2046 on 'Trafalgar' class submarines.

Contractor
Thomson Marconi Sonar Ltd, Templecombe.

UPDATED

Type 2082

Type
Passive intercept and ranging sonar.

Description
The Type 2082 is the replacement for Type 2019 mounted on Royal Navy submarines.

Under a contract awarded to PMES in 1993, eight systems have been produced as part of the 'Swiftsure' and 'Trafalgar' Class Update. Type 2082 is an advanced DSP-based intercept system offering enhanced sensitivity and analysis facilities, extended frequency coverage and intercept auto-tracking capability.

PMES has also conducted studies into advanced intercept and transient capture systems for the UK MoD and associated agencies.

Contractor
PMES Limited, Rugeley.

UPDATED

Lara

Type
Experimental towed array.

Development
Lara, an experimental large aperture research array, is being manufactured to carry out studies for future towed array development programmes. The array comprises a set of reconfigurable towed array modules. Both single and multiple array configurations will be used to evaluate acoustic performance in varying environmental conditions.

Operational status
A contract was awarded in late 1994 to Thomson Marconi Sonar to design and manufacture the array for the UK Defence Research Agency at Winfrith, Dorset. The array features many advanced techniques developed jointly with the DERA.

Contractor
Thomson Marconi Sonar Ltd, Templecombe.

VERIFIED

Flank array sonar

Type
Submarine flank array sonar system.

Description
The Thomson Marconi Sonar Ltd submarine flank array sonar offers the submarine operator an alternative form of linear array sonar. The flank array sonar provides a large acoustic aperture without directional ambiguity.

This flank array is a modular system enabling acoustic performance to be tailored to the requirements of a particular platform. Standard hydrophone modules which are 2 m long contain Thomson Marconi patented large-area hydrophones, embedded in a special polyurethane resin providing optimum decoupling from the flow noise. The spacing of the hydrophones can be varied during manufacture, depending on customer requirements, to a maximum of 12 hydrophones per module. The number of modules that can be fitted is dependent on the length of the parallel section of the submarine hull.

A special end fairing module provides a smooth contour at the end of the array to minimise flow noise. The array modules are fitted to prepared sites on the submarine pressure hull comprising studs or tapped bosses welded to the hull. Each hydrophone incorporates its own preamplifier which enables long cable lengths to be driven from the array without impairing performance. The preamplifiers are housed in the fairings of the module, enabling easy access for maintenance in dry dock without removing the module.

The low-frequency flank array can be interfaced to existing inboard narrowband and broadband processors, or can be supplied with a Thomson Marconi Sonar Ltd processor as a complete system.

Operational status
In service with the Swedish Navy on the 'Västergötland' class submarines.

Contractor
Thomson Marconi Sonar Ltd, Templecombe.

UPDATED

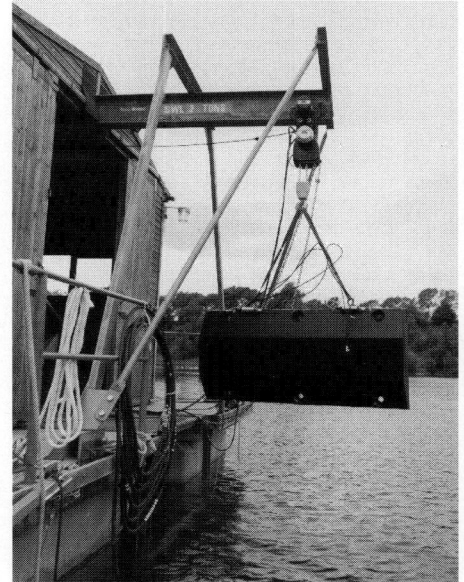

In-water testing of a submarine flank array

SUBTASS

Type
Passive low-frequency towed array sonar.

Description
Thomson Marconi Sonar Ltd, under contract to the Department of Supply and Service of Canada, has supplied a Submarine Towed Array Sonar System (SUBTASS) for the Canadian 'Oberon' class submarines. SUBTASS is a complementary array to the Type 2051 Triton attack sonar, and is required to interface with existing Type 2051 and 2007 sonars, the submarine fire-control system, log, gyro and data recording systems.

Based on Thomson Marconi Sonar Ltd modular Series 5 technology, the requirements for both narrowband and broadband processing, displays with provision for narrowband verniers, broadband trackers, TMA audio monitoring plus optional expansion including Demon analysis and a narrowband flank array facility, have been met using in-service modules. Commonality with CRISP acoustic analyser and CHIP concurrent high-speed integrated processor leads to reduced spares holdings and increased interchangeability.

Modular flexibility allows use of the indigenous designed and produced array modules from Hermes Electronics Ltd of Canada, as well as the inclusion of future enhancements and expansion. The array modules are housed in an opaque 6.03 cm polyurethane jacket. Six non-acoustic channels provide fore and aft measurements of the array's heading, depth and temperature.

The frequency range covers 1 Hz to 6 kHz and the 266 m array can be towed at depths down to 340 m.

Operational status
In service with the Royal Canadian Navy.

Contractors
Thomson Marconi Sonar Ltd, Templecombe.
Hermes Electronics, Dartmouth, Nova Scotia, Canada.

UPDATED

UNITED STATES OF AMERICA

AN/BQR-15 sonar

Type
Passive towed array sonar.

Development
The AN/BQR-15 towed array system was developed in the early 1970s for deployment on fleet ballistic missile submarines.

Description
The AN/BQR-15 high-frequency sonar system comprises a 2,200 ft tow cable with a 154 ft towed sonar array and was previously deployed by all US Navy Poseidon ballistic missile submarines. It provides hydrophone outputs to the shipboard electronics to facilitate target detection and classification. In addition, the sonar performs own ship noise monitoring and underwater communication. The sonar can simultaneously form 46 beams on fixed bearings and can form a beam steerable to any of 92 bearings.

Under optional conditions the array can be deployed in as little as 15 minutes.

Operational status
Currently in service with 'Ohio' class SSBNs and 'Sturgeon' class SSNs.

Contractor
Lucent Technologies Advanced Technology Systems, Greensboro, North Carolina.

UPDATED

Model 1110 sonar

Type
Passive flank array sonar.

Description
The Model 1110 uses a passive flank array to enhance the overall sensor capability of submarines. It is designed to complement an existing sonar, combat and fire-control systems already planned or installed. The Model 1110 provides for long-range detection, tracking and classification of surface and submarine targets, while the existing hull sonar can be optimised for close-in detection.

The Model 1110 is a single-operator system comprising an operator console, processor cabinet, signal conditioning electronics and acoustic sensor arrays.

The system provides capabilities for expanded target detection and analysis through use of low-frequency beam-forming, split-beam correlation, bearing determination techniques and frequency analysis. It has the capability to track more than 15 targets simultaneously. Target tracking is accomplished by broadband and narrowband signal processing interpolation. Track processing allows the Model 1110 to automatically track multiple targets with minimal supervision. Selected targets can be assigned to high-quality track processors for improved position solution accuracy. The trackers can maintain track even in a crossing-target environment without operator intervention. Tracking can be initiated by the operator using the cursor and soft key controls. An adaptive self-noise cancelling subsystem has been interfaced and tested at sea and has demonstrated greatly improved sonar performance. A towed array variant can be provided as an option.

The Model 1110 system single-operator console has two high-resolution colour displays. In existing submarines, the control unit can be located adjacent the host submarine sonar console and together will create an integrated sonar system. The microprocessor-based host submarine interface allows the Model 1110 to be integrated easily into the combat system of any new or existing submarine.

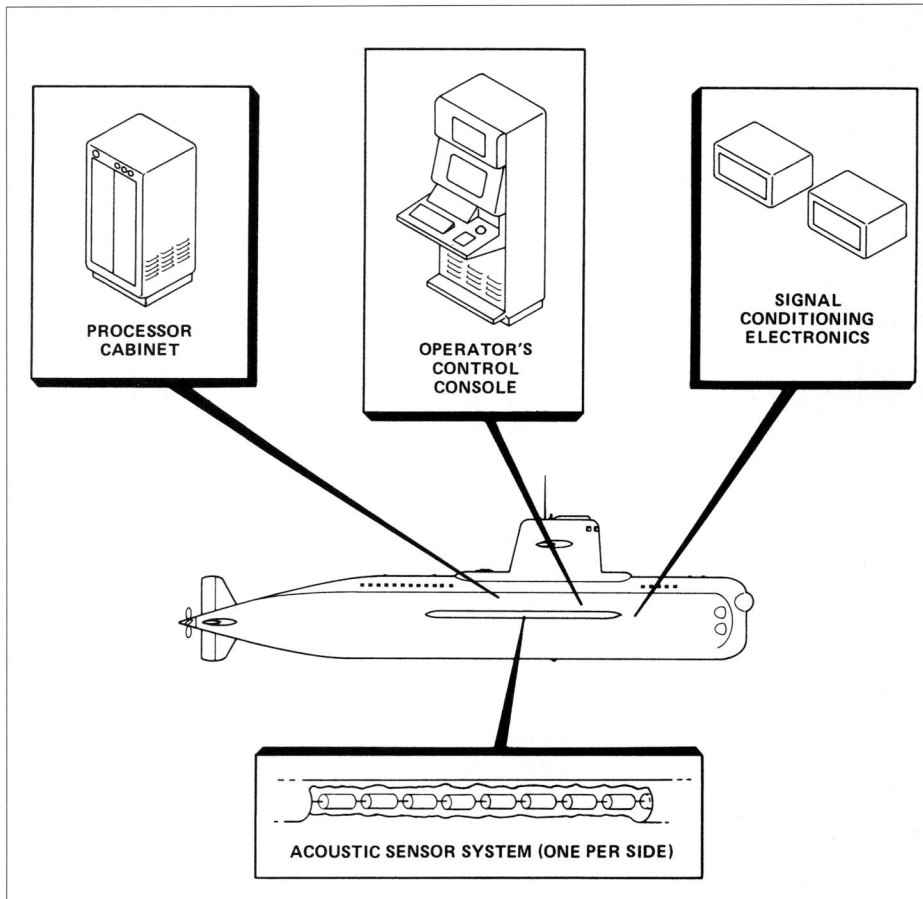

Model 1110 sonar layout

Operational status
Several systems are currently installed and operational with an international customer.

Contractor
EDO Corporation, Government Systems Division, College Point, New York.

UPDATED

Model 1120 series sonar

Type
Passive flank/hull array sonars.

Development
The Model 1120 Series is being developed by the EDO Corporation for sale on the international market. The systems being offered use the latest COTS/OSA hardware and software techniques. These modern sonars are new technology upgrades of the Model 1110 sonars. All variants can be provided with one or two flat panel type colour displays to further reduce the system footprint on small hulls. The use of standard digital interfacing standards simplifies the integration of the sonar with the submarine's combat system.

Description
The Model 1120 architecture is organised to allow the units to be programmed for operation in the desired mode. The multisensor variants are operator selectable for downloading in the desired configuration. Processing for simultaneous multisensor operation can also be provided.

The following models are available:
1121 flank array sonar
1122 hull cylindrical array sonar
1123 towed array sonar
1125 flank array with cylindrical hull array sonar
1126 flank/hull array with towed array sonar
1127 flank array with towed array sonar.

The Model 1121 is a single-operator system which uses a passive flank array to enhance the overall sensor capability of submarines. The new technology system is designed to match the performance capabilities of the operational Model 1110 systems. An adaptive noise canceller can be offered to reduce the effects of hullborne noise. The modernised flank array system retains the production Model 1110 outboard sensor mounting and fairing arrangement which greatly reduces installation risks.

The Model 1122 sonar uses a passive 1 m diameter forward hull-mounted transducer. The COTS variant sonar design is founded on EDO's experience with the operational BQR 2B and Model 1102 sonar equipments.

The EDO Model 1123 is a passive towed array sonar which uses a dual-nested hydrophone configuration in the array.

Contractor
EDO Corporation, Government Systems Division, College Point, New York.

VERIFIED

AN/BQQ-4

Type
Passive search and detection sonar.

Description
The AN/BQQ-4 MicroPUFFS system is a high-accuracy, long-ranging passive sonar designed for installation aboard submarines. This equipment is a derivative of the original US Navy PUFFS (Passive Underwater Fire-control Feasibility Study) hardware. The PUFFS designation was also applied to USN AN/BQG-1/2/3/4 anti-submarine warfare systems. It was developed for the detection and tracking of submarine and surface targets, and provides: automatic target tracking of up to four targets simultaneously, instantaneous and continuous ranging solution, fire-control range and bearing output, and secure operation. The basic technique is signal enhancement by cross-correlation of the signals received at the arrays with conversion of two measured time differentials into bearing and range data.

The system consists of three baffled hydrophone arrays mounted on each side of the submarine. The arrays are connected to a display control console, which utilises an analogue processor, digital processor and AN/UYK-26 computer. The console also includes built-in fault localiser equipment. The system is built to US military specifications and uses standard electronic modules throughout.

Operational status
MicroPUFFS systems are currently installed aboard 'Oberon' class submarines of the Royal Australian Navy and Royal Canadian Navy.

Lockheed Martin is currently developing an updated version, which uses all digital processing and COTS hardware. It will be available with an optional display for use as a stand-alone system or without the display for embedding in integrated combat systems.

Contractor
Lockheed Martin Tactical Defense Systems, Great Neck, New York.

UPDATED

AN/BQQ-5 sonar

Type
Hull-mounted and towed array.

Description
The AN/BQQ-5 sonar, the US Navy's first active/passive digital sonar system, is the principal sensor system of the US Navy's 'Los Angeles' class nuclear attack submarines. The AN/BQQ-5 is a digital, multibeam system employing hull-mounted, sphere-mounted and towed acoustic hydrophone arrays. The polyethylene covered towcable, which has a maximum length of about 800 m, is 9.5 mm in diameter, and the array at the end of the cable, where the hydrophones and electronics are located, is 82.5 mm in diameter. The array is tapered fore and aft to minimise flow noise. Drag is stated to account for a maximum reduction in speed of 0.5 kt, with no serious inhibition on submerged manoeuvres and little adverse effect on surface manoeuvres, with the one exception of those entailing going about.

During the late 1970s, the US Navy became concerned that the AN/BQQ-5 would not be able to handle the developing threats and embarked on a programme to update the capabilities of the system. An improved control display console has been developed and sonars that have been upgraded with the new consoles are known as the AN/BQQ-5B. A further modification to the system developed under this programme is the AN/BQQ-5C(V) Expanded

Directional Frequency Analysis and Recording System (DIFAR), which replaces the original signal processor with the AN/UYH-1 standard signal processor. The AN/BQQ-5D uses a thin-line array and utilises some of the technology common to the AN/BSY-1 system. It became operational in 1988. A contract for the development of the AN/BQQ-5E was awarded in December 1988. A new thin-line array for passive ranging has been integrated with the AN/BQQ-5E, and went to sea in 1993/4.

The US Navy has been developing a thin-line tow array and handling system and other modifications for the AN/BQQ-5 system (see Operational status). The thin-line towed array, known as the TB-23, forms part of the AN/BSY-1(V) combat suite. It is a lightweight system which can be reeled into the vessel's main ballast tank, instead of being housed on the side of the submarine, and will replace the TB-16 thick-line array currently employed in the AN/BQQ-5.

The AN/BQQ-5 provides improved detection, tracking and classification of low-frequency data. The latest development also provides processing capability for the new TB-29 towed array and interfacing to the new fire-control system CCS Mk 2. The combining of the AN/BQQ-5E, TB-29, and CCS Mk 2 will result in a system referred to as QE2 underwent technical evaluation in 1993.

The AN/WLR-9A acoustic intercept receiver, produced by Norden, has been successfully evaluated and forms an integral subsystem of the AN/BQQ-5 and AN/BQQ-6 on new attack and Trident missile

submarines. The subsystem features a CRT display, a digital readout and a remote unit for the submarine commander. A sensitivity improvement kit for the AN/WLR-9A has been developed by Norden and deliveries to the US Navy are in progress.

Operational status
The AN/BQQ-5 underwent extensive developmental testing during 1972 and 1973, and was approved for production later in 1973.

During the period 1973 to 1992, a number of companies, including Raytheon, Gould, Bendix, Norden and Tracor, has been awarded contracts for systems, subassemblies and technical support. In December 1988, IBM Federal Systems (now Lockheed Martin Federal Systems) was awarded a US$54 million contract for upgrade kits to bring the AN/BQQ-5 to its latest configuration.

To date, 100 ship sets have been delivered – 91 for submarines, four maintenance trainers and five engineering models. In addition, eight 21B64 versions are used as land-based operator trainers.

Operational on board 'Los Angeles' class submarines.

Contractors
Lockheed Martin Federal Systems, Manassas, Virginia.
Raytheon Company, Electronic Systems, Portsmouth, Rhode Island.
UPDATED

AN/BQQ-6 sonar

Type
Passive and active sonar.

Description
The AN/BQQ-6 is an advanced active/passive sonar set developed for the 'Ohio' class submarines and forms part of the Trident command and control system, which is an integrated complex of command, control, communications and ship defence equipment.

The AN/BQQ-6 is based on the AN/BQQ-5 and has many of the same parts. The primary detection group is a digital integrated system employing spherical array,

hull-mounted line array and towed array sensors, with an active emission acoustic intercept receiver and high-frequency active short-range sonar. In the passive mode, the system used is identical to that of the AN/BQQ-5. Support equipment has been added to provide for underwater communications, environmental sensing, magnetic recording and acoustic emergency devices. All other technical details are classified.

A version of the AN/BQQ-5E system is currently being procured as an upgrade for the Trident submarines. This subsystem will represent a major upgrade to the present AN/BQQ-6 sonar. By utilising to a great extent the current AN/BQQ-5 hardware, a large

improvement in logistics (provisioning, spares, training, and so on) is envisioned.

Operational status
To date 19 ship sets have been delivered – 15 for submarines, three training systems and one engineering development model.

Contractor
Lockheed Martin Federal Systems, Manassas, Virginia.
Lockheed Martin, Syracuse, New York.
VERIFIED

AN/BQR-15 9080 sonar

Type
Passive thin-line towed array sonar.

Development
The thin-line towed array is a new generation of digital towed sonar arrays which was developed to provide high-resolution target detection at greater ranges and at higher towing speeds.

Description
The 9080 thin-line towed array has an outside diameter of 1 in and features a wide acoustic aperture, low self-noise and reelable construction. Acoustic channels are digitised and multiplexed onto a coaxial signal path to inboard signal processing equipment. Sophisticated digital signal processing enables long-range detection, classification and bearing determination.

Operational status
In operational service aboard 'Ohio' class SSBN submarines.

Contractor
Lucent Technologies Advanced Technology Systems, Greensboro, North Carolina.
VERIFIED

AN/BQS-13 sonar (BQR-24)

Type
Passive/active sonar system.

Development
Acceptance testing of the AN/BQS-13 (BQR-24) multipurpose subsystem began in late 1973 and technical and operational testing were conducted aboard *Archerfish* during 1974. The AN/BQS-13 multipurpose subsystem (BQR-24) was service approved in late 1974. The extended operational test programme delayed procurement to FY75-76, when 11 systems plus support were procured for US$25.3 million. These subsystems provide interim improvement in sonar and fire-control capability for those submarines not scheduled to receive the

AN/BQQ-5 sonar until late in that programme. The BQR-24 has been installed only in submarines where it will have a useful life of at least three years before receiving an AN/BQQ-5 sonar system.

Description
The AN/BQS-13 sonar system is the primary sonar in the 'Sturgeon' SSN class nuclear attack submarines. The system was service approved in December 1971 and is in operation. A system addition, providing new functional capability, was service approved in late 1974. This addition, the AN/BQS-13 multipurpose subsystem (BQR-24), is being procured for a limited number of SSNs.

The system is also operational on the 'Ohio' class submarines where it forms the spherical array for the AN/BQQ-6 sonar.

Operational status
Commencing in October 1979, Raytheon was awarded a series of US Naval Sea Systems Command contracts calling for refurbishment and conversion of AN/BQS-11, 12, and 13 sonar transmission subsystems to AN/BQQ-5 transmission subsystem configuration. These subsystems are part of the AN/BQQ-5 system.

Contractor
Raytheon Company, Electronic Systems, Portsmouth, Rhode Island.
UPDATED

AN/BQS-14A submarine sonar

Type
Mine detection sonar.

Description
The AN/BQS-14A is an upgraded development of the Hazeltine AN/BQS-14 ice detection sonar developed during the late 1960s. In 1979, the US Navy initiated a

programme to upgrade the system to provide an improved mine detection and avoidance capability to the 'Sturgeon' class of nuclear-powered attack submarines. The improved BQS-14A system has been introduced into the fleet as an upgrade and is the primary sonar used for under-ice navigation. Later versions of the sonar array, named the submarine active detection system/high-frequency array, have been integrated as part of the AN/BSY-1 and AN/BSY-2.

Operational status
Operational aboard most US Navy submarines.

Contractors
Hazeltine Corporation, Greenlawn, New York.

Lockheed Martin Federal Systems, Manassas, Virginia.

VERIFIED

TB-23 towed array

Type
Thin-line submarine towed array.

Description
The TB-23 is a towed array component of the AN/BQQ-5D and AN/BSY-1 submarine sonar systems. It is a thin-line array which can be reeled into the boat's main ballast tank instead of being housed on the side of the pressure hull.

Operational status
In operational service.

Contractor
L-3 Communications Ocean Systems, Sylmar, California.

VERIFIED

TB-29 towed array

Type
Advanced thin-line array.

Description
The TB-29 is intended for use with the latest types of the AN/BQQ-5 sonar system and for the AN/BSY-1 and AN/BSY-2 combat systems. The array features increased length over earlier arrays. As is the case with the TB-23 towed array, the TB-29 can be reeled into the boat's main ballast tanks instead of being housed on the side of the pressure hull. The TB-29 will eventually replace the TB-23 on most submarines.

Operational status
In production.

Contractor
Lockheed Martin, Syracuse, New York.

VERIFIED

Model 1525 CTFM sonar

Type
Obstacle avoidance sonar system.

Description
The Model 1525 obstacle avoidance sonar provides rapid scanning, high-resolution display of subsea objects at ranges from 3 to 1,500 m, and a full 360° area coverage. It was designed for installation on conventional submarines and manned submersibles. The Model 1525 features a TV formatted colour display with memory circuits that provide a continuous presentation of the sector scan selected until updated by the next scan. The Model 1525 is based on the design of the 500A and uses the same outboard unit. Its size and connections are compatible with that system.

The Model 1525 features automatic sector modes and manual scan selection. Forward sectors of 60 and 120° are offset, while a 240° sector is centred on the display. In addition, a 60° sector can be positioned anywhere in 360° for higher information rate operation. Incorporated in the equipment is the latest digital analyser with 128 channels, providing 3.2 lines higher resolution than the 500A.

The system has a transponder mode for operation with CTFM transponders. The mode can be used alone or simultaneously with the regular sonar mode at the operator's choice. Transponders are used primarily as a navigation aid, but can also be dropped from

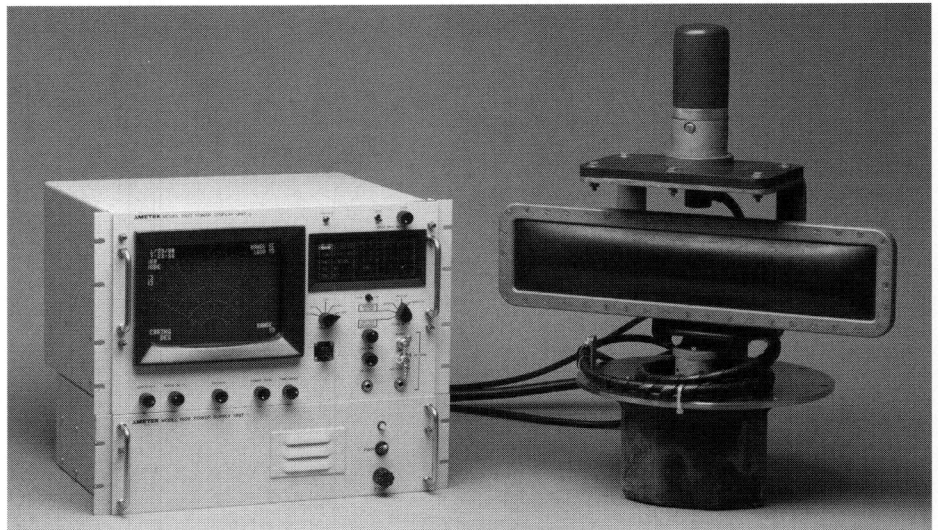

Model 1525 obstacle avoidance sonar

submarines in a submerged condition to mark positions of interest.

Operational status
In service with the US Navy, Norway and Japan.

Contractor
EDO Acoustic Products, Salt Lake City, Utah.

VERIFIED

TABLE IX

Submarine sonar equipments

System	Manufacturer	Country	Installed on	Navy	Units
410 A4	STN ATLAS Elektronik	Germany	Type 206	Indonesia	2
BQG-3	Sperry/Raytheon	USA	Guppy IIA	Turkey	4
BQG-4	Sperry/Raytheon	USA	Guppy III	Turkey	1
BQG-4	Sperry/Raytheon	USA	Tang	Turkey	2
BQG-501[1]	Sperry	USA	Oberon	Canada	3
BQQ-5	Lockheed Martin	USA	Los Angeles	USA	53
BQQ-6	Lockheed Martin	USA	Ohio	USA	18
BQR 2B	EDO	USA	Guppy II	Taiwan	2
BQR 2B	EDO	USA	Guppy IIA	Turkey	4
BQR 2B	EDO	USA	Guppy III	Turkey	1
BQR 2B	EDO	USA	Tang	Turkey	2
BQS 4	EDO	USA	Guppy IIA	Turkey	4
BQS 4	EDO	USA	Tang	Turkey	2
BQS 4C	Raytheon/EDO	USA	Guppy II	Taiwan	2
CSU 3	STN ATLAS Elektronik	Germany	Type 209/1,200 (Salta)	Argentina	1
CSU 3	STN ATLAS Elektronik	Germany	Type 209/1,300 (Thomson)	Chile	2
CSU 3	STN ATLAS Elektronik	Germany	Type 209/1,300	Ecuador	2
CSU 3	STN ATLAS Elektronik	Germany	Type 209/1,200 (Casma)	Peru	6
CSU 3	STN ATLAS Elektronik	Germany	Type 209/1,200 (Atilay)	Turkey	6
CSU 3-2	STN ATLAS Elektronik	Germany	Type 209/1,200 (Pijao)	Colombia	2
CSU 3-2	STN ATLAS Elektronik	Germany	Type 209/1,300 (Cakra)	Indonesia	2
CSU 3-32	STN ATLAS Elektronik	Germany	Type 209/1,300 (Cabalo)	Venezuela	2
CSU 3-4	STN ATLAS Elektronik	Germany	Santa Cruz TR 1700	Argentina	2
CSU 3-4	STN ATLAS Elektronik	Germany	Type 209/1,100-1,200 (Glavkos)	Greece	8
CSU 83	STN ATLAS Elektronik	Germany	Type 206A	Germany	12
CSU 83	STN ATLAS Elektronik	Germany	Type 209/1,500 (Shishumar)	India	4
CSU 83	STN ATLAS Elektronik	Germany	Type 209/1,200 (Chang Bogo)	Korea, South	8
CSU 83	STN ATLAS Elektronik	Germany	Ula (Type P6071)	Norway	6
CSU 83	STN ATLAS Elektronik	Germany	Kobben Type 207	Norway	6
CSU 83	STN ATLAS Elektronik	Germany	Vastergotland A 17	Sweden	4
CSU 83	STN ATLAS Elektronik	Germany	Type 209/1,400 (Preveze)	Turkey	3
CSU 83-1	STN ATLAS Elektronik	Germany	Type 209/1,400 (Tupi)	Brazil	3
CSU 90	STN ATLAS Elektronik	Germany	Type 209/1,100-1,200 (Glavkos)	Greece	8
CSU 90	STN ATLAS Elektronik	Germany	Oberon	Chile	2
CSU 90	STN ATLAS Elektronik	Germany	Dolphin	Israel	2
CSU 90	STN ATLAS Elektronik	Germany	Gotland A 19	Sweden	3
DMUX 20 (Eledone)	Thomson Marconi Sonar SAS	France	Rubis	France	6
DSUV 1	Thomson Marconi Sonar SAS	France	Hangor (Daphne)	Pakistan	4
DSUV 2	Thomson Marconi Sonar SAS	France	Daphne	France	1
DSUV 2	Thomson Marconi Sonar SAS	France	Albacora (Daphne)	Portugal	3
DSUV 2[2]	Thomson Marconi Sonar SAS	France	Daphne	South Africa	3
DSUV 2H	Thomson Marconi Sonar SAS	France	Hashmat (Agosta)	Pakistan	2
DSUV 22	Thomson Marconi Sonar SAS	France	Agosta	France	2
DSUV 22	Thomson Marconi Sonar SAS	France	Galerna (Agosta)	Spain	4
DSUV 62[3]	Thomson Marconi Sonar SAS	France	Galerna (Agosta)	Spain	3
DSUV 62A	Thomson Marconi Sonar SAS	France	Agosta	France	2
DSUV 62C	Thomson Marconi Sonar SAS	France	Rubis	France	6
DUUA 1	Thomson Marconi Sonar SAS	France	Hangor (Daphne)	Pakistan	4
DUUA 1D	Thomson Marconi Sonar SAS	France	Hashmat (Agosta)	Pakistan	2
DUUA 1D	Thomson Marconi Sonar SAS	France	Agosta	France	2
DUUA 2	Thomson Marconi Sonar SAS	France	Daphne	France	1
DUUA 2	Thomson Marconi Sonar SAS	France	Albacora (Daphne)	Portugal	3
DUUA 2[2]	Thomson Marconi Sonar SAS	France	Daphne	South Africa	3
DUUA 2	Thomson Marconi Sonar SAS	France	Delfin (Daphne)	Spain	4
DUUA 2A/2B	Thomson Marconi Sonar SAS	France	Hashmat (Agosta)	Pakistan	2
DUUA 2A/2B	Thomson Marconi Sonar SAS	France	Galerna (Agosta)	Spain	4
DUUA 2D	Thomson Marconi Sonar SAS	France	Agosta	France	2
DUUG 1B	Thomson Marconi Sonar SAS	France	Guppy II	Taiwan	2
DUUG 1D	Thomson Marconi Sonar SAS	France	Type 209/1,200 (Salta)	Argentina	1
DUUX 2	Thomson Marconi Sonar SAS	France	Type 209/1,300	Ecuador	2
DUUX 2	Thomson Marconi Sonar SAS	France	Agosta	France	2
DUUX 2	Thomson Marconi Sonar SAS	France	Daphne	France	1
DUUX 2	Thomson Marconi Sonar SAS	France	Type 206	Indonesia	2
DUUX 2	Thomson Marconi Sonar SAS	France	Type 206A	Germany	12
DUUX 2[2]	Thomson Marconi Sonar SAS	France	Daphne	South Africa	3
DUUX 2	Thomson Marconi Sonar SAS	France	Type 209/1,300 (Cabalo)	Venezuela	2
DUUX 2A	Thomson Marconi Sonar SAS	France	Hashmat (Agosta)	Pakistan	2
DUUX 2A/5	Thomson Marconi Sonar SAS	France	Galerna (Agosta)	Spain	2
DUUX 2C	Thomson Marconi Sonar SAS	France	Type 209/1,200 (Salta)	Argentina	1
DUUX 2C[5]	Thomson Marconi Sonar SAS	France	Type 209/1,200 (Casma)	Peru	n/k
DUUX 5	Thomson Marconi Sonar SAS	France	Santa Cruz TR 1700	Argentina	2
DUUX 5[4]	Thomson Marconi Sonar SAS	France	Romeo Type 033	China	n/k
DUUX-5	Thomson Marconi Sonar SAS	France	Type 209/1,500 (Shishumar)	India	4
DUUX 5	Thomson Marconi Sonar SAS	France	L'Inflexible	France	3
DUUX 5	Thomson Marconi Sonar SAS	France	Walrus	Netherlands	4
DUUX 5	Thomson Marconi Sonar SAS	France	Han	China	5
DUUX 5	Thomson Marconi Sonar SAS	France	Ming	China	15
Eledone	Thomson Marconi Sonar SAS	France	Delfin (Daphne)	Spain	4
Eledone 1102/5	Thomson Marconi Sonar SAS	France	Abtao	Peru	2
Eledone	Thomson Marconi Sonar SAS	France	Galerna (Agosta)	Spain	4
Eledone	Thomson Marconi Sonar SAS	France	Heroj	Yugoslavia	2

System	Manufacturer	Country	Installed on	Navy	Units
Feniks	RFAS	RFAS	Romeo Type 031	Korea, North	22
Feniks	RFAS	RFAS	Foxtrot Type 641	Libya	4
Herkules/Feniks	RFAS	RFAS	Foxtrot Type 641	Cuba	2
Herkules/Feniks	RFAS	RFAS	Foxtrot Type 641	India	4
Herkules[4]	RFAS	RFAS	Romeo Type 033	China	39
Herkules	RFAS	RFAS	Modified Romeo	China	1
Herkules	RFAS	RFAS	Foxtrot Type 641	Libya	4
Hydra	Plessey	UK	Sjoormen A 12	Singapore	3
IPD 70/S	Selenia Elsag	Italy	Sauro Type 1081	Italy	4
IPD 70/S	Selenia Elsag	Italy	Improved Sauro	Italy	4
Kariwara	Thomson Marconi Sonar Pty	Australia	Collins	Australia	1
MD 100	Selenia Elsag	Italy	Sauro Type 1081	Italy	4
MD 100S	Selenia Elsag	Italy	Improved Sauro	Italy	4
Mouse Roar	RFAS	RFAS	Akula I/II	RFAS	12
Mouse Roar	RFAS	RFAS	Delta I	RFAS	4
Mouse Roar	RFAS	RFAS	Delta III	RFAS	9
Mouse Roar	RFAS	RFAS	Kilo Type 877E	Algeria	2
Mouse Roar	RFAS	RFAS	Kilo Type 877E	Romania	1
Mouse Roar	RFAS	RFAS	Kilo Type 877EKM/636	China	4
Mouse Roar	RFAS	RFAS	Kilo Type 877EM	India	9
Mouse Roar	RFAS	RFAS	Kilo Type 877EKM	Iran	3
Mouse Roar	RFAS	RFAS	Kilo Type 877/877K/877M	RFAS	15
Mouse Roar	RFAS	RFAS	Oscar II	RFAS	12
Mouse Roar	RFAS	RFAS	Sierra I/II	RFAS	3
Mouse Roar	RFAS	RFAS	Tango	RFAS	6
Mouse Roar	RFAS	RFAS	Typhoon	RFAS	6
Mouse Roar	RFAS	RFAS	Victor III	RFAS	11
Mouse Roar	RFAS	RFAS	Yankee Notch	RFAS	1
PRS 3[4]	STN ATLAS Elektronik	Germany	Type 209/1,200 (Casma)	Peru	n/k
PRS 3-4	STN ATLAS Elektronik	Germany	Type 209/1,100-1,200 (Glavkos)	Greece	8
PRS 3-4	STN ATLAS Elektronik	Germany	Type 209/1,300 (Cakra)	Indonesia	2
PSU 83	STN ATLAS Elektronik	Germany	Type 207 (Tumleren)	Denmark	3
PSU 83	STN ATLAS Elektronik	Germany	Narhvalen	Denmark	2
Pelamida	RFAS	RFAS	Delta III	RFAS	9
Pelamida	RFAS	RFAS	Oscar II	RFAS	12
Pelamida	RFAS	RFAS	Typhoon	RFAS	6
Pike Jaw	RFAS	RFAS	Foxtrot	RFAS	2
Pike Jaw	RFAS	RFAS	Modified Romeo	China	1
Pike Jaw	RFAS	RFAS	Ming	China	15
SRS M1H	STN ATLAS Elektronik	Germany	Type 205	Germany	2
Shark Fin[4]	RFAS	RFAS	Kilo	RFAS	15
Shark Fin	RFAS	RFAS	Kilo	India	9
Shark Fin[4]	RFAS	RFAS	Tango	RFAS	6
Shark Gill	RFAS	RFAS	Akula I/II	RFAS	12
Shark Gill	RFAS	RFAS	Delta IV	RFAS	7
Shark Gill	RFAS	RFAS	Oscar II	RFAS	12
Shark Gill	RFAS	RFAS	Sierra I/II	RFAS	3
Shark Gill	RFAS	RFAS	Typhoon	RFAS	6
Shark Gill	RFAS	RFAS	Victor III	RFAS	11
Shark Hide	RFAS	RFAS	Delta III	RFAS	9
Shark Hide	RFAS	RFAS	Delta IV	RFAS	7
Shark Rib	RFAS	RFAS	Oscar II	RFAS	12
Shark Rib	RFAS	RFAS	Sierra I/II	RFAS	3
Shark Rib	RFAS	RFAS	Typhoon	RFAS	6
Shark Rib	RFAS	RFAS	Victor III	RFAS	11
Shark Teeth	RFAS	RFAS	Delta I	RFAS	4
Shark Teeth	RFAS	RFAS	Delta III	RFAS	9
Shark Teeth[4]	RFAS	RFAS	Kilo	RFAS	15
Shark Teeth	RFAS	RFAS	Kilo Type 877E	Algeria	2
Shark Teeth	RFAS	RFAS	Kilo Type 877EKM/636	China	4
Shark Teeth	RFAS	RFAS	Kilo Type 877EM	India	9
Shark Teeth	RFAS	RFAS	Kilo Type 877EKM	Iran	3
Shark Teeth	RFAS	RFAS	Kilo Type 877E	Poland	1
Shark Teeth	RFAS	RFAS	Kilo Type 877E	Romania	1
Shark Teeth	RFAS	RFAS	Kilo Type 877/877K/877M	RFAS	15
Shark Teeth	RFAS	RFAS	Tango (Som) Type 641B	RFAS	6
Shark Teeth	RFAS	RFAS	Yankee Notch	RFAS	1
Siass-2	Signaal	Netherlands	Hai Lung	Taiwan	2
Skat 3	RFAS	RFAS	Sierra I/II	RFAS	3
Skat 3	RFAS	RFAS	Victor III	RFAS	11
TSM 2225 (DUUX 5)	Thomson Marconi Sonar SAS	France	Collins	Australia	3
TSM 2233 (Eledone)	Thomson Marconi Sonar SAS	France	Collins	Australia	4
TSM 2253	Thomson Marconi Sonar SAS	France	Collins	Australia	4
TSM 2272 (Eledone)	Thomson Marconi Sonar SAS	France	Walrus	Netherlands	4
TSM 2285	Thomson Marconi Sonar SAS	France	Ula	Norway	6
Type 187	THORN EMI	UK	Oberon	Brazil	1
Type 187	THORN EMI	UK	Oberon	Chile	2
Type 2006	GEC-Marconi Avionics	UK	Walrus	Netherlands	4
Type 2007	Graseby Marine	UK	Oberon	Brazil	1
Type 2007	Graseby Marine	UK	Oberon	Canada	3
Type 2007	Graseby Marine	UK	Oberon	Chile	2
Type 2007[4]	Graseby Marine	UK	Trafalgar	UK	7
Type 2007	Graseby Marine	UK	Swiftsure	UK	5
Type 2019	Thorn	UK	Swiftsure	UK	5
Type 2019[4]	Thorn	UK	Trafalgar	UK	7
Type 2020[4]	Thomson Marconi	UK	Trafalgar	UK	7
Type 2026[4]	GEC Avionics/Plessey	UK	Trafalgar	UK	7

System	Manufacturer	Country	Installed on	Navy	Units
Type 2026	GEC Avionics/Plessey	UK	Walrus	Netherlands	4
Type 2040	Thomson Marconi Sonar SAS	France	Upholder	UK	4
Type 2043	Thomson Marconi	UK	Vanguard	UK	3
Type 2046	Thomson Marconi Sonar	UK	Swiftsure	UK	5
Type 2046[4]	Thomson Marconi Sonar	UK	Trafalgar	UK	7
Type 2046[4]	Thomson Marconi Sonar	UK	Vanguard	UK	3
Type 2051[5]	Plessey	UK	Oberon	Canada	3
Type 2054	Thomson Marconi Sonar	UK	Vanguard	UK	3
Type 2072[4]	Thomson Marconi Sonar	UK	Trafalgar	UK	7
Type 2074	Plessey	UK	Swiftsure	UK	5
Type 2074[4]	Plessey	UK	Trafalgar	UK	7
Type 2077	Thomson Marconi Sonar	UK	Swiftsure	UK	5
Type 2077	Thomson Marconi Sonar	UK	Trafalgar	UK	7
Type 2082	PMES	UK	Swiftsure	UK	5
Type 2082[4]	PMES	UK	Trafalgar	UK	7
Type 2082	PMES	UK	Vanguard	UK	3
Tamir	RFAS	RFAS	Whiskey V	Albania	4
Tamir 5[4]	RFAS	RFAS	Romeo Type 033	China	n/k
Whale series	RFAS	RFAS	Kilo Type 877E	Poland	1
ZQQ 5	Hughes/Oki	Japan	Yuushio	Japan	10
ZQQ 5B	Hughes/Oki	Japan	Harushio	Japan	7
ZQQ 5B	Hughes/Oki	Japan	Oyashio	Japan	2
ZQR 1	Hughes/Oki	Japan	Harushio	Japan	7
ZQR 1	Hughes/Oki	Japan	Yuushio	Japan	10
ZQR 1	Hughes/Oki	Japan	Oyashio	Japan	2
Passive TAS	Hermes Electronics/MUSL	Canada/UK	Oberon	Canada	3
NK	UEC	South Africa	Daphne	South Africa	3

[1] Micropuffs
[2] Being replaced by indigenous system
[3] In S 71-73
[4] In some units
[5] Also known as Triton

UPDATED

AIRBORNE DIPPING SONARS

FRANCE

HS 12 helicopter sonar

Type
Passive and active dipping sonar system.

Description
The HS 12 is an active/passive panoramic helicopter version of the SS 12 small ship sonar and uses the same electronics as the surface vessels version. It has similar capabilities for operation in shallow/noisy waters, and has a system weight of less than 240 kg, making it suitable for installation on light naval helicopters such as the Lynx. The HS 12 transducer is raised and lowered by a hydraulic winch at a high speed.

Operation in CW and FM modes is possible and digital signal processing is employed by the system's microprocessor. Automatic tracking of two targets and transmission of elements to an external equipment, such as a plotting table, are provided.

The system operates on 13 kHz in the active mode and in the 7 to 12 kHz range passively. A total of 12 preformed beams is employed giving 30° sectors, with a minimum range of about 10 km. The display consists of four quadrants, these being obtained by processing adjacent beams. The operator can select a CW mode which provides target range and Doppler. He can also select FM processing, and a sector mode is also provided.

Operational status
In series production. French and Chinese-built SA-321G Super Frelon helicopters have been equipped with HS 12 systems, to support the Chinese SSBN force

Typical helicopter installation of the HS 12 sonar on Dauphin naval helicopter

(four systems operational) together with about 10 Chinese-built Dauphin (Harbin) helicopters. It is also in operation in several other navies including the Indian Navy where it equips Sea King helicopters and the French Navy.

Specifications
Frequencies: medium frequency band
Transmission level: ≥=212 dB/µPa/m

Passive search: noise or pinger. Display of 12 preformed beams, each subject to adaptive signal processing. Other characteristics as for the SS 12
Total weight: 230 kg

Contractor
Thomson Marconi Sonar SAS, Sophia Antipolis.

UPDATED

HS 312 ASW system

Type
Passive and active ASW system.

Description
The HS 312 is an acoustic system for helicopters incorporating the facilities of the HS 12 system and the SADANG new-generation acoustic processor. The equipment functions in both passive and active modes (CW and FM) and has a longer range than the HS 12.

Only a single operator is required and the light weight and compactness of the system means that the HS 312 can be fitted to any type of light or medium-size

helicopter. The performances of the basic components have been improved by integration of the processing units, integration of a standardised keyboard and the use of only one display screen.

Operational status
In service with the Chilean Navy where it equips seven Nurtanio Cougar helicopters.

Contractor
Thomson Marconi Sonar SAS, Sophia Antipolis.

VERIFIED

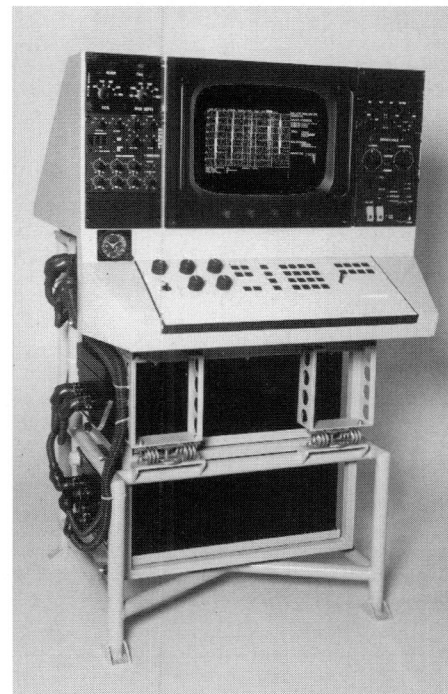

HS 312 helicopter acoustic system control/display

FLASH dipping sonar

Type
Helicopter dipping sonar system.

Description
FLASH (Folding Light Acoustic System for Helicopters) is a new low-frequency dipping sonar with sonobuoy processing capabilities developed by Thomson Sintra ASM for a number of applications. It operates at five

frequencies below 5 kHz, with 24 preformed beams. The winch is capable of 750 m immersion depth at a speed up to 10 m/s.

The system offers low frequency and large bandwidth in FM mode to enable detection of low-speed targets in shallow waters and strong reverberation conditions. With its light weight and volume, the FLASH can be fitted on board light ASW helicopters, such as the Super Lynx.

Operational status
Thomson Marconi Sonar SAS is teamed with prime contractor Thomson Marconi Sonar Ltd to supply FLASH (designated AQS 960 in the UK) for the long-range dipping sonar requirement for the Royal Navy's Merlin helicopter. This system will combine the French low-frequency array and winch with the British signal and data processing expertise. Thomson Marconi SAS has also teamed with Hughes Aircraft Company in the

USA for the US Navy's ALFS requirement in which configuration it is designated AN/AQS-22 (see below). The FLASH array and reeling machine have already been evaluated in trials for the US Navy.

The FLASH array, reeling machine, transmitter and new acoustic processor are being evaluated in flight trials by the French Navy for the NH 90 programme.

Acoustic in-water testing has been completed in France and in the UK, and hydraulic rig tests have been completed in Italy. During 1995, a comprehensive series of environmental and in-water tests were carried out. Thomson Marconi SAS has also been awarded a contract to supply FLASH to the UAE to equip their helicopters.

Contractor
Thomson Marconi Sonar SAS, Sophia Antipolis.

VERIFIED

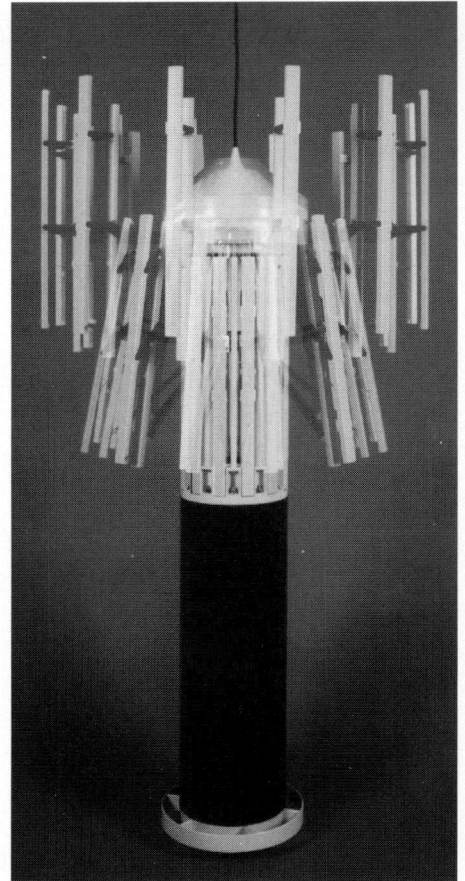

Wet end of the FLASH dipping sonar

DUAV-4 helicopter sonar

Type
Passive and active sonar system.

Description
The DUAV-4 is an active/passive directive sonar designed for submarine surveillance and location (azimuth, distance and radial speed). It is specially designed for use on board light, versatile, ship-based helicopters such as the Lynx.

The DUAV-4 differs from conventional sonar in its signal processing system, which is designed to give improved detection, especially in severe reverberation conditions such as shallow waters. The sonar can be operated in either the active or passive mode: in the active mode true bearing, range, and radial speed are measured; in the passive mode only true bearing is measured.

A combined display unit permits surveillance display (initial detection) or plotting display (precise azimuth determination). Total weight including electronic rack, cable and dome is 250 kg.

Operational status
The DUAV-4 sonar is in active service with the French Navy, the Royal Netherlands Navy (Lynx helicopter) and several other navies. Over 90 have been produced. The French Navy has opted to replace it in some of the Lynx helicopters with the HS 12 (see separate entry).

Contractor
Thomson Marconi Sonar SAS, Sophia Antipolis.

UPDATED

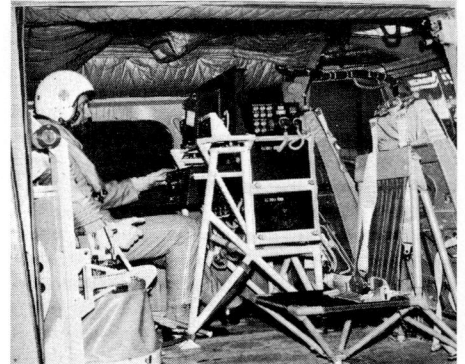

DUAV-4 helicopter sonar installation in Lynx

Upgraded DUAV-4 helicopter sonar

Type
Active and passive dipping sonar system.

Description
The upgraded DUAV-4 or DUAV-4 UPG differs from the DUAV-4 in its acoustic processor, which is now based on the SADANG new-generation acoustic processor. This state-of-the-art system offers more powerful signal and data processing, as well as improved man/machine interface.

The DUAV-4 UPG also offers specific software tools and operator aids designed to meet the requirement for detection in shallow, rocky bottom waters.

Operational status
A DUAV-4 UPG prototype passed factory tests and at sea trials and was supplied to the Swedish Navy in April 1992. Now operational at sea.

Contractor
Thomson Marconi Sonar SAS, Sophia Antipolis.

VERIFIED

INTERNATIONAL

AN/AQS-22 ALFS (Airborne Low-Frequency Sonar)

Type
Next-generation low-frequency dipping sonar.

Description
ALFS is a US Naval Air Systems Command project for a new-generation dipping sonar which is planned to replace the existing AN/AQS-13F, on board the SH 60R (the upgraded version of the SH 60F or B). No technical details have been announced, except that the associated processor will be the UYS-2 manufactured by Lucent Technologies. The UYS-2 has operator selectable full-band, eighth-band and quarter-band verniers, and constant resolution modes are available for up to eight omni or four directional sonobuoys over their full bandwidth. These processing options are available concurrently with the active or passive modes of the dipping sonar. The sonar, which is being developed by Hughes Aircraft and Thomson Marconi Sonar SAS, will use the expandable sonar array and reeling machine subsystem of the Thomson FLASH (Folding Light Acoustic System for Helicopter) system (see above) which will equip the Royal Navy EH101 helicopter.

The design integrates via the 1553B databus with existing display subsystems. The sonar is controlled by the operator using menus on the existing control display unit. A dedicated control panel provides fail-safe control of the reeling machine.

Operational status

In January 1992, the consortium of Hughes Aircraft Co and Thomson Marconi Sonar SAS was awarded a US$31.3 million contract to develop the ALFS system. The five year contract, awarded by Naval Air Systems Command, provides for full-scale development and a follow-on production option of up to 50 systems.

Contractors

Hughes Aircraft Company, Fullerton, California.
Thomson Marconi Sonar SAS, Sophia Antipolis.

UPDATED

AN/AQS-22 ALFS undergoing trials on an SH-60 Seahawk helicopter
***1998**/0006973*

RUSSIAN FEDERATION AND ASSOCIATED STATES (CIS)

Very little information on airborne submarine warfare equipment has been made public. There are at least two types of dipping sonar carried by helicopters. The photograph shows an Mi-14PW 'Haze-A' helicopter which is equipped with both MAD and dipping sonar. The dipping sonar used in a number of former Soviet corvettes, and NATO codenamed 'Rat Tail', equips the 'Hormone-A' helicopter. It apparently operates at a frequency of about 15 kHz and has a maximum range of about 5,000 m. The dipping sonar operated from the Mi-14 is almost certainly a medium-frequency type. The Kamov Ka 27 and 28 (Helix A/D) deploy the dipping sonar from the rear of the fuselage through open clamshell type doors.

The RFAS uses large numbers of sonobuoys, and airborne processing and analysis/display equipment must therefore be fitted to both helicopters and fixed-wing ASW aircraft, but no information is available.

The current dipping sonar deployed by the Ka 27 Helix helicopter is the VGS-3.

UPDATED

Mi-14PW of the Polish Air Force showing the towed MAD 'bird' stowed against the rear of the fuselage. A dunking type sonar is also fitted

UNITED KINGDOM

Type 2069 sonar

Type

Active helicopter dipping sonar.

Description

The Type 2069 sonar is the latest upgrade of the Type 195M. It employs solid-state transmitters and a longer cable. The sonar transducer has been re-engineered to permit operation at greater depth. The sonar is now integrated with the AN/AQS-902G-DS acoustic processing system. It is installed in the Royal Navy Sea King HAS Mk 6.

The sonar provides coverage in progressive 90° arcs. It can be programmed for automatic search, or the operator can manually select a particular search direction. The operator is provided with audio, Doppler and visual sector sonar information.

The sonar may be employed in either surveillance or attack control modes, or both simultaneously. Pulse length and detection ranges are operator selectable to optimise the system according to the prevailing conditions or tactical requirements. Effective range is understood to be in the region of 7.3 km.

The Type 2069 has been upgraded to provide enhanced performance at close ranges and greater immunity to interference.

Operational status

Type 2069 is currently in service with the Royal Navy and several other navies around the world including the Pakistan Navy.

Contractor

Thomson Marconi Sonar Ltd, Templecombe.

UPDATED

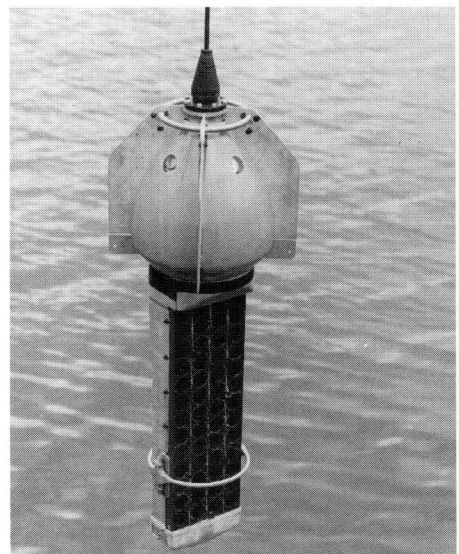

Type 2069 dunking sonar sensor

UNITED STATES OF AMERICA

AN/AQS-13 helicopter sonar

Type
Active and passive helicopter dipping sonar.

Description
The AN/AQS-13B system is a long-range, active scanning sonar which detects and maintains contact with underwater targets through a transducer lowered into the water from a hovering helicopter. Opening or closing rates of moving targets can be accurately determined and the system also provides target classification clues.

The sonar has significant advantages in operation and maintenance over earlier systems. To aid the operator, some electronic functions were automated to eliminate several controls. Maintenance was simplified by eliminating all internal adjustments and adding built-in test (BITE) circuits.

To enhance detection capability in shallow water and reverberation-limited conditions, while essentially eliminating false alarms from the video display, Bendix (now L3 Communications Ocean Systems), developed an Adaptive Processor Sonar (APS) for the system.

The APS uses digital processor Fast Fourier Transform (FFT) techniques to provide narrowband analysis of the uniquely shaped CW pulse transmitted in the APS mode. The display retains the familiar PPI readout of target range and bearing but APS adds precise digital readout of the radial component of target Doppler.

With APS, processing gains of greater than 20 dB with zero false alarm rates have been measured for target Dopplers under 0.5 kt.

The AN/AQS-13E was the first system to integrate sonar (APS) and sonobuoy processing in a common processor, the Sonar Data Computer (SDC). Improvements of this system led to the AN/AQS-13F.

The higher energy transmitted with the longer pulse APS mode, combined with the narrowband analysis, also substantially improves the Figure-Of-Merit (FOM) in the non-reverberant conditions more typical of deep water operation. Measured processing gains for APS under ambient, wideband noise-limited conditions exceeded 7 dB.

The AN/AQS-13F has been designed to provide rapid tactical response against the most advanced submarine threats. It is a sister equipment to the AN/AQS-18 and is identical in many respects. A new state-of-the-art transducer, when lowered to depths of up to 1,450 ft, permits instantaneous range improvements of over 100 per cent compared to previous systems. High-speed reeling allows a dip, cycled to maximum depth, to be completed in approximately 3 minutes. The powerful omnidirectional transducer providing 216 ± 1 dB source level is integrated with a sensitive directional receiver array providing azimuthal resolution in a small rugged unit. The sonar data computer offers digital matched filter processing for 200 and 700 ms sonar pulses, as well as sonobuoy control and processing. The azimuth and range indicator and receiver provides a video display for the operator.

Operational status
The AN/AQS-13 is widely used by American forces and over 1,000 sets have been supplied or ordered for naval helicopters of 15 foreign navies in Europe (Italy), Asia (Japan), the Middle East and South America. The AQS-13F was selected by the US Navy and is now in operation on the SH-60F carrier-based ASW helicopters.

Specifications
Frequencies: 9.25, 10, 10.75 kHz
Sound pressure level: 213 dB/µPa /yd (13B); 216 dB (13F)
Range scales: 1,000, 3,000, 5,000, 8,000, 12,000, 20,000 yds (0.9, 2.7, 4.6, 7.3, 11, 18.3 km)
Operational modes: active 3.5 or 35 ms, MTI, APS, passive, voice communications, key communications (13A only). The 13F has the following operational modes: active – 3.5 m/s and 35 ms rectangular pulse, 200 and 700 ms shaped pulse; MTI; passive – 500 Hz bandwidth 9-11 kHz; communicate SSB at 8 kHz (voice)
Visual outputs: (13A) range, bearing; (13B) range, range rate, bearing, operator verification
Audio output: (13A) single channel with gain control; (13B) dual channel with gain control plus constant level to aircraft intercom
Recorder operation: bathythermograph, range, aspect, MAD self-test
System weight: (13A) 373 kg; (13B) 282 kg; (13F) 280 kg
Operating depth: 1,450 ft at 50 ft hover (13F)

Contractor
L3 Communications Ocean Systems, Sylmar, California.

UPDATED

AN/AQS-18 helicopter sonar

Type
Active and passive helicopter dipping sonar.

Description
The AN/AQS-18 is a helicopter-borne, long-range active scanning sonar developed by Bendix (now L3 Communications Ocean Systems). The system detects and maintains contact with underwater targets through a transducer lowered into the water from a hovering helicopter. Active echo-ranging determines a target's range and bearing, and opening or closing rate, relative to the aircraft. Target identification clues are also provided.

The AN/AQS-18 consists of a small high-density transducer with a high sink and retrieval rate, a built-in multiplex system to permit use of a single conductor cable, a 440 m cable and compatible reeling machine and a lightweight transmitter built into the transducer package. The Adaptive Processor Sonar (APS), which provides enhanced performance in shallow water areas, is an integral part of the system.

The AN/AQS-18 offers a number of improvements over earlier dipping sonars. These include increased transmitter power output giving longer range, high-speed dip cycle time and reductions in weight of all units.

The APS increases detection capability in shallow water and reverberation-limited conditions, while essentially eliminating false alarms from the video display. The APS is a digital processor which uses Fast Fourier Transform (FFT) techniques to provide narrowband analysis of the uniquely shaped CW pulse transmitted in the APS mode. The PPI display retains the normal readout of target range and bearing.

AN/AQS-18 sonar deployed from German Navy Westland Lynx Mk 88 helicopter

The APS processing gain improvement over the normal AN/AQS-18 analogue processing is 20 dB for a 2 kt target and 15 dB for a 5 kt Doppler target. The higher energy transmitted with the longer pulse APS mode, combined with the narrowband analysis, also improves operation in the non-reverberant conditions more typical of deep water. The gain improvement outside the high-reverberation zone (10 kt or greater) under wideband noise-limited conditions exceeds 7 dB. The latest version is the AN/AQS-18(V) which is available with both 300 and 450 m length cables.

Operational status
In production and/or operational use with the Ecuadorian, Egyptian, German, Greek, Japanese, Italian, Portuguese, Taiwanese and Spanish navies, and in use by the US Navy in a similar version, the AN/AQS-13F, on the new SH-60F carrier-based ASW helicopter. Greece and Portugal have also ordered the system for their helicopters. The AN/AQS-18 is in production for and operates in the Lynx, SH-3, SH-60J and S70C(M)-1 helicopters.

Specifications
Operating depth: 440 m
Operating frequencies: 9.23, 10, 10.74 kHz
Sound pressure level: 217±1 dB/µPa/yd (0.9 m)
Range scales: 0.9, 2.7, 4.6, 7.3, 11, 18, 29 km
Operational modes: 3.5 or 35 ms pulse (energy detection) and 200 or 700 ms pulse (narrowband analysis)
Visual outputs: range, range rate, bearing, operator verification
Audio output: dual channel with gain control, plus constant level to aircraft intercom system
Recorder operation: bathythermograph, range, ASPECT, MAD, BITE
System weight: 252 kg plus 13.3 kg for APS

Contractor
L-3 Communications Ocean Systems, Sylmar, California.

UPDATED

AN/AQS-18A helicopter sonar

Type
Active and passive helicopter dipping sonar.

Description
The AN/AQS-I 8A system is a lightweight modern helicopter dipping sonar system (using L-3 Communications Ocean Systems' next generation dry-end processing equipment) designed for active, long-range search, localisation and attack of submarines in both shallow and deep water environments. The sonar detects and maintains contact with underwater targets through a transducer lowered into the water from a hovering helicopter. Active echo ranging determines the bearing, range and range rate of the target relative to the position of the helicopter. Aids in target identification are also provided.

The AQS-18A wet-end consists of proven AQS-13/18 components, which include a small high-density transducer assembly with a high sink and retrieval rate, a built-in multiplex system to permit use of a single conductor cable, a 440 m cable and compatible reeling machine and dome control. The new dry-end consists of the sonar interface unit, cable interface power supply, and sonar control unit. Optional dry-end sub-systems include cable payout indicator, bearing-range indicator and multifunction display.

The AQS-18A has significant performance improvements from the AQS-18 sonar. These enhancements include four additional CW pulse widths 800 ms, 1.6, 3.2 and 4.0 seconds: two FM pulses (Fmhi and Fmlo 625 ms): 16 beam signal processing; display integration; increased coverage range; simplified operator interface. The total ASW system improvement of the AQS-18A system is 14dB Figure Of Merit (FOM)

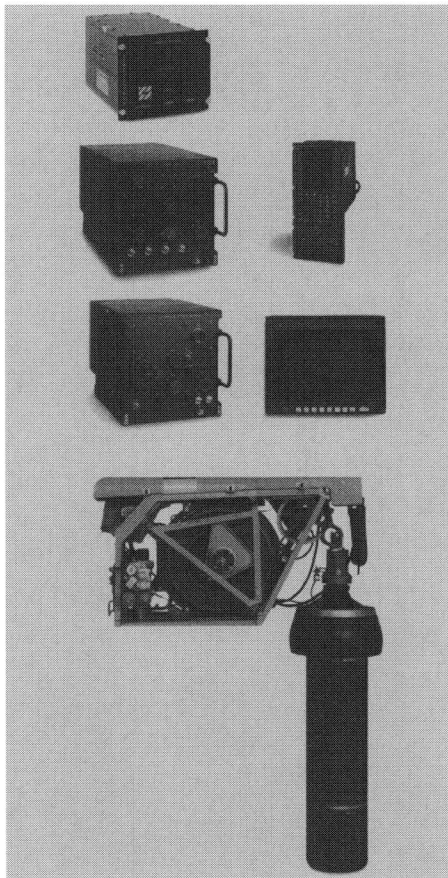

AN/AQS-18A sonar system (Courtesy of L-3 Communications Ocean Systems) **1999**/0024705

AN/AQS-1 8A sonar deployed from Egyptian Navy Kaman SH-2G (E) Super Seasprite helicopter (Courtesy of Kaman Aerospace Corp) **1999**/0024706

over current systems and can provide more than four times the area search rate of the AQS-18.

The system is fully compatible with MIL-STD 1553B databus architectures and has a built-in databus which can facilitate integration with aircraft subsystems and components. Interfaces are also provided to accept input data and provide graphic and data output to a sonics data recorder, as well as video for sensor data display.

The sonar interface unit has powerful signal processing algorithms specifically designed for the increased pulse lengths and spare processing space for additional processing features such as computer aided detection and classification, multisensor target fusion, embedded training and performance prediction based on environmental data collected during past or current missions.

The cable interface power supply provides the electrical coupling of the sonar cable to the uplink/downlink signals from the transducer assembly, develops high-voltage power for charging the transducer assembly battery pack between transmission periods and modulates and passes voice communication signals to the unit.

The sonar control unit provides the operator with

simple interface menus for operator command and control of the sonar system. These menus have been developed with the co-operation of sonar operators from various countries during flight trials and provide a simple interface that greatly reduces the operator's workload.

The new dry-end weapon replaceable assemblies (WRAs) of the AQS-18A have a significantly higher Mean-Time-Between-Failures (MTBF) than current systems. The WRAs also provide, through the systems internal 1553B databus, built-in-test data for each WRA that eases support and boosts the availability of the system.

Operational status

The AQS-18A is in production and in operational use with the Ecuadorian, Egyptian, Italian and Turkish navies, using the 412EP, SH-2G (E), SH-3D/H and 5-70 helicopters respectively.

Specifications

Operating depth: 440 m
Operating frequencies: CW 9.23, 10.003, 10.774 kHz; Fmlo 9.485 kHz, Fmhi 10.520 kHz
Sound pressure level: 2,171 dB/μPa/yd (0.9 m)
Range scales: 1, 1.5, 2.5, 5, 7.5, 10, 12.5, 15.20 nm
Operational modes: Active 3.5, 35 ms rect; 2, 7, 1.6, 3.2, 4 s shaped: 0.625 s FM; passive; UQC
Weight: 88.9 kg (transducer assembly); 122.7 kg (dome control, reeling machine, cable and reel); 51.8 kg (sonar interface unit, cable interface power supply and sonar control unit); 264 kg (total)
Optional components: 0.5 kg (cable payout indicator); 0.6 kg (bearing and range indicator); 14 kg (multifunction display)

Contractor

L-3 Communications Ocean Systems, Sylmar, California.

NEW ENTRY

Helras

Type

Helicopter long-range dipping sonar.

Description

L-3 Communications Ocean systems has developed the Helras (Helicopter Long-Range Active Sonar). The sonar is a deployable low-frequency volumetric array based on the company's experience with previous dipping sonars arrays. The wet end comprises a receiver array of eight hydraulically driven arms which expand to a diameter of 2.6 m when deployed, and a transmitter array of eight projector elements. These projectors form a vertical array 6.2 m long that descends from within the wet end central body during deployment. The projectors radiate at high power levels over 360° in azimuth and the transmitted beam can be vertically steered from −15 to +15°. A range of some

Helras deployed from a Sea King helicopter of the Italian Navy (Courtesy of L-3 Communications Ocean Systems) **1999**/0024703

75 km has been demonstrated in favourable weather conditions.

Helras employs a unique transducer design in the wet end to provide very high acoustic power levels in a lightweight array. The high efficiency vertical line projector radiates in a narrow beam to reduce boundary interaction and couple efficiently into long-range propagation modes. The highly directional receive array discriminates against noise interference and provides maximum sensitivity of the return echo for high detection ranges.

The sonar is capable of operating at depths down to 500 m and has Figure-Of-Merit sufficient to achieve convergence zone detections in deep water, and transmission/receive characteristics optimised for extremely long ranges in shallow water. The low frequency offers multiple boundary interactions and reduced reverberation contamination of the received signals. The use of high resolution Doppler processing and shaped pulses achieves detections of targets even at speeds below 1 kt. Extended duration FM pulses are available to detect the near-zero Doppler target as well.

The sonar interface unit has powerful signal processing algorithms specifically designed to take advantage of the wide area search capability of Helras. The unit performs demultiplexing, beam-forming, and signal processing functions, and provides an output for presentation on the aircraft display.

The advanced reeling machine is derived from the AQS-13/18 series and has considerable commonality of parts with the series. It allows automatic deployment and retrieval of the wet end to depths substantially below the deepest mixing layers for detection of the deepest diving submarines with minimum dip cycle time.

The dome control commands remote deployment and retrieval, and continuously monitors the wet end status. The automatic control ensures flight safety by varying the speed during raising and lowering operations. Cable angle and cable payout signals are provided for the automatic flight control system to assist helicopter stabilisation during hover.

The system is fully compatible with MIL-STD 1553B databus architectures. Interfaces are provided to accept data input and provide graphic and data output to a sonics data recorder, as well as video for sensor data display. The power for the wet end electronics is derived from a battery pack within the

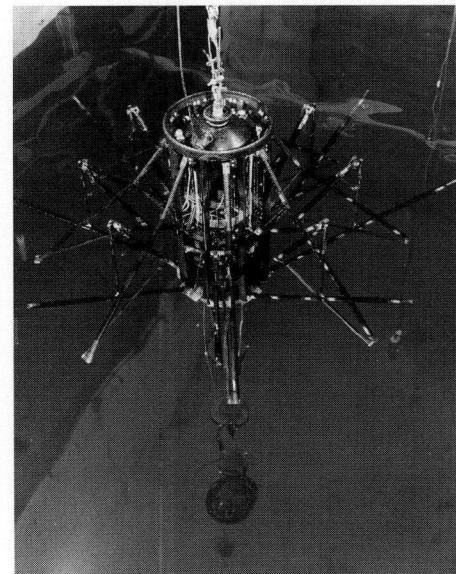

Array arms of Helras deployed (Courtesy of L-3 Communications Ocean systems) **1999**/0024704

body which is recharged between transmission periods.

Operational status

Advanced development has been completed and flight trials have been conducted with the system by the US Navy Australian, Italian and Norwegian navies.

The Helras system is in production for and operates in the Italian EH101 helicopter. The sonar has also been selected by NAHEMA (Netherlands, Germany, France and Italy) for the NH-90 helicopter.

Specifications

Operating depth: 500 m
Projector: 8 elements, array length 5.2 m
Operating frequencies: 1.31-1.45 kHz

Source level: 220 dB/µPa/yd
Range scales: 1, 1.5, 2.5, 4, 6, 10, 16, 25, 40, 60 nm
Receive array: 2.6 m diameter × 1.2 m high
Operational modes: active CW to 10 s pulses (FM to 5 s pulses); passive, UWT

Weight: 152 kg (wet end); 115 kg (dome control, reeling machine, reel and cable); 31.75 kg (sonar interface unit); 45 kg (sonar interface unit, cable interface power supply); 312 kg (total)

Contractor
L-3 Communications Ocean Systems, Sylmar, California.

UPDATED

AN/AQH-4(V)2 system

Type
Mission recording and playback system for fixed-wing aircraft.

Description
The AN/AQH-4(V)2 recording and playback system is designed for use on board ASW maritime patrol aircraft such as the P-3 Orion. In its standard configuration, the system records data transmitted from sonobuoys together with a time code to chronologically reference any given segment of data. The system can also be operated remotely and, with an optional interface unit, can communicate directly with a Central Processing Unit (CPU).

High-density digital data is recorded with bit densities up to 34 KBPI per track, at 7.5 ips. The recorder accommodates a mixture of FM, direct and digital inputs simultaneously or independently on data Printed Circuit Board (PCB) selections. With the availability of various multiplexing schemes and interfaces, the recorder offers considerable flexibility in choice of recording methods. The system records and monitors acoustic signals on 28 active tracks and records all tracks simultaneously. Hard-wire provision is made for monitoring four channels (including time code information) and one switchable output to monitor any of the 28 recorded tracks.

The unique tape transport assembly floats on its own shockmounts within the main frame, allowing hard installation without the need for additional vibration

isolation. The AQH-4(V)2 is designed to permit the interchanging of electronic data cards (PCBs) between systems without adjustments. This overcomes the need to normalise the electromagnetic characteristics inherent in the design of magnetic heads.

Operational status
Approximately 300 units have been sold to the US Navy and to other foreign navies, including Australia, Canada, Holland, Japan and Norway.

Contractor
DRS Precision Echo Inc, a subsidiary of DRS Technologies Inc, Santa Clara, California.

VERIFIED

AN/AQH-9 system

Type
Helicopter mission recording and playback system.

Description
The AN/AQH-9 Mission Recorder System records ASW data for contact validation, post-mission analysis and crew training. The equipment can be readily reconfigured for rotary-wing, fighter/attack and reconnaissance aircraft, and for Remotely Piloted Vehicle (RPV) missions.

The system consists of a mission tape recorder interface unit, a VHS video cassette data recorder, a remote-control unit and a video cassette. The video cassette links the airborne equipment to a ground station playback system.

As configured for the SH-60F helicopter, the AN/AQH-9 will record any mix of four channels of raw sonobuoy data, voice communications and time code data, processed acoustic display data and avionics information via a dual-redundant MIL-STD-1553B database.

To keep weight and volume to a minimum (37 lb and 0.65 cu ft) for critical applications, the equipment was configured to 'record only' using standard VHS tape cassettes. Other configurations are available to record and provide in-flight playback facilities. The unit is shockmounted near the sonar operator's console.

The rotary head data recorder is a helical scan unit which accepts a modified digital datastream for recording B/W video signals. The recorder incorporates two side tracks (time code and auxiliary voice) for expanded capabilities. User data rate using a modified RS-170 format is 3.3 Mbits/s for a 2 hour recording.

The ground element of the system is provided by the RD-591/AQH-9 Mission Data Playback System which is capable of reproducing all the data from the mission recorder system. It recalls acoustic, tactical, communications and avionics data from a standard 1553B databus. It contains an IRIG 'B' time code generator which allows the playback system to provide real-time, post-mission analysis reconstruction of mission data, as well as high-speed search for specific information.

The acoustic data is recorded from four sonobuoy receiver channels, maintaining precise phase

AN/AQH-9 mission tape recorder system (left to right) interface unit, remote-control unit, data recorder

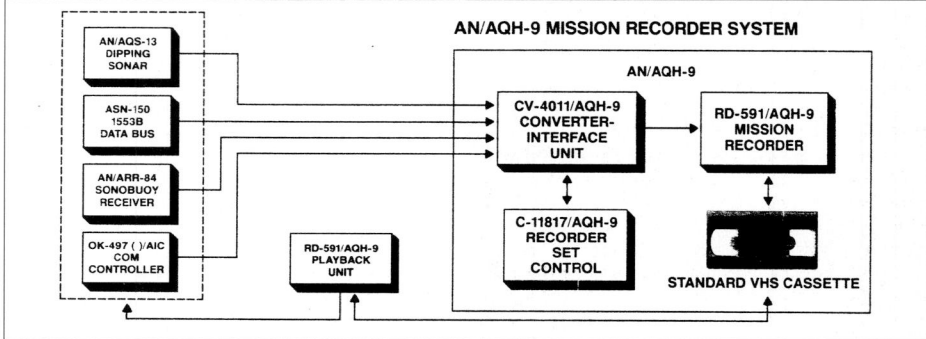

AN/AQH-9   *1996*

information, and presented in analogue form for analysis by the AN/SQR-17A. Intercom and UHF radio are reconstructed and made available both as analogue electrical outputs and in audible form through a front panel loudspeaker.

The man/machine interface consists of a high-resolution, full colour display, presented to the operator as three interactive windows. The operator interface is accomplished primarily via a trackball located next to the keyboard. The real-time graphics display window provides symbolic reconstruction of the mission's tactical scenario. Two supplementary windows are provided for additional data relevant to the current status of the aircraft and the significant events logged during the mission. Interactive data filtering is provided

to control the type and extent of the data displayed in these windows.

Operational status
The recorder system has been manufactured for operational use on board the US Navy SH-60F CV carrier-based helicopters and other rotary-wing aircraft, with over 100 sold. Playback systems have been provided to the US Naval Air Development Center.

Contractor
DRS Precision Echo Inc, a subsidiary of DRS Technologies Inc, Santa Clara, California.

VERIFIED

AN/AQH-11 system

Type
Helicopter high-density digital mission recorder system.

Description
The AN/AQH-11 compact, lightweight, high-density, rotary-head digital mission recorder system is designed for ASW helicopter operations. The AQH-11 is a variant of the AQH-9 system (see previous entry) currently deployed on the SH-60F CV inner zone ASW helicopter. The low-power, ruggedised recording system is

equipped with a dual-redundant MIL-STD-1553 databus interface. Using standard commercial S-VHS tape cassettes, the system is fully integrated with a portable digital mission data playback support system for post-mission analysis. When configured in a single-channel mode, the system captures 2 hours of data at a user rate of 8.3 Mbps.

The AN/AQH-11 can be configured to acquire a variety of datastreams, from wideband acoustics to raster video. The AQH-11 features four sonobuoy input channels from the AN/ARR-84 sonobuoy receiver with a frequency response of 5 to 17.5 kHz. Other inputs include four ICS/UHF voice audio channels and two

sonar audio signal channels which are provided to the sensor operator's headset.

Operational status
The system has been manufactured for the US Navy under a contract worth approximately US$8 million.

Contractor
DRS Precision Echo Inc, a subsidiary of DRS Technologies Inc, Santa Clara, California.

VERIFIED

SONOBUOYS

AUSTRALIA

AN/SSQ-801B Barra sonobuoy

Type
Passive directional sonobuoy.

Development
The new Barra sonobuoy design is the end result of two years' development by the former GEC-Marconi Systems Pty company working closely with the RAAF and RAN. The complexity and cost of the old AN/SSQ-801A design prohibited extensive improvements to performance. The company therefore decided to take a radical departure from the original Barra design concept by moving to a modular design. This has resulted in a much simpler method of construction leading to high reliability, reduced cost and a much more easily upgradable buoy, compared with the previous AN/SSQ-801A which has been in service with the RAAF and RAN since 1981.

Description
The AN/SSQ-801B has been designed for rapid and accurate localisation of submarine targets.

The buoy's performance is achieved by beam-forming the output of 25 hydrophones in a horizontal planar array which provides a highly selective steered gain in the direction of the targets, and suppression of noise interference. All hydrophone and compass data is transmitted by an FSK VHF transmitter providing a minimum of 1 W output power.

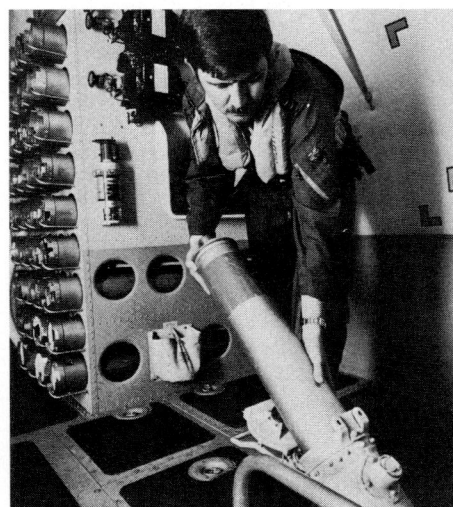

AN/SSQ-801 Barra sonobuoy about to be launched from an RAAF Orion ASW aircraft

RAAF P-3C Orion maritime patrol aircraft equipped with AN/SSQ-801 Barra ASW system

Performance improvements include: lower acoustic self-noise from the revised flotation and suspension systems in all sea states up to and including Sea State 5; rapid hydrophone array deployment; reduced payload weight; and Electronic Function Select (EFS) of RF channels, life and depth settings. Once deployed, changing the RF channels, life and data transmission (on/off) can be commanded from the aircraft using the Command Function Select (CFS) feature. The buoy is able to transmit on one of 99 channels, operates at 22 or 130 m and has life settings of 0.5, 1, 2 and 4 hours. These selections are made with a side-mounted push-button. A second button is used to verify stored selections. Selections are maintained in the buoy's memory until a new selection is made or for the five year life of the unit. This system is similar to that used in other Nato-approved sonobuoys.

Before launch, the operating life, array depth and RF channel number are programmed using the EFS. The AN/SSQ-801B may be launched from either helicopters or fixed-wing aircraft. Shortly after launch, a parachute is automatically deployed to slow down the descent and stabilise the trajectory. Upon water entry the depth selection mechanism is activated, the parachute is jettisoned and the unit automatically deploys, the RF transmitter unit surfaces while the array descends to the preselected operating depth. The unit is fully operational within 65 seconds of water impact. At the conclusion of the operating life, the RF is turned off and at 5 hours from water entry the unit automatically scuttles.

Operational status
Thomson Marconi Sonar Pty Ltd was awarded a contract for the supply of 13,000 Barra sonobuoys by the Royal Australian Air Force in May 1995, with options for a further 12,500 buoys. The A$50 million contract means that production of the AN/SSQ-801B sonobuoy will continue at least to the year 2001.

The AN/SQQ-801B began to enter service with the RAAF and RAN in December 1996.

Specifications
Weight: 8.2 kg
Transmitter power output: 1 W min
Operating frequency: 136-174 MHz
Frequency stability: ±15 kHz
Modulation: narrowband FSK
Number of channels: 99
Operating depth: 22 or 130 m
Operating life: 0.5, 1.2 and 4 h

Contractor
Thomson Marconi Sonar Pty Ltd, Sydney.

UPDATED

Advanced sonobuoy communication link receiver for AN/SSQ-801 Barra sonobuoy data exchange

LFA Sonobuoy

Type
Low-Frequency Active Sonobuoy (LFAS).

Description
Thomson Marconi Sonar Pty is now developing an active low-frequency sonobuoy to perform multistatic shallow and deep water operations in co-operation with standard passive sonobuoys. The buoy features a high transmission level within a narrow acoustic band. The array incorporates a new type of transducer, the Diabolo, developed by Thomson Marconi Sonar which provides a high Q at low frequency. Hence, the expected detection range is greatly increased when used in bistatic/multistatic modes.

Operational status
Under development.

Contractor
Thomson Marconi Sonar Pty Ltd, Sydney.

VERIFIED

FRANCE

TSM 8030 (DSTV-7) sonobuoy

Type
Passive omnidirectional sonobuoy.

Description
The one-third 'A' size TSM 8030 (DSTV-7) is an omnidirectional passive buoy which provides a number of additional facilities. These include the selection of the VHF frequency channel from the 31 or 34 available before the buoy is dropped, choice of 31 or 34 channels from those available between 136 and 173.5 MHz at 365 or 375 kHz spacing, a three-position selector to give a choice of hydrophone depth immersion and an increased maximum depth. Other improvements are reduction in overall weight and volume, and the use of a small balloon to act as a brake during the trajectory of the buoy in the air, and then as a floater on the water surface.

Three variants are available: TSM 8030B, C and D. All are airdropped from 150 to 10,000 ft within a speed range of 80 to 250 kt. They are NATO interoperable.

Operational status
In production and service.

Specifications
Frequency coverage: 10-20 kHz (optionally 5-20 kHz)
VHF emission: TSM 8030B – selectable among 31 channels spaced 375 kHz apart from 162.25-173.5 MHz; TSM 8030C – selectable among 34 channels spaced 375 kHz apart from 136-158.375 MHz; TSM 3080D – selectable among 34

channels spaced 365 kHz apart from 148.75-161.125 MHz
RF power: 1 W
Operating life: 1, 3 or 8 h (selectable)
Launch profile: airdrop from 150-10,000 ft within the speed range 80-250 kt
Operating depth: 20, 100 or 300 m (selectable)

Power supply: magnesium/lead chloride battery
Size: 'A'/3 ± 123.82 mm (diameter) ×304 mm (high)
Weight: 4.5 kg

Contractor
Thomson Marconi Sonar SAS, Sophia Antipolis.
VERIFIED

TSM 8060 DIFAR

Type
Directional passive sonobuoy.

Description
The 'A' size TSM 8060 DIFAR passive sonobuoy is designed for the detection and localisation of submarines from aircraft. VHF transmission is possible from one of 99 channels which are selectable from on board the aircraft. Channels are spaced at 375 kHz

ranging from 136 to 173.5 MHz. The buoy operates at one of three selectable submersion depths and its operational life is selectable from one of five periods.

Operational status
Interoperable within NATO.

Specifications
VHF emission: selectable from 99 channels spaced 375 kHz apart, from 136-173.5 MHz.
RF power: 1 W

Operating life: 0.5, 1, 2, 4 or 8 h (selectable)
Operating depth: 30, 120 or 300 m
Power supply: lithium cell; magnesium/lead chloride battery
Size: 'A' 123.82 mm (diameter) × 914 mm (high)
Weight: 9 kg

Contractor
Thomson Marconi Sonar SAS, Sophia Antipolis.
VERIFIED

Alkan launchers

Type
Sonobuoy launchers for aircraft.

Description
Alkan has designed a number of sonobuoy launchers for use in both fixed- and rotary-wing aircraft. They meet a range of operational requirements, such as in-flight

reloading and release or ejection from pressurised or unpressurised cabins. Various launcher sizes and capacities are available to allow for easy installation in the cabin or equipment bay.

Container-launcher
Each buoy is packed in a container-launcher which provides for its storage and ejection, and is equipped

with a pressurised vessel at the top and a releasable cap at the bottom. The power for ejection is provided by compressed gas stored in the vessel at the top of the buoy.

Type 8030/8031
These launchers have been designed specifically for the Dassault-Breguet Atlantique Mk 1 and Atlantique Mk 2 aircraft which can house up to four launchers. The capacity of each launcher is 18 size 'A' and 18 size 'F' buoys respectively.

Type 8050
This launcher designed for the launch of 'F' size sonobuoys from naval Lynx helicopters. It has a capacity of 18 'F' size buoys.

Operational status
The Type 8030 and Type 8050 are both in production.

Specifications (Type 8050)
Capacity: 18 'F' sized buoys
Ejection speed: 6 m/s
Installation: 14 in (standard lugs)
Dimensions: 2,500 × 576 × 412 mm
Weight: 80 kg (empty)

Contractor
R Alkan & Cie, Valenton.
VERIFIED

Sonobuoy launch pod for the Lynx

ITALY

BI series sonobuoys

Type
Passive omnidirectional sonobuoys.

Description
The Servomeccanismi organisation produces three types of passive sonobuoy:
BI T-3: this is an 'A' size omnidirectional passive sonobuoy for use at depths between 20 and 100 m,

with a selectable life of 1 to 3 hours, and transmitting data over a 31 channel IW link.
BI T-8: this is similar to the BIT-3 but has different frequency/sensitivity characteristics.
BI R: this is a miniature sonobuoy (500 × 100 mm, 3 kg) for use with helicopters. It is omnidirectional and can be deployed at depths to 20 m. Life is 1 hour.
Receivers for these sonobuoys are the REA-16 and REA-31, having 16 and 31 reception channels respectively. The BI series buoys have been designed

primarily for use in the Mediterranean, under conditions normally found in that sea area.

Operational status
In production for the Italian Navy.

Contractor
Servomeccanismi, Pomezia.
VERIFIED

MSR-810 sonobuoy

Type
Passive omnidirectional sonobuoy.

Description
The MSR-810 passive sonobuoy, for detecting and locating submarines from aircraft, is manufactured by Whitehead Alenia Sistemi Subacquei, under license from Thomson Marconi Sonar SAS for the Italian market. The sonobuoy transmits, by VHF/FM radio, the

underwater acoustic AF signals received from the hydrophone. The MSR-810 complies with NATO and Italian Navy specifications.

Operational status
Production and in service.

Specifications
Type: passive
Frequency: low
Weight: 8 kg

Operating life: 1.3 or 8 h
RF channels: 31
RF output: 500 mW (at antenna)
Antenna: ¼wavelength

Contractor
Whitehead Alenia Sistemi Subacquei, Genoa.
VERIFIED

RUSSIAN FEDERATION AND ASSOCIATED STATES (CIS)

Virtually no information on former Soviet sonobuoys had ever been made public until the US Department of Defense released the photograph shown here and published it in *Soviet Military Power*. Early sonobuoys were the RGB-56/64, the latter being a smaller version of the former. In the late 1960s, a passive omnidirectional buoy, the BM-1, was produced. It apparently has a 29 channel FM datalink at about 171 MHz.

The buoy shown here is the Type 75 which is similar to the US AN/SSQ-41B Lofar sonobuoy, although somewhat larger.

Other types of buoy known to exist are designated RGB-1 and RGB-2.

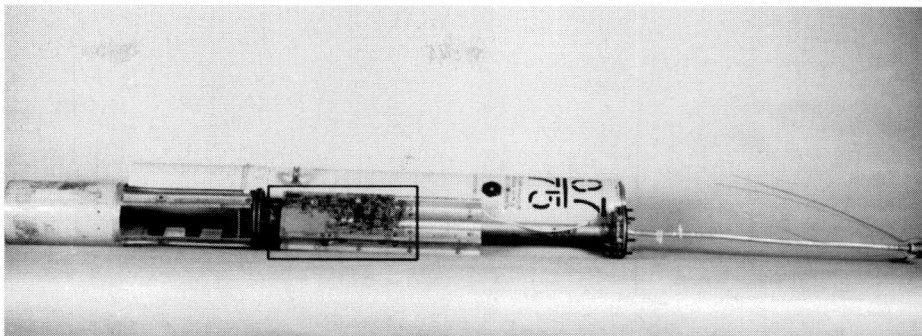

UPDATED   *Type 75 Lofar sonobuoy* (Soviet Military Power)

UNITED KINGDOM

SSQ-906/906A/907A Lofar sonobuoys

Type
Miniature passive omnidirectional sonobuoys.

Description
These 'F' size Jezebel sonobuoys are designed to be launched from aircraft or helicopters at airspeeds between 60 and 300 kt and at altitudes between 50 and 9,200 m. Once deployed, they detect and amplify underwater sounds to modulate the self-contained FM transmitter. The buoy is capable of operating with a range of airborne signal processors, including the AQS-901, AQS-902 and AQS-903.

The buoys feature an exceptionally low-frequency acoustic performance which results in a significant improvement in operational performance over previous Lofar sonobuoys. The 907A is a calibrated version of the 906A.

After launch a parachute is deployed to slow down the rate of descent. On impact with the sea, the parachute is jettisoned, the gas inflation system deploys the combined float and antenna system. As the sonobuoy canister descends to the preset depth it streams the lower unit and suspension system; thereafter the canister is discarded.

Once deployed, the lower unit hydrophone sensor detects and amplifies underwater sounds in a preamplifier to establish a satisfactory signal-to-noise ratio. The output from the preamplifier is further

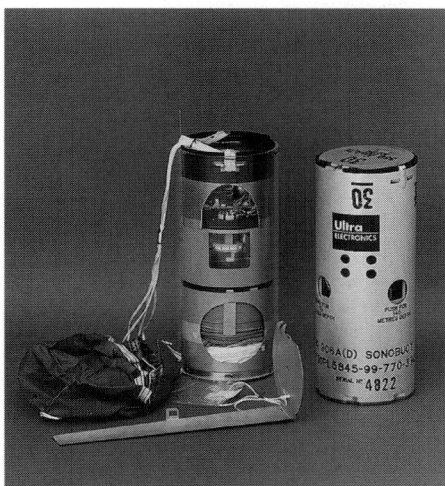

SSQ-906 sonobuoy   *1995*

amplified in the sonic amplifier, which consists of three IC active filter stages. The first two provide virtually all the shaping with the third stage giving HF roll-off, gain and symmetrical limiting. The amplifier is directly coupled through a buffer to the modulator to ensure efficient low-frequency performance.

The VHF FM transmitter consists of a crystal-controlled oscillator with a variable capacity diode in series with the crystal. The output from the sonic amplifier is applied across the variable capacitance diode and results in frequency modulation of the crystal oscillator. The second harmonic of the oscillator drives a doubler stage, the output of which drives a second doubler stage. Finally, a power amplifier is used to produce an output of 1 W. The radiation pattern is omnidirectional using a vertical quarter-wave antenna and ground plane.

Operational status
No longer in production. In service with the Royal Navy, Royal Air Force and with the French Navy.

Specifications
Length: 395 mm
Diameter: 125 mm
Weight: 4.5 kg
Frequency range: 4-3 kHz
Operating life: 1, 4 or 6 h
Operating depth: selectable for either 18 or 91 m; or 18 or 137 m
RF channels: 99
RF output: 1 W (min)

Contractor
Ultra Electronics Sonar and Communication Systems, Greenford.

VERIFIED

SSQ-954D DIFAR sonobuoy

Type
Miniature passive directional sonobuoy.

Description
The SSQ-954D DIFAR (Directional Frequency Analysis Recording) 'G' size sonobuoy is a fully automatic directional passive acoustic sensor which has been developed to meet specific ASW operational requirements of the Royal Navy and Royal Air Force.

The SSQ-954D can be launched from either helicopters or fixed-wing aircraft, using standard sonobuoy launch systems at speeds of between 50 and 300 kt and altitudes of between 15 and 9,200 m. After launch, a parachute is automatically deployed to slow down descent rate to approximately 30 m/s to stabilise the buoy's trajectory. On entering the water the parachute is jettisoned and separation is achieved. The surface unit float is automatically inflated to deploy the RF antenna, while the hydrophone assembly descends to its preselected operating depth.

The sonobuoy employs a high-performance Difar hydrophone system. Acoustic signals received by the omni and directional hydrophones are multiplexed, together with azimuth bearing reference data, onto the VHF carrier as a frequency modulated signal for

transmission to the aircraft. At the end of its selected life, the unit automatically scuttles.

The SSQ-954D includes an Autonomous Function Selection (AFS) system to select, set and verify electronically the fully synthesised VHF transmitter frequency and its operating life and depth.

Operational status
In service with the Royal Navy, Royal Air Force and the French Navy and is qualified for use by both the US and Canadian services.

Specifications
Length: 419 mm
Diameter: 124 mm
Weight: 6.8 kg
Operating life: 1, 2, 3, 4, 5 or 6 h (selectable)
Operating depth: 30/140/300 m (selectable)
Acoustic frequency range: 5-2,400 Hz
RF channels: 99
RF power: 1 W (min)
Frequency: 136-173.5 MHz (synthesised and programmable)

Contractor
Ultra Electronics Sonar and Communication Systems, Greenford.

VERIFIED

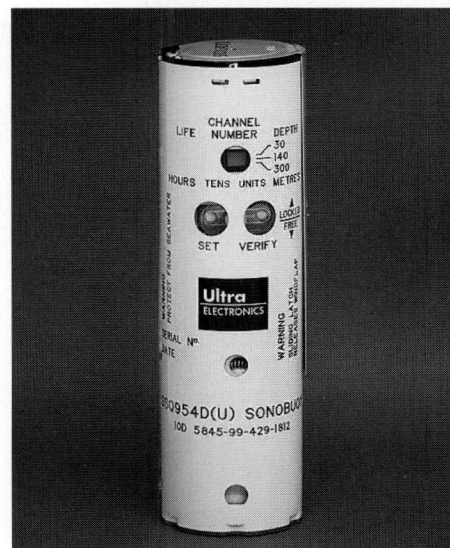

The SSQ-954D DIFAR sonobuoy   *1998*/0006974

SSQ-963C sonobuoy

Type
Command active multibeam sonobuoy.

Description
The SSQ-963C Command Active MultiBeam Sonobuoy (CAMBS) is an 'A' size sonobuoy which was developed in 1997 by Ultra Electronics under contract to the British Ministry of Defence. The latest version of Cambs, the SSQ-963C, entered service with the Royal Air Force at the end of 1997 and offers further significant operational enhancements.

Cambs is an advanced active directional sonobuoy which features a high data rate, and directional information which is obtained by the use of an omnidirectional transmitting acoustic projector combined with a directional receiving hydrophone array. Hydrophone depth is adjustable by radio command and there is a 98 channel radio telemetry link operating on standard NATO frequencies for communication with the parent aircraft. The antenna is a monopole with ground plane, erected by and integral with the buoy flotation bag. The design incorporates a flux-gate compass with very rapid stabilisation.

While in the passive mode, the buoy relays all signals detected by the receive array hydrophones together with magnetic compass information to the controlling platform via the telemetry link. In the active mode acoustic pulses are emitted from the omnidirectional

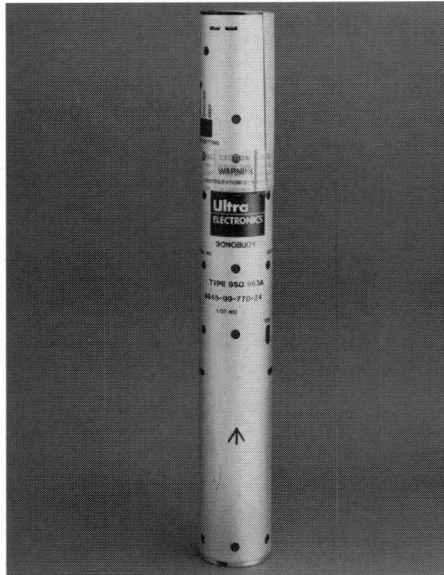

SSQ-963C Cambs sonobuoy **1995**

projector and the returned echoes are detected by the receive array and, along with compass information, are amplified and fed to a multiplexer. There they are combined to provide a multiplexed signal which is transmitted to the airborne platform where it is processed to derive a 'fix' of detected targets.

The power output of the transmitter is 0.5 W and this is fed to the antenna whose radiation pattern is omnidirectional in the horizontal plane with a single lobe typically between 5 and 25° in the vertical plane.

The SSQ-963C includes an Autonomous Function Selection (AFS) system to preset, before launch from the aircraft, the RF channel and initial depth settings.

Operational status
The SSQ-963C sonobuoy is part of the equipment developed for the Mk II version of the Nimrod ASW aircraft. In production.

Specifications
Length: 914 mm
Diameter: 124 mm
Weight: 11.3 kg
Operating life: 1 h, automatic or command scuttle
Operating depth: 4 selectable depths
RF channels: 98
RF power: 0.5 W (min)
Frequency: 136-173.5 MHz

Contractor
Ultra Electronics Sonar and Communication, Greenford.

VERIFIED

SSQ-981B Barra sonobuoy

Type
Passive broadband directional sonobuoy.

Description
The SSQ-981B represents a further evolution of the SSQ-981A Barra sonobuoy currently operational with the UK services. It is an advanced passive broadband sonobuoy sensor; 25 hydrophones mounted on five arms and arranged over a large acoustic planar array to achieve enhanced detection over conventional buoys. Spectral and broadband energy is digitally telemetered to the ASW aircraft. A Barra-compatible acoustic processor forms 60 beams providing 360° cover to give excellent signal-to-noise discrimination, resulting in good azimuth resolution and extremely accurate contact bearings.

The SSQ-981B is configured as a conventional 'A' size sonobuoy which may be launched from helicopter or fixed-wing aircraft using existing stowage and launching systems. It operates with an Autonomous Function Selection (AFS) microprocessor system to preset any one of the 50 RF channels and its operating life and depth settings.

Operational status
First production deliveries to the UK MoD commenced in 1998.

Specifications
Length: 914 mm
Diameter: 124 mm
Weight: 9 kg
Operating life: 1, 2, 3 or 4 h
Operating depth: 21.5 or 121.5 m (selectable before launch)
RF channels: 50
RF power: 1 W (min)
Frequency: 136-173.5 MHz

Contractor
Ultra Electronics Sonar and Communications Systems, Greenford.

UPDATED

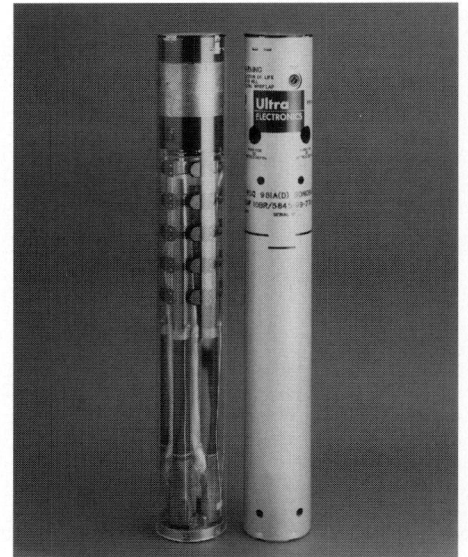

SSQ-981 Barra sonobuoy **1995**

ALST sonar transponder

Type
Helicopter-launched expendable sonar transponder.

Description
The air-launched sonar transponder is an expendable device for the alignment of ASW helicopter radar/dipping sonar equipment.

The transponder is housed in an 'A' size canister, enabling it to be launched from existing stowage and launch systems. The ALST can be adapted to operate with sonars in the medium ASW acoustic frequency band. The transponder is capable of several pulse transmission modes, CW, shaped CW, Doppler simulation frequency shifted pulses, and FM pulse form.

The radar target is a folding reflector of twin tetrahedral configuration, ensuring detection at up to 4 n miles range.

Contractor
Thomson Marconi Sonar Ltd, Templecombe.

VERIFIED

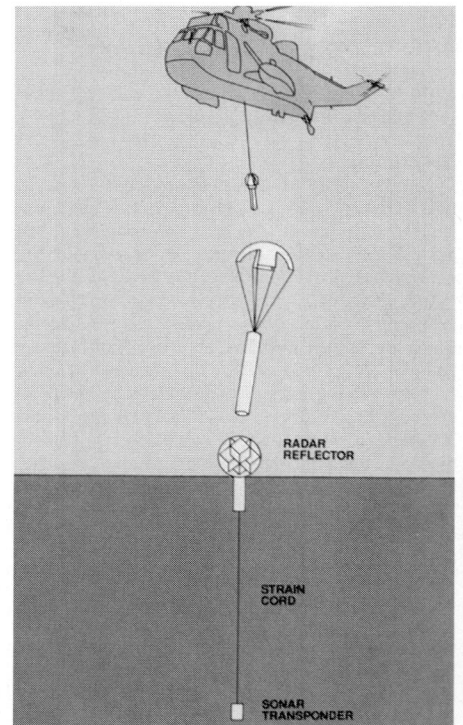

Schematic diagram of operation of the ALST transponder **1995**

UNITED STATES OF AMERICA

AN/SSQ-41B sonobuoy

Type
Passive omnidirectional sonobuoy.

Description
The AN/SSQ-41B is an 'A' size omnidirectional, passive sonobuoy with a multiple life and depth capability. In volume production since 1963, the series has been the mainstay of US Navy sonobuoy operations but is gradually being replaced by the newer directional AN/SSQ-53B.

The AN/SSQ-41B, extensively redesigned from the earlier AN/SSQ-41, provides an audio bandwidth expanded to 20 kHz and improved dynamic range capabilities. The depth selection is 60 or 1,000 ft and operating life selection is 1, 3 or 8 hours.

The AN/SSQ-41B may be launched from aircraft at airspeeds up to 370 kt and from altitudes up to 9,000 m.

After a controlled and stabilised descent and upon impact with the water, the termination mass and the hydrophone are released and descend to a preselected operating depth between 18 to 310 m. On contact with seawater, the battery is activated and the sonobuoy becomes fully operational within 60 seconds at shallow depth and 100 seconds at deep depth.

Modulation is accomplished in the crystal oscillator stage by a variable capacity diode. Two frequency doubling amplifiers multiply the RF oscillator-doubler frequency to the desired operating frequency of the sonobuoy, which is preset at time of manufacture to one of 31 VHF channel frequencies within the 162.25 to 173.50 MHz band or at one of 99 channels between 136.00 to 173.5 MHz. Available in both standard and low noise versions.

Operational status
Although large-scale production was contracted until 1982, the US Navy has discontinued further development. This unit will continue to be available. Over 500,000 units have been manufactured over the last 20 years.

Specifications
Audio frequencies: 10-20,000 Hz
Weight: 6 kg
Operating life: 1, 3 or 8 h
Operating depth: 18-310 m
Launch altitude: 0-9,000 m
Launch speed: 30-370 kt
RF channels: 31 o⁻ 99
RF power: 1 W
Frequency band: 162.250-173.5 MHz (31 × VHF channels) or 136-173.5 MHz (91 channels)

Contractor
Sparton Corporation, Electronics Division, DeLeon Springs, Florida.
VERIFIED

AN/SSQ-41N sonobuoy

Type
Passive omnidirectional sonobuoy.

Description
The AN/SSQ-41N sonobuoy is an 'A' size omnidirectional, passive buoy with a multiple life and dual-depth capability specifically designed for low-frequency, low noise operation. The buoy differs from the 41B model in various parameters, for example operating depth (just two selectable depths rather than multiple depth selectivity).

Upon launching, a parachute deploys to slow down descent and provide stability. After entering the water a seawater battery activates the electronics and initiates deployment of the hydrophone system to its preselected operating depth. Buoyancy for the transmitter electronic assembly is provided by a CO_2 gas filled urethane float. The sonobuoy becomes fully operational within 60 seconds of entering the water at the shallow operating depth, and after 100 seconds at the deeper operating depth.

Modulation is accomplished in the crystal oscillator stage by a variable capacity diode. Two frequency doubling amplifiers multiply the RF oscillator-doubler frequency to the desired operating frequency of the sonobuoy, which is preset at time of manufacture to one of 31 VHF channels within the 162.25 to 173.50 MHz band.

The buoy can be supplied at different operating depths and a variety of attenuated source levels for special applications.

Specifications
Audio frequencies: 5-2,400 Hz
Weight: 9 kg
Operating life: 1, 3 or 8 h
Operating depth: typically 20 or 120 m
Launch altitude: 12.2-15.2 m (in hover attitude), 30.5-9,000 m @ 45-370 kt
Launch speed: hover or 45-370 kt
RF channels: 31
RF power: 1 W
VERIFIED

AN/SSQ-47B sonobuoy

Type
Active omnidirectional sonobuoy.

Description
The AN/SSQ-47B is an active, non-directional, 'A' size sonobuoy. The operational function is that of detection, classification, tracking and location of submarines.

The AN/SSQ-47B provides an active sonar capability for fixed-wing aircraft. It can be operated from a minimum range of zero to 10 n miles at an altitude of between 500 and 10,000 ft and in sea conditions up to Sea State 5. A series of separate VHF channels permits interference free (RF or sonic) operation of up to six sonobuoys, either individually or simultaneously in a single sonobuoy field.

The buoy consists of three main sections: air descent retarding unit, surface unit, and subsurface unit. The retarding unit consists of a spring-loaded parachute release assembly and a parachute. The surface unit sends an automatic keying command to the subsurface unit and target information to the aircraft via a VHF transmitter. The subsurface unit transmits sonar pulses when keyed from the surface unit and receives sonar target echoes for transmission to the aircraft. The AN/SSQ-47B permits launching and operation of up to six sonobuoys, either individually or simultaneously, without encountering RF or sonar interference.

Operational status
In operational service. The AN/SSQ-47B, while no longer carried in the US Navy inventory, is in widespread use with other navies and is still in production for this purpose.

Specifications
Sonar modes: automatic keyed CW
Operating depth: 18 or 240 m
Sonar channels: 6 (HF)
RF channels: 12
RF power: 0.25 W
Weight: 10 kg

Contractor
Sparton Corporation, Electronics Division, DeLeon Springs, Florida.
VERIFIED

AN/SSQ-53D(2) sonobuoy

Type
Passive directional sonobuoy.

Description
The AN/SSQ-53D(2) DIFAR (DIrectional Frequency Analysis and Recording) low noise sonobuoy is the US Navy's primary sonobuoy sensor. Compared with earlier omnidirectional buoys, the SSQ-53D(2) provides target bearing as well as improved acoustical sensitivity, particularly in the low-frequency ranges.

The SSQ-53D(2) is the result of a US Navy-sponsored development programme to improve the operational capabilities of the SSQ-53. Changes include extension of the lower limit by one octave to 5 Hz. In addition, omni-acoustic performance has been improved by 10 dB across the frequency range, and the directional hydrophones have been improved by 16 dB at the 10 Hz point.

The sonobuoy may be dropped from an aircraft at indicated airspeeds of 30 to 370 kt, and from altitudes of 30 to 9,144 m. Descent of the buoy is stabilised and slowed by a parachute assembly.

Immediately after water entry, the saltwater activates the lithium battery. Buoy separation occurs, jettisoning the parachute assembly and erecting the VHF transmitting antenna. This allows the surface assembly to rise and separate from the sonobuoy housing. The housing serves as a descent vehicle and separates from the subsurface assembly at the operating depth. Before launch the buoy is programmed, via push-buttons, to one of 99 channels, life and depth. Readout of these is on an LED display located on the side of the buoy. After 8 hours in the water the buoy automatically scuttles itself.

Development of the SSQ-53E DIFAR is currently underway at Hermes for the US Navy. Various features are being incorporated, aimed at increasing operational performance. These include: command function select; automatic gain control; constant shallow omni-directional hydrophone at 14 m; and an additional fourth operating depth.

Operational status
In production for the Canadian Forces since 1992 with more than 100,000 delivered. The buoy is also in service with the US Navy.

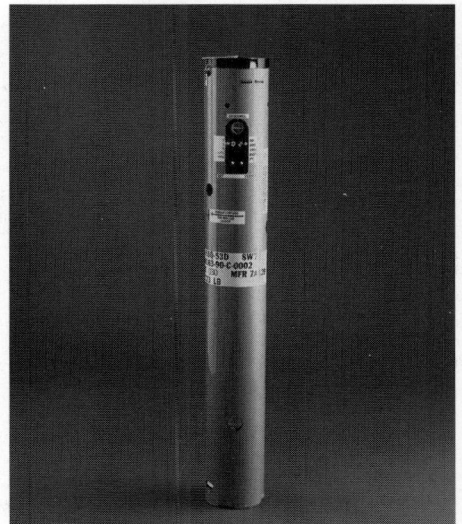
AN/SSQ-53D directional sonobuoy

Specifications
Transmitter frequency: 136-173.5 MHz
Acoustic frequency: 5-2,400 Hz
RF power output: 1 W (min)
RF channels: 99 VHF
Sensitivity-directional: 122 ± 3dB rel to 1μPa at 100 Hz

Sensitivity-ominidirectional: 122 ± 3dB rel to 1μPa at 100 Hz
Operating life: 0.5, 1, 2, 4, 8 h (preselectable)
Operating depth: 30, 120 or 300 m (100, 400 or 1,000 ft) (preselectable)
Dimensions: 91.44 × 12.4 cm (diameter)
Weight: 10 kg

Contractors
Sparton Corporation, DeLeon Springs, Florida.
Hermes Electronics, Dartmouth, Nova Scotia.
Raytheon Electronic Company, Fort Wayne, Indiana.

UPDATED

AN/SSQ-57B sonobuoy

Type
Passive intelligence gathering sonobuoy.

Description
The AN/SSQ-57B sound reference sonobuoy features a hydrophone that affords exceptional performance stability under temperature and pressure extremes. The hydrophone is a piezoelectric ceramic segmented cylinder with a smooth frequency response to 40 kHz. The sonobuoy is calibrated to allow determination of underwater acoustic sound pressure levels over this wide frequency range.

The AN/SSQ-57B can be air-launched from altitudes between 100 and 30,000 ft (30 and 9,000 m) into sea conditions of up to Sea State 5 and is effective at ranges up to 10 n miles.

On launching, a parachute deploys to provide descent retardation and stability. Immediately after entering the water the seawater-activated battery system is energised to activate the electronics and initiate deployment of the hydrophone system to its preselected operating depth. CO_2 gas filled cylinders inflate the sonobuoy float, jettisoning the parachute assembly and erecting the VHF transmitting antenna. This permits the surface assembly to rise and separate from the sonobuoy housing. The sonobuoy housing serves as a descent vehicle and separates from the subsurface assembly at the operating depth.

Operating life is 1, 3 or 8 hours, selected before launch. At the end of the operating life, the transmitter is shut off and the buoy scuttles itself using an electronic timing circuit to activate a burn device in the tip of the float.

The AN/SSQ-57A (XN-5) is a version of the AN/SSQ-57A that has a fixed operating depth of 1,000 ft (305 m) and is available in both standard and low noise versions.

Operational status
The AN/SSQ-57A is in limited production. The AN/SSQ-57B has been produced by both Sparton and Hughes. It is in widespread operational use in the US Navy and a number of other navies.

Specifications
Type: passive, calibrated audio
Audio frequencies: 5-40,000 Hz
Size: 'A'
Weight: 9 kg
Operating life: 1, 3 or 8 h
Operating depth: 20 or 120 m
Launch altitude: 30-12,200 m
Launch speed: 45-370 kt
RF channels: 31
RF power: 1 W

Contractors
Sparton Corporation, Electronics Division, DeLeon Springs, Florida.
Raytheon Electronic Company, Fort Wayne, Indiana.

VERIFIED

AN/SSQ-57M sonobuoy

Type
Shallow water, moored, passive sensor buoy.

Description
The AN/SSQ-57M is a two-thirds 'A' size, air-deployable, low-cost long-life, moored, passive omnidirectional (Jezebel) sensor buoy. After launch from aircraft, or ship, the buoy relays, by FM VHF radio transmission, in the audio frequency range, acoustic signals initially received by the ominidirectional hydrophone system.

The AN/SSQ-57M features the same Sparton-designed hydrophones as used in the AN/SSQ-57 buoy and has similar acoustic performance characteristics. It has a mooring capability similar to the AN/SSQ-58 moored buoy.

The buoy can be launched at altitudes between 12 to 15 m from a stationary ship or between 30.5 to 9,144 m altitude at speeds between 45 and 370 kt. Water entry is facilitated by a parachute. The hydrophone and mooring system are released on water entry. Buoyancy is obtained from a CO_2 gas inflation system. The buoy can operate in surface currents up to 2 kt and in up to Sea State 5.

A saltwater battery provides 12 hours of operation. The buoy is capable of being anchored in water depths from 25 to 120 m. Hydrophone depths are 15, 30 or 60 m. Operating life is 1, 4 or 12 hours. Scuttle is automatic at the end of the selected life period, or may be overridden, depending on circumstances. An Electronic Function Selector (EFS) provides prelaunch selection for the RF channel, operating life, hydrophone depth, and scuttle on/off.

The frequency range, hydrophone operating depth, attenuated sensitivity level and operating life of the buoy can be modified by Sparton, according to customers requirements. System calibration data to allow measurement of underwater acoustic sound pressure levels for special applications can also be supplied.

Operational status
Available for the US and other navies. Currently being used by the Swedish Navy for shallow water, coastal deployment.

Specifications
Audio frequencies: 4-20,000 Hz
Sensitivity: 116 ±3 dB/μPa= ±25 kHz deviation
RF power output: 1 W
RF channels: 99 selectable
Operating life: 1, 4 or 12 h
Operating depth
　Hydrophone: 15, 30 or 60 m
　Anchoring: 25-120 m
Weight: 9.3 kg
Launch altitude: 12-15 m (in hover altitude), 30.5-9,000 m @ 45-370 kt

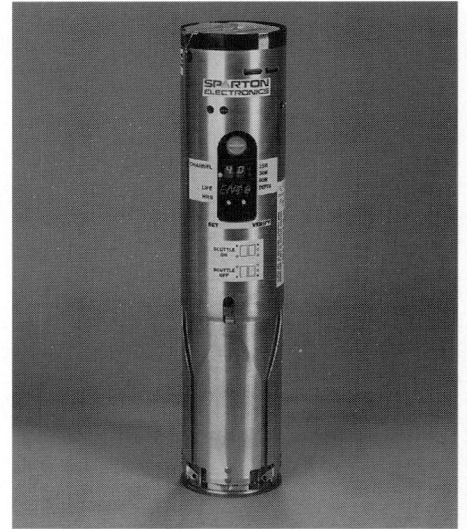

AN/SSQ-57M sonobuoy

Contractor
Sparton Corporation, Electronics Division, DeLeon Springs, Florida.

VERIFIED

AN/SSQ-57SPC

Type
Passive, omnidirectional sonobuoy.

Description
The 'A' size AN/SSQ-57SPC passive directional sonobuoy is the latest version of the SSQ-57 that has been in service for over 25 years. The buoy has been updated to both a 31 and a 99-channel version. Both versions allow channel selection via an electronic function selector. The buoy is factory calibrated to allow operation over its entire frequency range. The unit's VHF transmitter is rated at 1 W.

The buoy can be air launched across the same envelope as the previous SSQ-57 buoys and can also be deployed from the deck of a research vessel.

The unit is powered by a seawater battery which is activated when the buoy enters the water. Operating life is selectable at either 1, 3 or 8 hours with an automatic scuttle at the end of 8 hours. Other optional life setting selections are available within the 8-hour period.

Using the electronic function selector the buoy can be set to operate at one of two operating depths (either 60 or 400 ft) selected before launch. Other optional depth suites are available, with up to three depth choices in a suite.

The buoy is also available with an optional command function selector which allows the operator to command the unit to change VHF frequencies, to change to a longer life than originally set, and to toggle the RF on and off to conserve life. Command function selection commands are detected via a 291.4 MHz UHF receiver.

The buoy can also be modified for special application to include fixed calibrated attenuators up to a maximum of 80 dB.

Operational status
On order for Australia with deliveries commencing in 1998.

Specifications
Audio frequencies: 5-20,000 Hz
Weight: 10 kg
Operating life: 1, 3 or 8 h
Operating depth: 18-122 m
Launch profile: 12-15 m (in hover altitude), 30.5-9,000 m @ 45-370 kt
RF channels: 31 or 99
RF power: 1 W

Contractor
Sparton Corporation, Electronics Division, DeLeon Springs, Florida.

NEW ENTRY

AN/SSQ-58A/B moored sensor buoy

Type
Passive/active sensor buoy.

Description
Sparton Electronics has developed the moored sensor buoy, a multipurpose building-block system for at sea measurements. Two separate versions of the buoy are available: a passive system for harbour surveillance, and an active system for use in test ranges and so on. Other applications of the buoy include a variety of oceanographic and environmental requirements. The buoy is retrievable and reusable and can also be configured in a free-floating version. Powered by commercially available batteries, or extended life alternative power sources, it has a long operational life. It is available with 31 VHF (AN/SSQ-58A model) selectable frequency channels and is compatible with all sonobuoy receivers and processors.

The buoy consists of: a foam filled glass fibre float, from which is suspended its mooring and hydrophone array; a suspended hydrophone assembly; sonobuoy electronics package integrated with a 31 channel radio transmitter; ruggedised radio antenna; radar reflector;

and navigation warning light that can be switched on or off, depending on the tactical situation. It is normally deployed from a small vessel, usually in depths of approximately 12 to 35 m. A deep mooring version is also available. A VHF antenna and a radar reflector are mounted on top of the float.

Operational status
Available for the US and other navies.

Specifications
Frequency range: 50-10 kHz
Operating life: 100 h (between battery recharging)
Operating depth: 6 m
Dimensions of float: 1 m (diameter) × 60 cm (deep)
RF Channels: 31
RF power: 1 W

Contractor
Sparton Corporation, Electronics Division, DeLeon Springs, Florida.

VERIFIED

The Sparton AN/SSQ-58 Moored Inshore Undersea Warfare (MIUW) buoy

AN/SSQ-62 series DICASS sonobuoy

Type
Active directional 'A' size sonobuoy.

Description
This series is the sonobuoy component of the Directional Command-Activated Sonobuoy System (DICASS). A high-performance sonobuoy, it detects the presence of submarines using sonar techniques under direct command from ASW aircraft. The AN/SSQ-62C can also determine the range and bearing of the target relative to the sonobuoy's position.

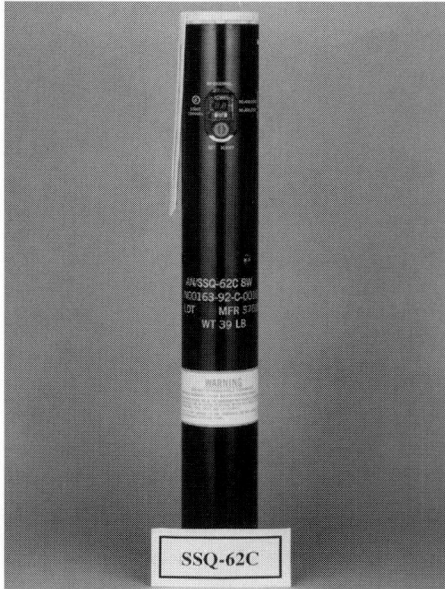

The AN/SSQ-62C DICASS sonobuoy **1995**

The DICASS sonobuoy is composed of three main sections: a parachute, a surface unit and a subsurface unit. The parachute slows down descent immediately after launch. The surface unit receives commands from the controlling aircraft, via a UHF receiver and sends target information to the aircraft, via a VHF transmitter. The subsurface unit transmits sonar pulses in the ocean upon command from the aircraft and receives sonar target echoes for transmission to the aircraft.

Command signals are received by the sonobuoy and are accepted if the correct address code is identified by a decoder. The command capability includes depth selection, scuttle and selection of transducer (sonar) transmission signals. The echoes from the selected activating signal are multiplexed in the subsurface unit before being transmitted to the receiving station or aircraft. There are two deep depth settings (460 and 760 m) one of which may be chosen via EFS selection before launch.

The buoys can be launched from fixed- or rotary-wing aircraft within the parameters specified in the launch envelope.

Upon impact with the water, the transducer is released for descent to its shallow operating depth. Immediately after entry, the float is inflated and the VHF-transmitter/UHF-receiver antenna is erected. When the float surfaces, the VHF transmitter begins emitting a continuous FM carrier signal. The sonobuoy is now operating in the passive mode and is ready to receive commands.

The main power source for the sonobuoy comes from either a lithium sulphur dioxide or thermal battery pack, instead of the more costly silver chloride batteries commonly used in sonobuoys. The sonobuoy is designed for economical volume production without compromising performance or reliability. Electronic design is exclusively solid-state with maximum use of multifunction integrated circuits.

Operational status
The older AN/SSQ-62B with three immersion depths and single channel 0.25 W VHF transmitter is still

available in limited production. The SSQ-62C is now in full production.

Specifications
AN/SSQ-62C
Sonar modes: pulse CW or linear FM
Operating life: 1 h
Operating depth: commandable, 27, 120, 460/760 m
Weight: 15.5 kg
Sonar channels: 4
RF channels: 87 EFS selectable
RF power: 1 W
Launch profile: altitudes up to 3,048 m and airspeeds up to 300 kt.

AN/SSQ-62B
Operating life: 1 h
Operating depth: commandable, 27, 120, 460 m
Weight: 14.5 kg
Sonar channels: 4
RF channels: 31 EFS selectable or fixed
RF power: 1 W
Launch profile: altitudes up to 3,048 m and airspeeds up to 300 kt.

AN/SSQ-62D
Operating life: 1 h
Operating depth: shallow, 15, 45, 90 m
Deep, 27, 120, 460 m
Weight: 14.5 kg
Sonar channels: 4
RF channels: 87 EFS selectable
RF power: 1 W
Launch profile: altitudes up to 9,000 m and airspeeds up to 370 kt.

Contractors
Sparton Corporation, Electronics Division, DeLeon Springs, Florida.
Raytheon Electronic Company, Fort Wayne, Indiana.

VERIFIED

AN/SSQ-77B (VLAD) sonobuoy

Type
Passive directional search and surveillance 'A' size sonobuoy.

Description
The SSQ-77B Vertical Line Array DIFAR (VLAD) is an 'A' size passive, tactical search and surveillance sonobuoy designed to improve detection and tracking capability for the DIFAR system in a noisy, high-traffic environment. The VLAD concept utilises a vertical line array of omnidirectional hydrophones in place of the single omnidirectional hydrophone used in the standard DIFAR unit. A directional hydrophone, similar

to the DIFAR, is mounted at the array phase centre to provide target bearing data. Improvements include multiple depth settings, horizontal or vertical selectable beam patterns and an increased number of omnidirectional hydrophones in the vertical line array.

The electronic configuration for the hydrophone outputs is the same as that of the AN/SQQ-53D sonobuoy and no modifications are required to airborne DIFAR equipment to process SSQ-77B signals. All beam-forming functions are accomplished with the sonobuoy, with provision for sea noise equalisation and omnidirectional phase tracking.

The buoy includes Electronic Function Select (EFS) and multichannel transmitter functions. It incorporates a frequency synthesised FM transmitter selectable over

the entire range of 99 VHF channels. Selection of transmitter channel, depth, beam-forming and sonobuoy life is accomplished using a push-button mounted on the side of the buoy. A second button is used to verify stored selection.

The buoy is deployed as other buoys and is fully operational within 4 minutes of entering the water and automatically scuttles after 8 hours.

Operational status
In quantity production and in service with the US Navy.

Specifications
Acoustic frequency: 10-2,400 Hz
Transmitter frequency: 136-173.5 MHz

Weight: 11.3 kg
Operating life: 1, 4 or 8 h (selectable)
RF channels: 99 selectable (136 to 173.5 MHz)
RF power: 1 W
Launch envelope: Altitudes up to 9,144 m and airspeeds up to 370 kt.
Operating depth: Dual depth, 150 or 300 m

Contractors
Sparton Corporation, Electronics Division, DeLeon Springs, Florida.
Raytheon Electronic Company, Fort Wayne, Indiana.
Hermes Electronics, Dartmouth, Nova Scotia.

UPDATED

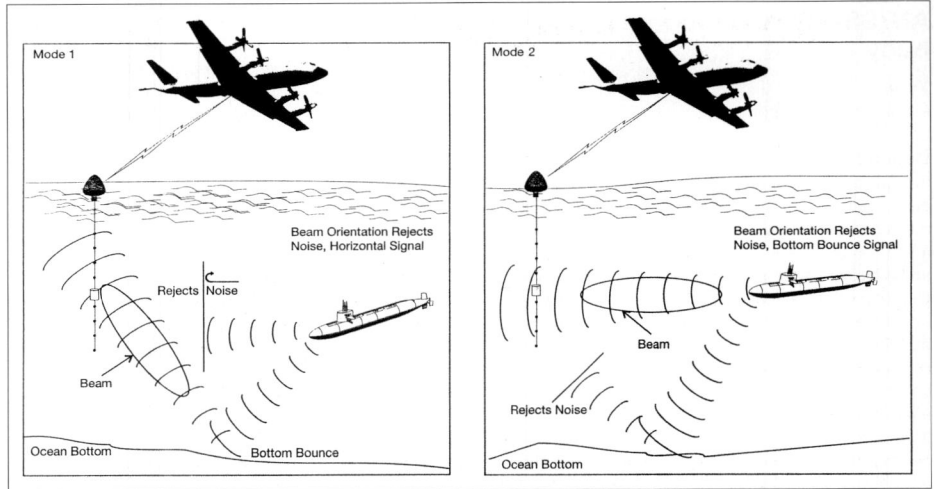

Operating geometries of the AN/SSQ-77B
1998/0006975

AIRBORNE ACOUSTIC PROCESSING AND RECEIVING SYSTEMS

CANADA

AN/UYS-503 acoustic processing system

Type

Multisonobuoy acoustic processing system.

Description

The AN/UYS-503 acoustic processor is a small, lightweight acoustic processing system designed for use in a variety of surface and subsurface surveillance platforms. It employs a new concept in processor architecture that exploits the advantages of high-density digital technology and the rapid pace of its evolution, providing processing for all types of sonobuoy and growth capability for current and future sonobuoys.

This feature enables the AN/UYS-503 to remain ahead of the ever changing threat, meeting the detection, localisation and attack challenges posed by new-generation nuclear submarines and diesel submarines operating in shallow and littoral waters. It functions as a complete system providing all the required input signal conditioning, signal processing and analysis, post-detection processing and control and display processing.

A typical maritime patrol aircraft configuration would feature concurrent processing of 16, 32 or 64 sonobuoys, while a helicopter configuration could include simultaneous processing for eight sonobuoys and a low-frequency dipping sonar. Unique algorithms

AN/UYS-503 acoustic processing unit   **1995**

provide acoustic data fusion functions that combine available data to compute fixes and automatically track targets of interest. Other unique features include proprietary algorithms that provide consistent and reliable detection and localisation capability for broadband, swathes and transient emissions. A new colour display is an intrinsic part of the signal processing and greatly reduces operator workload, while dramatically enhancing detection and tracking performance.

The processor can handle analogue or digital data from a wide variety of sources including: Omni, DIFAR/VLAD, VLAD, DICASS, CAMBS, CODAR, BARRA, dipping sonar, bathythermal and ambient noise.

Operational status

Currently in service with Australian, Canadian, Japanese, Swedish and US forces. The RAF ordered 25 AN/UYS-503 processing systems in 1997 with delivery to commence in October 1998. The systems will equip the upgraded Nimrod MPAs. Deliveries will continue until 2001.

Specifications

Frequency range: full-band Difar

Input channels: any standard sonobuoy receiver

Control input: MIL-STD-1553B or RS-232C/RS-422

Tactical data output: MIL-STD-1553B, RS-232C/RS-422, or other as specified

Video output: RS-343A composite video colour or monochrome

Size: 26.2 × 39.6 × 24.6 cm

Weight: 22 kg (8-DIFAR system) 135 kg (64-DIFAR system)

Contractor

Computing Devices Canada Ltd, Ottawa, Ontario.

UPDATED

FRANCE

SADANG

Type

Sonobuoy signal processing and display system.

Description

The two-operator SADANG system is adaptable to all types of ASW aircraft. It is capable of processing Lofar (up to 64 simultaneously), DIFAR, DICASS and Barra sonobuoys. The basic module, which can be fitted on ASW fixed- or rotary-wing aircraft, handles mission tactics and processes sonobuoy data on one console manned by a single operator. Automated functions relieve operators from repetitive tasks, particularly during the search phase. Operators can choose to display sonobuoy detection or classification images.

As a processing system, the SADANG can be integrated with new-generation data management systems, or can operate in a stand-alone mode, handling complex information, including tactical data and fire control.

The SADANG is designed both to accommodate a large number of configurations and to offer a large potential for evolution, which will enable it to process data from advanced digital sensors such as sonobuoy networks and new active arrays. Additionally, a general purpose beamformer has been developed and tested in co-operation with Australian affiliate Thomson Marconi Sonar Pty Ltd.

Using the latest processing technologies coupled with current graphics displays and data fusion techniques the systems are designed for the sanitising and survey of both deep and shallow waters. Up to 16 DICASS can be processed simultaneously enabling a typical coverage of about 200 n miles.

Advanced processing covers the full spectrum of noise including narrowband frequency lines, broadband signals, swaths, transients and Demon signals.

The SADANG 2000 offers a major boost in performance using modern, proven technology. It

SADANG version 1C during sea trials with the French Navy on board an ATL 2 maritime patrol aircraft   **1995**

features an impressive range of operator aids, including colour-coded software tools for improved classification and data fusion. It also provides lightweight, optimised operations, designed to meet a wide range of threats and tactical situations.

Operational status

The first of the SADANG 2000 series, version 1A, equips 28 French Navy Dassault Atlantique 2 Maritime Patrol Aircraft. The French Navy has also awarded Thomson Marconi Sonar SAS a contract to supply the new-generation SADANG version 1C. A prototype 1C was installed on board an Atlantique 2 in September 1992 and successfully tested during 1993.

SADANG 2000 1C was selected by the Pakistan Navy and completed successful operational evaluation sea trials during 1995.

Contractor

Thomson Marconi Sonar SAS, Sophia Antipolis.

VERIFIED

TMS 2000

Type

High capability COTS-based processors.

Description

The TMS 200 family of acoustic processors uses sophisticated graphic displays and data fusion techniques to efficiently survey deep, as well as shallow waters.

The passive functions of the high-capacity model, performed on up to 32 channels, rely on a number of processors devoted to wideband, swaths and narrowband frequency lines, transients and DEMON signals. Double DEMON processing ensures full coverage of both the carrier noise bandwidth and

modulating spectrum to achieve permanent detection of transient cavitation phenomena. The active functions of TMS 2000 can process up to 16 DICASS in monostatic/multistatic mode, over a 200 n miles area, with an overlap of 15 per cent between sonobuoys. The MMI makes extensive use of the Geographic Energy Plot (GEP) to provide a plot which results from data fusion. This GEP can be superimposed on an acoustic localisation plot (sonobuoy position, coastlines and sea bottoms) enabling the operator to easily fulfil the mission.

A simple COTS-based processor, whose capabilities make it possible to process four Barra/ four DIFAR/two DICASS, is also offered incorporating PC components housed in a ruggedised package. It offers conventional functions (beam-forming, spectrum analysis, power

versus bearing display) and tools (harmonic division, cursors for frequency and bearing measurement). The processor is suitable for navies owning small, radar equipped, maritime surveillance aircraft, whose performance would be much enhanced by the use of acoustics.

Operational status

Currently under development with ground-based prototypes undergoing evaluation by the UK MoD. Airborne components have been trialled on Nimrod Mk2 aircraft of the RAF.

Contractor

Thomson Marconi Sonar SAS, Sophia Antipolis.

UPDATED

Calcutta

Type

Acoustic data processing and display unit and control unit for RR 229 sonobuoy receiver.

Description

Calcutta is the acoustic processing and display system associated with the RR 229 sonobuoy receiver (see following). The unit performs a number of different functions which are selected by a cursor within scrolling menus by activating either a trackball or a mouse. Control of the display is carried out on the front panel screen under Windows.

System functions include:

(1) Lofar, DEMON, CODAR, DIFAR, DICASS and RO processing and display of four sonobuoys (a

maximum of eight is possible) with the following characteristics – 50 kHz to 10 Hz analysis bandwidth with averages and zoom; amplitude, frequency, time and bearing measurements; tactical measurements

(2) early warning 8-way Lofar switched analysis which boosts the Calcutta VHF receiver system capability to 48 sonobuoys

(3) control and checking of one (or two) RR 229 sonobuoy receivers for reception parameters set up and analysis of: sonobuoy VHF channel number; parameters of the digital sonobuoy in use; triggered Built-In Test (BIT) request and report; free channel search and display (scanning).

To carry out processing, Calcutta is fitted with a 25 MHz 68EC030 microprocessor and TMS 320C31

signal processor with 16-bit acquisition at variable sampling rates. Information is presented on a 640 × 400 pixel electro-luminescent screen.

Interface components include serial RS 422 or RS 232 port and 1553B dual port.

As an option, Calcutta can be fitted with an 8-buoy radio data relay control for remote processing on board a surface vessel.

Contractor

DCN International, Paris.

VERIFIED

RR 229 sonobuoy receiver

Type

Sonobuoy receiver.

Description

The 1/2 ATR RR 229 sonobuoy receiver is able to simultaneously receive VHF signals emitted by eight analogue or digital sonobuoys within the 136 to

173.5 MHz frequency range. Incorporating an internal amplifier, the receiver is compatible with all known sonobuoys. The receiver transmits on one of 99 analogue or 50 digital channels, with an audio frequency response between 2 Hz to 50 kHz. The output per acoustic channel provides three independent audio standard outputs for analogue signals or one coded NRZL data output and the related clock for digital signals. Eight IF outputs are available

for simultaneous retransmission. The receiver sensitivity is: S/N >–40 dB for 1μV VHF input level for analogue signals and BER >10^6 for a VHF input level of >–85 dbm for digital signals.

Contractor

DCN International, Paris.

VERIFIED

ASRC

Type

Active sonobuoy remote control equipment.

Description

ASRC is a 1/2 ATR processing Line Replacement Unit (LRU) providing active buoy control signals. The unit is controlled through a serial datalink (RS 422) by a PC or a host computer. This datalink continuously provides

system configuration and sonar emission triggering messages. Buoys to be controlled are selected through their VHF channel and the system provides: remote control of active directional buoys (DICASS SSQ 62 type) for depth and scuttle process; and remote control (DICASS SSQ 62) with single tone pulse emission (short medium or long pulses) or FM pulse emission (up or down linear modulation). Pulse repetition may be either automatic (8, 16, or 32 seconds) or manual.

The ASRC can independently control eight buoys at

the same time. A processor unit manages: emitted frequency coherence; emission rhythm and buoy power handling coherence; and correct signal multiplexing to feed a UHF emitter.

The unit is connected to an onboard UHF emitter receiving multiplexed control signals.

Contractor

DCN International, Paris.

VERIFIED

UNITED KINGDOM

AQS-901 processing system

Type

Sonobuoy processing and display system.

Description

The AQS-901, which first entered service in 1979, processes data from all NATO inventory sonobuoys and exploits the considerable advantages gained from advanced buoys such as BARRA, CAMBS and VLAD. It is claimed to be the most powerful and comprehensive system in service today.

The aircraft contains two systems, each with two CRT displays, (one being colour) and two hard-copy displays. A fifth CRT, which is able to show data from either system, is provided for system management. Information is displayed via both hard-copy and CRT display media, and is controlled and interrogated by the operator using a keyboard, trackerball and cursor.

AQS-901 sonobuoy processing and display system

Operational status
In service. The AQS-901 is fitted to over 30 RAF Nimrod MR Mk 2s and 20 RAAF P-3C Orions. Deliveries were completed in Autumn 1986. The current software and hardware product improvement programme will continue for several years. The latest upgrade involves integrating Barco ruggedised colour displays into the system for the RAF Nimrod which was installed in 1995.

Contractor
Thomson Marconi Sonar Ltd, Rochester.

UPDATED

AQS-902/AQS-920 series processing systems

Type
Sonar processing and display systems.

Description
The AQS-902 and its export variants, the AQS-920 series systems, have been designed and developed for installation in ASW helicopters, maritime patrol aircraft and small ships. This range of lightweight and flexible systems will handle data from all current and projected NATO inventory sonobuoys and will interface with a wide range of 31 or 99 RF channel sonobuoy receivers. A maximum of eight sonobuoys may be processed, while for ASW helicopters these systems offer the additional advantage of simultaneously displaying and processing dipping sonar data.

The modular structure of the AQS-902/920 series acoustic processors provides the capacity for every configuration to be tailored to the aircraft, operational role and specific requirements of each customer. As a result the equipment is suitable for both mid-life updates of existing ASW aircraft and equipment fits in aircraft having no current sonobuoy capability. Installation of AQS-920 on a small ship provides a flexible, cost-effective, variable depth ASW capability without requiring the modifications dictated by hull-mounted sonar systems.

The installation comprises four basic units, each of which contains the dedicated modules which allow construction of the selected system configuration at minimum weight and cost. These allow subsequent expansion and modification of the system to suit changing requirements.

The processor unit converts the received sonobuoy signals into digital form, carries out the necessary filtering and analysis, and processes the data into a form suitable for display. It is packaged into a short 1 ATR case.

The display unit may be in the form of a CRT display or a chart recorder, with configurations using either or both of these as required. All CRT and hard-copy displays are fully annotated. The operator uses a simple control panel to specify processing modes and parameters, and a trackerball or stiff stick control and cursor to interrogate data on the CRT display. AQS-902 provides numerous standard processing and display facilities, including: simultaneous passive and active processing; wide and narrowband spectral analyses; broadband correlation; CW and FM active modes; passive and active auto alerts and auto calculation and tracking of line frequency; target bearing and range; signal-to-noise ratio; Doppler fixing; CPA analysis; bathythermal processing and ambient noise measurement.

A unique feature of the system is the acoustic localisation plot which presents the operator with a geographical plot of sonobuoy and target positions.

AQS-902 G/DS systems equip Royal Navy Sea King ASW helicopters. The systems feature new CRT display facilities and also control, process and display data from the 2069 dipping sonar.

Operational status
Configurations of this range of equipment are in service with the Royal Navy, the Swedish Navy and the Indian and Italian navies. Further variants are installed in Grumman S-2 Turbo Tracker aircraft. Further system enhancements and developments, including the incorporation of colour display processing techniques, are continuing. The Swedish Navy has ordered a number of AQS-928 systems which will process data from up to eight passive and active sonobuoys in any combination simultaneously.

Contractor
Thomson Marconi Sonar Ltd, Rochester.

VERIFIED

AQS-902G/DS installation in the RN Sea King Mk VI

Integrated mission avionics suite in an Indian Navy Sea King incorporating AQS-920 series system (foreground) and TPS (left)

AQS-903/AQS-930 series processing systems

Type
Sonar processing and display systems.

Description
The AQS-903 and its export variants, the AQS-930 series, are powerful lightweight acoustic processors for helicopter and fixed-wing applications. These systems process data from eight to 32 sonobuoys in any combination. This flexibility is derived from the design based around a number of processing 'pipelines', each able to handle a certain quantity of passive and/or active sonobuoys. Four of these pipelines will provide a processing power eight times that of the AQS-901, at a fraction of the size and weight.

AQS-903/930 series make extensive use of distributed processing, with the systems incorporating the latest component technology to provide the operator with a wealth of information. The system provides a wide range of processing options to enable the operator to smoothly progress from initial contact

RN Merlin ASW helicopter

detection to classification, localisation and tracking to achieve an effective mission. Analysis and display cues are in plain language, requiring a minimum of operator training. Control of the twin CRT displays is achieved by the operator using a simple control panel and multifunction keys. AQS-930 systems will interface to other systems responsible for sonobuoy launching, navigation, tactical tracking and weapon release.

Operational status
Manufacture of 44 AQS 903 systems for the Royal Navy Merlin helicopter has commenced. The basic configuration will process data from the complete range of NATO inventory sonobuoys.

Contractor
Thomson Marconi Sonar Ltd, Rochester.

VERIFIED

AQS-903 acoustic processing rack

Type 843 sonar receiver

Type
Sonobuoy signal receiver for helicopters.

Description
The Type 843 is a multichannel sonar receiver of modular construction for use in ASW helicopters. It simultaneously receives four VHF channels which are selected by serialised digital commands from a processor or control unit. Advanced filter technology, extensive use of hybrid circuits and wideband techniques ensure high availability. An antenna preamplifier unit gives a low noise figure and the system's high dynamic range and wide bandwidth enable high-speed telemetry data to be received.

The receiver provides its full RF performance when installed at up to 50 m from the preamplifier. All interfaces are to ARINC specification 404A. Outputs are compatible with existing aircraft data processing systems and with more advanced systems under development. A built-in test signal generator provides extensive check facilities for the receiver and associated processing equipment, and is designed to facilitate servicing by ATE systems.

Operational status
In service aboard ASW Sea King helicopters of the Royal Navy.

Specifications
Frequency: 134.5-174.625 MHz

Frequency stability: ±3 PPM (at −26 to +45°C ambient temperature)
No of channels: 4
Channel coverage: 99 RF channels plus 2 test channels
Channel spacing: 375 kHz
Input impedance: 50 Ω
Modulation: FM/FSK
Weights: 1.3 kg (preamplifier); 2.5 kg (control unit); 11.4 kg (receiver)

Contractor
GEC-Marconi Radar and Defence Systems, Command and Information Systems Division, Camberley.

VERIFIED

ARR 901 sonar receiver

Type
Multichannel sonar receiver.

Description
The ARR 901 is a multichannel sonar receiver of modular construction for use in fixed-wing maritime patrol aircraft. It simultaneously receives eight VHF channels which are selected by serialised digital commands from a processor or control unit. Advanced filter technology, extensive use of hybrid circuits and wideband techniques ensure high availability. An antenna preamplifier unit gives a low noise figure and the system's high dynamic range and wide bandwidth enable high-speed telemetry data to be received.

The receiver provides its full RF performance when installed at up to 50 m from the preamplifier. All interfaces are to ARINC specification 404A. Outputs are compatible with existing aircraft data processing systems and with more advanced systems under development. A built-in test signal generator provides extensive check facilities for the receiver and associated processing equipment, and is designed to facilitate servicing by automatic test equipment.

Operational status
In service aboard Nimrod MPA aircraft of the RAF.

Specifications
Frequency band: 134.5-174.625 MHz

Frequency stability: ±3 PPM (at −26 to +45°C ambient temperature)
No of channels: 8
Channel coverage: 101 RF channels plus 2 test channels
Channel spacing: 375 kHz
Input impedance: 50 Ω
Modulation: FM/FSK
Weights: 1.3 kg (preamplifier); 23.6 kg (receiver)

Contractor
GEC-Marconi Radar and Defence Systems, Command and Information Systems Division, Camberley.

VERIFIED

Type R605 receiver

Type
Sonobuoy data receiver.

Description
The R605 receives, demodulates and amplifies FM sonobuoy signals in the frequency range from 136 to 173.5 MHz over 99 channels. Simultaneous operation can be carried out on four channels and a demodulated output for analysis and display is provided. The equipment is DIFAR and DICASS compatible.

The R605 is a much improved version of the AN/ARR-75 and uses many of its mechanical and electrical design features to simplify retrofit applications.

Control options include MIL-STD-1553, RS-422, ARINC 429 or a manual control unit. MTBF is quoted as better than 2,000 hours, and the life of the system is better than 20,000 hours of continuous operation.

Extensive BITE is provided. An optional remote preamplifier allows receiving sets to be mounted at a distance from the system antenna.

Operational status
In production for the Royal Navy's Merlin EH 101 helicopter.

Specifications
Dimensions: 276 × 260 × 194 mm
Frequency range: 136 -173.5 MHz
Channels: 99 RF, 4 acoustic
Noise figure: 5 dB at 50 Ω
Stability: ± 3 PPM
Weight: 11 kg

Contractor
Ultra Electronics Sonar and Communications, Greenford.

UPDATED

R605 sonobuoy receiving set

Type R612 receiver

Type
On-top position indicator sonobuoy receiver.

Description
The Type R612 is a modular 99 channel, on-top position indicator which covers the frequency range 136 to 173.5 MHz. It is suitable for use on both fixed-wing aircraft and helicopters. The equipment is fully compatible with all known sonobuoys and a variety of antenna systems. When used in conjunction with an ADF system it enables the operator to home on and locate sonobuoys on any of the 99 channels. The R612 is interchangeable with the R-1651/ARA without changes to existing wiring. Control options include RS-422, MIL-STD-1553B (via R605), or a manual control box.

The R612 functions as an amplitude demodulator for the direction-finding system, and produces an indication that an adequate signal exists when the RF signal input is sufficient for reliable direction-finder operation. The presence of the adequate signal is indicated in three ways; a 400 Hz tone to the operator's headset, a light on either side of the control box, or as a discrete signal to the R605 sonobuoy receiver.

Operational status
In production and in operational service.

Specifications
Sensitivity: 10 dB S/N ratio at 2.5 μV input
Dimensions: 126 × 135 × 78 mm
Weight: 2.16 kg
MTBF: 3,000 h

Contractor
Ultra Electronics Sonar and Communication Systems, Greenford.

VERIFIED

Type T613 receiver

Type
Sonobuoy data receiver.

Description
The T613 is an improved version of the T843 receiver which it replaces. The lightweight system is encased in an aluminium chassis, offering higher stability and a lower noise signature. The receiver operates on four simultaneous channels and has a full 99 channel capability but without the high level of sophisticated BITE provided with the other receivers. A dedicated control unit and separate RF preamplifier for increased signal-to-noise performance is also incorporated.

Contractor
Ultra Electronics Sonar and Communication Systems, Greenford.

VERIFIED

AA34030 sonobuoy command transmitter

Type
Sonobuoy command transmitter.

Description
The AA34030 is a UHF transmitter developed specifically to function as the sonobuoy command transmitter for the RF downlink of a sonics data processing system, allowing selection of the sonobuoy mode, uplink channel, ping and so on. The transmitter provides AM or FM analogue or digital communications over the frequency range 282 to 292 MHz with 50 kHz channel spacing throughout the band. This can be reduced to 25 kHz increments if desired.

The design of the transmitter is related to the AD3400 multimode VHF/UHF airborne communications transmitter/receiver in using the same 'sliced' modular construction, with five of seven modules making up the transmitter being common to both units. An additional Transmitter Control Interface Module (TCIM) unique to the sonobuoy provides an interface between the AQS 903 acoustic Radio Control Module (RCM) and the sonobuoy.

Operational status
The AA34030 is being installed in the Royal Navy's Merlin helicopter.

Specifications
Frequency range: 282-292 MHz
Channel spacing: 50 kHz
No of channels: 200
Frequency accuracy: 2 ppm
Transmitter power: 50 W FM, 15 W AM
Weight: 6.5 kg

Contractor
GEC-Marconi Electro-Optics Ltd, Airadio Division, Basildon.

UPDATED

AA34030 sonobuoy command transmitter

UNITED STATES OF AMERICA

LAMPS Mk III

Type
Integrated helicopter ASuW/USW system

Development
The US Navy's LAMPS (Light Airborne MultiPurpose System) Mk III has been designed to extend and enhance the capabilities of surface combatants particularly in anti-surface (ASuW) and undersea warfare (USW), and increase the effective operational range for the weapons systems fitted on the helicopter and in surface combatants of the US Fleet. This system extends the electronic and acoustic sensor ranges and provides a weapon delivery capability for cruisers, destroyers and frigate classes of ship by the deployment from them of armed helicopters. Sensors, processors and display capabilities aboard the helicopter, coupled via secure broadband digital datalink with the sensors, processors and display capabilities aboard the surface combatant, enable the seven-man crew (three in the aircraft; four on the ship) to extend the LAMPS Mk III capable surface combatant's tactical decision making and weapons delivery capabilities. Classic line of sight limitations for surface ships and limitations to underwater acoustic detection, classification, and localisation are mitigated through the use of a manned LAMPS aircraft and thus the surface combatant's sphere of influence is increased tenfold.

In 1974, the USN selected Lockheed Martin Federal Systems-Owego (at that time IBM Federal Systems Division) as the system prime contractor to work with the Naval Air System Command in developing LAMPS Mk III. In 1976, the USN and IBM-FSD completed proof of concept testing at sea. The results far exceeded

SH-60B helicopter (Courtesy of Lockheed Martin)   *1999*/0017498

expectations. In 1978, the Department of Defense and the Navy authorised IBM-FSD to proceed with full-scale engineering development with Sikorsky Aircraft as the air vehicle contractor and General Electric as the engine manufacturer.

Description
The LAMPS Mk III Weapons System embodies the integration of the host surface combatant (cruiser, destroyer, frigate) and the manned aircraft (SH-60B Seahawk) operating from that surface combatant for both ASuW and USW missions. As an adjunct to the sensor and attack systems of the host or similar equipped surface combatants, the aircraft extends the detection, classification, location and attack capabilities where needed to increase line of sight and range capabilities of the host surface combatant. Radar/FLIR and ESM sensors extend the surface combatant's line of sight surveillance and engagement range on surface threats, whereas, sensors that operate relative to the ocean medium (such as sonobuoys and magnetic anomaly detectors) enable the surface combatant to engage the subsurface threats.

In an ASW mission the aircraft is deployed from the host surface combatant when a suspected threat has been detected by the surface combatant's systems or offboard cueing. It proceeds to the estimated target area, where sonobuoys are dropped in the water in a pattern designed to trap the target. The acoustic signatures detected by the sonobuoys are transmitted over a radio frequency link to the aircraft where they are analysed, codified and retransmitted to the surface combatant for interpretation and analysis. When the location of the threat is determined with adequate precision, the aircraft descends near the ocean's surface for final confirmation. This may be accomplished using active sonobuoys, passive directional sonobuoys or by trailing a Magnetic Anomaly Detector (MAD) behind the aircraft. On final confirmation by any of these methods, a torpedo attack can be initiated. The extension of the surface combatant's sensor, tactical control and attack capabilities is achieved by using a secure duplex digital datalink that transfers acoustic (ASW mode) or radar (ASuW mode), electromagnetic and command data and voice from the Seahawk aircraft, and command data and voice to the aircraft.

Data transmitted to the host surface combatant (or other LAMPS capable surface combatant) are processed by digital computers and specialised processors, distributed and evaluated by the surface combatant's operators: the LAMPS Air Tactical Control Officer (ATTACO); Acoustic Sensor Operator (ASO); Remote Radar Operator (REMRO), and ESM Operator (ESMO). The combat information centre evaluator uses

the data in context with the overall tactical situation to determine what actions are to be taken, and tactical instructions and weapon delivery information are linked back to the SH-60B Seahawk. Operation of the LAMPS Mk III Weapon System was described in detail in earlier editions of *Jane's Weapons Systems*.

The LAMPS Mk III system has been updated to better address current threats and includes the incorporation of a GPS navigation system, a 99 channel sonobuoy receiver, Mk-50 lightweight torpedo and Penguin anti-ship missile armament capability and a UHF-VHF radio capability.

Operational status

Introduction to the US Fleet of the SH-60B LAMPS Mk III Weapon System began in 1984, and these

helicopters are now deployed aboard US Navy surface vessels including CG-47 cruisers; DD-963, DD-993, and future DD-52 Flight IIA destroyers; and FFG-7 frigates. A total of 181, SH-60Bs have been delivered to the US Navy, completing the planned production. A Block II update is currently in design, which will add the Advanced Low-Frequency Sonar (ALFS) and an Inverse Synthetic Aperture Radar (ISAR). This capability is planned to be operational by 2002.

Contractors

Lockheed Martin Federal Systems-Owego, New York, (system prime contractor).
Sikorsky Aircraft (helicopter contractor).
General Electric Company (engine contractor).

UPDATED

AN/AYA-8B processing system

Type
ASW data processing system for the P-3C.

Description
Since 1968, Lockheed Martin (formerly GE) has been producing the P-3C Data Processing System (DPS). The DPS equipment constitutes a major portion of the P-3C anti-submarine aircraft avionics.

The purpose of the AN/AYA-8B DPS is to provide an interface between the central computer and other aircraft systems. The main aircraft systems consist of the TACtical Co-Ordinator (TACCO) station, non-acoustical sensor station, NAVigation COMmunications (NAV/COM) station, acoustical sensor stations, pilot station, radar interface unit, armament/ordnance system, navigation systems, ARR-72 receivers, submarine anomaly detector, and OMEGA. The DPS consists of four major logic units, as well as a selection of manual keysets and panels.

Logic Unit 1 provides an interface to four types of peripheral information systems:
(a) manual entry
(b) system status
(c) sonobuoy receiver
(d) auxiliary readout display.

The manual entry subsystem provides the communication between the operator stations and the central computer of the man/machine interface. Each operator has a complex of illuminated switches and indicators by which he communicates with the central computer.

System status is received and stored by the status logic sub-unit of Logic Unit 1 and transmitted to the central computer. The status words are transmitted whenever any status bit changes or upon interrogation by the computer.

The sonobuoy receiver logic provides for the digital tuning of the sonobuoy receiver subsystem by the central computer. Each of the 20 receiver processor channels can be tuned to one of 31 RF cells by the central computer or manually by the operator.

The auxiliary readout display logic provides the digital interface between the computer and the combat information displays located at the TACCO station and the NAV/COM station.

Logic Unit 2 provides digital communications to three major aircraft subsystems:
(a) navigation
(b) armament/ordnance
(c) magnetic tape transports.

Logic Unit 3 provides the interface between the computer and three display subsystems:
(a) TACCO multipurpose display
(b) sensor multipurpose display
(c) pilot display.

Logic Unit 4 provides for the expansion of the computer input/output capability by means of the Data Multiplexer Sub-unit (DMS), furnishes an increased memory capacity by means of the Drum Auxiliary Memory Sub-unit (DAMS) and provides an interface between the computer and two aircraft subsystems:
(a) OMEGA
(b) auxiliary display.

Keysets and panels
Three universal keysets provide the navigation/communication operator and the two acoustic sensor operators with the capability of entering information into and receiving information from the computer program. The pilot keyset is used by the aircraft pilot to control information presented on the CRT display, to enter navigation stabilisation information into the computer, to drop or cause the dropping of smoke floats and weapons, and to enter information on visual contacts.

AN/AYA-8B modernised logic units for P-3C aircraft

AN/AYA-8B pilot keyset (front) with (left to right) ordnance panel, universal keyset and armament/ordnance panel

The ordnance panel is used to display commands from the computer to the ordnance operator concerning search stores, that is, bin and chute number, and status information.

The armament/ordnance test panel provides the capability of monitoring each output to the aircraft armament and ordnance systems from the armament output logic unit and ordnance output logic sub-units of Logic Unit 2.

Operational status
In production.

Contractor
Lockheed Martin, Utica, New York.

VERIFIED

BIRADS

Type
ASW processing and display system.

Description
BIRADS (BIstatic Receiver And Display System) is a portable equipment which has been developed to upgrade existing ASW electronics suites. By using off-the-shelf computer and display components, the BIRADS system integrates matched filtering, pulsed energy detection and constant wave energy detection. It is intended to work alongside existing hardware to improve real-time, active sonar data analysis and interpretation. It can be used as an aircraft or a towed array active ASW receiver.

BIRADS uses a Concurrent (MASSCOMP) 6650 computer to digitise and process 32 channels of

acoustic sensor data from different sensor types, for example: 32 omnidirectional sonobuoys; 10 DIFAR or VLAD buoys; or 32 beams from a towed array sonar. For coded active pulses, the system can store the signals in raw form, perform optional beam-forming, and conduct matched filtering. For CW signals the system executes a high-resolution frequency analysis through a user-defined configuration file.

Processed signals are presented on a Sun 4 Sparcstation display as either A-scan traces or colour/intensity modulated sonograms. Operators can pick out signal peaks manually, using a mouse or cursor control, or set a threshold for automatic detection. On the display a split-screen approach allows the operator to see not only the signals themselves, but also observe the detected peaks mapped out on a geographic display. These displays provide information on bathymetric contours, convergence zones, sensor

fields, planned source and target tracks, and transmission loss coverage. The data for all peak detections are stored for subsequent track processing.

For post-mission analysis BIRADS can play back multiple platform sensor data, reanalyse processed data and reprocess raw data as required. Special features include analysis of echo features and manipulation of the measurement database. BIRADS can also be used in a simulation mode for training purposes.

Operational status
In production.

Contractor
BBN Technologies, Arlington, Virginia.

UPDATED

OL-320/AYS

Type
Acoustic processing system.

Description
The system incorporates the CP-1584/AYS post-processing and display elements and the C-11262/AYS

monitoring control unit. The latter provides switching between sonobuoy audio processing and the integrated communications system on the S-3B Viking ASW aircraft.

Operational status
The computer is operational in all S-3B Viking aircraft.

Contractor
Sanders, a Lockheed Martin Company, Manchester, New Hampshire.

VERIFIED

AN/AQR-185

Type
Sonobuoy receiving system for the P-3C Update IV.

Description
The AN/AQR-185 99 channel sonobuoy receiver performs the functions of analogue and digital sonobuoy reception. Each receiver will tune the 99 RF channels between 136 and 173 MHz. The system is designed for applications where weight and size are important and the sonobuoy field area is limited.

The receiver system is divided into a set of modular units, a dual RF preamplifier, a control unit, and from one to five receiver units. Each receiver unit will accommodate up to 20 interchangeable receiver modules, providing a growth potential to monitor 100 different sonobuoys simultaneously.

The receiver system accepts optional plug-in assemblies that perform the functions of RF spectrum scanning to determine channel activity and on-top position indication in conjunction with a direction-finding antenna and a heading indicator. Phase angle measurement equipment supports sonobuoy location determination in conjunction with sonobuoy reference system antennas and software, and the receiver provides acoustic test signal generation and the provision of acoustic audio signals to operators' headsets and display.

Sonobuoy receiver for Update IV electronics suite to be fitted to P-3C ASW aircraft

Operational status
In development for the US Navy P-3C Update IV.

Contractor
Dowty Avionics, Arcadia, California.

VERIFIED

R-1651/ARA

Type
On-top position indicator for sonobuoy receivers.

Description
The R-1651/ARA radio receiving set is a 31 RF channel On-Top Position Indicator (OTPI) used aboard rotary- or fixed-wing aircraft to provide bearing and on-top position indication of deployed sonobuoys. When used in conjunction with a suitable DF system, it enables the operator to locate and verify the position of deployed sonobuoys operating on any of 31 channels.

The R-1651/ARA is form and fit interchangeable with the R-1047 A/A OTPI. This radio receiving set is compatible with ARA-25 and ARA-50 antennas and it is also available in a modified configuration which is compatible with the DF-301E antenna system. The R-1651/ARA is controlled by the optional C-3840/A control box.

The R-1651/ARA is a single-conversion, 31 channel, superheterodyne VHF receiver which is designed to receive signals from deployed sonobuoys over the frequency range of 162.25 to 173.50 MHz.

The antenna input is switched by a relay to the

receiver or the UHF receiver, on external command. The RF signals (R-1651/ARA) pass through bandpass filters, are amplified, and converted to an IF of 25 MHz. After IF amplification and detection, the resultant audio signal is amplified to the level required by the ADF system. The AGC level is developed at the detector, amplified and used for gain control of the IF and RF stages and to operate the adequate signal strength circuits.

The local oscillator provides one of 31 possible LO signals. Thirty-one crystals are appropriately selected using external controlled code lines containing a 6-bit binary code. Line receivers, decoding gates and matrix drivers connect one of 31 crystals to the local oscillator circuits.

The optional C-3840/A radio receiver control is used to switch the antenna system, by energising an external coaxial relay, from the UHF/ADF mode of operation to the sonobuoy OTPI mode of operation. This enables the selection of any one of the 31 RF channels and indicates adequate signal strength when the R-1651/ARA receives an RF signal of -86 dBm or greater.

Operational status
In production for the US Navy.

R-1651/ARA sonobuoy OTPI radio receiver

Contractor
Flightline Electronics Inc, Fishers, New York.

UPDATED

AN/ARR-75

Type
Sonobuoy receiving system.

Description
The AN/ARR-75 sonobuoy receiving set is a 31 RF channel FM receiver designed for use on ASW fixed-wing aircraft, as well as shipboard applications. Independent receiver modules provide four simultaneous demodulated audio outputs each capable of selecting one of 31 RF input channels. The AN/ARR-75 is used on the LAMPS Mk III, LAMPS Mk I and SH-3H, as well as various other surface combatant platforms.

The AN/ARR-75 receiving set is composed of two units. The first is the OR-69/ARR-75 receiver group assembly, which in turn consists of a power supply (PP-6551/ARR-75), four receiver modules (R-1717/ARR-75) and the chassis (CH-670/ARR-75). The receiver group assembly contains the majority of the system's electronics.

The second unit is the radio set control (control box), C-8658/ARR-75 or C-10429/ARR-75. This control unit provides independent RF channel selection and signal strength, monitoring of any of 31 RF channels for each of the four receiver modules.

Flightline Electronics also manufactures a maintenance kit for the AN/ARR-75. This optional equipment is designated MK-1634/ARM maintenance

kit (also known as the suitcase tester) and is designed to facilitate interim maintenance of the AN/ARR-75 at the intermediate level. The MK-1634 contains module extenders, adaptors, cables and a test bench panel for interconnecting with and testing the AN/ARR-75.

Operational status
In service with US Navy SH-2, SH-3 and SH-60 helicopters.

Contractor
Flightline Electronics Inc, Fishers, New York.

VERIFIED

AN/ARR-76

Type
Fixed-frequency sonobuoy receiver system.

Description
The AN/ARR-76 sonobuoy receiver system receives up to 31 VHF/FM channels in the frequency range of 162 to 174 MHz, and provides 16 separate video output channels for monitoring. A general purpose digital computer or a manual computer simulator is used to control the unit relative to assignment of receiver

channels to the video output channels at any one time. Communications between the AN/ARR-76 and the computer are carried out by 'Manchester Code' at a 6 MHz rate.

Operational status
In service with US Navy S-3A and Canadian CP-140 aircraft.

Contractor
Dowty Avionics, Arcadia, California.

VERIFIED

AN/ARR-76 sonobuoy receiver

AN/ARR-78(V)

Type
Sonobuoy receiver for operation and management of buoys.

Description
The AN/ARR-78(V) radio receiving set, also known as ASCL (Advanced Sonobuoy Communication Link), is for the operation and management of anti-submarine sonobuoys by the crew of ASW aircraft, such as the P-3C Orion (Update III) maritime patrol aircraft and S-3B Viking. The equipment comprises five main items (described below) and is employed with an associated onboard ASW signal processor. The main units of the AN/ARR-78(V) are:

(a) the AM-6875 RF preamplifier, an optimised high-performance, low noise VHF preamplifier which provides amplification and prefiltering of received RF signals
(b) the R-2033 radio receiver, which contains 20 fully synthesised high-performance receiver modules (16 acoustic and 4 auxiliary), and one each of the following modules: RF/ADF amplifier multicoupler; reference oscillator, Proteus digital channel, I/O processor, clock generator, BITE, and DC power supply module. Each single conversion receiver module includes mixer conversion, frequency synthesised LO, demodulator and output interface circuits. Each of the acoustic receiver modules processes FM/analogue or FSK digital signals at any of the 99 channels in the extended VHF band. Each of the four auxiliary receiver modules processes signals at any of the 99 channels and provides: one channel for selection and processing of the On-Top Position Indicator (OTPI) signals; two channels for the operator to monitor acoustic information; and one channel to monitor the RF signal level in any of the RF channels. Common receiver modules are interchangeable

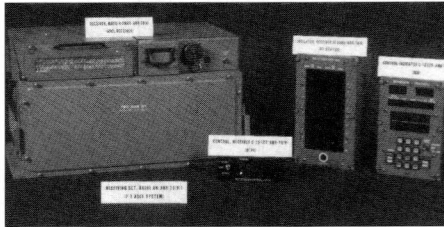

AN/ARR-78(V) advanced sonobuoy communications link equipment (P-3C configuration)

(c) BITE circuits provide comprehensive end-to-end evaluation of each receiver from the VHF preamplifier to the receiver output interface circuits. BITE is initiated automatically by the computer (such as Proteus) and/or by the operator through the Indicator Control Unit (ICU). Performance status is displayed on the ICU and routed to the computer
(d) the C-10126 Indicator Control Unit (ICU), provides the operator with manual control over the assignment of each receiver channel frequency, receiving mode and self-test. It also displays status of the receiving set and operator entry information
(e) the ID-2086 receiver status indicator, which continuously displays the control mode setting, the RF channel number and the received signal level for each receiver
(f) the C-10127 receiver control unit, which provides the operator with control over the OTPI channel frequency.

Surface Acoustic Wave (SAW) IF filters are incorporated to provide high selectivity, coupled with good linear phase characteristics, which yield low levels of distortion. Microprocessors are used in the ICU and I/O module to process the commands to, and display status data from, the receivers. Versatile

programs accommodate changes without requiring aircraft modification.

The auxiliary receivers permit the operator to aurally monitor signals in any of the 99 RF channels without interrupting data processing in that channel.

The ASCL PLUS incorporates enhancements which doubles the receiver complement by incorporating two digital receivers in the current AN/ARR-78 (V) receiver module to provide up to 40 acoustic channels for processing. Each digital receiver is easily commanded to any frequency in the 99 channel receiver band. The demodulised output of each receiver is fully compatible with existing ASW signal processors.

Operational status
By the end of 1991, 408 systems had been delivered for P-3C and S-3B aircraft. An additional 91 receivers are on order. The receiver is used by the Norwegian, Japanese and South Korean defence forces.

Specifications
Frequency: extended VHF
Receivers: 20 (16 acoustic/4 auxiliary)
Channels: 99 per receiver
Audio output
Analogue: 2 V or 4% V RMS (balanced)
Power input: 115 V AC ±10%, 380-440 Hz, 3-phase 450 W; 18-32 V DC, 7 W; 26.5 V AC, ±10%, 400 Hz, 50 W
Dimensions
AM-6875: 7.6 × 14.6 × 10.8 cm; 0.95 kg (weight)
R-2033: 30.9 × 54.1 × 38.9 cm; 45.9 kg (weight)
C-10126: 22.9 × 14.6 × 16.5 cm; 3 kg (weight)
ID-2086: 26.7 × 14.6 × 12.7 cm; 1.9 kg (weight)
C-10127: 5.7 × 14.6 × 8.3 cm; 0.6 kg (weight)

Contractor
GEC-Marconi Hazeltine Corporation, Greenlawn, New York.

UPDATED

AN/ARR-84

Type
Airborne and surface vessel sonobuoy receiver.

Description
The AN/ARR-84 represents the latest generation of 99 channel sonobuoy receivers. It is designed to receive signals from all current and planned future US and allied sonobuoys. This receiver features four acoustic channels capable of receiving signals from up to four deployed sonobuoys simultaneously on any of 99 RF channels. It is suitable for installation on a variety of rotary- and fixed-wing aircraft, as well as large surface vessels and fast patrol craft. It is a form and fit

replacement for its predecessor, the AN/ARR-75 (see earlier entry).

Additional features of the AN/ARR-84 include dual selectable IF bandwidths for optimum performance, with a wide variety of sonobuoy types. It provides high AM rejection and mechanical vibration immunity which severely reduces interference from propeller/rotor multipath and platform vibration. The receiver has a very low susceptibility to conducted and radiated energy, allowing the set to be used near strong onboard emitters and shipboard search radars.

An optional radio set control, intended for use when online computer control is not available, provides power control and BIT operation for the receiver group. RF channel selection, sonobuoy type selection and RF

level readout is provided independently for each of the four receivers in the group. Operator input is through a multifunction keypad with signal strength provided as a histogram display and, entry readback/failure data provided by a 16 character message display.

Operational status
In production and in use on board US Navy SH-60B, SH-60F, LAMPS Mk III, Royal Australian Navy S-70B and Taiwanese S-2T tracker fixed-wing aircraft.

Contractor
Flightline Electronics Inc, Fishers, New York.

VERIFIED

AN/ARN-146

Type
On-top position indicator for sonobuoy receivers.

Description
The AN/ARN-146 radio receiving set is a 99 RF channel on-top position indicator used on board ASW rotary- and fixed-wing aircraft to provide bearing and on-top position indication of deployed sonobuoys. When used in conjunction with a suitable DF system, the equipment enables the operator to locate and verify the position of deployed sonobuoys operating on any of 99 channels.

The AN/ARN-146 has a modular, solid-state design and it is designed to be used with operator control, either via RS-422 directly, or when connected with the AN/ARN-84 sonobuoy receiver. The AN/ARN-146 is form and fit interchangeable with the R-1651/ARA and

R-1047 A/A OTPI receivers. It is compatible with ARA-25, ARA-50 and OA-8697/ARD (DF-301E) DF antenna systems.

The AN/ARN-146 set consists of the R-2330/ARN-146 radio receiver and the C-11699/ARN-146 radio set control. The R-2330/ARN-146 is a VHF, AM receiver, which receives signals from sonobuoys operating on any of 99 RF channels, in the frequency range from 136 to 174 MHz. This receiver houses all the electronics, including internal transfer circuits for the switching of RF, baseband, power and phase compensation circuits for sharing the DF system between the OTPI and an associated UHF receiver system for the DF. The R-2330/ARN-146 also contains a PLL synthesised local oscillator, digital address decoder, voltage tuned BP filters, AGC and BIT circuitry.

The C-11699/ARN-146 radio set control is an optional manual control box which is used in place of

the RS-422 bus or the MIL-STD-1553B dual bus of the AN/ARN-84. This control box provides the capability to apply power to the AN/ARN-146 system, to select any one of 99 RF channels, to activate the BIT circuitry and to display adequate signal strength and results of the BIT via the adequate signal strength indicator.

The equipment is compatible with all known or planned sonobuoys, including DIFAR, LOFAR, Ranger, BT, CASS, DICASS, VLAD, CAMBS, BARRA, HLA, ATAC and SAR.

Operational status
In production for the US Navy.

Contractor
Flightline Electronics Inc, Fishers, New York.

UPDATED

AN/ARR-502

Type
Acoustic receiving set.

Description
The AN/ARR-502 is a 495 channel RF radio receiver,

primarily used for reception of sonobuoy signals that provides simultaneous reception of 16 acoustic channels in the 136 to 174 MHz frequency band. Additional optional features provide for digital output, end-to-end test, On-Top Position Indicator (OTPI), and sonobuoy referencing/positioning all in one lightweight

package that substantially reduces power consumption.

It is capable of receiving and simulating all types of sonobuoys, including Lofar, DIFAR, BT, VLAD, ambient noise, range only, and DICASS. Additionally, capability is provided to operate with CAMBS, BARRA and ATAC buoys, as well as future developmental sonobuoys such as ADAR and EEC. Capability is also provided for receipt and demultiplexing of GPS signals from all types of existing sonobuoys being modified to incorporate GPS.

The AN/ARR-502 can be controlled either by a MIL-STD-1553B or RS-422. Multiple bandwidths and channels are provided for improved performance and compatibility with existing and future sonobuoys.

Operational status
The AN/ARR-502 is in production for the Canadian Air Force and is currently installed on Sea King, CP-140 Aurora and RAF Nimrod aircraft.

Contractor
Flightline Electronics Inc, Fishers, New York.

UPDATED

STATIC DETECTION SYSTEMS

CANADA

CSAS-80 sonar

Type
Active seabed or platform-mounted sonar system.

Description
The CSAS-80 surveillance sonar is a high-performance omnidirectional sonar designed for the detection of underwater intruders such as divers using open or closed breathing apparatus, swimmers, Swimmer Delivery Vehicles (SDVs), mini-submarines and so on. Surveillance over a large area (2,000 m radius) is provided, and the sonar can either be used as a single unit or combined in a series of underwater units to provide coverage over a much larger area. The system may be integrated with other sensors such as radar, IR camera, motion sensors, and so on, to provide a complete security system.

The system can be used in a number of applications including: harbour surveillance; perimeter surveillance for naval bases or other coastal facilities; protection of offshore platforms or other high value assets; and to deny access to strategic waterways.

The sonar comprises an underwater unit (with transducer and transmitter/receiver subassemblies), sea cable (2,000 or 3,000 m optional), and a shore-based control/display formatter unit, operator unit and power control unit. It provides surveillance coverage of 360° with sector operation for restricted channel or cluttered archipelago areas. Electronic tilt (±24° from the horizontal) provides surveillance coverage of the entire water column in deep water. The system provides automatic detection and tracking of new and moving targets at varying depths. Target history track is provided for all targets on the display. A constant update of target range, bearing, depth, speed and course is also provided as a digital readout on the screen. To aid classification the sonar incorporates an operator-activated zoom capability. The system uses adaptive processing techniques to suppress noise and provide a clear image of targets in the operating area. Stable static background targets within the sonar's area of coverage are suppressed allowing changing target echoes to be easily identified. The non-fading display presentation is in eight colours, each colour representing a certain echo strength. An alphanumeric display presents key sonar operating parameters, as well as time and date. The system can be optimised to suit the specific conditions of any particular site.

Operational status
In operation with the Swedish Navy.

Contractor
C-Tech Ltd, Cornwall, Ontario.

UPDATED

CSS-80AS sonar

Type
Active seabed or platform-mounted sonar.

Description
The CSS-80AS OMNI surveillance sonar is designed for use in very shallow coastal waters for applications such as harbour surveillance, perimeter surveillance for coastal facilities and coastal platforms.

The sonar comprises an underwater unit, a 2,000 m long sea cable, a transmit/receive unit and an operator unit with display.

The sonar provides 360° surveillance coverage with a 3° vertical beam. Electronic tilt may be selected ±24° from the horizontal. An automatic tilt step mode is also available.

Automatic target detection and tracking is provided with automatic update of bearing, range and depth. An automatic alarm indicates when a new target is detected. Adaptive processing techniques suppress static targets and reduce noise interference, enabling new and moving targets to be easily detected. The source level is 211 dB/µPa/m.

Operational status
In use by the Swedish Navy.

Contractor
C-Tech Ltd, Cornwall, Ontario.

UPDATED

FINLAND

Sonac PFA

Type
Passive static fixed-sonar array.

Description
Moored on the seabed, the passive Sonac PFA can detect surface vessels as well as underwater targets in restricted waterways, such as straits, estuaries, harbours and so on. The system can also be used for surveillance in a large water area. The requirement to identify and locate the target can be fulfilled by using two or more arrays. These are connected by an underwater fibre optic cable to one operating unit on the shore.

The received signals are digitised in the array and transmitted to the processing system via the fibre optic cable in digital form, which allows the arrays to be installed far away from the shore. The shore-based equipment incorporates a display unit in one cabinet and a signal conditioning unit housed in another cabinet. The processing system, which is organised in a parallel multiplexed configuration, provides high computing capacity in order to utilise adaptive algorithms. The results are presented on four independent colour video monitors by means of clear waterfall displays. Sonac PFA provides the operator with easy access to all essential information. The five main functions are:
(1) broadband surveillance information; signals from all directions are presented showing both source direction and signal power
(2) narrowband surveillance information; spectral estimates of signals from different directions are presented
(3) more detailed information on targets being tracked is analysed and displayed
(4) tactical display monitoring, target movement and surface traffic situation on the map.
(5) audio information from a selected direction is generated for headphone and/or for classification purposes
(6) alarm and replay of unexpected events, transients, and so on is performed.

Operational status
In service with the Finnish Navy.

Specifications
Total length of array: 45 m
Operating depth: 600 m (max)
Length of optical cable: 40 km (max)

Contractor
Patria Finavitec Oy, Tampere.

VERIFIED

FHS-2000

Type
Hydroacoustic surveillance system.

Development
The shallow water and many islands which comprise much of the coastal areas of the Baltic Sea present an acoustic environment which provides a severe test for any ASW system. The frequency domain hydroacoustic surveillance system (FHS-2000) has been developed specifically to operate reliably in these waters. The first underwater surveillance systems were installed in Baltic waters in 1984. Since then over 100 systems have been delivered.

Description
Optimised for operation in water from 20 to 300 m depth, the FHS-2000 system uses arrays of passive sensors to detect and track surface and subsurface targets. A single sensor unit comprises a triangular array of three hydrophones, which form an equilateral triangle with sides of 2 m. Each sensor unit is connected by cable to a shore-based processing station where the data is processed together with that obtained from other sensor units in the sensor field. The cable length between the sensor units and the processing station can be up to 30 to 40 km without repeaters. The sensor array responds to the acoustic spectrum from 0.5 Hz to 100 kHz. In addition to the tracking of continuous radiators, special provision is made for the detection of transients and high-frequency active sonar transmissions.

The use of triangular hydrophone array sensors; the transmission of analogue sensor data; a unique frequency domain approach to target bearing

A typical chart display of FHS-2000 (Courtesy of Elesco Oy) *1999*/0024713

determination; the use of commercially available hardware where possible; and the easy location and deployment of sensor units offers a high-performance system which is easy to expand and upgrade.

The processing station simply requires a normal office environment to house its processing equipment. The hardware consists of a Pentium PC as the main processor, with a dedicated PC for DSP-based spectrum analysis. Target bearing analysis is performed by purpose designed microprocessor systems operating in parallel on the outputs of each sensor unit. Peripheral hardware includes the Frequency Shift Keying (FSK) data transmission system between each sensor unit and the processor, audio monitoring and recording facilities, a printer/plotter, and a modem connection to other workstations.

A PC-hosted DSP system processes the information required to generate the Lofar displays and spectrum analysis of targets. Data is transferred by LAN from this subsystem to the main Pentium-based processor. The system software uses the SCO/open desktop UNIX operating system with X-Windows graphical user interface. Two high-resolution 20 in displays per workstation are standard. These normally show a chart display of the sensor area showing sensors, current targets, and their track histories on one, and a set of Lofar and bearing analysis displays on the other.

Several enhancements for data processing and MMI have/are being made.

Operational status
The delivered systems are in operation.

Contractor
Elesco Oy, Espoo.

UPDATED

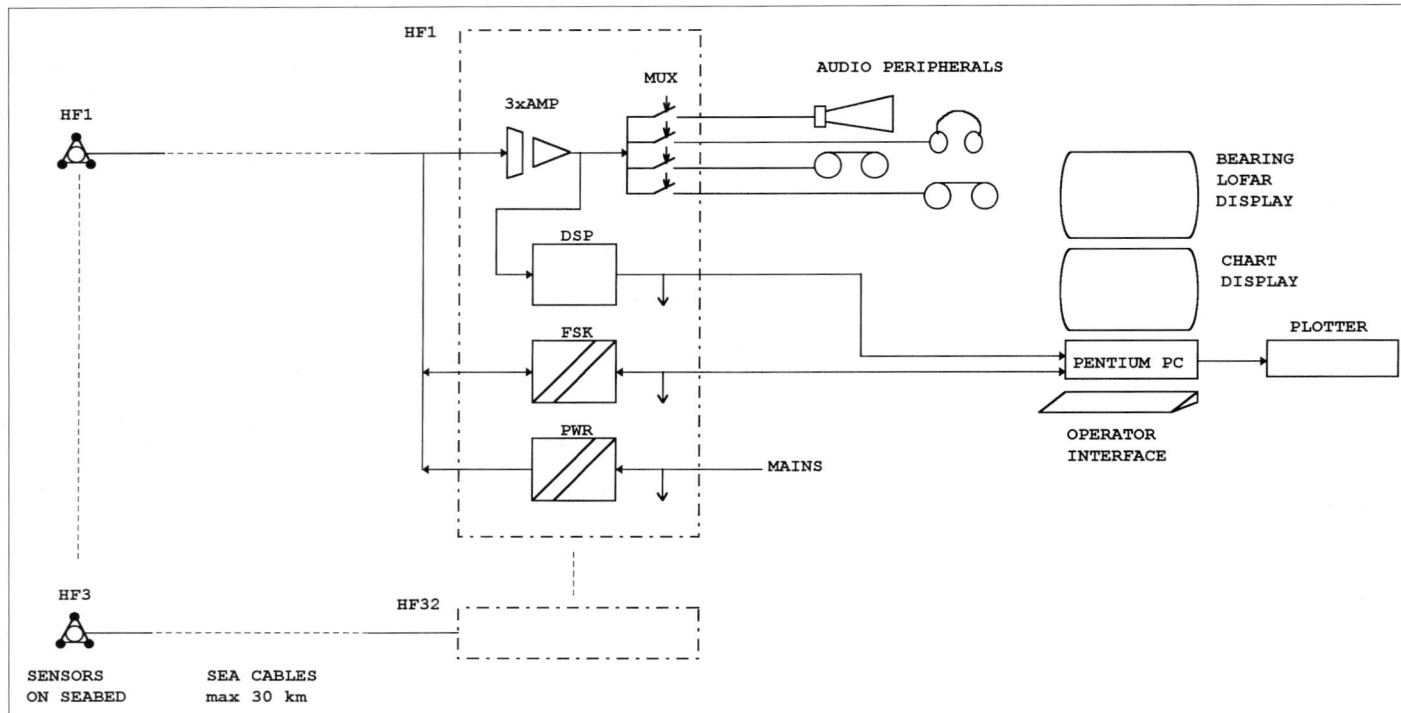

Schematic overview of the FHS-2000 hydroacoustic surveillance system

GERMANY

Hydroacoustic Coastal Range

Type
System for detection, collection and analysis of underwater sounds and signals by means of fixed arrays in defined coastal or sea areas connected to relevant shore stations, surveillance of coastal and sea areas as well as choke points.

Description
The Hydroacoustic Coastal Range has been developed and manufactured to enable naval forces not only to control the surface and airspace of a certain shore or sea area but also to detect and classify/identify subsurface targets to provide continuous underwater surveillance.

For this task the subsurface system performs:
(1) continuous control of the entire hydroacoustic frequency spectrum, in order to detect acoustic

phenomena and characteristic sounds and signals of passing vessels and objects
(2) recording, displaying, classification/identification of the received sounds and signals; analysis and storage of data in order to steadily improve the capabilities of the system; correlation of data with information gained from other sources
(3) transmission of received and evaluated data to:
    a) intelligence centres for further evaluation and as input for establishment of databases
    b) operational authorities for tactical assessment and early warning.

The results gained by the Hydroacoustic Coastal Range improve with increasing significance the vital support of: surveillance and control of specific sea and coastal areas in any aspect and application; data collection for ASW and mine warfare (target information); training of sonar operators; and early warning.

Passive sensors are placed at suitable distances in the area to be surveyed. The arrays have directional sensitivity to allow directional resolution, rangefinding and tracking of individual targets.

If active sensors are integrated they can guarantee independence from the noise of the target in extreme situations.

Manning of the system is kept to lowest requirements.

Operational status
In operational service.

Contractor
STN ATLAS Elektronik GmbH, Bremen.

VERIFIED

ITALY

GUARD D101

Type
Underwater surveillance and reaction system.

Description
The GUARD D101 (Global Underwater Alert Reaction and Dissuasion) system is designed for underwater surveillance and defence of maritime borders, littoral

sea routes and strategic maritime areas, especially when located in disputed waters. In combination the system provides a wide range of capabilities from surveillance barrier to non-lethal dissuasion and close-/ long-range underwater reaction systems, offering a cost-effective tool to counter the silent threat posed by covert, low-speed minelaying midget submarines, chariots and underwater intruders such as frogmen.

The system can be used in a number of different

applications including: harbour surveillance and defence; oil rig surveillance and defence; point surveillance and defence; area surveillance and defence; surveillance barriers; controlled access; and the surveillance of strategic straits.

The system comprises: a control console; a unit to manage automatic underwater surveillance and reaction control; active/passive underwater sensor units to provide omnidirectional and/or sectorial coverage; automatic alert system; dissuasion system; inner layer defence capability; short-range reaction system; mid-layer defence capability; long-range reaction system; and outer layer defence system.

The system's operational phases are: detection and tracking with time-space coherence within a sector; TMA providing amplitude and bearing, speed assessment; alert capability in which the system switches to active mode with automatic sectorial scanning and target tracking; a non-lethal reaction decided by the console operator which offers a deterrent effect through the use of non-lethal devices; a short-range reaction capability involving the deployment of lightweight depth charges with an effective range of 80 m in the proximity of the target; and a long-range reaction capability involving the firing of A202 mini-torpedoes with an effective range up to 2,000 m against the target by the Medusa units.

Contractor
Whitehead Alenia Sistemi Subacquei SpA, Genoa.

VERIFIED

RUSSIAN FEDERATION AND ASSOCIATED STATES (CIS)

Fixed sonar systems

Type
Static sonar surveillance systems.

Description
The RFAS has developed an extensive number of fixed and moored sound surveillance systems which have been deployed at a number of locations around the world. Very little information is available on this sphere of operations.

It is known that the former USSR has deployed Cluster Lance planar acoustic arrays in Pacific waters near the Russian landmass where broad area surveillance is required. Barrier arrays have been deployed at points of egress and ingress from SSBN operating areas, such as in the Barents, Greenland and Kara Seas. These are placed in or near trenches at 'choke points' which would serve as 'tripwires' more than long-range surveillance systems. One system developed by the former Soviet Union and now available on the open market is Searchlight. This is a bistatic system with a range of only a few hundred kilometres.

The USA considers that acoustic sensors have considerably increased their detection ranges and will include improved hydrophone arrays and acoustic processing capability. There is even a suggestion that long-range, low-frequency arrays could be mounted on the permanent Arctic ice pack.

The former Soviet Union is also believed to have been developing non-acoustic surface signature detection systems which sense the disturbance on the surface caused by a passing submarine. Wake detection systems that sense turbulent wake, internal wave wake or contaminant (chemical or radioactive) wake have also been under development.

VERIFIED

UNITED KINGDOM

Sea Guardian

Type
Static or mobile surveillance and swimmer detection sonar system.

Description
The Thomson Marconi Sonar Sea Guardian is designed

Sea Guardian Swimmer Detection Sonar and deployment system

to counter the threat of terrorism and sabotage, offering comprehensive protection to naval bases, commercial ports and other economically vital assets such as offshore oil platforms, oil refineries and nuclear power stations. It can be supplied either as a fixed system or in a mobile form.

The system will continuously detect, classify and track intruders at ranges that allow successful interception. It is fully automatic. When a target is detected a series of detection, classification and tracking algorithms are initiated before an alarm is activated. This level of interaction with the possible intruder provides very low false alarm rates. The auto-detection process is programmable to meet the requirements of target threat versus environmental conditions. The special characteristics, or signatures, of swimmers and small vehicles ensure that they can be accurately tracked while similar sized harmless debris is eliminated electronically.

Having confirmed the presence of an intruder, the system will predict its track and estimate the time of interception. Using the sonar beam, a police launch or similar craft fitted with a transponder can be directed to a position immediately above the intruder.

Due to the high degree of automation, it is possible for an unskilled person to interrogate the system and, by means of the tactical display, to identify quickly the type of intrusion, its range, speed and bearing.

Sea Guardian uses a mirror array as the sensor. The array provides near perfect beam-forming with a minimal number of components, in a robust, low-cost configuration. The fixed installation consists of a 120° sonar unit, giving an azimuth resolution of 1.3° over the total field of view. The trainable mobile installation has a 40° azimuthal field of view. The electronically scanned sonar heads have no moving parts, thus ensuring high reliability and ease of maintenance.

The company also offers complete intruder detection system packages comprising radar, IR sensors, acoustic fences and physical barriers. Used in combination, these systems offer protection against intruders from up to 30 miles away right up to the inner harbour zone.

A flexible communications system with facilities for point-to-point, broadcast and group communications is available at all displays. Interfaced external communications provide up-to-date data for naval forces and harbour authorities. Full command and control facilities allow optimum co-ordination and prevent any possibility of undetected escape or reinforcement.

Operational status
Sea Guardian has successfully undergone extensive proving trials in a variety of demanding environmental conditions worldwide.

Contractor
Thomson Marconi Sonar Ltd, Templecombe.

VERIFIED

Sea Witness

Type
High-resolution ultrasonic holographic imager.

Description
Sea Witness is a family of high-frequency imagers utilising an orthogonal implementation of acoustic mirror sonars, similar to the Mills Cross. The acoustic system outperforms an optical system because it fully operates in total darkness and highly turbid conditions, and allows the operator solid three-dimensional (3-D)

presentation of the object, plus a level of detail of its internal structure. The mirror technology is based upon experience gained in operational minehunting sonars.

The system comprises an outboard module with an acoustic transmit mirror assembly, a receive mirror assembly and associated electronics. The design of the sonar gives low sidelobes so reducing the image clutter. Inboard processing uses data visualisation techniques to convert the raw sonar data into recognisable solid 3-D images.

The mirror technology allows a family of sonars to be configured for different ranges, resolution and sizes, for

installation on ships, harbour walls, ROVs or for diver operations.

Operational status
In development. Successful prototype trials have been completed.

Contractor
Thomson Marconi Sonar Ltd, Templecombe.

VERIFIED

TG-1 HF sonar system

Type
HF static surveillance sonar system.

Description
Thomson Marconi Sonar Ltd, has designed, developed and produced a new high-frequency sonar system which offers electronic scanning at a cost which, the company states, makes it competitive with mechanically scanned equivalents. It has been given a provisional nomenclature of TG-1. The equipment, which is understood to have a range of some 700 m, is built to withstand continuous operation down to depths of 500 m. Targets are identified in both azimuth and depth.

The modularity of the system makes it suitable for a diversity of applications. These include harbour surveillance, offshore platform defence, mine

identification and classification, ROV navigation, and oil/gas pipeline profiling.

Typically a single unit comprises two 64 element transducer arrays each connected to its own underwater pod, and using a digital beam-forming technique. An umbilical cable provides both power and datalink to the pods from the surface control unit. A standard VDU presents high-resolution data which include the bearing, range and depth of sonar contacts. An auto-alarm system causes an audible alarm to sound when a possible sonar return is received.

In more general surveillance applications up to 255 pods can be linked to form a system. The control system then acts as a master and accesses each of the slave pods in turn.

Operational status
The equipment has been tested successfully.

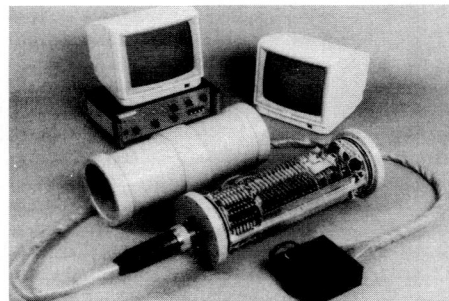

Units of Thomson Marconi TG-1 sonar system

Contractor
Thomson Marconi Sonar Ltd, Templecombe.

VERIFIED

Sea Shield

Type
Diver detection system.

Description
Sea Shield is designed to protect ships and waterside installations from underwater incursion through the active detection of diver targets. The system forms a protective ring around the potential target by deploying fast-scan sonar sensors on the seabed. Any intruder is quickly monitored and an accurate calculation of range and bearing is displayed on a screen. In an extensive research and development programme, using both closed-circuit and open circuit breathing apparatus, Sea Shield was able to plot several diver targets at the same time.

The detection system is based around a dual-frequency sonar which consists of two mechanically scanned imaging sonars in a single subsea pressure housing. The first is a 325 kHz sonar with a true operational range in excess of 250 m for long range target acquisition and the second ranges between 625 to 975 kHz as an ultra high-definition micro stepping imaging system. Other frequency options are available upon request. Each module has a scan rate of up to 180

per second, depending on the range and selected resolution.

Using digital signal processing technology Sea Shield produces subsea images of great clarity, which go a long way towards crossing the divide between underwater cameras and traditional sonar systems.

The unit is controlled by a multitasking sonar processor which can control up to 250 dual-frequency sonars. Using a personal computer, Sea Shield is very simple to operate and can also control and display information from other equipment such as video cameras, pipetrackers and survey computers.

Each device connected to the processor runs in real time in its own screen window. The display on the monitor may comprise multiple windows showing, for example, sonar, video and profiler with pipetracker. Alternatively, any one of these may be expanded to full-screen at the touch of a button. An access terminal enables control to be exercised over the various devices connected to the system. The access terminal may be detached from the main processor allowing it to be either hand-held or desk mounted. A built-in trackball allows for quick and accurate range and bearing calculations to be made.

All sonar and profiler data may be logged onto the built-in disk drive for replay and post-mission analysis.

Also linked with Sea Shield is Sea Sentry, a high-powered diver deterrent which can prevent intruders approaching any ship or underwater installation. The unit, which operates on a combination of frequencies and modulation, is designed to cause distress and disorientation to an intruding diver. The signal produced by the unit provides an acoustic barrier or exclusion zone around any underwater installation or platform.

Two high-powered, low-frequency sources combine to provide a high acoustic emission. Research and development, using ex-navy divers, both tested and proved the unique modulation scheme applied to these high-power transmitters.

The surface unit consists of two 1 kW power amplifiers mounted in 19 in enclosures. These rugged units may be fitted onto a ship or shore installation and can be powered from a convenient AC power supply. Any incursion monitored on the Sea Shield screen will trigger an immediate response.

Contractor
Intech Corporation Ltd, West Molesey.

VERIFIED

UNITED STATES OF AMERICA

IUSS (Integrated Undersea Surveillance System)

Type
Large area, ocean basin ASW detection and reporting system.

Description
The US Navy's IUSS programme is of high priority. The system uses much COTS software to provide basic building blocks based on open system architecture.

The IUSS comprises a number of sensor systems including SOSUS, SURTASS, FDS (Fixed Distributed

System), SDS (Surveillance Direction System) and the ADS (Advanced Deployable System) involving a number of major contractors.

(See entries on individual elements of IUSS below).

UPDATED

Fixed Distributed System (FDS)

Type
Static passive underwater surveillance systems.

Development
The FDS is being developed in response to the rapidly changing submarine threat environment and is designed to detect and track quiet threat targets as well as surface targets in the deep ocean and littoral

environment. The system features advanced signal and information processing, operator focused operational concept and modular distributed architecture. The system will significantly reduce operator workload.

Description
This system is the first in the Integrated Undersea Surveillance System (IUSS) programme that employs fibre optic cable and data transmission technology. The FDS system has orders of magnitude greater data

transmission capabilities, and on a per channel basis is much more cost effective than previous systems. In addition, FDS detection capabilities are far superior to current operational systems and intended to operate in either a stand-alone configuration or in conjunction with the existing SOSUS and SURTASS systems (see separate entries).

FDS is designed to detect and track extremely quiet SSNs and modern SSKs operating in deep and shallow water. The shore-based element SSIPS (Shore Signal

Information Processing Segment) features a flexible operational concept with open architecture to commercial industry standards based around a high-performance fibre optic LAN.

Operational status
The programme, which is managed by the US Navy's Space and Naval Warfare Systems Command, is currently deployed at four sites. In addition to the larger EDM system, a smaller rapidly deployable version (FDS-D) was manufactured and deployed in 1994.

FDS-D successfully demonstrated rapid deployment capability, as well as successfully proving its ability to perform its mission in a shallow water, high noise environment.

General Dynamics Advanced Technology Systems is responsible for development, manufacture, deployment and support for the underwater hardware part of the system while Lockheed Martin/BBN is responsible for the development, installation and support for the shore signal and information processing part.

Contractors
General Dynamics Advanced Technology Systems, Arlington, Virginia.
Lockheed Martin Federal Systems, Manassas, Virginia.
BBN Systems and Technologies, Cambridge, Massachusetts.

UPDATED

ADS (Advanced Deployable System)

Type
Subsystem of IUSS.

Description
The ADS comprises a family of wet end hardware and

dry end hardware/software subsystems, designed to meet a broad range of short- and long-term IUSS mission requirements. The system provides detection, localisation and tracking capabilities designed to deal with both nuclear and diesel-electric submarine threats in potentially hostile, near shore areas in adverse conditions of shallow water and strong currents.

Contractors
Lockheed Martin Federal Systems, Manassas, Virginia.
Raytheon Naval & Maritime Systems, Mukilteo, Washington.

VERIFIED

SDS (Surveillance Direction System)

Type
Command, control and communications (C3) system element of IUSS.

Description
The SDS will provide the Integrated Undersea Surveillance System (IUSS) with advanced C3 and intelligence functions, with a command and control toolset that ensures that all IUSS sensors (FDS, SOSUS and SURTASS) are optimised for threat detection and reporting, as well as a solid communications suite for rapidly reporting threat targets to strategic authorities and tactical forces.

Lockheed Martin is developing and integrating SDS Systems using commercial technology as well as application software in Ada, C, and C++.

The system features: a number of powerful workstations which interact equally with acoustic, geographic and textual message data; superior data processing technology for correlating, fusing and localising contacts between the sensors; improved intersite communications for transfer of data using commercial bridge-routeing systems and wide area networking; and robust external communications connectivity for distribution to strategic and tactical users.

The system will provide the cornerstone for the ASW portion of the US Navy's C3 architecture (COPERNICUS) which uses the same open system

architecture and transparent user-pull communications nodes as SDS.

Operational status
The programme, which is managed by the US Navy's Space and Naval Warfare Systems Command, is well into the manufacturing and deployment phase. The SDS is currently deployed with the FDS (see above) at four operational sites.

Contractor
Lockheed Martin Federal Systems, Manassas, Virginia.

UPDATED

Pilot Fish

Type
Static seabed sonar system.

Description
Pilot Fish is understood to be an underwater detection and position locating system in development, which

involves the use of a transmitter placed on the ocean bottom to send data to submarines. The transmitter would have sufficient self-contained power to transmit sonar signals. The echoes would be reflected from a hostile submarine, and would reach a listening submarine which would be far enough away so that echoes rebounding off it would not in turn be strong enough to disclose anything to the enemy.

Operational status
The programme is highly classified and no official information has been released. It is understood, however, that the programme remains a high-priority item for the US Navy.

VERIFIED

SOSUS

Type
Fixed sound surveillance system.

Development
The current US philosophy regarding defence against the principle ASW threat (hostile submarines) can be summarised by relying on engaging these submarines in forward areas and at barriers before they can get within attacking range of US forces. To do this, reliance rests primarily on American SSNs and long-range P3 patrol aircraft supported by underwater surveillance systems.

The ability to locate enemy submarines within broad ocean areas is essential to the task of containing the large hostile submarine force. The Sound Surveillance System (SOSUS) which is now part of the Integrated Undersea Surveillance System (IUSS) has played a major role in this effort for several years. This 40 year old programme has been recognised as nationally important by the Joint Chiefs of Staff.

Description
SOSUS consists of fixed undersea acoustic networks of passive hydrophone arrays deployed in the Atlantic and Pacific oceans to provide significant detection capabilities. SOSUS functions as an operational network of shore sites that collect, analyse, display and report acoustic data. Strings of hydrophones, preferably laid at the depths of the deep sound channel are located at intervals along a linking cable connected

to the shore stations. Submarines were detected using narrow band frequency analysis, using the same LF acquisition and ranging processors common to modern sonobuoys. Over the years the network has evolved as increments were added to expand coverage. In 1992, the US Navy embarked on a SOSUS Consolidation Programme to address the evolving threat, and take advantage of advances in technology. The specific details of this programme remain classified; however, it is known that the improvements include workstation display and communication elements and complementary improvements to the signal and data processing elements, thereby facilitating greater operational utilisation of the data obtained by the undersea array sensors.

With the collapse of Soviet Russia and declining budgets, the SOSUS Consolidation Programme also focused on reducing manning requirements by consolidating processing facilities through the evolution of technology and utilising the SOSUS system assets for Dual Uses programmes that included whale monitoring and detection. The whale detection system, called the Marine Mammal Acoustic Tracking System (MMATS) centres around a hydrophone monitoring system that automatically identifies calls from several whale species, correlates signals between hydrophones and produces positions and tracks. The Dual Uses programme also makes the considerable assets of SOSUS available to the scientific community for other purposes, to include: acoustic thermometry of ocean climate to study global warming; seismic

monitoring; and acoustic tomography for mapping the ocean floor.

To reduce operational costs, more processing and analysis are being accomplished by fewer personnel and with greater utilisation of COTS-based computer systems. In addition, the number of shore facilities and personnel required to support those facilities have been significantly reduced by remoting the signal processing function to distant facilities. As a result, the station forces have reduced staffing to approximately one half the 1992 level. The utilisation of COTS equipment has evolved from about 1 per cent before the SOSUS Consolidation Programme to 60 per cent after its implementation.

Operational status
The SOSUS network is fully operational and continues to play a key role in Integrated Undersea Surveillance.

Contractors
General Dynamics Advanced Technology Systems, Greensboro, North Carolina (research, engineering and integration).
TRW, McLean, Virginia (engineering and technical assistance).
Simplex Wire and Cable, Portsmouth, New Hampshire (coaxial cable).
Lockheed Martin Federal Systems, Manassas, Virginia (signal processing).

VERIFIED

Swimmer detection system

Type
Rapidly deployable active acoustic sonar barrier.

Development
The SAIC active acoustic sonar barrier was specifically designed for the protection of valuable commercial and military assets in shallow water and port areas. This barrier is ideal for harbour security and is intended for use against a variety of target types and sizes including scuba divers, free swimmers, Swimmer Delivery Vehicles (SDVs) and mini-submarines. Lightweight active acoustic elements attached to a cable are arranged along the ocean bottom to form a 'tripwire' zone. The upward-looking active elements form a 'fence' resulting in the display of a target in real time as it moves through the barrier (see photo of the PC display and the transducer node). This simply designed PC-based system is made up of readily available commercial-off-the-shelf (COTS) components which make maintenance and parts replacement quick and easy. Overall system cost is well below that of a fixed conventional system. In addition, the ability to deploy and retrieve this system in a matter of minutes is ideal for the unpredictable and ever changing operational scenario of modern limited warfare and counter-terrorist situations.

Originally conceived as an internal research and

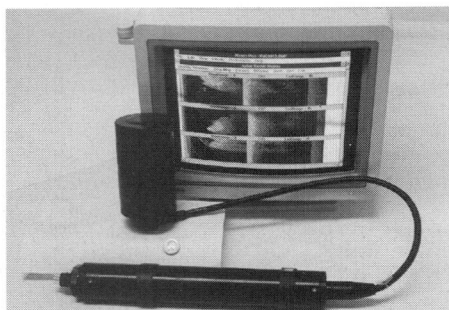

Swimmer detection system   *1996*

development project in 1990, the active acoustic barrier was first demonstrated in the waters off San Diego in July 1991. In this instance, the system showed excellent detection capability by picking up divers crossing the barrier in 30 ft of water. After another successful demonstration in January 1992, the system was modified to handle greater water depths. During the subsequent July 1993 test, the system successfully detected divers in 60 ft of water with an increased detection range. Maximum water depths of 100 m are practical.

Operational status
The demonstration system consists of three upward-looking active transducer nodes, 1,000 ft of underwater cable, and a shore-based PC-based signal processing and display system. An operator monitoring the colour

waterfall display can clearly see divers or other targets passing through the barrier as brightly coloured swaths against a blue background. Enhancements to the system's target detection and classification algorithms and efforts to add more active nodes to increase the barrier's length (up to 2 km) are continuing. The enhanced design is being considered by the US Navy, US Customs Service, and the Defense Special Weapons Agency.

Contractor
Science Applications International Corporation (SAIC), Ocean and Remote Systems Division, San Diego, California.

VERIFIED

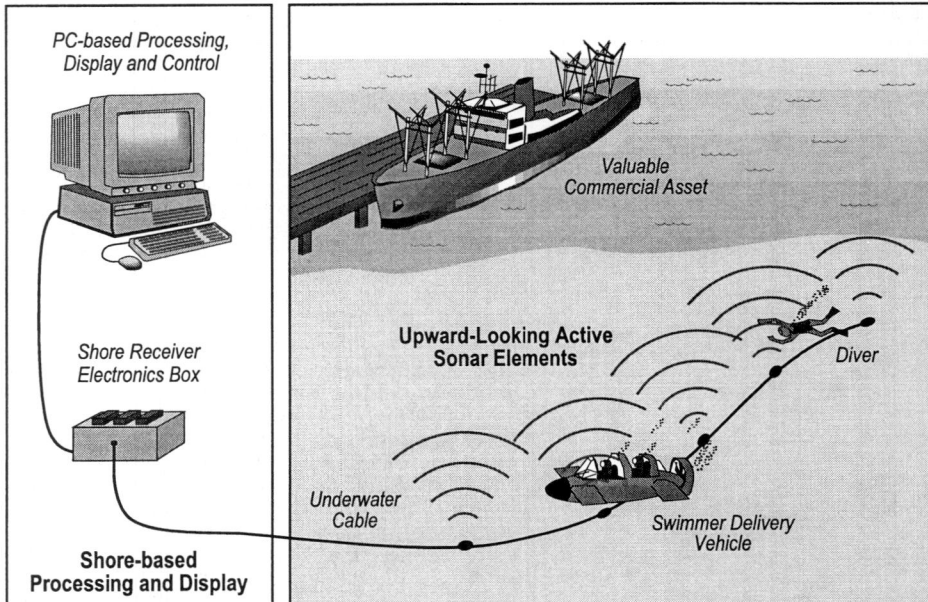

Swimmer detection system (rapidly deployable active acoustic sonar barrier)   *1995*

Underwater Sentry

Type
Rapidly deployable passive acoustic boat detector.

Development
The Underwater Sentry is designed to detect and track small vessels entering a restricted area or heading in a suspicious direction on lakes, rivers or coastal waterways. The area under surveillance, the boundary line, and system hours of operation are first preset by the user. Once the system is set up, it will automatically provide an RF-transmitted alert or visual/audio indications when a boat is detected crossing into the restricted area. This covert, rapidly deployable underwater system is designed to operate autonomously in the field for several weeks at a time. An additional option is the provision of a bearing versus time track that can be displayed on a PC or laptop computer, providing the user with a real-time observation tool or monitoring capability. The key features of this system are: autonomous operation – no operator required; auto alert; position tracking capability on a 2-dimensional (2-D) geoplot (using two systems); special digital processor-based signal

Underwater Sentry small boat detection system
1996

processing to eliminate false alarms; and rapid deployment capability (within 30 minutes).

Operational status
The current demonstration system consists of two

passive hydrophones, 400 m of underwater cable, a shore-based signal processing module and a battery pack. The cable and hydrophones are wound onto a portable cable reel for easy, rapid deployment and recovery. The signal processing modules and battery pack are housed in an all-weather environmentally sealed container that can be buried or concealed for covert operation. This basic demonstration system can be adapted or configured to specific customer needs. US and foreign law enforcement agencies have shown interest in the application of this system to detect small boats engaged in illegal smuggling or poaching activities along international border waterways and in marine sanctuaries.

The system has been demonstrated operationally with the US Border Patrol.

Contractor
Science Applications International Corporation (SAIC), Ocean and Remote Systems Division, San Diego, California.

UPDATED

CSDS (Coastal Surveillance Display System)

Type
Real-time, multisensor correlation and graphical display system.

Description
The CSDS was developed to allow operators to monitor surface and subsurface contacts in shallow coastal waters. The system displays large quantities of sensor data from a variety of sensors including radar, sonar,

visual and thermal imaging sensor systems using easily recognised symbols and formats overlaid on maps of the operational area. The operator also has the capability to control and calibrate multiple sensor systems in the configuration. In addition, the system provides automatic sensor input recording, replay and permanent archive capabilities, as well as a video camera image capture feature. The CSDS can track as many as 500 active targets simultaneously. Potential applications for the CSDS include coastal area surveillance, use as a vessel tracking management system, near shore shallow water surveillance, port and

harbour security, and offshore environmental monitoring. The software and hardware can be customised to meet specific requirements.

The system typically comprises a number of COTS elements including: Hewlett-Packard J210 workstation; 20 in high-resolution colour monitor; CD-ROM and digital audio tape drive; radar track processor; control/ video processor; interface units (radar, video and thermal imaging, underwater acoustic sensor systems).

Operational status
Customised system on order for a foreign navy for

harbour surveillance applications. US and foreign government agencies have shown interest in both the coastal water and land-based surveillance configurations of the system. The US Navy's Mobile Inshore Undersea Warfare (MIUW) units incorporate a Graphical Data Fusion System (GDFS) (a forerunner of the CSDS) developed jointly by SAIC and CEA Pty Ltd.

Operational with the US Navy under the description GDFS.

Contractors

Science Applications International Corporation (SAIC), Ocean and Remote Systems Division, San Diego, California.

CEA Pty Ltd, Canberra, Australia.

UPDATED

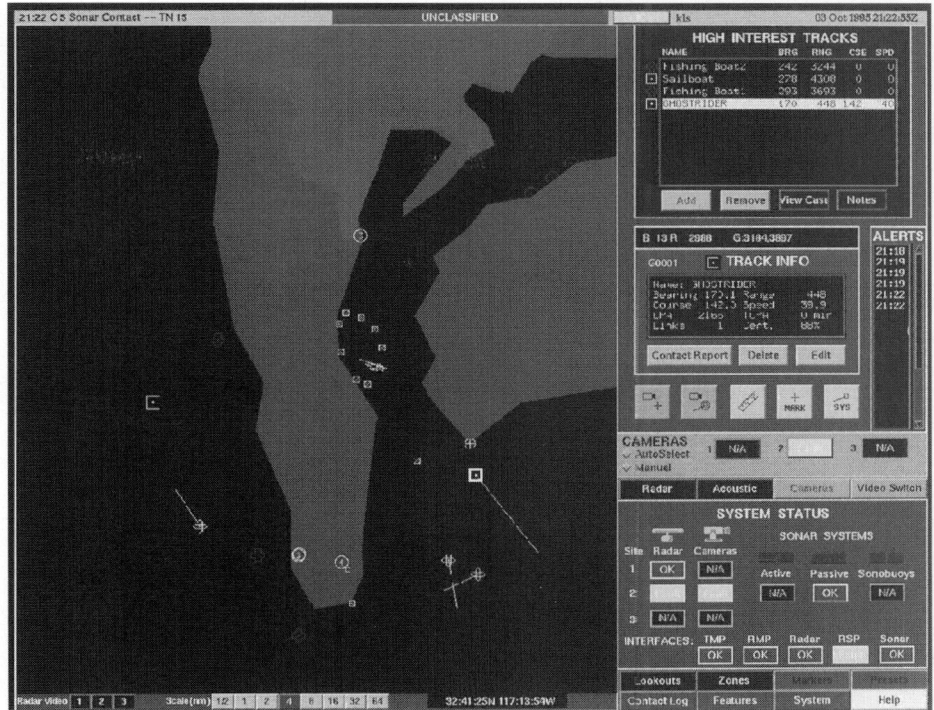

Basic CSDS screen
1997

Undersea Coastal Surveillance System (UCSS) Sea Sentinel

Type

Surface and undersea coastal detection and tracking system.

Description

The Undersea Coastal Surveillance System (UCSS) Sea Sentinel is an undersea detection tracking system which uses ruggedised acoustic sensors on ocean floors to perform uninterrupted surveillance and detection of surface or submerged intruders. Potential targets for the system include: surface and subsurface vessels; patrol boats; mining activity; illegal fishing; illegal immigration; terrorists (frogmen); or pirates.

The UCSS quickly and accurately identifies and locates threats, without the need for repeated aircraft or surface vessel patrols or traditional aircraft, surface vessel, or sonobuoy methods. The surveillance system's shore-based command centre operates on a modular, open architecture system, incorporating a single or multiring workstation configuration interconnected by industry standard fibre optic LAN that can be tailored to meet specific surveillance requirements. The command centre includes acoustic, geographic, and system control and maintenance displays. Acoustic displays present visual images of sound energy detected by underwater sensors. Geographic displays show contacts on built-in world and national maps. The UCSS system control and maintenance display enables operational commanders to configure the system displays and workstations for specific tactical situations and continually monitor performance of the system. Monitoring is carried out via individual displays or on large screen format. The system is built entirely from COTS hardware. Software is based on modern object oriented programming techniques. Virtually all operator interface programming is done in C++. Each object is individually configurable.

The UCSS provides more efficient use of existing tactical systems, including ships and aircraft, by timely alert and accurate localisation of threats. UCSS can be implemented for a small fraction of the cost of traditional aircraft, surface vessels or sonobuoy

UCSS Command Centre

methods. The underwater components sensors by Raytheon Naval and Maritime Systems comprise hydrophones with internal electronics telemetry components mounted in watertight housings. Sensors are connected to the shore station via fibre optic cable. General Dynamics Advanced Technology Systems provides for long haul telemetry multiplexing and power distribution.

Advanced spectrum analysis converts signals into frequency components making detection and classification more precise. Adaptive beam-forming reduces interfering sources of noise. Workstations use the same hardware as the signal processing computer. Logistics support and spares are compatible for low cost and ease of maintenance, while COTS hardware and open system architecture reduce installation and maintenance costs.

Contractors

Lockheed Martin Federal Systems, Manassas, Virginia, (shore processing and systems integration).

Raytheon Naval and Maritime Systems, Mukilteo, Washington (underwater sensors and components).

General Dynamics Advanced Technology Systems, Greensboro, North Carolina (long haul telemetry and power).

Cable and Wireless (Marine), Southampton, UK (installation and maintenance of underwater cables and arrays).

UPDATED

Coastal Surveillance System

Type

Coastal and harbour surveillance system.

Description

The system is designed to provide surveillance and

protection for inshore applications including coastlines, straits, harbours and approaches and choke points from intrusion by ships, small boats, submarines and swimmer delivery vehicles operating offshore. The relocatable system can be deployed on short notice for defence, law enforcement or intelligence gathering missions.

The system is deployed in vans for relocation, or in fixed-site installations using radar, acoustic and optical sensors for detection. All sensors are man-deployable in the field, and acoustic units can be fully deployed from a small 22 ft boat.

The Coastal Surveillance System uses commercially available components which allow for easy upgrading and expansion to provide systems customised to meet specific requirements. The radar, video and modular acoustic units offer optimum performance by allowing site-specific system configurations.

The sonar subsystem consists of commercially available processors and displays integrated in a modular open architecture that makes use of industry standard interfaces such as VME, SCSI, Ethernet, RS-232, RS-343, RGB and commercially available software such as X-Windows. The system is also UNIX/POSIX compliant. The underwater unit uses passive acoustic sensors such as sonobuoys, arrays and omnidirectional string barriers to acoustically detect and track surface and subsurface targets. Sonar contacts are correlated with sonar data and overlaid on site-specific geographic maps to simplify the surveillance assessment. The modular string sensor design allows the customer to site the passive sensors in the optimum configuration dependent on the geographic requirements of the area to be placed under surveillance. The passive string sensors are powered from the van and use fibre optic telemetry for data transmission from sensor to shore.

The radar subsystem comprises one or more commercially available surface search radars modified to improve detection and tracking performance. This subsystem detects and tracks a large number of surface contacts, which it overlays on site-specific geographic maps. This allows the operator to monitor surface vessel positions and movements with geographically referenced real-time tracking data. A radar warning receiver is also provided to gather additional information on surface and airborne emitters.

Command and control is achieved using a combination of voice and record datalinks. HF/VHF/UHF and SATCOM voice links are available in a variety of configurations to meet customer specific requirements.

An embedded training system provides multisensor team training, while acoustic sensor positional training is available as an option. Tactical situation data is readily available in a standard US format for evaluation by the appropriate national authorities.

In addition to the sensor specific displays a single tactical console is used to provide the operator with multisensor contact information.

Contractor

DRS Electronic Systems Inc, Gaithersburg, Maryland.

VERIFIED

COUNTERMEASURES

The threats facing both submarines and surface ships are many and varied, but one that neither can afford to ignore at their peril is the torpedo. For surface ships in particular, the torpedo threat is very real. However, most navies have tended to ignore it and pay closer attention to the anti-ship missile. But while attack from the torpedo has not produced such spectacular results as the anti-ship missile during recent conflicts (apart from the sinking of the Argentine cruiser *Belgrano* in the South Atlantic War of 1982), the threat is, nevertheless, very real. It is in fact possibly more dangerous than the anti-ship missile in that, unlike the latter, there is little available in the way of countermeasures to counter the latest generation of torpedoes.

For both submarines and surface ships there has been little real progress in the development of anti-torpedo measures since the Second World War. For both, the main method of defence remains a towed or expendable noisemaker.

Modern torpedo defence systems comprise three main elements:
- Intercept sonar
- automated torpedo attack warning processor
- countermeasures

On board submarines, a dedicated sonar provides the intercept capability. On surface ships, torpedo warning is provided by a specific array which forms an adjunct to an existing sonar (these systems are covered in the section on sonars).

The torpedo warning processor is designed to carry out detection, classification, localisation and threat evaluation. In addition it provides tactical advice to the command team on the most effective evasive manoeuvre to conduct, linked with the tactical deployment of countermeasures.

The countermeasure stores themselves currently fall into two categories. Both are noisemakers, but one category is designed to decoy a hostile sonar as it searches for the submarine – either confusing the hostile torpedo's homing sensor as it seeks to find its target, or the parent platform's sonar if it is 'locked-on' to the submarine preparatory to firing a torpedo. Decoy noisemakers are designed to emulate the target's signature in such a way that the hostile sonar chooses the noisemaker as the preferred target on which to home, rather than the submarine.

The jammer is used when the torpedo's on board sensor has homed on to its target and the weapon is commencing its terminal homing phase. Exact sequence of deployment of the decoy or the jammer may obviously vary according to the tactical situation.

In addition, both types of decoy may be of the static or mobile type, and can be deployed in a mix – again the method of deployment depending on the tactical situation.

The primary method of launching expendable stores from a submarine is by the Submerged Signal Ejector (SSE) which is fitted inside the pressure hull, allowing the system to be replenished from the boat's internal stores. For very deep diving boats, such as strategic submarines, a launch method using an external stores

launcher is fitted. The main disadvantage with this system is that, once the store has been expended, the launcher cannot be reloaded until the boat returns to base. External launchers deploy their stores using explosive charges, which enables them to be used at greater depths and at greater speed than the SSE.

It is essential that, whatever method of deployment is used, the store is launched noiselessly and positioned correctly at all operating depths and through the full range of manoeuvres in which the submarine is likely to engage, while trying to evade torpedo attack. Launching a decoy from a moving platform poses enormous problems operationally, both for the store and for the platform. The primary concern with the store is that it is not damaged on launch. The store can easily be damaged as it leaves the launch tube, particularly if the submarine is carrying out a violent evasive manoeuvre at high speed. Hydrodynamic forces can also damage the store. Inert stores are particularly susceptible to such damage when they leave the boat under their own power. As they are ejected, they may be damaged by collision with the boat as it manoeuvres. The most delicate part of the store is the tail fin assembly that ensures that the store is maintained at the correct altitude and depth once launched. It is essential that this vital part is not damaged. Equally important is the need to protect the internal electronic assemblies of the store from damage that can be caused through mechanical or hydrodynamic forces as the store is launched.

Developments are currently under way to develop an effective hard kill system. Various concepts are under investigation such as electromagnetic guns and high-speed projectiles powered by chemical propellants, as well as mini-torpedoes.

Other countermeasures used to disguise a target's signature are covered in the section on signature management.

But what is really required for the future, for both submarines and surface vessels, is a torpedo defence system that is fully integrated into the platform's combat system and which, like the surface ship's anti-missile soft/hard kill countermeasures, can be almost entirely automatic in response. The difficulty in developing such a system, however, is that it must be highly sophisticated in order to be 100 per cent effective. If it is not, then the submarine is 'dead' and with it its highly trained crew. Few navies seem willing, even in collaborative ventures, to devote the necessary funds to developing such a system.

However, the entry into service of the latest generation of torpedoes, using highly sophisticated homing systems and embodying counter-countermeasures, is giving rise to increased efforts to enhance torpedo defence, and both soft and hard kill systems are currently being developed, albeit at a very slow pace.

The other main countermeasure fitted in the submarine is the ESM system. Its primary role is to detect radar emissions from surface and airborne platforms, identify them, and warn the boat if it is under threat from a hostile platform or weapon as it comes to

the surface. Using this data, together with other sensor data, enables the boat to rapidly draw up a tactical picture of the above water scenario.

The sensor detects emissions, measures their amplitude and analyses the parameters so that the transmission can be identified against a stored library of emitter characteristics. This is combined with an indication to the command that to raise a radar or other sensor mast above the surface would expose the submarine to detection.

Additionally identification of the potentially hostile emitter through signal analysis of its characteristics and comparison with the stored emitter library also provides alert for the initiation of self-protection measures.

In some submarines the ESM can also be used to provide passive targeting information for long-range weapons such as anti-ship torpedoes or missiles. For this function a highly selective antenna is required (for example a high gain reflector). Because this antenna is necessarily much larger than the standard ESM antenna (which is usually mounted on the search periscope) it has to be mounted on a separate mast which is only used for the short duration required to gather data for weapon deployment.

Finally the ESM can be used in the ELINT role providing high intercept capability, limited onboard display, analysis and recording functions so that gathered data can undergo detailed analysis and evaluation ashore.

In the past most ESM systems have been configured for open ocean warfare. Now, however, the littoral threat presents a much more demanding scenario for which many ESM systems were not conceived. Signal density is very high and covers a much wider range of targets including naval platforms, commercial shipping, shore-based emitters as well as airborne emitters, both commercial and military. In addition, electromagnetic interference from TV and cellular radio, as well as military emissions from datalinks all add to the confusion of signals. Furthermore all these potential emitters are in much closer proximity to the submarine than they might have been in the open ocean.

To cater for this complex scenario, ESM systems require channelised receivers to provide for the intercept and measurement of several simultaneous signals while narrow band receivers are needed to search for individual low power radars that would be hidden in the dense environment. In addition the measurement of unintentional modulation within transmitted radar pulses, together with the use of more capable signal sorting algorithms, are necessary to provide greater confidence in signal analysis and identification.

Finally, all this data must be analysed and processed to provide the command with a real time situation report to allow appropriate measures to be undertaken for successful operation.

NEW ENTRY

ACOUSTIC DECOYS

FRANCE

Salto

Type
Anti-torpedo warfare system.

Description
Salto is a fully integrated defensive system designed to provide complete and automatic defence against torpedo attack for surface vessels. It consists of two main components, the anti-torpedo warning system

Alto and the anti-torpedo countermeasures system Contralto.

Alto consists of a detection, classification and warning receiver providing 360° coverage in azimuth. The receiver is mounted in the main towed array antenna, but can be supplemented by an optional complementary antenna mounted in the bow of the surface ship. Due to its modular architecture, Alto can be fitted to various types of array, towing system and receiver and can fit, on request, any surface ship or

submarine. Automatic torpedo warning and classification is provided, with the possibility of operator confirmation if required.

Contralto consists of two soft kill countermeasures systems which are thrown or dropped from the ship. One is fired from a launcher and descends to the sea surface by parachute where it is then suspended from a float, transmitting a signal which masks the acoustic signatures of the ship, allowing the ship to manoeuvre out of danger. The ship can also deploy a mobile countermeasure. The acoustic decoys/jammers operate over an extensive broadband with high emission levels. Control is exercised from a two-screen console or two single-screen consoles.

As well as providing decoys/jammers, Salto also suggests the most effective evasive manoeuvres. Salto can be installed as a stand-alone system or integrated into a ship's combat management system.

Contractor
DCN International.

VERIFIED

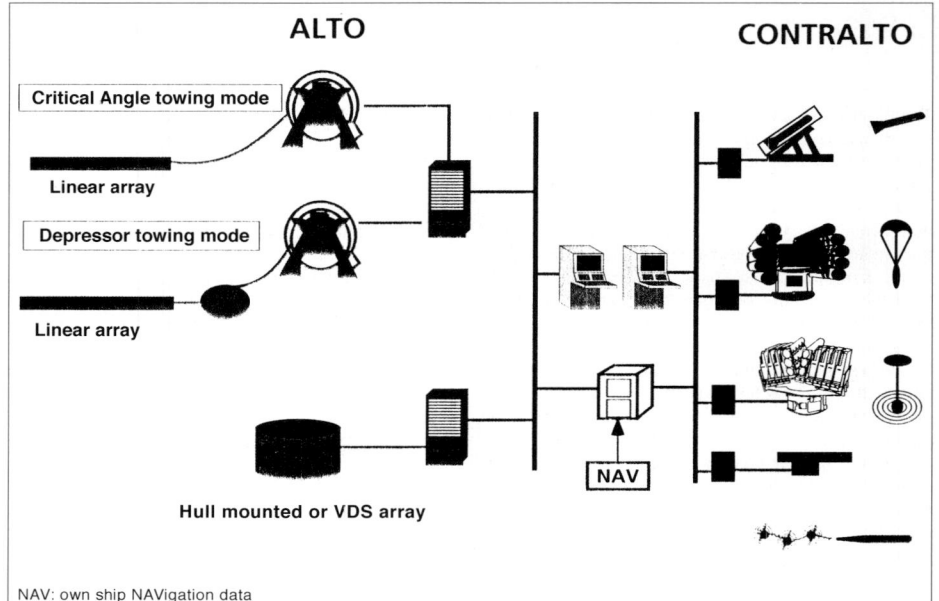

ALTO   CONTRALTO

Critical Angle towing mode

Linear array

Depressor towing mode

Linear array

Hull mounted or VDS array

NAV

Salto architecture
1996

NAV: own ship NAVigation data

INDIA

Toted

Type
Towed torpedo decoy.

Description
Toted is an acoustic countermeasure system developed for deployment from surface ships. Its primary role is to decoy attacking active and passive homing torpedoes away from the ship towards its remotely controlled decoying signal.

The decoy signal is electronically generated in the main electronics unit. The signal is amplified by a high-gain wideband amplifier and fed to the piezoelectric transducer housed inside the towfish via an

electromechanical cable wound on a cable drum on the winch deck. The towfish is deployed approximately 400 m away from the ship. The high-decibel acoustic signal produced is designed to divert passive homing torpedoes astern and confuse or jam the steering mechanisms of active homing torpedoes.

Provision is made to eliminate interference with 'own ship's' sonar or with sonars of other ships nearby through front panel selectable filters.

Facilities are available to monitor the depth and roll of the towed fish.

Operational status
In service with the Indian Navy. It is believed that up to six systems have been delivered.

Specifications
Operating modes: noise – passive homing torpedoes; CW – active homing torpedoes; noise/CW – both passive and active homing torpedoes
Beam pattern: horizontal >140°; vertical >50°
Transmission source level: 85 dB (min – ref 1 micro bar) in noise mode

Contractor
Bharat Electronics Ltd, Bangalore.

UPDATED

INTERNATIONAL

Joint Surface Ship Torpedo Defence (J-SSTD)

Type
Development programme for anti-torpedo defence.

Description
For some years the US Navy and the Royal Navy have been studying various methods of anti-torpedo capability, under the Surface Ship Torpedo Defense (SSTD) and Talisman programmes respectively. These studies culminated in a Memorandum of Understanding (MoU) signed in October 1988 which eventually became known as the Joint Surface Ship Torpedo Defence (J-SSTD) Programme. The objective was to collaborate in the three-phase development of an anti-torpedo defence system to protect surface ships against attack from the increasingly sophisticated wire-guided and wake homing torpedoes.

The J-SSTD programme was to be developed to counter any projected torpedo threats in the future, the priority being to provide surface vessels of all categories with defence against 70 km range, wake homing torpedoes which lock onto the turbulence created by a passing warship. Detection of such a

threat is difficult, since the attacking weapon needs to be picked up within the turbulence created by the ship – an area which represents a poor detection environment for sonar. To overcome this problem a specialised variable depth sonar or towed array may be employed.

Following detection, the torpedo threat will be categorised and countermeasures selected to provide the most appropriate defence. Various hard kill and soft kill options were examined.

Hard kill options under study included the use of a modified lightweight torpedo, such as the US Mk 46 as a development model with a view to adapting the US Lightweight Hybrid Torpedo (LHT). Other possibilities included adapting the Franco/Italian MU90 as an anti-torpedo torpedo, and development of a 6.25 in weapon to provide a launch system for vessels lacking space and weight for 324 mm torpedo tubes.

Soft kill options under consideration included the use of nets and anti-torpedo decoy countermeasures. A combination of both hard and soft kill options could eventually form a complete anti-torpedo package. Since the MoU was signed, the programme focused on developing a combined soft/hard kill solution.

Operational status
The US Department of Defense and the UK Ministry of Defence completed the concept evaluation phase of the project and two binational teams received parallel contracts for demonstration and validation.

The team led by Northrop Grumman in the USA, with Lockheed Martin (USA) and Ultra Electronics Sonar and Communication Systems (UK) were selected for the soft kill; Thomson Marconi Sonar Ltd (UK) and BAeSEMA for a particular form of countermeasure was selected by the US Naval Sea Systems Command early in 1995 to carry out the demonstration/validation phase.

The programme is currently in the final stages of an extended demonstration/validation phase. However, the US has since decided not to assign further funds to the project for the engineering and manufacturing development phase. Further US funding may not be made available until FY99 and hence the future of the programme is in some doubt.

UPDATED

SSTDS

Type
Surface Ship Torpedo Defence System (SSTDS).

Development
The SSTDS project is aimed at providing an integrated self-defence system for the new generation of frigates

that are to be acquired by a number of nations over the next few years. It is designed to carry out detection, classification, localisation, evaluation and deployment of countermeasures against torpedo threats well into the next century.

The companies involved in the feasibility study of the SSTDS project are Empresa Nacional Bazan of Spain, DCN and Thomson Marconi Sonar SAS of France,

Whitehead Alenia Sistemi Subacquei (WASS) of Italy, and Kongsberg of Norway. Within the feasibility study STN ATLAS Elektronik of Germany, WASS and Eurotorp addressed both soft kill and hard kill solutions, while all six companies carried out integration studies.

One of the most likely systems to be chosen for the SSTDS project is the Franco-Italian SLAT (Système de

Lutte Anti-Torpille) on which development began in 1993 and which was due to complete its project definition phase in mid-1998. The project involves WASS (countermeasures and launchers), Thomson Marconi Sonar (detection) and DCN (classification and tactical software).

The SLAT is being developed for installation on board the carriers *Charles de Gaulle* and *Giuseppe Garibaldi* and the 'Horizon' type frigates of France and Italy. Italy will also install such a system on the navy's more vital support vessels and auxiliaries. The proposed in service date for SLAT is 2002.

SLAT is a fully integrated torpedo defence system encompassing a number of vital subsystems including acoustic sensors (ALERTO), reaction subsystem (RATO) and countermeasures (CMAT). The system is fully automatic and is said to offer a high probability of detection and classification with a low false alarm rate. SLAT incorporates an assessment capability that determines the mix of jammers and decoys and their most effective deployment together with a subsequent alert to the command of the need to launch further

countermeasures should the initial deployment fail to engage the target. Included in this capability is an advice programme that will indicate to the command the most effective evasive manoeuvres for the platform to conduct while the countermeasures are being deployed.

SLAT consists of a linear towed array that features up to 100 beams and in which left/right ambiguity is resolved and also an LF hull-mounted array. Automatic and semi-automatic and operator classification of targets is based on specific broadband and narrowband signature analysis algorithms while real-time computing is used to determine ship manoeuvres and timing, mix and positioning of countermeasures. Engagement is followed using a dedicated intercept array and special purpose signal processing while fusion techniques are employed to integrate data from the TAS and hull-mounted arrays with the intercept array.

Operational status
The SSTDS programme commenced under an MoU

between the six countries (France, Germany, Italy, the Netherlands, Norway and Spain) in April 1996. The feasibility study was completed and submitted towards the end of 1997 and preliminary discussions concerning the next stage of the programme are now underway. This will involve a 24-month project definition phase due to commence in the latter part of 1999.

Contractors
DCN, Paris, France.
Eurotorp, Sophia Antipolis, France.
Thomson Marconi Sonar SAS, Sophia Antipolis, France.
STN ATLAS Elektronik, Bremen, Germany.
Whitehead Alenia Sistemi Subacquei (WASS), Genoa, Italy.
Kongsberg Gruppen ASA, Kongsberg, Norway.
Empresa Nacional Bazan, Cartagena, Spain.

NEW ENTRY

TAU-2000

Type
Torpedo countermeasure for submarines.

Description
TAU-2000 is a rapid reaction, multi-effector, soft kill concept developed to meet the requirement for a torpedo countermeasures system for the new Type 212 submarines for the German Navy. It is designed to counter modern and advanced lightweight torpedoes and also wire-guided and non wire-guided, acoustic homing heavyweight torpedoes, by effectively decoying them using threat-matched jam and decoy functions.

TAU-2000 is designed to operate in a similar way to missile decoys, seducing the torpedo and causing it to break contact with its intended target, deceiving the weapon's sonar by one or more decoys, and exhausting its energy in carrying out ineffective re-attack procedures to regain contact with the target, which meanwhile carries out evasive manoeuvres. The essential functions will be carried out by autonomously deploying effectors that work in jam and/or decoy mode. The deployment of a number of effectors creates a multitarget scenario which confuses the weapon's intelligence and its reattack capabilities. To build up a multitarget scenario the mobile effectors are fitted with powerful acoustic functions for jamming and decoying the torpedo.

For jamming, the transmitted signal from an effector must be significantly higher than the incoming echo signal from the submarine, so that the latter is hidden by the decoy signal and torpedo contact with it is broken. For decoying the deceptive signal generated by the effector is as similar as possible to the signal

which would be created by the real target. It is, however, transmitted from a different location and therefore attracts the torpedo away from the submarine.

To counter passive homing torpedoes or torpedoes operating in active plus passive mode, the jammers can assume the characteristic of a passive target.

In order to carry out its designed functions the TAU-2000 consists of multiple acoustic transducer arrays which provide omnidirectional transmitting and receiving characteristics in an appropriate broad frequency range. State-of-the-art signal processing in the acoustic section allows both jammer and decoy functions to be combined in one unit of hardware.

Torpedo threat detection and alarm are generated by the submarine sonar providing bearing details which are continuously fed to the TAU-2000 which is maintained in a constant standby condition ready to respond to any alert. After launch some of the effectors operate in jamming mode to confuse the torpedo and cause it to break contact, after which other effectors operate in decoy mode to simulate an alternative target. After a preset time delay some of the decoy jammers automatically revert to decoy mode. The movement and false target signatures presented by the effectors follow precisely the speed profile and noise signature of the intended target, causing the torpedo to reattack various decoys and so be drawn away from the real target.

The TAU-2000 effector is a compact, self-propelled, self-controlled, underwater device comprising: afterbody; transducer array; battery unit; receiver and electronics; and control unit. A single propeller driven by an electric motor propels the effector. A thermal battery provides propulsion energy.

Short reaction time is one of the design features, and the system is flexible enough to launch effectors having various performance parameters and meeting the customer's specific requirements. About 10 discharge units with 127 mm effectors are placed in one launch container that is integrated to form a flush surface with the surrounding structure. One or two containers can be provided on each side of the boat.

Main features of the system are:
(1) Short time of preparation for discharge
(2) Discharge of several effectors in a short time
(3) Discharge without bubbles and swell
(4) High degree of functionality and safety
(5) No adverse effect on hull signatures
(6) Compatibility with submarine systems
(7) Easy loading and unloading of launch containers
(8) Possibility of retrofit
(9) Design according to rules and regulations for submarine.

Operational status
Full-scale experimental effectors have been built and tested and acoustic devices have been prepared for sea trials. Further trials were conducted during 1997 and extended trials with the final version are due to be conducted in 1998-99. Delivery is scheduled for 2001.

Contractors
HDW, Kiel, Germany.
Whitehead Alenia Sistemi Subacquei, Livorno, Italy.

UPDATED

ISRAEL

ATC-1 torpedo decoy

Type
Acoustic torpedo decoy.

Description
ATC-1 (Acoustic Torpedo Countermeasures-1) is a towed decoy, developed to counter modern, high-speed acoustic-homing active/passive torpedoes. It generates and transmits strong acoustic signals which are accepted as a legitimate target by attacking torpedoes which home in on the decoy and attack and reattack until its power source is exhausted and it sinks.

The ATC-1 does not restrict the speed or manoeuvrability of the towing ship. Lightweight, compact and reliable, it is simple to install and requires a minimum of deck and below deck space.

Compatible with a wide range of both naval and merchant vessels, from fast missile boats, corvettes, frigates and destroyers to cargo ships and tankers, the ATC-1 has been designed, built and tested to MIL-E-16400. The decoy does not interfere with other systems on board, and is not affected by them. Optimal

vessel protection is ensured by a frequency range that covers all existing homing heads.

The ATC-1 features a towed body 120 cm long and 30 cm in diameter, weighing 25 kg, together with winch, launch and recovery units weighing a total of 1,325 kg. Also on board the ship is an electronics cabinet weighing 190 kg and a remote-control unit weighing 10 kg, neither of which requires much space. The manufacturer claims that a single operator is required for the system.

Specifications
Length: 1,200 mm
Diameter: 300 mm
Weight: 25 kg

Operational status
In production and in service. Operational aboard the Israeli 'Eilat' class corvettes.

Contractor
Rafael, Haifa.

UPDATED

Rafael ATC-1 towed body

ATDS

Type
Advanced Torpedo Defence System.

Description
This concept has been developed to provide surface ships with a defence system to counter modern torpedoes fitted with advanced counter-countermeasures subsystems.

The system comprises a Torpedo Detection Towed Array (TDTA) for long-range sonar detection of torpedoes, the ATC-2 (an upgraded ATC-1) towed decoy which simultaneously transmits acoustic signals and provides close-range localisation of the torpedo and Scutter – an expendable, intelligent decoy which is deployed from the ship to finally divert and decoy the torpedo until its fuel is exhausted.

The TDTA and ATC-2 are towed in tandem by a cable carrying the electronics signals between the surface ship and the sensors. Typically the TDTA and ATC-2 are towed at a distance of about 500 m behind the ship, being deployed before the vessel enters a danger zone.

The first phase of the operation involves the TDTA providing a 360° coverage alert and bearing data of the launch of a torpedo and noise signals radiated by the weapon at detection ranges of several kilometres. Onboard electronics process the signal data from the TDTA received from the torpedo's propulsion system to provide classification and identification of the attacking weapon. This intelligence provides the defending vessel with sufficient time to engage the second level of defence to effectively counter the threat.

The second level of defence requires the processed data from the TDTA to be fed to the ATC-2 which responds by transmitting a signal tailored to decoy the torpedo. ATC-2 can synthesise a signal with a coverage of all known torpedo frequencies from 17 to 85 kHz and with a variety of profiles. The torpedo then attacks the towed ATC-2 and even modern torpedoes will make their initial attack on the first target of preference, the ATC-2. The ATC-2 can transmit sufficient power (up to 187 dB/µPa/m) to assure this process. Simultaneously, while transmitting, the ATC-2 also listens for the radiated signature of the torpedo as it passes at close range. Detection of the torpedo enables it to be localised in time and space, thus providing the ship with time to manoeuvre while initiating deployment of the Scutter decoy.

Originally designed for submarine defence against homing torpedoes, Scutter is able to receive, classify and identify active acoustic transmissions from homing torpedoes and to respond accordingly with its own decoy transmission. A passive homing torpedo will correspondingly receive transmissions from Scutter which are acceptable on board computers. The torpedo then attacks the Scutter. Depending on the type of threat, a sequence of Scutter decoys can be deployed while the torpedo attacks each one in turn until its fuel supply is exhausted.

The ATDS comprises a 'wet end' integrated towed array equipment and on board hardware and software. An operator's console contains two 19 in colour monitors and MMI tools.

The 'wet end' features a 500 m towcable, a towed body housing the TDTA front end electronics, depth, vibration and direction sensors and the ATC-2 transducers, the TDTA mini-towed line array and a tail stabilising module.

Onboard hardware and software features the electronic console containing: TDTA analogue front end, amplifiers, filters, digital/analogue and analogue/digital circuitry, beam-forming algorithms, trackers and signal processors; ATC-2 close-range detection circuitry, low power unit and synthesiser and power amplifiers.

Target detection is affected in two modes: broadband and narrowband. Tracking and classification is based on broadband, narrowband and Demon data. The principle features include: broadband detection with automatic normalisation and jammer rejection; narrowband detection with 4 Hz resolution; eight trackers using broadband, narrowband and Demon data; and long-range torpedo detection and alert.

The broadband detection process provides the operator with all frequency bands and a sector coverage of nearly 360°. The broadband detector features true linear beam-forming with a full dynamic range. It reduces the masking effects of low level sources by high-level sources and beam-forming of patterns with very low sidelobes. Other features include automatic narrowband jammer rejection and true bearing stabilisation to allow long time integration. Spatial normalisation provides a clear and legible picture on the display by suppressing sea noise on each beam and compressing sea noise variance over the 360° coverage. In addition, integration time is selectable from 1 to 128 seconds.

The narrowband detector features true linear beam-forming and panoramic filtering with a fine spectral analysis performed on each beam. Frequency resolution is matched to expected tonal widths and offers a preclassification capability right from the time of detection. Frequency normalisation improves detection and enhances the display presentation. Integration time is selectable to optimise the detection range with contact dynamics and signal-to-noise ratios. A complementary integration gain is attained through visual integration provided by the display which presents a picture of the frequency lines.

Eight targets can be tracked simultaneously and automatically following operator initiation. Additional targets can be tracked manually using a cursor. The automatic tracker uses up to seven data sources to obtain each target's bearing. The sources include broadband information, three narrowband trackers and three Demon trackers. Data sources are integrated to form one common bearing. The tracking module can handle a large dynamic range, high bearing rates, crossing targets and end-fire crossings.

An alert is sounded immediately a torpedo is detected. Simultaneously, torpedo bearing and identification data are provided to facilitate activation of anti-torpedo self-defence countermeasures.

Contractor
Rafael-Armament Development Authority, Ordnance Systems Division, Haifa.

VERIFIED

Scutter

Type
Submarine and surface ship-deployed expendable torpedo decoy.

Description
Scutter is a self-propelled expendable torpedo decoy designed to respond to two different types of torpedo simultaneously and against the threat from acoustic homing active and/or passive torpedo threats. On receipt of a torpedo alert, Scutter is launched manually or from any standard submarine signal ejector or external hull-mounted launcher. System operation is automatically initiated after launch, the decoy propelling itself automatically to a preset depth between 10 and 300 m where it listens for torpedo transmissions.

Active acoustic transmissions detected are analysed and the decoy selects and generates the appropriate deception signals for transmission, including Doppler effect simulations and simulation of noise radiated by the launch platform. This allows the launch platform to perform evasive manoeuvres. If no torpedo transmissions are detected, Scutter incorporates generic responses which are transmitted to counter passive homing weapons. The decoy's computer 'library' contains threat correlations and threat identification and its programs provide tailored responses to defeat various logics incorporated in a modern torpedo such as range gates, Doppler shifts, target discrimination and so on. The decoy can be programmed for specific known threat characteristics.

The decoy can operate for up to 8 minutes at depths down to 300 m providing 360° coverage. The decoy self-destructs and sinks on completion of the mission thus preventing retrieval by hostile forces.

Scutter can be deployed by all types of submarine and has no special installation requirements. Due to its light weight and compact size it takes up very little space.

The decoy's hydrodynamically shaped body is manufactured from composite materials which house the acoustic transducer, electronic boards, thermal batteries, electric motor, propeller and an external waterproof test connector.

Operational status
Scutter has completed final acceptance trials by the Israel Navy and is in full-scale production.

Specifications
Length: 1,000 mm
Diameter: 100 mm
Weight: 8 kg
Frequency range: 17-85 kHz
Transmission source level: >180 dB/µPa/m
Operating life: 8 min

Contractor
Rafael-Armament Development Authority, Ordnance Systems Division, Haifa.

UPDATED

ITALY

C303/310 Torpedo countermeasures

Type
Submarine anti-torpedo countermeasures system.

Description
This is a series of high-speed, high-endurance, lightweight, submarine anti-torpedo countermeasures systems, designed to counter attacking acoustic homing, active/passive torpedoes using expendable jammers and decoys. The devices can emulate any type of acoustic signature through the use of multiple frequency operation in wide bandwith using a multibeam transmit/receive pattern. The systems feature low secondary lobes, important when attempting to decoy a high-speed target.

The countermeasures system consists of:
(a) jammers and decoys
(b) launching system, mounted externally on the pressure hull
(c) control panels and computer which can be integrated with the torpedo detection system.

The jammer features a highly efficient transducer, covering the whole bandwidth with a switching power amplifier and high-energy density lithium battery. The jammer generates a very high amount of energy spread over the entire reception band of the acoustic head of the torpedo. To achieve the decoy effect the unit uses a sophisticated transponder to simulate a real target, generating in real time an acoustic signature to any multifrequency coded pulse emitted by the torpedo.

The modular design and commonality with existing devices combined with simple functional architecture enables the decoy to be easily interfaced with the majority of existing platforms.

The devices are deployed from a dedicated launching system comprising a multibarrelled module fitted externally to the pressure hull immediately abaft the sail. The number of barrels per module and the number of modules can be tailored to meet customer requirements. Generally, a configuration of three identical modules each comprising seven launch tubes is proposed, with three modules sited to port and three to starboard. Each barrel would be loaded with appropriate devices containing its own air bottle, valve and launching tube, thus remaining independent of all the other launch barrels. Such an arrangement would provide defence against up to six torpedo engagements. The C303 system is designed specifically for mounting on submarines and carries six decoys and 24 jammars. The C310 system is designed for surface ships carrying six decoys and 12 jammars.

Development of the devices began in 1989. The C303 version features improved performance enabling the countering of two torpedoes simultaneously using self-propelled devices. The vectors of the mobile target emulator are derived from the A200 mini-torpedo developed by Whitehead Alenia Sistemi Subacquei, within the frame of the Low-Cost Anti-submarine Weapon (LCAW).

Specifications

	C303	C303/s	C310
Jammer	stationary	stationary	stationary
Length	1,150 mm	1,150 mm	750 mm
Diameter	76.2 mm	76.2 mm	76.2 mm
Weight	approx 10 kg	approx 10 kg	approx 6 kg
Emulator	stationary	mobile	mobile
Length	1,150 mm	1,150 mm	1,150 mm
Diameter	76.2 mm	123.8 mm	123.8 mm
Weight	approx 10 kg	approx 14 kg	approx 14 kg

Operational status
In development.

Contractor
Whitehead Alenia Sistemi Subacquei, Genoa.

UPDATED

UNITED KINGDOM

Bandfish

Type
Expendable torpedo countermeasures system.

Description
Bandfish (UK Royal Navy designation Type 2066) is an expendable countermeasures system for use against acoustic torpedoes. It is designed to be launched from a submarine's submerged signal ejector or from a surface ship launcher. Once launched, the device operates independently of the launch platform.

In operation, Bandfish hovers in mid-water and transmits a high-intensity broadband acoustic signal. The operating depth and acoustic pattern can be designed to meet the customer's requirements. The buoyancy control system and hover are preset before launch. It has a minimum shelf life of five years and requires no user or depot maintenance. Simple tests are provided to ensure that it can be launched with maximum confidence.

Operational status
In full production and currently in use aboard Royal Navy submarines.

Specifications
Length: 995 mm
Diameter: 102 mm
Weight: <12 kg

Contractor
Ultra Electronics Sonar and Communications, Greenford.

UPDATED

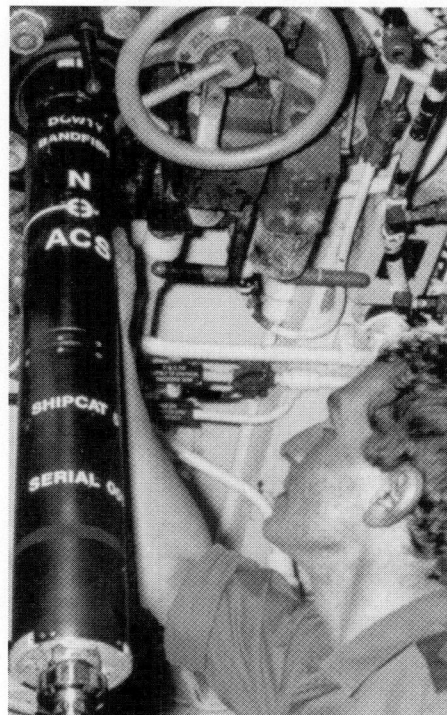

Bandfish expendable torpedo countermeasure

G 738 decoy

Type
Towed torpedo decoy equipment.

Description
The G 738 is an improved and re-engineered version of the Royal Navy Type 182 towed torpedo decoy, now being manufactured for several other navies. The equipment uses modern solid-state signal generation equipment and compact, lightweight deck machinery.

G 738 is designed to decoy both active and passive homing torpedoes. Decoy signals are electronically generated within the ship and fed via the towing cable to electro-acoustic transducers within the towed body. The signals thus produced divert passive homing torpedoes astern and confuse or jam the steering mechanisms of active homing torpedoes. It can also be used as a high-frequency noise source for trials purposes.

The G 738 simultaneously produces two independently controllable output signals which are radiated from the towed decoy.

An amplitude modulated noise signal is provided to simulate the ship's propeller noise and to cause passive homing torpedoes to be seduced away from the ship towards the towed decoy.

A swept CW signal is provided to simulate sonar echoes to confuse or jam the guidance system of active homing torpedoes, and cause them to miss the ship.

The frequency bands covered by both types of output are switchable, as required for tactical scenarios and to enable concurrent operation of other sonar and communication equipments.

A variety of deck handling arrangements are produced, and dual systems may be fitted. Remote control selection of signal characteristics provides for instantaneous changes from operations room or bridge. The equipment can incorporate handling systems for open deck on extended flight deck ships.

The operating characteristics remain confidential between the manufacturer and the purchaser.

Operational status
No longer in production. In operational service with the Royal Navy and a number of other navies including those of Argentina and India.

Specifications

Dimensions and weights	Height	Depth	Width	Weight
Electronics cabinet	1,725 mm	640 mm	585 mm	375 kg
Remote-control unit	152 mm	190 mm	305 mm	2.3 kg
Deck connection box	203 mm	115 mm	265 mm	13 kg
Resistance box	444 mm	115 mm	381 mm	14.3 kg
Towed decoy	533 mm	2,005 mm	457 mm	74 kg
Cable reel	1,245 mm	2,340 mm	1,390 mm	2,540 kg
Gantry	203 mm	5,918 mm (retracted)	440 mm	810 kg
Davit	3,500 mm (Jib raised)	3,400 mm (max)	1,270 mm	818 kg
Triple control unit	357 mm	240 mm	440 mm	188 kg

Towing cable
Length: 411 m
Diameter: 14.48 m
Weight: 269 kg

Contractor
Graseby Dynamics Ltd, Watford.

UPDATED

Thomson Marconi Sonar acoustic countermeasures

Type
Range of acoustic countermeasures units.

Description
The company has designed and produced a range of acoustic countermeasures units. These are high-power, high-efficiency devices for deployment against acoustically guided weapons to protect both surface vessels and submarines. Both spatial targets and jammers have been designed and can be tailored to individual requirements in submarine or surface (air-launched) configurations. These devices are available in formats which can be deployed by most standard launch systems. Special requirements, such as overall frequency bands, source levels and noise structures can be made available as required.

An anti-torpedo acoustic countermeasure known as ATAAC is a noisemaker launched from a standard chaff launcher. There are both 76 mm and 100 mm versions and the rounds are programmed before launch to produce broadband noise. A 203 mm version, for discharge from a submarine countermeasures tube, is also available.

Operational status
In production.

Contractor
Thomson Marconi Sonar Ltd, Templecombe.

UPDATED

Submarine Signal and Decoy Ejectors (SSDE)

Type
Reloadable submarine discharge systems.

Description
Submarine signal and decoy ejectors provide a small reloadable discharge system, capable of launching cylindrical stores approximately 100 mm in diameter and 1,250 mm in length, over the complete submarine operating envelope. These ejectors can also be designed and manufactured to meet particular requirements.

Existing ejectors currently in service are used to deploy communications buoys, bathythermal probes, exercise and emergency signalling stores, decoys and countermeasures, including programmable stores and stores requiring datalinks. The ejectors can be configured for purely manual operation, or they can be provided with an automatic operating capability, enabling discharge to be initiated either locally or from a remote position.

The submarine signal and decoy ejector contains two main assemblies built into a common hull insert; the launch tube assembly, into which the store is loaded prior to launch, and a ball valve closure outboard. Mounted alongside is the water/air ram assembly. The hull insert is normally positioned radially in the pressure hull, discharging across the ship's flowfield. It can be positioned with its exit flush with the pressure hull, or it can be moved radially outboard if required to bridge the space between the pressure hull and the free-flooding casing.

The stores are loaded into the ejector which is then automatically flooded and equalised at depth pressure. Launch is effected by admitting high-pressure air to the air ram causing it to move and discharge water into the tube, thereby ejecting the store. Ejectors of this type combine a virtually instantaneous discharge capability

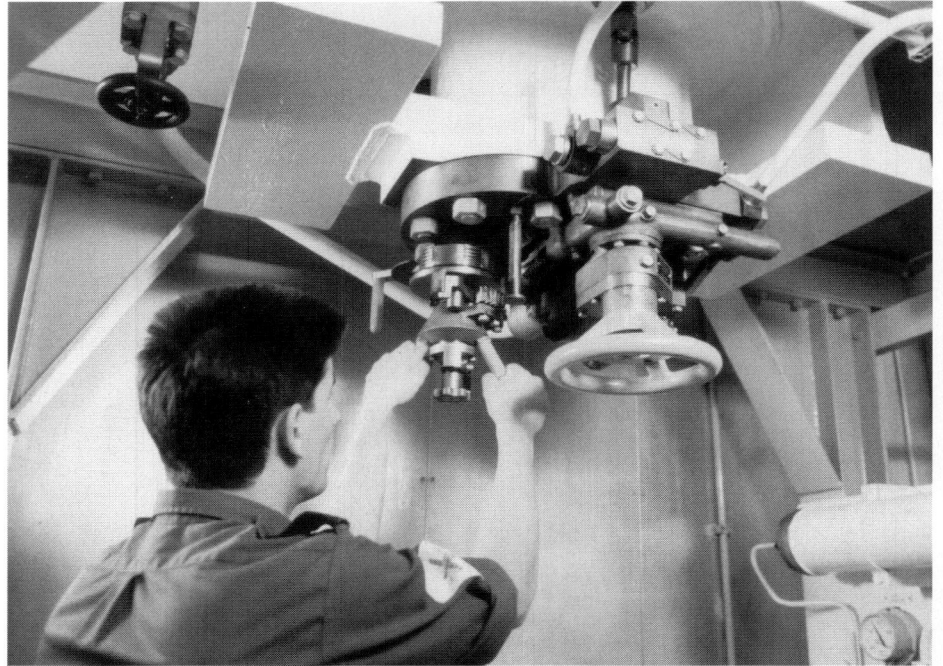

Submerged signal ejector trainer

with a very rapid cycle time, and have a proven record of high reliability and low maintenance.

In the submarines of the Royal Navy two SSDEs are fitted, one forward and one aft. Three variants are currently fitted: Mk6 ('Swiftsure' class), Mk 8 ('Trafalgar' class) and Mk 10 ('Vanguard' class). All three variants can launch both Type 2066 and Type 2071 decoys and also communications buoys ECB-608 and ECB-699.

Operational status
In operational service. Strachan & Henshaw is currently the design authority for the Royal Navy.

Contractor
Strachan & Henshaw, Bristol.

UPDATED

UNITED STATES OF AMERICA

Countermeasures command and control unit – an advanced submarine countermeasures controller unit with an expandable capability for countermeasures device inventory management, processing tactical solutions, target data management, and launch sequencing of all externally configured countermeasures launchers. The equipment was an early design with engineering development models due in 1993-94. It is believed that about 50 sets are likely to be procured by 1998.

UPDATED

ADC Mk 1 Mod 0

Description
127 mm diameter expendable acoustic countermeasures device produced by Conax Corporation. Procurement quantities have exceeded 2,000 units.

UPDATED

ADC Mk 1 Mod 1

Description
152 mm diameter expendable acoustic countermeasures device.

UPDATED

NAE Beacon Mk 3

Description
127 mm diameter sonar countermeasures device

Mobile multifunction device

Description
152 mm diameter expendable submarine-launched advanced countermeasures device.

CSA Mk 2 Mod 0/Mod 1

Description
152 mm diameter launching system for acoustic countermeasures stores.

UPDATED

AN/SLQ-25A countermeasures

Type
Torpedo countermeasures transmitter (Nixie).

Development
Development of the AN/SLQ-25A Nixie was accomplished under a US Navy project in the mid-1970s. This project was involved with the development of systems to enable ASW and non-ASW ships to counterattack. The first production contract was awarded in 1975.

Description
The AN/SLQ-25, also known as Nixie, is a towed electro-acoustic device that provides surface ships with an effective countermeasure against homing torpedoes. Together with a command/display and information processing system, the AN/SLQ-25A comprises the Ship Acoustic and Torpedo Countermeasures (SATC) system. In addition to detecting and recognising threats posed by hostile acoustic sensors and acoustic homing torpedoes, the Nixie device attempts to jam the acoustic sensor and transcribe necessary data so that appropriate action can be taken.

Nixie consists of two small lightweight towed bodies (TB14A) which are streamed and recovered by two winches. Below deck equipment consists of a number of electronics cabinets and a control console sited either in the command centre or near the sonar consoles. The towed bodies transmit acoustic signals, intended to complement the ship's acoustic signature and decoy the torpedo away from the ship. The decoys on Canadian ships are deployed by an all-weather handling system developed by Indal Technologies. A single operator is able to quickly launch, retrieve and stow the towed body at the touch of a button.

The towed bodies are streamed in tandem in case one should be hit by a decoyed torpedo, and thus leave the ship defenceless against a salvo attack. Among the decoy's features are those of simulating specific noises created by the towing vessel, such as machinery and propellers, or generating noises of a specific nature at different frequencies, depending on the nature of the anticipated threat and its mode of operation.

produced by Pique Engineering. Procurement quantities scheduled have exceeded 3,000 units.

UPDATED

Operational status
In early development.

Contractor
ESCO Electronics Corporation.

UPDATED

Frequency Engineering Laboratories have also developed a torpedo defence controller unit which can interface to a Mk 36 launcher to control the firing of decoys based on data fed to the controller from the ship's sonars.

Operational status
The AN/SLQ-25A is still in production. The device is in operational use in the US Navy and a number of other navies, including those of Australia, Belgium, Canada, Egypt, Germany, Greece, Israel, Italy, Japan, South Korea, Netherlands, Portugal, Spain, Taiwan and Turkey. The Royal Navy has acquired four sets for evaluation and trials purposes.

Contractor
Frequency Engineering Laboratories, Farningdale, New Jersey.

UPDATED

AN/SLQ-25B

Type
Towed array sensor module of surface ship torpedo defence system.

Description
The SLQ-25B delivers improved torpedo detection capabilities in a flexible configuration that provides rapid alert and situation assessment, recommended towed and deployed countermeasure responses and manoeuvres to torpedo attacks. The system incorporates and integrates multiple sensors and provides multiple decoy solutions including (but not limited to): AN/SLQ-25A Towed Acoustic Countermeasure (TAC): AN/SLX-1 Multisensor Torpedo Recognition and Alertment Processor (MSTRAP); AN/SLQ-25B Towed Array Sensor (TAS); and Launched Expendable Acoustic Device (LEAD).

Specifications

	TB-14A towed body	AN/SLQ-25B (TAS)	LEAD Rocket round Mk 12	LEAD Rocket round Mk 13
Length	1,193.8 mm	64 m	1,232 mm	1,193.8 mm
Diameter	152.5 mm	50.8 mm	130 mm	130 mm
Weight	263 kg	1,247.5 kg	249.5 kg	215.5 kg

Operational status
The SLQ-25B has been proven in US Navy service and is available.

The LEAD device is a ship-launched rocket or mortar-propelled system designed to decoy an incoming torpedo away from its target. The rounds are designed to be launched from the standard AAW Mk 36 launcher.

Contractor
Northrop Grumman Corporation, Annapolis, Maryland.

UPDATED

AN/SLX-1 MSTRAP

Type
Multisensor torpedo recognition and alertment processor.

Description
Designed for installation aboard surface ships and submarines, the SLX-1 is a COTS-based system providing automatic alert of incoming torpedo threats on an easy-to-read display. This multifunction system provides information, signal processing, and the controls necessary to detect, classify and localise threat torpedoes. It also offers command and control functions, TMA threat evaluation and tactical advice.

Displays with automatic visual and audio alarm of torpedo threats, allow sufficient time to make tactical decisions and effective deployment of countermeasures, and appropriate ship evasive manoeuvres. Data is received simultaneously from either an existing ASW hull sonar, towed array or sonobuoys, or the AN/SLQ-25B towed array sensor module. MSTRAPs open system architecture ensures easy installation, hardware/software upgrades, and future improvements.

Currently MSTRAP is designed to be integrated with the AN/SQQ-89(V) ASW combat system, receiving acoustic data from the AN/SQR-19 TACTAS, AN/SQQ-28(V) SSPS and AN/SQS-S3C hull-mounted sonar via the AN/SQQ-89(V) system.

Operational status
The first unit of MSTRAP was delivered 14 months after receipt of the contract. MSTRAP has been operationally tested aboard the USS *Arleigh Burke* (DDG-51) and proven at sea. Fourteen units have been delivered with six more in production. Options can be exercised to take the total production to 60 units. Currently in process of being provided with options available for up to a total of 60 units. (See also section on Surface ship sonar systems.)

Contractor
Northrop Grumman Corporation, Annapolis, Maryland.

UPDATED

ADC Mk 2 Mod 1 countermeasure

Type
Expendable acoustic torpedo countermeasure.

Description
The ADC (Acoustic Device Countermeasure) Mk 2 Mod 1 is a submarine-launched expendable decoy designed to counter acoustic torpedoes. It is a 76 mm diameter device which hovers vertically at a preselected depth, emitting an acoustic signal.

In the lower electronics section the signals are generated and amplified, while the upper section consists of ceramic transducers and impedance-matching networks for the acoustic transmitter. The bottom section contains a seawater-activated battery which powers a pressure-controlled motor, driving a small, shrouded propeller which keeps the decoy hovering.

Operational status
In production since 1978, more than 10,000 ADC Mk 2

units have been produced by Emerson Electric Co (now ESCO Electronics Corporation). Follow-on contracts were awarded in 1992 and 1994 to Hazeltine for ADC Mk 2 production.

Contractor
GEC-Marconi Hazeltine Corporation, Greenlawn, New York.

UPDATED

ADC Mk 3 countermeasures

Type
Submarine countermeasures equipment.

Description
ADC Mk 3 is a wideband 152 mm diameter unpowered jammer device ejected from submarines, designed

primarily to jam torpedo sonar homing heads. Information on this device is restricted and no other technical details have been released.

Operational status
In production. Over 500 systems have been manufactured.

Contractor
L-3 Communications Ocean Systems, Sylmar, California.

UPDATED

ADC Mk 4 countermeasure

Type
Expendable acoustic countermeasure.

Description
The ADC (Acoustic Device Countermeasure) Mk 4 is a

submarine-launched acoustic countermeasure. No other details are available.

Operational status
Completed development in FY93, and commenced initial production in FY95.

Contractor
GEC-Marconi Hazeltine Corporation, Greenlawn, New York.

UPDATED

ELECTRONIC WARFARE

CHINA, PEOPLE'S REPUBLIC

921-A ESM equipment

Type

Shipborne radar ESM receiver.

Description

The 921-A is a wideband pulse radar direction-finding receiver and is installed on submarines to detect emitters of airborne, shipborne and shore-based radars. It provides coarse measurement of azimuth, frequency band and operational state of the hostile emitters. The equipment is derived from the Russian Stop Light EZM system.

Operational status

In service with the Navy of the People's Liberation Army (PLA-N) on 'Xia', 'Han', 'Song', and North Korean 'Romeo' class submarines.

Specifications
Frequency: 2-18 GHz in four bands
Sensitivity: 1.5×10^{-3} to 10^{-4} W/M^2
Bearing accuracy: Better than ±30°
Dimensions and weights
Antenna unit: 560 mm (diameter) × 515 mm (high); 80 kg
Receiver and display: 450 × 468 × 124 mm; 40 kg
Distribution unit: 145 × 214 × 291 mm; 6 kg

Contractor
China National Electronics Import and Export Corporation, Beijing.

VERIFIED

921-A ESM equipment

FRANCE

DR 2000 U ESM receiver

Type

Submarine radar threat warner.

Description

The DR 2000 U is the submarine variant of the DR 2000 series of ESM systems. The system incorporates a crystal video receiver and has some analysis capability. When coupled with the Dalia analyser, it can be converted into a basic ESM suite.

The receiver is built up of six DF antennas and one omni-antenna and a processing control/display console. It provides a virtual 100 per cent probability of detection over the complete 360° of azimuth. The ESM system carries out passive search and detection of all pulse and CW signals and gives an instantaneous audible and visual alert. The DR 2000 U Mod 3 is an improved version with better sensitivity.

The Dalia analyser provides alarm analysis and identification facilities through a library of 1,000 radar modes and parameters and is easily reprogrammable.

Operational status
In service.

Contractor
Thomson-CSF, Radars and Contre-Mesures, Elancourt.

VERIFIED

DR 3000 U ESM system

Type

Submarine ESM system.

Description

The DR 3000 series is the latest family of ESM systems from Thomson-CSF. Available in a number of variants (the submarine version is the DR 3000 U French Navy designation ARUR 13), the surface ship configuration has been selected by the French Navy to equip the 'La Fayette' class frigates and the nuclear-powered aircraft carrier *Charles de Gaulle*, and several foreign navies.

The DR 3000 is a fully automatic system providing 360° of coverage over an elevation arc of −10° to + 45° in the C to J frequency bands. It uses GaAs technology and radar identification algorithms developed in an expert system, supplemented with a high- IFM receiver. The lightweight (150 kg) DR 3000 uses autonomous air cooling of low-power consumption units. Features include a high probability of interception of modern radars, high sensitivity (said to be up to −68dBmi with a dynamic range of 60dB) providing range advantage compared with other systems, accurate direction-finding, and reliable instantaneous identification of

radars and platforms. The DR 3000 U can operate in a high pulse density of up to 1 million pulses per second. The emitter library can store data on up to 4,000 radar models related to 192 platforms.

DR 3000 missions include radar threat warning, tactical situation build-up and updating, target designation, ECM control and ELINT.

The DR 3000 is offered in a variety of configurations. Different types of antennas are available, including a lightweight periscope miniature aerial, small standard DF antennas, and a high-accuracy DF unit. The associated console is a high-definition colour TV display. A bus interface is provided for integration of the DR 3000 U unit with multipurpose consoles in different submarine combat systems.

Operational status
In production. In service aboard French nuclear attack submarines and will equip the new 'Agosta-90B' submarines for Pakistan.

Contractor
Thomson-CSF, Radars and Contre-Mesures, Elancourt.

UPDATED

DR 3000 U submarine ESM system **1994**

DR 4000 U ESM system

Type

Submarine ESM surveillance system.

Description

The DR 4000 U is a variant of the DR 4000 series of surface ship and submarine ESM systems. It is a radar

detection, analysis and identification system which has an extremely fast reaction time and a high instantaneous data collection capability. High-sensitivity IFM receivers allow 100 per cent intercept probability, even on a single pulse. The highly automated system has an instantaneous 360° DF

capability with bearing accuracies of about 5°. Frequency coverage is over the C- to J-bands with the possibility of extension to the range as an option.

The highly sensitive antenna system includes an omnidirectional unit for frequency measurements and two concentric six-port assemblies, one for C- to

G-bands and the other for H- to J-bands for DF. The information is presented in a variety of forms to the operator on advanced colour consoles.

Operational status
In service aboard the 'Tupi' class Type 209 boats of the Brazilian Navy.

Contractor
Thomson-CSF, Radars and Contre-Mesures, Elancourt.

VERIFIED

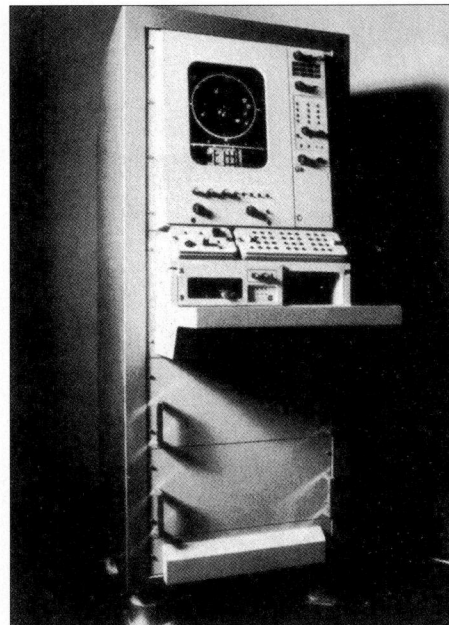

DR 4000 U IFM ESM system console

GERMANY

FL 1800 U

Type
Submarine ESM/DF system.

Periscope antenna OD/DF sensor USK 800/4 for Type 212
1998/0006976

Description
The fully automatic, modular FL 1800 U ESM system incorporates the direction-finding sensor USK 800/4 for submarines. The system provides interception,

FL 1800 U operator console for Type 212
1998/0006977

automatic analysis, automatic DF, classification and display on all signals within its coverage of the radar frequency range. The processed, analysed and classified signals are presented on a tactical picture of the radar environment. Emitters are classified against a stored library and automatic visual and external warnings are given of signals that exhibit certain types of radar characteristics, such as lock on.

The display features a snapshot mode in which emitter track data can be frozen and reviewed after the ESM sensor has been lowered. Finally the FL 1800 U transfers all selected data to the main combat system.

The compact sensors of the USK 800 series can be integrated easily in all common periscopes or optronic masts. In one sensor head the OD/DF antennas can be combined with any electronic modules of ESM, navigation (GPS) and communications (VHF/UHF) systems according to customer requirements. This overcomes the necessity of having a dedicated ESM mast and the ESM is available immediately after the periscope penetrates the surface, allowing simultaneous optical and electromagnetic search and correlation of detected targets.

Operational status
The system has been selected for the new 'U212' class of submarines to be built for the German Navy.

Contractor
Daimler Chrysler Aerospace, Ulm (Donau).

UPDATED

USK 800 series

Type
Modular family of ESM antenna for submarines and surface vessels.

Description
The USK 800 is a modular family of DF systems offering the following capabilities: USK 800/1: ESM omnidirectional reception; USK 800/2: ESM omnidirectional reception and GPS; USK 800/3: ESM omnidirectional reception, GPS and VHF/UHF omnidirectional receive; USK 800/4: ESM omnidirectional reception, GPS and ESM direction-finding reception; USK 800/5: ESM 4-port omnidirectional reception, GPS, VHF/UHF

omnidirectional transmit/receive and ESM direction-finding reception; USK 800/6: ESM omnidirectional receivers, GPS and 6-port DF reception.

Sensor electronics in the USK 800/1-3 are housed in the head, while in the case of the USK 800/4-6 the electronics can be accommodated in a remote housing, its position depending on the mast version. Interfaces from the various sensors are kept to a minimum, different channel functions being combined or separated in the sensor head through frequency separating networks. Data is passed to the control room via a single RF coaxial cable (with rotary joint) in the periscope mast.

All antenna elements are protected by radomes and to reduce radar cross-section, the metallic side surfaces of the sensor element are covered with RAM.

All elements are designed to withstand pressures of 60 bar.

ESM reception is achieved using a biconical antenna with the modelling of the cone contour designed for wide elevation coverage and the greatest possible acquisition range (>40 km). To permit the acquisition of vertically, horizontally and circularly polarised EM radiations, the aperture of the antenna is provided with a 45° polariser. A built-in high-pass filter suppresses any disturbances and interferences caused by strong VHF/UHF signals, such as might be experienced from TV and radio broadcast stations near the coast. Additional limiters prevent damage to the input branch when receiving extremely strong radar pulses.

ESM DF information is derived from signal levels received from the four/six special broadband

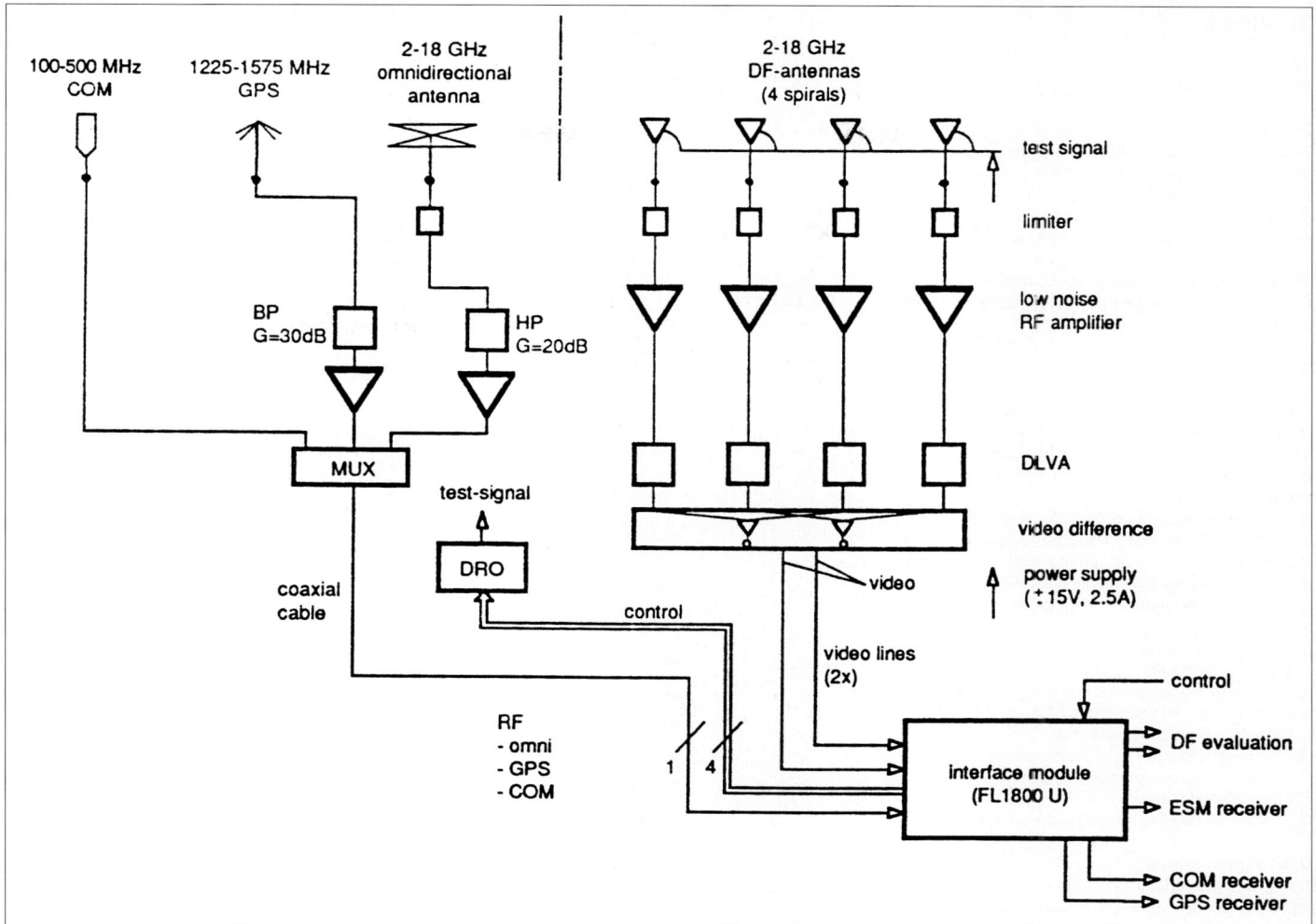

Block schematic of the USK 800/5 system   *1995*

directional receiving antennas, together covering the 0 – 360° azimuth range. The signals are processed through a low noise RF preamplifier and detector log video amplifier to a video differential network, which combines two opposite antenna branches. This reduces the number of video lines in the sensor mast to two (USK 800/4).

Bearing is calculated by a DF processor of the ESM unit. Different levels of the two DF video channels and of the pattern data of the antennas permit determination of the incident angle of the signal in azimuth.

Operational status
A variant of the USK family has been selected for the Type 212 submarines to be built for the German Navy.

Contractor
Daimler Chrysler Aerospace, Ulm (Donau).

UPDATED

Periscope antenna DF sensor USK 800/5
1998/0006978

ISRAEL

NS-9003A-V2/U ESM/ELINT system

Type
Submarine radar warning and ESM system.

Description
The NS-9003A-V2/U is a compact and sophisticated ESM/ELINT radar warning system for submarines. It features instantaneous high-sensitivity frequency measurement combined with accurate direction-finding. It will carry out accurate and rapid detection and analysis of all emitters in dense electromagnetic environments. The system uses an ESM mast covering the 2 to 18 GHz frequency range. An omnidirectional array also could be mounted on the periscope mast. DF accuracy is 2 to 3° RMS. Elevation coverage is from –10 to +30°. A variety of data display facilities is available.

The NS-9003A-V2/U can be installed in new submarines or retrofitted to existing vessels.

Operational status
In production.

Contractor
Elisra Electronic Systems, Bene Beraq.

UPDATED

Timnex II

Type
ESM/ELINT system.

Description
Timnex II is an advanced IFM channelised ESM receiver for ESM/ELINT operations. It provides omnidirectional detection, intelligence gathering, and direction-finding in the 1 to 18 GHz band, performing in-depth tactical analysis on detected emitters. It offers a choice of eight monopulse DF ports across two bands (1 to 8 and 8 to 18 GHz). Bearing accuracy is believed to be in the region of 1.5°, and the emitter library holds up to 500 parameters.

Detected emitters are identified and data on them displayed on a sophisticated graphic and alphanumeric console; processed radar pulses are digitally recorded for replay; selected radar video signals are recorded in analogue. The fully automatic system provides instantaneous frequency coverage and instantaneous direction-finding with real-time signal processing.

The system is fully automatic in operation but incorporates a manual override. It assures short response time and high omnidirectional detection probability in dense environments.

Various configurations ranging from simple to sophisticated can be tailored to meet specific operational requirements.

Timnex systems can be configured for both surface and subsurface applications and airborne and ground applications can also be configured.

Two main variants are available:
(1) an integrated system in which one of the control consoles of the command system is allocated to the ESM function or
(2) a stand-alone system using a special ESM console.

Operational status
Operational in submarines of several navies. It has been reported that the system is installed on the Chinese 'Han' class submarines. The system is also being fitted to the 'Dolphin' class submarines of the Israel Navy. The system is currently fitted to the Israeli 'Gal' and Taiwanese 'Hai Lung' classes of submarine.

Contractor
Elbit Systems Ltd, Haifa.

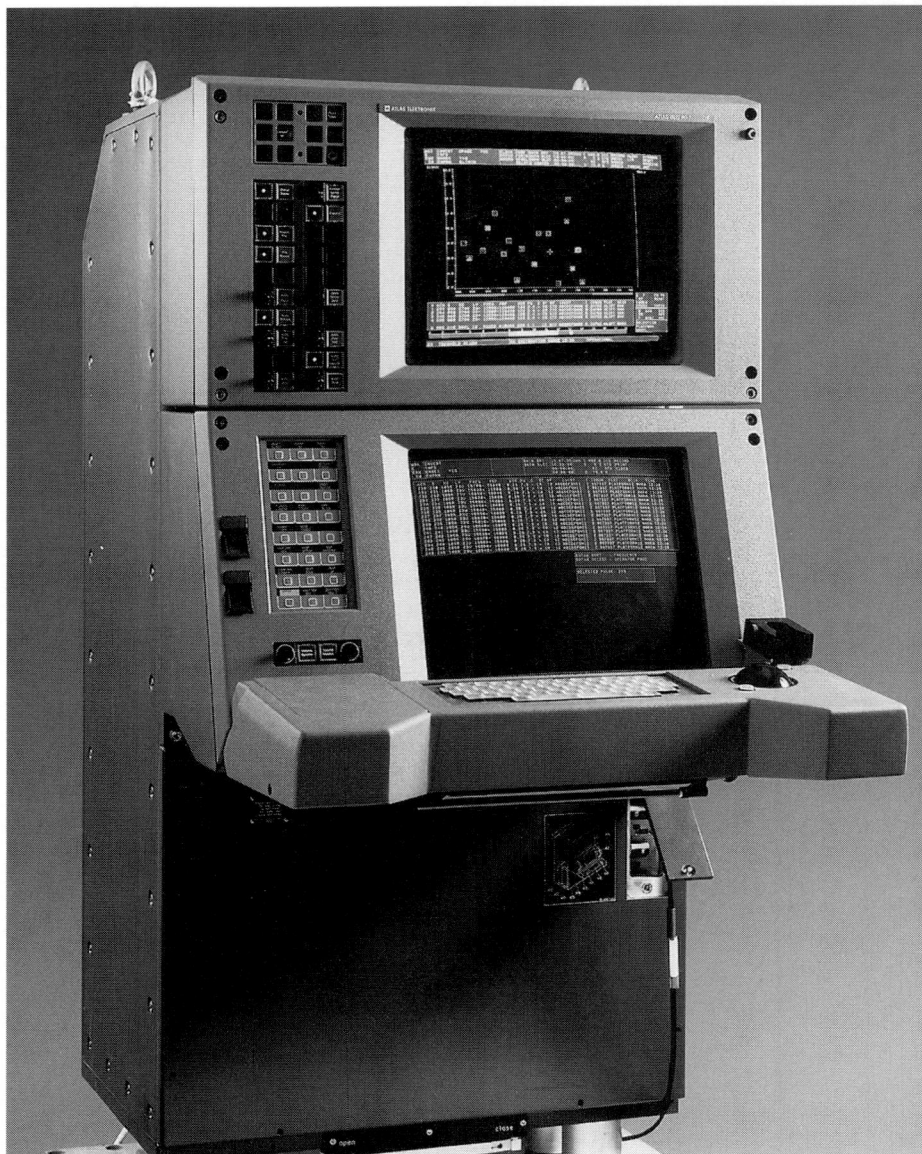

VERIFIED   Operator console for Timnex II ESM/ELINT system   *1998*/0006980

ITALY

Thetis ESM system

Type
Submarine ESM surveillance system.

Description
Thetis is a modular ESM system designed to fulfil the EW requirements of a submarine. The system is produced in two configurations. The first and simpler of these, the ELT/124-S, performs high-sensitivity instantaneous threat warning and DF functions, while the second, the ELT/224-S, extends these functions to those of a full ESM system or tactical ELINT. Both variants use the same DF antenna as the sensor, and both can be fitted with an additional and separate RF preamplified omnidirectional antenna as an option. This latter antenna is very small and can be installed on top of the submarine's attack or search periscope for primary warning.

The ELT/124-S variant consists of a threat warning and DF receiver system. The hardware comprises a DF antenna and receiver, a processor unit and a warning display. It provides instantaneous, high-sensitivity,

warning of threats programmed into the processor and continuous surveillance of the radar environment with 100 per cent intercept probability. A dedicated operator is not required.

The ELT/224-S consists of the units of the ELT/124-S plus an IFM receiver and an ESM display. It provides instantaneous DF with 100 per cent intercept probability of pulse and CW emissions, display of all intercepted emissions, IFM, automatic technical analysis and identification of emitters and their platforms. It will also provide warning that the vessel is entering an area under hostile radar surveillance. A single operator is required for this system.

The antenna assembly consists of one conical spiral omnidirectional antenna plus eight plane spiral DF antennas assembled in a single casing. It has a very low radar cross-section and is also covered with microwave absorbing material. It is light but strong enough to be mounted on the search periscope. The small, lightweight optional antenna can be mounted on the attack periscope and will provide the submarine commander with a first alarm facility without DF facilities.

Thetis submarine ESM system (display/control console and two antennas)

Operational status
In production and operational service aboard Italian Navy submarines.

Contractor
Elettronica SpA, Rome.

VERIFIED

ULR-741 ESM/ELINT system

Type
Submarine ESM/ELINT automatic system.

Description
An improved version of the Thetis ESM system is being commercialised under the designation ULR-741.

The system features fully automatic ESM operation with an environment 'snapshot' capability and fast, high-quality ELINT analysis and data recording.

The system uses the same antenna unit of the ELT/244-S, with minor circuit changes, while the internal units are derived from the latest generation of airborne automatic ESM/ELINT systems, manufactured by Elettronica SpA.

Operational status
Ready for series production.

Contractor
Elettronica SpA, Rome.

VERIFIED

UNITED KINGDOM

Manta ESM system

Type
Submarine radar detection and analysis system.

Description
Manta is a range of advanced ESM systems optimised for submarine applications. Each of the Manta systems intercepts, analyses, classifies and identifies enemy radars operating in the 2 to 18 GHz frequency band.

Instantaneous detection of radar threats at long range provides submarine commanders with maximum warning of the presence of enemy surface ship or airborne ASW systems. In addition, Manta provides information for tactical battle management, targeting and information gathering. Instantaneous warning enables the submarine commander to maintain a covert posture.

Manta is designed as a modular equipment that enables the most cost-effective system to be built up to match the operational requirement for a particular class of submarine.

Different antenna configurations provide options for extended frequency coverage and high-resolution targeting. The ESM can be interfaced with the central command system and the output data may be recorded on either magnetic tape and/or hard-copy.

Using wide open ESM techniques, the systems provide 360° coverage and 100 per cent probability of intercept. Intercepted signals are automatically analysed and identified by reference to a comprehensive library of known radar types covering simple pulse, frequency agile, PRF agile, CW high duty cycle, FMOP and PMOP flags. Processed data is immediately displayed to the operator in both Cartesian and alphanumeric formats on two interchangeable colour displays. Manta uses a central management computer controlling subsystems, each containing advanced microprocessors carrying out local processing.

A version of Manta, with DF accuracy down to 2° is available for use in long-range weapon targeting. Manta variants are available for ocean-going and patrol submarines, and the Manta antennas can be mounted on a dedicated EW mast, or elsewhere as requested.

Specifications
Frequency range: 2-18 GHz (optional extensions either end are available)
Azimuth coverage: 360°
Elevation coverage: −10 to +30°
Signal polarisation: horizontal, vertical and either RH or LH circular
Bearing accuracy: 6° RMS
PRF: 50 ns to 100 µs
Scan: circular, sector, steady (simple and complex)
Scan periods: 0.02-40 s
Mission library: 2,000 emitter modes

Operational status
In production. Sea trials of a variant of Manta, on an 'Oberon' class submarine, were completed in August 1988 as part of a programme to update the Royal Navy's submarine ESM system. The Manta E variant has been supplied to Inisel to retrofit the 'Agosta' and 'Daphné' class submarines of the Spanish Navy. The Manta S outfit also equips the Swedish 'Gotland' class submarines.

Contractor
Racal Radar Defence Systems Ltd, Crawley.

UPDATED

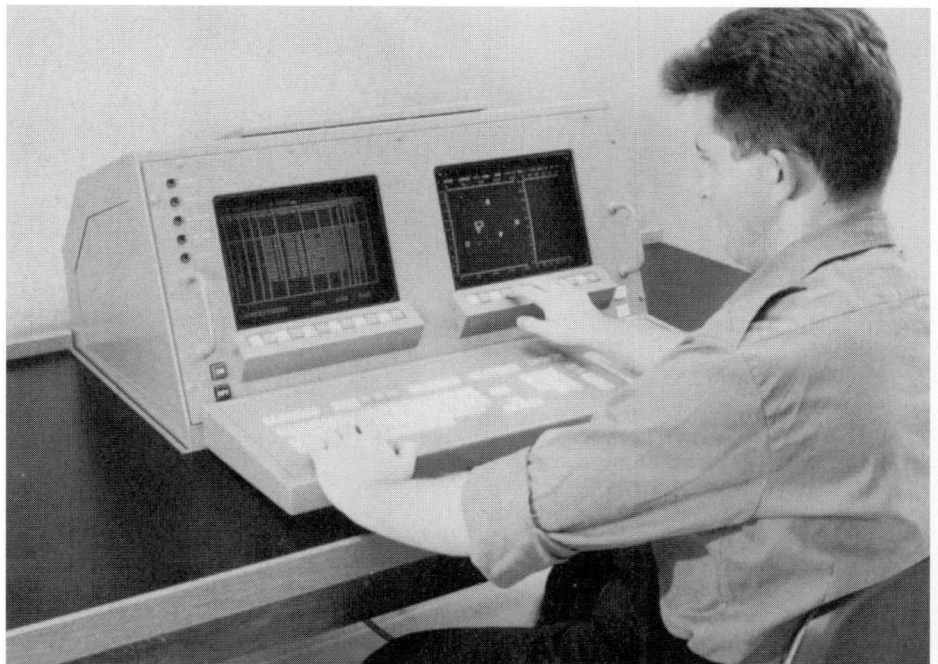
Operator's console for Manta ESM system

Sealion ESM system

Type
Submarine ESM search and warning system.

Description
The Sealion ESM system, is configured to address a wide range of submarine EW requirements including: threat warning; tactical surveillance; assistance in targeting, and intelligence gathering.

The system operates in dense pulse environments providing accurate instantaneous bearing and frequency measurement over a full 360° azimuth, giving a Probability Of Intercept (POI) of 100 per cent in the frequency range 2 to 18 GHz. The advanced system architecture is modular in design with a variety of options available to extend its performance to satisfy future operational requirements.

Key system features include:
(1) near instantaneous parameter measurement. The wide open receiver architecture uses a highly reliable omnidirectional channel
(2) accurate DF; the eight element bearing measurement system, using Racal Radar Defence System's advanced proprietary quadratic fit algorithm combined with squint correction and gain calibration, provides exceptional DF accuracy

(3) state-of-the-art pulse train deinterleaving system
(4) a modern user-friendly Windows-based operator interface
(5) extensive BITE
(6) compact size, high reliability and low power consumption.

The system consists of an omnidirectional antenna, a receiver processor unit and a low volume/lightweight operator console. The antenna may be installed in a penetrating or non-penetrating mast configuration. The receiver incorporates an advanced ESM processor using VLSI integrated circuits to provide fast analysis of complex RF environments. Real-time comparison with the preprogrammed library of 10,000 radar emitter modes provides rapid emitter recognition with the processor automatically tracking over 200 emitters simultaneously. An auxiliary library of 100 emitter modes is provided to hold operator loaded data. The compact operator console is fully ruggedised to meet the demands of the naval environment and has a high resolution colour display. This provides the operator with a tabular or graphic display of emitter identity,

Sealion antenna
1998/0006979

bearing and threat significance. The processor provides a continuous update on emitter status. Data may be recorded onto magnetic disc for subsequent detailed analysis. The system runs on Windows NT using a variety of COTS flat panel displays.

Operational status

In service on board the Royal Danish Navy 'Tumleren' and 'Narvhalen' class submarines. Sealion has also been selected in a hull-penetrating mast configuration for installation in a number of submarines currently undergoing modernisation.

Specifications

Frequency range: 2-18 GHz
Bearing accuracy: 3° rms
Sensitivity: –60 dBmi
Dynamic range: 55 dB
Elevation coverage: –30 to +30°
Azimuth coverage: 360°
Pulse density: 1,000,000 pulse/s

Contractor

Racal Radar Defence Systems Ltd, Crawley.

UPDATED

Operator's console of the Sealion ESM system

Porpoise ESM equipment

Type

Submarine radar surveillance system.

Description

A submarine version of the Cutlass ESM equipment, Porpoise is a fully automatic ESM system operating throughout 360° of azimuth. The equipment receives signals in the 2 to 18 GHz frequency range, measures their parameters and compares these with those contained in a preprogrammed radar threat library. Processing of the radar emitter signals is carried out against the library to give the operator an alphanumeric or graphic display of identification and threat significance. Up to 2,000 radar modes are held in the library and the operator can feed in data on up to 100 other emitters.

Intercepted signals are preamplified in the mast unit before being passed to the processing and analysis equipment inside the hull. Bearing data is extracted using amplitude comparison. Amplitude, pulsewidth, frequency and time are combined in a single digital word before being passed to the processor unit in the operator's console. There the pulse trains of the different radars are deinterleaved and identified from the library information.

The system is capable of integration with the vessel's fire-control and communications systems and may also be integrated with periscope-mounted radar warning equipment. Porpoise also has the ability to give an alert warning when prime threats, such as helicopter or maritime surveillance radars, reach a preprogrammed danger level.

The Porpoise antenna is a compact six-port system giving good bearing accuracy and may be mounted on either hull penetrating or non-hull penetrating masts. It is built of titanium to reduce weight and overcome corrosion, and is pressure resistant to 60 bar. With the exception of the antenna system, the primary subassemblies of Porpoise are fully compatible with those of Cutlass and so enable common logistic facilities.

Porpoise console unit

Operational status

In production and in service with several navies both in and outside NATO including the Turkish Navy ('Preveze' class Type 209 and some 'Atilay' class Type 209s) and the Chilean Navy where they may fit the new 'Scorpene' class.

Contractor

Racal Radar Defence Systems Ltd, Crawley.

UPDATED

Porpoise antenna unit

UAP 1/3 Family of ESM systems

Type

ESM systems for submarines of the Royal Navy.

Description

Racal Radar Systems Ltd is the sole supplier of submarine radar ESM systems to the Royal Navy. With the removal from service of the older UAP(2) systems in mid-1996, the UAP family of equipment will consist of outfits UAP(1) in the 'Swiftsure' and Trafalgar' class SSNs and UAP(3) in the four 'Vanguard' class SSBNs.

Outfit UAP(1) is a state-of-the-art submarine radar ESM system designed, developed and manufactured by Racal Radar Defence Systems Ltd. The equipment has successfully completed the full range of MoD (UK) specified harbour and sea acceptance trials, exceeding its declared performance specification in a number of key areas, such as DF accuracy. As a result, UAP(1) has achieved Fleet Weapon Acceptance status and has been transferred to the Naval Support Command (NSC). UAP(3) has also successfully completed all mandated trials and was transferred to the NSC late in 1995.

Outfit UAP(1) has a frequency range covering all the required radar bands and a frequency extension capability is already in place, in anticipation of future operational requirements.

The UAP(1) and UAP(3) equipment fulfil the four principal submarine ESM requirements of self-protection, surveillance, data gathering and, in the case of UAP(1) only, over-the-horizon targeting for weapon release. UAP(1) and UAP(3) also incorporate Sadie, a high-performance signal analysis processor. They also benefit from a number of operational and technical enhancements as follows:

(a) automatic alarm levels – the calculation of the alarm level, which indicates that the submarine is at risk of detection by radar, has been automated. Factors considered in this calculation include the number of masts raised, the sea state, command risk factor, frequency band, and library identity of the intercepted radar

(b) disposition of antennas – the minimising of mast exposure is a critical requirement of submarine

UAP(1) operator's console unit showing two colour displays, advanced MMI, and an onboard training unit
1996

ESM. The disposition of the UAP(1) antenna elements allows the command to carry out in excess of 95 per cent of the radar ESM role using the search periscope only. The main ESM mast is used only for high-accuracy direction-finder targeting, or as a reversionary mode in the event of search periscope failure. The search periscope antenna assembly Outfit AZE(2) also includes GPS satellite navigation and tactical communication facilities, thus further obviating the requirement to raise additional masts

(c) Man/Machine Interface (MMI) – UAP(1) and UAP(3) are single operator systems employing a number of special function keypads, which together with a rollerball allow the operator to achieve the majority of his interaction with the system by using a single keystroke. The MMI features twin high-resolution colour displays which present unprocessed and processed data to the operator

(d) onboard training – UAP(1) and UAP(3) will be the first submarine ESM systems to have an integral onboard trainer. This will allow prerecorded scenarios to be played into the system while the submarine is deep. The onboard trainer is capable of simulating the most complex scenario with full environmental representation. Realistic onboard training is a critical asset in maintaining an operator's professional skill both at sea and in harbour

(e) data gathering – UAP(1) and UAP(3) allow the operator to log to disk both parametric and pulse-by-pulse data of any selected radar intercept. The recordings can be reviewed on board before submission for post-patrol analysis.

Operational status

The UAP(1) submarine ESM system which has been proposed for, and is compliant with, the requirements of the Royal Navy 'Astute' class, is the subject of a mid-life enhancement programme which commenced in late 1997. Selected enhancements will also be incorporated into UAP(3) where appropriate and cost effective.

Contractor

Racal Radar Defence Systems Ltd, Crawley.

UAP(1) ESM mast antenna

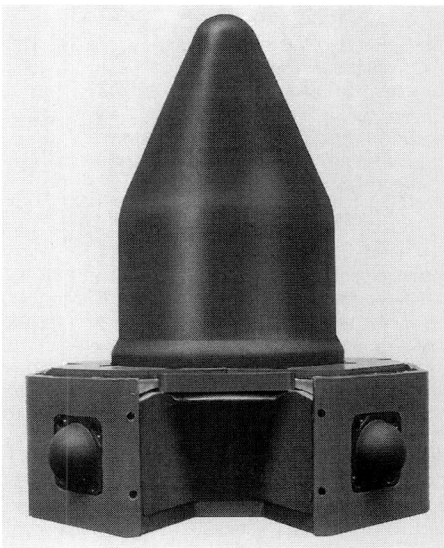

The AZE antenna   *1996*

UPDATED

UNITED STATES OF AMERICA

AN/BLD-1 DF system

Type

Submarine ESM direction-finding system.

Description

The AN/BLD-1 is a submarine-based precision ESM system, installed on the US Navy's 'Los Angeles' class attack submarines, which passively detects and tracks airborne, sea surface and land-based radar threats. It employs a mast-mounted antenna that is raised just above the sea surface for operation, and delivers

precise threat bearing information that is integrated with other sensor data for tactical surveillance and over-the-horizon targeting.

No technical details are available, although it is understood that Litton Systems has developed a number of complex algorithms required to solve the multipath problems caused by the close presence of surface waves.

Operational status

In service with the US Navy. Some 38 systems have been delivered or are on order. Three systems have

been ordered for the 'Seawolf' class where they form part of the AN/WLQ-4(V)1 EW system.

Contractor

Amecon Division of Litton Systems Inc, College Park, Maryland.

UPDATED

AN/WLQ-4 SIGINT system

Type
Submarine SIGINT detection and analysis system.

Description
The AN/WLQ-4 performs SIGINT missions aboard 'Sturgeon' class submarines of the US Navy. It is an automated, modular signal collection system which allows for the identification of the nature and sources of unknown radar emitter and communications signals. It incorporates a network of mini-computers and microprocessors, and data from these computers is correlated with information received from satellite sensors. The system is part of Sea Nymph, a highly classified US Navy NAVELEX programme.

The AN/WLQ-4 system has a number of key features, including:
(a) automatic search, acquisition and signal processing

(b) automatic logging, book-keeping and reporting
(c) semi-automatic correlation of real-time measured data with input from an external system
(d) 400,000 lines of AN/UYK-44 source code
(e) 50,000 lines of executable code in 40 microprocessors
(f) a significant growth capability to handle new threats.

Operational status
In service on board the US Navy's 'Sturgeon' class nuclear-powered attack submarines, ordered for the 'Seawolf' class and proposed for the NSSN.

Contractor
GTE Electronic Defense Division, Mountain View, California.

VERIFIED

AN/WLQ-4(V) skeletal system in a US Navy 'Sturgeon' class submarine

---

AN/WLR-8 EW receiver system

Type
Submarine tactical radar detection and analysis system.

Operating console of AN/WLR-8 tactical EW receiver

Description
The AN/WLR-8 is a tactical electronic warfare and surveillance receiver designed for fitting in both surface ships and submarines of the US Navy. The system is of modular construction and provisions are made for operation in conjunction with numerous types of direction-finding or omni-antennas, and a wide range of optional peripheral equipment, to provide comprehensive ESM (Electronic Support Measures) facilities. The WLR-8 is compatible with NTDS (Navy Tactical Data System) and similar action information automation systems. The system can be expanded in frequency or signal handling capability by means of simple additions and/or software changes. Four versions are available: V(1) for submarines, V(2) for 'Los Angeles' class (SSN-688) submarines, V(4) for large surface ships and V(5) for 'Ohio' class Trident submarines. Frequency coverage is from 50 MHz to 18 GHz.

Two digital computers are incorporated: a Sylvania PSP-300 for system control, automatic signal acquisition and analysis, and file processing; and a GTE PSP-200 microcomputer for hardware level control functions. Digital techniques are employed throughout the WLR-8 system, which is all solid state.

Operational facilities provided include:
(a) automatic measurement of signal direction of arrival
(b) signal classification and recognition
(c) sequential or simultaneous scanning over a wide frequency range

(d) signal activity detection for threat warning
(e) analysis of signal parameters such as frequency, PRF, modulation, pulsewidth, amplitude and scan rate
(f) logging signal parameters for display to operator (s), and printout of hard-copy to teletype or printer
(g) extensive built-in test equipment
(h) directed priority searches of specific frequency segments.

Direct reporting to onboard computers, such as NTDS, permits response times in the millisecond range with minimal operator involvement. A two-trace CRT is provided for display purposes and this can be supplemented by an optional five-trace panoramic display for presentation of signal activity data. Another CRT display is incorporated if the WLR-8 is used with automatic or manual DF antenna systems.

Operational status
Operational with USN and possibly other navies. More than 45 sets have been produced. The system is in use on board SSN-688 'Los Angeles' class submarines, two 'Benjamin Franklin' class and the V(5) version is fitted to 'Ohio' class ballistic missile submarines.

Contractor
GTE Electronic Defense Division, Mountain View, California.

UPDATED

Phoenix (AR-700) ESM system

Type
Submarine radar ESM system.

Description
Phoenix is a range of ESM systems which has been designed for use in submarines to provide automatic identification and bearing of intercepted radar emissions in the frequency range 2 to 18 GHz. The system comprises an antenna array and a highly sensitive IFM receiver, which provides precise

measurement of the threat radar operating frequency. A detailed display is provided to the operator, giving all significant parameters of the incoming signals. The system is modular in concept and can be configured to meet individual customer requirements. Both manual and fully automatic systems and analysis capabilities are available. The fully automatic version is known as Phoenix IV.

Operational status
In production. AR-700 Series systems have been supplied to India (Type 209 submarine), Sweden

('Vastergotland' and 'Nacken' class) and the Netherlands ('Walrus' class). The system is also installed on the Egyptian 'Romeo' and Greek Type 209 class submarines.

Contractor
ARGOSystems Inc, Sunnyvale, California.

VERIFIED

AR-700(S)

Type
Submarine ESM/DF system.

Description
The AR-700(S) ESM/DF modular electronic warfare system provides threat warning and DF capabilities over the frequency band 2 to 18 GHz. The system features a nominal sensitivity of –60 dBm, automatic signal/data processing and monopulse direction-finding. The DF element of the system features an accuracy of 5° rms, high sensitivity (–60 dBm), high-frequency accuracy using digital IFMs and advanced software technology including tactical graphics display.

Four other levels of extended performance are available for the ESM element, which allows the system to be upgraded to meet a variety of missions.

Five versions of the AR-700 are available covering the

2 to 18 GHz band and are wide open in frequency, bearing and time (no band stepping). All configurations include the ASP-32 advanced distributed signal multiprocessor specifically designed for operating in dense, complex naval electronic environments.

All configurations feature a 200 active signal handling capability, automatic signal and platform identification, high processing speed, a cassette recorder for loading/storing/changing library data while at sea, and an optional hard-copy printer for logging signal data. A weapon control system interface is also included.

The AR-700 consists of two subassemblies: an antenna unit with omnidirectional antenna channel for the frequency receivers and a 6-element monopulse spiral array for DF (a rotating DF is optional); a control/display assembly with two high-resolution digital IFM receivers, bearing digitiser unit which processes direction-finding data from the spiral array, and ASP-32 advanced signal processor, digital cassette recorder,

operator's tactical data display, and keyboard with function control keys.

Nine operator display formats are included with five alphanumeric formats which are used for reviewing signal intercept and library data. A graphics format summarises the overall tactical situation.

Operational status
The AR-700(S)5 configuration has become one of the standard fits for the German HDW Type 209 submarines, and remains in production. The AR-740 offers numerous performance extensions and improvements to the basic AR-700(S)5 for specialised applications.

Contractor
ARGOSystems Inc, Sunnyvale, California.

UPDATED

AR-900

Type
ESM/DF system.

Description
The AR-900 is the latest ESM system from ARGOSystems which is designed for electronic warfare and ELINT applications. The system consists of three subassemblies: the antenna, receiver/processor unit and operator workstation. The AR-900 provides a very high probability of intercept across the 2 to 18 GHz frequency band.

The antenna system features a new lower radar cross-section wideband omni-antenna and an 8-element wide-open amplitude monopulse DF array with a bearing accuracy better than 3° rms. The receiver subassembly consists of two wideband digital IFM receivers, a monopulse bearing receiver and a VME-band Electronic Signal Processor (ESP). The ESP can process up to one million pulses per second receiving digitised high-resolution frequency measurements for each received radio frequency pulse and continuous wave signals from the IFM receivers. Parameters measured include frequency, pulse repetition, pulsewidth, scan and amplitude. The monopulse bearing processor provides direction of arrival and amplitude of each pulse. Types of signal detected include conventional pulse train, agile frequency, staggered pulse repetition (2 to 16 positions), frequency/phase/amplitude on-pulse, jittered PRI, continuous wave, FMCW and pulse Doppler. The ESP processes both sets of data in parallel, compares the pattern with patterns held in a signal library to identify the emitter and its platform and alerts the operator to a high-threat signal within one second of acquiring the signal. Emitter libraries can be programmed with up to 10,000 emitter modes that can reference as many as 500 radar names and associated platforms.

The operator workstation comprises: a keyboard with integrated trackerball; high-resolution, flat panel colour display; a floppy disk drive; a printer; and an embedded computer. The display provides the operator with tactical and intelligence information using simple, comprehensive displays incorporating Windows technology. The 'activity', 'tactical', 'graphics' and 'tactical summary' pages give the operator threat warning information, while the frequency/azimuth, frequency/PRI, and frequency/amplitude pages are for analysis. An intercept report generator is provided to prepare intelligence reports.

The system is available with or without an operator console and can be integrated easily into a submarine combat system via any standard interface. As an option, the system can be provided with integrated GPS or SATCOM antennas.

Operational status
In production.

Specifications

Frequency band	2-6 GHz		6-18 GHz
System sensitivity	−65 dBm		−65 dBm
DF measurement accuracy	3.5°		2°
Dynamic range		>70 dB	
Receiver recovery time		<100 ns	
Azimuth coverage, instantaneous		360°	
Displayed resolution	1 MHz		1 MHz
Accuracy, rms	3 MHz		6 MHz
Range		0.1 to 99.9 µs	
Resolution		0.1 µs	
Amplitude measurement			
Range		>60 dB	
Resolution		0.5 dB	
PRI measurement			
Range		2 to 10,000 µs	
Resolution		0.1 µs	

Contractor
ARGOSystems Inc, Sunnyvale, California.

UPDATED

Guardian Star shipborne EW system

Type
Shipborne and submarine radar detection and surveillance systems.

Description
Guardian Star is a family of EW systems designed to cover requirements from threat detection to ELectronic INTelligence (ELINT). Various configurations allow the system to be configured to the specific requirements of surface ships and submarines.

The Mk 1 system offers early warning to small surface craft or patrol boats, with an average bearing accuracy of ±10° provided by octave frequency measurements. The Mk 2 system is designed to meet the basic ESM needs of surface ships and submarines. It provides YIG tuned frequency measurement and emitter average bearing accuracy of ±5°. The Mk 3 is an ELINT system for surface ships and submarines. It carries out instant threat warning and accurate frequency measurement by IFM devices in the receiver front end. Average bearing accuracy of less than ±5° is achieved.

The systems consist basically of an antenna assembly, an RF/Digital Interface Unit (RFDIU) and a Display/Controller Unit (DCU). The antenna assembly comprises an omnidirectional and six spiral DF antennas covering the frequency range from 2 to 18 GHz. All the necessary preamplifiers and preprocessing components are included in the assembly. The RFDIU processes all signals before transfer to the main digital processor in the DCU. All signal conversion electronics, auxiliary outputs and power supplies are contained in this RFDIU. The DCU includes the input/output section, main processor, magnetic tape unit, keypad and all required operator interfaces. The operational program and library file data are loaded into the processor memory via the magnetic tape unit.

In the surveillance mode the system operation is broadband from 2 to 18 GHz. This mode is entirely automatic and the operator has only to view the situation summary display. The display will provide data on up to 50 emitters (10 per page). The displayed emitter characteristics are frequency, PRF/PRI, pulsewidth, pulse amplitude, true bearing (instantaneous and averaged), scan rate, emitter name (if matched in the library), and threat level (if matched in the library). The library file handles up to 2,000 sets of emitter parameters. The operator can store previously known emitter data together with preassigned threat levels in the file. The system matches incoming emitter data with the file and issues an immediate alert on high-interest threats. Library data can be changed or entered via the display/controller unit.

Operational status
In service aboard Canadian 'Oberon' class submarines and Venezuelan Lupo type frigates.

Specifications
Frequency range: 2-18 GHz (optional expansion down to 0.5 GHz and up to 40 GHz)
Library: 2,000 emitter parameters
Dynamic range: 55 dB
Sensitivity: −60 dBmi
Bearing accuracy: 2-8 GHz <10° RMS; 8-18 GHz <5° RMS
Dimensions: antenna assembly 360 mm (high) × 15 cm (diameter) (including radome); RFDIU 430 × 360 × 350 mm; DCU 250 × 430 × 530 mm
Weights: antenna 5 kg; RFDIU (Mk 3) 33.5 kg; DCU 27 kg

Contractor
Sperry Corporation, Arlington, Virginia.

UPDATED

TABLE X

Electronic warfare systems[1]

System	Manufacturer	Country	Installed on	Navy	No units
921-A	CNEIEC	China	Xia	China	1
921-A	CNEIEC	China	Han	China	5
921-A	CNEIEC	China	Song	China	1
921-A	CNEIEC	China	Romeo	North Korea	21
AR-700	ArgoSystems	USA	Walrus	Netherlands	4
AR-700[5]	ArgoSystems	USA	Type 209/1500 (Shishumar)	India	4
AR-700	ArgoSystems	USA	Type 207 (Kobben)	Norway	6
AR-700	ArgoSystems	USA	Romeo	Egypt	4
AR-700	ArgoSystems	USA	Hai Lung	Taiwan	2
AR-700-S5	ArgoSystems	USA	Type 209/1100-1200 (Glavkos)	Greece	8

System	Manufacturer	Country	Installed on	Navy	No units
AR-700-S5	ArgoSystems	USA	Vastergotland A 17	Sweden	4
AR-700-S5	ArgoSystems	USA	Nacken A 14	Sweden	3
AR 740	ArgoSystems	USA	Collins	Australia	1
Arud	Thomson-CSF	France	Agosta	France	3
Arud[3]	Thomson-CSF	France	Hashmat (Agosta)	Pakistan	2
Arud	Thomson-CSF	France	Hangor (Daphne)	Pakistan	4
Arud[4]	Thomson-CSF	France	Daphne	South Africa	3
Arur/DR 3000	Thomson-CSF	France	Agosta	France	3
Arur/DR 3000	Thomson-CSF	France	Albacora (Daphne)	Portugal	3
Arur/DR 3000 U	Thomson-CSF	France	Le Triomphant	France	2
Arur/DR 3000 U	Thomson-CSF	France	L'Inflexible	France	3
Arur/DR 3000 U	Thomson-CSF	France	Rubis	France	6
BLD-727	Elettronica	Italy	Improved Sauro	Italy	4
BLD 727	Elettronica	Italy	Sauro Type 1081	Italy	4
BRD-7		USA	Los Angeles	USA	58
Bald Head	RFAS	RFAS	Alfa	RFAS	1
Brick Group	RFAS	RFAS	Kilo Type 877E	Algeria	2
Brick Group	RFAS	RFAS	Kilo Type 877E	Poland	1
Brick Group	RFAS	RFAS	Kilo Type 877E	Romania	1
Brick Group	RFAS	RFAS	Charlie II	RFAS	2
Brick Group	RFAS	RFAS	Delta IV	RFAS	7
Brick Group	RFAS	RFAS	Delta III	RFAS	12
Brick Group	RFAS	RFAS	Delta I	RFAS	5
Brick Group	RFAS	RFAS	Victor III	RFAS	18
DR 2000	Thomson-CSF	France	Type 209/1200 (Salta)	Argentina	1
DR 2000	Thomson-CSF	France	Type 209/1200 (Atilay)	Turkey	6
DR 2000U	Thomson-CSF	France	Type 206	Germany	2
DR 2000U	Thomson-CSF	France	Type 206A	Germany	12
DR 2000U	Thomson-CSF	France	Type 206	Indonesia	4
DR 3000U	Thomson-CSF	France	Hashmat (Agosta)	Pakistan	3[6]
DR 4000	Thomson-CSF	France	Type 209/1,400 (Tupi)	Brazil	3
FL 1800U	Daimler Chrysler	Germany	Type 212	Germany	4[6]
Guardian Star	Sperry Marine	USA	Oberon	Canada	3
Manta	Racal-Thorn Defence/Inisel	UK/Spain	Galerna (Agosta)	Spain	4
Manta	Racal-Thorn Defence	UK	Delfin (Daphne)	Spain	4
Manta S	Racal-Thorn Defence	UK	Gotland A 19	Sweden	3
Mavis ODU	AWA	Australia	Oberon	Australia	3
Porpoise	Racal Radar Defence	UK	Type 209/1400 (Preveze)	Turkey	2
Quad Loop	RFAS	RFAS	Kilo Type 877 EKM	Iran	3
Quad Loop	RFAS	RFAS	Kilo Type 877E	Poland	1
Quad Loop	RFAS	RFAS	Kilo Type 877E	Romania	1
Quad Loop	RFAS	RFAS	Kilo Type 877/877K/877M	RFAS	24
Rim Hat	RFAS	RFAS	Oscar II	RFAS	12
Rim Hat	RFAS	RFAS	Typhoon	RFAS	6
Rim Hat	RFAS	RFAS	Akula I/II	RFAS	13
Rim Hat	RFAS	RFAS	Sierra I/II	RFAS	4
Sarie 2	THORN EMI	UK	Type 206A	Germany	12
Sealion	Racal-Thorn Defence	UK	Type 207 (Tumleren)	Denmark	3
Sealion	Racal-Thorn Defence	UK	Narhvalen	Denmark	2
Sealion	Racal-Thorn Defence	UK	Ula (P 6071)	Norway	6
Sea Sentry[5]	Kollmorgen	USA	Type 209/1500 (Shishumar)	India	4
Sea Sentry III	Kollmorgen	USA	Santa Cruz TR 1700	Argentina	2
Squid Head[2]	RFAS	RFAS	Kilo Type 877EKM	China	3
Squid Head	RFAS	RFAS	Kilo Type 877EKM	Iran	3
Squid Head[2]	RFAS	RFAS	Tango Type 641B	RFAS	10
Squid Head[2]	RFAS	RFAS	Kilo Type 877/877K/877M	RFAS	24
Stop Light	RFAS	RFAS	Foxtrot Type 641	Cuba	2
Stop Light	RFAS	RFAS	Foxtrot Type 641	India	6
Stop Light	RFAS	RFAS	Foxtrot Type 641	Libya	4
Stop Light	RFAS	RFAS	Foxtrot Type 641	Poland	2
Stop Light	RFAS	RFAS	Charlie II	RFAS	2
Stop Light	RFAS	RFAS	Sava	Yugoslavia	2
Stop Light	RFAS	RFAS	Heroj	Yugoslavia	2
Timnex	Elbit	Israel	Gal	Israel	3
Timnex	Elbit	Israel	Dolphin	Israel	3[7]
Timnex	Elbit	Israel	Hai Lung	Taiwan	2
UAP 1	Racal-Thorn Defence	UK	Trafalgar	UK	7
UAP 1	Racal-Thorn Defence	UK	Swiftsure	UK	5
UAP 3	Racal-Thorn Defence	UK	Vanguard	UK	3
WLR-1/3		USA	Guppy II	Taiwan	2
WLR-8	GTE	USA	Benjamin Franklin	USA	2
WLR-8(V)2	GTE	USA	Los Angeles	USA	58
WLR-8(V)5	GTE	USA	Ohio	USA	18
WLR-10		USA	Benjamin Franklin	USA	2
WLR-10		USA	Los Angeles	USA	56
WLQ-4	GTE	USA	Sturgeon	USA	13
WLQ-4	GTE	USA	Narwhal	USA	1
WLQ-4(V)1	GTE	USA	Seawolf	USA	1
ZLR 3-6	Japan	Japan	Harushio	Japan	7
ZLR 3-6	Japan	Japan	Yuushio	Japan	10
ZLR 7	Japan	Japan	Oyashio	Japan	1
n/k	ArgoSystems	USA	Chang Bogo 209/1200	Korea, South	5

[1] According to information available   [4] Being upgraded with domestic design   n/k not known
[2] Or Brick Pulp   [5] In some units
[3] Unmodernised units   [6] Building   *UPDATED*

UNDERWATER COMMUNICATIONS

This section covers those systems that are primarily devoted to underwater communications. Information on these is frequently sparse because of security restrictions. Equipments described cover communications between surface ship/submarine and vice versa, aircraft/submarine and vice versa, and submarine/submarine.

Radio communications between submerged submarines and ships, aircraft or shore-based stations pose many problems.

Waveband communication problems

Generally speaking, surface vessels communicate over long ranges using the 3 to 30 MHz HF band, since neither the 0.3 to 3 MHz MW nor the 30 to 300 MHz VHF bands offer adequate range. The possibility of detecting communications in this waveband is impeded by the use of burst transmission and wideband modulation. This is because the signal energy is distributed over a wide frequency band. For satellite communications the 3 to 300 GHz band is used.

However, submarines have difficulty in using any of these wavebands, as they are unable to penetrate the water. To establish two-way radio communication for urgent priority messages the submarine must therefore either surface or deploy an antenna while remaining submerged. Both options are unsatisfactory because they lower the indiscretion rate of the submarine. Any transmitter can be rapidly and accurately detected by modern electronic DF equipment, even if the message is sent as a digitised burst lasting only a few seconds. From this, the ASW aircraft can immediately determine the likely area of operation of the submarine and plan an effective search and attack. Submarines therefore have to use alternative methods to communicate from greater depths without direct physical contact with the surface.

Transmitter buoys

One method that enables the submarine to put itself at some considerable distance from its transmitter is the towed buoy. Tethered to the submarine by a datalink cable, the buoy houses an antenna, its own power supply and a wideband amplifier. However, given the high resolution of modern airborne search radars, this solution too is vulnerable to detection. Smaller buoys are therefore required which the submarine can launch at depth from its SSE. In addition to the transmitter and power supply, these free floating communications buoys feature a pre-set timer, an extending antenna and a recorder. When the buoy reaches the surface the timer triggers the deployment of the antenna and starts

to count down the delay period pre-set by the submarine to allow it to clear the launching area. At this point the message stored on the recorder is transmitted. On completion of the transmission to a satellite the buoy scuttles itself.

ELF communications

The only way in which a submarine can receive messages from its national command centre, without placing itself in a position where it is vulnerable to detection, is to use the 3 to 30 Hz ELF waveband. Transmissions in this wavelength can travel down to a depth of around 122 m at ranges of over 6,000 n miles, enabling the submarine to remain discrete.

The major problem with using ELF is the size of the required antenna system. For example, an ELF transmitter can only operate to its maximum efficiency at 75 Hz using an antenna array covering several square kilometres and in an area with very low conductivity characteristics. The power requirements are also heavy: sending an ELF signal of a few watts through the air demands a transmitter input to the aerial of several megawatts. Furthermore, data transmission speed is even slower than in the VLF band due to the limited bandwidth. This implies that the messages to be transmitted must be kept to a bare minimum using short, pre-established code groups. It has been established that over a 15-minute period no more than three characters can safely be transmitted. In order to guarantee that the message is properly received, it is continuously repeated until a new message or a station standby tone is transmitted. This means that the ELF station has to remain continuously on air. Finally, to achieve the best possible reception requires the submarine to tow a very long antenna on an absolutely straight course in both horizontal and vertical planes.

Another drawback of the ELF band is the extreme vulnerability of the large ground-based antenna, which would obviously become a priority target in a major conflict. Although ELF signals are inherently resistant to the electromagnetic pulses that would arise from a nuclear explosion, the American system is not designed to survive a nuclear strike, and since the station transmits continuously, it acts as a 'fail-safe' system. This means that should the station cease to transmit, submarines would assume that war had broken out. In this case the submarine would rise to upper levels to communicate with its assigned command centre using the HF network, or contact a defence communication satellite for further orders.

Today, ELF transmitters are used by the RFAS and the United States for the transmission of commands to their strategic missile submarines.

Satellite laser communications

Although the sophisticated VLF and ELF communications systems help submarines to remain undetected, they are hardly compatible with even the most basic requirements of current command and control concepts. The need for a two-way voice and/or data communications system for deep-diving submarines is paramount. Since over short distances sonar-based underwater telephony is feasible, surface ships can serve as relay stations to submarines while aircraft can drop buoys containing sonar-based data transmitters. Submarine hunter-killer groups that comprise all three elements of naval warfare (surface, sub-surface and airborne units) employ such methods. However, the use of sonars for longer-range two-way contact is, for a number of technical and tactical reasons, neither feasible nor desirable.

One possibility examined during the 1970s was the use of blue-green laser radiation in the visible light spectrum. This is capable of penetration to depths of more than 90 m. Towards the end of the decade experiments were underway to develop a submarine laser communication system. To obtain the widest possible coverage it was decided to use satellites as relays between ground stations and submarines, the messages being first directed to the satellites as microwave signals and then converted into modulated blue-green laser light. The concept called for a system comprising up to four satellites, several ground stations and laser receivers in virtually all US Navy ballistic missile and attack submarines. The scientists faced enormous problems in developing such a system, and although early trials showed that such a system would confer enormous advantages, particularly in trials with a system mounted in an aircraft, both budgetary constraints and the ending of the Cold War led to a reappraisal of the need for such a highly complex and costly system. However, the successful trials using a blue-green laser communication suite mounted in an aircraft led to a shift away from satellite communication for submarines to aircraft, with the intention of adapting the developing technology for use in aircraft engaged in tactical ASW scenarios.

For submarines to have to surface to establish a UHF/VHF link with a hunter-killer task force has never been regarded as an ideal means of communication. Neither has sonar telephony ever been rated as very reliable. Airborne laser communications, on the other hand, could radically change this unsatisfactory situation and increase the effectiveness of ASW teams.

UPDATED

SUBMARINE COMMUNICATIONS SYSTEMS

FRANCE

MCA 30/MCA 45 cable antenna

Type

Buoyant cable antenna for diesel and nuclear submarines, allowing VLF/LF/HF reception.

Description

The buoyant cable antenna is a dispensing and retrieval system for submarine VLF/LF/HF reception and navigation. Acting as an underwater antenna, the cable is paid out swiftly, according to the speed and depth of the submarine, so that the end of the antenna rises to the surface of the sea and close enough to receive electromagnetic signals. The MCA 30 (300 m version)

or MCA 45 (450 m version) is a compact system consisting of a buoyant cable dispensed by a winch equipped with a cable pusher and a pressure hull throughway.

The buoyant cable is an insulated conductor, terminated with a short electrode in direct contact with the seawater. For HF communications the cable is fitted with an online HF preamplifier.

The cable pusher unit drives the cable in and out and has a cable length detector which will automatically dispense the cable to the required length. The pressure hull throughway is a complete subassembly equipped with a series of pneumatic and manual redundant safety devices. This subassembly ensures

watertightness in the different phases of operation, which are: cable in, cable out and cable in motion.

Operational status

The buoyant cable is standard equipment on the French Navy's submarines and has been installed on submarines of a number of other countries.

Specifications

Diameter: 16 mm
Length: 300, 450, 600 m
Buoyancy: 0.7
Cable deployment speed: 0.8 m/s
Operating depth: down to 300 m

Inflated seal pressure: 50 bars
Maximum towing speed: >15 kt
Radio channels: VLF 14-23 kHz; LF 60-80 kHz; HF 500 kHz-30 MHz

Contractor
Société Nereides, Les Ulis.

UPDATED

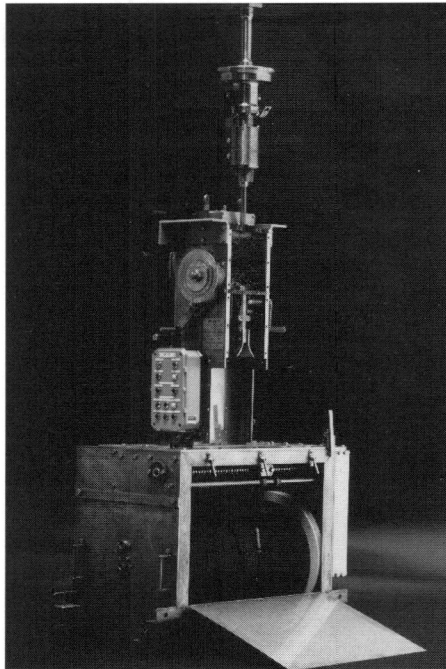

MCA 30/MCA 45 buoyant cable antenna

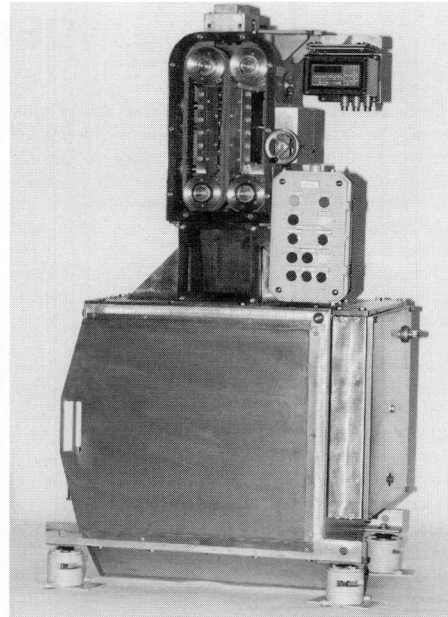

MCA buoyant cable antenna for submarines
(Nereides)   *1999*/0017502

GERMANY

UT 2000

Type
Underwater communication system.

Description
The UT 2000 is a small, compact microprocessor-controlled, multifunction underwater acoustic system with graphic EL display and keyboard. In addition to telephony and telegraphy modes, various other additional functions and features are available. It fulfils the frequent demand of a wider and variable frequency range, adjustable output power and sector or omnidirectional operation. Depending on the acoustic propagation conditions, these features allow far- or close-range communication with the shortest possible intercept range, which is especially essential for submarines.

Operating modes and parameters are menu-guided and selected from the keyboard so that untrained personnel can use the system. The EL display always shows the whole frequency range from 1 to 60 kHz as well as parameters according to the selected mode (for example – output power, carrier frequency, modulation control and so on).

Control is simplified by an automatic switchover from receive to transmit.

Use of upper and lower sideband respectively with suppressed carrier (SSB-operation) guarantees a high signal-to-noise ratio at a high transmission bandwidth. In addition, while in telegraphy mode, the transmission bandwidth is reduced so that an optimum signal-to-noise ratio and consequently greater range can be achieved.

To cover the frequency range from 1 to 60 kHz it is necessary to operate with different sets of transducers for the preferred frequency bands; low frequency for long-range communication, high frequency for reduced range of intercept or for communication with divers.

Operational status
More than 130 units have been sold worldwide in the last 10 years equipping both submarines and surface vessels of most categories.

Specifications
Frequency: 1-60 kHz (adjustable in steps of 50 Hz)
Power output: LF 300 W (max); HF 50 W (max)
Audio output: 1 W, 4 Ω;

Contractor
L-3 Communications ELAC Nautik, Kiel.

UPDATED

ISRAEL

SACU 3100

Type
Stand-alone digital communications unit.

Description
The SACU 3100 provides two-way, secure, high-speed transmission/reception for messages in standard formats or free text.

Each terminal can handle three communication channels simultaneously, connected directly to standard radio communication equipment (FM and AM, VHF, UHF and HF) as well as two-wire telephone lines.

The system features manual and/or automatic operation, and can serve as an automatic communication controller for command and control systems. A variety of network communication features is provided such as: encryption (optional), automatic and/or manual acknowledgement, multinetwork relay capability and broadcast messages.

Messages can be transmitted in a wide range of baud rates, while error detection and correction algorithms enable communication in adverse conditions. The system supports single- or dual-language operation using a 2000 character graphic display. The software supports special purpose processing such as ballistic computations, navigation

The SACU 3100 stand-alone communications unit

1995

and so on. Messages are transmitted in short bursts while received messages are stored in a memory and displayed at operator request.

The hardware configuration is based on an 80386 microprocessor CPU, with optional GPS integrated receiver, crypto unit and HF high-speed modem. Optional features include a command and control

software package to enable tactical picture processing and display in addition to message display. A smaller version (Mini SACU) with 320 character display, touch panel keyboard and FSK (75 to 1,200 baud) modem is available.

A ruggedised laptop (LT 2000) version with HW and SW extension is also available as an option.

Operational status
SACU is currently in service with several navies throughout the world.

Contractor
Elbit Systems Ltd, Haifa.

VERIFIED

ITALY

Integrated communication system

Type
Submarine integrated radio communication system.

Description
Main features of Elmer's integrated communication system for submarines include a complement of wideband and tunable antennas for transmission and reception covering the LF to UHF frequency bands, the use of MF/UHF multicouplers and antenna filters, the assembly of equipment in preconfigured racks, centralised system control and supervision, and the use of serialised data transfer.

Communication facilities available include those for ship-to-ship, ship-to-shore and ship-to-air.

Operational status
Manufactured in a number of variants, the systems have been installed on a number of naval units belonging to Italy and other countries.

Specifications
Frequency range: LF to UHF
Power output: 30 or 400 W
Types of service: analogue and digital voice data
Antenna types: magnetic loop, whip. UHF section of UHF/IFF antenna
Number of users: up to 3

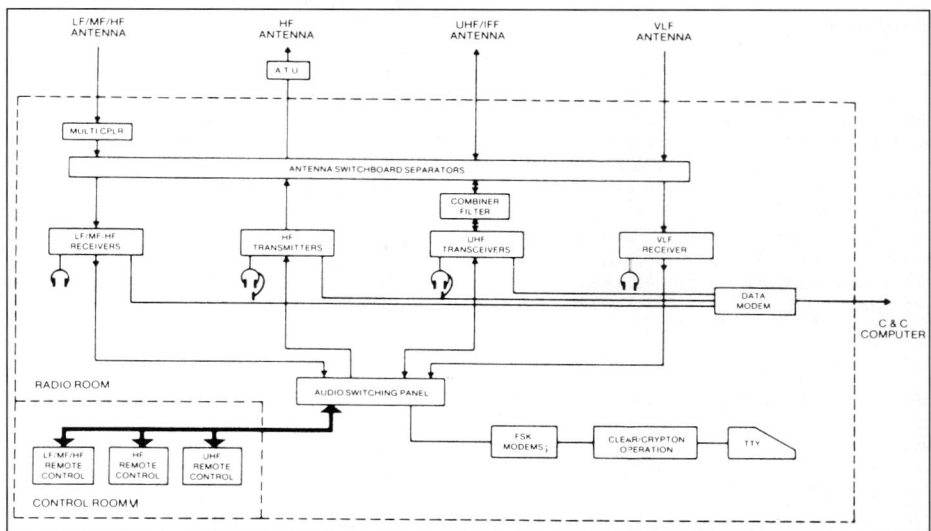

Typical integrated communication system for submarines

Contractor
Elmer, Pomezia.

VERIFIED

UNITED KINGDOM

Seafox

Type
Submarine integrated communications system.

Description
Seafox is an integrated system designed for submarines and light naval forces. An alternative range of equipment is also available which is designed to suit vessels ranging from patrol boats, mine countermeasures vessels and offshore patrol vessels up to large corvettes and light frigates.

Each system provides for intercom – for combat systems co-ordination and control – plus ship management; tactical radio communication with other ships and aircraft, and strategic radio communication with the command on shore.

The Seafox range of equipment includes the H1073 amplifier, H4640 transceiver, H6701 control outfit, H6702 internal communication system and the H1473 automatic tuning unit.

H1073 250/400 W MF/HF amplifier
When used with the H1473 automatic antenna matching unit and a suitable drive (for example, ICS3 or H4640), the H1073 provides a 250/400 W transmission system. Tuning is fully automatic over the 1.5 to 30 MHz frequency range. The frequency range can be extended down to 240 kHz as an option. Tuning time is typically 1.5 seconds.

H4640 transceiver
The H4640 is a fully synthesised drive/receiver designed for naval shipborne use. The basic equipment comprises four units mounted in a single cabinet. These are the drive, receiver, synthesiser and a common power supply unit. This basic configuration allows for simplex or two-frequency working; other configurations provide for duplex working and for the

use of pre/post-selectors where improved out-of-band performance is required.

H6701 control outfit
The H6701 control outfit has been designed to provide remote control of radio communications equipment. The system consists basically of a central Control Selection Unit (CSU) and a number of remote-control units situated at remote operating positions, together with radioteletype, data terminals and loudspeakers as required.

Time division multiplex techniques are employed to achieve switching between the various user positions and the radio equipment. The controls necessary for selecting the user equipment combinations, HF transmitter power level and HF transmit/receive muting, are centralised. Frequency channel and audio characteristics are selected remotely by the operator.

The standard system has a capacity for a maximum of 16 user lines and 16 equipment lines but can be sub-equipped for a smaller number of lines. For larger systems, two CSUs can be interlinked by cross-connecting two or more user and equipment lines according to system requirements.

H6702 tactical internal communication system
The H6702 system is designed to provide tactical internal communication (voice) between operational users. Forms of communications available are: group, in which a number of users is connected in conference on a single speech channel; interphone, for single user-to-user connection; and broadcast, for a single user transmitting to multiple users on receive only. The basic system is designed around a central Internal Communication Exchange (ICE) and provides for up to 16 user positions. There are 16 separate speech channels available for voice communication between any combination of user-to-user, conference groups and general broadcast. Each user position can have up

to 16 keys available (depending on the number of internal CSUs fitted) which can select particular connections to other users or combinations of users via the switching exchange.

The system can be expanded to accommodate up to 32 user positions by adding a second ICE. Facilities are provided for interconnection to the external communication circuits.

H1473 automatic antenna tuning unit
The H1473 antenna tuning unit automatically matches the HF output from an H1073 amplifier (or the 500 W combined output from a pair of such amplifiers) to whip antennas 7 to 12 m long, or equivalent wire antennas, over the 1.6 to 30 MHz frequency range.

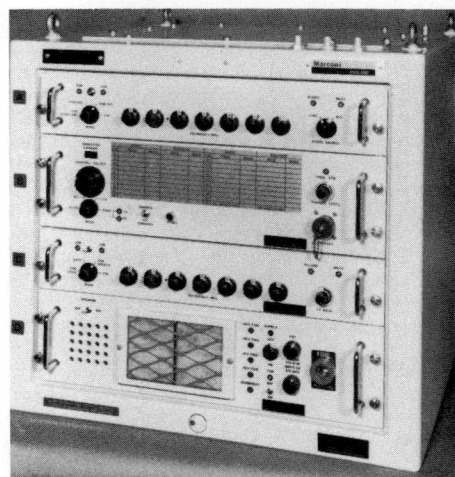

H4640 transceiver

Operational status

The Seafox equipment was introduced in 1983. The system is in service with the Royal Navy and the navies of Greece and Indonesia.

Specifications

H1073 Amplifier

Frequency range: ed 01 – 1.5-30 MHz; Ed 02 – 1.5-30 MHz and 240-1,496 kHz

Power output: 6-400 W mean or PEP into 50 Ω; load with level adjustable in 7 steps

Contractor

GEC-Marconi Communications Ltd, Chelmsford.

VERIFIED

GS7210 receiver

Type

VLF/LF multichannel receiver.

Description

The GS7210 multichannel MSK receiver is an enhanced version of the GS7110 (see page 143 of 1996-97 edition). The receiver guarantees interoperability within NATO and exceeds many of the requirements of the NATO STANAG 5030.

The receiver retains all the key functions and capabilities of the GS7110 including FSK/MSK multimode demodulation and traffic format processing requirements of STANAG 5030. Operating over a frequency range of 10 to 200 kHz (as in the GS7110), the receiver provides 50 and 70 baud FSK/MSK reception.

The performance features of the 7210 receiver result from research undertaken by the manufacturer for NATO and covering critical aspects of the VLF submarine communications environment including propagation, atmospheric noise and spectrum occupancy. Optimal performance is achieved in atmospheric noise and adjacent channel rejection to enable message reception in many difficult locations throughout the world. Performance features include S/AN (Vd = 10 dB): to –12.7 dB, depending on mode, and S/AN (Vd = 14 dB): to –16 dB, depending on mode.

The fully modular system is mounted in a standard 19 in rack unit and existing GS7110 receivers are upgradable to GS7210 standard.

Specifications

Frequency range: 10-200 kHz in 5 Hz steps

Performance in atmospheric noise (CER 0.1%):
S/AN (Vd = 10 dB): FSK-8.2 dB; MSK2 –12.7 dB; MSK4 –6.4 dB
S/AN (Vd = –14 dB): FSK –10 dB; MSK2 –16 dB; MSK4 –8 dB

Noise: 13 dB max

Sensitivity: 50 Ω; input –130 dBm for (S + N)/N of 10 dB

RF intermodulation: third order intercept +18 dBm

IF rejection: 90 dB

Image rejection: 90 dB

Contractor

GEC-Marconi Radar and Defence Systems, Command and Information Systems Division, Camberley.

VERIFIED

VLF shift keying system

Type

Communications and management control of submarine forces.

Description

The GEC-Marconi VLF shift keying control system has been designed for management and control of submarine forces. Sophisticated broadcast preparation increases the capacity, efficiency and security of processing command traffic, while networked communications to the broadcast transmitters provide high interconnectivity and information exchange across the entire theatre of operations, thus permitting effective use to be made of the new Minimum Shift Keying (MSK) technology.

This system, developed for NATO, and complying with the latest NATO standards, equips two broadcast preparation stations and one broadcast transmitter in each of three countries to cover the Atlantic and Mediterranean NATO fleets.

Each equipped broadcast preparation centre can accept up to 24 simultaneous incoming circuits from anywhere in NATO, and provides electronic editing and collation of messages for the generation of up to eight broadcasts. These are then re-encrypted and sent via a complex line switching matrix which permits full interconnectivity between all sites to the broadcast transmitters, where they are transmitted using MSK equipment.

Computing and processing and software-controlled automatic switching are essential features of the VLF system which employs Ada software throughout. To complete the upgrading of the broadcast stations' off-the-air monitoring, error detection and correction systems are provided, while the associated electronics automatically tune the broadcast variometers to ensure optimum broadcast quality.

Operational status

In operational use.

Contractor

GEC-Marconi RDS, Portsmouth.

VERIFIED

MSK 5030 demodulator

Type

VLF/LF MSK demodulator.

Description

The compact, fully automatic MSK 5030 demodulator provides a standard VLF/LF receiver with NATO STANAG 5030 MSK compatibility.

The demodulator is connected to the output of any suitable receiver by means of a receiver headphone socket, or other audio interface. The output of the demodulator interfaces to the decrypting equipment or teletype. When connected to a radio receiver the device demodulates a received MSK transmission, regains the timing frame and demultiplexes the data to the original four individual channels.

Suitably modified, the MSK demodulator can also be used to provide redundancy coding.

Operational status

In use by NATO VLF/LF ground stations. Also in service with submarines from two NATO navies.

Contractor

Thomson Marconi Sonar Ltd, Templecombe.

VERIFIED

SMBPS

Type

SubMarine Broadcast Processing System (SMBPS).

Description

The SMBPS provides a fully integrated VLF/LF communications management facility to Royal Navy submarine operations staff. It comprehensively updates and combines the roles of operations and communications management onto a single, highly secure and available platform.

The system is based on the security-enhanced variant of the highly reliable VMS operating system from Digital and the E3 assured OpenIngres database from Computer Associates. A client-server design takes advantage of the high availability characteristics of a VAX cluster, while optional site-to-site replication affords a further level of protection.

SMBPS provides complete separation and protection of secure information (based on the Bell and La Padula model) without compromise to the systemic usability.

A comprehensive application suite provides Security Officers with the capability to mix and match security clearances of users, workstations, message and broadcast channels and submarine clients. This provides F-B1 functionality and is evaluated to ITSEC E3 assurance level.

SMBPS provides interconnectivity to ACP 127 message networks and a complete messaging service for the operations and communications staff using the system.

The integrated vetting and scheduling applications applied to message traffic and the collation of VLF and satcom broadcasts which SMBPS provides, offers a significant manpower advantage over manual methods. Automatic vetting of traffic according to user-selectable criteria provides particular benefits, for example, in ensuring that the latest intelligence traffic is routed directly to the earliest appropriate broadcast. The multilevel secure design of SMBPS will prevent routeing of a message to a broadcast that is not cleared to carry that traffic.

The system supports the planning of transmitter usage and crypto allocation up to two years ahead.

Detailed planning facilities include individual channel plans on MSK assets, the allocation of broadcast schedules to channels and the allocation of submarines to schedules. These plans are used by the system automatically to compile and key MSK VLF/LF and satcom broadcasts.

Full implementation for SSIXS protocol satellite communications is provided, supporting pre-planned satcom broadcast and random access query-response mode operation. The system is not dependant on transmitter hardware and is DAMA-ready.

Operations staff is provided with full Waterspace Management (WSM) tools using the proven Nauticus system from INRI UK. WSM is fully integrated into the SMBPS system, providing benefits in the speed and integrity of WSM communications.

SMBPS integrates geographic applications from INRI with the messaging and broadcast facilities to provide auto vetting of up-to-the-minute track information. Of particular value is the capability to filter information in user-definable areas of interest (AOI) onto specific submarine broadcast via the SMBPS vetting system.

SMBPS workstation (British Aerospace Defence Systems)
1999/0017500

Operational status
SMBPS provides the core operations and communications functions for the UK Royal Navy's CTF 345 Command Centre project.

Contractor
British Aerospace Defence Systems Limited.

NEW ENTRY

Submarine communications mast

Type
Integrated non-hull penetrating communications mast for submarines.

Description
The mast forms part of an integrated submarine communications system comprising HF, VHF and UHF aerials which are supported on a non-hull penetrating low radar cross-section GRP mast.

The integrated communications system performs multiple functions, from MF, HF, VHF and UHF communications through JTIDS and IFF, to SATCOM and SATNAV (both Transit and NAVSTAR), using only one mast. The system is capable of simultaneous transmission and reception for most functions, and reception only for others. Spatial coverage is from low-angle line of sight communications to high-angle SATCOM.

The hydrodynamic shape of the mast enables deployment while submerged, at high submarine speeds and in severely adverse weather conditions to provide high operational availability for radio communications.

The low drag, streamlined mast profile exhibits excellent vortex-shedding characteristics and provides extreme resistance to vibration while extended, thus conferring stability to the masthead payload at high speeds. With hydraulically actuated extension and retraction, the mast is reliable in operation under all conditions. Being non-hull penetrating, the mast allows flexibility in submarine design and saves valuable space inside the pressure hull. The mast can be supplied as a single drop-in module which is ideal for retrofit programmes.

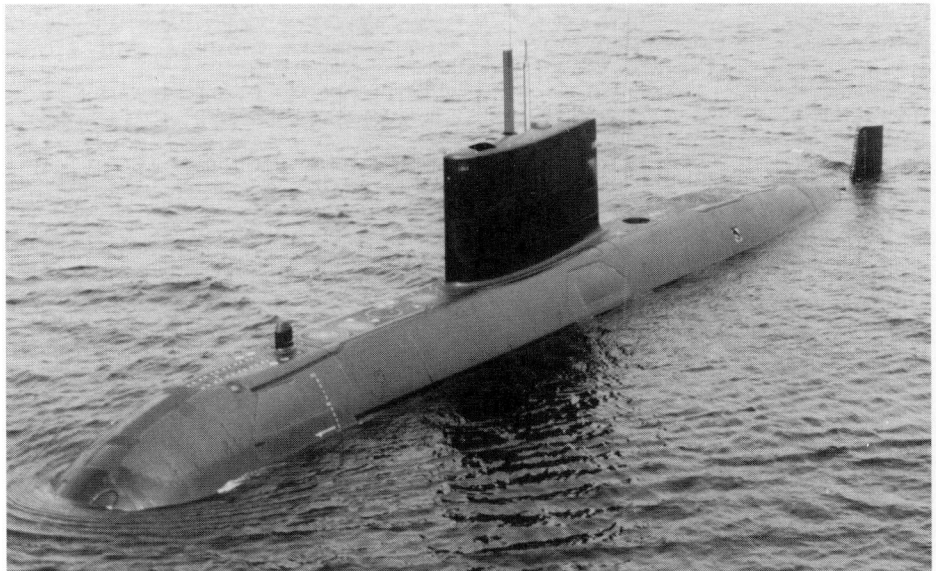

AJU integrated communications mast fitted in HMS Upholder

To minimise radar signature the upper portion of the mast incorporates multiband RAM in the GRP laminate.

The submarine communications mast meets all the requirements of the Royal Navy AJU communications mast.

Operational status
Thomson Marconi Sonar communications masts are supplied to the Royal Navy and Royal Australian Navy.

In 1994, the company was awarded a contract by the US Navy to carry out a feasibility study into the design of a communications mast for the 'Los Angeles' class.

Contractor
Thomson Marconi Sonar Ltd, Templecombe.

VERIFIED

AVXD

Type
Submarine SATCOM multifunction antenna

Description
Thomson Marconi Sonar Ltd has developed a range of omnidirectional antennas, many of which can simultaneously accommodate transmit (Tx) and receive (Rx) over different frequencies. The Tx/Rx information, together with the facility selection, can be provided on a single RF coaxial conductor.

The Submarine SATCOM multifunction antenna consists of a pressure-tight broadband antenna system containing a head amplifier unit, together with an inboard control unit.

The compact, unobtrusive antenna affords excellent spatial coverage, from low angle Line-Of-Sight (LOS) to high angle SATCOM. The single antenna element is capable of performing many simultaneous communications functions, covering VHF, UHF JTIDS, IFF, UHF, SATCOM, SATNAV and GPS.

AVXD has been developed specifically to provide the performance essential to operate DAMA UHF SATCOMS. Within the conical envelope a vertically polarised dipole antenna provides VHF and UHF LOS

communications, a conical logarithic spiral antenna provides a circularly polarised hemispherical pattern for UHF SATCOMS and GPS reception, and a monopole antenna is provided for IFF, JTIDS and optionally, terrestrial cellular telephone operation.

The antenna control unit provides all necessary RF and control interfaces between the submarine communication system and the antenna and is fitted with a 250 W power amplifier for UHF SATCOM transmissions.

The antenna outfit has been fully tested under extreme depth conditions, and meets all the requirements of the Royal Navy AVDI communications

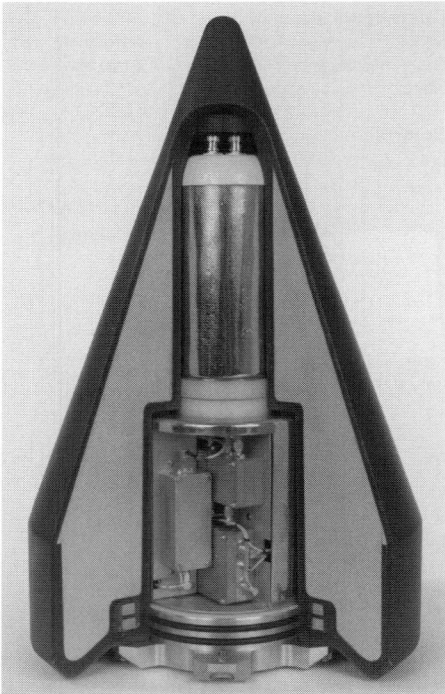

Submarine communications multifunction antenna

antenna. Dimensions are 580 mm high with a maximum overall diameter of 356 mm, and a weight of less than 20 kg for the complete outfit of antenna, flange and electronics package. The single coaxial connection and universal adaptor flange permits easy retrofitting on existing masts.

Operational status
In service with the UK and a number of foreign navies.

Contractor
Thomson Marconi Sonar Ltd, Templecombe.

VERIFIED

Submarine communications and NAVSTAR antenna

Type
Submarine broadband periscope-mounted antenna for communications and navigation.

Description
The compact submarine communications and NAVSTAR antenna consists of a pressure-tight broadband antenna system containing a head amplifier, together with an inboard control unit. The unit offers excellent spatial coverage, the antenna/head amplifier combination giving a signal-to-noise ratio at the receiver which meets the NAVSTAR/GPS requirements for worldwide coverage.

The single element antenna is capable of performing communication functions covering VLF, LF, MF, HF, VHF, UHF and D-band GPS reception (with simultaneous capability), and V/UHF line of sight transmission.

The antenna presents a very low visual profile, and its special water-shedding coating ensures quick response communications. It has been designed specifically for operation in highly reflective environments, thus ensuring full operational low-angle NAVSTAR coverage. It has been fully tested under extreme conditions.

A universal flange fitted to the base of the antenna provides a general purpose mounting suitable for a wide range of submarine periscopes. Size of the antenna is 323 mm high by 197 mm maximum overall diameter. Weight of the complete outfit, including the electronics package, is less than 6 kg.

Contractor
Thomson Marconi Sonar Ltd, Templecombe.

VERIFIED

Communications and NAVSTAR antenna

R800B receiver

Type
VLF/LF receiver system.

Description
The R800B receiver offers full coverage of the VLF and LF frequency bands, and is designed for fully automatic reception on MSK, FSK, LW and FDM. Full operating parameters can be programmed for up to 63 channels, 10 of which can be protected from accidental reprogramming.

The R800B is suitable for use in submarines, surface ships and shore stations, and incorporates Built-In Test Equipment (BITE). It can be used with normal Crypto

equipment, and a number of spread spectrum and frequency-hopping techniques can be included.

Operational status
The R800B is the specified receiver of, and is in service with the Royal Navy, and is also in service with the Portuguese Navy and an Asian navy. Latest sales include further units to the Royal Navy (1995) and to an Asian navy (1996).

Contractor
Redifon MEL Ltd, Crawley.

VERIFIED

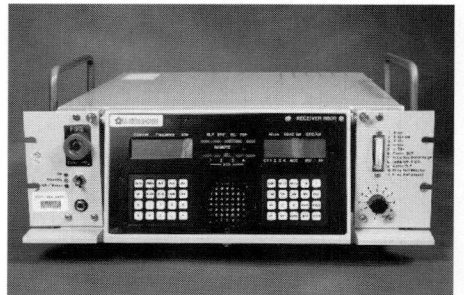

R800B VLF/LF receiver

RICE 10

Type
Internal communication system.

Description
The RICE 1O (Rationalised Internal Communications Equipment), offers an extremely flexible internal communications capability and provides all the facilities required by a modern submarine. Facilities available include ship-wide broadcast, alarms, point-to-point communication and conference. These can be combined with the external communication components to form a fully integrated communications system.

The RICE 1O is based on a digital mesh network that offers high survivability, by reconfiguring message paths automatically in the event of damage. Operation of the system is via a voice terminal outstation.

A numeric keypad allows the user to dial any other outstation on the system. The keypad is also used to re-

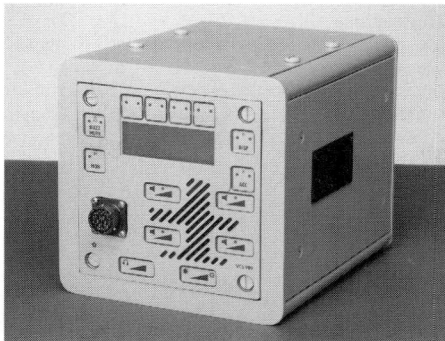

A typical outstation of the RICE 10 system (Redifon MEL Ltd)   ***1999*/0017499**

programme the outstation in the event of system damage or to programme a user specific facility. This programmability allows the programmed outstation

facilities to follow the individual user to other outstations. An incorporated dynamic tallying facility allows the operator to immediately ascertain the available functions. Two basic outstations are available and these, with their variants, cater for a wide range of operator requirements. A 16-channel outstation, which provides the full RICE 1O capability, is coupled with a smaller five-channel unit which offers limited functionality for those areas not requiring the full range of services.

Operational status
In production. Supplied to the Royal Navy in 1998 for the 'Swiftsure' and 'Trafalgar' class submarine refit programmes.

Contractor
Redifon MEL Ltd, Crawley.

NEW ENTRY

UNITED STATES OF AMERICA

Extremely Low-Frequency (ELF) communications programme

The concept of using extremely low-frequency radio signals to communicate with submerged submarines was first suggested over 30 years ago. However, the extremely large antenna size and environmental worries prevented its introduction until recently.

ELF signals can travel great distances with low loss and can penetrate seawater to considerable depths. In practice, an ELF consists of one or more shore-based transmitters, operating at around 40 to 80 Hz, connected to long horizontal wire antennas (either just above ground or buried for additional security) that are earthed at each end. Orthogonal antennas are used to provide omnidirectional radiation patterns. The transmitted signals are sensed by an antenna on the submarine and decoded by a sophisticated, computer-based receiver. Because the bandwidth is low at ELF, the message transmission rate is necessarily very slow but, even by employing a simple three letter code system, any of a great number of messages can be transmitted in reasonable time.

The most favourable site in the US for the installation of the system is in northern Michigan or northwestern Wisconsin, where the low-conductivity bedrock formations in the Laurentian Shield greatly enhance propagation.

Development
In 1969, the US Navy constructed an experimental transmitter in Wisconsin to study propagation and environmental effects of ELF. The antenna consisted of two 22.5 km long pole-mounted lines at right angles. In 1976, a message handling capability was added and a small quantity of shipboard receivers was built and installed to prove conclusively that an ELF system would perform as anticipated.

Encouraged by this trial, the navy planned to construct an operational ELF system with a completely buried antenna and redundant transmitters. Codenamed Sanguine, this vast system would have been highly resistant to blast overpressure and could have absorbed a moderate number of direct nuclear hits. However, in 1975, a defence analysis group reached the conclusion that the increasing accuracy and number of Soviet nuclear warheads could neutralise the system. Sanguine was, accordingly, downgraded and eventually cancelled.

The navy, undeterred by this setback, persisted and developed a more modest system with above ground transmitters and a pole-mounted antenna 45 km long. This system, called Seafarer and installed at Clam Lake, Wisconsin, immediately ran into opposition from the residents of Wisconsin and the environmentalist lobby, who were concerned about the environmental impact of the system and health issues. Despite numerous navy-sponsored biological studies showing no adverse environmental or ecological effects from ELF, and a

1977 study carried out by the National Academy of Sciences giving ELF a clean bill of health, the project was cancelled in 1978.

In 1981, President Reagan ordered that the Wisconsin transmitter be reactivated and upgraded to operational status and, at the same time, ordered the Department of Defense to conduct a study of ELF requirements. That study resulted in the conclusion that ELF would enhance the US strategic C³ posture. It recommended that in addition to the upgrade of the Wisconsin system, a supplementary 90 km system should be constructed at the nearby KI Sawyer Air Force Base on the Upper Peninsula of Michigan. Congress approved funds for the project in 1982 and research and development resumed. Construction commenced at both sites in 1983 but, in January 1984, a court injunction halted further work pending the preparation of a supplemental Environmental Impact Statement (EIS) to evaluate health effect studies of ELF fields performed since the original EIS was filed in 1977. A study was carried out by the American Institute of Biological Sciences which reaffirmed the results of previous investigations and the EIS was filed in 1985. Construction restarted following the cancellation of the injunction by the US Court of Appeal. The Supreme Court subsequently upheld that decision.

By the end of 1986, both stations were completed. Prototype submarine receivers were delivered in April 1985. The first test in May 1985 aboard a submarine of the Pacific Fleet was successful. In subsequent tests, submarines in the Mediterranean, the western Pacific and on patrol under the North Polar ice cap have successfully received signals from the Wisconsin station.

Description
There are four segments to the Wisconsin/Michigan ELF communications system: the Broadcast Control Segment (BCS), the Message Input Segment (MIS), the Transmitter Segment (TS) and the Receiver Segment (RS).

The main input port is the BCS, which is controlled by the Commander, Submarine Forces Atlantic (COMSUBLANT) in Norfolk, Virginia. The MIS, the secondary input port for ELF messages, is located at KI Sawyer AFB and can take over from the BCS should that become disabled. It can if necessary pre-empt the BCS.

The TS comprises the ELF stations at Wisconsin and Michigan which normally operate simultaneously but can operate independently when required. Two frequency bands are used, 40 to 50 Hz and 70 to 80 Hz and each transmitter facility uses commercial prime power from its local utilities companies. In the event of power failure, each site has back-up diesel generators. Each transmitter facility has installed two spare back-up power amplifiers and an uninterruptible power system to ensure necessary reliability to continue transmitting

in case mechanical or power failures occur. The TS is a soft, surface-deployed subsystem with ECCM and electromagnetic pulse protection. However, it is not expected to withstand a hostile physical attack.

Receiver Segments (RS) are fitted on all US Navy submarines, and the Broadcast Control Segment (BCS) and Message Input Segment (MIS) also incorporate receivers to perform monitoring functions. Messages are normally fed into the system at the BCS, although the MIS can also input messages if required. Messages are then sent to the Transmitter Segment (TS) via dedicated communication lines. Both the BCS and MIS include a message entry element consisting of a data terminal set (teletype Model 40), a link encryption device (KG-84) and a datalink selector panel. This arrangement allows the operator to enter messages into the message queue at the master transmitter facility via telephone company lines. The BCS and MIS also include an order-wire for operator-to-operator communications between facilities. The master transmitter facility is in contact with the BCS and maintains the message queue and automatically updates the back-up queue at the slave facility.

At each transmitter facility a processor element converts the encrypted ELF message from the message processor into drive signals for the power amplifiers, monitors antenna current and controls the transmitter facility master/slave protocol.

The antenna arrays at each transmitter facility consist of antennas oriented north-south and east-west. Two pairs of power amplifiers exist at each facility, one pair for each antenna configuration. Only two power amplifiers are used at any one time, one for each antenna, and each is rated at an output level of 660 kW.

Signals which are picked up via the OE-315 towed antenna are then fed to the ELF receiver terminal group via an in-line amplifier. The receiver performs analogue and digital signal processing to detect any message that may be present. The receiver terminal group (OR-279(XN-1)/BRR) consists of five main units: the preamplifier, receiver timing and interface unit, the combined processor and key generator unit, time and frequency junction box and navigational interface junction box. To extract the ELF message, the receiver has first to filter out atmospheric and ocean noise and eliminate interference caused by the submarine's onboard power systems. This is achieved by a UYK-44 computer that sorts and decodes the ELF bit-stream and monitors signal strength.

The ELF system is synchronous requiring that the receiver know accurate time information relative to the transmission, and time compensation has to be allowed for the ELF signal propagation delay. A band spreading key stream is generated and removed from the message, which is then decrypted. By using an embedded AN/UYK-44 militarised reconfigurable processor, many of these functions are performed digitally.

Operational status
The system is in service with the US Navy. The base at Michigan became operational in 1986, and at Wisconsin in 1991. By the end of the 1980s some 90 receivers had been installed in US Navy submarines. All US Navy submarines are now fitted with ELF receivers as standard equipment.

Contractors
GTE Government Systems Corporation, Needham, Massachusetts (prime).
Illinois Institute of Technology, Chicago, Illinois.
Computer Sciences Corporation, Falls Church, California.
Mitre Corporation, McLean, Virginia.

R M Vredenburg Company, McLean, Virginia.
Spears Communications Group, Sippican Inc, Marion, Massachusetts.
Booz-Allen and Hamilton Company, Washington DC.

UPDATED

AN/WQC-2A communication set

Type
Sonar underwater communication set.

Description
The AN/WQC-2A is a sonar underwater communication set for surface ships, submarines and shore installations. The system provides an SSB general purpose voice and CW communication set consisting of a control station, remote-control station, receiver/transmitter and LF and HF transducers.

Control and remote-control stations for AN/WQC-2A system

The system transmits and receives voice, audio and low-speed telegraphy in two frequency bands: high (8.3 to 11.1 kHz) for close range, and low (1.45 to 3.1 kHz) for long-range underwater communication. In addition, the system can be used to amplify and transmit signals from external sources in a frequency range of 100 Hz to 13 kHz.

The receiver/transmitter contains the main electronic assemblies of the sonar communication system. It consists of a rack-type cabinet with three removable drawer assemblies containing the power supply and test panel assembly, the receiver/transmitter assembly and the final amplifier assembly. The primary functions of the receiver/transmitter are to develop the high-powered, single sideband transmission from voice, audio and external source signals to drive the LF or HF transducers and to receive and demodulate LF, HF and CW signals received from the transducers. The receiver and transmitter outputs are made available to external tape recordings and monitoring equipment. The circuits are also capable of muting, or being muted by, other external equipment.

The control station contains the required controls, indicators, microphones and so on. The remote-control station is a secondary operating position, if required.

The two transducers (LF and HF) have a horizontal

(radial) omnidirectional beam pattern. The electromechanical energy conversion is accomplished by piezoelectric ceramic elements which are totally encapsulated and covered by an acoustically transparent neoprene boot.

Operational status
In production.

Specifications
Frequency bands: 1.45-3.10 kHz (LF); 8.3-11.1 kHz (HF); 100 Hz to 13.0 kHz (auxiliary mode)
Output power: 600 W (LF); 450 W (HF); 1,000 VA (auxiliary mode)
Weights
Receiver/transmitter: 220.5 kg
Control station: 5.8 kg
Remote control: 3.1 kg

Contractor
SeaBeam Instruments Inc, East Walpole, Massachusetts.

VERIFIED

SUBTACS

Type
Submarine ELF tactical communications system.

Description
The Submarine Tactical Communications System (SUBTACS) is designed for land-based transmission, as well as tactical battle group communications, to

submarines with the installation of shipboard transmitters. With a data rate 300 times faster than the original GTE Extremely Low-Frequency (ELF) systems, SUBTACS functions at ranges in excess of 1,000 n miles, and at submarine operating depths and speeds.

The system is more jam-resistant than a VLF system, is less expensive to install, has a greater range and far greater water penetration. The ELF system operates at frequencies between 30 and 300 Hz. (See also the

Extremely Low-Frequency (ELF) communications programme entry.)

Contractor
GTE Government Systems Corporation, Needham, Massachusetts.

UPDATED

Mast antenna

Type
Erectable communications mast antenna.

Description
This equipment provides a submarine with communication reception and transmission over a wide range of frequencies. The antenna can be erected above the seawater surface from its normal storage area within the submarine fin. Once clear of the water,

communications in various bands, IFF transpond and GPS reception can be accomplished. The system comprises an antenna/radome, mast raising equipment, outboard cable, hull penetrator, junction box, and antenna control unit. The outboard cable, hull penetrator, and junction box provide connectivity between the antenna/radome and the antenna control unit in radio. The antenna control unit provides the interface point between the antenna/radome and the shipboard transceivers and transponders.

Operational status
In service aboard US Navy submarines.

Contractor
Spears Communications Group, Sippican Inc, Marion, Massachusetts.

NEW ENTRY

VLF/LF antennas

Type
VLF/LF antennas.

Description
The ability to effectively copy VLF/LF broadcast, while submerged, has long been a requirement for submarines. The high performance, crossed-loop VLF/LF antennas needed to accomplish this have been designed and manufactured by Spears/Sippican for many years. These military quality units are designed to

operate in the submarine environment and to interface directly with standard VLF/LF communication receivers. The antennas can be configured in varying sizes and shapes, which makes for easy mounting to the submarine. All the VLF/LF loop antennas offered contain integral low noise preamplifiers and are provided with an antenna coupler (located in the ship's radio room) which interfaces the active antenna to a standard 50 ohm receiver. These systems are also offered with the outboard antenna transmission line, hull penetrator, and inboard antenna transmission line.

The antenna/preamplifier covers the range from 10 to 160 kHz, with extended coverage up to 300 kHz.

Contractor
Spears Communications Group, Sippican Inc, Marion, Massachusetts.

NEW ENTRY

Communication antenna

Type
Emergency HF communication antenna.

Description
The emergency HF communication antenna is used when the submarine's normal HF antenna systems are inoperable. The system consists of a manually erectable antenna, an antenna base unit, an RF transmission line, and a hull penetrator. The telescoping antenna is approximately 11 m in length

when fully extended and is capable of progressive adjustment and being locked to any desired length. The antenna collapses to approximately 2 m in length that allows for easy stowage inside the submarine when not in use. The antenna base unit is a pressure-proof assembly that is permanently located outside the hull of the submarine in the vicinity of the bridge. This unit contains broadband matching networks that allow for direct connection of the ship's HF transceiver to the antenna base unit, thereby eliminating the need for radio personnel to operate an antenna tuner. These broadband networks provide a match to the antenna

such that the nominal SWR at the inboard end of the transmission line is less than 3:1. The antenna base unit also houses integral electromagnetic pulse (EMP) and lightning surge protection and is totally compatible with the seawater environment.

Contractor
Spears Communications Group, Sippican Inc, Marion, Massachusetts.

NEW ENTRY

BWA

Type
Buoyant Wire Antenna (BWA) system.

Description
Buoyant wires are long, towed antennas that provide a submarine with the ability to communicate while remaining deeply submerged. The system consists of a buoyant wire antenna, a reeling machine that deploys, tows, and retrieves the antenna, reeling machine controls, a transmit/receive switch, and an antenna coupler. When the submarine wishes to communicate, the buoyant wire antenna is deployed via the reeling machine which can be mounted either inboard or outboard of the pressure hull. A portion of the antenna floats at or near the sea surface and receives radio signals. The antenna is 610 m long with a diameter of 16.5 mm. Shorter lengths are also available. An antenna that allows both transmit and receive in the HF band is also available. Signals received on the BWA are filtered and amplified in the antenna coupler located in the radio room. This coupler is a broadband device that provides the interface between the special antenna and the standard submarine radio receivers. Because the system is broadband, it is LINK 11 compatible.

Contractor
Spears Communications Group, Sippican Inc, Marion, Massachusetts.

NEW ENTRY

Towed buoys

Type
Towed communication buoy systems.

Description
The major submarine navies of the western world have employed towed communication buoys for many years. These systems allow for maximum operational flexibility by placing minimal restrictions on the boat's speed and depth while offering a wide range of communications capabilities. Today's towed buoys offer a stable platform that can be outfitted with communications antennas and electronics that will allow a wide range of frequency coverage.

Non-communication type sensors can also be mounted in the buoy. Several system configurations are available to suit any size and type of submarine and can offer a full range of communication/navigation capabilities.

The buoys are tethered to the submarine by a multiconductor tow cable, and are deployed and retrieved using either an inboard hydraulic winch, or an outboard palletised submersible electric winch.

Operational status
Spears/Sippican manufactures all of the towed buoy systems used by the US Navy submarine fleet.

Contractor
Spears Communications Group, Sippican Inc, Marion, Massachusetts.

NEW ENTRY

MFM

Type
Multifunction mast antenna.

Description
The standard MFM system is a high performance, mast-mounted, communication and navigation antenna that is intended for new construction or as a replacement upgrade for antennas on existing submarines. The system supports communications in several operating bands which include VLF/LF receive, HF transceive, VHF transceive, UHF transceive and IFF transpond. GPS, Omega and Loran C navigation capabilities are also included. The individual antennas of this integrated communication and navigation system are housed in a hydrodynamically faired, non-penetrating mast assembly.

The MFM system is capable of simultaneous transmission and reception for all functions. The wide spatial coverage allows communication from low angle, line of sight assets to high angle satellite assets. The external mast is an extremely rugged hydrodynamically optimised assembly, allowing use from a submerged submarine operating at high speeds and in severe weather and sea state conditions. The small volume of the deployable mast allows impressive communications capabilities to be offered in a compact mast.

The standard MFM system comprises six units:
Unit 1. The antenna/radome, is a pressure-proof radome that houses all antenna elements and associated outboard electronics; Unit 5. The Antenna Control Unit (ACU), provides control of the antenna/radome and RF input/output ports for external receivers and transceivers; Units 2, 3 and 4. These provide connectivity between the antenna/radome and the ACU. Unit 6. Incorporates the mast raising equipment which includes the fairing, structural module and all related hydraulic and mechanical components.

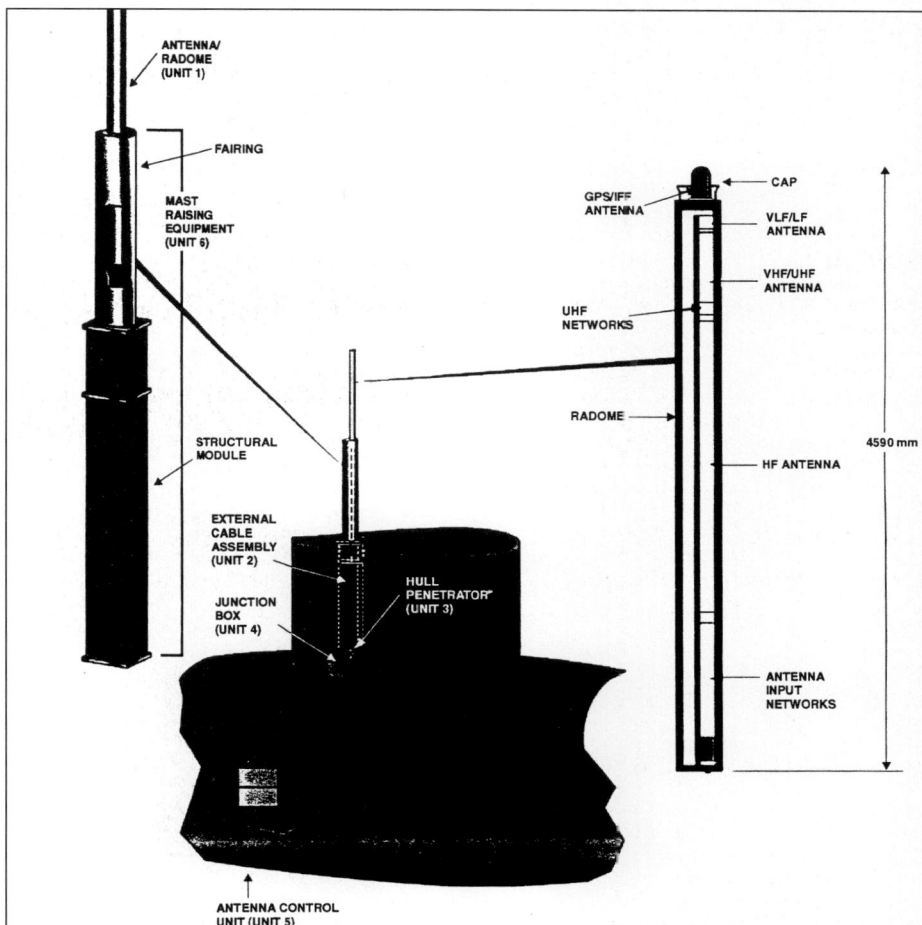

Multifunction mast antenna

1998/0006981

Specifications

Range	HF	UHF(LOS)	VHF	UHFSATCOM
Frequency range (MHz)	2-30	225-400	150-174	240-320
Antenna impedance	50 Ω	50 Ω	50 Ω	50 Ω
Power handling (max)	500 W CW	100 W CW	50 W CW	100 W CW
Polarisation	Vertical	Vertical	Vertical	Circular
Radiation pattern				
Horizontal	Omni	Omni	Omni	Omni
Vertical	Equivalent to a half wave dipole	Equivalent to a half wave dipole	Equivalent to a half wave dipole	Hemispherical

Range	IFF	VLF and LF	GPS
Frequency range (MHz)	950-1,150	0.01-0.17	1,227/1,575
Antenna impedance	50 Ω	50 Ω	50 Ω
Power handling (max)	1 KW (peak)	Receive only	Receive only
Polarisation	Vertical	Vertical	Circular
Radiation pattern			
Horizontal	Omni	Figure-eight	Hemispherical
Vertical	Equivalent to a half wave dipole	Equivalent to a half wave dipole	Hemispherical

Contractor
Spears Communications Group, Sippican Inc, Marion, Massachusetts.

VERIFIED

UNDERWATER TELEPHONES

AUSTRALIA

Acute 9000

Type
Underwater telephone unit.

Description
The UnderWater Telephone (UWT) is designed to perform underwater voice and telegraphic communications from surface ships and submarines. It can also ensure communications with a diver carrying autonomous equipment with compatible characteristics.

The Acute 9000 UWT has been developed and tested to operate on naval vessels where compliance to military standards is required. The telephone is normally supplied with an omnidirectional transducer. However, sectorial operation is available.

The UWT unit consists of a transmitter and receiver, which are mounted in a common cabinet, and is supplied with an accessories set (headset, handset and morse key).

The underwater telephone unit operates in a number of modes (listed below) which the operator selects from the front panel. These modes operate in conjunction with transmit power selection, receiver sensitivity and volume controls to enable the operator to select the setting best suited to the conditions of the day. Manual/automatic AGC control of the receiver enables the operator to adjust the receiver gain to enhance the quality of the received signal.

Watch: silent mode – allows listening for incoming communications, with transmitter inhibited.
Phone: intercommunication by voice.
Transp: automatic reply to a CW call.
CW & Reception: two-way morse communication.
CW: one-way morse transmission with receiver inhibited.
Test: self-test of underwater telephone unit.
Auto: automatic transmission of CW pulses, with reception between pulses.

Operational status
The UWT is currently in service in the Royal Australian Navy and Royal New Zealand Navy on board 'Anzac' class frigates.

Specifications
Electrical
Nominal frequency: 8,087 Hz (carrier frequency)
Maximum input (Pmax): 1,000 VA
Input voltage: 115 VAC, 60 Hz, 3-phase; (115 V between phases)
Maximum output voltage: 140 V rms
Mechanical
Width: 546 mm
Depth: 428 mm
Weight: 54 kg

Contractor
Thomson Marconi Sonar Pty Ltd, Rydalmere, New South Wales.

UPDATED

FRANCE

ERUS-3

Type
Emergency underwater telephone.

Description
The ERUS-3 ultrasonic telecommunications equipment has been designed for telephone and telegraphic communication between a surface ship and one or several submerged craft, diving bells or submarines, or between several submerged craft.

The unit operates with the ship's mains power supply or with an autonomous battery power supply. A selector switch enables the operator to choose the required power supply.

The ERUS-3 incorporates an automatic transmission device, for use in an emergency, and a responder, enabling ships to estimate the distance between them (this device is also used for standby operation). The range is greater than 5 n miles under normal propagation conditions with the ships running at silent speed.

The ultrasonic signals are transmitted and received by a type 2Z9A omnidirectional transducer. Signals are transmitted with SSB AM.

In the passive mode, when a signal with a particular call frequency is received, the receiver is unlocked to enable it to receive telephone signals. In the responder mode, reception of a signal which has a particular call frequency initiates retransmission of a signal of the same frequency.

The power supply batteries are placed in special battery containers. Autonomous operating time on transmission is 3½ hours, intermittently transmitting for 3 minutes per hour for 72 hours, for example. Autonomous operating time on reception is 72 hours for continuous operation. These times relate to a single battery container and are multiplied up according to the number of battery containers installed.

The equipment can operate either on a classified NATO frequency for military use or on 10.5 kHz for commercial use. For particular needs, the two frequencies can be used on the same equipment with a changeover switch.

Operational status
In production.

ERUS-3 transmitter/receiver unit

Contractor
Safare Crouzet, Nice.

VERIFIED

ERUS-6

Type
Emergency underwater telephone.

Description
ERUS-6 meets the STANAG 1074 requirements, which are to have available in each manned, watertight compartment of a submarine, a self-powered underwater telephone using the NATO standard channel.

A number of operating modes is available including: voice transmission and reception; passive (squelch mode where the receiver remains silent until a call signal is received); pinger; and transponder/responder for distance measurement from a rescue vessel. Transmission/reception is omnidirectional.

The ERUS-6 is powered from batteries housed in a container delivered with the equipment, but it can also be connected to an emergency 24 V DC supply network when available.

The system is simple to use in emergency conditions and easy to install.

Operational status
Fully developed.

Contractor
Safare Crouzet, Nice.

UPDATED

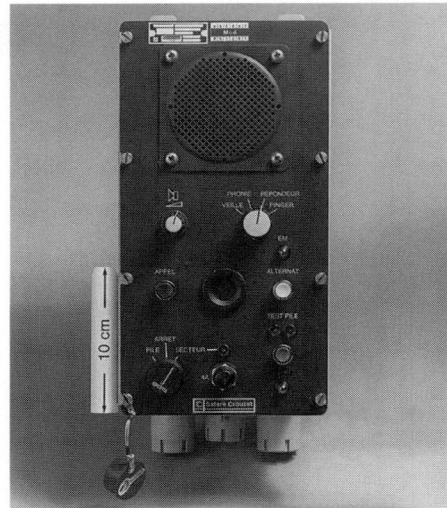

ERUS-6 receiver unit
1996

TUUM-4A/B underwater multichannel wireless telephone

Type
Underwater multichannel telephone.

Description
The advanced multichannel underwater communication system TUUM-4A/B allows communications between surface ships, submarines, divers, and diving bell, for normal and rescue operations.

The TUUM-4A/B is based on a powerful synthesised transceiver that covers the frequency range 1.45 to 50 kHz with preset channels facilitating switching.

On each standard channel, the following modes are available: voice; passive (squelch mode); key/TTY (transmission and reception of telegraphic signals, manually or automatically from a serial link); transponder/distance measurement between two ships with direct display of the distance; pinger; recording of the received message and playback (very useful in bad reception conditions); dual-channel monitoring (permanent monitoring of the NATO standard frequency in parallel with the selected channel).

The equipment can be remotely controlled through the integrated internal communication system or from a dedicated remote station.

The operational range is over 7.5 n miles in the NATO mode but can be extended up to 10.75 n miles by the addition of an optional AM5300 amplifier. It can be limited by reducing the output power.

There are 20 standard preset channels for normal communication, tailored to customer requirements and/or programmable in depot. There are 10 specialised preset channels, allowing automatic repetitive transmission of a prerecorded message with a programmable period (for instance Victor-Delta-Sierra when a VDS is in use); and improvement of intelligibility, particularly when used for long-range communication in bad propagation conditions (patented process). In addition, the NATO channel is permanently monitored.

Usually one or more omnidirectional transducers, according to the required operational frequency range, are used on surface ships; one or more sets of directional transducers are used on submarines.

The equipment is powered from 115/230 V AC – 50/60 Hz, or 28 V DC (used as standard power supply or as automatic back up of AC mains).

Operational status
In service in the French Navy from the early 1990s.

Manufacturer
Safare Crouzet, Nice.

UPDATED

TUUM-4A/B underwater telephone

AN/WQC-501(V) underwater telephone

Type
Acoustic underwater communications telephone.

Description
The AN/WQC-501(V) is an underwater telephone system operating on NATO frequencies and designed for communications between surface ship and submarine. Normal operation is in the passive mode, the receiver remaining silent until a call signal is received. An automatic encoder input and an audio output can be used for transmission and reception of a slow-speed teleprinter transmission.

The surface equipment consists of a transceiver, remote-control unit, power amplifier and an omnidirectional transducer. Submarine equipment comprises a transceiver, power amplifier, four directional transducers and a transducer switching unit. The directional transducers are installed forward, aft, starboard and port.

Transmission and reception from the surface vessel is omnidirectional. The submarine transmission and reception can be either directional or omnidirectional as selected by the operator, using one transducer or four simultaneously.

Operational status
In operational service with the Canadian Navy.

Specifications
Operating modes
Voice: voice transmission and reception

Keying: transmission and reception of telegraphic signals
Responder: automatic transmission on reception of call signal
Surface ship output power: 5.5, 45, 55 or 450 W as selected by the operator
Submarine output power: 5.5 or 55 W (directional); omnidirectional transmission is 5.5, 22, 55 or 220 W (1.4, 5.5, 14 or 55 W per transducer respectively)

Contractor
Safare Crouzet, Nice.

VERIFIED

TSM 5152A, 5152B

Type
Transmitter/receivers for underwater telephony.

Description
The TSM 5152 transmitter/receivers (TSM 5152A for submarines and TSM 5152B for surface vessels) allow two-way telephony and telegraphy between submerged submarines, between surface ships and submarines, and between surface ships.

The system has a range of operating modes which include: a watch mode for silent survey when the receiver is normally silent but which is activated by reception of a call; a phone mode for communication; a CW transmission mode controlled with the signalling key; a CW and reception mode with the equipment automatically switched to reception during breaks of transmission; an automatic transmission mode used to obtain a first contact; a transponder mode in which a pulse is automatically transmitted on receipt of a CW transmission; and a test mode to allow the operator to test the equipment automatically.

Operational status
In production for the French and other navies.

Specifications
Mode: USB transmitted-call FM
Carrier frequency: in accordance with NATO or other specifications
Power output: 400 W for 0.33 form factor with AC supply
Power supply: 115 V AC ± 5%, 48-63 Hz, 3-phase

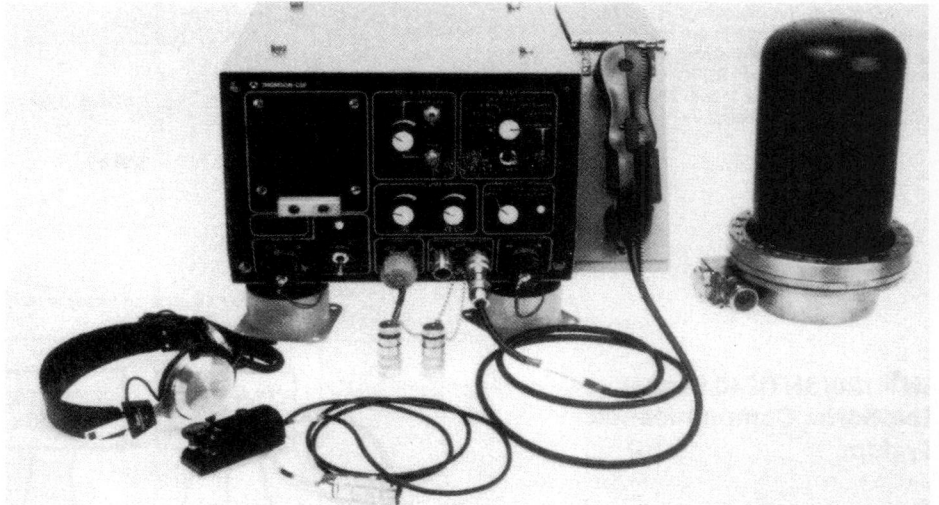
TSM 5152 transmitter/receiver and associated projector

Consumption: 800 VA (max)
Weight: 33 kg
Projector
Type A
Each 1 of 3 groups includes 2 independent and identical hydrophones
Bearing aperture: 120° (3 dB)
Elevation aperture: 50° (3 dB)
Operating depth: 300 m (max)

Type B
Omnidirectional in bearing
Elevation aperture: 50° (3 dB)
Range: 10.75 n miles (approx)

Contractor
Thomson Marconi Sonar SAS, Sophia Antipolis.

VERIFIED

ITALY

TS-200 system

Type
Underwater telephone system.

Description
The TS-200 underwater telephone set has been designed, following Italian Navy specifications, to be

installed on surface vessels and submarines. By means of the TS-200, both phonic and telegraphic underwater communications between surface and underwater vessels are available. This equipment can transmit and receive in either omnidirectional or directional modes (with three selectable bearings) following NATO standards concerning signal modulation and operating frequency.

Transmitted power is such as to allow communications at distances up to several tens of kilometres in optimum sound propagation conditions. The emitted power can be reduced to half or quarter of maximum for shorter range communications.

The equipment consists of a set of piezoelectric transducers, a cabinet containing all the electronics required, and a main and auxiliary control panel from which the various operating modes can be selected by the operator.

In addition to standard IFF applications, the equipment can also integrate online, an automatic communications coding system by means of an optional interface.

Operational status
Currently installed on the *Giuseppe Garibaldi* helicopter carrier, 'Maestrale' class frigates and 'Minerva' class corvettes.

Specifications
Frequency range: 8.3-11.1 kHz
Modulation: SSB/SC and CW
Carrier suppression: >40 dB
Carrier frequency: NATO standard
Horizontal beamwidth: 360° (omni); 30° (directional)
Vertical beamwidth: 90° (omni) (approx); 28° (directional) (approx)

Contractor
Whitehead, Alenia Sistemi Subacquei, Livorno.

UPDATED

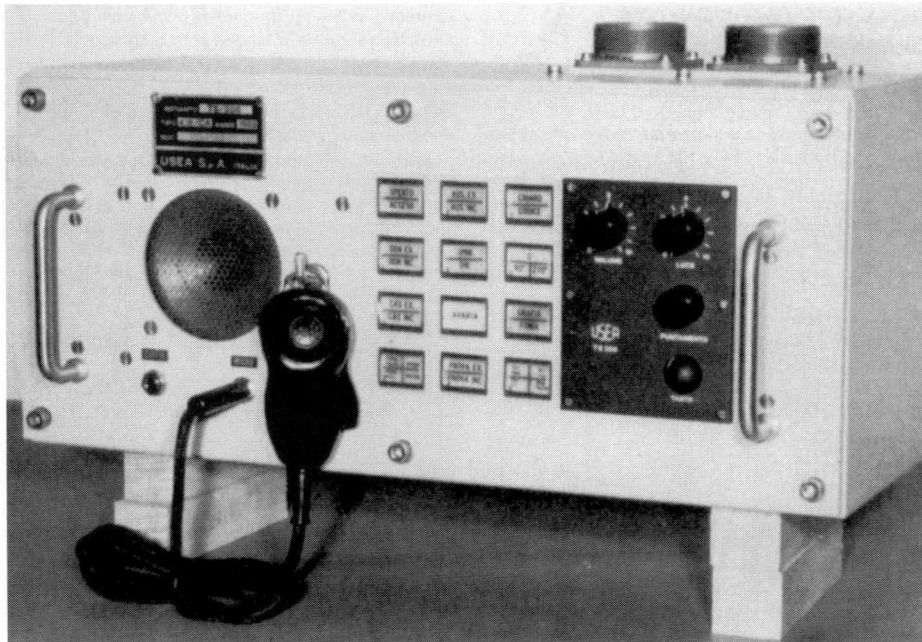
TS-200 auxiliary control panel

UNITED KINGDOM

G732 Mk II underwater telephone

Type
Acoustic throughwater communications system.

Description
The G732 Mk II is an improved underwater telephone system for use in submarines and surface vessels. It provides communications facilities for speech, morse, digital and coded messages, plus an emergency pinger mode of operation.

The system operates on the standard NATO frequency of 8.0875 kHz using upper single sideband transmission. Alternative operational frequency bands can also be supplied and a choice of transducer installations is available to suit customer requirements. The transmit direction, which can be omni, port, starboard, ahead and above, is user selectable, as also are the transmit source levels to maintain security. A narrowband is also available via a suitable transducer.

The receiver/transmitter electronics are housed in a ruggedised bulkhead-mounted unit which connects to one or more transducers to produce directional transmissions as required by the user. The system normally operates from the main power supply but a battery-powered unit can be supplied to maintain communication in the event of complete failure of the vessel's own electrical power supplies.

A combined lower and upper sideband version of the G732 Mk II is available to communicate with ex-Russian ships and submarines.

Operational status
In production.

Specifications
Operating frequency: 8.4-11.3 kHz (nominal)
Mode: upper SSB (NATO compatible)
Transmitter source level: 186 ± 2 dB/µPa/m (max) (port, starboard, ahead, upper); 183 ± 2 dB/µPa/m (max) (omni)
Dimensions: 400 × 300 × 200 mm
Weight: 18 kg (max)

G732 Mk II underwater telephone

Contractor
Graseby Dynamics Ltd, Watford.

UPDATED

Type 2073 underwater telephone

Type
Emergency underwater telephone.

Description
The Type 2073 is an emergency underwater telephone/pinger which has been designed to assist in locating and communicating with a submarine or submersible in distress. In an emergency, the equipment can operate as a pinger producing low-frequency (10 kHz) tone pulses, with the receiver operational at the same time. As a rescue vehicle approaches, the pinger can be changed to a higher frequency (37.5 kHz) to enable the vessel to home in on the submarine or submersible escape hatch. For naval use, a 43.5 kHz frequency is included for communication with divers.

The equipment can also operate as a throughwater communication system, on any of three standard channels to communicate with rescue vessels, divers and other ships.

Operational status
In production.

Specifications
Underwater telephone
Frequency: 10, 27 and 43 kHz
Range (min): 3,600 m (low frequency); 2,200 m (mid-frequency); 685 m (high frequency)
Pinger
Frequency: 10 and 37.5 kHz
Range (to a sonar): 7,300 m (10 kHz); 1,800 m (37.5 kHz)

Contractor
AB Precision (Poole) Ltd, Poole.

UPDATED

ABLO36

Type
Emergency portable underwater telephone.

Description
The ABLO36 portable underwater telephone has been designed for communication from a ship or helicopter to a submarine, submersible or diver.

The equipment can operate on any of three standard frequencies compatible with other in-service equipment allowing communication in a search and rescue role to divers or other ships. The equipment also has a high power mode for longer distance communication to other sonars.

A robust and efficient multifrequency omni-directional transducer is supplied which can be deployed from the ship or helicopter by suspension from its own strain relieved cable.

The ABLO36 underwater telephone is built to withstand the severe conditions in which it will have to operate. It can operate from its own integral battery, DC or AC supplies for maximum versatility.

Operational status
In production.

Specifications
Frequency: Voice 10, 27 and 43 kHz
Pinger: 10 or 35.5 kHz

Contractor
AB Precision (Poole) Ltd, Poole.

NEW ENTRY

UNITED STATES OF AMERICA

Model 5400

Type
Acoustic underwater telephone system.

Description
The Model 5400 underwater telephone features extensive communications and navigation functionality in a modern, lightweight, compact design. Primarily intended for voice communications, the equipment is based on a synthesised transceiver and is compatible with all commonly used underwater telephone modes, including WQC, UQC and BQC modes of operation. The unit covers the frequency band 5 to 45 kHz in 1 Hz steps plus the AN/WQC and ARD8000 frequencies. Up to 50 frequency combinations can be stored in memory for easy recall at any time. Upper and lower sideband modulation is operator selectable, as is output power in five steps: 2, 10, 50, 100 and 200 W.

The system includes auxiliary modes of operation which support submerged navigation of manned submersibles and submarines, with frequencies selectable from 5 to 45 kHz in all modes. The unit is fitted with TIPE (Transponder, Interrogator, Pinger/Echo-sounder). The transponder/interrogator functions provide automatic ranging between two Model 5400s (or any other compatible device, such as ATM-504A equipped with TIPE). The pinger and echo-sounder mode offers acoustic tracking, depth sounding and echo ranging. Used with a vertically directed transducer, the Model 5400 can provide digital readout of depth below the surface or distance to the seabed. Transducers are available for operation to full ocean depth. The operating frequencies for the TIPE mode are operator selectable within the system's entire frequency band.

The Model 5400 is controlled by an operator using a touch front panel, hand microphone and operator headset. Squelch control enables background noise to be reduced or eliminated.

Remote control is available using separate Model 5400 remote-control units, or via an RS-232/RS-422 interface.

Operational status
The system is in service in the US Navy, NATO and other foreign navies.

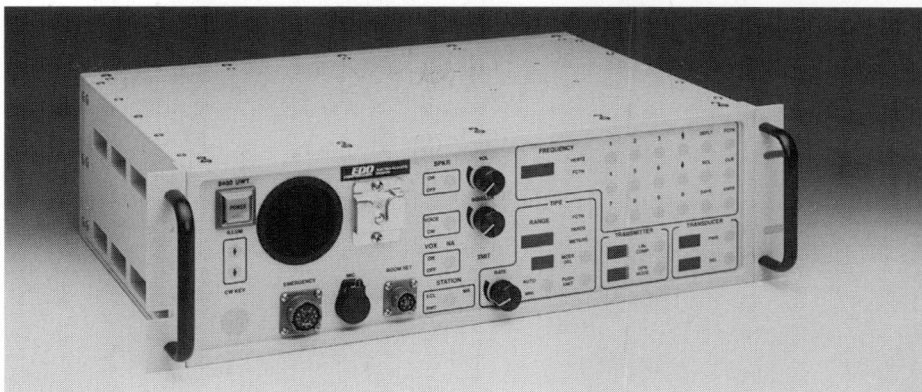

Model 5400 underwater telephone   *1995*

Specifications
Height: 132.5 mm
Width: 431 mm
Depth: 352.5 mm
Weight: 16 kg

Contractor
EDO Acoustic Products, Salt Lake City, Utah.

VERIFIED

Model 505

Type
Emergency underwater telephone.

Description
This is a compact, rugged, self-contained emergency underwater communications system featuring single sideband, suppressed carrier operation. The unit provides both voice and keyed CW operation at three power levels for both optimum performance and for conserving battery power. Power can be supplied by batteries or by ship's DC power. The Model 505 is designed for use with the vessel's installed UQC/BQC transducers. The unit is housed in a portable case suitable for bulkhead, overhead or under-seat mounting which provides easy access to the telephone handset, system controls and self-contained batteries.

Operational status
The system is in service in foreign navies in deep submergence/submarine rescue programmes.

Specifications
Height: 386 mm
Width: 342 mm
Depth: 172.5 mm
Weight: 18 kg

Contractor
EDO Acoustic Products, Salt Lake City, Utah.

VERIFIED

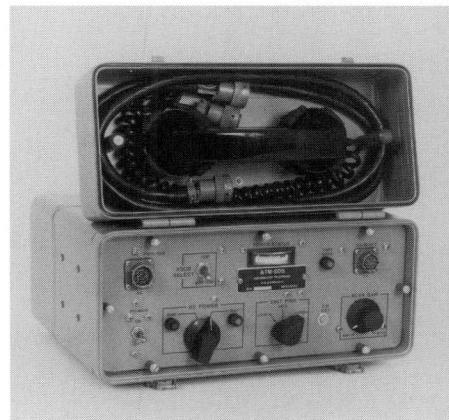

Model 505 underwater telephone
1995

COMMUNICATIONS BUOYS

UNITED KINGDOM

ECB-680(1) buoy

Type
Expendable communications buoy.

Description
The Expendable Communications Buoy ECB-680(1) system has been developed by Thomson Marconi Sonar Ltd for Royal Navy operations to provide a one-

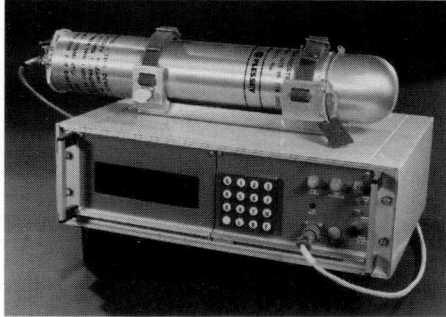

ECB-680(1) expendable communications buoy

way line of sight VHF/UHF radio communications relay between a submerged submarine and surface receivers.

The ECB-680(1) is designed for launching from either forward or aft Submerged Signal Ejector (SSE) tubes at all operational depths and up to 15 kt. Before launch, the ECB is programmed by means of an inboard control interface unit, with its operating parameters and plain or encoded messages as dictated by the tactical situation. Alternatively, in an emergency, the ECB-680 (1) can be launched with no programme input; in this mode of operation a preprogrammed distress SARBE signal is transmitted.

On release, the buoy ascends to the surface after a preset transmission delay has elapsed, allowing the submarine to leave the area discreetly. The aerial unit with its flotation collar is then deployed and message transmission is initiated. Transmissions are made in the VHF/UHF band between 168 MHz and 310 MHz; modulation may be programmed for either AM or FM transmissions.

The message cycle allows up to 3 minutes of message with a 1 minute gap and may be repeated up to 60 times, allowing the buoy 4 hours' transmission time. For the emergency role, an unlimited number of

cycles may be selected, in which case the buoy will transmit in excess of 8 hours. On completion of the preset number of message cycles, the buoy scuttles automatically.

The buoy's aluminium hull is capable of withstanding the extreme water pressures experienced during normal submarine operations, and is reinforced to permit launch at high speed for suitably equipped submarines.

The ECB-680(1), operating in conjunction with the submarine's communication system, provides a comprehensive capability for emergency distress and tactical communications. By employing existing submarine SSE launch facilities, the ECB-680(1) is easy to install, requiring minimum maintenance and support.

Operational status
Widely fitted to submarines of the Royal Navy and other navies.

Contractor
Thomson Marconi Sonar Ltd, Templecombe.

VERIFIED

ECB-699 communication buoy

Type
Expendable buoy for satellite communications from submarines.

Description
The ECB-699 UHF Satellite Communications System provides a digital communications link from submarines to shore.

A Control Interface Unit (CIU) is used to load messages, frequency and timing information for subsequent transmission. The CIU also provides built-in test facilities.

The ECB-699 buoy is launched from the submarine's Submerged Signal Ejector (SSE) at normal operating depths and speeds. After launch, the buoy rises to the ocean surface. Once the programmed delay has elapsed, the buoy automatically erects its antenna, and transmits, via satellite, to shore. To prevent interference with established submarine broadcast periods, transmissions are automatically inhibited except in the emergency mode. If the submarine is in distress, the

buoy can be used to transmit a distress signal on 406 MHz via the KOSPAS/SARSAT system, which provides reliable, rapid, worldwide indication of the buoy's identity and location.

On completion of transmission the buoy automatically scuttles itself within 20 minutes.

Operational status
In production for the Royal Navy since the early 1990s.

Specifications
Buoy
Weight: 4.4 kg
Dimensions
Length: 502 mm
Diameter: 101 mm
Communications mode: digital data preset for 2,400 bits/s. Selectable for 75, 300, 600, 4,800 and 9,600 bits/s
SSE holding and transmission delay: selectable up to 8 h
Message cycles: 1-15 programmable messages each repeated once

Buoy scuttle: after message cycle
Transmit power output: 100 W ERP
RF transmission modes: dual-phase Frequency Shift Keying (FSK)
Operating frequency range: 290-320 MHz
Channel selection/spacing: multiple programmable frequencies/25 kHz channel increments
Alert mode: COSPAS/SARSAT tone at 406 MHz
CIU
Weight: <10 kg
Dimensions
Height: 889 mm
Width: 482 mm
Depth: 350 mm

Contractor
Ultra Electronics Sonar and Communication Systems, Greenford.

UPDATED

UNITED STATES OF AMERICA

AN/BRR-6

Type
Submarine towed buoy communications antenna.

Description
The AN/BRR-6 is a towed buoy RF receiver communications antenna system which is installed on the 'Ohio' class Trident submarines.

Operational status
In production and in service.

VERIFIED

AN/BRT-6 communications system

Type
UHF satellite communications system.

Description
The AN/BRT-6 UHF satellite communications system was developed by the Hazeltine Corporation, in conjunction with the US Naval Underwater Systems Center and the US Naval Electronics System Command, as a digital communications link from

submarines at operating depth to shore communications stations and operating naval forces. The expendable UHF SATCOM buoy provides reliable, high-fidelity, one-way communications using a uniquely designed Hazeltine right circular polarised floating antenna. The SATCOM buoy is manually or automatically launched from standard 3 in signal ejectors on fleet submarines and can be operated in conditions up to Sea State 5.

The expendable buoy is a transmit-only digital communications device. After launch it rises to the surface at a speed in excess of 3 ft/s. After a suitable

delay period the buoy automatically extends its antenna and transmits to the Submarine Satellite Information Exchange System (SSIXS) receiving system via the operational satellite communications channels.

The buoy operates on multiple programmable frequencies between 290 and 315 MHz in channel increments of 25 kHz. The modulation is compatible with the AN/WSC-3 transceiver. Messages are digitally formatted and are selectable for 1 to 15 transmissions. Data is transmitted at a preset rate of 2,400 bits/s and selectable for 75, 300, 600, 1,200, 4,800 and 9,600

bits/s. The delay between message transmissions is selectable in 5 minute increments.

The buoy incorporates a self-scuttling device which is activated within 30 minutes of the last transmission.

Operational status
In production.

The AN/BRT-6 buoy deployed
1994

Contractor
GEC-Marconi Hazeltine Corporation, Greenlawn, New York.

UPDATED

AN/SSQ-86(XN-1) DownLink Communication (DLC) sonobuoy

Type
One-way surface-to-submarine communication equipment.

Description
The AN/SSQ-86(XN-1) DownLink Communication (DLC) sonobuoy is a one-way communication device designed to transmit a preprogrammed message to a submerged submarine. Developed to meet the need for a reliable means of transmitting a message without revealing the receiving submarine's position, the DLC is compatible with existing sonobuoy launch platforms. The unit is packaged in a standard 'A' sized sonobuoy envelope and can be launched from any properly equipped fixed-wing aircraft or helicopter, or from a surface ship.

The desired message is programmed into the DLC using a single push-button switch and four seven-segment LED displays. Four groups of three digits make up the message. The message can be verified and corrected if necessary with a second push-button, provided to ensure proper data insertion. Once programmed, all remaining operational functions are automatic. The buoy is tunable to one of 99 RF channels.

When air-launched, the DLC is slowed and stabilised by a small parachute. For launches from surface

AN/SSQ-86(XN-1) downlink communication sonobuoy

vessels, it is merely thrown overboard. Upon water entry, the DLC immediately deploys. As soon as the subsurface unit reaches the shallow operating depth,

the message, coded into appropriate tones, is acoustically transmitted. The first transmission is followed by a 5 minute pause while the subsurface unit deploys to the greater depth. The message is repeated at the greater depth; after a second 5 minute pause, the message is transmitted a third time. At the end of the final transmission the DLC automatically scuttles. The nominal life of the DLC from water entry to scuttle is 17 minutes.

The omnidirectional projector is a simple resonant bender element designed to operate efficiently at the desired bandwidth and power is supplied by a lithium battery.

Operational status
No longer in production but still in service. During 1996, Sparton refurbished over 1,000 buoys for the US Navy, thus extending their useful life for another five years.

Specifications
Depth: shallow and deep
Size: 'A' (123 × 910 mm)
Weight: 11.4 kg
Operating life: 17 mins (nominal)

Contractor
Sparton Corporation, Electronics Division, DeLeon Springs, Florida.

VERIFIED

SUS Mk 84

Type
Air-to-submarine communications device.

Description
The SUS Mk 84 Mod 1 underwater sound signal device is an expendable electro-acoustic device which provides one-way acoustic communications with submarines. It may be dropped or deployed from fixed-wing aircraft or helicopters as well as by over-the-side ship-launched methods. It can also be used to simulate

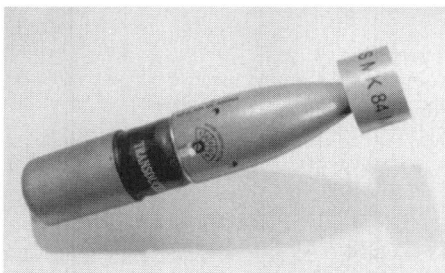

SUS Mk 84

the drop of an ASW weapon during a tactical exercise. The low centre of gravity and spin stabilising design of the tail allows launches from altitudes up to 10,000 ft and speeds up to 380 kt without parachutes or other descent stabilising systems. The device activates within 2 seconds after entry into the water to provide a source level of 160 dB 11 µPa throughout the life of the units.

The SUS Mk 84 is able to transmit any one of five prelaunch-selected coded acoustic signals, each of which may convey a predetermined message to the submarine. The device transmits four acoustic tonals, one to four at between 3.3 and 3.5 kHz, each of which may be pulsed for either ½ or 1½ seconds. The fifth code setting generates a continuous 3.5 kHz tone. Third harmonics of the fundamental frequencies are also generated at levels slightly less than those of the fundamentals. The coded sequences are selected using a five-position switch on the side of the unit. The submerged submarine may use either passive sonar or underwater telephone to receive the SUS output.

The SUS is a compact device, measuring 38 cm in length by 7.6 cm diameter and weighing 2.7 kg. A heavy zinc nose houses a seawater battery, a hermetically sealed inlet port to the battery compartment and most of the electronics.

Operational status
In production and in service with the navies of Australia, Canada, France, Greece, Italy, Netherlands, New Zealand, Norway, Portugal, Spain, Turkey, UK and USA.

Specifications
Dimensions: 38 × 7.6 cm
Weight: 2.7 kg
Launch altitude: 0 – 3,000 m
Launch speed: 0 – 300 kt
Operating life: 70 s
Frequencies: 3.3, 3.5 kHz (plus third harmonics)
Source level: 160 dB
Code table: (long 1.5 s, short 0.5 s)

Code	F1 (3.3 kHz)	F2 (3.5 kHz)
1	Long	Long
2	Short	Long
3	Short	Short
4	Long	Short
5	Off	Continuous

Contractor
Sparton Corporation, Electronics Division, DeLeon Springs, Florida.

VERIFIED

ELECTRO-OPTICAL SENSORS

Although the sonar suite is the submarine's main sensor system, the periscope, and its successor the optronic mast, will remain major sensors for some time to come. These sensors provide amplifying information to that already gathered by the sonar suite and, for this reason, will remain important for the foreseeable future.

The modern periscope is highly sensitive and sophisticated, generally integrating image intensifier, infra-red detection and ranging, camera recording equipment, together with gyrostabilised fixed lines of sight and so on. Other features common to modern periscopes are split beam binocular viewing, heated windows, alphanumeric data display in the ocular, data transmission to the combat information system, power drive and other optional extras which all help to improve the effectiveness of the system. In addition, the mast often carries various other sensors such as ESM and communications antenna. With the advent of new command systems and multifunction consoles and new software technologies including Windows, operators are able to display both tactical data and visual images from the periscope/optronic mast on a single console display, which considerably enhances the command team's ability to assess a tactical situation.

Periscopes are either of the hull penetrating or non-penetrating type. Non-penetrating masts have the advantage that they leave the control room floor clear of obstruction and offer much improved positioning for the operator. In this arrangement the ocular remains in a fixed position, overcoming the need for the operator to follow the periscope lens round. Hull-penetrating masts, on the other hand, require the operator to follow the periscope lens round, which takes up more space. Modern hull penetrating periscopes no longer require the operator to adopt awkward crouching positions when using the ocular at the base of the mast. This problem has been overcome by mounting the periscope inside a large rotating tube to which is fixed the ocular, which remains at a constant level irrespective of the position of the periscope, which slides up and down as normal inside the outer tube. The field of view is reflected off a prism mounted in the lower part of the tube, up to the ocular sited at eye level, affording the operator a continuous view during the full vertical movement of the periscope.

There are at present two main types of periscope, each with its own particular ocular arrangement. The main periscope is the search periscope with a large (in relative terms) lens designed to gather the maximum amount of light to provide the best possible means of detection, even in conditions of low light. These periscopes incorporate extra optical systems for use at night and often include a tilt mechanism which allows zenith observation for navigation. The much smaller attack periscope is designed to create minimum disturbance of the water and present the minimum radar cross-section to reduce the possibility of detection.

As soon as a mast penetrates the water surface it is subject to enormous forces which, when it is a certain height above the surface and moving above a certain speed, cause vibration in the mast. Vibrations occur due to the resonance between the natural bending frequency of that part of the periscope submerged in the water and the frequency of the vortex shedding from that part of the periscope above the water as it passes through the water. The vibration can seriously affect the clarity of the image viewed in the ocular. This vibration is overcome by the use of hydrodynamically designed fairings around the mast, by corrective optical lenses and through stabilisation of the line of sight. Careful design enables periscopes to provide accurate vibration-free viewing at speeds up to 12 kt.

Periscopes are now undergoing a revolutionary change. This has been made possible by the latest developments in optronics, fibre optics and allied technologies. The direct optical channel is being made redundant, to be replaced by high-resolution, broadband Electro-Optical (EO) sensors which will have a major impact on the operational capabilities of the platform. These new technologies will enable the periscope view to be displayed on a large screen in the control room, obviating the need for the operator to view the scene through the fixed ocular system.

Because the operator views the horizon from a fixed position, he has no immediate feel for the direction of view and bearings to objects will have to be displayed against a fixed scale presented on the screen.

These revolutionary developments will result in future submarine control room layouts which may well differ considerably from current designs.

VERIFIED

AUSTRALIA

'Collins' class periscope system

Type
Search and attack periscopes.

Description
British Aerospace Australia is the Australian Industry partner to Pilkington Optronics (Barr and Stroud Ltd), UK, for the provision of the periscope systems for the 'Collins' class submarines on behalf of the Royal Australian Navy. As the primary above water sensor, the periscope systems embody state-of-the-art optronics which interface with and update the submarine's combat system during surveillance and strike missions. The 'Collins' class submarines are equipped with two periscopes which collectively possess: still camera; LLTV; passive rangefinder; image intensification and thermal capability.

British Aerospace Australia is responsible for the manufacture, assembly, system testing, onboard installation and setting to work of the periscope systems for the 'Collins' class submarines. Additionally, the company undertook the design and manufacture of the LLTV sensor which is also being supplied to the Royal Navy for service with the UK 'Vanguard' class submarine periscope systems.

Operational status
Periscope systems have been installed on board HMAS *Collins* and HMAS *Farncombe*.

Contractor
British Aerospace Australia, Salisbury, South Australia.

UPDATED

FRANCE

SFIM periscopes

Type
ST 5 attack periscope.

Description
The ST 5 attack periscope has recently been upgraded to provide it with enhanced transparency, a magnification of ×12, one-axis stabilisation, night vision capability with image intensification, improved stealth and a reduction in noise signature. The image intensifier camera is sufficiently compact to be mounted in the small head of the attack periscope. To improve stealth the head is specially shaped and coated with RAM to reduce the radar cross-section (RCS).

Operational status
In service with the French Navy.

Type
Modernised J-type search periscope.

Description
Upgrades to the J-Type search periscope include magnification of ×1.5 and ×6, elevation arc of −10 to +90°, image intensification and improved stealth.

Operational status
Being delivered for the 'Agosta 90B' boats for the Pakistan Navy.

Contractor
SFIM Industries, Massy.

UPDATED

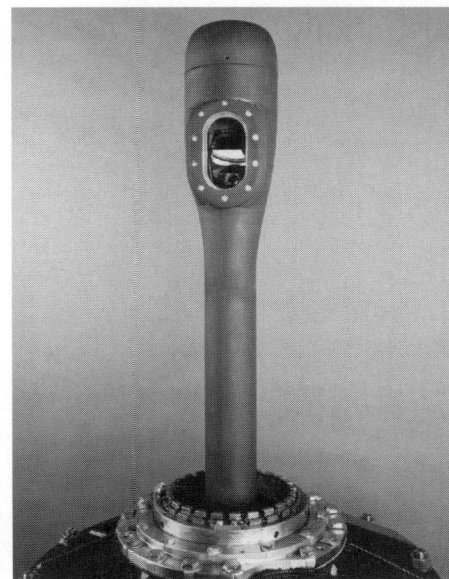

ST 5 attack periscope head
1996

APS

Type

Small diameter attack periscope.

Description

The small diameter (140 mm) attack head includes a single-axis gyrostabilised line of sight for both optical and low-light level TV channels. Four magnifications (×1.5, ×3, ×6, ×12) are available for the optical channel. The LLTV (optional black and white or colour CCD) camera is mounted in the head. The head is covered with RAM for low RCS. The ocular can either be fixed or rotating.

Optional features include a laser rangefinder replacing the LLTV camera, a built-in ESM antenna or GPS antenna on top of the mast, and operation from a combat system multifunction common console.

Operational status

Developed from periscopes operational on board French submarines.

Contractor

SAGEM SA, Defence and Security Division, Paris.

VERIFIED

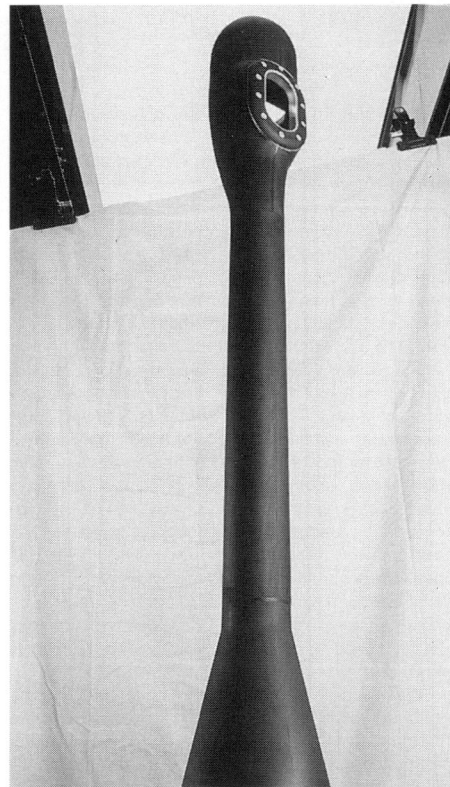

Attack PeriScope (APS)
1996

SPS

Type

Submarine visual and infra-red periscopes.

Description

The optronic search periscope Pivair is built around three fundamental characteristics: two-axis gyrostabilisation of the line of sight, built-in vertical reference, and integration of infra-red vision and an optical channel. The head is 320 mm in diameter. The head and upper fairing are coated in RAM. The fairing is of new design optimised to reduce wake and head vibration which results from vortex shedding.

A two-axis gyroscope located in the head overcomes problems from mast vibration and improves visual and IR observations by gyrostabilising the line of sight in elevation and bearing.

Associated with the built-in vertical reference, very accurate sextant measurements can be obtained, providing a reliable position fix.

The use of 8 to 12 μm band infra-red extends the operational capabilities, both at night and in poor meteorological conditions. The periscope can be operated in an auto look-around mode for a quick panoramic view above the water and in a real panoramic IR search mode providing long-range airborne and surface detection.

A number of options is also available, including:
(a) choice of magnification (optical channel) ×1.5 to ×12
(b) rangefinder
(c) built-in ESM antenna on top of the mast
(d) built-in GPS Navstar receiver antenna on top of the mast
(e) RAM coatings
(f) low-light level TV camera
(g) display and recording of video and infra-red images on remote console.

Operational status

Operational on board all French Navy nuclear submarines (SSBN and SSN).

Contractor

SAGEM SA, Defence and Security Division, Paris.

UPDATED

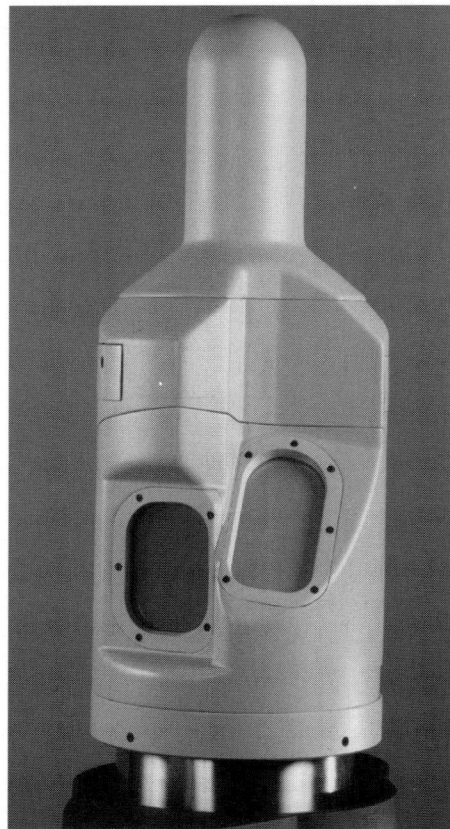

Search head
1995

SMS

Type

Submarine optronic non-penetrating mast.

Description

The optronic mast combines the advantages of SAGEM's search optronic periscopes with the advanced safety of non-penetrating masts. With the mast fully located outside the pressure hull, control room space is increased and more comfortable operating conditions are provided. The optronic non-penetrating mast is thus more easily accommodated in submarine designs and facilitates retrofit upgrades. The diameter of the head is 320 mm.

The SAGEM search optronic mast includes:
(a) dual field of view IRCCD thermal imaging system (conventional manual mode, auto look-around mode, panoramic search mode)
(b) high-resolution ×4 magnification (×1.5, ×3, ×6 and ×12) TV system
(c) two-axis gyrostabilisation of the line of sight
(d) interface for ESM antenna and/or GPS antenna

(e) electrical rotation
(f) range estimation
(g) RAM coatings.

The SMS mast can be fitted on any type of non-penetrating hoisting device. A remote-control and display console includes a 19 in CRT display, VCR, joystick and keyboard as well as electronic PCBs and power supply.

Furthermore, it can be operated from a multifunction common console in the command and control system of the combat system.

Operational status

Available for production. Trials were undertaken on board a French 'Daphné' class submarine (*Psyche* (in 1992)), the Swedish submarine *Västergötland* (in 1993), a South Korean Type 209/1200 (in 1995) and a Norwegian 'Kobben' class (*Svenner* in 1995). In production for several export customers.

Contractor

SAGEM SA, Defence and Security Division, Paris.

UPDATED

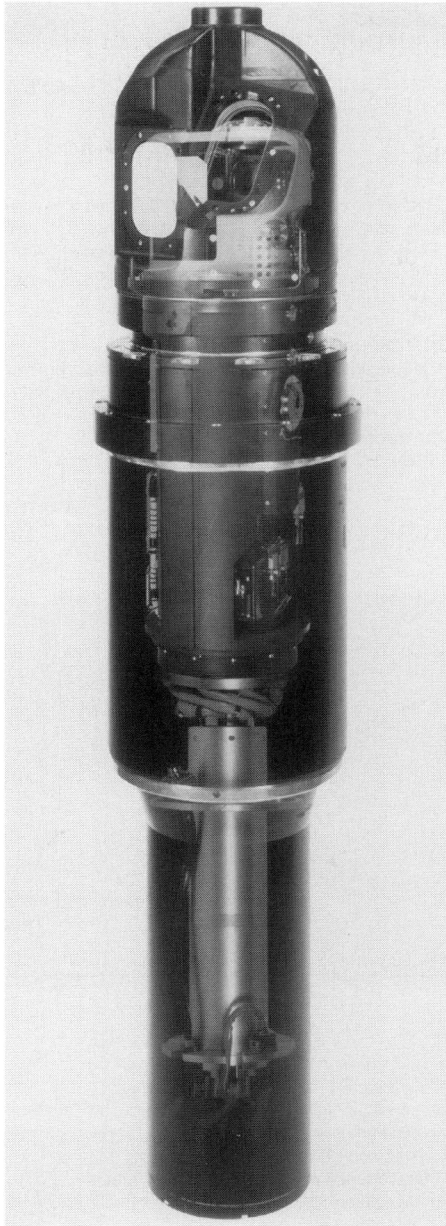

Search optronic mast

OMS

Type

Submarine optoradar non-penetrating mast.

Description

The optoradar mast combines the capabilities of SAGEM's optronic mast and the integration of a navigation radar. The optoradar mast allows elimination of separate masts for these functions, thus improving the discretion of the submarine. The non-hull penetrating system facilitates submarine retrofit upgrades. The head is 360 mm in diameter.

The SAGEM optoradar mast includes:

(a) thermal imaging system (conventional manual mode, auto look-around mode, panoramic search mode)

(b) high-resolution ×2 magnification TV system

(c) single-axis gyrostabilisation of line of sight

(d) X-band navigation radar

(e) electrical rotation

(f) interface for ESM antenna and/or GPS antenna

(g) range estimation

(h) RAM coatings.

Stabilised azimuth surveillance can be presented on one of four range scales between 4 and 32 km. Up to five targets can be tracked and automatic target acquisition is provided. Separate control and display consoles are provided to operate the radar and the optronic system. The radar console includes PPI display, software keys and a trackerball. The optronics

OMS non-penetrating optoradar mast
1995

The OMS mast as installed on the 'Le Triomphant' class SSBNs of the French Navy **1998**/0007233

console includes a 19 in CRT display, VCR, joystick and keyboard as well as electronic PCBs and power supply.

Operational status

Entered service on board French Navy SSBNs in 1993.

IMS

Type

Submarine infra-red non-penetrating mast.

Description

The infra-red mast combines the capabilities of SAGEM's non-penetrating masts while including a unique IR channel (8 to 12 μm). Reduced dimensions of the part emerging from the water improves discretion while providing a real day and night capability. The head is 210 mm in diameter.

The infra-red mast includes:
(a) thermal imaging system (conventional manual mode, auto look-around mode, panoramic search mode)
(b) two-axis gyrostabilisation of line of sight
(c) interface for ESM antenna or GPS antenna
(d) electrical rotation
(e) range estimation
(f) RAM coatings.

Contractor

SAGEM SA, Defence and Security Division, Paris.

UPDATED

The infra-red mast can be fitted on any type of non-penetrating hoisting device. A remote-control and display console includes a 19 in CRT display, VCR, joystick and keyboard as well as electronic PCBs and power supply.

Operational status

In production for export customers. Installed on two 'Narhvalen' class submarines of the Royal Danish Navy.

Contractor

SAGEM SA, Defence and Security Division, Paris.

UPDATED

IMS infra-red non-penetrating mast
1995

GERMANY

SERO 40 STAB

Type

Family of search and attack periscopes.

Description

SERO 40 STAB family comprises a stabilised attack periscope and stabilised modular search periscope with electronic control. The system features two-axis line of sight stabilisation, binocular viewing with geometrical beam splitting plus secondary eyepiece with magnification having click stop magnifications of ×1.5 and ×6, highly efficient visual integrated optical range finding system, remote-control capability, and connection facilities for a wide variety of antennas and camera systems, and installation into a hoisting device with streamlined fairing.

The line of sight prism has an elevation range of −15 to +75° (restricted to +60° if an antenna is fitted) and with a wide field of view at ×1.5 magnification (36° azimuth, 28° elevation).

The ocular provides a digital readout in the eyepiece of true and relative bearing, elevation angle, target height and range.

The following options are currently available:
(a) Thermal camera (in search periscope = SERO 40 STAB IR)

(b) Low-light level CCD TV camera (at auxiliary eyepiece)
(c) Video unit (monitor and VCR)
(d) Radar early warning system (omnidirectional/DF)
(e) Navigation (GPS) and communication (UHF/VHF) antenna
(f) Digital and roll film camera (35 mm)

Operational status

Currently in production. In service aboard Greek Type 209/1200 ('Poseidon' class), Indonesian Type 209/1300 ('Cakra' class) and Peruvian Type 209/1200 ('Casma' class), as the SERO 40 and on board the Chilean Type 209/1300 ('Thomson' class), South Korean Type 209/1300 ('Chang Bogo' class), Taiwanese 'Hai Lung' and Venezuelan Type 209/1300 ('Cabalo' class) as the SERO 4O STAB.

Contractor

Zeiss-Eltro Optronic GmbH, Oberkochen.

UPDATED

The SERO 40 STAB periscope (Zeiss-Eltro)
1999/0017503

SERO 14

Type
Search optronics periscope.

Description
The SERO 14 stabilised modular search optronics periscope features an integrated IR camera and continuous zoom magnification, two-axis line of sight stabilisation, binocular viewing with geometrical beam splitting plus secondary eyepiece with magnification having click stop magnifications of ×1.5, ×6 and ×12, highly efficient thermal channel, remote-control capability, and connection facilities for a wide variety of antennas and camera systems, and installation into a hoisting device with streamlined fairing. The unit incorporates an 8 to 12 μm thermal camera with monitor and recorder.

Other features are similar to the 40 STAB series.

Operational status
The SERO 14 optronics periscope is operational in the Norwegian 'Ula' class submarines and is in production for the German and Italian navies' new Type 212 submarines.

Contractor
Zeiss-Eltro Optronic GmbH, Oberkochen.

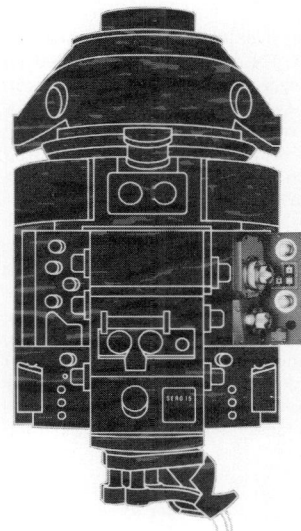

UPDATED

The SERO 15 (foreground) and SERO 14 (behind) aboard a Norwegian 'Ula' class submarine
1998/0007234

SERO 15

Type
Modular attack periscope.

Description
The SERO 15 stabilised modular attack periscope with integrated laser rangefinder features two-axis line of sight stabilisation, binocular viewing plus secondary eyepiece with zoom system magnification fitted with click stops at ×1.5, ×6 and ×12, remote-control capability and connection facilities for a wide variety of camera systems at the auxiliary eyepiece. Installation into a hoisting device with streamlined fairing. As optional extras the unit can be fitted with a TV/LLTV camera or a thermal imager (3 to 5 μm).

Operational status
The SERO 15 optronics periscope is in service aboard Norwegian 'Ula' class submarines and is in production for the German and Italian navies' new Type 212 submarines.

Contractor
Zeiss-Eltro Optronic GmbH, Oberkochen.

UPDATED

The SERO 14 (left) and SERO 15 (right) periscopes (Zeiss-Eltro)
1999/0017501

OMS 100

Type
Optronics mast system.

Description
The rotatable sensor head of the optronics mast can be raised above the bridge fin of the submarine using a hoisting device. The head features a TV channel with a three-chip colour CCD camera with the field of view changed either in steps or zoomed, and an IR channel with a second-generation thermal camera operating in the 7.5 to 10.5 μm band. The TV camera field of view is either zoomed or changed in steps from 30 × 22.7° to 3.5 × 2.6°. The digital IR camera is equipped with IRCCD 96 × 4 or IRCCD 288 × 4 detectors with a field of view from 12.4 × 9.3° to 4.1 × 3.1° changed in steps. The sensor head has an elevation from −15 to +60°.

Digital interfaces allow data to be transmitted to the control and display unit via a fibre optic link. The daylight and IR images can be displayed on monitors and recorded on videotape.

The lines of sight can be smoothly and continuously set using a brushless DC motor drive. They are stabilised in elevation and azimuth to ensure optical observation even under pronounced movements of the submarine.

The sensor head can be fitted with a GPS and/or ESM omni/DF antenna and with a RAM coating.

Operational status
The mast is in service with the German Navy and in production.

Contractor
Zeiss-Eltro Optronic GmbH, Oberkochen.

UPDATED

ITALY

Riva Calzoni non-penetrating masts

Type
Range of optronic, ESM and radar masts for submarines.

Description
Riva Calzoni produces a range of submarine masts for various applications. All masts are non-penetrating into the control room; there are no translating parts and sliding seals through the hull boundary. Only a short fixed head is sometimes fitted inside for hydraulic connections. This completely frees the control room and provides the crew with an unobstructed view and access.

Each hoist consists of three main parts:
(a) guide fixed to the main sail structure
(b) streamlined mast which slides, by means of sliding shoes, inside the guide
(c) hydraulic cylinder, inside the mast and fixed to the hull.

Electrical connections to the antennas on the mast top are generally of flexible cables located inside the mast and forming a loop housed in a side recess, making the hoist units compact and self-contained.

The integration of hydraulic, electrical and mechanical parts ensures reduced size and weight, and also fast and easy installation. The masts are hydrodynamically shaped in order to reduce wake and vibration. They are fully free-flooding except for the hydraulic cylinder, where the actuating rod has double seals for water and oil tightness.

Riva Calzoni non-penetrating masts

Masts are applicable to a number of configurations, including optronic, radar, radio, ESM/ECM and other special purposes. These configurations refer to single-stage and double-stage masts.

Available are compact solutions of two or more masts integrated within one modular structure.

The modular hoisting mast of non-hull-penetrating design features a rectangular guide trunk housing a faired mast in one or two sections. It is raised and lowered by a non-penetrating hydraulic cylinder. The mast is designed to withstand speeds well in excess of 12 kt at periscope depth with minimum wake and plume.

Operational status
Riva Calzoni has supplied or is supplying non-penetrating masts to the Italian Navy and a number of other countries' navies, including Argentina, Netherlands, Norway, Germany and Australia.

An integrated solution of three masts in one modular hoisting mast structure has been installed in the new Norwegian 'Ula' class submarines. In late 1989, Riva Calzoni received a contract from Kollmorgen Corporation in the USA to develop a universal modular non-hull-penetrating mast. The programme, funded by DARPA, completed successful sea trials aboard the USS *Memphis* in 1992. The new mast design will interface a new range of optronic sensors for the US NSSN.

Contractor
Riva Calzoni, Bologna.

UPDATED

RUSSIAN FEDERATION AND ASSOCIATED STATES (CIS)

Standard periscope

Type
Standard submarine periscope systems.

Development
Lomo Plc has been involved in developing periscope systems since 1939. The main types manufactured are attack, search, celestial correctors and high-precision optical reference systems. Celestial correctors are used to provide high-precision measurements of celestial reference objects' co-ordinates to determine the position, course correction and weapon reference for large submarines.

Description
The standard periscopes with tube diameters of 180 or 260 mm are designed for installation in large and small displacement submarines. Facilities available include day and night target acquisition and classification, measurement of range and azimuth angles, height and azimuth measurement of celestial objects, satellite navigation and preliminary acquisition of radio signals. A video recording facility is also incorporated in the system. All mounts are fitted with vibration suppressors.

The periscopes feature a wide range of information channels which are provided according to customer requirements. The two-axis stabilised systems feature optically matched channels and are fully autonomous in operation. Laser rangefinders are fitted as standard. The optical channel is fitted with two magnifications ×2 (or ×4) and ×8, with fields of view of 40° (or 20°) and 10°. Elevation is from −10 to +60°. The TV day and LL channel is provided with a field of view of 18° with elevation from −10 to +30°. The laser rangefinder has a wavelength of 1.54 mcm (or 1.06 mcm according to customer requirements) with a maximum range of 10 n miles and minimum range of 60 m. Maximum training speed is 20°/s.

Operational status
In service in most Russian submarines and in 'Kilo' class submarines exported overseas.

Contractor
Lomo Plc, St Petersburg.

UPDATED

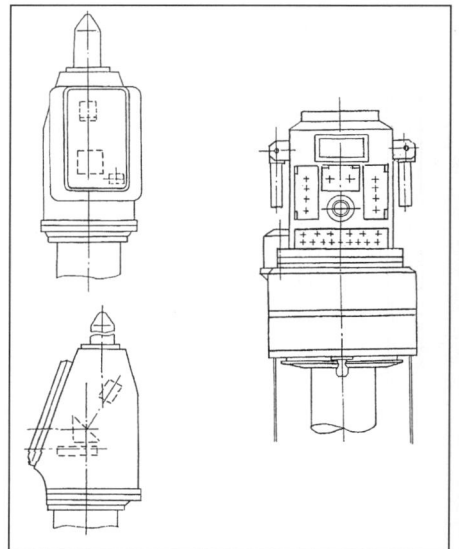

General arrangement of standard periscope
1998/0006982

Non-retractable periscope

Type
Single-tube periscope for installation in small displacement submarines.

Description
This periscope has a tube diameter of 260 mm and is designed for installation in small displacement (midget) submarines. Facilities available include day and night target acquisition and classification, measurement of range and azimuth angles, height and azimuth measurement of celestial objects, satellite navigation and preliminary acquisition of radio signals. A video recording facility is also incorporated in the system. All mounts are fitted with vibration suppressors. The system is available in single or double tube mounts with an adjustable optical system. The ocular can be either fitted as a stationary mount (which does not retract into the hold) or trainable.

The periscope features a wide range of information channels which are provided according to customer requirements. The two-axis stabilised system features optically matched channels and is fully autonomous in operation. A laser rangefinder is fitted as standard. The optical channel is fitted with two magnifications ×2 (or ×4) and ×8, with fields of view of 40° (or 20°) and 10°. Elevation is from −10 to +60° and maximum training speed is 20°/s. An optional TV day and LL channel with a field of view of 18° can be provided with optional equipments being thermal imaging channel, satellite navigation antenna, communication antenna and laser for use in bad light conditions.

Operational status
In service.

Contractor
Lomo Plc, St Petersburg.

VERIFIED

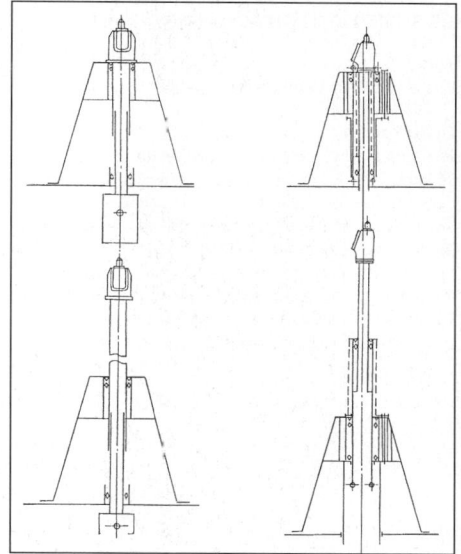

General arrangement of non-retractable periscope
1998/0006983

UNITED KINGDOM

Compact periscopes

Type
Compact periscopes for small submarines.

Description
This range of periscopes is optimised for small platforms of between 50 and 500 tonnes. With a range of seven models (CK032, CK037, CK039, CK041, CK044, CK044S and CK060) using modular designs and 127 mm diameter tubes, solutions are available to meet all requirements. Offering high-performance optics coupled with image intensification for night vision, stabilisation, electronics weapons system interface, still camera, elevation of line of sight (−15 to +60°) facility and dual magnification, the periscopes can be further enhanced by the addition of ESM warning and GPS sensors. The full range of compact periscopes uses proven, in-service technology providing the optimum solution for small submarines without compromising performance.

Operational status
The CK032, CK037, CK039 and CK041 periscopes are in service.

Contractor
Pilkington Optronics, Glasgow.

VERIFIED

Compact search periscope
1998/0006984

190 mm electronic periscopes

Type
Search and attack periscopes.

Description
The search periscope, CK038, and attack periscope, CH088, are both fitted with 190 mm diameter tubes. They are compatible with medium-sized diesel-electric submarines for either new build or retrofit applications and can also be offered in search/attack or stand-alone configurations.

The CH088 attack periscope features a low-profile top stem and top hood. Offering exceptional reliability, both the CK038 and CH088 periscopes can be fitted with image intensification, thermal imaging, LLTV, video recording, stabilisation, torque drive or torque assist and an electronic weapons system interface. Further upgrades including ESM warning, GPS and communications sensors can be provided on the CK038 search periscope. All functions on the periscopes are controlled by electronic controls.

The modular design of the periscopes offers the lowest cost upgrade path, low maintenance costs and minimal platform modifications.

Operational status
The CK038 periscope is in service on board the Swedish 'Västergötland' class submarines.

Contractor
Pilkington Optronics, Glasgow.

VERIFIED

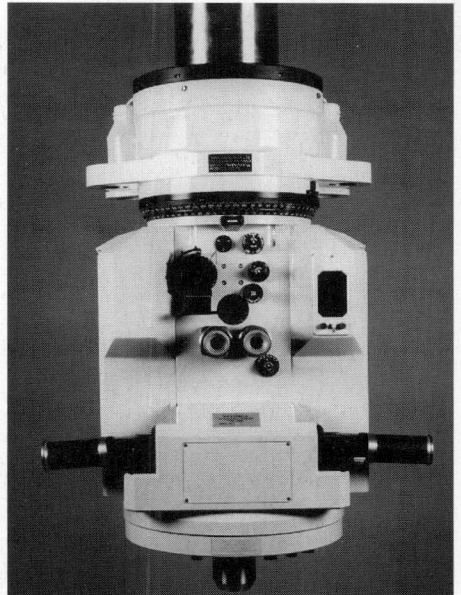

190 mm electronic periscope
1998/0006985

254 mm optronic periscopes

Type
Search and attack periscopes.

Description
The most advanced periscope systems currently in service, the CK043 search and CH093 attack periscopes offer the optimum combination of day and night vision and electronic surveillance sensors. The periscopes are capable of being operated totally by remote control from either a dedicated console or multifunction console and are designed to allow full integration with the platform's combat system.

Fitted with thermal imaging, image intensification,

LLTV, ESM warning and DF, GPS and communications sensors, all contained within a 254 mm diameter main tube, the periscopes are ideally suited for large platforms.

Operational status
All versions of the optronic periscope are in service. The CK043 and CH093 have been selected to equip the Australian 'Collins' class submarines.

Contractor
Pilkington Optronics, Glasgow

VERIFIED

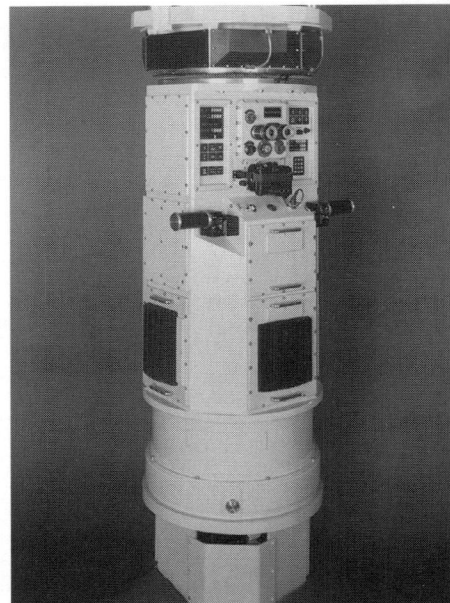

254 mm optronic periscope as fitted to the 'Collins' class submarines

CM010 series

Type
Non-hull penetrating optronic masts.

Description
The CM010 series of optronic masts combines innovative design with the latest proven technology to provide exceptional performance and operational advantage under all conditions. All CM010 masts are non-hull-penetrating and can be fitted with a wide range of modular sensor heads. Sensor technologies available include thermal imaging (8 to 12 µm or 3 to 5 µm), high-definition monochrome TV, LLTV, colour TV, and support for high-sensitivity broadband ESM, communications and GPS sensors. Images captured by the system are complemented by advanced image manipulation and image processing capabilities which further enhance the operational advantages of the system.

The optronic mast systems are controlled and operated from dedicated remote-control consoles equipped with high-resolution monitor displays, allowing the command team to obtain a complete above-water picture. Alternatively, should the submarine require, the system can be controlled from suitably equipped multifunction consoles. The common mast raising equipment facilitates integration with other payloads such as dedicated ESM, radar, satcom, and communications antenna packages.

Operational status
The CM010 optronic mast is in production. Two masts are being delivered to the Royal Australian Navy to meet its non-hull-penetrating periscope requirement. They will be fitted on board 'Collins' class submarines to replace either the CH 093 periscope or the radar mast.

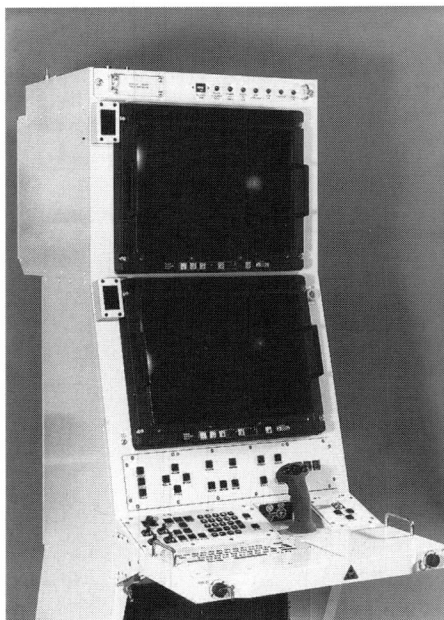

Optronics mast console   *1997*

The UK Royal Navy 'Trafalgar' class submarine *Trenchant* has also been fitted with a CM 010 mast for trials. It temporarily replaces the CH 084 attack periscope.

Contractor
Pilkington Optronics, Glasgow.

UPDATED

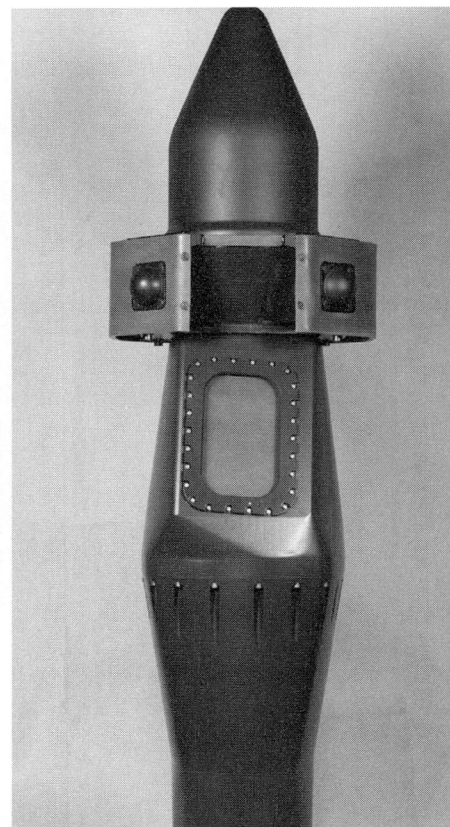

Optronics mast sensor head fitted with AZE-2 ESM head   *1998*/0006986

FD 5000 periscope camera

Type
Periscope-mounted video camera.

Description
This lightweight (135 g) monochrome video camera has been designed to be mounted within a periscope head to assist in periscope observation. It measures 37 × 52 × 55 mm and features a 24 mm lens and offers a resolution of 400 lines.

Submarines equipped with the system would come to periscope depth long enough for a scan of the horizon. The submarine can then dive while the images

are examined on a TV monitor linked to the camera by a screened cable.

The camera has no moving parts, giving high reliability.

Contractor
GEC-Marconi Avionics, Edinburgh.

VERIFIED

FD 5000 periscope video camera

Optronic sensors

Type
Submarine periscope optronic sensors.

Description
The increasing need to supply accurate information day and night in all weather conditions requires high-performance electro-optic sensors and processing. The GEC-Marconi electro-optical systems for periscopes and non-hull-penetrating masts incorporates fully stabilised, high-performance daylight, low light, colour and thermal imaging sensors. Processing includes automatic target detection, image enhancement, and other processing features. The data from the multiple sensors may then be correlated to provide optimum information. Fibre optics and signal multiplexers are used to minimise the system complexity.

The TICM mini-scanner and electronic modules are utilised, operating over the 8 to 13 μm wavelength. Staring focal plane array camera technology can be employed to provide a miniature IR sensor in the 3 to 5 μm range. In the visible spectrum, use is made of high-resolution CCD cameras for colour, monochrome or low light.

Operational status
In production.

Contractor
GEC-Marconi Electro-Optics Ltd, Sensors Division, Basildon.

VERIFIED   *Optronic sensors equip submarine periscopes*

OE 0285

Type
Submarine ice operation camera.

Description
Developed for use aboard submarines operating in ice conditions, the OE 0285 is an ISIT camera capable of viewing objects in ultra-low light level conditions in the region of overcast starlight, enabling thinner patches of ice to be identified. This provides an important aid to submarines surfacing through ice when engaged on Arctic operations. The cameras, which are the first to be developed specifically for under ice operations where low-light levels present a major obstacle to obtaining well-defined images, are connected via suitable cabling and hull gland penetrators to the control equipment inside the submarine's hull. The control equipment comprises a dual-redundant camera power supply, data overlay generator, display monitor and high-quality video monitor. The cameras are housed in nickel aluminium bronze castings and are positioned on the submarine's fin, allowing both forward and overhead images to be obtained.

Operational status
In production and operational aboard Royal Navy submarines.

Contractor
Simrad Ltd, Aberdeen.

VERIFIED

Model OE 0285 ice cameras in bronze housings

Model OE 0285 combined display monitor, video monitor, power supply and control unit

UNITED STATES OF AMERICA

Model 76 periscope

Type
Modular attack and search periscope.

Description
The Kollmorgen Model 76 is a modular periscope system with common components for the attack and search versions. The basic difference is that the attack periscopes have a smaller profile head, while the search periscopes have a larger head to act as multipurpose reconnaissance platforms.

The system consists of a mast unit with optical train, a display and control unit including a split-beam binocular eyepiece, a 35 mm camera and training handles. In addition to the mast unit, there is a hoisting yoke, a control unit, and a junction box unit.

The basic Model 76 periscope incorporates numerous features in the basic configuration including: elevation line of sight stabilisation; integral electric torque drive; microphone; three magnifications; mechanical bearing dials; eyepiece data display for range; true/relative bearing and elevation displays; heated head window; 35 mm photo camera kit; optical stadimeter; fail-safe election line of sight; image intensification; and LLLTV-CCD TV cameras.

Features offered as options include RAM, ESM/DF antennas (2-18 GHz), video tape recorders, GPS antenna, voltage probe antenna, thermal imaging capability (3-5 µm or 8-12 µm), SATCOM antenna, remote-control console, sextant capability and laser rangefinder.

Operational status
In production. The Model 76 is fitted in a number of submarines including Brazilian Type 209 submarines and in the Improved 'Sauro' class submarines of the Italian Navy. Also operational in Swedish 'Nacken' and Israeli 'Dolphin' class and others (see Table XI).

Specifications
Diameter: 190.42 mm
Elevation: −10 to +74° (attack); −10 to +60° (search)
Magnification: ×1.5, ×6, ×12
Fields of view: 4°, 8°, 32°

Contractor
Kollmorgen Corporation, Electro-Optical Division, Northampton, Massachusetts.

UPDATED

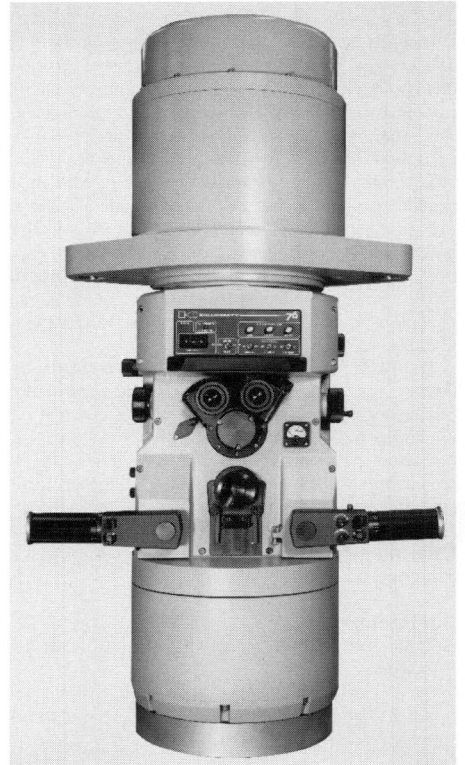

Model 76 periscope

Model 86 optronic mast series

Type
Submarine non-hull-penetrating optronic masts.

Description
Capabilities which previously required a hull-penetrating optical periscope can now be included in one of these sensor packages. The Model 86 optronic mast series includes a sensor unit, a hydraulically operated streamlined mast including connection cabling external to the hull, and electronic interface unit and a control/display console internal to the hull. Electronic processing and data transmission permit all sensor information to be processed and displayed in a dedicated operating console, or incorporated into the main combat consoles to further reduce space requirements and enhance operation.

Features of the optronic mast include:
(a) thermal imaging for day/night/adverse weather viewing without major hull penetration
(b) colour or black and white television for daylight, low-light level and 'quick-look' viewing
(c) two-axis line of sight stabilisation to eliminate ship's motion and mast vibrations
(d) ESM (2-18 GHz) warning to detect radar threats
(e) unique, rotating sensor package (sealed statically) with quick response and low power consumption
(f) manual or automatic mast control with a 'quick-look' mode.

Operational status
In production. The Model 86 has been fitted in US Navy 'Los Angeles' class submarines.

Specifications
Fields of view: 3°, 4°, 12°, 16° (visual); 3°, 4°, 6.75°, 9°(IR)
Elevation: −10 to +72° (visual); −10 to +55° (IR)
IR camera: 3-5 µm or 8-12 µm

Contractor
Kollmorgen Corporation, Electro-Optical Division, Northampton, Massachusetts.

Remote-control console for Model 86 optronic mast series

UPDATED

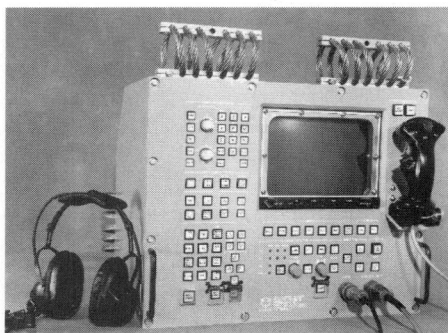

Model 86 tactical optronic mast

Model 90 optronic periscope system

Type
Submarine optronic periscope system.

Description
The optronic periscope system has been developed to allow the operator to search the sea surface during day and night utilising a thermal imaging subsystem and, at the same time, to supply a direct viewing visual channel. The periscope system combines a wide range of sensors in one periscope; a thermal imaging camera, (3-5 µm or 8-12 µm) CCD TV camera, 35 mm photographic camera, laser rangefinder as well as passive TV and visual stadimeter, omniradar early warning antenna, a radar direction-finding antenna and GPS. The periscope provides high optronic performance by utilising accurate stabilisation to compensate for induced vibrations and platform motion to the visual and thermal lines of sight. Additionally, in combination with this function, the operator is provided with a periscope rotation and line of sight elevation rate control which allows fast direction and target tracking. The operator has a direct view of the scene in addition to a video display and eyepiece data display of target range, target bearing and line of sight elevation angle.

A remote-control station is supplied as part of the optronic periscope system, in addition to a complete control datalink to the submarine fire-control system.

Operational status
In production. The mast is operational.

Specifications
Periscope tube diameter: 190.42 mm
Line of sight elevation: −10 to +74° (visual and TV);
−10 to +55° (thermal imaging)
Magnifications: ×1.5, ×6, ×12, ×18
Fields of view: 2.5°, 4°, 8°, 32° (visual) 4.4°, 10° (IR)

Contractor
Kollmorgen Corporation, Electro-Optical Division,
Northampton, Massachusetts.

UPDATED

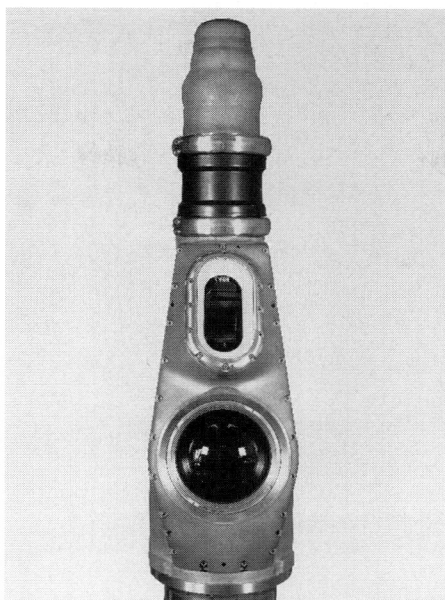

Model 90 optronic masthead unit

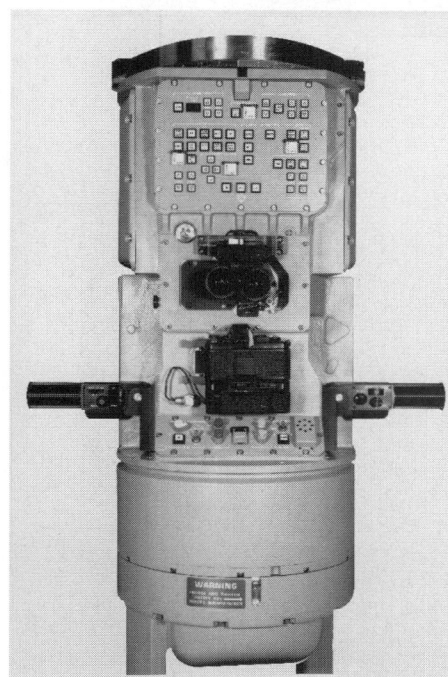

Model 90 optronic mast eyepiece unit

TABLE XI

Operational periscope systems[1]

System	Manufacturer	Country	Installed on	Navy	No units
Trident mast	Riva Calzoni	Italy	Ula (Type P 6071)	Norway	6
CH 074 attack	Pilkington Optronics	UK	Oberon	Australia	2
CH 074 attack	Pilkington Optronics	UK	Oberon	Brazil	3
CH 074 attack	Pilkington Optronics	UK	Oberon	Canada	3
CH 074 attack	Pilkington Optronics	UK	Oberon	Chile	2
CH 074 attack	Pilkington Optronics	UK	Walrus	Netherlands	4
CH 081 attack	Pilkington Optronics	UK	Sauro (Type 1081)	Italy	4
CH 093 attack	British Aerospace Australia	UK	Collins	Australia	1
CK 024 search	Pilkington Optronics	UK	Oberon	Australia	2
CK 024 search	Pilkington Optronics	UK	Oberon	Brazil	3
CK 024 search	Pilkington Optronics	UK	Oberon	Chile	2
CK 024 search	Pilkington Optronics	UK	Walrus	Netherlands	4
CK 024 search	Pilkington Optronics	UK	Oberon	Canada	3
CK 030 search	Pilkington Optronics	UK	Kobben (Type 207)	Norway	6
CK 031 search	Pilkington Optronics	UK	Sauro (Type 1081)	Italy	4
CK 034 search	Pilkington Optronics	UK	Type 207 (Tumleren)	Denmark	3
CK 038 EO search	Pilkington Optronics	UK	Vastergotland A 17	Sweden	4
CK 043 search	British Aerospace Australia	UK	Collins	Australia	1
IMS	SAGEM AS	France	Narhvalen	Denmark	2
Model 76	Kollmorgen	USA	Dolphin	Israel	1
Model 76	Kollmorgen	USA	Nacken	Sweden	3
Model 76 search	Kollmorgen	USA	Type 209/1400 (Tupi)	Brazil	5
Model 76 search	Kollmorgen	USA	Improved Sauro	Italy	4
Model 76 attack	Kollmorgen	USA	Type 209/1400 (Tupi)	Brazil	5
Model 76 attack	Kollmorgen	USA	Improved Sauro	Italy	4
Model 76 search	Kollmorgen	USA	Type 209	India	5
Model 76 attack	Kollmorgen	USA	Type 209	India	5
Model 76 search	Kollmorgen	USA	Walrus	Netherlands	4
Model 76 attack	Kollmorgen	USA	Walrus	Netherlands	4
Model 76 search	Kollmorgen	USA	TR 1700	Argentina	5
Model 76 attack	Kollmorgen	USA	TR 1700	Argentina	5
Model 76 search	Kollmorgen	USA	Type 209/1400	Turkey	4
Model 76 attack	Kollmorgen	USA	Type 209/1400	Turkey	4
n/k	Kollmorgen	USA	Dolphin	Israel	1
OMS	SAGEM AS	France	Le Triomphant	France	1
SERO 40 STAB	Zeiss-Eltro	Germany	Type 209/1300 (Thomson)	Chile	4
SERO 40	Zeiss-Eltro	Germany	Type 209/1200 (Poseidon)	Greece	4
SERO 40	Zeiss-Eltro	Germany	Type 209/1300 (Cakra)	Indonesia	2
SERO 40	Zeiss-Eltro	Germany	Type 209/1200 (Casma)	Peru	6
SERO 40 STAB	Zeiss-Eltro	Germany	Type 209/1200 (Chang Bogo)	South Korea	7
SERO 40 STAB	Zeiss-Eltro	Germany	Hai Lung	Taiwan	2
SERO 40 STAB	Zeiss-Eltro	Germany	Type 209/1400 (Preveze)	Turkey	3
SERO 40 STAB	Zeiss-Eltro	Germany	Type 209/1300 (Cabalo)	Venezuela	2
SERO 14	Zeiss-Eltro	Germany	Ula	Norway	6
SERO 15	Zeiss-Eltro	Germany	Ula	Norway	6
SPS	SAGEM AS	France	L'Inflexible	France	3
SPS	SAGEM AS	France	Rubis	France	6

*Details are given for periscope fittings where known.   n/k not known.

UPDATED

MAGNETIC ANOMALY DETECTION SYSTEMS

CANADA

AN/ASQ-504(V) AIMS system

Type
Magnetic Anomaly Detection (MAD) equipment.

AN/ASQ-504 MAD mounted on a sponson on a Lynx helicopter. (Photograph by Robin A Walker via Anchor Consultancy)
1998/0006987

A CAE MAD detector head being inserted into the tail of a CP-140 Aurora maritime aircraft

Description
The AN/ASQ-504(V) Advanced Integrated MAD System (AIMS) is an inboard system designed to provide optimum performance when used on helicopters and fixed-wing aircraft. AIMS is a fully automatic Magnetic Anomaly Detection (MAD) system which improves detection efficiency while significantly reducing operator workload. For helicopter installation, the detecting head is mounted inside the aircraft. This provides true 'on-top' contact when over a target by eliminating the delay inherent in a towed detecting system.

The AN/ASQ-504(V) system combines sensitivity and accuracy with ease of operation, eliminates aircraft-generated interference fields and delivers continuous, automated feature recognition. Contact alert, both visual and audible, is provided. Slant range is displayed allowing the operator to determine if he is within target acquisition range.

AIMS eliminates hazards associated with towed systems and permits high-speed surveillance and manoeuvrability, thereby increasing patrol range, improving detectability and substantially reducing false alarms. When used with dipping sonar, transition from one system to the other can be performed quickly and effectively.

The AIMS package weighs 26.3 kg (49 lb) and consists of an optically pumped magnetometer, a vector magnetometer, an amplifier computer and an optional control indicator. The system operates from a single-phase, 115 V, 400 Hz power source. AIMS has the flexibility either to perform independently or to accept and execute commands from common control/ display units via a 1553 databus.

Operational status
Production quantities have been delivered for requirements on the RAN Sea Hawk, Grumman Turbo-Tracker, RAF Nimrod and RN Lynx and Sea King and to the Canadian Forces for their Sea Kings. CAE is under contract to deliver MAD systems to the Australian Air Force for its P-3C Update Programme. The system is also in service in the Republic of South Korea (Lynx helicopter) and Taiwan.

Specifications
Detecting head: caesium, optically pumped, oriented
Sensitivity: 0.01 γ; (in-flight)
Features: automatic target detection. Operator alert, visual and audible. Slant range estimate to 854 m (2,800 ft)
Outputs
Visual: contact alert and estimated slant range
Analogue: adaptive MAD signal
Digital: control and display interface to MIL-STD-1553B
Audio: ICS alert
Dimensions: 178 mm (diameter) × 813 mm (long) (detecting head); 193 × 257 × 559 mm (amplifier/ computer); 152×152×152 mm (vector magnetometer); 190 ×145 ×145 mm (control indicator)
Weights: 6.1 kg (detecting head); 17.7 kg (amplifier/ computer); 0.7 kg (vector magnetometer); 1.8 kg (control indicator)

Contractor
CAE Electronics Ltd, Saint Laurent, Montreal, Quebec.

VERIFIED

FRANCE

MAD Mk III detector

Type
Magnetic anomaly detection system.

Description
The MAD Mk III is derived from several years' experience with the Mk I and II systems and is specifically designed for inboard use on fixed-wing aircraft and helicopters to detect the presence of a submersible by measuring the disturbance to the earth's magnetic field. In the helicopter version the MAD body is towed at the end of a 70 m cable to eliminate disturbances. It is an integral, digital, solid-state and highly reliable airborne system which becomes operational at switch on without any warm-up time. Target parameters are automatically delivered in real time on a CRT control and display unit.

The basic equipment can be broken down into three seperate units; a detection unit, a computer and a control/display unit. A 1553B databus interface card is included in the system. A graphic recorder is available as an option.

The sensor operates on the nuclear magnetic resonance principle and uses the precession of protons in a liquid, the frequency of which, measured by the pick-up coils, is proportional to the magnetic field to be measured. Compensation is employed to eliminate from the received signal all disturbances created by the magnetic element in the aircraft and their movement in the earth's field. The compensator uses a 16-terms model, representing the magnetic components of the aircraft. The MAD Mk III includes rapid aircraft identification modes.

The target is detected and located automatically. This is a fundamental role, allowing the MAD system to give a high-detection probability with a very low false alarm rate in extracting the target signal from the background noise. The Sextant Avionique MAD system is based on a mathematical comparison between the current MAD signal and an analytical model of the target signals, as opposed to the more conventional

method using threshold detection in several frequency bands. All computing tasks are performed by the ALPHA 732 proprietary 1 MOPS digital computer operating in Pascal.

Operational status
No longer in production but remains in service with the French Navy (Lynx and Atlantique 2), and Argentine Navy (Fennec helicopter). The MAD Mk III has been specified for the French-designed AMASCOS (Airborne Maritime Situation Control System) series developed by Thomson-CSF RCM.

Specifications
Dimensions: 1,250 mm (long) × 125 mm (diameter) (detection element); ½ ATR (long) (computer); 190 mm (long) × 146 mm (wide) × 162 mm (high) (control/display unit)
Weights: 5.5 kg (detection element); 10 kg (computer); 3.5 kg (control/display unit)

Contractor
Sextant Avionique, Vélizy Villacoublay.

UPDATED

UNITED STATES OF AMERICA

AN/ASQ-81(V) ASW magnetometer

Type
Magnetic anomaly detection system.

Description
The AN/ASQ-81(V) Magnetic Anomaly Detector (MAD) system was developed for the USN for use in the detection of submarines from an airborne platform. The system operates on the atomic properties of optically pumped metastable helium atoms to detect variations in total magnetic field intensity. Changes in the Larmor frequency of the sensing elements are converted to an analogue voltage, which is processed by bandpass filters before it is displayed to the operator.

Two configurations of the AN/ASQ-81(V) are available: one for installation within an airframe and one for towing behind an aircraft. The USN uses the AN/ASQ-81(V)-1 inboard installation with carrier-based S-3 aircraft as well as with the land-based P-3C ASW aircraft. For towing, the configuration is the AN/ASQ-81(V)-2, which is employed by USN SH-3H and SH-2D helicopters. The towed version is also fitted in the USN's latest helicopter, the SH-60B Seahawk.

The equipment has been upgraded with a form, fit and function replaceable digital implementation version, which incorporates microprocessor technology to achieve automatic aircraft compensation and to perform all signal processing and built-in test. The upgraded version features improved performance and can eliminate up to 13 Weapons Replaceable Assemblies (WRAs).

Operational status
The AN/ASQ-81(V) is in service aboard USN P-3, S-3 and SH-60 aircraft and a variety of international platforms such as the WG-13, HSS-2, 500D and SH-60J. Among navies operating the system are Australia (Orion), Egypt (Seasprite), Japan (Orion), Netherlands (Orion) Norway (Orion), Portugal (Orion), Spain (Orion), and Taiwan (Seasprite).

Contractor
Raytheon Systems Company, Sensors & Electronic Systems, McKinney, Texas.

VERIFIED

AN/ASQ-208(V) MAD system

Type
Digital magnetic anomaly detection system.

Description
The AN/ASQ-208(V) is a derivative of the AN/ASQ-81 magnetometer that operates on the same principle of the atomic properties of optically pumped metastable helium atoms to detect variations in total magnetic field intensity. The AN/ASQ-208(V) is a digital implementation which incorporates microprocessor technology to achieve aircraft compensation, multiple channel filtered display and threshold processing. The system can be controlled either through an online 1553B digital interface, or offline through a control indicator unit. For ease of integration there is also an RS-232 interface which can be connected directly to a standard PC for displaying and/or archiving data. Two system configurations are available; one for inboard installations and one for towed installations. The system has been configured for ease of installation on either existing or new aircraft and uses current AN/ASQ-81(V) wiring where applicable.

This system is form, fit and function replaceable with the AN/ASQ-81(V) and eliminates up to 13 ancillary WRAs. Digital signal processing yields a 25 per cent improvement in open water detection range and a 50

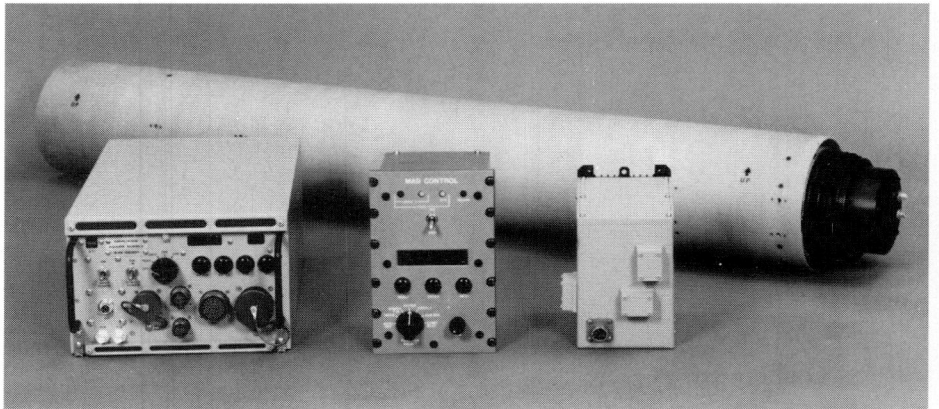

ASQ-208 digital MAD system   *1994*

per cent improvement in shallow water detection range compared to the AN/ASQ-81(V). The reduced number of WRAs has a significant advantage in initial price and has an even greater advantage in reduced maintenance costs.

Operational status
In service aboard USN P-3 aircraft.

Contractor
Raytheon Systems Company, Sensors and Electronic Systems, McKinney, Texas.

VERIFIED

Stationary Helium-3 ASW magnetometer

Type
Magnetic anomaly detection system.

Description
The He³ magnetic anomaly detection system was developed to have characteristics ideal for missions in which the sensor is stationary and the target is moving, such as MAD buoys, arctic ASW arrays, underwater arrays for shallow water or coastal surveillance and sensors for advanced mines. It derives its signal from the free precession of He³ nuclei, polarised by an optical pumping process, and is therefore an absolute measure of magnetic field. This provides the stability required for detecting the extremely low-frequency (1 to 50 mHz) signals generated at typical target speeds.

The 100 MW sensor has 1 pT/√H₃ sensitivity in the

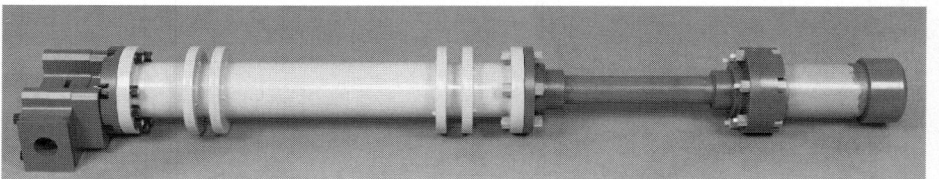

Stationary Helium-3 ASW magnetometer   *1994*

1 to 10 MHz passband. Array processing techniques have achieved 80 dB background noise suppression in undersea and Arctic array tests. Each sensor unit consists of two subassemblies, the basic sensor, and the associated electronics package. The sensor is a 4 in diameter cylinder 10 in long and the electronics telemetry package is a cylinder 5 in in diameter and 25 in long. The total weight of a sensor and the electronics package, with batteries installed, is 10 kg.

Operational status
The system has been used in sea tests at depths over 600 ft and on the Arctic ice cap to record the passage of submarines and surface vessels.

Contractor
Raytheon Systems Company, Sensors and Electronic Systems, McKinney, Texas.

VERIFIED

UNDERWATER WEAPONS

Strategic and cruise missiles

Guided ASW/ASUW weapons

Torpedoes

Mines

ASW rockets

Depth charges

STRATEGIC AND CRUISE MISSILES

Although the overall numbers of Russian Federation and Associated States (CIS) and American strategic missiles are being reduced, the diversity and deployment of such weapons throughout the world is on the increase. The one category of system which remains static is the submarine-launched ballistic missile, with the RFAS, the United States, France, the United Kingdom and China currently being the only operators of such systems. Both the RFAS and the United States have unilaterally agreed to withdraw submarine-launched tactical nuclear cruise missiles currently in service and to scale down the numbers of operational strategic nuclear weapons.

While the RFAS and the United States have agreed to reduce their inventories of such weapons, France and

the United Kingdom do not appear to be willing to reduce their stocks, and China is actively increasing its stock, albeit at an extremely slow rate. However, all three countries possess very few numbers of such weapons when compared to those held by the RFAS and the United States.

Of increasing concern is the developing capability among smaller nations to develop nuclear weapons. Although there does not seem to be any likelihood of either India or Pakistan developing submarine-launched ballistic missiles in the very near future, this must, in the long term, remain a distinct possibility. India, especially, is devoting resources to develop nuclear weapons, probably including submarine-launched weapons.

India is said to have an SLBM called Dhanush at the conceptual design stage.

Of greater concern must be the possibility that ballistic missiles might be armed with biological or chemical weapons, a much easier, and in some ways more deadly, threat to develop and deploy.

Another use for ballistic missiles, which is currently being studied in the United Kingdom, for instance, is the possibility of deploying conventional high-explosive warheads as an alternative to NBC warheads.

UPDATED

CHINA, PEOPLE'S REPUBLIC

JL-1 (CSS-N-3)

Type
Intermediate-range, submarine-launched ballistic missile.

Development
Development of China's first submarine-launched ballistic missile, the Julang-1 (Giant Wave-1 NATO designation CSS-N-3), began in 1967. Designed for deployment from the 'Xia' class submarine, the JL-1 is thought to embody some technology from the CSS-2 missile. Difficulties experienced in mastering the technologies associated with solid propellant fuels, together with delays in the nuclear warhead and submarine programmes, resulted in an extended development time for the missile. Initial test firings were carried out in 1982, reportedly from a submarine pontoon, and subsequently from a modified 'Golf' class submarine. The first launching took place on 30 April 1982 and the second from the 'Golf' submarine on 12 October 1982.

The basic missile is currently being used as a test vehicle for the follow-on JL-2 missile.

Description
The missile is a two-stage, solid-propellant, ballistic

The CSS-N-3 submarine-launched missile

1995

missile employing inertial guidance. The first stage is ignited after the missile has emerged from the water. The warhead carries either 3/4 MIRV of 90 kT or a single nuclear charge of 250 kT.

Operational status
The missile became operational in 1983 and featured in the October 1984 parade in Beijing. The missile is deployed in the 'Xia' class submarine which has 12 missile launch tubes. A successful submerged launch of the missile from the *Xia* was accomplished on 27 September 1988.

Specifications
Length: 10.70 m
Diameter: 1.40 m
Launch weight: 14,700 kg
Warhead: 3/4 90 kT MIRV or 1 × 250 kT nuclear.
Range: 917 n miles
Accuracy: c 700 m CEP

Contractor
Not known.

UPDATED

JL-2 (CSS-NX-5)

Type
Intercontinental, submarine-launched ballistic missile.

Development
This missile, believed to have the Chinese designation Julang-2 (Giant Wave-2), is a second-generation missile with a longer range than the JL-1. The 'Xia' class submarine commenced a refit in 1995 to enable her to deploy the new missile. She may not emerge from the refit to deploy the new missile until towards the end of the decade. A new class designated Type 094 is under development to deploy the JL-2. Four boats are to be built, each carrying 16 missiles.

Description
JL-2 is a three-stage solid-propellant missile which may carry some form of accurate delivery system within a 700 kg payload. The payload weight would seem to indicate a single warhead, probably with a yield similar to that used in the JL-1. However, nuclear tests in 1994-95 could indicate the capability to deploy a 3/4 MIRV type warhead. A range of 4,300 n miles has been reported.

Operational status
The first underwater launch was reported in 1985, but there is as yet no evidence that the missile has been operationally deployed. A new class of submarine,

Type 094, is under design and may deploy the new missile. Anticipated date of entry into service is 2002.

Specifications
Diameter: 2 m
Launch weight
Warhead: 700 kg
Range: 4,300 n miles

Contractor
Ministry of Aerospace Industries.

UPDATED

FRANCE

M-4

Type
Intermediate-range, submarine-launched ballistic missile.

Development
The M-4 is the fourth in the MSBS (Mer-Sol-Balistique-Stratégique) family of submarine-launched missile systems and is similar in form to the US Polaris and Poseidon systems. Initial development work for the M-4 was carried out using the submarine *Le Gymnote*,

using twin tubes for the tests. The first test launch took place in November 1980 from the Landes land-test range and the final test firing from the submarine was conducted on 24 February 1984. The first operational launch, from *Le Tonnant*, was carried out on 15 September 1987.

The M-4 was developed to carry multiple re-entry vehicles, and two versions are reported, the M-4A with a range of 4,000 km and the M-4B with a range of 5,000 km. An improved version, the M-45, is being developed with a range of 6,000 km.

Description
The M-4 missile is a three-stage, intermediate-range missile using inertial guidance. The first stage burns for 62 seconds, the second for 71 seconds and the third for 43 seconds. The three-stage solid-propellant motors contain 20,000 kg, 8,015 kg and 1,500 kg of propellant respectively. Each motor has a single flexible nozzle for control. The missile carries six thermonuclear MRV (Multiple Re-entry Vehicles) which are believed to have some independent targeting capability. This implies

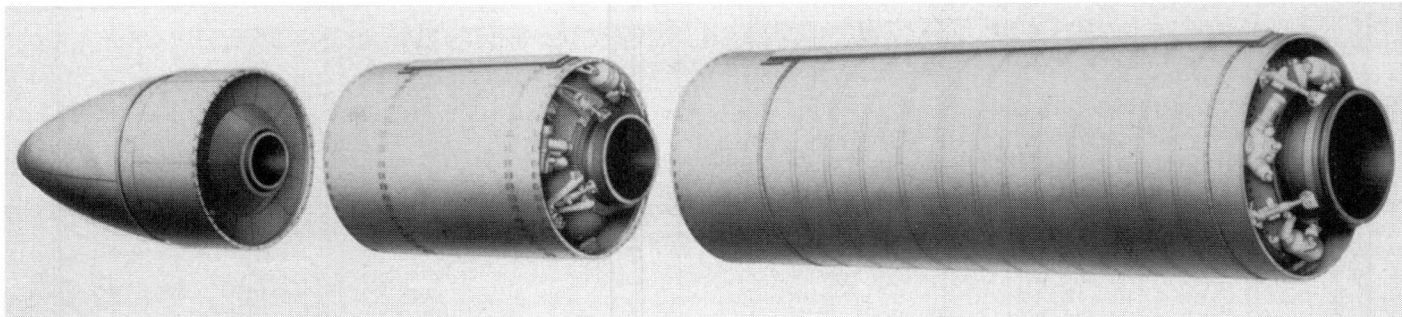

A diagram of the main assemblies of an M-4 missile

1995

that there is some additional guidance system incorporated within the delivery system.

The TN-71 warheads have a yield of 150 kT and each re-entry vehicle is believed to weigh about 250 kg. This would suggest a total payload capability of approximately 1,700 kg. An estimated accuracy of the M-4 missile system is 500 m CEP.

The M45 carries a new warhead with improved stealth and penetration aids. Control is exercised through a single flexible nozzle in each stage.

Operational status
The M-4 entered service in 1986 and is operational aboard the three SNLE submarines – *L'Inflexible*, *Le Tonnant* and *L'Indomptable* – each of which is fitted with 16 missile launch tubes. It is believed that there are 16 M-4A missiles and 48 M-4B missiles in service.

Test firings to qualify the M45 variant were successfully conducted by the new SSBN *Le Triomphant* in February 1995. The missile entered service in spring 1997 aboard *Le Triomphant*.

Specifications
Length: 11.05 m
Diameter: 1.93 m
Launch weight: 35,000 kg
Payload: 6 re-entry vehicles in MRV configuration
Warhead: 150 kT per vehicle
Range: 2,000 n miles (M-4A), 2,700 n miles (M-4B), 3,200 n miles (M-45)
Accuracy: 500 m CEP

Contractor
Aerospatiale, Space and Strategic Systems Division, Les Mureaux, (prime contractor)

UPDATED

The M-4 SLBM being fired from L'Inflexible *during an operational test*
1995

M-51

Type
Intercontinental-range submarine-launched ballistic missile.

Development
The M-51 is the latest development of the MSBS family and is planned to become operational in the 'Le Triomphant' class SNLE-NG boats. Full-scale development of the M5 began in 1993. In February 1996, the President of France confirmed the programme would continue with some changes in specifications to conform with budgetary constraints. The missile was renamed M-51, but retains the essential characteristics laid down for the M-5.

Description
The three-stage solid-propellant M-5 missile is equipped with modern penetration aids capable of matching the perceived upgrades to the Russian anti-ballistic missile system. Guidance is by inertial system. The solid stages are equipped with flexible nozzles made by filament winding of carbon fibre/epoxy material.

The payload has been reported as being between six and 10 MIRV armed with the TN 75 nuclear warhead with an expected yield of 100 kT.

Operational status
The missile is planned to enter service in 2010 aboard the 'Le Triomphant' class as a replacement for the M-4 and M-45 SLBM.

Specifications
Length: 12.00 m
Diameter: 2.30 m
Launch weight: >50,000 kg
Payload: 6-10 MIRV
Warhead: 150 kT per vehicle
Range: >5,400 n miles

Contractor
Aerospatiale, Space and Strategic Systems Division, Les Mureaux, (prime contractor).

UPDATED

RUSSIAN FEDERATION AND ASSOCIATED STATES (CIS)

SS-N-8 Sawfly (RSM-40 Vyosta)

Type

Intercontinental, submarine-launched, ballistic missile.

Development

Development of the SS-N-8 began in 1961 and was considered to be a new generation because it was the first Russian SLBM to use stellar sensing techniques to enhance its accuracy. In addition, its increased length required a new submarine design for deployment.

Description

The two-stage SS-N-8 liquid-propelled intercontinental missile has been developed in two versions, the first carrying a single 800 kT warhead and the Mod 2 carrying two MIRV re-entry vehicles of 500 kT each.

The use of stellar inertial guidance greatly improved the accuracy over the previous generation of missiles. The missile uses unsymmetrical dimethyl hydrazine and nitrogen tetroxide liquids with a maximum first-stage burn time of 150 seconds and a second-stage burn time of 80 seconds.

Operational status

The SS-N-8 became operational in 1973 and is deployed aboard the 'Delta I' and 'Delta II' SSBNs, the former fitted with 12 missile launch tubes and the latter with 16. To comply with the START 1 and START 2 agreements the RFAS are decommissioning the SS-N-8 and all missiles will be out of service by about 1998/9. Only four Delta I boats remain in service and these will have been decommissioned by 1998-99.

Specifications

Length: 13.9 m
Diameter: 1.80 m
Launch weight: 33,300 kg
Payload: single Re-entry Vehicle
Warhead: Mod 1 – 1 MT nuclear; Mod 2 – 800 kT nuclear
Range: Mod 1 – 4,200 n miles; Mod 2 – 4,900 n miles
Accuracy: Mod 1 – 1,500 m CEP, Mod 2 – 900 m CEP

Contractor

Not known.

UPDATED

SS-N-8 'Sawfly' being carried on a transporter vehicle during a parade

1995

SS-N-18 Stingray (RSM-50 Volna)

Type

Intercontinental submarine-launched ballistic missile.

Development

Development of the SS-N-18 began in 1968 and was first tested from land-based launch sites in 1975 and from a submarine in November 1976. Long-range tests were carried out in the Pacific in late 1978.

The missile has many similarities with the SS-N-8, from which it may be derived; it does, however, have a more sophisticated guidance system and an MIRV warhead dispensing system.

Description

The SS-N-18 is a two-stage, liquid-propellant, intercontinental ballistic missile fitted with a Post Boost Vehicle (PBV). Range is 6,500 km for the MIRV variants and 8,000 km for the single RV version. The Mod 1 carries a payload of three RVs in MIRV configuration; the Mod 2 a single RV and the Mod 3 up to seven RVs again in an MIRV configuration. The provisional START agreement allocated seven RVs to the SS-N-18, but in 1991, the Russians announced that they would download all missiles to three RVs in MIRV configuration.

The accuracy of the MIRV versions is around 900 m,

indicating continuing use of the stellar inertial guidance system used in the SS-N-8.

Operational status

The SS-N-18 Mod 1 entered service in 1977, the Mod 2 in 1978 and the Mod 3 in 1979. The missile is deployed aboard 12 'Delta III' class SSBNs, each fitted with 16 launch tubes. Following the START 1 and 2 agreements it is anticipated that most SS-N-18s will have been removed from service by the year 2000. Units of the Delta III type are starting to decommission.

Specifications

Length: 14.60 m
Diameter: 1.80 m
Launch weight: 35,300 kg
Payload: Mod 1 – 3 RVs in MIRV configuration; Mod 2 – single RV; Mod 3 – 7 RVs in MIRV configuration
Warhead: Mod 1 – 3 × 200 kT nuclear; Mod 2 – single 450 kT nuclear; Mod 3 – 7 × 100 kT nuclear
Range: Mods 1/3 – 3,500 n miles; Mod 2 – 4,300 n miles
Accuracy: 900 m CEP

Contractor

Not known.

UPDATED

A 'Delta III' class submarine, which carries 16 SS-N-18 'Stingray' missiles
1995

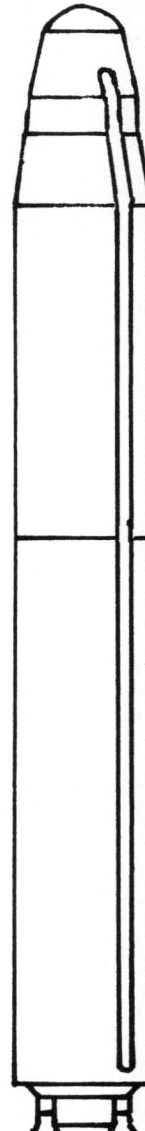

A diagram of the SS-N-18 missile
1995

SS-N-20 Sturgeon (RSM-52)

Type
Intercontinental, submarine-launched ballistic missile.

Development
Development of the three-stage solid-propellant SS-N-20 began in 1972, the first test flight taking place in January 1980. Following a long run of failures, successful tests were finally conducted in 1981, and in October 1982 a simultaneous launch of four missiles was reported. A further improved variant of the missile (SS-N-28) is said to be undergoing flight testing but the development programme is progressing very slowly.

Description
The SS-N-20 solid-propellant ballistic missile carries 10 re-entry vehicles in an MIRV configuration. US reports suggest that an early version only carried six vehicles,

but this was not confirmed in START 1 information supplied by the RFAS in 1991. Guidance is inertial with stellar reference update. Two missiles can be launched within 15 seconds.

Operational status
The SS-N-20 entered service in 1982 and is currently deployed in the six 'Typhoon' class submarines, each of which is fitted with 20 launch tubes located forward of the boat's sail. This is the first SSBN to carry its ballistic missiles in such a configuration. It is reported that the RFAS are proposing to download SS-N-20 missiles to six or four RVs under the START 2 agreements. A programme is underway to modify the 'Typhoon' class to deploy the SS-N-28 missile, indicating that the SS-N-20 will slowly be withdrawn from service.

A total of 120 missiles remained in service in 1995.

Specifications
Length: 16.00 m
Diameter: 1st stage – 2.40 m; 2nd stage – 2.30 m
Launch weight: 84,000 kg
Payload: 10 RVs in MIRV configuration
Warhead: 10×100 kT nuclear
Range: 4,500 n miles
Accuracy: 500 m CEP

Contractor
Not known.

UPDATED

A 'Typhoon' class submarine, platform for the SS-N-20 'Sturgeon'   *1995*

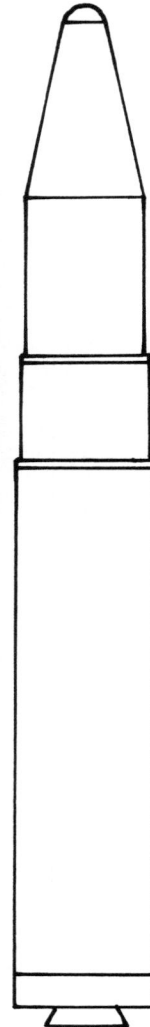

A diagram of the SS-N-20 missile   *1995*

SS-N-23 Skiff (RSM-54 Shetal)

Type
Intercontinental, submarine-launched, ballistic missile.

Development
The SS-N-23 is the successor to the SS-N-18 and began testing in 1983. Development of this missile indicates that the RFAS continue to prefer liquid-propelled missiles over solid-fuelled ones, perhaps on the grounds of extensive experience and confidence in their safety. This may be the result of disappointing past experiences with the solid-propelled SS-N-17 and 20.

Description
The SS-N-23 MIRV-capable missile, like the RFAS' other ballistic missiles, uses inertial guidance with stellar reference for correcting in-flight trajectory. The missile is Russia's first three-stage liquid-propelled SLBM.

The missile carries between four and 10 warheads in an MIRV configuration, but the START agreement only allocates four RVs and hence it is assumed that the 10 RV version has never been tested.

The 'Delta IV' submarine. Platform for the SS-N-23, the newest of the Russian SLBMs   *1995*

Operational status

The SS-N-23 entered operational service in 1985 and is deployed aboard the seven 'Delta IV' class SSBNs fitted with 16 launch tubes.

It is reported by US Intelligence that a liquid-propelled follow-on to the SS-N-23, the SS-NX-27, is to be fitted to a modified 'Typhoon' class SSBN and a possible follow-on design to the 'Delta IV' class boats.

A total of 112 missiles remained in service in 1996.

Specifications

Length: 15.3 m
Diameter: 1.90 m
Launch weight: 40,300 kg
Payload: 4 RVs in MIRV configuration
Warhead: 100 kT nuclear each
Range: 4,500 n miles
Accuracy: 500 m CEP

Contractor

Not known.

UPDATED

A diagram of the SS-N-23 missile
1995

SS-N-19 Shipwreck (P-500 Granit)

Type

Submarine-launched cruise missile.

Development

The SS-N-19 is the third-generation cruise missile and follows on the SS-N-3 Shaddock and SS-N-12 Sandbox missiles. It was developed in the late 1970s and was Russia's first vertically-launched ship or submarine deployed cruise missile.

Description

Very little detail is available concerning the missile, but it is believed to be similar, but much smaller than, the

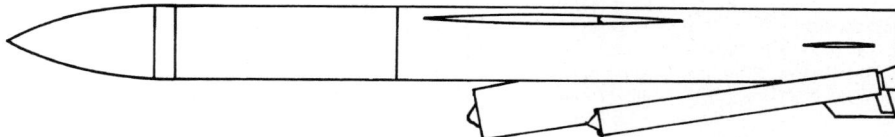

A line diagram of an SS-N-19 missile   *1995*

SS-N-12 with a lower flight profile. Mid-course guidance is inertial, with command update provided by aircraft or helicopter. Terminal homing is provided by an active radar system with an IR option as an interchangeable terminal seeker.

Vertical launch thrust is provided by a tandem solid-propellant boost motor, which is probably jettisoned after use, after which a turbofan engine takes over to provide power for the cruise phase of the trajectory. The missile is reported to have a high-altitude cruise phase of around 24,000 m, which is followed by a steep dive on to the target. The payload is either nuclear or conventional HE. It is also believed that Russia could have developed a fuel-air explosive warhead, which is thought to be particularly effective against ship targets.

The launch tubes on the 'Oscar' class submarines are inclined at 40° and mounted in six sets of pairs on either beam outside the pressure hull, each pair of tubes being sealed with a single hatch.

Operational status

The SS-N-19 is thought to have entered service in 1980, and is currently operational on board the 12 'Oscar II' class attack submarines equipped with 24 launch tubes.

It is estimated that around 265 missiles are in service.

Specifications (approx)

Length: 10.00 m
Diameter: 0.85 m
Launch weight: 4,000 kg
Warhead: 500 kT nuclear or 750 kg conventional HE
Range: 300 n miles
Speed: M2.5

Contractor

Not known.

The Russian battle cruiser Admiral Ushakov, *photographed in 1980, showing the 20 hatches on top of the vertical launch tubes for SS-N-19 'Shipwreck' missiles, with the smaller 12 hatches on the right of the picture for the SA-N-6 'Grumble'*   *1995*

UPDATED

SS-N-21 Sampson (RK-55 Granat)

Type
Submarine-launched land-attack cruise missile.

Development
It is believed that the SS-N-21 was first test flown in 1982 and became operational later in the decade.

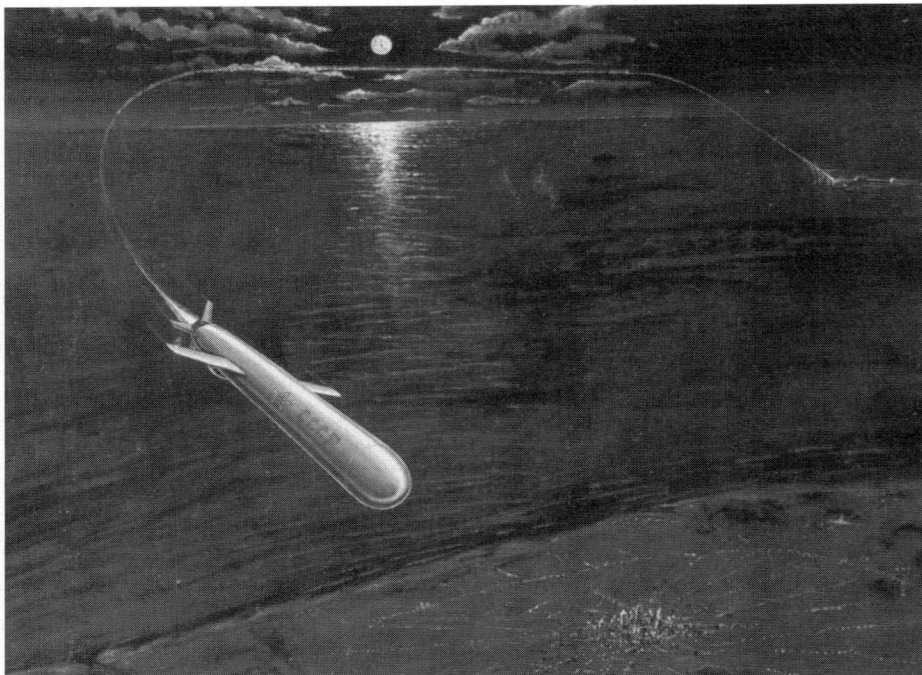

Description
Very little is known about the SS-N-21, but it is assumed that it is similar to the SSC-X-4 ground-launched cruise missile. Details of the latter were disclosed at the time of the exchange of data in connection with the INF Treaty.

Inertial guidance is used during the cruise phase of the trajectory, with some form of terrain matching for updates and terminal guidance. The missile has a radar altimeter and flies at around 200 m altitude. The missile is launched from a standard 533 mm torpedo tube and the wings unfold on leaving the water.

It is reported that the missile uses a rear-mounted expendable solid-propellant booster to achieve cruise speed, while cruise speed is maintained by a turbofan engine whose air inlet is mounted underneath the after body of the missile.

Operational status
The missile is believed to have become operational around 1987 and is deployed from the 11 'Victor III', 13 'Akula' and three 'Sierra I/II' class SSNs and one 'Yankee Notch' converted SSBNs (from six additional 533 mm torpedo tubes with 35 missiles). In 1992, as part of the START 2 agreement, the RFAS indicated that the SS-N-21 would be removed from service and production terminated.

Specifications
Length: 8.09 m
Diameter: 0.51 m
Launch weight: 1,700 kg
Payload: single warhead
Warhead: 200 kT nuclear
Range: 1,620 n miles
Accuracy: 150 m CEP
Speed: M0.7

Contractor
Not known.

UPDATED

An artist's impression of the SS-N-21 'Sampson' submarine-launched cruise missile
1995

Alfa

Type
Submarine-launched cruise missile.

Description
Alfa is a new design supersonic cruise missile which was first seen in 1993. The design may be a scaled down version of the AS-19 Koala strategic nuclear cruise missile programme which is thought to have been terminated in 1992.

The missile is powered by a turbojet engine with rectangular air inlet below the centre body while flight control is exercised through two delta wings mounted at the mid-point of the missile's body. The missile can be launched from land, ship, submarine or aircraft, and can be targeted against land or ship targets.

Specifications (approx)
Length: 6.00 m
Diameter: 0.55 m
Weight: 1,500 kg

Warhead: 400 kg
Range: 325 n miles

Contractor
Novator NPO.

UPDATED

UNITED STATES OF AMERICA

UGM-96 Trident C-4

Type
Intercontinental submarine-launched ballistic missile.

Development
The Trident C-4 missile was designed to succeed the Poseidon system, being lighter and smaller than Poseidon and with improved range capability, which allows SSBNs more ocean in which to manoeuvre.

Two basic fire-control systems are used: the Mk 88 Mod 2 for retrofitting to Poseidon-converted boats and the Mk 98 Mod 0 for the new 'Ohio' class SSBNs. A new digital fire-control computer has also been developed to handle the increased information flow in providing presetting on the larger submarines. A new Mk 4 re-entry vehicle was developed for the C-4 and a Mk 500 Evader, manoeuvre capable (MARV) system was put in development. The C-4 uses the full volume of the Poseidon missile tube and extends the range without penalty in payload.

Description
The three-stage solid-propellant C-4 MIRV-capable missile carries eight Mk 4 RVs fitted with W76 nuclear warheads. It features an optional MARV package for ABM penetration, in keeping with its second strike retaliatory role. A feature of note is the use of an extendable 'spike' which is deployed from the missile's nose after launch to reduce aerodynamic drag during the early phases of flight.

The missile can be launched from either a submerged or surfaced submarine. It is ejected from the launch tube using a gas generator, and once the missile is clear of the surface the first stage ignites. Guidance is inertial with stellar reference update.

Operational status
The missile entered service in 1979 and was deployed from 'Benjamin Franklin', 'Lafayette' and 'Madison' class SSBNs each fitted with 16 launch tubes and latterly from the first eight 'Ohio' class fitted with 24 launch tubes. The C-4 missile will be phased out of

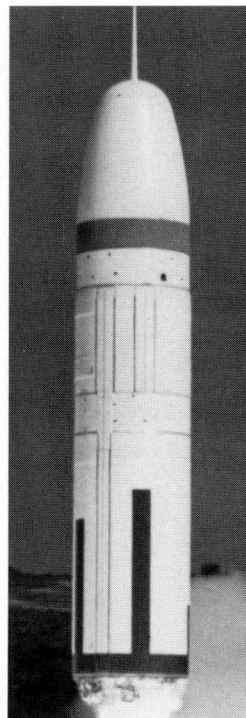

The first test launch of a Trident C-4 missile, at Cape Canaveral
1995

service and four of the first eight 'Ohio' class deploying the C-4 will be modified to deploy the D-5 while the remaining four may be modified for D-5 deployment, or converted to another role. Modification commenced in 1998. Following the agreement to limit SLBM numbers agreed under START 1 and 2, a rationalisation of boats and missiles is being made and 'Benjamin Franklin', 'Lafayette' and 'Madison' class boats have been paid off. If any C-4 missiles are retained, these will probably be downgraded to carry only four RVs.

By 1996 a total of 192 missiles remained in service.

Specifications
Length: 10.39 m
Diameter: 1.88 m
Launch weight: 32,850 kg
Payload: 8 × Mk 4 RVs in MIRV configuration
Warhead: W76 100 kT nuclear each
Range: 4,000 n miles
Accuracy: 450 m CEP

Contractor
Lockheed Missiles and Space Corporation Sunnyvale, California, (prime contractor).

UPDATED

UGM-133 Trident D-5

Type
Intercontinental submarine-launched ballistic missile.

Development
The D-5 Trident has been developed together with the 'Ohio' class SSBNs to provide the US Navy with a capability that extends range, payload and inventory of its missile force and to introduce a hard target capability in the submarine-based missile. The high accuracy is achieved through sophisticated improvements to the boat's navigation system and to the missile. A wide variety of mid-course updating techniques and terminal guidance systems are being considered to further enhance accuracy. The use of the Global Positioning System (GPS) satellite navigation system is an example of the former and there is a number of terminal guidance techniques that might be applied. It is reported that an earth penetrator warhead is being developed for the missile to attack underground hardened command centres and initial tests were conducted in 1993.

The first D-5 was launched from a land-based test site in January 1987 and the first sea launch, which proved to be a failure, in March 1989. The first successful submerged launch took place from the USS *Tennessee* in December 1989 and the missile was first deployed operationally in March 1990. The full test programme has now been completed with a total of 49 firings recorded by 1993.

A limit of eight RVs was set in 1991 and it is anticipated that the D-4 may be downgraded to carry four RVs during the late 1990s as part of the implementation to limit SLBMs under the START 2 agreements. Further reports indicate that a single warhead and even conventional HE warhead versions might be developed in the future.

Description
The three-stage solid-fuelled D-5 MIRV-capable ballistic missile uses an inertial guidance system with stellar sensor update. Like the C-4, the D-5 is fitted with an aerodynamic 'spike' which deploys from the nose early in flight to improve the drag characteristics of the missile.

The missile's payload is from eight to 12 MIRVs which are either the Mk 4 RV with W76 nuclear warhead, or the Mk 5 RV with W88 nuclear warhead.

The propulsion system manufactured under a joint venture by Alliant Techsystems (now Hughes Aircraft Naval and Maritime Systems) and the Thiokol

Corporation uses lightweight graphite motor cases and high-energy solid propellants.

Operational status
The D-5 became operational in 1990 and is successively equipping the 'Ohio' class SSBNs from the ninth boat onwards. Currently some 40 to 50 missiles are being built each year, but it is not clear if the planned full production total of 434 missiles, due to be built by 2000, will be achieved, due to the reductions demanded under START 1 and 2. A total of seven missiles per year are scheduled to be ordered under FY97 to FY99, with 12 per year in FY00 and FY01. The plan is to build up an inventory of 434 missiles sufficient to support a force of 14 'Ohio' class SSBNs all armed with the D-5 missile.

In 1980, an agreement was reached with the UK for the sale of Trident D-5 missiles. The missile will be fitted with the US-designed MIRV capability with US Mk 4 RVs, but with UK-built warheads believed to be similar to the US-designed W76 warhead. Under plans announced in November 1993 each 'Vanguard' class submarine carries a maximum of 96 warheads (6 MIRVs). The UK version is to be carried by the four boats of the 'Vanguard' class.

The first UK missile was successfully launched from HMS *Vanguard* on 26 May 1994 and its first operational patrol undertaken early in 1995. On 24 July 1995, the second 'Vanguard' class SSBN, HMS *Victorious*, successfully launched a D-5 missile.

The third boat *Vigilant* has also successfully launched two D-5 missiles.

In October 1997, it was announced that the UK would buy an additional seven missiles, making a total of 58 D-5 missiles held in the UK inventory.

Specifications
Length: 13.42 m
Diameter: 2.11 m
Launch weight: 59,090 kg
Payload: 8 × Mk 4 or Mk 5 RVs in MIRV configuration
Warheads: 8 × W76 at 100 kT nuclear or 8 × W88 at 475 kT nuclear
Range: 6,500 n miles
Accuracy: 90 m CEP

Contractor
Lockheed Missiles and Space Corporation, Sunnyvale, California (prime contractor).

UPDATED

Launch of the second development test Trident D-5 missile from Cape Canaveral in March 1987
1995

BGM-109 Tomahawk

Type
Intermediate-range submarine-launched cruise missile.

Development
Development of surface-launched cruise missiles began in 1972 and the US Navy programme was planned to provide both surface ship and submarine-launched variants for attacking ships and land targets. In the early days it was planned that the weapon would have a nuclear warhead to provide an additional survivable nuclear force. The first underwater test firing was conducted in 1976 and General Dynamics (now Hughes Missile Systems) was awarded the development contract.

Originally there were three versions of the ship and submarine-launched cruise missile:
(a) the BGM-109A TLAM-N (Tomahawk Land Attack Missile-Nuclear)

(b) the BGM-109B TASM (Tomahawk Anti-Ship Missile) with a conventional HE warhead
(c) the BGM-109C TLAM-C (Tomahawk Land Attack Missile-Conventional) with a conventional HE warhead.

A fourth version, BGM-109D, a conventional submunition warhead land-attack missile, TLAM-D, entered service in 1989.

Several developments and improvements are funded for the SLCM programme including:
(a) laser radar guidance
(b) ability to select the most important target from a group of ships
(c) updates to the Digital Scene Matching Area Correlation (DSMAC) terminal guidance system
(d) time of arrival control
(e) new warheads
(f) GPS updates to improve mid-course guidance
(g) improved performance turbofan engines

(h) mission planning resource improvements to reduce reaction times and possibly provide mission planning aboard ships.

The Tomahawk Block 3 upgrade programme (which has only been applied to the land-attack versions BGM-109A/C/D) includes a GPS receiver, time of arrival control, and improved guidance computer, warhead and propulsion systems. The first test flight was completed in February 1991. Range in this variant has been increased by more than 30 per cent to 620 n miles. Some earlier missiles will be retrofitted with these improvements. Test flights of the Block 3 began in 1991 and production commenced in early 1992.

Three further development programmes are under discussion and include an ASW variant armed with a torpedo, an advanced sea-launched cruise variant and a conventional cruise missile. These developments may be met by improvements to the basic Tomahawk design. A Block 4 upgrade is already planned for

A trials BGM-109B in flight   *1995*

Tomahawk and might incorporate propfan engines and stealth technologies, together with a damage assessment capability and datalink, a new penetrator warhead for use against reinforced concrete targets, laser radar, synthetic aperture radar and GPS options mentioned above, and with an in-service date of about 2000.

Description

The missile carries a single warhead. Initial launch thrust to take the missile up to cruise speed from its launch container is provided by a boost motor weighing 250 kg, which on burnout is succeeded by a Williams International 65 kg F107-WR400 turbofan producing about 272 kg thrust. The boost motor is jettisoned during cruise flight. Missiles are ejected vertically from a steel launch canister or from torpedo tubes. Development programmes are in progress which will result in the canisters being built of lightweight composite materials, rather than steel.

Guidance varies between the various versions. The BGM-109A uses inertial navigation with terrain contour matching (Tercom). The BGM-109B uses inertial plus active radar/anti-radiation homing. The BGM-109C/D versions have inertial and Tercom mid-course guidance (TAINS – Tercom Aided Inertial Navigation) used with DSMAC for terminal guidance. Tercom is achieved by storing digital terrain profile map information in the missile before launch and comparing this with radar altimeter measurements of ground elevations below the missile during a set number of sections en route to the target. The Block 3 upgrade programme will include a GPS receiver linked to 24 satellites circling the earth to supplement or replace the Tercom system. The terminal phase DSMAC system uses a stored digital representation of the target area and compares this with the scene viewed below the missile by a TV camera. This latter system is claimed to be extremely accurate for attacking land targets, with a CEP of 10 m being reported.

The BGM-109A is fitted with a W80 135 kg nuclear warhead, the BGM-109B/C with old Bullpup B 454 kg unitary HE warheads and the BGM-109D with 166 combined effects bomblets BLU-97B each weighing 1.5 kg and with shaped charges, fragmentation and incendiary capabilities. These submunitions can be dispensed in groups against up to four separate targets. The BGM-109C Block 3 missiles will have a smaller WDU-36B 320 kg warhead carrying a unitary HE charge with a selectable fuze delay to increase warhead penetration before detonation. Both the BGM-109C and D versions have a programmable terminal dive attack mode option.

There have been unconfirmed reports that some Tomahawk missiles used during the 1990-91 Gulf War were fitted with high-power microwave (EMP) generators in place of warheads to disrupt electronic circuits. Others are reported to have unrolled spools of carbon fibre to cause shorting between electrical power supply cables.

Range capabilities between the various versions differ because of the weight of the guidance systems and the warheads used. The range of the Block 3 missiles will be increased by at least 30 per cent due to increased fuel capacity and more efficient turbofan engines.

Operational status

The BGM-109 Tomahawk entered service in 1983 and it is planned to acquire nearly 4,000 missiles consisting of 637 BGM-109A, 593 BGM-109B, 1,486 BGM-109C, 1,157 BGM-109D. With an annual production rate of around 200 missiles per year, the planned total will probably be reached about 1998.

In 1982, a second source contract was placed with McDonnell Douglas (now the Boeing Company), which has competed with General Dynamics (now Hughes Missile Systems) since FY85, and with Teledyne CAE competing with Williams International from FY89 for the engine.

Most submarines carry reloads of Tomahawk, up to eight missiles being carried in a 'Sturgeon' class submarine and 45 (TLAM-N) in a 'Seawolf' class SSN.

The BGM-109D entered service in 1989 and the Block 3 upgrade versions first went to sea in 1993.

During the 1990-91 Gulf War, a total of 264 TLAM-C and 27 TLAM-D variants was fired, 15 of them from two submarines. Following the START 2 agreement it is expected that all BGM-109A nuclear warhead missiles will be removed from service. The UK government has announced that it will procure 65 109C Block 3 Tomahawk missiles for the 'Trafalgar' class SSNs. It has been reported that up to 200 missiles may be acquired. The first of these missiles is now operational aboard the submarine *Splendid*.

Contractors

Hughes Missile Systems, Tucson, Arizona, (prime contractor).
Boeing Company, St Louis, Missouri, (second source).

USS Merrill, *a 'Spruance' class US Navy destroyer, with two Mk 44 armoured box quadruple BGM-109 Tomahawk launchers located forward of the superstructure on either side of the RUR-5 ASROC launcher*   *1995*

Specifications

	BGM-109A (TLAM-N)	BGM-109B (TASM)	BGM-109C (TLAM-C)	BGM-109D (TLAM-D)
Length	6.25 m	6.25 m	6.25 m	6.25 m
Diameter	0.52 m	0.52 m	0.52 m	0.52 m
Launch weight	1,452 kg	1,452 kg	1,452 kg	1,452 kg
Payload	1 warhead	1 warhead	1 warhead	1 warhead
Warheads	W80 200 kT	454 kg HE	454 kg HE*	166 bomblets
Range	1,400 n miles	250 n miles	490 n miles	490 n miles
Accuracy	80 m CEP	N/A	10 m CEP	10 m CEP

* 318 kg shaped charge and range of 620 n miles in Block 3.

UPDATED

TABLE XII

Submarine-launched strategic and cruise missiles

Weapon	Type	Country origin	Launch weight	No of RVs	Warhead	Range
Alfa	Cruise	RFAS	1,200 kg	–	400 kg	325 n miles
BGM-109 Tomahawk	Cruise	USA	1,452 kg	–	454 kg	250-1,400 n miles
JL-1	IRBM	China	14,700 kg	1	250 kT	917 n miles
JL-2	ICBM	China	n/k	1	700 kg	4,300 n miles
M-4	IRBM	France	35,000 kg	6	1,000 kg	2,000 n miles
M-51	ICBM	France	48,000 kg	6 to 10	100 kT ea	5,400 n miles
SS-N-8 Sawfly (Mod 2)	ICBM	RFAS	33,300 kg	1	800 kT	4,900 n miles
SS-N-18 Stingray	ICBM	RFAS	35,300 kg	3	200 kT ea	3,500 n miles
SS-N-19 Shipwreck	Cruise	RFAS	4,000 kg	–	500 kT	300 n miles
SS-N-20 Sturgeon	ICBM	RFAS	84,000 kg	4/10	100 kT ea	4,500 n miles
SS-N-21 Sampson	Cruise	RFAS	1,700 kg	–	200 kT	1,620 n miles
SS-N-23 Skiff	ICBM	RFAS	40,300 kg	4	100 kT ea	4,500 n miles
SS-NX-26	ICBM	RFAS	In development			
SS-NX-27	ICBM	RFAS	In development			
UGM-96 Trident	ICBM	USA	29,500 kg	4 to 8	100 kT ea	4,000 n miles
UGM-133 Trident	ICBM	USA	59,090 kg	1 to 8	100/475 Kt	6,500 n miles

UPDATED

GUIDED ASW/ASUW WEAPONS

In the Western world submarine-launched anti-ship guided weapons tend to be variants of existing surface-launched anti-ship missiles such as Exocet and Harpoon. To some extent the same is true of submarine-launched anti-ship missiles developed by Russia under the Communist regime. However, they have also developed a number of anti-ship weapons designed specifically for submarine launch. The latest

country understood to be developing an indigenous submarine-launched cruise missile is India. Under development by the Defence Research and Development Organisation, the missile called Sagarika is in the early development phase and may possibly arm the nuclear-powered submarine under development by India.

In addition to anti-ship weapons, a number of guided weapons has been developed for deployment in the anti-submarine role. These are long-range ASW weapons designed primarily to deploy lightweight, homing torpedoes.

UPDATED

CHINA, PEOPLE'S REPUBLIC

CSS-N-4 Sardine (YJ-1/C-801)

Type
Submarine-launched SSM.

Development
Development of the Ying-Ji (Eagle Strike) is believed to have started in the early 1970s, but it was not until 1984 that the missile was first revealed by the Chinese. Initially the design was developed for surface ship deployment, but the weapon can also be launched from surfaced submarines.

Description
Similar in appearance to the French Exocet, the missile features four clipped delta wings at the mid-point of the body and four small clipped-tip triangular moving control fins at the rear. Propulsion is by a tandem-mounted solid-propellant boost motor which is jettisoned after use. The YJ-2 is powered by turbojet with an air inlet scoop beneath the missile body between the central wings. Mid-course guidance is inertial, with a monopulse active radar (probably operating in the J-band) in the terminal homing phase. Cruise altitude is believed to be around 20 m, followed by a terminal homing phase altitude of around 5 to 7 m. Altitude is controlled by a radio altimeter.

Specifications
Length (oa): 5.81 m
Diameter: 0.36 m

A launch of a Ying-Ji-1 missile from a modified Chinese-built 'Romeo' class submarine **1995**

Wing span: 1.18 m
Launch weight: 815 kg
Warhead: 165 kg HE
Range: 20 n miles
Speed: M0.9

Contractors
Chinese State Factories (manufacturer).
CPMIEC, Beijing (marketing).

UPDATED

Operational status
In production, entering service about 1984. Operational on board a Chinese modified 'Romeo' class submarine and the last three 'Han' class submarines.

Ying-Ji-1 (C-801 export version) surface-to-surface missile mounted beside its container, showing the jettisonable tandem boost motor assembly at the left-hand (rear) end **1995**

CY-1

Type
ASW surface-launched missile.

Description
The CY-1 single-stage ASW missile was first revealed in 1987. It would appear to be similar to the American RUR-5 ASROC system. Designed to be deployed by surface ships from multicelled launchers, CY-1 has been seen mounted on a 'Luda III' class destroyer in four inclined twin-launcher canisters. It is estimated that the missile has a range of about 15 km and is armed with a torpedo warhead. The missile is 5.5 m long and is estimated to weigh 700 kg.

Specifications
Length: 5.50 m
Weight: 700 kg
Range: 8 n miles (approx)
Operating depth: 300 m

Contractor
China Precision Machinery Import and Export Corporation (CPMIEC), Beijing.

VERIFIED

FRANCE

Malafon

Type
Surface-to-subsurface torpedo missile system.

Description
Malafon is a shipborne weapon consisting of a radio command-guided winged vehicle carrying an L4 acoustic homing torpedo. It is intended primarily for use from surface vessels against submarines, but may also be used to attack surface targets.

Malafon has the appearance of a small conventional aircraft with short, unswept tapered wings and a tailplane fitted with endplate fins.

The missile is ramp-launched and propelled by two solid-fuel boosters for the first few seconds of flight. Subsequent flight is unpowered. A radio altimeter is fitted to the missile to maintain a flat trajectory at low level. On reaching the target area, approximately 800 m from the target's estimated position, a tail parachute is deployed to decelerate the missile. The homing torpedo is ejected from the remainder of the vehicle and enters the water to complete the terminal guidance phase of the attack by acoustic homing.

Target detection and designation in the case of submerged targets is by means of sonar and by radar in the case of surface targets. These sensors, as appropriate, are used during the flight of the missile to provide data on the target for the generation of command guidance signals which are sent, via radio command link, to guide the missile. Missile tracking is aided by flares attached to the wingtips.

Operational status
Malafon was deployed in an interim form on French Navy vessels while full development trials were still in

Malafon anti-submarine missile on its mount on the French destroyer Aconit (Stefan Terzibaschitsch)

progress, these installations being updated as development continued. These fittings have been referred to as Malafon Mk 1 systems. Deployment of the system includes installations in the two 'Suffren' class destroyers. Malafon is now obsolescent and non-operational.

Specifications
Length: 6.15 m
Diameter: 0.65 m
Wing span: 3.30 m
Weight: 1,500 kg (launch)
Range: 7 n miles
Speed: 450 kt
Payload: L4 torpedo

Contractor
Société Industrielle d'Aviation Latécoère, Paris.

UPDATED

SM 39 Exocet

Type
Subsurface-to-surface anti-ship missile.

Description
The French Navy deploys a submarine-launched version of Exocet, known as the SM 39.

Like the other Exocet missiles, with which it has many components in common, the SM 39 is an all-weather, fire-and-forget, sea-skimming weapon. Target acquisition and tracking is carried out by an active electromagnetic homing head.

The SM 39 missile consists of a watertight, highly resistant, powered and controlled underwater capsule, housing an aerial missile before the latter's ejection following surface broach.

Known as the VSM (*Véhicule Sous-Marin*), the capsule can be discharged from a submarine's 533 mm standard torpedo tube at any speed. At a safe distance from the submarine, the solid-propellant motor of the VSM is ignited and the VSM is piloted up to ejection of the aerial missile by jet deviation of gases; this deviation is performed by four interceptors actuated by electromagnets placed to the rear of the nozzle outlet section.

The guidance orders for pitch and yaw, which slave the VSM to a preset bearing and attitude depending on the launch depth and offset angle, are elaborated by the inertial system and guidance computer in the aerial missile.

The missile breaks the surface at an angle of 45° in the direction of the target, whatever the sea state. The nosecone is jettisoned. Jet interceptors enable the VSM attitude to be stepped down to limit the culmination altitude (30 m). After unlatching, the aerial missile is ejected, its wings and control surfaces are deployed, the boost motor is ignited and then it flies towards impact with the target like all other Exocets. All VSM's components sink rapidly to avoid detection.

The submarine-launched SM 39 Exocet missile, shortly after broaching the water and leaving the launch capsule (VSM) after a submerged firing
1995

Mid-course guidance is inertial, using one axial and two vertical gyroscopes operating in conjunction with three accelerometers. Coupled with the radio altimeter, the inertial system allows the missile to follow a sea-skimming trajectory to the target before the active radar takes over for the terminal phase of the flight.

Operational status

Development of the SM 39 was completed in 1984. Successful launches have been carried out by the diesel-electric submarine *La Praya*, the nuclear-powered submarine *Saphir* and the nuclear-powered ballistic missile submarine *L'Inflexible* from periscope to maximum launching depth at the Centres d'Essais de la Mediterranée (CEM) or des Landes (CEL). SM 39 is now operational in the French 'Le Triomphant', 'L'Inflexible', 'Rubis' and 'Agosta' class submarines and has also been exported to Pakistan for her 'Agosta' class submarines.

Specifications
Length (launch vehicle): 5.80 m
Diameter: 0.35 m
Launch weight: 652 kg

Range: 27 n miles
Warhead: 165 kg HE shaped charge fragmentation
Speed: M0.9

Contractor
Aerospatiale Missiles, Châtillon.

UPDATED

INTERNATIONAL

Milas

Type
Surface-launched ASW torpedo-carrying missile system.

Development
Milas, first revealed at the Le Bourget Naval exhibition in 1986, is being developed by Matra BAe Dynamics of France and Alenia Oto Sistemi Missilistia of Italy to meet a joint French and Italian naval requirement for an air flight vehicle which will be carried by frigates and be capable of carrying a lightweight anti-submarine torpedo.

The first two development firing trials of Milas were conducted at the San Lorenzo range in Sardinia on 21 and 27 June 1989. The successful firings validated the separation of the torpedo from the carrier.

Description
Milas is derived from the Otomat missile, but with the warhead and terminal seeker replaced by a lightweight torpedo. Attached to the missile's body are four folding delta-shaped wings which are mounted mid-way along the length of the body. The command receiver is mounted at the wingtip. At the rear of the missile are four small clipped triangular moving control fins.

The missile is launched using two 75 kg jettisonable boost motors and on reaching cruise speed the missile is powered by a 400 kg thrust turbojet engine.

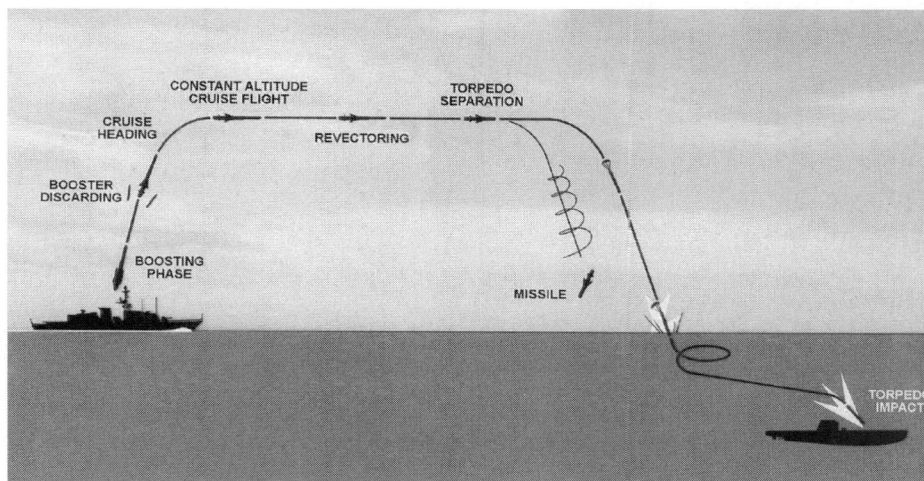

Flight trajectory of MILAS   *1996*

Target co-ordinates are preprogrammed into the missile's inertial navigation system before launch and several in-flight mid-course updates are passed by datalink from the launch vessel. The missile also features an automatic in-flight retargetting capability. During the terminal flight phase, the turbojet is shut down and the torpedo separates from the missile's body and descends to the water by parachute. On entering the water the torpedo's motor is activated and the active sonar homes the weapon onto the target. Milas is designed to carry the new MU 90 Impact lightweight torpedo and design studies are under way to examine the possibility of mounting other lightweight torpedoes such as the US Mk 46, Stingray and Mk 50. The time between launching and torpedo release at 32 n miles is 3 minutes.

Operational status
Under development with first production delivery now taking place. Early in 1993 successful launch trials were carried out with a dummy Impact torpedo. A test launch was carried out from the Italian trials ship *Carabiniere* in April 1994.

The missile turned through 180° and launched the torpedo at the correct speed and angle at a range in excess of 10 km.

Although originally planned to equip the three French 'Tourville' class destroyers and the 'Georges Leygues' class, the programme has now been dropped by the French government. Development is continuing, however, for the Italian Navy, and Milas will equip the two destroyers of the 'De la Penne' class. *De la Penne* conducted acceptance trials during 1996. The latest test firing took place in September 1998.

Specifications
Length: 6.00 m
Diameter: 0.46 m
Span: 1.06 m
Launch weight: 800 kg
Speed: M0.9
Range: 32 n miles
Warhead: 59 kg

Contractor
GIE Milas, Paris, France.

The Milas ASW missile under development by Matra and Otobreda

A trials firing of a MILAS anti-submarine missile, launched at San Lorenzo in Sardinia in 1989   *1995*

UPDATED

RUSSIAN FEDERATION AND ASSOCIATED STATES (CIS)

SS-N-9 Siren (P-50 Malakhit)

Type
Submarine-launched anti-surface missile.

Development
The missile was developed together with the SS-N-7 in the 1960s.

Description
The SS-N-9 is believed to achieve better performance than the SS-N-7 as it includes provision for autopilot mid-course updates from airborne platforms. Terminal homing is either by active radar (J-band, 10 GHz) or an IR seeker. The warhead is either conventional HE or nuclear. Propulsion is provided by a solid-propellant motor.

Operational status
The SS-N-9 is believed to have entered service in 1972, but is no longer in production. Since the one remaining 'Charlie II' class boat has now been withdrawn from service it is assumed that the missile is no longer operational.

Specifications
Length: 8.84 m
Diameter: 0.50 m
Launch weight: 3,300 kg
Warhead charge: 500 kg HE or 200 kT nuclear
Range: 60 n miles
Speed: M0.9

Contractor
K B Machinostroyenia, Kolomna.

UPDATED

SS-N-14 Silex

Type
Missile-carrying ASW torpedo system.

Description
The SS-N-14 'Silex' is thought to be somewhat similar in concept to the French Malafon or the Australian Ikara anti-submarine weapons in which a subsonic winged vehicle carries a homing torpedo to the position of the target submarine and can have its course corrected using command guidance techniques during flight so as to allow for submarine movements and make an accurate drop. The solid fuel-powered SS-N-14 deploys a 400 mm parachute-retarded homing torpedo or alternatively, can carry a nuclear warhead, with a weight of about 150 kg and a yield in the low kiloton range.

It is also believed possible that the SS-N-14 has an anti-ship capability and may be able to drop a homing torpedo outside a ship's normal defensive cover. Control of the weapon is exercised through a Head Light or Eye Bowl radar director.

Operational status
Operational since 1968 on board 'Kirov', 'Krivak I/II' and 'Kara' class ships. Later classes noted with this

A Russian Navy 'Udaloy' class destroyer with the starboard quadruple SS-N-14 'Silex' launcher just forward of the bridge (US Navy)

weapon system include the 'Udaloy' class of anti-submarine destroyers which are fitted with two quadruple launchers. The total number of missiles in service is considered to be about 250.

Specifications
Length: 7.20 m
Diameter: 0.55 m
Launch weight: 2,500 kg
Warhead: conventional HE or 5 kT nuclear
Max range: 30 n miles
Payload: nuclear warhead or Type 40 or E53-72 homing torpedo
Trajectory: programmed/radio command flight to target area under autopilot control with command override capability. Speed is about M0.95 at 750 m above sea level.

Contractor
Raduga OKB.

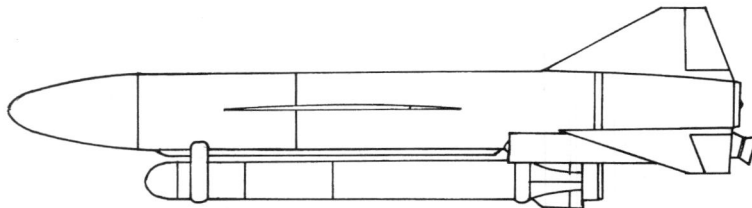

A line diagram of the SS-N-14 missile, with a torpedo attached underneath   **1995**

UPDATED

SS-N-15 Starfish

Type
Submarine and surface ship-launched ASW missile.

Development
This missile is reported to be of the same general type as the US SUBROC.

Description
The missile is horizontally launched from a standard 533 mm torpedo tube to follow a short underwater path before broaching the surface to follow an airborne trajectory with inertial guidance for the major part of the flight to the target area. On reaching the target area, a depth charge is released to continue on a ballistic trajectory until it enters the water near the target, sinking to the optimum depth before detonation. This type of weapon relies upon accurate localisation of the target in the first instance, followed by rapid launch and flight to the target area before the target has time to travel far from its last known position.

It was originally estimated that the solid-propellant motor gave the missile a range of about 18 n miles, but later official US reports have credited the missile with a range of between 24 and 27 n miles.

Operational status
The missile is reported to have entered service in 1973 and is currently operational aboard 'Typhoon', 'Charlie II', 'Delta IV', 'Oscar II', 'Victor III', 'Sierra I and II' and 'Akula' class submarines as well as 'Kirov' and 'Neustrashimy' class surface vessels.

Specifications
Length: 6.50 m
Diameter: 0.53 m
Launch weight: 1,800 kg
Warhead charge: 300 kg depth charge or 200 kT nuclear
Range: 24-27 n miles
Payload: Type 40 torpedo in place of charge

Contractor
Not known.

UPDATED

SS-N-16 'Stallion'

Type
Submarine and surface ship-launched ASW missile.

Development
Believed to be a further development of the SS-N-15 to provide a torpedo delivery vehicle instead of a depth charge. Considered to be similar to the US ASROC in concept.

Description
The SS-N-16 is believed to carry a nuclear-tipped Type 65 or ET-80 torpedo, although a conventional HE-armed torpedo could be fitted in place of the nuclear weapon.

The missile is horizontally launched from a standard 650 mm torpedo tube to follow a short underwater path before broaching the surface to follow an airborne trajectory with inertial guidance for the major part of the flight to the target area. On reaching the target area the solid-propellant motor is jettisoned and the torpedo descends to the water by parachute. On entering the water the torpedo's motor is activated together with its active sonar homing head to carry out a preprogrammed search and attack pattern. This type of weapon relies upon accurate localisation of the target in the first instance, followed by rapid launch and flight to the target area before the target has time to travel far from its last known position.

Operational status
The missile is reported to have entered service in 1979 and may still be in limited production. The SS-N-16 is deployed on 'Oscar II', 'Victor III', 'Sierra I and II', 'Akula' and 'Severodvinsk' class submarines and the 'Neustrashimy' class surface vessels. It has been estimated that each submarine carries, on average, four ASW nuclear weapons, either SS-N-15 or SS-N-16.

Specifications
Length: 8.17 m
Diameter: 0.65 m
Launch weight: 2,445 kg
Warhead: 200 kt nuclear or 150 kg HE or Type 40 torpedo
Range: 65 n miles

Contractor
Not known.

UPDATED

Yakhont

Type
Anti-ship missile.

Description
The Yakhont ship- or submarine-launched missile was first exhibited in 1993. Launched from a canister, the ramjet-powered missile is believed to use active radar terminal guidance. Yakhont has a distinctive shape with a nose-mounted ramjet air intake assembly, four clipped delta wings and four smaller clipped delta wings probably located on a solid-propellant boost motor. Range is stated to be 160 n miles with a launch weight of 1,300 kg.

Operational status
Reports suggest that development began in 1985.

Contractor
Novator NPO.

UPDATED

YP-85

Type
Surface ship-launched ASW missile.

Description
The YP-85, first seen in 1993, has a body similar to the SS-N-2 Styx and appears to be an enlarged SS-N-14. It is designed to carry a large torpedo underneath the missile's body. Cruise speed is achieved by solid propellant boost motors, the cruise speed then being maintained by a liquid-propellant motor, the missile having a maximum speed in the region of M0.95.

Operational status
The missile is thought to have been developed for the export market.

Specifications
Length: 7.20 m
Diameter: 0.55 m
Launch weight: 4,000 kg
Range: 250 km
Speed: M0.95

Contractor
Not known.

UPDATED

UNITED STATES OF AMERICA

RUR-5 ASROC

Type
Short-range ASW rocket.

Development
Development of ASROC began in 1956 to meet a USN requirement for a quick reaction ASW system. Since the original design entered service in 1961, it has undergone several upgrades, with five major variants (5A, 5B, 5C, 5D and 5E) entering service. The latest variant to be developed is a vertical launch version to equip both surface ships and submarines (see following entry).

Description
The RUR-5 is basically an all-weather unguided rocket system powered by a solid-propellant motor and armed with either a depth charge (which can be nuclear) or a Mk 46 acoustic homing torpedo. The warhead is enclosed in a split shell flared into the front end of the rocket. The rocket motor and the front end shell are fitted with four stabilising fins at their rear ends.

After launch, the missile follows a ballistic trajectory before the rocket motor is jettisoned at a predetermined point and the missile follows a ballistic flight to the point of airframe separation. At this point, the steel band that holds the front shells together is split by an explosive charge, after which the payload continues towards the target. In the case of the

Eight-round ASROC launcher aboard USS David R Ray (W Donko)

torpedo, a parachute is deployed which affords the weapon a certain degree of control during its descent and allows it to enter the water at the correct angle where its protective nose cap is shattered, the parachute discarded, the weapon's motor ignited and its seeker head activated. The depth charge descends in free flight and sinks to a predetermined depth before detonating.

The rocket is fired from the Mk 112 eight-cell launcher (which is also used to fire Harpoon), the Mk 26 launching system, or from the Mk 10 Terrier (SAM) missile launcher system. Other major components of the system include a fire-control computer and an underwater sonar detector.

Operational status
ASROC first entered service in 1961 and it is believed that limited spares are still being manufactured. The weapon is in widespread service throughout the US Navy in many classes of ship in the cruiser, destroyer and frigate categories. In 1991, the US government announced that the nuclear warheads for ASROC would be withdrawn from service. The conventional HE warhead is in operational service with many other countries including Brazil, Canada, Egypt, Germany, Greece, Italy, Japan, South Korea, Mexico, Pakistan, Spain, Taiwan, Thailand and Turkey. It is expected that the RUR-5 variant will be phased out of service in the late 1990s, although limited numbers of spares are still being manufactured to support the missiles still in service.

Specifications
Length: 4.57 m
Diameter: 0.325 m
Launch weight: 435 kg
Payload: Mk 46 torpedo (Mod 5 Neartip in US Navy service)
Range: 5.5 n m les

Contractor
Lockheed Martin Tactical Defense Systems, Akron, Ohio.

UPDATED

RUM-139 Vertical Launch ASROC (VLA)

Type
Short-range vertical launch ASW rocket.

Development
Initially, plans were developed to provide separate ASW standoff capabilities for surface ships and submarines. These were subsequently suspended on the grounds of cost and a requirement to merge the project with the

ASW-SOW standoff weapon programme. However, changes in the USN's overall ASW policy led to a reversal of this decision in 1982 and development of the RUM-139 Vertical Launch ASROC (VLA) began in 1984. It will be deployed from the Mk 41 vertical launch system using the Mk 116 Mods 6.7/8 ASW combat systems.

Description
The VLA features a boost motor fitted with thrust vector controls which are used to turn the missile over onto a

ballistic trajectory after its vertical launch phase. The booster is jettisoned at a predetermined time, and the payload flies on a ballistic path to the splashpoint. A parachute is deployed to slow the descent of the Mk 46 torpedo and ensures that it enters the water at the correct angle after which the parachute is discarded, and the torpedo's boiler fired and its search initiated. All VLA systems (including export models to Japan) will be armed with the Mk 46 Mod 5 Neartip torpedo.

Operational status
Development commenced in 1984 and VLA was initially deployed in 1993. Contracts for 510 missiles have been awarded to Lockheed Martin Tactical Defense Systems, Akron, and orders have been placed by Japan. In US service, VLA will arm the 'Ticonderoga' class cruisers and 'Arleigh Burke' and 'Spruance' class destroyers. In the Japanese MSDF, VLA equips the 'Kongo' and 'Murasame' class destroyers.

Specifications
Length: 5.08 m
Diameter: 0.35 m
Payload: Mk 46 or Mk 50 torpedo
Launch weight: 635 kg
Warhead (torpedo): 45 kg
Range (missile only): 9 n miles

Contractor
Lockheed Martin Tactical Defense Systems, Akron, Ohio.

A Japanese Maritime Self-Defence Force 'Chikugo' class frigate with a conventional HE RUR-5 ASROC octuple launcher aft of the funnel   **1995**

UPDATED

UGM-84 Harpoon

Type
Short-range submarine-launched anti-ship missile.

Development
The USN commenced studies for an anti-ship missile in 1965, at about the same time as McDonnell Douglas was carrying out private studies to meet a similar requirement. In July 1971, McDonnell Douglas won the development contract and in May 1973, the US Navy selected the Harpoon as its prime anti-ship missile system for deployment from aircraft, submarines and surface ships. All these versions entered service in 1977, since when various improvement programmes have been initiated. Block 1B was introduced in 1982 and featured ECCM improvements, while 1C featured an improved seeker and increased range, and entered service in 1985. Block 1G, which incorporates re-attack and other guidance improvements entered service in 1997.

Description
Harpoon has four clipped-tip triangular wings sited midway along the body and four similar in-line clipped triangular moving control fins at the rear. Initial thrust is provided by a jettisonable tandem boost motor with in-line tail fins attached, with a turbofan sustainer motor (the Teledyne CAE J402 turbojet) maintaining cruise speed. Maximum range is 70 n miles.

The missile is carried in a capsule which is launched from the torpedo tube. As the capsule leaves the

torpedo tube, stabilising fins unfold to establish a proper glide angle to enable the weapon to reach the surface. A sensor detects when the missile broaches the surface and initiates release of the capsule nose and tail sections, followed by ignition of the boost motor. As the missile exits the capsule, its wings and fins unfold and the weapon is propelled into a flight trajectory similar to that of surface launches.

After booster separation, the missile descends to a low-cruise altitude determined by its altimeter and flies towards the target by heading reference guidance, under the power of its turbofan engine. The Block 1C and 1G versions can execute mid-course waypoints based on preset data to avoid obstacles in the trajectory, approach the target from a desired direction or provide multiple-missile simultaneous arrival on target. At a point preset by the launch platform the J-band, frequency-agile, two-axis active radar seeker is activated into its search and acquisition mode. The missile is usually launched in this preset range and bearing launch mode, turning on the radar seeker at the last moment to acquire the target. The radar can be set for large, medium or small acquisition windows that determine the range-to-target at which the seeker is activated. The smaller the window, the more precise the initial target data must be, and the less chance that the missile will succumb to ECM. An alternative launch technique is the bearing-only launch in which the missile is fired along the target bearing and seeker is activated early in the trajectory to scan a 45° search sector either side of the missile's course. If no target is acquired after a suitable time on the initial bearing then

the missile switches to a preset search pattern, and if it then fails to acquire the target will self destruct. In either launch mode, once the target is detected and the seeker is locked on its tracking mode the Block 1A missile climbs rapidly at about 1,800 m from the target in a pop-up manoeuvre before diving down onto the target at an angle of about 30°. The later Block 1B, 1C and 1G missiles have a sea-skimming terminal attack profile. The Block 1C and 1G have an optimal shallow 'pop-up' manoeuvre. The Block 1G, instead of self-destructing at the end of the search phase, will turn and execute a reattack manoeuvre if it missed the target on its first pass.

Operational status
The missile entered service in 1977 and is operational in the submarine forces of a number of navies including Australia, Egypt, Greece, Israel, Japan, Netherlands, Turkey, UK and the US.

Specifications
Length: 4.64 m
Diameter: 0.34 m
Launch weight: 690 kg
Warhead: 227 kg, HE blast penetration
Range: 70 n miles

Contractor
The Boeing Company, St Louis, Missouri.

UPDATED

TORPEDOES

The torpedo is the primary ASW weapon used by surface ships, aircraft and submarines. The weapon dates back to 1866 when Robert Whitehead developed the first real torpedo in his factory at Fiume. Driven by cold compressed air at a speed of about 7 kt, it was successfully demonstrated to a number of interested countries and by the early 1880s had been developed to run at speeds of up to 30 kt for distances of about 1 km. This method of propulsion continued to be developed until the early part of this century when experiments with heated air in the UK produced a dramatic increase in performance. In this system, paraffin, water and air were mixed, the result being a steam-air mix driving a radial engine. In the United States a turbine-powered weapon was developed using alcohol as the fuel.

By the middle of the First World War, the torpedo had become a much more effective weapon, despite problems with the fuzing. Indeed, the British Mk 4 torpedo could run at distances of around 5 km at speeds of up to 40 kt, figures that, with the exception of a Japanese torpedo in 1942, were not exceeded until a few years ago.

Between the First and Second World Wars a great deal of development continued in the field of warhead, fuzing and propulsion systems. One of the most remarkable and certainly the most long-lived products of this period was the British Mk 8 which was developed in the early 1930s and remained in service with the Royal Navy until 1986, more than 50 years of service. It was however, Germany, which revolutionised the torpedo by introducing an electronically propelled weapon in 1939. Powered by lead/acid batteries this weapon could travel at 27 kt over ranges up to 8 km. The German Navy also produced the first acoustic homing torpedo in 1943, another major milestone in torpedo development, as well as the wire guidance system that is used on all modern heavyweight torpedoes, and also wake following techniques. Post-war the latter had primarily been developed in the former Soviet Union, but other countries are now increasingly developing this technology.

By this time it had also been realised that not only were torpedoes essential for use against surface shipping but also they were even more necessary to combat hostile submarines. Indeed, this latter capability has now become the main requirement for the modern torpedo.

After the end of the Second World War, development continued apace, mainly in the field of electrically propelled weapons with acoustic homing systems, until the increasing speed and depth capabilities of submarines necessitated the return to a thermally powered torpedo. The two latest torpedoes are the thermally powered US Mk 50, which has now reached its end of nearly 20 years of development, and the Swedish Torpedo 2000. Torpedo development takes a very long time.

In the present economic climate it is proving increasingly difficult to justify funds for development of a weapon which may take years to perfect. It is not, therefore, surprising to find increasing interest in international collaboration. Once such venture is that between France and Italy who have developed a new lightweight weapon based on experience with national programmes involving the Murene and A290 weapons. Post-war advances in submarine design have laid down the most important requirements for ASW torpedoes as:
(a) stealth
(b) speed (sufficient to chase and overtake high-speed underwater targets)
(c) range and endurance (sufficient to carry out search, attack and, if required, re-attack procedures)
(d) onboard homing
(e) depth
(f) guidance.

In addition to these features, the modern torpedo must be able to distinguish between real and false targets and to disregard countermeasures. It must also be effective in shallow water as well as at the maximum operating depth of submarines. Finally the submarine-launched torpedo must have a dual capacity to operate effectively against both surface targets as well as

submarines. Space on board submarines is too precious for two types of torpedo with different roles to be carried.

Torpedoes have undergone a major transformation over the last two decades. Today's torpedo is a highly sophisticated, guided, homing missile. To achieve the features outlined above, a vast span of technologies have had to be encompassed covering such diverse disciplines as hydrodynamics; mechanical; electrical; thermal propulsion; electro-acoustic transducer design; signal processing and correlation; beam-steering; self-generated noise suppression; computers employing very advanced algorithms and incorporating signature libraries used for comparison; computer architecture; flow noise techniques; simulation and warhead design.

With weapons embodying such sophisticated technologies, it is no longer necessary to employ salvo firing to attain a high-hit probability. A single wire-guided or fully autonomous torpedo in the hands of expert operators is sufficient to assure at least a 95 per cent hit probability. The problem is being able to place the weapon in exactly the right position to ensure a 100 per cent kill probability. This in turn demands that the weapon carries even greater sophisticated computation capabilities, combined with even more effective warhead charges, in many cases of the shaped directed type. While deploying torpedoes in salvos is a tactic of the past, two modern sophisticated weapons can be deployed in certain circumstances against a target which is using highly sophisticated evasive manoeuvres. Using this tactic the target can be attacked from different directions, both weapons homing on to the target at the most vulnerable spot. Used in this way the modern torpedo can enable a submarine to considerably increase the number of targets which it can effectively attack.

An area which has been the subject of considerable development in recent years has been propulsion. Today, new and powerful thermal propulsion systems are under development which will enable weapons to achieve much greater ranges and higher speeds, enabling high-speed manoeuvring targets to be attacked and reattacked until either the target is destroyed or eventually the weapon runs out of fuel.

The development of effective torpedo warning capabilities in the latest generation of sonars, linked with the development of sophisticated countermeasures, is beginning to pose a major problem for the torpedo. To remain effective and achieve a high-hit/kill probability, torpedoes must now incorporate an increasingly sophisticated degree of stealth technology combined with a much more effective onboard sonar and processing capability. Together these will enable the next generation of weapons to 'burn through' the countermeasures, detect the real target behind them and effectively home on to it.

Guidance involves the resolution of the fire-control equation by the combat information system and control of the weapon until its own onboard sensor can take over and home the weapon on to the target. Because of size constraints, torpedoes can only be fitted with small transducer arrays, which limits them to operating in the L-band with only a relatively short detection range capability. Hence the need to guide the torpedo to the vicinity of the target before its own onboard sonar can take over and complete the terminal homing phase.

Initial control of the weapon from the submarine command system is achieved using a very fine wire, some tens of kilometres in length. This fine wire is paid out from both submarine and torpedo. This ensures that the wire remains stationary in the water and unaffected by any strain resulting from manoeuvres by either the submarine or torpedo. The wire is used to carry commands from the submarine to the weapon and for response signals from the weapon to be fed back to the combat information system.

The possibility of using fibre optics in the future with even finer connections between submarine and torpedo will permit the weapon to be guided for much longer periods and out to greater ranges than hitherto. Furthermore, the much wider bandwidth will allow a much greater volume of data to be passed between

weapon and launch platform. Linked to the use of fibre optics are developments designed to give the torpedo much greater tactical autonomy. However, improved operator control could be an advantage when the torpedo is used in an ASV role.

The need outlined above for higher speed creates severe problems at the front end of the weapon where the sonar transducers are mounted. To be effective, the weapon must be able to 'hear' its target, but the target must not be able to 'hear' the torpedo. The torpedo must, therefore, be a very silent weapon, employing the latest stealth technologies. This requires very careful design with detailed attention being paid to noise reducing techniques. To achieve detection at the maximum possible range the torpedo uses an LF sonar system. For terminal homing, torpedoes are fitted with an onboard signal processor and sophisticated algorithms which are capable of detecting the correct target, ignoring false contacts and countermeasures, and steering the weapon to the exact position and inclination of the target for the warhead to detonate with the maximum effect.

Modern software enables the weapon to operate effectively at high speed, the enormous signal processing capacity being able to sort out any problems arising from adverse hydrodynamic and self-generated noise. This enables the weapon to overcome any acoustic degradation at high speeds. To achieve this, fast Fourier transform is used for spectral analysis, together with correlation techniques which are used to cancel out false returns and other ambiguities.

One area of development which has not been addressed to any major degree, except by the RFAS, is wake homing. Wake homing, when it can be effectively conducted, is immune to current countermeasures. Wake homing torpedoes are fitted with a single, upward-looking sensor which is used to detect and localise a ship's wake which may extend for some miles behind the vessel. The weapon is fired at an angle which anticipates crossing the wake and after registering the initial disturbance, and the exit from it, the weapon then continues to execute a zig-zag pattern across the wake following the narrowing width of the wake as it gets closer to the ship. This homing method offers a number of advantages. It is almost impossible to replicate with a decoy, by using an upward looking sensor it overcomes to a large degree the self-generated noise of the torpedo. Finally, the wake sensor can also be used to detonate the torpedo when it senses the disappearance of the wake. The tactics can be further enhanced by the integration of satellite sensors used initially to detect a ship's wake, thus ensuring better positioning of the torpedo. Wake following technology will become really effective with the introduction of high-speed, long-range thermal torpedoes, for the weapons currently suffer generally from too low a speed – the speed of advance of the weapon being considerably reduced due to its zig-zag movement across the wake and hence the extended time required to reach the target, currently limited by the weapon's endurance. Wire-guidance and satellite information used by the launching platform's command system can, to some extent, overcome this disadvantage.

The torpedo tube has now become a general delivery system for missiles and mines as well as the torpedo, and most boats today ship a mix of missiles and torpedoes. As a general rule two or three mines can be substituted for each torpedo and submarine mines are available from numerous international sources. A feature now available for submarines is an external add-on container for mine stowage so that there is no reduction in the number of missiles/torpedoes carried.

Owing to the increased space demanded by bow sonars, there is a tendency to resite torpedo tubes from their traditional bow position to a position slightly further aft, and angled out from the centreline.

The type of torpedo carried depends, to some extent, on the boat's mission. Torpedoes are optimised for anti-submarine or anti-surface vessel missions. However, it is both a restriction on mission capability and a waste of valuable space for a submarine to carry two types of torpedo. Modern torpedoes are thus optimised for use against both surface and submerged targets.

Weapons stowage is an area of considerable importance and shock requirements, position of tubes, shape of compartment, size of weapon and handling arrangements all dictate the number and availability of weapon reloads that can be carried.

Two prime methods of discharge have been developed: positive discharge and swimout. In the swimout tube the weapon is discharged using its own propulsive system. In the positive discharge system, energy is imparted to the weapon by an external source to push it out of the tube either by pneumatic water pulse or mechanical means. Positive discharge systems allow a submarine to deploy tube-launched weapons at a wide range of speeds and depths.

While the swimout tube requires less space it does require large volumes of water to be transferred during weapon loading, which requires large capacity tanks. The swimout system is of low weight and does not require great energy. It does suffer from a disadvantage that the torpedo propulsion makes a small noise, but with modern torpedoes this is fairly insignificant.

Four types of positive discharge are available. The compressed air type can be independent of depth, but at greater depths it requires greater energy to accomplish discharge. The piston pump, turbine-driven rotary pump and telescopic ram methods are all independent of depth.

The turbine rotary driven pump system is perhaps the most effective for it requires less space than a piston pump, weighs less and the discharge cycle is quicker. The system can also be adapted to suit various types of weapon (for example, mines, torpedoes and missiles).

One of the most important aspects of weapon discharge is the safety interlocks which must be incorporated to ensure that tubes are closed and drained of water before loading occurs. The number and siting of torpedo tubes, the discharge system selected and its design requirements, together with the type and number of weapons carried have a significant impact on the design of a submarine.

The latest developments underway in torpedo design relate to new propulsion systems which will drive torpedoes faster and deeper, autonomy will be further developed to parallel this development with increased onboard computing power, improved signal processing and much more sensitive homing heads. Modular weapon design will enable parts of a weapon to be upgraded without the need to design a completely new weapon (leading to reductions both in cost and in time of entry into service). Other areas that are currently the subject of study include the development of quieting and stealth techniques, improved acoustic performance to overcome submarine quieting techniques, and counter-countermeasures (both soft and hard kill) and improvements in initial guidance.

UPDATED

FRANCE

L3

Type
Submarine/surface ship-launched acoustic torpedo.

Description
This is a conventionally shaped, ship- or submarine-launched, anti-submarine torpedo with a strong body in light alloy and a laminated nosecone with the following five compartments: AS-3 active acoustic self-guidance

and electromagnetic firing device; explosive charge and impact fuze; secondary battery; air tank and automatic pilot; electric motor for propulsion. Location in bearing and elevation is used to guide the torpedo along a pursuit curve. As the torpedo nears the target, the pulse rate increases to improve the accuracy of pursuit and of the acoustic detonator. If after a computed time the target has not been detected at the previous position, the torpedo commences a circular search (in shallow water) or a helical search (deep

water). The torpedo can attack vessels proceeding at speeds up to 20 kt and down to a maximum of 300 m depths.

Operational status
The weapon is no longer in service with the French forces but is still in service in other navies (Portugal and Uruguay). Approximately 600 weapons were manufactured. The weapon is no longer in production.

Specifications
Length: 4,300 mm
Diameter: 550 mm
Weight: 910 kg
Warhead: 200 kg Tolite A1
Propulsion: electric
Speed: 25 kt
Range: 3 n miles
Operating depth: 300 m
Guidance: acoustic, active, range approx 600 m with favourable inclination of the target submarine. Type AS3T
Fuze: impact/acoustic

Contractor
Direction des Constructions Navales (DCN), Saint Tropez.

UPDATED

L3 acoustic torpedo launch

L5

Type
Surface/submerged launch dual-role torpedo.

Description
The L5 is powered by silver-zinc batteries which are activated at launch and provide the power for the motor, which drives twin contrarotating propellers.

There are three models: L5 Mod 1; L5 Mod 3; L5 Mod 4.

L5 torpedo about to enter the water (DCN)

The Mod 1 and Mod 3 have been in operational service with the French Navy for a number of years. The Mod 4, derived from the Mod 1, is carried by surface ships of the French Navy and several other navies. All models are designed to attack both underwater or surface vessels. The Mod 3 is submarine-launched and the Mod 1 and Mod 4 are launched by surface ships and submarines. All models are fitted with a passive/active homing head which has various operating modes, including direct attack or programmed search. Maximum operating depth is 550 m. The compact size of this weapon makes it ideal for operation from small displacement submarines.

Operational status
Operational with the French Navy and other navies (Belgium, Pakistan, Spain). The weapon is no longer in production.

Specifications
(L5 Mod 4)
Length: 4,400 mm
Diameter: 533.4 mm (can be increased to 550 mm if required)
Weight: 1,300 kg (Mod 3)
Warhead: 150 kg HE
Propulsion: electric
Speed: 35 kt
Range: 5 n miles
Guidance: passive/active homing
Fuze: impact/magnetic
Depth: 555 m

Contractor
Direction des Constructions Navales (DCN), Saint Tropez.

UPDATED

E14

Type
Submarine/surface ship-launched acoustic anti-ship torpedo.

Description
This is a conventionally shaped, submarine- or surface ship-launched, anti-ship (or anti-submarine against noisy targets close to the surface) torpedo with a strong body in light alloy, and a laminated nosecone with the following five compartments: acoustic/passive self-guidance and electromagnetic firing device; explosive charge and impact fuze; secondary battery; air tank and automatic pilot; and electric motor for propulsion. As many parts as possible have been commonalised with those in the L3 torpedo. The torpedo can attack surface vessels proceeding at speeds up to 20 kt and submarines at shallow depth. The depth setting is variable between 6 to 18 m. The E14 has the same geometry and mechanical features as the L3. Some E14 torpedoes have been upgraded to E15 Mod 2 standard.

Operational status
The E14 is no longer in service with the French forces, but is still in service in some other navies (Portugal, South Africa, Spain). The weapon is no longer in production.

Specifications
Length: 4,279 mm
Diameter: 550 mm
Weight: 927 kg
Warhead: 200 kg Tolite
Propulsion: electric
Speed: 25 kt
Range: 3 n miles
Depth: 18 m
Guidance: acoustic/passive, average range 500 m
Fuze: impact/magnetic

Contractor
Direction des Constructions Navales (DCN), Saint Tropez.

UPDATED

E15

Type
Submarine/surface ship-launched acoustic anti-ship torpedo.

Description
A conventionally shaped, submarine- or surface ship-launched, anti-ship (or anti-submarine against noisy targets close to the surface) torpedo with a strong body in light alloy and a laminated nosecone, the E15 is a lengthened version of the E14 Mod 1 torpedo. It has the following five compartments: acoustic/passive self-guidance and electromagnetic firing device; explosive charge and impact fuze; secondary battery; air tank and automatic pilot; electric motor for propulsion. As many parts as possible have been commonalised with those in the L3 torpedo. The torpedo can attack surface vessels proceeding at speeds up to 20 kt and submarines at shallow depth. Depth setting is variable between 6 to 18 m. The self-guidance system is the same as the E14, but geometry, range and explosive charge are different.

The E15 has been upgraded to Mod 2 standard. This constitutes fitting a new homing head of the AH-8 type, and replacing the batteries with silver-zinc primary batteries.

Operational status
The E15 is no longer in service with the French forces, but remains operational in some other navies (Pakistan, Portugal, South Africa, Spain, Uruguay). The weapon is no longer in production.

Specifications
(Mod 2)
Length: 5,900 mm
Diameter: 550 mm
Weight: 1,387 kg
Warhead: 300 kg Tolite A1
Propulsion: electric silver-zinc battery
Speed: 31 kt
Range: 6.5 n miles
Detection range: 2,000 m
Search depth: 25 m
Attack depth: 6-18 m presettable
Max depth rating: 300 m
Guidance: passive homing, medium-range
Fuze: impact/magnetic

Contractor
Direction des Constructions Navales (DCN), Saint Tropez.

UPDATED

F 17

Type
Submarine/surface ship-launched wire-guided torpedo.

Description
The F 17 torpedo family comprises the following models: F 17 Mod 1, F 17P, F 17S and F 17 Mod 2.

These torpedoes are deployable from surface vessels or submarines and are able to destroy surface ships and submarines (except for the F 17 Mod 1 which can only be used against surface ships or noisy submarines close to the surface).

Launching by submarine may be performed in one of the following three modes:
(a) pneumatic ram
(b) water ram (including turbo-pump set)
(c) swimout (this mode has been qualified with a MAK tube).

The F 17 torpedo carries an extensive warhead charge which enables it to operate successfully against all naval vessels. The sonar homing system can operate either in passive or active mode (except F 17 Mod 1, which operates only in passive mode) giving a high-detection performance. An internal counter-countermeasures system enables the F 17 Mod 2 and F 17S to strike their targets under the most adverse conditions. These performance features, associated with a long-range, high-speed, great operating depth, plus the ability to operate with considerable stealth give the F 17 torpedoes a formidable capability.

The F 17 Mod 1 and F 17P can be updated to the Mod 2 or S standard. (All the F 17 Mod 1 torpedoes in the French Navy have been updated to F 17 Mod 2.)

Maintenance time interval has been increased to 18 months, and also simplified.

Operational characteristics

	Target[2]	Speed	Range	Counter-countermeasures capabilities	Maximum attack depth	Search depths[3]
F 17 Mod 1	ss	35 kt	18,500 m	low	>500 m	Za, 30, 100 or 200
F 17P	ss/sub	35 kt	18,500 m	low	>500 m	Za, 30, 60
F 17 S	ss/sub	35 kt	18,500 m	high	>600 m	Za, 30, 100 or 200
F 17 Mod 2	ss/sub	24/40 kt	20,000 m (at 40 kt)	very high	>600 m	Za, 30, 100 or 200

Notes:
[2]ss = surface ship; sub = submarine
[3]Za = depth against surface ship attack (Za = 6 to 20 m)

Physical characteristics of the torpedo and its tube container for 'Agosta' class submarines

	Diameter[1]	Length	Weight
F 17 Mod 1	533 mm	5,914 mm	1,431 kg
F 17P	533 mm	5,914 mm	1,431 kg
F 17 S	533 mm	6,003 mm	1,539 kg
F 17 Mod 2	533 mm	5,406 mm	1,428 kg

Note: [1]can be increased to 550 mm if required

The F 17 Mod 2 is the latest version of the F 17 torpedo family which is in current operational service. The Mod 2 has been upgraded, the main improvements relating to the propulsion, homing head and the counter-countermeasures capabilities. The weapon features a new all-digital control and guidance unit. At high speed the F 17 Mod 2 electric heavyweight torpedo has sufficient range capability to enable it to overtake most escaping targets.

Although the F 17 Mod 2 is quite stealthy at high speed, to take into account new detection performances of some targets, it now features a two-speed capability. The ability to command a slow speed (24 kt) enables the torpedo to carry out a very stealthy approach phase, but with the possibility of switching to high speed (40 kt) via wire-guidance commands. The homing head can operate in active or passive mode, or both simultaneously.

A special version of the F 17 Mod 2 has been designed for the A 209 or TR 1700 submarines (mechanical interfaces and adaptation for the swimout discharge). In addition, interoperability with different fire-control systems has been examined and is possible for most with few modifications.

Operational status
The F 17 Mod 1 and F 17P are operational with several navies (Pakistan and Saudi Arabia). F 17 Mod 2 has been in service with the French Navy since 1989 and the Spanish Navy since 1993.

Specifications
(Mod 2)
Length: 5,406 mm (with 'Agosta' class tube container)
Diameter: 533 mm (can be increased to 550 mm if required)
Weight: 1,428 kg (with 'Agosta' class tube container)
Warhead: 250 kg HBX3
Propulsion: electric
Speed: 24/40 kt
Range: 10.75 n miles at 40 kt (15.5 n miles if the torpedo runs at 24 kt during the whole wire-guidance phase)
Guidance: wire and automatic homing
Fuze: magnetic proximity and impact

Contractor
DCN International, Paris. *UPDATED*

F 17 Mod 2 torpedo

GERMANY

SUT

Type
Anti-surface vessel and anti-submarine torpedo.

Description
The SUT (Surface and Underwater Target) torpedo is the latest and most versatile member of the Seal,

Surface and Underwater Target (SUT) torpedo

Seeschlange and SST 4 family of torpedoes. It is a dual-purpose wire-guided torpedo for engaging both surface and submarine targets. The SUT can be launched from submarines and surface vessels, from fixed locations or mobile shore stations. Its electrical propulsion permits variable speed in accordance with tactical requirements, silent running and wakelessness. The wire guidance gives immunity to interference with a

two-way datalink between vessel and torpedo. The acoustic homing head has long acquisition ranges (active – 1 n mile; passive – 3 n miles) and a wide search sector for active and passive operation. After termination of wire guidance, SUT continues operation as a highly intelligent homing torpedo, with internal guidance programmes for target search, target loss and so on. The large payload with combined fuze systems ensures the optimum effect of explosive power. The SUT operates at great depths as well as in very shallow waters. Consort operation permits exploitation of the full over-the-horizon range of the SUT. The body is made of plastic or aluminium. SUT at present exists in three different versions with slight differences in the internal guidance programs and the extent of data transferred via the guidance wire. The latest version is SUT Mod 2 with its special feature of additional data from the homing head being signalled back to the vessel including an 'Audio Channel'.

Operational status
In service and in production for several navies (NATO – Greece; South America – Chile; Asia – India, Indonesia, South Korea, Taiwan). In the early 1980s Indonesia signed a contract for indigenous manufacture of the SUT within a long-term programme still running.

Specifications
Length: 6,150 mm (6,620 mm with guidance wire casket)
Diameter: 533 mm
Weight: 1,420 kg (without casket)
Warhead: 260 kg
Speed: selectable, max 35 kt
Range: 15 n miles at 23 kt or 7 n miles at 35 kt
Fuze: magnetic proximity and impact
Operating depth: 2-460 m

Contractor
STN ATLAS Elektronik GmbH, Bremen/Hamburg.

UPDATED

Seal (DM2/DM2A1) and Seeschlange (DM1)

Type
Anti-surface vessel and anti-submarine torpedo respectively.

Description
These two wire-guided, heavyweight, 21 in (533 mm) torpedoes were purpose-developed for use in the German Navy: Seal for use against surface ships and the smaller Seeschlange against submarines. There is a high degree of equipment commonality between the two weapons. Major differences are that the anti-submarine model has half the propulsion battery capacity of the Seal, but is fitted with three-dimensional (3-D) sonar. The following main features apply to both types.

All essential data are sent to the torpedo throughout its run via a dual-core guidance wire. Similarly, actual torpedo running data are simultaneously transmitted to the ship. An active/passive homing head with a steerable transducer array. Attack options following acquisition are either by manual or computer control from the launch ship, or by self-homing by the torpedo.

Provision is made for a programmed run after

guidance wire pay out or loss of signals from the onboard fire-control system. Different programmes adapted to various tactical situations are available and can be selected via the guidance wire.

Launch arrangements include compressed-air firing from surface ships and swimout from submarines. There are no limitations on ship movements during the launch and guidance phase. The electric propulsion system employed provides long running distances, permitting launch from beyond target defence area. Torpedo speed is selectable.

Other features are: combined impact and proximity fuze; full performance in shallow or deep water; 3-D internal stabilisation; identification of different targets and high-hit probability; and automatic system check before firing.

Operational status
Both torpedoes are in service with the German Navy aboard '206' class submarines. The weapon has been replaced by the Seehecht (DM2A3) on the modernised Type 206A boats. Seal is also deployed on surface ships, in particular on the Type 142 and 143 fast patrol boats. Seal has been modified to form the special surface target torpedo SST 4 and the dual-purpose surface and underwater target torpedo SUT.

Specifications
Seal
Length: 608 cm (655 cm with guidance wire casket)
Diameter: 533 mm
Weight: 1,370 kg
Warhead: 260 kg
Speed: 18-35 kt
Range: >16 n miles
Guidance: active/passive
Fuze: magnetic proximity or impact

Seeschlange
Length: 415 cm (462 cm with guidance wire casket)
Diameter: 533 mm
Warhead: 100 kg
Speed: selectable, max 33 kt
Range: 6.5 n miles
Guidance: active/passive
Fuze: magnetic proximity or impact

Contractor
STN ATLAS Elektronik GmbH, Bremen/Hamburg.

UPDATED

Seehecht (DM2A3)

Type
Anti-surface and anti-submarine torpedo.

Development
This wire-guided, heavyweight, 21 in (533 mm) acoustic homing torpedo has been developed to replace the Seal and Seeschlange torpedoes in service with the German Navy and NATO countries.

In a further development it is planned to modify the DM2A3 into the DM2A4 configuration incorporating a new propulsion system leading to increased speed and range, and improved electronics.

Based on the DM2A3 a new generation of torpedoes for a specific export market designated Seahake has been developed in accordance with the regulations of the German MoD. There will be some differences in hard and software in order to meet the export regulations and the special requirements for worldwide

application. However, the main characteristics of the DM2A3 (Seehecht) will be incorporated in the new design.

Description
Seehecht is a dual-purpose torpedo which can engage surface and submarine targets and can be launched from both surface vessels and submarines. The main features of the weapon are: extremely long guidance

distance; silent running resulting from the greatly improved propeller design (5-bladed skew propellers in GRP) and other special measures (elastic motor suspension and special shaft bearings); impact and improved magnetic proximity fuze with high ECCM capability; improved communication system for two-way transmission of extended volume of data, including the complete acoustic panorama and wideband noise samples; stabilisation system with highly accurate sensors and regulator systems optimised for multiple operational requirements; multistage guidance concept including highly intelligent internal guidance programs; new TOSO acoustic homing head developed by WASS using PFB technology and covering a wide panorama; operation in active and passive modes in several frequency bands with pre-formed beam technology; ability to handle multiple target situations with numerous features capable of identifying or suppressing jammers and decoys; the latest technology using microprocessors connected via a MIL-BUS system and fibre optic conductors; and extensive BITE (Built-In Test Equipment).

The DM2A4 will feature an improved battery, new propulsion system driving a contra-rotating propeller and a redesigned after section. Speed will be increased by ×1.5 compared to the DM2A3 and propulsion energy will be increased by more than ×3. The high-speed electric motor of the DM2A4 will be powered by a Zn-AgO battery. The new motor will be a stepless permanent magnet DC motor developing 275 kW.

Operational status
DM2A3 – In service in the German and Norwegian navies. The last DM2A3 was delivered to the German Navy at the end of 1997.
DM2A4 – In full scale development. Technical evaluation scheduled for 2000/2001, with first full production weapons due to be delivered in 2003 to arm the U212 boats.

Prototypes and series torpedoes of Seahake have been proven successfully in some hundreds of wet firings.

Specifications
Length: 615 cm (662 cm with guidance wire casket)
Diameter: 533 mm
Speed: 18-35 kt
Range: 7-15 n miles

Contractor
STN ATLAS Elektronik GmbH, Bremen/Hamburg.

Seehecht torpedo on test-stand

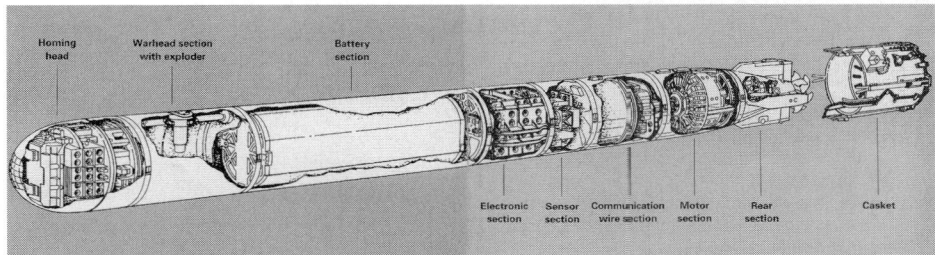

UPDATED   *Cutaway of Seehecht* (STN ATLAS Elektronik)   *1999*/0038527

SST 4

Type
Special anti-surface vessel torpedo.

Description
The SST 4 (Special Surface Target) torpedo is comparable in its dimensions, construction and capabilities to the Seal weapon, except for those features which can only be applied within the operational area of the German Navy. In this respect the SST 4 has been adapted to a standard international version. It has an operating depth down to 100 m.

The basic design has been continually improved over the years, especially in the homing head functions and in the stabilisation and related control system.

Further improvements covering additional return signals (actual course, speed and depth) and magnetic proximity fuze are incorporated in the SST 4 Mod 1 configuration. A modification kit is available for conversion of SST 4 weapons to SST 4 Mod 1 configuration.

Operational status
In service on '209' class submarines, fast patrol boats of the 'Combattante I', 'Combattante II' and 'Jaguar II' classes; introduced into various NATO (Greece and Turkey) and South American (Argentina, Ecuador and Venezuela) navies.

Specifications
Length: 608 cm (655 cm with guidance wire casket)
Diameter: 533 mm
Weight: 1,263 kg
Warhead: 260 kg
Propulsion: electric
Speed: selectable, 23/35 kt
Range: 6.5/15 n miles
Guidance: wire-guided, dual-core with two dispensing systems; active/passive homing sonar
Fuze: impact (SST 4 Mod 1 proximity also)

Contractor
STN ATLAS Elektronik GmbH, Bremen/Hamburg.

UPDATED

INTERNATIONAL

Impact MU90

Type
Lightweight ASW homing torpedo.

Development
Following the French and Italian governments' MoU (Memorandum of Understanding) concerning the future of their two lightweight torpedo programmes, the Murène and A 290, work commenced on merging the

two national programmes to develop a new-generation lightweight torpedo.

Description
The Impact uses an advanced acoustic homing head developed by Thomspon Marconi Sonar SAS with the Mangouste digital signal and data processor which is capable of achieving high-detection ranges against anechoic-coated submarines in both deep and very shallow water using both active and passive detection

and homing. Multiple lobing, using constant or modulated waveforms, is used to overcome reverberation and multiple false echoes. The torpedo incorporates a high ECCM capability which will make it immune to most types of countermeasures. The weapon features very low self and radiated noise and is armed with a powerful warhead based on a directed energy charge.

The torpedo s powered by a high-efficiency, low-noise propulsion system based on an STN Atlas

Elektronik electrically controlled stepless electric motor with a variable speed from 29kt and beyond. It is powered by a high-energy/weight ratio silver oxide-aluminium battery with a potassium hydroxide electrolyte driving a pump jet. The torpedo is highly manoeuvrable and has a high-endurance and high-operating depth. The weapon can be armed either with a directed-energy hollow-charge or a semi-directional HE warhead.

Other features include a short reaction time realised through the integrated automatic system, covert attack capability because it is only detectable at very short range and wide interoperability with other lightweight torpedoes. The guidance system features a strapdown inertial guidance unit. The weapon can be launched from any type of platform including surface ships, missiles, fixed-wing aircraft, helicopters, submarines and continental shelf mines in depths as shallow as 25 m.

Operational status

The Reconfiguration Programme has been completed. Sea trials were concluded in 1996 and the weapon received formal qualification from the French and Italian governments at the end of 1996. The main production contract for France and Italy was signed at the end of 1997, as was a German contract. The three countries will then organise production over a 12 year period both for domestic use and for export. In 1997 the French DGA placed an order for 600 MU90 torpedoes. Initial production deliveries are due in late 2000.

Specifications
Length: 3,000 mm
Diameter: 324 mm
Weight: 300 kg
Speed: 29->50 kt
Operating depth: >1,000 m
Range: >10,000 m

Contractors
Eurotorp, Sophia Antipolis, France.
Production – DCN St Tropez & Thomson Marconi Sonar SAS, Sophia Antipolis, France.
Whitehead Alenia Sistemi Subacquei, Livorno, Italy.

Impact MU90

1999/0024775

UPDATED

A 244/S Mod 1

Type
Lightweight active/passive self-homing torpedo.

Development
Developed for the Italian Navy, the A 244 is a lightweight torpedo designed to replace the older US Mk 44 type. The A 244 Mod 1 was designed to provide surface units (including small ships), fixed-wing aircraft and helicopters with an ASW weapon system which could meet operational requirements facing submarine technical and tactical evaluation.

Description
The A 244/S Mod 1 is a lightweight torpedo based on an advanced computerised homing system with reprogrammable software enabling search and/or homing behaviour to be adjusted without any hardware changes. The acoustic homing head is capable of active, passive or mixed modes for closing on to its target. It can discriminate between decoys and real targets in the presence of heavy reverberations by specially emitted pulses and signal processing. The head has a large search volume covered by multiple preformed beams following a number of self-adaptive search patterns. The computerised homing system also provides for presettable combinations of signal processing, spatial filtering and tactical torpedo manoeuvring to match the torpedo's performance in the ever changing operational situation as dictated by the threat and ASW tactics.

Active acquisition range is in excess of 2,100 m and the weapon is capable of shallow water operation in depths of 30 m with a maximum operating depth in excess of 600 m.

The propulsion system uses a saltwater battery to power a DC counter-rotating motor providing direct drive to a two-blade propeller shaft.

The A224/S Mod 3 (earlier variants can be upgraded using modification kits) features a dual speed motor (30/36 kt) and increased range (13,500 m/10,000 m).

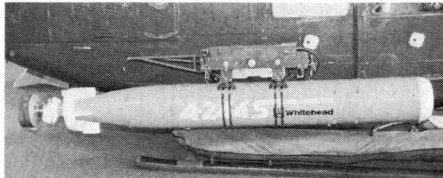

Type A 244 lightweight torpedo on helicopter

Other improvements include enhanced ECCM and reduced self-noise signature.

Operational status
In production. In service with over 14 navies and air forces including Argentina, Croatia, Greece, Indonesia, Morocco, Pakistan, Peru, Singapore, Sweden, Venezuela and Yugoslavia. The United Arab Emirates also ordered the weapon in 1997.

Specifications
Length: 2,750 mm
Diameter: 324 mm
Weight: 221 kg (approx) (warshot version)
Warhead: HBX-3 HE
Propulsion: electric
Guidance: active/passive sonar, self-adaptive programmed patterns
Speed: 37 kt (max)
Range: >7,000 m

Contractor
Eurotorp, Sophia Antipolis, France.
Whitehead Alenia Sistemi Subacquei, Livorno, Italy.

UPDATED

B 515

Type
Surface vessel torpedo launching system.

Description
Developed by Whitehead Motofides of Italy, the B 515 torpedo launching system is designed for the operation of lightweight anti-submarine torpedoes such as the A 244/S, A 244/S Mod 1, Mk 44, Mk 46, Sting Ray and Impact MU90.

The torpedo launcher is designed to be installed on the ship's deck. Normally the launchers are mounted in pairs, one at the starboard and the other one at port side of the ship, both being arranged with the muzzle facing forward.

The tubes are mounted on a base which can be manually trained by means of a suitable retractable handle. The torpedo is fired after the tube has been trained into the designated sector. The tubes are provided with an electrical de-icing system, meant to ensure suitable temperature for the torpedo loaded into the barrel. Similarly heaters are installed on the rotating base. Moreover, the launcher is equipped with an alarm circuit which signals overheating in the tubes.

Each tube is also provided with manual safety lock of the pneumatic circuit and a series of electric interlocks to prevent anomalous and/or unwanted firing operations.

Firing is achieved using compressed air stored in air bottles secured to the tube's outer surface and refilled via the air charging station (also made by Whitehead Alenia Sistemi Subacquei).

The torpedo presetting is carried out electrically via a connecting wire (snap connector) connected to a plug mounted on each tube. Presetting is normally remote controlled but in case of a failure of the launching network or/and of the remote-control panel the presetting and firing sequence can be performed by means of a portable presetter which has to be directly connected to the aforementioned plug. Torpedo launching can be controlled either electrically, by means of the Shipborne Remote-Control Panel (Whitehead Alenia Sistemi Subacquei production), or manually, acting on a special push-button located on each tube.

Operational status
In production. In service with the Italian and 13 other navies.

Specifications
Length: 3,400 mm
Width: 1,200 mm
Height: 1,285 mm
Weight: 1,050 kg

Contractor
Eurotorp, Sophia Antipolis, France.
Whitehead Alenia Sistemi Subacquei, Livorno, Italy.

B 515 torpedo launcher

UPDATED

ITALY

A 184 torpedo

Type

Submarine- or surface-launched heavyweight torpedo.

Description

The A 184 is a compact, dual-purpose, wire-guided, electrically-propelled torpedo equipped with an AG 67 panoramic homing head, controlling both course and depth. It is suitable for use against both submarines and surface vessels. It is a dual-speed weapon, carrying a high-explosive charge and capable of operating to considerable depths. Dedicated fire-control systems exist in different versions for submarines and surface vessels. On board a number of non-Italian built submarines, the A 184 is controlled by existing fire-control systems. The launchers are Whitehead B 512 and B 516 for submarines and surface ships respectively, although the A 184 may be used also with a large variety of 533 mm discharge systems using both swimout and positive discharge techniques.

Commands carried by the guidance wires include: course, depth, acoustic mode (active, passive and combined), enabling range, stratum allowed, speed, impact and influence fuze setting, torpedo stop. Replies from the weapon include: course, distance, depth, acoustic mode, speed and other data on interrogation. The fire-control system displays the tactical scenario, acquiring data from onboard sensors,

Type A 184 dual-purpose torpedo

and allows the underwater weapon selection, check-presetting, start and guidance against the designated targets. The fire-control system is modular, each module being capable of being used independently in case of failure of the others.

The latest development is the A 184 Mod 3 capability upgrade and life extension package which will enable the weapon to remain effective up until around 2025. The upgrade package features: wake homing and improved ECCM. A digital fire-control interface is being added, a new Ag-Zn battery and a low-noise skewed direct drive, contra-rotating propeller which will reduce self-radiated noise. The ASTRA (Advanced Sonar Transmitting and Receiving Architecture) acoustic housing head and classification software is also being integrated into the weapon.

Operational status

In production. The A 184 is in service with the Italian and Peruvian navies.

Whitehead Alenia Sistemi Subacquei was one of four companies bidding for a US contract for 2,000 torpedoes. Only the Whitehead Alenia Sistemi Subacquei contender reached the initial evaluation sea trials phase, which has been completed successfully. However, the US Navy has not requested funding for the Anti-Surface Warfare (ASuW) torpedo. Progressive upgrading of A 184 torpedoes to Mod 3 standard commenced in 1998.

Specifications

A 184 Mod 1
Length: 6,000 mm
Diameter: 533 mm
Weight: 1,265 kg
Warhead: 250 kg HBX-3 HE
Propulsion: electric, silver-zinc battery 24-36 kt
Range: 13.5 n miles at 24 kt (under wire guidance) or 9 n miles at 38 kt

Contractor

Whitehead Alenia Sistemi Subacquei, Livorno.

UPDATED

A200 LCAW

Type

Mini torpedo.

Description

The A200 Low Cost Anti-submarine Weapon (LCAW) is a self-homing, self-propelled, fire-and-forget mini-torpedo. The weapon is considered to be complementary to existing ASW weapons such as lightweight torpedoes and has been specifically designed to operate in severe environmental conditions with high levels of performance and high probability of success.

Its small size enables it to be launched from existing size 'A' sonobuoy dispensers and it can be carried and used in quantities from surface vessels and the great majority of fixed- and rotary-wing aircraft.

The LCAW is available in various configurations including the basic system, a rocket-assisted ship-launched system, air delivered system and practice round.

The A200 LCAW comprises an active seeker optimised for search, classification and homing; a directed energy charge warhead; exploder; guidance and electronics units; thermal propulsion battery;

saltwater mechanical activator; direct drive DC propulsion motor; electrically controlled fins; and a single, skew-bladed propeller.

The main features of A200 are its long range; active acquisition; high target discrimination and classification capabilities; double-hull penetration capability; and high-kill probability. In addition, it has long endurance, very shallow water operational capability and can be deployed in salvos if circumstances demand.

The ship-launched configuration features a low-cost ballistic rocket designed to deliver the weapon to the desired range by means of existing rocket launchers (if compatible), or dedicated launchers. For short-range applications the weapon can be deployed by means of compressed air tubes.

The air-launched version comprises an A200 LCAW coupled to a rotary-type air stabiliser. The device is deployed from sonobuoy dispensers or dedicated canisters. Launch altitude is between 45 and 600 m at speeds of 50 to 200 kt. Alternatively the device can be deployed from a hovering helicopter at an altitude of 15 m or more.

The practice round incorporates a suitable acoustic pinger for the localisation of the mini torpedo at the end of exercise firings; a recovery system based on an

inflatable bag; a data acquisition system based on non-volatile memories; and a synchronised acoustic transmitter designed to enable the device to be tracked within a 3-D tracking range during exercises.

Operational status

Flight trials undertaken and initial production units due to enter service 1999.

Specifications

Length (basic torpedo): 883.4 mm
Length (ship-launched): 2,036 mm (approx)
Length (air-launched): 914.4 mm
Diameter: 123.8 mm
Weight (basic torpedo): 12 kg (approx)
Weight (ship-launched): 32.7 kg (approx)
Weight (air-launched): 12.5 kg (approx)
Speed: 17 kt
Operational depth: 15-300 m

Contractor

Whitehead Alenia Sistemi Subacquei, Livorno.

UPDATED

Toso improved

Type

Heavyweight torpedo sonar.

Description

This is an improved version of the heavyweight Toso sonar, and is designed to handle multitarget situations, in an environment which features high-reverberation characteristics, high anti-torpedo countermeasures such as jamming and decoys, and capable of carrying out detailed analysis on target characteristics.

The improved Toso incorporates highly compact processing units, high-integration technology, very high

computational power, MIMD (Multiple Instructions, Multiple Data) structure, flexible motherboard technology, and standard electrical and optical interface for handling and transferring command and data information.

The main operational features are: passive and active operation modes, high ACCM capability (logical codes) search, acquisition, classification, target data extraction, extensive acoustic panorama definition (achieved through very wide horizontal and vertical coverage), and online and offline auto test.

Compared to the earlier version, the improved Toso offers an increased acquisition range in the passive mode, an increase in processing gain of more than 10

dB and improved detection capability in the presence of high noise. More than 30 per cent of the analogue technology used in the previous version has been replaced by digital technology. FFT processing technology considerably improves target discrimination and detection.

Contractor

Whitehead Alenia Sistemi Subacquei, Livorno.

VERIFIED

JAPAN

Type 73

Type
Lightweight anti-submarine torpedo.

Description
Formerly known as the GRX-4, the Type 73 is a short-range, air-launched ASW torpedo that is replacing the US Mk 46 in Japanese service. The weapon is said to be comparable to the US Mk 50, and is fitted with a SCEPS propulsion system powering a propulsor. The weapon is armed with a shaped charge warhead.

Operational status
In service with the Japanese armed forces.

Contractor
Mitsubishi Heavy Industries.

UPDATED

Type 89

Description
Under the Medium-Term Defence Programme (FY86-90) another torpedo, the Type 89 (formerly the GRX-2), has been developed as an equivalent to the US Mk 48 ADCAP weapon. Powered by a thermal propulsion system with a maximum speed said to be in the region of 70 kt, the wire-guided Type 89 uses active/passive homing and has a range of about 50,000 m.

Operational status
Entered service in the early 1990s. Operational aboard 'Yuushio', 'Harushio' and 'Oyashio' class submarines.

Specifications
Diameter: 533 mm
Speed: 55 kt
Range: 27 n miles (at 40 kt), 21 n miles (at 55 kt)
Warhead: 267 kg
Operating depth: 900 m (max)

Contractor
Mitsubishi Heavy Industries.

UPDATED

RUSSIAN FEDERATION AND ASSOCIATED STATES (CIS)

Type 53 Series

Type
Submarine/surface vessel-launched heavyweight torpedoes.

Description
The Type 53 is a long-running series of weapons that have been continually updated over the years with new technologies, as they have become available. Type designations appear to indicate (within a year or two) the date of entry into service of the weapon, thus 53-68 indicates that this weapon entered service about 1968.

The standard 533 mm torpedo of the Second World War was the Type 53-38 Type 3. This 7.27 m long weapon weighed 1,610 kg (dry) and had a maximum range of 6.5 n miles. Maximum speed was 46 kt and the torpedo was armed with a 300 kg TNT/tetryl warhead.

Post Second World War torpedo development in the former Soviet Union has generally proceeded along similar lines to that in the UK, France and the USA. This was primarily derived from experience gained with wartime German pattern running and homing torpedoes, together with intelligence acquired concerning developments in the West.
Around the late 1950s variants of this torpedo began to be deployed armed with a nuclear warhead. This was subsequently confirmed in 1981 when a Soviet 'Whiskey' class submarine ran aground in Swedish waters. The former Soviet Union regarded it as normal

Specifications

	53-56V/VA	53-65	53-66	53-68
Length (mm):	7,000	7,800	n/k	n/k
Diameter (mm):	533	533	n/k	n/k
Weight (kg):	n/k	2,100	n/k	n/k
Warhead:	400	305	n/k	20 kT
Speed (+) (max):	51	55	n/k	n/k
Operating depth:	n/k	n/k	2-14 m	n/k
Range:	2 n miles @ 51 kt	7.5 n miles @ 55 kt		
	4.5 n miles @ 41 kt	13 n miles @ 40 kt		

practice to fit alternative nuclear warheads to its torpedoes. Swedish sources estimated a warhead yield of about 15 kT.

The Type 53-56 is a straight or pattern running weapon powered by reciprocating air/steam propulsion and is mainly deployed from FACs and submarines.

The Type 53-65 is an anti-surface vessel wake homing torpedo powered by high-test peroxide (HTP). It is understood that this weapon has, at times, been armed with a nuclear warhead. The Type 53-65KE weapon is deployed in coastal defence systems covering ranges from 17 to 20 n miles within a 0° to ± 170° sector. The weapon is fully automatic in operation and is fitted with an acoustic homing system and dual electromagnetic contact inertia exploder. A turbine engine powered by a mixture of gaseous oxygen, kerosene and seawater provides propulsion. Four variants of the weapon exist including a warshot,

practice, mine carrier and simulator for training personnel on construction, maintenance and repair.

The Type 53-66 is an electrically powered, straight and pattern running weapon.

The Type 53-68 straight running torpedo is derived from the Type 53-65 and is armed with a 20 kT nuclear warhead. It is said to be capable of operating down to depths of 300 m.

The Type 53-83 HTP powered, wake-homer is the latest variant of this torpedo.

Operational status
The Type 53/56 and 53-68 are now probably withdrawn from service. Other Type 53 weapons remain in production in Russia, China and India.

NEW ENTRY

Type E40 (E40-79)

Type
Air-launched lightweight ASW torpedo.

Description
This weapon is thought to have been developed as an equivalent to the US Mk 46 based on intelligence acquired regarding the Mk 46. The weapon consists of four main sections: a nose-section housing the active/passive sonar seeker (which is said to operate on a frequency of 12 kHz) and the guidance electronics; the warhead; a centre section housing the seawater-activated battery; and a rear section with the electric

motor and fin actuators. The rear of the weapon contains a parachute retarding pack and the shrouded propulsor. The weapon is understood to be capable of acquiring a target at a range of 1,200 m. The parachute pack is discarded just before the torpedo enters the water.

The export variant of the Type 40 is the APSET-95. This weapon is said to have an unusual pointed nose shape with a seawater scoop on the underside and four clipped-tip delta fins,

Operational status
Stated to be operational on board some small destroyers and deployed by some large helicopters

and from SS-N-14, SS-N-15 and SS-N-16 missiles. The APSET-95 was offered for export in 1994.

Specifications
Length: 4,500 mm (3,840 mm APSET-95)
Diameter: 406 mm
Weight: 720 kg (APSET 95)
Warhead: 60 kg HE
Fuzing: impact and proximity
Speed: 50 kt
Range: 8 n miles @ 50 kt
Operating depth: 15-500 m

UPDATED

Type 45

Type
Helicopter/aircraft/missile deployed lightweight anti-submarine torpedo.

Description
Development of post Second World War air- and small ship-launched torpedoes started during the 1950s and 450 mm weapons, powered either by steam or solid propellant motors, emerged.

The Type E45-75A is designed to be launched by helicopters or the SS-N-14 or SS-N-16 ASW rockets. The torpedo is fitted with an active/passive homing head and is believed to be powered by an electric motor

Operational status
In service with the Russian Navy. Carried by SS-N-14 and SS-N-16 missiles.

Specifications
Length: 3,900 mm
Diameter: 450 mm
Weight: n/k
Warhead: 100 kg HE
Speed: 30 kt
Range: 8.1 n miles @ 30 kt

UPDATED

Type E53

Type
Air-launched lightweight ASW torpedo.

Description
The Type E53 is designed to be launched from fixed-wing aircraft or the SS-N-14 ASW rocket system and is powered by an electric motor. The terminal homing phase is carried out using an active/passive seeker.

Operational status
In service with the Russian Navy and deployed from fixed-wing aircraft.

Specifications
Length: 4,700 mm
Diameter: 533 mm
Weight: n/k
Warhead: 150 kg HE
Speed: 40 kt
Range: 5.5 n miles @ 40 kt

NEW ENTRY

Type 65

Type
Submarine-launched anti-surface vessel 650 mm torpedo.

Description
Development of this long-range wire guided, wake-homing weapon probably began in the early 1960s. Designed primarily for attacking large surface targets the torpedo was scheduled to arm the 'Victor II' class SSNs and entered service in 1972.

The propulsion compartment with the fuel tanks and turbine occupies most of the midships section of the weapon. Propulsion is said to comprise a closed-cycle thermal system (HTP) powering contra-rotating propellers that are not protected by a shroud. The weapon uses a passive/active seeker mounted in the nose section while the guidance section is sited to the rear of the weapon in front of the drive shaft and actuators which form the rear compartment. Guidance is carried out using the wire datalink to the launch platform with depth and course commands being continuously fed to the weapon. A re-attack capability is incorporated in the weapon's on-board guidance section. The warhead mounted behind the nose section is either a 450 kg (approximate) conventional charge detonated by impact and/or proximity fuze, or a low-yield nuclear warhead.

The export variant of the Type 65 is the DST 92. This is said to be powered by a gas turbine propulsion system using HTP, kerosene and compressed air fuel. The sensor operates in a vertical direction to detect the edge of the target's wake and then sweeps to either side until, in the terminal phase, the distance to the target can be gauged from the wake's width.

Operational status
The torpedo first became operational in 1981 and equips the later generations of Russian submarines fitted with 650 mm tubes such as the 'Victor III', 'Akula' and 'Sierra' classes.

Specifications
Length: 11,000 mm
Diameter: 650 mm
Weight: 4,500 kg
Warhead: 450 kg
Speed: 50 kt
Range: 27 n miles at 50 kt; 54 n miles at 30 kt

UPDATED

Test-71 Series

Type
Submarine/surface vessel-launched ASW torpedo.

Description
The torpedo consists of five main compartments running from fore to aft: seeker section, warhead, battery compartment (batteries are probably of the AgO-Zn type), guidance section and electric motor and actuators in the aft section together with twin propellers.

The Test 71ME export variant is a submarine-launched ASW weapon fitted with active/passive homing and with an external dispenser for the wire guidance link. This is used for depth and course control. The warhead is equipped with both impact and proximity fuzes. The weapon is stored and carried in a watertight container with nitrogen. Reported to be wakeless.

An upgraded variant of the Test-71, the USET/TE2 has been developed. This wire-guided weapon is fitted with an active/passive seeker and can be deployed by both submarines and surface ships. The weapon carries a conventional warhead of 250 kg HE. Length has been increased to 8,200 mm.

Specifications
Length: 7,930 mm
Diameter: 533 mm
Weight: 1,820 kg
Warhead: 205 kg
Fuzing: impact and proximity
Range: 10 n miles @ 20 kt or 8 n miles @ 40 kt

NEW ENTRY

Test-96

Type
Submarine/surface vessel-launched dual-role torpedo.

Description
The wire-guided Type-96 torpedo has now superseded the Test-71ME weapon for export. It is a dual-role (ASW/ASuW) weapon equipped with active/passive/wake homing sensors; the latter for use in the AsuW role.

The weapon consists of five main compartments: seeker section, warhead, guidance section, battery section, electric motor and actuators. The rear section may also internally house the wire dispenser.

Depth and course are controlled by the launch platform, with commands being sent to the weapon via the wire link. The weapon is said to be capable of attacking a surface target at an angle of 90°.

Specifications
Length: 8,000 mm
Diameter: 533 mm
Weight: 1,800 kg
Warhead: 250 kg
Fuzing: impact and proximity
Range: n/k

NEW ENTRY

APR-2E

Type
Air-launched lightweight active/passive anti-submarine torpedo.

Description
The APR-2E lightweight torpedo, said to have been developed specifically for airborne ASW, is designed to destroy current and advanced submarines at depths of 1,000 m travelling at speeds up to 62 kt.

The torpedo is of standard shape featuring a long cylindrical body with flattened rounded nose and four long stabilising fins at the rear. It incorporates a shroud at the rear, but there is no propeller as it is rocket propelled (APR). A retarding parachute pack is fitted at the rear, which is discarded on entering the water.

The weapon consists of six major elements: sonar and guidance electronics; warhead, fuzing and arming circuits; control system; rocket motor; control surface actuators; parachute brake.

The weapon is designed to strike its target outside the pressure hull in the bow section housing the sonar sensor and torpedo tubes, in the region of the propellers, or in the region of the control room.

The torpedo is fitted with a new sonar system described as a hydro-acoustic correlation-phase guidance system. This features a multi-channel detection and direction finding system, which is capable of detecting underwater targets in active mode at ranges up to 1,500 m (surface target range is stated to be 1,000 m). The sensor uses a search pattern of 90° ×45° and has a signal-to-noise resolution claimed to be 0.2 to 0.3 dB maximum and bearing accuracy 1.5 to 2°. In passive mode the sensor has a detection range of 500 m and two search patterns are available – a flat spiral search pattern for shallow water operations between 50 to 250 m and a downwards spiral search trajectory for deep water operations between 150 to 1,000 m.

Following launch the parachute brake opens and stabilises the weapon's descent until it enters the water at an angle of 17° with a minimum of noise. The parachute pack is then discarded and the weapon sinks to a depth of 20 m before the various safety devices are de-activated, the fuzing system armed and the passive search mode activated. If the torpedo reaches a depth of 150 m without detecting the target the active sonar is switched on and the weapon continues the search in active mode for a further 15 to 20 minutes. If, after this time, the target has still not been detected, the weapon self-destructs.

When the target is acquired the rocket motor is ignited and the sonar switches to active mode and locks on to the target. The torpedo then homes on to the target at speeds up to 62 kt and down to a depth of 600 m.

An upgraded variant, the APR-3 has been developed. This weapon is powered by a waterjet propulsion system with a controlled speed. The motor has a running time of 113 seconds.

Operational status
In service with the Russian navy aboard shipborne helicopters since 1992. The weapon is also being exported to countries believed to be: Poland, Romania, Syria and Serbia. India may also have acquired the APR-2E

Specifications
Length: 3,700 mm
Diameter: 350 mm
Weight: 575 kg
Warhead: 100 kg of Trotyl equivalent
Speed: 62 kts
Range: 2 n miles
Operating depth: 800 m

UPDATED

VA-111 Shkval

Type
Submarine-launched anti-submarine torpedo.

Description
Shkval was developed as a short-range anti-submarine torpedo to provide former Soviet submarines with a rapid reaction defence system against US submarines, in case the latter remained undetected by the Soviet submarine's sonar system. The concept is to provide a short-range defensive weapon which cannot be avoided by a hostile submarine and which is immune to countermeasures.

The rocket-powered torpedo is deployed from a standard 533 mm torpedo tube using a standard torpedo propulsion motor to propel it at 50 kt. When it reaches a safe distance from the submarine the rocket is fired propelling the weapon to a speed of some 200 kt. It has been reported that the exhaust gas from the rocket is fed internally to the weapon's nose where it is used to create a supercavitating bubble along the torpedo's body. Guidance is provided by a gyroscopic system.

Operational status
In service aboard Russian submarines and exported to China in 1998.

Specifications
Length: 8,200 mm
Diameter: 533 mm
Speed: 50/200 kt
Range: 5 n miles

UPDATED

SOUTH AFRICA

Advanced A44

Type
Air-launched, lightweight homing torpedo.

Development
In 1993 the South African Institute for Maritime Technology (IMT) commenced two private venture upgrade packages for the ageing US Mk 44 lightweight ASW torpedo. Although the weapon remains in widespread service, particularly with smaller navies, it is no longer sufficiently effective against modern submarines and is becoming increasingly difficult to support. The broad aim of the programme was to develop an effective upgrade for roughly 30 per cent of the cost of a new torpedo.

The initial upgrade resulted in an electronics replacement kit for the old vacuum tube electronics. While this improved reliability, maintainability and availability, it did not improve functional capabilities.

The second upgrade, referred to as the Advanced Mk 44 (A44) lightweight torpedo, has resulted in a virtually new weapon.

In 1994 it was reported that the South African government had funded the programme for two years.

Description
The A44 is essentially a Mk 44 body fitted with a new directed energy warhead, a new guidance system and new sonar sensor.

The body consists of three major compartments: nose section containing the guidance system and warhead with its firing system; centre section with battery and priming elements; and tail section with electric propulsion unit and the steering mechanism.

To ensure a smooth entry into the water the torpedo is fitted with a retarding parachute aft which stabilises the weapon during its aerial flight and which is discarded on entry into the water. The torpedo may also incorporate a cap over the nose to protect the delicate homing sensor until the weapon is in the water, when it is discarded.

The 300 mm shaped charge warhead has been designed to penetrate a 40 mm pressure hull behind a 1,500 mm water-filled void space. Prior to release from the launch platform the torpedo guidance system is fed with target distance, course to steer, initial search depth, maximum depth, attack mode and optional target depth and speed parameters. It can run circular, direct or sector attack patterns. Countermeasures against decoys include spatial filtering, a decoy-triggering pulse, and a multi-frequency sonar with multiple modes (FM, Doppler, and short pulse).

On entering the sea the saltwater battery is activated and the weapon is driven to its initial search depth. The sonar scans 98° horizontal and 19° vertical. Acquisition range is said to be better than 1,000 m in deep water, and better than 700 m in shallow water. When the target is detected the torpedo locks on and homes onto the target. The weapon probably uses a dual fuzing system with acoustic proximity and impact fuzing.

The A44 is credited with an endurance of 6 minutes.

Operational status
The 1996 trials programme is reported to have included pre-programmed guidance, target acquisition, long-range homing and target-lost algorithms, as well as static seeker sea trials for evaluating target signatures, environmental influences and shallow water effects. There is no reported production order for the South African Navy, and no known export orders.

Specifications
Length: 2.57 m
Body diameter: 324 mm
Launch weight: 196.8 kg
Warhead: 45 kg HE shaped charge
Speed: 32 kt
Range: 6,000 m
Operating depth: 10-500 m

Contractor
Institute for Maritime Technology (IMT), Simonstown.

NEW ENTRY

SWEDEN

Tp 45 (43X2)

Type
Multipurpose lightweight torpedo.

Description
This series of torpedo has been developed to meet the challenge of ASW in shallow waters. The Tp 45 (Tp 43 ×2 for export purposes) is the fourth generation of Swedish ASW torpedoes. Like previous 43-series torpedoes, the Tp 45 is wire-guided, using a hydro-acoustic homing system for the terminal guidance phase. The weapon can be launched from submarines, surface ships and helicopters (the latter flying at up to 70 kt), and can be wire-guided from a flying or hovering helicopter, being deployed without the use of a parachute.

The weapon is fitted with an onboard computer system which monitors the comprehensive volume of data which it both gathers itself and which is fed to it via the wire communication link and which makes highly accurate inertial navigation possible. The wire link enables more than 80 different types of message to be transmitted in both directions. This information controls the weapon's parameters and targeting, supervises homing procedures and so on. In the event of breakage in the wire communication, the computer takes full control of the weapon using the latest data received as

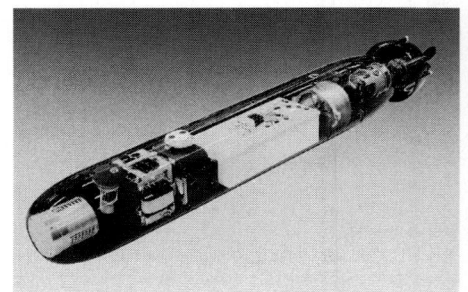

Arrangement of the 43X2 torpedo

well as computed search patterns incorporating safe/attack zones and so on.

The seeker is of new design (almost identical to that fitted in Torpedo 2000) and specifically developed for use in shallow water using advanced signal processing carried out by the computer housed in the electronics module of the weapon. Three selectable homing modes are possible: active, passive or simultaneous active/passive. The weapon can track several targets simultaneously, classifying target signals and rejecting all false signals from the environment and from acoustic countermeasures. Data generated includes among others: target presence in one or more of the preformed beams; the sign and value of the target's position in both azimuth and depth; plus the data necessary for countermeasures action.

The main computer software is written in Pascal/Ada high-level language, which simplifies software maintenance.

The torpedo is fitted with a multifrequency hydro-acoustic proximity fuze operating on several frequencies, and which can also detect the presence of targets overhead. An impact fuze is also incorporated. The transducers are positioned in a recess in the front end of the hydrodynamically shaped nose, which is designed for low self-noise at high speed. Advanced analysis of the received echoes enables waves, wakes and countermeasures from the hull of a ship to be discriminated.

Propulsion is by a secondary silver-zinc oxide 4.2 kW h battery feeding a DC electric motor with gearbox. The torpedo uses three selectable speeds (managed through a thyristor battery switching unit and believed to be 15, 25 or 35 kt) in order to optimise hit probability and minimise time from launch to strike.

The torpedo is launched from a helicopter without the use of retarding parachutes. The weapon enters the water at an angle of 15 to 35°. On entering the water the propulsion system is activated and the weapon steered towards the target from data processed on the launch platform and fed to the torpedo via the wire datalink. In the vicinity of the target the torpedo is commanded to activate its housing system for the final attack phase.

The body of the weapon is made from aluminium alloy castings and comprises interchangeable modules.

Launching a 43X2 torpedo from a Swedish Navy corvette

Work has now commenced on the development of the next-generation lightweight torpedo, together with Denmark. It is envisaged that the weapon will be deployed by the new YS 2000 surface effect ship, the next-generation submarine – submarine 2000 – and Danish 'Flyvaefisken' class vessels. The torpedo will incorporate an improved propulsion system with a wide speed range and fit a new homing system which will remain effective at great depths.

Operational status

The weapon has completed its final development phase for the Swedish Navy and a production contract worth in the region of SKr80 million was awarded to Swedish Ordnance (subsequently renamed Bofors AB) in the autumn of 1991 for series deliveries to commence in 1993. In May 1992 the Swedish FMV placed a contract worth approximately SKr100 million for a second batch of Tp 45 weapons. The torpedo is deployed from Swedish Navy Boeing Vertol 107 helicopters and the 'Göteborg' class corvettes. In the

spring of 1994 Pakistan placed an order for the 43X2 for use on the Type 21 frigates acquired from the UK.

A test firing in the autumn of 1995 using a warshot weapon against a target similar to a submarine proved the ability of the weapon to both strike and destroy a target.

Specifications
Length: 2,800 mm
Diameter: 400 mm
Launch weight: 310 kg
Warhead: 45 kg shaped charge
Speed: 15, 25 or 35 kt
Range: (approx) 10.75 n miles at 15 kt

Contractor
Bofors Underwater Systems AB, Motala.

UPDATED

Tp 46 Grampus

Type
Surface ship/submarine launched lightweight torpedo

Development
The Swedish FMV signed a project definition contract with Bofors Underwater Systems on July 1, 1996 for work to begin on the development of the next-generation lightweight torpedo, the Tp 46 Grampus (also known as Torpedo Weapon system 90). This project is being undertaken in collaboration with Denmark whose own FMV is providing part of the funding for the project. It is envisaged that the weapon will be deployed by the new 'Visby' class YS 2000 surface effect ship, the next generation submarine – Submarine 2000 – and the Danish 'Flyvefisken' class vessels.

Description
The wire-guided torpedo will incorporate an improved propulsion system with a wide speed range and fit a new homing system that will remain effective at great depths. The requirement for the weapon is to be deep diving, and be powered by a variable speed drive using a thermal battery system and AC brushless motor (with stepless speed control) and a pumpjet propulsor to achieve a maximum speed of 45 kt. Manoeuvrability is a primary objective in the development of the new

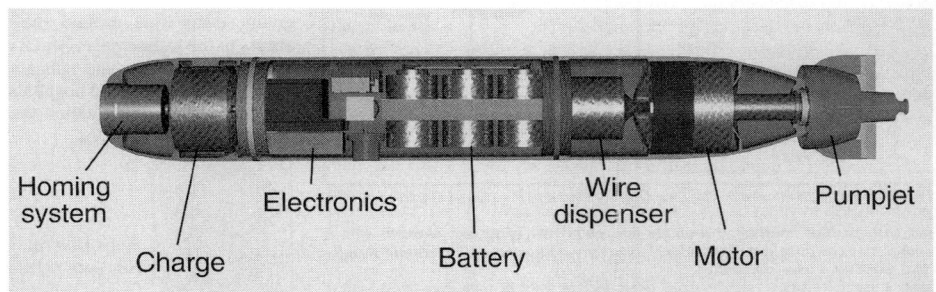
Block schematic arrangement of the Tp 46 (Bofors Underwater Systems) *1999*/0038526

weapon with emphasis being placed on slow speed, hovering and finally high-speed terminal phase. It is intended that the weapon will be able to detect submarines on the seabed and report their presence to the launch platform. Data will be transmitted via a fibre optic two-way communications link while a radio buoy will be used to relay data for air-launched weapons. The torpedo will offer very high detection and classification performance in both shallow water and open ocean environments.

Operational status
Currently under development. Full development/initial production is anticipated for early 1999 with prototypes

undergoing sea trials in 2001 and entry into service around 2005.

Specifications:
Length: c 3,000 mm
Diameter: 400 mm
Weight: c 300 kg
Warhead: PBX blast (shaped charge available as option)

Contractor
Bofors Underwater Systems AB, Motala.

NEW ENTRY

Tp 617

Type
Anti-surface vessel heavyweight torpedo.

Description
The Tp 617 is a homing, wire-guided, long-range torpedo for launching from submarines and surface ships against surface targets. It is based on the

well-proven Tp 61, which entered service in 1966 and retains the thermal propulsion system with hydrogen peroxide/alcohol/water as propellants.

The Tp 617 is divided into five main sections comprising: nose with acoustic sensor; electronics guidance section with fuzes; fuel compartment with tanks containing ethyl alcohol and HTP; motor; and afterbody with the course and depth control fins electric alternator and wire dispenser.

The electronics guidance section houses a programmable digital computer that controls the homing system, communication with the launching fire-control system and the torpedo navigation system. The guidance wire communication link allows the launch platform to transmit orders to the torpedo controlling speed, depth, course, and target data. The torpedo reports its position, speed, course and depth, homing system parameters and target noise. In the event of

Tp 617, latest version of Swedish Type 61 long-range torpedo

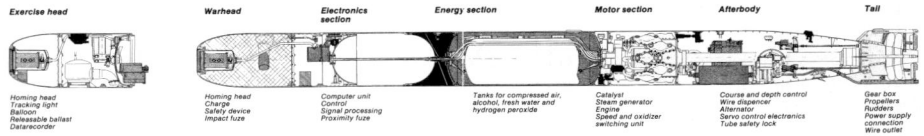

Tp 617

communications being interrupted, the torpedo computer calculates the expected target position, guides the torpedo to the predicted point and initiates one of several possible search patterns.

The fuzing system consists of a dual-impact/ computerised proximity fuze, both of which can be controlled from the launch platform's fire-control system.

The homing system is designed for use against surface targets and is mounted in a recess in the nose with the acoustic sensor using preformed beams for horizontal homing.

The 12-cylinder, double star, single-stroke steam engine is powered by superheated steam produced through the combustion of the alcohol with the oxygen from the decomposed HTP. Water is added to cool the process and to produce the superheated steam. The torpedo can be driven at one of two speeds which are preselected before launch, and can be changed according to the tactical situation during the torpedo's search and homing phase. Exhaust and steam leave the weapon through the propeller shaft leaving no visible trace as it contains 20 per cent carbon dioxide and 80 per cent steam.

Operational status
The Tp 617 has been in operational service with the Royal Swedish Navy since early 1984 and is also in service with the Danish and Norwegian navies.

Specifications
Length: 6,980 mm
Diameter: 533 mm
Weight: 1,860 kg
Warhead: 250 kg
Speed: 25 kt
Range: 10 n miles

Contractor
Bofors Underwater Systems AB, Motala.

UPDATED

Tp 62 Torpedo 2000

Type
Heavyweight torpedo.

Description
The 533 mm dual-purpose heavyweight Tp 62 is being developed for the next generation of Swedish submarines and surface vessels for use against underwater and surface targets. The weapon is powered by a new, high-powered thermal pump jet propulsion system with a very high energy content based on Bofors' long experience of High-Test Peroxide (HTP) systems. This results in a wakeless system giving the weapon high speed and long range down to considerable depths with low radiated noise level. The motor is an axial two-stroke piston engine with seven cylinders and bore and stroke of 70 mm. Admission temperature is 800°C and the engine uses a two-stage compressor with compression pistons linked to the engine pistons. The propellant comprises 85 per cent HTP and paraffin together with hydrogen peroxide (H_2O_2), the engine developing between 25 and 300 kW of power. A condenser is built around the engine and the whole unit is mounted in shock-absorbing elements in the weapon's body. The pump jet is based on the system used in the British Spearfish and developed by GEC-Marconi.

The weapon is equipped with an advanced active/ passive homing head and a wire guidance communication link. Sonar operating modes cover active, passive, simultaneous active/passive using multiple frequency and wide bandwidth. The guidance and control system is set up as a neural net processor offering advanced control and signal processing

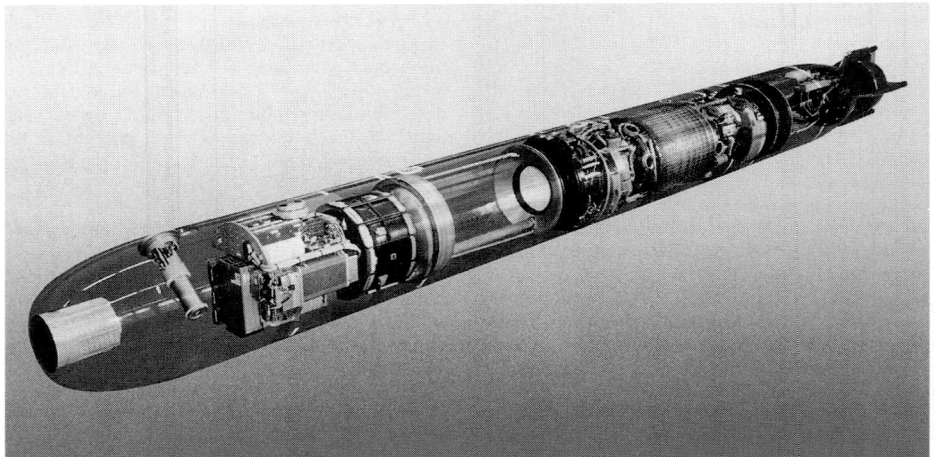

Schematic of Tp 62 (Bofors Underwater Systems)   *1999*/0038525

algorithms, which allows several targets to be tracked simultaneously and their signals classified. In the event of a communication breakdown between platform and weapon, the torpedo's onboard computer takes full command of the weapon and calculates the target's anticipated position and guides the weapon to the predicted point of impact initiating one of several possible search patterns programmed into the computer.

Operational status
An initial contract for Tp 62 was signed on 17 December, 1997.

Specifications
Length: 5,990 mm
Diameter: 533 mm
Weight: 1,314 kg
Warhead: c 200 kg
Speed: <50 kt
Range: <27 n miles

Contractor
Bofors Underwater Systems AB, Motala.

UPDATED

Torpedo propulsion system

Description
Bofors Underwater Systems has developed a new torpedo propulsion system to meet future torpedo requirements which will demand a minimum speed of not less than 25 kt, a running depth of 500 m and very low noise levels. A number of propellants has been considered and Bofors has elected to develop a system based on a bipropellant using a combination of hydrogen peroxide (HTP) and kerosene fuel. This combination has been chosen because of its high-energy content, high-density, the fact that the propellants remain liquid at prevailing temperatures and the combustion reaction products (steam and carbon dioxide) can be fed directly into a heat exchange engine, the exhaust leaving no visible wake as CO_2 is soluble in water.

By using a mixture of HTP and kerosene fuel a number of working modes can be chosen for the engine. The semi-closed cycle method has important advantages in that engine performance is not significantly influenced by the ambient pressure. The

HTP is decomposed by a catalyst prior to feeding into the steam generator, where the HTP reacts with the aviation fuel to generate a large amount of energy. By adding fresh water the temperature of the reaction products is reduced to 700 to 800°C, which is acceptable to the engine.

The exhaust from the engine mainly consists of steam and carbon dioxide at a temperature of 200 to 300°C. The steam is condensed in a condenser and most of the condensation is recycled back into the system. The remaining part of the exhaust gases is compressed in a compressor before being released into the surrounding sea.

The HTP is stored in a flexible bag located in a special tank which is directly connected to the water and kerosene fuel tanks, so that all tanks are the same pressure. A water pump supplies pressurised seawater to the water tank to provide feed pressure. The capacity of this pump is controlled by the engine speed.

The engine is a seven-cylinder axial piston engine (cylinder bore 70 mm, stroke 70 mm), the engine pistons acting against a sinusoidal double rise cam. With the selected shape of the cam curve a complete

dynamic balance of the inertial forces is achieved. The engine is designed to run at low speed so that the cam rollers of the pistons are always in contact with the cam. This design reduces engine vibration and improves stealth capabilities.

The admission gas is supplied via a central connection, distribution being controlled by inlet and outlet valves. The engine pistons are linked to the compression pistons with compression being achieved in two stages. The condenser is built around the engine which assists in dampening any engine noise, and the whole unit is mounted in shock-absorbing elements in the torpedo shell.

It is estimated that the prototype engine will produce a power of 25 to 300 kW at 600 to 1,500 rpm at an admission pressure of between MPa 2-8 and with a fuel consumption of between 3.5 and 5 kg/kWh.

Contractor
Bofors Underwater Systems AB, Motala.

UPDATED

UNITED KINGDOM

Tigerfish Mod 2 torpedo

Type
Submarine-launched torpedo.

Development
Development of Tigerfish Mod 2 was completed in 1985 incorporating homing improvements based on techniques developed in the Sting Ray lightweight torpedo and the wire guidance system is developed for the new Spearfish heavyweight development. Performance of the Mod 2 was demonstrated successfully in a series of trials which culminated in the sinking of a decommissioned Type 12 frigate. A modification kit has been developed which enables Mod 1 torpedoes to be upgraded to Mod 2 standard.

Description
Tigerfish is a 533 mm wire-guided acoustic homing torpedo capable of engaging surface and submarine targets in all operational scenarios from deep water to the littoral. The weapon is designed for minimum radiated noise and is used in passive sonar mode whenever possible to take full tactical advantage of covert operation. Tigerfish can either swim out or be ejected with a positive discharge system, making it compatible with all classes of submarine.

Wire guidance is used in the initial stages of an engagement up to the point where the torpedo's automatic three-dimensional passive/active acoustic homing system can control the run into the target. Wire is dispensed from both torpedo and submarine so as to avoid any wire stress due to their relative motion.

The torpedo carries its own computer which is connected through the guidance wire to the computer of the submarine's torpedo fire-control system. During the wire guidance phase the torpedo's computer responds to the demands of the submarine computer and during the homing run it interprets the data from the homing system sensors and calculates and commands the appropriate course.

The sonar can operate in either passive or active mode. During the attack the interrogation rate is progressively increased, as the torpedo nears the target, so as to improve system accuracy. The interrogation rate is controlled by the onboard computer, which performs several functions during this phase: interrogation control, sonar data computation, torpedo steering control and data transmission to the submarine to update its own computer memory.

Tigerfish is propelled by a powerful two-speed electric motor driving a pair of high-efficiency, low-noise, contrarotating propellers. Power for the motor is

derived from two high-capacity silver zinc primary batteries using potassium hydroxide as the electrolyte. Fast run-up gyros for directional and attitude stability provide a very quick reaction time. High or low speed is selectable at all times.

The torpedo is fitted with a dual-action fuzing system comprising an impact fuze and an all-round proximity fuze, designed to operate at the point of closest approach to the target.

Operational status
Tigerfish Mod 2 is in service with the Royal Navy. Data given below relate to the warshot torpedo; there are also exercise and dummy (handling) versions. The exercise version is similar to the warshot but has rechargeable batteries, becomes buoyant at the end of the run, and has an instrumentation pack for data analysis in place of the warhead.

Tigerfish has been released for export with sales of Mod 1 to Brazil for the Type 209 'Tupi' class submarines and sales of Mod 2 to Turkey for the Type

209 'Preveze' class. Tigerfish has been successfully integrated into a number of fire-control systems including all UK Royal Navy equipments, the KAFS system and ISUS 83.

Specifications
Length: 6,464 mm
Diameter: 533 mm (21 in)
Weight: 1,550 kg (in air)
Propulsion: electrically driven contrarotating propellers; 2 speeds
Speed: dual high/low selectable at all times, 35/24 kt
Range: 13 km at 35 kt (active), 29 km at 24 kt (passive)
Fuze: impact and proximity
Warhead: 135 kg PBXN 105 HE

Contractor
GEC-Marconi Underwater Systems Group, Waterlooville.

UPDATED

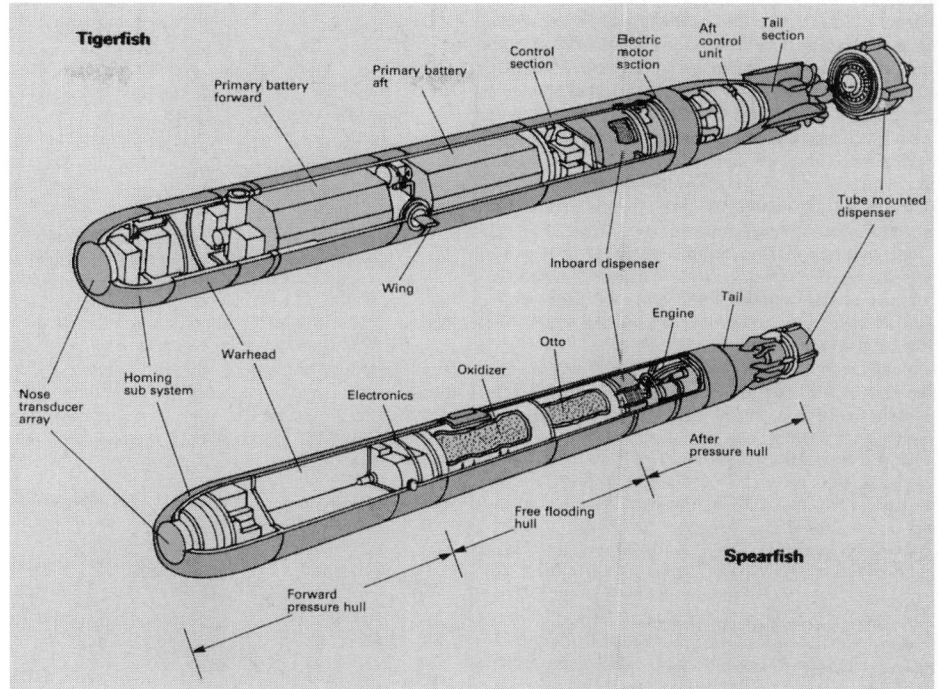
Tigerfish (top) and Spearfish (below) heavyweight torpedoes

Spearfish torpedo

Type
Submarine-launched torpedo.

Development
Following a competition between Spearfish and the proposed US Mk 48 ADCAP carried out in 1979-80, the UK MoD awarded a fixed-price contract to GEC-Marconi for full development and initial production of

Spearfish. The contract is understood to have been worth about £350 million.

Description
Spearfish is a submarine-launched heavyweight torpedo designed to operate totally autonomously from all Royal Navy submarines and interfacing with a minimum of modification to all weapon handling and fire-control systems. Spearfish meets the Royal Navy staff requirement to counter faster, deep diving, quieter and stronger-hulled submarines, as well as dealing with surface targets. Spearfish is now in full production and will form the main fleet armament of all Royal Navy submarines.

Spearfish is equipped with a Sundstrand 21TP01 1,000 hp turbine thermal engine powered by an advanced liquid fuel and oxidiser contained in separate tanks. The maximum speed, endurance and diving depth significantly exceed those of all other torpedoes, either thermally or electrically propelled and double that of other electrically propelled heavyweight torpedoes. The propulsor itself is a pump jet consisting of a rotor and stator using design techniques from the successful Sting Ray development. The thermal propulsion system has been designed to be very quiet in operation to provide the covert operation needed to protect the firing submarine from detection on torpedo launch and subsequent counterattack. Small control surfaces mounted in the efflux from the pump jet make Spearfish extremely agile.

Spearfish production warshot torpedo and associated test equipment

The torpedo's advanced sonar and homing system enable it to operate primarily in a passive mode. However, when required to operate against a very quiet target, or in the final stages of attack, the active mode is used. In this mode, the powerful transmitters give Spearfish a long detection range and enable it to burn through enemy countermeasures. The detection capabilities are further enhanced by an array offering a large search volume with frequency-agile transducers which enable salvo firing to be carried out.

Communication between submarine and torpedo is via a guidewire link. The torpedo normally operates autonomously, relaying data back to the submarine, but the command team can assume control at any time. The wire is dispensed from two reels, one in the torpedo and one in the submarine launch tube, through a tube-mounted dispenser which allows discharge at high submarine speeds as well as complete freedom of manoeuvre for the submarine after launch.

Spearfish contains a number of homing and tactical computers to control the weapon, enabling it autonomously to select search, detection and attack modes, to classify signal returns, to decide on

appropriate tactics including re-attacks on the target if necessary, and to classify, track and overcome countermeasures and decoys. The torpedo's capability can be enhanced by software updates as target and countermeasure characteristics evolve in the future.

Spearfish carries a large warhead to allow both double-hulled submarines and major surface units to be effectively engaged. Against submarines, the homing system guides torpedo to the optimum point on the target's hull. Against surface targets the torpedo detonates the warhead under the hull creating a whipping effect which breaks the target's back.

Data given below relate to the warshot torpedo; there are also exercise and handling and discharge versions. The exercise version is identical to the warshot except that the warhead is replaced by a Recovery and Instrumentation (R and I) section. This section carries a tracking system and an advanced data recorder with a large data capacity, together with a recovery system. At the end of run the torpedo, which is heavier than water, is floated to the surface by means of a toroidal bag contained in the R and I section. Developed from the

Sting Ray system, it uses cold gas (CO_2) to inflate the bag and is clean, effective and reliable.

Operational status
Spearfish is in production for the Royal Navy following the award of a £600 million contract to GEC-Marconi in December 1994. Full production is likely to be 200 to 300 weapons.

Specifications
Length: 6,000 mm
Diameter: 533 mm
Weight: 1,850 kg
Warhead: 300 kg shaped charge PPX-104 HE
Speed: >70 kt
Depth: >900 m
Range: 12.5 n miles at 60 kt

Contractor
GEC-Marconi Underwater Systems Group, Waterlooville.

UPDATED

Sting Ray torpedo

Type
Anti-submarine lightweight torpedo.

Development
Development was started in 1977 by GEC-Marconi and the first fixed-price production contract was placed in November 1979. First production weapon was delivered in August 1981 and acceptance trials began in April 1982, resulting in the issue of the design certificate in December 1982. The torpedo was officially handed over to the RAF and RN in September 1983. Fleet Weapon Acceptance trials were successfully completed in 1985 when the weapon's complete performance envelope was checked out. Subsequent to these trials an RAF Nimrod successfully launched a Sting Ray and actually sank a submarine target in the open sea. The torpedo will be updated to Sting Ray Mod 1 standard and will remain capable of countering submarine threats effectively until well into the next century.

Description
Sting Ray is an advanced lightweight torpedo designed for launching from helicopters and fixed-wing aircraft as well as from surface ships. It has a multimode, multibeam sonar, quiet, high-speed propulsion system and a fully programmable onboard digital computer, which together give the weapon a high performance not only in deep water, but also in shallow water where sonar conditions are difficult. A very considerable effort was put into gathering acoustic data in shallow water with many hundreds of trials in order to optimise the

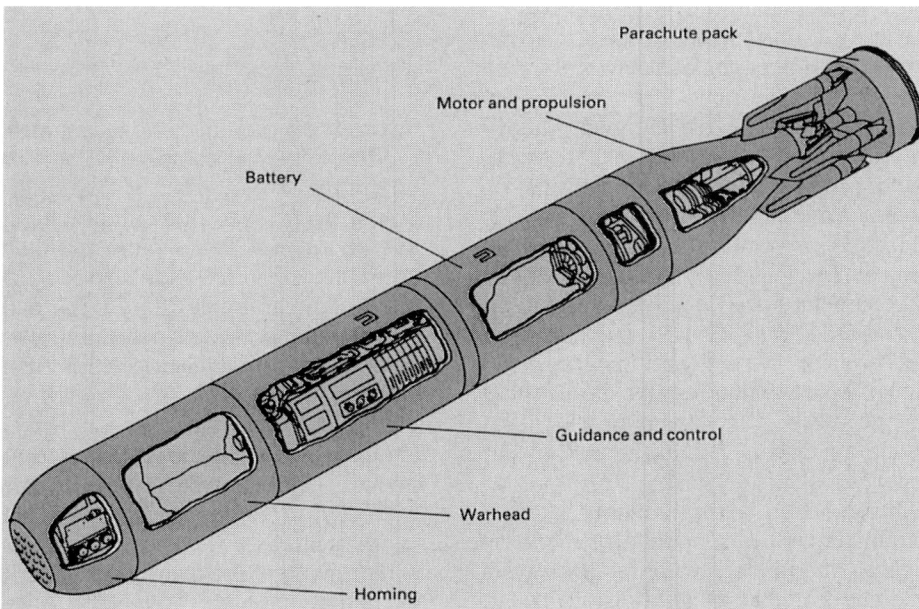

Sting Ray torpedoes on a Sea King helicopter of the Royal Navy

torpedo's tactics and software algorithms under such conditions. Its computer system enables it to make tactical decisions during an engagement to optimise the various homing modes to suit the environment and target behaviour. The computer can be reprogrammed 'through the skin' whenever updated software, or programs tailored to a particular customer's operational requirements are required. These programs then adapt themselves to the particular tactical and environmental conditions in which the torpedo operates. The high-speed performance of

Sting Ray comes from its pump jet propulsion system. Vehicle hydrodynamics are accomplished by an electrohydraulically driven proportional control system. The four control surfaces are mounted aft of the propulsor. The advanced seawater battery gives extended endurance and no performance degradation with depth. The battery uses silver chloride and magnesium alloy electrodes. A carefully controlled flow of seawater through the battery removes heat sludge and hydrogen gas, while electrolyte recirculation offsets the effects of cold water or lack of salinity.

Sting Ray carries a directed energy warhead required to counter modern submarine design. This type of warhead demands high-accuracy guidance to ensure the torpedo strikes the most vulnerable part of the target.

In operation, when the torpedo enters the water and frees itself from its parachute, it sets up a preprogrammed search pattern designed to maximise the chances of target detection. On acquiring the target, the computer and active/sonar sensors make it almost impossible for the target to evade attack. The advanced software of the guidance/homing system allows it to filter out background noise and decoys and to make an interception rather than perform a conventional tail-chase attack. It is believed to have a maximum speed of about 45 kt with an estimated endurance of more than 8 minutes at that speed. Maximum operating depth is believed to be about 1,000 m.

Data given below relate to the warshot torpedo; there are also exercise and dummy (handling, carriage and release) versions. The exercise version is identical to the warshot except that the warhead is replaced by a novel design of recovery and instrumentation section. This 'R and I' section carries a tracking system, a data recorder and a recovery system. At the end of the run the torpedo, which is heavier than water, is floated to the surface by means of a toroidal bag contained in the R and I section which is inflated by a solid-propellant

Parachute pack

Motor and propulsion

Battery

Guidance and control

Warhead

Homing

Sting Ray outline

hot gas generator. The system has been demonstrated to be effective and reliable.

Over 1,700 in-water runs have now been completed both in development and operational usage, mainly against targets, not only making it one of the most tested torpedoes in the world but also ensuring an ever expanding database of in-water and environmental data to enable continuous upgrading of its software and algorithms to be made. The upgraded Mod 1 version will incorporate new tactical and acoustic processing hardware incorporating a new digital correlator. The weapon will also be given a new sonar with new active sensor and a solid-state autopilot.

Operational status
In full production. In service with RN and RAF and the Brazilian, Norwegian, Thai and Egyptian navies. A £400 million production order for 2,500 torpedoes was signed in January 1986 by UK MoD.

On 15 December 1989 Marconi signed a contract with the Norwegian Navy and Air Force for a substantial number of Sting Ray torpedoes with a contract value in excess of £25 million. Delivery commenced early in 1991 together with the work necessary to modify the Norwegian Air Force P-3C Orions and Mk 32 launch tubes to take Sting Ray.

The UK MoD has awarded GEC-Marconi a £109 m contract to enhance Sting Ray capabilities and extend the life of the weapon. The resulting weapon, to be known as Sting Ray Mod 1, will have a new signal processing system and enhanced computer hardware and software. Improvements will also be made to the control sensors, actuation system (digitally controlled electric actuators) battery (new Mg-AgCl batteries) and motor controller including the possibility of replacing the existing mechanical correlator with a digital unit, together with a multispeed propulsion system. With these improvements shallow water performance will be enhanced. Following design and development, between 50 and 100 preproduction weapons will be manufactured prior to the main production programme. The upgrade will extend the life of Sting Ray for another 25 years from 2003.

Specifications
Length: 2,597 mm
Diameter: 324 mm
Weight: 265 kg
Warhead: 45 kg Torpex shaped charge HE
Propulsion: electrically driven, contrarotating pump jet
Speed: 45 kt
Range: 6 n miles at 45 kt
Depth: 750 m

Contractor
GEC-Marconi Underwater Systems Group Division, Waterlooville, (prime contractor).

UPDATED

Torpedo steering actuation systems

Type
Integrated steering actuation systems for torpedoes.

Description
Typically the source of power for a long-endurance control system necessary in torpedoes such as Spearfish is a hydraulic pump with a return oil system.

To cover essential control activity in the immediate post-launch period and prior to the run-up speed of the propulsion engine, a stored-pressure system gives a smooth hand-over to pump-driven operation.

The following subassemblies are required:
High-pressure pack
Low-pressure pack
Ring main
Fins (with rudder and actuator)
Electronic controller.

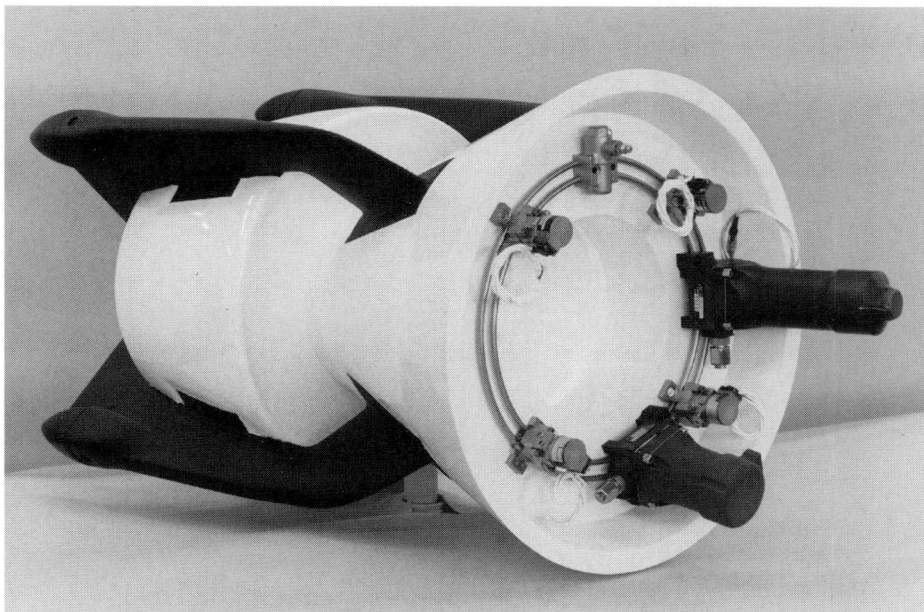

Part of FHL's electrically signalled, hydraulically powered control system for the rudders of the Spearfish heavyweight torpedo

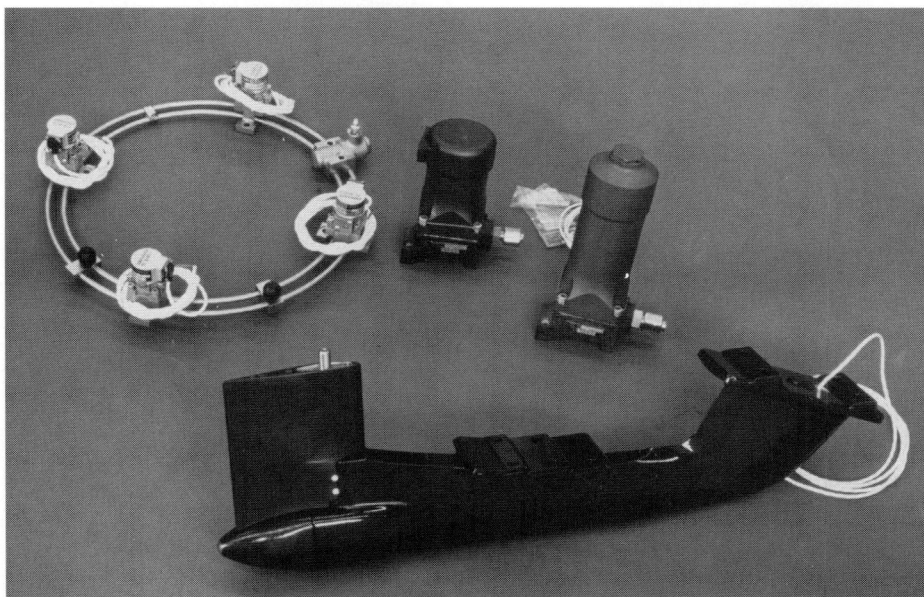

FHL hydraulic systems are specifically configured to suit the space envelope constraints in a torpedo tailcone. The ring main integrates the following essential components in an electrohydraulic system: pump, hoses (pressure and return) with quick-disconnect couplings, pressure and return ring-main pipes, relief valve, accumulator with gas cartridge release valve and filter, pressurised reservoir, electrohydraulic servo valves (four), rotary actuator with position feedback (four), and the control system's electronic package.

In electrohydraulic systems a pump, usually near the torpedo motor and remote from the ring main, is connected via hoses and quick-disconnect couplings. Particular attention is paid to pump noise levels and to mounting structure stiffness for maximum noise attenuation.

Quick-disconnect couplings enable a modularised torpedo to be broken down easily for servicing without draining oil from the system.

A ring main of concentric pressure and return rigid pipes links mounting pads to which are attached accumulator; reservoir; relief valve; and four electrohydraulic servo valves. These components are arranged around the propeller shaft giving an optimum use of space. A low-hysteresis relief valve limits the pressure peaks in the hydraulic system and discharges from the high- to the low-pressure ring main. An accumulator provides a back-up source during the starting period and when actuator demands momentarily peak above the pump output flow.

On torpedo release an electrical signal actuates a frangible valve in a sealed gas bottle. This gas immediately pressurises accumulator fluid to provide hydraulic power to the actuators via the electrohydraulic valves.

A spring-loaded reservoir maintains a positive return-line pressure at all times and absorbs fluid displacements due to accumulator discharge and fluid expansion.

Four two-stage electrohydraulic servo valves mounted on the ring main control four rotary actuators. Each valve receives an electrical signal from the torpedo guidance system via an electronic control loop-closure package.

Four rotary actuators directly coupled to the control rudders are fitted in the fins. The actuator/control rudder position is monitored through a mechanically linked rotary potentiometer which provides the position feedback signal to the electronics package. The fins are individually attached to the torpedo tailcone.

Test sets are provided to put the four-axis control system through its paces by injecting various electrical demand input signals, including steady-state, sinusoidal and other forms. Hydraulic power is used to energise the ring main of the control system under test; and an electronic section simulates the torpedo control. A microprocessor is programmed to take each control system through the complete test routine.

Contractor
FHL, Bristol.

VERIFIED

FHL system components for the Spearfish heavyweight torpedo showing a fin with rudder

Torpedo launchers

Type
Deck and magazine launchers for lightweight torpedoes.

Description
The family of Mk 32 launchers together with STWS 1 and STWS 2 comprise deck-mounted or magazine-mounted triple or twin GRP torpedo tubes. Deck tubes are mounted on a training mechanism, allowing the tube assembly to be trained outboard. Muzzle closures and anti-condensation heaters provide an all-weather storage capacity for the torpedoes. The Mk 32 family comprises a triple-barrelled (Mod 5) or twin-barrelled (Mod 9) variant.

The family of Mk 32 launchers can be supplied to fire the US Mk 46 (all variants), Sting Ray (all variants) and Italian A 244/S, torpedoes. Weapons are fired by the release of compressed air stored in a breech assembly air reservoir. Air reservoir charging terminals are installed at each launcher position and connected to the ship's high-pressure air supply.

The launchers can be interfaced with a variety of launch controllers and presetters giving the option of either automatic or local control.

Various configurations are available through the use of modular construction, including the ability to fit onto small ships which lack a combat information system. Various interfaces can be fitted allowing the system to integrate with the ship's other ASW systems including sonar.

Operational status
GEC-Marconi launchers are in service with 19 navies worldwide and can be adapted to launch all current standard lightweight torpedoes.

Contractor
GEC-Marconi Underwater Systems Group, Waterlooville.

UPDATED

Weapon handling and discharge system

Type
Positive discharge system for submarines.

Description
The design can accommodate various weapon types including:
(a) Mk 24 'Tigerfish' wire-guided torpedo
(b) Spearfish heavyweight torpedo
(c) Royal Navy anti-ship missile Sub-harpoon
(d) Mk 48 US wire-guided torpedo
(e) mines
(f) Tomahawk Land Attack Missile.

The equipment takes the weapon through embarkation, stowage, loading and launching from four-, five- or six-tube boat configurations, incorporating a variety of stowage requirements, including hull-mounted dependent or independently shockmounted stowages and integral raft structures.

The equipment provides for all the handling systems, the torpedo tubes and electronic firing equipment necessary for a positive discharge launch method, utilising one of the following processes:
(a) Air Turbine Pump (ATP)
(b) water/air ram discharge
(c) high-pressure air discharge system.

There is a growing demand for positive discharge systems brought about by the adoption of tube-launched missiles with no 'swimout' capability and the increasing need for quiet discharge.

Operational status
Positive discharge systems are in use in overseas navies, and most recently a positive discharge system has been chosen by Kockums of Sweden and the Royal Australian Navy to equip the new 'Collins' class submarine. Systems are operational with the Royal Navy, Royal Australian Navy, Canadian Forces, Brazilian Navy and Chilean Navy aboard the following classes of submarine:
(a) 'Oberon' Type SSK in service with RAN, RCN, Brazil and Chile
(b) 'Swiftsure' class SSN
(c) 'Trafalgar' class SSN
(d) 'Upholder' class SSK
(e) 'Vanguard' class SSBN
(f) 'Collins' class SSK.

A system of this type will be incorporated in the Royal Navy's new 'Astute' class SSN.

Contractor
Strachan & Henshaw Ltd, Bristol.

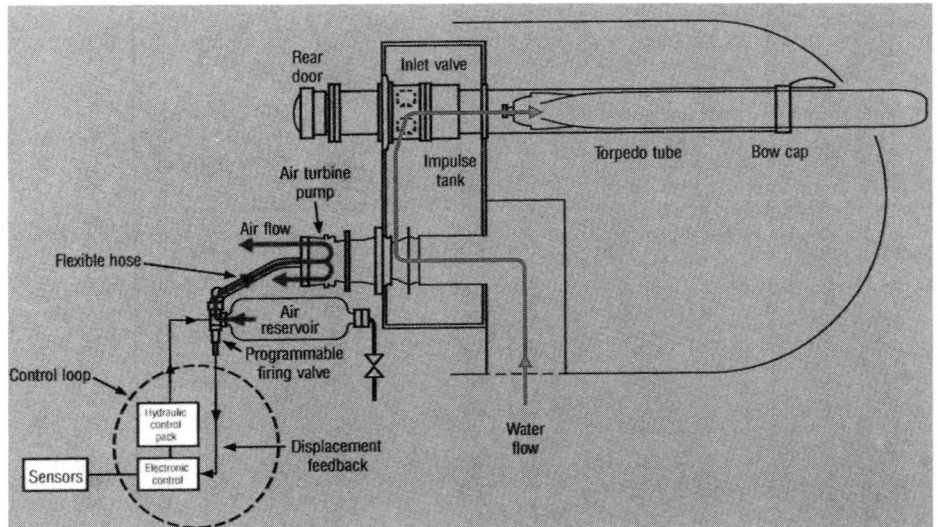

Diagrammatic arrangement of weapon discharge system

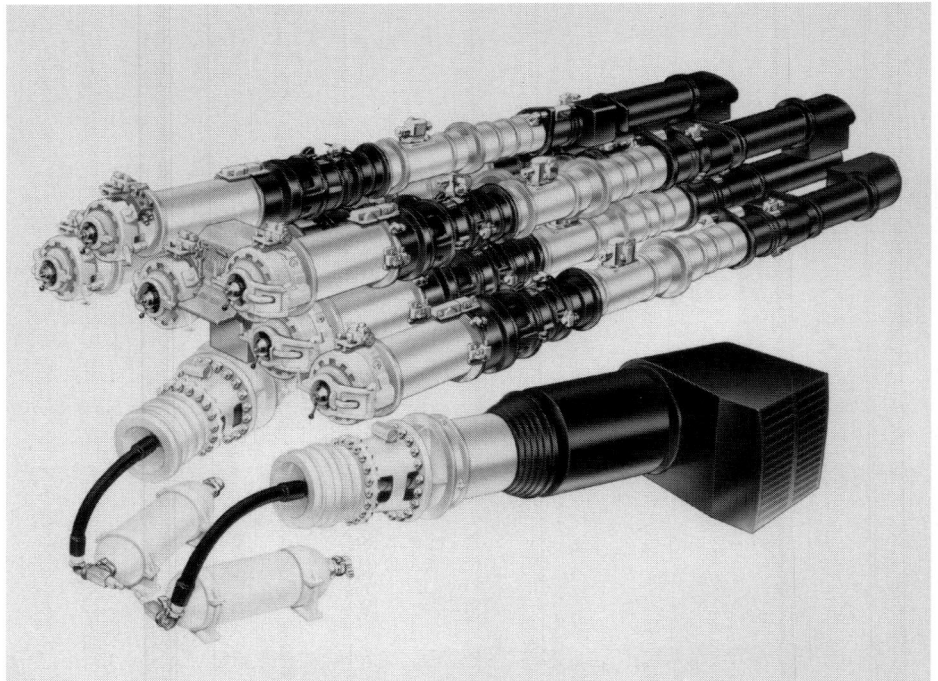

VERIFIED   *Submarine weapon discharge system*

Torpedo recovery system

Type
Torpedo recovery system.

Development
Originally designed for the recovery of the Sting Ray torpedo in a vertical floating recovery position the system has since been adapted to recover both Mk 44 and Mk 46 lightweight torpedoes, including the horizontally floating Mk 46 exercise variant. The system has been engineered for use from Sea King (RN), Lynx (RN), Puma (RAF), Sikorsky SH-3 (USN), Agusta 109, Dauphin and Bell LongRanger helicopters. The system has also been cleared by the Joint Air Transport Establishment in the UK to fly under any suitable UK military aircraft.

Description
The system comprises two elements: a recovery net and a landing platform. The conical net arrangement is suspended by a 9 m cable from the standard cargo hook on a helicopter. A remote SACRU (Semi-Automatic Cargo Release Unit) is attached to the end of the cable and is activated from the aircraft cabin by a completely independent power supply. The SACRU supports three colour-coded lines whose relative lengths are changed by SACRU release, to effect weapon capture. Auxiliary lines consisting of a rubber bungee and weak link arrangement also assist in this automatic procedure. The polyester braided net is kept in shape by means of two aluminium hoops which are coated to prevent damage to the torpedo.

The transportable landing platform is a sectioned octagonal frame of tubular aluminium across which is stretched a braided net which is kept under tension by

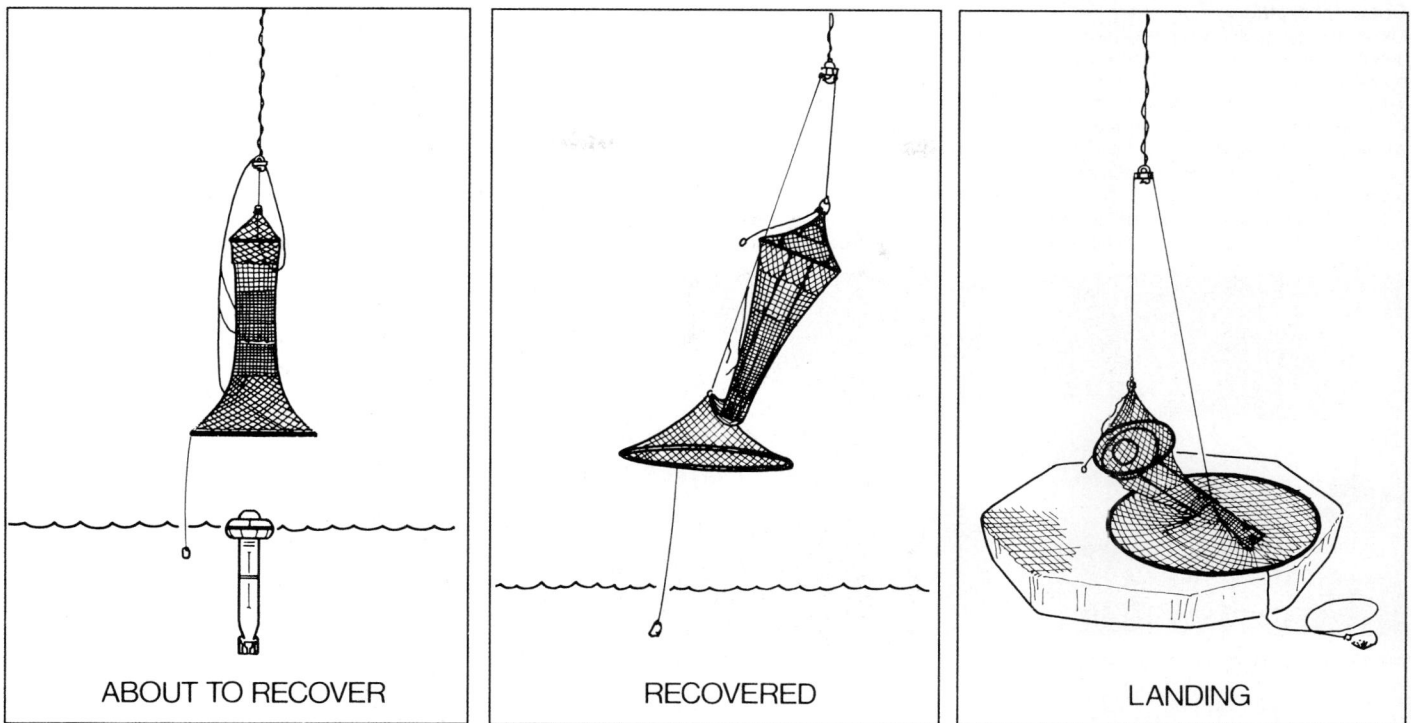

ABOUT TO RECOVER   RECOVERED   LANDING

Principle of operation for the torpedo recovery system

means of ratchet tensioners. This strong but soft platform all but eliminates the risk of damage to the weapon. Attached to the inside of the frame is a PVC sump to catch any effluent.

Weapons can be recovered by the system in up to Sea State 6. Turn round time on weapons is kept to a minimum and the requirement for personnel in the water during recovery has been eliminated.

Operation
To capture a floating weapon the aircraft hovers over the weapon until the catchment area of the bottom

hoop is approximately centred over the weapon. The aircraft and net are then lowered until the nose of the weapon is seated in the apex of the net. The remotely operated cargo hook is then released allowing the net to drape fully over the weapon. The pilot then slowly increases hover height, allowing the net's trapping lines to operate thus enclosing the weapon with the bottom hoop folded under and to one side of the net. The aircraft then lifts the net complete with the weapon from the water and proceeds to the landing site where the net is lowered onto the landing platform and released from the aircraft.

Specifications
Length of net: 5,480 mm (from apex to bottom hoop)
Bottom hoop diameter: 3,000 mm
Top hoop diameter: 1,200 mm
Weight: 94 kg

Contractor
Bridport Aviation Products, Bridport.

VERIFIED

UNITED STATES OF AMERICA

Mk 44

Type
Air/surface/missile-launched anti-submarine torpedo.

Description
This is a lightweight torpedo designed for launching from aircraft or helicopters, from surface vessels (using Mk 32 tubes), or by the ASROC rocket system.

Two models have so far been produced but the differences in dimensions are trivial. Both torpedoes are electrically propelled and their calibre is 12.75 in (324 mm). Approximate length is 2.56 m and the

torpedoes weigh about 233 kg with a 34 kg warhead. Active acoustic homing is used. Depth and course settings are entered by umbilical cable. Estimated maximum submersion depth is approximately 300 m. A range of about 5,000 m at a speed of 30 kt has been reported. Arming is by seawater scoop.

The battery uses silver chloride and magnesium electrodes with seawater as the electrolyte.

Operational status
Obsolescent. Licence production was initiated in a number of foreign countries. Replaced by Mk 46 in most navies. A number of navies still deploy the Mk 44.

Specifications
Length: 2,560 mm
Diameter: 324 mm
Weight: 233 kg
Warhead: 34 kg HE
Speed: 30 kt
Range: 2.5 n miles
Operating depth: 300 m

UPDATED

Mk 46

Type
Surface-, ship-, or air-launched lightweight torpedo.

Development
Developed as a successor to the Mk 44 in the early 1960s. Deliveries of first Mod 0 model were made to US Navy in 1965. The Mod 1 was introduced into US Navy in April 1967 followed by Mod 2 in 1972. The latest version is the Mod 5 resulting from the Near-Term Improvement Programme (Neartip), which was aimed at improving acoustic performance, countermeasure resistance, guidance and control system and fire-control system. This latest version is in full production and was introduced to combat the adoption of anechoic coatings by former Soviet submarines (Codename Clusterguard).

Description
This is a deep-diving, high-speed torpedo fitted with active/passive acoustic homing and is intended mainly for use against submarines. After water entry it starts a helical search pattern, acquires and attacks its target; if it misses the target it is capable of multiple re-attacks. Two search modes are available: snake or circle. It has a maximum speed of about 40 kt. The Mk 46 Mod 0 used a solid-fuel motor whereas the Mod 1, which is slightly lighter, uses a five-cylinder liquid monopropellant (Otto fuel II) motor. This latter propulsion system was introduced because of the maintenance problem with the original solid-fuel motor.

The latest version is the Mod 5A(S) which features a new passive/active sonar capable of detecting most types of underwater submarine target including anechoically coated hulls. The sonar offers improved target acquisition capabilities in all types of acoustic

environment, including shallow water. The Mod 5A(S) is fitted with the Mk 103 Mod 1 conventional warhead with proximity fuze. Guidance and control systems have been upgraded and feature reattack logic. The propulsion system features a two-speed capability and range and endurance have been increased. A current Service Life Extension Program (SLEP) for the Mod 5A(S) will further improve shallow water performance (improved shallow-running countermeasures and bottom avoidance) and maintainability, enabling the weapon to remain effective until about 2017. This weapon is designated Mk 46 Mod 5A(SW). This upgrade will concentrate on addressing the shallow water multiple threat scenario.

The Mk 46 can be launched by surface vessels, rocket-assisted ASROC and vertical launch ASROC (VLA), as well as by fixed-wing aircraft and helicopters.

Operational status

Mk 46 Mod 0 weapons, which were produced in limited quantities, were in use as air-launched torpedoes only by the US Navy.

By early 1975, most Mod 1 torpedoes were converted to Mod 2. Currently, Neartip implementation is being applied to all existing Mod 1 and Mod 2 torpedoes in the form of modification kits applied on a retrofit basis.

Since 1965 more than 25,000 Mk 46 torpedoes have been produced for the USN and other navies. The current SLEP programme will ensure maintainability and full support through to 2017.

Users of Mk 46 torpedoes include: Australia, Brazil, Canada, Chile, People's Republic of China, Ecuador, Egypt, France, Germany, Greece, Indonesia, Iran, Israel, Italy, Japan, Mexico, Morocco, Netherlands, New Zealand, Norway, Pakistan, Portugal, Saudi Arabia, South Korea, Spain, Taiwan, Thailand, Turkey, UK and USA.

In 1995 Alliant Techsystems (now part of Hughes Aircraft Naval and Maritime Systems) was awarded a contract with the US Navy for production of 204 torpedoes and 70 Mod 5 upgrade kits for foreign users. The first production units of the Mk 46 upgrade were due to be delivered about the end of 1996.

Hughes Aircraft has signed an MoU with Delex Systems Inc of Vienna, Virginia to provide computer simulated tactical training for the Mk 46 torpedo. This will be offered to all users of the weapon.

The Mk 46 Mod 5A(SW) is being introduced in a series of upgrades which commenced in 1996. The weapon currently features a new stern gland to correct a seawater leakage problem and a bottom avoidance capability.

Specifications

Length: 2,590 mm
Diameter: 324 mm
Weight: 231 kg
Warhead: 44 kg PBXN-103 HE
Propulsion: monopropellant
Speed: 40 kt
Range: 11,000 m (max)
Operating depth: >370 m

Contractor

Raytheon Systems Company, Mukilteo, Washington.

UPDATED

Mk 46 Mod 5 (Neartip) torpedo produced by Hughes Aircraft which succeeded the Mk 46 Mods 1 and 2

Mk 50

Type

Advanced lightweight torpedo.

Development

The Advanced LightWeight Torpedo (ALWT) (now designated the Mk 50) was developed as a successor to the Mk 46 Mod 5 (Neartip) torpedo to counter the advancing and sophisticated Soviet submarine threat. It was designed for delivery by surface ships, submarines and aircraft, and to be interoperable with existing Mk 46 launch platforms. Development commenced in the late 1970s and is now completed.

Description

The Mk 50 has similar dimensions to the Mk 46 but has improved speed (in excess of 50 kt), greater endurance, greater operating depth (1,000 m) plus improved shallow water and terminal homing, signal processing and greater destructive effect. It also features greater search volume and increased ECCM. In addition to the ability to operate against fast, deep-diving submarines there is also a requirement for capabilities against shallow, slow-moving submarines and surface ships.

For maximum effect the directed energy blast must strike normal to the submarine hull amidships. The US Navy has established a programme to improve the performance of the warhead, but for the present there will be no major upgrades.

From the two original competing contractors under US procedure A-109, Alliant Techsystems (formerly Honeywell Defense & Marine Systems and now part of Hughes Aircraft Naval and Maritime Systems) was selected in 1981 as the prime contractor for Mk 50 development and production, with the Garrett Division of AlliedSignal Corporation being responsible for propulsion. Northrop Grumman Naval Systems became second source in 1987 and is currently completing the delivery of the last purchase of 212 torpedoes.

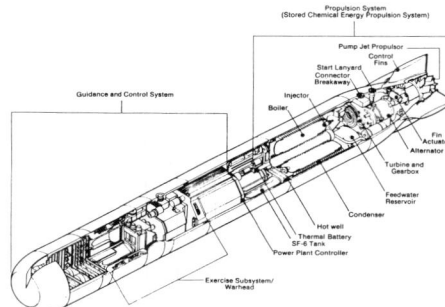

Block schematic of Mk 50

Propulsion is achieved by a closed-cycle steam turbine engine in a Stored Chemical Energy Propulsion System (SCEPS) with the energy being supplied by a chemical reaction in which solid lithium is ignited to produce an oxidant with an extreme high-energy density. This is then injected into the boiler and mixed with lithium to generate the very high temperatures needed to produce superheated steam for driving the turbine. The steam is then converted back to water for recycling through the boiler. The propulsion unit develops full power at all depths and is capable of multiple speed settings as required by the tactical situation.

Development testing of the Mk 50 terminal homing system has been carried out by using seven modified Mk 45 torpedoes as high-speed mobile torpedo targets (ADMATT). These run at 41 kt and tow a hydrophone/echo repeater array to simulate the spatial extent of a real submarine. A three-dimensional instrumented shallow water range has also been developed for Mk 50 evaluation.

The active/passive sonar comprises a low-noise nose array assembly, transmitter unit, receiver and two digital signal processing units. The sonar features a multiple beam/multiple ping capability with acquisition ranges in excess of 2,700 m. It operates with multiple, selectable transmit and receive beams. Acoustic returns are analysed in real time.

Mission control, including commands to the sonar and propulsion subsystems, autopilot navigation and target detection, classification, tracking and countermeasures, is carried out by an AN/AYK-14 standard US Navy computer. Guidance is carried out using two digital signal processors, processing sonar data to generate a range map for the AYK-14 missile computer.

Operational status

Airdrop tests were carried out in February 1987, followed by operational tests in March 1987. In June 1987, Northrop Grumman Naval Systems won a US$7.5 million initial contract to act as a second manufacturing source. Full-scale development of the Mk 50 concluded in 1992 with completion of operational evaluation trials. The Mk 50 entered service in late 1992. A total of 1,063 Mk 50 torpedoes were ordered. Production was terminated in FY96.

Block Upgrade 1 software upgrades currently in progress will enhance shallow water capability.

Specifications

Length: 2,896 mm
Diameter: 324 mm
Weight: 363 kg
Warhead: 45 kg shaped charge HE
Propulsion: closed-cycle
Speed: >50 kt at all depths
Range: >12,000 m at >50 kt 15,000 m at normal speed

Contractors

Raytheon Systems Company, Mukilteo, Washington (prime contractor).
Northrop Grumman Corporation, Electronic Sensors and Systems Division, Naval Systems, (second source) Cleveland, Ohio.

UPDATED

Mk 37 Mods 0 and 3

Type
Submarine-launched anti-submarine torpedo.

Description
Designed primarily as a submarine-launched anti-submarine torpedo, but suitable for deck-launching by the Mk 23 and Mk 25 torpedo launchers, the Mk 37 torpedo is a weapon that has been described as the first successful high-performance anti-submarine torpedo. The 483 mm diameter of all versions of this torpedo was chosen to enable the torpedo to swim out from a standard 533 mm launch tube, an arrangement with obvious operational advantages.

Mods 0 and 3 are free-running torpedoes. After one has been launched on a target interception course it maintains course until, at a preset range, it initiates a process which arms the warhead and switches in the attack logic circuits. The latter includes various preselectable options such as depth limits, search pattern and type of homing. The final attack is by sonar auto-homing which can be active, passive, or both combined. Power is provided by Ag0-Zn 36 kW batteries and a DC electric motor.

Mod 3 is an updated version of Mod 0, the updating consisting of the incorporation of a large number of minor modifications based on operational experience.

Operational status
No longer in production. The Mk 37 has been largely replaced by the NT 37 and Mk 48, although quantities of the Mk 37 remain in service with Argentina, Greece, Peru, Taiwan and Turkey.

Specifications
Length: 3,520 mm
Diameter: 483 mm with guides to fit 533 mm tubes
Weight: 645 kg (warshot); 540 kg (practice)
Warhead: 150 kg H3X-3
Speed: 24 kt (Mod 3)
Range: 11.5 n miles
Operating depth: to 270 m (Mod 3)

UPDATED

Mk 37 Mods 1 and 2

Type
Submarine-launched anti-submarine torpedo.

Description
These versions of the Mk 37 torpedo differ from Mods 0 and 3 in their size and method of guidance; in other respects they are substantially similar.

Both versions are wire-guided, whereas the Mk 37 Mods 0 and 3 are free-running torpedoes. Both are 4,099 mm long and weigh 766 kg (warshot) or 657 kg (practice).

Just as Mod 3 is an updated version of Mod 0, so is Mod 2 a version of Mod 1, updated by the incorporation of a number of minor modifications.

Operational status
Operational with the Brazilian Navy.

Specifications
Length: 4,099 mm
Diameter: 483 mm with guides to fit 533 mm tubes
Weight: 766 kg (warshot); 657 kg (practice)
Warhead: 150 kg HBX-3

Speed: 24 kt (Mod 2)
Range: 11.5 n miles
Submersion: to 270 m (Mod 2)

Contractor
Raytheon Systems Company, Mukilteo, Washington.

UPDATED

NT 37 C/E

Type
Anti-surface vessel/anti-submarine torpedo.

Development
The NT 37 heavyweight torpedo is a total redesign of the original US Navy Mk 37 torpedo. Only the original torpedo hull remains common to both. The Mk 37 torpedo subsystems, including propulsion, sonar and guidance, have been completely replaced by new hardware and software.

Additional performance enhancements include a new proximity fuze, two-speed engine and increased torpedo speed.

User navy exercises on the NT 37D model have demonstrated long-range active and passive homing attack on acoustic and live targets. The tests also demonstrated the following improvements: increased dynamic control and accuracy, complete wire guidance capability, ease of tactical software, low sonar background noise levels, low acoustic false alarm rate, long-range active and passive acoustic attacks on acoustic targets, increased torpedo reliability and reduced maintenance. A semi-automatic test set has been designed to replace manual test equipment.

Description
The NT 37 torpedo is a dual-purpose ASW and anti-ship torpedo that can be fired both from submarines and surface ships. Interfacing to a standard Mk 37 fire-control interface or a standard RS-422 bus digital fire control, the torpedo is compatible with the following platforms and fire-control systems: German 205/206/MSI-70U; German 209/HSA-SINBADS (later version); RNIN Zwaardvis/HSA-M8; or any class equipped to fire Mk 37 torpedoes.

Complete NT 37 torpedoes are available as well as upgrades to earlier Mk 37 configurations. The design is modular, allowing the upgrade to be installed in various configurations. The upgrade consists of three major subsystems:

(a) an Otto fuelled thermochemical propulsion system. When compared to the performance of an unmodified Mk 37 silver/zinc oxide battery and electric motor, the new propulsion system demonstrates a 40 per cent increase in speed, a 150 per cent increase in range and an 80 per cent increase in endurance. The propulsion system is compatible with existing Mk 46 fuelling and engine maintenance facilities, engine tools and engine refurbishment kits. A fuel bag contained in a sealed hull provides double sealing protection for the Otto fuel

(b) a solid-state acoustic system and a noise reduction laminar-flow nose assembly replaces the Mk 37's vacuum-tube acoustic panel and hemispherical nose. The new sonar substantially improves the passive detection range against high-speed surface targets and active detection range against small silhouette submarine targets; in most cases, target acquisition range has been doubled. The new self-noise reduction nose assembly increases transducer isolation while reducing flow noise effects, reducing the likelihood of self-decoying at all depths

(c) a computer-based control system upgrade eliminates the original relay logic, resulting in greater system stability, reliability and accuracy. The new system utilises an Intel 80186 embedded microprocessor, allowing fully programmable tactics. Should future naval tactics require modified torpedo command and control, changes can be implemented by revision of software and replacement of programmable read-only memory units (PROMs).

Operational status
The US Navy Mk 37 family of torpedoes was purchased by at least 17 navies. The NT 37 E version was delivered to the Egyptian Navy in 1994-95. The NT 37E (range 20 km at 35 kt) is in service with Israel, the NT 37C with Norway and the NT 37D with the Netherlands. Various running patterns are available including: (1) straight run/salvo anti-ship; (2) straight run with acoustic miss indicator to initiate acoustic reattack; (3) active snake and circle – ASW; passive snake and circle – anti-ship

Specifications
Length: 4,505 mm (Mod 2); 3,846 mm (Mod 3)
Diameter: 485 mm (fits 533 mm launch tubes)
Weight: 750 kg (Mod 2); 642 kg (Mod 3)
Warhead: 150 kg HE
Fuze: contact and proximity
Propulsion: thermochemical rotary piston cam engine with Otto fuel
Speed: 32 kt (Mod 2)
Range: 13.5 n miles (Mod 2)

Contractor
Raytheon Systems Company, Mukilteo, Washington.

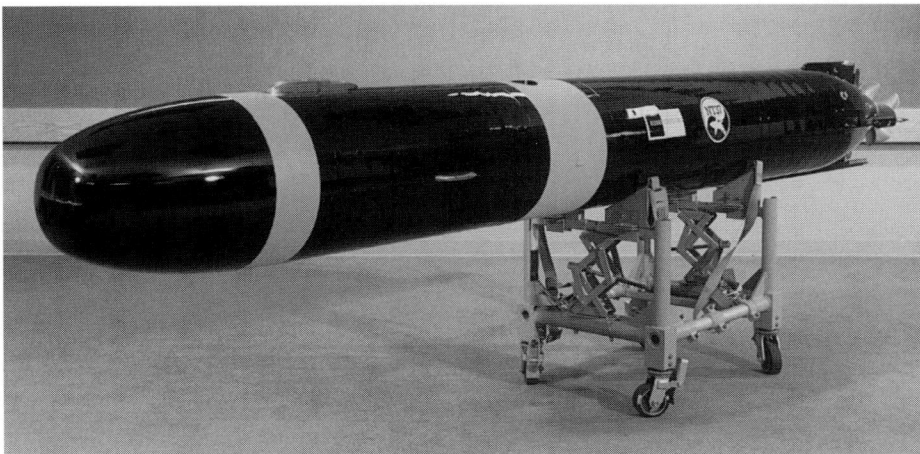

The NT 37 torpedo

UPDATED

NT 37G Seahuntor

Type
Anti-surface vessel/anti-submarine heavyweight torpedo.

Development
The next-generation heavyweight torpedo to succeed the NT 37F is the NT37G Seahuntor, developed by Alliant Techsystems (subsequently Hughes Aircraft Naval and Maritime Systems and now Raytheon Systems Company). The Seahuntor uses a two-speed Otto engine, programmable digital control logic, improved signal processing and an enlarged warhead with a new STN ATLAS Elektronik proximity fuze.

Description
The dual-role ASuW/ASW Seahuntor torpedo is designed to be deployed from the launch tubes of all known existing submarines. Being dual role eliminates the need for a submarine to carry two types of torpedo.

The torpedo consists of six main sections: nose assembly with seeker, fuze assembly, warhead, Otto fuel tank (near the centre of the weapon), processor and guidance system electronics, guidance wire dispenser (the guidance wire is dispensed through a tube at the rear of the weapon), and propulsion system.

The enlarged warhead with the advanced proximity fuze ensures a high kill probability against all known targets both surface and submerged.

The weapon is equipped with an improved thermal-chemical rotary piston cam two-speed engine driven by Otto fuel to provide the long range. The Seahuntor retains the self-noise reduction nose assembly of the NT-37F with solid-state ceramic acoustic panel (similar to the one used in earlier NT 37 and Mark 44 upgrades) for the active/passive seeker. The detection range in either active or passive mode is reported to be 200 per cent better than the Mark 37. The noise-reducing laminar-flow nose assembly reduces acoustic interference when operating at high speed or near the sea surface. The system includes adjustable detection sensitivity to compensate for high sea states and range gating and control logic to avoid decoys.

The modular software-controlled design featuring plug-in subassemblies is designed for easy and inexpensive maintenance. An automated test set can completely test the torpedo electronics in less than one day while a self-test capability ensures that the weapon is operable prior to launch.

Exercise torpedo operations are inexpensive using only Otto fuel as the propellant. A digital data recorder provides a complete record of the torpedo exercise operation.

Weapons control compatibility with existing and new submarines is ensured by the availability of both digital and analogue interfaces with the weapons control system. The torpedo interface was designed to be compatible with most versions of the German Type 209 submarines.

Operational status
Initial production deliveries began in 1996-97.

Specifications:
Length: 4,763 mm
Diameter: 433 mm (fits 533 mm launch tubes)
Weight: 868 kg
Warhead: 150 kg HE (optional 250 kg HE)
Range: 15 n miles

Contractor
Raytheon Systems Company, Mukilteo, Washington.

UPDATED

Seahuntor (S)

Type
Anti-surface vessel/anti-submarine short-length torpedo.

Description
The Seahuntor (S) has the same operating characteristics as the standard-length Seahuntor.

Repackaged electronics and smaller sensors provide a 29 per cent reduction in length. The Seahuntor short-length torpedo doubles the firing capacity of submarines originally designed to fit two Mk 37 Mod 3 torpedoes in each launch tube. Seahuntor (S) also increases the number of torpedoes that can be stored in the submarine.

Specifications
Length: 3,200 mm
Diameter: 483 mm
Weight: 650 kg

Contractor
Raytheon Systems Company, Mukilteo, Washington.

UPDATED

Mk 48 ADCAP

Type
Submarine-launched dual-purpose torpedo.

Development
Anticipation by the US authorities of impending advances in submarine technology by the former USSR, noted in the 1960s and 1970s, led to studies of the Mk 48's capabilities against likely threats and, in 1975 this resulted in an Operational Requirement issued by the Chief of Naval Operations for a programme to develop appropriate modifications to the Mk 48 torpedo to keep pace with anticipated submarine threat developments. The origins of the Mk 48 ADCAP (advanced capabilities) programme lay in this requirement, but the extent and rate of Soviet submarine technology advance hastened both ADCAP progress and another Mk 48 improvement programme.

Recognition (by the USA) of the impressive operational characteristics of the former Soviet 'Alfa'

class submarine in late spring 1979 resulted in a decision, taken in September 1979, to accelerate the ADCAP programme. It was also responsible for an intensive test and analysis programme to determine the true limits of the then current Mk 48 in terms of depth, speed and acoustic capabilities. This was known as the expanded operating envelope programme, and showed that the Mk 48 was structurally reliable at the depth needed to engage 'Alfa' class submarines, the target speed recognition capability required could be achieved, the vertical coverage was adequate as was the self-noise at higher speeds with the existing nose and array, and additional speed could be achieved. Laboratory modifications were made to a few torpedoes for tests and these changes were implemented in the form of a programme to update fleet Mk 48 torpedoes to what is now the Mk 48 Mod 4 standard.

Of the performance requirements demanded by the ADCAP programme, the most important are:
(a) sustained long acquisition range
(b) minimised adverse environment and countermeasure effects
(c) minimised shipboard tactical constraints
(d) enhanced surface target engagement capabilities.

Hardware changes involved in ADCAP entail replacing the entire nose of the weapon housing the acoustics and beam-forming circuits, and replacement of the signal processing by the latest electronics. The latter will also incorporate the current command and control electronics. Warhead sensor electronics will be improved.

Application of the expanded operating envelope programme findings to ADCAP has resulted in the upgraded ADCAP, which incorporates: upgraded acoustics and electronics; an expanded operating envelope (depth, target speed, weapon speed options); increased fuel delivery rate and capacity for optimum speed and endurance; improved surface target capabilities.

Description
The torpedo is wire-guided through a two-way communications link in the current Mod 3 version.

The Mk 48 torpedo is propelled with an axial flow pump jet propulsor driven by an external swashplate combustion gas piston engine. This engine, like that in the Mk 46 torpedo, is a Gould design. The fuel for the

engine is a monopropellant: Otto fuel II which uses nitrogen ester and an oxidant.

The Mk 48 torpedo is capable of operation in wire-guided active or passive, acoustic and non-acoustic modes. The acoustic modes of operation allow active or passive target detection capabilities. The seeker has an active electronically-steered 'pinger' which enables the torpedo to avoid having to manoeuvre as it approaches its target. A complete description of the torpedo will be found in *Jane's Weapon Systems 1987-88*.

Operational status
The Mk 48 has been in production since 1972 and is used aboard USN attack submarines and strategic submarines for self-defence. By early 1980, more than 1,900 torpedoes of this type had been delivered to the USN and an estimated 800 plus were in the production and procurement line. It was then estimated that another 1,050 might be required to meet inventory objectives and to allow for peacetime training and testing.

In August 1979, Hughes Aircraft Company received a contract for development of digital guidance and control electronics for the Mk 48 ADCAP programme.

The first test run was carried out by the USN at Nonoose Bay in early 1982, using the inertial guidance system developed for the ADCAP programme. About 240 more runs were programmed before completion of this phase of the programme, after which it was expected that entry into service would take place in 1983-84. However, the 1985 report by the US Secretary of Defense amended this to indicate anticipated deployment of the system in the mid- to late-1980s.

In the US Secretary of Defense's Annual Report to Congress in February 1985, it was stated that following completion of a successful test programme in 1984, it had been decided to accelerate production of the ADCAP torpedo. The five year programme called for production of 1,890 ADCAP units and included 123 in 1986 at a cost of US$433 million and 280 units in 1987 costing US$671 million. Hughes began production of the ADCAP version in 1985 and it became operational in 1988. In 1986 Gould (subsequently taken over by the former Westinghouse company) was named the second ADCAP source.

Hughes Aircraft Company was awarded contracts for 370 plus Mk 48 ADCAP torpedoes and related test

Data processor cards which are part of digital guidance and control system being developed by Hughes for Mk 48 torpedo ADCAP programme

equipment, and the former Westinghouse Electric Corporation a contract for 96 Mk 48 ADCAP. In June 1989, the US House Armed Services Seapower subcommittee cut US$331 million from the US Navy's US$493 million procurement request for Mk 48 ADCAP for 1990, reducing the proposed buy from 320 to 140 for 1990. At the beginning of 1989 Pentagon had approved ADCAP for full-rate production. In mid-1992 Hughes Aircraft Company was awarded a US$183 million contract to manufacture 324 Mk 48 ADCAP weapons over a five year period. This contract eliminated the second source supplier, leaving Hughes as the sole supplier of the Mk 48 ADCAP. Final

deliveries of Mk 48 ADCAP were conducted under the FY94 programme. In mid-95 the then Westinghouse company won a four year contract to retrofit 450 Mk 48 ADCAP torpedoes. The upgrade includes COTS microprocessors, improved receivers and improved quieting of the propulsion system involving the fitting of a muffling device to the motor which will be activated prior to launch. This will improve the weapon's performance in littoral regions. The Mk 48 ADCAP is to undergo a Block IV upgrade commencing in 1999.

The only known foreign users are Australia, Canada, Israel and the Netherlands.

Specifications
Length: 6,100 mm
Diameter: 533 mm
Weight: 1,814 kg
Warhead: 267 kg
Max speed: 55 kt
Range: 38 km at 55 kt or 50 km at 40 kt
Max depth: 800 m

Contractor
Raytheon Systems Company, Mukilteo, Washington.

UPDATED

HOTTorp

Type
Training torpedo.

Description
The aluminium-hulled HOTTorp has been developed to provide a low-risk device to train military personnel in the handling, preparation, launching and recovery of lightweight torpedoes. The system is compatible with all existing lightweight torpedo handling systems, launch platforms, fire-control systems and recovery equipment. The system contains no explosives or hazardous materials and requires no additional personnel, equipment, tools or power supplies. The system incorporates launch platform interfaces and hookups.

The external dimensions, weight and centre of gravity are the same as a Mk 46 lightweight torpedo, and enable HOTTorp to simulate an actual lightweight torpedo up to and including water entry characteristics.

The system is programmable at any time prior to launch, enabling it to provide fire-control preset training. A self-contained, hull-mounted liquid crystal display provides a readout of the fire-control presets for verification after recovery. The fire-control module also monitors hull integrity and battery life.

The system features a dual weight release mechanism which ensures recovery. Water impact is the primary mechanism for weight release. Upon entering the water a mechanical plunger-piston mechanism releases six 11.5 kg drop weights from the system's nose. As a back-up, HOTTorp uses the increasing static water pressure to release the weights at a depth of 11 m. The release of these weights allows HOTTorp to return to the surface and float vertically 25-30 cm out of the water. A dye container is seawater-activated to help locate the position of HOTTorp. If it is not recovered within 24 hours of launch, HOTTorp automatically scuttles to avoid becoming a shipping hazard. An optional pinger is available to aid in tracking HOTTorp on the ship's sonar.

Turnaround time on board ship is less than 60 minutes. This enables personnel to be trained more quickly and with greater frequency. On board the turnaround cycle includes a complete washing with fresh water, removal and replacement of the flooding valve and dye container, testing of the vacuum, releasing the piston and installing new drop weights, recaging the gyro, inserting a seawater lanyard and installing any launch accessories.

Designed for a minimum of 150 launches, HOTTorp requires minimal maintenance, with preventative

maintenance based on an annual basis or after 10 launches.

The system's shallow plunge depth enables its use in all operating environments. It will not exceed 23 m when launched from a torpedo tube, or 37 m when launched from rotary- or fixed-wing aircraft.

Operational status
Believed to be in limited production for the US and other navies.

Specifications
Length: 2,591 mm
Diameter: 325 mm
Weight: 230 kg
Weight (in water): 52.36 kg

Contractor
Raytheon Systems Company, Mukilteo, Washington.

UPDATED

Mk 54 LHT

Type
Lightweight surface ship, aircraft-launched ASW torpedo.

Development
As a result of its reassessment of undersea ASW weapon requirements for the next century, the US Navy has noted a significant shift in the most likely threat scenarios. The current generation of lightweight torpedoes were specifically designed and built to counter high-speed Russian SSNs operating in blue water, open ocean environments. Studies now indicate that the main threat in the future will come from SSKs operating in shallow water, littoral areas.

To face this threat the US Navy is embarking on a new programme referred to as the Lightweight Hybrid Torpedo (LHT now designated as Mk 54, formerly Mk 46 Mod 8). This programme is designed to maximise on the technologies developed for the Mk 50 and Mk 48 ADCAP programmes together with proven Mk 46 warhead, the Mk 103, and propulsion subsystems (modified to incorporate the thermal battery and dual-winding alternator). This is designed to provide an effective low-cost weapon system optimised for littoral warfare.

Description
The LHT will integrate components from various existing weapons (Mk 46, Mk 50 and Mk 48 ADCAP)

with COTS signal and tactical processors and components. It will use the Mk 50 array and transmitter; a new digital receiver, signal processor, tactical processor and depth sensor using COTS technology with open system architecture software; the Mk 46 warhead and propulsion system; the Mk 50 forebody; Mk 48 ADCAP and Mk 50 software. On the propulsion side the variable speed fuel control valve used in the Mk 48 ADCAP will be adapted for the LHT, together with the same OTTO fuel-based motor and the dual-winding alternator from the Mk 50, together with the Mk 50 thermal battery for quick start-up.

The primary new components will be the receiver, signal processor and tactical processor, which will be housed in the nose assembly. The new COTS-based processors will provide memory capacity and processing speed necessary to run extremely complex software algorithms. This is necessary to overcome the problems of distortion in return sonar signals caused by the turbulence from surface waves, bottom scattering and incidental acoustic interference from marine life and surface craft. A further complication in detecting SSKs is the fact that it is extremely difficult to make use of Doppler signals because of their very slow speed (6 to 8 kt). Detecting slow-moving submarines requires extremely sophisticated processing techniques such as Frequency Modulation (FM) to provide a verifiable acoustic image of the target.

Successful in-water tests conducted between January and November 1994 by Alliant Techsystems (now part of Hughes Aircraft Naval and Maritime

Systems) and the US Navy have validated the LHT concept. These trials, using a Control and Acoustic Test Vehicle (CATV), demonstrated the feasibility of integrating the Mk 50 nose array, Mk 46 propulsion system and ADCAP fuel valve. Mechanical electrical and software interface between the Mk 46 afterbody and the Mk 50 forebody were verified. Three CATV systems were delivered to the US Navy for in-water tests.

Operational status
In June 1996, Hughes Aircraft was awarded a competitive contract to develop and deliver 31 LHT engineering development models to support development and operational testing of LHT beginning in mid-1998. Production is scheduled to commence around 2001 with entry into service scheduled for 2003. A total inventory of 1,000 warshots is projected.

Specifications
Length: 2,590 mm
Diameter: 324 mm
Acquisition range (in deep water): 2,000 m
Sonar frequency: 25 kHz (approx)
Bandwidth: 10 kHz
Speed: 28 and 36-43 kt

Contractor
Raytheon Systems Company, Mukilteo, Washington.

UPDATED

TABLE XIII

Operational torpedoes[12]

Model	Manufacturer	Country	Role	Dimensions[1]	Guidance	Weight[2]	Charge[2]	Range[3]	Speed[4]	Depth[1]	Operators
A 184	Whitehead	Italy	Dual	6.00 ×0.533	wire-active/ passive	1,265	250	13.5/9	24/36	n/k	Italy, Peru
A244/S	Eurotorp	Internat	ASW	2.75 ×0.324	active/passive	221		>3.5	37		Singapore, Sweden+others
APR-2E	RFAS	RFAS	ASW	3.7 ×0.35	active/passive	575	100	2	62	800	RFAS, Poland, Syria, Romania, Serbia
DM 1[5]	STN ATLAS	Germany	ASW	4.15 ×0.533	wire-active/ passive		100	6.5/3.3 [13]	23/35 [13]		Germany
DM 2[6]	STN ATLAS	Germany	ASV	6.08 ×0.533	wire-active/ passive	1,370	260	15/7 [13]	23/35 [13]		Germany
DM 2A3[7]	STN ATLAS	Germany	Dual	6.08 ×0.533	wire-active/ passive	n/k	260	15/7	23/35	n/k	Germany, Norway
DM 2A4	STN ATLAS	Germany		6.08 ×0.533							Germany
E14	DCN	France	ASV	4.27 ×0.55	passive	927	300	3	25	18	Portugal, Spain, South Africa
E15	DCN	France	Dual	5.9 ×0.55	passive	1,387	300	6.5	31	25	Portugal, South Africa, Pakistan, Spain, Uruguay
F17 Mod 2	DCN	France	Dual	5.4 ×0.533	wire-active/ passive	1,428	250	15.5/10.8[13]	24/40[13]	600	France, Pakistan, Spain, Saudi Arabia
Impact	Eurotorp	Internat	ASW	3.00 ×0.324	active/passive	300	n/k	>10	29/50	>1,000	France, Italy, Germany
MU90											
L3	DCN	France	ASW	4.3 ×0.55	active	910	200	3	25	300	Portugal, Uruguay
L5	DCN	France	Dual	4.4 ×0.53	active/passive	1,300	150	5	35	555	Belgium, France, Pakistan, Spain
Mk 37/1-2		USA	ASW	4.09 ×0.483	wire-active/ passive	766	150	11.5	24	270	Argentina, Brazil, Turkey, Peru, Venezuela
Mk 46	Raytheon	USA	ASW	2.59 ×0.324	active/passive	231	44	11	40	>370	Australia, Brazil, Canada, Chile, China, Ecuador, Egypt, France, Germany, Greece, Indonesia, Iran, Israel, Italy, Japan, Mexico, Morocco, Netherlands, New Zealand, Norway, Pakistan, Portugal, Saudi Arabia, South Korea, Spain, Taiwan, Thailand, Turkey, UK, USA
Mk 48 ADCAP	Raytheon	USA	Dual	6.1 ×0.533	wire-active/ passive	1,814	267	43/33 [13]	40/55 [13]	800	Australia, Canada, Israel, Netherlands
Mk 50	Raytheon	USA	ASW	2.896 ×324	active/passive	363	45	10	>50		USA
NT37		USA	Dual	4.50 ×0.485	wire-active/ passive	750	150	13.5	32		Canada, Egypt, Norway, Netherlands, Israel
SAET-40	RFAS	RFAS	Dual	4.5 ×0.40	active/passive[8]		100	5.4	30		
SAET-60	RFAS	RFAS	ASV	7.8 ×0.533	passive			400	8.1	40	Albania, Bulgaria, China, India, North Korea, Libya, Poland, Syria
SET 53	RFAS	RFAS			active/passive			400	8.1	45	Egypt
SET-65E[10]	RFAS	RFAS	ASW	7.8 ×0.533	active/passive			205	8.1	40	Cuba, India, Libya, Poland, RFAS, Yugoslavia
SET-92K[11]	RFAS	RFAS	ASW	7.8 ×0.533	active/passive						
SST 4	STN ATLAS	Germany	ASV	6.08 ×0.533	wire-active/ passive	1,263	260	6.5/15[13]	35/23[13]	n/k	Argentina, Turkey, Venezuela
SUT	STN ATLAS	Germany	Dual	6.62 ×0.533	wire-active/ passive	1,420	260	15/7[13]	23/35[13]	n/k	Chile, Greece, India, Indonesia, South Korea, Taiwan
Seahuntor	Raytheon	USA	Dual	4.76 ×0.433	active/passive	868	150	15			
Shkval	RFAS	RFAS	ASW	8.00 ×0.533		n/k	n/k	5	50/200	n/k	RFAS, China
Spearfish	GEC-Marconi	UK	ASW/ASV	6 ×0.533	wire-active/ passive	1,850	300	12	60	>900	UK
Sting Ray	GEC-Marconi	UK	ASW	2.597 ×0.324	active	265	45	6	45	750	Egypt, Norway, Thailand, UK
Test-71[9]	RFAS	RFAS	ASW	7.9 ×0.533	wire-active/ passive[14]	1,820	205	8	40		Algeria, China, India, Iran, Poland, Romania, Yugoslavia

Model	Manufacturer	Country	Role	Dimensions[1]	Guidance	Weight[2]	Charge[2]	Range[3]	Speed[4]	Depth[1]	Operators
Test 96	RFAS	RFAS	Dual	8 ×0.533	wire/wake active/passive	1,800	250				
Tigerfish	GEC-Marconi	UK	Dual	6.46 ×0.533	wire-active/ passive	1,550	135	7/15.7[13]	35/24[13]		Brazil, Turkey, UK
Type E 40	RFAS	RFAS	ASW	3.8 ×0.4	wire-active/ passive	720	60	8	50		RFAS
Type 41	Bofors	Sweden	ASW		passive	45	10.8	25			Denmark
TP 45	Bofors	Sweden	ASW	2.85 ×0.40	wire-active/ passive	310	45	10.75	15/25/35		Sweden
Type E 45	RFAS	RFAS	ASW	3.9 ×0.45	active/passive		100	8.1	30		RFAS
Type E53	RFAS	RFAS	ASW	4.7 ×0.533	active/passive		150	5.5	40		RFAS
Type 53	RFAS	RFAS	Dual	7 ×0.533	pattern-active/ passive	400	10.8	4.5	51		RFAS
Type 53-65	RFAS	RFAS	ASV	7.8 ×0.533	passive wake	2,100	305	13	55	2-14	Algeria, India, China, Iran, Libya, Poland, Romania
Type 65	RFAS	RFAS	ASV	11 ×0.65	wire-passive wake	4,500	450	54/27	30/50		RFAS
Type 61	Bofors	Sweden	ASV		wire-passive		240	13.7	45		Denmark, Norway
Type 613	Bofors	Sweden	ASV		wire-passive		250	8.2	45		Sweden
TP 617	Bofors	Sweden	ASV	6.98 ×0.533	wire-active/ passive	1,860	250	10	25		Denmark, Norway, Sweden
Type 89		Japan		×0.533	wire-active/ passive		267	27/21 [13]	40/55[13]	900	Japan
TP 62	Bofors	Sweden	Dual	5.99 ×0.533	wire-active/ passive	1,314	200	c32.5	c50	n/k	
n/k	IPqM	Brazil						9.7	45		Brazil

ASV Anti-Surface Vessel.
ASW Anti-Submarine Warfare.
[1] Dimensions of length and diameter and operating depth are in metres.
[2] Total weight and explosive charge in kilograms.
[3] In nautical miles.
[4] In knots.

[5] Seechlange.
[6] Seal.
[7] Seehecht.
[8] Export version Uset-95 may have ASV wake-homing option.
[9] Export version Test-71 superseded by torpedo promoted for export as Test-96.

[10] Domestic designation ET80-67.
[11] Supersedes ET80-67 as an export variant.
[12] Details provided where available. Operators listed only where confirmed.
[13] Active/passive.
[14] Wire guidance optional.
n/k not known

UPDATED

MINES

Mines

One of the most cost-effective forms of naval warfare is the mine. Mines are small, easily concealed, cheap to acquire, require virtually no maintenance, have a long shelf life, are easy to store in considerable numbers and can be laid easily and simply from almost any type of platform, which need not necessarily be a military platform. They can be used both strategically and tactically to deny waters to hostile forces and to defend high-value targets such as ports, anchorages and offshore structures from amphibious or seaborne attack, and can very quickly wipe out or very seriously impair the effectiveness of surface forces. To counter and neutralise the mine requires an effort out of all proportion to its size. In short, the mine is probably one of the most deadly weapons that any navy can deploy in its armoury.

Moored mines

The moored mine has changed little from its forbears of the First and Second World Wars. Although a relatively unsophisticated weapon, it remains, nevertheless, an extremely effective one. Being simple to manufacture and relatively easy to lay (all that is necessary is a set of rails and a reasonably accurate navigation system which can be installed quickly on a wide variety of available ships), an extensive minefield will require a large force of minesweepers to sweep the field, an exercise which may take considerable time. The moored mine is therefore an ideal weapon for defensive purposes. A moored minefield can be used to protect extensive areas of vulnerable coastline and important port areas and anchorages. It can be used to create specified navigation channels leading to important areas such as ports; channels which can be rapidly closed using ground mines in face of impending attack.

Because of the nature of the weapon, the modern moored mine can be laid in much deeper waters than the ground mine (that is, beyond the limits of the continental shelf). The moored mine is therefore eminently suitable for use in a barrier system, denying passage in open waters. Although post-Second World War developments concentrated on the ground or influence mine, the moored mine has again come into prominence in new forms for use in submarine barriers in deep waters and to control choke points.

While many moored mines are of the older horned variety which are detonated by physical contact, the new moored mines for deep water barriers are detonated by magnetic or acoustic influence. These new types of moored mine can be countered using deep sweeps towed between two minesweepers. Naturally the mines incorporate anti-sweep devices such as time delays and snag lines.

The normal contact mine is usually triggered in one of three ways:
Contact – hydrostatic
Contact – mechanical
Contact – chemical.

The moored mine comprises the mine itself coupled to an anchor box for laying. Mine and anchor are laid together by the minelaying ship. On reaching the seabed the mine is released from the anchor mechanism and rises on a cable attached to the anchor to a preselected height, determined in advance from operational requirements. The body of the mine is made of steel or GRP and contains the explosive charge, fuzes, detonators and other elements (such as influence devices, batteries and security devices) required for the correct functioning of the weapon.

Seabed mines

Seabed mines are detonated using one of three influences created by a target: acoustic, magnetic or pressure signature, or a combination of these influences. Influence mines are technologically the most sophisticated of all types of mine, and their reactivity to the specific type of influence on which they operate is infinitely variable. So sophisticated is the technology associated with these types of mine that they literally can be programmed to react not only to a specific class of ship, but, where sufficient data is available, even to a named ship within that specific class.

Unlike the moored mine, the seabed mine houses all its operating equipment within the steel or GRP casing, which is usually cylindrical in shape. A cylindrical shaped mine has obvious advantages in that it is compatible to laying from a wide variety of platforms, ranging from submarine torpedo tubes to surface vessels of all types and sizes to aircraft. However, cylindrical shaped objects reflect fairly definite sonar patterns or shadows which can be identified by a minehunter. To improve stealth capabilities and make the task of mine neutralisation more difficult, some seabed mines employ specific 'shaping' which diffuses the acoustic shadow or return to an indefinable pattern. Sonar patterns are further confused by the use of anechoic coatings.

For use on soft or sandy beds, some modern ground mines have the capability to bury themselves in the seabed, and are virtually undetectable to all but the most sophisticated sonars. Such mines, however, are small in size, with a small explosive charge and are better suited to very shallow water operation such as in estuarial areas or in areas where hostile submarine activity is anticipated fairly close to the coast or in specific navigable channels.

The two principal types of influence are the acoustic and magnetic. The acoustic influence can be programmed to react to specific types of ship's signature such as propeller cavitation, engine noise, hull cavitation or shaft revolutions and even imperfections in this regular sound pattern, when these are known.

Magnetic influence mines react to changes in the ambient magnetic field and can be fuzed to react to magnetic polarity in either a vertical plane or fore and aft or athwartships in a horizontal plane on a target, and to specific fields and strengths, reacting to specific parts of a ship such as the machinery area or stern area.

In addition to the multitude of fuzing refinements available, triggering mechanisms can also be controlled with activate/deactivate devices, and various types of delay, counters and time responses.

The other principal type of influence used is the pressure influence. As a ship moves through the water an area of low pressure is created beneath the hull which varies according to the speed and draught of the vessel. This pressure change can be used in a number of ways to operate the pressure sensing device in the mine, the decrease in pressure opening and completing an electrical circuit to detonate the mine. Because of the method of operation, mines incorporating this type of influence certainly cannot be swept using normal methods (as can some magnetic and acoustic influence mines) and have to be 'hunted' using sonar and neutralised using either clearance divers or remote-control submersibles.

Tethered mines

Tethered mines are specifically reactive to the approach of the target, which is principally the submarine, as distinct from the normal type of moored mine which remains completely passive until struck by the victim.

The most common form of tethered mine is known as the 'rising mine'. Both the US and the former Soviet Union have developed such weapons. Most tethered systems rely on passive detection of an approaching target to activate a powered, mobile homing system carrying the explosive charge.

Tethered mines are very much a barrier weapon to prevent passage of hostile shipping and to control major choke points.

The other major type of tethered system comprises a mobile homing system such as an acoustic torpedo tethered to the seabed in depths below 3,000 m. The system is linked to a static passive acoustic sensing system (such as the US SOSUS system) which detects, analyses and classifies underwater sounds, and in response to preprogrammed target signatures activates the release of the homing torpedo which then reacts in its normal way to the approach of the target.

Controlled mines

These were, in fact, the very earliest type of mine ever devised. They simply consisted of a series of large explosive charges laid on the seabed and connected by electric cables to a position on the shore overlooking the minefield. As the target was observed passing over the position of the mines they were detonated by an observer in the shore station.

Such systems using modern explosives and technology are still valid today and indeed some navies do operate such controlled minefields.

However, more sophisticated means are being developed to provide the remote activation of minefields. Experiments in this area have been geared towards the control of deep water minefields controlling principally acoustic influence mines. This is necessary due to the fact that while a mine remains in water its casing acts as home to a variety of marine life which affects the sensitivity of the acoustic transducers. Consequently, the longer a mine remains in water, the less sensitive the transducers.

Furthermore, there is a finite limit to the life of the batteries which power the mine's various electronic systems. If the mine can be remotely activated at intervals well beyond the capacity of any inbuilt timing device in the mine then it could remain dormant in the water for considerable lengths of time, its onboard sensors all being activated by a simple remote command. Naturally any such remote activating system must be proved to be almost infallible before any system relying on this method is deployed. This is essential if any such system is to be deployed before active hostilities.

Remote control of such minefields will, in all probability, rely on coded VLF transmissions which are the only satisfactory means of long-range underwater communications at depth. Such a communications system, however, requires very high power and would most probably be shore-based.

Sensors

Traditionally, the main sensors used to detonate mines (apart from the contact variety) have been magnetic and acoustic with, to a lesser degree, pressure. The pressure sensor has not quite achieved the same prominence as the acoustic and magnetic sensors, as it is far more difficult to control and creates enormous problems in clearance, not only for hostile forces but for friendly forces too. However, as greater control is exercised over platform signatures through the incorporation of modern stealth techniques, so mine designers have sought to employ newer forms of sensor; both to make the mine more effective and to reduce the effectiveness of mine countermeasures forces. Hence the development of electric field (UEP and ELFE) and seismic sensors.

Acoustic

Acoustic signatures are notoriously unreliable influences to use in mine warfare for they are subject to the vagaries of multipath interference and high ambient noise resulting from the existing sea state, marine life and the increasing volume of marine traffic now using the oceans; all of which mask the actual acoustic signature of the target. The sensor is also dependant on the propagation conditions existing in the area where the mine is laid. These factors create more of a problem for mine countermeasures forces than they do for the mine's sensor when used to detect an oncoming target. However, they do affect the effectiveness of the mine. This is usually overcome by combining the acoustic sensor with a magnetic sensor which is less susceptible to outside interference. Modern acoustic sensors for mines are usually constructed of piezoelectric materials which detect the acoustic signature which is then fed into a low noise amplifier with a response ranging from a few Hz up to several kHz. The problems associated with detecting ship acoustic signatures are overcome by the signal processing incorporated in the mine, in which the minute electric signals are changed into the frequency domain using a Fast Fourier Transform engine before undergoing processing from which tonal signal information can be extracted for analysis.

Magnetic

The traditional magnetic sensor is based on an induction coil (usually referred to as the coiled rod) which responds to the rate of change in the magnetic field surrounding the mine rather than the actual magnetic field. It usually operates in a single axis of detection with a low sensitivity. Modern ground mines, on the other hand, usually incorporate a three-axis fluxgate magnetometer which incorporates a filter which separates into several components the earth's static field, the target's static field and alternating fields in the immediate vicinity of operation. The magnetometer is much smaller and more highly sensitive than the induction coil, with a much lower self-noise signature. Unlike the acoustic sensor, the magnetic sensor is less influenced by outside interferences for their strength is less than the earth's magnetic field. This makes it easier to filter out the unwanted magnetic anomalies thus greatly assisting in the signal processing to reduce the effect of non-coherent signals.

Generally static fields of targets are detected in the 5 mHz to 5 Hz band and for undegaussed platforms the signal-to-noise ratio is usually of sufficient strength for targets to be detected at medium range. However, with degaussed vessels the signal-to-noise ratio is lower and the form of the signature is very different, making it much more difficult to detect a target with consequent reduction in range of detection. With degaussed platforms, therefore, it is necessary to use a different type of magnetic field for detection, and this is usually the alternating magnetic field created by a target's onboard electric generating and other electrical systems. Other alternating fields are created through corrosion currents (that is, a rusty ship would exhibit a higher alternating field in this region than a newly built ship) eddy currents created by rotating machinery and magneto-hydrodynamic fields generated by the wake of the ship. These fields are all of low intensity, but as they are at a higher frequency, typically ranging from 5 to 500 Hz, the modern sensor with its high sensitivity is more easily able to detect them against the lower background magnetic noise.

Pressure
The modern pressure sensor is based on piezoelectric materials which are used to detect a reduction in pressure caused by the passage of a vessel over or near a mine. These signals are extremely small, but modern electronics used in signal conditioning are able to detect the small target signals against the much larger signal created by tidal movement and the static water head above the mine. Modern signal processing enables the system to compare short and long term averages of the signal and to measure the duration of pressure pulses created by the passage of the bow and stern of a target through the water, thus further enhancing target discrimination against other extraneous signals. Other pressure sensors operate on a strain gauge principle in which the deformation of a diaphragm exposed to pressure signals is measured. However, these latter demand higher power than the piezoelectric sensor and are lower in resolution resulting from hysterisis and non-linear problems.

Electric field
Two types of electric field can be used to detect a target, the Underwater Electric Potential (UEP) and the alternating field or Extra Low-Frequency Electric field (ELFE). Sensors used to detect these fields operate in the 5 MHz to 1 kHz band with a sensitivity of 10 nV per metre per root Hertz. The UEP static field of a platform is usually created by the platform's cathodic protection system. These fields propagate for some considerable distance beyond a vessel and with sufficiently sensitive sensors, such fields can be detected at reasonable ranges. The ELFE field results from the modulation of the target's static field by moving parts such as the propeller, machinery, generators, shafts and so on. The ELFE field is not so well defined as the UEP field,

particularly in deep water and propagation falls off quickly as frequency increases. On the other hand, sensors have a higher sensitivity in the ELFE band compared to the UEP band, while the background noise in the ELFE field is much lower than that associated with UEP signals, which results in a much higher probability of detection in the ELFE band than the UEP band.

Seismic
When acoustic wave fronts emanating from a target at below 10 Hz strike the seabed, some of the energy is changed into velocity or seismic energy, which is then more easily retransmitted into the surrounding environment. It is this seismic energy which is used to detect the presence of a target. Because of the form of the energy it is not necessary for any part of the sensor to be in contact with the sea through which the energy passes. This means that seismic sensors are not unduly affected by mines which become buried in sediment or mud on the seabed.

Seismic sensors are either based on the piezo-electric accelerometer, or on a pendulous mass which incorporates a permanent magnet and a fixed sensor coil. Both have advantages and disadvantages. The accelerometer uses electronic processing to derive a velocity signal, while the pendulous mass has to be mounted in such a way that the mechanism is maintained in the correct plane of operation using a gimbal type mounting. This requires a certain delicacy of construction which is not required by the accelerometer. On the other hand, the pendulous mass provides a direct reading signal proportional to the velocity and does not require any signal processing.

VERIFIED

BRAZIL

MCF-100

Type
Moored contact mine.

Description
The MCF-100 moored contact mine is designed for use against submarines and surface ships at depths between 3 and 50 m in strong currents. It can be laid at bottom depths between 10 and 100 m. The mine can be programmed to remain inert on the bottom for a fixed period of time and then to self-release and anchor itself at the desired depth.

The newly developed installation procedure allows the mine to be moored with an accuracy in anchoring level of 0.50 m independently of currents and surface waves.

During 1992, Consub, in co-operation with the Brazilian Navy Research Institute, completed development of an influence (magnetic and acoustic) sensor which will increase the efficiency of the MCF-100. The sensor replaces the contact head assembly in the mine. This sensor can also be used in combination with up to three explosive modules, a priming charge and fuze, resulting in a powerful ground influence mine which could also be deployed from submarine torpedo tubes.

Three different versions of this mine are available: combat (MCF-100C), exercise (MCF-100E) and handling simulation (MCF-100M).

The MCF-100 can also be supplied with modular skids allowing:
(a) smaller storage area
(b) easier and safer transportation
(c) the skids can also be mounted and aligned on almost any COOP vessel to create a modular rail deployment system.

Operational status
The mine has been exhaustively tested by the Brazilian Navy. It has also been designed to meet US Navy military standards. A contract for 100 mines was placed by the Brazilian Navy in August 1991. Since then the Brazilian Navy has purchased the MCF-100 on a regular basis.

Specifications
Length: 1,400 mm
Width: 1,020 mm
Height: 1,500 mm
Weight: 770 kg
Charge: 160 kg Trotyl
Operating depth: 3-50 m
Sinker depth: 10-100 m

Contractor
Consub SA, Rio de Janeiro.

VERIFIED

MCF-100 stored in modular skids reducing storage space requirements

MCF-100 mine on launching rail   **1997**

Skid mounted on COOP to form minelaying rail   **1997**

MCF-100 in skid facilitating handling   **1997**

CHINA, PEOPLE'S REPUBLIC

Chinese naval mine upgrading system

Description

The Chinese Navy, which largely relies upon former Soviet-designed naval mines, has developed an 'intelligent' sea mine actuation system for what are described as 'large' and 'medium' moored mines. The Chinese also possess a rising mine, the EM 52, which closely resembles the first Russian rising mine.

The system features a programmable central processor which can accept inputs from acoustic and magnetic sensors and, optionally, pressure sensors as well. It incorporates a ship counter system which can permit up to 15 actuations before detonation, a delay mechanism of up to 250 days before arming and a self-destruction timer for up to 500 days. There are eight operating modes which are believed to be mixtures of fuze and logic settings to meet different operational or environmental conditions.

Contractor

Dalian Warship Institute, Dalian.

VERIFIED

DENMARK

MTP-19

Type

Cable controlled mine.

Description

The MTP-19 is a fully remote-controlled mine, designed to block channels in uncontrolled minefields, the entrances to harbours and protected anchorages, and so on. The mine system comprises a portable mains-/battery-powered weapon control unit, distribution box and the mine itself which consists of two sections – the flotation unit and the weapon unit which incorporates a microcomputer, enabling settings to be changed to meet particular needs and situations, and the explosive charge. A number of mines can be linked together via the distribution box and each can be independently controlled from the control unit. The weapon can be laid from minelayers, ferries or any other suitable vessel. It is mounted on a trolley which is designed to fit the transport rails in Danish ferries.

After launching, the mine is turned to the correct attitude within the first 3 m of water and the weapon's electronics actuated. After laying, the two parts of the mine separate and the buoyancy unit is retrieved enabling the cable connecting the buoyancy unit to the mine to be hoisted and reconnected to another cable or the distribution box. The control unit retains in its memory the exact location of every mine it controls and enables the acoustic and magnetic sensors to be set in four steps of sensitivity. The control unit also indicates the complete status of each mine. Mines can be armed/detonated manually from the control unit ashore, which can be linked to an automatic sensor system indicating the approach of targets. In the auto-alarm mode the mine uses its own sensors to actuate the detonator. The final arming mode can be set in the event that should the cables be cut, damaged or disconnected, the mine automatically activates itself, after which it remains fully active all the time.

In view of the range of automatic modes available, the mine incorporates a very high degree of safety. In an emergency, the system can be sterilised by short-circuiting the batteries and disconnecting the detonator, which can be done from the shore. The mine can also initiate self-sterilisation under certain conditions, for example, when battery power drops below a certain sensitivity. The normal operating depth of the mine is 20 m, but it can operate at greater depths. Weapon control can be exercised at distances up to 12 km, but this too can be extended if required. The charge is 300 kg of high explosive.

Specifications

Height: 1,128 mm (max)
Length: 1,000 mm
Width: 1,090 mm
Weight: 800 kg
Charge: 300 kg
Operating depth: 3-20 m

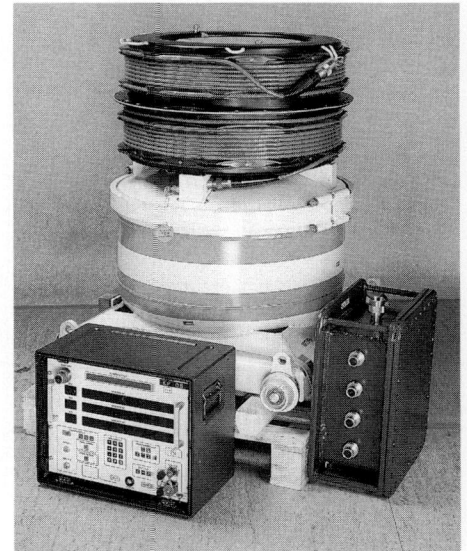

MTP-19 mine with sinker unit, drum of cable, portable weapon control unit and distribution box

Contractor

Danish Aerotech, Ballerup.

VERIFIED

GERMANY

SM G2

Type

Ground influence mine.

Description

The SM G2 heavyweight, non-magnetic ground mine is designed for blocking shipping lanes, and for laying defensive minefields in coastal waters and sea areas.

The mine comprises an equipment section incorporating acoustic, magnetic and pressure influence sensors and signal processing electronics and an explosive charge section with detonator and safety devices.

The sensors detect target signatures which are then analysed by the signal processor and compared with preprogrammed data held in the computer's programmable memory. When certain preset parameters are recognised the detonator is actuated and the mine explodes. The analysis programs and parameters are fed into the onboard computer via the testing and programming units; particular tactical parameters can still be programmed into the weapon just before laying.

Operational status

Serial production has been authorised under a MoU between Denmark and Germany and assembly will be carried out in German and Danish naval arsenals. For training purposes an exercise mine is available, and for data acquisition of ships' signatures a special version is also available.

Specifications

Length: 2,000 mm (approx)
Diameter: 600 mm (approx)
Weight: 750 kg (approx)
Operating temperature range: −2.5 to +38°C

Contractor

STN ATLAS Elektronik GmbH, Bremen/Hamburg, (prime contractor).

VERIFIED

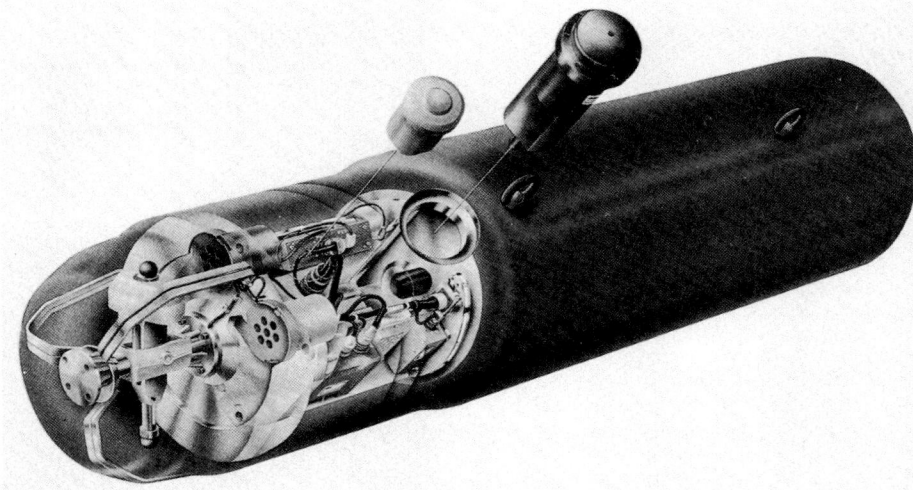

Cutaway of SM G2

SAI/AIM

Type
Anti-invasion mine.

Development
STN ATLAS Elektronik has developed the SAI (Seemine Anti-Invasion) mine under a bilateral programme in conjunction with the German MoD and another NATO country.

Operational status
SAI has been produced for the German Navy and another NATO country, and is generally available for NATO navies.

Contractor
STN ATLAS Elektronik GmbH, Bremen/Hamburg.

UPDATED

DM 211 and DM 221

Type
Anti-frogman depth charge and underwater signal charge.

Development
The anti-frogman depth charge is designed to provide an effective weapon for the protection of ships and harbour installations against frogmen. The underwater signal charge is used for encoded submarine-to-surface ship communications. Both charges incorporate safety features providing for safe handling and use by untrained personnel.

Operational status
In production and used by the German, Canadian and other navies.

Specifications
DM 211
Length: 268 mm
Diameter: 60 mm
Weight: 1,400 g
Charge: 500 g
Operating depth: 6 m
Throwing distance: 15 m

DM 221
Length: 145 mm
Diameter: 60 mm
Weight: 800 g
Charge: 50 g
Operating depth: 6 m
Throwing distance: 20 m

Contractor
Rheinmetall W & M, Ratingen.

VERIFIED

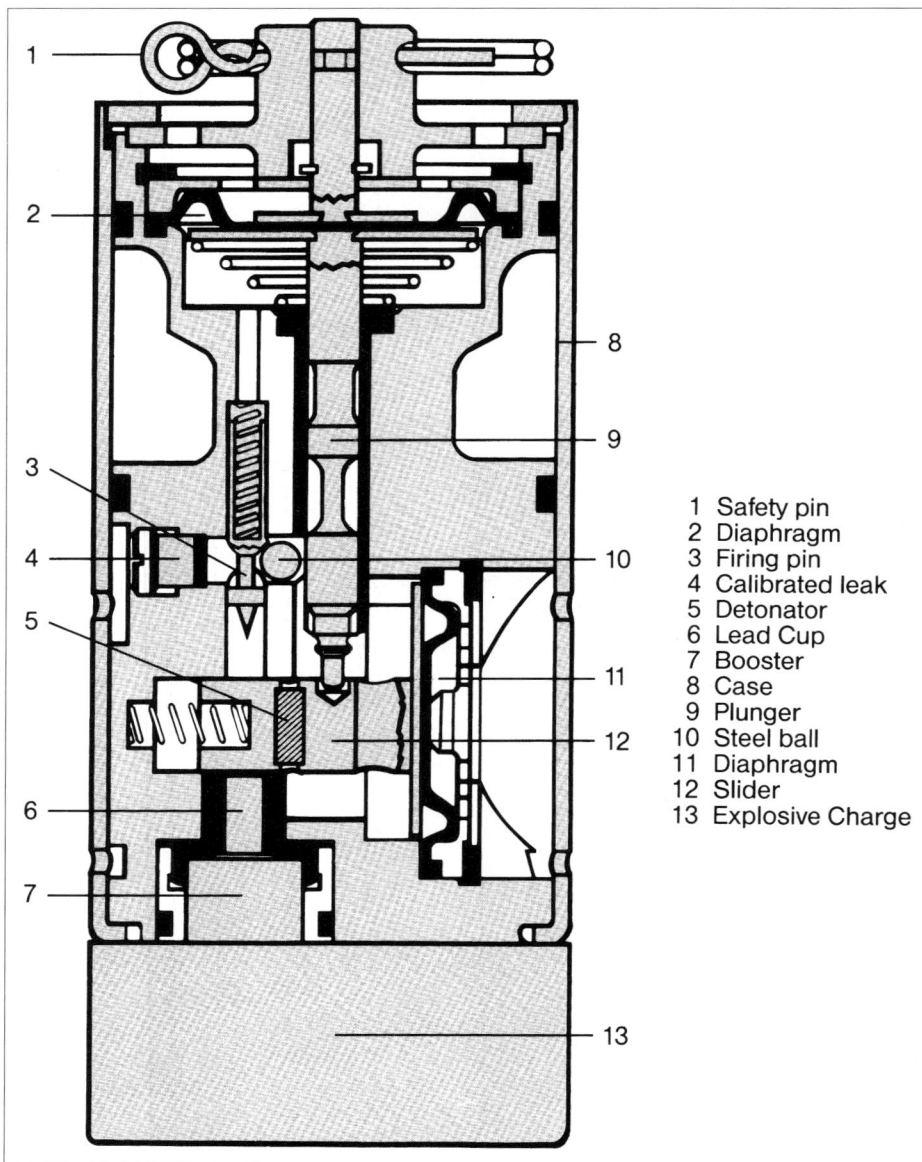

1 Safety pin
2 Diaphragm
3 Firing pin
4 Calibrated leak
5 Detonator
6 Lead Cup
7 Booster
8 Case
9 Plunger
10 Steel ball
11 Diaphragm
12 Slider
13 Explosive Charge

DM 211 and DM 221

IRAQ

Al Kaakaa/16

Type
Floodable submersible mine.

Description
The Al Kaakaa is possibly the largest mine in the world. It is a floodable mine designed to destroy large offshore structures such as drilling platforms, bridges and under seabed pipes. The shape and structure of the charge has been designed in such a way that detonation achieves its maximum effect, even in very deep water.

Specifications
Dimensions: 3,400 × 3,400 × 3,000 mm
Weight: 16.1 t
Charge: 9 t (equivalent to 13 t of TNT)
Detonation: by timer or remote control

VERIFIED

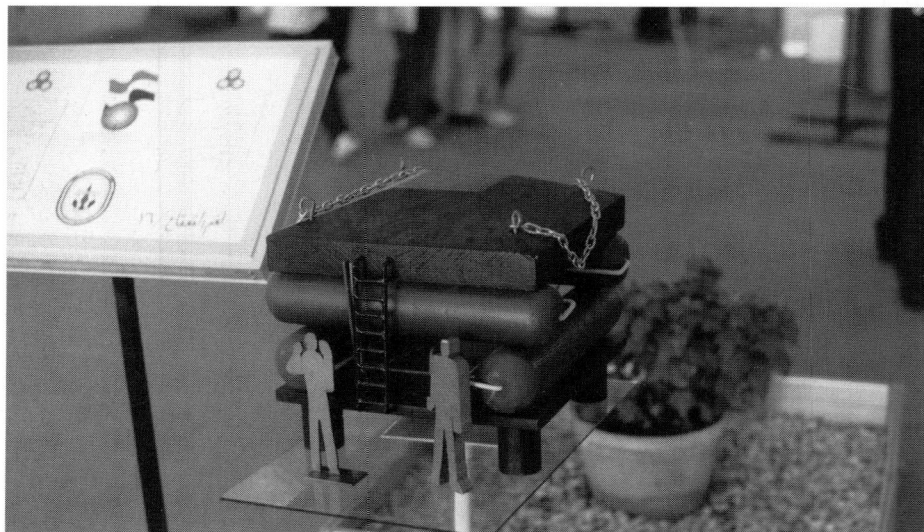

Model of Al Kaakaa/16 mine

Sigeel/400

Type
Ground mine.

Description
This seabed mine which has been developed from the Italian Manta mine by Chile, of which Iraq possessed a number, is designed for laying in both deep and shallow water for use against medium- and large-sized surface targets. It can be launched from ships and helicopters. The mine is supplied with both safety and sterilising devices. A number of these weapons were cleared from off Kuwait at the end of the Iraq/Kuwait war.

Specifications
Height: 850 mm
Diameter: 700 mm (upper); 980 mm (lower)
Weight: 535 kg
Charge: 400 kg

UPDATED

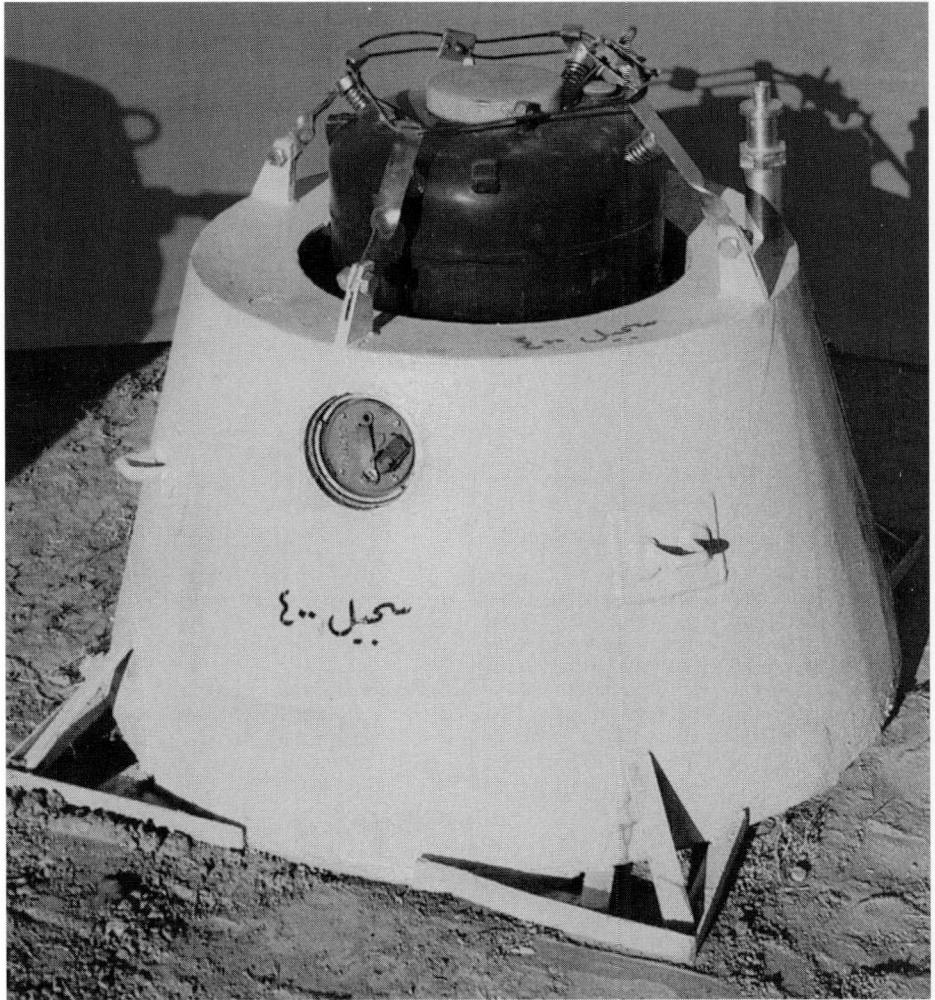

Sigeel/400 mine

ITALY

MR-80 sea mine

Type
General purpose seabed influence mine.

Description
The MR-80 cylindrical seabed mine, which can be laid from any type of platform, is actuated by a combination of influences (magnetic, pressure, low-frequency acoustic, audio frequency acoustic) from the target and is able to damage or sink surface vessels and submarines of all types, either conventional or nuclear-powered. The body is of epoxy resin and glass fibre, which renders the mine resistant to sea corrosion and makes it lighter than previous generation mines. The main body contains the explosive charge, while the tail section contains all the actuation, priming and operating devices. It is closed by means of a cover, which houses the acoustic, magnetic and pressure sensors together with the safety and arming device.

All influence devices are connected to a central unit, where all delay functions, safety intervals, sterilisation, influence combinations, ship counting, anti-removal and firing control are located.

The sole plate supporting all the electronic modules and the magnetic sensor is linked to the cover by slides that run inside the tail section. By unscrewing the bolts of the crown, the tail cover is easily removed with all the devices linked to it; in this way every part of the mine is accessible for assembly and adjustment operations.

The MR-80/1 is an exercise version of the warshot mine incorporating a data transmission device or smoke signals. As such, it can be used for training, research, analysis or evaluation.

Operational status
In service with the Italian Navy but no longer in production. Replaced by Murena (MN 102).

Specifications
Length: 2,750 mm (Mod A); 2,096 mm (Mod B); 1,646 mm (Mod C)
Diameter: 533 mm
Weight: 565-1,035 kg (according to model and type of explosive)
Charge: TNT, HBX-3 or similar HE; 380-865 kg (according to model)
Service depth: 5-300 m (for surface targets and for submerged submarines)
Power supply: dry lithium or alkaline-manganese batteries
Arming and sterilisation delay: variable up to 999 days. Electronic (quartz) clock
Working temperature: −2.5 to +35°C
Storage life: 30 years with a maintenance mean time of 5 years
Operational life: 500-1,000 days

Contractor
SEI SpA, Ghedi (Brescia).

UPDATED

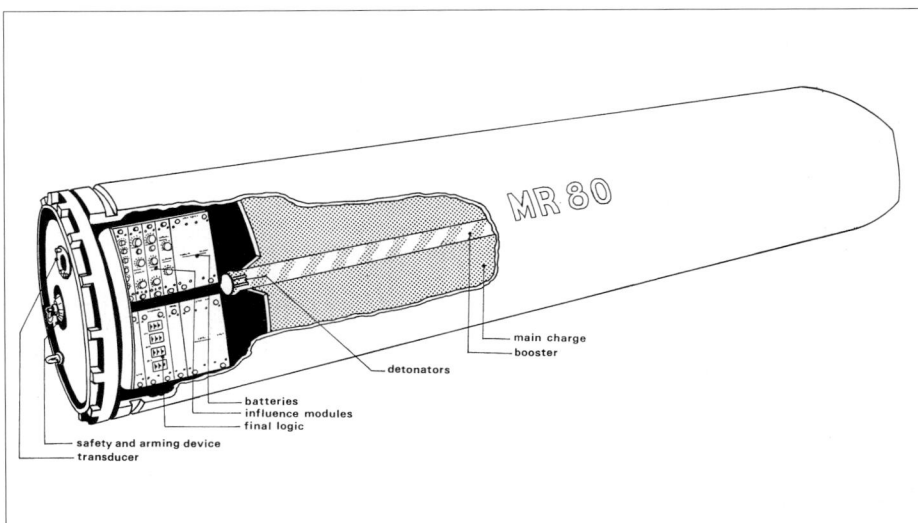

Diagram of MR-80 general purpose seabed influence mine

MP-80

Type

General purpose seabed mine.

Development

Designed in co-operation with the Italian Navy to meet a requirement for a general purpose seabed mine capable of meeting the latest concepts on mining doctrine.

Description

The MP-80 uses the same general mechanical structure as the previous MR-80, but with an updated firing mechanism and the addition of sophisticated data processing to improve performance. The target detection device uses four microprocessors to provide digital processing of target signatures and comparison with an onboard threat library. The system can discriminate between targets and countermeasures and, when presented with multiple choice targets, selects the most suitable.

Activation is by combined magnetic, acoustic and pressure sensors controlled by the programmable microprocessor. Arming and neutralisation delay time is also programmable.

Operational status

In service with the Italian and other navies. No longer in production. Replaced by Murena (MN 102).

Specifications

Length: 2,096 mm
Diameter: 533 mm
Weight: 780 kg
Service depth: 5-300 m

Contractor

SEI SpA, Ghedi (Brescia).

VERIFIED

Murena

Type

General purpose seabed influence mine.

Description

Derived from the MRP mine, the Murena uses a microprocessor firing device with increased computing capability, so achieving the maximum effectiveness and discrimination of the target countermeasures. Its mechanical, structural and explosive characteristics are the same as those used in the MR-80 general purpose seabed mine. The mine consists of a cylindrical shaped casing constructed of glass fibre impregnated with epoxy resin which houses the explosive, a target detection device, and a safety and arming unit.

The Mine Programming Unit (MPU) is used to compile the operator settings, mine programming and testing.

Detonation is microprocessor-controlled by pressure, acoustic and magnetic influences with a wide range of selectivity on the threshold, times, combination logistics and localisation logic.

Operational status

In production and in service with the Italian Navy. Murena customers include India.

Specifications

Length: 2,096 mm
Diameter: 533 mm
Weight: 780 kg (approx)
Charge: 630 kg HBX-3
Service depth: 5-300 m

Contractor

SEI SpA, Ghedi (Brescia).

UPDATED

Manta mine

Type

Shallow water seabed influence anti-invasion mine.

Description

The Manta mine can damage, sink or destroy amphibious and landing craft, small- and medium-sized surface vessels, and submarines. It operates at depths of between 2.5 and 100 m, and is a dual-influence (acoustic and magnetic) mine. It remains effective under water for more than one year. Its weight and shape are such that it rests firmly on the sea bottom even in running or tidal waters. Total weight is about 220 or 240 kg depending on the explosive.

The mine consists of two units: the body containing the explosive, and the igniter which comprises all the safety, target detection and firing devices. The mine is equipped with all the safety devices needed for handling and transport. The priming device keeps the detonator away from the explosive until the operating depth is reached so as to prevent any explosion due to a casual ignition.

Electronic circuits include:

(a) delay circuit: started by the priming device, this enables an actuation delay from 0 to 127 days, adjustable by steps of half a day, to be set

(b) prealarm device: this detects the noises caused by the passage of the targets and energises the sensing circuit

(c) sensing circuit: activated by the prealarm circuit, this actuates the firing circuit

(d) sterilising circuit: this sterilises the mine after an adjustable preset time.

The mine is laid by surface vessels or by frogmen. The mine is also available in a remotely controlled configuration.

Operational status

In production and in service with the Italian Navy. A number of Manta mines acquired by Iraq were cleared from Kuwaiti waters following the war in the Gulf.

Specifications

Diameter: 980 mm
Height: 470 mm
Weight: 225 kg
Charge: 150 kg TNT or 180 kg HBX-3
Service depth: 2.5-100 m
Firing system: prealarm and influence firing circuit

Contractor

SEI SpA, Ghedi (Brescia).

UPDATED

Manta mine

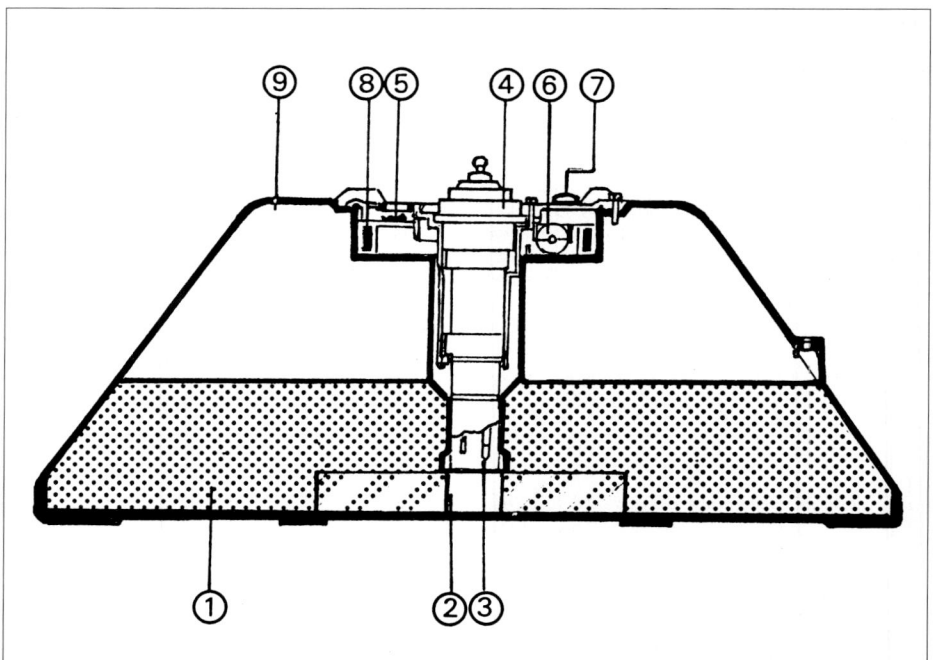

Manta shallow water seabed influence mine. Key to diagram: (1) *main charge* (2) *booster* (3) *detonator* (4) *priming device* (5) *firing mechanism* (6) *battery* (7) *acoustic transducer* (8) *magnetic sensor* (9) *calibrated holes*

Seppia

Type

Moored influence mine.

Description

Seppia is a programmable moored influence mine which can be laid by submarines, surface vessels equipped with rails (manual or automatic in operation), or by aircraft. It is designed to operate on any type of bottom against all types of target, both surface and submerged.

Using its built-in microprocessor, the mine is fully programmable, enabling it to select specific targets and discriminate against countermeasures.

Before laying, the mine's various parameters are programmed according to operational requirements to set depth of operation, sensitivity of influence firing mechanism (acoustic and magnetic), arming delay and life time, and ship counter. The programming unit can also be used to remotely control the mine.

Operational status
Believed to be in production and possibly in service with the Italian Navy.

Specifications
Length: 1,560 mm
Diameter: 533 mm
Weight: 870 kg (approx)
Charge: 200 kg HBX-3
Storage life: 30 years
Operating depth: 20-300 m

Contractor
SEI SpA, Ghedi (Brescia).

UPDATED

Diagram of Seppia moored influence mine

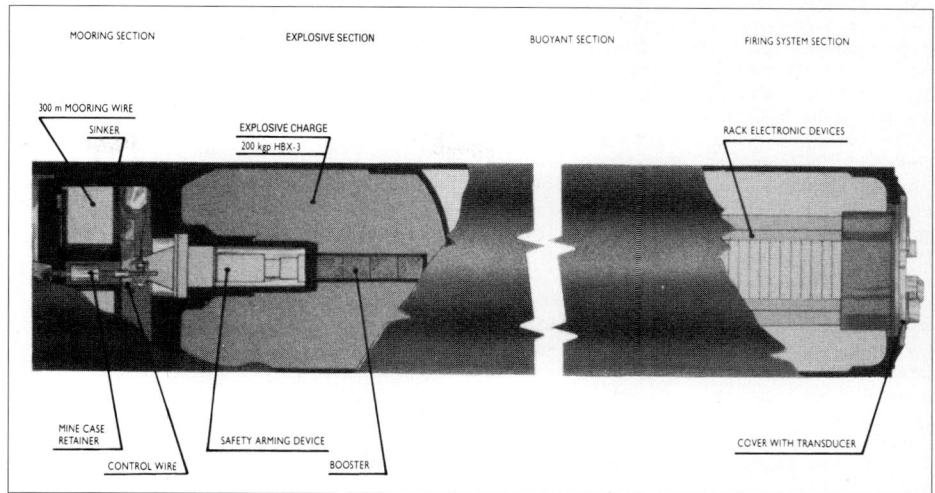

Exercise mines

Type
Exercise seabed and assessment mines.

Description
Exercise variants of all SEI mines are available, designed to support all mining activities both in workshops and at sea for training and trials in minelaying and minesweeping/hunting. The exercise mines can be used for minehunting/sweeping and minelaying and counter mining exercises, maintenance training, collecting data on the influence signatures of surface vessels and submarines, simulating hostile mines, and the study of new MCM techniques. The exercise variants are almost identical to the warshot versions, which enables operators to become thoroughly familiar with all aspects of the weapon and to provide a very effective, realistic and accurate method of obtaining data.

All exercise mines are remotely controlled via cable and/or acoustic link. Using these links, various functions relating to data collection, hardware and software tests and programming can be performed according to the type of mine being simulated.

Exercise variants available include Seppia, which incorporates a microprocessor-controlled target detecting device. The Murena exercise mine is supported by a complete range of optional equipment which allows operators to carry out programming exercises on the mine both in the workshops ashore and at sea, testing of hardware and software and the gathering of influence data, as well as simulation activities. Manta is a shallow water exercise mine which, like the other exercise mines, differs only in that it does not contain any explosive. For exercises and trials, actuation is indicated either by cable or by a pyrotechnic signal. Use of the cable enables a more accurate study to be undertaken of the firing mechanism in different environmental conditions and against various types of target. From this it is possible to determine the most effective mechanism settings according to the type of target.

Contractor
SEI SpA, Ghedi (Brescia).

VERIFIED

Updating mines

Type
Influence mines.

Description
Large quantities of old sea mines, manufactured shortly after the end of the Second World War, still remain in many inventories around the world. Both technically and operationally, these mines are now considered obsolete and are unable to comply with the modern requirements of naval mine warfare. In addition, their firing devices, designed and using technology of 40 years ago, are no longer reliable and are thus often unusable. Modern targets present operational, constructional and passive countermeasure techniques completely different from those extant when these mines were designed. To be effective, such mines require a modern, sophisticated and intelligent firing mechanism.

To meet these requirements a refit package, based on the Target Detection Device (TDD) which can be fitted inside the original mine casing and allowing the mine to retain its main charge, is available. The TDD can operate over a wide range of pressure and magnetic sensors, as well as different acoustic frequencies. TDD can be used to update any type of mine, both ground and moored. The refit TDD package is also available in exercise versions.

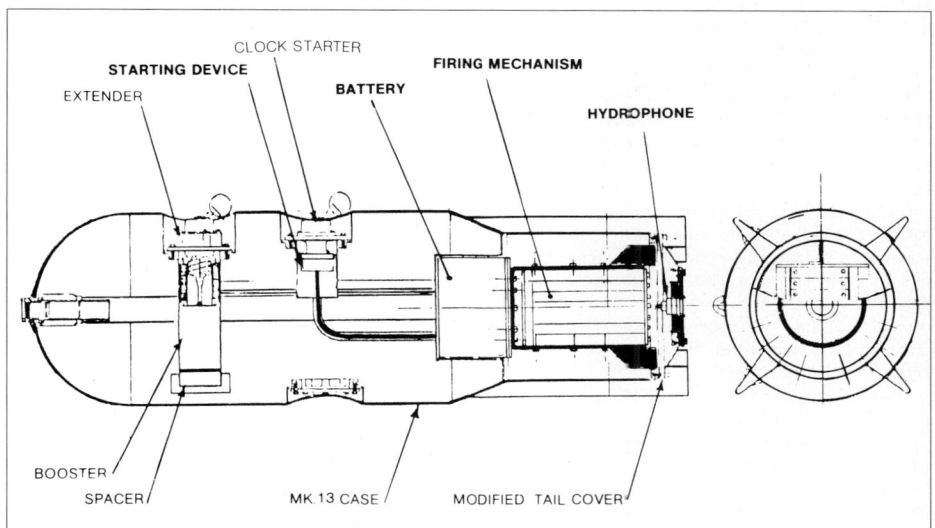

Typical elements incorporated in an updated Mk 13 Mod 6 mine

Operational status
The TDD is currently in production.

Contractor
SEI SpA, Ghedi (Brescia).

VERIFIED

RUSSIAN FEDERATION AND ASSOCIATED STATES (CIS)

Introduction
In 1987, the US DoD estimated that the former Soviet Union possessed an inventory of 300,000 naval mines. According to US intelligence, the inventory includes about 100,000 moored contact mines. In addition, mines have been provided to other former Warsaw Pact navies and also exported to many other countries. Among countries outside the Warsaw Pact that are believed to have received stocks of former Soviet mines are China, Egypt, Finland, Iran, Iraq, Libya, North Korea and Syria. It is also likely that other Third World navies possess mines of former Soviet design. At least, some of these countries have set up production lines to manufacture mines based on these designs.

Moored mines
The RFAS is known to have at least 11 different types of contact mine. Some of these are no longer in production, but many of them could be encountered.

The main type is the basic contact moored mine using an inertia type firing mechanism which can be either galvano-contact, mechanical-contact or electrical-contact.

The two smallest mines are designated MYaRM and MYaM. Both have a conventional spherical shape with horns and small sinker units. The MYaRM has an explosive charge of 3 kg and is used in rivers and lakes. The MYaM has an explosive charge of 20 kg and is used in lakes and shallow coastal areas to protect them from small boats and landing craft. The MYaM is thought to have entered service immediately after the Second World War and was first encountered in 1952 by the US Navy. It has been used by both the Iranians and Iraqis in the Persian Gulf.

The medium-sized moored contact mine is largely confined to the extensive stocks of the M-08 series used for coastal defence barriers. The M-08 was developed in 1908 and remains in widespread service in various navies. The robust, reliable, spherical mine is filled with 115 kg of explosive and the firing mechanism consists of 5 Hz horns. Intended for use against surface ships, the M-08 can be laid in up to 110 m of water. The M-08 mines used in the Persian Gulf were, according to US intelligence reports, assembled by the Iranians, probably from components supplied by the North Koreans, who apparently manufacture the M-08 themselves.

There are two improved versions of the M-08: the M-12 and M-16, which are identical to the M-08 except that they have longer cables connecting the mine case to the anchor, enabling them to be laid in deeper waters. In addition, the RFAS has devised a version of the M-16 in which two of them are linked together in tandem, with only one mine case being released at a time. If the first mine is swept or detonates, the second mine is released.

The M-26 is totally different from the M-08, using a PLT inertial firing mechanism. When armed, a shock causes the trigger shaft to move out of position, thus allowing the trigger to strike percussion caps that detonate the main charge. The entire mine comes to rest on the bottom before the mine case is released from the anchor. Although this makes it possible to use a simple anchor, it also makes it impossible to use the M-26 in deep water, where the case might be crushed by high pressures. The M-26 has been used by the North Koreans.

The M-KB is a large mine with a 230 kg explosive charge intended for use in deep waters down to 300 m. It operates in much the same way as the M-08 and even uses some of the M-08's components. The M-KB is known to have been supplied to a number of countries including Egypt and North Korea.

Several variants of the M-KB mine have appeared. The case of the M-KB-3 is 6 in shorter with a corresponding reduction in the size of its explosive charge. The M-AG is almost identical to the M-KB except that it also has an antenna firing device and can be laid in considerably deeper water. The mine has a 35 m long upper antenna and a 24 m long lower antenna. The antenna operates by generating an electric current when the copper wire of the antenna comes into contact with the steel hull of a surface ship or submarine. The mine is intended primarily for use as an ASW weapon as it increases the vertical area covered by the mine. Antenna mines tend to be unreliable, however, and most countries now use influence mines for ASW.

The AMG-1 is a version of the M-KB adapted for air delivery. Not being fitted with a parachute, it must be laid at low altitudes and slow speeds. However, because of its high-impact velocity it can penetrate several inches of ice. Because of this feature it is possible that this otherwise obsolescent mine might still be used operationally by the RFAS.

Two contact mines exist which can be laid by submarines, the PLT and PLT-3. The large PLT can only be laid by specially designed submarines fitted with an internal minelaying mechanism. As there are no such submarines operational, the PLT is likely to be encountered only if laid by surface ships. The newer PLT-3, however, can be laid through submarine torpedo tubes.

It is likely that the RFAS has additional types of moored mines in her inventory. According to one report there is an acoustic moored mine, and it would seem probable that a magnetic influence moored mine has also been developed. These could be versions of existing moored mines fitted with influence sensors. It is also probable that the RFAS has developed a new aircraft-delivered moored mine. The AMG-1 is a primitive weapon not suited for deployment from modern high-speed aircraft. Hence, the appearance of a new moored mine fitted with either a contact or an influence firing mechanism would not be surprising.

Mine protectors

Because of the ease with which moored mines can be swept, the RFAS has developed the MZ-26 mine

Egyptian KMD mines on display in 1977

Iranian M-08 mines captured by the US Navy in the Persian Gulf

An Egyptian KMD mine. The AMD mine is similar

protector. Attached by a cable to an anchor is a magazine with a buoyancy chamber and four floats. The magazine is normally set to float at a depth of 18 m, but it can be up to 46 m deep. Each of the floats is attached to the chamber by a cable usually 12 m long. The floats contain a small explosive charge which is detonated when a sweep cable comes into contact with it, cutting the sweep cable, and halting minesweeping

Specifications

Type	Mine designation	Firing mechanism	Overall weight	Explosive charge	Minelaying depth		Case depth	
					Min	Max	Min	Max
Bottom	AMD-1000	Influence	987 kg	782 kg	4 m	200 m	0 m	54.9 m
	AMD-500	Influence	–	299 kg	4 m	70 m	0 m	24.4 m
	KMD-1000	Influence	987 kg	782 kg	4 m	200 m	0 m	54.9 m
	KMD-500	Influence	500 kg	300 kg	4 m	70 m	0 m	24.4 m
	Mirab	Influence – magnetic	279 kg	64 kg	2 m	–	0 m	9.1 m
Mobile	?	Influence	–	–	40 m	70 m	0 m	–
Moored	AMG-1	Contact – chemical horn	1,034 kg	262 kg	13 m	100 m	2 m	9 m
	M-08	Contact – chemical horn	–	115 kg	6 m	110 m	0 m	6 m
	M-12	Contact – chemical horn	–	115 kg	6 m	147 m	0 m	6 m
	M-16	Contact – chemical horn	–	116 kg	6 m	366 m	0 m	6 m
	M-26	Contact – inertial	–	240 kg	6 m	139 m	1 m	6 m
	M-AG	Antenna	1,089 kg	230 kg	80 m	454 m	0 m	91 m
	M-KB	Contact – chemical horn	1,089 kg	230 kg	0 m	300 m	6 m	9 m
	M-KB-3	Contact – chemical horn	1,061 kg	200 kg	0 m	273 m	0 m	9 m
	MYaM	Contact – chemical horn	175 kg	20 kg	3 m	50 m	1 m	3 m
	MYaRM	Contact – chemical horn	–	3 kg	3/4 m	50 m	–	2 m
	PLT	Contact – impact-inertial	839 kg	230 kg	9 m	139 m	0 m	9 m
	PLT-3	Contact – chemical horn	998 kg	100 kg	0 m	128 m	0 m	9 m
	UEP ?	Influence – electrical	–	227 kg	0 m	490 m	0 m	–
Obstructor	MZ-26		413 kg	1 kg	24 m	46 m	0 m	34 m
Rising	'Cluster Bay'	Influence – acoustic	–	230 kg	80 m	200 m	0 m	609.6 m
	'Cluster Gulf'	Influence – acoustic	–	230 kg	80 m	2,000 m	0 m	–

operations. Although the float is destroyed, a replacement float is automatically released when one is destroyed. Thus each MZ-26 has the potential for cutting up to four sweep wires. The protectors can be used in conjunction with M-16 tandem mines.

Seabed mines

Relatively little is known about Russian influence mines. The first seabed mine available in any quantity was the KMD developed in the late 1940s. The AMD is an aircraft-delivered version of the KMD. It was first identified in the late 1950s but may have entered service earlier in the decade. The mines can be fitted with one of four different types of sensor: acoustic (using either or both low-frequency and high-frequency noise generated by the target), magnetic (relying on either the intensity of the horizontal or vertical component of the target's magnetic field or the rate of change of its field), pressure, or a combination firing mechanism using two or all three different influences in conjunction. The KMD mine is fitted with a timer that can be set to activate the mine after a delay of up to 10 days, and it can also be fitted with a ship counter that requires up to 11 activations before the mine actually detonates. Both mines can carry either 500 kg or

M-AG antenna mine

1,000 kg of explosives, and both have been exported to other former Warsaw Pact navies and to countries in the Third World.

The former Soviet Union is known to have developed additional influence mines in the past 30 years. The first of these was identified in 1985 after being used by the Libyans in the Red Sea. According to published reports the mine was 533 mm in diameter, suggesting it was designed for submarine deployment. The mine is believed to have been of modular design with replaceable circuit boards, which enabled the sensitivity to be altered according to the type of target to be countered. The mine carried an explosive charge of 680 kg.

It is probable that several different types of ground mine are available, including an air-deployed version of the submarine-launched mine. As that particular mine was large, it is probable that smaller anti-invasion ground mines also exist. The latest types of ground mine probably follow Western practice in that they are programmable microprocessor-controlled types, whose mechanisms can also be retrofitted into existing warstocks.

While the RFAS has exported its newer mines to countries as diverse as Libya and Finland, it is unlikely that they will incorporate firing devices as sophisticated as those fitted in the former Soviet Navy's inventory.

Specialised mines

Several special purpose mines intended for a deep water ASW role have been developed. Two such weapons are the torpedo shaped, rocket-propelled, tethered rising mines, NATO designation 'Cluster Bay' (operating in water 80 to 200 m deep) and 'Cluster Gulf' (which can be laid in 2,000 m of water) which first entered service during the 1970s. The speed of rise of this rocket-propelled mine is understood to be about 155 kt.

Targets are initially detected by a passive acoustic sensor and located by transmissions from an active

acoustic sensor. If the target is confirmed as being within the vertical attack zone, the tether is cut and the rocket ignited. The fast upward speed allows very little time for the target to evade the device if its launch has been detected.

The RFAS is also believed to have developed an Underwater Electric Potential (UEP) mine. Nothing is known about this mine, except that the firing mechanism operates by detecting the electrical field generated by a target.

It is possible also that the RFAS has developed a mobile mine similar to those used by the US Navy. Mobile mines are attached to torpedoes, making it possible for a submarine to lay a mine at a distance from the location where the mine is released into the water.

Nuclear mines

The RFAS is believed to have a small stockpile of nuclear mines with yields ranging between 5 and 20 kT for use against high-value surface units and base targets. Laying of these mines is almost certainly assigned to specially selected SSK/SSN units.

UPDATED

Firing mechanism for KMD mine

MDM series

Type

Seabed mines.

Description

This family of seabed influence mines is designed for delivery from a wide range of platforms. The MDM-1 and MDM-6 variants are designed for laying from the torpedo tubes of submarines and surface ships, while the MDM-2 is designed for launching from surface ships. MDM-3 and MDM-5 are designed to be laid by aircraft. These mines differ from most normal air-launched mines in that the parachute is dispensed with, enabling the mine to be laid much more covertly and from a much lower altitude. While this is advantageous from the tactical point of view, it does give rise to severe technical design problems with regard to damage resulting from high-speed impact with the water, the possibility of ricochet, and the possibility of deep penetration of soft seabed. This is overcome by the use

Specifications

	MDM-1	MDM-2	MDM-3	MDM-4	MDM-5
Diameter	533 mm	790 mm	450 mm	650 mm	630 mm
Length	2,860 mm	2,300 mm	1,580/1,530 mm	2,790/2,300 mm	3,060/2,400 mm
Weight	1,120 kg	1,413 kg	525/635 kg	1,370/1,420 kg	1,500/1,470 kg
Charge	960 kg	950 kg	300 kg	950 kg	1,350 kg
Operating depth	12-120 m	12-125 m	15-35 m	15-125 m	15-300 m

of soft stabilisers which also help to ensure accuracy of placement.

The MDM-1 features a two-channel influence exploder of the acoustic-induction type and conventional devices, a timing device and ship counter. It can be laid from surface ships with rails and stern ramps or from submarine torpedo tubes. Laying speed from surface ships is between 4 and 15 kt and from submarines 4 to 8 kt.

The MDM-2 is laid from surface ships at speeds of between 4 and 10 kt. It features a three-channel acoustic/magnetic/pressure influence fuze and a set of

protective and functional devices including a time delay, ship counter and self-destruct unit.

The MDM-3/5 are fitted with a three-channel fuze using acoustic, magnetic and pressure influences and carries a set of protective and functional devices as the MDM-2.

Contractor

Rosvooruzhenie, Moscow.

UPDATED

SMDM series

Type
Self-propelled seabed mines.

Description
These mines are a development of the basic MDM seabed influence mine, combining a mine into the body of a torpedo. They are designed to be laid from submarines as defensive mine barriers in the entrances to naval bases, harbours and restricted waterways, forming part of an integrated defensive system with other weapons such as coastal defences and torpedoes.

The SMDM-1, after covert launch, travels up to a distance of about 10 n miles before descending to the seabed. The trajectory and distance are preset before launch and the weapon is believed to incorporate various devices such as timing circuits for activation/deactivation, counters, target identifying algorithms and so on. The SMDM-2 variant prototype is believed to have an extended range in the region of about 25 n miles.

Standard torpedoes can be adapted into SMDM.

Specifications
SMDM-1
Diameter: 533 mm
Length: 7,900 mm
Weight: 1,980 kg
Charge: 480 kg
Operating depth: 4-100 m

SMDM-2
Diameter: 650 mm
Length: 11,000 mm
Weight: 5,500 kg
Charge: 800 kg
Operating depth: 8-150 m

Contractor
Rosvoorouzhenie, Moscow.

VERIFIED

PMK-1

Type
Underwater rocket-powered torpedo.

Description
The PMK-1 features a combined launcher and mine based on a torpedo. Designed for attacking submarines, the onboard sensor and computer detects and identifies the target, establishes the course and speed and computes the required intercept trajectory before activating the torpedo motor. Detonation is achieved using a combined influence, contact and time fuze which operates either independently or in combination.

Specifications
Diameter: 533 mm
Length: 4,500 mm (torpedo)
7,830 mm (overall)
Weight: 1,850 kg
Warhead: 324 mm diameter
Charge: 350 kg
Speed: 30 or 60 kt
Deployment depth: 200-400 m
Radius of action: 1,000-1,500 m

Contractor
Rosvoorouzhenie, Moscow.

VERIFIED

MSHM

Type
Continental shelf mine.

Description
The MSHM is designed for use against submarines and surface ships in continental shelf waters at depths between 60 and 300 m. The weapon is fitted with an acoustic sensor which detects and identifies the target, establishes the course and speed and computes the required intercept trajectory before the underwater rocket is fired. The rocket then homes on to the target using its acoustic homing device.

The mine is constructed with a double-casing with a large air gap in between. This effectively reduces the impact velocity performance of a penetrator by some 25 per cent.

Specifications
Diameter: 533 mm
Length: 4,000 mm
Weight: 1,500 kg
Charge: 250 kg
Operating depth: 60-300 m

Contractor
Rosvoorouzhenie, Moscow.

VERIFIED

SPAIN

MO-90

Type
Moored influence mine.

Description
Developed by SAES in close co-operation with the Spanish Navy, the MO-90 moored mine is an intelligent, multi-influence mine designed for use against submarines or surface ships. It is designed for 25 years' shelf life with up to two years in-water life.

The mine consists of a GRP buoy which contains the sensors, target detection device, safety and arming device, and the explosive charge. The sinker and reel containing Kevlar mooring rope also form part of the mine. More than 10 parameters of the mine can be programmed into the weapon's computer just before launch without opening the casing. Among the features incorporated in the design are anti-minesweeping countermeasures. The depth setting mechanism of the buoy is claimed to be extremely accurate, being in the region of ±1 m.

Operational status
In production and in service with the Spanish Navy.

Specifications
Length: 1,180 mm
Width: 1,090 mm
Height: 1,690 mm
Weight: 1,060 kg
Charge: 300 kg
Buoy depth setting: 5-340 m

Contractor
SAES, Cartagena.

UPDATED

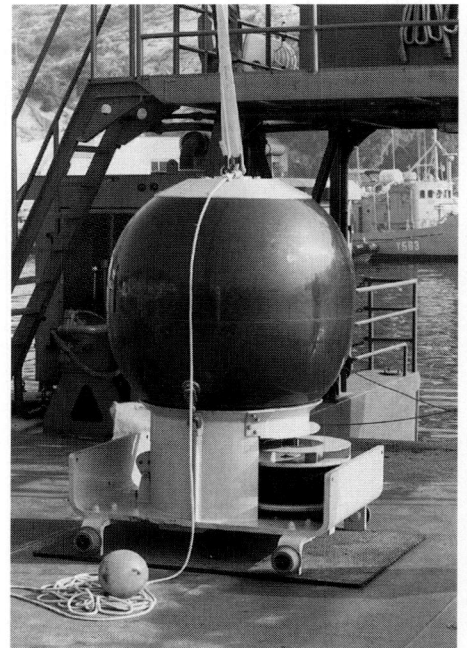

Moored mine MO-90
1996

Mila-6B

Type
Naval limpet mine.

Description
Mila is a time-fuzed tactical underwater limpet mine that can also be used as a demolition charge.

The limpet mine can be attached to all kinds of underwater structures by either of two different methods of attachment: magnetic or mechanical.

The main features of the limpet mine are safe transport and handling, simple operating procedures, countermeasures to prevent removal from the target, enhanced robustness and high reliability and low maintenance.

The limpet mine comprises two main subassemblies. The upper assembly houses the electronic fuze and arming device, while the main section contains the relevant hardware for attaching the mine to its target, the explosive charge and an anti-removal device.

Mila incorporates a large number of safety devices covering transportation and arming. Arming safety methods include explosive train out of line, unpowered firing circuit and detonator short circuit.

A portable test set is provided for testing the mine without requiring disassembly. A Mila Exercise Version (MEV) is available for operator training (above and underwater).

Operational status
Final stage of development.

Specifications
Height: 150 mm
Diameter: 350 mm
Weight: 65 kg (air)
Fuze delay: up to 48 h (programmable)
Operating depth: down to 40 m

Contractor
SAES, Cartagena.

NEW ENTRY

A Mila-6B limpet mine (SAES)
1999/0024774

SWEDEN

BGM 100 (Rockan)

Type
Anti-invasion ground influence mine.

Description
This ground influence mine is intended for use as an anti-invasion deterrent against small and medium tonnage vessels in coastal waters, inlets, and confined waters such as harbours. It may also be deployed in deeper waters against submarines to protect friendly routes and bases. A feature of the design is a shape which enables minelaying over a wide area while the minelayer covers the minimum distance. This is achieved by shaping the mine in a casing that enables it to plane or glide in the water for a distance of up to twice the depth of the water it is being laid in. This also enables mines to be sown in pairs from each side of the ship, or for mines to be spread out directly from the quay.

Another advantage of this shape is its low profile, which makes it difficult to detect by minehunting sonars and other anti-mine techniques. The mine is constructed of an outer shell of reinforced plastic containing an explosive charge, a fuze with arming unit,

and a sensor/electronic unit. The outer shell has sliding runners to enable the mine to be moved on a mine rail, orientated either along or across the rails. The shell can be opened easily for maintenance or for adjustments and so on.

New multisensors and SAI units have been developed.

Operational status
In production.

Specifications
Length: 1,015 mm
Width: 800 mm
Height: 385 mm
Weight: 190 kg
Charge: 105 kg
Service depth: 5-100 m
Glide speed: 2 m/s (approx)
Minimum distance between mines: 25 m

Contractor
Bofors SA Marine, Landskrona.

VERIFIED

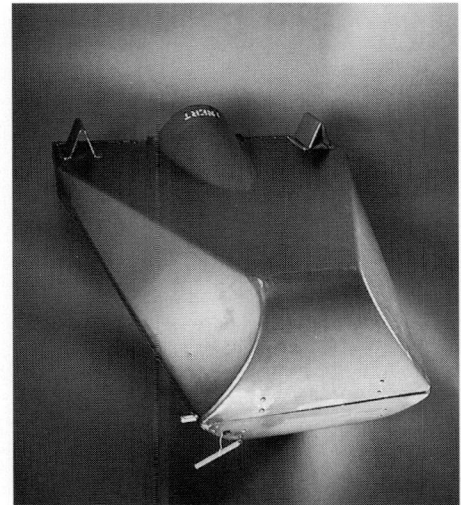

After resting on the seabed for a time the Rockan ground mine becomes very difficult to detect both visually and by minehunting sonar **1997**

BMM 80

Type
Moored influence mine.

Description
This influence mine is designed for use against small and medium tonnage surface ships and submarines. It is suitable for laying in water 20 to 200 m deep. The mine itself consists of a buoyancy section of cellular plastic, which also contains the explosive charge, the fuze and the sensor/electronics package (which is similar to that of the BGM 100, but adapted to the moored mine role).

The BMM 80 is programmed to anchor itself automatically at the desired depth, which can be set before sinking. A sinker unit is employed with the BMM 80 and the mine can be laid from rails.

Operational status
In production.

Specifications
Length: 1,125 mm
Width: 660 mm
Height: 1,125 mm
Total weight: 450 kg
Sinker weight: 240 kg
Charge: 80 kg
Bottom depth range: 20-200 m
Mine depth range: 8-75 m
Minimum distance between mines: 25 m

Contractor
Bofors SA Marine, Landskrona.

VERIFIED

BMM 80 mine on launching trolley **1997**

BGM 601 (Bunny)

Type
Ground influence mine.

Development
Development ended in May 1990 and was followed by a production contract.

Description
The BGM 601 ground influence mine has been developed as part of a new submarine weapon which will be carried in a mine girdle attached to the outer hull

of submarines. The mine will also be able to be launched from any type of surface ship.

The mine incorporates a buoyant upper section which gives the weapon an underwater weight of 200 kg. Against surface targets the mine will have an operating range of 40 to 50 m but can be laid in deeper water down to 150 m for deployment against submarines.

The mine is fitted with the CelsiusTech 9SP 180 programmable naval mine sensor unit, which incorporates multiple sensors using sophisticated logic, making it resistant to countermeasures. It will be able to discriminate specific types of target and initiate

detonation at the lethal range for that particular target. The sensor unit is powered by a lithium battery pack and/or external power source. The sensor unit can remain in water for up to 12 months. The PCHS-1 combined hydro-sensor unit senses depth, dynamic pressure and sound in separate channels, while the three-axis magnetometer measures variations from the locally determined magnetic field. The measured signals are filtered and evaluated against preprogrammed criteria by the microprocessor. The logics unit reports the evaluation result through a standard electrical interface to the firing device or corresponding unit in non-mine applications.

The sensor unit can also be retrofitted to existing mines.

The volume of explosive carried is 400 litres.

A moored version of the mine has been successfully tested for submarine deployment.

The exercise variant VEM 90 can be both active and passive and is equipped for all types of training from laying to simulation of hostile attack.

Operational status
In production for the Swedish Navy since 1992. The mine is also being offered as an option for the new submarines for Australia.

Specifications
Length: 2,000 mm (approx)
Width: 750 mm (approx)
Height: 750 mm (approx)
Total weight: 800 kg (approx)
Sinker weight: 240 kg
Charge: 80 kg
Bottom depth range: 20-200 m
Mine depth range: 8-75 m
Min distance between mines: 25 m

Contractor
Bofors SA Marine, Landskrona.

VERIFIED

Training version of the BGM 601 ground influence mine shown fitted with a recovery unit (Anchor Consultancy Photo Library)   *1996*

BGM 600

Type
Cable controlled mine.

Development
The BGM 600 advanced cable controlled mine has been developed for the Swedish Coastal Artillery's amphibious battalions to provide an efficient and rapid means of deploying a controlled minefield, featuring on/off arming capabilities which enable a navy's own forces to pass over the minefield in safety for rapid deployment.

Description
The BGM 600 is designed for remote control using a slim, flexible cable and the entire mine and cable system can be deployed at high speed. When required it may also be used in autonomous mode without cables.

The mine features a modular main charge consisting of two, three or four charge modules each containing over 100 kg of high explosive and a central instrumentation tube containing the sensors and control electronics. The bottom plate may optionally be fitted with wheels to enable it to be handled on mine rails. The electronics include a multisensor system and advanced signal processors which are programmable for different targets, the sensitivity being changed from the shore base. The mine is capable of operating in different environments and against different target characteristics, and can be upgraded and reprogrammed.

Operational status
Developed for the Swedish Coastal Artillery and in full production.

Specifications
Length: 1,700 mm (max)
Width: 600 mm (max)
Height: 700 mm (max)
Weight: 700 kg (max) (approx)
Charge: 2, 3 or 4 modules each containing over 100 kg explosive
Sensors: multiple sensors defined by customer

BGM 600 mine showing modular charges and central instrumentation unit   *1997*

Contractor
Bofors SA Marine, Landskrona.

VERIFIED

UNITED KINGDOM

Sea Urchin

Type
Programmable influence mine.

Description
Sea Urchin is an intelligent seabed mine which utilises the mine modernisation components developed by BAeSEMA for the Royal Navy. It can be programmed to detonate on a range of influence characteristics either singly or in combination, including a ship's acoustic or magnetic influences. Its advanced microprocessor control ensures that detonation occurs at the closest approach point of the target within the damage radius of the mine.

Sea Urchin seabed mine exercise variant

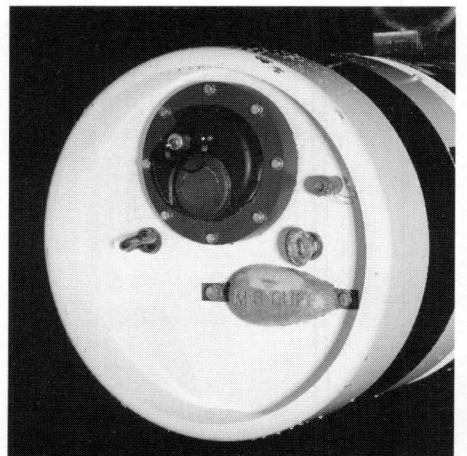

Sensing and processing unit installed in Sea Urchin exercise mine

Mine setting and self-testing procedures are simple and secure, being achieved by means of a small plug-in setting box weighing less than 5 kg.

Equipped with a 350 kg warhead (an additional warhead can be added), Sea Urchin can be deployed in water depths from 5 to 200 m; various launch attachments are available for the nose and tail of the mine to permit deployment by surface vessels, submarines or aircraft. An exercise version is also available.

Stonefish mines

Type
Surface-, air-, submarine-launched ground influence mine.

Development
GEC-Marconi Underwater Weapons Division continues to develop its family of advanced mines comprising warstock, exercise, training and assessment versions. Stonefish, as this family is known, is designed on the modular concept enabling mines to be made up to meet any tactical, exercise or training requirement.

Description
The warstock variants are provided with an advanced design plastic bonded explosive warhead. The warhead, together with the appropriate launch kit and a standard tail unit containing the sensors, signal conditioning and processor package, can be configured to cover the range of launch vehicles, depths and targets. The basic sensor pack comprises acoustic, magnetic and pressure sensors. The signal processing includes signal conditioning and information processing, the setting of thresholds, mine logic and sterilisation delay times. The mine programme can be modified by means of a portable presetter, changing threat and operational conditions before the mine is deployed. Different sizes of warhead can be fitted according to water depths, minefield pattern and expected target size.

The Stonefish warstock versions have a storage life of over 20 years and an in-water or laid life of over 700 days. They are intended to be laid at depths of between 10 and 90 m for surface targets and down to depths of 200 m for submarine targets. The refurbishment life for stored warstock weapons is nominally six years. To provide training in mine countermeasures, Stonefish is available as an exercise variant.

The exercise variant is capable of assessing minehunting and minesweeping effectiveness in its true environment. When laid it is available for data gathering relevant for target selection, together with a facility to emulate known mine characteristics. The main features are:
(a) command activated recovery system
(b) acoustic command system, enabling mine states to be interrogated or changed
(c) digital signal processing and large solid-state data store
(d) it is available as cable controlled or a seabed test platform for sensor/processing development.

Using the command link the exercise variant can accurately measure target range, change its operational status and recover itself, even if buried in deep mud on the seabed, by means of an inflatable flotation collar.

It can be refurbished at sea and recharged within 4 hours ready for relaunch.

The assessment variant is used to gather mine and target data for use by the tactical planner so that the optimum minefield can be planned. The variant uses a modified version of the warstock version's electronics system and is connected by underwater cable to a shore-based computer.

The training variant is inert but can accept all the available equipment options to facilitate classroom instruction, handling, programming and test trials.

Mines of the Stonefish family are up to 2 m in length, with a diameter of 533 mm and all-up weight varying up to 770 kg depending on configuration.

The Stonefish presetter, designed as a self-contained unit, employs a microprocessor and

Operational status
Development has been completed but no information is available on production or orders.

Specifications
Length: 1.44 m (single warhead), 2.54 m (two warheads)
Diameter: 533 mm
Weight: 569 kg

Specifications

	Mk II Warstock mine	Exercise mine	Assessment mine
Length	2,500 mm	1,912 mm	506 mm
Diameter	533 mm	520 mm	530 mm
Weight	990 kg	440 kg	120 kg
Warhead	600 kg PBX	inert	
Operating depth	10-200 m	10-200 m	

Stonefish mine

incorporates built-in data checking during input and automatic test facilities.

Operational status
In production. GEC-Marconi has received orders for the exercise variant of Stonefish from the Royal Australian Navy and has sold warstock units to Finland and Pakistan. Assessment mines are in service with six countries.

Contractor
GEC-Marconi Underwater Systems Group, Waterlooville.

UPDATED

Charge: 350 kg HE
Operating depth: 5 to 200 m

Contractor
BAeSEMA, Bristol.

UPDATED

Mine modernisation

Description
BAeSEMA has designed and developed a new Target Information Module (TIM) which can be installed quickly and easily in older mines to convert them cost-effectively into modern weapons, thus considerably extending their useful service life.

The TIM enables modernised mines to be configured to meet a user's operational requirements. Maximum use may be made of existing internal components. For example, subsystems such as safety and arming units could be retained. The TIM houses acoustic and magnetic influence sensors and an electronics unit employing the latest in digital microprocessing technology.

If required the flexibility exists to add other types of sensors. A new safety and arming system is available, designed to cater for all types of ground mine, including submarine-launched ground mines. High-energy batteries ensure the modernised mines have a long storage and in-water life.

Mine preparation is simple and rapid using a hand-held plug-in setting box. The minefield planning officer programmes the unit with the appropriate mine mission parameters, which are then downloaded into the mine immediately before laying. Before transferring the mission parameters, the setting box automatically tests the mine's electronics, and only if they are functional will the information be downloaded. Mine testing and setting takes approximately 30 seconds. The safety and arming unit, TIM and battery module have been installed in mines of the Royal Navy and are suitable for installation in most mine types.

Operational status
Production equipment delivered.

Contractor
BAeSEMA, Bristol.

VERIFIED

TIM installed in updated Royal Navy mine

Advanced Versatile Exercise Mine System (VEMS)

Type
Exercise ground and buoyant mines.

Description
The Advanced Versatile Exercise Mine System (Advanced VEMS) is designed for training and for assessment of the effectiveness of mine countermeasures equipment and tactics. The recently developed Advanced version has much greater capability than the original system with a wider range of sensors, improved processing capability and a much larger data recording capacity. With its improved sensors and signal processor, it provides a system which can emulate the firing characteristics of the most sophisticated ground mines currently available. The enhanced underwater acoustic link is used to signal mine fires, to measure ship to VEM range and to command release. The Advanced version can also transmit recorded data while laid and permits change of emulations during exercises.

The VEM is available in a variety of forms: the standard generic ground mine, moored buoyant mine and in stealth mine shapes. It is capable of processing target acoustic and seismic influences, three independent magnetic axes (plus rate of change and total field) and pressure influences. An electric field sensor, capable of detecting static and alternating

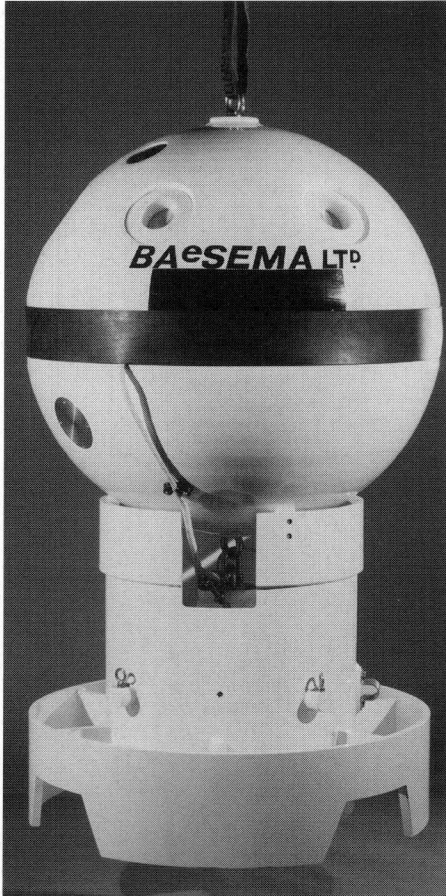

Buoyant variant of VEMS　　　*1995*

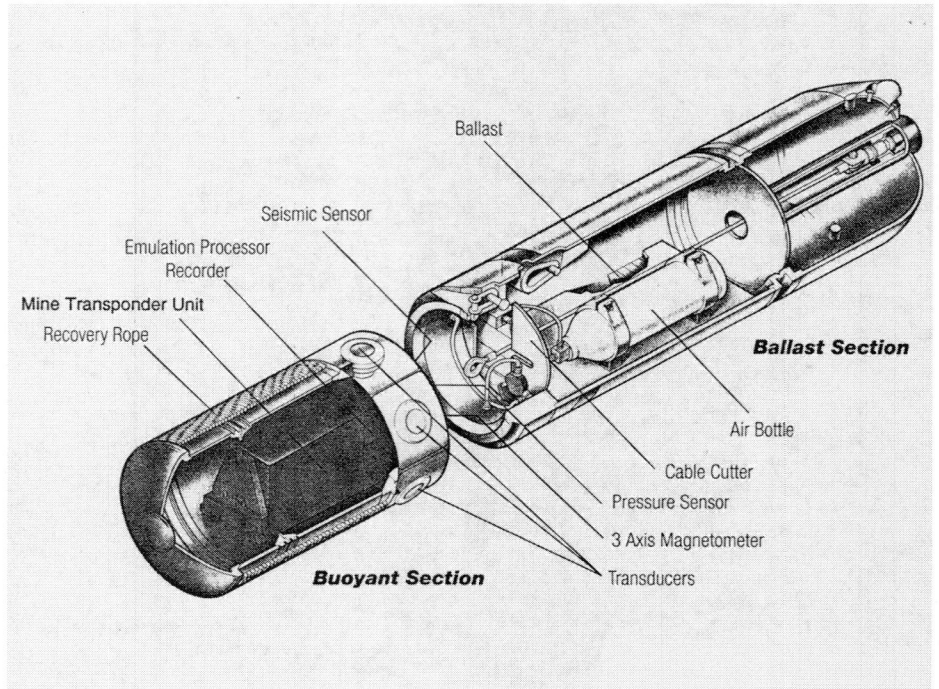

Diagram of the Versatile Exercise Mine System (VEMS) Ground mine variant　　　*1994*

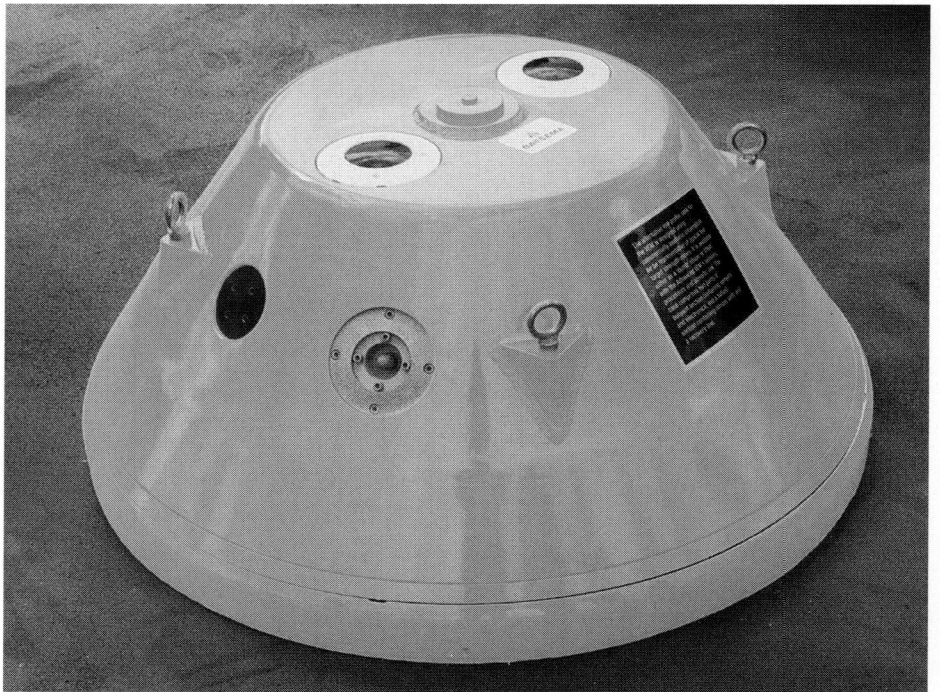

VEM Type S – the low-profile VEM (Anchor Consultancy Photo Library)　　　*1996*

fields is also available. The processor uses digital signal processing techniques to implement spectral analysis of acoustic and seismic signals. This permits accurate programming of frequency response to emulate all threat mine types including those which have a resonant response. Raw spectral data is also recorded for detailed analysis. The system can store and execute multiple emulations and these can be selected during an exercise by means of the acoustic link system. The incorporation of improved memory devices has led to a major increase in stored data capacity during exercise and the use of a low power design has led to a longer operational life.

The ship transponder system is used on the exercise vessel for two-way communication between the ship and exercise mines. When a mine 'fires', coded acoustic signals are transmitted which are detected by the shipborne transducer. This transmits the information to the Exercise Support Unit (ESU) in the ship's Operations Room which signals an alarm and displays mine identity and range. Data recorded in the mine during the exercise can be transmitted back to the ship to permit 'quick look' analysis during the exercise. The acoustic link system can interrogate mine status and is also capable of surveying VEM positions and of acoustically tracking the ship relative to the VEMS. Each VEM can also be programmed to signal additional events, such as individual influence channel detections and ship counts.

The ESU is a ruggedised laptop computer which can be used on shore, on the deck of the vessel or in the ship's Operations Room. It is used to prepare mine programs and to download those programs into the VEM before laying. During an exercise the ESU is connected to the shipborne transducer via a ship transponder unit in the Operations Room and is used to control the VEMS acoustic link. After the mine has been recovered the ESU is used to download data recorded in the mine via an umbilical cable. The analysis facility of the ESU allows an MCM exercise to be analysed in detail in order to provide an assessment of the MCM performance and risk, and threat to target vessels.

Improvements for applications in the minehunting role currently in development include: fitting of a sonar insonification detector; provision of anechoic jackets to vary the target strength of VEMS; and reconfiguration of VEM sensors, processor and transponder into threat mine shapes to provide more realistic sonar targets during classification and identification operations.

Enhancements can also be provided to allow use of the system as a mobile multi-influence range. These include an item umbilical cable, additional sensors and a full suite of influence analysis and modelling software. In this role, VEMS can provide many influence range configurations, from a fixed semi-permanent range to an onboard self-check facility. Advanced VEM can also be fitted with an Electric Field Sensor (EFS). This is a broadband three-axis sensor capable of detecting the static and alternating electric field signatures of vessels. The unit is fitted to the end of the buoyant section of the VEM.

Displayed for the first time in October 1994 was a stealth variant, designed to have a low sonar signature representative of a number of currently available warshot mines. Known as VEM Type S (low-profile VEM) it features a 'dumb' shape for use in simple sonar exercises or as a fully equipped variant with a full range of influence sensors and VEM electronics. Standard sensors include acoustic and magnetic, while seismic, pressure and electric field sensors are available as options. The case consists of two sections: a ballast section containing release units and a recovery line, and a buoyant section with sensors and electronics.

Mines supplied to the Canadian Navy are provided with innovative anechoic jackets which enable the target strength of the mines to be varied. Also incorporated are modifications to allow the recording of diver influences coupled with a diver warning device which further enhances the training provided to explosive ordnance disposal divers.

Operational status

VEMS is currently in production and in service with the Royal Navy, the US Navy and the Royal Thai Navy. Current orders will make the US Navy the largest operator of the VEM. VEMS is also operated by the Royal Saudi Navy and has recently been purchased by the Belgian and Canadian navies.

Contractor

BAeSEMA, Bristol.

UPDATED

Limpet mines

Type
Underwater demolition devices and exercise/training variants.

Description
BAeSEMA in collaboration with MP Compact Energy produce a family of naval limpet mines which includes live mines and training mines which are in service with naval diving units throughout the world.

The live limpet mine contains a modular plastic explosive charge and is equipped with electronic timer, with manual override and anti-tamper features, and a safety and arming system, built into a modular self-contained timer unit. A firing delay of up to 24 hours can be set. Explosive charge options range from 1 to 6 kg. Slave explosive units, attached by special couplers, are available to increase explosive power.

Magnetic and strap attachment methods are standard and other options are available. Inbuilt buoyancy is standard. The training mine is equipped with a high-intensity light to simulate detonation.

Operational status
Range of variants in service with navies worldwide.

Contractors
BAeSEMA, Bristol.
MP Compact Energy.

VERIFIED

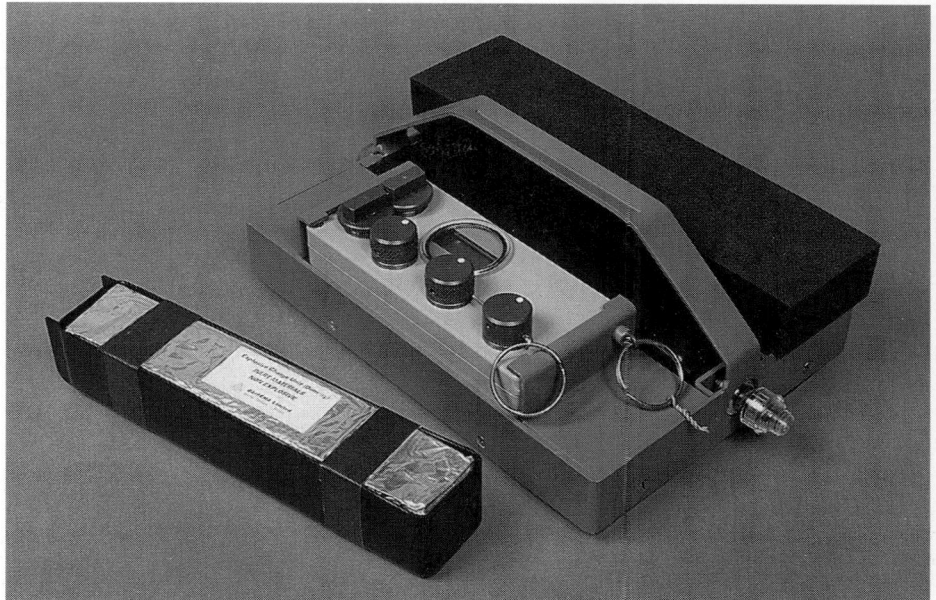

Limpet mine   *1998*/0007236

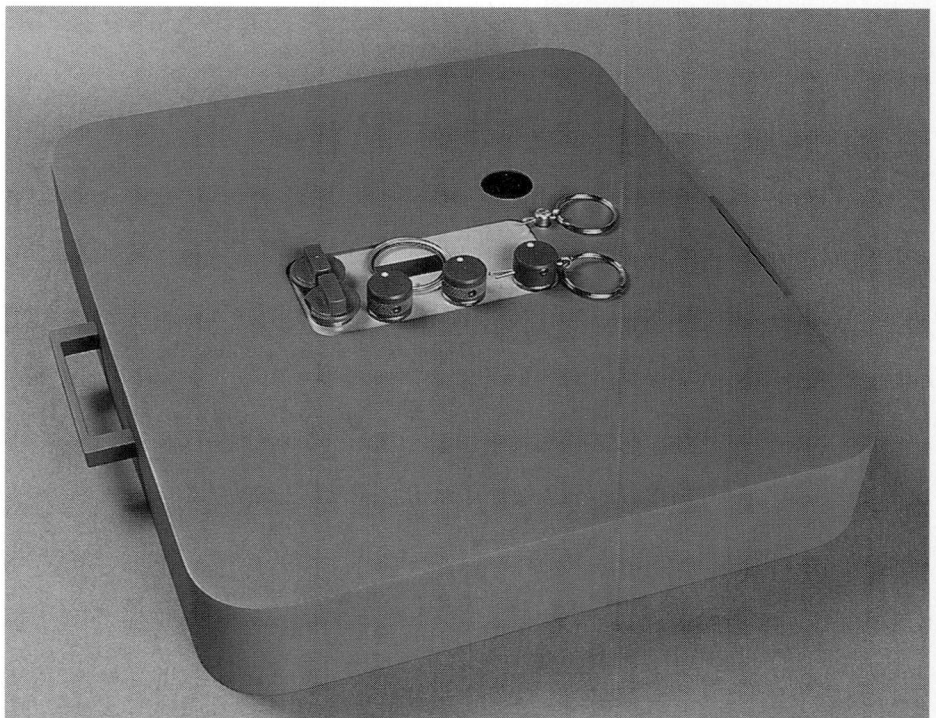

Limpet mine variant
1998/0007237

LNS4-3 magnetic sensor

Type
Three-axis magnetic anomaly detector.

Description
The PMES LNS4-3 is a three-axis magnetic sensor which detects small magnetic anomalies or measures low-frequency magnetic fields. Although designed primarily for use in sea mines, other applications include geophysical surveys, surveillance, security screening, and monitoring for unwanted ferrous matter. The salient feature of the LNS4-3 is that it incorporates a fast reset facility, enabling multiple target detection.

LNS4-3 can detect magnetic anomalies as low as 2.4 nT in the earth's static field and requires less than 5 mW of power from a ±5 V DC supply. Two low-impedance outputs are provided for each axis; one is AC coupled and the other is DC coupled. Both outputs are suitable for interfacing directly to a multiplexer and an analogue to digital converter. The device is compact (100 × 690 × 46 mm), robust, totally self-contained and encapsulated. Reliability is quoted as 300,000 h MTBF.

Operational status
In service.

Contractor
PMES Ltd, Rugeley.

VERIFIED

UNITED STATES OF AMERICA

Mk 52, 55, 56 and 57

Type
Seabed mines.

Description
The current stock of sea mines consists of the Mks 52, 55, 56 and 57 types.

Operational status
Used extensively in the Vietnam War, this family of mines is now obsolete, although a number may still remain in the US inventory.

Specifications
Type: Mk 52, submarine/aircraft laid seabed mine
Length: 2,250 mm
Diameter: 844 mm
Weight: 542 kg (Mod 1); 567 kg (Mod 2); 572 kg (Mod 3); 570 kg (Mod 5); 563 kg (Mod 6)
Charge: 300/350 kg HBX-1
Service depth: 45.7 m (Mod 2, 183 m)
Actuation: acoustic (Mod 1); magnetic (Mod 2); pressure/magnetic (Mod 3); acoustic/magnetic (Mod 5); pressure/acoustic/magnetic (Mod 6)

Type: Mk 55, submarine/aircraft laid seabed mine
Length: 2,890 mm
Diameter: 1,030 mm
Weight: 580 kg (Mod 2); 992 kg (Mod 3); 992 kg (Mod 5); 996 kg (Mod 6); 995 kg (Mod 7)
Charge: 576 kg HBX-1
Service depth: 45.7 m (Mods 2/7, 183 m)
Actuation: magnetic (Mod 2); pressure/magnetic (Mod 3); acoustic/magnetic (Mod 5); pressure/acoustic/magnetic (Mod 6); dual-channel magnetic (Mod 7)

Mk 36, 40, 41 and 115A

Type
Air laid naval mines.

Description
Also employed by the US Navy is a range of air-deployed munitions based on modified general purpose low drag bombs and which can be released without requiring a parachute. The modification involves the use of a Mk 75 Mod 0 Destructor Modification Kit which can be added to 500, 1,000 and 2,000 lb Mk 80 series bombs to form the Service Destructors (DST) Mks 36, 40 and 41, respectively. These are mostly intended for use in shallow waters such as estuaries, against typical coastal targets. There is also the DST 115A, which can be employed with either aircraft or surface craft for use against surface targets.

Operational status
Replaced from the 1980s onwards by the Quickstrike mines. Some may remain in servce.

Quickstrike series

Type
Air launched seabed mines.

Description
The Quickstrike seabed mine development programme embraced a family of mines using different size cases but with common target detection and classification mechanisms. The four members of the Quickstrike family are the Mks 62, 63, 64 and 65. The last of these (Mk 65 Mod 0) is in the 2,000 lb (900 kg) class and is in full production by Aerojet TechSystems in Sacramento, California. The Mk 64 is also in the 2,000 lb (900 kg) class, based on a Mk 84 2,000 lb bomb and measuring 3.8 m long with a 633 mm diameter.

Quickstrike mines are for shallow water deployment (to approximately 100 m) and targets will have to approach to within a few hundred feet for it to act. It will use existing Mk 80 series GP bomb cases as well as a new mine case. Quickstrike mines are deployed

Mk 60 Captor

Type
Encapsulated torpedo.

Description
Captor, a contraction of 'encapsulated torpedo', is the name given to an anti-submarine system comprising a Mk 46 torpedo inserted into a mine casing.

Deployment is in deep water, generally in the vicinity of strategic routes travelled by enemy submarines, and submarines are the intended targets. US officials have stated that Captor has the ability to detect and classify submarine targets, while surface ships are able to pass over a Captor field without triggering the Mk 46 Mod 4 torpedo which carries the warhead (43.5 kg PBXN103 explosive). This capability is reported to have been tested.

The Detection and Control Unit (DCU) that performs these functions is the most costly subassembly of the complete Captor weapon and accounts for about 45 per cent of the total unit cost – approximately US$130,000. DoD statements imply that the DCU incorporates facilities for turning itself on and off, in addition to its principal operational functions of

Type: Mk 56 Mod 0, aircraft laid moored mine
Length: 3,500 mm
Diameter: 1,060 mm
Weight: 1,010 kg
Charge: 500 kg HBX-3
Service depth: 366 m
Actuation: total field, dual-channel magnetic, acoustic/magnetic

The Mk 56 first became operational in 1966. Specifically designed for use against high-speed deep operating submarines.

Specifications
Type: Mk 36, aircraft laid seabed mine
Length: 2,250 mm
Diameter: 400 mm
Weight: 240 kg (with fixed conical fin); 261 kg (with tail retarding device)
Charge: 87 kg H-6
Service depth: 91.4 m
Actuation: magnetometer (Mods 0/3); magnetic/seismic (Mods 4/5), dual-channel magnetic, acoustic

Type: Mk 40, aircraft laid seabed mine
Length: 2,860 mm
Diameter: 570 mm
Weight: 447 kg (with fixed conical fin); 481 kg (with tail retarding device)
Charge: 204 kg H-6
Service depth: 91.4 m
Actuation: magnetometer (Mods 0/3); magnetic/seismic, dual-channel magnetic, acoustic

Type: Mk 41, aircraft laid seabed mine
Length: 3,830 mm

primarily by aircraft, but surface ships or submarines can also launch the weapons.

This family of mines is based primarily on conversion of existing ordnance (bombs and torpedoes). An exception is the Mk 65 mine which is not a bomb conversion. It has a thinner case than the equivalent bomb and contains the effective underwater PBX explosive. It is 3.25 m long and 533 mm in diameter. The Mk 65 has now been deployed with the US Navy and is being evaluated by the Italian Navy.

Operational status
In service with US forces.

Specifications
Type: Mk 62, aircraft laid seabed mine
Length: c 2,200 mm
Diameter: 273 mm
Weight: 227 kg
Charge: 89 kg
Actuation: magnetic and pressure

detecting possible targets, classifying them by their sound signatures, and initiating release of the Mk 46 torpedo when the target is within range of its homing head. It is probable that the turn on/turn off system is quite sophisticated in the interests of power conservation to ensure the maximum operational life for deployed Captor mines. Factors which are likely to be taken into account include the levels of traffic (surface and submarine), ambient conditions, sea state and so on.

Both active and passive sensor modes are employed and the system first operates in a listening (passive) mode, which continues for a certain length of time sufficient to identify the target as a submarine and not a surface vessel. The system then switches to an active mode, during which it is assumed target ranging is carried out to determine the optimum release time for the homing torpedo.

The DCU is gated to ignore surface traffic and has an estimated range of about 1,000 m on submarine targets. There is no IFF capability and friendly units must be warned of Captor minelaying and positions of deployed mines.

There are presumably provisions for some method of self-deactivation or self-destruct for those Captors

Type: Mk 57 Mod 0, submarine or ship laid moored mine
Length: 3,000 mm
Diameter: 510 mm
Weight: 934 kg
Charge: 154 kg HBX-3
Service depth: 350 m
Actuation: total field, dual-channel magnetic, acoustic/magnetic

UPDATED

Diameter: 630 mm
Weight: 926 kg (Mods 0/3); 921 kg (Mods 4/5)
Charge: H-6
Service depth: 90 m
Actuation: magnetometer (Mods 0/3); magnetic/seismic (Mods 4/5), dual-channel magnetic, acoustic

Type: 115A, aircraft laid surface mine
Length: 450 m
Diameter: 620 mm
Weight: 61 kg
Charge: 24 kg HBX-3
Service depth: 45 m
Actuation: magnetic/seismic

UPDATED

Type: Mk 63, aircraft laid seabed mine
Length: c 3,000 mm
Diameter: 350 mm
Weight: 454 kg
Charge: 202 kg
Actuation: magnetic and pressure

Type: Mk 64, aircraft laid seabed mine
Length: 3,800 mm
Diameter: 460 mm
Weight: 908 kg
Actuation: magnetic or magnetic/pressure

Type: Mk 65, submarine/aircraft laid seabed mine
Length: 3,250 mm
Diameter: 533 mm
Weight: 908 kg
Charge: PBX
Actuation: magnetic and pressure and combined magnetic/pressure

UPDATED

which are life-expired and not capable of retrieval, and other measures to prevent unauthorised salvage or interference may be expected.

The current deployed life of a Captor mine is thought to be in the region of six months, but it is not known if this is the USN's target, although it is a fact that considerable effort has been placed on ensuring maximum shelf and operational life.

Captor mines can be sown by surface ships, submarines and by aircraft. In the first delivery mode, mine rails or other delivery systems are not required, the chosen technique being by means of an over-the-side boom (or yard and stay) with a capacity of 1,045 kg. The Captor is brought to a point about 10 m above the surface before release, which has to be at an angle to ensure proper entry into the water. Any submarine equipped with standard 21 in (533 mm) torpedo tubes can lay Captor mines. Aircraft employ a parachute technique for delivery of Captor.

According to USN statements, there are two main options for Captor delivery: one utilises P-3 maritime aircraft, nuclear strike submarines and surface ships; the other envisages a combination of USAF B-52 bombers and P-3 aircraft, plus surface ships and nuclear strike submarines. Captor has been tested for delivery on P-3C, A-6, A-7 and B-52 aircraft, LKA and other cargo ships, and aboard 10 different classes of submarine, including conventionally powered boats.

Most Captor minefields will be barriers located at some distance from possible enemy defences, and in the more highly defended areas Captor mines would be delivered by aircraft or submarines. The unique capability of submarines to plant mines covertly and under ice would be employed selectively.

Operational status

Initial production commenced in March 1976, although the US Secretary of Defense's report of January 1981 stated the procurement had been at a low level while development and testing were conducted to correct performance deficiencies. In December 1978, reduced performance against shallow water targets had been indicated. Nevertheless, initial operational capability was achieved in September 1979. Further procurement was cancelled after FY80 by a decision taken in December 1980, and in January 1981 a Captor improvement programme was approved. The weapon was granted approval for service use in February 1981, but no procurement funds were sought in the FY81 budget 'because Captor failed to provide the high level of effectiveness we had sought'. Subsequent testing has demonstrated that recent modifications have corrected its performance deficiencies.

Despite this somewhat chequered history, it was disclosed that about 630 Captors had been procured. The system is no longer in production, although weapons remain in the inventory.

Specifications
Length: 3,700 mm (overall)
Diameter: 533 mm
Weight: 1,075 kg (with torpedo and mooring)
Warhead: 45 kg Mk 46 torpedo
Destructive range: 1,000 m
Operating depth: >305 m

Contractor
Lockheed Martin, Akron, Ohio.

UPDATED

Mk 67 Submarine-Launched Mobile Mine (SLMM)

Type
Guidance and control system upgrade for SLMM.

Development
The Submarine-Launched Mobile Mine (SLMM) Mk 67 is intended to provide the US fleet with a capability for planting mines in shallow water (to approximately 100 m) by submarine, using a self-propelled mine to reach water inaccessible to other vehicles. It is also meant for use in locations where covert mining would be particularly desirable from a tactical standpoint.

The Mk 67 SLMM is based on a Mk 37 torpedo with a Mk 13 mine warhead, an auxiliary controller that occupies the previous acoustic panel location, and a standard Mk 37 electromechanical guidance system with a few modifications. The tail section is a standard Mk 37 with modifications to the enabler. The current Mk 67 guidance and control system comprises multiple relays and vacuum tubes that can no longer be supported and require frequent and extensive maintenance. To overcome these difficulties in supporting the Mk 67, Hughes Aircraft Naval and Maritime Systems (formerly Alliant Techsystems) began development in 1992 of a software-controlled guidance and control system to replace the obsolete electromechanical system. The guidance and control system successfully passed in-water production verification tests in 1993.

A shortened version of the guidance and control system offers a mine of reduced length that allows stowage of two mines per storage rack on 'Los Angeles' class submarines.

The various versions use either magnetic/seismic or magnetic/seismic/pressure target detection devices.

Description
The upgrade programme retains the existing Mk 13 mine section and replaces the auxiliary controller and original Mk 37 guidance system with a microprocessor-controlled, solid-state guidance system that incorporates the auxiliary controller functions. The cumbersome Mk 37 electromechanical enabler is replaced by a simpler rpm counter. The new guidance and control system enables current inventories of Mk 67 mine to be upgraded or it can be incorporated in new conversion programmes.

The new system replaces unreliable electromechanical relay logic and vacuum tube control circuitry with software-implemented logic and control algorithms. As a result, the maintenance overhaul cycle for the guidance and control system has been increased from 18 months to six years.

The new hardware easily adapts to the Mk 67, fitting inside the afterbody chassis envelope using existing mounting provisions.

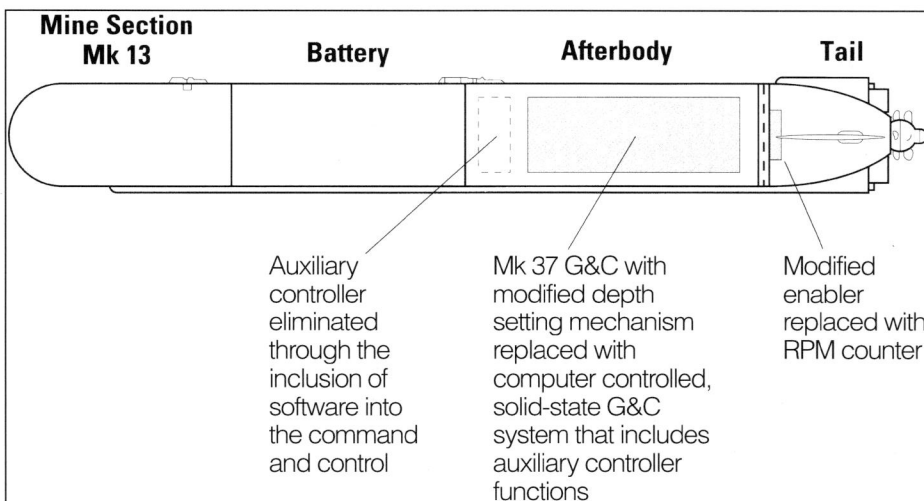

Mine Section Mk 13 — **Battery** — **Afterbody** — **Tail**

Auxiliary controller eliminated through the inclusion of software into the command and control

Mk 37 G&C with modified depth setting mechanism replaced with computer controlled, solid-state G&C system that includes auxiliary controller functions

Modified enabler replaced with RPM counter

Mk 67 SLMM upgrade　　*1997*

Specifications
Length: 4,090 mm
Diameter: 485 mm
Weight: 745 kg

Contractor
Raytheon Systems Company, Mukilteo, Washington.

VERIFIED

YUGOSLAVIA, FEDERAL REPUBLIC OF

M66

Type
Diversionary underwater mine.

Description
This practically non-magnetic mine is designed for the destruction of vessels, harbour installations and other fixed offshore installations and in rivers and lakes. The weapon incorporates a pyrotechnical safety device which enables full safety in handling and preparation of the mine. Various clockwork fuze settings from a minimum of 20 minutes up to a maximum of 10 hours are possible. Fuze timing is set during weapon preparation. Once the mine has been attached to its target and the mine safety and pyrotechnical safety elements removed, the weapon cannot be removed from the target without risking detonation.

Specifications
Length: 670 mm
Diameter: 320 mm
Width: 430 mm
Weight: 50 kg
Charge: 27 kg TNT (approx)
Operating depth: 30 m

Contractor
Yugoimport SDPR, Belgrade.

VERIFIED

M70

Type
Acoustic influence seabed mine.

Description
The M70 acoustic influence seabed mine is designed to destroy or seriously damage warships up to 5,000 tonnes and merchant ships up to 20,000 tonnes and over. Highly sensitive sensors and a large explosive charge make the weapon extremely effective and applicable for either offensive or defensive roles.

The mine can be laid from submarine torpedo tubes in depths suitable for anti-submarine (150 m) and anti-surface vessel (50 m) targets. Targets and laying depths can be preselected.

The detonating system comprises both mechanical and fully transistorised electrical devices using state-of-the-art printed board circuit technology. The detonation system is compact, robust and resistant to vibration/shock and stable under any climatic conditions for long periods of time.

Specifications
Length: 2,823 mm
Diameter: 534.4 mm
Weight: 1,000 kg
Charge: 700 kg
Operating depth: 50-150 m

Contractor
Yugoimport SDPR, Belgrade.

VERIFIED

M71

Type
Limpet mine.

Description
The M71 limpet mine is designed for use against both ships and submarines. Detonation of the weapon is designed to rupture the underwater part of the hull plating. It is attached to the target by means of magnets, and to wooden structures by special screws fitted on the mine.

The weapon is fitted with a time fuze and clockwork mechanism with time settings ranging from 0.5 to 10 hours and with anti-removal devices. Fuze activation is performed after the mine has been fixed to the target.

Specifications
Diameter: 345 mm
Height: 245 mm
Weight: 14 kg
Charge: 3 kg pressed TNT
Operating depth: 30 m

Contractor
Yugoimport SDPR, Belgrade.

VERIFIED

ASW ROCKETS

Medium-range rocket launchers
One of the most widely adopted developments in the years since the Second World War has been the medium-range anti-submarine rocket launcher. Brief details of the French system, and Swedish system on which the French system is based, are given. Although it differs in some important respects from the French and Swedish systems described below, the Norwegian Terne system is also in this general weapon category. Many similar developments have also taken place in recent years in the former USSR.

Other entries give details of the widely used Bofors four-tube launcher and the more recent two-tube launcher.

Depth charge mortars
Towards the end of the Second World War a more streamlined type of depth charge with a higher rate of sinking was introduced by the RN and subsequently by other navies. The intention was to project charges ahead of an attacking ship so that they could be fired more accurately while still in sonar contact, which was otherwise lost when the attacking ship ran over the submarine to deliver conventional depth charges. Such mortars have been developed in several countries, Squid and Limbo in the UK were two examples, the Italian Menon being a third.

Another weapon in this category is the French four-barrelled mortar. This is mounted in a turret and is automatically loaded. It fires a heavier projectile and has a longer range (about 2,750 m) than any of the others. All these mortars fire 12 in (305 mm) depth charges. Sweden has developed the small grenade launcher ELMA which is now operational in some Scandinavian navies.

It is believed that the former Soviet navies have not adopted the streamlined form of depth charge. They do, however, use depth charge mortars, but it is thought possible that they use compressed air to propel the charge whereas other countries use an explosive cartridge.

VERIFIED

FRANCE

Mk 54

Type
Surface-to-subsurface anti-submarine rocket.

The French Bofors ASW rocket launcher on the frigate Lieutenant de Vaisseau le Hénaff (Stefan Terzibaschitsch)

Description
This system comprises an anti-submarine rocket launcher associated with a sonar and a computer. The rocket launcher is remotely controlled, aiming and rocket fuzing being determined by the computer, which in turn receives input data from the sonar.

The launcher is made by Mécanique Creusot-Loire under licence from Bofors and is a six-tube device with automatic reloading from a magazine. It fires single rockets or salvos as required and will accept any of the range of Swedish Ordnance 375 mm rockets, thereby giving a choice of ranges from approximately 655 to 3,625 m. Rate of fire can be up to one round per second.

The computer calculates ballistic data for initial velocities of 100, 130, 165 and 205 m/s for the different rockets that may be used with the systems.

Input data from the sonar comprise the location and rate of change of position of the target.

Mécanique Creusot-Loire is also licensed for the manufacture of the twin-tube rocket launcher.

Operational status
The rocket system equips the 12 'Type A69' frigates of the French Navy and three 'E71' class escorts of the Belgian Navy. The system is currently being removed from the French frigates.

Specifications
Warhead: 107 kg
Range: 1,600 m

Contractor
Mécanique Creusot-Loire, Immeuble Ile de France, Paris La Défense (launcher).

UPDATED

NORWAY

Terne III

Type
Anti-submarine rocket-propelled depth charge.

Description
Terne III is a short- to medium-range anti-submarine weapon system, ripple-firing rocket-propelled depth charges, in salvos of six, against submarine targets. It is designed to be the principal anti-submarine weapon for ships up to frigate size, and a secondary weapon for ships carrying long-range anti-submarine armament. Salvo patterns can be formed according to the tactical situation up to the moment of firing. Recoverable practice missiles are available. The remote power-controlled six-tube launcher is automatically loaded within 30 seconds from a manually loaded magazine hopper. The launcher is specially protected against arctic conditions.

The main features of Terne III are:
(a) a unified and balanced design
(b) low weapons weight and small space requirements. A complete installation weighs less than 10 tonnes
(c) the weapon part can be interfaced with existing sonars and other target data sources. Norwegian frigates have a new modular fire-control unit to which the Terne III system is interfaced via databus STANAG 4156
(d) freedom of choice in the position on board
(e) remote-control firing at elevation angles between 47 and 77°, giving variable ranges throughout 360° of bearing
(f) depth charge is fitted with proximity/time/percussion fuze
(g) high rate of fire.
The system has been upgraded to the Mk 10 version.

Operational status
In production since 1960. Fitted in four Royal Norwegian Navy frigates of the 'Oslo' class.

Specifications (Rockets)
Length: 1.97 m
Diameter: 20 cm
Weight: 120 kg
Warhead: 70 kg TNT equivalent
Fuze: proximity, time and impact
Range: 400-5,000 m (Mk 10 version)

Contractor
Kongsberg Defence and Aerospace AS, Kongsberg.

VERIFIED

RUSSIAN FEDERATION AND ASSOCIATED STATES (CIS)

ASW rocket launchers

There is a wide variety of short-range anti-submarine rocket launchers in the units of the RFAS fleet, although all appear to operate on much the same principle of firing charges in a pattern ahead of an attacking ship from a mounting which can be trained and elevated under remote control. Two different calibres of 250 mm and 300 mm exist and the weight of the bombs in the 250 mm versions has been estimated as 180 to 200 kg. The RBU associated number is understood to mean the maximum range possible in metres. Some of the systems are also understood to have an anti-torpedo capability. The projectiles are thought to be fitted with impact fuzes.

RBU 2500: a 16-barrelled 250 mm calibre system introduced in 1957 and fitted with two horizontal rows each of eight launching tubes about 1.6 m long. The trainable launchers can elevate up to about 85°. Depth bombs are loaded by hand. The system is fitted in the older cruisers and destroyers although also in the more modern but smaller 'Petya I'. Production appears to have ceased.

RBU 6000: probably of 300 mm calibre, and fitted very widely this system entered service around 1967. Twelve launch tubes are arranged in a circular fashion and it is believed that they are reloaded automatically by bringing them to the vertical and then indexing one by one while depth bombs are loaded from below. Range is 3.25 n miles and the launch barrel length has been extended to about 1.8 m and can be elevated to about 50°. The system remains in production.

RBU 1200: a five-barrelled system reportedly introduced in 1957 and which was fitted normally on the quarters of larger ships such as the *Kara* and *Kresta*. Almost certainly it has a range of no more than 1.5 n miles. The system, because of its size and mounting almost certainly employs automatic reloading. It is probably intended to deal with torpedoes if they can be detected in the closing stages of an attacking run. The tubes are on a fixed mounting and can elevate up to about 65°. Production continues, but possibly only for export.

RBU 12000: A 10-tube system deploying a much heavier rocket than other variants. The system equips the latest carriers of the Russian Navy. Like the other ASW rocket launchers, the RBU 12000 is designed to form part of a layered defence. It is able to deploy two types of munition: a decoy round and an ASW rocket round. In Russian service the system is referred to as UDAV-IM.

The system is thought to still be in production.

Specifications

	RBU 12000	RBU 6000	RBU 2500	RBU 1200
Length	n/k	1.8 m	1.6 m	1.4 m
Range	6.5 n miles	3.25 n miles	1.5 n miles	0.8 n miles
Warhead	80 kg	31 kg	21 kg	34 kg
Diameter	n/k	300 mm	250 mm	250 mm
Weight	232.5 kg*	100 kg	n/k	70 kg
Elevation	n/k	50°	85°	65°

Note: * ASW rocket, the decoy round weighs 201 kg

UPDATED

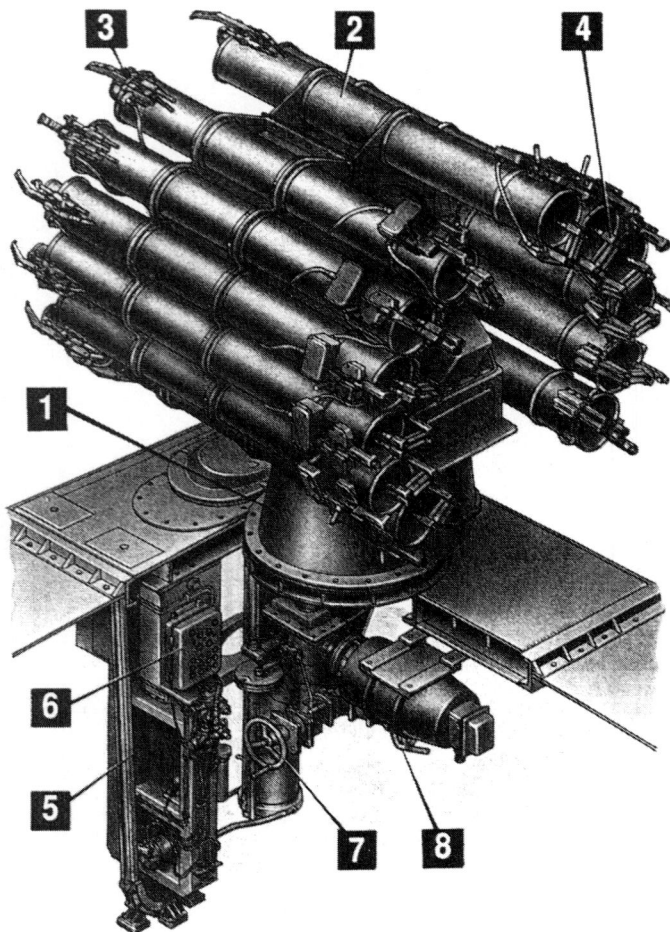

RBU 6000 anti-submarine rocket launcher: **(1)** *launcher pedestal;* **(2)** *tube;* **(3)** *front clip;* **(4)** *rear clip;* **(5)** *rocket hoist;* **(6)** *control panel;* **(7), (8)** *traversing and elevating handwheels*   *1998*/0007238

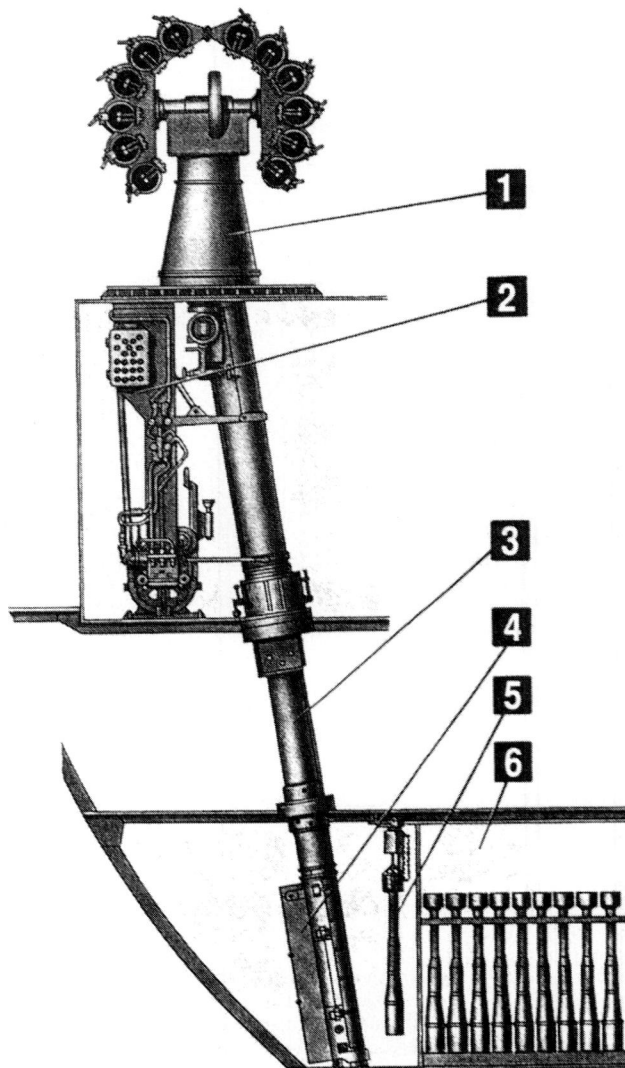

RBU 6000 anti-submarine rocket launcher with rocket hoist and elevator: **(1)** *rocket launcher;* **(2)** *rocket hoist;* **(3)** *elevator;* **(4)** *elevator feeder;* **(5)** *anti-submarine rocket RGB-60;* **(6)** *rocket magazine*   *1998*/0007239

SWEDEN

Type 375

Type
Shipborne surface-to-surface short- and medium-range ASW rocket system (two-, four- and six-tube arrangements).

Description
The ship's sonar provides submarine position for prediction of Type 375 launcher elevation and bearing data. The launcher has either two or four tubes and can fire single- or multiple-shot salvos. A special design ensures a predictable and accurate underwater trajectory. The launcher is reloaded by automatic means from the magazine, which is disposed of directly below the launcher. The twin launcher has an integral motor-driven twin hoist, for loading both tubes at once, and a rotating loading table. The laying mechanism is electrohydraulic with a choice of local or remote operation. Fuze setting is with the rocket in the tube. The rocket table holds four rounds and the total number of rockets in the operating room is 24. The shortest time between successive firings is 1 second. The time for firing six ready rockets (four on the table

Bofors twin-tube ASW rocket launcher

and two in the tubes) is 1 minute, and the firing rate for continuous fire is two rockets every 45 seconds. The manning complement during continuous firing is three men.

Rockets have two types of motor giving differing range brackets. The rocket trajectory is flat, thus giving a short time of flight to minimise target evasive action. Fuzes are fitted with proximity, time, and DA devices.

The 80 mm MI type is launched from a special holder that is handled as a live rocket in the operating room, hoist and launcher.

Operational status
The twin-tube launcher and anti-submarine rockets are in service with a number of navies including: the Mk 9 corvettes of the Nigerian Navy; 'Nilgiri' class ('Leander' type) frigates of the Indian Navy; the 'Descubierta' class frigates of the Egyptian, Spanish and Moroccan navies; the 'Niteroi' class of the Brazilian Navy; the 'Fatahillah' class frigates of the Indonesian Navy and the 'Kastori' class frigates of the Malaysian Navy.

Specifications (twin-tube launcher)
Weight, excluding rockets but including flame guard and deck plate: 3.8 t
Traversing speed: 30°/s
Elevating speed: 27°/s
Traverse limits: unlimited
Elevation limits: 0 to +90° (mechanical); 0 to +60° (for firing)
Power supplies: 440 V, 60 Hz, 3-phase
Power consumption: tracking 6 kW (mean)

Contractor
Bofors AB, Motala.

Specifications

375 mm ASW	Weight with fuze	w/out fuze	Length time fuze	proximity fuze	Charge	Max velocity 1 motor	2 motor	Final sinking speed	Range to 30 m depth
Erika	250 kg	236 kg	2,000 mm	2,050 mm	107 kg hexotonal	100 m/s	130 m/s	10.7 m/s	655-1,635 m
Nelli	230 kg	216 kg	2,000 mm	2,050 mm	80 kg hexotonal	165 m/s	205 m/s	9.2 m/s	1,580-3,625 m
Practice Type H	100 kg	86.5 kg	–	–	–	135 m/s	–	–	980-1,570 m
80 mm MI 15.5	–	–	1,120 mm	–	–	130 m/s	–	–	980-1,600 m

VERIFIED

ASW-600

Type
Lightweight multibarrel ASW system.

Development
The Saab ASW-600 was developed to meet the threat of small- and medium-sized submarines operating in coastal waters in the Baltic.

Description
The ASW-600 system consists of four launchers, each with nine barrels for grenades, control electronics and an operator's fire and launcher selection panel.

The launcher is loaded with grenades from the muzzle and propulsion charges from the rear. The grenade has a shaped charge specially designed to penetrate the submarine's form and pressure hull. A new grenade, designated M90 and with enhanced efficiency, is currently replacing the basic M83 grenade in the Swedish Navy.

Upon target detection the operator designates the target and selects the number of grenades for the salvo. Target location and predicted impact area of the grenades are presented on the sonar display in the CIC. This information is used to steer the ship towards the firing point. Firing is then normally initiated at a distance of 300 to 400 m from the ship to the impact area.

A carpet of up to 36 grenades can be laid in a selectable pattern over the submarine's located position. The size of the carpet can be varied from about the size of a football field to a tennis court.

The latest generation of the ASW-600 multibarrel

ASW system is the ASW-601 which comprises two trainable platforms each fitted with four launchers carrying nine barrels. The ASW-601 is specially designed for operation in shallow waters. Using data from sonar or other target designators the system calculates an impact point and trains the platform to the predicted angles. These are then continuously updated to track the target until the decision is made to fire. The computer in the fire-control unit calculates and recommends the best firing instant, salvo size and number of grenades. A carpet of up to 36 grenades can be laid in a selectable pattern optimised to achieve the highest hit probability. Salvos are alternated between launchers. The non-firing launchers aim continuously at the target area guided by the fire-control units, thus optimising the grenade pattern. The ASW-601 can cover an area out to a range of 150 to 450 m. The ammunition used is the M-90E grenade fitted with a shaped charge specially designed for underwater penetration. For the ASW-601 system there is also chaff and IR ECM ammunition.

The ASW-604 is designed to fit almost any type of helicopter. Part of the concept involves integrating the grenade launcher with the Hawkeye and using a simple sighting system. The grenades are then fired from two multibarrel launchers mounted on either side of the helicopter, each containing 20 grenades. Grenades are either launched in batches of two to four or all together with a short timed delay between each grenade launching.

Operational status
More than 40 systems of the ASW-600 have been delivered and are operational on all corvettes, FACs and MCMVs of the Swedish Navy. The system is also operational on board Finnish patrol boats.

Specifications
System
Number of grenades: ASW 600 – 9, 18, 27 or 36 per salvo; ASW 601 – max 36 per salvo
Range: ASW 600 – 400 m (approx); ASW 601 – 150-450 m (approx)
Dispersal: adjustable

Grenade
Calibre: 100 mm
Length: 478 mm

ASW-601 grenade system undergoing trials on board a Swedish 'Kaparen' class FAC   **1997**

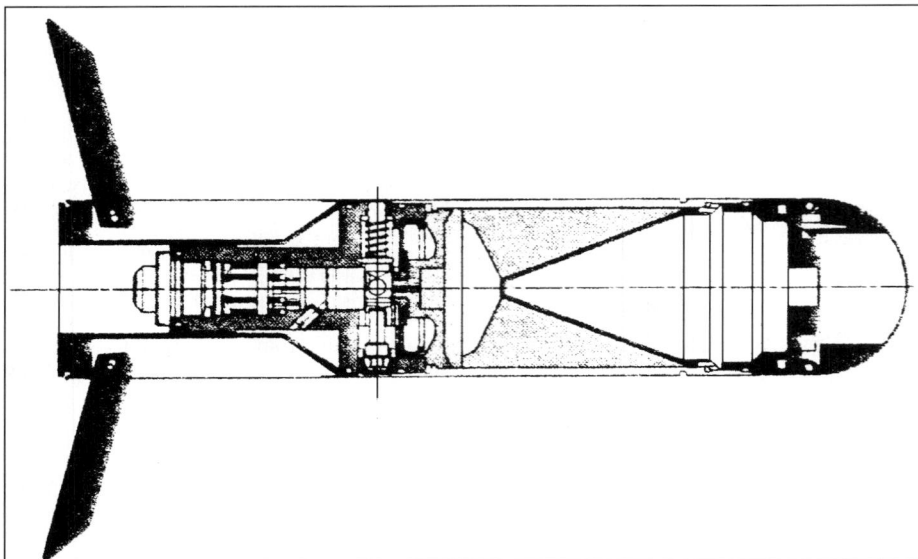

Weight: 5.7 kg
Equipment weights and dimensions
Launcher: 103 kg, 460 × 920 × 620 mm
Relay unit: 7 kg, 330 × 230 × 110 mm
Firing unit: 6 kg, 330 × 230 × 148 mm
Rectifier: 34 kg, 610 × 315 × 210 mm

Contractor
Saab Dynamics, Linkoping.

M90 grenade
1996

KAS 2000

Type
Active acoustic homing ASW Weapon System.

Description
The KAS 2000 is a new-generation ASW weapon based on terminally guided 127 mm grenades. Using target designation data from sonars, the system calculates the predicted point of impact and trains the gyrostabilised launcher. The system then locks on and tracks the target until firing. Target data is fed into the grenade immediately before launching. After firing, the grenade achieves a maximum speed of 120 m/s and follows a ballistic path to the target area. Descent is stabilised and slowed to 80 m/s by a parachute assembly, which ensures the correct angle of entry into the water. On entering the water the parachute is jettisoned and the active acoustic homing system commences seeking the target using a medium sized number of preformed beams. On impact, a specially shaped charge detonates to punch a hole in the submarine's hull. Firing can be generated from a general purpose console or from a separate dedicated operator console.

The launcher can be trained in both elevation and azimuth providing a range of between 100 to 1,200 m and the grenade is effective in both extremely shallow and deep waters.

The system can be used in a multipurpose role, deploying active and passive ECM decoys both above and below the surface, as well as the ASW ammunition.

Operational status
The system is designed to be installed on the new 'Visby' YS 2000 class corvettes now under construction, but is awaiting further official funding.

Contractor
Saab Dynamics, Linkoping.

UPDATED

UNITED KINGDOM

Limbo AS Mk 10

Type
Surface-launched anti-submarine mortar.

Description
This is a shipborne surface-to-subsurface medium-range anti-submarine mortar system. Mortars are stabilised in pitch and roll by a metadyne system referenced to the ship's stable platform. For details of this operation see *Jane's Weapon Systems 1987-88*.

Weight of the projectile is about 177 kg with a 92 kg HE warhead and range is about 1,000 m.

Operational status
Only a few systems remain in service: on three 'Nilgiri' ('Leander' type) class frigates of the Indian Navy; the three ex-UK 'Tribal' class frigates in the Indonesian Navy; three 'Alvand' class frigates of the Iranian Navy; and the frigate *Rahmat* of the Malaysian Navy.

Specifications
Weight of system: 35 t
Weight of bomb: 177 kg
Warhead: 92 kg
Range: 1,000 m

Contractor
Manufactured to MoD (Navy) designs by several contractors.

UPDATED

DEPTH CHARGES

Depth charges
Relatively simple cylindrical depth charges that can be rolled or catapulted into the sea or dropped from aircraft are the longest established anti-submarine weapons; they were first used by the Royal Navy in the First World War. They are generally depth-fuzed and have a low sinking rate to give the launching vessel time to get clear. They are still extensively used by many navies: a typical weight is 150 kg and a launcher can project the charge up to about 150 m.

Nuclear depth charges
Also introduced by the US Navy is the nuclear depth charge. To take this clear of the launch vessel a rocket is required, and these depth charges have so far been associated only with ASROC and SUBROC. In the UK, it is understood that Royal Navy Sea King helicopters can be equipped with 'Bomb 600 lb special ordnance'. Based on the US B-57 depth charge, this is intended to attack deep-diving submarines, which are running too

deep or too fast for homing torpedoes. These nuclear devices are of UK manufacture. It is believed that nuclear depth charges are being removed from the inventory of NATO navies.

UPDATED

CHILE

AS-228 depth charge

Type
Air/surface-launched depth charge.

Description
The AS-228 depth charge is an anti-submarine weapon with a hydrostatic pressure-activated fuze that permits its use against targets at depths from 100 to 1,600 ft (30

to 490 m) and the detonation depth can be preset to any one of 19 depths between these limits. The detonator also incorporates three safety measures for

Cardoen anti-frogman underwater grenades

Cardoen AS-228 anti-submarine depth charge

handling and transportation, inertia, and submarine action. The charge itself can be launched by conventional methods from naval vessels or from aircraft, including helicopters. The fuze, manufactured by Industrias Cardoen, is also supplied as a separate unit as a replacement for outdated fuzes in depth charges and bombs of other manufacture.

This company also manufactures underwater hand grenades for use as an anti-diver weapon for the protection of ships moored or anchored in insecure waters. These are operated by hydrostatic fuzes and can be set to explode at depths between 4 and 12 m. Thrown overboard at intervals alongside warships they offer a defence against the attentions of frogmen.

Operational status
In production.

Contractor
Industrias Cardoen Ltda, Santiago.

VERIFIED

ITALY

SLD-302-SAF and DC-103-HAF

Type
Anti-intruder defence system.

Description
The SLD-302-SAF (Scattering Anti-Frogman) system has been designed to protect vessels and maritime installations against underwater attack. The system comprises a launching unit, expendable launching tubes and control panel.

The standard system consists of four to eight tubes, each containing two or three depth charges and the electric activation device. The tubes can be mounted on the launching unit with different orientations to cover a wide sector with ranges up to 80 m from the launch site.

The charges contained in the same tube can be factory set for different operating depths. The control panel allows sequential or single ignition of the tubes.

After discharge the tubes can be rapidly replaced which allows the firing of approximately one salvo per minute.

The system can be integrated with the hand-launched DC-103-HAF (Hydrostatic Anti-Frogman charge) which is designed to provide a safe and reliable hand weapon for the security personnel responsible for the protection of vessels and maritime installations against underwater attack.

Operational status
In production.

Specifications (launch tube)
Length: 210 mm
Width: 400 mm
Height: 520 mm
Weight: 18 kg
Operating depth: 6-30 m
Charge
Diameter: 66 mm
Weight: 0.4-0.9 kg (HAF charge)
Charge: 0.2-0.5 kg HBX3

Contractor
SEI SpA, Ghedi.

UPDATED

DC101

Type
Smart anti-submarine charge.

Description
The DC101 has been developed to replace the obsolete Mk 54 anti-submarine depth charge under the Italian Navy programme MS500.

The low-cost charge is an intermediate type of weapon between the depth charge and the lightweight ASW torpedo.

The DC101 is torpedo shaped with an elliptic head and tail with four fins and stabilisation ring. A parachute is fitted in the tail to ensure the correct angle of entry and speed into the water.

The DC101 is fitted with an influence fuze which is controlled by an active sonar mounted in the nose of the weapon which ensures the optimum possibility of the charge striking the target without the need for pre-setting the detonation depth. At a depth of 5 m the system is armed and the active sonar commences searching for the target. All processing is carried out within the charge using sophisticated algorithms in order to localise the target and filter out false echoes and background noise. On acquiring a definite target the range is continuously monitored and the warhead activated when the charge achieves minimum distance from the target (less than 3 m) or a programmed depth is reached or when the charge is 3 m above the seabed. The charge can be launched at a maximum air speed of 250 kt and at a minimum launch height of 30 m

The weapon is easily handled on any type of platform

using a standard carrier and no electrical link is required.

Operational status
Under development for the Italian Navy.

Specifications
Length: 2,170 mm
Diameter: 324 mm
Weight: 225 kg
Charge: 105 kg desensitised CBX
Operating depth: >5 m

Contractor
SEI SpA, Ghedi.

NEW ENTRY

SWEDEN

BDC 103

Type
Underwater signal/anti-frogman charge.

Description
The hand-launched BDC 103 is designed to fire a 55 or 500 g explosive charge when it has sunk to a preset detonation depth. It incorporates a hydrostatic safety device and can be stored and transported ready for use.

Normally used in peacetime as a user-safe signalling device during underwater exercises, for instance to simulate a depth charge attack, the BDC 103 is equally useful as a frogman deterrent in a threat scenario. A complementary 500 g charge can be screwed onto the standard 55 g charge for enhanced impact.

Operational status
The BDC 103 is in service with the Swedish Navy.

Contractor
Bofors SA Marine, Landskrona.

VERIFIED

*The BDC 103 anti-frogman charge
(Courtesy Bofors SA Marine)*
***1999**/0038679*

BDC 204

Type
Air-launched depth charge.

Description
The BDC 204 depth charge is designed for use against submarines operating in shallow waters or at periscope depth. The weapon can be deployed in patterns, weapons being set to detonate at different depths to achieve greatest shock and damage effect against submarines.

The charge comprises a steel case with fixings for standard NATO helicopter bomb launchers. The fuze

Type BDC 104 is of unique design and is a self-contained unit operating on the pressure sensing principle for depth control. It is fitted with devices which eliminate the effect of shockwaves and also make it insensitive to inertia forces in any direction. Because the weapon is not subject to sympathetic detonation from nearby exploding charges it can be deployed in patterns.

Four variants of the weapon are available in different weight categories and with different sinking speeds ranging between 5.2 and 6.8 m/s.

A training version of the weapon is available.

Operational status
The depth charge entered service with helicopters of the Swedish Navy in 1985.

Specifications
Length: 988 mm
Diameter: 240 mm
Weight: 61 kg
Charge: 50 kg
Operating depths: 10-70 m

Contractor
Bofors SA Marine, Landskrona.

VERIFIED

BDC 204 depth charge   **1997**

*BDC 104 depth charge fuze with 9 settings
between 19-72 m*   **1997**

UNITED KINGDOM

Mk 11 depth charge

Type
Air-launched depth charge.

Description
The Mk 11 depth charge quick-reaction air-launched ASW weapon is ideally suited for shallow water operations against submarines on the surface or at periscope depths. It is fully compatible for carriage and release from a wide range of ASW helicopters and fixed-wing maritime patrol aircraft. The Mk 11 depth charge tolerates the harsh vibration levels associated with helicopter operations. The Mod 3 version in service with the Royal Navy has the charge case and nose section strengthened to withstand entry into the water at high velocities without distortion. The charge is fitted with a modern fuze comprising pistol unit, detonator and primer assembly which can withstand severe vibration and shock to ensure accurate detonation at the set depth. On impact with the sea the hydropneumatic arming system is activated and the tail assembly is detached.

The Mk 11 depth charge

The Mk 11 is currently supplied with a Torpex filling. However, a Polymer Bonded Explosive (PBX) filling was developed as an alternative in the early 1990's. This uses long-chain molecules to bind the explosive together forming a relatively insensitive filling enabling it safer to handle in rough conditions.

Operational status

The Mk 11 has been operational with the Royal Navy for a number of years and has been supplied to many overseas navies including the French Navy in 1989 and the Pakistan Navy.

Specifications

Length: 139 cm
Diameter: 27.9 cm
Weight: 145 kg
Warhead: 80 kg HE

Contractor

BAeSEMA, Bristol

UPDATED

MINE COUNTERMEASURES FORCES

Mine countermeasures forces

Mine countermeasures vessel designs

MINE COUNTERMEASURES FORCES

Albania
The three former Soviet T43 and three T301 type minesweepers transferred in the 1950s-60s are obsolete.

Algeria
Plans exist to acquire two new mine warfare vessels and tenders are expected to be invited sometime in 1998-99.

Angola
Two former Russian 'Yevgenya' class minehunters were acquired in 1987. Neither are ever likely to be serviceable again.

Argentina
Two former UK 'Ton' type purchased in 1967 remain in service, although obsolescent. However, it is doubtful if the current economic situation and military budget will permit replacement.

Australia
The May 1991 Force Structure Review recommended the urgent acquisition of a proven design of coastal minehunter. The review and subsequent Minehunter Coastal Project, established in June 1991, projected a force of six vessels and, in 1994, a consortium of Australian Defence Industries and Intermarine of Italy were awarded the contract for the construction of the vessels. The first of class is now in service.

Two 'Bay' class inshore minehunters remain in service, together with a number of auxiliary minesweepers.

Bangladesh
In 1994, four former Royal Navy 'River' class minesweepers were acquired. The navy has also acquired a new T 43 minesweeper ordered from China.

Belgium
The main strength of the current MCM force consists of seven 'Tripartite' minehunters. A capability upgrade programme is now underway. An international project to build a new class of GRP-hulled minesweepers collapsed, but Belgium decided to continue with the project on her own and authorisation for the acquisition of four units was given in July 1994. Work on the first unit is due to commence in 1998-99 and it is expected to be completed in 2000. The ships will be equipped with the French Thomson Sintra ASM Sterne M magnetic sweep equipment.

Brazil
Class of six minesweepers acquired in the 1970s are in urgent need of major modernisation, and this may be undertaken towards the end of the decade, but funds are limited. In view of coastal littoral a larger type of vessel is needed. Numbers are also inadequate in view of size of the navy, number of bases and so on. If SSNs are eventually acquired, an effective MCM capability including modern minehunters, will be essential. It is thought that plans exist for a new class of minesweeper.

Bulgaria
A number of former Soviet minesweepers, all of which are obsolete and need replacing, remain in service. The most recent units to enter service are six 'Olya' class built in Bulgaria between 1988 and 1996. A force of about 20 minehunters/sweepers is planned with an in-service date around the year 2000.

Burma
It is anticipated that two new-build Chinese Type T 43 minesweepers may be acquired in due course.

Canada
There is a priority to build up an effective MCMV force, and a class of 12 MCDVs is now operational. The units will have a dual role of MCM and general patrol, operating with one of three modular payloads; enabling them to adapt quickly to different tasks.

China, People's Republic
Although there is an MCM force of considerable size,

Egyptian T 43 type minesweeper Sinai (H & L van Ginderen)   *1997*

this does not appear to be very effective. Plans for a new class of minehunter are thought to have been drawn up. A force of up to 38 Italian 'Lerici' type minehunters may possibly be planned.

Croatia
Two inshore minehunters are on order and the first has entered service.

Cuba
A number of former Soviet units exist, many of which are probably non-operational.

Denmark
Modern MCM capability is being developed based on the Stanflex 300 'Flyvefisken' multirole design. Units use 'plug in' systems so force can be rapidly expanded or reconfigured for other roles. Six MCM modules are available. Six SAV drones are operational and four more are currently being delivered to a modified larger design to enable them to operate more effectively in a more open ocean environment.

Egypt
A modern MCM capability is being developed with three MHC ordered from Swiftships in America in 1991, the first unit was delivered in June 1994. Apart from this, the navy operates a mixed force of 10 former Soviet units which are now obsolescent. Swiftships has also delivered two route survey vessels and it is planned to acquire two more.

Estonia
Two ex-German Type 394 'Frauenlob' class vessels were transferred in 1997 and may be upgraded with new ROVs.

Finland
A small MCM force of 13 inshore minesweepers exists and plans are thought to exist for the construction of a new class of minehunters. The four 'Kuha' class have been refitted with an additional 6 m long section added to afford greater versatility in MCM. The vessels recommissioned in the autumn of 1998.

France
A major construction programme to update the MCM force was cancelled in 1992. All five 'Circe' class have been withdrawn from service. The bulk of the MCM force, 13 'Eridan' class 'Tripartite' type, is stationed at Brest to protect the SSN and SSBN force, but overall MCM forces are insufficient to provide effective cover for all bases and shipping routes.

Germany
A powerful, modern MCM capability is retained. A major programme of new construction to replace units built in the 1950s and 60s is almost complete. The Type 343 minesweepers are being modernised and five will be upgraded to include a minehunting capability, while five will be converted to include a drone control capability. These plans are currently in the definition phase. In addition, the navy is in the development phase of the MA 2000 programme which is an integrated plan for upgrading mine countermeasures equipment. Under this programme, disposable ROVs are currently being evaluated.

Greece
The MCM force is obsolescent and desperately in need of replacement. The eight former US 'MSC 294' class have recently completed a major upgrade, and may also be given a new sonar in the near future. Acquisition of modern units and equipment is now an urgent priority. Plans exist to acquire a force of three vessels, and a 'Tripartite' may be acquired from Belgium. Two old 'Adjutant' class vessels were acquired from Italy in 1995.

Hungary
Hungarian forces have an important role in patrolling the river Danube. Six 'Nestin' class river minesweepers are in service together with a number of small road-transportable craft.

India
With an extensive coastline and many major ports and naval bases, not to mention a large merchant fleet, the current MCM capability is inadequate for the task. The

'Natya I' class minesweeper Bedi *of the Indian Navy* (H & L van Ginderen)   *1997*

force needs to be expanded and existing units either need urgently replacing or modernising with new systems. The need for at least 10 minehunters has been recognised, and construction of six units at the Goa shipyard is projected, but no decision has yet been made concerning this programme.

Indonesia
A modest programme of two 'Tripartite' minehunters was completed, but more ambitious plans to build up to 12 units have been shelved due to budget restrictions. The MCM force urgently needs building up, and the acquisition of nine former East German patrol vessels partly meets this need, although their priority task is Economic Exclusion Zone (EEZ) patrol.

Iran
A small force of two minesweepers exists but may be non-operational. No effective minehunting force exists.

Italy
A major programme of modernisation is now complete with the construction of 12 'Lerici' and 'Gaeta' class minehunters. An upgraded 'Gaeta' design to encompass a multirole configuration is under consideration and a class of six vessels may be ordered before the end of the decade.

Japan
The mine threat has been clearly recognised and a major programme of modernisation has been completed. Three new ocean-going MCMVs have been built, together with 27 coastal MCMVs. The first two units of a new class of coastal minehunter ('Sugashima' class) were ordered in 1995 and a third in 1996. Two Swedish SAM drones were ordered in 1997.

The Royal Malaysian Jerai, *an Italian 'Lerici' type MCMV* (H & L van Ginderen)   *1996*

North Korea
A modern force of coastal minesweepers has been built.

South Korea
A major mine threat exists and a new MCM programme is under way, with six GRP-hulled minehunters of the 'Kang Keong' class completed. A new 750 tonne combined minehunter/sweeper design is in hand and seven were ordered in December 1996 with sea trials due to be undertaken in 1999. A total force of seven vessels is planned.

Kuwait
There are long-term plans to acquire three minehunters.

The Polish 'Krogulec' class minesweeper Czajka (H & L van Ginderen)   *1996*

A Russian 'Lida' class minesweeper (H & L van Ginderen)   *1996*

Latvia
Two former East German 'Kondor' class vessels were transferred in 1993.

Libya
Eight former Soviet 'Natya' units delivered in the early to mid-1980s remain operational.

Malaysia
A major programme of construction of four minehunters has been completed. Plans for a further four units have been delayed in favour of coastal units.

Netherlands
The current force comprises 15 'Tripartite' minehunters. Modernisation of this class commenced in 1996, and as part of the programme, units are to be converted to the drone control role each controlling four drones. Four other units will additionally be modified to deploy propelled variable depth sonars.

Nigeria
Two Italian 'Lerici' units exist.

Norway
Construction of nine units under a major new programme to replace the obsolete former US minesweepers has now been completed.

Pakistan
Three French 'Tripartite' MCMVs are being acquired to replace obsolete units. The first of these was a unit transferred directly from the French Navy. The second unit has been built in France and outfitted in Pakistan and commissioned in 1996, while the third unit was assembled and fitted out in Pakistan under a technology transfer agreement and has now commissioned.

Philippines
Acquisition of three MCMVs is projected for 2002.

Poland
A mixed fleet of MCMVs exists, with 13 Polish-built 'Notek' class completed between 1981-91, and three older 'Krogulec' class. Four coastal minehunters based on the 'Notec' design entered service in 1992-94 but plans to build more of this class appear to have been abandoned.

Portugal
Currently no MCM force exists and plans to acquire four new coastal minesweepers in conjunction with Belgium have been dropped. Development of an MCM force is now an urgent requirement.

Romania
The MCM force was modernised in 1986-89 with four locally built 'Musca' class.

Russian Federation and Associated States (CIS)
Russia currently operates the largest MCM force in the world, but much of it is obsolete. A number of new designs has been tried in recent years, but not proved successful. The latest class to enter service are the two 'Gorya' class. Apart from this the most modern units are the 27 vessels of the 'Natya' class, but early boats are now paying off. An uprated minehunting version of this class is expected to complete sometime during 1999.

Construction of the 'Lida' class coastal minehunter continues. Glass Reinforced Plastic (GRP) construction has not proved a success and the main material used is still wood, sometimes sheathed in GRP.

Saudi Arabia
An effective MCM force is being built up with three 'Sandown' class units being delivered, and three more projected.

Singapore
Four Swedish 'Landsort' type MCMVs have been acquired to establish an MCM capability.

South Africa
Four 'Ton' class units have been substantially rebuilt and are now considered to have another 20 years of life. Four coastal minehunters of the 'River' class have been built to a design based on the German 'Schutze' class. These are being upgraded with a domestically developed tactical data system.

Spain
Plans have been drawn up to acquire eight minehunters to be built by Bazan based on GRP technology. The first unit commissioned in July 1998.

Sweden
A modern MCM force exists based on the seven 'Landsort' class, and a new class of four inshore minesweepers the 'Styrso' is now operational. A new class of drone for remote control by the 'Styrso' class is under consideration.

Syria
For a small country, a sizeable MCM capability exists (six units), although much of it must be considered obsolete. All units are of ex-Soviet design.

Taiwan
Existing forces are obsolete and ineffective and the force is being modernised. Four MWV 50 offshore oil support vessels were built by Abeking & Rasmussen in Germany in 1990-91 for the Chinese Petroleum Corporation (CPC), and on arrival in Taiwan were converted to coastal minehunters. A new class is planned, and tenders are expected to be issued when funds are made available.

Al Kharj   *1998*/0007413

Thailand
Two M 48 MCMVs have been built in Germany, no more units of this type will be acquired. An order for two new coastal minehunters was placed with Intermarine of Italy on 17 September 1996. Built to a modified 'Gaeta' design, the first unit commissioned in December 1998. Six more units may be acquired, inshore minehunters may also be acquired.

Turkey
Turkey has concluded an agreement with France to purchase all five French Navy 'Circe' class which will be refitted with a new tactical system Mintac developed by DCN. All five units were delivered before the end of 1998.

Ukraine
Two 'Sorya' class vessels were transferred from Russia in 1996 together with a 'Yevgenya' class unit.

United Arab Emirates
In December 1997, an ITT for three MHCs was issued. This now forms the main priority of the UAE.

United Kingdom
A major programme of MCM construction continues with the acquisition of the 'Sandown' class, and orders for a further seven units have been placed. Two units are now in service. A cut back mid-life update for the 13 'Hunt' class is scheduled. A remote-controlled minesweeping drone is also to be acquired.

United States of America
A major programme has been defined to replace obsolete units and build up a modern MCM force. A programme of 14 'Avenger' class and 12 'Osprey' class MCMVs is now complete. In addition, the use of MCACs for use in littoral mine countermeasures is being evaluated.

Uruguay
Four former East German 'Kondor' class minesweepers have been acquired.

UPDATED

TABLE XIV

MCMV building programme 1994-97

Name (Pt N)	Type	Builder	Date	Stage
Australia				
'Huon' class				
Huon (M 82)	MH	ADI	Dec 1993	Cd
Hawkesbury (M 83)	MH	ADI	23 Apr 1998	L
Norman (M 84)	MH	ADI	Mar 1999	L
Gascoyne (M 85)	MH	ADI	13 Sep 1997	LD
Diamantina (M 86)	MH	ADI	4 Aug 1998	LD
Yarra (M 87)	MH	ADI	12 Aug 1994	O
Bangladesh				
Sagar (M 91)	MS	Guangzhou	27 Apr 1995	Cd
Belgium				
Four unnamed (M 926-929)	MS		1998	O
Canada				
'Kingston' class				
Kingston (700)	MS	Halifax	21 Sep 1996	Cd
Glace Bay (701)	MS	Halifax	26 Oct 1996	Cd
Nanaimo (702)	MS	Halifax	10 May 1997	Cd
Edmonton (703)	MS	Halifax	21 June 1997	Cd
Shawinigan (704)	MS	Halifax	14 June 1997	Cd
Whitehorse (705)	MS	Halifax	17 Apr 1998	Cd
Yellowknife (706)	MS	Halifax	18 Apr 1998	Cd
Goose Bay (707)	MS	Halifax	May 1998	Cd
Moncton (708)	MS	Halifax	June 1998	Cd
Saskatoon (709)	MS	Halifax	Sep 1998	Cd
Brandon (710)	MS	Halifax	Feb 1999	Cd
Summerside (711)	MS	Halifax	Apr 1999	Cd

Name (Pt N)	Type	Builder	Date	Stage
Croatia, Unnamed	MSI	n/k	1998	O
Denmark				
'Flyvefisken' class				
Skaden (P 561)	MRB	Danyard	10 Apr 1995	Cd
Viben (P 562)	MRB	Danyard	15 Jan 1996	Cd
Soloven (P 563)	MRB	Danyard	28 May 1996	Cd
Slave minesearchers				
Four unnamed (MRD 3/6)	Slave MS	Danyard	1996	Cd
Four unnamed (MRD 7-10)	Slave MS	Danyard	1998-99	Cd
Egypt				
Yard No 0423	MH	Swiftships	13 July 1997	Cd
Yard No 0424	MH	Swiftships	13 July 1997	Cd
Yard No 0425	MH	Swiftships	13 July 1997	Cd
France				
'Eridan' class				
Sagittaire (M 650)	MH	Lorient DY	2 Apr 1996	Cd
'Altair' class				
Aldebaran (M 772)	M-S	Socarenam	10 Mar 1995	Cd
Germany				
'Frankenthal' class				
Dillingen (M 1065)	MH	Abeking & Rasmussen	25 Apr 1995	Cd
Homburg (M 1069)	MH	Kroger	26 Sep 1995	Cd
Sulzbach-Rozenburg (M 1062)	MH	Lürssen	23 Jan 1996	Cd
Fulda (M 1058)	MH	Abeking & Rasmussen	28 May 1998	Cd
Weilheim (M 1059)	MH	Lürssen	Nov 1998	Cd
India				
Ten unnamed (M 89/M 98)	MH	Goa Shipyard	1993	P
Italy				
Chioggia (M 5560)	MH	Intermarine	19 May 1996	Cd
Rimini (M 5561)	MH	Intermarine	26 Nov 1996	Cd
Japan				
'Uwajima' class				
Tobishima (678)	MH	NKK	10 Mar 1995	Cd
Yugeshima (679)	MH	Hitachi	11 Dec 1996	Cd
Nagashima (680)	MH	NKK	25 Dec 1996	Cd
Type 510				
Sugashima (681)	MS	NKK	25 Aug 1997	L
Notojima (682)	MS	Hitachi	3 Sep 1997	4
Unnamed (683)	MS	Hitachi	Aug 1998	L
Unnamed (684)	MS	NKK	Aug 1999	L
Kuwait				
Three unnamed	MH		1994	P
Norway				
'Oksoy' class				
Maaloy (M 342)	MH	Kvaerner-Mandal	24 Mar 1995	Cd
Hinnoy (M 343)	MH	Kvaerner-Mandal	8 Sep 1995	Cd
'Alta' class				
Alta (M 350)	MS	Kvaerner-Mandal	12 Jan 1996	Cd
Otra (M 351)	MS	Kvaerner-Mandal	8 Nov 1996	Cd
Rauma (M 352)	MS	Kvaerner-Mandal	2 Dec 1996	Cd
Orkla (M 353)	MS	Kvaerner-Mandal	4 Apr 1997	Cd
Glomma (M 354)	MS	Kvaerner-Mandal	1 July 1997	Cd
Pakistan				
'Tripartite' type				
Muhafiz (163)	MH	Lorient Dockyard	15 May 1996	Cd
Mujahid (164)	MH	Karachi Shipyard	June 1998	Cd
Philippines				
Three unnamed	MH	n/k	2002	P
Saudi Arabia				
'Sandown' class				
Al Kharj (424)	MH	Vosper Thornycroft	7 Aug 1997	Cd
Onaizah (426)	MH	Vosper Thornycroft		P
Al Raas (428)	MH	Vosper Thornycroft		P
Al Bahan (430)	MH	Vosper Thornycroft		P
Singapore				
'Landsort' class				
Bedok (M 105)	MCMV	Karlskronavarvet	7 Oct 1995	Cd
Kallang (M 106)	MCMV	Singapore SB	7 Oct 1995	Cd
Katong (M 017)	MCMV	Singapore SB	7 Oct 1995	Cd
Punggol (M 108)	MCMV	Singapore SB	7 Oct 1995	Cd
Spain				
'CME' Type				
Segura (M 51)	MH	Bazan	July 1998	Cd
Sella (M 52)	MH	Bazan	Aug 1998	L
Tambre (M 53)	MH	Bazan	Mar 1999	L
Turia (M 54)	MH	Bazan	Nov 1999	L
Sweden				
'YSB' Type				
Styrso (M 11)	MCMV	Karlskronavarvet	20 Sep 1996	Cd
Sparo (M 12)	MCMV	Karlskronavarvet	21 Feb 1997	Cd
Skafto (M 13)	MCMV	Karlskronavarvet	13 June 1997	Cd
Sturko (M 14)	MCMV	Karlskronavarvet	19 Dec 1997	Cd
Taiwan				
Eight unnamed	MCMVs		1993	P
Thailand				
'Gaeta' type				
Lat Ya (633)	MH	Intermarine	Dec 1998	Cd

Name (Pt N)	Type	Builder	Date	Stage
Tha Din Daeng (634)	MH	Intermarine	Dec 1999	Cd
United Arab Emirates				
3 unnamed	MHC	n/k	10 Dec 1997	OTT
UK				
'Sandown' class				
Penzance (M 106)	MH	Vosper Thornycroft	May 1998	Cd
Pembroke (M 107)	MH	Vosper Thornycroft	6 Oct 1998	Cd
Grimsby (M 108)	MH	Vosper Thornycroft	10 Aug 1998	L
Bangor (M 109)	MH	Vosper Thornycroft	Mar 1999	L
Ramsey (M 110)	MH	Vosper Thornycroft	July 1994	O
Blythe (M 111)	MH	Vosper Thornycroft	July 1994	O
Shoreham (M 112)	MH	Vosper Thornycroft	July 1994	O
USA				
'Osprey' class				
Oriole (MHC 55)	MH	Intermarine, Savannah	16 Sep 1995	Cd
Pelican (MHC 53)	MH	Avondale	18 Nov 1995	Cd
Kingfisher (MHC 56)	MH	Avondale	26 Oct 1996	Cd
Black Hawk (MHC 58)	MH	Intermarine, Savannah	27 Apr 1996	Cd
Falcon (MHC 59)	MH	Intermarine, Savannah	8 Feb 1996	Cd
Cormorant (MHC 57)	MH	Avondale	12 Apr 1997	Cd
Raven (MHC 61)	MH	Intermarine, Savannah	5 Sep 1998	Cd
Shrike (MHC 62)	MH	Intermarine, Savannah	Feb 1999	Cd
Robin (MHC 54)	MH	Avondale	11 May 1996	Cd
Cardinal (MHC 60)	MH	Intermarine, Savannah	18 Oct 1997	Cd

Type: AMS – Auxiliary minesweeper, MCMV – Mine countermeasures vessel, MH – Minehunter, MRB – Multirole boat (can be equipped as a minelayer/minehunter/minesweeper as required), MS – Minesweeper, M-S – Minesearcher
Stage: A – Authorised, B – Building, C – Completed, Cd – Commissioned, L – Launched, LD – Laid down, O – Ordered, OTT – Out-to-tender, P – Planned/projected

UPDATED

TABLE XV

MCMV age 1970-97[1]

Country	1994-98 0-5 years	1989-93 6-10 years	1984-88 11-15 years	1979-83 16-20 years	1974-78 21-25 years	Pre-1974 over 25 years	TOTAL[1]
Albania						6	**6**
Argentina						2	**2**
Australia	1		2				**3**
Bangladesh	1 + 4[2]						**5**
Belgium		4	3			1	**8**
Brazil					2	4	**6**
Bulgaria	1	4	1 + 2[2]	4[2]	4[2]	2 + 2[2]	**20**
Canada	12						**12**
China	4	3	<		40[3]	>	**47**
Croatia	1						**1**
Cuba			6[2]				**6**
Egypt	3					10[2]	**13**
Estonia	2[2]						**2**
Finland				7	6		**13**
France	1		12				**13**
Germany	8	14				21[4]	**43**
Greece	2[2]					8 + 4[2]	**14**
Hungary				6		45	**51**
India		1[2]	8[2]	7[2]	2[2]		**18**
Indonesia		9[2]	2			2[2]	**13**
Iran						5[2]	**5**
Italy	4	4	4				**12**
Japan	6	8	10	3			**27**
North Korea	1	<	23	>			**24**
South Korea	1	4	1		2[2]	6[2]	**14**
Latvia		2[2]					**2**
Libya			3[2]	5[2]			**8**
Malaysia			4				**4**
Netherlands		2	10	3			**15**
Nigeria			2				**2**
Norway	9						**9**
Pakistan	2	1					**3**
Poland	3	5	8	3			**19**
Romania		3	3	14	5	7	**32**
RFAS	14	27	13	14	30	32	**130**
Saudi Arabia	1	2		4			**7**
Singapore	4						**4**
South Africa				4		4	**8**
Spain	1					12[2]	**13**
Sweden	3	2	6		1	12	**24**
Syria			5[2]			1[2]	**6**
Taiwan	4[2]	4				4[2]	**12**

Country	1994-98 0-5 years	1989-93 6-10 years	1984-88 11-15 years	1979-83 16-20 years	1974-78 21-25 years	Pre-1974 over 25 years	TOTAL[1]
Thailand	1		2			2	5
Turkey	5[2]			1[2]	5[2]	15[2]	26
Ukraine	3[2]						3
UK	3	5	6	7			21
Uruguay		4[2]					4
USA	10	10	5				25
Vietnam		2[2]	3[2]	3[2]			8
Yemen		4[2]	1[2]	1[2]			6
Yugoslavia	1			3	3	3[2]	10
TOTAL	**97 + 19[2]**	**102 + 22[2]**	**117 + 28[2]**	**68 + 21[2]**	**87 + 13[2]**	**146 + 64[2]**	**784**

[1] Figures do not cover drones or miscellaneous craft such as small minesweeper auxiliaries or container modules (for example for Denmark).

[2] Transferred – hulls may be considerably older.

[3] Many of these were built over a large number of years and actual dates are not readily available. The majority however, are in need of replacement

[4] Being replaced by 'Hameln' and Frankenthal' classes.

UPDATED

MINE COUNTERMEASURES VESSEL DESIGNS

AUSTRALIA

'Huon' class

Type
Minehunter.

Description
The GRP 'Huon' class coastal minehunters now under construction for the Royal Australian Navy are being built to a modified Italian 'Gaeta' design at a purpose-built GRP construction yard in Newcastle, by ADI Ltd, in partnership with Intermarine of Italy. The heart of the ships will be the GEC-Marconi Nautis-II M minehunting command and control system. Data will be fed to the Nautis by a comprehensive range of sensors including: a Thomson Marconi Sonar Type 2093 VDS minehunting sonar; a Kelvin Hughes Type 1007 I-band navigation radar; Link 11; and British Aerospace Australia Prism ESM system with two ML Aviation SuperBarricade chaff/IR decoy rocket launchers. For mine disposal, the ships will carry two Bofors Double Eagle Mk II ROVs carrying Danish DAMDIC charges. In addition to their primary minehunting role, the ships will also be fitted with an ADI double oropesa mechanical sweep and will also be able to deploy the Mini-Dyad influence sweep. Weapons control is exercised through a Radamec 1000N optronic surveillance system.

Propulsion is provided by a Fincantieri GMT B230 8M 1,460 kW diesel engine driving a Lips cp propeller. For precise manoeuvring during minehunting, the vessels are fitted with three Italian Riva Calzoni retractable and rotatable thrusters powered by three electrohydraulic motors driven by three Isotta-Fraschini 1312.TI.ME V-form 350 kW generating sets.

Self-defence is provided by a single MSI Defence Systems DS30B 30 mm gun.

The vessels are also fitted with a recompression chamber together with a Rigid Inflatable Boat (RIB) and inflatable diving boat to support the six-man diving team.

Operational status
Under construction for the Royal Australian Navy. The lead ship *Huon* is scheduled to commission in December 1998 Sea trials of the second unit, *Hawkesbury*, were commenced in Spring 1998.

Contractors
ADI Ltd, Newcastle, New South Wales.
Intermarine SpA, Sarzana, Italy.

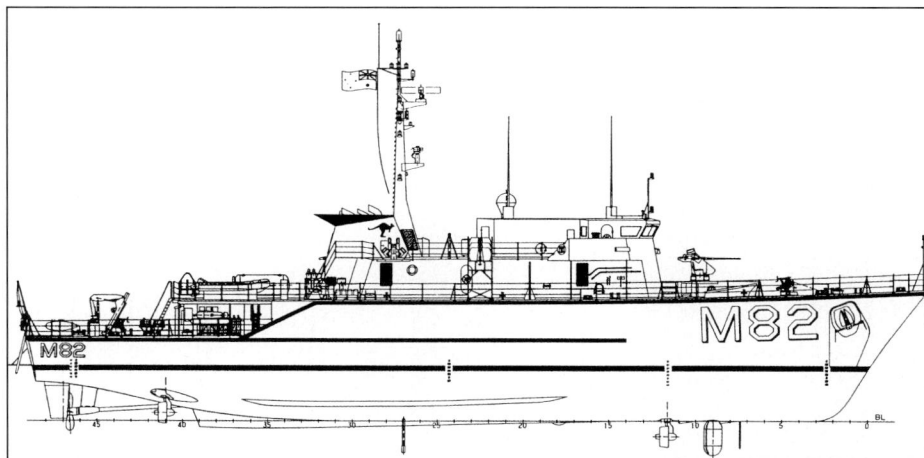

Outboard profile of the 'Huon' class **1998**/0007401

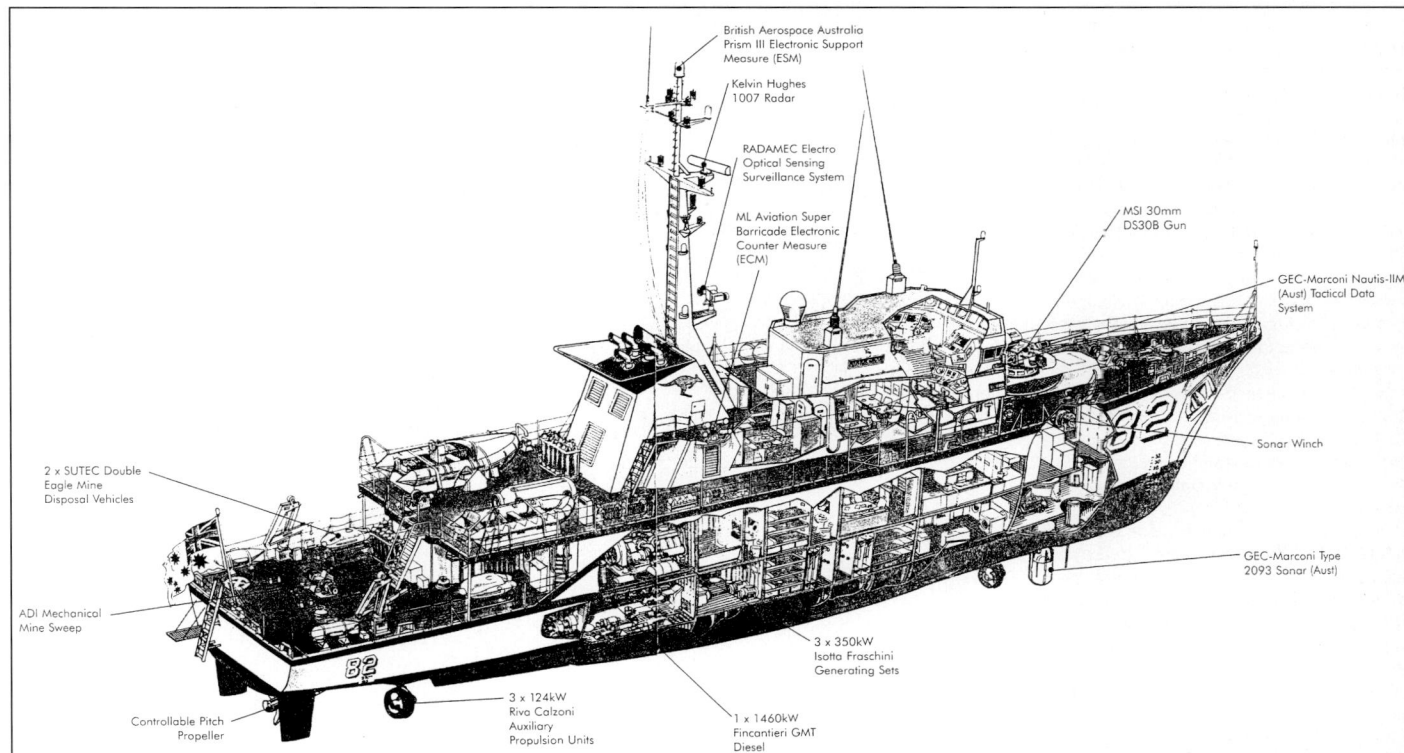

General layout of the 'Huon' class **1998**/0007400

BELGIUM

Minesweeper project

Type
Minesweeper.

Description
This project is based on a design prepared by Beliard Polyship and van der Giessen-de Noord in 1990. The vessels will be equipped with the new Sterne M minesweeping system developed by Thomson Marconi Sonar SAS. The sophisticated multi-influence sweep gear will be capable of simulating different ship influences including magnetic, acoustic, Extremely Low-Frequency Electromagnetic (ELFE) and Underwater Electrical Potential (UEP). In addition, the ships will carry a Mk 9 wire sweep system. All sweep equipment will be handled by an A-frame centreline crane on the after deck. Main propulsion will be by two 1,088 bhp Brons/Werkspoor diesel engines driving twin shafts with cp propellers through a reduction gearbox. Hotel services will be provided by three

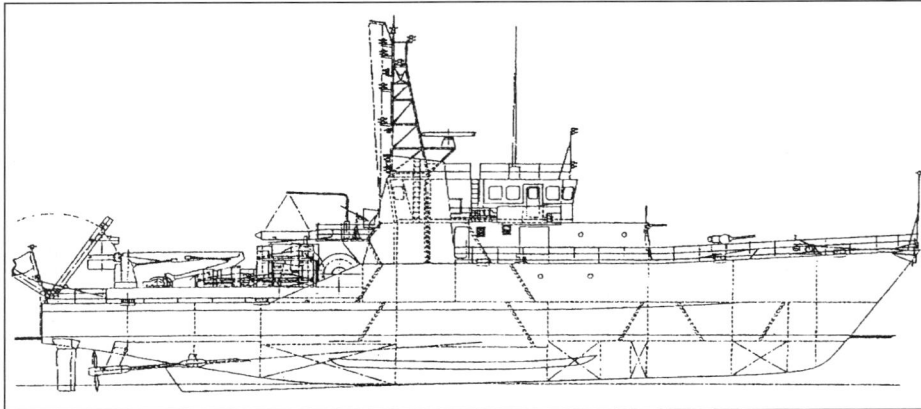

300 kW diesel alternators. The vessels will be armed with a 25 mm AA gun on the foredeck, and a machine gun on each side of the aft end of the long fo'c'sle deck.

In addition to their mine sweeping equipment, the vessels will be fitted with a high-definition sonar to enable them to undertake rapid route survey tasks. Conversion to the route survey role will take about 6 hours.

Hull construction is based on that used in the 'Tripartite' vessels.

Operational status
The contract for the class was signed on 3 October 1996 and construction has begun.

Contractor
SKB Shipbuilders, Antwerp.

UPDATED

CANADA

'Kingston' class

Type
Multirole mine countermeasures vessel.

Description
The 12 steel-hulled vessels of the multirole MCDV programme have been designed to combine general patrol duties with MCM tasks. The ships are being built primarily to commercial standards using COTS equipment wherever practicable, with military standards being applied to stability, flood control zones, doors, turning/stopping distances and ammunition spaces. The design allows the vessels to accommodate various modular MCM payloads including: an Indal Technologies SLQ38 Mechanical Minesweeping System (MMS) including single and double oropesa sweeps; route survey system; and mine disposal deploying ROVs. Payloads are contained in standard ISO containers which can be changed over in 12 hours. The heart of the MCM system is the integrated command and control geographic database developed by MacDonald Dettwiler, which incorporates their geocoded sonar imagery system. The primary task is route survey, and using the geocoded database, considerable savings in time can be achieved in minehunting tasks, as only new seabed targets need to be examined at any one time. For route survey the vessels deploy a remotely operated survey inspection craft and tow a side scan sonar to maintain the up-to-date database of seabed objects. The surface search radar is a Kelvin Hughes 6000 E/F-band, while a Kelvin Hughes I-band radar is fitted for navigation. Communications facilities are the Thomson Systems TS 700 integrated system, which includes two each of VHF, UHF and HF, as well as secure voice. The ships are fitted with an automated message processing system.

Propulsion is provided by four Wartsila-SACM UD 23V12 S5D 2,450 bhp diesels directly driving two Jeumont 715 kW alternators to power two 2,000 hp (1,150 kW) Jeumont CI 560L DC electric motors, each turning a Lips FS1000-234 podded Z-drive azimuth thruster fitted with a five-bladed fixed-pitch propeller. Electrical services are provided by a Wartsila-SACM UD19 diesel powering a 300 kW alternator and generator. For emergency use, a Wartsila-SACM UD60 105 kW diesel prime mover is fitted.

Self-defence is provided by a 40 mm Bofors AA gun and two 12.7 mm machine guns.

Minesweeping equipment includes the SLQ-38 wire sweep which is capable of being used in single or double orpesa or team sweep configuration.

Currently only seven MCM payloads are budgeted, four route survey, one ROV and two mechanical minesweeping systems.

Operational status
Twelve units are now operational.

Contractor
Halifax Shipyard Ltd, Halifax.

Glace Bay the second of the 'Kingston' class (Canadian Forces photograph)   **1998**/0007403

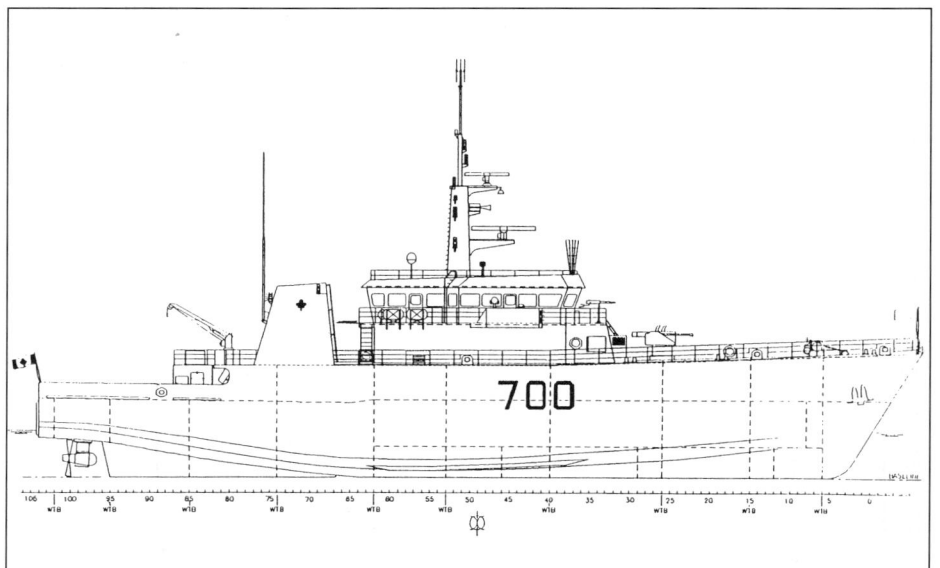

CHINA, PEOPLE'S REPUBLIC

Type 312

Description
The Type 312 is a small, remotely controlled mine countermeasures system believed to be for use in harbours and estuaries. It uses a 20.94 m long drone with a displacement of 39 tonnes which is controlled by radio signals from a shore station or mother ship up to 3 n miles (5.55 km) away.

During transit the drone is powered by a 12V 150C 500 hp supercharged diesel engine, giving a maximum speed of 11.6 kt. During minesweeping operations the drone is powered by an electric motor. It has a range of 108 n miles at 9 kt. Acoustic and magnetic sensors are used but it is not clear how located mines are destroyed.

Operational status
In service with PLA navy.

Specifications
Length: 20.9 m
Beam: 3.9 m
Draught: 1.3 m
Displacement: 47 t

Chinese remote-control minesweeper Type 312. Key to diagram: **(1)** *steering engine room;* **(2)** *special instrument compartment;* **(3)** *wheelhouse;* **(4)** *main engine compartment;* **(5)** *material compartment, this is within the engine room compartment and is not separated from it by a watertight compartment;* **(6)** *fuel compartment;* **(7)** *auxiliary engine compartment;* **(8)** *acoustic generator compartment*

Contractor
China Shipbuilding Trading Company, Beijing.

UPDATED

DENMARK

'Flyvefisken' class

Type
Multirole vessel capable of undertaking a mine countermeasures role.

Description
This multipurpose design can be fitted out as a missile boat, patrol boat, minelayer, minehunter, or minesweeper as requisite. The weapon fit is accommodated in four standardised containers – one forward and three aft – each measuring 3.00 × 3.50 × 2.50 m. Container slots not utilised are closed by watertight hatches and displacement, draught and trim vary depending on the role. A passive tank stabilisation system has been installed for use at low minehunting/sweeping speeds, when hydraulic drive is used for propulsion in order to minimise the noise signature. At high speeds rudder roll control is used.

Regardless of role, the machinery, sensor fit and command system is standardised. Main propulsion is by a CODAG arrangement of a General Electric LM-500 gas turbine (on the centreline) and two MTU 16V 396 TB94 diesel engines driving three shafts fitted with Stone Vickers fp (centre shaft) and cp (wing shafts) propellers through Allen double/single-reduction gearboxes. A speed of 12 kt is attained on one diesel engine, 20 kt on two diesel engines, and over 30 kt with all engines online.

Auxiliary slow-speed propulsion is provided by a 500 bhp General Motors 12V 71 diesel engine driving three Rexroth hydraulic pumps powering a 240 shp Rexroth hydraulic motor on each shaft line, and to the transverse thrust unit forward. When operating in the diesel mode, the centreline shaft is windmilled by the hydraulic motor to reduce drag, with power supplied by the hydraulic pumps driven from the diesel engine gearboxes. All main and auxiliary machinery is remotely operated and monitored from central control stations on the bridge and from a compartment forward of the engine room.

The sensor fit comprises: TRS-3D/16 multimode G/H-band phased array surveillance radar and integrated GEC-Marconi IFF interrogator; Terma I-band navigation radar; two CelsiusTech tracking radars fitted with TV cameras and laser rangefinders; and Racal Sabre ESM and Cygnus ECM jammer; all linked to a CelsiusTech 9LV-200 Mk3 AIO/weapon control system.

For minehunting, a Thomson Marconi Sonar SAS TSM 2054 side scan sonar is streamed either from the multipurpose boat itself, or from two remotely controlled 18 m, 30 tonne GRP unmanned slave craft

The Danish minehunter drone MRD 2 (H & L van Ginderen)   *1996*

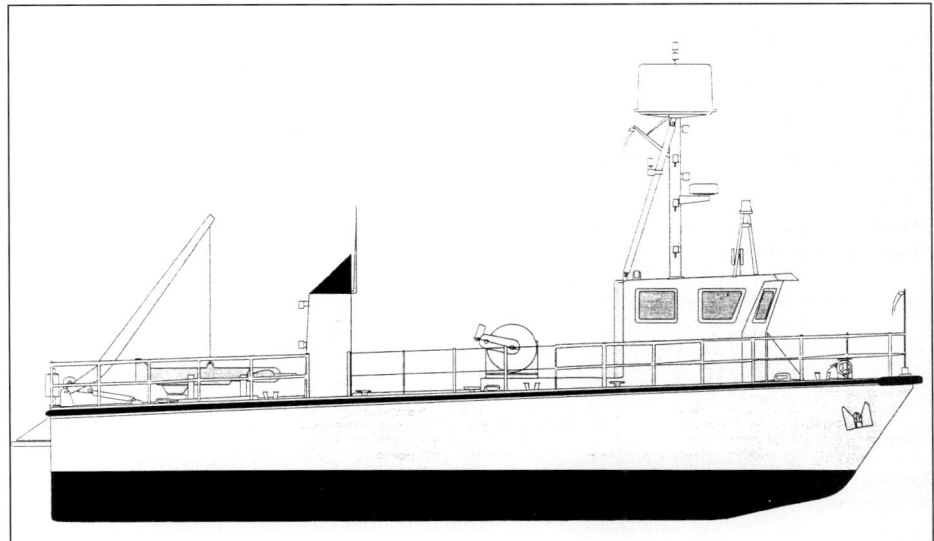

Elevation of the minehunter drone   *1996*

(MRD), and Double Eagle ROV for classification and disposal; together with the associated Thomson-Sintra IBIS 43 minehunting command and TSM 2061 tactical systems. Wire, acoustic and electric sweeps – all handled by a centreline crane – will be provided for minesweeping.

An Otobreda Super Rapid 120 rpm 76 mm gun is mounted forward.

Operational status
Fourteen units are now operational.

Contractor
Danyard A/S, Aalborg.

UPDATED

The Danish 'Flyvefisken' class vessel Stoeren, *outfitted with an MCM module*
(H & L van Ginderen)
1996

SAV class

Type
Remote controlled minesweeping drone.

Description
Built by Danyard, these Surface Auxiliary Vessels (SAV) are controlled in pairs by the 'Flyvefisken' class as configured for MCMV operations. The GRP-hull design is based on the design of the Swedish 'Hugin' class patrol vessels. Special attention has been paid to reducing the acoustic signature to an absolute minimum. The drones can deploy a Thomson Marconi Sonar SAS TSM 2054 side scan sonar from the stern gantry. The drones are controlled from the parent vessel via a Terma radio link, which is used to relay back to the parent vessel the sonar picture and precise position information. The radio link is also used to carry manoeuvring control signals for the drone and control of the side scan sonar. The drones are fitted with a Furono I-band navigation radar.

A new larger drone, able to operate effectively in more open ocean conditions, is now under construction. The new 26.5 m GRP drone displaces 125 tonnes and is driven by water-jets at a speed of 12 kt.

Operational status
Six units in service.

Contractor
Danyard A/S, Aalborg.

VERIFIED

MSF class

Type
Remote-controlled mine countermeasures drone.

Description
A new 26.5 m mine countermeasures drone, built by Danyard Aalborg, is carrying out sea trials with the Danish Navy. Four vessels have been ordered and an eventual force of 10 units is projected. The drone, which is scheduled to enter operational service during 1999, is based on a GRP hull, displacing 125 tonnes. Propulsion is provided by waterjets to give the vessel a speed of 12 kt. This drone is remote-controlled for unmanned operation, and can carry out minehunting as well as minesweeping or can be operated in autonomous mode with a crew of three or four.

The drone is filled with datalinks, relaying position and control orders as well as sonar data from the Thomson Marconi Sonar SAS TSM 2054 side scan sonar.

In order to comply with the Standard Flex concept, the drone can accommodate a $3.0 \times 3.5 \times 2.5$ m Standard Flex container as utilised in the Flyvefisken class multirole vessels.

Propulsion is provided by two 500 bhp diesels each driving a water-jet propulsor for a maximum speed of 12 kt.

Operational status
Four units now in service.

MSF1 undergoing sea trials (Courtesy of Naval Team Denmark)   *1999*/0024813

Contractor
Danyard A/S, Aalborg.

NEW ENTRY

GERMANY

'Hameln' class

Type
Minesweepers.

Description
The 10 Type 343 minesweepers of the 'Hameln' class built by Lürssen, Abeking & Rasmussen and Kröger are designed for minesweeping operations in the Baltic with Troika control capability, and with mechanical, acoustic and magnetic sweeps. Hull and superstructure are completely constructed from non-magnetic and non-corrosive austenitic steel, fully welded. Plates 4 to 6 mm thick are used for the hull, welded to web frames at intervals of about 1 m with longitudinals at 300 mm. The longitudinal stiffenings are offset bulb plate or flat-bar sections. The transverse web frames are built of girders with single sided web plate. Partitions are made from non-magnetic sandwich construction, elastically mounted, for reason of shock resistance. Although aluminium also has non-magnetic

properties, eddy currents cause problems, creating significant magnetic signatures when moving in the earth's magnetic field. By using thinner non-magnetic steel, which has a higher resistivity than aluminium and ordinary steel, this problem has been overcome. The non-magnetic steel does, however, possess sufficient conductivity to achieve earth bonding and electromagnetic shielding, a problem which has been experienced with completely non-conductive materials such as wood and GRP.

The sensor fit includes: a Signaal WM-20/2 surveillance and tracking radar; Raytheon SPS-64 I-band navigation radar; Palis datalink; and Thomson-CSF DR-2000 ESM. The ships are also fitted with the STN ATLAS Elektronik DSQS-11M minehunting sonar.

All systems are integrated into the minehunting control and M-20 weapon control system (the latter removed from the 'Zobel' class torpedo boats).

Main propulsion is by two MTU 16V 396 TB84 diesel engines, each rated at 3,070 bhp at 2,000 rpm and driving twin shafts fitted with Escher Wyss cp propellers through Renk single-reduction gearboxes. Electric power at 400 V 3-phase 60 Hz is supplied by three Siemens 230 kW alternators, each powered by an MWM TBD-601-6S auxiliary diesel rated at 639 bhp at 1,800 rpm.

The craft are armed with two Bofors SAK 40L/70 (2 × 1) guns, two Stinger SAM launchers, and two Silver Dog chaff rocket launchers, and can stow about 60 mines.

Under the MA2000 Programme, the class will be converted into minehunters (MJ 343 five units) and upgraded Troika control vessels (HL 343 five units), the latter each controlling up to four remote-control minesweeping drones.

The MJ 343 conversion will achieve the same minehunting capabilities as the Type 332 class. This will include fitting the DSQS-11M minehunting sonar, a tactical data system and a new ROV concept. This will be the Seafox expendable mine destructor, a one-shot mine destruction drone. Work is due to commence in 1998.

The HL 343 conversion will include fitting a modified DSQS-11M Mod 3 mine avoidance sonar for detection and classification of moored mines, the Seafox ROV, and MCM command and control system to control up to four drones. The command and control system is to be jointly developed by the German and Netherlands navies, the latter using the system to update their 'Tripartite' minehunters. Development of the system began in 1996 and will conclude in 1998 when the class is to commence conversion.

Operational status
In January 1998, the German Navy began its conversion programme. Five of the class are being converted to minehunters, to be designated as Type MJ 333. The remaining vessels are to be converted to advanced Troika control ships designated HL 352.

Contractors
Lürssen Werft, Bremen.
Abeking & Rasmussen, Lemwerder.
Krögerwerft, Rendsburg.

The German 'Hameln' class minesweeper Siegburg (H & L van Ginderen)   **1997**   **UPDATED**

'Frankenthal' class

Type
Minehunters.

Description
The 10 Type 332 minehunters of the 'Frankenthal' class, built by the same builders as the 'Hameln' class, have the same hull and sensor fit (less the M-20 element) as the 'Hameln' class, but with the addition of the STN ATLAS Elektronik MWS 80-4 minehunting command system and two Pinguin ROVs for mine disposal in place of the sweep gear. Main propulsion machinery is identical to the 'Hameln' class, but with the addition of an electric slow-speed drive for minehunting. Only a single 40 mm gun is mounted forward.

Under the MJ 2000 Programme configured to address precisely minehunting and future aspects of MCM scenarios relevant to the German Navy) current limitations in minehunting capability – specifically the problem of detecting buried mines – will be addressed. These will centre on developing a sediment sonar (SEDIS), the Seawolf expendable mine destructor, a launch and recovery system to handle the towed sonar from the Seahorse surface drone, shock resistance of the remote-control drone Seahorse, a sonar datalink between the control ship and the surface drone and a location/relocation capability using GPS.

Development phase is due to commence in 1999

The German Type 332 minehunter Frankenthal (Spearhead Exhibitions)   **1996**

with conversion of five Type 332 vessels or construction of a new platform to commence in 2005.

Operational status
Class of 2 units is now operational. Project definition for MJ 2000 commenced in 1997.

Contractors
Lürssen Werft, Bremen.
Abeking & Rasmussen, Lemwerder.
Krögerwerft, Rendsburg.

UPDATED

Lürssen M48 design

Type
Multirole mine countermeasures vessels.

Description
Lürssen has also built two Type M48 MCMV for the Royal Thai Navy. These are multipurpose vessels whose design is based on experience with the Type 343. The hull is of laminated wood, with non-magnetic steel frames, tanks and superstructure. As far as is technically and economically possible, all equipment is built of non-magnetic materials. Items which do contain ferromagnetic materials are all compensated and wherever necessary stray magnetic fields caused by electrical equipment and machinery (including cabling) are compensated for by additional magnetic loops. The ships are fitted with a permanent degaussing system

comprising horizontal, vertical and athwartships coils, the fields created by the coils being automatically controlled and adjusted to minimise the actual components of the permanent (induced) and eddy (stray) current magnetic fields.

Acoustic signature reduction has also been carefully studied, and the ships are fitted with a five-bladed cp propeller with highly skewed back profile. A cp was chosen in order to meet the requirement for different loads created by high and low speeds. For slow-speed minehunting operations the ships are powered by an auxiliary electric motor driving the cp propeller at very low rpm via a tooth belt drive through gearboxes. All machinery and moving equipment is resiliently mounted, to reduce noise and provide shock resistance, the main engines being double-elastically mounted via an intermediate frame. Additional sound insulation for the engine rooms is arranged to prevent

transmission of airborne noise into the structure and to improve conditions for the crew.

The design features a long forecastle deck with a continuous main deck to provide the best longitudinal strength. The high freeboard forward with moderate flare ensures a dry ship with good sea-keeping qualities. A large clear deck aft gives good access and handling for the MCM gear, while a large A-frame hydraulic gantry at the stern greatly eases the job of handling heavy gear of all kinds. In particular it can be used to handle the minesweeping gear or oceanographic equipment. A foldable hydraulic crane is also provided for handling mine disposal vehicles, the inflatable dinghies, marker buoys and provisions. A decompression chamber on the aft deck and diving equipment in a dedicated divers cabin are arranged to support mine clearance diving.

The twin funnel arrangement was chosen together

with the amidships position of the wheelhouse and operations room, to provide the best all-round view, in particular of the sweep deck, and to reduce the effects of motion experience by operators in the operations room during heavy weather.

The ships are powered by two MTU 12V 396 TB83 diesel engines, each developing 1,610 bhp driving two KaMeWa cp propellers through a single reduction gearbox. Twin Becker rudders are installed for manoeuvrability.

Onboard systems include an STN ATLAS Elektronik MWS 80 integrated mine warfare control system and DSQS-11H sonar. Precise positioning is achieved using a Motorola MRS III Miniranger navigational positioning system. Two Gaymarine Pluto ROVs are carried for mine disposal duties.

Three 20 mm guns are carried for self-defence. The ROVs are stored and maintained in an interchangeable container between the funnels, being moved out on to the open deck on rails ready for handling by the crane.

Operational status
Two units operational with the Royal Thai Navy.

Contractor
Lürssen Werft, Bremen.

VERIFIED

Troika

Type
Remote-control mine simulation sweeping system.

Description
The problem of sweeping mines without endangering the parent MCMV and her crew can be overcome by using the Troika system. This system is designed to carry out remote simulation sweeping of influence mines and consists of a parent MCMV and three remotely controlled minesweeping drones called Simulation Sweeping Craft (SSC). The parent MCMV acts as a control ship and has at her disposal the command and control facilities required to direct and control the unmanned SSC for influence simulation sweeping in a defined mine danger area.

The SSC (German Navy designation HFG-F1) is a self-propelled craft specifically designed to perform simulation sweeping with high reliability and a very high sweep rate. The overall design is characterised by extremely high shock resistance and very good seaworthiness. The main body of the SSC is a heavy unframed steel cylinder which houses most of the machinery and sweeping equipment. The propulsion system consists of a diesel-driven hydraulic power transmission and a combined rudder-propeller unit. Maximum speed is about 10 kt. For optimum sweeping operation, even in rough seas, an autopilot is provided.

The magnetic sweeping field is generated by two magnetic coils, mounted forward and aft on the cylindrical steel body. For acoustic simulation sweeping two medium-frequency broadband noise generators are positioned in the forward area and for low-frequency transmission a third noise generator is mounted in the aft area of the SSC. The respective magnetic fields and acoustic signatures to be generated by the SSC are programmed by the MCM control officer on board the parent MCMV.

The Troika system of the German Navy uses a converted coastal minesweeper ('Lindau' class, Type 351) as a guidance and control ship with a mine warfare operations centre and associated remote-control facilities. The latter comprises: a horizon-stabilised X-/C-band precision navigation radar; a digital target extractor; a master console with three control displays; and a UHF datalink transceiver. Control of SSC operations is exercised from each control display which is able to greatly magnify the swept channel so that a high degree of precision is achieved in guiding the SSC along the predetermined tracks in the mine danger area.

Control commands are sent to the SSC in the form of multiplexed data messages using the UHF datalink transceiver with up- and downlink capability. Datalink messages may refer to the SSCs autonomous navigation, equipment control and magnetic/acoustic simulation subsystem. The responses from the SSC concerning their status and functions are monitored automatically by an integrated monitoring system on board the control ship. The last main feature is the reference buoy, which provides a geographically stabilised radar image and constitutes an important component for precise navigation in this Troika configuration.

Operational status
The system has been delivered since 1979 and operational with the German Navy since 1980. Its effectiveness was proven during allied MCM operations

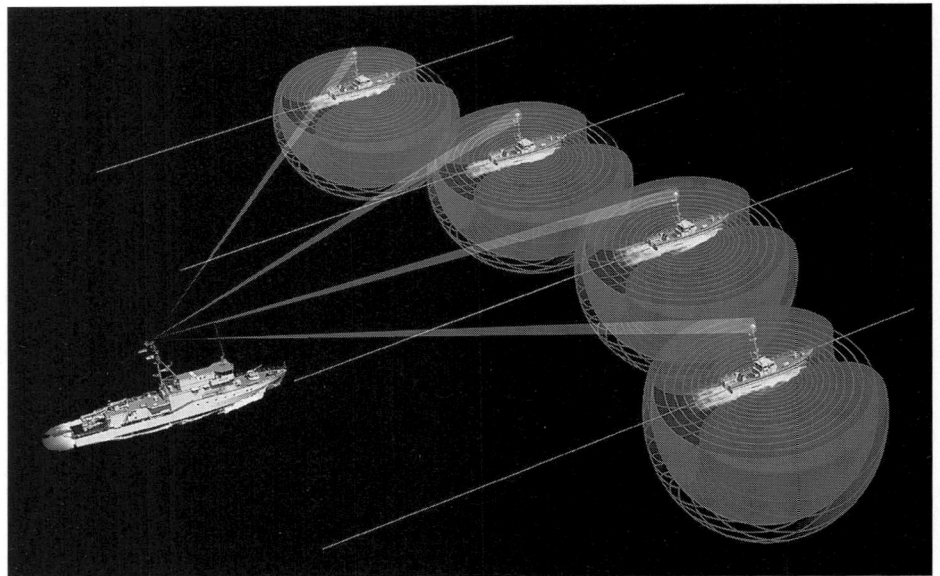

Operating principles of Troika remote minesweeping system   *1998*/0007405

Seehund 5 drone of the Troika system (H & L van Ginderen)   *1997*

in the 1990-91 Gulf War. A modified version is being studied by the Royal Netherlands Navy with the aim of installing it on three of their 'Tripartite' minehunters to be converted to Troika control vessels by 1998. In January 1998, the German Navy began converting five 'Hameln' class minesweepers to Troika control ships designated HL 352. These vessels will be equipped with an advanced command and control system for remote control of Simulation Sweeping Craft (SSC), a mine avoidance sonar MHS 90 and a number of one-shot Seafox mine disposal vehicles for countering moored mines. The advanced Troika C² system will comprise a single multifunction console for the control of up to four SCCs; one multifunction console for the

supervisor; GPS on both the control ship and the SSC for accurate positioning and position control; improved UHF datalink for enhanced data exchange and control of the SSCs; and Loran-C equipment as back-up navigation facility. The advanced C² system also provides the capability of MCM mission planning, supervision of execution, mission recording and evaluation, and onboard operator training.

Contractors
Lurssen Werft, Bremen.
STN ATLAS Elektronik, Bremen (C² system).

UPDATED

INTERNATIONAL

'Tripartite' design

Type
Minehunter.

Description
The 'Tripartite' design is a collaborative venture between France, Belgium and the Netherlands. The hull, decks and partitions are constructed of a single GRP skin stiffened by trapezoid section formers. Hull and former joints are reinforced with glass fibre pins. Hull reinforcements are principally transverse, but longitudinal ribs and binding strakes provide resistance to buckling.

Two independent propulsion systems are fitted: a main conventional diesel and propeller shaft (supplied by the Netherlands); and an auxiliary propulsion system for minehunting operations which consists of active rudders (supplied by Belgium). The main engine is a supercharged 1,860 bhp (1.37 MW) Brons Werkspoor A-RUB 215 V12 diesel driving, via a flexible coupling, a Rademakers epicyclic reduction gearbox and a Lips five-bladed cp propeller. Under auxiliary power the propellers are maintained in a 'feathered' position. The auxiliary propulsion system comprises two Acec active rudders with six-bladed fixed-pitch propellers. In addition, a Schottel thrust unit comprising two electric motors each driving a propeller transversely mounted in a tunnel is fitted near the bows. The active rudders and bow thrust unit are driven by three Astazou gas-turbine-powered alternators mounted high up in the ship. When minehunting, one gas-turbine unit powers the auxiliary propulsion system, a second provides electrical power for ship's services and the third remains on standby. A fourth diesel-driven alternator provides electric power for ship services when in port. The machinery can be controlled from the bridge in manual mode, or from the operations room in either manual or automatic mode. In emergency the machinery can be controlled from a soundproofed machinery control room located above the engine room.

The minehunting system comprises the Thomson Sintra DUBM 21B sonar integrated with a Thomson CSF Evec command and control display system. Navigational systems include a Racal Decca 1229 I-band radar, Loran and Syledis and Decca Hifix. Mine disposal is carried out using the two ECA PAP 104 ROVs carried on board. Up to six clearance divers are also carried. Light mechanical and acoustic sweep gear (the AP-4) can be deployed.

The vessels are fitted with comprehensive communications facilities comprising: an SNTI internal communication system; two sets of Signaal HF radio receivers; and a UHF receiver supplied by Belgium. Self-defence armament comprises a Giat 20F2 20 mm gun and a single 12.7 mm machine gun.

Operational status
Vessels in service with the Belgian, Royal Netherlands and French navies, and units have also been exported to Indonesia and Pakistan.

If funds are forthcoming, the Belgian minehunters will be given an upgraded capability between 1998 to 2003.

Three of the Royal Netherlands units began converting to Troika control vessels in 1996. They will receive the new C² system being jointly developed by Germany and the Netherlands. The remaining operational units will be upgraded with command and sonar system improvements, a new ROV, and propelled variable depth sonar system (in four units).

Contractors
DCN, Lorient, France.
Beliard Polyship, Ostend, Belgium.
Van der Giessen-de Noord, Netherlands.

VERIFIED

French 'Tripartite' minehunter La Croix du Sud.
Note the variation abaft the main superstructure, where the decompression chamber is mounted
(Spearhead Exhibitions)
1996

Belgian 'Tripartite' minehunter Narcis (H & L van Ginderen)   *1996*

Netherlands 'Tripartite' minehunter Hellevoetsluis (Spearhead Exhib tions)   *1996*

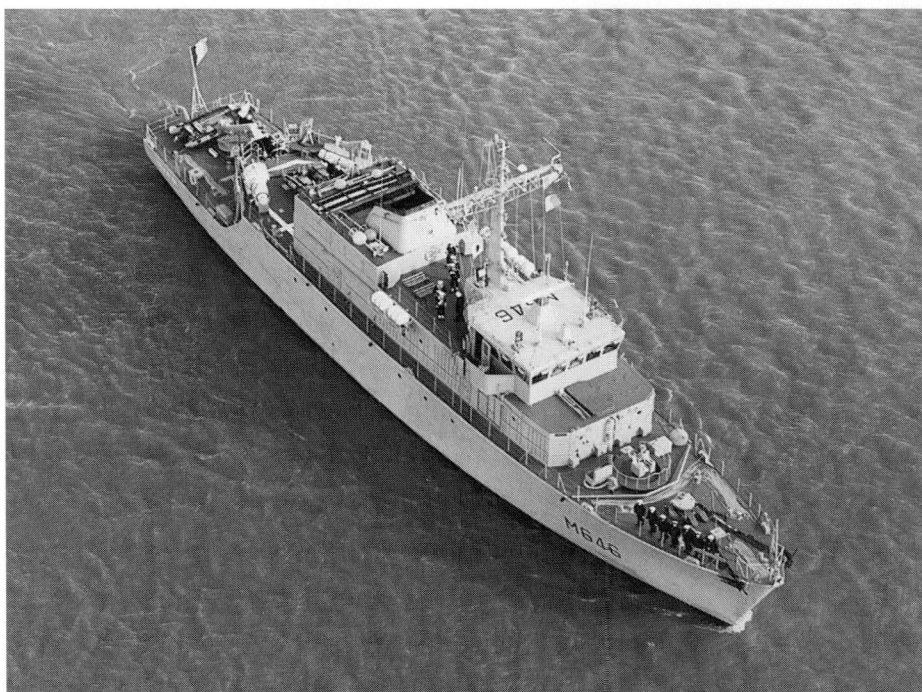

ITALY

'Gaeta' class

Type
Minehunter.

Description
This is a slightly lengthened version of Intermarine's 'Lerici' class minehunter and is equipped with: SMA MM/SPN-728 radar; FIAR 2072 (an Italian variant of the US SQQ-14 with digital processor) minehunting sonar; GEC-Marconi Speedscan side scan route-mapping

sonar; one SMIN Constorium MIN Mk 2 ROV and one Gaymarine Pluto Plus ROV for mine disposal; and a single Mk 4 mechanical wire sweep. Precise positioning is exercised through the POSYS-Miniranger III system. Command and control is exercised through a Datamat MM/SSN-714 command system. A twin 20 mm AA gun is mounted forward for self-defence.

Main propulsion is by a Fincantieri-GMT B230-8M diesel engine driving a single shaft fitted with a KaMeWa cp propeller through a Tosi single-reduction gearbox. Electric power at 440 V 3-phase 60 Hz is

supplied by three Ansaldo 420 kW alternators, each driven by a 550 bhp Isotta-Fraschini ID36SS-6V auxiliary diesel engine. For slow-speed propulsion there are three 120 shp electrohydraulic motors, each driving a Riva Calzoni retractable/rotatable hydraulic thruster, positioned one forward and two aft. All machinery is resiliently mounted, and constructed of amagnetic material.

To prevent buckling, Intermarine developed a monocoque hull of relatively thick single skin whose thickness decreases from keel to deck sides. The material used is Intermarine's patented unsaturated isopthalic resin system which achieves twice the design-lethal limit specified by the Italian Navy and five times the shock resistance of wooden minesweepers. Machinery, equipment and crew are physically isolated from the hull bottom by means of cradles suspended from bulkheads which minimises shock effect on the deck and bulkheads and considerably reduces self-generated noise and vibration, much of which is dissipated before it reaches the hull.

Operational status
The six unit class is now operational. A modified design has been developed for the Royal Australian Navy (see 'Huon' class entry). Two units are also building for the Royal Thai Navy the first, *Lat Ya*, was launched in March 1998 and the second in June 1998.

Contractor
Intermarine SpA, Sarzana.

'Gaeta' class minehunter Viareggio. *Note the decompression chamber mounted on the upper deck aft* (Intermarine)
1998/0007406   **UPDATED**

JAPAN

'Yaeyama' class

Type
Ocean-going minehunter/minesweeper.

Description
These 1,000 tonne wooden-hulled ocean minesweepers are powered by two Mitsubishi 6NMU-TAI diesel engines, each developing 1,200 bhp (1.76 MW), driving twin shafts fitted with cp propellers through single-reduction gearboxes for a speed of

18 kt. Precise manoeuvring is aided by a 350 hp (257 kW) bow thruster unit. The ships are being retrofitted with an integrated tactical command system. They deploy the Type S-7 deep water minehunting system and the Type S-8 (SLQ 48) deep water mechanical minesweeping system and are equipped with the Fujitsu OPS9 I-band surveillance radar; Raytheon SQQ-32 variable depth minehunting sonar and Klein Associates AQS-14 side scan sonar. Armament consists of a single General Electric 20 mm Sea Vulcan gatling-type gun.

Operational status
Three units operational, but plans for a further three abandoned.

Contractors
Hitachi, Kanagawa.
NKK, Tsurumi.

VERIFIED

'Uwajima' class

Type
Coastal minehunter/minesweeper.

Description
This is a slightly lengthened (by 2.7 m) version of the 'Hatsushima' class MCMV, the increased length being used to provide improved accommodation facilities. The sensor fit includes a Fujitsu OPS-18B I-band surveillance radar (OPS 39 in later units) and NEC/

Hitachi ZQS 3 high-frequency hull-mounted minehunting sonar.

Mine destructor equipment is based on the remote-controlled Type S-7 equipped with a counter mining charge. The ships are armed with a General Electric 20 mm Sea Vulcan gatling-type gun mounted forward.

Main propulsion is by two Mitsubishi 6-NMU-TAI diesel engines driving twin shafts fitted with KaMeWa cp propellers through Mitsubishi single-reduction gearboxes; electric power is provided by one 1,450 kW gas-turbine DC generator (for the electric sweep),

and two 160 kW AC diesel-alternator sets (for ship services).

Operational status
Nine units operational.

Contractors
Hitachi, Kanagawa.
NKK, Tsurumi.

VERIFIED

'Sugashima' class

Type
Coastal minehunter/sweeper.

Description
This new class of coastal minehunters has been designed to replace the obsolete 'Takami' class, most of which have now been decommissioned. The design differs markedly from the 'Uwajima' class and will be much more capable than the current mine countermeasures vessels in service in the JMSDF. The

upper deck has been extended aft to provide additional stowage for mine disposal equipment. Following Japanese experience in minesweeping operations in the Gulf during the Iraq/Kuwait war, the ships will be fitted with an integrated mine countermeasures command system, the GEC-Marconi Nautis M command system, and a version of the Type 2093 VDS minehunting sonar, and French ECA PAP 104 Mk 5 ROV. The vessels will also deploy the Australian Dyad minesweeping system. It is believed that the Japanese decision is based on a desire to compare the latest generation UK and US command and control and sonar

systems. Other systems to be fitted include Fujitsu OPS 39B I-band surface search radar and a 20 mm Sea Vulcan gatling-type gun. Propulsion will be provided by two Mitsubishi 6 NMU-TAI diesels.

Operational status
Programme authorised under the FY95 programme. Four units under construction.

UPDATED

KOREA, SOUTH

'Swallow' class

Type
Minehunters.

Description
The 'Kang Keong' (Swallow) design features: a Racal Mains 500 integrated AIO; Thomson Marconi Sonar Type 193M minehunting sonar; Type 2048 Speedscan sonar; Raytheon I-band navigation radar; and two Italian Gaymarine Pluto ROVs. The ships are also fitted with a single wire sweep system. Self-defence is provided by an Oerlikon 20 mm light AA gun and two 7.62 mm machine guns. Propulsion comprises two MTU 1,020 bhp diesels driving two Voith-Schneider vertical cycloidal propellers, together with a 102 hp bow thruster for precise manoeuvring.

Construction of a slightly larger (600 tonnes) variant began in December 1996. This is thought to be the first of a class of seven units designed to deploy a VDS of the Type 2093.

Operational status
Six units operational in South Korean Navy.

Contractor
Kangnam Corporation.

UPDATED

NORWAY

'Oksoy/Alta' classes

Type
Split class minehunter (four units) minesweeper (five units)

Description
One of the main features of this design is the high transit speed requiring less power than comparable conventional designs, which will facilitate deployment over the lengthy Norwegian coastline. By using a catamaran design built in advanced composite sandwich structure, it is claimed that lower magnetic and acoustic signatures will be exhibited due to the small wet area of the hull, leading to safer operation and less susceptibility to shock, and reduced disturbance of the water leading to improved sonar response, particularly in shallow water. Other advantages claimed are improved manoeuvrability, reduced draught, improved view from the bridge – an important element in minesweeping operations, increased deck area (approximately 70 per cent more compared to a monohull) and greater comfort and improved sea-keeping qualities due to a much lower roll acceleration compared to other catamaran-type designs and a much reduced roll angle compared to standard displacement hulls (approximately 2 to 3° in Sea State 3).

The minehunter version is equipped with an integrated minehunting sonar comprising the Thomson Sintra TSM-2023N classifier and Simrad detection sonar, the combined system being carried on a retractable cantilever amidships inside the air cushion. Two Gaymarine Pluto ROVs are carried for mine disposal, being handled by two deck cranes. Minehunting operations will be controlled by the Simrad/Thomson Sintra Micos integrated MCM system. The minesweepers are fitted with the Simrad SA 950 mine avoidance sonar, and the customary wire/ acoustic/magnetic sweeps. Extensive navigation equipment is fitted including a Racal Decca radar and Doppler log, GPS, wind sensors and so on. These are linked to the Norcontrol Databridge 2000 ARPA and SeaMap electronic chart system.

An Aeromaritime Systembau integrated computerised distributed communications system is fitted, incorporating a comprehensive message handling system. The company is also responsible for the EMI environmental control on the craft. The ships are armed with a six-tube Sadral SAM launcher firing Mistral missiles, and two Rheinmetall 20 mm AA guns for self-defence.

Bridge layout of the Oksoy (Kvaerner)

1996

Oksoy *undergoing shock trials* (Kvaerner)
1999/0024794

General arrangement layout of the Norwegian Oksoy *(Kvaerner)*

1996

Main propulsion is by two raft-mounted MTU 12V 396 TE94, 1,920 bhp (1,400 kW) diesel engines powering twin Kvaerner Eureka water-jets through ZF BW465 single-reduction gearboxes. The water-jets are controlled by the dynamic positioning system. Two MTU 8V 396 TE54, 950 bhp (700 kW) auxiliary diesel engines power the lift fans for a draught of 0.90 m when on-cushion. Machinery and ship control is exercised through a Norcontrol Automation Ship Technical Control system (STC) and a Surface Surveillance System (SSS).

Operational status
Four 'Oksoy' class and five 'Alta' class.

Contractor
Kvaerner Mandal, Mandal.

UPDATED

The Norwegian MCMV Oksoy *(Kvaerner)*
1998/0007407

RUSSIAN FEDERATION AND ASSOCIATED STATES (CIS)

'Gorya' class

Type
Ocean-going minehunter.

Description
The hull is probably of aluminium construction. Main propulsion is by two diesel engines driving two shafts for a speed of 18 kt. It is armed with: eight (2 × 4) SA-N-5/SA-N-14 (Grail/Gremlin) SAM missile launchers; one 76 mm DP gun; one six-barrelled 30 mm AA gatling-type gun; and two 16-barrelled chaff rocket launchers. The associated sensor fit includes: Palm Frond I-band surveillance radar; Bass Tilt H/I-band fire-control radar; Nayada I-band navigation radar; Salt Pot IFF interrogator; HF minehunting sonar; passive ESM (Cross Loop and Long Fold); and a Kolonka optronic director. Accurate position fixing systems may be fitted.

Mechanical, acoustic and magnetic sweeps are provided for minesweeping, and HF sonar and ROV for minehunting. The ROV is deployed from behind a large sliding door sited on 02 deck at the after end of the superstructure to port and starboard beneath the gatling-type gun. The sweep deck aft is very cramped for space.

Operational status
Two units operational.

Contractor
Kolpino Yard, Leningrad.

VERIFIED

'Lida' class

Type
Coastal minehunter.

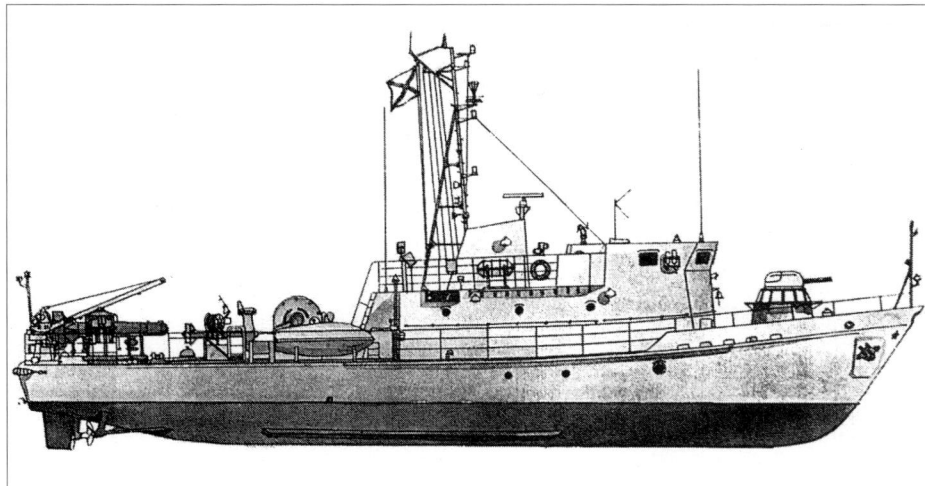

Description
This class of coastal minehunter was designed as a follow-on to the 'Yevgenya' class, to which it is similar in appearance. The ships are powered by three 900 hp (690 kW) diesels driving three shafts for a speed of 12 kt. The vessels are armed with a 30 mm six-barrel gatling-type gun and minesweeping equipment comprises the AT-6 acoustic, PEMT-4 and ST-2 magnetic and GKT-3MO wire sweeps, the latter with a swept path of 50 to 120 m and a depth of 2 to 120 m. In addition, the vessels carry the IT-3 ROV equipped with underwater TV camera for mine identification. A Karbarga I hull-mounted high-frequency minehunting sonar is carried. The radar outfit comprises a Pechora I-band surveillance radar which may be complemented by Bass Tilt fire-control radar in due course.

The hull is of GRP with low magnetic steel and aluminium-magnesium alloy in parts. Transverse framing is employed with frames spaced at 400 mm. All equipment is shockmounted with flexible connections on all pipework.

Operational status
Twenty three units are in service and construction is continuing.

UPDATED

Outboard profile of the 'Lida' class
1998/0007408

'Natya' class

Type
Ocean-going minesweeper/hunter.

Description
The low magnetic steel-hulled Project 266ME (export 'Natya' class) ocean-going minesweeper continues to be built in small numbers. The ships deploy: the GK-2 wire sweep with a sweep width of 260 to 280 m and maximum sweep depth of 200 m and minimum sweep depth of 10 m; the AT-2V acoustic sweep (with a towing speed of 5 to 12 kt in up to Sea State 4 and a minimum depth of 15 m; and the TEM-3 coil sweep with a towing speed of up to 10 kt at Sea State 4 and a minimum depth of 15 to 20 m with a length of 455 m. Control of influence sweeping operations is carried out by two sets of Micron control equipment.

For mine detection, the ships are fitted with the MG-89 minehunting sonar with a detection range of 1,000 to 1,200 m against moored mines and 350 to 400 m against seabed mines. Angular resolution is 1.5°. The ship is also fitted with the MG-35E underwater sound communications sonar used to maintain non-directional HF/LF telephone and telegraph communications.

Self-defence is provided by two AK-306 30 mm six-barrel automatic gun mounts sited at the ship's bow and stern. Two Igla-1 portable air defence missile systems are also carried; located on either side of the ship and two RBU-1200 ASW mortars mounted to port and starboard just aft of the forward 30 mm mount. The ship is also capable of minelaying, a total of eight being carried and secured on special mine rails aft.

The ships are powered by two M503B-37 four-stroke diesels with a total output of 5,000 bhp. These drive two VR226 AE-TM three-bladed cp propellers with a diameter of 2.1 m through a reduction gearbox. The export variant differs from the Russian Navy version in having the main engines and cp propellers electrically controlled instead of pneumatically.

Degaussing equipment consists of three coils (horizontal main, frame course and keel course) and control equipment.

Operational status
In widespread service around the world.

VERIFIED

'Sonya' class

Type
Coastal minesweeper.

Description
The Project 1265-E ('*Sonya*') coastal minesweeper features a wooden hull sheathed in GRP. The ships are equipped to handle the GKT-2 wire, AT-2 acoustic, PEMT-4 and ST-2 magnetic sweeps. It is designed to detect and mark seabed mines. For minehunting a MG-89 minehunting sonar is also fitted. The ship is also fitted with an IT-3 mine detector exploder with a hunting swath of 20 to 30 m and an operating depth of 10 to 60 m and a towing speed of 3.5 to 6 kt. The exploder is towed at a distance of 70 m astern of the ship. In addition to the MG-89, the ship is also fitted with the MG-35E underwater communication system which maintains non-directional HF and LF telephone and telegraph communications and also measures ranges using the interrogations-response methods. The operating range is 10 km for telephone communications and 20 km for telegraph communications.

For self-defence the ships are fitted with two AK-306 30 mm automatic six-barrel gun mounts.

Propulsion is provided by two 12ChPN 18/20 four-stroke diesel engines rated at 1,000 bhp each (at 1,550 rpm) driving two three-bladed VR1265E-TM5 cp propellers with a diameter of 1.7 m. Electrical supplies are provided by three diesel generators each rated at 110 kW and one 50 kW diesel generator.

Degaussing equipment consists of three coils (horizontal main, frame course and keel course) and control equipment.

Operational status
Numerous units in service around the world.

Contractors
Petrozavodsk shipyard.
Vladivostok shipyard.

VERIFIED

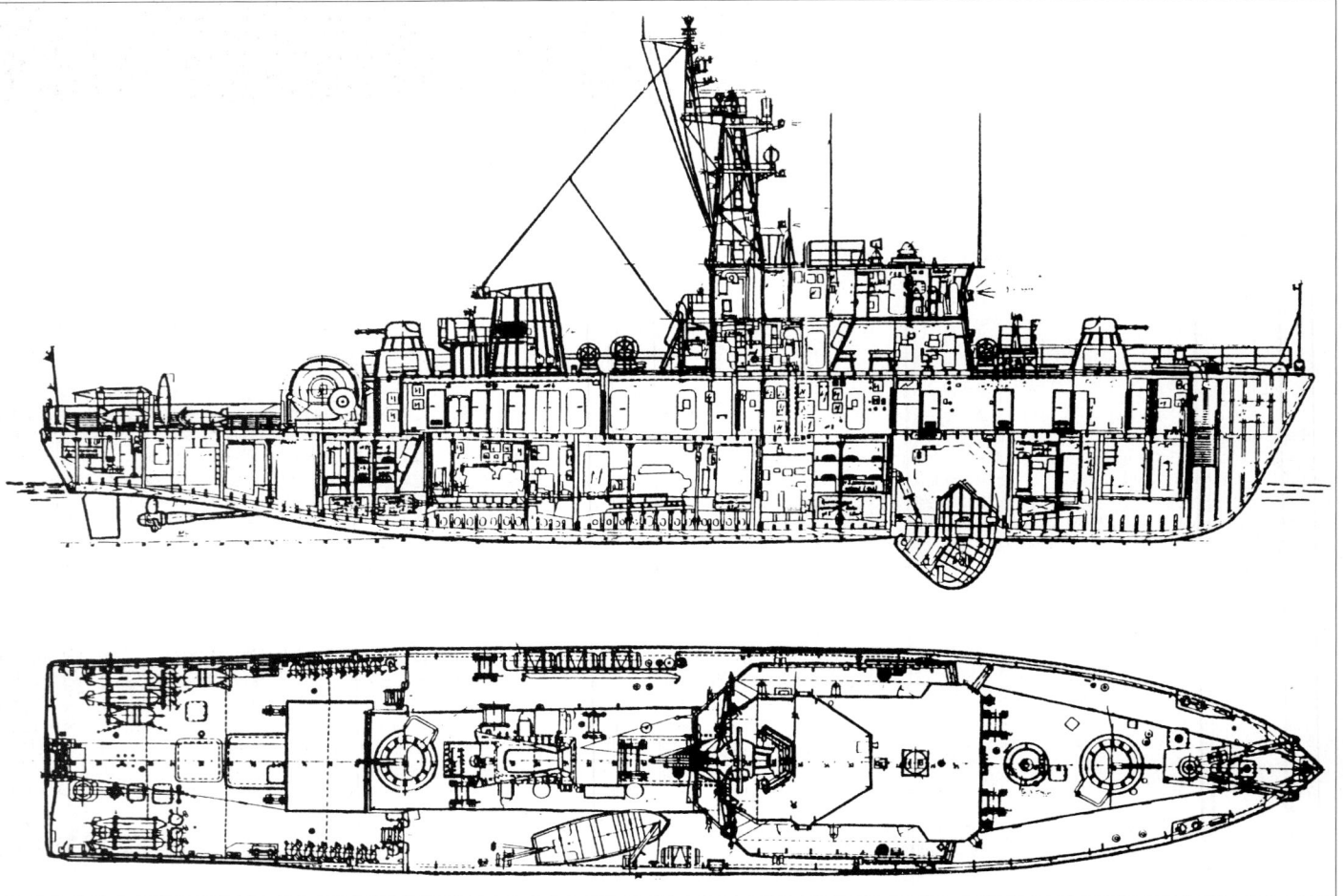

General arrangement of the 'Sonya' class   *1998*/0007409

SPAIN

'Segura' class

Type
Minehunter.

Description
On 4 July 1989, Bazan signed a technology transfer agreement with Vosper Thornycroft for the shipyard to design a new MCM vessel based on the British 'Sandown' design. This technology transfer agreement was followed by another agreement signed in November 1993 between DCN and Bazan which provides for personnel at Bazan to be trained in GRP technology. A class of 12 vessels is envisaged, eight minehunters and four minesweepers.

Collectively known as the CME (Contra Minas Espanol) the new design will be fitted with a minehunting system manufactured by FABA-Bazan in co-operation with Inisel and based on the British GEC-Marconi Nautis minehunting command system. The minehunting sonar will be a multifunction high-frequency VDS SQQ-32 sonar manufactured by ENOSA under licence from Raytheon. For mine neutralisation the ships will carry two Gaymarine Pluto Plus ROVs.

The vessels will be powered by two MTU-Bazan 6V 396 TB83 diesels (1,523 bhp) built by Bazan and driving two Voith Schneider cycloidal propellers for a speed of 14 kt. Two 55 kW bow thrusters will also be mounted for precise manoeuvring during minehunting operations. Slow speed propulsion for minehunting will be provided by two 125 kW electric motors. Hotel services will be provided by three 270 kW diesel alternators.

Armament will comprise a single Oerlikon 20 mm GAM-BO1 gun and one or two Pluto Plus ROVs will be carried.

Operational status
First unit under construction. The four units are scheduled to complete as: *Segura* May 1998, *Sella* May 1999, *Tambre* January 2000 and *Turia* September 2000.

Contractor
Bazan.

UPDATED

SWEDEN

'Landsort' class

Type
Multirole mine warfare vessel.

Description
This is an extremely compact multipurpose minelaying/minehunter/minesweeper design. For minesweeping duties the class is equipped with wire, acoustic and electric sweeps; and for minehunting with the Bofors Sea Eagle ROV for mine disposal. With the addition of a radio command link, the ships can control two slave 18 m SAM GRP catamarans fitted with acoustic and electric sweeps. In the minelaying role the vessels are fitted with rails laid along each side of the aft deck, but merging at the stern to form a single track over the transom.

The sensor fit includes: Terma navigation radar; Thomson Marconi Sonar SAS TSM 2022 minehunting sonar; TVT-300 optronic director fitted with TV camera and laser rangefinder which feed the CelsiusTech 9MJ-400 AIO system and the 9LV-100 weapon control system. The armament comprises a Bofors 40 mm DP gun, two 7.62 mm GP (2 × 1) machine guns and four nine-barrelled Saab Missiles Elma ASW rocket launchers. It is possible that the 40 mm will be replaced in time by the Bofors Trinity 40 mm CIWS system which has been trialled on the *Vinga*. In addition, the ships carry the Saab Dynamics RBS-70 MANPADS SAM.

Main propulsion is by four SAAB-Scania DSI-14 diesel engines driving two Voith-Schneider cycloidal propellers with the engines coupled in pairs by an especially quiet clutch cone belt transmission system.

MCMV Landsort (Kockums Naval Systemst)

1998/0007410

Electric power at 440 V 3-phase 60 Hz is supplied by two 180 kW alternators, each driven by a 337 bhp SAAB-Scania DSI-11 auxiliary diesel engine; and a 108 kW alternator driven by a 183 bhp SAAB-Scania DN-11 auxiliary diesel engine. All diesel-alternator sets are mounted on the upper deck to reduce the noise signature; while slow-speed propulsion is achieved by using only one main engine per shaft.

Operational status
Seven units operational with the Swedish Navy and four units with the Republic of Singapore Navy.

Contractor
Kockums Naval Systems, Karlskrona.

UPDATED

SAM

Type
Remote-control minesweeping drone.

Description
Developed by Karlskronavarvet, the remote-controlled SAM uses a diesel-powered rudder propeller to drive the GRP catamaran hull. The craft is fully equipped for power supply, sweeping of acoustic and magnetic mines, navigation and remote control from a 'Landsort' class MCMV. The MCMV, or any other ship, can control a number of the unmanned sweeps, providing remote control of the diesel engine, steering, minesweeping equipment, control and monitoring. The hull and superstructure are built in saltwater-resistant aluminium with the catamaran pontoons of GRP sandwich material.

The hull of the SAM is fitted with a number of coils in the pontoons which are used for sweeping magnetic mines, and a towed acoustic transmitter is used to counter acoustic mines. A number of buoys is carried to mark the swept passage. A digital Global Positioning System (GPS) and microplotter is also installed to monitor the sweeping track.

Onboard power is provided by a Volvo Penta TAMD 70D diesel engine with a continuous output of about 159 kW at 2,200 rpm, to give the craft a speed of 8 kt.

Power is fed to a Schottel propeller unit via reversing gear couplings and a shaft. The diesel and reversing gear are mounted on a girder framework which is

Remote-controlled sweeping drones SAM 03 and SAM 05

shockmounted in an aluminium structure on to the main platform.

Although the machinery is normally remotely controlled from the MCMV, it can also be controlled at the engine or from the operating platform. Remote-control functions include diesel engine control, steering control, minesweeping equipment control and monitoring.

Operational status
In service with the Swedish and US navies. Four SAM drones have also been delivered to the Japanese Maritime Self Defence Force.

Specifications
Length: 18.2 m
Beam: 6.2 m
Draught: 1.6 m
Weight: 25 t
Displacement: 27 m^3
Speed: 8 kt
Range: 330 n m les at 7 kt

Contractor
Kockums Naval Systems, Karlskrona.

UPDATED

'Styrso' class

Type
Inshore mine countermeasures vessel.

Description
In 1994, Karlskronavarvet and Erisoft AB were awarded a contract to build a new class of MCMV capable of controlling two SAM drones. These ships are also designed for use in an inshore surveillance role.

'Styrso' class MCMV Skafto
(Kockums Naval Systems)
1998/0007411

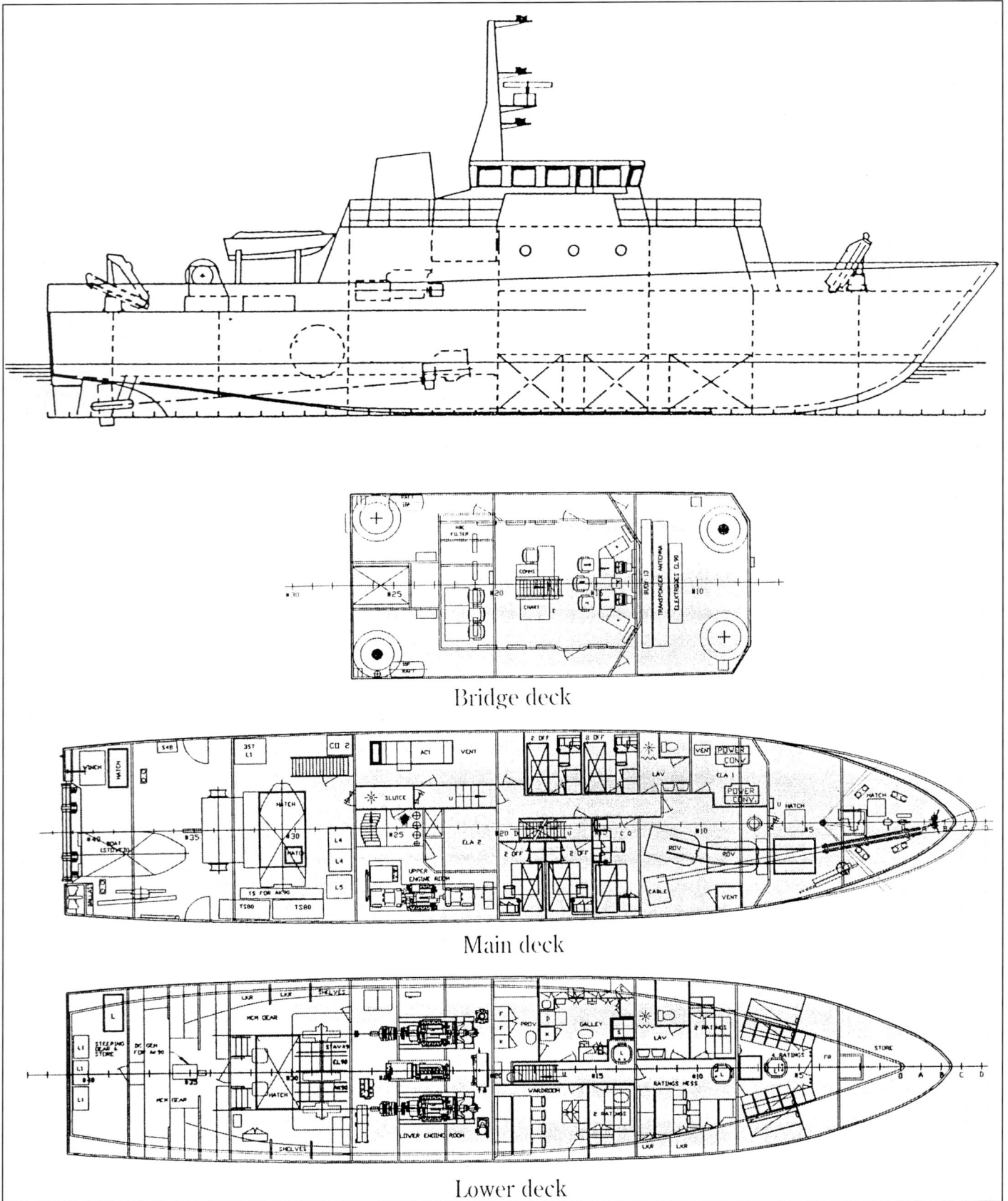

Bridge deck

Main deck

Lower deck

Outboard profile and general arrangement plan of 'Styrso' class

1998/0007412

The main MCM systems comprise two Double Eagle ROVs which are intended for clip-on sonar and mine disposal charges. The ships themselves are fitted with a Reson mine avoidance sonar and an EG & G side scan sonar system for route surveying tasks. The navigation radar is a Racal Bridgemaster I-band system. All systems integrate with an Ericsson tactical data system. For minesweeping the ships carry an AK-90 acoustic, EL-90 magnetic and mechanical sweeps. The main combat system is based on COTS PC technology integrating the various mine countermeasures equipments via a LAN system.

The vessels are powered by two SAAB-Scania DSI 14 diesels of 1,104 bhp, driving two shafts through reduction gears for a speed of 13 kt. A bow thruster unit is also installed.

There is no main armament, but the ships carry two 12.7 mm machine guns.

Operational status
Four units in service with the Swedish Navy.

Contractor
Kockums Naval Systems, Karlskrona.

UPDATED

UNITED KINGDOM

'Sandown' class

Type
Minehunter.

Description
The 'Sandown' class, which is also being acquired by the Royal Saudi Naval forces, is the third generation of Fibre Reinforced Plastic (FRP) minehunters designed by Vosper Thornycroft to counter the increasingly sophisticated mine threat. To face anticipated future shock levels, new concepts have been realised. Shockwaves affect FRP hull structures in two ways. Firstly, primary shockwaves cause the hull, and particularly the bottom, to undergo high accelerations which can result in delamination and breaking of bonded fixtures. Secondly, bubble pulse pressures cause the hull to whip through, which means the bottom and main deck panels can be severely buckled, possibly leading to the breaking of the ship's back. Any structural connections and their designs therefore play a major part in ship survivability in the face of severe shockwaves.

New design techniques have enabled the builders to dispense with the need for through bolting of hull frames previously required to meet the exacting shock standards demanded in MCMV construction. The FRP skin of the hull bottom is longitudinally stiffened, with stiffeners laminated to the hull using flexible resin fillets. The main deck is also longitudinally stiffened, with transverse stiffeners fitted on the sides of the hull. This has eliminated the need for through bolts and crossovers of structural members resulting in a lighter structure, while at the same time achieving the required strength to meet the exacting shock standards. The main watertight and transverse bulkheads are of corrugated FRP, while all other bulkheads are of FRP sandwich construction with a balsa wood core. The stern is specially designed to reduce slamming and improve conditions for the launch and recovery of the ROV and results in a much more stable platform.

For the second batch of 'Sandowns' to be built for the Royal Navy, Vosper Thornycroft is introducing new construction techniques, including the SCRIMP (Seeman Composite Resin Infusion Moulding Process). In this process, the resin is drawn into a sealed mould under a vacuum. The traditional construction process involves laying up the FRP by hand, with resin applied from a bucket. This laying up process results in a quality of FRP far surpassing existing FRP construction and will result in a lighter and stronger laminate.

From the safety aspect, the new process will virtually eliminate dangerous styrene fumes. Other improvements in construction are being applied to the modular construction techniques, resulting in enhancements to three modules, in which outfitting will be maximised before module installation in the ships.

Other changes to the design include an increase in the size of the propulsors in the second batch (from size 16: 1.6 m diameter to size 18) which will enable the vessels to maintain speeds currently attained by operational units, but with more equipment on board. Accommodation arrangements are being remodelled to cater for the inclusion of female personnel in the ship's crew. Other changes to the design include tropicalisation to enable the units to conduct operations in out-of-area regions such as the Gulf. This will primarily affect habitability standards, and include air conditioning. Finally, a new telephone exchange will be installed together with a two-man decompression chamber. Apart from the new construction techniques, all the changes are relatively minor, but will considerably improve the effectiveness of the ships.

The minehunting system comprises the Thomson Marconi Sonar 2093 variable depth minehunting sonar (which is integrated with the ship's position control system) and the Nautis command and control system combined with highly sophisticated navigation equipments. The sonar can be operated either as a Variable Depth Sonar (VDS), or locked into the hull for operation in shallow water conditions. Navigation equipment comprises a Kelvin Hughes Type 1007 navigation radar with bright raster display for bridge viewing, and a Kelvin Hughes echo-sounder. Other navaids include gyrocompass, AGI log which combines

Bridge layout of HMS Sandown (Vosper Thornycroft) *1996*

Engine room of a 'Sandown' class minehunter (Vosper Thornycroft) *1996*

HMS Penzance *the first of the second batch of 'Sandown' class minehunters for the Royal Navy* (Vosper Thornycroft) *1999*/0024796

Operational Spaces

Machinery Spaces

Accommodation

Galley and associated spaces

RADAR

BRIDGE TOP

01 DECK

OFFICERS (3)

OFFICERS (2)

WARDROOM

GEN. ROOM (S)

OFFICERS (2)

C.O.'s CABIN

I DECK

FAN CHAMBER

S.R.'s (6)

J.R.'s

ACCOMMODATION

DINING & REC. SPACE

GALLEY

OFFICE

SCHOTTEL MOTOR ROOM

HAWSER STORE

OFFICE

BATHROOM

SPARE CABIN (3)

J.R.'s

ACCOMMODATION

BATHROOM

J.R.'s

DINING SPACE

LAUNDRY

2 DECK

VOITH SCHNEIDER COMPARTMENT

ENGINE ROOM

COLD ROOM

NO.1 PUMP ROOM

SCHOTTEL

CHAIN LOCKER

NO.2 PUMP ROOM

COLD ROOM

AUX. MACHINERY COMPARTMENT

3 DECK

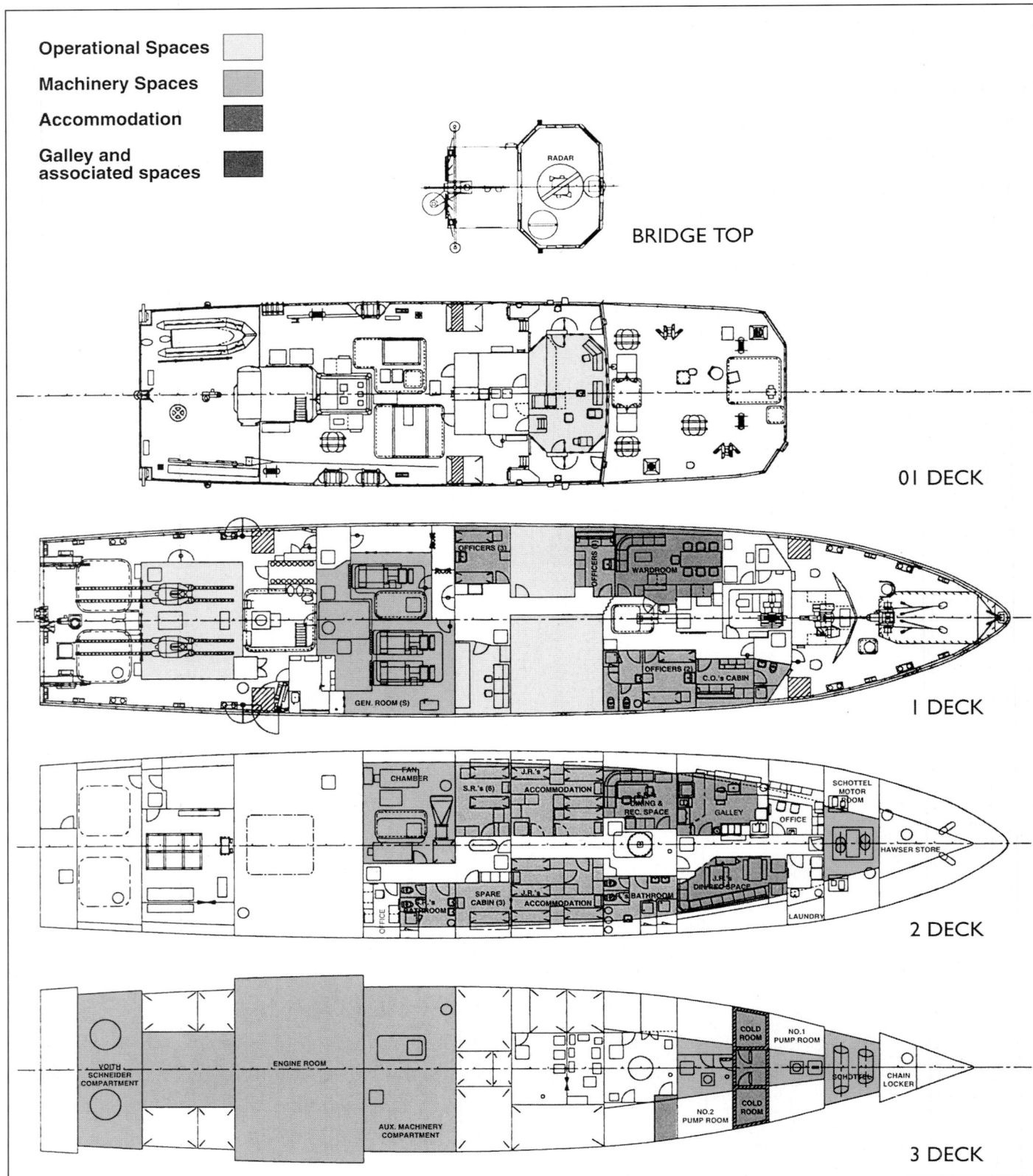

General arrangement layout of a multirole variant of the 'Sandown'

1996

EM log with ground correlation, and Evershed Aeolus wind measuring system which are linked into the command system. Radio fixing aids comprise the Racal Hyperfix and QM14 systems and Decca Navigator Mk 21 and Navstar GPS.

All systems feed to the command system via a PEMS PDM3 dual-redundant serial data highway and embedded microprocessors in each of the equipments. The system architecture devolves data processing functions to individual systems rather than through a central processing unit. This enables each system to exercise effective full local control but at the same time allowing the command to more effectively control and integrate the general functions and operations of the various systems through the command system. This allows various system

operators to concentrate their efforts on the immediate task in hand.

Communications comprise HF transmitter, LF/MF/HF receivers and V/UHF transceivers and V/UHF FM transceiver.

Main propulsion is by two Paxman Valenta 6-RP-200EM diesel engines each with a continuous propulsion rating of 630 kW at 1,450 rpm driving two Voith-Schneider 18GS cycloidal propellers through fluid couplings and GEC Marine & Industrial Gears single-reduction gearboxes. The low rotational speed of the Voith-Schneider propellers exhibits little or no cavitation and very low noise, providing infinitely variable thrust all round. Electric power at 400 V 3-phase 60 Hz is provided by three Mawdsley 250 kW alternators, each powered by a 335 bhp Perkins V8-

250G air-cooled auxiliary diesel engine. For auxiliary slow-speed propulsion, each propeller is belt-driven by a 135 shp Mawdsley electric motor, and draw their power from any diesel alternator set. Manoeuvrability is further aided by two Schottel electric-powered bow thruster units. To reduce the acoustic signature, which was a design priority, the main engines are raft-mounted, the electric plant is resiliently mounted on the upper deck with air-cooled engines, and hydraulic transmission has been replaced by electric drive in the minehunting mode.

Finally, the ship is armed with an MSI Defence Systems Ltd 30 mm gun mounted on the foredeck and two ECA PAP 104 Mk 5 ROVs for mine classification and disposal. These are carried on a trolley in the hangar and moved out on rails. A single hangar and

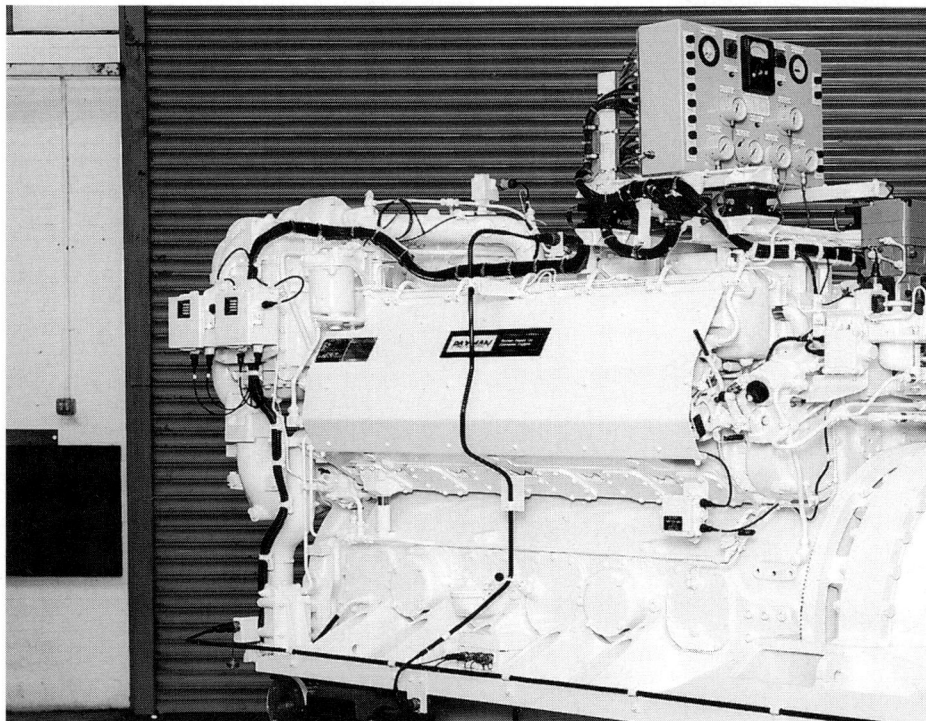

wire sweep winch can replace the double hangar fitted in the UK ships, together with an extra crane for handling the minesweeping equipment.

The units ordered by the Royal Saudi Naval forces are very similar to the UK vessels, except that as they are also to be utilised as patrol vessels the main engines are each uprated to 845 kW for a slightly higher speed, and an optronic director fitted to control the light AA gun on the fore deck.

Operational status
Five units of Batch 1 operational. The first unit of the second batch, *Penzance*, was commissioned in May 1998. The second ship in Batch II, *Pembroke*, was launched on 15 December 1997. The third ship, *Grimsby* was launched in August 1998. Three Royal Saudi Naval Force ships are in service.

Contractor
Vosper Thornycroft, Woolston.

UPDATED

A Valenta 6EM diesel for HMS Sandown
(Alstom Engines Ltd)
1999/0024795

UNITED STATES OF AMERICA

'Avenger' class

Type
Ocean-going minehunter/minesweeper.

MCM 2, USS Defender *of the 'Avenger' class. Note the large winch forward of the bridge for lowering the SQQ-32 minehunting sonar* (H & L van Ginderen)   *1996*

Description
The 14 units of the 'Avenger' class are equipped with: wire and SLQ-37(V)2 acoustic/magnetic influence sweeps for minesweeping; two Hughes Aircraft Naval and Maritime Systems SLQ-48 ROVs for mine classification and disposal; General Electric SQQ-30 sonar (being replaced by Raytheon SQQ-32) for minehunting; Mk 116 minehunting control system; Cardion SPS-55 surface search radar, and an IFF interrogator; and two 12.7 mm GP (2 × 1) machine guns mounted for self-defence. The vessels are gradually being refitted with a GEC-Marconi Nautis M system integrated with the Lockheed Martin SYQ-13.

The first two units of the class are powered by four Waukesha LN-1616 diesel engines driving twin shafts fitted with KaMeWa cp propellers through DeLaval twin-input/single-output single-reduction gearboxes. Three more Waukesha LN-161 diesels drive Tech Systems 375 kW alternators supplying electric power at 440 V 3-phase 60 Hz. In the third and all subsequent units, 655 bhp Isotta-Fraschini ID36SS-6V-AM diesel engines are substituted. In addition, there is a Siemens alternator powered by a Solar gas turbine for the electric sweep. For auxiliary slow-speed propulsion, Hansome 200 shp electric motors are coupled to each shaft line.

Operational status
Fourteen units operational.

Contractors
Peterson Builders Inc, Sturgeon Bay.
Marinette Marine Corporation.

VERIFIED

'Osprey' class

Type
Ocean-going minehunter.

Description
The 'Osprey' class is an enlargement of the Italian 'Lerici'. The vessels are fitted with the Raytheon/Thomson Marconi Sonar SQQ-32 minehunting sonar and Hughes Aircraft Naval and Maritime Systems SLQ-48 ROVs for mine classification and disposal; the

Raytheon SPS-64(V)9 radar and are armed with two 12.7 mm GP (2 × 1) machine guns. Command is exercised through a Lockheed Martin SYQ-13 command and control system. Precise positioning is controlled by a Microfix system.

Main propulsion is by two Isotta-Fraschini ID36SS-V8 diesel engines driving two Voith-Schneider cycloidal propellers through single-reduction gearboxes; with electric power at 440 V 3-phase 60 Hz provided by three 300 kW alternators, each driven by an Isotta-

Fraschini IF 36558V-AM auxiliary diesel engine. These latter units also power hydraulic pumps for the two 180 shp hydraulic motors for auxiliary slow-speed propulsion, and a 180 shp transverse thrust unit forward.

The 'Osprey' combines a relatively stiff hull structure with flexible decks and bulkheads which are connected with flexible attachment details. These have been designed and tested for underwater explosive shock loads. There are no stiffeners to cause stress

concentrations on the hull and the attachment of equipment is minimised and located in low-deflection areas.

Operational status
All 12 units now in service.

Contractors
Intermarine, Savannah.
Avondale Industries, New Orleans.

UPDATED

MHC Osprey. *Note the large winch forward of the bridge for handling the AN/SSQ-32 variable depth minehunting sonar* (Intermarine)
1998/0007414

TABLE XVI

Mine warfare vessel designs – specifications and electronics

Class	Navy	Displacement[1]	Dimensions[2]	Command	Sonars	Radar	EW	Hull
MINESWEEPERS								
K8 (M/S boat)	Vietnam	n/k/26	16.9 × 3.2 × 0.8	None	None	None	None	
MSB (M/S boat)	Thailand	21/25	15.3 × 4 × 0.9	None	None	None	None	Wood
MSB 07 (M/S boat)	Japan	50/n/k	22.5 × 5.4 × 1	None	None	OPS-29D	None	Wood
KMV (new minesweepers)	Belgium	n/k/644	52.4 × 10.4 × 3.1		Mine avoidance	ARPA	ESM	GRP
MSC (river minesweepers)	Romania	n/k/97	33.3 × 4.8 × 0.9		None	Nayada		
MSC 268	South Korea	320/370	43 × 8 × 2.6	None	UQS-1 or TSM 2022	Decca 45	None	Wood
MSC 268	Pakistan	330/390	43.9 × 8.5 × 2.6	None	UQS-1D	Decca 45	None	Wood
MSC 268 & 292	Iran	320/384	44.5 × 8.5 × 2.5	None	UQS-1D	Decca 707	None	Wood
MSC 289	South Korea	315/380	44.3 × 8.3 × 2.7	None	UQS-1 or TSM 2022	Decca 45	None	Wood
MSC 294	Greece	320/370	43.3 × 8.5 × 2.5		UQS-1D	Decca		Wood
M 15	Sweden	70/n/k	27.7 × 5 × 2	None	None	Terma	None	
M 301	Yugoslavia	n/k/38				Racal Decca		
PO 2 Type 501	Bulgaria	n/k/56	21.5 × 3.5 × 1					
T 43	Albania	500/580	58 × 8.4 × 2.1	None	Stag Ear	Ball End, Neptun	None	Steel
T 43 (China built)	Bangladesh	520/590	60 × 8.8 × 2.3	None	Tamir II	Fin Curve	None	Steel
T 43 Type 010	China	520/590	60 × 8.8 × 2.3	None	Tamir II	Fin Curve/Type 756	None	Steel
T 43	Egypt	500/580	58 × 8.4 × 2.1	None	Stag Ear	Don 2	None	Steel
T 43	Indonesia	500/580	58 × 8.4 × 2.1	None	Stag Ear	Decca 110	None	Steel
T 43 Type 254	RFAS	520/590	60 × 8.4 × 2.3	None	Stag Ear	Ball End & Don 2 or Spin Trough		Steel
T 43	Syria	500/580	60 × 8.4 × 2.1	None	Stag Ear (MG 11)	Ball End, Don 2	None	Steel
T 301	Albania	146/170	38 × 5.7 × 1.6	None	None	None	None	Wood
T 301	Romania	145/170	38 × 5.7 × 1.6	None	None	None	None	Wood
Adjutant & MSC 268	Spain	355/384	43.9 × 8.5 × 2.5	None	UQS-1D	TM 626 or RM 914	None	Wood
Adjutant & MSC 268	Taiwan	320/375	43.9 × 8.5 × 2.4		UQS 1D	Decca 707		Wood
Adjutant, MSC 268 & MSC 294	Turkey	320/370	43 × 8 × 2.6	None	UQS-1D	TM 1226	None	Wood
Aggressive	Belgium	720/780	52.6 × 10.7 × 4.3	None	SQQ-141T	Type 1229	None	Wood
Aggressive	Spain	720/817-853	52.6 × 10.7 × 4.3	None	SQQ-14	TM 1226 & Decca 626	None	Wood
Aggressive	Taiwan	720/780	52.6 × 10.7 × 4.3		SQQ-14	SPS-53L		Wood
Alta	Norway	n/k/375	55.2 × 13.6 × 2.5	Micos	SA 950	Racal Decca	None	FRP
Arko	Sweden	285/300	44.4 × 7.5 × 3	None	None	Skanter 009	None	Wood
Baltika Type 1380	RFAS	210/235	25.4 × 6.8 × 3.3	None	None	Spin Trough	None	Wood
Bluebird	Denmark	350/376	45 × 8.5 × 2.6	None	UQS-1D	NWS 3	None	Wood
Bluebird	Thailand	317/384	44.3 × 8.2 × 2.6	None	UQS-1D	Decca TM 707	None	Wood
Cape	Iran	200/239	33.9 × 7 × 2.4	None	None	Decca 303N	None	
Cove	Turkey	180/235	34 × 7.1 × 2.4	None	None		None	
Dokkum	Netherlands	373/453	46.6 × 8.8 × 2.3	None	None	TM 1229C	None	Wood
Frauenlob Type 394	Germany	n/k/246	38 × 8.2 × 2	None	None	Kelvin Hughes 14/9	None	
Gassten	Sweden	120/135	24 × 6.5 × 3.5	None	None	Skanter 009	None	GRP
Gilloga	Sweden	110/130	22 × 6.5 × 3.5	None	None	Skanter 009	None	Wood
Ham	Yugoslavia	120/159	32.5 × 6.5 × 1.7	None	None	Decca 45	None	Wood
Hameln	Germany	590/635	54.4 × 9.2 × 2.5	MWS-80	DSQS-11M	WM20/2, SPS-64	DR 2000	Steel
Hisingen	Sweden	130/150	24 × 6.5 × 3.5	None	None	Skanter 009	None	Wood
Kiiski	Finland	18/20	15.2 × 4.1 × 1.2	None	None	Type 1229	None	GRP
Kondor II Type 89	Indonesia	n/k/310	56.7 × 7.5 × 2.4	None	AQS 17	TSR 333	None	
Kondor II Type 89.2	Latvia	n/k/310	56.7 × 7.5 × 2.4	None	None	I-band	None	
Kondor II	Uruguay	n/k/310	56.7 × 7.5 × 2.4	None	None	TSR 333 or Raytheon 1900	None	

Class	Navy	Displacement[1]	Dimensions[2]	Command	Sonars	Radar	EW	Hull
Krogulec Type 206F	Poland	n/k/503	58.2 × 7.7 × 2.1	None	MG 11 & SHL 200	TRN 823	None	
Kuha	Finland	n/k/90	26.6 × 6.9 × 2	None	None	Type 1229	None	GRP
Leniwka Type 410S	Poland	195/269	25.8 × 7.2 × 2.7	None	None	SRN 311	None	
Lienyun	Vietnam	n/k/400	40 × 8 × 3.5					
Musca	Romania	660/790	60.8 × 9.5 × 2.8		Hull	Krivach, Drum Tilt		
Natya I Type 266M	India	650/804	61 × 10.2 × 3	None	MG 69/79	Don 2, Drum Tilt	None	Aluminium/ steel
Natya I Type 266ME	Libya	650/804	61 × 10.2 × 3	None	Hull-mounted	Don 2. Drum Tilt	None	Aluminium/ steel
Natya I Type 266M	RFAS	650/804	61 × 10.2 × 3	None	MG 79/89 or Tamir	Drum Tilt & Don 2 or Low Trough	None	Aluminium/ steel
Natya II Type 266DM	RFAS	650/804	61 × 10.2 × 3		MG 79/89	Drum Tilt & Don 2 or Low Trough		Aluminium/ steel
Natya I Type 266M	Syria	650/804	61 × 10.2 × 3	None	MG 79/89	Don 2, Drum Tilt	None	Aluminium/ steel
Natya I Type 266ME	Yemen	650/804	61 × 10.2 × 3	None	MG 69/79	Don 2	None	Aluminium/ steel
Nestin (river)	Hungary	66/72.3	27 × 6.5 × 1.2	None	None	Decca Type 101	None	
Nestin (river)	Iraq	65/72	27 × 6.5 × 1.2	None	None	Decca Type 101	None	
Nestin (river)	Yugoslavia	65/72	27 × 6.3 × 1.6	None	None	TM 1226		
Notec Type 207P	Poland	208/225	38.3 × 7.2 × 1.8	None	MG 79/89	SRN 302	None	GRP
Olya Type 1259	Bulgaria	44/64	25.8 × 4.5 × 1	None	None	Pechora	None	Wood
Olya Type 1259	RFAS	44/64	25.8 × 4.5 × 1	None	None	Don 2	None	Wood
River	Bangladesh	n/k/890	47.5 × 10.5 × 2.9	None	System 880	TM 1226C	None	Steel
River	UK	n/k/770	47.5 × 10.5 × 2.9	None	System 880	TM 1226C	None	Steel
Schutze	Brazil	230/280	47.2 × 7.2 × 2.1	None	None	ZW06		Wood
Sonya Type 1265	Bulgaria	350/450	48 × 8.8 × 2	None	MG 69/79	Kivach	None	Wood & GRP
Sonya Type 1265	Cuba	350/450	48 × 8.8 × 2	None	MG 69/79	Don 2	None	Wood & GRP
Sonya Type 1265	Syria	350/450	48 × 8.8 × 2	None	MG 69/79	Don 2	None	Wood & GRP
Sonya Type 1265	Vietnam	350/400	48 × 8.8 × 2	None	MG 69/79	Nayada	None	Wood & GRP
Styrso	Sweden	n/k/175	36 × 7.9 × 2.2	Ericsson	RESON & EdgeTech	Bridgemaster		GRP sandwich
Ton	South Africa	360/440	46.6 × 8.8 × 2.5	None	Type 193[9]	Type 978 or 1006	None	Wood
Vanya Type 257D	Bulgaria	220/245	40 × 7.3 × 1.8	None	MG 69/79	Don 2	None	Wood & GRP
Vanya	Syria	220/245	40 × 7.3 × 1.8	None	MG 69/79	Don 2	None	Wood & GRP
Vegesack	Turkey	362/378	47.3 × 8.6 × 2.9	None	Simrad	Decca 707	None	
Wosao	China	320/n/k	44.8 × 6.8 × 2.3		Hull-mounted	China Type 753		Steel
Yevgenya Type 1258	Bulgaria	77/90	24.5 × 5.5 × 1.4	None	MG 7	Spin Trough	None	GRP
Yevgenya	India	77/90	24.6 × 5.5 × 1.5	None	MG 7	Spin Trough	None	GRP
Yevgenya	Syria	77/90	24.6 × 5.5 × 1.5	None	MG 7	Spin Trough	None	GRP
Yukto I	North Korea	n/k/60	24 × 4 × 1.7	None	None	Skin Head	None	Wood
Yukto II	North Korea	n/k/52				Skin Head		Wood
Yurka Type 266	Egypt	400/540	52.4 × 9.4 × 2.6	None	Stag Ear	Don 2	None	Aluminium/ steel
Yurka Type 266	RFAS	400/540	52.4 × 9.4 × 2.6	None	Stag Ear	Drum Tilt & Don 2 or Spin Trough	Watch Dog	Aluminium/ steel
Yurka Type 266	Vietnam	400/540	52.4 × 9.4 × 2.6	None	Stag Ear	Don 2 & Drum Tilt	None	Aluminium/ steel
MINEHUNTERS								
Segura	Spain	n/k/530	54 × 10.7 × 2.2	Nautis	SQQ-32	I-band		GRP
MHC (Swiftships type)	Egypt	178/203	33.8 × 8.2 × 2.3	SYQ-12 mod	TSM 2022	Sperry	None	GRP
Bay	Australia	100/178	30.9 × 9 × 2	MWS-80-5	DSQS-11M	Type 1006	None	GRP sandwich
Circe	France	460/510	50.9 × 8.9 × 3.4	None	DUBM 20B	Type 1229	None	Wood
Flyvefisken	Denmark	n/k/450	54 × 9 × 3		Ibis 43			GRP
Frankenthal	Germany	590/650	54.4 × 9.2 × 2.5	MWS-80	DSQS-11M	SPS-64	DR 2000	Steel
Gorya Type 1260	RFAS	950/1,130	66 × 11 × 3.3		Hull	Palm Frond, Nayada Bass Tilt	Cross Loop, Long Fold	Wood
Huon (Gaeta)	Australia	n/k/720	52.5 × 9.9 × 3.0	Nautis 2M	Type 2093	Type 1007	Prism	GRP
Lida Type 10750	RFAS	n/k/135	31.5 × 6.5 × 1.6		Karbarga I	Pechora		GRP/ aluminium/ steel
Oksoy	Norway	n/k/375	55.2 × 13.6 × 2.5[3]	Micos	TSM 2023N	Racal Decca	None	FRP
Osprey	USA	750/918	57.3 × 11 × 2.9	SQQ-13 & SYQ 109	SQQ-32	SPS-64(V)9		GRP
Sandown	Saudi Arabia	450/480	52.7 × 10.5 × 2.1	Nautis M	Type 2093	Type 1007	Shiploc	GRP
Sandown	UK	450/484	52.5 × 10.5 × 2.3	Nautis M	Type 2093	Type 1007		GRP
Swallow	South Korea	470/520	50 × 8.3 × 2.6	Mains 500	193M Mod 1 or 3	Raytheon; I-band		GRP
Tripartite	Belgium	562/595	51.5 × 8.9 × 2.5	Evec	DUBM 21B	Type 1229C		GRP
Tripartite	France	562/595	51.5 × 8.9 × 2.5	Evec	DUBM 21B or 21D	Type 1229C	None	GRP
Tripartite	Indonesia	502/568	51.5 × 8.9 × 2.5	Ibis V	TSM 2022	Type 1229C		GRP
Tripartite	Netherlands	562/595	51.5 × 8.9 × 2.6	Sewaco IX	DUBM 21A	Type 1229C		GRP
Tripartite	Pakistan	562/595	51.5 × 8.9 × 2.9	TSM 2061[4]	DUBM 21B or 21D[5]	Type 1229C	None	GRP
Yevgenya Type 1258	Angola	77/90	24.5 × 5.5 × 1.4	None	MG-7	Don 2	None	GRP
Yevgenya Type 1258	Cuba	77/90	24.6 × 5.5 × 1.5	None	MG-7	Don 2	None	GRP
Yevgenya Type 1258	RFAS	77/90	24.6 × 5.5 × 1.5	None	MG-7	Spin Trough or Mius	None	GRP
Yevgenya Type 1258	Vietnam	77/90	24.6 × 5.5 × 1.5	None	MG 7	Spin Trough	None	GRP
Yevgenya Type 1258	Yemen	77/90	24.6 × 5.5 × 1.5	None	MG 7	Spin Trough	None	GRP

Class	Navy	Displacement[1]	Dimensions[2]	Command	Sonars	Radar	EW	Hull
MINEHUNTERS/ SWEEPERS								
Sugashima	Japan	n/k/510	57.7 × 9.4 × 4.2	Nautis-M	Type 2093	OPS-39		Wood
MSC 322	Saudi Arabia	320/407	46.6 × 8.2 × 2.5	None	SQQ-14	SPS-55	None	Wood
MWV 50	Taiwan	n/k/500	49.7 × 8.7 × 3.1	Ibis V	TSM-2022	I-band		
Adjutant	Greece	330/402	44.2 × 8.5 × 2.4	None	SQQ-14 or UQS-1D	Decca or 3RM 20R	None	Wood
Avenger	USA	1,145/1,312	68.3 × 11.9 × 3.5	Nautis M[6]	SQQ-30[7]	SPS-55 & SPS-66(V)9		Wood & GRP
Bang Rachan	Thailand	390/444	48 × 9.3 × 2.7	MWS-80R	DSQS-11H	STN 8600 ARPA		Composite
Gaeta	Italy	592/697	52.45 × 9.9 × 2.9	Mactis	SQQ-14(IT)	SSN-728V(3)	SPN-714V(2)	GRP
Hatsushima	Japan	440/510	55 × 9.4 × 2.4		ZQS 2B	OPS-9		Wood
Hunt	UK	615/750	60 × 10 × 3.4	CAAIS DBA 4	Type 193M Mod 1	Type 1006 or 1007	Matilda UAR 1	GRP
					Type 2059 & Mil Cross		Mentor A (in some)	
Kingston	Canada	713/962	55.3 × 11.3 × 3.05		Towed side-scan	Type 6000		Steel
Landsort	Singapore	270/360	47.5 × 9.6 × 2.3	Ibis V	TSM 2022	Terma	None	GRP
Landsort	Sweden	270/360	47.5 × 9.6 × 2.3	9MJ400 + Mains	TSM-2022	Terma	Matilda	GRP
Lat Ya (Gaeta)	Thailand	n/k/650	52.45 × 9.9 × 2.9	MWS 80-6	DSQS-11M	n/k	n/k	GRP
Lerici	Italy	470/572	50 × 9.9 × 2.6	Mactis	SQQ-14(IT)	SSN-714V(2)	SPN-728V(3)	GRP
Lerici	Malaysia	470/572	51 × 9.9 × 2.8	Ibis V	TSM 2022	Decca Type 1226	None	GRP
Lerici	Nigeria	470/540	51 × 9.9 × 2.8	Ibis V	TSM 2022	Decca Type 1226	None	GRP
Lindau Type 331	Germany	388/463	47.1 × 8.3 × 3	None	DSQS-11 or 193M	14/9 or TRS N	None	Wood
Lindau Type 351	Germany	390/465	47.1 × 8.3 × 2.8	None	DSQS-11	TRS N	ESM	Wood
Notec II Type 207M	Poland	208/225	38.3 × 7.2 × 1.6	None	SHL 100/200	SRN 401XTA or RN 231	None	GRP
River	South Africa	n/k/380	48 × 8.5 × 2.5	None	Simrad, Klein	Racal Decca		Wood
Sonya Type 1265/1265M	RFAS	350/450	48 × 8.8 × 2	None	MG 69/79	Don 2 or Krivach or Nayada	None	Wood & GRP
Ton	Argentina	360/440	46.6 × 8.8 × 2.5		Type 193[8]	Decca 45		Composite
Uwajima	Japan	490/586	58 × 9.4 × 2.9		ZQS 3	OPS-39		Wood
Vanya Type 257D/DM/DT	RFAS	200/245	40 × 7.3 × 1.8	None	MG 69/79	Don 2 or Don Kay	None	Wood
Vukov Klanac (Ton)	Yugoslavia	365/424	46.4 × 8.6 × 2.5	None	TSM 2022	DRBN 30		Wood
Yaeyama	Japan	1,000/1,275	67 × 11.8 × 3.1		SQQ-32	OPS-39		Wood
DRONES								
MSD	Australia		7.3 × 2.8 × 0.6	None	None	None	None	GRP
MSF	Denmark	n/k/125	26.5 × n/k × n/k	None	TSM 2054	n/k	None	GRP
SAM	Sweden	n/k/20	18 × 6.1 × 1.6	None	None	None	None	GRP
SAM II	Sweden	n/k/56	22.5 × 11.7	None				
SAV	Denmark	32/38	18.2 × 4.75 × 1.2	None	TSM 2054	Furuno	None	GRP
Futi Type 312	China	47/n/k	20.9 × 3.9 × 2.1	None	None	None	None	
Futi Type 312	Pakistan	47/n/k	20.9 × 3.9 × 2.1	None	None	None	None	
Ilyusha Type 1253	RFAS	n/k/85	26.4 × 5.9 × 1.4	None	None	Spin Trough		
Tanya Type 1300	RFAS	n/k/73	26.5 × 4 × 1.5	None	None	Spin Trough		
Troika	Germany	99	26.9 × 4.6 × 1.4	None	None	None	None	Wood

[1] Standard/Full Load tonnes.
[2] Length (overall) × beam × draught in metres.
[3] 0.9 m on cushion.
[4] In last pair

[5] TSM 2054 in last pair.
[6] In last two ships and to be retrofitted to rest of class.
[7] SQQ-32 in MCM 10 onwards and being retrofitted to rest of class.

[8] In minehunters.
[9] Only in Kimberley.
n/k not known

UPDATED

TABLE XVII

Mine countermeasure vessel designs – mine warfare systems

Class	Navy	ROVs	Sweep gear	Guns[1]	Missiles	Mines[2]
MINESWEEPERS						
K8 (M/S boat)	Vietnam					
MSB (M/S boat)	Thailand					
MSB 07 (M/S boat)	Japan					
KMV (new minesweepers)	Belgium	2 ROVs	Sterne & Mk 9 wire	1-25		
MSC (river minesweepers)	Romania					6
MSC 268	South Korea			1 × 2-20 or 2 × 1-20		
MSC 268	Pakistan			1 × 4-23 or 1-20		
MSC 268 & 292	Iran		Wire + M, A	1 × 2-20		
MSC 289	South Korea			1 × 2-20 or 2 × 1-20		
MSC 294	Greece			1 × 2-20		
M 15	Sweden					
M 301	Yugoslavia			2-20		
PO 2 Type 501	Bulgaria					
T 43	Albania			2 × 2-37		16
T 43 (China built)	Bangladesh		MPT-3 wire M, A	2 × 2-37 & 2 × 2-25		12-16
T 43 Type 010	China		MPT-3 wire M, A	1 or 2 × 2-37 1-85 in some		12-16
T 43	Egypt			2 × 2-37		20
T 43	Indonesia			2 × 2-37		

Class	Navy	ROVs	Sweep gear	Guns[1]	Missiles	Mines[2]
T 43 Type 254	RFAS			2 × 2-37		16
T 43	Syria			1 × 2-37		16
T 301	Albania			2-37		18
T 301	Romania			2 × 1-37		18
Adjutant & MSC 268	Spain			1 × 2-20		
Adjutant & MSC 268	Taiwan			1 × 2-20		
Adjutant, MSC 268 & MSC 294	Turkey			1 × 2-20		
Aggressive	Belgium					
Aggressive	Spain	Pluto[3]		1 × 2-20		
Aggressive	Taiwan		SLQ-37 wire, M & A			
Alta	Norway	None	Wire + M, A	1 or 2-20	1 × 2 Sadral	
Arko	Sweden			1-40		
Baltika Type 1380	RFAS					
Bluebird	Denmark			1 Bofors 40		
Bluebird	Thailand		Mk 4 & Mk 6 Type Q2 M	1 × 2-20		
Cape	Iran		Wire, M, A			
Cove	Turkey					
Dokkum	Netherlands			1 or 2-20		
Frauenlob Type 394	Germany			1-40		n/k
Gassten	Sweden			1-20		
Gilloga	Sweden					
Ham	Yugoslavia			1 × 2-20		
Hameln	Germany		SDG-31 wire	2 × 1-40	2 × 4 Stinger	60
Hisingen	Sweden			1-20		
Kiiski	Finland					
Kondor II Type 89	Indonesia		Dyad influence	3 × 2-25		2 rails
Kondor II Type 89.2	Latvia	None	Temp. removed	3 × 2-23		
Kondor II	Uruguay		MSG-3 variable depth	1-40		2 rails
Krogulec Type 206F	Poland			3 × 2-25	2 × 4 SA-N-5	16
Kuha	Finland	1 × Pluto	M, A, P	1 × 2-23		
Leniwka Type 410S	Poland		Wire + M, A			
Lienyun	Vietnam					
Musca	Romania			2 × 2-30		
Natya I Type 266M	India		GKT-2 wire AT-2 A TEM-3 M	2 × 2-30 & 2 × 2-25	2 × 4 SA-N-5[3]	10
Natya I Type 266ME	Libya		GKT-2 wire AT-2 A TEM-3 M	2 × 2-30 & 2 × 2-25		10
Natya I Type 266M	RFAS		1 or 2 GKT-2 wire 1 AT-2 A 1 TEM-3 M	2 × 2-30 or 2 × 6-30	2 × 4 SA-N-5/8 & 2 × 2-25	10
Natya II Type 266DM	RFAS			2 × 2-30 or 2 × 6-30	2 × 4 SA-N-5/8	10
Natya I Type 266M	Syria			2 × 2-30 & 2 × 2-25	2 × 4 SA-N-5	
Natya I Type 266ME	Yemen		Wire, M, A	2 × 2-30 & 2 × 2-25		10
Nestin (river)	Hungary	None	M, A Kram wire	1 × 4-20, 2 × 1-20	24	
Nestin (river)	Iraq		Wire + M, A	1 × 3-20 & 2 × 1-20		24
Nestin (river)	Yugoslavia		Wire + M, A	1 × 3-20 & 2 × 1-20		24
Notec Type 207P	Poland		Wire + M, A	1 × 2-23		24
Olya Type 1259	Bulgaria		AT-6, SZMT-1 3 PKT-2			
Olya Type 1259	RFAS			1 × 2-25		
River	Bangladesh		Wire	1-40		
River	UK		Wire	1-40		
Schutze	Brazil		Wire + M, A	1-40		
Sonya Type 1265	Bulgaria			1 × 2-30, 1 × 2-25		5
Sonya Type 1265	Cuba			1 × 2-30 1 × 2-25		8
Sonya Type 1265	Syria			1 × 2-30 or 2 × 1-30 & 1 × 2-25		8
Sonya Type 1265	Vietnam			2-30 & 1 × 2-25		8
Styrso	Sweden	2 × DE	AK-90 A EL-90 M & wire			
Ton	South Africa			1-40		
Vanya Type 257D	Bulgaria			1 × 2-30		8
Vanya	Syria			1 × 2-30		8
Vegesack	Turkey			1 × 2-20		
Wosao	China		Wire, M, A	2 × 2-25		6
Yevgenya Type 1258	Bulgaria			1 × 2-25		
Yevgenya	India			1 × 2-25		
Yevgenya	Syria			1 × 2-25		
Yukto I & II	North Korea			1-37 or 1 × 2-25		8
Yurka Type 266	RFAS		Wire + M, A	2 × 2-30	2 × 4 SA-N-5/8	10
Yurka Type 266	Egypt	? ROV		2 × 2-30		10
Yurka Type 266	Vietnam			2 × 2-30		10
MINEHUNTERS						
Segura	Spain	2 Pluto Plus		1-20		
MHC (Swiftships type)	Egypt	Pluto				
Bay	Australia	2 PAP 104				
Circe	France	PAP	None	1-20		
Flyvefisken	Denmark	DE	None	1 × 76		n/k
Frankenthal	Germany	2 × Pinguin		1-40	2 × 4 Stinger	
Gorya Type 1260	RFAS	1 ROV	Wire + M, A	1-76 1 × 6-30	2 × 4 SA-N-5	
Huon (Gaeta)	Australia	2 × DE	Oropesa wire Mini-Dyad influence	1-30 mm		
Lida Type 10750E	RFAS		1 wire + M, A	1 × 6-30		
Oksoy	Norway	2 Pluto	None	1 or 2-20	1 × 2 Sadral	

Class	Navy	ROVs	Sweep gear	Guns[1]	Missiles	Mines[2]
Osprey	USA	SLQ-48	SLQ-53			
Sandown	UK	2 PAP 104		1-30		
Sandown	Saudi Arabia	2 PAP 104		1 × 2-30		
Swallow	South Korea	2 Pluto	Wire	1-20		
Tripartite	Belgium	2 PAP 104	Mechanical	1-20		
Tripartite	France	2 PAP 104	AP-4 + wire	1-20		
Tripartite	Indonesia	2 PAP 104	OD3 oropesa wire F-82 M & AS 203 A	2 × 1-20		
Tripartite	Netherlands	2 PAP 104	OD 3 wire	1-20		
Tripartite	Pakistan	2 PAP 104	Wire + MKR 400 A & MRK 960 M	1-20		
Yevgenya Type 1258	Angola		M	1 × 2-25		
Yevgenya Type 1258	Cuba					
Yevgenya Type 1258	RFAS			1 × 2-25		8 racks
Yevgenya Type 1258	Vietnam			1 × 2-25		
Yevgenya Type 1258	Yemen			1 × 2-25		
MINEHUNTERS/SWEEPERS						
Sugashima	Japan	PAP 104	Dyad influence	1 × 3-20		
MSC 322	Saudi Arabia		Wire + M	1-20		
MWV 50	Taiwan	2 Pinguin		1-40		
Adjutant	Greece			1-20		
Avenger	USA	2 SLQ-48	SLQ-37M/A Oropesa SLQ-38 wire ALQ 166 M, M/S vehicle			
Bang Rachan	Thailand	2 Pluto	Wire + M, A	3-20		
Gaeta	Italy	1 MIN Mk 2 1 Pluto	Oropesa wire	1 × 2-20		
Hatsushima	Japan		S4	1 × 3-20		
Hunt	UK	2 PAP 104	MS 14 M MSSA Mk 1 A Mk 8 oropesa wire	1-30 2-20		
Kingston	Canada	ROV module	M/S module	1-40		
Landsort	Singapore	2 PAP 104	SAM wire	1-40		2 rails
Landsort	Sweden	2 × DE	Wire + M, A	1-40		
Lat Ya (Gaeta)	Thailand	2 Pluto Plus		1 × 2 – 20		
Lerici	Italy	1 MIN 77 1 Pluto	Oropesa wire	1-20		
Lerici	Malaysia	2 PAP 104	'O' Mis-4 wire	1-40		
Lerici	Nigeria	2 Pluto	'O' Mis-4 wire	1 × 2-30 & 2 × 1-20		
Lindau Type 331	Germany	2 × PAP 104		1-40		
Lindau Type 331	Germany	None	Wire	1-40		
Notec II Type 207M	Poland		Wire, + M, A	1 × 2-23	2 SA-N-5	6-24
River	South Africa	2 PAP 104		1-20		
Sonya Type 1265/1265M	RFAS			2 × 6-30 or 1 × 2-30 & 1 × 2-25	2 × 4 SA-N-5	8
Ton	Argentina		Mk 3 wire Mk 2 M, AD Mk 3 A	1 or 2-40		
Uwajima	Japan		S7	1 × 3-20		
Vanya Type 257D/DM/DT	RFAS			1 × 2-30 & 1 × 2-25		8-12
Vukov Klanac (Ton)	Yugoslavia	PAP 104[4]		2-20		
Yaeyama	Japan	S7	S8 (SLQ-48)	1 × 3-20		
DRONES						
MSD	Australia					
MSF	Denmark					
SAM	Sweden		M, A			
SAM II	Sweden		A, M +?E			
SAV	Denmark					
Futi Type 312	China		M, A			
Futi Type 312	Pakistan		M, A			
Ilyusha Type 1253	RFAS					
Tanya Type 1300	RFAS					
Troika	Germany		M, A			

[1] Machine guns not listed
[2] Depth charges are not listed
[3] In some
[4] In minehunters

DE Double Eagle
M Magnetic sweep
A Acoustic sweep
P Pressure sweep

E Electric influence
n/k not known

UPDATED

TABLE XVIII

Mine warfare vessel designs – machinery

Class	Navy	Engines[A]	Shafts	Thrusters	Auxiliary	Propellers	Speed	Range	Crew
MINESWEEPERS									
K 8 (M/S boat)	Vietnam	2/300	2				18		6
MSB (M/S boat)	Thailand	1/165	1				8		10
MSB 07 (M/S boat)	Japan	2/480	2				11		10
KMV (new minesweepers)	Belgium	2/2,176	2	n/k	None	cp	16	3,000/12	26
MSC (river minesweepers)	Romania	2/870	2				13		
MSC 268	South Korea	2/880	2				14	2,500/14	40
MSC 268	Pakistan	2/880	2				13.5	3,000/10.5	39
MSC 268 & 292	Iran	4/696³	2				13	2,400/10	40
MSC 289	South Korea	4/696	2				14	2,500/14	40
MSC 294	Greece	2/1,760	2				13	2,500/10	39
M 15	Sweden	2/320	2				12		10
M 301	Yugoslavia	2/n/k	2				12		
PO 2 Type 501	Bulgaria	1/300	2				12		8
T 43	Albania	2/2,000	2	None			15	3,000/10	65
T 43 (China built)	Bangladesh	2/2,000	2				14	3,000/10	70
T 43 Type 010	China	2/2,000	2				14	3,000/10	70
T 43	Egypt	2/2,000	2				15	3,000/10	65
T 43	Indonesia	2/2,000	2				15	3,000/10	77
T 43 Type 254	RFAS	2/2,000	2				15	3,000/10	65
T 43	Syria	2/2,000	2				15	3,000/10	65
T 301	Albania	3/900	3	None			14	2,200/9	25
T 301	Romania	3/900	3	None			14.5	2,200/9	25
Adjutant & MSC 268	Spain	2/880	2				14	2,700/10	39
Adjutant & MSC 268	Taiwan	2/880	2				13	2,500/12	35
Adjutant & MSC 268 & MSC 294	Turkey	4/696	2				14	2,500/10	35
Aggressive	Belgium	4/1,760	2			cp	14	2,400/12	40
Aggressive	Spain	4/2,280	2			cp	14	3,000/10	74
Aggressive	Taiwan	4/2,280	2			cp	14	3,000/10	86
Alta	Norway	2/3,700		2²	2/1,740		20.5	1,500/20	32
Arko	Sweden	2/1,360	2				14		25
Baltika Type 1380	RFAS	1/300	1			cp	9	1,400/9	10
Bluebird	Denmark	2/880	2				13	3,000/10	31
Bluebird	Thailand	2/880	2				13	2,750/12	43
Cape	Iran	4/1,300	2				13	1,200/12	21
Cove	Turkey	4/696	2				13	900/11	25
Dokkum	Netherlands	2/2,500	2				16	2,500/10	27-36
Frauenlob Type 394	Germany	2/2,200	2				12+	700/14	25
Gassten	Sweden	1/460	1				11		
Gilloga	Sweden	1/380	1				10		
Ham	Yugoslavia	2/1,100	2				14	2,000/9	22
Hameln	Germany	2/6,140	2		Electric	cp	18		37
Hisingen	Sweden	1/380	1				10		
Kiiski	Finland	2/340				2²	11	260/11	4
Kondor II Type 89	Indonesia	2/4,408	2			cp	17	2,000/14	31
Kondor II Type 89.2	Latvia	2/4,408	2			cp	17		31
Kondor II	Uruguay	2/4,408	2			cp	17	2,000/15	31
Krogulec Type 206F	Poland	2/3,750	2				18	2,000/17	48
Kuha	Finland	2/600	1			cp	12		15
Leniwka Type 410S	Poland	1/570	1				11	3,100/8	16
Lienyun	Vietnam	1/400	1				8		
Musca	Romania	2/4,800	2				17		
Natya I Type 266M	India	2/5,000	2			cp	16	3,000/12	82
Natya I Type 266ME	Libya	2/5,000	2			cp	16	3,000/12	67
Natya I/II Type 266M/DM	RFAS	2/5,000	2			cp	16	3,000/12	67
Natya I Type 266M	Syria	2/5,000	2			cp	16	3,000/12	65
Natya I Type 266ME	Yemen	2/5,000	2			cp	16	3,000/12	67
Nestin (river)	Hungary	2/520	2				15	860/11	17
Nestin (river)	Iraq	2/520	2				12	860/11	17
Nestin (river)	Yugoslavia	2/520	2				15	860/11	17
Notec Type 207P	Poland	2/1,874	2				14	1,100/9	24
Olya Type 1259	RFAS	2/471	2				12	500/10	15
Olya Type 1259	Bulgaria	2/471	2				12	300/10	15
River	Bangladesh	2/3,100	2	None			14	4,500/10	30
River	UK	2/3,100	2	None			14	4,500/10	30
Schutze	Brazil	4/4,500	2			2 × cp	24	710/20	39
Sonya Type 1265	Bulgaria	2/2,000	2				15	1,500/14	43
Sonya Type 1265	Cuba	2/2,000	2				15	3,000/10	43
Sonya Type 1265	Syria	2/2,000	2				15	3,000/10	43
Sonya Type 1265	Vietnam	2/2,000	2				15	3,000/10	43
Styrso	Sweden	2/1,104	2	1 bow			13		17
Ton	South Africa	2/3,000	2				15	2,300/13	27
Vanya Type 257D	Bulgaria	1/2,502	1				16	2,400/10	36
Vanya	Syria	1/2,500	1				16	1,400/14	36
Vegesack	Turkey	2/1,500	2			cp	15		33
Wosao	China	4/4,400	4				25	500/15	40
Yevgenya Type 1258	Bulgaria	2/600	2				11	300/10	10
Yevgenya	India	2/600	2				11	300/10	10
Yevgenya	Syria	2/600	2				11	300/10	10

Class	Navy	Engines[A]	Shafts	Thrusters	Auxiliary	Propellers	Speed	Range	Crew
Yukto I	North Korea	2/	2				18		22
Yukto II	North Korea	2/	2				18		22
Yurka Type 266	RFAS	2/5,350	2				17	1,500/12	45
Yurka Type 266	Egypt	2/5,350	2				17	1,500/12	45
Yurka Type 266	Vietnam	2/5,350	2				17	1,500/12	45
MINEHUNTERS									
Segura	Spain	2/1,523		2	2/150	2 × VS	14	2,000/12	40
MHC (Swiftships type)	Egypt	2/1,068		1/300		2 × Schottel	12.4	2,000/10	25
Bay	Australia	2/650	2			2 × Schottel	10	1,500/10	14
Circe	France	1/1,800	1				15	3,000/12	48
Flyvefisken	Denmark	1/5450 GT 2/5800	3	1 bow	1/500	cp (outer)	30	2,400/18	19-29
Frankenthal	Germany	2/5,550	2		1	cp	18		37
Gorya Type 1260	RFAS	2/5,000	2				17		70
Huon (Gaeta)	Australia	1/1,986	1	3	3/506	cp	14	1,600/12	36
Lida Type 10750E	RFAS	3/900	3				12	650/10	14
Oksoy	Norway	2/3,700		2[2]	2/1,740		20.5	1,500/20	38
Osprey	USA	2/1,600		1 bow/180	2/360	2 × VS	12	1,500/10	51
Sandown	Saudi Arabia	2/1,500	2	2 bow		2 × VS	13	3,000/12	34
Sandown	UK	2/1,500	2	2 bow		2 × VS	13	3,000/12	34
Swallow	South Korea	2/2,040		1 bow/102		2 × VS	15	2,000/10	44
Tripartite	Belgium	1/1,860	1	2 bow	2/240	2 × cp	15	3,000/12	46
Tripartite	France	1/1,860	1	2 bow	2/240	2 × cp	15	3,000/12	46
Tripartite	Indonesia	2/2,610	2	2 bow/150	2/240	2 × Schottel	15	3,000/12	46
Tripartite	Netherlands	1/1,860	1	2 bow	2/240	2 × cp	15	3,000/12	29-42
Tripartite	Pakistan	1/1,860	1	2 bow	2/240	2 × cp	15	3,000/12	46
Yevgenya Type 1258	Angola	2/600	2				11	300/10	10
Yevgenya Type 1258	Cuba	2/600	2				11	300/10	10
Yevgenya Type 1258	RFAS	2/600	2				11	300/10	10
Yevgenya Type 1258	Vietnam	2/600	2				11	300/10	10
Yevgenya Type 1258	Yemen	2/600	2				11	300/10	10
MINEHUNTERS/SWEEPERS									
Sugashima	Japan	2/1,800	2				14		40
MSC 322	Saudi Arabia	2/1,200	2				13		39
MWV 50	Taiwan	2/2,180	2				14	3,500/14	45
Adjutant	Greece	2/880	2	.			14	2,500/10	38
Avenger	USA	4/2,400	2	1/350	2/400	cp	13.5		81
Bang Rachan	Thailand	2/3,220	2		1 electric motor	cp	17	3,100/12	33
Gaeta	Italy	1/1,985	1	3 × 506	3/1,481	cp	14	2,000/12	51
Hatsushima	Japan	2/1,440	2				14		45
Hunt	UK	2/1,900	2	1 bow	1/780		15	1,500/12	45
Kingston	Canada	4-2/3,000[4]		2 Z drive		2 fp	15	5,000/8	37
Landsort	Singapore	4/1,592				2 × VS	15	2,000/12	38
Landsort	Sweden	4/1,592				2 × VS	15	2,000/12	29
Lat Ya (Gaeta)	Thailand	1/1,985	1	3 × 506	3/1,481	cp	14	2,000/12	n/k
Lerici	Italy	1/1,985	1	3 × 506	3/1,481	cp	14	1,500/14	47
Lerici	Malaysia	2/2,605	2		3/1,481	cp	16	2,000/12	42
Lerici	Nigeria	2/3,120				2[2]	15.5	2,500/12	50
Lindau Type 33I	Germany	2/4,000	2			2	16.5	850/16.5	43
Lindau Type 35I	Germany	2/5,000	2			2	16.5	850/16.5	44
Notec II Type 207M	Poland	2/1,874	2		2/816		13	790/14	24
River	South Africa	2/4,515				2 × VS	16	2,000/13	40
Sonya Type 1265/1265M	RFAS	2/2,000	2				15	3,000/12	43
Ton	Argentina	2/3,000	2				15	2,500/12	27-36
Uwajima	Japan	2/1,400	2				14		40
Vanya Type 257D/DM/DT	RFAS	1/2,502	1				16	1,400/14	36
Vukov Klanac (Ton)	Yugoslavia	2/1,620	2				15	3,000/10	40
Yaeyama	Japan	2/2,400	2	1 bow/350			14		60
DRONES									
MSD	Australia	2/300					45		
MSF	Denmark	2/500	2	n/k	n/k	pump jet	12	n/k	
SAM	Sweden	1/210				1 × Schottel	8	330/8	
SAV	Denmark	1/350				1[1]	12		4
Futi Type 312	China	1/300				cp	12	144/12	3
Futi Type 312	Pakistan	1/300				cp	12	144/12	3
Ilyusha Type 1253	RFAS	2/500	2				12	300/10	10
Tanya Type 1300	RFAS	1/270	1				10		
Troika	Germany	1/446	1				10	520/9	3

A number/horsepower
[1] Pump jet propulsion
[2] Water-jets
[3] Except MSC 268-2/880

[4] Diesel electric drive, 4 diesels, 2 motors
[5] VS Voith Schneider propellers
n/k not known

UPDATED

MINE WARFARE

Command and control and weapon control systems
Combat information systems
Positioning and tracking systems

Mine warfare sonar systems
Hull-mounted minehunting sonars
Variable depth minehunting sonars
Side scan minehunting route survey sonars
Self-propelled sonar systems
ROV sonars

Mine warfare laser systems

Mine disposal vehicles
Remotely operated vehicles
Remotely operated vehicles' ancillary equipment

Minesweeping systems

Divers' systems
Diving equipment
Diver vehicles

MINE WARFARE

There is no doubt that the mine threat is real. One has only to consider the evidence from recent history to see that this is so. Three US ships in the early 1980s were severely damaged by mines in the Gulf. The USS *Roberts* had her keel broken and engines thrown off their mounts; the USS *Princeton* suffered damage and the USS *Tripoli* had a large hole blown in her hull which resulted in a large proportion of her fuel leaking out into the sea.

These incidents highlight the difficulty facing naval forces today as to where the threat is coming from and what the strategic and tactical requirements are for mine countermeasures to handle that threat.

The key to mine warfare, therefore, is the ability to define the threat. The more that is known about the threat, the easier it is to counter. There is little that is not already known about the present day threat covering the moored mine, various types of ground influence mine and the volume threat posed by devices such as rising mines and so on. What is not so clear, however, is where and how such weapons will be used, and in what circumstances.

During the Cold War emphasis on MCM was placed on the need to ensure a safe passage for SSBNs exiting and returning to their base. While this will remain an important strategic requirement in the future for nations operating SSBNs, other equally important roles for MCM have begun to emerge since the ending of the Cold War.

Primary among these has been the realisation by some that expeditionary forces operating in support of an international coalition are extremely vulnerable to the mine threat, especially in the littoral region. Amphibious forces in particular now demand protection and immunity from the mine threat not only during their transit of the littoral but also during the actual assault through the surf zone and on to the beach itself. While previously ensuring that the beach landing area is free of mines has been very much the responsibility of the army/marines, it is now becoming much more the responsibility of naval forces. The US is devoting much effort to developing ways and means of ensuring that the surf zone and immediate beach landing area is free of all types of mines before the infantry go ashore. The experience and lessons of Normandy June 1944 are being relearned.

Within this type of scenario it has been further realised, following the 1990-91 Gulf War that MCM operations must form part of the overall tactical plan and that any MCM units must be fully integrated within the Task Force (TF). This demands that adequate protection be given to the MCM units which will, in all probability, be operating in advance of the main TF. Hence, they will, in addition to other forms of protection, require air cover.

Such detailed integration of small and large units within a TF requires adequate and secure communications at all levels and between all units.

Mine countermeasures techniques fall into two main areas: active and passive.

Active MCM

Active countermeasures can also be divided into two main spheres of action: minehunting and minesweeping.

Minehunting has proved to be the only relatively safe and effective method of dealing with modern sophisticated influence mines. If the minehunter is to be effective in its task then it must be equipped with highly sophisticated equipment. In order to be able to pinpoint the exact location of a mine and to record its position precisely requires accurate navigation systems of the highest order. Next, in order to detect the mine and carry out accurate classification, the vessel must be equipped with an efficient sonar system able to detect the smallest targets under the most adverse conditions.

Having detected and classified contacts, the next task is to neutralise those classified as mines. The main task of mine destruction is now carried out by small

submersibles called Remotely Operated Vehicles (ROVs) deployed from the minehunter. These either position a remotely detonated countermining charge next to the mine, or they are equipped with powerful cutters which sever the tethering wire of moored mines so that they float to the surface where they can be destroyed.

With the advent of the Continental Shelf Mine, which is extremely difficult to deal with and laid in much deeper waters, minesweeping has taken on a new lease of life. The new role of the minesweeper is to deploy the deep-armed sweep designed to operate at much greater depths than the normal oropesa sweep. To achieve this the minesweepers have to operate at least in pairs. This operation also requires extremely accurate navigation and station-keeping, and such vessels must be equipped with the most up-to-date navigational aids available if they are to perform their task effectively.

Precise positioning

MCM requires accurate navigation and precise positioning. It is often necessary for vessels to subsequently return to the position of previously defined contacts to re-examine them or, in the case of confirmed mines, to carry out countermining. Precise positioning and plotting are absolutely essential elements if an MCMV is to carry out this task effectively. With such a system a vessel can return to the precise position where a contact was previously found with considerable saving in time and with the sure knowledge that the original contact will be found again.

Route surveying

The most effective form of MCM is that of route surveying. In peacetime this is a primary task not only of the MCM force but also of the hydrographic service. To be effective, route surveying requires that the whole of a proposed wartime shipping route be surveyed extremely carefully to produce bottom contour charts which show in the minutest detail the composition of the seabed and precise data on all objects on it. By regularly surveying and updating the charts, accurate pictures of proposed routes can be maintained which will enable safe ones to be selected in wartime. Furthermore, these routes can be checked by the minehunters much more rapidly and accurately as the vessel then only has to check objects not already marked on the chart.

Command and control

With such volumes of data now being made available to the command, a co-ordinated, effective means of correlating and presenting this wealth of information is essential if the MCMV is to carry out its task effectively and within a reasonable time span. Modern tactical plan displays frequently using colour graphics indicate to the command the route to be swept or surveyed, the ship's position, all objects on the seabed, the format of the seabed with integrated contour lines, the direction of the sonar beam, danger and specified circles around objects, map overlay and alphanumeric tote display. As they are of modular construction and use distributed processing and preprocessing techniques, these systems can integrate with a wide variety of minehunting sonars now available and allow the command to accept only valid data required for the task in hand.

Among features now being incorporated into the latest command systems are window techniques which allow the display of information such as raw sonar data, contact data and track data and the possibility of correlating and extracting radar positional data, ESM data and so on. Other features under study include the capability to automatically control the ship in various attitudes such as hovering and heading using autopilot and propulsion control. The problem with propulsion control is integrating the thrust developed by the various manoeuvring control systems. Development of such an automatic control system will greatly ease the strain on the crew during minehunting operations.

Finally, the Man/Machine Interface (MMI) has to be of the highest order. The presentation of the vast amounts of data required in minehunting demand displays of high resolution and clear definition, the careful use of colour techniques and the possibility of presenting more specific data from various sensors using window techniques.

Passive MCM

Passive MCM involves such techniques as noise reduction (reducing cavitation to a minimum, reducing machinery noise and so on), degaussing (to reduce a ship's magnetic signature to a minimum) and optimising the hydrodynamic shape of the hull with a minimum displacement (to reduce pressure signature to a minimum). Alternatively, it may be possible to modify a ship's signature so that it no longer appears to be what it actually is. Such techniques, however, require that one can precisely define the threat. Mine avoidance is another passive technique which can be employed, but again requires prior knowledge of the precise nature of the threat.

The future

MCM will continue to lag behind the development of the mine. The gap is narrowing, however, and there are limits to the extent to which mines can be developed. The two main developments in the future will be the self-propelled mine and mines which can be laid in much deeper waters; both of these will pose enormous problems for MCM forces. Other likely developments include increasing the mine's effectiveness against specific targets, in particular minehunters and submarines, and seeking to capitalise on influences so far not developed.

For example, it may be that some mines will be influenced to react to a minehunter's sonar, or for tethering cables of moored mines to react to the operations of an ROV and then to set off a chain reaction of other mines in an attempt to destroy the minehunter.

On the MCM side, more efficient mine detecting sonars, command and control systems will be developed to deal with the mass of data accumulated. Efforts will be made to reduce reliance on human operators, for minehunting can be an extremely tedious task and, when operators become bored or tired, errors can be made that prove fatal. Greater efforts will be made to make MCMVs more immune to mines, and new generations of submersibles will be developed with much greater capacities for dealing with more sophisticated mines, longer mission times, increased speed and range and deeper diving capabilities and above all much greater autonomy. Also, smaller ROVs will become attractive as complementary systems to the larger vehicles.

The use of datalinks to relay information between vessels on task and the shore-based MCM headquarters is now becoming essential to the speeding up of MCM operations and ensuring that shipping is allowed to use cleared channels at the earliest opportunity.

Sonar prediction techniques, already widely used in ASW operations, will also become more important as a means of improving effectiveness and safety with the ability to define safety circles more carefully.

Finally, the requirement to keep the MCM platform as far removed as possible from danger zones, particularly in areas which have not previously been route surveyed, will lead to the further development of remote-controlled, autonomous surface and subsurface platforms for minehunting and disposal. Already major developments are underway to develop autonomous small platforms which can be deployed not only from MCMVs but also from larger surface platforms to provide advance warning of a mine threat to an expeditionary force.

UPDATED

COMMAND AND CONTROL AND WEAPON CONTROL SYSTEMS

COMBAT INFORMATION SYSTEMS

DENMARK

MCM system

Type
Mine countermeasures control system.

Description
Erisoft in Sweden has subcontracted Eiva of Denmark responsibility for the development and supply of the mine countermeasures system for the new 'Styrso' class mine countermeasures vessels of the Swedish Navy.

The system comprises integrated functions for mine detection, minesweeping as well as navigation and positioning. The system is based on a common network and UNIX multifunction workstations with Windows-based MMI.

Eiva supplied all the sensors required to meet the navigational and operational requirements, developed and supplied software for navigation data handling and positioning, and the design and supply of integrated ROV and MCM operators consoles, as well as installation of the complete MCM system.

Operational status
Four systems and a training system have been supplied to the Swedish Navy. The system is installed on the 'Styrso' class.

Contractor
Eiva A/S, Hasselager.

UPDATED

FRANCE

EVEC 20

Type
Data handling system.

Description
The data handling system developed for the 'Tripartite' minehunter is the Thomson-Marconi EVEC 20. The EVEC 20 samples and processes data from the sonar, radio navigation systems, navigation radar, gyrocompass and Doppler log to present the operator with a continuously updated display of the minehunting area showing position of own ship, located targets, radar contacts, tracks and lines representing specified zones. All or part of the displayed images can be recorded on cassette for subsequent recall, updating

and maintaining an up-to-date permanent record of operations relating to a specified zone. A repeater on the bridge displays point co-ordinates from the minehunting sonar derived by the EVEC 20 and transmitted to the repeater.

The computer in the EVEC 20 also assists the automatic pilot, providing command inputs and error generation, and helps to control the automatic radar navigation systems providing location computations, error corrections, location of tracking windows and so on.

The automatic pilot is used to keep the ship on a predetermined track during minehunting operations, receiving necessary data from the gyrocompass and Doppler log.

The navigation radar and automatic tracking system

integrate with the EVEC 20 to provide a buoy location system.

The horizontal plotting table uses an 80 cm^2 screen on which up to 200 sonar contacts can be displayed. The table receives data from the central processing unit which has a 20 kbyte random access memory.

Operational status
Operational in French 'Eridan' class MCMVs and the three 'Tripartite' units being acquired by Pakistan.

Contractor
Thomson-Marconi Sonar SAS, Sophia Antipolis.

UPDATED

IBIS systems

Type
Mine countermeasures data handling system.

Description
IBIS III integrates the TSM 2021B (DUBM 21B) dual-antenna sonar with a TSM 2060 NAVIPLOT tactical plotter for navigation, plotting the precise location of detected targets and the recording of all relevant data. The NAVIPLOT uses the 15M 05 computer.

Data are displayed on a large four-colour CRT which is mounted at an angle to enable the operator to work in a sitting position, giving greater safety in the event of a mine exploding close to the ship. For operations in the vicinity of complex geographical formations (such as archipelagos and fjords) the raw radar video can be overlaid on the graphic symbology. In all cases, coastlines are presented in the synthetic mode.

The IBIS III system provides a continuous simultaneous display of the ship's planned and actual track, fully labelled display of up to 256 contacts, complete recording of all operational data and storage in the computer's memory. This enables a continuous comparison to be made between previously recorded data and the current situation with regard to the area being surveyed. New underwater contacts are thus instantaneously identified and located, providing increased safety for the ship and resulting in considerable saving in time on minehunting operations.

Associated subsystems of IBIS III are the ship's Doppler sonar log (Thomson-CSF 5730), radio navigation system (for example Thomson-CSF TRIDENT III), navigation radar (for example Racal/Decca 1229), compass, log, SATNAV, echo-sounder,

radio datalink and SWEEPNAV, an acoustic location system for mechanical sweeps. As optional extras the IBIS III can be integrated with the ship's autopilot for automatic track following course to a point and hovering.

The IBIS III Mk II is an upgraded version of IBIS III which comprises the TSM 2021D (DUBM 21D) sonar and the TSM 2061 tactical system. The TSM 2021D is a fully digitised minehunting sonar equipped with Computer-Aided Detection (CAD)/ Computer-Aided Classification (CAC) sonar processing.

The IBIS V is a lightweight system suitable for retrofit, as well as new construction. The sonar associated with the IBIS V is the TSM 2022 single-array system, which takes up less space in the hull than the TSM 2021. The TSM 2026 NAVIPLOT is the second main component of the IBIS V. Whereas the 2021 can provide simultaneous detection and classification in different directions, the 2022 performs detection and classification in sequence and requires only a single operator.

IBIS V weighs 1.5 tonnes, the array having a span of about 1.5 m but pivoting to retract vertically into the sonar well, which is 75 cm in diameter. IBIS V Mk II is an upgraded version of IBIS which comprises the TSM 2022 Mk II sonar and the TSM 2061 tactical system. The TSM 2022 Mk II uses a new 19 in high-resolution colour console with sonar processing such as CAD, CAC and PI (Performance Indicator).

The latest system to be developed in the IBIS series is the IBIS 43. Using sophisticated computing equipment and the TSM 2054 side scan sonar, with antenna incorporated in a towfish, IBIS 43 produces high-quality images of the seabed with a high resolution, enabling operators to identify underwater

mines. The computer system ensures real-time optimisation of sonar operating parameters depending on the speed of the towed body, its height above the seabed and the automatic compensation for roll, pitch and yaw. Computerised aids are used in conjunction with an image management system to ensure real-time detection and classification of unknown objects and to compare images obtained in successive missions in a given sector.

The towfish can be navigated manually or automatically at constant depth between 6 and 200 m with an altitude in relation to the seabed from 4 to 15 m, programmable from the sonar console keyboard. It can automatically follow seabed terrains with gradients of more than 15 per cent with high precision. It also has an obstacle avoidance capability for obstacles up to 10 m high at maximum speed.

Operational status
The IBIS V is in service on board Malaysian and Nigerian 'Lerici' class minehunters, Indonesian 'Eridan' class minehunters and the Swedish 'Landsort' class minehunters and the four 'Yung Feng' class minehunters of the Republic of China Navy.

The IBIS V Mk II is installed on the four 'Landsort' type minehunters built for the Singapore Navy.

The IBIS 43 has been purchased for use with the STANFLEX 300 multirole vessels of the Royal Danish Navy.

Contractor
Thomson Marconi Sonar SAS, Sophia Antipolis.

UPDATED

GERMANY

MWS 80

Type
Minehunting weapons system.

Development
MWS 80 was developed for the German Navy's 'MJ 332' class minehunters by STN ATLAS Elektronik.

Description
STN ATLAS Elektronik has developed the MWS 80 as an integrated multirole MCM system which performs search, detection, classification, identification and neutralisation of ground and moored mines, precision navigation and vessel control, full control and co-ordination of all MCM operations, and maintains an accurate record of all areas searched, targets located and their geographic position and details of classification results.

MWS 80 integrates the DSQS-11M minehunting sonar, the NVC (Navigation and Vessel Control equipment), the TCD tactical command and control system for co-ordinating and documenting MCM operations and the mine disposal system and the AIS 11 active identification sonar installed on the ROV.

The DSQS-11M provides independent and simultaneous detection and classification modes through 360° azimuth. The stabilised sonar beams simultaneously cover a 90° horizontal sector and a 60° vertical sector for three-dimensional target location. Functions include performance prediction, computer-aided detection, classification and tracking with full-colour display to improve detection, performance and discrimination between targets. The man/machine interface features function keys and interactive, colour-coded control displays.

The TCD assists the operator in controlling all phases of an MCM operation by providing means for display, plotting, storage and retrieval of tactical data. The equipment employs a comprehensive database which is continuously updated during operation.

The navigation sensors such as GPS, radio location systems, gyro or inertial platform and so on, and the

MWS 80 mine countermeasures system

integrated Doppler log DLO 3-2 enable the NBD to compute with a very high degree of accuracy the various navigational parameters such as ship's geographical position, groundspeed, speed through water, course, heading and drift.

All data handled are recorded on hard disk, tape and printer to enable missions to be stopped and started at will, provide recall of data for comparison and evaluation, maintenance of a permanent record of the seabed for MCM survey and for training purposes ashore.

The new integrated minehunting weapon system

MWS 90 is functionally based on MWS 80 and additionally includes the Self-propelled Variable Depth Sonar SVDS 90.

Operational status
In operational service. The MWS 80 is being installed in the new Thai minehunters/sweepers building in Italy.

Contractor
STN ATLAS Elektronik GmbH, Bremen.

UPDATED

TCD (Tactical Command and Documentation) equipment

Type
Action information system.

Description
The TCD forms an integral part of the MWS 80 minehunting weapon system (see previous entry).

The TCD provides comprehensive facilities to the commanding officer and the minehunting director to prepare, control and document all phases of MCM operations.

A variety of functions is available on the two consoles including:
(a) track planning for the selected search area, taking

into account tactical considerations such as sonar performance data and environmental conditions
(b) presentation of tactical, navigation and status information, including radar video (on commanding officer's console) and live sonar displays
(c) presentation of manoeuvre data to the helmsman
(d) control of sonar and navigation equipment
(e) contact management with correlation of detected and stored sonar targets
(f) documentation of all results for evaluation and follow-on missions
(g) data logging of selected parameters including BITE messages relating to all connected MWS 80 subsystems.

Virtually all these functions are linked by the TCD's comprehensive tactical database management system.

The TCD console includes two high-resolution raster scan displays and keypads for direct system operation. Additionally, the operation is supported by easily accessible menu guidance on both graphic displays. The peripheral equipment includes the mass memory unit, printer, plotter, the manoeuvre display and the minehunting sight.

Operational status
In operational service.

Contractor
STN ATLAS Elektronik GmbH, Bremen.

VERIFIED

NVC (Navigation and Vessel Control) equipment

Type
Integrated navigation and vessel control equipment.

Description
The NVC (abbreviation in German Navy MWS 80 configuration is NBD) forms a part of the MWS 80 minehunting weapon system. It combines the functions of precise navigation, vessel control and precise speed measurement, performed by an integrated Doppler log. Data is accepted from a wide range of navigational sensors such as GPS, radio location system, Decca, inertial platform, gyro, echo-sounder, anemometer and

radar. From these inputs the NVC computes optimal values for position, heading, course, longitudinal and transversal speed over ground and through the water, set drift and water depth by Kalman filtering. Sensor data quality is continuously monitored. In case of sensor failures degradation modes are automatically selected. By providing datum transformations the NVC will accept and output navigational data in most of the commonly used reference systems. Using track and waypoint data from the tactical database or entered directly at the control and display unit, the NVC automatically steers the vessel along the planned search pattern. A manual vessel control mode via joystick is available.

The NVC equipment comprises the electronic cabinet containing all interfaces, the control and display unit, and the transducer for the Doppler log.

Operational status
In operational service.

Contractor
STN ATLAS Elektronik GmbH, Bremen.

VERIFIED

INTERNATIONAL

MWSC

Type
Mine Warfare System Centre (MWSC).

Description
Developed by CelsiusTech of Sweden in association with Computing Devices Company Ltd of the UK for the Royal Australian Navy, the MWSC is designed to undertake responsibility for the conduct of all aspects of planning, tasking, modelling and evaluation for the entire MCM fleet and its tasks. The operational methodology is based on NATO standards.

The MSWC is established at Waterhen, Sydney and is capable of conducting all operational mine warfare activities in wartime and will provide extensive tactical development and training facilities in peacetime. The centre is able to comprehensively undertake the modelling of equipments (for example sonar), of assets (minehunters) and tactics. The entire system is based on a large network of open system workstations based on rugged commercial computers and commercial software supplemented by specially developed program functions in a Windows environment to provide an effective Man/Machine Interface (MMI).

Operational status
In service.

Contractor
CelsiusTech Naval Systems AB, Järfälla, Sweden.
Computing Devices Company Ltd, St Leonards-on-Sea, UK.

UPDATED

ITALY

MACTIS MM/SSN-714(V)2

Type
Minehunting data processing system.

Development
Under the designation MM/SSN-714(V)2, SMA and Datamat have developed a digital navigation and plotting system for the Italian Navy's new 'Gaeta' class of mine countermeasures ships.

Description
The main functions of the MM/SSN-714(V)2 MACTIS (minehunting action information subsystem) are:
(a) operations planning
(b) automatic computation and presentation of the ship's current position; navigation control (through autopilot interface)
(c) display of the tactical situation
(d) location of surface and underwater targets
(e) analysis and presentation of target characteristics
(f) guidance of surface and underwater craft
(g) event recording.

The system is based on a computer of the same or similar type to that employed in the SACTIS submarine combat information system, namely a Rolm MSE 14 machine. In the MM/SSN-714(V)2 system this is interfaced with recording units, display units, controls, printer/plotter and ship's sensors. The principal items in the last of these categories are radar, sonar, compass, log and various navigation aids. The

operator's display has a vertical screen CRT, an alphanumeric display, a keyboard for communications with the system, supplementary data readout display units and associated input controls on the bridge and in the operations room.

The following units comprise MACTIS: an operator station with one 16 in graphic video screen; a functional keyboard and a trackerball; two display units (bridge and CIC/OPs room) for data display and navigation; processing signal distribution; and power supply units.

Analysis and presentation of target characteristics is achieved using the FIAR SQQ-14 fully solid-state mine-detection and classification sonar. The sonar is a dual-frequency, variable depth, beam-steering sonar enabling simultaneous detection and classification of targets. Separate search, classification and memory display consoles are provided. Comprehensive navigation equipment, in addition to the MM/SPN-703 navigation radar, is provided to ensure precise positioning of the ships and accurate plotting of underwater objects encountered.

Operational status
In service with Italian Navy 'Gaeta' class minehunters.

Contractor
GF Galileo SMA, Florence.
Datamat Ingegneria dei Sistemi, Rome.

VERIFIED

Graphic display for MM/SSN-714(V)2 system

Mine Data Centre

Type
MCM mission planning centre.

Description
The Mine Data Centre (MDC) has been developed to support the operational activities of MCM vessels.

It enables operators to plan MCM missions through the analysis of data stored in the system files and the extraction of information pertinent to the missions. MCM missions can also be evaluated through the analysis of information acquired during the MCM mission. It also enables MCM missions to be fully exploited by the updating of information stored in the system files.

The input to the centre consists primarily of data recorded by MCMVs during missions using their

combat information system (in the Italian Navy the MM/SSN-714(V)2).

Input data are recorded both manually and automatically on magnetic support. Various types of outputs can be obtained including reports on missions, statistics, content of the database and planning information for future missions. Various output formats can be used including map display on graphic workstations or large screen projector, plots and listings.

Planning data are delivered to the MCMV on magnetic support, the data being directly input to the combat information system.

The system is based on a DEC mainframe (VAX 8250) equipped with disk and tape units and running the ORACLE Rdbms. User interaction is via graphic workstations (VAX stations) connected through Ethernet LAN, PCs and terminals. Peripherals include

colour plotter, printers, digitiser and large screen projector.

All the application software is written in Ada, and has been designed with the HOOD methodology; database design used the Entity-Relationship approach.

Operational status
The system has been in use at the Italian Navy mines and mine countermeasures centre since July 1991 and was used to support Italian MCMVs deployed in the Persian Gulf.

Contractor
Datamat Ingegneria dei Sistemi, Rome.

VERIFIED

NORWAY

MICOS

Type
Minehunting command and control system.

Description
The MICOS command and control system has been

developed for the Norwegian mine countermeasures programme. MICOS is an integrated mission control system which can be configured for either minehunting or minesweeping. The configuration on board the 'Oksoy' class minehunters comprises: an integrated bridge system; an MCM command and control system; the sonar system; and ROV and mine disposal system.

The configuration on board the 'Alta' class minesweepers comprises: an integrated bridge system; an MCM command and control system in the operations room; the sonar system; interfaces to the sweep gear.

The integrated bridge system comprises a Simrad ADP 701 navigation and dynamic positioning system for high-precision navigation and automatic control of the vessel, and a Norcontrol DB 2000 officer-of-the-watch ARPA system for surface surveillance. The MCM command and control system comprises two Simrad ATC 900 multifunction tactical consoles, two operator consoles for the Simrad/Thomson Sintra TSM 2023 detection and classification minehunting sonars and a control console for the ROV operator. The system components are all integrated via a dual-redundant Ethernet LAN. Micos is based on an open architecture design using a high degree of COTS.

The command and control system has a wide range of functions for planning, operation and reporting and documentation of an MCM mission. The system comprises a database containing two-dimensional electronic charts, three-dimensional bottom terrain models, previous observations of bottom objects, navigation routes and so on. Also available are functions for performing MCM calculations for coverage and system performance analysis, including updating of the MCM database.

The dynamic positioning system is based on Simrad's ADP 701 which is in service with a great number of offshore vessels. It includes functions for automatic control of the vessel comprising a full three-axis manual mode, autopilot (for transit), hovering mode, auto-track and auto-sail modes. The two latter modes are used to steer the vessel automatically on a planned course with minimum cross-track error.

The dynamic positioning system is interfaced to the sweep gear for automatic calculation of the gear's position and coverage and for measuring and counteracting the forces induced by the gear on the vessel.

Operational status

A total of nine MICOS systems has been ordered for the Royal Norwegian Navy's surface effect ships of the 'Oksoy' class (minehunters) and 'Alta' class (minesweepers). The minehunting systems are in service.

Contractor

Kongsberg Defence & Aerospace AS, Kongsberg.

VERIFIED

Console arrangement of the MICOS MCM system 1995

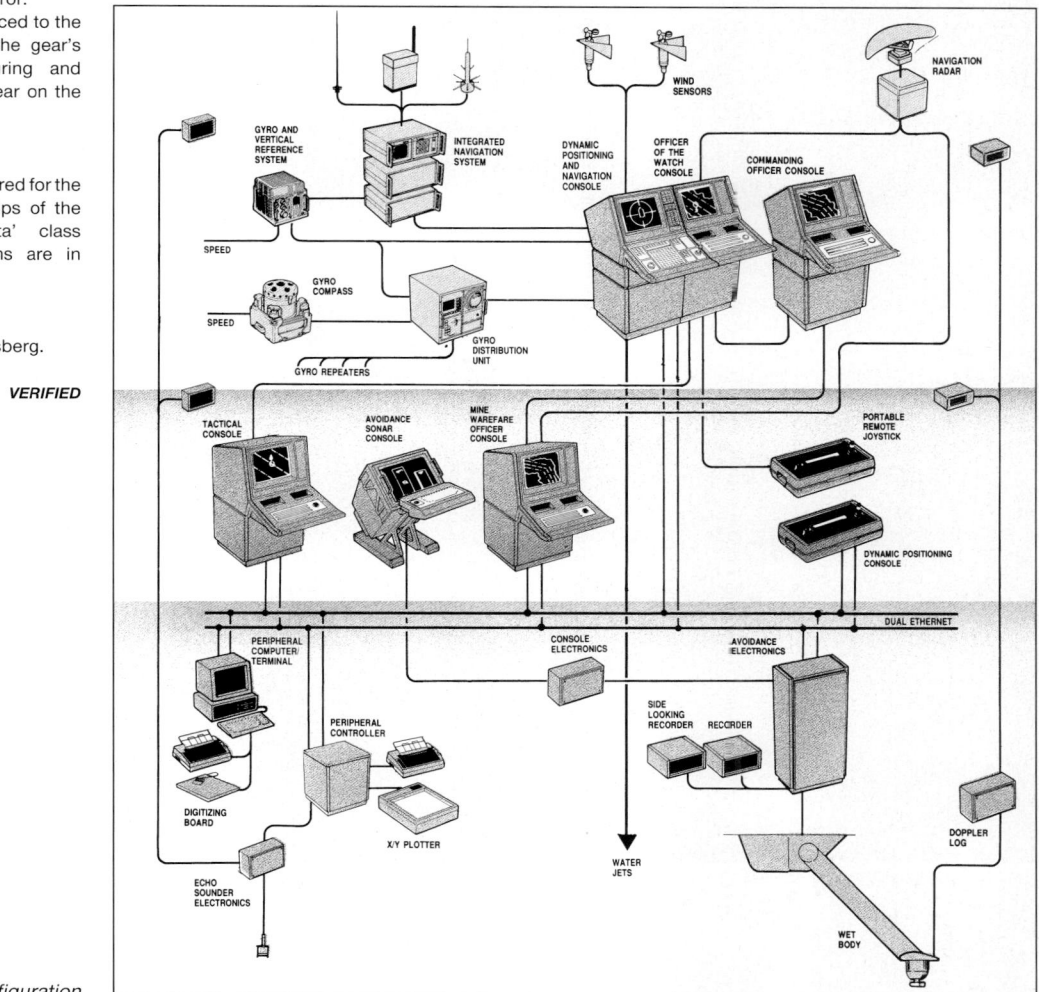

MICOS minesweeper configuration

Satyr

Type

Minefield inspection, control and maintenance command and control system.

Description

The Satyr system is an adaptation of the MICOS (see previous entry) command and control system developed by Simrad for the Norwegian mine countermeasures programme. It is an integrated mission control system comprising: a Simrad tactical command and control system; a Simrad ADP 701 dynamic positioning system; a Simrad HPR 410 hydroacoustic reference system; a Simrad EA 300 echo-sounder; the Artemis surface position reference system; and a windspeed/direction sensor.

The system components are all integrated via a dual-redundant Ethernet LAN. Satyr is based on an open architecture design using a high degree of COTS equipment.

The command and control system has a wide range of functions for planning, operation and reporting and documentation of missions for minefield laying, inspection and maintenance. The system comprises a database containing two-dimensional electronic charts and high-resolution three-dimensional bottom terrain models.

The DP system is based on a Simrad design currently in service on a large number of vessels worldwide. The ADP 701 system installed in the *Tyr* includes, in addition to the standard DP functions, a 'Follow ROV' mode where the vessel automatically follows an ROV, maintaining an operator-defined distance between the ROV and the vessel.

Operational status

In service on board the Norwegian Navy KNM *Tyr*.

Contractor

Kongsberg, Defence & Aerospace AS, Kongsberg.

VERIFIED

IDA

Type
Command, weapon control and information system for cable controllable minefields and torpedo batteries.

Development
In 1990 it was decided to undertake a comprehensive modernisation of existing minefields and torpedo batteries which form part of Norway's coastal defences. These comprise a series of fortresses which form an essential part of the country's anti-invasion defence system. The fortresses consist of a combination of cable-controllable minefields, fortified torpedo batteries and artillery installations.

Description
The modernised minefields and torpedo batteries will have identical and completely new weapon control systems and sensors. The existing torpedoes and mines will be upgraded and new mines are to be developed.

The central part of the system is the fire-control centre with the operator console and weapon control panels. The console accommodates two operators – one fire-control operator and one sensor operator.

Operational functions such as target tracking, track correlation, threat evaluation and weapon allocation are fully automatic but provide for operator interaction at certain preselected stages.

Target detection is based on radar and Electro-Optic (EO) sensors, which may be located at a distance from the weapons control room and linked to it by fibre optic cable. The EO sensors, which include a TV camera, laser rangefinder and thermal camera, are protected by a hydraulically elevated, camouflaged armoured steel cupola.

Mines can also be equipped with sensors that send information to the control room. Consequently, maximum operational efficiency is obtained through this extensive system of surface surveillance that contributes threat evaluation and target data for engagement analysis.

The minefields consist of both ground and moored mines. Supervision and control are executed via a new concealed distribution system. The mines are operated individually or in groups. Extensive safety measures are employed in order to prevent unintended firings including new safety and arming units in each mine. The system must also ensure that friendly shipping can traverse these minefields without risk during a crisis and, of course, that absolute safety is assured during peacetime.

The torpedo system consists of wire-guided torpedoes that are fired against targets from an underground torpedo hall via sloping tubes that protrude below the waterline. The use of new technology will significantly improve the equipment, systems and procedures used in the preparation and loading of torpedoes, as well as the overall working environment in the torpedo hall.

The torpedo system employs the same radar and EO sensors as the minefield system. Data signals between the weapons control room and the torpedo hall are conveyed via fibre optic cables. Before engagement, torpedo initialisation data is sent from the fire-control system to the torpedo control system. During the engagement phase, command messages are sent to the torpedo hall and guidance data to the torpedo. Messages on the status of the control system, ramps and torpedo are then returned to the fire-control system. After firing, the torpedo sends back course data on its way to the target.

Components and units are standardised so that they may be used in any of the modernised fortresses. The fire-control system will also be standardised so that it is suitable for both the controllable minefield and torpedo systems. This will result in lower operative and maintenance costs and a simplified logistics system.

The coastal fortresses will also be equipped with integrated tactical and administrative radio communications, a new digital telephone exchange, and an integrated alarm/command system.

Operator console of the cable-controllable defence system IDA   **1994**

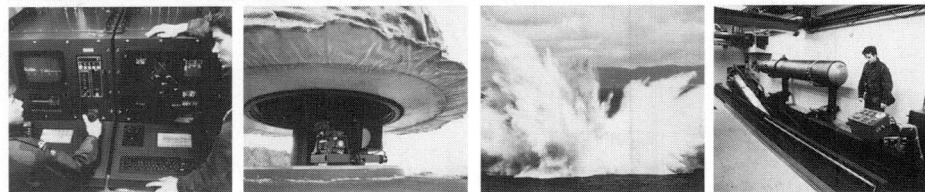

Artist's impression of the cable-controllable defence system IDA   **1996**

Operational status
IDA is under contract for the Royal Norwegian Navy. Kongsberg has completed basic system development. The first two systems became operational in March 1995. Delivery of systems will end in March 1999.

Contractor
Kongsberg, Defence & Aerospace AS, Kongsberg.

UPDATED

SOUTH AFRICA

MTDS

Type
Minehunting Tactical Data System (MTDS).

Development
Altech Defence Systems of South Africa has developed the MTDS as an upgrade to replace the TSM 2060 system which the South African Navy can no longer economically support aboard its 'River' class minehunters.

MTDS is being developed from core technology derived from Project Nickels (see separate entry) and Project Diamant (an AIO system developed for the South African strike craft force). The pilot system was fitted aboard SAS *Umtoli* in 1995, with a second prototype fitted in SAS *Urnzimkulu* in 1996.

Description
The MTDS is destined to integrate sensor information, compile a real-time tactical picture and facilitate

co-ordination and control of minehunting operations. As part of its operating capabilities MTDS is designed to assist the command team in the compilation and maintenance of a mission plan, defining search track parameters for a predefined search area. As part of this process the MTDS manages own ship navigation data (it is planned to incorporate differential GPS in the future) to provide the helmsman with tactical directions as to courses to be steered. Other capabilities include: direction of own ship sensors during search operations; tactical picture record and replay for post-mission analysis; and a hard-copy printout for mission planning and debriefing.

In its basic configuration, MTDS consists of a single-operator console with a graphics printer, together with a remote display for the commanding officer and an optional helmsman's indicator display. The console houses two 20 in colour raster displays. One of these displays the tactical situation with the mine plot, the other is a tabular display for totes and stateboards and for function menu access and data entry.

Applications software is executed inside the console chassis using Intel Pentium processor cards on a Multibus II backplane.

Operational status
Prototypes of MTDS have been installed aboard two 'River' class vessels and development of the system is continuing. Sea trials of a fully ruggedised standardised console with operational software were undertaken in 1997.

Contractor
Altech Defence Systems.

VERIFIED

SWEDEN

9 MJ 400

Type
Integrated navigation and combat information minehunting system.

Development
Developed as a joint project between CelsiusTech and Racal (UK) the 9 MJ 400 integrated navigation and combat information system provides a wide range of navigation and MCM functions together with sensor interfaces.

Description
The 9 MJ 400 enables the operator to plan in minute detail all manner of MCM tasks. The system computes and displays the navigation plan to and from a search area, and during the MCM task displays computerised search tracks, taking into account the type of sonar fitted and environmental conditions.

The sonar search is planned using two types of search plan displayed on the conference-type combat information console. The operator then defines an area or route to be searched and enters the anticipated sonar coverage and required percentage overlap. The system then computes and displays a search plan, the sonar being controlled in one of two remote-control modes (search or target indication). The search plan and actual tracks, together with the position of located targets, are automatically drawn on the X-Y plotter.

During a search the operator can track other ships and aircraft operating in the area, the system tracking 16 targets automatically or 30 targets semi-automatically.

Data on mine contacts are stored in the direct-access memory of the computer (up to 100 contacts) and 'dumped' on to magnetic tape, as well as being fed to the plotter and/or the printer. The magnetic tape can store data on 1,000 mine contacts, which can be read into the direct-access memory when planning the operation. All data recorded on tape or printer are available as a permanent record for subsequent comparison when carrying out searches of specific channels.

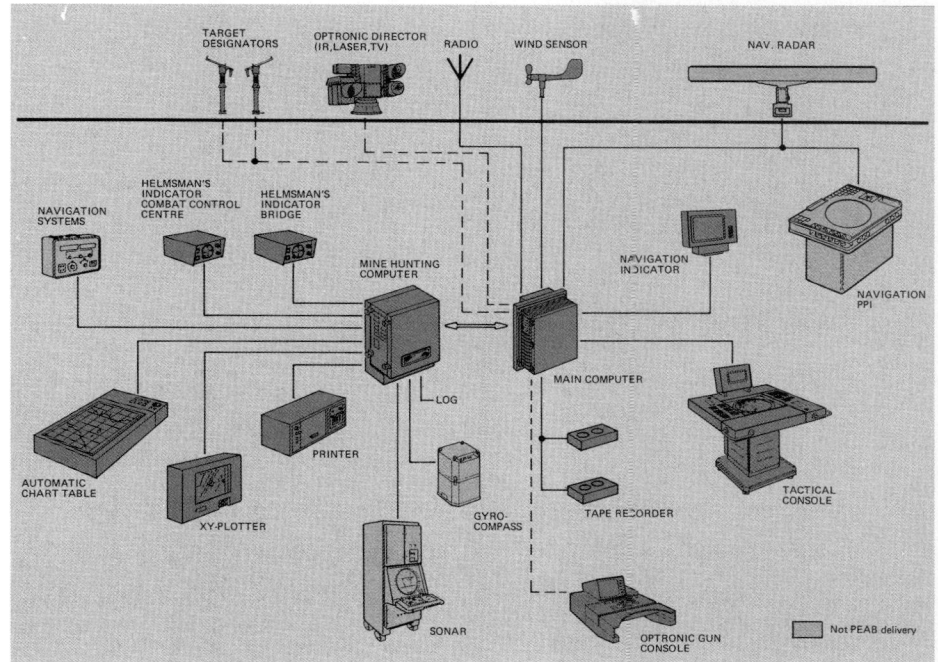

CelsiusTech Type 9 MJ 400 minehunting system

'Own ship' position and true speed are continuously calculated from each sensor input which includes Decca Navigator, transponders, navigation radar and dead reckoning. The operator selects the most reliable source for input to the system. The computer can store up to four range-measuring beacons when using a microwave transponder system.

In addition to its MCM function, the 9 MJ 400 system provides an ASW capability and weapon control function for depth charges and mortars. When linked with the CelsiusTech 9 LV 100 FCS, the 9 MJ 400 offers an upgraded self-defence capability for the ship as well as improving MCM tracking functions.

The 9 MJ 400 also features an integrated datalink for the onward transmission to shore-based control centres of important MCM data, as well as to other units in the operating area.

Operational status
In service with the Swedish 'Landsort' class MCMV.

Contractor
CelsiusTech Naval Systems AB, Järfälla,

VERIFIED

UNITED KINGDOM

System 880

Type
Minesweeping/minehunting control system.

Description
The Racal System 880 minesweeping package is available for precise navigation in minesweeping and co-ordination between ships and has been fitted to the 'River' class minesweepers of the Royal Navy. The system has also been sold to the US Navy and is recommended for Craft Of Opportunity (COOP). A minehunting version of the System 880 is also available using a comprehensive tactical display, in addition to the equipment available on the minesweeping package. The system can integrate with a variety of

sonars (an early version in the Royal Navy 'Ton' class minehunters integrated with the Plessey 193M sonar) and has provision for control of ROVs. Based on the modular concept and using distributed processing, the system can be reconfigured easily to meet different requirements and to interface with different equipments. A concept currently under study is the possibility of incorporating an automatic hovering capability using the autopilot and propulsion control. Other possibilities include feeding the command and control system with extracted positional radar data, and the display of extracted ESM data.

System 880 is a third-generation, low-cost MCM system comprising two major subsystems: the Racal Integrated Navigation System (RINS) and Racal Action Display System (RADS).

RINS is designed for minesweeping vessels, while for minehunters the addition of RADS provides a full MCM capability. System 880 automatically gathers information from sensors and navigational aids and presents it to the command on a ground-stabilised 19 in colour CRT display. The display uses colour graphics to identify route, ship's position, objects, direction of sonar beam, alphanumeric tote display, and danger and safety circles prescribed around the object under examination, all presented on a plan display. Associated with the main display is a control panel, alphanumeric keyboard and smaller alphanumeric display for online communication with the computer system.

RINS provides a minesweeper with mission planning capability, accurate navigation display data for the

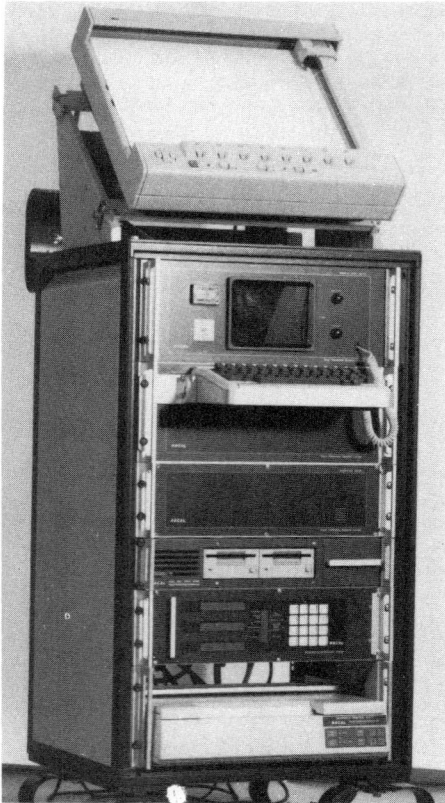

System 880 MCM system

helmsman to aid accurate track-keeping, hard-copy printout on a plotter or automatic chart table, and autopilot control. For team sweeping a special purpose version of RINS is available called RAFTS (Racal Aid For Team Sweeping). This provides all the facilities noted above plus a communications link between the RINS systems on the ships involved in the team sweep to ensure that the wing ship is automatically maintained on station with respect to the lead ship.

RADS can be added to RINS to provide integrated sonar, radar, MCM database and a tactical display for minehunting. RADS also provides an acoustic navigation capability enabling inputs from an underwater acoustic positioning system such as Racal's Aquafix 4 to navigate a remotely controlled submersible.

Operational status
No longer in production. In service on board 'River' class minesweepers of the Bangladesh Navy.

Contractor
Racal Survey Ltd – Products, New Malden.

VERIFIED

MAINS

Type
Minehunting Action Information and Navigation System (MAINS).

Description
MAINS provides the following main functions:
(a) accurate navigation
(b) ship control guidance
(c) integration of minehunting sonar
(d) combined surface/subsurface tactical display
(e) automatic hard-copy tactical plot with detailed data printout
(f) patrol navigation/combat information operating characteristics as an optional second role.

The main features of the equipment are: simplified operating procedures; flexible characteristics; compact equipment; low cost; and reliable and easy maintenance. Being of modular construction MAINS is capable of interfacing with a wide variety of equipments suitable to meet the individual requirements of any navy.

'Own ship's' position is fixed using one or more high-accuracy radio and microwave ranging systems such as the Racal Hi-Fix 6 (large area coverage extending to 150 km) or Hyperfix, or the portable line of sight Trisponder with a range of 80 km. Other navaids include the statutory navigational radar which has sufficient resolution to track marker buoys, GPS, Decca, Loran C, the Doppler sonar, gyrocompass, and conventional EM or Doppler/acoustic log.

The MAINS computer can also be used for ship control and steering guidance.

An automatic plotter and printer provides a hard-copy record of operations necessary for subsequent operations covering the same area.

The interactive PPI display is the Racal ED1202 which combines both radar surface situation and own vessel position relative to sonar search plan together with underwater contacts as a single presentation. The displayed picture has multilevel brilliance and correlated data and is presented in alphanumeric form.

The computer software is modular and is capable of accepting a variety of languages.

One of the major features of the MAINS is the compact size of the equipment, the control and display unit being tabletop mounted. Being modular and compact means that such a system can be fitted rapidly to suitable vessels to build up a mine warfare capability.

Operational status
In service on board the Swedish 'Landsort' class MCMVs where it integrates with the 9 MJ 400 system.

Racal Marine Systems action information system

The MAINS 500 system is operational on board the 'Swallow' class MCMVs of the South Korean Navy.

Contractor
Racal Radar Defence, Crawley.

UPDATED

NAUTIS-M

Type
Third-generation integrated command, control and navigation system for shipborne mine countermeasures.

Development
NAUTIS is a modular family of naval command and control systems for air, surface and underwater requirements.

NAUTIS-M was originally developed for the Royal Navy's new single-role minehunters, the 'Sandown' class. Requirements included the integration of the new Type 2093 variable depth sonar and remotely controlled mine disposal system, multiple navigation and environmental sensors, radar and ship control as an integrated combat system. Functional requirements for NAUTIS-M cover all phases of minehunting operation including mission planning, route surveys, minehunting, classification, disposal and mission reporting.

Since the award of the development contract to Plessey Naval Systems (now GEC-Marconi Combat Systems) in 1984, NAUTIS-M systems have successfully undergone rigorous environmental testing and software proving by the UK MoD (Navy). The first production system successfully completed Royal Navy acceptance trials on HMS *Sandown*, and five systems are now operational. A further seven systems are in production for follow-on ships of the class.

A 1988 contract from the UK MoD (Navy) has produced NAUTIS consoles with high-resolution shock-hardened colour raster displays with radar video superimposed from a scan converter within each console. The US Navy has ordered a number of these consoles as part of the AN/SSN-2 Phase 3 evaluation programme. Under this programme NAUTIS has been given the US Navy nomenclature of AN/SYQ-15.

NAUTIS-M is installed in all 14 'Avenger' class MCMVs of the US navy and three 'Sandown' class minehunters of the Royal Saudi Naval Forces, and is being built under licence by Bazan – FABA for Spain's CME programme. Systems are also in development for the Australian 'Huon' class coastal minehunters.

Variants are currently proposed worldwide for a range of other MCM combat systems and ship types.

Description
The basis of each NAUTIS system is a new technology autonomous intelligent console. The console includes interfaces for sensors, weapons and, if required, a datalink, an integral radar autotracker, processors and memory to handle an extensive command system database. High-level system application software and a selection of interfaces determine the functionality of the console for air/surface/underwater warfare roles. A high-definition colour raster display presents a labelled radar picture with tactical graphics superimposed.

A NAUTIS system comprises a number of consoles networked via a dual-redundant digital highway. The highway is used to maintain automatically a replica of the command system database within each console. Each console user has independent access to the database and to interfaced sensors, weapons and peripherals according to his task requirements. User facilities enable tasks to be readily changed to give functional interchangeability between consoles of a system. Should any console be out of use then the other consoles remain fully operational as a system, using duplicated interfaces, and can take up the additional tasks.

NAUTIS-M systems can be configured with one or two networked consoles in a ship's Operations Room (CIC) and, if required, with a further console on the bridge (as in the 'Sandown' class). The number of consoles depends on operational requirements, the performance of the overall combat system and the number of operators required to handle the system data.

The NAUTIS-M online database covers an area of up to 2,000 sq n miles. The data include route plans, above and below water geographic and environmental data, known seabed contacts and other supporting data that will enable a mission to be planned and carried out with best use made of the ship and its combat system. Examples of database content are: at least 200 radar and sonar tracks, threat evaluations and weapon assignments, route/search plans and navigation status, six user-designated tactical maps, 15 synthetic charts, 200 labelled reference points, 5,000 fully detailed sonar contacts for MCM, 32 labelled bearing lines and operational warning messages. This database is online for access by the operator: radar, labelled graphics, totes and an interactive main area clearly displayed in one viewing area; two different display compilations maintained for alternate selection by single-key action; display scales typically from 0.125 n miles to the limits of the database; off-centring anywhere within the database area; true/relative motion stabilisation; labelled electronic range and bearing lines which can be hooked onto fixed points and moving tracks. The database is retained in each console, with the system operating program, in non-volatile memory. It can be loaded and retrieved using a magnetic tape cartridge. The man/machine interface also includes a typewriter keyboard, up to 32 assignable special function keys, a trackerball and electroluminescent panel. There is a comprehensive library of characters, symbols and lines. Size and brilliance can be selected by the operator.

Operational effectiveness of the ship's systems is maximised by full integration of sonar, navigation, collision avoidance and ship control systems through NAUTIS-M together with the online presentation of the MCM database.

Advanced display presentations, simplified user controls and operating procedures enable NAUTIS-M to be fully utilised with minimum training.

No specialised technical skills or test equipment are required for onboard maintenance. Comprehensive built-in tests are performed automatically at start-up and during run-time and any faults are indicated at plug-in module level.

The modular architecture of NAUTIS-M enables it to be readily adapted for MCM combat system applications with various forms of variable depth and hull-mounted sonars, influence and mechanical sweeps, sweep monitoring equipment, navigation systems, radars, mine neutralisation systems and ship control systems. NAUTIS-M typically comprises one, two or three consoles.

Operational status

In service with the Royal Navy and the Royal Saudi Naval Forces ('Sandown' class) and the US and Australian navies and being installed in the new MCMV classes of the Spanish and Japanese navies.

Contractor

GEC-Marconi Radar and Defence Systems, Command and Information Systems Division, Camberley.

UPDATED

NAUTIS-M command and control system as fitted to HMS Sandown

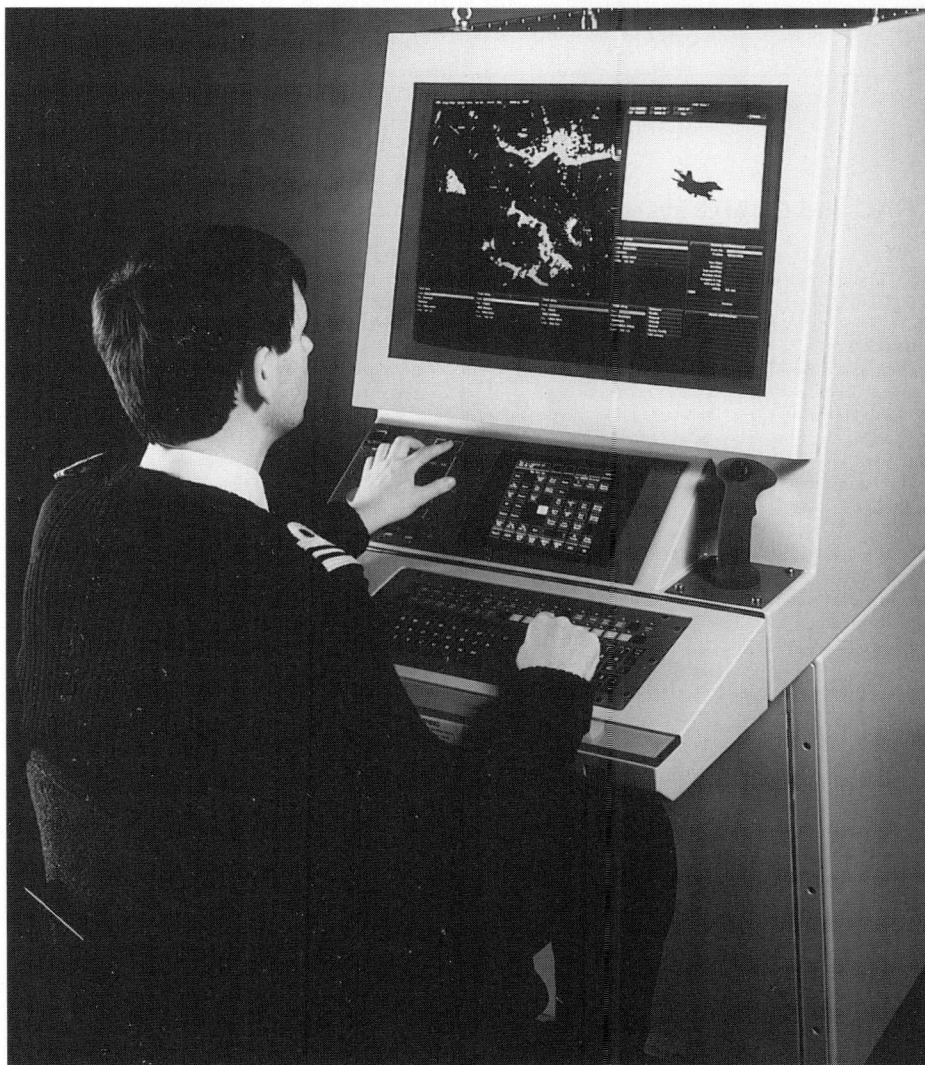

The latest technology NAUTIS console with colour raster display

POSITIONING AND TRACKING SYSTEMS

AUSTRALIA

ASK 4000 series

Type
Vessel dynamic positioning systems.

Description
The ASK 4000 series has been designed to provide manoeuvrability and control of all types of vessel. In the most basic system control is exercised using a simple joystick while the most complex system is a triple-redundant system. The various models available are:

ASK 4000JS – joystick/autopilot

ASK 4001 – single-dynamic positioning system

ASK 4002 – dual-redundant dynamic positioning system

ASK 4003 – triple-redundant dynamic positioning system.

Each system can interface with all known positioning sensors. Control is achieved simply by using clearly identified function keys and comprehensive graphic displays. Expanding systems is accomplished by simply adding one or more modular hardware units which are easy to install.

The basic design of each system comprises a standard modular hardware unit which simplifies procurement and eliminates the need for lengthy specification and custom engineering. The hardware requirement is much reduced as the console unit contains all the system elements – no bulkhead-mounted equipment is required. Most operational features are supplied as standard, keeping customisation to a minimum. The use of a multidrop data highway for system interconnections reduces both labour and material costs for installation.

The multifunction console is housed in a single cabinet and features a large, ruggedised 20 in diagonal high-resolution, full-colour display with IBM VGA compatibility. Signal processing is carried out using an industry standard processor – a ruggedised VME bus-compatible IBM 80486 digital processor running at high-clock frequency. Multiple RS-232/422/485 serial interfaces with optical isolation are fitted as standard.

The processing software control algorithms are programmed in high-level language C. Software development can be accomplished on board. If required, signal processing can be relocated to a remote location to reduce signal cabling costs. Signal modules communicate with digital processors over a high-speed RS-485 data highway.

Station-keeping performance data is always shown by using a split screen display. The selected graphics page is shown on the left side of the screen while the permanent display of station-keeping is always shown on the right. Ten information displays are accessed through dedicated panel control keys.

The summary page provides a numeric summary of the control process, with heading and position setpoints and actual positions, environmental data such as wind and sea current and a review of the control system values. The sensor page displays individual measurements reported from each sensor. Raw input data is also available for environment and velocity sensors. The alarm page provides full-colour-coded descriptive alarm messages to indicate alarm conditions and current alarm status. The plot page provides a graphic 'strip chart' display of up to four internal parameters simultaneously. The operator may select any variable for recording together with an appropriate display scale and bias.

Thruster Page No 1 reports in graphic format command and feedbacks for individual propulsors under control of the dynamic positioning system. Mode and alarm status are also provided. Thruster Page No 2 provides numeric values for the commands and feedbacks reported on the graphics page. In addition, a summary is provided in tabular form of the composition of the total command to each axis. Waypoint Page No 1 addresses the detailed heading and position setpoints and actual vessel location. Course-keeping values are also reported. For trackline operations, a set of eight waypoints may be specified along with required turn radius and track leg speed. Waypoint Page No 2 illustrates in graphic form the geographic locations of all waypoints identified. When the trackline is traversed, the actual vessel trajectory is plotted against the commanded locations.

Operational status
ASK 4000 JS systems are installed in the Australian inshore minehunters *Rushcutter* and *Shoalwater,* and are being supplied to the US Coast Guard's new 'Juniper' and 'Keeper' class vessels being built by Marinette Marine.

Contractor
Nautronix Ltd, Fremantle, Western Australia.

VERIFIED

UNITED KINGDOM

TRAC and CHART systems

Type
Integrated navigation, hydrographic survey and data logging system.

Description
The TRAC and CHART systems are a family of automated data acquisition and processing systems used extensively by navies and commercial organisations for a wide variety of operational tasks such as mine warfare, maritime patrol, hydrography, trials and evaluations, fisheries protection and offshore engineering.

TRAC provides integrated navigation, data logging, and precise and dynamic positioning for vessel control. It can form an important part of a naval Action Information Organisation (AIO) by providing real-time data in a variety of formats directly from the vessel's sensors.

TRAC is a self-contained system which uses an advanced position computation algorithm to handle up to 20 position lines which may be any combination of hyperbolae, ranges, bearings or latitude/longitude lines. The system can operate on any spheroid and projection and takes into account the various geodetic reference systems.

Control of TRAC is through a keyboard, rollerball and colour display. The system supports a wide range of peripherals, including remote colour displays, plotters and printers.

CHART is a complete data processing system which complements TRAC. It may be supplied as a stand-alone unit or networked to TRAC. The system design supports multiple TRAC and CHART workstations networked together to create an integrated multi-user system. CHART may be used afloat or ashore for data analysis and the production of fair sheets, and contains all the necessary processing facilities to carry out these functions.

Operational status
TRAC and CHART systems are in service with the Royal Navy, Royal Australian Navy, Italian Navy, Royal Navy of Oman, Republic of Singapore Navy, Turkish Navy and Venezuelan Navy. TRAC and CHART systems equip the survey boats of the Italian Navy's hydrographic survey vessel *Ammiraglio Magnaghi* and the Antarctic survey vessel HMS *Endurance*. TRAC is also fitted to the NATO ASW research vessel *Alliance*.

Contractor
Kelvin Hughes Ltd, Hainault.

UPDATED

PIMS

Type
Navigation and information system.

Description
PIMS (Polaris International Missing System) is one of a family of integrated navigation and information systems with applications covering minelaying and minesweeping as well as hydrographic and other applications. In the minelaying role, PIMS provides powerful tools for planning minefields, arranging mine cargo plans for minelaying vessels, supporting and controlling the deployment of mines at sea and producing reports and maps of the operation.

The minefield plan contains the number of mines, their planned positions and the mine types. Planning is undertaken on the screen in a graphical interactive way. NOAA charts and the user's own chart information, such as operation area boundaries, can be displayed as the reference on the background within the map window.

Mine laying lines can form an independent sailing plan or the mine laying lines can be included as sail lines in a given sailing plan. Operator options include the definition of the number of mines on the line and the spacing between adjacent mines. It is also possible to lay a single mine on a given point as in an exercise scenario. The planned positions are presented on the map windows as graphics representing the individual mine types.

There is a separate database that contains data for each mine type in use. It holds information such as type name, class, weight, length, width, height, centre of gravity, security radius and security times. Other desirable data can be added and the database can easily be restructured to contain other data when needed. There are tolls provided within the system allowing the operator to create new mine types and manipulate existing ones in the database.

The mine cargo editor (a spreadsheet window on the screen) allows the operator to change the track number and the mine type data of the mines in the cargo plan. After each editing action the deck diagram is updated. When the mine cargo plan has been completed, the loading lists are printed out. When the mines are physically loaded on deck, the serial numbers and other identification information is entered into the mine cargo plan using the mine cargo editor,

During deployment the PIMS system serves as the control centre. The integrated positioning provides continuously updated position and movement status for minelaying control. The system calculates the distance and bearing to the next deployment point and issues the command to drop the mine when that point has been reached. The ETA (Estimated Time of Arrival) to the next deployment point is calculated and the digital display panel on the mine deck is updated.

The PIMS system also controls the 'traffic lights' fitted at the end of each mine track. Th drop command can be issued electronically to hydro-pneumatic valves or audibly for manual operator control. Microswitches

fitted on the track give information to the system when the mine locates on the dropping platform and when the mine is deployed. The true drop position and the water depth are recorded when the deployment signal from the microswitch is received. The true drop position appears immediately on the map window as a graphical symbol.

The deck diagram is visible during deployment and is updated at each drop showing the latest situation on the mine deck. The current status of traffic lights and ETA are presented on the screen.

Deployment reports can be obtained both during and after the deployment operation.

Operational status
Under development.

Contractor
Polaris International Ltd, Woodrow Way, Gloucestershire.

NEW ENTRY

Sonar 2059

Type
Acoustic tracking system.

Description
The Sonar 2059 ultra-short baseline acoustic tracking system has been specifically designed to fulfil a number of military applications, including minehunting. The system interrogates an acoustic source and measures the time it takes for a reply to be received. The direction from which the signal came is calculated via the phase differences sensed by an array of receiving hydrophones. The software programme then

calculates the range and bearing to the source, simultaneously measuring and compensating for the pitch and roll motion of the ship. The information is then displayed by the control unit in various user-defined configurations. In MCM operations, this information can be transferred to the plotting table via a suitable interface.

The system can simultaneously track and display data for up to five targets. Each acoustic source is interrogated sequentially, with a different interrogation frequency dependant on the transponders in use.

The system can also be employed with underwater responders; these provide an acoustic reply to an electric signal triggered via an umbilical cable.

Alternatively, the system can also track a free-running acoustic pinger.

The 2059 uses MIL-SPEC components to ensure maximum reliability of operations.

Operational status
In service with the Royal Navy on board 'Hunt' class MCMVs to track PAP ROVs and submarine escape and rescue services.

Contractor
Kongsberg Simrad Ltd, Bordon.

UPDATED

UNITED STATES OF AMERICA

PINS

Type
Integrated navigation system for MCMs.

Description
Magnavox has developed the AN/SSN-2(V) PINS (Precise Integrated Navigation System) for MCMs. The heart of the PINS is a powerful military computer interfaced with an array of commercial and military sensors, control units and data display and recording systems. The PINS uses a Kalman Filter routine to smooth received data and apply appropriate

corrections for known error factors. The system integrates and compares data from multiple navigation aids and other sensors to perform precise navigation and position determination, mission planning, plotting and data recording, target location and positioning and post-mission analysis. The TRANSIT SATNAV system is used to provide navigation fixes, interfaced via PINS with the ship's Doppler sonar operating in either groundspeed or water speed modes. Other interfaced systems include LORAN C and Hyperfix.

All contacts and their positions are recorded and displayed, together with the ship's track, on a vertical plotter in the operations room. The high-speed vertical

belt-fed plotter provides a hard-copy printout using standard navy charts if required.

Operational status
PINS is in service with the US Navy on board the 'Avenger' class MCMVs.

Contractor
Hughes Aircraft, Torrance, California.

VERIFIED

TABLE XIX

Command and tracking systems

System	Manufacturer	Country	Class installed on	Navy	Units*
9MJ400	CelsiusTech Systems	Sweden	Landsort	Sweden	7
ASK 4000	Nautronix	Australia	Shoalwater	Australia	2
MWS 80	STN ATLAS Elektronik	Germany	Frankenthal	Germany	12
MWS 80	STN ATLAS Elektronik	Germany	Bang Rachan	Thailand	2
MWS 80	STN ATLAS Elektronik	Germany	Bay	Australia	2
MWS 80	STN ATLAS Elektronik	Germany	Lat Ya	Thailand	2
NVC	STN ATLAS Elektronik	Germany	Frankenthal	Germany	12
Type 2059	Simrad	UK	Hunt	UK	
PINS	Hughes	USA	Avenger	USA	14
TCD	STN ATLAS Elektronik	Germany	Frankenthal	Germany	12
System 880	Racal	UK	River	Bangladesh	4
Chart	Kelvin Hughes	UK			
EVEC 20	Thomson Marconi Sonar	France	Tripartite	France	10
EVEC 20	Thomson Marconi Sonar	France	Tripartite	Netherlands	15
EVEC 20	Thomson Marconi Sonar	France	Tripartite	Pakistan	3
IBIS V	Thomson Marconi Sonar	France	Tripartite	Indonesia	2
IBIS V	Thomson Marconi Sonar	France	Lerici	Malaysia	4
IBIS V	Thomson Marconi Sonar	France	Lerici	Nigeria	2
IBIS V	Thomson Marconi Sonar	France	Landsort	Sweden	7
IBIS V	Thomson Marconi Sonar	France	Landsort	Singapore	4
IBIS V	Thomson Marconi Sonar	France	MWV 50	Taiwan	4
IBIS 43	Thomson Marconi Sonar	France	Flyvefisken	Denmark	4
MACTIS	Datamat	Italy	Gaeta	Italy	8
MAINS	Racal	UK	Landsort	Sweden	7
MAINS	Racal	UK	Swallow	South Korea	6
MICOS	Kongsberg Defence	Norway	Oksoy	Norway	4
MICOS	Kongsberg Defence	Norway	Alta	Norway	5
NAUTIS-M	GEC-Marconi	UK	Sandown	Saudi Arabia	3
NAUTIS-M	GEC-Marconi	UK	Segura	Spain	4
NAUTIS-M	GEC-Marconi	UK	Sandown	UK	12
NAUTIS-M	GEC-Marconi	UK	Avenger	USA	14
NAUTIS-M	GEC-Marconi	UK	Huon	Australia	6
NAUTIS-M	GEC-Marconi	UK	Sugashima	Japan	2
Satyr	Kongsberg Defence	Norway	Tyr	Norway	1
TRAC	Kelvin Hughes	UK			
TRAC	Kelvin Hughes	UK	Alliance	NATO	1

* In service or on order

UPDATED

MINE WARFARE SONAR SYSTEMS

Many of the problems associated with sound transmission in Anti-Submarine Warfare (ASW) apply equally to Mine CounterMeasures (MCM), although there are considerable differences between the sonars used in the two forms of warfare. While ASW is essentially a long-range detection problem requiring low frequency, MCM is a very short-range problem which has generally required the use of high frequencies. The use of acoustics in minehunting is further compounded by the problem of multipath reflections from the bottom, experienced in shallow waters and more recently the need to penetrate the top soft layer of a sea bottom to detect buried mines.

Conventional mine detection and classification is carried out using high-frequency side scan imaging sonars. The usual frequency range for these sonars is in the order of 35 to 350 kHz with bandwidths which spread by some 10 to 20 per cent either side of the centre frequency. Within this frequency range it is possible to achieve signals and high-gain arrays in a relatively small package.

However, to detect and provide a high-definition image of small objects these high-frequency systems are practically limited to a range of about 100 m, well inside the danger zone of most modern mines (the danger zone being in the region of 200 m from the mine). Beyond this, the signal is strongly attenuated and the angular resolution is no longer adequate to resolve small objects. In order to carry out search and classification outside the danger zone, that is at ranges much greater than 100 m, a substantial increase in array gain and source level is required, both of which are prohibitive. As well as this limitation, sonars operating at these frequencies cannot penetrate the sediment to detect buried or partially buried mines, or penetrate the anechoic coatings now fitted to some modern mines.

Another limitation is the inherent inability of sonars operating at these frequencies to fully classify objects, even at close range. The acoustic images used for classification are fundamentally two-dimensional (for example range and azimuth), and thus no information about the shape of the object in three dimensions is obtained. Although in some circumstances height can be estimated from the length of an object's acoustic shadow, fundamental shape characterisations such as spherical, flat or cylindrical cannot be carried out using this approach.

The approach to defining requirements for MCM sonars is therefore quite different from that of ASW, and very specific. As mines become more stealthy the use of low frequency is becoming even more important as these signals stand a better chance of penetrating anechoic coatings which are being increasingly used on mine casings as well as the mud and sediment in which mines may be buried.

To provide as much detail and data over a very small insonified area (the area covered by a sonar) as possible, high-frequency sonars are currently used. This means either that a Mine CounterMeasures Vessel (MCMV) with an onboard sonar probably has to approach within the mine's danger zone in order to detect, identify and classify it, with the possibility of detonating it, or that the sonar must be displaced from the MCMV in some way so allowing the MCMV to remain well outside the danger radius of any possible mine detonation. In the past the use of hull-mounted minehunting sonars has required a trade-off between the high frequency required for high definition, and the lower frequency required to ensure that the MCMV can standoff at a safe distance.

To overcome this problem most MCM sonars incorporate a dual-frequency capability, using a lower frequency to survey a specific area, before moving in closer to use the high-definition sonar to classify and identify the target. However, to ensure ship survivability against modern mines, most MCMVs now carry an offboard system (the Remotely Operated Vehicle (ROV)) as well, carrying either a camera or small, high-definition sonar (or both), or use a diver to positively identify and neutralise the detected target.

Standoff sonars for MCM

There are two methods of deploying minehunting sonars in a standoff form. One is to use a side scan sonar towed astern of the platform. Alternatively the sonar may be fitted in an ROV, of which numerous examples now abound, or it may be deployed as a variable depth sonar. The latest systems, now the subject of intense development and investigation, include autonomous offboard ahead swimming vehicles capable of carrying out a fully integrated minehunting operation against both shallow and deep water laid mines.

Route surveillance

One of the most important factors in mine warfare is route surveillance. This requires that areas susceptible to mining must be constantly surveyed well in advance of hostilities. Detailed records of the seabed and all objects on it must be maintained by constantly surveying the area and this will enable routes to be quickly checked for new targets in time of rising tension and hostilities. This will enable routes to be cleared more quickly, or alternative routes to be assigned. Surveillance such as this requires a high speed of advance and high-coverage rate. Route surveying thus needs to be carried out at speeds up to 15 kt with a path cover about 2,000 m wide. This requires a sonar range of about 1,600 m. In good propagation conditions a sonar with an azimuth resolution of 3° and a range resolution of 20 n miles would be sufficient to carry out such a route survey.

Classification, on the other hand, demands a much slower speed of advance with much greater resolution. Thus speeds of 5 kt are the norm, and to achieve good classification results using acoustic shadow and echo to allow evaluation of the shapes of detected objects at a range of at least 275 m (to ensure platform safety), requires a bearing resolution of 23° with 5 n miles resolution. In addition, classification is further aided by employing sector-scan techniques. For close-range identification an even higher resolution capability will be required.

As targets become smaller and more elusive the ability to detect and classify accurately will demand even higher frequencies and smaller beamwidths, while identification will be aided by optics and laser systems.

The future

The key to the whole question of sonars, both for ASW and MCM, is technology.

One of the new technologies emerging which may overcome the problems of long-range detection and classification of mine-like objects is the low-frequency broadband sonar. This sonar exploits the full spectrum of low frequencies below 20 kHz, thus avoiding substantial attenuation at longer ranges and in the sediment, while achieving a bandwidth comparable to that of sonars operating at much higher frequency. Furthermore, the bandwidth of these signals is such that reduced reverberation and localisation can still be achieved while maintaining low frequency over the entire band. The use of multibeam multipulse imaging to detect and localise objects can thus be achieved using the most favourable incident angle.

Because of the fact that the acoustic reflection pattern of small objects is relatively broad in azimuth at low frequencies, this pattern can be unambiguously measured as a two-dimensional function of aspect angle and frequency, while still maintaining a relatively fast sweep rate. Such a sonar could provide key information on both the size and three-dimensional structure of the contact. Furthermore, for objects that have flat or cylindrical surfaces, this also insures that detection of a small, smooth reflecting surface will be achieved, increasing signal strength by 10 to 20 dB over random aspect angle detection.

Since this low frequency does not depend on high-azimuth resolution for classification the physical array length requirement, although greater than that for high-frequency systems, is still practical.

New and improved transducer materials are required to increase cavitation limits, synthetic aperture sonar processing would enable the range and coverage rate of sonars for MCM to be increased without degrading resolution and requiring a large underwater body. Parametric sonar techniques would enable sonars to penetrate the seabed to detect buried mines, a feature which is now of paramount importance in the face of the latest mine technology. The potential threat to MCMVs from modern sophisticated mines is such that previous generations of ROV fitted with HF sonars no longer provide the necessary safety radius for the parent vessel. The next generation of ROV must, in effect, be a self-propelled sonar – a vehicle carrying both HF and LF sonars which can swim well ahead of the parent vessel, sending back to it all the necessary data for detection, classification, and identification.

As with ASW, there is also the need for increased capabilities in the field of real-time data processing and post-processing techniques. Finally, the Man/Machine Interface (MMI) has to be improved, together with computer-aided detection and classification, which will greatly assist the operator.

UPDATED

HULL-MOUNTED MINEHUNTING SONARS

CANADA

CMAS-36/39

Type
Mine detection and avoidance sonar.

Description
Derived from the CTS-36/39 multipurpose sonar, the CMAS-36/39 provides all of the CTS 36/39 features as well as an enhanced small target detection feature (it can locate tethered, bottom, near surface mines and divers). Like the CTS-36/39, this sonar is a high-speed scanning, multibeam, hull-mounted sonar using narrow receiving beams and digital processing techniques to provide instantaneous detection and tracking of targets. Video and audio processing provides for target classification assistance.

High-resolution Mine CounterMeasures (MCM) mode operation provides a 180° sector which may be positioned through the 360° azimuth area. A high-resolution (0.25 m) 250 m window may be set through the full operator-selectable range scale.

Operational status
The CMAS sonar is in service with the navies of Australia ('Pacific' class AGS), Bangladesh ('Sagar' class MSO), Canada and Denmark.

Specifications
Frequency: dual-frequency 36 kHz/39 kHz, operator selectable in CW or FM mode
Beamwidth: receive – (horizontal and vertical) 6°; transmit – horizontal omni or selectable sector; vertical 6°
Transmitter: 270 channels; power output/channel 50 W, 13.5 kW omni total
Source level: 223 dB (at 1 μPa/1 m)

Contractor
C-Tech Ltd, Cornwall, Ontario.

UPDATED

FRANCE

TSM 2021 sonar

Type
Minehunting sonar system.

Description
The Thomson Marconi Sonar SAS TSM 2021B (service designation, DUBM 21B) is a modern minehunting sonar, the design of which applies experience with the DUBM 20A, also developed by Thomson Marconi Sonar SAS. The hull-mounted sonar is intended for use in conjunction with precision navigation equipment, and a Mine CounterMeasure (MCM) (destruction) system such as the PAP 104. A former version is the 2021A with different array housings and minor display variations. TSM 2021 functions are:

(a) the detection of mines at distances up to 600 m
(b) the classification of such targets by the study of the shape of their echo and acoustic shadow, at distances up to 250 m.

Each detection and classification sonar chain consists of an acoustic transmitter/receiver transducer, the associated transmission and reception electronics, and a display console. A subassembly performs the stabilisation, steering, and retraction of the detector and classifier arrays. This part of the system is the result of studies and development by ECAN at Ruelle.

The Electro-Acoustic subAssembly (EAA) includes:
(a) the classification sonar transmitter/receiver chain cabinet
(b) a display console which controls the transmission and reception functions and has Cathode Ray Tube (CRT) presentation of mine detection and location by the detector sonar chain
(c) a display console with CRT displays for presentation of mine locations within the classification chain
(d) bridge repeater display, showing range and bearing of designated targets.

The TSM 2021D (French Navy designation DUBM 21) is an upgraded version of the TSM 2021B. It comprises two 19 in high-resolution colour sonar consoles, an electronics cabinet and two acoustic antennas. The mechanical subassembly of TSM-2021D is the same as for TSM 2021B. This subsystem is produced by ECAN Ruelle.

Operational status
The TSM 2021 is in service with the French and other navies. Altogether some 60 systems have been ordered.

The latest fully digital version, DUBM21D, is in service on the French Navy MCMV, *Sagittaire*.

Specifications
Coverage: ±175° in sectors of 30, 60 and 90° for the detector chain; 3, 5 or 10° for the classification chain
Elevation: variable, between −5 and −40°
Stabilisation: ±15° (roll); ±5° (pitch)
Detection sonar
Frequency: 100 kHz (modulated ±10 kHz)
Pulse duration: 0.2 or 0.5 ms
Range scales: 400, 600 or 900 m
Max sound level: 120 dB
Channels: 20
Beam aperture: 1.5°
Display: PPI
Classification sonar
Range scales: 200 or 300 m
Max sound level: 122 dB
Channels: 80
Beam aperture: 0.17°
Display: CRT and storage magnifier tube

Contractor
Thomson Marconi Sonar SAS, Sophia Antipolis.

UPDATED

TSM 2021B minehunting sonar, showing the main electronics cabinet flanked by the classification and detection display console (left and right), with the transducer arrays in the foreground

TSM 2022 family

Type
Hull-mounted minehunting sonar system.

Description
The TSM 2022 is a lightweight minehunting sonar designed specifically for small- and medium-sized mine countermeasures vessels. The sonar can also be used for mine avoidance on minesweepers, or in civil applications such as bottom profiling and side scan surveys.

The TSM 2022 is based largely on experience gained with the DUBM 21B in the field of high-resolution beam-forming techniques. It is a hull-mounted equipment intended also for use in conjunction with precision display and navigation equipment and a mine countermeasures system.

The main feature of the TSM 2022 is its small-size retractable single-array assembly, used for both detection and classification, which enables easy installation and maintenance. The high-resolution features of the DUBM 21 are maintained and improvements are mainly in the field of digital processing technology.

The main functions of the sonar are: detection and classification of moored and bottom mines, detection at distances up to 600 m (2,000 m for submarines), and classification of targets by analysis of the shape of their echo or shadow at distances up to 250 m. In the classification mode, horizontal beamwidth is 7°; in detection mode either 14 or 28° beamwidth can be selected. In both modes the vertical beamwidth is 15°.

The main assemblies are:

(a) the hoisting and stabilisation system together with the retractable array which is installed in the sonar trunk (0.75 m diameter). The total weight is only 900 kg

(b) electronics cabinet

(c) operator's console.

The operator console provides display of sonar images in the various modes, memory and display of former sonar contacts, and includes interfaces with current mine disposal and navigation equipments.

In the Mk II configuration the TSM 2022 is equipped with a 19 in high-resolution colour console and specific sonar processing such as Computer-Aided Detection (CAD), Computer-Aided Classification (CAC) and Performance Indicator (PI).

The TSM 2022 Mk III, now being marketed, is equipped with the same electronic elements as the DUBM 21B. Classification and detection functions have been upgraded.

Operational status

Some 31 equipments have been ordered by seven navies. The Swedish Navy has seven TSM 2022s in operation, Malaysia four, Indonesia two, Nigeria two, Singapore four, Egypt three, and the former Yugoslavia two.

Contractor

Thomson Marconi Sonar SAS, Sophia Antipolis.

UPDATED

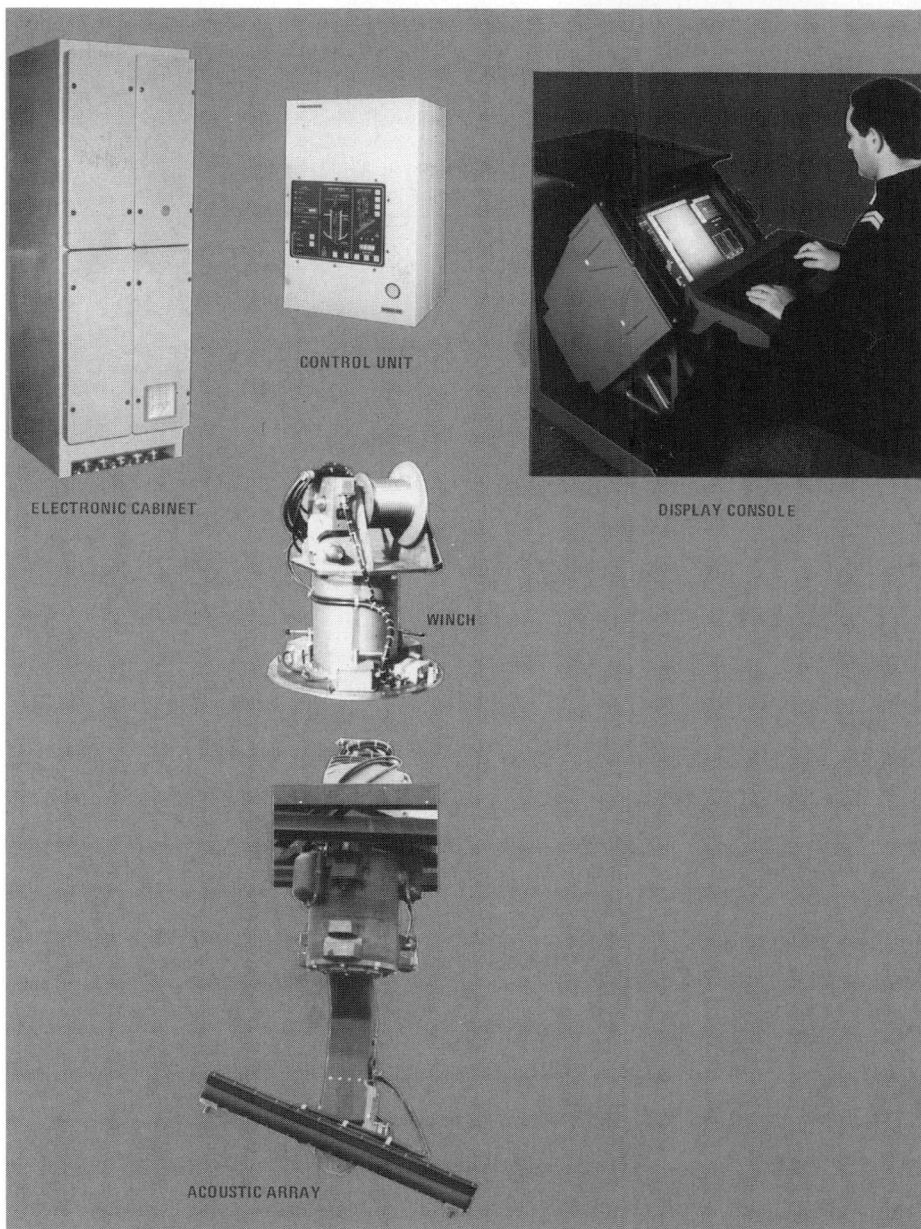

CONTROL UNIT

ELECTRONIC CABINET

DISPLAY CONSOLE

WINCH

ACOUSTIC ARRAY

Elements of the TSM 2022 Mk II minehunting sonar system, with acoustic array (far right)

TSM 2023 sonar

Type

Hull-mounted minehunting sonar.

Description

The TSM 2023 has been developed by Thomson Marconi Sonar SAS in collaboration with Simrad Subsea, with Thomson Marconi Sonar SAS as the main contractor and Simrad Subsea as subcontractor being responsible for the detection sonar.

The TSM 2023 performs mine detection and classification simultaneously, and comprises:

(a) a detection sonar which covers a bearing sector of 90° electronically steerable through 360°

(b) a classification sonar which can perform shadow analysis at three different frequencies. The three different frequencies allow the beam resolution and bearing sector to be adjusted according to the size of the detected object

(c) two sonar consoles, one for detection, one for classification, equipped with Computer-Aided Detection (CAD) and Computer-Aided Classification (CAC) processing and a performance indicator.

Operational status

The TSM 2023N was ordered by the Norwegian Navy in 1990 and four systems are in service aboard the 'Oksoy' class minehunters.

Contractor

Thomson Marconi Sonar SAS, Sophia Antipolis.

VERIFIED

GERMANY

DSQS-11 M

Type

Hull-mounted minehunting sonar.

Description

The DSQS-11 M minehunting sonar is a high-definition, dual-triple-frequency sonar which forms an integral part of the STN ATLAS Minehunting Weapon System MWS 80 and MWS 90

The sonar performs:

(a) detection and classification of all types of ground, short tethered and moored mines

(b) high-clearance performance by simultaneous detection and classification within a 90° sector

(c) simultaneous three-dimensional precision target localisation

(d) Computer-Aided Detection (CAD) and Classification (CAC)

(e) automatic tracking of stationary and moving targets

(f) integrated sonar performance prediction function

(g) integrated route survey mode (side scan function)

(h) video and data recording and playback

(i) comprehensive Built-In Test Equipment (BITE) to ensure system availability.

The system comprises an operator console, an electronics cabinet, the hoisting unit (with transducer arrays, signal processing units and stabilisation unit) and the hydraulics unit.

The DSQS-11 M console includes two high-resolution raster scan displays and keypads for direct system operation. Additionally, the operation is supported by easily accessible menu guidance on both graphic displays.

Operational status

The DSQS-11 M is in operational service in the German Navy's 'Frankenthal' class vessels and in the Royal Australian Navy's 'Bay' class catamarans. The 'Hameln' class of the German Navy are also being upgraded with a full DSQS-11 M system. The DSQS-11 M is also being installed in the Thai Navy's 'Lat Ya' class MCMVs as part of the MWS 80-6 minehunting system.

Contractor

STN ATLAS Elektronik GmbH, Bremen.

UPDATED

NORWAY

SA950 sonar

Type

Mine detection and avoidance sonar system.

Description

The Simrad SA950 is an active, high-resolution, multibeam sector scanning sonar designed for fast and accurate detection of moored and bottom mines. The design of the sonar ensures simple and accurate operation, a high degree of maintainability, and low size and weight.

The system consists of a hull unit (with transducer array), transceiver unit, servo unit, and an operator/display unit. Detection is achieved using 95 kHz frequency combined with high-source level and CW or FM signal processing. Transmission is sectoral, covering a sector of 45 or 60° with 32 beams of 1.7 or 1.9° each with a vertical beamwidth of 10°. The sector is mechanically trainable and tiltable covering ±190° in the horizontal plane and from +10 to −90° in the vertical plane. The centre beam is stabilised against pitch and roll.

The signal processing is performed in a fast digital system using the full-dynamic range of the echo information. The receiver has time variable gain and automatic gain control on the preamplifiers, background normalisation after the beam-forming, and ping-to-ping correlation.

The echoes are presented in mono or 64 colour high resolution on a 20 in monitor. There is sector presentation and time/range (echogram) presentation. In addition, PPI, relative or true motion is selectable and target or position can be tracked. The system is controlled by a single operator using designated hardware switches, joysticks and trackerball on the operator unit front panel and interactive with a menu on the display.

The SA950 Mk 2 can be used as a stand-alone system or form part of the MICOS integrated combat information system. It can also be used in combination with Simrad towed array sonars.

The sonar interfaces to Doppler log, gyro, radio position system, VRU, CTD sensor and tactical system.

Operational status

In January 1989 Simrad received an order from the Turkish MoD for SA950 systems to be fitted to six exGerman Navy minesweepers.

The five 'Alta' class minesweepers of the Norwegian Navy are fitted with the SA950 as part of their MICOS system. The SA950 also equips a number of the 'Hugin/Kaparen' class. The 'Stockholm' and 'Goteborg' class corvettes of the Swedish Navy are also fitted with the sonar.

Contractor

Simrad AS, Horten.

UPDATED

UNITED KINGDOM

Type 162M sonar

Type

Side-looking target detection sonar.

Description

Sonar Type 162M, detects and classifies both mid-water and seabed targets. It displays port and starboard recordings simultaneously on a single straight-line recorder, which has a maximum range scale of 1,200 yd. Operation is simplified by entirely automatic gain control. Reliability is enhanced with solid-state technology and maintenance is facilitated by ease of access and comprehensive built-in testing and monitoring features.

The three transducers are all similar and employ 49.8 kHz barium titanate elements. Their beam pattern is fan shaped, about 3° wide and 40° vertical angle; the side-looking elements have their axes 25° below the horizontal.

The recorder design uses a double helix (left and right hand) so that there are two points of contact with the moist electro-sensitive paper. As the helix rotates, the two points of contact move outwards from the centre. Port and starboard signals are fed respectively to the left- and right-hand points of contact so that they are recorded simultaneously. There are two zero lines near the middle of the paper, and the 18 mm gap between them is used for time marks every 5 minutes.

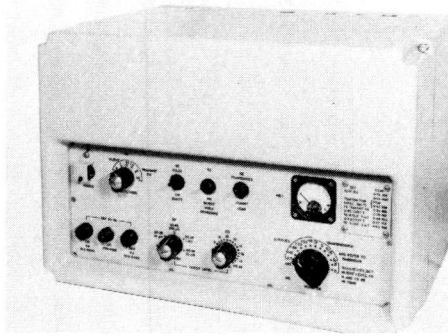

Type 162M sonar transmitter/receiver unit

The paper width is 286 mm, and each trace occupies 133 mm.

An electronic oscillator provides a controlled frequency supply for the driving motor and interval marks. Motor speed changes, for the three range scales, are made by frequency division so as to avoid the use of change speed gearbox, and a stroboscopic arrangement is included so that the helix speed may be easily checked.

A loudspeaker and a socket for headphones enable signals to be monitored aurally if required.

The three range scales are 0 to 300, 0 to 600 and 0 to 1,200 yds, and accuracy is better than two per cent assuming a sound velocity of 4,920 ft/s. The paper speed changes automatically when the range scale is selected, and the speeds are 6, 3 and 1.5 in/min (152, 76 and 38 mm/min) respectively. At a ship speed of 10.8 kt the display scales are the same across the paper and vertically. A take-up spool is fitted but its use is optional. A fix marker draws a line across the width of the paper when a button is pressed, and it can be operated in conjunction with a Type 778 echo-sounder or other equipment.

A new transducer array has been developed as a direct replacement unit. The new array incorporates redesigned elements and techniques using a low-voltage device which virtually eliminates insulation problems. Tuning boxes for the three transducers are no longer required since the transducer tuning coils are incorporated into the individual housing gland.

Operational status

In service in a number of ex-Royal Navy ships in different navies.

Contractor

Kelvin Hughes, Hainault.

UPDATED

Type 193M/193M Mod 1 sonar

Type

Sector scanning minehunting sonar.

Development

Development of the Type 193M was undertaken by Plessey (subsequently GEC-Marconi Naval Systems and now Thomson Marconi Sonar) in 1968. Since then a continuous programme of development and operational trials in conjunction with the UK MoD has resulted in further system improvements defined as the Type 193M Mod 1.

The system offers greatly improved performance in detection and classification by the use of digitally processed video and display systems, together with computer-aided target classification.

Description

The 193M is a short-range, high-definition sector scanning sonar, operating at two frequencies, providing detection and classification of both moored and ground mines. It has been developed from the Type 193 minehunting sonar; the adoption of solid-state electronics and other advances in technology has resulted in a reduction in installed weight to about 860 kg, which compares with a figure of some 2,100 kg for the older Type 193. Among the operational improvements are extensive use of digital displays and facilities for interfacing with computer systems.

The basic concept is to employ high-definition sonar to locate and classify mines which are then destroyed by explosive charges.

The sonar provides both bearing and range data; the fine range and bearing resolution enables the operator to assess accurately the shape of a target and hence its nature. Since the resolution depends both on operating frequency and pulse length, two selectable frequencies are provided, with a choice of pulse lengths in the classification mode.

Range and bearing data appear on two displays: in the search mode one display shows the total range covered; in the classification mode, a 27 m (30 yd) section of the search display is expanded to fill the second display screen and permit close examination of the target.

Two frequencies are employed by the 193M sonar: 100 kHz for long-range search and 300 kHz for short-range search and classification. The transducers for these signals are carried beneath the ship on a stabilised, steerable mounting, the whole assembly contained within a dome. A choice of either inflated fabric or GRP material is provided for the dome. The receiver uses a modulation scanning technique with 15 beams, 1° wide in LF and 0.33° wide at HF, giving azimuth coverage of 15 and 6° respectively. The searching of wider areas of the sea bottom is achieved by means of an automatic search sequence selected in accordance with the type of sea bottom. The returned echoes are presented to the operators on separate CRTs at the control console. One of these displays range and bearing of targets within the sector being scanned by the search transducer, while the other is used for expanded range presentation in search mode and also the presentation of the classification channel data. The controls for adjusting the transducer position, signal parameters such as frequency and pulse length and for co-ordinating with the rest of the minehunting and destruction systems, are also provided at the console. Type 193M data can also be fed into other ship systems.

The Speedscan system can be fitted to the Type 193M to allow it to be operated in the side scan mode and to generate a hard-copy printout of the seabed.

Range and bearing data outputs are provided in synchro and digital forms, allowing the sonar to be interfaced with various plotting tables, action information systems and remote indicators. The system is also fully compatible with the NAUTIS-M command system as chosen for the Royal Navy single-role minehunter.

Operational status
In service with the Royal Navy and many other navies, the Type 193M/193M Mod 1 is suitable for installation in all current minehunters.

Type 193M is fitted in the Royal Navy's 'Hunt' class MCMVs. In total, 12 navies have now purchased some 50 equipments, including both the 193M and the earlier version, 193.

The Type 193M Mod 1 enhancements have been retrofitted as a modification to RN ships fitted with 193M, while complete Type 193M Mod 1 sets are now being exported.

Additions to the system are a surface and near-surface mine detection mode, which also provides a mine avoidance capability while minesweeping.

Contractor
Thomson Marconi Sonar Ltd, Templecombe.

VERIFIED

Type 193M Mod 1 sonar operator's console

MS 58 minehunting sonar

Type
Hull-mounted minehunting sonar.

Description
The MS 58 is a new low-cost, lightweight hull-mounted dual-frequency minehunting sonar. Use of modern high-speed processing modules has created a flexible system architecture which allows the system to be offered in a number of forms:
(a) mine avoidance sonar for minesweepers or non-MCM vessels
(b) standard single-operator minehunting sonar
(c) standard dual-operator minehunting sonar.

All minehunting variants have facilities for search and classification of ground mines, and special array and processing features for detection and classification of moored mines, and surface and near-surface mines. A full route survey capability is incorporated as a standard feature.

A modern lightweight directing gear provides good platform stability characteristics with small hull aperture requirements.

MS 58 also offers a number of optional facilities. These include computer-aided detection, and multiping processing and computer-aided classification for the minehunting variants.

Operational status
The system has successfully undergone minehunting trials.

Contractor
Thomson Marconi Sonar Ltd, Templecombe.

VERIFIED

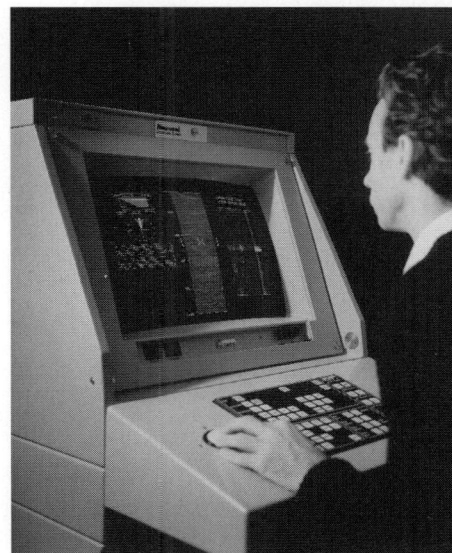

MS 58 operator console

Sea Scout

Type
Shallow water minehunting/avoidance sonar.

Description
Sea Scout is a lightweight, portable, rapidly deployed mine avoidance sonar which can also be used for detection and classification of seabed ordnance in shallow water conditions. It is designed to operate from small craft of opportunity offering rapid response in high-threat areas.

The sonar is based on Thomson Marconi Sonar Ltd's family of high-frequency mirror sonars, and carries out a continuous search of the seabed, detecting and classifying objects in the water column or on the seabed. Contact range and bearing are relayed to any command system fitted on the parent platform to geographically position contacts for further investigation.

The mirror array offers optimum in-water beam-forming with a minimal number of components. It also allows the system to be used at greater speeds than conventional MCM systems. The mirror provides a fixed-azimuthal field of view of 20° with a narrow azimuthal beam that is electronically scanned over the full field of view to provide an azimuthal resolution of nominally 0.6°. Sea Scout is capable of being mechanically scanned ±30° to provide an overall azimuthal field of view of at least 80°. The system has a fixed-vertical field of view of ±5° selectable within the total vertical range of +10° up to −45°, this being set as a function of the operational mode. When not in use the array is simply and rapidly removed from the bow and stowed on board.

The system is controlled using a simple MMI. In mine avoidance mode the operator sets the centre of the vertical field of view nominally around the sea surface to a specific depth in the water column, for the detection and classification of contact mines. In seabed search mode, the operator simply sets the depression of the sonar system to provide optimum cover of the seabed as a function of the water depth. The array can also be turned through 90° in order to provide a height-finding capability. In all operational modes, target data can be transferred automatically to the command system by the operator positioning a cursor over the detected target and pressing the 'mark' button.

Operational status
In service with US Explosive Ordnance Disposal teams.

Specifications
Frequency: 250 kHz
Azimuthal coverage: 20° fixed, 80° scanned
Azimuthal resolution: nominal 0.6°
Range: >300 m
Weight: 26 kg (in water)

Contractor
Thomson Marconi Sonar Ltd, Templecombe.

VERIFIED

PMS 75 Speedscan/Sonar 2048

Type
Side scan modification for minehunting sonars.

Description
The PMS 75 Speedscan is a self-contained equipment for ships already fitted with a compatible sonar system. It allows the sonar system to be operated in a side scan mode and generates a hard-copy printout of the seabed. It can be fitted to all types of minehunting sonar.

When the sonar is operating with Speedscan, its transducer is trained 90° to port or starboard of the ship's track. As the ship proceeds along a predetermined track, an area of seabed parallel to the track is interrogated by the sonar. The received sonar data and reference timing signals are fed to the Speedscan processor which forms a side scan beam and presents the processed data as hard-copy on continuous recording paper.

The Speedscan equipment consists of two main assemblies, each contained in a portable case. In use, the two cases are stacked and locked together with the end covers removed. The upper case houses the processor and the operator's control panel, the lower case the recorder unit.

Speedscan presents information on light-sensitive paper exposed in the recorder. On the record, the seabed is represented as a series of parallel narrow strips of selected length, perpendicular to the ship's track. The sonar data from the whole azimuth sector scanned by the sonar is extracted by Speedscan to represent a narrow strip of seabed along the sonar beam axis. This data is extracted by the beamformer to generate a parallel beam of sufficient width to overlap the strips covered by the two previous transmissions, so providing the three-ping sample of each point of the seabed along the range axis. The beam-forming system is programmed by ship's speed and sonar range to provide a controlled interrogation of the seabed. This information is used to intensify and modulate a light source in the recorder which produces a high-definition image of the seabed.

Operational status
The PMS 75 is designated Royal Navy sonar Type 2048 in mine countermeasures vessels. It is also used by other navies.

Contractor
Thomson Marconi Sonar Ltd, Templecombe.

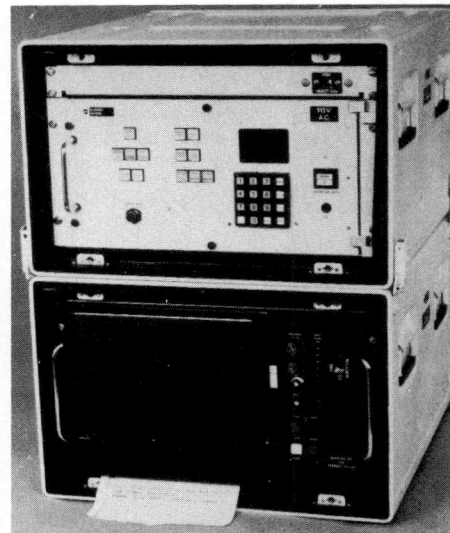
PMS 75 Speedscan

VERIFIED

UNITED STATES OF AMERICA

SH100 minehunting sonar

Type
Mine detection and classification sonar.

Description
The SH100 mine detection and classification sonar is an active, dual-frequency, high-resolution sonar designed for hull mounting on coastal or inshore minehunting or minesweeping vessels. The operational capability of this very compact sonar is from the amphibious landing zone out to depths of 100 m for detecting and classifying bottom mines and beyond that for detecting moored mines. The SH100 is an extension of the single-frequency SA950 mine avoidance sonar (see separate entry).

The transducer arrays are mounted on a fully retractable hull unit which is mechanically stabilised against roll, pitch and yaw. With the stabilised array, reliable detection and classification performance is maintained even in severe conditions.

The SH100 design uses hybrid preamplifier circuits to maximise performance and reliability while reducing the size, weight and cost of the total system. The signal processing, including beam-forming, is performed in a fast digital signal processing system using the full dynamic range of the signals.

The SH100 operates at 95 kHz (LF detector) and

335 kHz (HF classifier) and has a sectoral transmission and multibeam reception covering an LF sector of 45°, with 32 beams of 1.4°, and an HF sector of 16° with 64 beams of 0.25°. Vertical coverage is 10°. The sectors, which are aligned along the same axis, are mechanically trainable (±200°) and tiltable (+10 to −90°).

The selection of the two frequencies is optimised for the separate tasks of detection and classification. Extensive test and application results from a range of sea environments have proved this sonar's ability to consistently detect bottom and moored mines at a range of 600 m and classify at a range of 200 m. Mine detection with the SH100 has been demonstrated at ranges greater than 1,000 m.

The sonar can operate on both the LF detection and HF classification frequencies simultaneously. Signal enhancement and display techniques, such as shadow mode normalisation, FM pulse compression, ping-to-ping filtering, echogram detection and zoom, can be applied separately to each frequency. This allows the sonar operator to optimise system performance for the specific mission task. The following operational modes are available:
(a) wide sector search and target acquisition using only the LF mode

(b) high-resolution classification and precise localisation using only the HF mode
(c) combination of LF and HF on alternate transmissions
(d) vertical positioning of the transducer for moored mine classification and depth measurement.

The SH100 sonar consists of a hull unit, preamplifier, transceiver, servo control unit, servo transformer unit, and a control and display unit. The basic configuration includes software and hardware interfaces to Doppler speed log, gyrocompass, roll and pitch sensor, and surface navigation systems. Simrad can deliver the SH100 as a stand-alone system which includes these additional sensors and power units if required. Interfaces to buyer-supplied combat systems for transferring target data or the full sonar image data can be readily adapted.

The control and display console includes two high-resolution 20 in RGB monitors mounted one above the other. The lower monitor displays the sonar echo image while the upper one displays tactical information. The echo display presents the sonar image in B-scan and/or echogram modes. The tactical display presents a map-like image of the sonar situation including the ship symbol and past track, target positions and classifications, LF and HF sector coverage, and tilt and

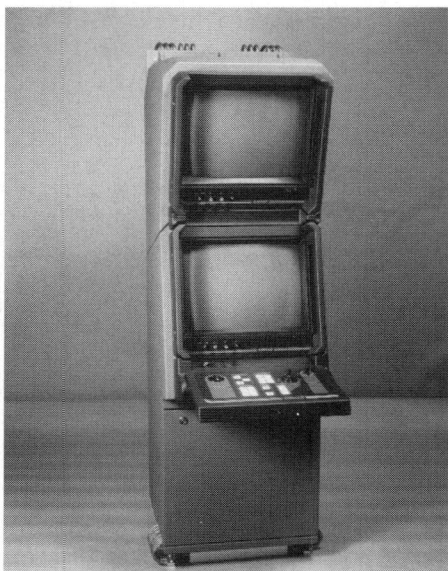
Control/display unit of the SH100 minehunting sonar

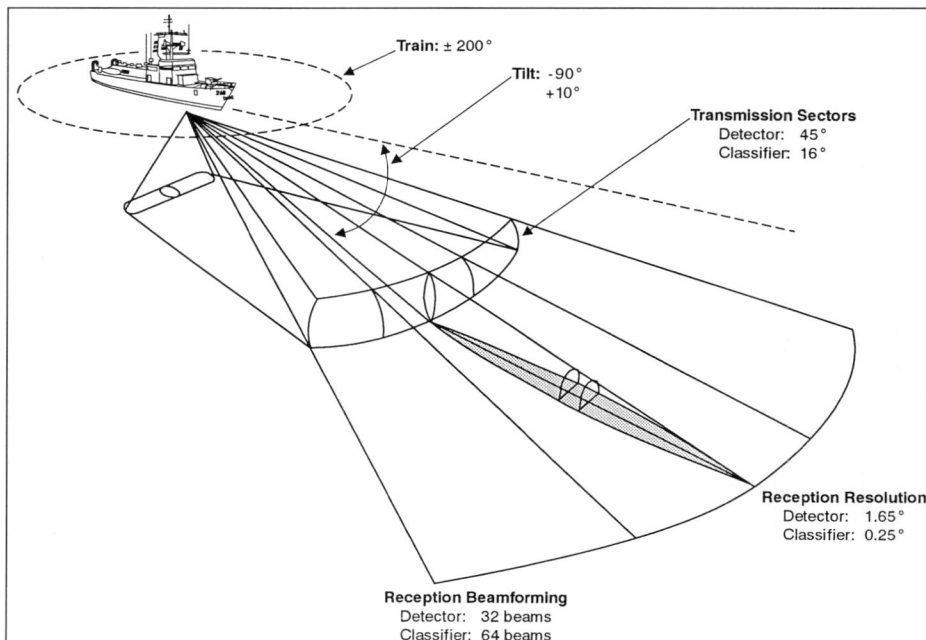
SH 100 mine detection and classification sonar

1995

train graphics. Additional tactical data displayed include the ship data such as position, course and speed, and target positions (both relative and geographical). The simple MMI provides access to all the primary controls with the buttons and joysticks on a control panel. Secondary commands are accessed through an on-screen menu.

Enhancements to the basic configuration are available providing data distribution to multiple locations on the vessel, recording of target database, sonar image printing, and more powerful tactical and navigational functions. The sonar can also integrate with other Simrad systems such as: vessel control system; Electronic Chart Display and Information Systems (ECDIS).

Operational status
Five systems in service.

Contractor
Kongsberg Simrad Inc, Lynnwood, Washington.

VERIFIED

VARIABLE DEPTH MINEHUNTING SONARS

INTERNATIONAL

AN/SQQ-32 sonar system

Type
Advanced minehunting search and classification sonar system.

Development
Technical and operational evaluation of the advanced minehunting sonar system (AN/SQQ-32), produced for the US Navy by Raytheon teamed with Thomson Sintra in France, has been completed. The system is installed in the US Navy's MCM ('Avenger') and MHC ('Osprey') class MCMVs.

Description
The system consists of two separate sonars: a search sonar for initial detection; and a high-frequency, high-resolution sonar for classification of the targets. The two sonars are partially housed in a hydrodynamically shaped vehicle and towed at various speeds by the minehunting ship. The latest technologies in beam-forming, signal processing, modular packaging and displays are used throughout the system.

The detection sonar, designed and manufactured by Raytheon, detects mines over a wide range of distances and bottom conditions. This system incorporates a computer-aided detection facility which is designed to help the sonar operator to discard non-mine objects detected by the sonar. Raytheon is also providing the computer facilities and display consoles and is responsible for overall system integration.

The classification sonar, designed and manufactured by Thomson Marconi Sonar SAS, is based on its experience with the DUBM 21 and TSM 2022 minehunting sonars. The classification sonar provides high-resolution transmission and reception of underwater signals to enable targets to be identified with near-picture quality.

The wet end, consists of a hydrodynamically shaped vehicle and a deployment/retrieval system, designed by Charles Stark Draper Laboratory of Cambridge, Massachusetts. The towed body is housed in a vertical trunk extending from keel to deck, just forward of the bridge. The body is roughly egg shaped, with a pair of boomerang shaped vane arms pivoted on either side. When stowed in the trunk, these lie alongside the body itself, once clear of the ship's keel they swing aft to stabilise the body as it is towed along in the direction normal to its axis. The body is slung from pivots on either side to a fork assembly that is connected to the towing cable, so that its axis remains vertical. Within it are the search sonar staves arranged as a 360° ring array belted around the body, together with immediate sonar electronics and associated equipment. Below the body is the rotating scanner for the classification sonar which can operate at three frequencies down to 300 kHz.

The AN/SQQ-32 consists of two identical operator consoles with high-resolution displays. Search and classification data can be displayed simultaneously or independently. A search display presents video patterns for the detection operator to examine, and computer-aided detection indicates objects likely to be mines. Search data is presented on six screen displays allowing the operator to determine if a mine-like object is a real target or a natural bottom feature. The system proceeds to long-range classification, with the second operator, by measuring the detected object's height above the seabed in order to assess the probability of it being a moored mine. The classification operator then uses the high-resolution classification sonar, which has very narrow beams with dynamic focusing, to examine objects detected on the bottom.

Automated detection and control functions are performed by two AN/UYK-44 computers.

Operational status
In February 1989, Raytheon received a US$46 million contract to build two production systems and to

AN/SQQ-32 operators' consoles

Simulation of the AN/SQQ-32 minehunting sonar system under way

refurbish two of the three full-scale development systems. In January 1990, Raytheon received a contract worth US$125 million for 11 systems (eight for the US Navy and three for Japan where it equips the 'Yaeyama' class).

In February 1992, Raytheon received a contract worth US$73 million for six systems, plus a set of shipboard electronic units to support shore-based training. In 1994-95 Raytheon was awarded the competitive follow-on production contract worth US$66 million for the balance of the systems required by the US Navy. AN/SQQ-32 production systems are installed

on six 'Avenger' class MCMVs. All 14 vessels in the class will be back-fitted with the system. The AN/SQQ-32 is being installed on all the new 'Osprey' class MHC vessels and will equip the new Spanish 'Segura' class.

Contractor
Raytheon Company, Electronic Systems, Portsmouth, Rhode Island.
Thomson Marconi Sonar SAS, Sophia Antipolis.

UPDATED

ITALY

SQQ-14IT

Type
Minehunting sonar.

SQQ-14IT minehunting sonar

Description
The aim of the SQQ-14IT programme has been to redesign the General Electric AN/SQQ-14 minehunting sonar, and completely new electronics have been developed for both the dry and wet ends. Only the mechanical parts of the old hoist system remain.

In line with the next-generation sonars, new functions such as real-time performance monitor, speed scan, memory display and remote display have been included, as well as Built-In Test Equipment (BITE) functions which provide self-diagnosis and the location of malfunctions down to the printed circuit cards.

The transfer of the transmitter unit into the towed body has provided an increase in the system's operational performance, as well as a redundancy of the principal functions, that is, console interchangeability and the possibility of operating with reduced efficiency in the case of malfunction in one or more of the transmitter sections. The adoption of the latest consoles for search and classification operations has greatly improved the man/machine interface, and simplified the use of the system and the training of operators. Recording and mission playback facilities are available, as well as training and simulation programmes.

Operational status
12 SQQ-14IT minehunting sonars are in service with the Italian Navy.

Contractor
Whitehead Alenia Sistemi Subacquei, Livorno.

UPDATED

UNITED KINGDOM

Type 2093 sonar

Type
Variable depth, multimode minehunting sonar.

Description
Designed by Thomson Marconi Sonar Ltd, Sonar 2093 is fitted in the 'Sandown' class single-role minehunter ships of the Royal Navy. The system is designed to operate in either hull-mount or variable depth mode, deployed through the ship's centre well, and will detect, localise and classify all current and future mine threats.

As the culmination of a six year development programme, the system is designed to carry out its mission at high speeds without compromising the ship's safety. A variable depth, dual-frequency search and classification capability enables the sonar to operate under all bottom and sea conditions at a range claimed to be twice that of hull-mounted sonar and with an increase of three times detection depth. The Type 2093 is integrated with the ship's command system to allow simultaneous operation in combinations of search, detection, classification and route survey modes. The area coverage rate capability will allow a significant improvement on current route survey operations.

The towed body of the 2093 is cylindrical with hemispherical ends. Stabilising vanes consist of fixed strips mounted a few centimetres away from the shell, largely parallel to its surface and curved so as to be concentric with its axis, and with the centres of the hemispherical ends over an arc of 45°. The sonar arrays within the body consist of two 360° rings for LF and VLF operation, hydrophones for HF and VHF reception, and separate HF and VHF projectors. Below is a cylindrical VLF projector and finally a VLF depth sounder. LF and VLF are used for search, survey and moored mine classification; HF and VHF for classification and survey.

Sonar 2093 employs the unique approach of multiple operating modes, which offer the MCM commander the choice of a number of operational configurations to match the prevailing threat and environment. These multiple operating modes comprise: VLF and LF search and moored mine classification, VLF and/or LF dual search, search and ground mine classification, also

Sonar Type 2093 consoles fitted in HMS Sandown

search either VLF or LF and route survey. Sonar 2093 can combine these modes to enable concurrent operation.

Operational status
In production, with 15 systems having been ordered for the Royal Navy and three systems for Saudi Arabia. The Type 2093 entered service with the Royal Navy in 1992.

A further six systems have been ordered by Australia for the 'Huon' class MCMVs and serial production is underway for the navies of South Korea and Japan.

Contractor
Thomson Marconi Sonar Ltd, Templecombe.

VERIFIED

UNITED STATES OF AMERICA

AN/SQQ-14 sonar

Type

Variable depth towed mine detection sonar.

Description

The AN/SQQ-14 system is essentially a variable depth, dual-frequency sonar for detecting and classifying bottom mines in shallow water. It utilises a towed body in the shape of an elongated sphere towed through a centre well on a US Navy minesweeper.

The AN/SQQ-14 can operate down to 45 m. It transmits at 80 and 350 kHz respectively for search and classification, scanning over azimuths of 100 and 18°, and through an elevation of 10° in each case. Azimuth and range resolutions are 1.5° and 1 m for search, and 0.3° and 8 cm for classification.

A unique aspect of the AN/SQQ-14 design is the towing cable. This consists of discrete, 18 in (457 mm) sections of articulated struts with universal joints at each section, permitting the cable to flex in any vertical plane, but restraining it from torsional motion. This configuration imparts a constant heading to the towed body, thereby eliminating the need for a gyroscopic heading reference system.

A rubber-jacketed electric cable containing 35 shielded, coaxial and individual conductors passes through the centre of the strut sections. The cable terminates in a slipring in the towed body and in a winch with compound winding drum in lieu of a slipring installed on the 01 deck of a mine countermeasures ship.

The articulated strut type of towcable posed unique problems related to winching, storage on the winch drum, and ship's tow point construction. Struts, of which there are 117 per system, are stored in a single layer on a cylindrical drum equipped with specially designed pads serving as contact points for the articulated strut knuckles. Since a conventional level wind mechanism could not be used with the struts, the winch drum itself is designed to traverse axially by

AN/SQQ-14 projectors and hydrophone arrays

means of an Acme lead screw and rotating nut, thereby accomplishing the level wind function and maintaining a fixed fleeting point. All equipment for this application must be non-magnetic. Struts, bearings and knuckles are fabricated of high-strength Inconel alloy. The winch drum is cast aluminium; the supporting structure is aluminium and stainless steel.

Operational status

In operational service.

Contractor

Lockheed Martin, Syracuse, New York.

UPDATED

AN/SQQ-30 sonar

Type

Variable depth towed mine detection sonar.

Description

The AN/SQQ-30 is the successor to the AN/SQQ-14 and consists of two sonars: a search sonar for initial detection; and a high-frequency, high-resolution sonar

for classification of targets. The two sonars are separated to give the area coverage needed. They are housed in a hydrodynamically shaped vehicle and towed at various speeds by the minehunter. The towed body is streamed from a well in the forward part of the ship, using a winch driving a 3 m diameter cable drum on the foredeck. Two display consoles are provided for the search and classification sonars.

Operational status

Fitted to the first eight US Navy MCM-1 'Avenger' class minehunters but being replaced by SQQ-32.

Contractor

Lockheed Martin, Syracuse, New York.

VERIFIED

SIDE SCAN MINEHUNTING ROUTE SURVEY SONARS

FRANCE

DUBM 41B (IBIS 41) sonar

Type
Side scan minehunting sonar system.

Development
Development of the DUBM 41B was conducted in collaboration between the French Navy Mine Warfare DCN (GESMA) and Thomson Marconi Sonar SAS.

Description
The TSM 2050 (service designation DUBM 41B) system is a high-resolution side scan sonar system designed for the location and classification of targets lying on the seabed. The system consists of three towed sonar vehicles, two consoles, the necessary cables, shipboard hoists and handling gear. In normal operation two of the three sonar bodies are used, with the third kept as a ready spare. A permanent record of the sonar data gathered is made on a facsimile type recorder and a tape recorder. The information is also presented simultaneously on two CRT storage tubes, these console-mounted displays providing an image of the seabed. The system is designed for use by low-tonnage vessels and in waters of 100 m or deeper. The towed sonars operate at about 5.5 to 7.5 m above the seabed at speeds of 2 to 6 kt. The towcables are provided with deflector vanes which ensure that the bodies are towed at a distance to left and right of the ship's track, and marker floats are provided to indicate the line of travel for each sonar.

The sonar body is a streamlined vehicle fitted with the side scan sonars, plus additional sonar transducers for determining height above seabed and for obstacle detection. Servo-controlled fins are provided for depth control and roll stabilisation is incorporated by means of a separate set of four fins. Dimensions of each sonar vehicle are: length 3.725 m, diameter 36 cm and in-air weight 340 kg. An acoustic pinger is installed as an aid to recovery should the sonar body break its tow. Manual controls are provided at the operating console to permit manual piloting, either to override automatic control or as a standby mode.

Each of the two sonar bodies operates on its own

Towed sonar DUBM 41B on manoeuvres

frequency, the two being separated by 50 kHz, with the area of seabed between them being scanned by both sonars. A typical scanned area covered by the two bodies amounts to a total width of about 200 m, with maximum sonar range selected. There are two other range settings, 50 and 25 m, and the latter setting resolution is stated to be better than 5×10 cm in the lateral (scan) direction and in the line of travel.

Operational status
In service.

Thomson Marconi Sonar SAS was under contract to GESMA to carry out a feasibility study of an experimental panoramic sonar for the detection of buried mines, as part of an international research programme involving France, Netherlands and the UK. The contract was concluded in 1992 following trials of a

derivative of the DUBM 41 aboard the Netherlands Navy hydrographic vessel *Tydeman*.

The sonar and a navigation system are derived from the DUBM 41B towed sonar, and the new towed unit will house non-linear acoustic arrays and part of the processing system. Thomson Marconi Sonar is responsible for the towed arrays and transmission circuits, as well as the interface and display systems, and British Aerospace (Dynamics) Limited for the receiver section.

The DUBM 41B is deployed from the French Navy 'Antares' class route survey vessels.

Contractor
Thomson Marconi Sonar SAS, Sophia Antipolis.

UPDATED

DUBM 42 (IBIS 42) sonar

Type
Side scan mine detection sonar system.

IBIS 42 sonar

Development
Development of the DUBM 42 was carried out jointly by the French Navy (DCN Research Establishment specialised in mine warfare – GESMA) and Thomson Marconi Sonar SAS in association with Société ECA.

Description
The DUBM 42 is a side scan sonar able to detect and classify mines at ranges up to 200 m, with an area coverage of 2,000 m²/s. The system is designed for use at an operational speed of 10 kt and a height above the seabed of 30 m.

The basic system consists of one vehicle with three sonars (two lateral and one front-head) on the sonar body, one Colibri sonar display, one Colibri tactical display and one high-density magnetic recorder. The data collected by the three sonars are recorded for analysis in a land-based processing centre. In addition, a visual display of the seabed is provided on the operator's console. A magnifier provides a more accurate classification. All navigation functions such as piloting and automatic control are integrated on the same console.

The two lateral and single front-head sonars operate on their own individual frequency, the swept channel being 400 m wide.

Operational status
In production. IBIS 42 is the export nomenclature. In service aboard French Navy Survey vessels of the 'Laperouse' class, and the *Thetis*.

Contractor
Thomson Marconi Sonar SAS., Sophia Antipolis.
ECA, Boulogne-Billancourt.

UPDATED

IBIS 43

Type
Mine surveillance system.

Description
IBIS 43 consists of the TSM 2054 multibeam side scan sonar and the TSM 2061 tactical system. Baseline configuration data collected is analysed offline in a shore centre. In Mk II the configuration data is analysed in real time on board. To carry this out, the IBIS 43 Mk II comprises:
(a) TSM 2054 multibeam side scan sonar
(b) TSM 2061 tactical system
(c) detection and classification sonar console
(d) database
(e) CAD and CAC sonar processing.

Operational status
In service with the MCM modules of the Danish 'Flyvefisken' class.

Contractor
Thomson Marconi Sonar SAS, Sophia Antipolis.

UPDATED

TSM 2054 sonar

Type
Side scan multibeam minehunting sonar.

Description
The TSM 2054C (IBIS 43) is a high-resolution side scan sonar derived from the DUBM 42. The detection and classification of mine-like objects is performed at a maximum speed of 15 kt, and the swept channel width is 200 m.

The towfish sonar vehicle is 3 m long and its weight is less than 300 kg. The fish operates at depths down to 200 m and carries an obstacle avoidance sonar that allows it to 'overfly' 10 m high obstacles at its maximum towing speed. It is also equipped with a Doppler sonar for accurate speed measurement. Automatic correction of speed, yaw and pitch, and the precise location of the towed body, is computed aboard the ship to an accuracy of 9 m and less than 1 m in altitude.

Two operating concepts are available:
(a) data collected by the sonar are recorded for analysis in a land-based processing centre
(b) detection and classification of new mine-like contacts is carried out on board ship.

The system is based on a multifunction Colibri display console and is used for tactical detection and classification operations as the IBIS 43.

Operational status
The Royal Danish Navy has ordered a number of systems. A contract for a third batch of systems was placed in October 1996. The award covers two systems: four 2054 sonars and two real-time processing units.

Specifications
Manual or automatic towfish navigation at constant depth or altitude

The TSM 2054 side scan sonar towfish

Water depth: 6 to more than 200 m
Altitude in relation to the seabed: 3-15 m, programmable from the sonar console keyboard. Automatic, high-precision terrain-following of irregular seabeds with gradients of more than 15%
Obstacle avoidance capability: obstacles up to 10 m high at maximum speed
Speed in relation to water: 4-15 kt
Sonar range: 2×50 m; 2×100 m

Resolution: 2×10 cm or 2×20 cm, according to sonar range
Coverage: up to 5.4 km²/h

Contractor
Thomson Marconi Sonar SAS, Sophia Antipolis.

VERIFIED

UNITED KINGDOM

Type 3010/T (Minescan) sonar

Type
Mine detection and route surveillance sonar.

Description
Minescan (3010/T) is a variant of the 3010 sonar and is a low-cost route surveillance and mine detection side scan which is particularly suited to Craft Of Opportunity (COOP) applications, and for use by reserve or regular forces. The 3010/T version has a greater range and includes the Ferranti Trackpoint II towfish tracking system.

Key elements of the system are:
(a) heavy-duty, dual-frequency towfish, 100/325 kHz, operable to 300 m depth, with optional depressor
(b) dual-cabinet transportable system in shock-mounted, splashproof units
(c) single-cabinet system for permanent installation applications
(d) colour video display with image enhancement and target marking facilities
(e) range selection – 50, 100, 200 and 400 m port and starboard; total swathe width up to 800 m
(f) image correction – slant range and speed over ground
(g) high-density sonar data recorder with random access target image recall

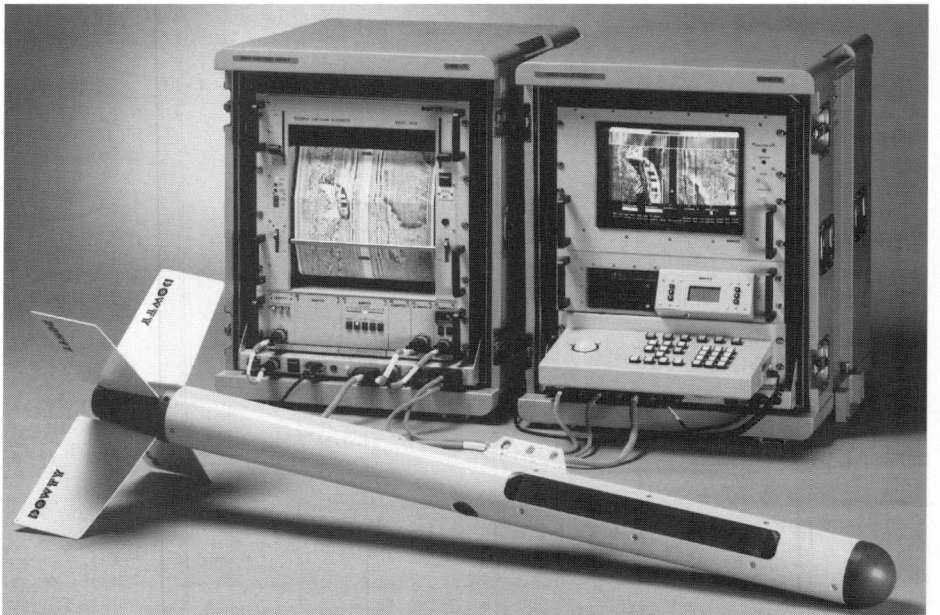

Minescan sonar in the dual-cabinet configuration for the US Navy COOP programme

(h) high-resolution thermal linescan recorder
(i) interfacing to integrated navigation processors.
    The manufacturer states that other elements of the complete COOP mine countermeasures, such as the winch, towfish tracking system, navigation aids, ROV, and shore planning and analysis, can be supplied as a turnkey operation.

Operational status
The Type 3010/T and subsequent variants, including the Type 3020 and 3030, are in service with several navies for route survey, route surveillance and hydrographic survey. More than 50 systems are in use with the US Navy, Royal Navy, navies in Asia and the Dutch Rijkwaterstaat organisation.

Contractor
Ultra Electronics Ocean Systems, Weymouth.

VERIFIED

UNITED STATES OF AMERICA

AN/AQS-14 minehunting sonar

Type
Helicopter or surface towed side scan sonar.

Description
This is a helicopter or surface towed side-looking multibeam sonar, with electronic beam-forming, all-range focusing and an adaptive processor, intended for use with the MH-53E Sea Dragon helicopters used by the US Navy for mine countermeasures. The US Navy has also operated the AN/AQS-14 from surface craft, remotely controlled drones and air cushion vehicles. The underwater vehicle is 3 m in length and has an active control system which enables it to be run at a fixed height above the seabed, or at a fixed depth beneath the water surface as chosen by the operator in the helicopter. The towcable is armoured and is non-magnetic.
    The sonar can scan up or down to picture objects moored above, or lying on the seabed. Two sonars are mounted to scan to port and starboard along the axis of the body.
    Controls in the helicopter (now updated to the Mod A configuration) include: a high-resolution monitor waterfall sonar display; underwater vehicle controls and status displays; system status indicators; and a magnetic tape recorder for sonar data recording. The system functions include all that is needed to locate, classify, mark, permanently record and review records of mines, mine-like objects and underwater features in the search area. Because the sonar employs multibeam techniques, a rapid search speed is possible. At lower tow speeds the towed sonar automatically reduces the

number of transmitted beams. The system is adaptable to surface platforms, especially hovercraft and remotely operated drone vessels for route survey missions.
    In 1993, the US Navy contracted three system enhancements. The first is a Post-Mission Analysis (PMA) system, which allows for portable side scan sonar tactical analysis. Designed primarily to support mine countermeasures missions, the PMA system incorporates computer-aided detection/classification of mines and mine-like objects, and automatic target logging. The PMA systems have been delivered. The second enhancement is a realistic AN/AQS-14 operator training system. The trainer utilises interactive simulation to train sonar operators, and evaluates each student's performance when the simulation is complete. Three AN/AQS-14 training systems have been delivered. Under the third enhancement Northrop Grumman developed, built, tested and delivered eight Mod A upgrades to the AN/AQS-14 system. Mod A updated all the helicopter-borne displays and vehicle control electronics (hardware and software).
    The AQS-14A also incorporates a digital recorder-reproducer, high-resolution 19 in colour video monitor and a new navigation and acoustic control processor which uses COTS boards to provide the operator with accurate, real-time target position.
    The AQS-14B features an improved and digitised wet end together with additional sensors (SM 2000 laser linescan) and a bottom penetrating sonar. The AQS-14B offers twice the resolution of the 14A and can operate at twice the depth. A new volume mine location facility provides the geographic position of moored mines.

A quadrature amplitude modulation telemetry system allows a coaxial tow cable to carry more than 40 MHz of signal bandwidth over lengths of 600 m or more.
    Also, Ocean Systems recently performed a feasibility demonstration of a mobile operations command and datalink from the MH-53C helicopter to a ground station. The US Navy has data linked to the sonar for operation for remotely operated vehicles.
    The US Navy plans to use the AN/AQS-14 sonar vehicle as the sensor for the Remote Minehunting System (V) 3, which is under development. First introduced in 1981, the AQS-14 was used successfully towards the end of Desert Storm (the Gulf War) in 1991.

Operational status
Introduced in 1984. Upgrades in production for the US Navy MH-53E ASW helicopters with over 30 systems delivered. Eight AN/AQS-14A systems were delivered to the US Navy in 1995 and additional systems are on order.
    Northrop Grumman has incorporated additional upgrades to the sonar towfish to improve performance, including adding its SM 2000 laser line scan into the system to provide paritive target identification.

Contractor
Northrop Grumman Corporation, Electronic Sensors and Systems Division, Oceanic Systems, Annapolis, Maryland.

UPDATED

AN/AQS-20

Type
Helicopter towed side scan sonar.

Description
The AN/AQS-20 has completed engineering and development and is ready for production. The system is a replacement for the AN/AQS-14, and features improved resolution and greater range. The system is designed to be towed at high speed by MH-53E Sea Dragon MCM helicopters of the US Navy. The towfish is about 5 m long, compared with the 3 m of the AQS-14. The AQS-20 is designed to provide total coverage of the seabed unlike the AQS-14, which leaves a gap

immediately beneath the tow body which then has to be covered by an additional pass. In addition to the two standard side-looking sonars, the AQS-20 incorporates a 'gap-filler' sonar which scans the seabed immediately below the towfish; a forward looking sonar; and a volumetric sonar which provides coverage upwards and around the towfish in a near 360° arc.
    In operation the AQS-20 towfish is linked by a fibre optic cable to the processing suite located in the towing helicopter. As the surveillance data is transmitted back to the aircraft it is processed using automatic detection algorithms and presented on a display screen and monitored by customised software to identify mine-like objects. Sonar contacts can be recorded and played back during mission analysis.

Operational status
Two engineering and manufacturing development models have been successfully tested by the US Navy. They will be subject to operational and technical evaluation prior to a production decision in late 1999. Delivery of the initial production system is anticipated for 2002.

Contractor
Raytheon Electronics Systems, Naval Systems Group, Portsmouth, Rhode Island.

UPDATED

MHSSS

Type
Mine Hunting Side Scan Sonar (MHSSS).

Description
The MHSSS is a high-resolution side scanning sonar for mapping and photographing the ocean bottom primarily for the detection of mines. It produces high-resolution, photo quality, and motion compensated real-time images of the bottom at tow speeds up to 10 kt and depths to 200 m. It operates in three range modes up to 200 m on each side with resolutions of 12.5 cm both along and across track. The towfish is actively stabilised and operates in either constant

depth or terrain-following modes. It is accompanied with a motion-compensated towfish-handling device that deploys and retrieves the towfish and dynamically decouples it from the ship's motion. The system is packaged in an ISO container for transportation and can be installed in less than four hours.

Operational status
On order for the Canadian Navy. The first unit was delivered in late 1998.

Specifications
Length: 3.4 m
Weight: 13,550 kg
Search rate: 1.8 n miles^2/h

Resolution: 12.5 cm (both along and across track)
Shadow contrast ratio: >12 dB
Range: 200 m (max on each side)
Tow speed: 10 kt (max)
Operating depth: 200 m

Contractor
L-3 Communications, Sylmar, California.

NEW ENTRY

5952 mine countermeasures sonar system

Type
Dual-beam, high-resolution side scan sonar using simultaneous dual-frequency insonification of the target for multispectral data collection and processing.

Description
The Type 5952 sonar system uses the latest technology available in lightweight side scan sonar systems to provide the highest resolution and target discrimination for the detection of current-generation bottom mines. The simultaneous dual-frequency insonification technique is used for target discrimination because targets reflect various frequencies in different manners, thereby providing the operator with the opportunity to observe the reflectivity of difficult targets with both high and very high frequencies.

The returns of both frequencies are displayed simultaneously on either hard-copy, very high-resolution thermal paper or on a high-resolution Video Display Unit (VDU). The sonar operator has a direct, real-time visual comparison of the target returns for optimum evaluation and target selection.

The sonar system uses a Variable Depth Sonar (VDS) transducer which can be selectively deployed to avoid interference from thermoclines. Combined 100/500 kHz frequency for simultaneous insonification of target areas, and 3.5 kHz for penetration of the sea bottom are available for use in the determination of bottom hardness for use in the assessment of a buried mine threat.

The sonar images are transmitted up the VDS towcable to a combined sonar transceiver and graphic recorder for processing and display on high-resolution thermal graph paper. Alternatively, the combined sonar image can be displayed on a high-resolution digital VDU. Performance enhancing accessories are available which permit advanced image processing of the sonar data, as well as data reduction. Fully integrated sonar systems, which interface with navigation and shipboard equipment, are also available.

The sonar system is lightweight, weighing less than 100 kg and is easily transported and installed aboard vessels of various sizes down to 5 m in length. The VDS transducer weighs less than 25 kg when operated in the simultaneous dual-frequency mode and is easily deployed by a single person without specialised handling equipment.

A full range of accessories is available, including towcables, towing winches, as well as various related systems including acoustic positioning, surface navigation and data management equipment.

Operational status
The 5952 is currently in production and in service with various navies worldwide for MCM applications.

Contractor
Klein Associates Inc, Salem, New Hampshire.

Elements of the 5952 side scan sonar system   *1994*   **VERIFIED**

System 2000

Type
Side scan MCM sonar system.

Description
System 2000 is a simultaneous dual-beam (100 kHz and 500 kHz) side scan sonar system which includes integrated target analysis and data storage for MCM applications. Using the latest digital technology it provides the highest resolution and target discrimination for the detection of current-generation bottom mines. The simultaneous dual-frequency insonification technique is very beneficial against modern mines that employ shapes and materials to make detection by sonar difficult.

The system has both a very high-resolution (300 dpi) hard-copy thermal printer and high-resolution video display. The sonar returns of both frequencies are displayed simultaneously on the hard-copy printer along with relevant navigation and status information. The high-resolution video displays the real-time sonar data and through the use of the integrated trackerball, true target zooms, mensuration, and position can be made. This target information/position can be logged off to a file on the digital tape storage unit, as well as output to an external printer. Selectable and programmable colour palettes are available to enhance bottom features and targets.

System 2000 uses a variable depth sonar deployed via a single coaxial towcable. The fully digital towed sensor with integral dual-frequency transducers features a built-in multiplexer and provides 12-bit data resolution. Programmable pulse length, tone burst transmitters coupled with a very low noise front end maximise range performance and noise immunity. This

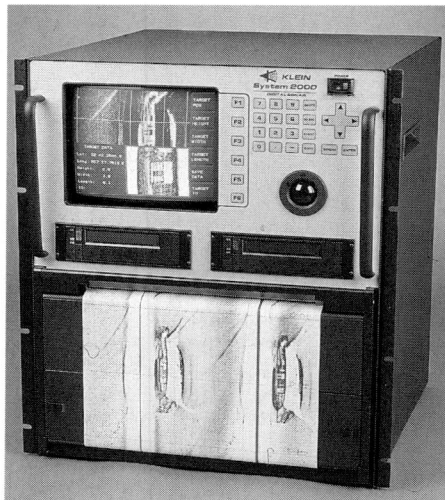

Sonar control, display and thermal printout of System 2000 (Klein Associates)   *1998*/0007416

design architecture results in a system performance which is suitable for very shallow water MCM applications as well as deeper operating depths. Optional sensors are available such as heading, depth, roll, pitch, responder and so on.

The lightweight sonar system weighs less than 40 kg for the surface processor displays and is easily transported and installed aboard vessels of various sizes down to 7 m. The VDS sensor weighs less than 25 kg in the simultaneous dual-frequency

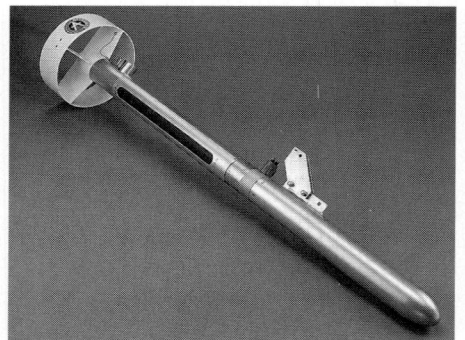

System 2000 towfish Model 2260NU fully instrumented with heading, pitch, roll, depth, temperature and responder (Klein Associates)   *1998*/0007417

configuration, and is easily deployed by a single person without specialised handling equipment.

A full range of accessories is available including towcables, winches, depressors, as well as related systems including acoustic positioning, surface navigation and data management equipment.

Operational status
System 2000 is currently in production and in service with navies worldwide for MCM applications.

Contractor
Klein Associates, Salem, New Hampshire.

VERIFIED

SELF-PROPELLED SONAR SYSTEMS

The task of mine countermeasures is becoming increasingly difficult as mines become more sophisticated. It is now recognised that when minehunting, MCMVs must remain well outside the danger zone of any potential mine. As the safety and efficiency of minehunting platforms is now regarded as a high priority, then the sensors used to detect mines must realise greatly improved performance characteristics. One of the technologies now being examined by a number of navies involves deploying a remotely operated vehicle fitted with a highly sensitive minehunting sonar some considerable distance ahead of the parent MCMV platform frequently referred to as the 'dog-on-lead' principle.

Ever since the Second World War, shallow water minehunting has been carried out using hull-mounted minehunting sonars mounted on the MCMV. These sonars were designed to scan ahead of the vessels searching for, detecting and classifying objects at what was then considered to be a safe distance ahead of the ship. However, because of the position of the sonar in relation to the seabed (the angle of incidence of insonification) and the problems of acoustic propagation near to the surface where thermal layers can create barriers to the sound propagation, together with the effects of reverberation in shallow water and a high ambient noise factor, these sonars were really only effective in depths down to about 100 m.

To overcome some of the problems associated with hull-mounted sonars, and in particular for operation in deeper waters where mines began to be laid, the Variable Depth Sonar (VDS) was developed. This allowed the sonar to penetrate the inhibiting thermal layers, and to be placed much closer to the level of the potential target. However, the VDS is limited by the fact that being a towed body it necessarily trails in the water, a characteristic which increases with the vessel's speed and the depth of immersion of the body. The towed body could therefore finish up quite some distance behind the towing platform, placing the latter in a vulnerable position in relation to the danger zone of any mine. To overcome this limitation necessitated the sensor having improved detection and classification ranges. These characteristics could only be achieved by use of a lower frequency, which in turn resulted in loss of resolution and contrast, features essential in minehunting.

To overcome the limitations of both the hull-mounted and variable depth sonars requires that a low-priority underwater vehicle be deployed to carry out a continuous search ahead of the parent MCMV platform. Such a vehicle is the self-propelled sonar, development of which is now reaching the stage where such systems can operate effectively, considerably reducing the danger to high-value MCM platforms.

VERIFIED

CANADA

Canadian minehunting system

Type

Remote minehunting system.

Development

Macdonald Dettwiler of Canada is developing the next-generation remote minehunting system which will have applications to all minehunting sonar sensors, as well as to shore-based facilities.

Canada has ordered 12 multirole Maritime Coastal Defence Vessels (MCDVs), one of whose key roles will be to provide the infrastructure for a remote minehunting system. Macdonald Dettwiler has developed the route survey module as part of the MCDV programme, having been awarded the contract for system integration. The concept involves the use of a sonar imagery map database of areas of the seabed which are susceptible to mining, showing the position of all known objects. Using this database as a reference, the MCDV can carry out a rapid survey of a selected area, comparing a new sonar image against one previously recorded, and from which any new object can be quickly detected and where necessary investigated. Such a concept will result in a significant reduction in the time required to monitor a route and neutralise any new mine-like objects which are detected.

In this system the MCDV will be able to accurately position objects in a continuous sonar map database, geocoding and mosaicking the data. The use of this geocoded sonar imagery will enable the MCM tasks of route survey, minehunting and neutralisation to be carried out independently, yet at the same time integrated by a single geographic database.

Description

The system elements comprise the MCDV, a shipboard mine warfare control system, the route survey sonar payload, the Remotely Operated Vehicle (ROV) and a mine warfare data centre.

For route surveying the MCDV will deploy a towed side scan sonar which will provide the necessary data to build up and maintain a sonar image of the seabed. During minehunting, a remotely controlled vehicle will tow the sonar and radio the sonar imagery to the ship's command and control centre. Mine neutralisation may be carried out by an ROV or expendable mine destructor.

The master databases will be located in two mine warfare data centres, one on the east and one on the west coast of Canada. Each centre will support a squadron of MCDVs, providing post-mission data analysis, mission planning and maintenance of the database for route survey and all mine warfare related data for that coast. Both the data centres and the MCDVs will be equipped with common hardware and software. Operator consoles in each facility will provide for mission planning, sonar data analysis and system/database management. The system also incorporates a

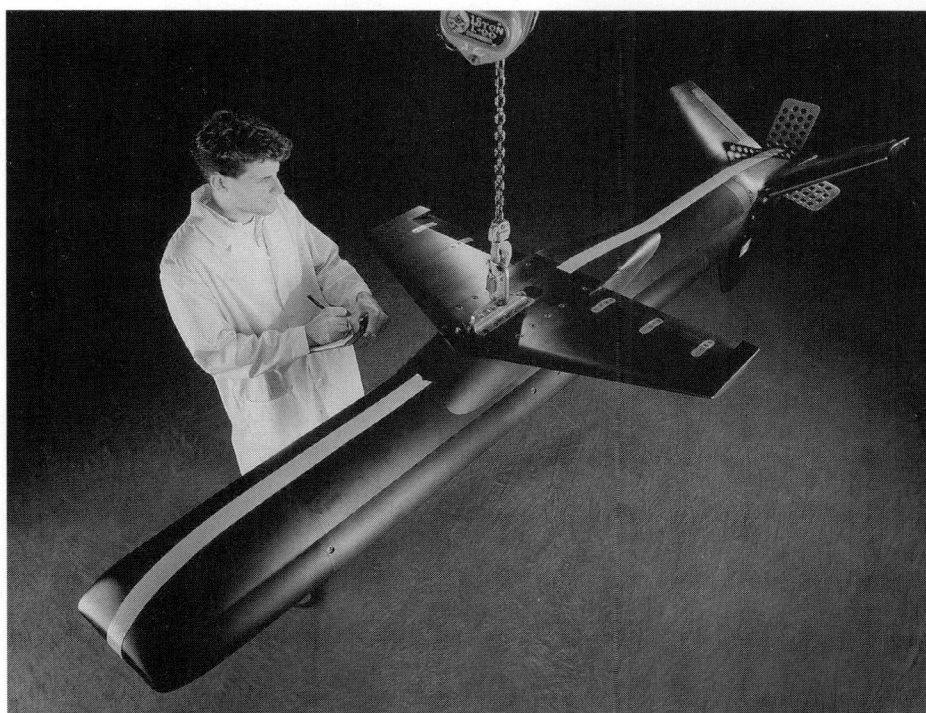

Development model of the towfish for the Canadian MCDV programme   **1995**

large format plotter (for printing route plans and electronic charts) and online and offline optical disc storage for the route survey database. Each MCDV will have a permanently installed mine warfare control system comprising tactical MCM sensor positioning, data analysis and DataBase Management Subsystems (DBMS).

The tactical subsystem will integrate all navigation sensor inputs and generate a local operations plot, based on electronic charts. These will display own ship's position, sonar towfish position, radar contacts and mine warfare contacts. The data analysis subsystem includes Digital Equipment of Canada multifunction detection and classification consoles for the display and analysis of real-time and historic sonar imagery. The DBMS retrieves all sonar imagery and contacts from previous missions for display by the data analysis subsystem. As new objects are detected and classified they are stored by the DBMS.

The primary requirement for the sonar was that it should be capable of detecting buried ground mines, mines with anechoic coatings and irregularly shaped mines in conditions of high background clutter. The sonar will operate at depths down to 200 m, close to the seabed to provide good object shadowing for accurate classification and avoiding thermal and salinity layers. Resolution in the classification mode is 12.5 × 12.5 cm and high-coverage rates dictate a speed of advance of 10 kt with a 400 m swath width.

To meet these requirements, a multibeam side scan sonar has been selected. The array is housed in a hydrodynamically stable towfish body, incorporating active control surfaces for bottom following, and comprehensive position monitoring instrumentation. In the route survey mode the sonar towfish is stored in a standard 6 m ISO container module incorporating its own deployment crane and winch. During both launch and recovery the crane maintains positive control of the towfish, release and recovery taking place below the surface. The module can be installed on the MCDV within 12 hours.

The route survey inspection payload comprises a towfish complete with launch and recovery system. The towfish is electrohydraulically powered through an umbilical which gives it unlimited endurance. Manoeuvrability in three axes is provided by four thrusters. Sensors include a side scan sonar, colour and LLTV cameras. The towfish is monitored and controlled from the ship's command centre while deployed, its maximum operating depth being 300 m at a range of 600 m.

The vehicle designed to tow the towfish sonar for minehunting and radio sonar imagery to the ship is the Dolphin which has undergone extensive field trials. The diesel-powered vehicle can operate for up to 24 hours at a speed of 12 kt. In water, both drone and probe refuelling is possible to provide for extended duration operations. The vehicle's rough water performance allows it to remain operating when most small surface drones would be forced to return to harbour. The Dolphin has been launched and recovered without any in-water manned assistance in Sea State 5. A shipboard handling system enables the vehicle to be easily transported to out-of-area operations in force projection missions.

To carry out mine neutralisation a number of options are available including clearance divers, ROVs and expendable mine destructors. However, ROVs are expensive, and the risk to divers from today's sophisticated mine is considerable. In view of these considerations, much effort is being devoted to developing effective mine destructors. These are relatively inexpensive self-propelled weapons which can be guided to their targets by sonar, detonating in close proximity to the mine with both high hard and soft kill effectiveness.

The sonar payload is designed to operate down to depths of 200 m, close to the seabed. Resolution in classification mode is 12.5×12.5 cm. High area coverage rates dictate a speed of advance of 10 kt with a 400 m swath width.

The multibeam side scan sonar is carried by the hydrodynamically stable towfish body which incorporates active control surfaces for bottom following and comprehensive position monitoring instrumentation.

The geocoding and mosaicking technology applied to the sonar imagery has been developed by Macdonald Dettwiler using its expertise in surveillance satellite technology. Algorithms have been developed

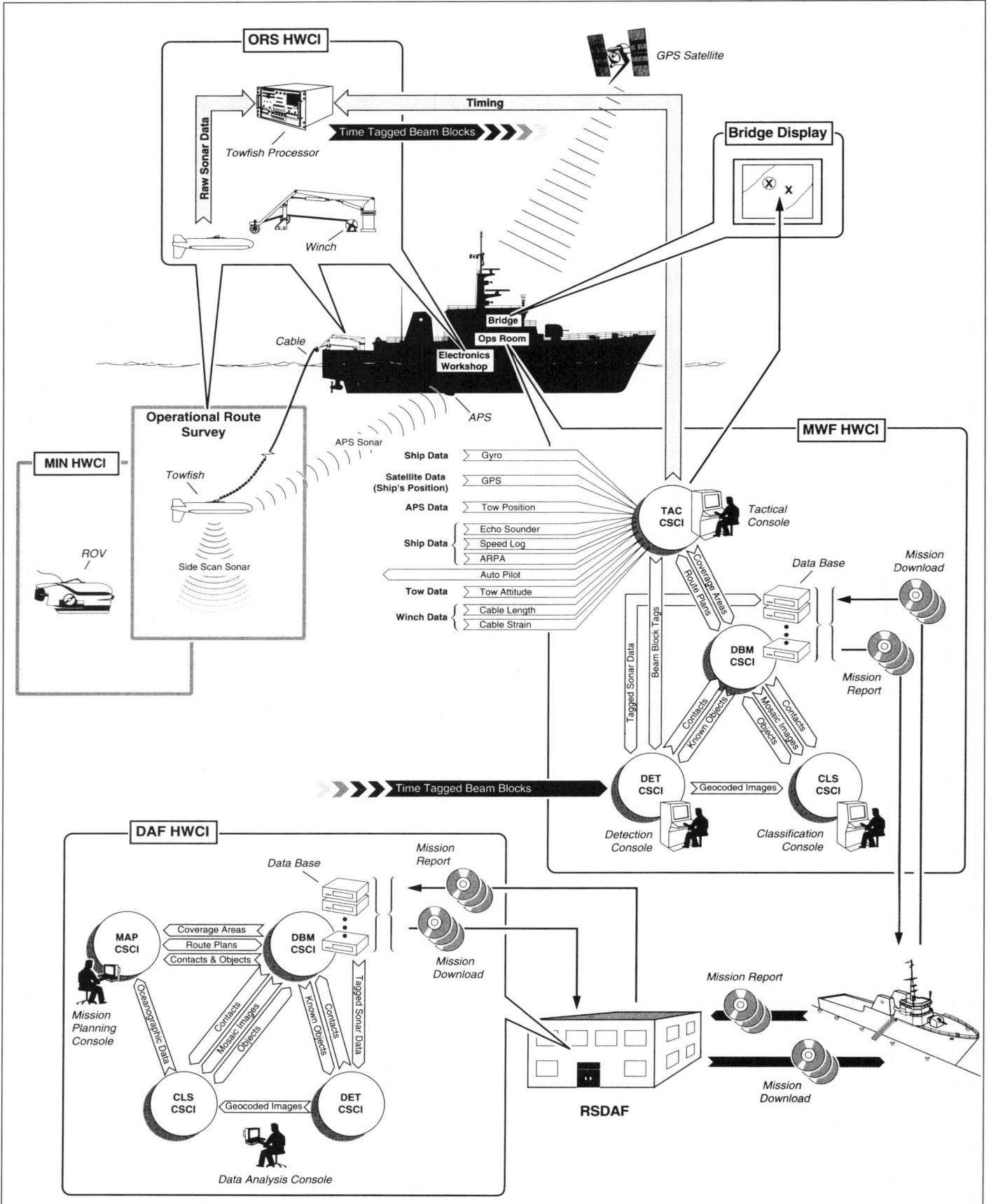

MCM sonar processing for the Canadian minehunting system

which automatically fuse geocoded imagery from sensors of different resolution into a single image. Geocoding is sensor independent and the technology can be used to enhance the performance of sector scan sonars and to integrate side scan sonar data with that of sector scan sonars.

Geocoding allows sonar imagery to be registered precisely to a chosen map projection and absolute geographic locations assigned to targets identified in the imagery. The sonar towfish ground speed, its position relative to the towing platform and the seabed, its roll, pitch and yaw, together with differential GPS and gyro navigational data are integrated into the geocoding process to position each pixel of sonar imagery in a geographic reference grid. Geocoding permits sonar images acquired during earlier missions to be compared with those of the present mission and changes detected semi-automatically.

Work in progress indicates that geocoded sector scan images may also be mosaicked, either for the purpose of building up a continuous map of the seabed or to smooth out background noise to improve the detection and classification range of the sonar.

Mosaicking is the process of combining swaths of geocoded imagery to create a continuous sonar image map database. Mosaicking fills in gaps in the sonar map with new mission data and discards duplicate imagery. The best data can thus be used to create the mosaic. Large areas which are in shadow from one aspect can be replaced by usable imagery acquired from a different aspect. The single mosaicked database is faster to access and the preparation of mission databases simplified.

Operational status
Four route survey modules are being supplied for the MCDVs.

Contractor
Macdonald Dettwiler, Richmond, British Columbia.

VERIFIED

INTERNATIONAL

PVDS

Type
Propelled Variable Depth Sonar (PVDS) system.

Development
The PVDS, developed by Thomson Marconi Sonar SAS of France in co-operation with Bofors Underwater Systems (SUTEC Division) of Sweden, is a third-generation minehunting sonar concept which seeks to meet the increased requirement for safety and efficiency in MCM. The aim of the PVDS concept is to provide a full minehunting capability covering survey, detection, classification and identification of all types of mine threat throughout the underwater environment, from a propelled underwater vehicle. It is intended that this will assure the complete safety of the MCMV by providing maximum range of detection and classification in front of the parent platform, without compromising the probability of detection and classification of the various types of threat. Being a modular concept, the PVDS has the flexibility to offer adaptation to all existing and new types of MCMV platform.

Trials off the Swedish coast in September 1995 demonstrated that this concept is particularly valid in areas where varying thermal layers create a mirror effect (in this instance at a depth of some 12 m, about 6 to 8 m above the seabed). In such conditions, the 'dog-on-lead' concept allowed the sonar to be navigated in the layer where the threat was anticipated, and well ahead of the parent MCMV, ensuring it remained outside the danger area of any mine.

The trials showed the PVDS was capable of being accurately navigated near the seabed, maintaining a satisfactory performance from the surveillance sonar. In these conditions, the PVDS provided a detection range on Manta and Rockan mines (two of the most difficult types to detect) of about 300 m at a distance of some 500 m ahead of the ship. In other words, an overall range between ship and mine of some 800 m. However, due to the proximity of the PVDS to the seabed, shadow classification was not as clear as in normal conditions due to the very shallow grazing angle and multipath propagation.

The production version of PVDS will feature improved performances including the incorporation of a third LF for long-range detection over a 90° sector; simultaneous detection and classification; computer-aided detection and classification; and a ×2 increase in resolution for VHF classification and HF detection.

Description
The PVDS concept is based on a self-propelled vehicle linked to the MCM by an umbilical cable. This cable provides power to the vehicle and also incorporates a fibre optic cable which is used to pass the full range of sonar data, which its sonar gathers, to the MCMV for further processing.

To assure full safety and efficiency the PVDS can navigate in combination with its parent platform at a speed of 5 kt, at depths between 5 and 300 m and at a distance of 200 m in front of the MCMV, closing to within about 10 m of the target for classification.

These characteristics have enabled the designers to mount a shorter-range high-frequency sonar in the vehicle offering improved detection and classification performances. Furthermore, the ability to manoeuvre in three-dimensions enables the vehicle to place the

Preparing to launch the PVDS Double Eagle vehicle from the Swedish MCMV Ulvon (Anchor Consultancy Photo Library)   **1996**

sonar array in the optimum position, irrespective of the speed of the MCMV.

Such features can offer a considerable increase in mission effectiveness, placing the vehicle close to the threat to provide a highly accurate classification.

In 1990 Thomson Marconi Sonar SAS began development of a sonar to meet the above requirements. The aim was to develop high integrated front-end electronics using hybrid technology which could be located inside the array beam without the need for an extra electronic container. In addition, dual-frequency transducers were developed to perform detection and classification, resulting in a slim acoustic array just 0.7 m in length exhibiting low drag. These developments led to a sonar weighing just 80 kg, small enough to be fitted to a medium-size ROV. The vehicle selected to carry the sonar was the Bofors Double Eagle, a fully proven and highly manoeuvrable, low-magnetic and acoustic signature, ROV.

The resulting PVDS design has realised a platform with much better performances than a VDS, at 500 m in front of the MCMV and using a detection frequency of 165 kHz (as opposed to the 30 to 80 kHz of a VDS) a bearing resolution of 3.6 m (compared to the 11 to 30 m of the VDS) is obtained.

The sonar is the TSM 2022 Mk 3 which is mounted at the front end of the Double Eagle with yaw and pitch mechanical stabilisation. Detection performance of the 165 kHz frequency sonar offers a search sector of 63° (mechanically steered) and a range on a sandy bottom of 500 m, an angular resolution of 0.7°, range resolution of 12.5 cm, and vertical beamwidth of 19°. The swept path is 300 to 400 m. The classification performance of the 405 kHz frequency sonar provides a 12° search beam which is electronically steered within the 63° classification beam operating to a range of 160 m, an angular resolution of 0.3°, range resolution of 6.25 cm and vertical beamwidth of 7°. For moored mine classification the vehicle is rolled through 90°. The Double Eagle weighs 400 kg (in air) and has dimensions of 1.9 × 1.3 × 0.8 m and a payload capability of 80 kg. Its maximum operating depth is 300 m and it can operate at a speed of 5 kt. The ROV is powered by eight thrusters (two forward, two lateral, four vertical) which can maintain the fish in any position. The umbilical cable is 600 m long and is very close to neutral buoyancy.

Operational status
The PVDS was demonstrated at Brest in October 1994. With the co-operation of the French Navy, and using the trials vessel *Thetis*, a series of demonstrations of the PVDS was given to a number of navies. Further demonstrations were carried out at Malmo, Sweden in September 1995 aboard the 'Landsort' class MCMV *Ulvon*.

Contractor
Thomson Marconi Sonar SAS, Sophia Antipolis, France.
Bofors Underwater Systems AB, Motala, Sweden.

VERIFIED

Target mines being prepared for launch from the Swedish MCMV U von   **1996**

SVDS 90

Type
Self-propelled Variable Depth Sonar (SVDS).

Description
The STN ATLAS SVDS 90 is part of the MWS 90 minehunting system. SVDS 90 is capable of fully autonomous operation and is based on the combined technologies of the DQS-11M sonar and PAP 104 Mk V ROV. The system is designed to operate at a speed of 8 kt and to provide sonar data down to depths of 300 m, deploying ahead of the parent vessel to counter ground, short-tethered, long-tethered and self-propelled mines.

The sonar is based on the STN ATLAS DSQS-11M family and comprises a high-resolution, dual-frequency (100 kHz for detection, 200 or 300 kHz for classification) system. The SVDS 90 vehicle houses the two sonar arrays as well as some of the signal processing electronics.

The 1.4 m linear array for horizontal detection and classification is mounted in a horizontally rotatable sonar wing mounted underneath the body of the ROV at the rear on a horizontal axis for 90° horizontal detection/classification. The depth array is mounted behind a sonar dome in the head of the vehicle and allows for simultaneous depth classification of contacts within the horizontal search sector of the linear array.

To ensure that the antenna orientation is aligned with the planned track, and to overcome current influences on the vehicle, the linear array can be operated in a stabilised mode. Normally the depth array automatically follows the orientation of the sonar wing, but can be freely trained within the detection sector if required.

The sonar is controlled from a single console with sonar data acquired by the parent vessel's hull-mounted sonar and the SVDS sonar displayed on the two displays of the console. The display includes the 90° horizontal detection sector with two windows for classification and depth evaluation. The operator's task is aided by CAD/CAC techniques and target data correlation between the hull-mounted and the SVDS sonar data.

The SVDS 90 self-propelled variable depth sonar undergoing sea trials   **1998**/0007418

Software performance evaluation tools to assess and optimise parent vessel missions and/or SVDS operation are integrated and can be used by the operator either for mission planning or during missions. By combining sonar and vehicle performance prediction under actual measured environmental conditions, mission effectiveness can be optimised including lap track planning, and vehicle operational depth and location. The integrated sensors and acoustic positioning system enable various vehicle control modes to be achieved, including constant height over the seabed, constant depth, automatic track keeping and manoeuvring and automatic winch control and tether management.

To provide for virtually unlimited mission duration the vehicle is powered via an umbilical cable from the parent vessel. The cable incorporates embedded fibre optic cores for the transmission of sonar data and vehicle control data.

Operational status
Under development. Factory acceptance tests were conducted in October 1995 and the sonar integrated into the body. Initial sea trials have been undertaken. In competition to meet requirement for the Netherlands Navy. The SVDS 90 was successfully tested aboard the STN ATLAS trial ship *Schall* in the Spring of 1997.

Contractor
STN ATLAS Elektronik GmbH, Bremen, Germany.
ECA, Boulogne-Billancourt, France.

UPDATED

TABLE XX

Sonar equipments

System	Manufacturer	Country	Installed on	Navy	Units[1]
AN/UQS-1		USA	MSC 268	South Korea	3
AN/UQS-1		USA	MSC 289	South Korea	1
AN/UQS-1D		USA	MSC 294	Greece	8
AN/UQS-1		USA	Adjutant/MSC 268	Spain	8
AN/UQS-1D		USA	Adjutant/MSC 268/MSC 294	Turkey	11
AN/UQS-1		USA	Bluebird	Thailand	2
AN/SQQ-14	Lockheed Martin	USA	Aggressive	Belgium	1
AN/SQQ-14	Lockheed Martin	USA	MSC 322	Saudi Arabia	4
AN/SQQ-14	Lockheed Martin	USA	Adjutant	Greece	6[2]
AN/SQQ-14	Lockheed Martin	USA	Aggressive	Spain	4
AN/SQQ-14	Lockheed Martin	USA	Aggressive	Taiwan	7
AN/SQQ-30	Lockheed Martin	USA	Avenger	USA	6
AN/SQQ-32	Raytheon	USA	Yaeyama	Japan	3
AN/SQQ-32	Raytheon	USA	Segura	Spain	4
AN/SQQ-32	Raytheon	USA	Avenger	USA	8
AN/SQQ-32	Raytheon	USA	Osprey	USA	12
CMAS 36	C-Tech	Canada			
DSQS-11H	STN ATLAS Elektronik	Germany	Bang Rachan	Thailand	2
DSQS-11M	STN ATLAS Elektronik	Germany	Bay	Australia	2
DSQS-11M	STN ATLAS Elektronik	Germany	Frankenthal	Germany	12
DSQS-11M	STN ATLAS Elektronik	Germany	Gaeta	Thailand	2
DSQS-11M	STN ATLAS Elektronik	Germany	Hameln	Germany	10
DSQS-11M	STN ATLAS Elektronik	Germany	Lindau	Germany	6
DUBM 20B	Thomson Marconi Sonar	France	Circe	Turkey	5
DUBM 41B	Thomson Marconi Sonar	France	Antares	France	3
DUBM 42	Thomson Marconi Sonar	France	Laperouse & Thetis	France	4
MG 7		RFAS	Yevgenya	Bulgaria	4
MG 7		RFAS	Yevgenya	India	6
MG 7		RFAS	Yevgenya	RFAS	24
MG 7		RFAS	Yevgenya	Syria	5
MG 7		RFAS	Yevgenya	Vietnam	2
MG 7		RFAS	Yevgenya	Yemen	5

System	Manufacturer	Country	Installed on	Navy	Units[1]
MG 79/89		RFAS	Natya I	Yemen	1
MG 69/79		RFAS	Sonya	Bulgaria	4
MG 69/79		RFAS	Sonya	RFAS	59
MG 69/79		RFAS	Sonya	Syria	1
MG 69/79		RFAS	Sonya	Vietnam	4
MG 69/79		RFAS	Vanya	Bulgaria	4
MG 69/79		RFAS	Vanya	RFAS	13
MG 69/79		RFAS	Vanya	Syria	2
MG 79/89		RFAS	Notec I	Poland	13
MG 79/89		RFAS	Natya I	RFAS	28
PMS 75	Thomson Marconi Sonar	UK	Swallow	South Korea	6
Q-MIPS	Triton Technology	USA			
SA950	Simrad AS	Norway	Alta	Norway	5
SA950	Simrad AS	Norway	Vegesack	Turkey	6
SA950	Simrad AS	Norway	Stockholm	Sweden	2
SA950	Simrad AS	Norway	Adjutant/MSC 268	Taiwan	5
SA950	Simrad AS	Norway	Goteborg	Sweden	4
SA950	Simrad AS	Norway	Hugin/Kaparen	Norway	6
SH 100	Kongsberg Simrad Inc	USA			5
SQQ-14IT	WASS	Italy	Gaeta	Italy	8
SQQ-14IT	WASS	Italy	Lerici	Italy	4
TSM 2021/DUBM 21B	Thomson Marconi Sonar	France	Tripartite	Belgium	7
TSM 2021/DUBM 21B/D	Thomson Marconi Sonar	France	Tripartite	France	13
TSM 2021/DUBM 21A	Thomson Marconi Sonar	France	Tripartite	Netherlands	15
TSM 2021/DUBM 21B/D	Thomson Marconi Sonar	France	Tripartite	Pakistan	3
TSM 2022	Thomson Marconi Sonar	France	MHC	Egypt	3
TSM 2022	Thomson Marconi Sonar	France	Tripartite	Indonesia	2
TSM 2022	Thomson Marconi Sonar	France	Landsort	Singapore	4
TSM 2022	Thomson Marconi Sonar	France	Lerici	Malaysia	4
TSM 2022	Thomson Marconi Sonar	France	Lerici	Nigeria	2
TSM 2022	Thomson Marconi Sonar	France	MSC 289	South Korea	4
TSM 2022	Thomson Marconi Sonar	France	Landsort	Sweden	7
TSM 2022	Thomson Marconi Sonar	France	MWV 50	Taiwan	4
TSM 2022	Thomson Marconi Sonar	France	Vukov Klanac	Yugoslavia	2
TSM 2023	Thomson Marconi Sonar	France	Oksoy	Norway	4
TSM 2050 DUBM 41B	Thomson Marconi Sonar	France	Antares	France	3
TSM 2054	Thomson Marconi Sonar	France	SAV	Denmark	6
ZQS 2B	NEC/Hitachi	Japan	Hatsushima	Japan	18
ZQS 3	NEC/Hitachi	Japan	Uwajima	Japan	9
Type 162M	Kelvin Hughes	UK	Type 42	Argentina	2
Type 162M	Kelvin Hughes	UK	Torrens	Australia	1
Type 162M	Kelvin Hughes	UK	Prat	Chile	4
Type 162M	Kelvin Hughes	UK	Leander	Chile	4
Type 162M	Kelvin Hughes	UK	Type 21	Indonesia	3
Type 162M	Kelvin Hughes	UK	Leander	Ecuador	2
Type 162M	Kelvin Hughes	UK	Godavari	India	4
Type 162M	Kelvin Hughes	UK	Nilgiri	India	5
Type 162M	Kelvin Hughes	UK	Type 21	Pakistan	6
Type 162M	Kelvin Hughes	UK	Leander	Pakistan	2
Type 162M	Kelvin Hughes	UK	Leander	New Zealand	2
Type 193	Plessey	UK	Ton	Argentina	2
Type 193M	Plessey	UK	Lindau	Germany	6
Type 193M	Plessey	UK	Swallow	South Korea	6
Type 193M	Plessey	UK	Hunt	UK	13
Type 2093	Thomson Marconi Sonar	UK	Huon	Australia	6
Type 2093	Thomson Marconi Sonar	UK	Sugashima	Japan	4
Type 2093	Thomson Marconi Sonar	UK	Sandown	Saudi Arabia	3
Type 2093	Thomson Marconi Sonar	UK	Sandown	UK	12
Stag Ear		RFAS	T 43	Albania	2
Stag Ear		RFAS	T 43	Egypt	6
Stag Ear		RFAS	T 43	Indonesia	2
Stag Ear		RFAS	T 43	RFAS	5
Stag Ear		RFAS	T 43	Syria	1
Stag Ear		RFAS	Yurka	Egypt	4
Tamir II		RFAS	T 43	Bangladesh	4
Tamir II		RFAS	T 43	China	34

* For route survey
[1] Building or on order
[2] Or UQS-1D

ROV SONARS

CANADA

MS 900/900D scanning sonar system

Type
ROV obstacle avoidance sonar, minehunting, diver support sonar.

Description
The MS 900 has been designed as a portable, cost-efficient processor to meet today's mechanically scanned sonar imaging and profiling needs. Its compact size, powerful capacity, and rugged construction means it is suited for both commercial and military applications.

Modes of operation include: polar, sector, linear, and profiling. Real-time acoustic imagery is available in VGA/RGB/NTSC and PAL video formats. A printer port is provided to enable hard-copy output, or Bitmap screen image transfer to a PC. External control and data acquisition of the MS 900 is possible via the serial port; NMEA serial data input is via the same port.

The MS 900 will operate approximately 40 different 971-series sonar heads (including 120/330/675 kHz, and 2.25 MHz frequencies). The sonar head shown is 6,000 m depth rated, 675 kHz with a 1.4 x 22° fan transducer.

The MS 900 is available in either analogue or digital telemetry.

Operational status
In full-scale production and in service.

Contractor
Kongsberg Simrad Mesotech Ltd, Port Coquitlam, British Columbia.

UPDATED

SM 2000

Type
ROV, hull-mounted, surveillance and obstacle avoidance multibeam sonar.

Description
The SM 2000's compact size, lightweight and versatile architecture makes it ideal for all acoustic-imaging applications. Windows 95-based operating software provides superior flexibility in sonar control, data output and interfaces. Data recording options include storage of raw or beam-formed data to JAZ or DAT drives with playback. The system's telemetry uses either electrical or fibre optic links. Electrical telemetry rates are variable and can accommodate a wide variety of cable

lengths and bandwidths. Selectable 'on-the-fly' features include variable beam-forming configurations, ping-to-ping averaging and preset or user-defined TVG and gain curves and maximum resolution/maximum speed settings.

The SM 2000 operating software includes a track plotter display, pinger receiver and passive tracking system modes of operation.

The SM 2000 processor is universal and can be used in conjunction with a variety of 200 kHz sonar heads including the 60, 120 and 180° fields of view versions, and the 90 kHz 90° coverage, long-range imaging multibeam (SM 900).

Applications for the SM 2000 include:
- obstacle avoidance
- inspection
- mine hunting
- underwater surveillance
- diver detection/diver support
- ROV/AUV sonar

The SM 2000 multibeam imaging sonar can be easily converted to a multibeam echo-sounder with the addition of a separate transmit transducer.

Operational status
In full-scale production and in service.

Contractor
Kongsberg Simrad Mesotech Ltd, Port Coquitlam, British Columbia.

NEW ENTRY

GERMANY

AIS 11 sonar

Type
Active identification sonar for minehunting.

Description
The AIS 11 is an active identification sonar intended for use with submersibles. Although the equipment can be used for a variety of underwater tasks, its prime role is that of mine countermeasures, for which purpose it is normally fitted to the PAP 104 or Penguin B3 mine disposal units (see separate entries).

The underwater parts of the AIS 11 consist of an

electronics unit and a transducer unit, the latter being fitted in the hull of the submersible. The parent surface ship is equipped with an evaluation unit, a control panel and a display. Weight of the overall system is 160 kg, that of the underwater part being about 20 kg. Data transfer between the underwater part and the surface takes place along a cable.

During operation, a 60° conical beam is transmitted ahead of the vehicle and the acoustic signals reflected by an object are received by a linear antenna. These signals are then processed in real time, by both analogue and digital methods, and transmitted to the surface vessel where the information is shown on a PPI

display in a sector-shaped format with correct angular representation. Fast real-time signal processing enables the underwater scene to be displayed at the frame repetition rate of a normal cine film.

Operational status
In operational service.

Contractor
STN ATLAS Elektronik GmbH, Bremen.

VERIFIED

ITALY

MIN Mk 2 sonar

Type
ROV sonar.

Description
This high-resolution, high-frequency sonar has been designed for the MIN Mk 2 ROV. The unit comprises a control and display unit which is integrated into the main shipboard console, and the underwater sonar head and electronic assembly mounted on the ROV.

The main features of the sonar are its high resolution in bearing and range and its ability to identify underwater objects using echo and shadow techniques. Displays are presented in either PPI or B-scan format. The sonar is mechanically sectorally scanned in the horizontal plane using a bidirectional motor. Vertical scan is achieved by tilting the head of the ROV.

The transducer consists of a horizontal transmit array, and horizontal and vertical receive arrays. The three arrays are mounted in the same plane, a configuration which provides for future implementation of sonar signal processing in order to give an acoustic image in pseudo three-dimensional form.

Operational status
In service with the MIN Mk 2 ROV in the Italian Navy.

Specifications
Range scales: 3/6/12/25/50 m
Scan rate: 15°/sec (min), 120°/sec (max)
Scan period: 0.6 s (min), 19 s (max) (depending on selected scan rate and range scale)
Pulse transmission modes: FM and CW
Max transmission bandwidth: 40 kHz with 0.5 ms pulse length (FM coherent processing)
Min pulse length: 0.5 ms (CW coherent processing)

Transmit beam: 1° horizontal, 40° vertical
Receive beam: horizontal array – 1.25° (horizontal) × 40° vertical; vertical array – 16° (horizontal) × 5° vertical
Source level: >=210 dB/µPa/1 m
Range resolution: between 2-8 cm (depending on selected full-scale range)
Max range: >25 m (bottom contact), >50 m (moored contact)
Angular resolution: better than 1.25°
Horizontal coverage: ±35° (with reference to longitudinal axis of ROV)
Mechanical vertical tilt: +60 to −90° (with reference to longitudinal axis of ROV)

Contractor
Whitehead Alenia Sistemi Subacquei, Livorno.

VERIFIED

UNITED KINGDOM

AS360 sonar

Type
Scanning sonar for ROV purposes.

Description
The AS360 is a series of mechanically scanned sonars of high-resolution capability designed specifically for installation on remotely operated vehicles. TV refreshed techniques offer a visual presentation in both monochrome and colour.

The AS360M and AS360 MS5 are derivatives of the basic AS360 equipment. The AS360M has a range of 100 m and employs a colour TV display. It is designed for use on large manned or unmanned submersibles, and can be used in a number of other applications.

Operational status
The Type AS360 MS1/1 has been selected for installation on the ECA PAP Mk 5 ROV to be used by the Royal Navy's single-role minehunters.

Specifications
Frequency: 500 kHz (AS360M and MS1A)
Beamwidth: 1.4° × 27° (AS360M and MS1A)
Depth rating: 300 m (AS360M); 1,500 m (AS360 MS1A)

Contractor
UDI-Wimpol Ltd, Aberdeen.

VERIFIED

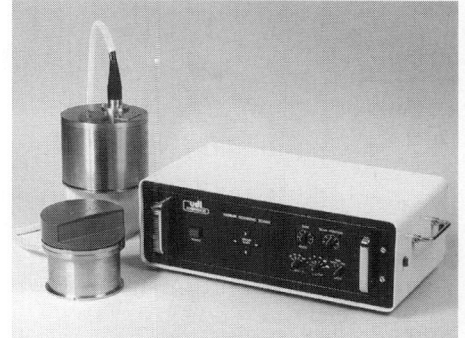
AS360 MS5 RCV sonar equipment   **1996**

UNITED STATES OF AMERICA

SeaBat 6012

Type
ROV scanning sonar.

Description
The SeaBat 6012 is a high-definition, real-time, electronically scanning, forward-looking sonar designed to be deployed on ROVs or small surface vessels (including inflatable boats). With 60 455 kHz beams in a forward-looking configuration, the SeaBat 6012 is used to locate objects such as mines and swimmers at distances up to 200 m.

The system includes a lightweight, low-magnetic sonar head; a surface processor; a trackball; and a high-resolution colour display. It is controlled by user-friendly 'point and click' on-screen menus, so that the operator never need avert his/her eyes from the sonar image. Commands include receive and transmit settings, five different colour ranges, grid on/off, distance measurement, zoom and freeze frame.

The display features 256 colours and can be recorded in PAL or NTSC VHS video format. An optional disk drive allows images to be imported into word processors for documentation.

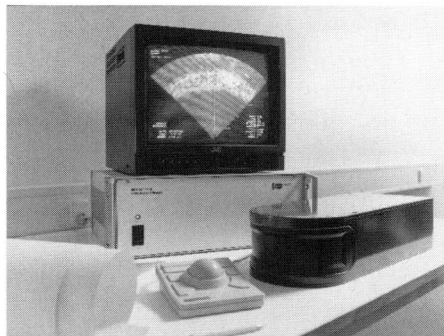
Elements of the SeaBat 6012 ROV sonar

The sonar scans a forward sector of 90 × 15° at 455 kHz using 60 individual 1.5° narrow sonar beams which update simultaneously up to 30 times per second. The result is high resolution with minimal noise and clutter. Seven range settings from 2.5 to 200 m are available, offering an actual increase in sonar display resolution. Undistorted images are produced in real time, even at speeds of up to 12 kt in water.

Operational status
Operational with the Coastal Systems Station of the Dahlgren Division, Naval Surface Warfare Center, USN; Carderock Division of NSWC where it is used in carrying out hull maintenance; Finnish and Danish navies.

Specifications
Operating frequency: 455 kHz
Bandwidth: 20 kHz
Range settings: 2.5, 5, 10, 25, 50, 100, 200 m
Range resolution: 5 cm
Horizontal beamwidth: 1.5° (receive, each beam); 165° (transmit)
Vertical beamwidth: 15° (receive); 15° (transmit)

Contractor
RESON Systems A/S, Slangerup, Denmark.
RESON Inc, Goleta, California, USA.
RESON Offshore Ltd, UK.

UPDATED

SeaBat 8100 mine avoidance sonar

Type
Obstacle avoidance sonar.

Description
The new-generation SeaBat 8100 is a beam-steered forward-looking multibeam sonar operating at 240 kHz. Optimised for use in shallow-water, high-reverberation environments, the SeaBat 8100 provides high-resolution, real-time displays out to 400 m, with a field of view that is configurable up to 360°. Updating up to 40 times per second and featuring a low magnetic signature, the SeaBat 8100 is ideal for fast MCM operations. Other applications include obstacle avoidance, static surveillance (against divers, chariots, ROVs and mini-submarines), marine mammal studies, and protection of offshore structures and vessels.

The SeaBat 8100 uses RESON's advanced 8100 Series processor architecture. This system combines advanced operating features (such as automatic beam steering, target location output, user-definable custom settings, and the ability to download new firmware through a communications port) with straight forward operation (including a point-and-click user interface, automatic calibration, and onboard self-diagnosis).

Specifications
Operating frequency: 240 kHz
Resolution: 1 cm
Number of beams: 100-240
Horizontal beamwidth:
    transmit – 170-360°
    receive – 1.5°
Vertical beamwidth:
    transmit – 1.5/3°
    receive – 20°
Range settings: 2.5, 5, 10, 25, 50, 100, 200, 400 m

Contractor
RESON Inc, Goleta, California.
RESON A/S, Slangerup, Denmark.
RESON Offshore Ltd, UK.

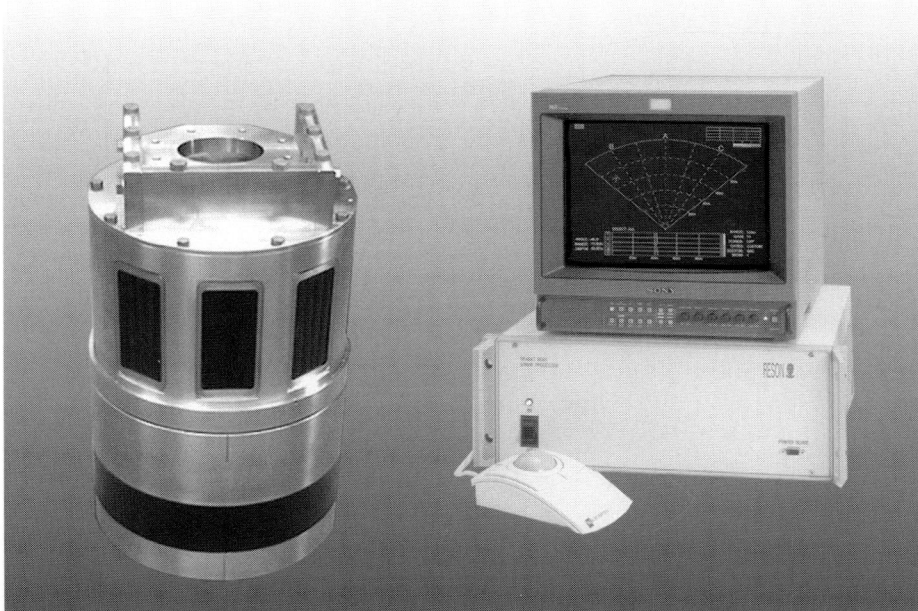
Sonar head and control unit and display of SeaBat 8100 (RESON Inc)   **1998**/0007419

UPDATED

MINE WARFARE LASER SYSTEMS

UNITED STATES OF AMERICA

Magic Lantern

Type
Laser-based minehunting system.

Development
The pod-mounted laser-based Magic Lantern minehunting system first underwent evaluation trials in the Persian Gulf in 1987 where it was used to detect, classify and locate mines during Operation Earnest Will. Subsequently, a prototype podded system was deployed aboard a LAMPS I helicopter system and used to detect mines in known mine lanes during Operation Desert Storm.

Kaman is currently proposing (unsolicited) four Advanced Development Models (ADMs) of Magic Lantern which the US Navy is evaluating. Two ADMs are to be installed on MH-53E Sea Dragon helicopters.

These could form the precursor of a full-development Advanced Laser Mine Detection System (ALMDS) which would be based on enhancements to the Magic Lantern prototype, and which is expected to be developed for fleet introduction after about 2003-04.

If linked to a GPS system the ADM could detect and precisely locate mines. For further enhanced capability the system requires an automatic target recognition capability. These two factors are now being addressed by Kaman in the ADM which will feature new computer hardware and software. This will allow the use of multiple receivers to focus on many depth levels simultaneously. Kaman is studying the incorporation of a GPS receiver and an improved data processing system.

Description
Magic Lantern uses a neodymium-YAG blue-green laser linked to six charge-coupled device cameras to detect mines at different depths. The generator pulses the laser into the sea and pulse-timing generators electronically open the shutters of the cameras to allow them to receive a return of laser energy from preselected depths. Objects that break the laser beam at those depths appear as reflections. Energy reflected from the mine is detected by any of six highly sensitive electronic image-intensified cameras and a scanner. The scanner allows the simultaneous measurement of several swaths of water perpendicular to the aircraft's flight path. The unique feature of the cameras is that they are able to distinguish reflected energy returned from below an object, such as a moored mine, and to produce a shadow image of that object. The received image is automatically catalogued, analysed and stored in the system's memory. It is also available for real-time analysis by an operator.

A third system was ordered to retrofit an SH-2G Seasprite helicopter of USN Reserve Squadron HSL-94 and was deployed in 1997.

The device is mounted in a 4 ft long pod mounted on the helicopter's fuselage.

An adaptation of Magic Lantern may form part of a

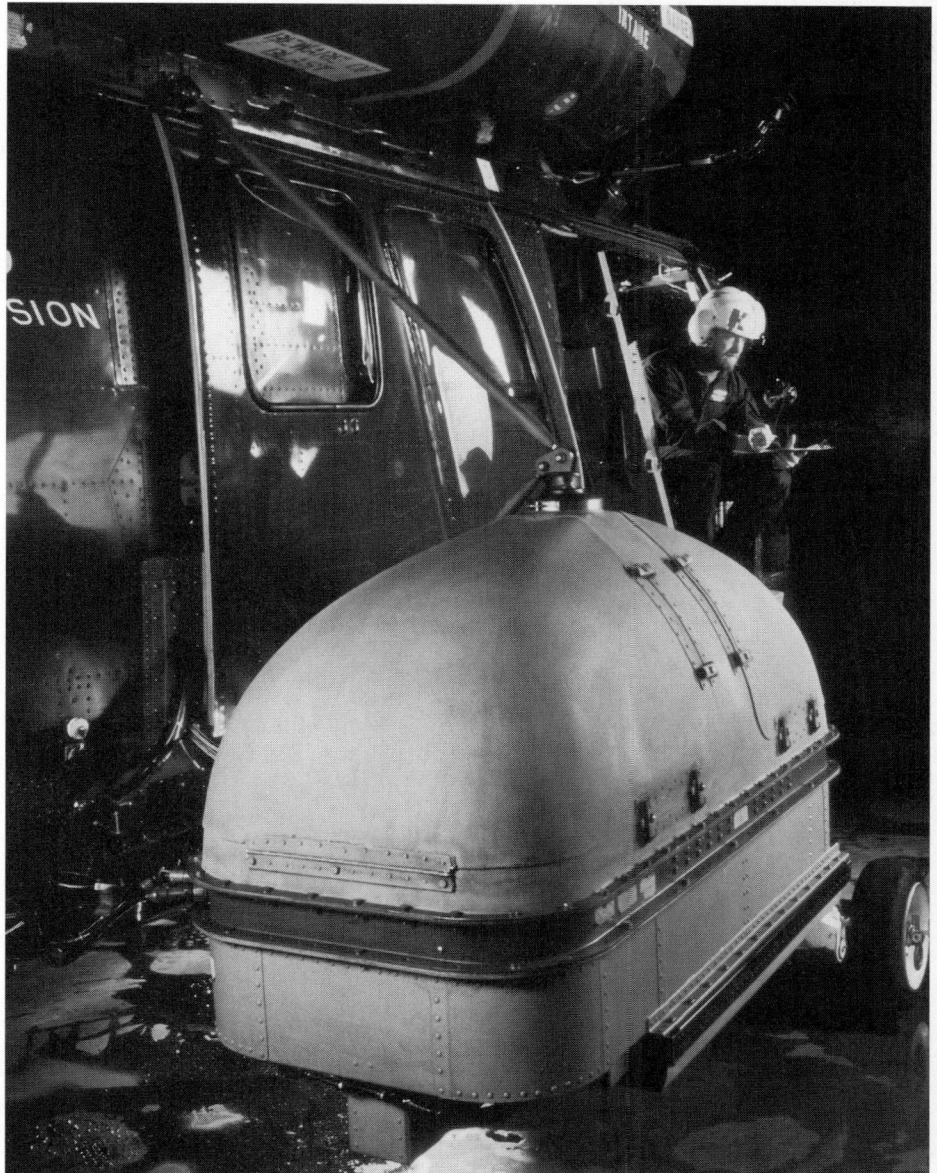

Magic Lantern pod mounted on a Super Seasprite (Kaman)   **1999**/0024793

US Navy programme for mine detection from the high water mark on the beach out to 12 m water depth in advance of amphibious operations.

Operational status
Under development. A prototype model has been tested in the Gulf. Three systems are now in service with the US Navy Reserve aboard SH-2G Super Seasprite helicopters of HSL-94 squadron.

Contractor
Kaman Electro-Optics Division, Tucson, Arizona.

UPDATED

SM 2000

Type
Laser line scan underwater imaging sensor.

Description
The SM 2000 laser line scan system utilises the latest high-power solid-state laser technology to provide high-resolution optical imaging while consuming less than 100 W continuous operating power. The system comprises an underwater unit and a command and control console mounted in the surface platform. The sensor unit consists of a bottle containing the Argon iou laser illuminator, scanning optics and photo multiplier tube sensor assembly. It provides an optical image of objects on the seabed, through water, with ranges up to 38 m, travelling at speeds of up to 10 kt. The laser scans the bottom with a narrow pencil beam as it moves forward, while a narrow instantaneous field of view receiver is synchronously scanned to detect light reflected from the spot and to provide a waterfall display on a high-resolution monitor. It can resolve objects as small as 0.3 cm at a distance of 7.5 m through turbid waters. The 532 nm wavelength of the laser is near the maximum bandpass for seawater and hence signal attenuation is low. For daytime operation nearly all the ambient light is filtered out while still allowing the laser light to reach the detector. The digital images can be pieced together with computer software to provide a comprehensive picture of a large ocean floor area. The system can be linked to a GPS to provide precise navigation and relocation of objects on the seabed. For daytime operations, virtually all ambient light is filtered out and rejected whilst still allowing the laser light to reach the detector.

The SM 2000 can be used for various commercial survey, environment, research or military applications such as minehunting and mine identification. Mine identification, determining which underwater objects detected by sonar are mines or other objects, can save considerable time during an MCM mission.

In the summer of 1996, the SM 2000 was used by the US Navy to identify sections of the aircraft flight TWA 800 that crashed in the Atlantic Ocean off Long Island, New York. The SM 2000 laser line scan system was deployed from a surface craft in about 30 m water depths. The laser was prepared and provided by the navy in only two days after Northrop Grumman was asked to assist in the search and salvage effort.

Northrop Grumman is integrating the SM 2000 into its AN/AQS-14A to provide positive target identification.

Operational status
In production as commercial or military undersea sensor.

Specifications
Dimensions (commercial)
Length: 1,752 mm

Diameter: 279 mm
Weight: 163 kg (dry)
Operating parameters
Operating depth: 1,525 m
Survey swath width: 3-61 m with 15-70° swath width (user adjustable)
Operating range: 2.5-38 m (depending on water turbidity)
Resolution: 0.3 cm at 7.5 m, 1.2 cm at 30 m
Laser wavelength: 200-400 mW, Nd:Yag @ 532 nm (green)

Power consumption: 100 W steady state (250 W max)
Speed of operation: 0.5-6 kt

Contractor
Northrop Grumman Corporation, Electronics Sensors and Systems Division, Oceanic Systems, Annapolis, Maryland.

UPDATED

MINE DISPOSAL VEHICLES

The task of Mine CounterMeasures (MCM) is becoming ever more difficult as mines become more sophisticated. It is now recognised that Mine CounterMeasures Vessels (MCMVs) must remain well outside the danger zone of any potential mines. Dealing with the modern influence mine is an extremely complicated affair requiring a very careful approach. Once an object has been detected and classified as a mine by a minehunting system it has to be neutralised. In the past this highly dangerous task was accomplished either by the MCMV itself – an operation of considerable danger to the ship – or by clearance divers operating from inflatables carried by the MCMV.

Ever since the Second World War, shallow water minehunting has been carried out using hull-mounted minehunting sonars on the MCMV. These sonars were designed to scan ahead of the vessel searching for, detecting and classifying objects at what was then considered to be a safe distance ahead of the ship. However, because of the position of the sonar in relation to the seabed (the angle of incidence of insonification) and the problems of acoustic propagation near the surface where thermal layers can create barriers to the sound propagation, these sonars were really only effective in depths down to about 100 m.

To overcome some of the problems associated with hull-mounted sonars, and in particular for operation in deeper water where mines began to be laid, the Variable Depth Sonar (VDS) was developed. This allowed the sonar to penetrate the inhibiting thermal layers, and to be placed much closer to the level of the potential target. However, the VDS is limited by the fact that being a towed body it necessarily trails in the water, a characteristic which increases with the vessel's speed and the depth of immersion of the body. The towed body could therefore finish up quite some distance behind the towing platform, placing the latter in a vulnerable position in relation to the danger zone of any mine. To overcome this limitation necessitated the sensors having improved detection and classification ranges. These characteristics could only be achieved by using a lower frequency, which in turn resulted in loss of resolution and contrast, features essential in minehunting.

Using technology originally developed for the offshore industry, a range of underwater Remotely Operated Vehicles (ROVs) has been developed to undertake mine neutralisation tasks. The advent of the ROV has given MCMVs used on MCM tasks a much greater flexibility of operation. Using the ROV, MCM operations can now be carried out under much more adverse conditions than would have been the case if only mine clearance divers from the MCMV were available. Moreover, the operation itself is completed more quickly and there is greater certainty of success without the risk of loss of life, or damage to the ship.

As the safety and efficiency of minehunting platforms is now regarded as a high priority, so the sensor used to detect mines must realise greatly improved performance characteristics. One of the technologies now being examined by a number of navies to overcome the limitations of both the hull-mounted and Variable Depth Sonar (VDS) and the limitation of current ROVs in the face of modern mines, involves deploying an ROV fitted with a highly sensitive minehunting sonar some considerable distance ahead of the parent MCMV platform. This is often referred to as the 'dog-on-lead' principle.

To carry out its role effectively the underwater vehicle must be capable of fulfilling a number of functions, the most important of which are: the ability to approach an object classified by the MCMV sonar as a possible mine target without the possibility of triggering the target and so endangering itself and the MCMV; to carry onboard sensors and relay data back to the MCMV for further evaluation of the target; on positive identification it must carry out a neutralisation attack under remote control (to lay a mine destruction charge near the mine).

In addition to these requirements, the vehicle must be capable of operating in a wide variety of environments ranging from shallow to deep water, sandy to rocky bottoms, clear water to very muddy water, and low or high sea states, in all areas of the world.

To develop an underwater vehicle to carry out the tasks outlined above in such a hostile environment requires the highest degree of technological development covering a wide range of disciplines such as mechanical and electrical engineering, metallurgy, computer technology, hydrodynamics and so on. It is not surprising, therefore, to find that firstly there are very few companies in the world involved in developing underwater vehicles for mine warfare and secondly that they and the equipment carried by the vehicles are at the forefront of technology.

An idea of the magnitude of the problems facing designers of underwater vehicles can be gauged from some of the following features which must be attainable in any such designs. Namely: they must be easy to handle, deploy and recover from the water, even in adverse weather conditions; the operator in the MCMV must be able to guide the vehicle to the target in the shortest time possible; the vehicle must be capable of allowing the operator in the MCMV to carry out positive identification of the target; to carry out its role effectively the vehicle must not generate any form of signature (magnetic, acoustic or pressure) which might trigger a mine; the vehicle must be extremely manoeuvrable, particularly at virtually zero speed in all planes, in order that the operator can carry out detailed inspection of the target.

To meet these criteria, underwater vehicles all exhibit similar characteristics but differ in specific details. Thus, all are constructed of non-magnetic materials with a hydrodynamic shape and are capable of carrying some form of countermining charge. Propulsion systems do differ but are usually either electric motor or hydraulic drive. Some form of sensor must be carried. In the past this has often been only a TV camera, but in the presence of limited visibility and for positive identification it is now recognised that to be fully effective the vehicle should also mount a small, high-resolution sonar.

The vehicle must be connected to the MCMV by some form of umbilical link through which the operator can control the movements of the vehicle, and via which the data gathered can be sent back to the MCMV. This requires that on the MCMV the operator's console should possess a TV monitor, PPI presentation of the vehicle's movements, some means with which to control the movement of the vehicle (usually a joystick), and presentation of the vehicle's sonar data. Alphanumeric tote display for data from the vehicle is also advantageous.

Generally, operation of underwater vehicles entails the operator guiding it down the beam of the MCMVs sonar to the vicinity of the target. When a satisfactory range is reached the operator switches on the vehicle's own TV camera and sonar and guides it onto the target. Once in close proximity to the target, the operator manoeuvres the vehicle around the target at very slow speed and close range. Having carried out a detailed examination of the target the operator manoeuvres the vehicle to a suitable position near the target and sends a coded signal to release the countermining charge. The vehicle is then recovered and the charge detonated by coded signal or time delay fuze.

Usually, underwater vehicles can only carry out one mission, as most are unable to carry more than one countermining charge. Furthermore, they usually only have sufficient onboard power for one mission.

Mine development continues to advance and new technologies are now being incorporated into future generations of underwater vehicles making them far more autonomous, enabling the parent vessel to stand off at much greater distances from danger areas, allowing the vehicle to carry out search, location and identification, tasks previously carried out by the MCMV.

Other autonomous vehicles will be used for rapid route survey while others will be deployed by high-value surface platforms to warn the vessel if it is heading into dangerous mined waters.

While much of the effort has concentrated on MCM, a developing role for ROV operation comprises the use and maintenance of fixed seabed installations such as deep ocean surveillance systems designed to detect and track deep-diving submarines. ROVs are increasingly being used in this area. The ROV offers an enhanced capability in operating fixed seabed ranges for determining a ship's acoustic and magnetic signatures, degaussing and in underwater weapon trials and so on. Here too the ROV can be an extremely valuable tool for weapon recovery. Finally, the ROV will become an invaluable tool for the hydrographer assisting in seabed surveys, and recovering seabed samples for analysis.

VERIFIED

REMOTELY OPERATED VEHICLES

CANADA

ARCS

Type
Autonomous Underwater Vehicle (AUV).

Description
Developed as a platform for autonomous vehicle research, ARCS is an alternatively programmable or remotely controlled vehicle which is similar in shape to a torpedo. The existing vehicle is 6 m long and 69 cm in diameter, but can be modified in length to accommodate different payloads.

ARCS is controlled using an onboard microprocessor which provides sufficient decision making ability to permit the vehicle to carry out relatively complex missions requiring good navigational accuracy. Depending on configuration and proximity to the support ship, it is also capable of being instructed remotely. In order to mitigate

unforeseen geographic problems, obstacle avoidance control is one of its onboard systems.

The vehicle has a maximum speed capability of 5 kt and is powered by a 20 kW Ni/Cd battery. It has also been operated using a developmental aluminium-oxygen fuel cell.

Range is variable, depending on overall mission power requirements and power source, but may approach 100 km in some configurations.

Operational status
The ARCS has participated in a variety of tasks, most recently as the testbed for the development of Theseus (see separate entry). It is currently being utilised in the continuing development of a submarine-compatible aluminum-oxygen fuel cell.

ARCS was developed by International Submarine Engineering (ISE) in 1984. It is now owned by the Canadian Department of National Defence and

operated by ISE. Under certain conditions, it is available for lease.

Specifications
Length: 6.40 m (nominal)
Diameter: 0.686 m
Displacement: 1,360.8 kg (nominal)
Payload: 136 kg
Speed: 4 kt (cruising)
Range: 36 km (with Ni/Cd batteries)
Operating depth: 300 m

Contractor
International Submarine Engineering (ISE), Port Coquitlam, British Columbia.

VERIFIED

Dolphin

Type
Deep ocean unmanned semi-submersible.

Description
The air-breathing, diesel-powered, semi-submersible Dolphin is a UHF radio link remote-controlled vehicle which is similar to a small snorkelling submarine. The vehicle can be preprogrammed to follow a set course with waypoints, running autonomously. When running autonomously, the control radio can monitor mission progress. The vehicle's configuration allows for easy integration of towed and hull-mounted sensor equipment, and it features low-noise signature and sea-keeping qualities. Current sizes are in the range of 8 m in length with 150 to 300 hp installed power providing speeds of 15 to 18 kt. The vehicle being trialled for the 'Kingston' class is powered by a 350 bhp Caterpillar 3116TA diesel, giving an operating speed in the region of 12 kt when towing a side scan sonar at a depth of 180 m. Under these conditions the vehicle has an endurance of 16 hours. Dolphin is a stable instrument/sensor/radiator platform. The vehicle is capable of fulfilling a variety of missions requiring stability in high sea states, long endurance and high speed. Two missions which have been demonstrated are hydrography and minehunting.

To demonstrate the hydrographic mission, a

multibeam hydrographic echo-sounder was installed in 1990. With accurate positioning from differential GPS, surveys can be conducted at substantially lower costs than those requiring the use of manned launches.

The second role for such vehicles is in MCM operations. These operations involve real-time transmission of water column and sea bottom mine-like contacts to a controlling platform, and provides battle groups and other seagoing formations with an organic minehunting capability. To demonstrate this mission, a cable winch, which could handle side scan sonar towfish, was developed in 1990-91.

Operational status
To date, 10 Dolphin vehicles have been built, four of which are operated by the Canadian Hydrographic Service as survey vehicles in areas where high sea states preclude operations from manned launches. One of these has been outfitted with multibeam echo-sounder and differential GPS positioning capabilities.

Two other vehicles known as SeaLions were delivered to the US Navy in 1986 for evaluation. These now reside at Naval Research Laboratory (NRL), Stennis Space Center, MS, and are used as tactical oceanographic platforms. For this purpose they have been outfitted with multibeam echo-sounders.

Two more vehicles were delivered to the USN in 1988, and have been outfitted with a variety of mine-hunting sensors for evaluation. These vehicles are

operated by Naval Coastal Systems Station (NCSS), Panama City, Florida. The most recent development was the conversion of one vehicle to the Remote Minehunting Operational Prototype (RMOP) in 1994. The RMOP vehicle was outfitted with a fleet MCM towed sonar, a cable winch complete with towfish handling/docking system, a high-bandwidth RF link for real-time transmission of sensor data, a precision GPS receiver, and hull-mounted forward-looking sonar. The command and control features were integrated with navy consoles, and installed into a standard military control van. The RMOP integration was a joint government/industry collaboration undertaken by NCSS, Rockwell AESD and International Submarine Engineering (ISE). It has been tested extensively and was operated from a combatant during an MCM exercise in March 1995, with good results.

The final two Dolphin vehicles are kept at ISER and are used for development and charter. They have been extensively tested for sea-keeping, towing capability, launch/recovery procedures, control strategies, and reliability. Much of this work has been conducted with Esquimalt Defence Research Detachment (formerly Defence Research Establishment Pacific – DREP) for development of a Canadian Remote Minehunting System (RMHS).

A Dolphin fitted with a SeaBat sonar and AQS-14 sonar operating from the 'Spruance' class destroyer *John Young* successfully supported a US amphibious exercise in March 1995, detecting target mines in the landing area. The vehicle was acquired as an advanced sensor suite to form part of the US Joint Countermine Advanced Concept Technology Demonstration (ACTD). This tested the capability of a joint task force to conduct land/amphibious mine detection and neutralisation in littoral warfare scenarios. The vehicle was used initially to tow advanced mine detection sensors, including a toroidal volume search sonar, side scan sonar, synthetic aperture sonar and a forward looking sonar.

Further trials of the Remote Minehunting System (RMS) were conducted from the US destroyer *Cushing* in the Persian Gulf early in 1997. The trials concentrated on testing the concept of detecting and classifying mines in waters ahead of navy battle groups. A production version of such a system is due to enter service in 2003 aboard 'Arleigh Burke' Flight IIA class destroyers. The vehicle used in these latest trials was a derivative of the Dolphin built by Boeing. Testing of a third variant being developed by Lockheed Martin and carrying the same sensor suites as earlier variants took place in 1998. A production decision on the third variant is due to be taken in FY99 and is planned for fleet service aboard DD 963s in limited numbers interfacing with the SSQ-89. Full series production of the fourth variant is due in FY03.

In October 1996 a vehicle based on the Dolphin began trials for the 'Kingston' class.

The Dolphin unmanned semi-submersible aboard the trials vessel Researcher *1996*

Specifications
(Base vehicle)
Length: 7.30 m
Diameter: 0.99 m
Weight: 3,265 kg
Speed: 12 kt continuous, 16 kt max

Endurance: 26 h at 12 kt
Sea State: capable of sustained stable operations in Sea State 5
Propulsion: Sabre 212 turbocharged, intercooled marine diesel engine (rated at 150 shp)

Contractor
International Submarine Engineering (ISE), Port Coquitlam, British Columbia.

UPDATED

Theseus

Type
Autonomous Underwater Vehicle (AUV).

Description
Designed as a multipurpose, modular vehicle, Theseus is an alternatively programmable, or remotely controlled, vehicle which is similar in shape to a torpedo. The 8,600 kg vehicle is 10.7 m long and 127 cm in diameter but can be modified in length to accommodate a variety of payloads. Designed operating depth is 1,000 m.

Theseus is controlled using an onboard MC68030 microprocessor which provides sufficient decision making ability to permit the vehicle to carry out complex missions requiring precision navigation over significant distances and under conditions where positioning information from external sources is infrequent or unavailable. Depending on configuration and proximity to a surface support station, it is also capable of being commanded with acoustic telemetry. Communication with the vehicle has been maintained, at various times, using a fibre optic cable being deployed by the vehicle, using acoustic telemetry, and using a radio link, via a surface-towed antenna.

Theseus is also fitted with a sophisticated obstacle-avoidance sonar (Sonatech STA-D13-1) design to detect, classify and avoid both mid-water and bottom objects.

The vehicle has a maximum speed capability of 5 kt and is powered, at present, either by a 78 kW Ni/Cd battery bank or by a 274 kW silver-zinc battery supply driving a single 61 cm diameter propeller. Future installation of an aluminium-oxygen fuel cell is considered to be viable.

Theseus autonomous underwater vehicle 1996

Range is variable, depending on overall power requirements of the mission and the selected power source, but may reach 700 km in some configurations.

Theseus is designed to be transportable to remote locations and can be broken down into modules for carrying by air or surface vehicles of more modest capacity.

Operational status
Theseus has achieved full operational status and has demonstrated deployment of fibre optic cable in ice-covered waters.

Upon completion of that task a series of varied tasks which benefit from the significant operational envelope of the vehicle are anticipated.

The Theseus AUV is owned by the Canadian Department of National Defence and operated by International Submarine Engineering (ISE), Port Coquitlam, British Columbia, Canada. It was developed by ISE in 1994.

Contractor
International Submarine Engineering (ISE), Port Coquitlam, British Columbia.

UPDATED

FRANCE

PAP Mark 5

Type
ROV mine disposal system.

Description
The PAP Mark 5, developed and manufactured by Société ECA, is the fifth generation of the PAP 104 family.

The Mark 5 has a capability to simultaneously hunt and destroy ground mines and tethered mines by either placement and remote detonation of a charge or by cutting the mooring rope of tethered mines. To achieve these aims, the PAP Mark 5 embodies both new features and also those retained from previous versions. Recently new options have been added to the PAP: an umbilical cable; CLAD – a new device to destroy tethered mines; and RECA – a remotely controlled ammunition for mine disposal.

The vehicle carries its own source of energy – a battery which can be recharged or changed on board between missions. It is wire guided and remote controlled from a control console situated in the operations room. The wire is not reusable and is dispensed by the vehicle.

The vehicle is horizontally propelled by means of two side thrusters.

By means of a guide rope, the vehicle can navigate at constant altitude.

The PAP Mark 5 is able to carry the 100 kg charge necessary to ensure the total destruction of the mine. Cutters can be carried with the charge.

The PAP Mark 5 employs modern new electronics, based on microprocessors, providing reliable, high-rate data transmission. It includes an ergonomic control console and provides the operator with different navigation aids.

The front section of the vehicle is devoted to an optional sonar of the customer's choice; it can be either a high-resolution nearfield sonar or a mid-range relocation sonar.

Identification is carried out by a tiltable low-light TV camera and/or by a tiltable colour TV camera.

Vertical thrusters are used which allow mid-water navigation. A new payload can be fitted in place of the charge, consisting of a manipulator with associated camera and projector.

Other features include: sealed lead-acid batteries; guide rope length variation to allow fine adjustments of vehicle altitude during a seabed mission; radio control for surface recovery in case of cable breakage; fibre optic wire.

The PAP Mark 5 has very low acoustic and magnetic signatures.

The latest variant is the PAP Plus. This is a modular variant based on the Mark 3 and 4 variants and designed to complement the more sophisticated Mark 5. With the same configuration as the Mark 3 and 4, the PAP Plus modular vehicle offers two alternative capabilities – one with monochrome, colour or LLLTV video and associated transmission electronics, and the other with amagnetic MG horizontal propulsor units.

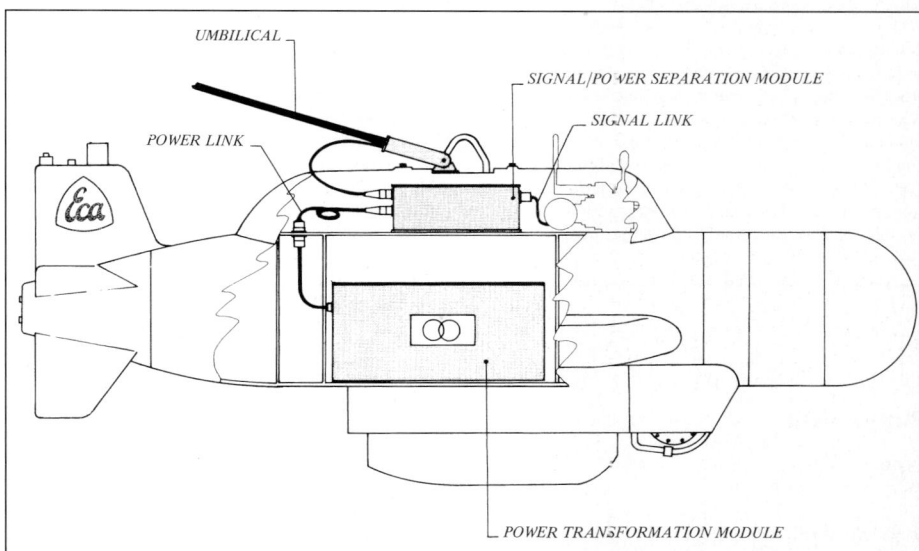

PAP Mark 5 umbilical cable option

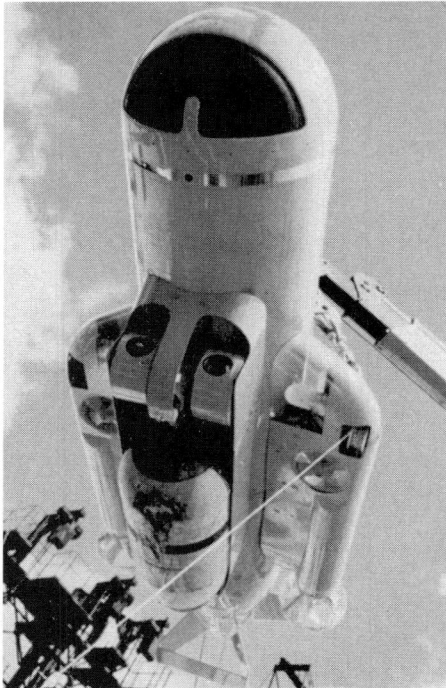

PAP Mark 5 mine disposal system

PAP Mark 5 control console

Additional options include a mid-water evolution capability using two vertical thruster units (fitted in the MG variant) and a high-definition sonar available for both versions. PAP Plus also features a new control console capable of controlling both PAP and RECA vehicles.

Operational status
More than 350 PAP vehicles have been sold and 14 navies are equipped with them. The latest navies to acquire the PAP Mark 5 are the Japanese MSDF, which will operate the system from its new 'Sugashima' class vessels and the Croatian Navy which has purchased the system for its new minehunter.

The PAP has carried out more than 30,000 runs including many successful combat missions in the Red Sea, the Falklands, the Persian Gulf and so on.

Specifications
Length: 3 m
Diameter: 1.2 m
Height: 1.3 m
Total weight: 850 kg (incl 100 kg charge)
Operating range: 2,000 m max (at depths of 300 m)
Speed: 6 kt (max)
Endurance: 90 min

Contractor
Société ECA, Boulogne-Billancourt.

UPDATED

ERATO

Type
Torpedo recovery vehicle.

Description
The ERATO (Engin de Ramassage de Torpilles – torpedo recovery vehicle) has been developed to recover torpedoes from the seabed at great depths (for example, the MU90 torpedo has a maximum operational depth far in excess of 1,000 m).

The recovery system comprises four units:
The ERATO underwater vehicle
The handling system
The power shelter
The operating and control shelter.

The handling system, power and control shelters are all carried on board the support ship.

The submersible is built of syntactic foam which is formed around a light-alloy tube framing. This structure gives the vehicle a slightly positive buoyancy. ERATO is propelled using four identical four-bladed thrusters (two longitudinal, one transverse and one vertical) which are driven at 800 rpm from a 2 kW power unit. The thrusters are electronically speed controlled using PWM inverters and are immersed in oil for pressure compensation. The sensor package comprises two monochrome video cameras for target observation, a

spotlight device, depth and immersion sensors, compass for course guidance, and an acoustic azimetry set tuned to detect the torpedo's pinger. A grab clip is provided for torpedo recovery. The vehicle is linked to the parent vessel by an umbilical coaxial cable 14 mm in diameter and 1,500 m long.

In operation ERATO homes on to the torpedo on the seabed using the torpedo's pinger, grasps the weapon with the grab and returns to the surface with it.

The handling system is used to transfer the vehicle from the parent vessel to the sea, recover it and hoist it back on board. It also handles and automatically feeds out the umbilical cable. The vehicle, umbilical cable and oscillating wire guidance system is carried on a motor-driven trolley running on two rails mounted on a support foundation. The oscillating arm provides compensation for heave motion as the vehicle is transferred between the parent vessel and the water.

The power shelter houses two alternators generating power for the units on board the parent vessel, and the ERATO vehicle, providing a completely self-contained recovery system. Surface power is provided by a 380/240 V, 50 Hz, 14 kW generator which powers the trolley, carrier cable hoist and parent vessel auxiliary units. A second generator delivers power at 208 V, 400 Hz, 15 kW to the ERATO vehicle. Power is supplied to the vehicle under high voltage (3,000 V) via the umbilical cable.

The operator control shelter is divided into two sections: one a high-voltage room incorporating safety devices; and a second room housing a control and monitoring station with a distribution panel and dedicated data processing system. The processing system handles and controls data using: a power circuit and monitoring video terminal; a console which provides status monitoring of the surface equipment; a real-time graphic display terminal for the TV picture; a data and ERATO command processing console; a keyboard; and a vehicle control console (manual controls or navigation commands).

Operational status
In service.

Specifications
Length: 2.3 m
Width: 1.8 m
Height: 2.4 m
Weight: 12,000 kg
Operating depth: 1,000 m

Contractor
DCN International, Paris.

VERIFIED

GERMANY

Pinguin B3

Type
ROV mine disposal system.

Development
The Pinguin B3 is a remotely controlled, reusable underwater vehicle designed for minehunting operations. With the input of target information (derived from minehunting sonars) Pinguin B3 is able to approach any mine-like object mainly automated. The vehicle can then identify the mine and, if necessary, drop one of two mine disposal charges carried to destroy it. The second charge capability enables the vehicle to continue the mission and approach a second target in only one mission. Thereafter Pinguin returns to the minehunter for recovery and rearming.

The Pinguin B3 provides different sonar options and is also fitted with a tilt TV camera. The vehicle is fitted with a 1,000 m reusable strong fibre optic cable for data transmission.

The propulsion system consists of two horizontal propeller thrusters, mounted laterally on the stern section of the vehicle. At low speeds the depth control is achieved by a vertical thruster mounted in the hydrodynamic centre point of the vehicle. At higher speeds (>2 kt) diving is dynamically controlled by flaps. The control system of Pinguin B3 provides ergonomic operator control of operation of the vehicle, only requiring sparse operator inputs for heading, depth and propeller speed.

The Pinguin B3 can be operated freely within the water column in the range of the minehunting sonars down to depths of 200 m. The vehicle carries two charges of up to 125 kg each. Equipped with a special moored mine disposal device, it can also counter moored mines very effectively by direct destruction of the mine casing (no cutting and thus floating mine results). The system offers a considerable growth potential for the adaptation of alternative sensors and of tools for salvage missions.

The Pinguin B3 remotely controlled and reusable mine disposal vehicle carries two mine destruction charges and sensors for vehicle navigation and target identification

Operational status

In service with the German Navy's 'Frankenthal' class minehunters and aboard the Republic of China Navy's 'Yung Feng' class.

Specifications

Length: 3.5 m
Hull diameter: 0.7 m
Width: 1.5 m
Height: 1.2 m/incl MDCs 1.5 m
Weight: 1,350 kg incl MDCs
Propulsion: 2 variable speed thrusters, 1 automatic controlled vertical thruster
Power supply: internal rechargeable battery systems
Operating depth: 200 m
Speed: 6 kt (continuous), 8 kt (max)
Endurance: 60/120 min
Payload: 255 kg (2 mine disposal charges or other payloads)
Sensors: TV camera, different sonar options

Contractor

STN ATLAS Elektronik GmbH, Bremen.

UPDATED

Pinguin B3 ROV mounted on its cradle on the poop deck of the MCMV Frankenthal. *The associated handling crane can be seen beside the ROV*
1994

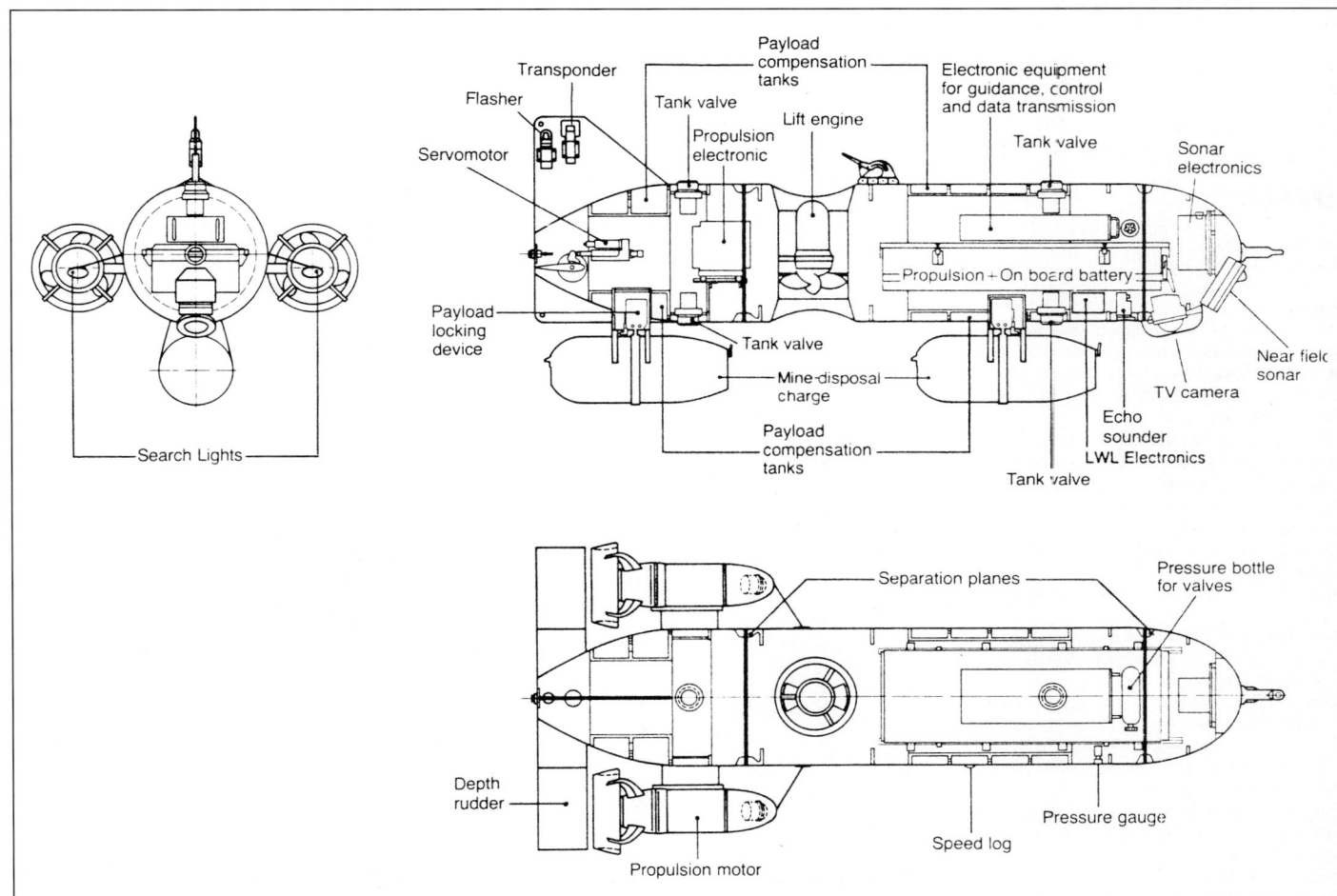

Pinguin B3 component arrangement

Seafox/Seewolf

Type

Family of one-shot mine disposal vehicles.

Development

This family of one-shot mine disposal vehicles is being developed as part of the German Navy's MA 2000 MCM improvement programme. The concept of one-shot mine disposal vehicles was developed to meet the requirements of the navy's advanced minehunting system under the MA 2000 programme. The MA 2000 concept is based on the use of remotely operated surface drones (Seahorses) towing different sonars and MCMVs controlling the drones and launching one-shot vehicles (the Seewolf) for mine destruction. Specific requirements that the Seewolf has to meet are that it should be capable of destroying all types of mines, including buried mines, and possess a relocation capability to enable a temporal split of reconnaissance and disposal operations.

Seafox is in the process of qualification and of type approval through the German Navy and Procurement Agency in a tight timescale as part of the scheduled upgrading of the current MCM force. There is a new role for the minesweepers of the 'Hameln' class. Within this conversion programme five boats will be converted into advanced minesweepers and five more boats into

minehunters similar to the MJ 332 boats, both for deploying the Seafox.

As a result of the navy's reduced requirement relating to the disposal of all types of mines except buried mines, Seafox was designed as a smaller, lighter and less expensive version of one-shot vehicles compared to Seewolf, but incorporates many of the components and subsystems of Seewolf. Full-scale development of Seafox began in 1993. MCMV upgrade programmes began in 1997-98.

Description

Seafox is a one-shot mine disposal vehicle capable of ensuring the destruction of moored mines at varying

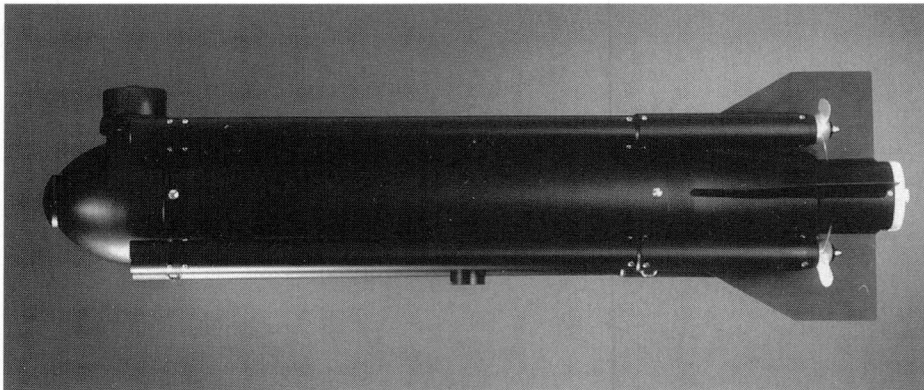

Seafox underwater vehicle   *1997*

depths and short-tethered mines using a shaped charge (Seewolf will deploy a larger blast charge to destroy buried mines through shockwaves). Seafox C is an electric powered, guided vehicle with onboard sonar and TV camera to identify targets and a shaped charge warhead. Seafox I is similar to Seafox C but has no warhead and is recoverable and reusable.

The Seafox vehicle is basically a self-propelled wire-guided (fibre optic cable with high bandwidth) ammunition with a homing capability. It is controlled from the MCMV from a console mounting four displays and normally operates within the range of minehunting sonars. It meets the following basic requirements:

(a) mine destruction in a very short time (less than 10 minutes overall)
(b) automatic guidance with manual override capability
(c) easy stowage and handling
(d) small size and weight
(e) low-cost expendable ammunition.

The vehicle is powered by four horizontal and one vertical thruster powered by lithium batteries. On board sensors include a panoramic sonar and TV camera with lights.

Seafox consists of five main subsystems:
(a) one-shot underwater vehicle
(b) control unit with adaptation to MCM platform
(c) stowage and launch facilities with adaptation to MCM platform
(d) fibre optic data transfer cable between control unit and vehicle
(e) data interfaces from the sonar system of the MCM platform to the control unit.

Operational status
Seafox is under development for the German Navy and is scheduled to enter service in mid-1999 aboard the converted 'Hameln' class minesweepers. Seafox was also ordered by the Swedish Navy for the 'Visby' class in March 1998. Seafox is also being incorporated into the US Lockheed Martin Airborne Mine Neutralisation System (AMNS) for deployment from a helicopter.

Specifications
(Seafox C)
Length: 1.3 m
Diameter: 0.2 m
Weight: 40 kg
Warhead: shaped charge 1.5 kg
Speed: > 6 kt (max)
Range: in excess of minehunting sonar range 900 m (approx)
Depth: continental shelf

Contractor
STN ATLAS Elektronik GmbH, Bremen.

UPDATED

ITALY

MIN Mk 2

Type
ROV mine disposal system.

Description
The MIN (Mine Identification and Neutralisation) system, developed and produced for the Italian Navy by Whitehead Alenia Sistemi Subacquei and Riva Calzoni, has the capability to identify and neutralise both bottom and moored mines.

The latest version, the Mk 2, has been developed from operational experience gained with the Mk 1, and is configured as follows:
(a) one self-contained, hydraulically powered, wire-guided submarine vehicle
(b) a main console in the operations room for vehicle guidance and control
(c) a tracking system for autonomous localisation of the vehicle
(d) one portable auxiliary console for guidance and visual control of the vehicle during launch and recovery operations
(e) one set of operational accessories including: an auxiliary battery charging station and a device for oleopneumatic power pack recharging.

The Mk 2 vehicle is powered by a closed-circuit oleopneumatic accumulator, enhancing the low-noise profile and non-magnetic characteristics. A steerable main propeller allows the vehicle to approach the target detected by the minehunter's search and classification sonar. In addition, the employment of horizontal or vertical thrusters gives the vehicle a high manoeuvrability and hovering capability when submerged and high maneuvrability when surfacing for the recovery phase. Vertical manoeuvring, even to the maximum depth of the vehicle, is controlled through water ballast tanks which are filled or emptied by means of pressurised air; hydraulic power is consequently used only during transfer from and to the parent ship and during identification of the target. The TV camera and sonar are orientable through 150° in the vertical plane allowing flexibility of observation during both bottom and moored minehunting. The vehicle's operational depth has been increased to over 300 m.

The system can use two types of weapons (bottom charge and explosive cutter) thus enabling the hunting of either bottom or moored mines. The main console is

MIN system Mk 1 underwater vehicle aboard a 'Lerici' class MCMV

of modular configuration and consists of separate processing units and an operator desk to minimise possible installation problems.

The desk on the main console is fitted with a control lever and a joystick for speed and direction control of the vehicle during the search and approach phases, and for fine positional adjustments of the vehicle when it is in close proximity to the target. The main console also incorporates the controls for the actuator of the high-resolution sonar in the MIN and for its tracking subsystem.

Two 9 in video display units are mounted on the operator's desk, a monochrome TV display and a high-definition display.

The sonar, which has been specially designed by Whitehead Alenia Sistemi Subacquei, is a high-resolution unit allowing both identification of targets in conditions of poor visibility and a reduction in mission time by means of its target relocation capability.

The sonar features very high resolution in both bearing and range, and is based on a transducer array specially developed using a diced-array technology. It is a mechanical sectoral scan sonar comprising two major sub-units: the control and display unit which is integrated in the main console, and the underwater unit which is mounted on the vehicle and which consists of a sonar head and electronic assembly.

The detailed sonar image is displayed in either PPI presentation or B (linear) presentation. A profiling inform function is also available. The MIN Mk 2 and its control console is integrated into the MCMV combat system via serial datalinks with the command centre, parent vessel's search/classification sonar and parent vessel's gyrocompass.

In addition, the MIN Mk 2 is fitted with an acoustic tracking system comprising an ultra-short baseline transducer mounted on the MCMV and a transponder on the vehicle.

The MIN Mk 2 system processes the data relating to the target under investigation, data from the combat information centre and the classification sonar, ship's data, vehicle data, vehicle position (from tracking and/or from the search sonar) in order to enable the operator to steer the vehicle towards the target following the optimum track and to manoeuvre the vehicle during transit from and to the MCMV.

MIN Mk 2 ROV on board an Italian minehunter
1995

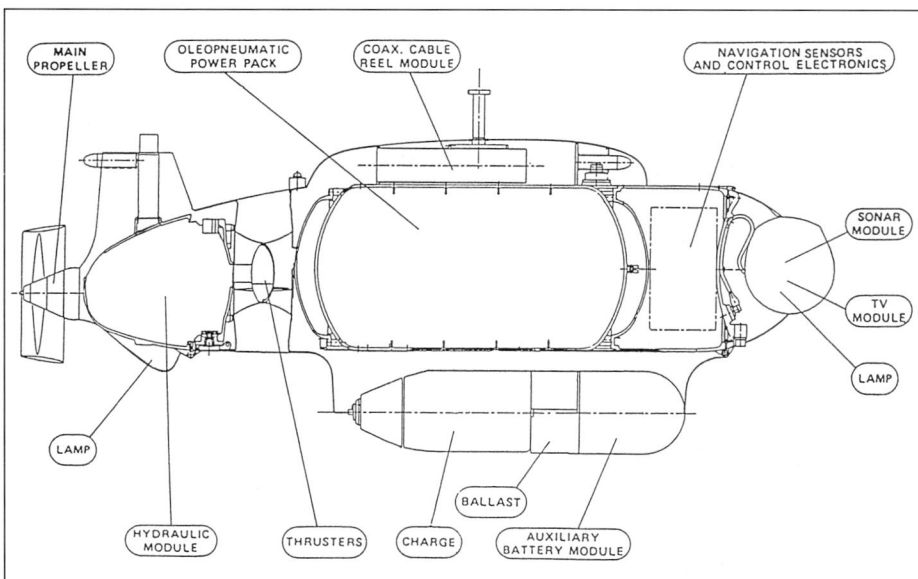

MIN Mk 2 vehicle arrangement
1994

Operational status

In series production. Four MIN Mk 1s are operational aboard Italian Navy minehunters and eight MIN Mk 2s are in service aboard the 'Gaeta' class minehunters of the Italian Navy.

Specifications
Weight: 1,150 kg
Max diving depth: 350 m
Speed: 6 kt
Endurance: <1 h

Contractor
Consorzio SMIN, Rome.
Whitehead Alenia Sistemi Subacquei (subcontractor), Livorno.
Riva Calzoni SpA (subcontractor), Bologna.

UPDATED

Pluto

Type
Remotely operated mine disposal vehicle.

Description
Pluto is available in three configurations: battery-powered with a 6 mm, 500 m long umbilical cable; battery-powered with a 3 mm, 2,000 m long fibre optic umbilical cable; or remote-powered of unlimited endurance with 8 mm, 500 m long umbilical cable.

The vehicle is powered by five thrusters: two horizontal for forward/reverse, two vertical for vertical and lateral shift, and a transversal thruster. The thrusters are controlled by two joysticks. The vehicle can maintain automatic depth control within ±10 cm. Forward maximum speed is 4 kt and vertical speed 1 kt.

Sensors are mounted in the forward tiltable section of the vehicle. Optional equipment includes black and

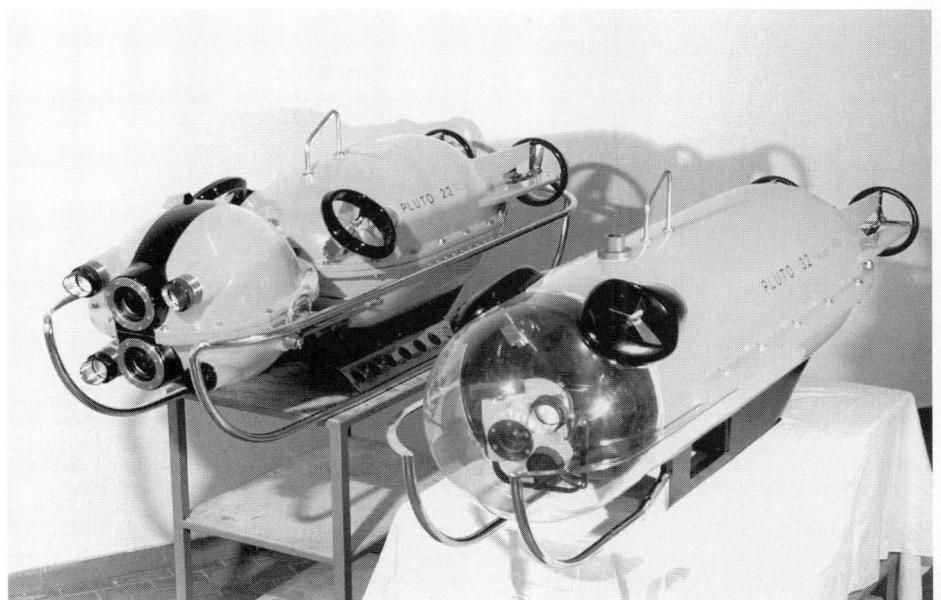

Pluto Plus (right) and standard Pluto (left)

white LLTV, colour TV or still camera, search or scanning sonar, acoustic pinger, strobe flash, measuring instruments, manipulators and so on. There are 10 free channels for remote control and two four-digit telemetry channels for measurements.

The console incorporates a 9 in TV monitor for display of information showing TV image, depth, compass, head tilt angle, elapsed time and sonar diagram. All displayed data can be video recorded.

Operational status
To date 70 units are in service with 10 navies.

Specifications
Length: 1.68 m
Width: 0.6 m
Height: 0.65 m
Weight: 130 kg
Operating depth: 300 m

Payload: 45 kg
Speed: 4 kt (battery), 5 kt (with power cable)
Endurance: 1-6 h (on batteries)

Contractor
Gayrobot, Milan.

VERIFIED

Pluto Plus

Type
Remote-control mine disposal vehicle.

Description
The basic configuration of Pluto Plus incorporates a number of improvements. Pluto Plus uses the latest fibre optic floating, reusable 2,000 m long cable and hydrodynamic design to provide improved observation capabilities, extreme standoff range, an increase in speed to 6 kt, an increase in endurance from 6 to 10 hours (battery capacity has been doubled) and the lowest possible drag factor.

The vehicle is fitted with special sonar sensors for navigation, search, obstacle avoidance and identification. The sensors are all mounted in a single package featuring ±100° tilt and ±80° pan. A single control console monitor gathers and displays the video picture, navigational data, sonar graphics and maps of the investigated area.

The vehicle exhibits low-magnetic and acoustic signatures and is resistant to shock and vibration to MIL-SPEC standards for minehunting.

The lightweight, compact Pluto Plus is designed to operate from all kinds of vessels without special or expensive handling equipment.

Operational status
To date 25 units are in service with four navies (South Korea, Norway, Spain and Thailand).

Specifications
Length: 2.1 m
Width: 0.6 m
Height: 0.61 m
Weight: 315 kg
Operating depth: 400 m
Payload: 100 kg
Speed: 6 kt
Endurance: 1-10 h (on batteries)

Contractor
Gayrobot, Milan.

VERIFIED

Pluto Gigas

Type
Remote-control mine disposal vehicle.

Development
Pluto Gigas has been designed and developed to meet increasingly stringent operational requirements such as the ability to operate in sea currents up to 5 kt at great depths (down to 1,000 m), and at long ranges.

Description
Pluto Gigas is the latest vehicle to be developed by Gayrobot and is capable of operating in extremely demanding conditions such as river estuaries where currents of 5 kt may be encountered and in zero visibility. It can also be used in MCM operations in the open ocean where it can carry out surveillance missions to depths of 1,000 m or greater.

The vehicle is based on the proven technology of Pluto Plus and incorporates some 90 per cent of the components used in Pluto Plus (see separate entry), thus improving logistics, as many of the parts are interchangeable.

Pluto Gigas features a redesigned body constructed of carbon fibre composites; a new propulsion system with double the power of Pluto Plus and capable of giving the vehicle a speed of 7.5 kt; increased endurance; greater payload (two charges); and greater depth (1,000 m).

The vehicle carries a long-range (up to 200 m) high-resolution electronically scanned sonar for search and navigation in addition to the Gayrobot SID identification sonar fitted in Pluto Plus. Navigation has been improved with the installation of a Doppler log operating in conjunction with a computer system for automatic 'hands off' mission operation. An improved colour TV camera is also fitted.

The umbilical link is a reusable, single-fibre, 2,000 m long optic cable, 4 mm in diameter with a breaking strain of 500 kg. The winch is similar to that used with Pluto Plus, but capable of generating an automatic constant pull up to 120 kg. The control console and

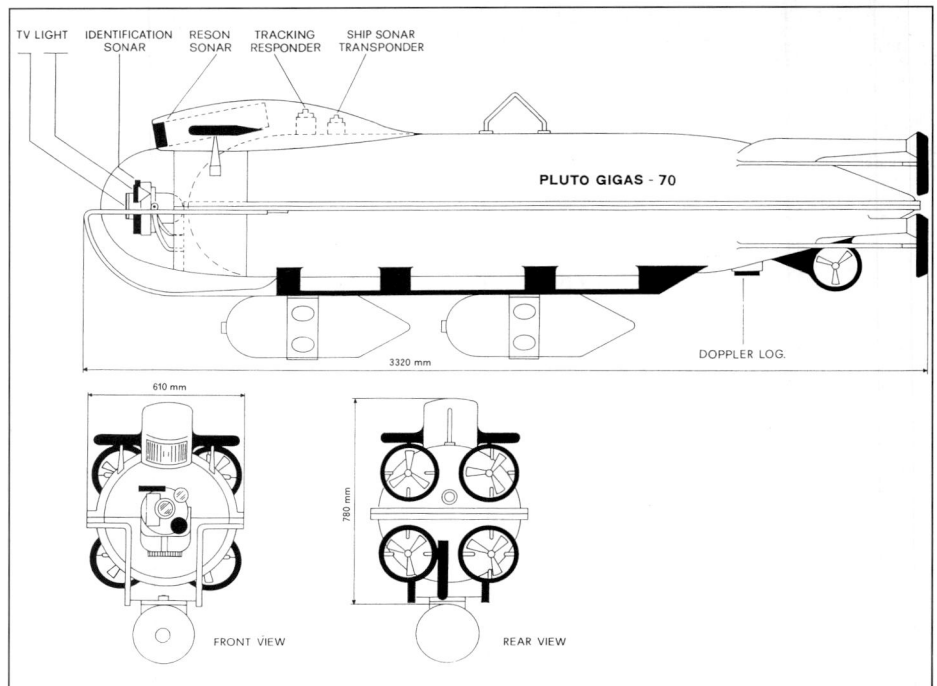

General arrangement of Pluto Gigas   *1997*

other equipment is identical to that used with Pluto Plus.

Operational status
Presented to the Italian Navy and currently undergoing sea trial demonstrations with other potential customers.

Specifications
Length: 3.32 m
Width: 0.61 m
Height: 0.78 m
Weight: 600 kg (approx)
Operating depth: 350, 600 or 1,000 m
Payload: 2 × 45 kg charges
Speed: 7-8 kt
Endurance: 8-9 h at 3 kt

Contractor
Gayrobot, Milan.

UPDATED

NORWAY

Minesniper

Type
Expendable mine destructor vehicle.

Development
Development commenced in 1990 as a technology demonstration project in co-operation between the Norwegian Defence Research Establishment and Bentech Subsea AS (acquired in 1993 by Simrad and now part of Kongsberg Gruppen). In 1992 the Norwegian Naval Material Command awarded a contract for a prototype system of five vehicles. The system was installed on the Norwegian minehunter *Tana* in May 1994 for tests and sea acceptance trials were undertaken in September 1994.

Description
The Minesniper is a lightweight mine clearance system utilising an expendable ROV-type weapon for mine verification and destruction.

The weapon is guided automatically towards the target by the Minesniper navigation computer on board the mother vessel. The navigation computer is linked to the minehunting sonar and the surface navigation system, and receives continuous updates of the target and weapon positions. Minesniper comprises a dedicated short baseline acoustic positioning system to bring the weapon rapidly into the sector covered by the sonar beam.

The communication cable to the weapon consists of a 0.25 mm optical fibre. The weapon is self-powered with a rated operation time of 60 minutes and a maximum travelled distance of 4,000 m. The maximum speed is 6 kt. The weight of the weapon is about 25 kg in the standard version which is depth rated to 500 m.

The weapon carries a short-range scanning sonar for relocation of the target, a camera for inspection/verification and a lightweight shaped charge for detonation of the mine. In the inspection phase, the weapon is controlled manually by use of a joystick. A dedicated navigation display in combination with the camera picture and a model-based navigation system enables the operator to manoeuvre the vehicle accurately during the final homing operation.

The existing version of Minesniper is fitted either with a 72 mm shaped charge of 300 g of Octol explosive or a 122 mm shaped charge of PBXN9 or PBXN110. A new warhead incorporating a semi-armour piercing gun under development by DERA in the UK is being studied for a Mk 2 variant of Minesniper. The Mk 2 version may also incorporate a new 360° sonar, a 6,000 m long fibre optic link and a new control surface to improve manoeuvrability.

A typical operation will take less than 10 minutes from the time the decision is taken to deploy the weapon until destruction of the mine is verified. A training vehicle is available along with a built-in simulator in the navigation computer.

Minesniper can use existing equipment for launching, positioning and display, and requires a small amount of explosives stored on board. Significant reductions in personnel risk combined with an increased system performance can thus be obtained in return for moderate investments in additional equipment.

Operational status
In development. Initial sea trials with Norwegian Navy in 1994.

Specifications
Length: 1.5 m
Diameter: 200 mm
Weight: 25-30 kg (depending on charge)
Speed: 6 kt
Operating depth: 500 m
Range: 4,000 m

Contractor
Kongsberg Defence & Aerospace AS, Kongsberg.

VERIFIED

Minesniper mine clearance system
1998/0009514

RUSSIAN FEDERATION AND ASSOCIATED STATES (CIS)

ANPA 6000

Type
Self-contained unmanned submersible.

Description
The ANPA 6000 is designed to carry out hydroacoustic and photographic surveys of the seabed. The vehicle is also used to measure hydrophysical and hydrochemical parameters of the water column. Onboard sensors include a gyrocompass, hydroacoustic and induction logs, fathometers, beacons and an electronic onboard computer which enables the position of the vehicle to be precisely fixed in relation to the mother ship.

The system can investigate the area according to a prearranged programme, survey a 1,500 m strip of seabed in the side scan mode, and carry out a photographic route survey (up to 3,000 stills with an image frequency of 8 seconds). The recorded data can be co-ordinated with the site being surveyed. The vehicle can also be used to collect and record data which can later be processed on board the mother ship.

Specifications
Operating depth: down to 6,000 m
Speed: 3 kt
Endurance: 10 h (from submersion to surfacing)

Contractor
Gidropribor, Scientific and Research Institute, St Petersburg.

VERIFIED

SWEDEN

Double Eagle Mk II

Type
Remotely operated mine disposal vehicle.

Description
The Double Eagle Mk II is the new improved version of the proven Double Eagle system. The ROV is a comparatively lightweight system with extremely good manoeuvrability fitted with powerful high-speed thrusters for use in currents up to at least 3 kt. The vehicle is driven by eight thrusters giving over 6 kt forward speed. Double Eagle Mk II is fitted with a computerised stabilisation control system and is extremely manoeuvrable, exhibiting unlimited movements in 6° of freedom (pitch ±180°, roll ±180°) which gives an operator the possibility to dive the maximum depth with maximum speed (>6 kt).

The display comprises a monitor (or several) displaying the picture from a colour camera. Digital data such as heading, depth, pitch and roll angle, cable twist, leakage warnings, real-time clock with date and time and diagnostics are superimposed on the TV monitor. High-quality video is assured through the fibre optic link in the umbilical cable the standard length of which is 1,000 m.

The vehicle can be equipped with several types of sensors and tools such as sonars (electronic scanning or conventional), echo-sounders, Doppler logs, tracking system, automatic navigation system and manipulators.

The vehicle uses a unique precision charge placement technique using an aiming sight to position the mine disposal charge with very high precision. The vehicle remains in an absolutely stable attitude as the charge is released which allows the operator to accurately control the operation and handling of the vehicle.

The system uses a tool which allows the vehicle to be launched and recovered from any ship equipped with a sea crane.

The vehicle has unlimited endurance which allows it to be moved from target to target for as long as necessary.

For the Australian Navy the Double Eagle has been equipped with brushless motors to meet the rigorous conditions in which it is required to operate.

Operational status
A Double Eagle Mk II vehicle has been used in the PVDS concept since 1994. The ROV is in service with the navies of Australia, Denmark, Finland and Sweden and in commercial versions in France and South Korea.

Specifications
Length: 2.1 m
Width: 1.3 m
Height: 0.5 m
Weight: 350 kg
Payload: >80 kg
Speed: >6 kt
Operational depth: 500 m

Contractor
Bofors Underwater Systems AB, Motala.

UPDATED

UNITED KINGDOM

Towtaxi

Type
Towed underwater vehicle.

Description
Towtaxi is an advanced, highly stable, actively controlled vehicle designed for normal operation at depths down to 100 m, but with a maximum depth capability of 250 m. Payloads and electronics may be housed in the body of the vehicle and an external payload, such as a transducer array, can be mounted underneath. The vehicle is controlled from on board the tow vessel by a command and monitoring system which is linked to the vehicle via the towcable.

Operational status
In use by the UK MoD Defence Research Agency and SACLANT Research Centre at La Spezia, Italy.

Contractor
BAeSEMA, Bristol.

VERIFIED

Towtaxi

NASP

Type
Non-Acoustic Sensor Platform (NASP).

Description
The NASP has been specially designed with extremely low-magnetic self-noise so that it can be used to mount magnetic anomaly detection sensors to detect mines and other seabed objects. To date it has only been possible to operate such sensors from towed bodies deployed from ships and submarines.

The flexible unmanned underwater test vehicle will enable the performance of several different types of non-acoustic sensors to be evaluated. The platform features radical new forms of construction and power. The modular vehicle incorporates a number of novel features in addition to its construction and power units.

Operational status
Under development. Sea trials of the sensor suite using the vehicle were undertaken in the summer of 1997, where the NASP and different types of sensor including magnetic gradiometers, electro-optic sensors and olfactory sensors were tested against mine-like objects.

Contractor
DERA, Gosport.

VERIFIED

Archerfish

Type
Self-propelled MCM munition.

Description
Archerfish is a low-cost, self-propelled, torpedo-shaped, one-shot mine disposal munition which has been developed to considerably reduce time on task, operator workload and risk. The vehicle is powered by twin propulsors which can operate in either transit mode for rapid movement (3 kt) between ship and target, or in hover mode during identification and destruction of the target.

The battery-powered munition is deployed from the MCMV and guided to the target using command guidance signals derived from the parent ship's sonar and passed to the munition via an optical fibre link. Alternatively, automatic waypoint navigation can be preset prior to the mission. The 1,000 m fibre optic communication link is dispensed from the rear of the vehicle. A sonar and LLTV camera are used to provide visual display from which the vehicle can be terminally guided on to the target using a joystick control. In the terminal approach, the vehicle's own short-range sonar and video link acquires the target and transmits more detailed information to the ship. The onboard sonar is dual mode and can be switched to imaging for use in turbid water.

The warhead is a directed energy charge of Plastic Bonded Explosive (PBX) developed by SNPE of France which is effective against the latest composite explosives in ground and moored mines.

The system includes a comprehensive PC-based simulation package for operator training or exercise simulation. Operators can customise and upgrade the software to meet their own requirements.

Operational status
Archerfish was to have been offered to meet the requirements for the UK 'Hunt' class mid-life update (HMLU) programme. However, since this programme has been scaled down in scope to a 'Hunt' Minimum Update (HMU), an 18 month feasibility study has concluded that the single-shot mine disposal technology was not sufficiently mature for the commencement of the HMLU, and will not be included in the HMU.

Archerfish has been offered into the US DoD for the Airborne Mine Neutralisation (AMNSYS) Programme.

Contractor
GEC-Marconi Underwater Systems Group, Waterlooville.

UPDATED

Hyball

Type
Remotely operated underwater vehicle.

Description
The Hyball is a small, portable vehicle designed principally for inspection of ships' hulls, harbour installations, buoys and moorings, seabed installations and so on. The vehicle's standard equipment fit includes a low-light CCD colour TV camera with wide-angle auto-iris lens mounted on a chassis which pitches through 360° and allows the camera to view up, down, forward or backwards through the unique patented Meridian viewport. TV pictures are relayed to a 15 in colour monitor in the surface platform.

For poor visibility conditions a low-light monochrome SIT camera may be mounted on the standard camera chassis together with the stills and colour cameras. The vehicle is also fitted with two fixed 75 W quartz halogen lamps pointing forward and two 75 W lights mounted on the camera chassis which follow the camera.

Vehicle status information is displayed along the bottom edge of the monitor screen and includes text lines for information such as customer name, job number or location which is defined by the operator at the beginning of each dive. The video overlay may be dimmed or switched off by the operator.

The Hyball is powered by four thrusters, two for forward, reverse and rotational motion and two for vertical and lateral movement. Power is derived from 0.5 hp 24 V DC brushed motors through 10.5:1 reduction gearboxes driving 5 in diameter propellers mounted in Kort nozzles. Thruster speed and direction are controlled by digital signals proportional to movements of the joystick. Vehicle movement is controlled by the single three-axis joystick and dive and surface buttons on the hand controller. Any combination of vehicle movement control may be held by the use of the trim button for hands-off operation or for recentering the joystick when operating in a cross-current. Vertical thrust power may be preset from the surface unit panel.

Operational status
In service with the navies of Germany, Philippines and UK.

Specifications
Length: 0.535 m
Width: 0.65 m
Height: 0.565 m

Weight: 42 kg
Speed: >2.0 kt
Operating depth: 300 m
Payload: 4.5 kg

Contractor
Hydrovision Ltd, Aberdeen.

VERIFIED

Offshore Hyball

Type
Remotely operated underwater vehicle.

Description
An alternative to Hyball, this ROV exhibits all the features of Hyball, but with twice the power, additional payload and interfaces provided for sonar and a manipulator.

Operational status
In service with the navies of Italy and Turkey.

Specifications
Length: 0.535 m
Width: 0.77 m
Height: 0.575 m
Weight: 45 kg
Payload: 7 kg

Speed: >2.5 kt
Operational depth: 300 m
Payload: 4.5 kg

Contractor
Hydrovision Ltd, Aberdeen.

VERIFIED

UNITED STATES OF AMERICA

AN/SLQ-48 MNS

Type
Remotely operated mine disposal system.

Description
The AN/SLQ-48 MNS (Mine Neutralisation System) developed by Alliant Techsystems Marine Systems (now part of Hughes Naval and Maritime Systems) and the US Naval Sea Systems Command, has completed production for the US Navy's new MCM and MHC ships. In operation since 1987, it is designed to detect, locate, classify and neutralise moored and bottom mines, using high-resolution sonar, low-light level TV, cable cutters and mine destruction charges. The tethered underwater vehicle carries these sensors and countermeasures and is controlled from the parent vessel out to a range of 1,000 m.

Initial target detection and vehicle guidance information is provided by the ship's sonar. Initial vehicle navigation is plotted and monitored within the MNS acoustic tracking system. Vehicle sonar is used during the mid-course search and final homing phases, and high resolution enables operations in poor visibility by sonar guidance alone. Low-light TV is used in conjunction with sonar during the precision guidance phase near the target. Underwater launch and recovery of the MNS vehicle assists operations in high sea states.

Vehicle power is provided via a neutrally buoyant umbilical cable, which also carries signal and control links between the vessel and the MNS vehicle. Aboard the parent vessel sonar screens on the control console display sensor and vehicle status information in alphanumeric form. Display facilities include vehicle sonar, vehicle TV, deck TV, vehicle control and navigation, and provision for monitoring system status.

MNS II is being developed with a smaller footprint and performance improvements. MNS II integrates the separate monitor and control consoles into a single console and reduces the size of the cable handling and power system. Intended for smaller MCMVs, MNS II also provides fibre optic telemetry and increased thrust that will raise vehicle speed to over 8 kt.

Using alternative sensors MNS II can extend the current MNS mission beyond bottom and moored mine neutralisation to the following applications: buried mine location and neutralisation, mine search and classification, bottom mine survey, search capability operating ahead of the parent platform and recovery or repair of underwater objects.

Operational status
In production since 1986. MNS was successfully used by USS *Avenger* and *Guardian* in the Persian Gulf

MNS vehicle showing handling arrangement   *1994*

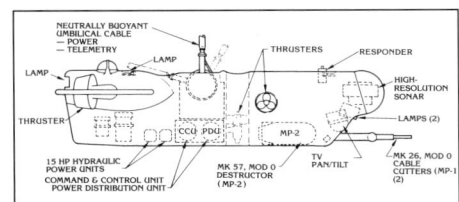

Umbilical cable-handling system provides constant tension to reduce cable catenary

Single console houses all monitor and control functions

Typical installation on minehunter vessel

Multipurpose crane handles vehicle launch and recovery

High-voltage transformer converts ship's power to vehicle power

MNS II vehicle for multiple payloads

Hydrophone provides independent tracking of vehicle

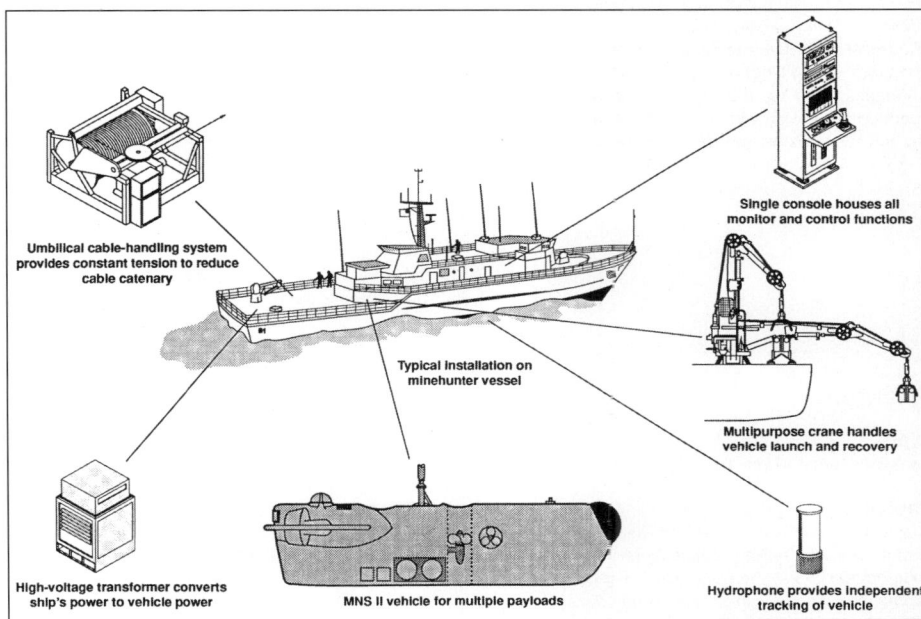

Arrangement of the MNS   *1994*

Elements of MNS II   *1999*/0024792

during Operation Desert Storm. In mid-1995 the total programme value exceeded US$225 million.

A total of 28 full systems including 67 vehicles have been delivered to the US Navy.

Specifications
Length: 3.8 m
Width: 0.9 m

Height: 1.2 m
Speed: 6 kt
Vehicle weight: 1,247 kg (in air)
Propulsion: hydraulic, twin 15 hp
Operating depth: 600 m
Thrusters: horizontal/vertical/lateral
Cable length: 1,070 m
Power requirements: 43 kVA

Contractor
Raytheon Systems Company, Mukilteo, Washington.

UPDATED

MUST Lab

Type
Underwater test laboratory vehicle.

Description
The Mobile Undersea Systems Test (MUST) Laboratory is designed to carry both wet and dry payloads. A versatile test facility, the MUST Lab provides rapid turnaround and multiple dives without a requirement to return to port. Rough weather does not impede the vehicle. An innovative hull design with unprecedented modularity and payload capacity supports a wide range of programmes.

The baseline vehicle is 9 m long and made up of five hull sections. Configurations with three to seven hull sections result in overall lengths of 7.5 to 10.5 m. The main pressure vessel arrangement permits the use of standard 19 in rack-mounted equipment. This eliminates the need to repackage each project. Nearly one tonne of payload capability and 1,500 litres of payload are available in the baseline five section vehicle. Maximum depth of the vehicle is 610 m. Vehicle endurance ranges from 8 hours at full propulsion and payload power to 24 hours at lower propulsion loads. Electrical power consists of 24 V DC and 120 V DC. Lead-acid batteries supply power.

A unique launch and recovery system allows UUV operations in sea conditions that were formerly considered unsafe. Minimal manpower and deck rigging are needed to maintain a positive control of the vehicle. The heart of the system is an articulated ramp and deck-mounted trunnion. Together they decouple

MUST underwater vehicle   **1994**

the ship's motion from the motion of the vehicle in the water. Typically the launch phase requires about 10 minutes and the recovery phase about 30 minutes or less. Ship speed during launch and recovery is 1 to 3 kt. Over 200 successful launches and recoveries have been performed in sea conditions up to and including Sea State 4.

Operation is monitored and piloted from the control/maintenance van containing the surface control console. Predive planning, software development, simulation, and post-dive analysis are also conducted in this van. The launch and recovery system, with vehicle hangar van, provide for vehicle handling, maintenance and storage during all phases of at-sea operations. MUST Lab's entire at-sea system is air-transportable to any location.

A complete UUV simulation and integration facility – Dry Lab – performs almost any preliminary work necessary for successful sea trials. This exact physical mock-up of the MUST Lab vehicle supports installation of rack-mounted electronics inside the pressure vessel, and sensor/actuator packages in the forward and aft fairings. The power distribution, vehicle operating system, and distributed input/output network duplicate the actual systems found in the vehicle. Dry Lab's capabilities allow independent verification and validation of MUST or payload software, operator training, mission planning and post-dive data analysis and mission reconstruction.

The vehicle is used to test a variety of underwater vehicle configurations, roles and payloads including autonomous mine surveillance, detection and destruction, route survey and autonomous clandestine surveillance roles.

Contractor
Lockheed Martin, Perry Technologies, Riviera Beach, Florida.

UPDATED

RSS

Type
Remotely operated vehicle.

Description
The Remote Surveillance System (RSS), a free-swimming, Remotely Operated Vehicle (ROV), offers cost-effective surveillance in a local or long-range system. Operating as a semi-submersible ROV, the RSS provides real-time video within a full sunlight to starlight capability. Assignments for covert surveillance, waterside security, drug interdiction and other data collection missions are readily accomplished with this portable and easily recoverable platform. The ability to monitor installations unobserved provides additional security for the RSS operators, who will not be exposed to danger.

Weighing 200 kg, the RSS can be deployed and operated by a team of three, independent of outside support. Operations can be conducted from a small

boat. This system minimises the need for large numbers of personnel and support facilities. The vehicle length is 3 m. The outer hull dimension measures 28 cm.

Vehicle endurance is 4 hours when running at 3 kt. A maximum speed of 6 kt can be attained. Power consists of 12 V DC lead-acid rechargeable batteries. The vehicle is controlled by Radio Frequency (RF) link with a range of 8 km.

Potential applications for the RSS vehicle include: base security of waterfront locations, boundaries and borders security, pollution and spillage patrol, intelligence and surveillance, anchorage security, forward observation post, wildlife survey, payload delivery, payload removal, sabotage, forward escort.

Contractor
Lockheed Martin, Perry Technologies, Riviera Beach, Florida.

VERIFIED

RSS underwater vehicle
1994

Phantom

Type
Remotely operated vehicle.

Description
The Phantom ROV has been utilised in a variety of commercial and military applications including mine countermeasures, harbour and port security, submarine rescue, ship hull inspections, bottom surveys, general diver support, and drug enforcement.

The fully streamlined Phantom HVS4 (deployed from

COOP) and Phantom HVS2 (for airborne deployment) were specifically developed for the US Navy as powerful MCM/neutralisation vehicles. As members of the Phantom Spectrum system, these vehicles use control consoles, umbilical cables and accessories that are interchangeable with nine other Phantom vehicle types. Other models that are frequently used in MCM applications include the Phantom S4, Phantom HD2 and Phantom 300.

All Spectrum-class Phantom ROVs feature a highly manoeuvrable vehicle platform protected within a full-perimeter stainless steel crash frame. High-resolution

colour and optional STI monochrome video cameras are mounted to tilt with a pair of high-intensity quartz halogen lamps. The system may be equipped with a high-definition scanning sonar, built-in acoustic navigation responder, manipulator, and a payload deployment mechanism. Other accessories include laser-scaling devices, video on-screen display, remote joystick control, fibre optic cables, and a cable cutter.

The vehicle is guided from the ship via a neutrally buoyant umbilical cable connected to the 19 in rack-mounted Spectrum control console. Umbilical cable lengths are available up to 1,200 m. The Spectrum

Phantom Ultimate **1997**

series is a hard-wired system, which minimises the complexity of subsea electronics and the cost of repair should the vehicle be flooded or lost.

Operational status

Since its introduction in 1985, more than 350 Phantom ROVs have been produced and are presently being operated worldwide in over 30 countries including nine navies. Phantom ROVs are now in service with military departments of Australia, Brazil, Egypt (six units for the Border Guard), New Zealand, Portugal, Singapore, Sweden (12 HD2), the UK, and the USA. US Navy Phantom operators include the Explosive Ordnance Disposal teams; Naval Facility Engineering Support Center, Naval R & D, Naval Surface Warfare Center, Navy Research Lab, and Naval Coastal System Center. The latest system to enter service is a 450 m rated Phantom Ultimate vehicle which completed acceptance trials with the Brazilian Navy in the summer of 1996. The vehicle will be used to handle equipment and tool packages for the rescue of crews from sunk submarines.

Contractor

Deep Ocean Engineering Inc, San Leandro, California.

VERIFIED

Phantom HD2+2 with associated umbilical and control and display unit **1997**

A Phantom HD2 vehicle similar to the type delivered to the Swedish Navy
1998/0009515

Super SeaROVER

Type

Remotely Operated Vehicle (ROV).

Description

The Super SeaROVER high-performance ROV can be customised to perform a variety of tasks including mine countermeasures, harbour security surveillance, inspections, surveys and so on.

The lightweight, compact ROV is fitted with 48 cm rack-mountable consoles, controlled via user-friendly controls, and with a sophisticated telemetry system with an RS-232 interface and a head-up video graphics overlay.

A number of special options can be fitted to the ROV including ultra-short baseline tracking system, high-resolution scanning sonar, altimeter, 1, 2 or 3 function manipulator arms, super thruster package or 35 mm cameras.

The vehicle is powered by two (optionally four) horizontal thrusters, and one vertical and one lateral powered by brushless DC drive motors. These give the vehicle a maximum speed of 5 kt and enable it to turn on its own axis, reverse, move laterally and vertically while maintaining heading, and hover motionless in

light to moderate currents. The ROV is rated to a depth of 300 m.

Standard equipment carried by the Super SeaROVER includes a high-resolution colour video camera and two high-intensity underwater lights. The camera has a wide field of view (90°) and can pan and tilt on command from the operator. Video signals from the camera and navigational data from the vehicle's sensors are transmitted via an umbilical cable to surface TV monitors, allowing the pilot to simultaneously view the underwater scene and the on-screen video graphics overlay showing heading, depth and other data.

A wide range of optional equipment can be fitted such as a positioning/navigation system, scanning sonar, manipulators, a low-light video camera and a tether management system.

Operational status

Originally ordered in 1988 by the US Navy as part of its effort to keep shipping lanes in the Gulf open during the Iran/Iraq war. Subsequently vehicles were used during Operation Desert Storm and later in the northern Gulf as part of the clean-up effort to identify and clear underwater mines laid by the Iraqis.

Contractor

Benthos, North Falmouth, Massachusetts.

UPDATED

The Super SeaROVER ROV with video camera and lights forward and thrusters aft

RCV-225

Type
Remotely Operated Vehicle (ROV).

Description
The RCV-225 system is designed to carry out detailed underwater tasks down to 400 m, such as the survey of underwater installations, search and salvage, remote monitoring of the laying of underwater installations, oceanographic surveys, placement of charges and arming. The vehicle is fitted with a low-light level SIT camera and optional colour TV and photographic systems. The SIT camera is equipped with a unique lens assembly that enables the operator to remotely pitch the angle of view ±90° from the horizontal. Two 45 W tungsten halogen lamps provide a viewing range of up to 10 m with no ambient light.

The ROV is driven by four oil-filled electric motors giving a speed of 3 kt in all directions. The motors are enclosed in a syntactic foam hull and the camera and electronics are contained in a pressure housing.

Full command of the vehicle is exercised from the control station with desired depth and heading automatically maintained by servo controls. Vehicle depth, heading and lens pitch angle are displayed in the TV picture and continuously recorded on videotape.

The vehicle is deployed using a protective RCV launcher with tether cable and winch, a deck winch with double-armoured cable mounted on a skid/A-frame assembly and a hydraulic power supply. The vehicle is carried to the working depth in the launcher, from which it is deployed for the operation, returning to the launcher for transport back to the surface.

Operational status
In service aboard the Japanese 'Futami' class survey vessels.

Specifications
Length: 5.1 m
Width: 6.6 m
Height: 5.1 m
Weight: 82 kg (in air)

Contractor
SEA Hydro Products Inc, San Diego, California.

UPDATED

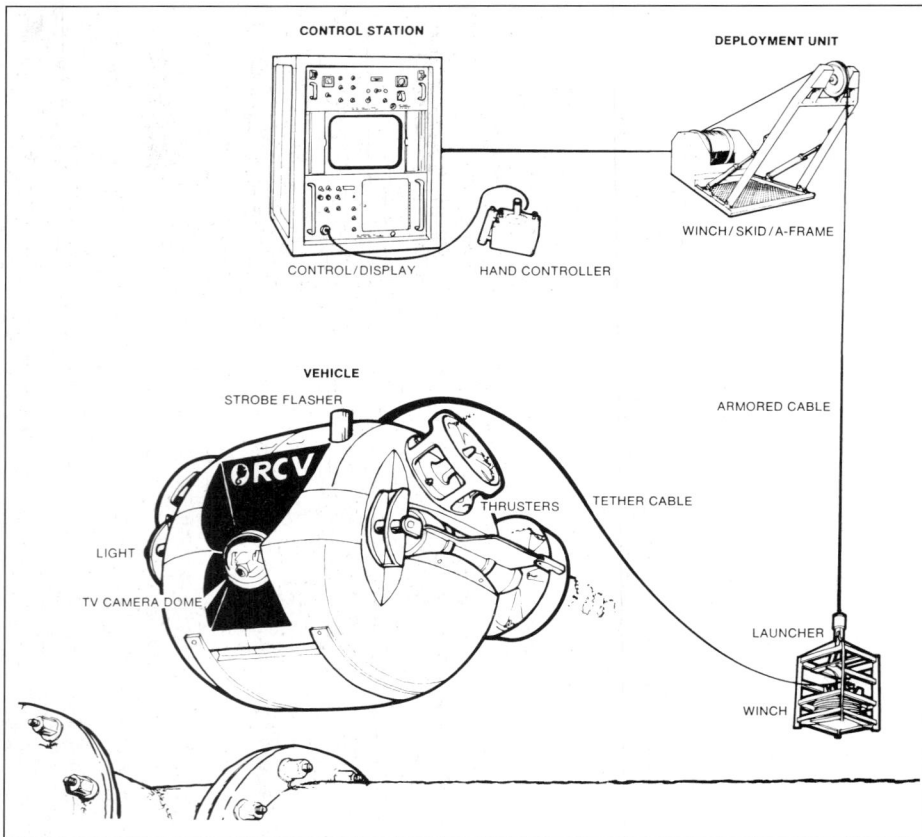

RCV-225 remotely operated vehicle

UUV

Type
Prototype Unmanned Undersea Vehicle (UUV).

Development
The US Defense Advanced Research Projects Agency (DARPA) together with the US Navy and Charles Stark Draper Laboratories has built two prototype UUVs to demonstrate the feasibility of developing a design for an ROV capable of undertaking covert high-priority missions. The vehicle is intended for deployment from submarines as well as surface ships. A third prototype is under construction and is due to be delivered next year. The US submarine *Memphis* has been adapted as a trials boat for testing the UUV concept.

The UUV will be fitted with a range of advanced acoustic sensors, communications and signal and data processing systems for trials which will demonstrate the vehicle's ability to carry out autonomous operations. Missions envisaged include MCM, surveillance and communications. Three mission systems are being developed for the UUV: a Tactical Acoustic System (TAS), Mine Search System (MSS) and Remote Surveillance System (RSS).

The hull is of titanium and pressure tests on the first hull were carried out in December 1990. An internal pressure hull contains the payload which occupies a section some 1.5 m in length. Power for the 12 hp electric motor will be provided by batteries housed in two 1.3 m long sections while the vehicle's control electronics will be contained in another 1.3 m section in the centre section. The electric motor and its control unit will be housed in a 3.5 m aft section.

Specifications
Length: 11 m
Diameter: 1.117 m

Contractor
DARPA.
Charles Stark Draper Laboratories.

UPDATED

XP-21

Type
Unmanned Underwater Vehicle (UUV).

Description
First deployed as a rapid prototype testbed platform, the XP-21 is a multimission UUV capable of performing operations similar to an ROV, with a power tether cable or a fibre optic only tether (the latter using an onboard power source). It can be operated either in a semi-automatic mode using an acoustic link for command and control and data transmission, or in a fully automatic mode.

The system is designed to carry a wide range of sensors for multimission applications including multibeam side scan sonar, forward-looking sonars for mine detection and classification, magnetometers and optical sensors such as the Applied Remote Technology LS-4096 laser imaging system which provides mine identification at up to six times the range and 10 times the resolution of conventional underwater LLTV cameras. One of the sensors incorporated in the

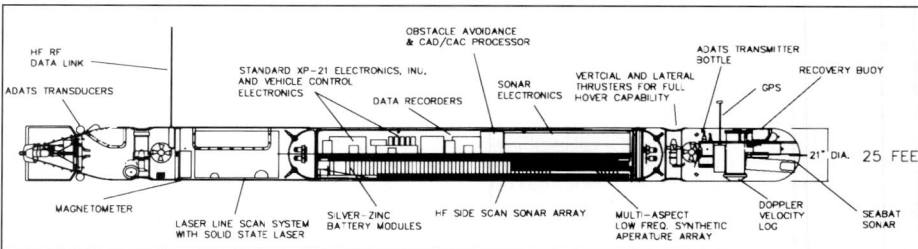

XP-21 survey and inspection platform   *1995*

XP-21 is the RESON Inc SeaBat 9001S switchable real-time multibeam echo-sounder and forward-looking sonar system. Navigation sensors include Doppler sonar, ring laser gyro Inertial Navigation Unit (INU) or GPS/ultra-short baseline system via the support ship. Onboard power is via either lead-acid or silver zinc batteries. A newly developed aluminium oxide fuel unit providing 10 times the range of lead-acid batteries is also available.

Lateral and vertical thrusters and the Doppler navigation enable XP-21 to transit from full hover to 6 kt.

Operational status
Operational since 1988. In service with the US Navy for mine surveillance, array development and search and recovery operations. Further demonstrations of the vehicle were carried out in UK waters during 1995.

Specifications
Length: 5-8.50 m (depending on payload)
Diameter: 0.533 m (submarine tube compatible)
Displacement (in air): 730-909 kg (depending on payload)
Payload: 0.1-0.75 m³ (dry)/0.1-0.45 m³ (wet)
Propulsion: DC brushless thrusters
Speed: 5 kt (capable of ROV-type hovering)
Operating depth: 600 m

Contractor
Applied Remote Technology Inc, a Raytheon company, San Diego, California.

VERIFIED

XP-21 UUV
1996

Hydra

Type
Cable repair vehicle.

Description
The Hydra-AT 1850 CRS has been designed with an extremely low magnetic signature for the location, tracking, excavation and recovery of buried cables for repair and reburial.

The vehicle is equipped with a powerful jet for blowing away mud and sand covering cables. The vehicle detects and tracks cables using the magnetic field developed around the cable, or by detecting superimposed AC tones or DC current generated through the cable.

The ROV is equipped with a range of tools including two seven function manipulators, the powerful jet system bolted to a skid on the bottom of the ROV, eight cameras (five colour), a short-range scanning sonar and the usual range of sensors for manoeuvring.

The five video cameras perform navigation, observation and control of jetting operations. Three of the cameras are on pan/tilt mountings for navigation, with another camera aft for observation. The system uses extensive graphic displays which enable the controller to manoeuvre the vehicle in conditions of zero visibility.

Power is provided by two 75 hp hydraulic packs driving seven thrusters; two twin units fore and aft, two lateral units and three vertical units.

Operational status
Delivered to the US Navy in 1988, operational at the beginning of 1990. A second system is on order for the US Navy and a third unit has been ordered by a commercial organisation.

Specifications
Length: 3.5 m
Width: 1.83 m
Height: 1.83 m
Weight: 3,775 kg (in air)
Speed: 3 kt (forward)
Depth: 1,850 m

Contractor
Ocean Systems Engineering.

VERIFIED

EMD

Type
Expendable Mine Destructor (EMD).

Description
The EMD is a small, self-propelled device for the destruction of seabed, moored and floating sea mines.

It features fully automatic operation from launch to target impact to minimise operator workload and training requirements and maximise speed to target.

Acoustic ultra-short baseline navigation or ship's sonar guides the EMD to an automatic terminal homing approach via a sophisticated acoustic seeker head. A bidirectional fibre optic link provides real-time video to the operator from two separate EMD onboard cameras. Sonar target image and ability to manually override EMD controls are also provided.

The device is fitted with a choice of warheads including modular, blast, shaped or advanced types including four shaped charge warheads which are symmetrically angled and designed to detonate the mine's explosive charge. In the event of this not being achieved, the detonation of the EMD is sufficient to disable the mine's triggering mechanism.

With its shrouded propulsor and manoeuvring surfaces the EMD is usable in currents up to 5 kt and in very shallow waters of only 2 m. The device is fitted with a seeker head which has a range of 30 to 50 m.

The EMD can be handled easily and deployed from ships, helicopters or drones. It is designed to be used with a mine-hunting sonar for kill verification, but can be used independently with expendable bottom transponders.

A recoverable reconnaissance/training version with a TV camera and optical fibre link is available.

Operational status
Under development as a future project.

Specifications
Length: 1.2 m
Diameter: 0.2 m
Weight: 32 kg
Speed: 10 kt
Operational depth: 300 m
Endurance: 15 min at 8 kt or 1 h at 4 kt

Contractor
L-3 Communications Ocean Systems, Sylmar, California.

Expendable mine destructor L-3 Communications Ocean Systems   *1999*/0024814   *UPDATED*

TABLE XXI

ROV inventory

ROV	Manufacturer	Role	Optional equipment	Status/In service
AN/SLQ-48	Hughes Aircraft	MH/mine destruction	Sonar, TV camera, cutter, MDC	Japan, US
ANPA 6000	Gidropribor	Survey	Sonar, TV camera	RFAS
ARCS AUV	ISE	Research	Various	Canada, Rockwell, Johns Hopkins Lab
Archerfish	GEC-Marconi	Mine destruction	Sonar, TV camera, shaped charge	Development
Dolphin AUV	ISE	Survey/research/MH	Sonar	Canadian Hydrographic Service, US, ISE
Double Eagle	Bofors Underwater Systems AB	MH/mine destruction	Sonar, TV camera, MDC	Australia, Denmark, Finland, Sweden
ERATO	DCN International	Weapon recovery	TV camera, manipulator	France
Hyball	Hydrovision	Inspection	TV camera, SIT camera, lights	Germany, Philippines, UK
Hydra	Ocean Systems	Cable operations	TV cameras, sonar, manipulator	US
MIN Mk2	SMIN Consortium	MH/mine destruction	Sonar, TV camera, MDC	Italy
Minesniper	Simrad	Mine destruction	TV camera, shaped charge	Development
MUST Lab	Lockheed Martin	Research	Various	
PAP 104	ECA	MH/mine destruction		Croatia, France, Germany, Indonesia, Japan, Malaysia, Netherlands, Pakistan, Saudi Arabia, South Africa, Turkey, UK, Yugoslavia
Phantom	Deep Ocean Engineering	MH/mine destruction	Sonar, TV camera, manipulator, MDC	Australia, Brazil, New Zealand, Portugal, Sweden, Singapore, UK, USA
Pinguin	STN ATLAS Elektronik	MH/mine destruction	Sonar, camera, MDC	Germany, Taiwan
Pluto (& Pluto Plus)	Gaymarine	MH/mine destruction	Sonar, TV camera, manipulator, MDC	Egypt, Finland, South Korea, Nigeria, Norway, Spain, Taiwan, Thailand
RC 225	SEA Hydro	Survey/research	LLLTV, colour TV	Japan
Seafox	STN ATLAS Elektronik	Mine destruction		Development
Super SeaROVER	Benthos	MH/mine destruction	Sonar, TV camera, manipulator	US
Theseus	ISE			
Towtaxi	BAeSEMA	Research	Various sensors	UK DERA, Canada, SACLANT
XP-21	Raytheon	Various	Various (laser, sonar)	US

Note: MDC – Mine Disposal Charge

VERIFIED

REMOTELY OPERATED VEHICLES' ANCILLARY EQUIPMENT

DENMARK

DAMDIC

Type
Mine disposal charge.

Description
The DAMDIC (Danish Mine Disposal Charge) is a medium-sized explosive charge developed for the destruction of mines and other ammunition detected on the seabed. The charge is carried to the vicinity of the target by an ROV where it is released to carry out a controlled descent to the location of the mine. The ROV then returns to the parent Mine CounterMeasures Vessel (MCMV) and the charge is detonated by cable control.

The fore end of the charge is a glass fibre-reinforced polyester housing, filled with Composition B explosive. At the core of the main explosive charge is an intermediate charge of 59.2 g of HNS. Mounted aft of the charge housing is the fuze mechanism which is of the interrupted type, with a tight-fitting aluminium blind of 16 mm. The detonator is EMS-Patvag ZP 71-2 55/60 with 55 mg AgN_3 and 60 mg PETN. The measuring current is 1 mA (maximum) for a maximum of 30 seconds. The maximum acceleration is 50,000 g. The lead charge is 600 mg Composition B, and the booster 21.6 g HNS.

Considerable attention has been paid to safety in the design of DAMDIC. One of the principle features is that the charge and its detonator mechanism do not contain an energy source which can activate the detonator.

Detonation is controlled from the MCMV by wire, in order to avoid dependency on acoustic signals (which can be unreliable in coastal waters with large variations in salinity through the water column). The operator can therefore move to a considerable distance (some 300 m) before connecting the detonation cable.

The fuze is an out-of-line mechanism with a depth (pressure-activated) arming mechanism, two independent locking mechanisms during operation, as well as an independent transport locking mechanism. The depth-based arming actuator requires a depth of 5 m to overcome spring pressure. The arming lock mechanisms require release of the charge from the ROV and one of the mechanisms requires additionally free space above the charge in order to float free.

Both locking mechanisms are held captive under a safety shield during transport and while the charge is attached to the ROV.

The charge can be made safe by decoupling the detonation wire, and can be restored to full out of line fuze safety on the seabed by externally clamping the fuze mechanism. Switch contacts prevent detonator activation if the fuze is not armed. The fuze mechanism is disarmed by a spring if the charge is brought up to the surface after arming.

Deployment of DAMDIC is carried out by the ROV which transports the charge to the location of the mine. The ROV then hovers over the mine at a relative height of about 2 m. The operator inspects the mine using the ROV's sensors (sonar, video camera and so on) and selects a suitable placement point for the charge. The command to release the charge from the ROV is given remotely from the MCMV to the ROV. A pneumatic piston removes the mounting rod from the DAMDIC support lug, and the charge is released.

Fins at the rear of the charge to provide guidance and stability during the charge's fall to the seabed. These fins ensure that the charge carries out a controlled descent. The actual release height and strength of the current must be taken into account when the point of release is selected. Extensive trials have resulted in a complete set of drop curves for relevant current speeds and heights above the seabed. Software based on this data has been integrated into the TV subsystem of the ROV and controls an aiming reticle on the operator's display. The ROV always heads into the current before releasing the DAMDIC.

On reaching the seabed the charge may roll, depending on the composition and contour of the seabed. Four short legs on the charge ensure that any rolling distance is short. Normally the charge does not roll at all, but stands firmly on the four legs on the seabed.

When DAMDIC is released from the ROV the lightweight detonation control cable is spooled out from its canister mounted behind the charge on the ROV. When the ROV has returned to the MCMV the detonation cable is connected to the exploder and detonation of the DAMDIC can be initiated.

Operational status
Currently being supplied to the Royal Danish Navy for deployment from the Stanflex 300 MCMVs when they are configured for the MCMV role. The Royal Australian Navy has ordered several hundred DAMDIC worth DKr 30 million. The charge will be deployed from the Swedish Sea Eagle ROV aboard the RAN's new MCMVs.

A total of around 200 have been manufactured, of which 60 have been used for trials and testing.

Specifications
Length: 816 mm (incl fins)
Diameter: 250 mm
Weight: 49 kg (in air)
Charge: 31 kg Comp B
Operating depth: 6-300 m

Contractor
Nordic Defence Industries A/S, Norresundby.

VERIFIED

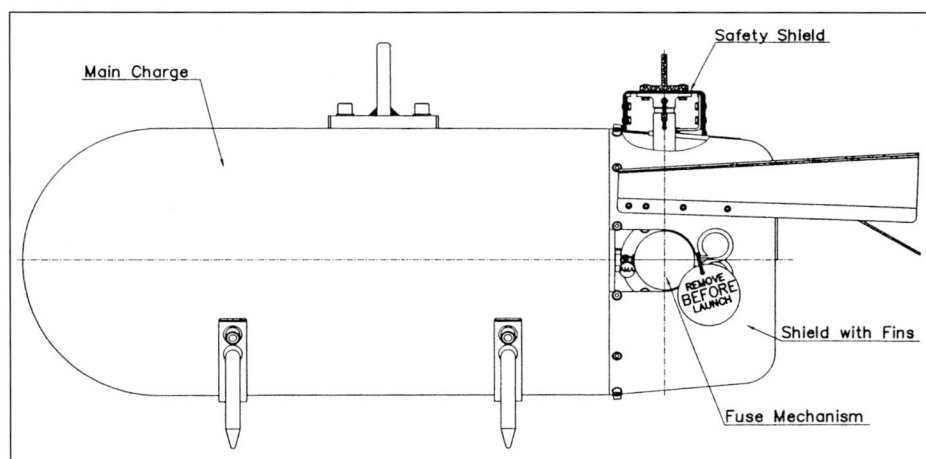

DAMDIC Mk I   *1998*/0009517

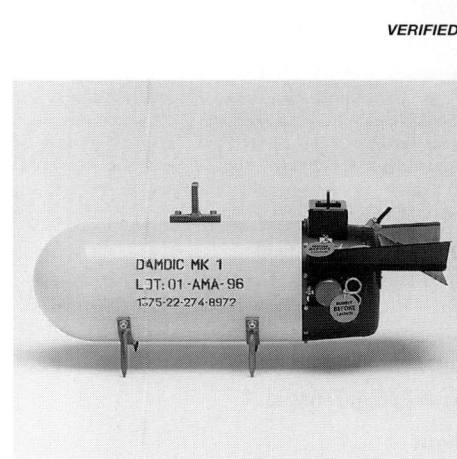

DAMDIC Mk I countermine charge   *1998*/0009516

FRANCE

RECA

Type
Remotely piloted mine disposal charge for use against moored and ground mines.

Description
Developed by ECA, the RECA is a new system designed to counter modern mines which are extremely sensitive to both acoustic and magnetic signatures. Carried by Remotely Operated Vehicles (ROVs) such as the PAP, the RECA is a very small, self-propelled remote-controlled mine disposal charge which is able to navigate towards the mine-like object using its autonomous acoustic tracking system, identify the mine using its onboard video and steer to the target by a simple command transmitted via its wire tether.

Like other standard mine disposal charges, RECA is fitted underneath the ROV. Once in the target area (up to 500 m from the target) the operator releases RECA from the ROV using the standard control console. The ROV then returns to the ship. The RECA's propelled section then detaches from the mooring frame which remains on the seabed to provide stability.

RECA is then remotely piloted by wire from the ship and can approach the target without endangering the ROV. Once the target is identified, RECA is piloted towards it and when contact is made the RECA explosive charge is remotely activated. If the target is identified as a non-mine, RECA is recovered on the surface.

RECA comprises two subsystems: the passive bottom mooring unit and the active unit comprising a power source, identification and tracking sensors (a 200 kHz sonar, two lights and a video camera) and control electronics, wire transmission line, vertical and horizontal propulsion thrusters and the small explosive charge.

RECA can carry a video camera and a 200 kHz sonar scanning through ±45°.

Operational status
Being developed and became available in 1997. Sea trials carried out during 1995 and 1996. RECA has been trialled by the French Navy.

Specifications
Weight (in air): 125 kg
Charge: shaped charge type or other
Operating depth (max): 100 m
Speed: >1.5 m/s horizontal, approx 0.4 m/s vertical
Autonomy: 10 min (at full speed)

Contractor
ECA, Boulogne-Billancourt.

UPDATED

GERMANY

DM 59/DM 69 to DM 119/DM 129

Type
Remote-Controlled Explosive Cutter (RCEC).

Description
Based in part on design features of the Rheinmetall explosive cutter for sweep wire attachment, this cutter is intended for use with underwater ROVs, such as the Société ECA PAP 104 system, STN ATLAS Elektronik Pinguin or Gayrobot Pluto.

The cutter's remote-control system is functionally identical to that used for detonating the mine disposal charge with the fuze DM 1002-3009 A1 to A3. With its hard foam housing, the RCEC is neutrally buoyant in water.

Operational status
In production and in service with the Japanese Navy, Royal Navy, Singapore and Royal Saudi Naval Forces.

Specifications
Cutting capacity
Stud link chains: 20 mm
Steel wires: 26 mm
Synthetic ropes: 40 mm
Operating range: 300-2,000 m
Operating depth: 5 to <300 m
Weight in air: 6.4 kg
Weight in water: neutral buoyant
Dimensions: 865 × 195 × 121 mm

Contractor
Rheinmetall W&M, Ratingen.

VERIFIED

*Remote-controlled explosive cutter
DM 59/DM 69
1995*

DM 1002-3009 A1-A3

Type
Remote-controlled fuze.

Description
The fuze is ignited by a coded acoustic signal only and deactivates automatically if no such signal has been received within a certain time span. The fuze is impervious to shocks from detonations or similar occurrences above or below the water, as well as to any other acoustic noise or signal.

Operational status
In production and used by the German Navy as well as by the Australian, Belgian, Dutch, Indonesian, Japanese, Norwegian, Royal, Saudi and Singapore navies.

Specifications
Length: 364.5 mm
Diameter: 200 mm
Operating range: <=2,000 m
Operating depth: 5-300 m
Weight: 8.0 kg
Weight of explosives: 0.92 kg
Self-deactivation: 45 min (after laying)

Contractor
Rheinmetall W&M, Ratingen.

VERIFIED

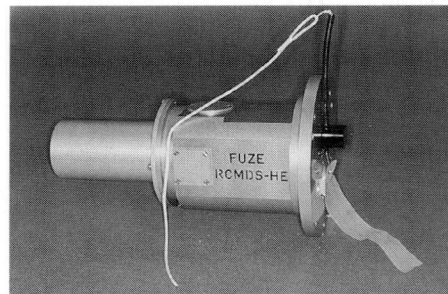

Remotely controlled fuze DM 1002 A1-A3   *1995*

E 67/E 67 Mod 1

Type
Firing transmitter.

Description
The firing transmitter generates a coded acoustic signal which is led via cable to the transducer in the water. The coded signal sets off the mine disposal charge via its remotely controlled fuze at operating distances of up to 2,000 m.

The transmitter exists in two versions. One is supplied in its own carrying case for use on the quarter-

deck, the other is for incorporation in a console below deck.

Operational status
In production and used by the German Navy as well as the Australian, Belgian, Dutch, Indonesian, Japanese, Norwegian, Royal, Saudi and Singapore navies.

Specifications
Firing transmitter E 67 capacity
Output: 14 W
Supply voltage for battery charger: 110 V
Batteries: 2 × 12V/500 mAh
Code setting: by two rotary switches
Weight of transmitter: 6.5 kg
Weight of cable with transducer: 10.8 kg
Length of cable: 33 m
Weight of storage container: 11.7 kg
Dimensions of transmitter: 230 × 200 × 130 mm
Dimensions of transducer: 200 mm (max diameter)
Dimensions of storage container: 440 × 600 × 250 mm
Firing transmitter E 67 Mod 1 capacity
Output: 14 W
Power supply: 115 V AC
Code setting: by two rotary switches
Weight transmitter: 3.5 kg

Weight of cable with transducer and cable drum: 15 kg
Length of cable: 100 m
Dimensions of transmitter: 340 × 153 × 153 mm

Contractor
Rheinmetall W&M, Ratingen.

VERIFIED

Firing transmitter E67 supplied in its own carrying case for use on the quarterdeck

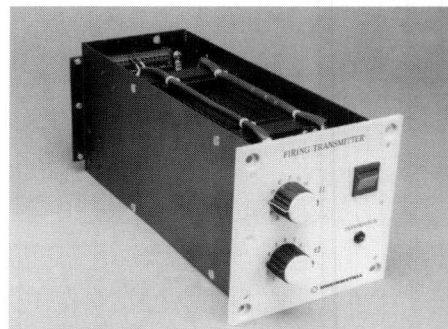

Firing transmitter E67 Mod 1 for incorporation in a console below deck

158 R/265 R

Type
Exercise fuze.

Description
The exercise fuze is functionally and dimensionally identical to the HE fuze DM 1002-3009 A1-A3 (see separate entry), except for the explosive train.

On receipt of the fuzing signal, the exercise fuze generates an electric impulse of 12 V and 1 A for a duration of 1 second. This impulse activates a relay or melts a wire which releases a buoy from an exercise charge.

Operational status
In production and use by the Royal Navy, Royal Australian Navy, Royal Saudi Navy and Singapore Navy.

Specifications
Operating range: <=2,000 m
Operating depth: 5-300 m
Total weight: 8 kg
Battery: 12 V/1,000 mAh
Relay impulse: 12 V, 4 A, 3.5 s
Reusable with same battery: 20-40 times
Dimensions: 364.5 mm (max length); 200 mm (max diameter)

Contractor
Rheinmetall W&M, Ratingen.

UPDATED

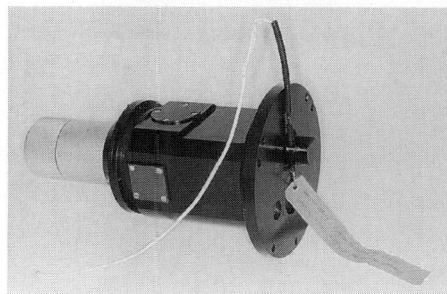
Exercise fuze 158 R   1995

ITALY

CM104

Type
Countermining charge.

Description
The CM 104 countermining charge operates through hydrostatic pressure acting on the safety arming device which turns a safety catch so that the two detonators automatically advance into a priming position. At the same time an electric switch is closed, while an additional electric circuit switches the detonators from the short circuit position to the working position. After the charge is released by the Remotely Operated Vehicle (ROV) the safety system of the charge is automatically deactivated.

When this occurs the electronic circuit is activated and an electronic arming delay of 16 minutes allows the operator to recover the ROV. During this time the charge remains in the safe condition.

An internal battery pack supplies electric power to activate the detonators. Remote charge activation is achieved by means of a coded acoustic signal generated by an ATX201 transmitter located on board the support ship.

The CM 104 can recognise 24 possible codes obtained with pulses in four different acoustic frequencies. Maximum activation range is 2,000 m.

Operational status
In production.

Specifications
Max diameter: 350 mm
Length: 940 mm
Height: 455 mm
Total weight (in air): 110 kg
Main charge weight: >80 kg CB 60/40
Operating depth: 6-300 m

Contractor
SEI SpA, Ghedi.

VERIFIED

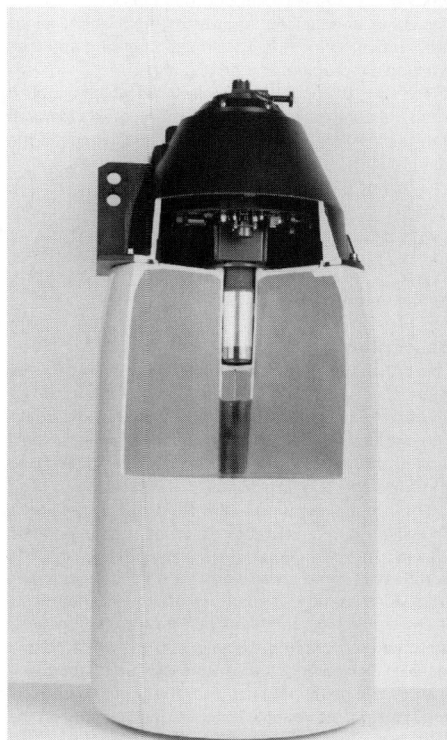
CM104

CAP (CM101)

Type
Countermining charge.

Description
This charge is particularly suitable for deployment by light remotely operated vehicles for mine destruction tasks. The charge is transported by the vehicle by means of a special attachment. When released, hydrostatic pressure activates the system and the two detonators automatically advance into a priming position. At the same time a contact is closed, completing the firing sequence while a second contact switches the detonators from the short-circuit position to the working position. The fire command is activated by dropping an explosive charge into the water at a safe distance or by a coded acoustic signal transmitted from the support ship. The charge can recognise a range of 24 possible codes contained in pulses in four different frequencies. Maximum activation range is 2,000 m.

Operational status
In service.

Specifications
Diameter: 250 mm
Length: 796 mm
Height: 367 mm
Weight: 44 kg (in air)
Main charge weight: >30 kg CB 60/40
Operational depth: 6-300 m

Contractor
SEI SpA, Ghedi.

VERIFIED

CAP underwater explosive charge

ECP (CM 102)

Type
Explosive cutter.

Description
The ECP is a specially designed explosive cutter device designed for mounting on the nose of Remotely Operated Vehicles (ROV) and is designed to cut the wire or chain of conventional moored mines. It is capable of cutting wires or chains up to a diameter of 30 mm. Two small guidance conveyors allow the device to be fitted on the wire by the pressure of the sensor against the wire or chain to be cut. At the same time the cutter device is disconnected from the vehicle which can then move back freely leaving the cutter locked in position on the wire. A switch is then activated to run an electrical circuit designed to start a time delay from the moment of the vehicle disconnection before the actual cutting operation.

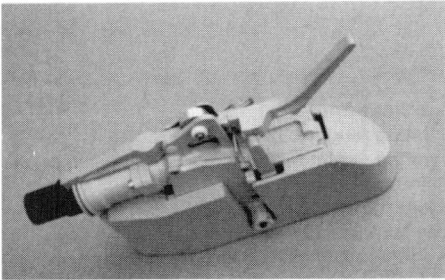

Operational status
In production and in service.

Specifications
Dimensions: 341 × 215 × 75 mm
Weight: 2 kg (in air)
Weight of explosive charge: 120 g (max)
Operating depth: 5-300 m

ECP explosive cutter device

Contractor
SEI SpA, Ghedi.

VERIFIED

CM 106

Type
Moored mine countermining charge.

Description
Currently under development to meet the Italian Navy ECMO project, CM 106 is designed to meet the requirement for an effective countermining charge dedicated to attacking and destroying moored mines without cutting them free from their mooring and thus creating the problem of a drifting mine.

The CM 106 incorporates interfaces which enable it to be transported by most current Remotely Operated Vehicles (ROVs) and to enable it to attach itself to the mooring cable or chain of a mine. It features a flotation system and a device which allows the charge to position itself against the mine in the most efficient position for mine destruction. Detonation is achieved via acoustic remote control.

After attaching the CM 106 to the mine's mooring cable, the ROV retires to a safe distance while an acoustic code is sent to the charge by remote control. This releases the charge to rise towards the surface while remaining attached to the mooring cable. On encountering the mine the charge is automatically positioned in the best attitude to assure close contact with the mine before detonating. After a fixed time, or upon acoustic firing command, the charge detonates causing destruction of the mine.

Operational status
Under development.

Specifications
Length: 430 mm
Width: 900 mm
Height: 190 mm
Weight: 27 kg (in air) (approx)
Charge: 8 kg (approx)
Operating depth: 150 m (max)

Contractor
SEI SpA, Ghedi.

VERIFIED

LARSC

Type
Launch And Recovery Smart Crane (LARSC).

Description
The LARSC has been specifically developed for the automatic launch and recovery of tethered or untethered underwater vehicles in high sea conditions. The system features reduced manpower requirements during the handling phase, improved safety and improved operational availability of vehicles.

The system comprises the ROV/AUV interface, a handling unit, crane and electronic remote-control system. The portable console allows the operator to exercise full control over the handling of the crane and its payload, using push buttons and control joysticks.

The electronic control unit performs all the functions of command, control, warning and system safety. It receives signals from the sensors located on the crane and the set points derived from the portable console and transforms them into output signals to operate the crane's booms.

The crane has two knuckle-type booms which automatically compensate for the relative motion of the ship and the sea. The ROV handling unit, fitted to the end of the boom, is designed to capture the ROV as it surfaces by clutching the vehicle's probe which can be interfaced on different vehicles.

LARSC overcomes the need for weighty and cumbersome hardware on board a small ship. The vehicle interface requirements are simplified to allow the system to be fitted on almost all existing vehicles without impairing performance.

Operational status
The system underwent sea trials in 1989 and has been operational on Italian navy 'Gaeta' class minehunters since 1994. A number of navies have shown interest in the system and sea trials with different ROVs have been proposed.

Contractor
Riva Calzoni, Bologna.

VERIFIED

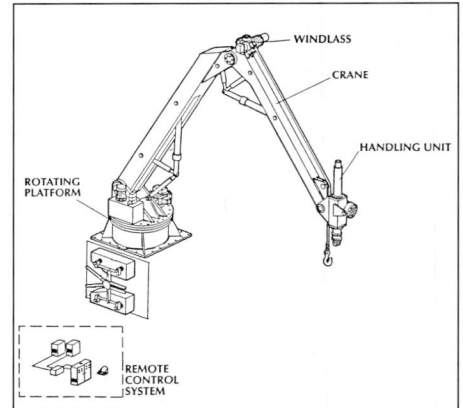

Launch and recovery system for ROV/AUVs **1994**

A LARSC on board the 'Gaeta' class minehunter Alghero *recovering a MIN ROV*   **1998**/0009518

Prototype LARSC in the factory prior to sea trials
1995

SWEDEN

MDC 605

Type
Mine disposal charge.

Description
The MDC 605 is designed primarily to detonate and destroy ground influence mines. It can also be placed on the sinker of a buoyant mine to sever its mooring, or be used for various underwater demolition tasks. The charge is normally deployed from an ROV, in particular the Sea Eagle or Double Eagle deployed from the Swedish 'Landsort' class MCMV. The MDC 605 can also be carried and placed in position by a diver as it is small enough to be handled and can be made weight neutral in water by fixing a detachable float.

The charge is detonated by acoustic telecommand using a coded (8-bit frequency shift) signal. For safety, firing commands are ignored for a fixed time after laying. The MDC 605 incorporates a hydrostatic device which precludes the charge being armed until it has reached a set depth. The detonator is not fitted until immediately before deployment and the MDC 605 contains only low-sensitivity explosives during storage and transport.

Operational status
In service with the Swedish Navy.

Specifications
Length: 380-615 mm
Diameter: 170 mm
Weight: 4 kg (excl of charge)
Charge: 3-10 kg HE (optional)

Contractor
Bofors SA Marine, Landskrona.

Mine disposal charge MDC 605   **1997**

VERIFIED

UNITED KINGDOM

Shaped charge mine disposal system

Type
Shaped charge mine disposal system.

Description
BAeSEMA has developed a small shaped charge capable of destroying both ground and buoyant mines *in situ*. This approach removes the need for cable cutting techniques and subsequent destruction of surfaced mines.

The weapon is deployed using a Remotely Operated Vehicle (ROV) with a limited function manipulator for placement and attachment and simple sensors such as low-light/TV for identification and placement. The attachment process is fully automatic, thus minimising the need for sophisticated operator interaction.

The disposal charge is virtually neutrally buoyant so that it does not impose significant demands in terms of ROV power and operability and the attachment mechanism ensures attachment to any mine surface.

The use of a small explosive charge ensures that the sonar environment remains undisturbed in the event of non-detonation of the target mine. A repeat engagement can be mounted quickly.

Operational status
In development.

Specifications
Length: >400 mm
Diameter: >170 mm
Weight (in air): 10 kg

Contractor
BAeSEMA, Bristol.

VERIFIED

ROV-Homer and Homer-PRO

Type
Relocation system for Remotely Operated Vehicles (ROVs) and divers.

Description
The ROV-Homer is a miniature range and direction guidance system for small or large ROVs. It is specifically designed for fast, efficient relocation of underwater targets such as lost diving bells, divers, seabed equipment or small objects. The system is particularly suitable for use in the marking of exercise mines and in EOD clearance. It enables points of interest to be marked either temporarily or permanently so that an ROV can be guided straight back to the target by the pilot, even in zero visibility, and therefore substantially reduces search time.

The system operates with a range of miniature battery-powered acoustic transponders, each being easily selectable by the ROV controller using the surface display unit. The transponders are small and lightweight so that they can be easily fitted to divers or equipment and have a long life battery pack making them ideal for use as permanent markers. Each transponder is individually encoded enabling many transponders to be used on the same site to unambiguously mark many targets.

The system comprises an ROV-mounted range and direction unit, a surface display for the operator and small, lightweight marker transponders. The operator uses the keypad on the surface display unit to select the marker transponder he wishes the ROV to home onto. The unit then interrogates the specified transponder and from the reply, determines both the range (distances up to 0.5 km are accurately measured) and direction of the marker transponder from the ROV. This data is relayed back to the surface and displayed, indicating to the operator the range to the target and the direction in which to steer the ROV towards the selected transponder.

The Homer-PRO is a diver hand-held relocation system which can be used to guide a diver directly towards a target. The diver's unit contains a Ni-Cd battery which must be charged before diving. A single push-button switch on the unit allows the 'address' of the transponder to which the diver wishes to 'home' onto to be selected while underwater. The unit then interrogates the transponder and indicates on a four-digit display the range and direction (right, left, ahead) to the transponder.

Homer-PRO operates to 500 m and can relocate objects in zero visibility from over 0.5 km distance.

The compact transponder beacons operate in the 35 to 55 kHz range and have a listening life of up to 24 months.

Contractor
Sonardyne International Ltd, Yately.

VERIFIED

ROV/Diver Trak

Type
Tracking system for Remotely Operated Vehicles (ROVs) and divers.

Description
ROV/Diver Trak has been developed for tracking the position of small ROVs and divers. It can track one surface and one underwater mobile target. A key feature of the system is its ability to mark waypoints which can be used to define arrays to represent sets of parallel search lines, tag underwater objects and features, mark transponder positions and tie-in known features such as jetties, shore markers and so on. The system is particularly applicable in explosive ordnance clearance, monitoring diver operations, search and salvage, vessel ranging and so on.

The number of waypoints is limited only by hard disc space. Additional display markers indicate the surface vessel and underwater mobile target. A display option provides a 'snail trail' for each of these two mobile targets. All marker and track data is automatically logged to disc, together with related information such as position, depth, heading and time of day, to provide a complete dive record.

Operational status
Being developed and was due for completion in 1996.

Contractor
Sonardyne International Ltd, Yately.

VERIFIED

Mini ROVNAV

Type

Long baseline acoustic interrogator for Remotely Operated Vehicles (ROVs).

Description

The Mini ROVNAV underwater long-baseline interrogator is specifically designed for fitting to ROVs or towfish. Its main function is to position the vehicle accurately and rapidly. It does this by simultaneously interrogating an array of seabed transponders, receiving their reply signals and measuring the time of each reply relative to its initial transmission. This information is then passed to a surface computer via the vehicle's umbilical, allowing the position of the vehicle to be determined relative to the transponder array. The system incorporates both pressure and temperature sensors to provide a measurement of the depth of the vehicle and estimate sound velocity.

The unit is small enough to be mounted in the most appropriate acoustic position on the vehicle, removing the need for remote transducers, and provides accurate tracking of vehicles for long-range search and salvage, explosive ordnance clearance, range support and so on.

Contractor

Sonardyne International Ltd, Yately.

VERIFIED

OE 1324

Type

Remotely Operated Vehicle (ROV) camera.

Description

The OE 1324 SIT low-light level image-intensified, monochrome camera is supplied as an upgrade for the PAP 104 Mk 3 camera in conjunction with other sensors supplied by CSIP/ECA. It provides higher resolution, longer viewing range and will operate at significant depths, even with ambient light levels. It is also fitted with a wide-angle lens thus providing the ROV pilot with greater vision enhancing obstacle avoidance and final positioning of the ROV-deployed mine disposal charge.

Operational status

In service with the Royal Navy.

Contractor

Simrad Ltd, Aberdeen.

VERIFIED

OE 2700

Type

ROV cameras.

Description

The OE 2700 system comprises a monochrome and a colour TV camera which are fitted to the ECA PAP 104 Mk V vehicle assembly. The cameras are mounted on cover plates which in turn are fitted with watertight housings and attached to a common tilt mechanism. Each camera views the underwater scene through a port which makes corrections for the refractive index of water, while the iris of each camera lens is automatically adjusted to suit the prevailing level of scene luminance. The cameras have the capability to lock to an external signal source – genlock.

Operational status

In service with the Royal Navy and Royal Saudi Naval Forces 'Sandown' class single-role minehunters.

Contractor

Simrad Ltd, Aberdeen.

VERIFIED

UNITED STATES OF AMERICA

Seaphire

Type

Underwater light.

Description

The Seaphire underwater two-light system is designed to handle the two most significant underwater illumination problems – common volume scattering and hot spots. The light allows the user to select the optimum beam pattern according to the application and operating conditions. The blue-green lens is designed to filter out the majority of red and yellow light above 600 nm wavelength, enabling optimum performance to be achieved from high-gain monochrome TV cameras (that is, SIT and ICCD).

Both long and short housing versions are available allowing direct replacement of existing lamp assemblies without changing mounting configurations. The internal reflector, glass domes and front cowls are also interchangeable, allowing superior beam patterns to be formed. The lights themselves are selectable for flood or spot patterns and run on 120 or 240 V supply with long-life tungsten-halogen lamps of 100, 150 or 250 W.

The lamps are housed in a lightweight aluminium mounting with a 1,000 m depth rating. As an option increased depth ratings are available using stainless steel or titanium housings. Optionally 20° spot and 90° flood beam patterns are available.

Specifications

Length: 216 mm (less connector)
Width: 89 mm
Depth: 146 mm
Weight: approx 1.8 kg in air
Depth rating: 1,000 m

Contractor

SEA Hydro Products, San Diego, California.

VERIFIED

Seavision

Type

Intensified CCD camera.

Description

Seavision is a low-profile, high-performance, intensified, solid-state TV camera with full hemispherical viewing and no external moving parts. This configuration provides a constant hydrodynamic profile and fixed electromechanical connections to the viewing system platform or ROV. When combined with the Seaphire range of underwater lighting systems, Seavision enables optimum viewing system performance in resolution, range and reliability.

The system incorporates the SEA Hydro Products Ultravision ICCD imaging system for extreme low-light (1×10^{-6} ft candles faceplate illumination) viewing. The camera is mounted on a high-speed pan and tilt mounting (45°/s) for optimum real-time viewing.

The camera is mounted in a lightweight aluminium housing rated to 1,000 m depth.

Optional features available include: increased depth rating using stainless steel or titanium housing; position feedback for annotation generation; various input/ output command and data formats; 6.5 and 12.5 mm lenses.

Specifications

Diameter: 258 mm
Length: 28 mm
Weight: 9.5 kg in air
Depth rating: 1,000 m

Contractor

SEA Hydro Products, San Diego, California.

VERIFIED

MINESWEEPING SYSTEMS

AUSTRALIA

AMASS

Type
Minesweeping system.

Description
The ADI Minesweeping And Support System (AMASS) is a systems approach to minesweeping, consisting of combined influence and mechanical sweeps, mission planning software, a portable magnetic and acoustic range and a minesweeping control system. The sweeps are self-contained and able to be deployed from both MCMV and auxiliary minesweepers (COOP).

The ADI system has enabled the Royal Australian Navy to utilise commercial craft such as fishing vessels as auxiliary minesweepers, providing an extremely flexible and cost-effective mine countermeasures capability.

Dyad influence sweep
The Dyad sweeps are self-contained modular systems consisting of distributed linear arrays of positively buoyant permanent magnets called Dyads, and water-driven acoustic generators. Signature manipulation is a function of the physical configuration of a particular sweep, and the system is unique in that there is no requirement for external sources of power or control. Consequently the sweeps can be deployed as clip-on arrays by a range of surface vessels or helicopters.

The ability to manipulate sweep signatures means that the system can be used in either a target emulation mode or a mine setting mode, and the resultant signature structure can be optimised to reflect the requirements of any geographic location.

The Dyads comprise mild steel cylinders magnetised by two ferrite magnet discs to produce a dipolar magnet with an extremely high magnetic moment independent of shipboard power supplies.

The Dyads are manufactured in two sizes: Mini Dyads which can emulate the influence signatures of degaussed vessels, and Maxi Dyads which can emulate the signature of undegaussed ships up to 100,000 dwt.

The sweeps are virtually maintenance free with an operational availability in excess of 98 per cent, and their manoeuvrability and ruggedness makes them ideal for use in confined waters and inland waterways as well as offshore operating areas. Significantly, the sweeps have been the subject of extensive shock testing and have survived levels of shock well in excess of those which would be experienced in an operational environment with no effect on their performance.

Mini Dyad influence sweep – deploying Dyads from transportation platform   *1998*/0009582

The sweeps, which are effective against even the most modern mines, are supported by specialised mission planning software.

Mission planning support system
This system includes sweep design and effectiveness software provided on a laptop computer. The software caters for both automatic calculation of sweep configurations to emulate ship signatures obtained through magnetic ranging and manual design. The software enables the effectiveness of sweep configurations to be established by modelling the sweep signatures against selected mine logics and sensitivity settings to establish actuation probability profiles.

A mission planning module is also available, which caters for tactical planning based on the unique characteristics of the Dyad sweeps and the deterministic nature of sweep-mine encounters.

Minesweeping control system
This is a portable minesweeping control system

integrated with differential Global Positioning System (GPS) and Syledis Remote Positioning System (RPS). The system can be fitted to commercial craft used to deploy the AMASS sweeps, and can provide them with the precise navigation and minesweeping control capability normally associated with purpose-built MCMVs.

Portable magnetic/acoustic range
This portable range can measure and record DC and AC magnetic, acoustic and pressure influences and can be deployed within an area of operations in a matter of hours. It provides the capability for check-ranging of MCMVs as well as the ranging of fleet units and commercial vessels used to deploy AMASS sweeps in the MCMV role. For nonferrous-hulled vessels, the range is complemented by a unique and highly effective permanent magnet degaussing system.

Mechanical sweeps
These are lightweight and portable sweeps fitted with explosive cutters. The team sweep is a traditional

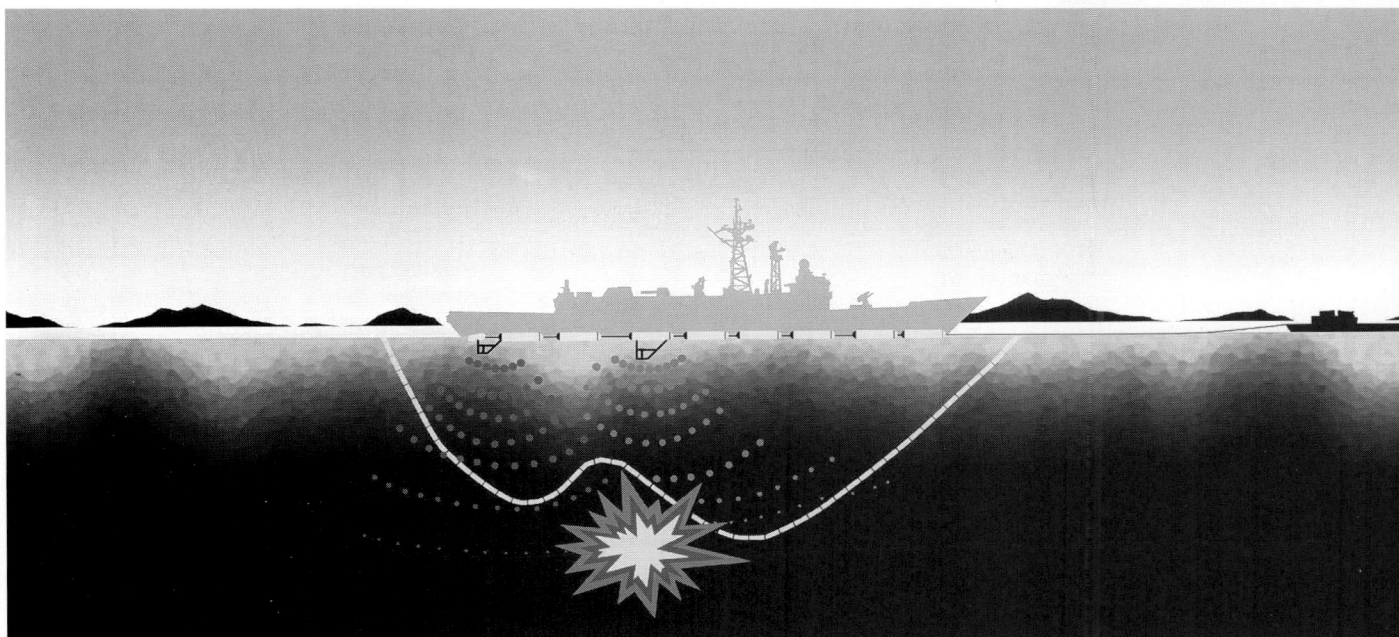

Dyad emulation sweep concept   *1998*/0009581

configuration with a number of innovative features that increase the effectiveness and efficiency of the sweep. A gravity kite is used rather than a depressor, providing simple deployment and a tight turning circle. A drogue minimises sag and ensures an optimum sweep wire angle.

The double oropesa sweep adapts modern fishing technology with lightweight otters and a kite that can be easily handled without special davits.

The sweeps are self-contained and can be effectively deployed from commercial craft with an optional winch and power pack.

Acoustic generator

This acoustic generator is a self-contained acoustic piston noise source powered by a water-driven turbine. The electrohydraulic acoustic module produces harmonically rich waveforms which can be manipulated to provide a broadband spectrum covering the infrasonic, audio and ultrasonic bandwidths with ship-like characteristics. Algorithms loaded to an onboard processor enable the programming of multiple tonals and the generator is independent of external sources for either power or control. Consequently, the system can be easily deployed by vessels of all types, including remote control drones, without any need for modification to the deploying platform. The generator is 2,000 mm long and 350 mm in diameter at the acoustic head. The trailing variable pitch turbine has a blade diameter of 730 mm. The weight in air is 200 kg and the generator has an operating speed range of 6 to 15 kt.

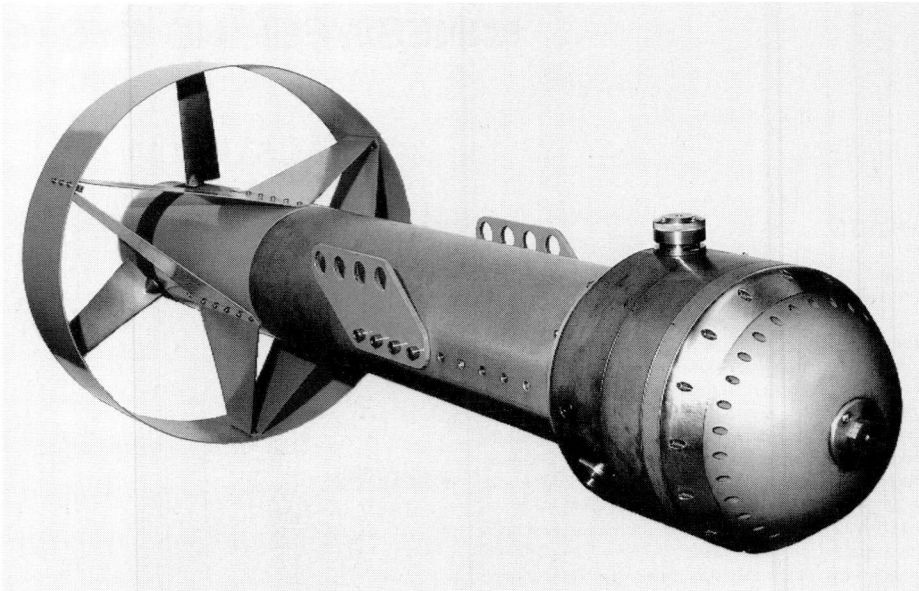

The programmable acoustic generator **1999**/0024837

Operational status

In service with the Royal Australian Navy. Procured by a number of countries including Denmark (eight units), Indonesia, Japan, Thailand and the Royal Navy.

Contractor

ADI Ltd, Bondi Junction, New South Wales.

UPDATED

FINLAND

FIMS

Type

Multi-influence sweep and combat control system.

Description

FIMS is an integrated package consisting of all components required for influence minesweeping operations on small- to medium-sized vessels. All tools required for a complete operation from initial planning to effectiveness assessment are provided in the package. The primary components of the system are:

(a) three electrode magnetic sweep cables with current capacities of 400 A DC, 1,000 A repetitive

peak. The sweep cable, with a diameter of 46 mm is buoyant and its low tow resistance enables operation at speeds of up to 10 kt

(b) various options of controllable and fixed output acoustic sources

(c) power generation units for both the magnetic and acoustic sweep systems. The magnetic sweep

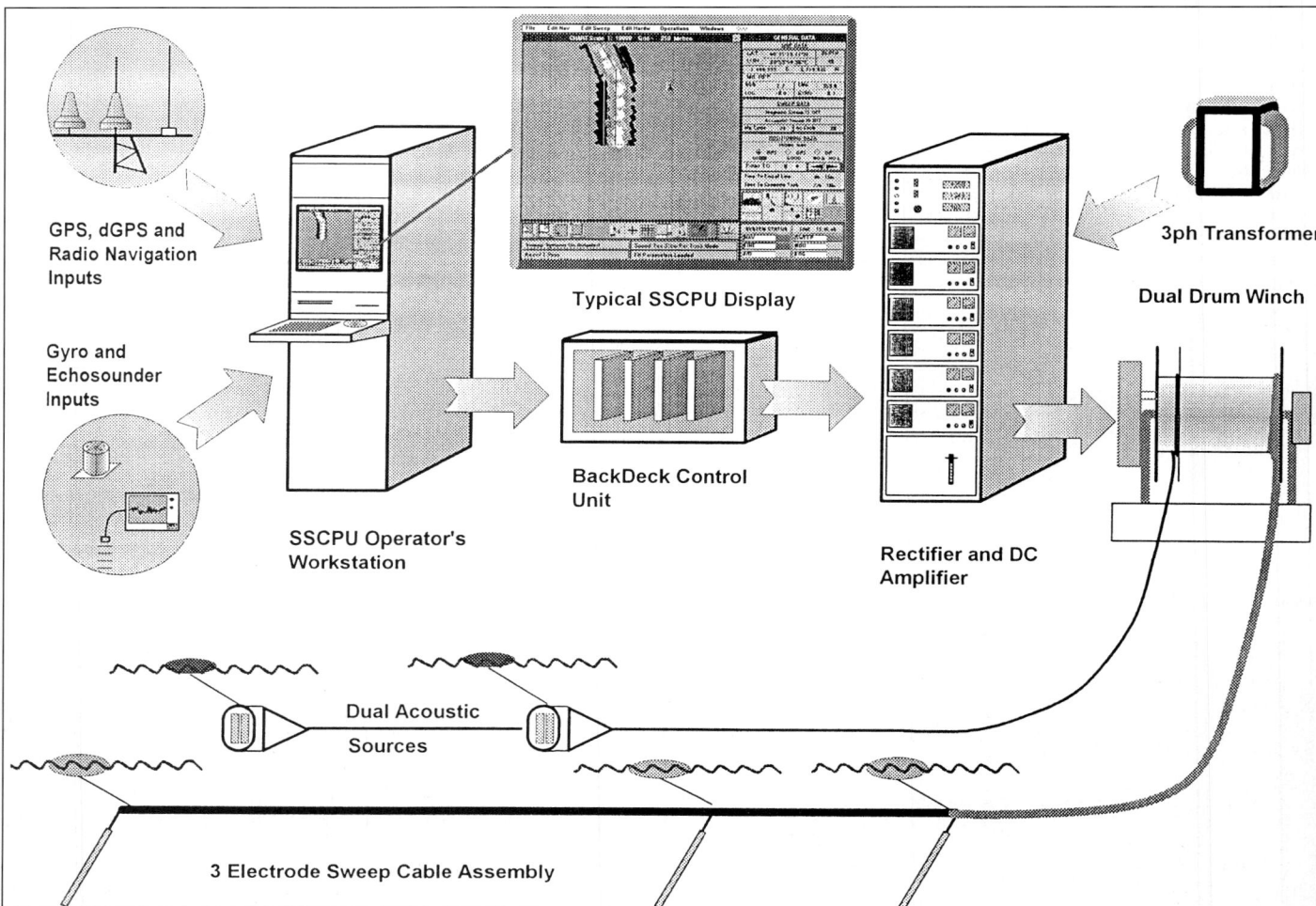

FIMS block schematic overview *1996*

current generator is normally powered from the ship's supply but can be supplied with its own diesel generator

(d) a powerful shipboard control computer system, with interfaces to the ship's positioning sensors.

The shipboard processing systems are at the heart of the FIMS concept. These systems are located on a 486 PC computer optimised for marine use, linked to a monitor and control outstation which utilises a processor Programmable Logic Controller (PLC) to interface the sweep systems to the PC. For single-ship operation the 486 computer is linked by cable to the PLC system, while for remote sweeping and multiship operations the units have a radio link. The complete shipboard system is termed the SSCPU (Shipboard Sweep Control and Positioning Unit), the PLC-based unit is termed the BDU (Back Deck control Unit), and the control software and interface is termed WINS (Window Integrated Navigation and Sweep control).

The SSCPU integrates control of the magnetic and acoustic sweep controller with the ship's positioning systems. This enables control of sweep line positions, displays of actual coverage, and position-activated generation of sweep footprints. Both the magnetic and acoustic sweep controllers can generate complex time varying signatures. These signatures are defined interactively, and need not be a precise mathematical or transcendental function. By interfacing to the ship's echo-sounder the variation of bottom cover with depth is determined. Also the estimated sweep signal strength beneath the ship is calculated. The control algorithms optimise coverage, while maintaining a safe operating environment for the ship. The positioning section of the system integrates inputs from GPS, radio positioning and log/gyro. Full navigation functions are provided, with the option of integral digital charting.

Graphic displays are provided to show both along-track and cross-track coverage, and all position and coverage data is logged for later analysis. Historical quality control information is provided for both positioning and sweep parameters. These can be used to provide both online and post-mission evaluations of sweep effectiveness. The WINS software includes a comprehensive training and simulation mode.

The sweep coverage subsystem is one of the key features of WINS. This consists of a database of sweep coverage information organised in small geographical boxes or bins. For each bin, information is stored on various sweep parameters such as type of sweep and number of over-runs. The information is designed to be viewed graphically, which quickly reveals the success of a sweep operation, or areas where further work is required. The database is updated on the completion of each sweep line.

Where possible, standard high-quality commercial products are used, modified where necessary by the addition of high-quality shockmountings. This results in an excellent price to performance ratio, and improved maintainability. The core technology is well proven and has a clear progress path. The shipboard equipment is small enough to be readily deployed on ships of convenience; in this mode the integration of the positioning function is particularly useful, as is the integral GPS option. Its ability to use several positioning sensors eases the problems of overseas deployment or non-availability of the primary sensor.

System options and variants include a containerised version, multiship variant, and integrated route management package, which adds a minehunting capability with the addition of a side scan sonar system.

Operational status
In service with the Pakistan Navy.

Contractor
Elesco Oy, Espoo.

UPDATED

MAINS

Type
Minesweeping And Integrated Navigation System (MAINS).

Description
MAINS is an integrated package consisting of all the components and functions necessary for influence minesweeping operations. The system integrates precision navigation and comprehensive sweep management with minesweeping signature generation/control implemented in a standard MS Windows NT environment.

The primary components of the system are:
a) Wet end units that include a 3-electrode magnetic sweep cable assembly (MRK-960) and acoustic sweep (SONAC AMS). The diameter of the buoyant sweep cable is 46 mm and its current capacity is 400 A DC continuous or 1,000 A repetitive peak. Its low tow resistance enables operation at speeds of up to 10 kt. The acoustic sweep (SONAC AMS) is programmable and fully controllable.
b) Shipboard units that include controllable power generation units for magnetic and acoustic sweeps and a powerful computer system network with optoisolated I/O interface.

The magnetic and acoustic sweep gears, MRK-960 and SONAC AMS enable the emulation of alternative overrun speeds independent from MCMV's towing speed. The low towing load and light weight of the equipment enables its installation on small MCMVs. The equipment has been shock tested.

The computer system comprises Sweep Management System (SMS) and complete navigation system (ProMare). The latter has interfaces to all ship's navigation sensors, such as electromagnetic log, gyro, differential GPS, radio navigation, echo-sounder, laser tracker (for accurate positioning of the sweeps), wind meter, navigation radar, autopilot and sonar. The displays can be mounted in control room, bridge or elsewhere as required. The number of displays is according to the user requirements and 20 in colour flat screens can be provided. Where possible, standard industrial grade COTS hardware is used with the addition of high-quality shock mountings.

The sweep coverage subsystem is one of the key features. This consists of a database of sweep coverage information organised in small geographical boxes or bins. For each bin, information is stored on various parameters, such as type of sweep and number of overruns. The information is designed to be viewed graphically, which quickly reveals the success of a sweep operation, or areas where further work is required. The combination of SMS and ProMare gives all the tools required for a complete minesweeping operation, from initial mission planning to ship's navigation, positioning of sweeping gears, sweep control, real-time monitoring and reporting of sweep results, and post mission effectiveness assessment using alternative sets of mine threat parameters when required. The system provides continuous monitoring and alarm facilities to ensure the safety of the towing vessel. The software package also includes a comprehensive training and simulation mode.

Operational status
Over 40 minesweeping systems in various configurations have been delivered since 1975. Two MAINS systems were delivered in 1998 and more are on order.

Contractor
Elesco Oy, Espoo, Finland.

NEW ENTRY

Sonac AMS

Type
Acoustic influence sweep.

Description
Sonac AMS is a high-efficiency fully controllable underwater sound source designed to sweep acoustic influence mines. It is specially designed for operation from small- and medium-sized vessels or drones.

The system comprises the Sonac AMS deck unit, wet end unit, 150 kg buoy and a buoyant towcable.

The deck unit consists of a control computer, signal generating unit, power amplifiers and matching transformers. The signal generating unit can create an unlimited number of acoustic signatures simulating any known surface vessel. Changing the signature is simply accomplished using the specially designed Sonac AMS user interface. The AMS system can be linked to external systems, for example multi-influence sweeps.

The wet end unit consists of a low-frequency sound source and several optional piezoceramic high-frequency transducers to cover the audible frequency band completely. The wet end unit has been designed to achieve extremely high shock resistance and has been successfully tested against an underwater explosion to provide the shock resistance capability.

The buoy is designed to maintain the wet end unit at a constant operating depth. Wet end body shape enables cross line offset towing.

The buoyant towcable is specially designed to achieve high breaking strength, and the wet end unit end of the cable has an armoured section 30 m in length.

Operational status
Entering service with the Finnish Navy.

Specifications
Height: 900 mm
Width: 630 mm
Length: 1,330 mm
Operating depth: 10 m
Weight: 500 kg (wet end unit)
Towcable length: 500 m
Frequency range: 17 Hz-25 kHz
Output: 190 dB at 20 Hz

Contractor
Patria Finavitec Oy Systems, Tampere.

UPDATED

FRANCE

Sterne 1

Type
Multi-influence minesweeping system.

Description
The Sterne 1 is a towed minesweeping system which simulates both the acoustic and/or magnetic and UEP/ELFE signatures of various types of ship.

The modular configuration of the Sterne system allows the output to be tailored to include magnetic (AC or DC); magnetic (AC or DC) and acoustic; or magnetic and electric (AC or DC) plus acoustic signature emulations.

In its basic configuration the system features a line of six towed magnetic bodies spaced at about 10 to 20 m apart with power provided from the towing vessel. Each body incorporates an M loop, and L loop, an ELFE loop (two of the six are active), and a UEP electrode (two of the six are active) Additionally, four of the bodies can incorporate an HF acoustic generator. The bodies can also be fitted with electrodes to generate an AC/DC field component. The entire sweep system is controlled from a three-cabinet system which incorporates a dedicated computer which is used to set up the

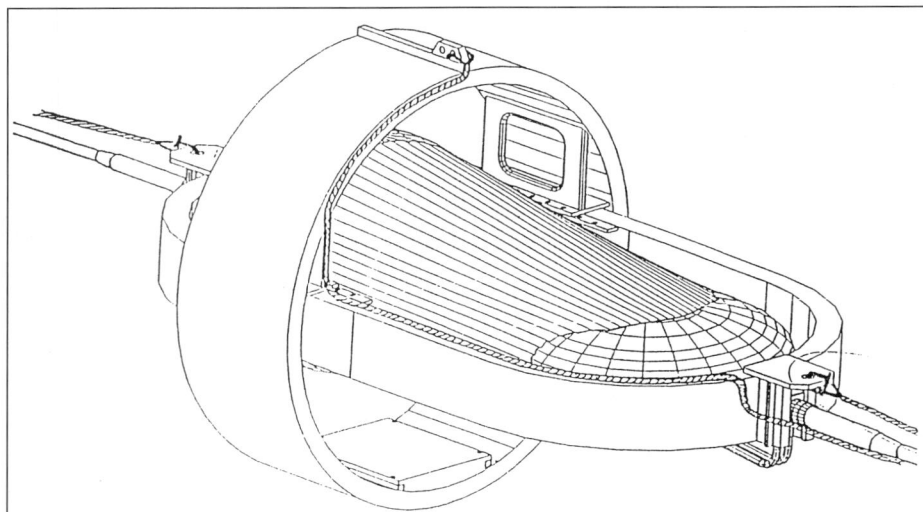

required signature profile. With the addition of a low-frequency acoustic generator the Sterne system can provide an improved low-frequency response. Onboard equipment includes the TSM 2061 tactical display, the

TSM 5722 sonar Doppler log, and navigation systems such as the Trident III radio location system and Decca 1226 navigation radar.

The system is towed at speeds of 6 to 10 kt at a minimum distance of 100 m and a depth of between 5 to 10 m. A mine-avoidance sonar, such as the Petrel, can be integrated into the system.

Operational status
On order for the new coastal minehunters for the Belgian Navy.

Specifications
Magnetic loop: 20 kW
ELFE: 1 kW

LF Transducer
Frequency: 10-2,000 Hz
Sound level: >160 dB/µPa/m

Contractor
Thomson Marconi Sonar SAS, Sophia Antipolis Cedex.

UPDATED

Towed magnetic body of the Sterne sweep system

AP 5

Type
Acoustic minesweep.

Description
The AP 5 acoustic sweep (the earlier AP 4 is similar, but with reduced frequency range) comprises: a signal generator which can generate up to 192 acoustic signatures and spectra for various types of ships, signals being preprogrammed within a wide frequency band (in both spectrum and modulation); a very high powered low-frequency amplifier with associated impedance matching circuit; an underwater vehicle which is towed by a combined power feed and towing cable and which is fitted with an underwater electrodynamic loudspeaker with two symmetrical diaphragms; and an optional piezoelectric transducer for high frequencies. The software programs are designed to meet operational requirements and can be easily developed or modified ashore, with the possibility of exchanging the preprogrammed electronics cards on board. These programs allow the generator to simulate acoustic signatures of various types of ship, to activate mines with known characteristics from a standoff position, to activate unknown mines by scanning over their probable acoustic spectrum, to inhibit mine detonating devices within an extended area, and to transmit acoustic spectrum in synchronisation with the transmission of the current pulse of a magnetic sweep. The magnetic signature of the shipborne equipment is very low, even when operating. A preprogrammed device enables the transmitted acoustic power to be adjusted according to acoustic propagation conditions over the swept path.

The underwater vehicle has successfully passed the impact resistance qualification tests against an underwater explosion of 1 tonne of TNT at 80 m abeam.

Minesweeping is performed at speeds between 3 and 12 kt (8 kt nominal) with a constant immersion of

AP 5 acoustic minesweep

the transmitting vehicle between 8 and 10 m; this can be extended as an option.

All frequency bands are covered by the sweep and electronically controlled by the computer.

Operational status
Development has been halted and only the AP 4 remains in service with the French Navy.

Contractor
DCN International, Paris.

VERIFIED

GERMANY

GHA

Type
Acoustic minesweeping unit.

Description
The GHA uses the flow of water through a turbine in the body of the vehicle to generate sound for sweeping acoustically sensitive sea mines. The GHA is towed – preferably together with a magnetic sweep – behind a minesweeper and supplies its two electrodynamic sound generators independently with power by means

of water flowing over its turbines. The beat frequency, sound pressure, operating and pause time, and wobble time of the sound generator are programmable. The beat frequency can be wobbled during towing (that is, the beat frequency fluctuates up and down periodically within the adjustment range). Once programmed, the operating mode is independent of the flow speed of the water and is thus also independent of variable towing speeds. The emission of sound via the electrodynamic system ensures largely wear-free reliability of operation for several thousands of operating hours.

Not requiring a power cable means that the buoy is

easily handled aboard ship, there being no need for generating sets, cable drums, control units and so on and the system is thus ideal for use on COOP as well as on purpose-designed minesweepers.

Operational status
The system has undergone extensive trials.

Contractor
IBAK, Kiel.

VERIFIED

DM 19

Type
Explosive minesweeping cutter.

Description
This proven explosive cutter for sweepwire attachment has been in naval service for many years. It uses a linear shaped charge and assures great operational safety even at high sweep speeds and offers a high degree of handling safety. The equipment is lightweight and capable of operation at considerable depths. It is a one-shot system and can be mounted onto sweepwires of different diameters. Neither onboard maintenance nor preparation for use are required.

No residual parts remain on the sweepwire after detonation. The sweepwire is sandwich-shield protected.

Operational status
In service with the German, Indonesian and Thai navies.

Specifications
Cutting capacity
Stud link chains: 20 mm
Steel wires: 26 mm
Synthetic ropes: 40 mm
Operating depth: 3 to <200 m
Weight in air: 4.4 kg
Weight in water: 2.3 kg
Dimensions: 410 × 470 × 85 mm

Contractor
Rheinmetall W&M, Ratingen.

VERIFIED

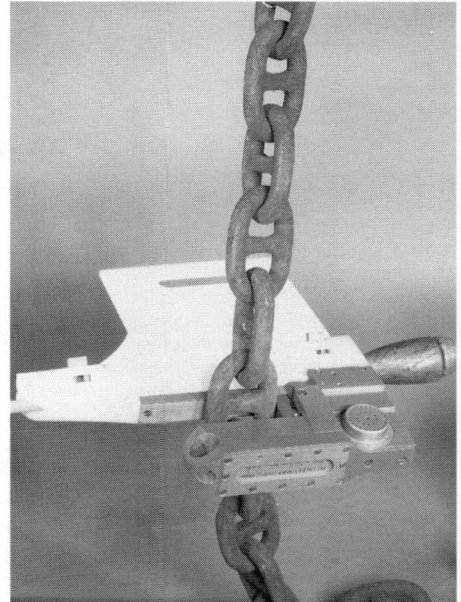

Explosive minesweeping cutter DM 19

ITALY

MDS

Type
Mine disposal system.

Description
The MDS is based on a one-shot expendable self-propelled self-homing mini-torpedo derived from the A200 LCAW. It is designed for deployment from surface vessels or helicopters to counter drifting, moored and ground mines. The device uses an advanced active acoustic seeker which allows it to considerably reduce the time taken to clear a minefield or open a route through a minefield.

The system is not adversely affected by meteorological conditions or by underwater currents

and can be deployed by day or night. The device consists of the Mine Disposal Torpedo (MDT), the control console and the launching system. The control console performs various computations to determine the best firing solution for the MDT, selects, presets and tests the weapon before firing, displays tube status and firing orders, tracks the target and MDT in real time, performs real-time cross correlation, and transmits orders for course, depth, activation of homing head and final attack procedure to the MDT via the guidance wire.

Specifications
Length: 1,250 mm
Diameter: 124 mm
Weight: 13 kg (approx)
Operating depth: 1-400 m

Speed: >15 kt
Range: 1,300 m @ (min)
Endurance: 3 min (approx)

Acoustic sensor
Detection range: 300 m
Mode of transmission: FM
Number of transmitted beams: 1
Coverage: 44° horizontal, 25° vertical
Number of received beams: 4
Coverage: 42° horizontal, 42° vertical

Contractor
Whitehead Alenia Sistemi Subacquei SpA, Genoa.

VERIFIED

NORWAY

Agate

Type
Acoustic sweep.

Description
Developed by the Norwegian Defence Research Establishment, Agate is designed for deployment from the Royal Norwegian Navy's 'Alta' class minesweepers. The system features a pneumatic air gun driven by high pressure trapped in a chamber. This results in a low-frequency sound suitable for use in minesweeping operations. Each acoustic impulse is created by the release of a controlled volume of air. The output level is

a function of the quantity of air that is released and the air pressure. The air guns incorporate two different firing chamber volumes and can be modulated to fire between one and six shots per second.

In addition to the air gun, Agate also incorporates Terfenol-D, a magnetostrictive material with a particularly high level of magnetostriction which converts electrical energy into mechanical energy. The CelsiusTech-developed Terfenol-D flextensional transducers develop a much higher power level than conventional piezoelectric transducers of about the same size and is well suited to the production of low- and medium-frequency sound.

The outputs from Agate include both noise and spectral lines. The entire system is controlled by a shipboard computer which contains a library providing data on a wide range of ship signatures.

Operational status
In service with the Norwegian Navy.

Contractor
GECO Defence, Kjorbokollen.

VERIFIED

RUSSIAN FEDERATION AND ASSOCIATED STATES (CIS)

SEMT-1 & ST-2

Type
Magnetic solenoid sweep.

Description
The SEMT-1 magnetic solenoid sweep is designed for use in shallow water and harbour clearance operations. The system comprises two magnets each 18.2 m long

which form a set and the entire system consists of two sets. The magnets can be used in waters as shallow as 5 m with a towing speed of 4 to 8 kt.

The ST-2 magnetic solenoid sweep is designed for sweeping operations in coastal waters using a current of 200 A. The sweep comprises a single magnet 29.1 m long and weighing 2.24 tonnes which can be towed at speeds between 5 and 12 kt in water depths below 10 m at a distance of 300 m behind the towing platform.

Both sweep systems are powered and controlled from the operating platform and the total weight of the shipboard elements is 1 tonne.

Operational status
In service aboard 'Lida' class minehunters.

UPDATED

PEMT-4

Type
Magnetic loop sweep.

Description
The PEMT-4 magnetic loop sweep consists of a 3-core cable 424 m in length with a 2-core feeder cable 227.5 m long attached. The 5.2 tonne sweep can be towed at speeds of 6 kt and depths between 7 and 35 m.

VERIFIED

IU-2, IT-3

Type
Ground mine detector/destructor.

Description
The IU-2 is a detector/destructor designed to counter ground magnetic influence mines in harbours and protected anchorages. It is a sophisticated system incorporating a mine detector towed 2 to 3 m above the seabed. When the magnetic field of the IU-2 is disturbed it releases either a marker or mine destructor which is towed in a container behind the detecting elements. The system is capable of operating at depths down to 60 m with a swept path of 16 m.

The IT-3 detector/destructor features a swept path of 20 to 30 m with a working depth of 10 to 60 m and a towing speed of between 3.5 and 6 kt. The device is deployed at a distance of 70 m behind the towing platform. It is designed to detect and mark ground mines. The equipment comprises an LVT3-2TM cable drum winch and a BBG-OT1 davit with a load capacity of 2.5 tonnes and a boom length of 1.37 to 4.73 m.

Operational status
In service aboard 'Sonya' and 'Lida' class MCMVs.

UPDATED

GKT-2, GKT-3, GKT-3MP

Type
Mechanical sweeps.

Description
The GKT-2 mechanical sweep is a deep sweep system in service with both coastal and ocean minesweepers. The system can be used in a stand-alone mode to deal with moored mines, or in combination with a sound generator to counter acoustic mines. In the latter case the cutters and cartridges are supplemented by corner reflectors. Swept path is between 260 and 280 m with a maximum sweep depth of 200 m and a minimum depth of 10 m. The sweep can be towed at speeds up to 10.5 kt depending on sweep depth.

The GKT-3 system is a deep sweep contact system for deployment from a single ship. It incorporates integral sonars to monitor the sweep position.

The GKT-3MP is a deep team sweep system incorporating mechanical sensors for bottom following.

Operational status
In service aboard 'Natya' class minesweepers around the world and Russian Navy 'Sonya' class vessels.

UPDATED

AT-2/3/6

Type
Acoustic sweep.

Description
The AT-2 and AT-6 are both wideband systems, with the AT-3 appearing to have three transducers powered by a water turbine. The AT-2 can be towed at speeds up to 12 kt at a distance of 550 m behind the towing platform. The system weighs 2.2 tonnes.

Operational status
In service aboard 'Natya' class minesweepers around the world.

UPDATED

SWEDEN

IMAIS

Type
Integrated magnetic and acoustic influence sweep.

Description
IMAIS generates both magnetic and acoustic influences with one sweep.

With the integrated acoustic transmitter, the IMAIS magnetic cable sweep can be streamed optionally as a straight-tailed electrode sweep or as a closed loop sweep. The straight-tailed configuration offers extremely low drag, while the closed loop variant permits a 100 per cent predictability of the generated magnetic field, regardless of variations in the seabed conductivity.

IMAIS provides safe magnetic sweeping, generating no more than about 10 nT within the minesweeper's damage radius. Simultaneously, mines within a 100 m broad path behind the ship will experience 1,000 nT or more, at 30 m depth.

The picture illustrates IMAIS deployed as a three-electrode sweep. The in-water cable arrangement features two buoyant cables, a short one feeding the front and middle electrodes, and a long one (660 m) feeding the rear electrode.

The long buoyant cable also serves as a tow/feed cable for the acoustic transmitter, positioned about half-way between the middle and the rear electrode. The electrodes, typically some 20 m in length, are supported by small floats to compensate for their weight.

The acoustic transmitter, Bofors SA Marine's type AT 205, is characterised by a 'ship-like' acoustic output covering frequencies from under 10 Hz up to hundreds of kHz. The entire dB versus Hz curve can be raised and lowered from the ship, for proper synchronisation with the pulses emitted by the magnetic sweep. The AT 205 is designed for good resistance to mine detonations.

IMAIS streamed as a straight-tailed electrode sweep with built-in acoustic transmitter   **1998**/0009583

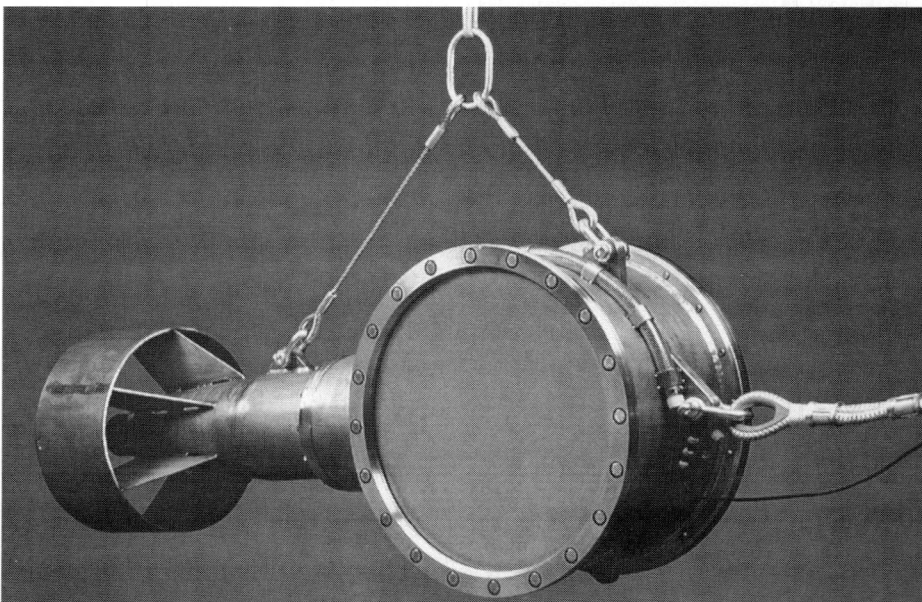

AT 205 acoustic transmitter of the IMAIS influence sweep system
1997

The arrangement with a long and a short buoyant cable lends itself to an easy conversion into a closed-loop configuration thus: removal of electrodes, joining of the outer ends of the two cables, and using the port or starboard mechanical sweep to open the loop. Thus the very same cable system is used in the two modes of the magnetic sweep with only some minor components needing to be removed or added when converting from one mode to the other.

When deployed as an electrode sweep IMAIS will require about 280 kW at 0.6 per cent salinity and 190 kW at 3.0 per cent salinity, while the closed loop will consume some 160 kW. (Power data refer to peak values; the rms values will be approximately half).

The tow loads, at 8 kt, amount to about 12 kN for an IMAIS deployed as an electrode sweep, against some 55 kN in the closed-loop configuration.

An IMAIS as described weighs about 3,000 kg, winch included. A full system also requires pulse generating equipment, including a suitable engine-generator set, weighing some 5,800 kg altogether.

Contractor
Bofors SA Marine AB, Landskrona.

VERIFIED

Mechanical sweep MS 103/104/106

Type
Mechanical minesweep.

Description
Similar to the Swedish Navy's wire sweeps M/58 and M/48, the MS 103 (double), MS 104 (single) and MS 106 (double) oropesa sweeps are compact, lightweight, of simple design and easy to handle. Their size means they are well suited to use as an ancillary sweep system for deployment from a minehunter. Each sweep is armed with six explosive cutters. The diagram illustrates the breakdown of the mechanical sweep MS 103.

Operational status
Delivered to various navies.

Contractor
Bofors SA Marine AB, Landskrona.

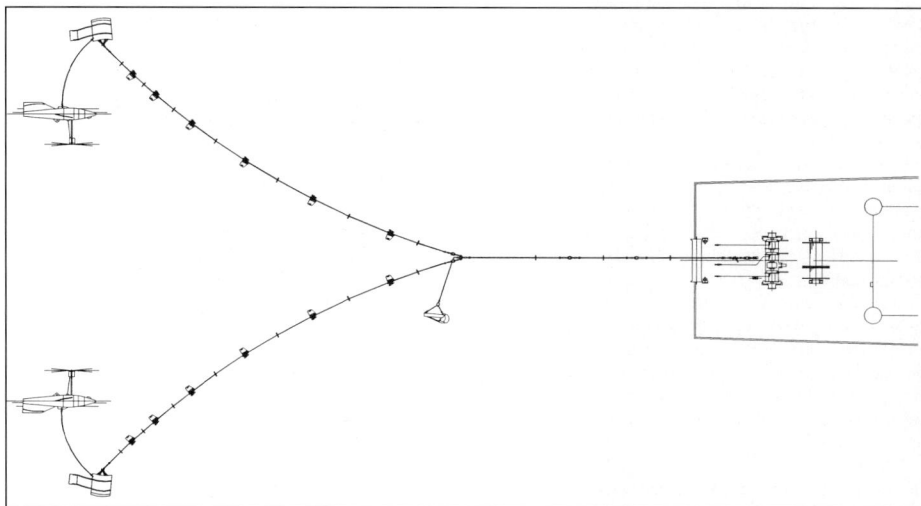

Arrangement of the MS 103 double oropesa mechanical sweep system **1997**

Specifications

	MS 104 (single oropesa)	MS 103 (double oropesa)	MS 106 (double oropesa)
Length of sweep wire	200 m	200 m	300 m
Diameter of sweep wire	12 mm	12 mm	13 mm
Swept path	75 m	150 m	200 m
Swept depth	8-25 m	8-25 m	
Swept speed (nominal)	7 kt	7 kt	
Towing load at nominal speed	13 kt	19 kt	
Explosive cutters	6	2 × 6	
Weight (total)	1.2 t	1.8 t	

UPDATED

Explosive cutter T Mk 9 Mod

Type
Explosive minesweeping cutter.

Description
A successor to the cutter T Mk 9 the T Mk 9 Mod, introduced in 1997, features a shaped charge instead of the former bulk charge. The result is a considerably enhanced cutting capacity, in spite of a reduced charge weight. The port/starboard independent cutter is made of stainless steel. Firing – which is possible only as long as the cutter is under water – results in only the fin being left on the sweep wire.

The cutter weighs about 7 kg and contains 80 g explosive material. It can cut chains up to 16 mm and steel wire rope up to 30 mm. Sweeping speed is 4 to 12 kt.

Operational status
The product is qualified and is in series production.

Contractor
Bofors SA Marine AB, Landskrona.

VERIFIED

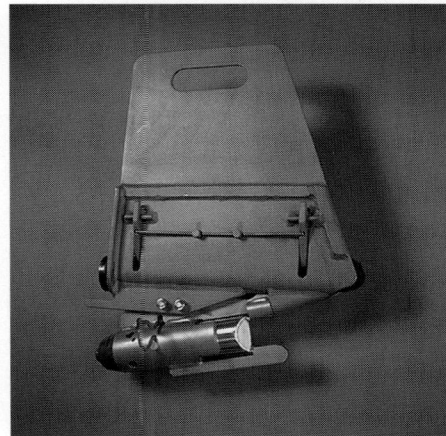

Explosive cutter T Mk 9 Mod
1998/0009584

UNITED KINGDOM

Wire sweep Mk 8

Type
Low-magnetic signature wire sweep system.

Description
The Mk 8 wire sweep system is a development of the wire sweep Mk 3 mod 2 used on the 'Ton' class minesweepers. It has been designed to have a low magnetic signature in order to be compatible with the overall low-magnetic characteristics of modern mine countermeasure vessels.

The Mk 8 is designed to sweep most types of buoyant mines that are moored to the seabed with sinkers by severing the mine mooring chain or wires with explosive cutters fitted to the sweep wires. The Mk 8 may be deployed in the following configurations: single oropesa, double oropesa and team sweep.

Operational status
The Mk 8 system is in service on the Royal Navy 'Hunt' class mine countermeasure vessels.

Contractor
BAeSEMA, Bristol.

VERIFIED

Wire sweep Mk 9

Type
Controlled, deep wire sweep system.

Development
With the advent of mines that could be laid in much deeper water and mines with much shorter mooring wires, work began, together with the Admiralty Underwater Weapons Establishment (AUWE), in 1974 on the development of a sweep system that could be worked much closer to the seabed and at far greater depths. To operate such a system efficiently the minesweeper would have to be able to control both the height of the sweep wire above the seabed and its flatness, so that it does not drag along the seabed and break or prematurely explode the wire cutters connected to it.

It was decided to adapt the existing UK Mk 3 wire sweep system but use a single wire with the sweep wire attached directly to the end of the kite wire. The kite would then only have to depress one wire instead of two, and allow the sweep to go much deeper with greater control.

Extensive use of computer modelling was followed by lengthy trials, which enabled tables to be drawn up comparing swept depth against length of kite wire and, when used in conjunction with Wire Sweep Monitoring Equipment (WSME), it proved possible to heave or veer the kite wire to the exact amount required for the sweep wire to follow the contour of the seabed.

Following extensive trials the system was accepted into Royal Navy service in the early 1980s.

Description
Wire sweep Mk 9 is an effective combat-proven countermeasure against buoyant moored mines in deep waters. The sweep's primary operational mode is team sweeping but it is also deployed as a double or single oropesa sweep. It is designed to be deployed from either purpose-designed minesweeping vessels

Wire sweep Mk 9

or ships taken up from trade – usually stern trawlers. Although usually operated with two ships, several ships have been linked together to form a multiship team sweep providing greater swept paths and allowing an odd number of vessels to be used.

Each vessel carries three lengths of wire on a winch. The outer barrels carry a sweep wire connected to a kite wire and the centre barrel carries a kite wire for oropesa sweeping.

Using WSME in conjunction with the ship's echo-sounder, the sweep can be adjusted to follow the seabed at a constant clearance by heaving or veering kite wire from the winch.

Operational status
Wire sweep Mk 9 is combat proven and has been in operation with the Royal Navy since the early 1980s on the 'River' class vessels. This system has also been delivered to Canada for the Canadian government Naval Reserves Mine Countermeasures Project and is now operational. A version of this system has also been sold to the Republic of Korea Navy.

Contractor
BAeSEMA, Bristol.

VERIFIED

Wire sweep Mk 105

Type
Compact, low-magnetic wire sweep system.

Description
The Mk 105 sweep configuration is a development of several Royal Navy wire sweep systems and is available in a range of sizes (including US size 1, 4 (and 5) to

enable a range of naval vessels to undertake minesweeping operations. The smaller variants of the sweeps may be fitted to small patrol craft or minehunting vessels to provide a dual-role mine

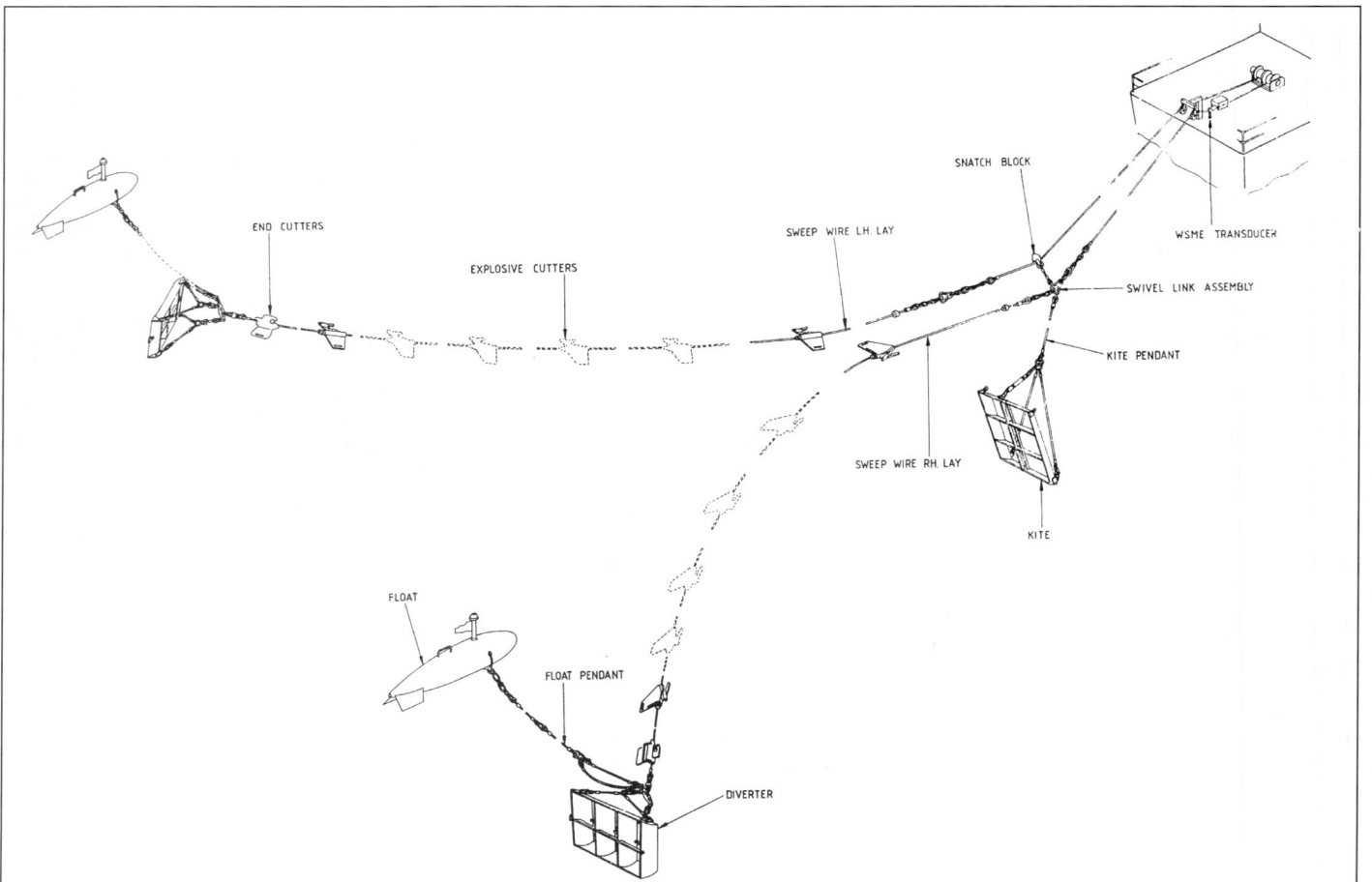

Mk 105 double oropesa sweep system

countermeasures capability. The larger variants are generally fitted to dedicated minesweepers or larger Craft of Opportunity (COOP) where the space and power available is sufficient for the larger equipment.

The design uses a simplified wire configuration enabling relatively small winches to be used to save deck space and in smaller sizes to reduce towing loads. The Mk 105 system may be deployed as either team, double or single oropesa sweeps and includes all items of minesweeping equipment necessary to fit out a vessel for minesweeping including the wires, winch, floats, cutters, kites and marker buoys. The complete systems are available in either low-magnetic or ferromagnetic materials. The system is used in conjunction with the BAeSEMA WSME system to achieve good bottom-following. The size 5 equipment incorporates a new float design which enables greater depths to be achieved in oropesa sweeping (25 m) together with a team sweep which is able to reach 120 m depth.

The Mk 105 sweep configuration is a very flexible system that has already been configured to suit several ship designs including minehunters, fast patrol craft, minesweepers and COOP.

Contractor
BAeSEMA, Bristol.

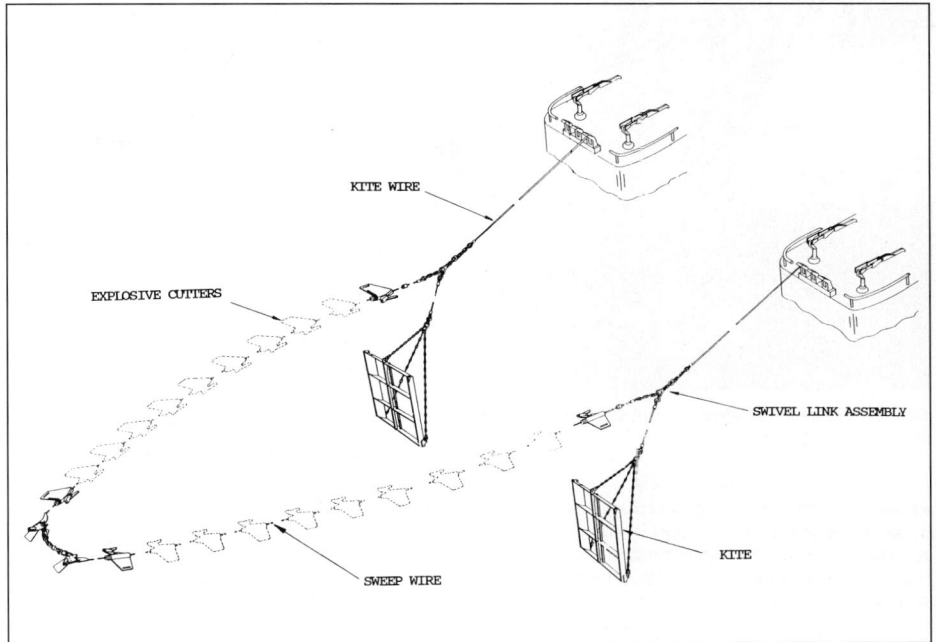

VERIFIED   *Team sweep with wire sweep Mk 105*

Wire Sweep Monitoring Equipment (WSME)

Type
Wire sweep monitoring equipment.

Description
WSME was developed primarily for use as part of the UK wire sweep Mk 9 to improve its accuracy of sweeping. WSME is fitted to purpose-built MCM vessels and vessels taken up from trade to allow effective minesweeping operations. The purpose of WSME is to provide a flat sweep profile at a steady seabed clearance.

For effective operations against all types of moored mines, it is important to tow the sweep wires just above the surface of the seabed otherwise mines on short moorings will not be swept. To achieve this position the sweep wire must be flat; if the wire sags the sweep will contact the bottom and damage the sweep, whilst if the wire hogs it may miss some mines.

Previous methods of minesweeping used vessel speed to indicate sweep flatness but the errors due to ships' logs, tidal flows and current profiles often lead to sweeps with significant hog or sag.

Sea trials have established that most of the tension measured as the wire passed through the fairlead on the ship to the water could be attributed to the drag of the sweep in the water and the forces acting on the kite. From measurement of this tension the speed of the sweep could therefore be calculated, and hence its flatness. This led to the development of the WSME which displays digital readings of average and peak load on the Sweep Monitor Console. The ship speed is then either manually or automatically adjusted to obtain the correct tow tension.

The WSME measures depth. Two options are available to the user. Firstly a direct measurement using an acoustic link to a depth sensor or, secondly, electronic measurement of the length of kite wire

passed through the tension meter to determine the depth of the sweep wire and displays a depth reading which can be directly compared with the digital output of the ship's echo-sounder. The winch is then either manually or automatically heaved or veered to maintain the required seabed clearance.

The WSME is also available with sweep location equipment for monitoring actual ground covered by the sweep. Links to the ship's command and control system are also provided.

Operational status
The WSME has been in service on the Royal Navy 'River' class minesweepers since 1984. In 1989 systems were ordered for the Canadian government's Naval Reserve Mine Countermeasures Project and the systems are now operational. Also operational on 'River' class minesweepers of the Bangladesh Navy and has been sold to the Republic of Korea Navy.

Contractor
BAeSEMA, Bristol.

VERIFIED

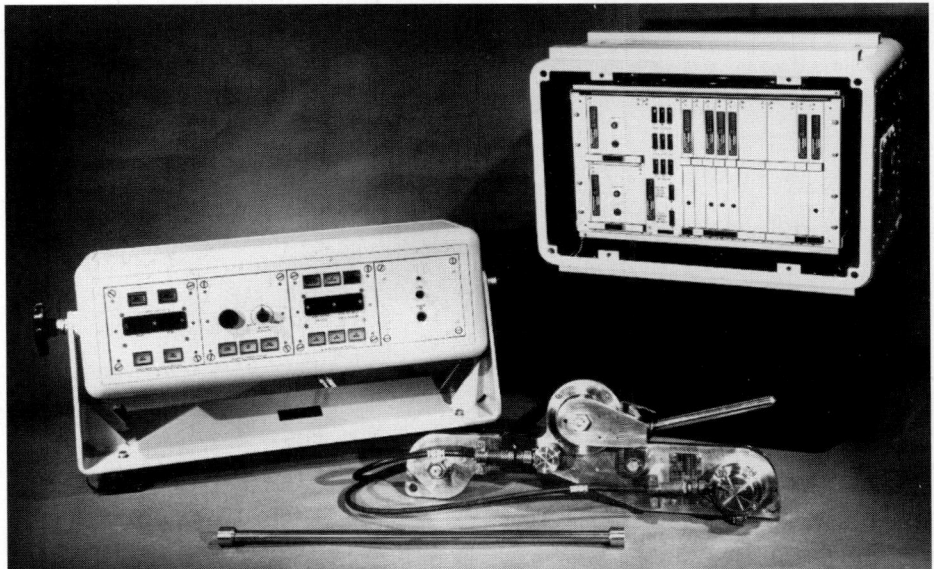

Wire Sweep Monitoring Equipment (WSME)

MSSA Mk 1

Type
Acoustic minesweeping system.

Description
MSSA Mk 1 is an advanced acoustic minesweeping system which can generate a wide range of target-like acoustic signatures. Its acoustic output is continually monitored and controlled to ensure that the required

amplitude at seabed level is maintained regardless of variations in acoustic propagation conditions. It provides a versatile sweeping system capable of very high power output. It is used to activate all acoustic mines, including those with frequency selective triggering characteristics which are designed to be actuated only by certain types or classes of vessels.

The system comprises a towed acoustic generator, a towed acoustic monitor with a towed hydrophone array, and an onboard control console.

Towed acoustic monitor of MSSA Mk 1 minesweeping system

Towed acoustic monitor (centre) and towed acoustic generator (left)

The system has been specially designed to withstand the repeated levels of explosive shock likely to be experienced in operation, and is normally deployed in association with a magnetic sweep to provide a very effective method of sweeping combined influence mines.

Operational status
MSSA Mk 1 is in service with the Royal Navy on the

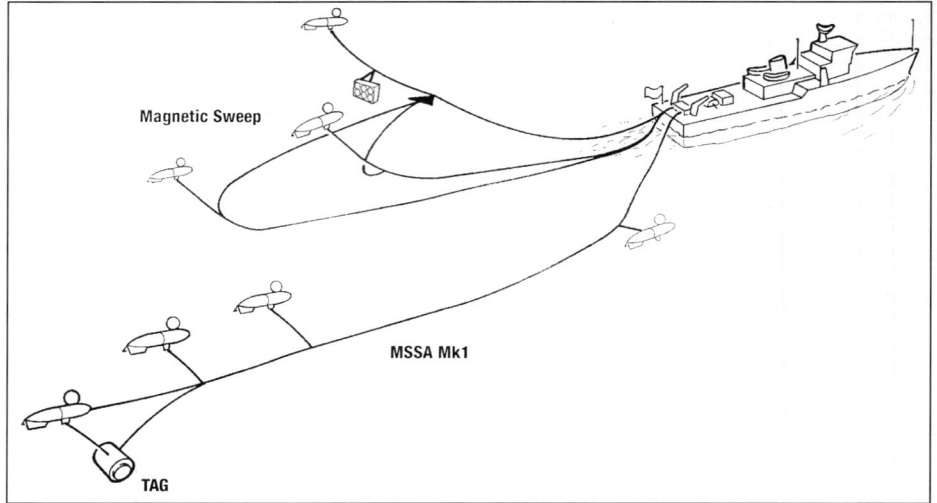

MSSA Mk 1

'Hunt' class MCMV, and has undergone evaluation trials with the United States Navy. The system has also been sold to the Republic of Korea Navy.

Contractor
BAeSEMA, Bristol.

VERIFIED

Combined Influence Sweep (CIS)

Type
Combined influence minesweeping system.

Description
By combining the MSSA Mk 1 towed acoustic generator (see separate entry) with an efficient

magnetic sweep, BAeSEMA offers a CIS system which is capable of sweeping the most modern microprocessor-controlled mines. In the CIS both the acoustic and magnetic signatures are microprocessor controlled, giving complete and total control of the generated signature. Both target setting mode and mine setting mode techniques may be employed readily with CIS.

Operational status
In service with the Royal Navy and has been sold to the Republic of Korea Navy.

Contractor
BAeSEMA, Bristol.

VERIFIED

MINENAV

Type
Acoustic minesweep monitoring system.

Description
The MINENAV system comprises a number of sensors used to measure the spread of the sweep using transponders, the swept depth using pressure

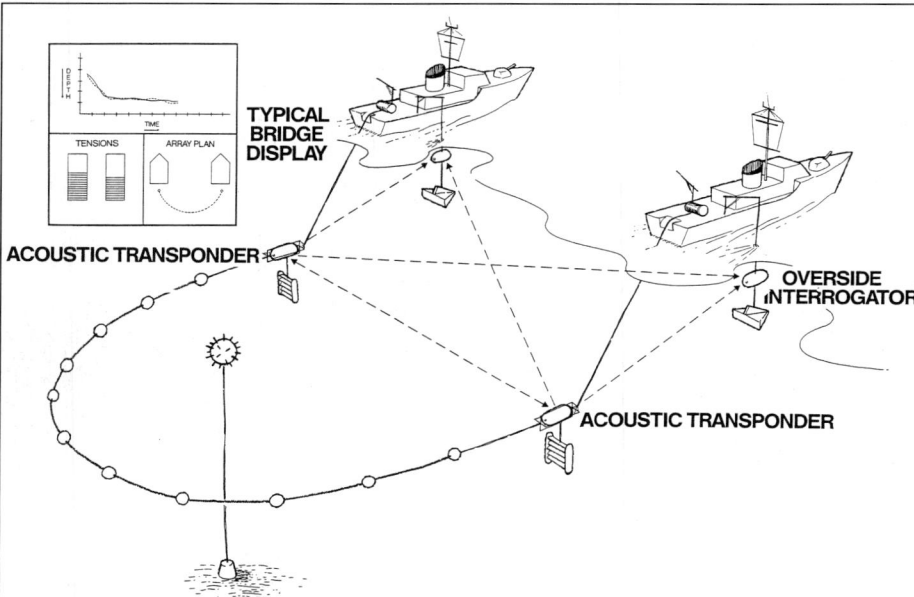

transducers and sweep height using echo-sounders, all of which are placed on the sweep wire. Subsea sensors are protected by an impact-resistant plastic outer layer and stainless steel inner cylinder. The receiver is deployed on the wire between the vessel and the sweep kite and is mounted in an attachment frame prior to deployment and suspended from a strain relieving cable, along which received signals are transmitted to the sweep vessel using an FSK acoustic link. It is protected against shock using four rubber universal joints. Computer processing allows graphic display of the information to provide the sweep officer with up-to-the-minute performance data. Data is logged to disk for further analysis.

Operational status
The system has been used annually since 1990 by the UK MoD DGUW (Director General Underwater Weapons) for trials of minesweeping equipment.

Contractor
Seametrix, Aberdeen.

VERIFIED

MINENAV system used in team sweep system

Limpet Mine Disposal Equipment (LMDE)

Type
Mine disposal equipment.

Description
The LMDE has been designed to destroy limpet mines which are difficult to remove from ships because of their powerful magnets and their anti-handling devices.

The system features a stainless steel barrel bomb disruption device similar to those used against terrorist bombs on land. The barrel is mounted in a movable sleeve and trunnion on a stand with flotation magnets to attach it to the ship's hull, and the whole device, which weighs less than 15 kg, has buoyancy blocks to reduce the effective weight in water.

When a limpet mine is discovered the diver assembles the LMDE out of the water, then takes it underwater and places it as close as possible to the mine. The barrel is carefully positioned so it faces the heart of the mine, then the diver returns to the surface from where he/she remotely fires a bridgewire cap cartridge. This propels a powerful slug of water at the mine, smashing open the thin casing and dispersing the main explosive charge.

Operational status
The system has been purchased by five customers.

Contractor
A B Precision (Poole) Ltd, Poole.

VERIFIED

Limpet Mine Disposal Equipment (LMDE)

UNITED STATES OF AMERICA

Mk 105 airborne minesweeping system

Type
Magnetic/acoustic influence minesweeping system.

Development
EDO Corporation has developed a highly mobile, helicopter-towed, unmanned MCM system that provides high-speed magnetic mine clearance and reduces the risks to personnel inherent in minesweeping operations. The Mk 105 is designed to be towed by the US Navy's MH-53E Sea Dragon helicopter and other heavy-lift helicopters of similar performance and size. Minesweeping operations may be either land-based or launched from various amphibious assault ships such as an LPH or LHA.

Description
The Mk 105 airborne minesweeping system is a helicopter-towed system carried on a hydrofoil that uses a gas-turbine engine generator set (power pack) to provide electrical power to an array of buoyant sweep cables. The electrical power is programmed and remotely controlled from the helicopter to simulate the magnetic signature of various ships. Magnetic influence mines are detonated by the magnetic field generated by the system. The Mk 105 is composed of a hydrofoil sled, a control programmer that is mounted in the towing helicopter, a towcable and an array of sweep cables.

The sled consists of a seaborne equipment platform and a generator set. The platform includes two tubular floats of light, metal-alloy construction, connected by an aerofoil wing assembly. Cavitating hydrofoils are attached to the forward and aft sections of each float. The surface piercing tandem foil configuration features two inverted V-shaped foils forward and two aft for

balance. High-riding pitch control subfoils of similar configuration are located ahead of the two bow foils. For landing and ground manoeuvres, the foils may be rotated up and parallel to the floats by a self-contained hydraulic system. Wheels are attached to the underside of the floats to facilitate ground or deck handling, and a retrieval system is mounted on the platform for helicopter recovery operations. Fuel for the turbine is carried in two centrally located tanks, one in each float. Towcable attaching points are provided on the forward inward faces of each float. Once launched, the sled is towed at about 25 kt by a 1,237.16 m towcable that is capable of carrying both electronic signals and fuel for the gas-turbine engine generator set.

Mounted on the wing assembly of the platform is the 350 kW, 2,000 A generator set that includes a gas turbine-driven AC generator, a rectifier, a controller containing the waterborne electronics and batteries to power the electronic system. The generator set provides energy for the generation of the magnetic field. The two electrodes (45 and 137 m long) are attached to the aft end of the sweep boom, which is located on the centreline of the sled, aft of and below the wing assembly. The sweep boom comprises an upper electrode attached to the end of a trailing cable and a lower electrode fitted to the boom fin. Using the water as a conductor, the potential between the electrodes produces a magnetic field that simulates the magnetic signature of a ship.

The control programmer, located in the helicopter, is the only manned station employed in the system. It contains the airborne electronics and all the controls and instrumentation necessary for operating the Mk 105. The control programmer console contains the fuel transfer control panel, turbine indicators, hydrofoil and boom actuators and the generator controls and indicators. From the control programmer, the operator can start and stop the turbine, raise and lower the foils

and control the magnetic influences generated through the sweep cables trailing behind the sled. The Mk 105 is to be upgraded with a new turbine offering improved reliability and endurance.

The Mk 106 configuration, also towed by the MH-53E, incorporates the Mk 105 magnetic influence sweep and the Mk 104 acoustic influence sweep. The Mk 104 comprises a venturi tube and a water-activated turbine that rotates a disc to reproduce a ship's acoustic signature.

Operational status
First used by the US Navy in 1973 during Operation End Sweep, the mine clearance operation in Haiphong Harbour, Vietnam. In continuous use since by the US Navy. Most recently used in the Persian Gulf during Operation Desert Shield/Storm. The system is currently undergoing a major upgrade.

Specifications
Length overall: 8.3 m
Beam
    Catamaran structure only: 3.5 m
    Across foils: 6.5 m
Height, foils extended
    To top of retrieval rig: 5.25 m
    To top of nacelle: 4 m
Height, foils retracted, to base of wheels: 3.5 m
Weight
    Gross: 3,800 kg
Speed: 25 kt

Contractor
EDO Corporation, Marine and Aircraft Systems, College Point, New York.

VERIFIED

Mk 105 sled foilborne

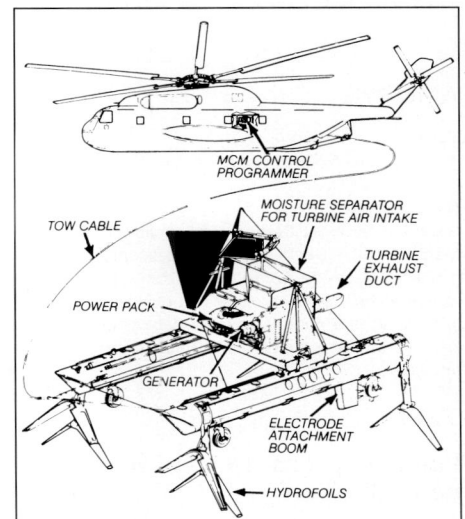

Schematic of Mk 105 system

SWIMS

Type
Shallow Water Influence Minesweep System (SWIMS).

Development
SWIMS is a US Navy shallow water magnetic/acoustic minesweeping system used to support amphibious operations, port breakout and riverine/brackish water operations. It is a small, high speed, influence sweeping system for all operating environments, deployable from helicopter or surface craft.

Description
SWIMS is currently operated from the MH-53B Sea Dragon helicopter, and is intended to support future organic MCM from smaller H-60-size helicopters.

It is totally contained within the helicopter during transit to and from the area of operation, thus providing the capability for high-speed transit from over the horizon.

The magnetic subsystem of SWIMS contains an electronically controlled dipole solenoid within a torpedo-shaped body. The magnetic output of SWIMS is preprogrammed to provide the operator with a variety of waveshapes to trigger the perceived threats. Demagnetisation is provided to allow for the magnetic body to be brought back aboard the helicopter upon mission completion.

The acoustic subsystem is towed aft of the magnetic, and can be either the Mk-104 or an Advanced Acoustic Source (AAS). This AAS produces speed-independent, operator-controlled, variable acoustic output which can emulate a number of target crafts.

Operational status
A prototype is currently being used by the US Navy in various national and international military exercises.

Specifications
Magnetic subsystem
Length: 3 m
Diameter: 50.8 cm
Weight
  Body: 453.5 kg
  Electronics: 181.5 kg
  Towcable: 181.5 kg
Advanced acoustic source
  Length: 1.5 m
Diameter: 40.5 cm
Weight
  Body: 90.75 kg
  Electronics: 317.5 kg
  Towcable: 22.5 kg

Contractor
EDO Corporation, Marine and Aircraft Systems, College Point, New York.

VERIFIED

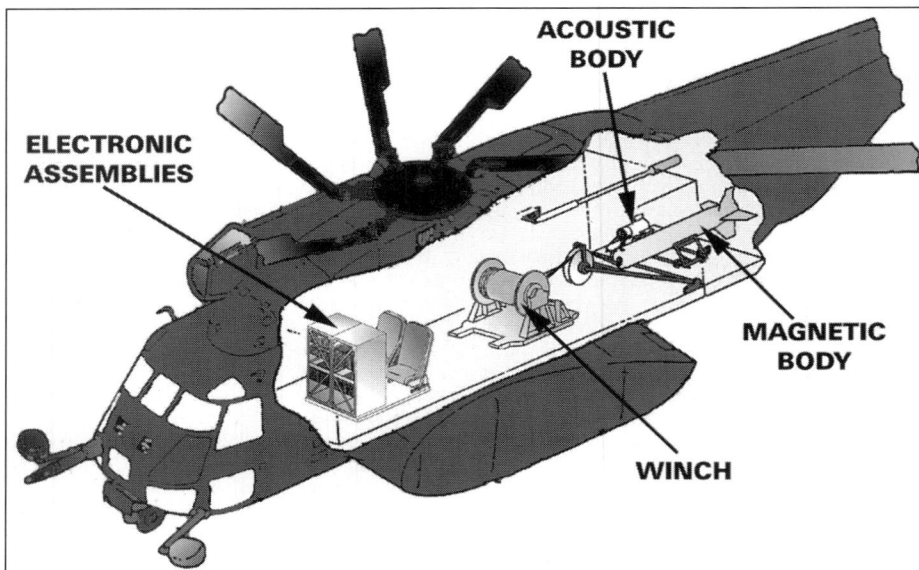

Helicopter-deployed SWIMS

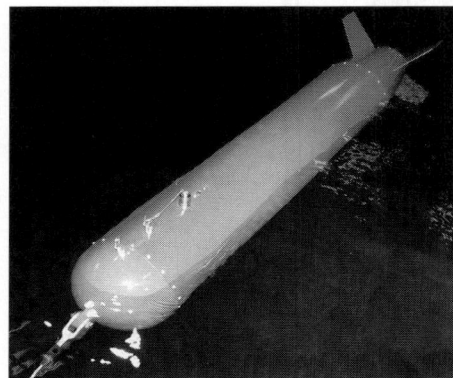

SWIMS under deployment

RAMICS

Type
Mine clearance gun system.

Description
RAMICS (Rapid Airborne Mine Clearance System) currently under development is based on a Light Detection and Ranging System (LIDAR) and a Gatling-type gun. The cannon fires a specially designed spin-stabilised 20 mm supercavitating discarding sabot projectile which can detonate surface or sub-surface mines, penetrating the casing down to depths in excess of 30 m providing a mine-free path in beach zones. The projectile enters the water at an oblique angle. On entering the water a small protrusion on the nose of the

projectile together with its speed produce a cavitation envelope in which the projectile travels in a straight low-drag trajectory. When the projectile strikes the mine it releases a highly reactive non-explosive lithium perchlorate oxidiser into the mine's explosive core causing it to detonate. The gun is aimed using a blue-green laser which penetrates the water column to survey the target from a number of directions to overcome any false alarms and provide a positive mine identification. Targets to be engaged by RAMICS are re-acquired by the LIDAR system from which the RAMICS controller is provided with the target's co-ordinates and automatically directs and holds the stabilised gun on the target firing a burst of between 20 and 50 rounds at a time.

Operational status
Hughes Aircraft has been awarded a contract worth up to US$12.5 million by the US Naval Surface Warfare Center Dahlgren Division's Coastal Systems Station at Panama City in Florida to carry out an advanced technology demonstration of RAMICS. The project is due for completion in September 2000.

Contractor
Raytheon Company, Mukilteo, Washington.

UPDATED

AMNSYS

Type
Airborne Mine Neutralisation System (AMNSYS).

Development
Raytheon Systems Company has teamed with GEC-Marconi in the UK to develop a helicopter-deployed multishot system to rapidly clear mines in intermediate water depths. Raytheon will be responsible for platform integration, fire control and the deployment system. GEC-Marconi will supply its Archerfish self-propelled munition.

Other systems offered to meet the request for proposals are the STN ATLAS Elektronik Seafox teamed with Lockheed Martin and the Kongsberg Defence Minesniper with an undisclosed American company.

Operational status
The US Navy evaluated off-the-shelf candidates for AMNSYS during 1998 and is now to select a system for developmental test and evaluation and operational test and evaluation during the second half of 1999. A contract for full production is due to be issued by the middle of 2000.

Contractors
Raytheon Systems Company, Mulkilteo, Washington, USA.
GEC-Marconi Underwater Weapons Division, Waterlooville, UK.
STN ATLAS Elektronik, Bremen, Germany.
Lockheed Martin.
Kongsberg Defence, Kongsberg.

UPDATED

Focused pressure wave mine neutralisation

Type
Mine neutralisation system.

Description
The focused pressure wave mine neutralisation system

is being developed jointly by United Defense and Tetra for rapid and reliable mine neutralisation tasks. This approach concentrates pulse power-based energy against mines from a safe standoff distance. It will enable forces to carry out rapid mine clearance operations without the need for detection, classification or explosive expendables.

The system uses intense pressure waves created by

pulsed electrical energy discharged into water. The pressure waves are generated from arc plasmas and neutralise the mine by rupturing its casing, rendering the sensing and fuzing systems useless, or causing the detonation of the mine.

The concept consists of an array of transducers which focus the high-energy pressure waves on the mine. The phased array generates high peak pressure at a focal point approximately 200 m from the array, with sufficient impulse to neutralise or detonate the target. The phased array will be capable of sweeping an access lane some 25 m wide by 8 m deep at a speed of 16 kt.

The concept is based on technology which has already been proven in underwater commercial applications used to drill and crush rock. Recent developments to efficiently load energy into underwater plasma sources and its conversion to pressure wave energy have significantly improved the efficiencies of these commercially based processes.

Such a system could be integrated in air-cushioned vehicle's surface craft or helicopter-deployed sleds for mine clearance operations in shallow and very shallow water and surf zone regions.

ALISS

Type
Advanced Lightweight Influence Sweep System (ALISS).

Development
The ALISS is being developed to meet the US Navy's need for a lightweight acoustic and magnetic minesweeping system that can clear a beach assault area in significantly less time than is currently possible. The acoustic subsystem generates high-power, low-frequency sound waves to detonate acoustic influence mines. It will replace fixed-frequency mechanical noisemakers used in the past with one system which is software configurable to the required frequency band. The acoustic system will be integrated with the magnetic system. Following an 18 month technology demonstrator programme, the US Navy is planning an engineering and manufacturing development programme to build an integrated acoustic and magnetic sweeping system.

Raytheon Systems Company (formerly Alliant Techsystems Marine Systems, subsequently part of Hughes Aircraft Naval and Maritime Systems), was awarded a US$4.6 million contract by the Coastal Systems Station, Dahlgren Division of the US Naval Surface Warfare Center to build and test the acoustic subsystem for ALISS.

The Coastal Systems Station will develop and test the spark gap acoustic transducers. Hughes Aircraft will integrate the entire acoustic subsystem of ALISS.

ALISS will be mounted in a QST-35 Craft. A super conducting magnet will be used to generate ship-like magnetic signatures while a spark gap generator will generate LF acoustic signatures.

Description
The sweep system uses advanced spark gap transducer arrays which provide the frequency agility necessary for spectral shaping of the acoustic output.

The magnetic sweep uses superconducting magnets and the coils are cooled using a specially developed, closed-cycle cryocooler.

The magnetic subsystem consists of a lightweight, high field, conductively cooled superconducting magnet housed in a low heat leak, vacuum insulated cryostat. The magnet is based on a solenoid wound with a niobium titanium superconductor, and operates at temperatures below 10K.

Operational status
Under development. The subsystem was scheduled to be delivered at the end of 1996 for demonstration testing.

Contractor
Raytheon Systems Company, Mukilteo, Washington.

UPDATED

Contractor
United Defense, Armament Systems Division, Minneapolis, Minnesota.

VERIFIED

DIVERS' SYSTEMS

This section covers many types of equipment associated with divers and their operations. Diving covers a vast field of activity ranging from sabotage, beach reconnaissance, disarming and removal of unexploded ordnance, to salvage and the inspection of underwater defences and so on. To support these varied underwater operations, a wide range of equipment including hand-held sonars, underwater telephones, communications equipment and cameras has been developed, some of which are described here. More of the equipment and systems used in these underwater operations will be covered in future editions of this book.

As this section expands it is hoped to be able to more sensibly divide entries into sections covering submarine escape and rescue, mine warfare (explosive ordnance disposal and so on), salvage and general military diving.

UPDATED

DIVING EQUIPMENT

BRAZIL

ARB

Type
Atmospheric Rescue Bell (ARB).

Description
The ARB is designed for the rescue of submarine crews. Eight submarine crew members plus two operators of the bell can be transported on each trip.

The bell comprises a control module, power supply, umbilical handling winch, umbilical cable and the bell itself.

Ancillary equipment fitted to the bell includes CCTV underwater cameras, an acoustic communication system, a closed-cycle life support system, two separate ballast systems for buoyancy and roll control, an emergency battery system, and an umbilical/lift cable release system.

Operational status
One unit has been delivered to the Brazilian Navy and is installed on board the NSS *Felinto Perry*. A second unit is under construction.

Contractor
Consub SA, Rio de Janeiro.

VERIFIED

ARB installed on board a rescue vessel
1997

CANADA

Scubaphone

Type
Diver underwater communications set.

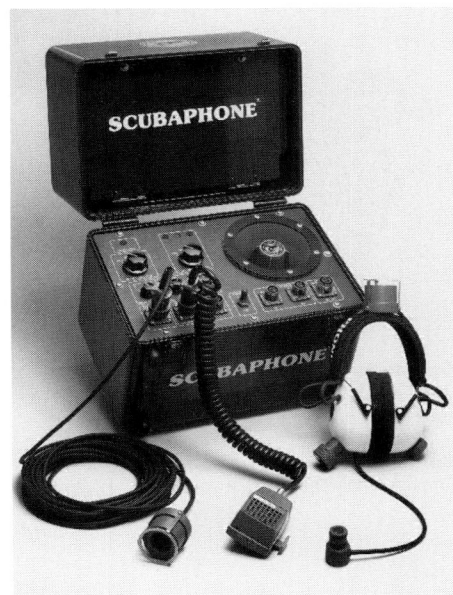

Scubaphone diver communications set

Description
The Scubaphone wireless underwater communication system is designed for both military and commercial use. The Scubaphone products enable diver-to-diver and diver-to-surface intercommunication. It employs a single-sideband suppressed-carrier transmission technique to achieve maximum power output and effective range. Special filters in the system process the incoming signal to remove background sounds such as the bubble noise caused by the exhalation of the diver.

With the Scubaphone, diver-to-diver communication is independent of the surface unit. Communication from the diver can be performed over ranges up to 1,200 m in saltwater, while that from the surface ranges up to 3,600 m. The diver's microphone and earphones are water- and pressure-proof to 100 m depth in saltwater. The set is powered by a 12 V rechargeable Ni/Cd battery.

Operational status
In service.

Specifications
Frequency: 30 kHz standard; others optional
Diver phone
Acoustic power output: 1 W
Audio output: 1 W RMS
Surface phone
Acoustic power output: 3 W
Audio output: 2 W RMS

Contractor
Orcatron Communications Ltd, Port Coquitlam, British Columbia.

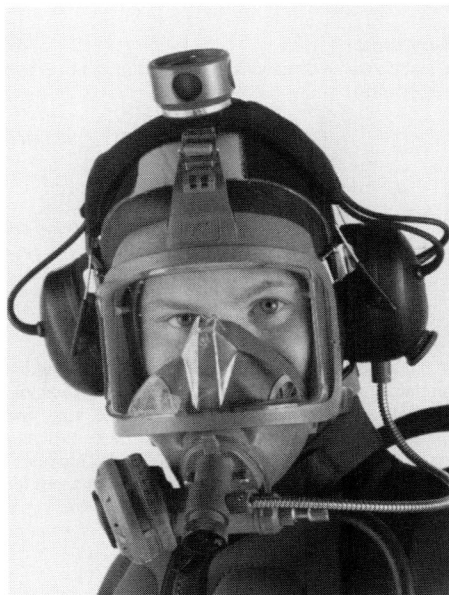

Diver unit of the Scubaphone

VERIFIED

Scubaphone Vox

Type
Voice-activated diver unit.

Description
The Scubaphone Vox (voice-activated) diver unit can either be belt-mounted or mounted on the air bottle. The unit offers hands free operation but can also, with a switch, be converted to push-to-talk mode to ensure 'silence' when needed. This system is fully compatible with the Scubaphone 2000D and 2000S series.

The Scubaphone Vox has a range of up to 3 km and is rated to operate at up to 100 m depth.

Operational status
In service.

Contractor
Orcatron Communications Ltd, Port Coquitlam, British Columbia.

VERIFIED

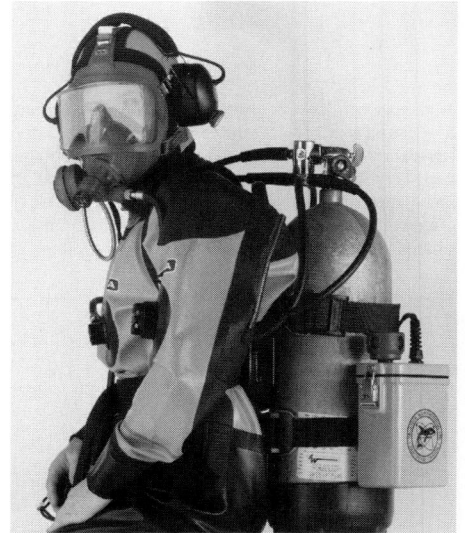

The Scubaphone Vox Model 1080 showing the unit mounted on a diver's air bottle

Subphone

Type
Submarine/submersible and surface vessel communications system.

Description
The Subphone is a single- or dual-frequency, single-sideband, selectable power, through-water communications system. It is a rack-mounted equipment for use in 1 atm submarines/submersibles or on surface vessels and can be mounted in a carrying case for surface support use.

The Subphone provides dual-frequency operation for two-voice frequencies and an internal pinger. It is intended to provide high-quality communications through saltwater at nominal ranges up to 10 km.

Subphone is also used as the surface station for the SDU (Special Diver's Unit) which allows clear communications between divers at nominal ranges of

10 km in saltwater. Model 2190 is a waterproof version of Subphone for use in a rubber dinghy for intrusive launch and recovery of commandos.

Operational status
In service.

Specifications
Frequencies: crystal controlled, user specified: 10, 27 and 30 kHz USB standard
Transmitter power: 50 W (0.5 W selectable)
Pinger: 15 ms
Nominal range: 6 km at 27 kHz; 10 km at 10 kHz

Contractor
Orcatron Communications Ltd, Port Coquitlam, British Columbia.

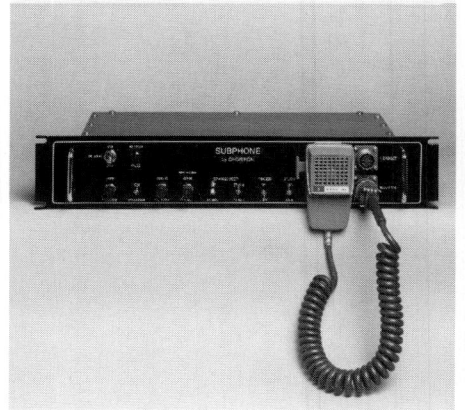

VERIFIED   *Subphone diver communications set*

Special Diver Unit (SDU)

Type
Long-range diver underwater communication set.

Description
The SDU is a single- or dual-frequency, single-sideband, selectable power, through-water communication system specially designed for long-range intrusive or anti-terrorist missions. The SDU is available in two configurations. The Mk I is designed to be worn on a backpack by divers equipped with chest-mounted rebreather apparatus. The Mk II is designed as a twin-tube, lower profile system which can be customised from the Mk I. Numerous customised features are available.

Low power allows diver-to-diver communication. High power allows up to 10 km communication between the divers and a submarine or surface ship equipped with a Subphone communication system.

The system can also be used for dry/wet swimmer delivery vehicles.

Operational status
In service.

Specifications
Frequency: crystal controlled, user specified: 10 kHz USB standard
Transmitter power: 50 W (0.5 W selectable)
Voice: Tx/Rx
Morse: Rx
Nominal range: 10 km at 10 kHz

Contractor
Orcatron Communications Ltd, Port Coquitlam, British Columbia.

VERIFIED

SDU communications set
1994

Video camera housings

Description
A variety of high-quality underwater video camera housings enables video cameras to be converted from hand-held models into units capable of operating at depths down to 100 m by plugging the housing to an underwater pluggable umbilical. This feature, coupled with high-quality images and low-lux capacity (low light required) make the units suitable for use in such diverse applications as reconnaissance, surveillance, the inspection of ships' hulls and so on.

Standard features on the housings include external amphibious microphone and fingertip electronic controls, which give the diver access to auto and manual focus, white balance, lens focus control and zoom control, as well as record and standby modes.

Amphibico/ has also developed an aspheric lens for its video housings which provides a 100° wide angle capability with no distortian either in or out of water or partially submerged.

Operational status
Units are in service with US Navy SEAL units and submarines and the Royal New Zealand Navy.

Contractor
Amphibico Inc, Dorval, Quebec.

UPDATED

Siva 55

Type
Diver rebreather apparatus.

Development
The Siva rebreather apparatus has been developed to provide divers with both shallow water oxygen and deep water mixed-gas capability.

Description
The Siva unit features a low-acoustic signature and, with a non-magnetic form available, the system is suitable for mine countermeasures duties as well as weapon recovery and other military tasks where minimal signature is important.

The modular Siva system accommodates two interchangeable gas control modules which offer the diver depth adaptability. The '55' gas control module permits the apparatus to be used with oxygen or premixed gases to a depth of 55 m. These are metered to the breathing circuit by a unique mass flow selector and control mechanism. The circuit for the control of premixed gases features four adjustable jets which can be calibrated by the user or supplied factory-set to suit the mixtures required for various depths down to 55 m. Once set, the user has the freedom to switch mixtures between dives without recalibrating the unit. In the self-contained mode, Siva has an endurance of up to 3 hours. The '+' gas control module delivers either 100 per cent oxygen or the dynamic mixing of oxygen and diluent gas using a pneumatic control system for dives exceeding 80 m in depth. These depths are made possible by significant gas conservation achieved through the control of the oxygen partial pressure. In the self-contained mode, Siva + provides enough gas and CO_2 removal capacity in 0°C water to support a working diver to a depth of 80 m. This includes a 20

minute bottom time and allows for in-water decompression. The maximum depth attainable is limited only by the quantity of stored gases.

The use of surface-supplied gases from an optional umbilical permits extended bottom time by employing Siva as a lightweight surface-supplied apparatus or by providing gas for in-water decompression following a free swimming dive. The design and the options extend depth and dive time to expand Siva's operational capabilities.

The ergonomically designed counterlung and breathing loop minimises flow resistance and provides for more comfortable breathing. The counterlung is positioned on the diver to reduce hydrostatic effects while optimising swimming actions. A diver-adjustable valve is employed to establish the best breathing characteristics in all positions. The lung is made of a tough, abrasive-resistant, reinforced polyurethane that will not mildew or rot and is unaffected by ultraviolet light. It comes fitted with integral and jettisonable weights and functions as a flotation device, maintaining a diver face up on the surface.

Specifications
Length: 58 cm
Width: 33 cm
Depth: 17 cm (excl breathing bag)
Weight: 30 kg (in air)
Gas storage volume: 1,177 litres
Breathing bag capacity: 8.0 litres
Endurance: 3 h (min) at moderate heavy work
Charging pressure: 297 bar

Contractor
Fullerton Sherwood Division, Carleton Life Support, Technologies Limited, Ontario.

VERIFIED

Siva rebreather apparatus

S-10

Type
Oxygen rebreather diving apparatus.

Description
The S-10 is a chest-mounted rebreather diving set designed for specific military and law enforcement agency work. It is a closed-circuit pure oxygen rebreather for use to depths of 8 m. Specific user rules and regulations on oxygen swimming operations may permit the use of the S-10, for brief intervals, to greater depths.

The S-10 may (optionally) be fitted with an adjustable flow metering valve, which allows the diver to alter the flowrate between 0 litres/min and approximately 2.5 litres/min via an adjustable valve having four detent positions which locate the intermediate flows. In addition, the set has a manually operated bypass valve.

The S-10 may also optionally be fitted with an automatic demand valve, located in the counterlung. This valve will operate and supply oxygen to the counterlung whenever the counterlung is breathed down to an empty, collapsed state.

Either the demand valve or the adjustable flow metering valve, or both, can be fitted in the S-10, depending on user preference.

The equipment is capable of being certified non-magnetic and acoustically safe in accordance with NATO requirements for shallow water MCM and EOD (Explosive Ordnance Disposal) operations.

Specifications
Length: 43 cm
Width: 30 cm
Depth: 17 cm
Weight: 12.5 kg
Oxygen flask volume: 1.3 or 1.9 litres
Breathing bag capacity: 5.5 litres
Endurance: 4 h
Charging pressure: 206 bar

Contractor
Fullerton Sherwood Division, Carleton Life Support Technologies Ltd, Mississauga, Ontario.

UPDATED

S-10 diving apparatus
1996

FRANCE

DC 55 UBA

Type
Underwater breathing apparatus.

Description
The DC 55 is a self-contained semi-closed circuit underwater breathing apparatus. It comes in the form

of a set comprising a soda-lime scrubber unit, two aluminium alloy mixture cylinders and a breathing bag protected by a cover. The diver breathes into the apparatus through a mouthpiece, maintained by a strap and connected to two corrugated hoses which lead to the inhalation and exhalation pipes containing the corresponding non-return valves.

The composition of the mixture breathed at different

depths is practically independant of the diver and the work performed. It is obtained by the release from the circuit of a certain volume of gas proportional to the breathing amplitude. In this manner, the mixture breathed is very stable. It is of demand type, comprising neither by-pass nor continuous flow. The reduction of the volume of gas in flow, due to gas release, oxygen consumption and carbon dioxide absorption, is automatically compensated by an addition to the mixture coming from the admission valve. The circulation of gases is ensured by the action of two non-return valves. The whole of the exhaled mixture passes through the absorber unit and there is complete absorption of the carbon dioxide.

Operational status
In service with the French Navy and in many foreign navies.

Specifications
Weight: 24 kg
Length: 0.68 m
Width: 0.39 m
Thickness: 0.17 m

Diving performance

	Without injector	With injector	With injector
Operating depth	0.25 m	20-40 m	30-55 m
Gas mixture	60% O_2/40% N_2	40% O_2/60% N_2	32.5% O_2/67.5% N_2
Duration of dive	180 m	90 m	75 m

Contractor
La Spirortechnique I.C, Carros.

NEW ENTRY

ERUS-2

Type
Underwater radio communication system.

Description
ERUS-2 equipment is designed to provide underwater radio communication, using HF ultrasonic waves, between divers, divers and surface craft, and between surface craft.

Transmission is omnidirectional. The average range in isothermal conditions is over 800 m and down to depths of about 100 m.

Two equipment models have been developed: the ERUS-2A4 for divers and the ERUS-2B4 for operation on surface craft.

The 2A4 equipment includes an enclosed transceiver unit, a full-face mask with a built-in microphone and an earphone. The transceiver unit is provided with a squelch device and a voice-operated transmit control (VOX).

The 2B4 equipment includes a portable case containing the transceiver unit, the loudspeaker, the battery charger (available also to load the battery of the 2B4 equipment). The equipment, provided with various adjustment and control features, allows the transmission of modulated telegraphic signals. A hand microphone and headphones are housed in a compartment provided in the case.

A 50 m connecting cable is available with a transducer housed inside the cable reel.

Operational status
Tested by the US Navy. Approved for navy use.

Specifications
ERUS-2A4
Power supply: rechargeable battery
Depth capability: 120 m
Diameter: 80 mm
Length: 250 mm
Weight: 2.5 kg

ERUS-2B4
Power supply: rechargeable battery
Transceiver case
Height: 250 mm
Width: 260 mm
Length: 350 mm
Weight: 12 kg
Cable reel
Diameter: 220 mm
Height: 240 mm
Weight: 6 kg

Contractor
Safare-Crouzet, Nice.

VERIFIED

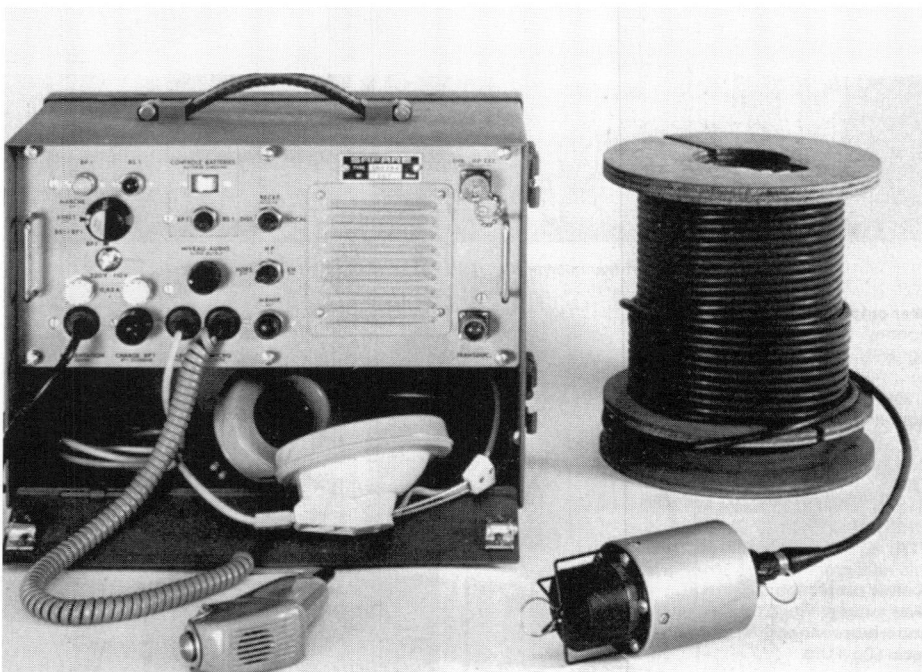

ERUS-2B4 surface craft equipment

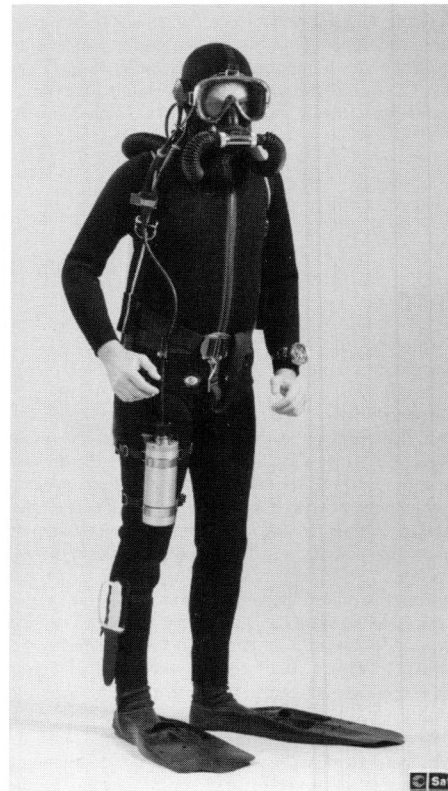

The ERUS-2A4 diver unit   *1994*

RUPG-1A

Type
Diver's homing receiver.

Description
The RUPG-1A receiver is a portable watertight equipment which allows divers either to home towards a submerged transmitter or to hear CW or phone communications from this transmitter.

Received signals are amplified by the equipment, which supplies the audio signal to a bone-conducting earphone. Directivity of the receiver informs the diver of the direction of the transmitter.

The shape of the equipment is similar to that of a portable torch. Metal parts are made from anodised aluminium alloy to withstand corrosion. A battery charger, designed for the RUPG-1A's storage batteries, can be operated from a 115 to 220 V 50/60 Hz mains or from a 24 V DC supply.

Operational status
In production for the French Navy in the early 1990s.

Specifications
Range: 2,500 m in normal propagation conditions with approx 10 W transmitter
Directivity: main lobe angle between ±10 and ±15° for 6 dB attenuation
Power supply: 9 V storage battery 15-20 h continuous operation

Pressure test: 10 bar
Max outer diameter: 191 mm
Overall length: 560 mm
Weight: 2.75 kg (in air)

Contractor
Safare-Crouzet, Nice.

UPDATED

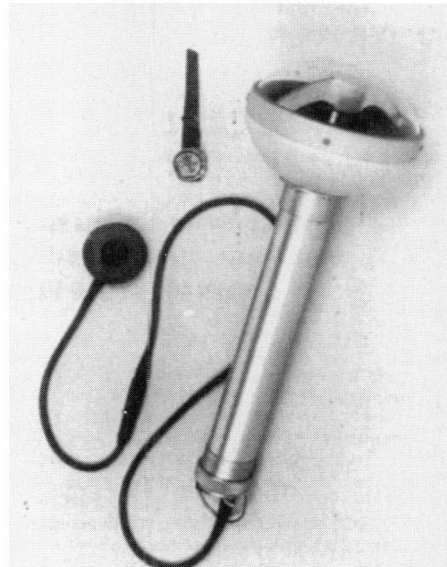

RUPG-1A diver's homing receiver

GERMANY

DSE 1

Type
Diver sonar camera.

Description
Mounted in a rugged, non-corrosive housing, the DSE 1 diver sonar integrates all sensors, electronics, controls and display in one common unit. The batteries are housed in a separate battery pack underneath the unit, which is easy to replace. The unit is nearly neutrally buoyant, which allows it to float up slowly to the surface if released. Different water densities are compensated for by means of trim weights.

The sonar is controlled by just two buttons – one each on the left and right hand side. The equipment is easily controlled by the diver's thumbs, even when wearing diving gloves. All functions are menu-controlled.

The sonar operates in active mode with a horizontal search sector of 64° and a vertical search sector of 12°. The search sector is divided into 32 channels for scanning, each of 2°. The active sonar returns are presented in an easily interpreted display. An audio channel enables acoustic observation over a frequency

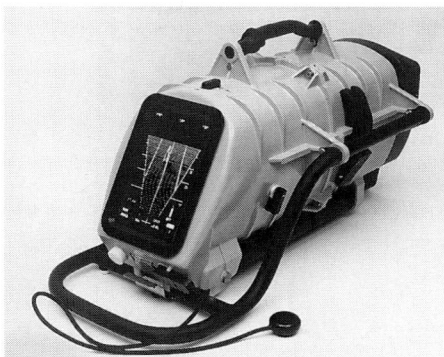

DSE 1 diver sonar camera   *1998*/0009569

range from 1 to 50 kHz. When using this mode the active sonar display is faded out. In addition to the sonar return, diving depth and a north reference enable the diver to determine position and obtain mission guidance.

The equipment is powered by replaceable batteries or with rechargeable Ni/Cd accumulators. An alarm

signal appears on the screen when the battery is close to discharge.

For special applications the DSE 1 can be split into two sub-units by means of two adaptor plates and a connecting cable (available as an option). This version allows for the separate operation of the sonar head under water while the monitor with display is carried on board an escorting vessel.

Specifications
Length: 732 mm
Width: 392 mm
Height: 396 mm
Weight: 31 kg
Field of view: active 64° (horizontal), 12° (vertical); passive 15° (horizontal), 30° (vertical)
Scale ranges: 12/25/50/100/200 m
Operating depth: 60 m

Contractor
L-3 Communications ELAC Nautik GmbH, Kiel.

VERIFIED

MW 1630/MW 1630B

Type
Underwater mine detectors.

Description
The MW 1630, specially designed for use by divers, is watertight down to depths of 60 m and is balanced for use underwater. It is a one-stick construction and comprises all accessories required for detection work in river environments as well as in shallow water. The equipment was introduced into US Navy service in 1992 and standardised as the Mk 29.

The successor model is the mine detector MW 1630B which is designed in two parts. The electronics unit is fixed on the leg of the diver and the search head with its light telescope carrying bar is hand-held. This is particularly suitable for use in strong currents.

The electronics are housed in a rigid cylindrical, amagnetic case with sensitivity adjustment being achieved by a single mechanical rotary switch and the volume control operated by a two-step toggle switch. Both detectors operate on the pulse principle. The advantage of this compared to detectors based on the eddy current principle is that, even in saltwater, detection work can be carried out without loss of sensitivity. The highly integrated electronics are designed using multilayer and SMD technologies.

The search coil is oval in shape providing a small (17 cm wide) swept path suitable for operation in areas of dense clutter such as rocks and so on. The detectors are powered by commercial batteries (MW 1630 – 4 × C size; MW 1630 B – 3 × D size) and incorporate an automatic battery voltage check and surveillance, which provides a permanent check of the batteries and functioning of the electronics without affecting the acoustic search signal. The unit provides constant sensitivity almost down to battery discharge level. An acoustic alarm sounds when the lowest voltage point is reached.

Operational status
Development completed in 1991 and 1994 respectively and in service with the US Navy, Belgium Army, German Army, France and Netherlands.

Contractor
Vallon GmbH, Eningen.

UPDATED

The MW 1630B mine detector in use
1995

Haux-Starcom

Type
Recompression chambers.

Description
This is a series of compact, modular diver recompression chambers for the treatment of divers, surface decompression and the training of divers. By incorporating the flat bottom principle and front-mounted control panel, chambers can be constructed in varying diameters. Complete utilisation of the entire inner length of the cylinder wall with optimum space can therefore be achieved. With minimised outer measurements the chambers can be easily transported and with front-mounted control panels they are especially suited for installation in containers, ships or trucks. The front panel houses all the instruments required for control and, with an adjacent observation window, supervision of the chamber occupants can be effected from a single point.

The aluminium alloy double-hinged doors are of pancake design and because of their symmetrical design can be operated from both sides.

The use of new Haux-Starvalve and Haux-Ventmaster valves and the integrated electronic Haux-Variomaster for the supervision of pressure changes offers a completely new way of supervising pressure changes enabling precise control over the build-up and maintenance of correct pressures.

With the use of the Haux-Scrubmaster IIIS CO_2 absorber, maximum operating time and gas consumption can be achieved. The system is low noise

Haux-Starcom in ISO container on ship's deck

in every phase of the operation and Haux-Phonkillers (inside and out) ensure minimal A-sound level. Newly developed Haux-Luxmaster/85D cold-light lamps provide the chamber with anti-dazzle illumination which can be gradually dimmed.

Contractor
Haux-Life-Support GmbH, Karlsbad.

VERIFIED

Haux-Super-Transstar

Type
Transportable one-man diver recompression chamber.

Haux-Super-Transstar. Note control panel with observation window to left

Description
The Haux-Super-Transstar pressure chamber is designed for emergency rescue operations by navy, police, rescue organisations, diving companies and so on. Due to its compact design and low weight, the chamber is easily transportable on trucks, ships and helicopters and can be carried by four people to the site of an emergency.

The pressure chamber is fitted with a standard bayonet flange enabling it to be connected to a treatment chamber.

For short-term operations, for example transfer of a patient to another transport system, mating with a treatment chamber and so on, the Haux-Super-Transstar is supplied with its own gas storage and supply systems. The cylindrical design of the pressure chamber in the region of the diver's trunk, and conical design in the leg region provides the occupant with the maximum amount of space. In the chamber the diver lies on a wide, comfortable stretcher with fabric covering and head rest.

The compact control panel is arranged with the observation window adjacent, enabling the operator to maintain permanent visual contact with the occupant. A precision manometer, semi-automatic fresh air ventilation, inlet and outlet valves for pressure control and an intercom system ensure a very high safety level during chamber operation.

Contractor
Haux-Life-Support GmbH, Karlsbad.

VERIFIED

Haux-Profi-Medicom

Type
Two-man diver recompression chamber.

Description
This unit is designed for diver rescue operations by, for example, navy, police, or diving companies. With its two occupants the chamber can be carried by six to eight people and is easily transportable. The newly developed structural shape allows the chamber to mate with any treatment chamber of any diameter equipped with a DIN/STANAG flange.

For short periods of operation the chamber is fitted with its own gas supply. The inside of the chamber provides the optimum amount of space for the injured diver who lies on a wide, comfortable stretcher which is covered with a hard-wearing canvas cloth. The assistant sits upright in the dome of the chamber.

The compact control panel and observation window are sited by each other enabling the operator to view the occupants. The chamber is equipped with a precision manometer, semi-automatic fresh air ventilation and communication system, which includes a battery-powered talkback speaker with volume control. The Medicom is supplied with a newly developed trolley which is easily manoeuvred into position by two people to mate with a treatment chamber.

Contractor
Haux-Life-Support GmbH, Karlsbad.

Haux-Profi-Medicom chamber on trolley ready to mate with large recompression chamber

Haux-Spacestar 1300/5.5 dome

Type
Amagnetic recompression chamber for MCMVs.

Description
The Haux-Spacestar recompression chamber is constructed of amagnetic stainless steel for the first-aid and treatment of divers on board MCMVs. The unit can also be used as a test and training facility.

The design features a dome mounted on the cylindrical pressure vessel which allows an occupant to stand upright inside the chamber without the need for increased floor space. This feature greatly improves the treatment which medical assistants can provide for injured divers. With neurological problems it is absolutely imperative that the injured diver can stand upright for diagnostic treatment. Increased freedom to move is desirable for the medical assistant and for the practice of a number of life-saving methods. For long-lasting treatment the ability to stand upright improves the physical and psychological condition of the

chamber's occupants. Finally, examination of eyes and ears is much easier if the patient and medical assistant are able to face each other.

The Haux-Spacestar consists of a main chamber and an antechamber. Under normal conditions accommodation is provided for two sitting and one lying occupant in the main chamber and one or two occupants in the antechamber. Pressurisation and fresh air ventilation are achieved using atmospheric air. The diagonal air flow in the chamber provides fresh air for the dome as well as the rest of the chamber. In order to improve decompression treatment, up to three oxygen breathing masks can be connected to a central point of attachment. Should pressurised gas storage begin to run low in long-lasting decompression treatments, safe operation can be assured for a number of days in the closed-circuit operation with the Haux-Scrubmaster CO_2 absorption unit and the addition of a small amount of oxygen from the supply.

The unit is equipped with an intercom system enabling communication between the main chamber, antechamber and control station. Three windows are

fitted in the dome and the cylindrical part of the chamber enabling continual observation of the occupants.

A Masterlock supply lock welded into the main chamber enables the occupants to be supplied with medication, food and drink without disturbing decompression in the main chamber.

For installation on MCMVs, recompression chambers have to meet stringent requirements and the Haux flat bottom design offering maximum economy of space and compact design with laterally sited operation centre, enables the unit to meet these requirements.

Operational status
A total of 10 of these chambers are being built to equip the 'MJ 332' class minehunters of the German Navy.

Contractor
Haux-Life-Support GmbH, Karlsbad.

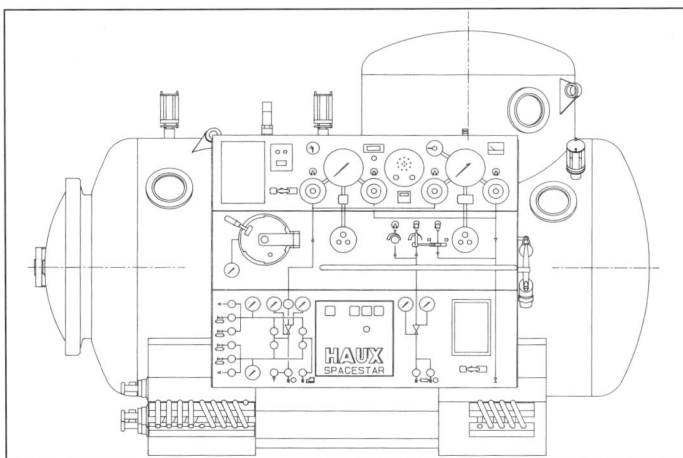

Spacestar recompression chamber

The Haux-Spacestar 1300/5.5 DOME decompression chamber for the 'MJ 332' minehunters of the German Navy
1994

Haux-Deepstar 80/C

Type
Containerised self-contained diving system.

Description
The Haux-Deepstar 80/C containerised diving system consists of: the Haux-Modulstar 1500/6 Dome, a four-man deck-decompression chamber with two compartments and top transfer coupling system for the diving bell; the Haux-Deepstar 1300/80 diving bell for

top transfer; the Haux-Divecontroller control panel for the diving bell; a handling system with two-guidewire bottom anchor; and a gas supply. The complete system is housed in two 20 ft containers. All containers are fully self-contained and operational except for the power supply which is fed from the ship's system or a generator.

The containers can be installed on a pontoon, a ship's deck or other suitable vessel. The containers are easy to transport and offer considerable flexibility for

multipurpose diving operations. The Haux-Modulstar 1500/6 Dome decompression chamber consists of a main chamber with an integrated dome to enable a man to stand upright inside and an entrance lock. On top of the main chamber is a mating flange which enables divers to transfer from the diving bell to the chamber without interrupting decompression.

The decompression chamber incorporates a wide range of safety devices and equipment including a CO_2 scrubber and external life support system.

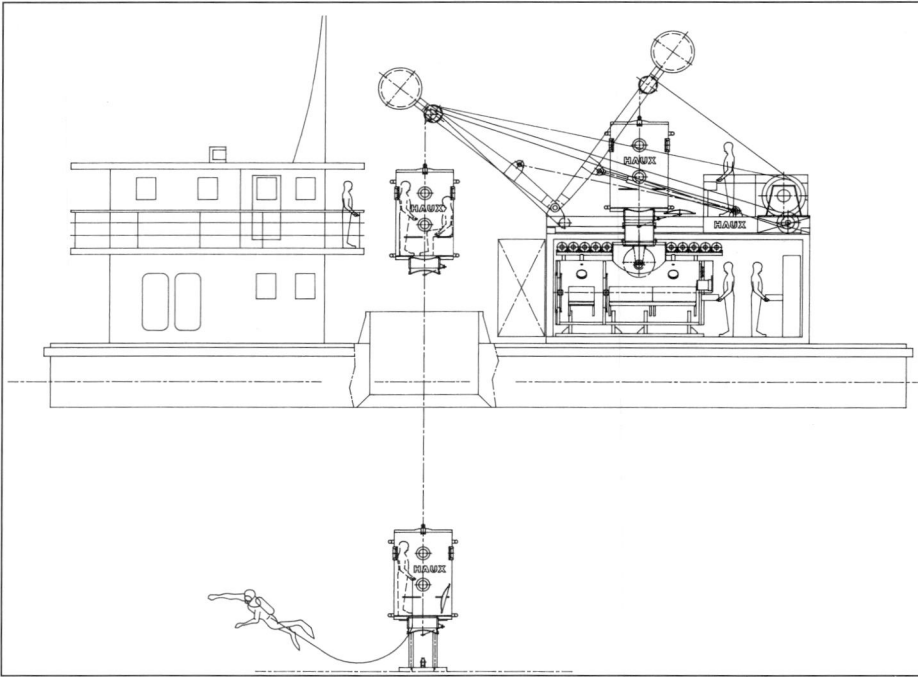

The diving bell is designed to carry up to three divers for bounce or heliox diving operations at a maximum depth of 80 m.

The bell is controlled from a separate control panel and incorporates various communications systems. Control and observation of divers can be effected from the bell or from the surface control station via TV and telephone.

Operational status
On order for the Indonesian Navy.

Contractor
Haux-Life-Support GmbH, Karlsbad.

VERIFIED

Haux-Deepstar 80 containerised diving system for the Indonesian Navy
1994

Hydra 2000

Type
Diving simulator.

Description
The Haux-Divestar Hydra 2000 is the first diving simulator to be built for the German Navy. The modular system comprises: a three-chamber complex of a simulation chamber, diagnostic chamber and therapy chamber operating at 20/10 bar; four separate control stations for each part of the decompression chamber; two control boxes which can also be used as fully operational emergency control stations; a Haux-Mixmaster computer-controlled mixer for breathing gases with preselectable but constant partial pressure; a supervision system for the operator under training which measures simultaneously under pressure such parameters as ECG, EEG, blood pressure, breathing frequency, transcutan PO_2 and PCO_2 on a separate observation stand; a colour TV monitoring system; gas supply with air, helium, oxygen and gas mixtures; and a water conditioner which automatically maintains water levels.

The diving and pressure simulator reproduces pressures that correspond to a diving depth of some hundred metres and can be simulated under secure conditions.

The horizontally mounted cylindrical diagnostic chamber is the largest. This can be almost completely filled with water and is divided into a wet and dry compartment by so-called buffalo walls. It is mainly used for the training of divers under realistic conditions, for testing diving and underwater equipment, diving methods, and underwater working procedures and for diver aptitude tests with several aspirants at the same time.

The simulation chamber is similar to the diagnostic chamber but can be pressurised only with breathing gas. It is mainly used for medical examinations of test persons, recompression of divers, and Hyperbaric Oxygen (HBO) treatments. The chamber consists of a main and an antechamber with room for up to 10 divers at a time.

The cylindrical pressurised therapy chamber is mounted vertically between the diagnostic chamber and the simulation chamber. It contains a complete operating theatre's equipment and is used for the intensive medical treatment of sick divers under pressure. The chamber is also fitted with a flange system that can interface with transportable chambers. This enables sick divers to be transferred from the portable chamber to the therapy chamber in a very short time and without suffering any further changes in pressure.

Operational status
In service with the German Navy.

Contractor
Haux-Life-Support GmbH, Karlsbad.

VERIFIED

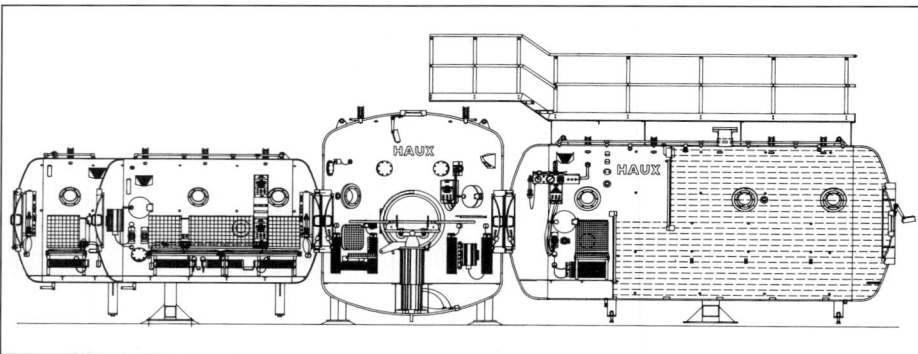

Hydra 2000 diving simulator for the German Navy
1994

Hydra 2000 diving simulator showing therapy chamber/simulation chamber/diagnostic chamber 1994

MEDUSA

Type
Diving simulator.

Description
MEDUSA is a new hyperbaric centre with integrated diving simulator. The system provides Hyperbaric Oxygen (HBO) treatment as well as tests and examination of divers and underwater equipment. The maximum working pressure of the system is 10 bar.

The hyperbaric centre comprises a treatment chamber and the actual diving simulator.

The treatment chamber is a horizontal cylinder with ante and main chambers where three and four persons respectively can be seated. The chamber is mainly designed for the treatment of sick divers, examination of students and hyperbaric treatment of civilians.

The main chamber of the diving simulator is a vertical cylinder (wet compartment) which can be equipped so as to function as an additional treatment chamber. With the antechambers, the diving simulator section and the treatment chamber section can be operated and used independently. The chamber is divided into wet and dry compartments by an intermediate bottom (grating). The wet compartment is mainly used for training divers and for tests of underwater equipment and working procedures and for diving aptitude tests.

The water column in the chamber is 3 m in height. The dry part of the main chamber can be fitted out with either eight seats or, if used as treatment chamber, four seats and two bunks.

Each chamber section has its own Haux-

Starcontroller control centre from where all proceedings are controlled and supervised. The system can be operated by just one person.

Operational status
The system is installed in the new building of the Mine department of the Royal Netherlands Navy at Den Helder. The treatment centre is available for both national and international military and civilian use on a daily basis.

Contractor
Haux-Life-Support GmbH, Karlsbad.

VERIFIED

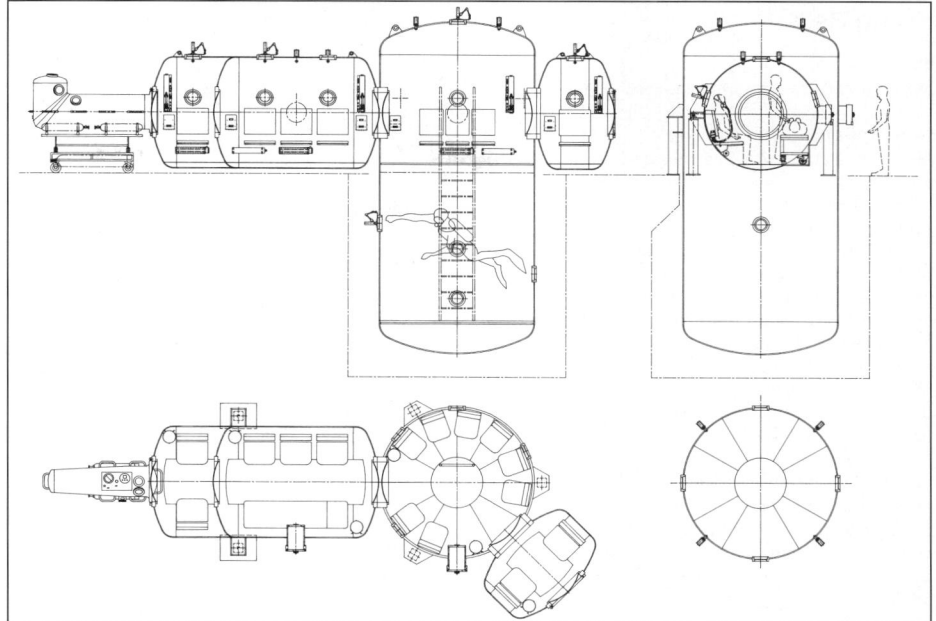

MEDUSA diving simulator with treatment chamber for the Royal Netherlands Navy
1994

Haux-Medistar 1250/5.5

Type
Two person diver recompression chamber.

Description
The Medistar is a newly developed modern transportable recompression chamber with a maximum working pressure of 5.5 bar (optionally 7 bar). The frusto-conical shell of the Medistar is constructed of saltwater-resistant aluminium alloy and is provided with several handgrips and three pad-eyes for handling by crane. Accommodation is provided for one lying injured diver and a seated attendant.

Maximum space has been provided for the diver and assistant, the assistant being supported in a seated position on a padded seat which enables him/her to provide adequate support for the injured diver who is supported on a wide, upholstered stretcher. Free access to the patient makes full medical treatment possible.

The low weight and small dimensions of the chamber enable it and its occupants to be transported in lorries, ships, aircraft or helicopters. It is easy to handle, and can be mated with any treatment chamber of any diameter which is equipped with a DIN/STANAG (NATO) flange.

The mating of the Medistar with the locking chamber Medilock offers a diver treatment chamber of considerable value. Injured personnel can be treated on site with minimum delay.

For short transit times between the site of the accident and a full medical centre for example, or for mating to another treatment chamber, the Medistar is able to use its own limited supply of gas. For long-duration transits or treatments, gas processing systems such as the Haux-Miniscrubber or other compatible systems ensure operation for hours even with limited gas storage (breathing air and particularly oxygen). The control desk is compact and clearly laid out in an ergonomic design with self-explanatory pictograms ensuring a higher degree of safety. The operator at the control console is able to maintain a constant watch on the chamber's occupants through observation windows in the chamber body. This constant visual contact of chamber occupants is particularly important for the psychological condition of the patient. Communication is available with a battery-powered intercom with volume control and headset microphone. Two oxygen breathing units are supplied as standard, with optional remote control of the chamber and for communication with the occupants via a control cable. Computer control is also available as an option, together with patient monitoring.

The Haux-Medilock is a one-lock chamber which is able to mate with the Medistar to enable the attendant to lock in and out of the chamber, and could enable a specialist doctor to replace the attendant while treatment continues.

The Haux-Medistar recompression chamber
1995

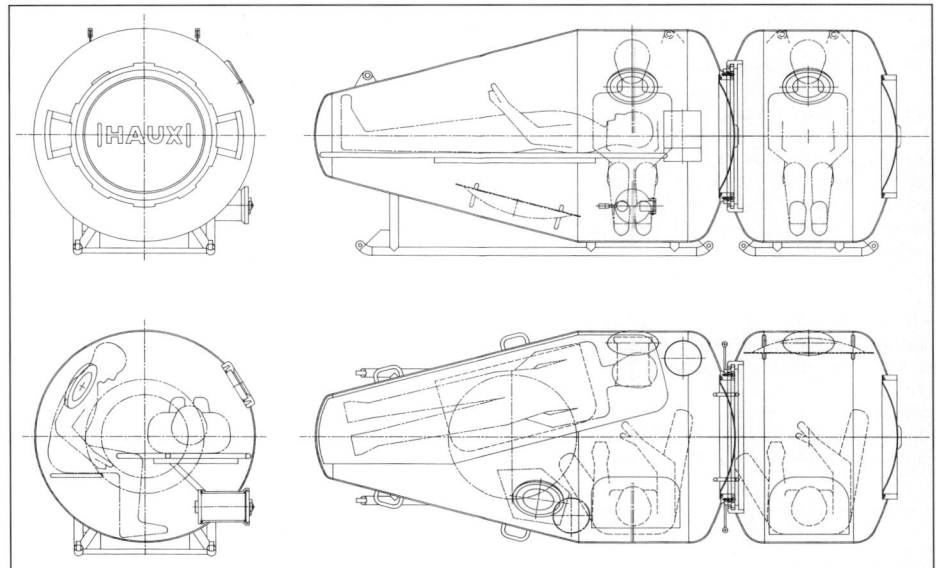

The Haux-Medistar with NATO flange/lock
1995

Depending on the loading capacity of a helicopter, it is possible to transport the Medistar or even the whole chamber system complete with lock to the nearest shore-based treatment chamber.

Contractor
Haux-Life-Support GmbH, Karlsbad.

VERIFIED

Flying Bell

Type
Mobile diving-, salvage- or rescue bell.

Description
Mobile diving bells have been designed to combine the advantages of freely swimming diver lockout submersibles with those of conventional, surface supplied diving bells.

The Flying Bell concept, as delivered and commissioned for saturation diving purposes, eliminates the limitations of the first with regard to energy and gas supply autonomy by provision of a connection to the surface via a specially designed umbilical. Due to it's variable buoyancy, propulsion and navigation systems the mobile bell achieves full three-dimensional manoeuvrability within a range only limited by the length of the supply cable. Furthermore, when launched and recovered through a moon pool a mobile diving bell may be less weather dependant than a rescue submersible handled over the ship's stern.

The pressure resistant body consists of a vertical cylindrical section with hemispherical caps on top and at the bottom. The cylindrical lockout trunk with double sealing hatch is inserted into the lower hemisphere. In the standard configuration the chamber accommodates up to three fully equipped saturation divers. The maximum operating depth of the system is 450 m (a rescue bell would be of spherical shape to typically accommodate 9 rescuees plus one operator, the mating skirt with bayonet door inserted into the lower part of the sphere. The rated depth could be extended to 600 m and more).

The chamber is fitted out with a number of flat acrylic viewports looking in almost all directions for optimum observation of the surroundings.

The vehicle's ballast system consists of two pressure resistant Hard Ballast Tanks (HBT) arranged on either side of the bell, plus a number of soft tanks on top. The HBTs can be emptied by means of compressed air or the HP drain pumps. The soft ballast tanks mainly serve for emergency surfacing and are blown by compressed air only.

Six large spherical, high-pressure gas containers are used for storage of air and breathing gas. This is a safety precaution in case the connection to the surface should fail. Should the umbilical become entangled it can be mechanically released from the connector installed on top of the bell. The umbilical contains electrical power and control cables, gas and hot water supply lines, gas control lines plus an integrated aramid fibre stress member. The latter is capable of carrying the weight in air of the fully equipped bell. The power cables end in a pressure resistant electrical power distribution module fitted out with terminals, circuit breakers, transformer and so on. The control cables are connected to a similar electronic module containing the bell units of the control computer and related interfaces. The mobile diving bell's propulsion system consists of six hydraulic thrusters, two located either side in a fore-aft direction plus two inclined vertical thrusters installed on top of the bell in a transverse position. The thrusters are fitted out with variable speed control. The two electrically driven hydraulic pumping aggregates feed into the hydraulic circuits either individually (at reduced overall power output) or in parallel. The hydraulic circuit and the hydraulic equipment installed on the bell are pressure compensated. The mobile diving bell is also equipped with north-looking gyro compass, autopilot, echo-sounder, hard wire and through-water communication systems, searchlights, underwater TV cameras plus sonar system.

Most of the functions of the bell, including manoeuvering, operation of ballast systems, claw, extendable pedestal, unlocking from the depressor and

Flying Bell   *1998*/0009570

so on, can be controlled by the operator from within the bell. However, under normal operating conditions the system is remotely operated from a control console installed on board the surface support vessel, which also contains the sonar, TV, echo-sounder and DP-system monitors, the manoeuvering controls with auto-pilot unit, the handling system controls and the electric power distribution instrumentation. The bell and related systems are protected by a tubular outer frame basically cylindrical in shape with integrated fairings made of glass fibre.

The mobile diving bell is preferably handled through a moon pool situated in the ship's centerline. A typical handling system consists of a two-wire main lifting winch, umbilical winch, trolley for moving the bell horizontally between moon pool and decompression chamber, cursor to guide the bell through the moon pool plus depressor to push the system through the air/water interface when launching and for lifting the bell out of the water. The depressor, which can be lowered by 100 m or more below surface, is preferably fitted out with a mechanically driven power sheave for the umbilical.

Contractor
Haux-Life-Support GmbH, Karlsbad.

VERIFIED

NORWAY

MoDive

Type
Dive monitoring system.

Description
The MoDive air diving monitoring system is an online system offering a total dive overview of all important diver and decompression chamber parameters.

The system is user friendly and easy to operate but at the same time fulfils all specifications for an offshore and inshore dive data recording system.

The system features advanced dive log, surface

Diving depth instrument module **1999**/0024835

A MoDive installation showing monitoring panel **1999**/0024836

interval timers, dive reports and multifunction control buttons using advanced software. The dive exposure matrix shows the diving activities during the last 18 days for each diver and is meant to be an aid for the supervisor to see who the next diver should be to carry out a dive. The digital value in each square in the matrix tells at which hour that particular dive ended. In the case of a repetitive dive this hour indication will be displayed in a red colour. The matrix updates itself automatically.

The online saturation diving monitoring system records diver, bell and habitat parameters which are presented both digitally and with graphical profiles. Additionally, a dive log records events and activities related to the dive. Alarms and manual inputs are also recorded in the log. To gain total control of time consumption during the dive, the bell and each diver have their own timers. All operations of the timers are registered in the dive log. To simplify use of software, all active objects are colour coded.

As a supplement to the standard diver and bell monitoring, the software is also able to monitor the valve positions in the bell. With colour codes it is possible to see if the valves are closed (red) or open (green). The tubes in the bell are also colour coded to monitor if they are connected to a closed or open valve. Monitoring the valve positions requires a minimum of additional equipment.

The MoDive offline dive monitoring system is a professional data acquisition system for monitoring depth and temperatures with very high accuracy. The system consists of three units: an offline dive recorder with carrying case; a battery charger and interface unit; and software for evaluating the collected data.

Fitted to a diver, the dive recorder measures depth, time, internal temperature and up to six external sensor parameters (for example temperatures). After the dive has been completed the recorder is inserted into the data socket in the charging/data interface unit and the collected data downloaded into the computer.

The offline software makes it very simple to evaluate the collected data, both on graphical profile diagrams and as an ASCII file for export to a spreadsheet or word processor.

As well as the MoDive data monitoring system, PAG has developed a diving depth instrument module with digital displays and necessary connectors for depth sensors and data logging. This system is specially made for military use and gives the user the possibility of connecting a mobile data monitoring system (for example a laptop computer).

A further development of MoDive is the monitoring and control system for hyperbaric treatment, called MoMed. This system is installed at NUI AS in Bergen, Norway and is also used by the parties involved in hyperbaric medical research such as the university in Bergen, Haukeland Hospital, Nutec, the Norwegian Navy and the State Diving School.

Operational status
In service with the US Navy, Norwegian Navy, French Navy and many commercial contractors like Stolt Comex Seaway, Rockwater, Global Industries, McDermot and so on.

Contractor
PAG Automasjon A/S, Moi Rana.

UPDATED

SOUTH AFRICA

RD 2300 Hyperbaric chamber complexes

Type
Hyperbaric compression chamber complexes for naval diver training establishments.

Description
RD 2300 chambers are designed for training establishments which require a multipurpose system capable of providing diving simulation and training, recompression treatments, clinical hyperbaric oxygen treatments and medical research. Occupant capacities range from 8 to 16 persons and are customised to suit each application. Available configurations for RD 2300 chambers include a main living chamber, transfer chamber wet chamber and treatment chamber. The chamber is supplied with the following systems and equipment:

* Remote-control console
* Rotating or fixed stainless steel NATO bayonet flange
* Wide-angle viewports which permit 95 per cent visibility into the chamber
* Quick-acting stainless steel clamp-type service lock with interlock
* Stainless steel bunks and seats with DNV-approved fire-retardant mattresses

The South African Navy's RD 2300 Hyperbaric chamber complex **1995**

- Easily removable deck plates
- Heavy-duty chemically stable epoxy paint system
- Dual-pressurisation system with selection manifold for air and four mixes
- Exhaust and pressurisation fine control system
- High-performance stainless steel venturi-type mufflers
- Oxygen Built-In Breathing (BIB) system with overboard dump back-pressure regulators and gas ejector system to reduce exhalation resistance at depths shallower than 5 m
- Mixed-gas BIB system with selection manifold for two gas mixes
- Dual, high-integrity ASME-approved safety valves in each compartment
- 300 mm diameter precision depth gauges
- Environmental conditioning system
- Emergency carbon dioxide scrubbers
- Metabolic oxygen injection system
- Dual oxygen and carbon dioxide analysis system
- Digital temperature monitors
- Humidity monitors
- Emergency power supply
- Alarm panel
- CCTV

- DNV-approved hyperbaric lighting
- High-quality communication system with digital signal processor helium voice unscrambler
- Sound-powered emergency communication system
- Water deluge type fire suppression system in each compartment
- Comprehensive operating and maintenance manuals in hard-copy and magnetic format for customer configuration management.

Available options include:

- Classification society certification
- Other design standards for certification authorities
- Other maximum depth ratings
- Hyperbaric medical lighting which permits detailed medical examinations and minor procedures to be performed in the chamber
- Hyperbaric patient ventilator
- Medical suction system
- Oxygen treatment hoods
- Intensive care monitoring system
- Sanitary system with toilet, hand basin and shower.

Pressure vessels are fabricated from carbon steel; stainless steel is used extensively for all components subjected to corrosion or wear including service locks,

door sealing surfaces, door hinges, bunks, lugs, mounting brackets and shell penetrators.

Specifications

Design standard: DNV or Lloyd's Rules for Diving Systems
Inspection authority: DNV or Lloyd's Register
Maximum depth rating: 200 msw
Occupant capacities: Min 8 to max 16 persons
Inside diameter: 2,300 mm
Manway ID: 800 mm
Viewport ID: 280 mm
Power supply: 440/380 V 6050 Hz 3-phase

Operational status

An RD 2300 multipurpose compression chamber is in service at the South African Navy's diving school at Simon's Town.

Contractor

Southern Oceanics (Pty) Ltd, Cape Town.

UPDATED

Type TC 665 transport chamber

Type

Diver's transport chamber.

Description

The Type TC 665 is a lightweight two-person aluminium transport chamber designed for the administration of emergency on-site recompression treatments and for transporting divers under pressure to larger treatment

chambers. With two occupants under pressure, the chamber may be hand-carried by 10 people.

An attendant sits in the vertical section of the chamber and a removable rail-mounted stretcher is provided in the horizontal section for the recumbent patient, allowing the attendant access to the patient's head and torso for the administration of basic life support. Three large-diameter viewports are provided for observing the occupants and a medical lock with safety interlock allows small medical and personal

items to be passed in and out of the chamber. A NATO bayonet flange is installed at the manway end for transferring the divers under pressure into the treatment chamber.

The TC 665 is equipped with onboard air, oxygen and power supplies. For therapeutic treatments or extended transport applications, connection points for external auxiliary supplies are provided for supplementing the onboard supplies. A safety feature built into the chamber is a permanent ventilation system which reduces oxygen and carbon dioxide build-up inside the chamber, while a sampling system allows the chamber atmosphere to be monitored for oxygen and carbon dioxide content. Standard equipment includes a nylon web-type lifting sling set, overboard dump-type oxygen masks, intercom system, waterproof panel-mounted stopwatch and carbon dioxide scrubber.

An optional non-magnetic version, the Type TC 665-MCM Transport Chamber, is also available for use aboard MCMVs.

Operational status

Type TC 665 and Type TC 665-MCM transport chambers are currently in use by the South African Navy.

Specifications

Depth rating: 55 msw
Inside diameter: 665 mm
Overall length: 2,580 mm
Overall height: 1,200 mm
Overall width: 900 mm
Weight: 265 kg

Contractor

Southern Oceanics (Pty) Ltd, Cape Town.

UPDATED

Type TC 665 transport chamber   **1999**/0024821

LM series recompression chambers

Type

Master recompression chamber.

Description

LM series Recompression Chambers (RCCs) are designed for military applications requiring a comprehensive specification and high level of operational reliability.

They are equipped with transfer under pressure flanges and can be used as master therapeutic treatment chambers for naval medical facilities or for routine operational recompression procedures. They are available with inside diameters of 1.5, 1.8 or 2.0 m

and occupant capacities ranging from 6 to 12 people. Standard configurations are:

- Type LM 1500/6 RCC: 1.5 m diameter – 6 occupants
- Type LM 1800/8 RCC: 1.8 m diameter – 8 occupants
- Type LM 1800/10 RCC: 1.8 m diameter – 10 occupants
- Type LM 1800/12 RCC: 1.8 m diameter – 12 occupants
- Type LM 2000/12 RCC: 2.0 m diameter – twelve occupants

LM Series RCCs are available with aluminium alloy, carbon steel or stainless steel pressure vessels. Standard equipment includes a rotating or fixed NATO

bayonet flange, 200 mm diameter viewports, large-diameter bayonet-type service lock with safety interlock, stainless steel bunks and seats with DNV-approved fire retardant mattresses, easily removed aluminium deck plates and epoxy paint system.

Standard life support systems include dual pressurisation, oxygen Built-In Breathing (BIB) system with overboard dump-type masks, back-pressure regulators and exhaust gas ejector system to reduce exhalation resistance at shallow depths, dual high-integrity ASME-approved safety valves, large-diameter precision depth gauges, oxygen and carbon dioxide monitoring system, temperature and humidity monitoring system, waterproof digital panel-mounted stopwatches, 80 A/h emergency power supply, alarm

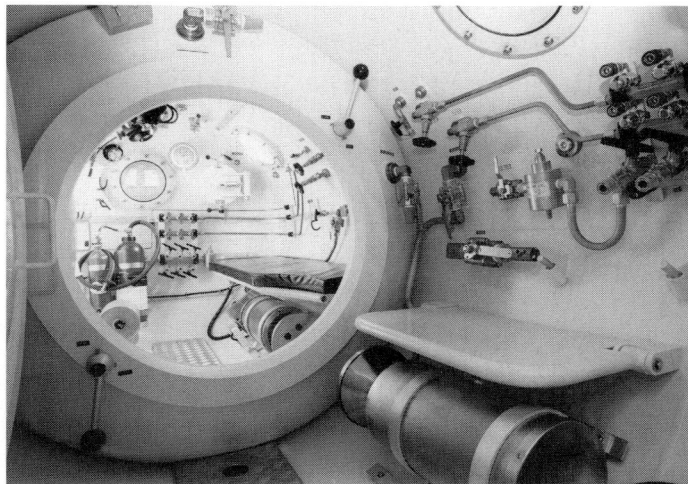

LMA 1500/6 RCC interior view   **1999**/0024833

LMA 1500/6 recompression chamber with aluminium alloy pressure vessel
1999/0024820

panel, DNV-approved hyperbaric lighting, high-quality duplex communication system and hyperbaric fire extinguishers.

Comprehensive operating and maintenance manuals are provided in hard-copy and magnetic format for customer configuration management.

A wide range of optional equipment is available, including mixed-gas BIB system, atmosphere conditioning system, carbon dioxide scrubbers, metabolic oxygen injection, remote-control console, CCTV monitoring, sound-powered emergency communication system, external acrylic light-pipe illumination, water deluge fire suppression system and toilet, hand-basin and shower.

A select range of medical equipment is also available, including hyperbaric patient ventilator, suction system, high-intensity medical lighting, HBO therapy hoods and intensive care monitoring system.

The LM Series RCCs are also available in standard ISO container packages for transportable applications. The chamber and it support equipment are installed in standard 6 or 12 m ISO containers according to requirement.

Operational status
Eight LM Series Recompression Chambers are currently in service with the South African Navy and Air Force.

Specifications
Design standard: ASME PVHO-1/Lloyd's Rules for Diving Systems
Certification authority: Lloyd's Register
Maximum depth rating: 100 msw

Contractor
Southern Oceanics (Pty) Ltd, Cape Town.

UPDATED

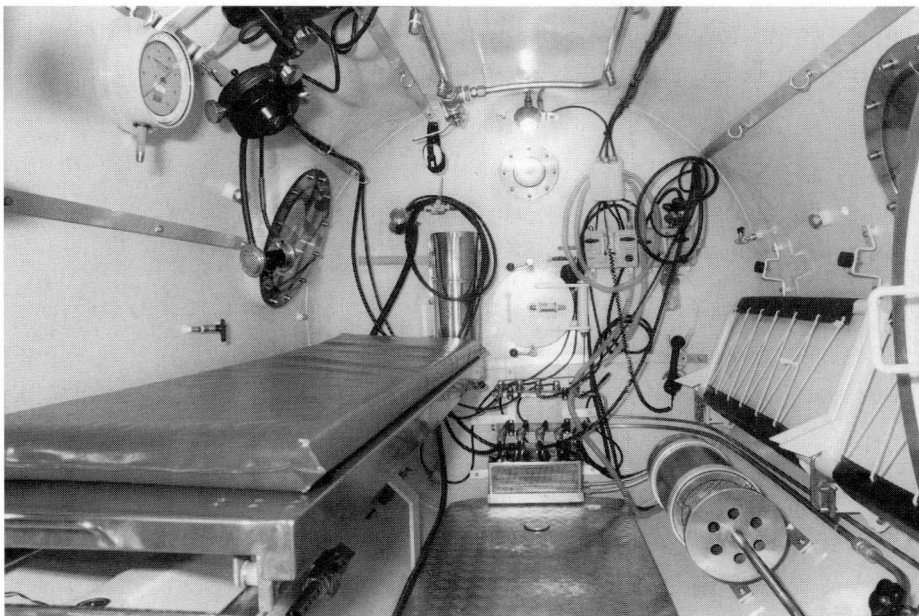

LMA 1800/6 medical hyperbaric chamber   **1999**/0024832

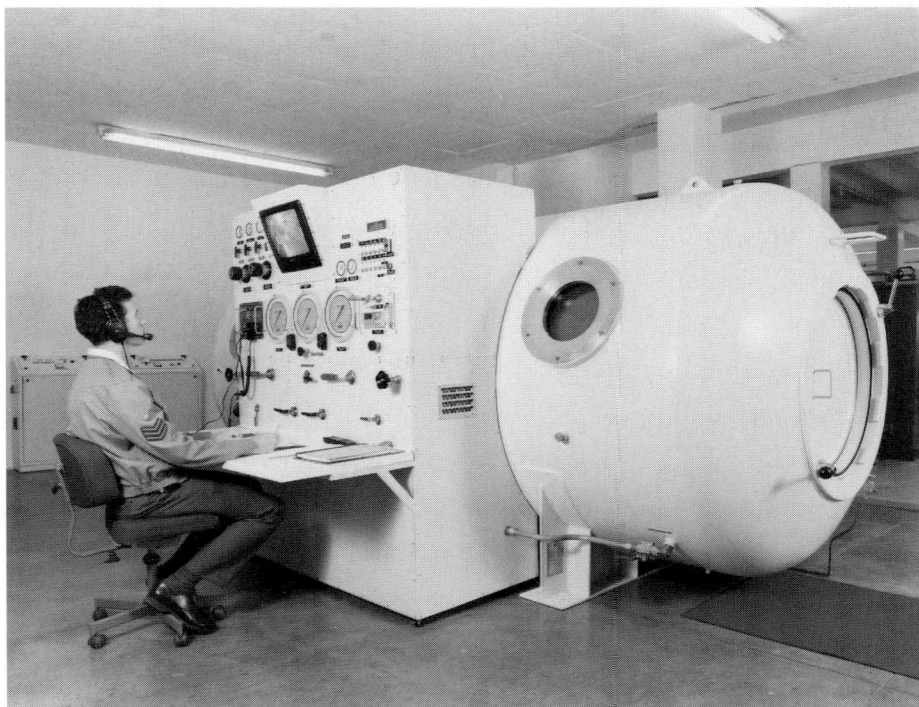

Interior view of LMC 1800/6 medical hyperbaric chamber (courtesy Southern Oceanics)   **1999**/0024831

LS 370 wet bell dive system

Type
Wet bell diving system.

Description
The LS 370 wet bell diving systems provide a safe and cost-effective vehicle for deep, surface-supplied diving to a maximum depth of 100 m.

Standard configurations are:

- LS 370-A: equipped for air diving to a maximum depth of 50 m
- LS 370-B: equipped for mixed gas diving to a maximum depth of 100 m

The system is self-contained, transporting only divers breathing gas and electrical power supplies. It comprises the following main equipment:

- skid-mounted bell handling system with folding A-frame, bell safety latch, hydraulically powered umbilical sheave, guide wire system, bell and guide wire winches and hydraulic power pack.
- wet bell with acrylic dome, emergency onboard gas supply, lighting and main umbilical.
- control van with divers' air and gas supply panels, diver communications, divers' gas analysis panel, bell handling system control panel, main and emergency electrical switchboards, alarm panels and standby diver station.

A lightweight air-transportable version is also available for rapid intervention applications. This system is packaged in standard 6 m ISO containers and is quickly and easily mobilised.

Numerous options are available, including diver heating, recompression chambers, LPIHP breathing air compressors and gas storage banks.

Specifications
Design standard: DNV Rules for Diving Systems
Maximum depth rating: 100 m
Main power requirement: 35 kW 440/380 V 60/60 Hz 3-phase
Emergency power requirement: 25 kW 440/380 V 60/50 Hz 3-phase

Contractor
Southern Oceanics (Pty) Ltd, Cape Town.

UPDATED

The LS 370-A wet bell dive system   *1996*

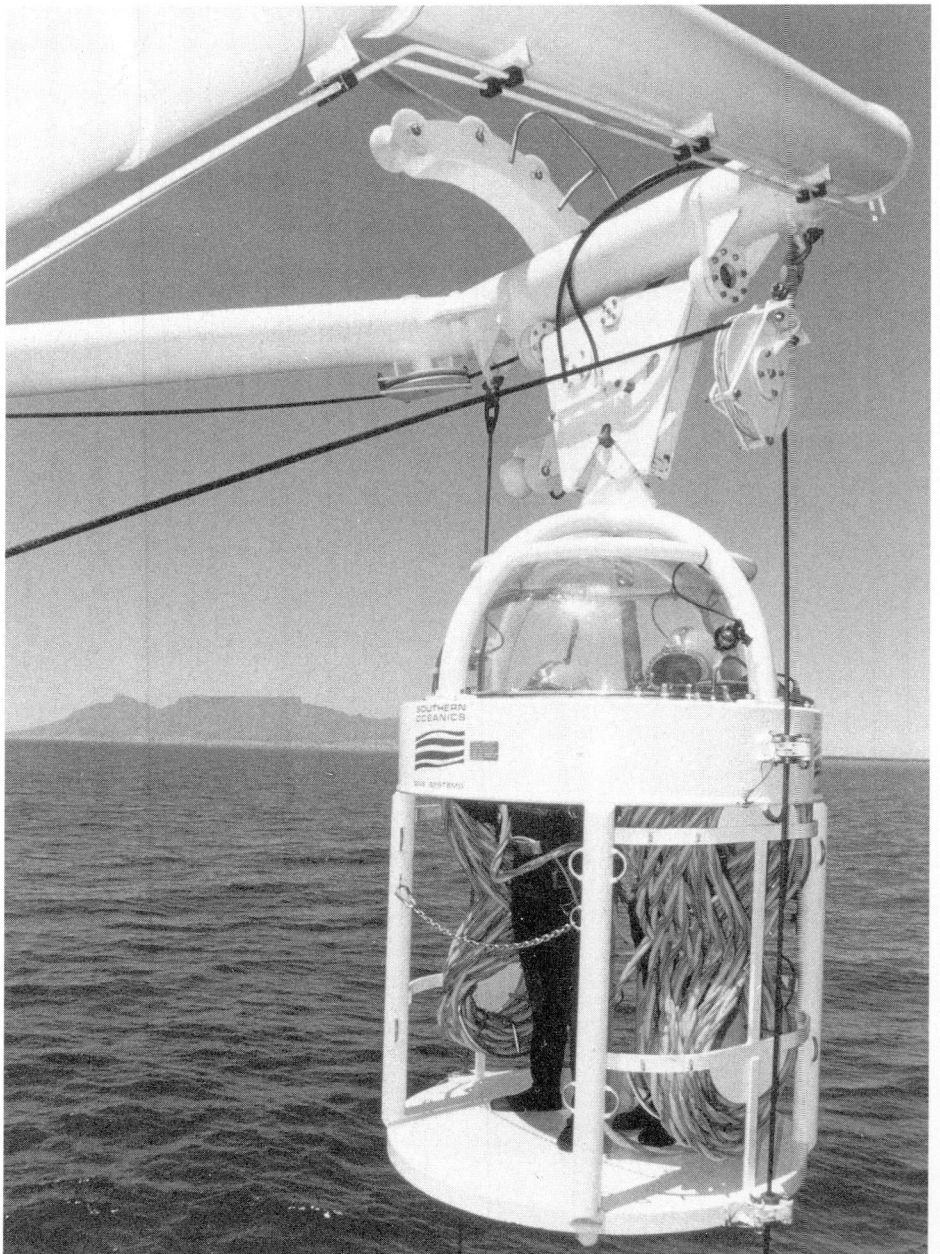
Launching an LS 370-A wet bell
1999/0024830

Offshore series recompression chambers

Type
Twin-lock diver recompression chambers.

Description
Offshore series recompression chambers are designed for routine operational decompression procedures. Occupant capacities range from two to six persons. They are robust, high quality commercial-style twin-lock chambers, available in five different diameters ranging from 1,10 m to 1,80 m configured as follows:

- Type 1100 Offshore RCC: 1.10 m diameter – two occupants
- Type 1220 Offshore RCC: 1.22 m diameter – three occupants
- Type 1372 Offshore RCC: 1.37 m diameter – four occupants
- Type 1500 Offshore RCC: 1.50 m diameter – six occupants
- Type 1800 Offshore RCC: 1.80 m diameter – six occupants

Stainless steel is used extensively in the construction of the pressure vessels for all components subject to wear or corrosion, including service locks, lifting lugs, bunks, door hinges, mounting brackets and shell penetrators. Standard equipment includes a weatherproofed control panel, precision depth gauges, Built-In Breathing (BIB) system, stainless steel service lock with safety interlock and three-way communication system.

A wide range of optional equipment is available, including NATO transfer under pressure bayonet flange, stainless steel door sealing surfaces, mixed-gas BIB system, oxygen and carbon dioxide analysis system, hyperbaric lighting, carbon dioxide scrubbers, hyperbaric fire extinguishers and a selection of approved hyperbaric medical equipment.

Operational status
Several units in service in Australia, South East Asia and South Africa.

Specifications
Design standard: ASME PVHO-1
Certification authority: Lloyds register
Maximum depth rating: 60 msw

Contractor
Southern Oceanics (Pty) Ltd, Cape Town.

UPDATED

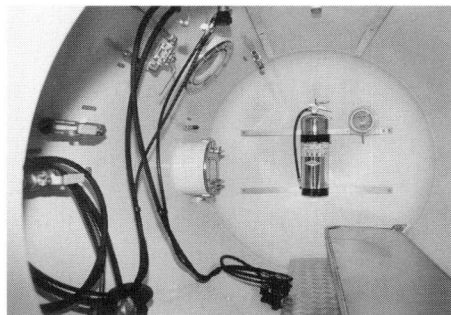

Interior view of Type 1372-D offshore RCC (Courtesy Southern Oceanics)   *1999*/0024828

Type 1372-D offshore RCC with stainless steel rotating NATO layout flange (Courtesy Southern Oceanics)   *1999*/0024829

Type TC 1150 transport chamber

Type
Diver's transport chamber.

Description
The Type TC 1150 transport chamber is designed for the administration of emergency on-site recompression treatments and for transporting divers under pressure to larger treatment chambers. It is equipped with an integral, wheeled handling system with steerable tow-bar which allows it to be manoeuvred by one person and is also fully transportable by helicopter, fixed-wing aircraft or light pick-up truck. A nylon web-type lifting sling set is included.

The TC 1150 will accommodate two divers or a recumbent patient and a seated attendant. An easily removed, rail-mounted stretcher with restraint straps is provided for the patient. The attendant has unobstructed access to the patient for administering CPR and basic life support.

The chamber is equipped with a NATO bayonet flange which allows it to lock-on to larger treatment chambers for transferring the divers under pressure. The stainless steel handling system is height-adjustable, allowing it to bayonet with treatment chambers of varying diameters and centreline heights. Two large-diameter viewports provide the operator with 95 per cent visibility of the chamber interior and a bayonet-type medical lock with safety interlock allows medical supplies and small personal items to be passed in and out of the chamber.

Life support for the occupants is provided by onboard air, oxygen and power supplies. Connection points for external auxiliary supplies are provided to supplement onboard supplies for extended therapeutic treatments and transport applications. All controls are housed in a weatherproof glass fibre control panel. Standard equipment includes an intercom system, overboard dump-type oxygen masks, chamber atmosphere sampling system, waterproof panel-mounted stopwatch, ECG penetrator and medical drip hook.

Optional equipment includes a detachable entrance lock, hyperbaric fire extinguisher, oxygen and carbon

Type TC 1150 transport chamber (Courtesy Southern Oceanics)   *1999*/0024827

Interior view showing patient stretcher, CO₂ scrubber, attendant seat and hyperbaric fire extinguisher (Courtesy Southern Oceanics)   **1999**/0024826

dioxide monitoring system, carbon dioxide scrubber, light-pipe chamber illumination system, mixed-gas built-in breathing system, 70 Ah onboard power supply, caisson depth gauge and various others.

The chambers are made of aluminium alloy or stainless steel.

Operational status

TCA 1150-B specification is currently in service with the Royal New Zealand Navy.

Specifications

Design code: ASME PVHO-1
Certification authority: Lloyd's Register
Maximum depth rating: 55 m (seawater)
Inside diameter: 1,150 mm
Length: 2,750 mm
Height: 1,600 mm
Width: 1,560 mm
Weight: 585 kg (unoccupied)

Contractor

Southern Oceanics (Pty) Ltd, Cape Town.

NEW ENTRY

Rear view showing the height-adjustable handling system and onboard air, oxygen and power supplies (Courtesy Southern Oceanics)
1999/0024825

600 Series diver's heaters

Type
Diesel-fired diver's hot water machine.

Description
The 600 Series diver's heaters provide heated seawater for up to three divers for cold water or extreme exposures. The heaters are equipped with a diesel-fired burner and fresh water boiler that circulates heated fluid through a titanium heat exchanger. Cold seawater is pumped through the titanium heat exchanger where it is heated and then supplied to the divers through a hose incorporated in the divers umbilical. A temperature control valve provides accurate and reliable temperature control to within ± 1°C.

Operational status
Currently in service with the South African Navy and various diving contractors in South and West Africa, the Middle East and Russia.

Contractor
Southern Oceanics (Pty) Ltd, Cape Town.

Type 635 divers' heater with deck enclosure
(Courtesy Southern Oceanics)
***1999**/0024834*

Specifications

	Type 620 Diver heater	**Type 630 Diver heater**	**Type 635 Diver heater**
Hot water delivery	21 litres/mm	42 litres/mm	21/42 litres/mm switchable
Hot water supply pressure	10 bar	10 bar	10 bar
Temperature range	35-65°C	35-65°C	35-65°C
Temperature control	±1°C	±1°C	±1°C
Boiler thermal power output	82 kW	157 kW	157 kW
Absorbed electrical power	3 kW	4 kW	4 kW
Power supply	on request	on request	on request
Other voltages	220 V 50 Hz 1 ph	380 V 50 Hz 3 ph	380 V 50 Hz 3 ph
Length	150 cm	190 cm	190 cm
Width	125 cm	140 cm	140 cm
Height	141 cm	135 cm	135 cm
Weight	880 kg	1,050 kg	1,085 kg

NEW ENTRY

BH Series carbon dioxide scrubbers

Type
Hyperbaric chamber carbon dioxide scrubbers.

Description
The BH Series carbon dioxide scrubbers are designed for use as primary or emergency scrubbers in diving chambers. They comprise a blower housing and bracket to which is attached a large, radial-flow CO_2 absorbent canister with dust filter. The absorbent canister attaches to the blower by quick-release toggle catches for replenishment of the absorbent chemical. The blower motor is a brushless DC unit equipped with thermal overload and reverse polarity protection.

Two sizes are available:
- The Type BH 500 CO_2 scrubber is designed for use in main or living compartments and will provide carbon dioxide removal for four people.
- The Type BH 300 CO_2 scrubber is smaller and designed for use in transport chambers and entry locks. It will provide carbon dioxide removal for three people.

Operational status
Currently in service with various military and commercial users in Australia, New Zealand, South Africa and the UK.

Specifications
Type BH 500
Airflow (measures at 1 bar absolute): 0.6 m³ /min actual

Absorbent canister capacity: 5.5 kg
Maximum working depth: 200 msw
Power requirement: 24 V DC 1 A
Dimensions: 210 mm (diameter) × 320 mm (height)

Type BH 300
Airflow (measures at 1 bar absolute): 0.45 m³ /min actual
Absorbent canister capacity: 2.5 kg
Maximum working depth: 200 msw
Power requirement: 24 V DC 1 A
Dimensions: 160 mm (diameter) × 320 mm (height)

Contractor
Southern Oceanics (Pty) Ltd, Cape Town.

NEW ENTRY

UNITED KINGDOM

BASAR

Type
Underwater breathing apparatus.

Description
BASAR (Breathing Apparatus Search and Rescue Mod A) has been designed to provide the diver with maximum breathing comfort (even under the harshest working conditions underwater or on the surface), to be as light as possible out of water, and to be always available for immediate use. The apparatus incorporates a stabilising jacket as an integral part of the diving set so that surface buoyancy can be increased, after rescue operations, while waiting for recovery.

A separate swim mask is used with the apparatus, which allows the diver to pinch his/her nose for ear clearing. The mask may be donned before jumping, or after entering the water.

The free mouthpiece allows the diver to conserve air by not having to insert it until the last moment before jumping; to communicate with a survivor in the water by removing the mouthpiece; and to dispense with the mouthpiece when supported by the stabilising jacket while waiting to be picked up.

Using BASAR the diver can jump from a helicopter door (from up to 15 m) without impediment and can sit comfortably whilst wearing the set when the helicopter is airborne.

The breathing circuit of the apparatus is of the open circuit type. High-pressure air is supplied from two steel cylinders, through a manifold to the first-stage regulator which reduces the pressure. Air, at a reduced pressure, passes to the diver via the intermediate hose, the second-stage demand valve and the mouthpiece.

Exhaled air from the diver is discharged through the second-stage exhaust valve to atmosphere/water.

The main cylinder is consumed as per the Certificate of Clearance for Use (CCU) at which point the diver opens the reserve cylinder valve. Air flows from the reserve cylinder to the main cylinder to equalise the pressure in the two cylinders. This equalisation of pressure takes about 10 seconds and can be distinctly heard by the diver.

Equalisation of pressure is carried out a second time

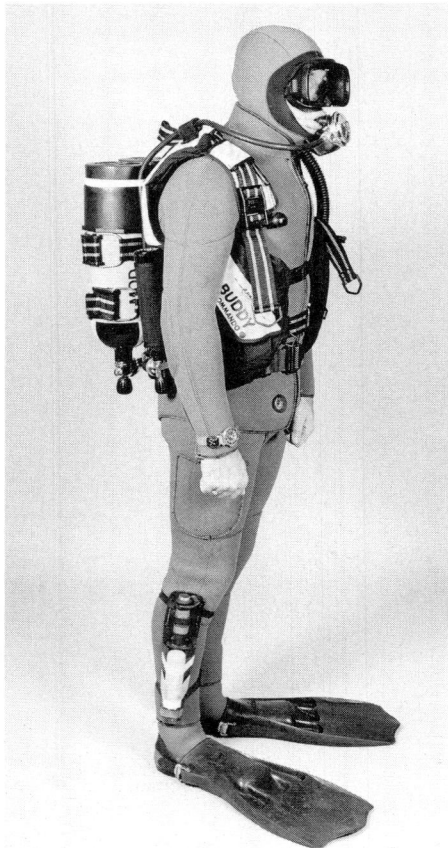

BASAR breathing apparatus
(Courtesy MSI – Defence Systems)   *1999*/0024822

Diver wearing BASAR breathing apparatus
(Courtesy MSI – Defence Systems)
1999/0024823

when the diver again finds difficulty in breathing. The necessity for the second equalisation must be taken as a signal to surface.

The BASAR commando stabilising jacket can be inflated either by mouth with the automatic mouthpiece on the corrugated breathing tube, or from an emergency cylinder. It provides: buoyancy for ascent in an emergency; buoyancy compensation at depth; underwater breathing in emergency; surface buoyancy before and after dives; high visibility for the pick-up crew.

A buoyancy ring and spherical float inflated by CO_2 from a small cylinder are stowed in pouches attached to the weight belt. These provide additional buoyancy if required during a rescue operation.

The apparatus is based on reliable, commercially available, proven components.

Operational status
Developed specifically for the Royal Navy's search and rescue organisation.

Contractor
MSI-Defence Systems Ltd, Norwich.

UPDATED

Compression chambers

Type
Military diving comprehensive chambers.

Description
Based on its experience over the last 10 years as in-service support contractor to the UK MoD for the Royal Navy's diving systems, MSI-Defence Systems has developed a range of compression chambers particularly suited to MCM/EOD operations. The range of configurations available includes: deck decompression chambers; recompression and treatment chambers; transportable chambers; and containerised compression chamber packages.

Military compression chambers supplied are designed to the application requirements. For example, they can be single- or two-compartment units with service locks; two-man or more occupancy; rated to the required depth; have low-magnetic signatures; be supplied complete with full controls, instrumentation, communications, Built-In Breathing Sets (BIBS); offer analysis where required; and Transfer Under Pressure (TUP) facilities.

Operational status
MSI-Defence Systems has won a contract in excess of £6 million from the UK MoD to supply transportable, two-compartment recompression chambers to the Royal Navy for MCM operations. The chambers, fabricated in aluminium with a forged personnel entrance can accommodate two patients and a seated medic in the main compartment and one person in the entry compartment. They are rated to 80 msw and are fitted with full controls, instrumentation, built-in breathing systems, service lock and communications.

Contractor
MSI-Defence Systems Ltd, Norwich.

VERIFIED

A transportable Delta, two-compartment recompression chamber rated to 80 msw, similar to more than 40 chambers being supplied to the Royal Navy
1997

Combat drysuit

Type
Combat drysuit.

Description
The Divex combat drysuit is manufactured from an easy-glide material which is water repellent and smooth; offering the lowest resistance whilst swimming. High-frequency welded seams fuse the suit to one piece of material offering maximum durability to withstand the shock of parachute drops or rapid water entries. The suit is ideal for wearing with closed-circuit and semi-closed circuit rebreathers.

A heavy-duty, low-magnetic zipper is fitted as standard, with high-frequency welded seams. Internal support straps ensure the suit remains supported in the crotch area. In addition, the suit has a direct fitting neck seal and integral ankle straps. Inlet and exhaust valve ports are positioned on the left chest and left arm.

The suit has great chemical and abrasion resistance and hence is ideal for use in commercial and professional diving environments. Other likely users include police diving teams, fish farmers and harbour divers.

For use in colder climates a 6 mm attached neoprene hood is available.

Contractor
Divex, Aberdeen.

VERIFIED

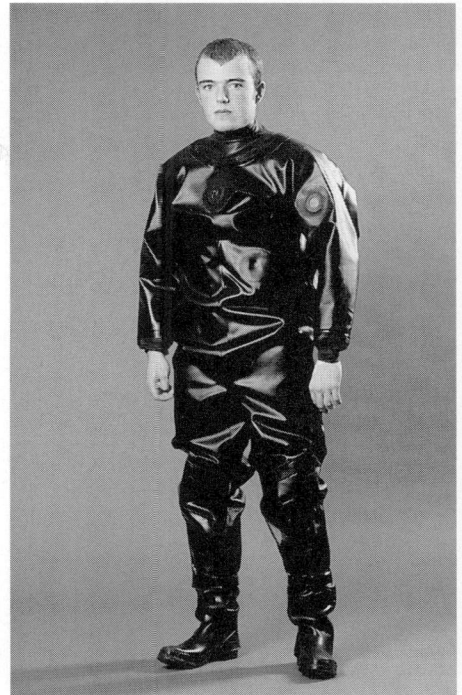

Combat drysuit
1998/0009574

SAR drysuit

Type
Diving suit.

Description
The Divex SAR drysuit is a very flexible, lightweight, drysuit comprising a trilaminate membrane ideally suited for rescue divers, scientific divers, professional divers and as a lightweight military suit.

The drysuit is constructed from the best quality, double-textured, high-tenacity nylon trilaminate with an interply of butyl rubber.

Every care is taken to ensure the security of the diver – seams are twin needle sewn, glued and hand taped with 19 mm Aquatape. The suit has a bellows-type neckseal, latex bottle-type wrist seals, and a heavy-duty zipper. Vlar pads are fitted to the knees for extra protection. To finish, the suit is fitted with full rubber boots with heel and toe reinforcements. These are, however, fully flexible and make an ideal swimmer's boot.

A 6 mm attached neoprene hood is available. In colder climates professional divers often require the hood to be an integral part of the suit.

Contractor
Divex, Aberdeen

VERIFIED

Stealth™ Combat and Stealth™ EOD

Type
Closed-circuit mixed-gas rebreathers.

Description
This military diving apparatus has been designed by Divex to meet the requirements of Explosive Ordnance Disposal (EOD) and special forces, incorporating modern materials, microprocessing power and current thinking on underwater breathing apparatus design.

The primary objective has been to create a very small package integrating primary and secondary (bail out) breathing systems complete with a buoyancy control facility in a swimmer-optimised configuration.

The Stealth™ rebreather is an electronically controlled closed-circuit mixed-gas rebreather available as two options for military diving. It features a buoyancy compensator and can be fitted with a bailout system.

Closed-circuit breathing systems function by maintaining a breathable gas mixture for the diver at depth through removing the CO_2 from exhaled gas via the scrubber unit, and injecting pure oxygen to replace the oxygen metabolised by the diver. The aim is to replace the oxygen in the quantity at which it is consumed – whatever the breathing rate and metabolic rate. Stealth™ achieves this through a well-proven electronic control system and microprocessor monitoring system packaged in a pressure-proof housing.

The oxygen partial pressure (ppO_2) can be factory set to customer requirements, but is normally preset to 1.3 or 1.5 bar. This maximising of safe ppO_2 levels minimises the amount of diluent (helium or nitrogen) uptake by the body's tissues and hence reduces decompression time.

Three independent oxygen sensors located in the scrubber outlet chamber continuously monitor the oxygen level in the gas and provide signals to the Electronic Oxygen Controller (EOC) and microprocessor monitoring module. The EOC reviews the three signals, creates an average and compares this to the desired set point. It then switches the oxygen addition valve on or off within a tolerance band of ±0.05 bar of the set point.

A Light Emitting Diode (LED) is fitted to the side of the diver's full-face mask. This warns the diver of any malfunction in the system, by changing the normal continuous green status LED to a flashing green for problem but not serious failure, to a flashing red if a dive abort is required.

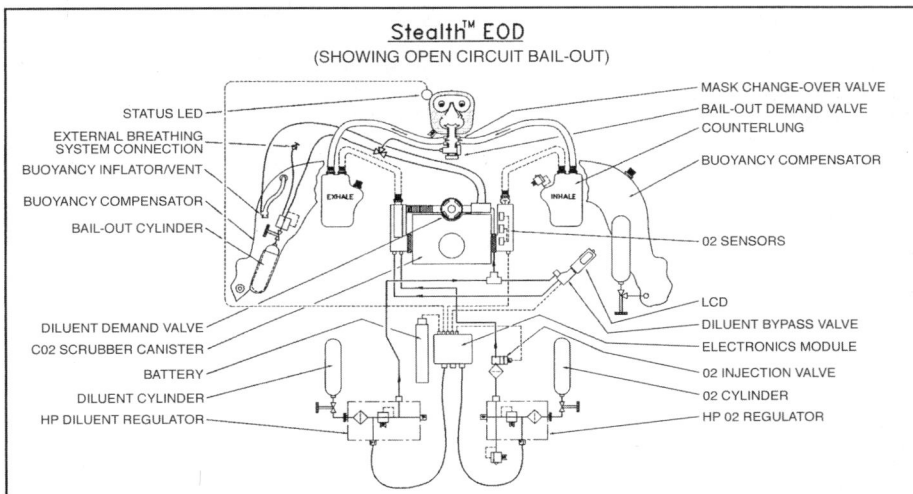

Stealth™ EOD
(SHOWING OPEN CIRCUIT BAIL-OUT)

- STATUS LED
- EXTERNAL BREATHING SYSTEM CONNECTION
- BUOYANCY INFLATOR/VENT
- BUOYANCY COMPENSATOR
- BAIL-OUT CYLINDER
- DILUENT DEMAND VALVE
- CO2 SCRUBBER CANISTER
- BATTERY
- DILUENT CYLINDER
- HP DILUENT REGULATOR
- EXHALE
- INHALE
- MASK CHANGE-OVER VALVE
- BAIL-OUT DEMAND VALVE
- COUNTERLUNG
- BUOYANCY COMPENSATOR
- 02 SENSORS
- LCD
- DILUENT BYPASS VALVE
- ELECTRONICS MODULE
- 02 INJECTION VALVE
- 02 CYLINDER
- HP 02 REGULATOR

EOD showing open circuit bail out
1998/0009576

Stealth™
1998/0009575

An open circuit bail out breathing system is offered as standard but a unique semi-closed circuit system is available that provides additional breathing gas for ascent and long decompression requirements.

The microprocessor also drives a back-lit Liquid Crystal Display (LCD) module located on a hanging pendant on the diver's right side. Attached to this is also a diluent bypass injection valve to enable rapid

gas injection of a safe breathable gas into the counterlungs.

The display gives such information as: current average ppO_2, current depth, maximum depth, elapsed time of dive, oxygen content, diluent content, battery status, alarm/abort description. With the Stealth™ rebreather a diver can stay at a depth of 100 m for up to 20 minutes.

Operational status
30 Stealth™ Combat systems were ordered by the Royal Navy early in 1998.

Contractor
Divex, Aberdeen.

UPDATED

Oxymax 3

Type
Closed-circuit oxygen diving set.

Description
Oxymax 3 is a compact oxygen rebreather set with up

to 3 hours duration – dependant upon conditions. It is inherently non-magnetic in construction (but can be ratified as fully 'non-magnetic' to special order) and is characteristically low in operating noise.

A streamlined glass fibre carapace is provided, both to assist swimming and to protect the breathing bag. This is, nevertheless, easily removable for

maintenance. A cylinder contents gauge is standard equipment and allows both diver and other divers to monitor gas available. This is especially useful during training.

The set features a demand-driven admission system for oxygen 'make-up' in the diver's gas. This is not only acoustically very quiet and consistent but also economical in use, giving adequate breathing gas for a given workload without wastage or under-supply. Depending upon activity, therefore, long endurance may be achieved – especially in warmer waters where problems of diver heat-loss do not arise.

Oxymax 3 features a crossover valve within the mouthpiece allowing the unit to be turned into an open-circuit system. This patented valve allows the diver to have the set ready for an emergency dive in open-circuit mode without water entering the bag.

Additional features include: quick-release neck and waist straps for ease of ditching; transparent filling and emptying ports for ease of filling and sodalime inspection; hygienic and simple cleaning; and easy to read contents gauge.

Operational status
In service with a number of navies worldwide.

Contractor
Divex, Aberdeen.

VERIFIED

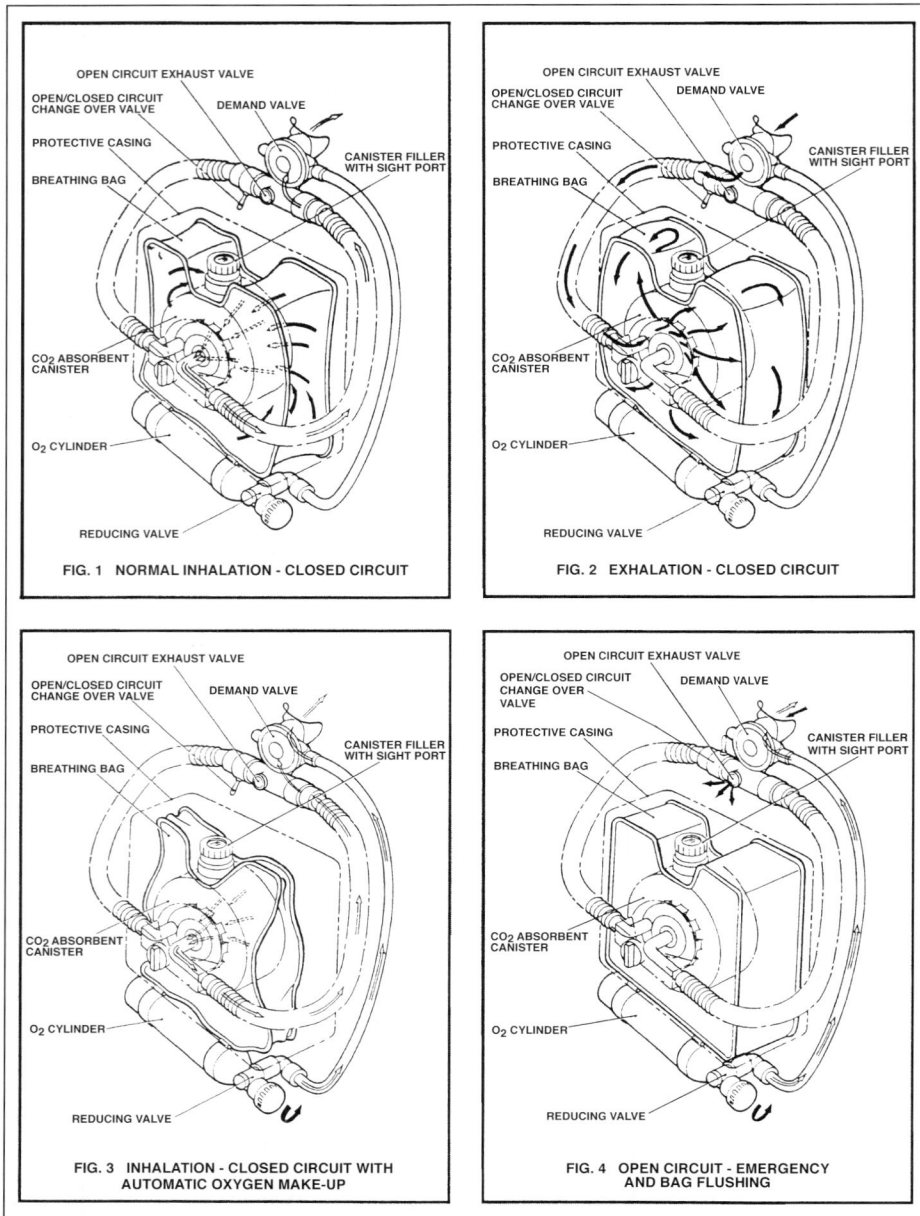

FIG. 1 NORMAL INHALATION - CLOSED CIRCUIT

FIG. 2 EXHALATION - CLOSED CIRCUIT

FIG. 3 INHALATION - CLOSED CIRCUIT WITH AUTOMATIC OXYGEN MAKE-UP

FIG. 4 OPEN CIRCUIT - EMERGENCY AND BAG FLUSHING

Oxymax 3 closed-circuit set   **1998**/0009578

Oxymax diving set   **1998**/0009577

Divator Mk II

Type
Breathing apparatus.

Description
The Divator Mk II breathing apparatus incorporates shoulder straps which have lockable buckles to prevent them from slipping. On the right shoulder strap there is a neat manifold to which the pressure gauge, the breathing hose, the extra air connector and the reserve air valve are fitted. The cylinder valve and the regulator are easy to reach on the back, and there is no risk of

hitting the head on them when diving. The back pad distributes the weight of the cylinders evenly across the back and keeps the apparatus in the correct position. Breathing and pressure gauge hoses are placed in a position close to the body and there are no hose loops behind the head which can get caught or damaged during dives.

With the large-capacity regulator breathing resistance is very low, even when the cylinder pressure is low. When the air is turned on the reserve air valve control lever will automatically turn to the correct position. The reserve air cannot be activated until the pressure has dropped close to the preset reserve air

pressure. An outlet on the regulator cover can be used for connection of a variable volume diving suit or a life jacket.

The apparatus incorporates a full face mask (developed jointly with the US Navy) which provides the diver with constant slightly positive pressure to prevent contaminated water from leaking in. For diving in clean water, a breathing valve without positive pressure can be supplied. Flushing the inside of the visor with dry air keeps the visor free from condensation. Airways are designed to provide very low breathing resistance and minimise dead space volume which prevents a build up of carbon dioxide.

The mask features an integral adjustable pressure equaliser and mounts for spectacles.

A divers' telephone can be integrated with the mask and a breathing valve with non-return valve minimises the risk of ice being able to form in the inlet port. A purge button makes it possible to drain any free water out of the valve.

The low-profile twin cylinders are built of special steel alloy and filled at 300 bar pressure giving a light, compact apparatus with long duration times.

The breathing valve can easily be changed from the full-face mask to a mouthpiece, which is easy to wear and is not tiring to the jaw during a long dive.

Operational status

In service with a number of organisations worldwide, including the US Navy, UK MoD, Swedish military, Royal Netherlands Navy, Indian Navy, Oman Navy and Indonesian Navy.

Contractor

Interspiro, Telford.

VERIFIED

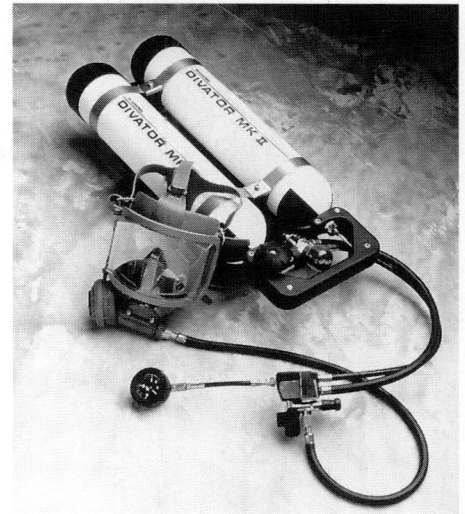

Divator Mk II
***1998**/0009579*

Oxydive

Type

Closed-circuit breathing apparatus.

Description

The Oxydive set consists of the following components: oxygen cylinder, regulator unit, bypass valve; dosage unit; relief valve; buoyancy control unit; breathing bag, absorber unit, mouthpiece and harness.

The oxygen cylinder is placed between the absorber unit and the breathing bag in such a way that it will protect the main part of the breathing bag and the regulator unit. The 1.5 litres (standard) steel cylinder has a charging pressure of 200 bar (3,000 psi). Other cylinders for non-magnetic versions are available.

The regulator unit consists of four main parts: bypass valve; dosage unit; relief valve; buoyancy control unit.

With the bypass valve the diver can always add oxygen into the breathing circuit by activating the valve manually. The dosage unit measures the consumed oxygen volume and injects an equal volume into the breathing circuit at each inhalation. In this way the diver will stay neutrally buoyant during the whole dive irrespective of depth (except if the diver ascends so fast to the surface that the expansion rate of the oxygen volume becomes greater than the consumption). The dosage unit operates without causing or being activated by any breathing resistance.

An automatic relief valve is connected to the regulator unit. Also, a relief valve for the breathing circuit which is independent of the diver's swimming position will open if the pressure in the breathing circuit should exceed 200 mm water column.

The buoyancy control unit allows the diver to change buoyancy during a dive by actuating the buoyancy control unit. This makes the diver become a little heavier before breaking the surface to make an observation. With this unit the diver can select four different settings for more or less buoyancy compared to the individual adjustment made before starting the dive. The original buoyancy can be restored at will.

The accurate and automatic regulation of the buoyancy at each inhalation is of great importance as it enables the diver to remain in desired swimming position without effort. The waste of energy on constant

Oxydive ***1998**/0009580*

adjustments of body position to compensate for an inadequate buoyancy is eliminated.

The breathing bag is placed on the diver's chest close to the lungs to achieve as low a breathing resistance as possible both during inhalation and exhalation. The small difference in hydrostatic pressure between the breathing bag and the lungs is mechanically eliminated when the diver swims in normal position. The breathing tag is uncomplicated and made of durable material.

The absorber unit consists of an absorber housing and an absorber cartridge. The absorber housing is made of glass fibre-reinforced plastic protected and insulated with rubber. The absorber cartridge is designed in such a way that there is always gas between the absorber housing and the sodalime. By insulating the sodalime from the surrounding cold water with rubber glass fibre-reinforced plastic and gas the duration time of the Oxydive set is almost independent of water temperature. The radial gas flow through the absorber cartridge has been proved to give superior utilisation of the sodalime and a low breathing resistance. The absorber cartridge can be prepacked

and stored for a long time in a small plastic box to be instantly available when needed. For an extremely long-time mission an extra prepacked absorber unit in a plastic box and an extra oxygen cylinder can be brought by the diver. The design is such that if some water should leak into the breathing circuit it will be held in the absorber housing and not go to the diver's lungs.

Non-return valves in the mouthpiece ensure low rebreathing of carbon dioxide. The assembly can be shut off with a valve to prevent ingress of water.

The harness is simple and uncomplicated. The whole unit can be quickly released by slipping off the neck collar and opening waist buckles.

As optional equipment a lifejacket for integration with the harness is provided. When not in use the lifejacket is protected in a zippered cover. The lifejacket is inflated by means of a CO_2 cartridge or an air cylinder. The zipper of the cover opens automatically and the jacket inflates in to correct position when the knob is pulled to activate the CO_2 cartridge or the air cylinder.

For increased safety the lifejacket can also be furnished with a hydrostatic valve for automatic inflation in case the diver should accidentally descend to a dangerous depth. However, If the diver deliberately goes deeper than the depth preset for the hydrostatic valve; he/she should close the air cylinder in advance. Failing that, the diver will definitely hear when the hydrostatic valve starts operating and then he/she should close the air cylinder valve, return to a safer depth, reset the mechanism of the hydrostatic valve by pressing the preset knob and open the air cylinder valve again. To restore a correct buoyancy the diver must again adjust the buoyancy control unit to compensate for the buoyancy of the lifejacket. No bubbles will leave the apparatus during the whole operation.

Operational status

The Oxydive set was tested by the US Navy Experimental Diving Unit (NEDU) in July 1982.

Contractor

Interspiro, Telford.

VERIFIED

DCSC

Type

Mixed-gas diving system.

Description

DCSC is a development of the well proven ACSC which has been in operation for over 10 years.

The equipment is extremely simple to prepare and operate. The breathing work is minimal due to the large-diameter hoses. The equipment is neutrally buoyant in water giving ease and exactness in diving procedures. The system has been dived under active service conditions to 150 m and in simulators to 450 m.

The set features very low magnetic and acoustic signatures (which can be adjusted to suit customer requirements) and the safety aspect has been given high priority. A warning device, bypass and separate harness lifejacket system ensure comfortable handling and safe diving.

DCSC is a mechanical system, that is simple and easy to use. The mechanical dosage system keeps the percentage of CO_2 constant in the breathing circuit, independent of depth and ventilation. The design allows a two gas mixture to be used, which simplifies logistics with regard to back up and charging.

The diving apparatus comprises a gas cylinder, control unit and breathing circuit.

The gas cylinder contains the gas mixture which is dosed for different diving depths.

The control unit comprises:
- Dosage cylinder, to be used in combination with the selected mixture for the various depths
- Dosage unit and warning system
- Hydrostatic valve
- Bypass valve.

The breathing circuit houses the following components:
- Absorber canister
- Breathing bellows, of which the spirometric function measures the breathing volume of the

diver and doses the amount of fresh gas to be supplied to the breathing circuit

- A diffusor, through which the surplus gas passes before it reaches the surrounding water
- Face mask (or mouthpiece).

Safety systems are built into the apparatus and include:

- Built-in warning of low breathing gas supply (approximately 10 minutes)
- Built-in warning of malfunction of gas dosing system and manual bypass.

Other features include: automatic buoyancy control via patented hydrostatic valve; single breathing gas charge which ensures the diver always receives breathable gas should a failure occur in the regulator or

The DCSC mixed gas diving system (Interspiro)
1998/0009873

dosage system; mask capable of accepting most available diving communication systems.

Specifications
Length: 745 mm
Width: 345 mm
Height: 265 mm
Weight: 35 kg
Gas volumes
Breathing bellows: 4.3 litres
Absorber unit: 2.6 litres
Breathing mixture: 5.0 litres (at 200 bar)
Suit inflation (air): 1.0 litres (at 200 bar)

Contractor
Interspiro, Telford.

NEW ENTRY

Pro-Am/Pro-HD

Type
Rubber dry diving suits.

Description
The Pro-Am dry diving suit is designed to keep divers dry, warm and comfortable under all environmental conditions. Developed over a period of four years, the Pro-Am diving suit has been subjected to intensive laboratory and field tests under varying conditions. Each suit is individually tank tested in addition to the many other tests undertaken as part of stringent quality control.

Designed to cope with the most rigorous of underwater conditions, the suit is manufactured from a double layer of rubber/EPDM blend on a two-way stretch nylon lining to provide maximum mobility and toughness combined with easy dressing.

The suit is manufactured to 1,050 g/m², which has been specified as meeting the requirements of some military forces. The seams are fully vulcanised while the high EPDM content increases resistance to 'weathering'. High-risk areas such as shoulder, waist, knees and shins are provided with pyramid reinforcement. Heavy-duty latex wrist seals ensure a longer operational life for the suit. A latex or neoprene hood is also available. The suit is quick and easy to repair on site if punctured and a repair kit is supplied with each suit.

The Pro-HD 1,550 g/m² suit has been developed for heavier-duty military applications. The suit features heavier reinforcements than the lighter Pro-Am suit, while the interior lining is a heavy-duty woven nylon which is extremely hard-wearing and also adds extra warmth to the suit. Reinforcements are of a shallow dimple pattern for ease of cleaning in case of contamination. The boots are also slightly larger to

allow for more underwear which could be necessary on an extended dive.

The suit can be supplied with a variety of neck options such as standard latex neck seal, latex neck yoke for use with neck ring and clamping band, or rubber neck yokes to attach to air breathing helmets.

Operational status
Suits are sold worldwide and are in service with a number of navies, including the French and Netherlands navies.

Contractor
The Gates Rubber Company Ltd, Dumfries.

VERIFIED

PRO-TDX 350

Type
Nylon trilaminate diving suit.

Description
An easy to dress 350 g/m² suit with an exterior nylon

6.6, butyl interply, lightweight interior nylon suit. The suit features unique Vulca-Seam technology which creates a virtually seamless suit.

Operational status
Suitable for military applications.

Contractor
The Gates Rubber Company Ltd, Dumfries.

VERIFIED

PRO-VSN 1100

Type
Hybrid stretch membrane dry suit.

Description
The Pro-VSN is a new unique stretch suit with the flexibility of a rubber dry suit and the toughness and durability of a nylon suit. The VSN has vulcanised

seams with a lifetime guarantee. The boots are constructed from 5 mm neoprene and over-dipped with latex for flexibility and durability. Braces are included with the suit.

Operational status
Currently being used by several Police Underwater Units.

Contractor
The Gates Rubber Company Ltd, Dumfries.

NEW ENTRY

New Concept

Type
Drysuit.

Description
The 'New Concept' range of drysuits combine regular fabrics with breathable panels which allow sweat to escape while still keeping the wearer completely dry from external water. The suits are lightweight, exceptionally comfortable and highly

mobile, allowing tasks to be carried out as efficiently wearing a drysuit as without.

Operational status
Developed and available.

Contractor
Typhoon International Ltd, Redcar.

VERIFIED

Special operations drysuit
1996

Ulvertech D100 diver's ranging unit

Type
Diver's echo-sounder sonar for target location.

Description
The D100 hand-held diver's ranging unit is intended for both naval and commercial diving applications and, operating at a frequency of 500 kHz, allows ranging and target location at up to 100 m range. It locks on to the strongest target on its bearing. A unique LED dot matrix display gives the equipment a sonar capability by allowing the diver to scan the target zone manually. This displays multitarget data present in a sector, dependent on the rate of manual scan, clearly indicating the range of the target. The rear of the unit contains both a digital display of range in either metres or feet, and a unique rolling display to give a graphic image of the surroundings.

The equipment is manufactured from low-magnetic signature materials and is internally powered by two PP3 batteries. The handle base has a fitting for use mounted on a tripod.

Operational status
In service.

Specifications
Range: 100 m
Resolution: 1 cm
Beamwidth: 4° conical
Depth rating: 100 m
Weight: 1.3 kg (in air); 0.32 kg (in water)

Contractor
Hyspec Systems Ltd, Barrow-in-Furness.

VERIFIED

D100 scanning sonar head

DIPS

Type
Diver Information and Positioning System (DIPS).

Development
DIPS is being developed to support Mine Counter-Measures (MCM) and Mine Investigation Evaluation (MIE) operations, particularly in conditions of zero visibility, both at depth and in the shallow water environment.

Future developments include target marking and real-time reconnaissance map generation, use of MCM/MIE databases, automatic generation of search patterns and integration into the new Closed Diver Breathing Apparatus (CDBA) life support equipment. DERA has developed the DIPS in conjunction with Systems Engineering to produce a technology demonstrator capable of testing new sensors and techniques for MCM operations. This system is being used to produce realistic specifications for equipment to fulfil the operational requirements of the UK armed forces.

Description
The function of DIPS is to provide a diver with a means of locating targets, viewing text, drawings and position information, at depth and in zero visibility. This is achieved by linking together a hand-held sonar and integrated computer, compass, depth transducer and either a GPS receiver or a communications link to a surface unit.

A wide range of capabilities and expertise from DERA have been used in developing DIPS, including: diving engineering; acoustic and magnetic signature reduction and assessment; life support equipment; diving physiology; sonar systems; ergonomics; a fully equipped dive team; various trials and test sites; and close links with the Royal Navy and Royal Marine dive teams.

The hand-held sonar is an SE500 high-resolution, electronic scanning sonar with an operating frequency of 500 kHz, the sonar picture being presented on a head-up display attached to the diver's mask. The sonar scans a horizontal sector 60° wide with 64 beams with a width of 1° with a maximum range of 50 m. Sonar controls are located at the rear of the unit and allow the diver to adjust the mode of operation to suit the operational conditions. Connections are provided to the power supply (a Ni/Cd battery pack or a 16 to 28 V DC surface supply), the head-up display and access to PC functions such as a Visual Graphics Array (VGA)

monitor, serial ports and floppy disc drive when operated on the surface.

The processor is a 33 MHz Intel 486sl with 12 Mbyte RAM and 256 Kbyte ROM and a 40 Mbyte Kittyhawk disc drive. The unit has two serial and one parallel ports and four analogue inputs. The depth rating is 150 msw. The diver's head-mounted display is a red LED-based system with 720 × 280 pixel resolution.

The system features an RS232 datalink which enables the diver to be connected to either a surface PC for supervised diver mode or a floating GPS antenna enabling stand-alone diver navigation.

Operational status
The equipment has been sold to the UK and Singapore navies.

Contractor
DERA, Gosport Systems Engineering, Lancaster.

VERIFIED

Buddyphone

Type
Through-water communications system.

Description
The Buddyphone has been developed by Interspiro, in co-operation with Ocean Technology Systems.

The system allows for diver-to-diver and/or diver-to-surface communication.

Specifically designed for use with the Divator Mk II full face mask, the Buddyphone is a small, easy to use, compact unit that is attached to the head harness and

sits directly over the diver's left ear. A single wire, running close to the face mask, from the ear unit, connects the 'push to talk button' and 'microphone'. These are located in the front port of the face mask to give good clear direct speech. The transducer is a ceramic cylinder.

Buddyphone is an ultrasonic transmitter/receiver unit which enables communication between users. Transmissions are carried using upper single sideband.

Specifications
Reference frequency: 32.768 kHz

Audio bandwidth: 300-3,000 Hz
Automatic gain control: >80 dB
Power supply: 9 V alkaline, 9 V lithium, or 9.6 V rechargeable Ni/Cd
Normal range: 50-500 m (dependent on sea conditions and noise levels)
Operating depth: 50 m

Contractor
Interspiro, Telford.

VERIFIED

Sea Piper

Type
Diver communications system.

Description
The Sea Piper diver communications system is fully modular and allows underwater communications between divers, between divers and a control platform and between a submersible and surface vessel. The equipment caters for divers working with compressed air or oxy-helium breathing apparatus, and provides line or through-water communications as required. A new design of helium speech converter, which gives increased clarity of speech together with simplified control, is a feature of the system.

In the tethered mode of operation, a four-wire neutrally buoyant cable is used. This doubles as a safety line and gives duplex or round-robin communications which, for divers using oxy-helium mixtures, feeds back an unscrambled version of the diver's own speech.

The through-water mode enables operators to communicate with one or more divers via a simplex link. This facility provides communications in the simplex mode between the diving tender, submersible and divers. In this mode, communications between two or more divers can be achieved without resorting to surface or submersible-carried equipment. For diver-to-surface communication, an adaptive headset is used by the surface operator; whilst the diver uses bone conduction earphones and microphones.

All parts of the system are designed for adverse environments. The control unit can be supplied as a backpack. The diver's module, constructed from plastic, is pressure-proofed to 100 m. The control unit carries its own internal rechargeable battery which has 8 hours duration but, in addition, external power from a ship's 115/240 V AC or 12 V DC supply can be used to power the control unit.

The system uses single-sideband, suppressed carrier HF transmitters and receivers. A two-way simplex communication is available, each transmission being preceded by a short tone-burst. The operation of the press-to-talk switch causes the changeover from receive to transmit.

When the electronic module is selected for long-range operation a minimum range of 1 km can be

Sea Piper diver communications system

achieved. A short-range facility (less than 100 m) is available, range depending on propagation conditions.

Operational status
Developed under contract from the UK Ministry of Defence with 150 units purchased by the Royal Navy; also in use with Qatari, Dutch, Spanish and other special forces.

Contractor
GEC-Marconi Radar and Defence Systems, Command and Information Systems Division, Camberley.

VERIFIED

Through-water communicator

Type
Diver's untethered through-water communicator.

Description
The system provides a compact communication system offering diver to diver and/or diver to surface through-water communications.

The communicator provides clear and reliable underwater communication and comprises a Diver's Electronic Module (DEM) headset and attachment belt. Communication is achieved using HF acoustic waves transmitted through water between acoustic transducers attached to the DEMs. Each DEM uses a single-sideband suppressed carrier, transmitter and receiver. A two-way simplex communication is available, each transmission being preceded by a short tone-burst. The operation of the press-to-talk switch causes the changeover from receive to transmit.

To provide diver-to-surface communication an adaptive headset is used for the surface operator, whilst the diver uses bone conduction earphones and microphones.

The sealed DEM has been designed for use down to 100 m, and a minimum range of 1 km can be achieved when the DEM is selected to 'long' range. For use in complex missions a facility exists to reduce to 'short' range (less than 100 m), dependent on prevailing propagation conditions. Divers using gloves can carry out simple battery changes.

Any number of divers can be involved with the controlling surface station.

Operational status
Developed under contract to the UK Ministry of Defence.

Contractor
GEC-Marconi Radar and Defence Systems, Command and Information Systems Division, Camberley.

VERIFIED

EL050 (2056), EL057, EL067 series

Type
Deep-phone underwater telephones.

Description
These multifrequency units provide voice communication and pinger indication in the military submarine and diver bands. They are employed as both portable and installed units in submarines, surface ships and submersibles, from helicopters, and as secondary/emergency units in submarines. They are also used for parachute delivery with SUBSUNK emergency assistance teams and are installed in the Royal Navy's crew rescue KR5 submersible.

The units provide diver-to-diver and diver-to-surface communication over ranges of 1,000 m. The diver's Ultra Short Baseline (USB) station operates at a carrier frequency of 40.2 or 42 kHz with a bandwidth of 3,400 Hz. It measures 78 × 300 mm (excluding helmet or hood microphone and earphones) and its weight in water is 800 g. An on/off switch is water-activated. Powered by Ni/Cd batteries, the unit provides 8 hours of operation at a 10 per cent transmit cycle. It is rated to a depth of 450 m.

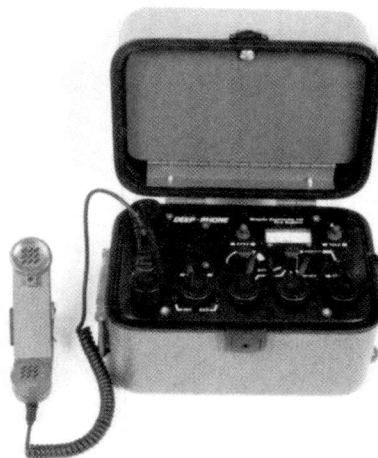

Deep-phone surface unit

The surface unit measures 200 × 300 × 280 mm (excluding handset) and weighs 8.65 kg. Its lead-acid battery provides 18 hours of operation at a 15 per cent

2056 unit

transmit cycle, and it also incorporates a battery charger, a built-in loudspeaker and a tape recorder.

Operational status
In use with the Royal Navy, the Swedish Navy and several other customers.

Contractor
Slingsby Engineering Ltd, Kirbymoorside.

UPDATED

Model 3220C two-diver telephone

Type
Underwater telephone for diver communications.

Description
The Model 3220C two-diver telephone is a two- or four-wire intercom unit. Using two wires to each diver the tender controls communications with one or two divers, or between two divers. Using four wires to each diver, up to five divers and the tender can participate in a 'conference call' bypassing the press-to-talk and cross-talk switches. A recording device can be connected when a complete record of all communications is required.

The Model 3220C is a heavy-duty telephone which has a wide range of applications for use with working divers, medical compression treatment chambers, and both naval and commercial operations. Its wide-frequency response provides good sound quality for helium atmosphere use to depths of over 60 m.

The unit is powered by easily replaceable 'C' cells, which are removed through the front panel. A rechargeable Ni/Cd battery unit is available as an option.

Operational status
In production.

Contractor
Nautronix Ltd, Aberdeen.

VERIFIED

Optical microphone

Type
Diver communications system.

Development
The Sea Systems Environmental Science unit at DRA, Alverstoke is developing a fully encapsulated optical microphone to complement the optically powered earphone. The system is designed for divers involved in mine countermeasures operations.

The DRA has carried out a series of underwater tests to investigate the frequencies at which a diver is best able to perceive underwater sound. From the tests an underwater audio response curve has been drawn which shows that the human ear responds to waterborne sounds in the frequency range 25 Hz to 16 kHz.

The reason for developing such a communication system is that modern influence mines are exceedingly sensitive to any electromagnetic signature, and can be programmed to react to the very low signatures exhibited from equipment carried by a diver, and in particular, normal communications systems. It is therefore essential that any communications are conducted using systems which exhibit no electromagnetic signature. Currently, the only safe method is by pulling on the diver's lifeline. Hence the desire to develop an audio system which would considerably improve efficiency and safety between the divers and the surface.

The current development work on the microphone has led to a fully encapsulated system which uses fibre optic technology. The optical microphone operates as an interferometric device in which the diaphragm of the microphone reacts to variations in pressure caused when the diver speaks. The movement of the diaphragm is then translated into a varying light intensity, using the interferometric process, for transmission via the fibre optic link to the surface.

The optical earphone uses a photo-diode with a ceramic sounder powered by a laser diode at the surface end. The output of the laser is modulated by the controller's voice.

Contractor
Defence Research Agency (DRA), Alverstoke.

VERIFIED

Model 3151 Hellephone

Type
Underwater communications unit.

Description
The Model 3151 Hellephone is a single-sideband underwater telephone which allows the operator to communicate with divers using the diver Hellephone. The unit can also be fitted in a submersible allowing communication to take place between submersibles, surface and divers in all directions. The unit comes complete with acoustic transducer on 15 m heavy-duty coaxial cable, rechargeable battery and battery charger.

The Model 3151 can also be provided in encrypted form.

Contractor
Nautronix Ltd, Aberdeen.

VERIFIED

Buddy

Type
Through-water communicator.

Description
The Buddy diver communications unit is a compact, multifrequency (10, 27, 30, 37, 41 or 43 kHz) through-water voice communication system for diver-to-diver, diver-to-surface or diver-to-submersible. The diver's unit is designed to interface with the Interspiro full-face mask but other masks can be interfaced if required.

The surface unit is powered from an external mains supply or an internal battery pack (typical life 8 hours) and incorporates a built-in loudspeaker and connections for a headset, microphone, dunking transducer and a tape recorder.

Communication is carried out using single upper sideband with a diver's unit having a typical power output of 1 W into the transducer. The maximum range of the diver unit is 1,500 m and of the surface unit 3,000 m.

Contractor
Graseby Dynamics, Watford.

VERIFIED

Under Hull Diver Location System (UHDLS)

Type
Diver locating system.

Development
Detailed predocking in-water surveys of the hulls of US Navy ships to assess damage is essential to accurately estimate time required in dry dock and to draw up work specifications. This enables time in dry dock to be reduced with considerable savings in cost. If such a survey shows only minor repairs are necessary it may even save the necessity of docking the ship at all.

The need to conduct these accurate underwater surveys of the hulls of US Navy ships has led to Oceaneering Inc in the USA developing ShipShape, a database of all US Navy ship hull profiles from which a map can be drawn against which the diver can carry out an inspection.

However, to carry out such a precision inspection requires the positions of the diver to be tracked with great accuracy. To meet this requirement the UHDLS system has been developed.

Description
UHDLS is an underwater acoustic positioning system which can track a single diver with accuracies of 150 mm at ranges of over 500 m. Three sets of positional data have to be obtained and computed to obtain the diver's position: the diver relative to the seabed; the ship relative to the seabed; and the diver relative to the ship. It is necessary to carry out these measurements because no ship remains absolutely stationary, even when it is tied up alongside, and any movement will introduce positional errors.

To position the diver relative to the seabed, an array of four or more Compatt (computing and telemetering transponder) units is deployed in a long baseline configuration on a grid. These 'intelligent' units communicate with each other to establish the baselines between them. This data is telemetered to the surface where it is received by an in-water transceiver and displayed and computed on board the ship on a 486 computer running on DOS.

The diver is fitted with a transceiver and his position relative to the seabed array is displayed on board.

The seabed array of Compatts forms the basis for positioning the ship and diver relative to each other and the seabed. In addition, three Minirov transceivers are attached to the vessel's hull at known locations – usually two on the keel and one on the side. This enables the position of the hull to be determined in three dimensions relative to the seabed array.

The position of the diver is similarly determined and his co-ordinates transformed into co-ordinates in the vessel's frame of reference. All positional information is updated once a second.

The system operates in conjunction with an online conductivity/temperature/depth probe which provides continuous updates on the speed of sound through water.

Operational status
On trials off Charleston the UHDL system successfully tracked a ship and dive team to the extent that a diver found a 4 in pipe inlet on the bottom of a frigate in the time it took him to swim to it. Handover of the system to the US Navy – with the added capability to track an ROV – was completed in December 1994.

Contractor
Sonardyne International, Blackbushe.

UPDATED

Model 6280

Type

Diver-held pinger receiver.

Description

This sensitive, directional acoustic receiver enables a diver or small boat operator to precisely locate underwater pingers. The hand-held, battery-powered ('C' size alkaline batteries) receiver can be used to locate any pinger on a frequency from 8 to 50 kHz. The unit is constructed of corrosion-resistant anodised aluminium and can operate down to 305 m depth. The unit is fitted with a parabolic reflector with a hydrophone mounted at the focal point forming a very narrow listening beam for the precise location of pingers.

A frequency synthesiser allows error-free frequency selection; once the pinger's frequency is selected, fine tuning is unnecessary. Since the frequency controls are detented, the diver can change the receiver frequency by 'feel', even in zero visibility.

For operation the sensitivity is set to maximum and the frequency controls adjusted to match the frequency of the pinger. The area is scanned by moving the unit through the water and listening for the pinger's acoustic signal. When a ping is heard the sensitivity is reduced and the operator searches for the strongest signal. By continuing to reduce sensitivity while listening to the ping the operator can easily determine the direction to the pinger by homing in on the strongest signal.

Contractor

Nautronix Ltd, Aberdeen.

VERIFIED

Model 6280 diver-held pinger receiver

Model 6281

Type

Surface and diver pinger receiver.

Description

The Model 6281 battery-powered pinger receiver is capable of locating any pinger in a frequency range from 8 to 50 kHz. The unit can be used from a small boat and can then be rapidly converted to a diver-held receiver, allowing the diver to swim directly to the located pinger.

The complete system comprises: the Model 6515 antenna; Model 6280 diver-held pinger receiver; a headset for surface operation; and an adaptor to convert the system from surface to diver use. The entire system is housed in its own waterproof carrying case.

For small boat operation the antenna sections are fitted together with the directional hydrophone on one end. The cable threads through the connected sections of the antenna and its electrical connector is connected to the adaptor mounted on the receiver.

For operation the sensitivity is set to maximum and the frequency controls adjusted to match the frequency of the pinger. The operator scans the area by rotating the antenna through the water and listening for the acoustic signal. When a 'ping' is heard the sensitivity is reduced and the operator searches for the strongest signal. By continuing to reduce the sensitivity while listening to the ping, the operator can easily determine the direction to the pinger by homing in on the strongest signal.

For diver use the hydrophone is connected to the end of the receiver and the headset replaced by the waterproof earphone. The diver operates the unit in the same way as described above.

Contractor

Nautronix Ltd, Aberdeen.

VERIFIED

Model 3342B

Type

Helium speech unscrambler.

Description

This two-diver helium unscrambler unit is a two- or four-wire intercom unit with two different unscrambler modes. Using two wires to each diver, the surface tender controls communication with one or two divers. Using four wires to each diver, a duplex connection allows the tender and up to five divers to communicate, bypassing the press-to-talk and cross-talk switches.

The two-diver unit uses an audio amplifier to allow communication between the tender and the divers. The unscrambler uses two complex forms of digital techniques to filter and lower the frequency of a diver's voice resulting in intelligible speech. Mode One is designed for use at normal depths with a normal microphone, Mode Two is intended for use in depths to 610 m using a high-quality microphone. A record jack allows a recording device to make complete records of all communications. Separate microphone and speakers improve communication. A remote press-to-talk capability is available.

Operational status

In use by military forces, deep sea research organisations and for medical hyperbaric chambers.

Contractor

Nautronix Ltd, Aberdeen.

VERIFIED

Model 3500

Type

DSP-based modular helium speech unscrambler.

Description

The Model 3500 helium speech unscrambler provides the highest possible speech intelligibility – 99 per cent at normal diving depths. It is designed in a modular form to allow full flexibility and expandability for saturation diving, hyperbaric training and hyperbaric medical facilities.

Each module provides separate volume control along with a limit indicator, and on wet diver modules, built-in breathing noise reduction. Modules are also available for record, telephone and emergency communications.

All modules in the system can be selected to be in one of six communication groups via panel-mounted, illuminated switches. Once selected the control of communication groups is from the tender module which allows the supervisor to select which groups he/she wishes to talk to and listen to separately, thus eliminating the traditional requirements for a separate switching panel.

Operational status

In use commercially.

Contractor

Nautronix Ltd, Aberdeen.

VERIFIED

Tender module of the Model 3500 DSP-based speech unscrambler
                                                    1997

OE 9020 Seahawk

Type

Diver TV system.

Description

The low-cost Seahawk underwater colour television system consists of a wide-angle Charge Coupled Device (CCD) camera complete with underwater lamp and pistol grip assembly, 50 m of cable and a surface control unit. The control unit incorporates a colour TV monitor, camera and lamp power supplies and a stowage bag for the camera/lamp assembly.

Operational status

In service with the Royal Navy and several navies worldwide.

Contractor

Simrad Ltd, Aberdeen.

VERIFIED

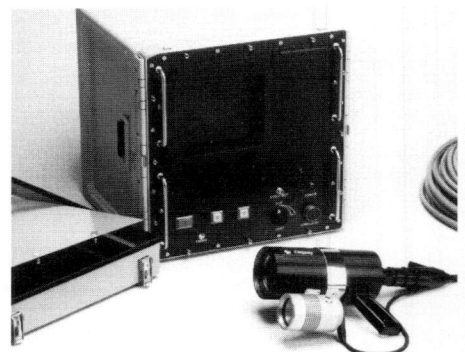

Seahawk diver TV system

Life Support Systems Laboratory

Type
Life Support Systems Laboratory

Description
The Life Support Systems Laboratory provides facilities for determining, quantitatively, the limits of performance of breathing apparatus under simulated conditions.

Breathing apparatus using oxygen, air or mixed gas can be tested dry, or immersed in water, within a temperature range of 2 to 40°C. The hyperbaric chamber and breathing simulator are rated to 1,000 msw with a capacity of 1.2 m³.

Instrumentation includes a computerised data acquisition system, real-time display and temperature control.

Equipment tested has included respirators, firefighting equipment, deep diving life support systems and submarine escape breathing apparatus.

The main features of the Life Support Systems Laboratory are:
(a) breathing simulator with a capability of 5 to 60 breaths per minute and tidal volumes 0.6 to 4.0 litres
(b) a second breathing simulator, linked electronically to the main breathing simulator, for testing equipment fitted with a second mask
(c) respiratory gas exchange system and gas analysis
(d) breath by breath CO_2 analysis
(e) internal video camera and recorder
(f) a secondary hyperbaric chamber that can be used for surface demand equipment, simulating divers operating at different depths
(g) an ambient pressure environmental chamber with 0.61 m³ capacity, temperature range of −75 to 100°C with relative humidities of 45 to 98 per cent. The chamber can be used for control panels connected to breathing apparatus in the hyperbaric chambers and testing of breathing apparatus.

Contractor
DERA, Gosport.

VERIFIED

UNITED STATES OF AMERICA

Mk 16 UBA

Type
Underwater Breathing Apparatus (UBA).

Description
The Mk-16 UBA is designed for EOD operations such as mine countermeasures or recovery and disposal missions against magnetic and/or acoustically sensitive ordnance. It is an electronically controlled, fully closed circuit, mixed gas, constant partial pressure breathing apparatus. The set permits completely autonomous diver operation with recirculation of 100 percent of the mixed-gas breathing medium resulting in bubble-free operation. With electronic control of the partial pressure, the Mk-16 allows divers to modify missions at will. These capabilities, combined with the equipment's extremely low magnetic and acoustic signatures, make the Mk-16 UBA well suited for mine clearance operations and special warfare operations, requiring minimum sound production and low-magnetic influence.

The rebreather is capable of mixing either nitrogen or helium depending on the depth of the dive, the limitations being 58 m with nitrox and 91.5 m with heliox.

Use of special alloy metals, tightly controlled processes and high-strength plastics result in a low-magnetic signature. A low-acoustic signature is obtained through careful design of high-pressure gas passages and by using a custom-designed piezoelectric oxygen addition valve that provides quiet operation, minimal power consumption and low-magnetic influence that is not possible with a solenoid valve. The utilisation of high-reliability electronics to maintain a constant partial pressure regardless of depth allows the diver to concentrate on the mission and not on the equipment. Other advantages of the system are efficient gas usage, extended dive duration, superior depth capability, and reduced breathing resistance.

As a closed-circuit rebreather, the Mk-16 was

Block schematic of Mk-16 UBA   **1999**/0024824

designed to: remove carbon dioxide produced by the metabolic activity of the body, add pure oxygen to the breathing gas to replace the oxygen consumed, maintain oxygen content at safe levels and recirculate the breathing gas for reuse.

The recirculation system consists of a closed loop (including inhalation and exhalation hoses), a lightweight mouthpiece or full-face mask (for use with communication gear), a carbon dioxide scrubber, and flexible breathing diaphragm. The diver's breathing gases are recirculated to remove carbon dioxide and permit reuse of the unused breathing gas. Inhalation and exhalation check valves in the mouthpiece assembly ensure the unidirectional flow of gas through the system.

Movement of recirculating gas through the circuit is normally achieved by the natural breathing action of the diver's lungs. As the lungs are capable of only small pressure differences, the entire circuit has been designed for minimum flow restriction and extremely low breathing resistance.

Oxygen is maintained at a constant level through the use of a micro-electronic control system, that analyses the input from three oxygen sensors, and signals the piezo-electric valve to add the precise amount of oxygen into the breathing loop. The removal of CO_2 is accomplished by passing the exhaled gas over a bed of CO_2 absorbent, either Sodasorb or Sofnolime.

The principal components of the Mk-16 include: two high-pressure gas reservoirs with shut-off valves, high-pressure regulators, piezoelectric oxygen addition valve, carbon dioxide scrubber, oxygen sensors, electronics, counter-lung, mouthpiece assembly and two independent displays. The assemblies are contained in a rugged ABS plastic housing and cover. LED readouts and a head-up display provide effective monitoring of the system's operation throughout the dive.

Operational status
In use with US Navy EOD and special operations forces since the mid-1930's. Also in service with the Royal Australian Navy.

Specifications
Weight: 29 kg
Operating depth: 90 m (max)
Duration of dive: 20 minutes (at maximum depth)

Contractor
Carleton Technologies Inc, Tampa, Florida.

UPDATED

AN/PQS-2A sonar

Type
Hand-held active/passive sonar.

Description
The AN/PQS-2A is a continuous transmission, frequency-modulated, hand-held non-magnetic sonar set intended for use by divers to locate and close on submerged targets such as mines, lost equipment, downed aircraft and other targets to depths of 91 m. Because of its non-magnetic construction, the set can be used safely when working near magnetically influenced underwater explosive devices. The equipment can be used by divers using only one hand.

The equipment has an active (continuous transmission) mode and a passive (listening) mode for detecting underwater pingers. The diver adjusts the frequency until pulses or a constant tone are heard in the headset. In the active mode, one of three range scales is selected (18, 55 or 110 m) and an acoustic signal is transmitted over a 30 kHz bandwidth, swept from 115 to 145 kHz. When a submerged object is detected, a ringing tone, proportional in frequency to the distance of the object, is produced in the earphones. The sonar has a 6° beamwidth and can detect a 30.5 cm diameter air-filled sphere at 110 m. In the passive mode, the unit detects active sound sources (pingers) in the 24 to 45 kHz range. The sonar can detect a 39 kHz acoustic beacon at a range of at least 1,800 m.

Operational status
In operation with the US Navy (over 200 units). The DLS-1 (Diver Locator Sonar 1), a commercial variant of the AN/PQS-2A, is currently available and in operation.

Specifications
Transmission: Continuous Transmission Frequency Modulated (CTFM) 145 down to 115 kHz (30 kHz bandwidth)
Active range: 18, 55 or 110 m
Passive range: up to 1,800 m

AN/PQS-2A hand-held sonar

Passive detection: pingers from 24-45 kHz
Output: audio tone (or pulse) in earphones. Frequency variable with range
Power source: two rechargeable gelled electrolyte lead-acid batteries
Operating depth: to 91 m
Operating temperature: u2 to +30°C
Weight in water: 0.23 kg (positive buoyancy collar included in accessories)
Size: 31.75 cm long × 11.43 cm diameter

Contractor
SeaBeam Instruments Inc, East Walpole, Massachusetts.

VERIFIED

DHS-100

Type
Diver held sonar.

Description
The DHS-100 diver sonar is a portable, self-contained sonar designed for use by divers to locate submerged objects. The system provides bearing and range information to the target and can also be used to locate pingers or other sound sources.

The unit is suitable for use in recovering submerged objects such as aircraft, vehicles and so on.

The sonar operates in a Continuous Transmission Frequency Modulated (CTFM) mode with an active mode for locating objects within 18, 55 or 110 m ranges. A continuous signal of varying frequency is transmitted and range information is determined by comparing the transmitted frequency to the echo frequency. When the target is located an audible tone is generated in the headset. A low tone indicates a close target while a high tone indicates a distant target. Moving in on the target causes the tone to gradually lower until contact is made. Experience with the system allows an operator to determine range and target characteristics from the echo signal. In the passive mode, the sonar provides bearing information to a pinger or other sound source.

Specifications
Range
active: 18, 55 or 110 m
passive: up to 1.5 km

Beam Width
active: 15° at -3 dB points
passive: 30° at -3 dB points
Operating frequency
active: 115-145Khz (30 Khz band)
passive: 25-45 Khz
Audio frequency: 250-2,500 Khz
Operating depth: 91 m

Contractor
Datasonics (manufacturer).
RJE International Inc (distributor), Irvine, California.

VERIFIED

Hydrocom

Type
Diver speech processor.

Description
Hydrocom is the US Navy's standard helium speech unscrambler system. Using a full-duplex connection, Hydrocom allows communication between three divers and a surface tender without the need for a press-to-talk function. The unit operates with either powered or non-powered diver microphones. The power for the microphone can be multiplexed onto the diver's microphone lines and the tender has the option to isolate each of the divers' inputs or outputs in the event of cable failure.

The system provides quality communications in a helium environment using a single audio printed circuit board which uses a differential input stage for each diver to minimise noise and cross-talk picked up in the umbilical cable. Low-noise amplifiers are used to amplify divers' voices and a ground plane is incorporated to minimise noise. An optional helium speech unscrambler printed circuit board is available which can be bypassed from the front panel of the control unit for air operation. The unscrambler uses analogue and digital techniques to convert distorted helium speech transmission into intelligible audio. The unscrambling is achieved with audio high-frequency boost and time domain sampling and is suitable for operation at depths down to 610 m. At depths of 305 m better than 75 per cent intelligibility has been achieved.

Hydrocom also incorporates inhalation noise attenuation circuitry to reduce breathing noise if an oral/nasal cavity is included in the mask.

Operational status
In service with the US Navy.

Contractor
SEA Hydro Products, San Diego, California.

VERIFIED

AN/UQN-5 SANS

Type
Swimmer Area Navigation System (SANS).

Description
The AN/UQN-5 SANS is used by the US NSWG SEAL Team divers to perform grid searches of large areas of the sea bottom in conditions of reduced visibility. The system employs a diver-held receiver unit synchronised with two acoustic beacons placed at extreme corners of the desired pattern. The diver is given a real-time readout of position and heading and the ability to record the location of significant objects on the seabed. On completion of the dive, the receiver is attached to a printer which produces a hard-copy record.

Operational status
In operational service.

Contractor
Allen Osborne Associates, Westlake Village, California.

VERIFIED

SeaLites

Type
Underwater lighting systems.

Description
This range of underwater illumination devices has been developed for a variety of underwater applications.

The standard lightweight SeaLite range offers lamp voltages from 12 V DC to 240 V AC with wattages ranging from 50 to 500 W. All lamps have positive locking bayonet mounts to reduce the possibility of vibration-related failures. Lamps are replaced simply by unscrewing the front bezel assembly from the light body. The glass-to-metal seal of the pyrex (or quartz) envelope to the bezel is never disturbed during lamp replacement. Modular interchangeable wet reflectors provide beam patterns ranging from a medium spot to a wide angle for almost any conceivable application with both diffuse and specular reflectors being available. The body and bezel are constructed of hard block anodised aluminium. The black finish minimises stray reflections which cause flare in video images. The standard depth rating is 6,000 m.

The Deep-SeaLite is an 11,000 m bezel-type light, similar to the SeaLite. It offers 12 V DC to 240 V AC lamps ranging from 50 to 1,000 W. The aluminium housing is extremely lightweight for a light of its depth rating. Three reflectors are offered: spot, medium and flood.

The MultiSeaLite is available in a variety of configurations using three different reflectors (spot to even, hot spot-free medium, wide-angle flood beam patterns), 120 or 240 V AC or DC, multiple wattages and two types of connectors. The light is available in aluminium or stainless steel housings. An external dome-retaining cowl offers protection and acts as a baffle to prevent stray light from entering the water column, which minimises backscatter. The light is rated to 1,000 m depth.

The Deep MultiSeaLite is a 6,000 m version of the MultiSeaLite. It has the same voltage, wattage, reflector, and connector options as the MultiSeaLite. Identical in outward appearance to the MultiSeaLite, the Deep MultiSeaLite has a different internal construction that allows for the greater depth rating.

The wet/dry MultiSeaLite is based on the MultiSeaLite design but has cooling fins that allow it to operate in either air or water. It can also be operated in air for prolonged periods and immediately be immersed in water while still operating. Lamps come in 12 V DC to 240 V AC and 50 to 100 W.

The Mini-SeaLite is a narrow-body light used for restricted access applications. The light body comes in either aluminium or stainless steel. Lamps for the Mini-SeaLite range from 12 V DC to 240 V AC and 50 to 250 W.

The Micro-SeaLite is a smaller version of the Mini-SeaLite, but like the wet/dry MultiSeaLite, can be operated in air or water and can survive immersion in water even after prolonged operation in air.

Contractor
DeepSea Power & Light, San Diego, California.

VERIFIED

SeaCams

Type
Underwater video camera systems.

Description
This range of underwater video camera devices has been developed for a variety of underwater applications.

The SeaCam 2003 is a hand-held or fixed-mount colour camera rated to 1,000 m. It offers a high-quality colour 12.7 mm Charged Coupled Device (CCD) video camera housed within a Delrin pressure housing with a scratch-resistant acrylic port. The fixed-focus 3.7 mm wide-angle lens produces crisp, high-resolution images from 10 cm to infinity.

The Micro-SeaCam 1001 is a monochrome camera available in a 6,000 m rating. The entire camera is contained in one small housing – there is no separate electronics bottle. Its 3.6 cm width makes it ideal for ROV or limited access use. The Micro-SeaCam 1001 offers 550 lines of resolution through a 12.7 mm CCD sensor, with fixed focus and a depth of field ranging from 10 cm to infinity.

The Micro-SeaCam 2000 is a colour camera rated to 6,000 m. Its incredibly small dimensions (2.5 cm wide by 21.6 cm long) make it ideal for ROV use. In addition to fixed-focus, wide-angle optics, the depth of field can be set to three positions. It has 470 lines of resolution with optional Y/C (S-VHS) output.

The Multi-SeaCam is a versatile camera designed to meet all deep ROV or shallow diver applications. One housing and mounting system accepts two different video camera cartridges, either black and white or colour. This enables all applications to be standardised with one electrical and mechanical interface. The ROV version is made from titanium and is rated to 6,000 m. The diver version is made from black Delrin and is rated to 300 m. Both versions use a scratch-resistant sapphire port. Identification of camera features in the field is simplified by the use of a label under the sapphire port that indicates such features as model type, depth rating, video format, and pinout.

The SeeSnake™ 1000 is a camera designed specifically for pipe, bore hole, or restricted access inspection in areas as small as 76.2 mm in diameter. The design incorporates both camera and lighting in one small housing measuring just 6.4 × 5.4 cm. Utilising 40 red Light Emitting Diode (LED) lights in a hardened stainless steel housing with sapphire crystal window, the SeeSnake™ is almost indestructible. It has fixed-focus, wide-angle optics that provide a tremendous depth of field. The SeeSnake™ is rated to a depth of 1,000 m.

The SS-1370 is designed to withstand the rigours of day-to-day inspection in areas where environmental and mechanical shock are expected and unavoidable. The SS-1370 has applications in pipe inspection, bore hole inspection, water wells, fixed-mount monitoring (such as underwater valves), and tank and void inspection. It is suitable wherever a compact camera with built-in lighting is the best solution. Features include: 35 adjustable bright red LEDs mounted in the camera face for lighting; stainless steel housing and a sapphire crystal lens port; individually tested and certified waterproof to 750 m; and fixed focusing with a consistently clear picture with a minimum 2.54 cm depth of field.

Contractor
DeepSea Power & Light, San Diego, California.

UPDATED

DIVER VEHICLES

CROATIA

R-1

Type
Swimmer Delivery Vehicle (SDV).

Description
The R-1 is a single-seat underwater SDV designed for covert tasks such as reconnaissance, harbour protection, minefield surveillance and so on.

The monohull vehicle is built of aluminium alloy and comprises light bow and stern sections that can be flooded. The vehicle can be transported in a submarine torpedo tube and used in both freshwater and seas of specific gravity of 1.000-1.030 t/m³ without reserve updrift.

Navigation instruments are housed in a watertight container and comprise a gyromagnetic compass, sonar, echo-sounder, electric clock and other measuring systems.

Operational status
Most units which have survived the civil war are believed to be in service with Croatian forces. Other units are in service with RFAS, Libya, Sweden and Syrian forces.

Specifications
Length: 3.72 m (oa)
Breadth: 1.05 m (max)
Height: 0.76 m (max)
Diameter: 0.52 m (max)
Displacement: 145 kg (less payload)
Payload: 40 kg
Propulsion: DC electric motor, 1 kW; 24 V silver-zinc battery
Speed: 3 kt (max); 2.5 kt (cruising)
Range: 6 n miles (at max speed); 8 n miles (at cruising speed)
Operating diving depth: 60 m

Contractor
Brodosplit, Split.

UPDATED

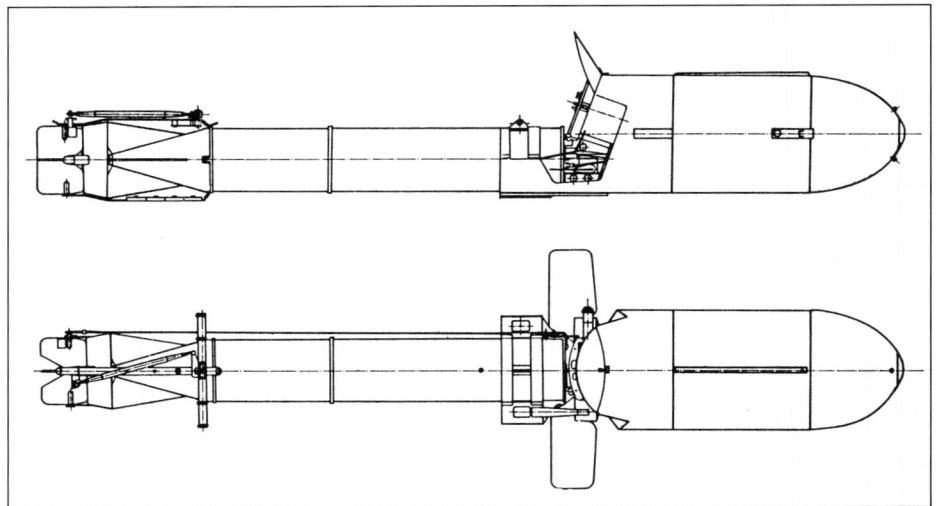

Frogman submersible R-1

R-1 submersible
1998/0009588

R-2

Type
Two-man Swimmer Delivery Vehicle (SDV).

Description
The R-2 is designed for the transport of frogmen and underwater limpet mines (up to 250 kg payload), for carrying diversionary equipment and for underwater reconnaissance. The hull is constructed of aluminium alloy resistant to saltwater corrosion. The front upper part of the submersible is made of Plexiglas. The spindle-shaped hull can be fully flooded, except for the cylinders housing the storage battery, propulsion unit, navigation equipment, compressed air and ballast tanks.

The built-in ballast system is used for blowing and flooding the ballast tanks to secure static diving in seas of specific gravity 1.01-1.03 t/m³ and at depths of 0-60 m, as well as for compensation of the air consumed by the frogmens' respiratory system.

The vehicle is powered by a DC electric motor which derives its power from a storage battery. The three-bladed propeller is driven via an electromagnetic clutch.

Navigation equipment comprises an aircraft-type gyromagnetic compass, a magnetic compass, depth

R-2M (dry delivery)
1998/0009589

gauge with scale 0 to 100 m, echo-sounder, sonar, two searchlights and so on. All navigation equipment is housed in a waterproof container.

Control is exercised through fore and aft hydroplanes and a conventional cruciform tail arrangement with rudder mounted behind the screw.

Operational status

In service with Libya, RFAS, Sweden and Syria. One or two units may be in service with Croatian forces.

Specifications

Length: 4.90 m (oa)
Diameter: 1.22 m

Breadth: 1.40 m (max over hydroplane)
Height: 1.32 m (max over appendages)
Displacement: 1,400 kg
Propulsion: 3.3 kW (DC electric motor); 24 V 192 Ah (storage battery)
Speed: 4.4 kt (max); 3.7 kt (cruising)
Range: 18 n miles at max speed, 23 n miles at cruising speed of (with lead-acid batteries); 38 n miles at max speed, 46 n miles at cruising speed (with silver-zinc batteries 310 Ah)
Max diving depth: 100 m
Armament: two 50 kg underwater mines

Contractor
Brodosplit, Split.

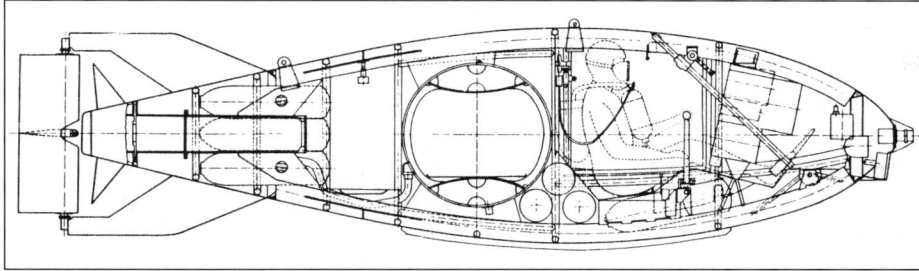

Schematic of the R-2 two-man submersible

UPDATED

FRANCE

535 UM

Type

Submersible inflatable craft.

Description

The 535 UM submersible inflatable is designed to carry either five fully equipped men or a payload of 1,000 kg and a pilot with 2,000 litres of fuel to a surface range of 100 n miles at a cruising speed of 20 to 25 kt, depending on the load. For submerged propulsion the craft is powered by two 5 kW thrusters mounted forward giving an endurance of up to 4 hours. The thrusters are powered by lead-acid batteries giving an underwater speed of 3 kt. With silver batteries and higher powered thrusters an underwater speed of up to 8 kt can be achieved. The maximum diving depth is 50 m. The craft can be transported in a standard 40 ft container.

Specifications

Length: 5.35 m
Beam: 2.3 m
Height: 1.5 m
Weight (unloaded): 1,000 kg

Contractor

Sillinger SA, Paris.

VERIFIED

ITALY

Chariot CE2F/X100T

Type

Small chariot.

Development

The Chariot or SDV (Swimmer Delivery Vehicle) as it is now referred to, is the result of an ongoing upgrading project of the famous Italian Maiali of World War II. Because of their endurance, chariots are normally transported to within their operating range by a parent ship such as a SWAT, patrol submarine, helicopter or even innocent-looking fishing vessel. The Chariot is designed to navigate undetected in hostile waters and carry two combat divers to their target (harbour, oil rig, coastal installation and so on) and then back to the parent ship after completion of the mission.

Description

The CE2F/X100T is a sturdy, two-man wet submersible vehicle designed for covert operations. The vehicle is characterised by its simplicity of construction, small dimensions and low weight, making it easy to handle and operate. Although new technologies are continuously developed and applied in order to improve the design, capabilities and maintenance of the Chariot, its fundamental operating principles remain the same.

Offensive capability includes limpet mines, which can be secured to targets such as ships hulls or underwater structures and depth charges laid on the seabed near the target. Both weapons are fitted with electronic time fuzes, which in the case of limpet mines are provided with an anti-removal device. Recently, mini-torpedoes with their launcher have been added to provide the craft with standoff capabilities.

The latest improvements include a digital Control Module where all navigation and platform parameters are displayed as well as a fully integrated autopilot.

Operational status

In service with a number of navies.

Specifications

Length: 7.00 m (overall)
Diameter: 0.80 m
Height: 1.50 m
Displacement: 2.10 t
Operational depth (max): 100 m

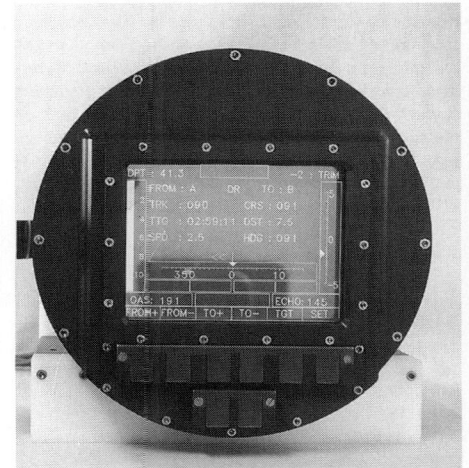

Control module dial for Chariot
1997

Chariot Mod CE2F/X100T with relative armaments
1998/0009590

Chariot underwater

1997

Speed: 5 kt (max); 4 kt (cruise)
Range: 50 n miles (at cruise speed)
Crew: 2
Armament: 1 × Mk 31 (230 kg explosive) or 2 × Mk 41

(105 kg explosive) or 1 × Mk 41 and 5 × mini-torpedoes, 12 × Mk 414 limpet mines (7 kg explosive each) or 10 × Mk 430 limpet mines (15 kg explosive each)

Contractor
Cosmos SpA, Livorno.

VERIFIED

UNITED KINGDOM

SSK96

Type
Surface/underwater rigid inflatable boat.

Description
The SSK96 Subskimmer is an unusual boat designed for divers, especially combat frogmen. The basic hull is that of a rigid-hull inflatable, fitted with a 2-stroke, 3-cylinder 90 hp Yamaha outboard engine but which is also capable of operating submerged. Like all rigid inflatables it is fast, reliable and good in heavy seas and can carry up to 600 kg to a considerable distance.

Conversion to submersible mode is quick and simple. The pilot stops the outboard engine, operates the controls which seals off the engine compartment, then switches on the electric pump to deflate the side tubes. When these are empty, the boat is supported awash by the buoyancy control chamber, a large box amidships, which is flooded to alter the buoyancy of the craft and cause it to submerge. This operation takes approximately 90 seconds. Alternatively, the hull tubes can be deflated completely whilst the craft is travelling on the surface so that immediately the outboard motor is stopped and pressurised the boat can be dived. With this method, conversion from surface to underwater operation takes 14 seconds.

Underwater the craft is very manoeuvrable and easy to control. Electric propulsion units flexibly mounted near the bows provide forward motion that is achieved by simply pointing the unit in the desired direction. The electric propulsion units are powered by a bank of lead-acid accumulators in a sealed tube that runs the length of the craft, increasing its rigidity and stability both on the surface and underwater. As all the dry systems are in water- and pressure-tight housings, the vehicle may be left parked on the seabed for hours or even days as operational needs dictate.

Underwater its endurance is over 2 hours on standard batteries.

To resurface from a dive, the electric propulsion units are inclined upwards and air is blown into the buoyancy chamber. Once the snorkels are above the surface, the dive procedure is reversed: air is first pumped into the side tubes, then the surface instruments are brought into operation and the outboard engine restarted. Such is the stability of the craft that if the pilot wishes to make

SSK 96 travelling in awash mode (KSA (Underwater)) *1999*/0024839

a fast getaway, he/she can motor away as soon as the snorkels are above the surface, reinflating the side tubes as he/she speeds off.

Subskimmer may also be motored awash with the body of the craft submerged and the outboard engine snorkelling. This mode is excellent for reconnaissance, when the boat is extremely difficult to detect either by naked eye or by surveillance sensors.

The craft is equally manoeuvrable on the surface, cruising at up to 20 kt with a maximum speed in excess of 25 kt in favourable conditions. The hull design ensures excellent seakeeping and as all systems are waterproofed, the craft are reliable even in rough weather. Surface range with standard fuel tanks is approximately 70 n miles. Adding extra tanks may increase this.

The usual payload is pilot, one crew and 300 kg stores, but four people can be carried up to a total of 600 kg. Although the craft is designed for a crew of two, it can be operated single handed.

Operation training of suitable personnel can normally be completed within a few weeks.

Operational status
In service in Thailand and a number of other countries.

Specifications
Length (inflated): 5.80 m
Length (deflated): 5.40 m
Width (inflated): 1.90 m
Width (deflated): 1.60 m
Height: 1.20 m
Draft (max/min): 0.70/0.30 m
Displacement: 1,520 kg
Payload (crew, fuel, and stores): 600 kg

Contractor
KSA (Underwater) Ltd, Alston.

NEW ENTRY

Subtug

Type
Diver tug.

Description
Developed with the assistance of former UK Royal Marine divers, the Subtug is designed to fulfil the needs of naval and special forces' divers for a robust, speedy vehicle with sufficient endurance to tow two divers and their equipment over a considerable distance.

The vehicle consists of a torpedo-shaped body, with rear-mounted propellers carried in a tubular steel frame. The propulsion system is a development of the thruster used in the Subskimmer vehicle. The frame offers protection from accidental damage and enables two divers to carry the vehicle easily between them. It also provides extra handholds and equipment fixation points so that divers may remain unencumbered during their mission. Also incorporated into the unit are permanent buoyancy pads and ballast attachment points so that neutral buoyancy may be achieved in any water conditions. The main body of the vehicle houses the batteries which are easy to recharge. Alternatively a spare set permanently maintained fully charged can be swiftly exchanged for the depleted set, keeping turnaround time to a minimum. The whole unit is streamlined and faired to reduce underwater drag.

To operate the tug the first diver holds the rear of the tubular steel frame and operates an on/off button with his/her right hand. If the driver should accidentally let go, the vehicle stops automatically. The second diver either holds a removable trapeze which tows behind the first diver; or he/she may hold the steel frame so that the two divers travel side by side allowing them to communicate.

Contractor
KSA (Underwater) Ltd, Alston.

NEW ENTRY

UNITED STATES OF AMERICA

OWL™ Mk II

Type
Remotely controlled marine vehicle.

Description
The remotely controlled OWL™ can perform a wide range of tasks including among others: surface surveillance, special operations, covert missions, mine countermeasures, coastal defence, drug interdiction.

The system consists of two sub-units: the vehicle and the control console. The flexible design of the vehicle allows it to accept a variety of custom or standard features and subsystems. A wide range of payload options up to 204 kg can be carried including video, underwater, night vision and thermal cameras; sonar; ECM/ESM equipment; station-keeping and other sensors.

The unique hull design delivers a high degree of stability and extremely low drag when operating at high speeds.

The vehicle is propelled by an internal combustion engine (petrol or diesel) with a jet drive giving it a speed in excess of 35 kt. An optional high-speed, high-power configuration can propel the vehicle at speeds in excess of 50 kt. Slow-speed operation is achieved using an electric motor which also reduces noise output, making the craft suitable for covert missions. With a very shallow draught of 4 in, the craft is unaffected by shallow water conditions.

At a cruise speed of 10 to 12 kt and with 35 gallons of fuel the OWL™ has an endurance of approximately 20 hours. Basic RF control range is about 10 n miles. An optional long-range, high-endurance configuration will increase the straight line range to between 250 and 300 n miles.

The vehicle can be remotely controlled from the ground, aircraft or satellite link at distances up to several hundred miles.

Operational status
In service with the US Navy.

Contractor
Navtec Inc, West Palm Beach, Florida.

VERIFIED

ASSOCIATED UNDERWATER WARFARE SYSTEMS

Environmental measuring systems
Environment measuring probes
Environment measuring buoys
Prediction systems

Oceanographic/hydrographic survey systems
Single-beam echo-sounders
Side scan sonars
Multibeam mapping systems
Deep towed survey systems
Miscellaneous survey systems
Ocean survey systems

Signature management
Acoustic ranges
Magnetic ranges
Degaussing systems
Acoustic control

Training and simulation systems
Command team trainers
Mine warfare trainers
Submarine equipment trainers
Aircraft equipment trainers
Sonar equipment trainers
Targets

General navigation equipment

Submarine radar

Consoles and displays

Transducers

Miscellaneous equipment

ENVIRONMENTAL MEASURING SYSTEMS

All forms of warfare are governed by the surrounding environment, and nowhere is this more true than in the case of underwater warfare. The ocean is the largest of all environments and covers some 70 per cent of the earth's surface. At its deepest it reaches 11,000 m, with an average depth of 3,850 m. The ocean environment is the most difficult of any in which to conduct warfare, and its vagaries are such that even today there is a great deal that still remains unknown about how the sea behaves and how it is affected by the land and air within and above it. Every aspect of underwater warfare – submarine operations, anti-submarine warfare, mine warfare, diver operations and so on – is subject to the behaviour of the sea. It is for this reason that the science of oceanography has become so important to many of the world's navies.

The underwater environment differs in many ways from the land and air environments and is subject to quite different forces. For example, in seawater, radiation of certain parts of the electromagnetic spectrum is severely limited. This greatly restricts the use of visual, radar or other detection systems which use that part of the spectrum.

Conversely, sound waves can travel and be detected

over very long distances underwater and are therefore the most important factor relating to underwater warfare, for they form the primary means of detection. Certainly this is true at long range, although at a relatively short range other phenomena such as magnetism and electric potential phenomena are also very important, particularly in mine warfare.

The physical properties of the sea, and how they affect radiation, play a vital role in underwater warfare. For example, geographical considerations, local atmospheric conditions, the season, and the composition of the seabed (is it smooth or rough, does it undulate and by how much, what is its underlying structure, is it rock or is it a sediment of mud or sand?) all affect very considerably the nature of the environment and how it behaves. Knowledge of how the oceans are conditioned by the atmosphere above, how the masses of ocean water move, and how temperature and salinity are affected, for example, is essential to assess how the propagation of sound in water will be affected. Other factors affecting sound propagation in water relate to boundary layers, both in water movement (in different parts of the ocean, currents running in opposite directions can be found

one above the other) and temperature boundaries (as opposed to gradients), scattering effects, and ambient noise, as noise from shrimps, whales, dolphins and so on, is a constant source of interference.

Knowledge of how sound propagation is affected in the ocean environment can be used to exploit methods of detection, or to avoid detection and to limit the effects of uncontrollable noise generation. Using such knowledge, it is also possible to predict how sound propagation will be affected in a given set of circumstances, hence the probability of detection under those conditions can be predicted.

In mine warfare other phenomena such as magnetic fields, electric potential and so on are also subject to variations in the surrounding natural environment. Again the effects caused by the natural environment need to be known in order for sensors operating in these domains to be effectively used.

Various methods and equipment are available with which to investigate and measure the various physical factors which affect the underwater environment.

UPDATED

ENVIRONMENT MEASURING PROBES

BRAZIL

CTD probe

Type
Multiparameter oceanographic probe.

Description
The CTD probe is a microprocessor-based oceanographic probe fitted with a variety of sensors suitable for employment in a range of oceanographic applications.

With a case of cathodic-protected aluminium with redundant sealing, the CTD/M can be deployed from ships/platforms or in moored systems or mounted in special buoys. It is compatible with PC/XT/AT microcomputers.

The buoy features 10 W power CMOS technology and uses either a real-time cable data transmission system or a solid-state (static RAM) data storage

computer with 512 kbytes of memory. The buoy has a maximum operating depth down to 2,000 m. Additional sensors can be fitted at customer request.

Typical power consumption is 1.5 mA and the probe has a typical autonomy of one year.

The conductivity (C) sensor is an inductive cell with a range of 0 to 65 mmho/cm and an accuracy of ±0.02 mmho/cm. The temperature (T) sensor is of linearised platinum with a range of u5° to 45° C and an accuracy of 0.02°C. The depth (D) sensor is a semiconductor strain gauge with a range of 0 to 2,000 m and an accuracy of 0.5 per cent FS.

Contractor
Consub SA, Rio de Janeiro.

VERIFIED

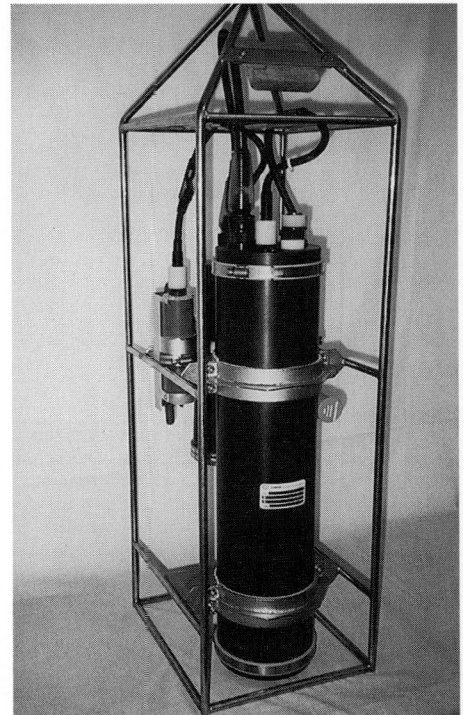

CTD probe in deployment frame
1995

COR

Type
Current meter.

Description
The COR 2000/M is a microprocessor-based current

meter using an electromagnetic sensor and fluxgate compass.

The case is manufactured in cathodic-protected aluminium with redundant sealing. It can be deployed from ships/platforms or in moored systems and special buoys. The COR 3/300 is compatible with PC/XT/AT microcomputers through an IEEE-485 interface.

The meter incorporates low-power CMOS technology and uses real-time cable data transmission or solid-state (static RAM) data storage.

The meter can operate down to 2,000 m depth in a temperature range of u5°C to 55°C, offering a measurement range from 0 to 300 cm/s current speed and through 0 to 360° for current direction. Accuracy is 3 per cent FS and ±1.5°. The maximum sample rate is 10 Hz (electromagnetic) and typical power consumption is 1 mA.

Contractor
Consub SA, Rio de Janeiro.

VERIFIED

Current meter COR 2000/M
1995

CANADA

AXOTD/XOTD

Type
Expendable optical temperature depth probes.

Description
The AXOTD/XOTD offer a low-cost solution for acquiring near-synoptic data for suspended particle concentrations and temperatures in the sea. The XOTD is fully deployable from underway vessels and the AXOTD is aircraft deployable. Both probes are fully functional for day/night deployments and data are easily collected and processed using the SOC multifunction PC expansion card.

The AXOTD is contained in a standard 'A' size sonobuoy housing with a parachute/surface float and transmitter electronics at one end, and the optical sensor/SOC temperature sensor at the other. The XOTD is a standard XBT canister.

The light scattering output is linearly correlated with suspended particle concentration in water and has high accuracy for the measurement of very low particle concentrations. Applications include pollution investigations, oceanographic research, visibility studies, laser communication experiments, turbidity measurements and so on.

Operational status
The AXOTD/XOTD are in development.

Specifications
Depth capability: 500 m
Fall rate: 6 m/s (in water) (approx)
Ship-launch speed: 15 kt
Temperature range: u2 to +35°
Particle concentration range: 5 µg to 3g
Sample rate: 12.5 Hz
Depth resolution at 6 m/s: 0.5 m
Operating depth: 500 m

Contractors
Sparton of Canada, London, Ontario.
Sea Tech Inc, Corvallis, Oregon, USA.

UPDATED

STD-12 plus

Type
Miniature conductivity, temperature, depth probe.

Description
The STD-12 Plus is a self-contained, intelligent CTD probe designed for precision measurement of conductivity, temperature and pressure in both fresh and saltwater. Its small size and low weight make it ideal for hand-hauled profiling from small vessels, while its rugged, hard anodised aluminium pressure casing allows deployment to depths of 5,000 m.

The STD-12 Plus features microprocessor-based, low-power, CMOS circuitry; a digital converter; its own set of programming commands as well as 128 kbytes of battery-protected, solid-state, non-volatile RAM memory. Two optional RAM modules of 512 kbytes are available for a potential total of 1.1 Mbytes of memory.

Designed to be used with an IBM-compatible computer, output from the probe is standard ASCII RS-232. The baud rate is automatically selected with a maximum of 19,200. Data output may be configured to display either unprocessed integers or computed engineering values.

The STD-12 Plus is supported by an MS-DOS-compatible software package, Total System Software (TSS) which allows the user to program the instrument for deployment, monitor the data in real time, download the memory, process and then column print and/or plot the data.

TSS consists of a DOS interface as well as the system software. The DOS interface is a program which allows an IBM-compatible computer to communicate directly with the STD-12. Once TSS is installed, the STD-12 can be accessed as if it was another disk drive.

TSS requires an IBM AT 286 or 386 computer or compatible, with a minimum of 640 kbytes of memory, DOS 3.0 or later, 20 Mbytes hard drive, a serial port and an EGA, VGA or monochrome monitor. A mouse and maths co-processor are optional but will greatly improve the performance and ease of use of the package.

The STD-12 is programmable for either continuous operation or for extended (moored) deployment of up to one year (in on-off mode). Programming options allow data to be logged continuously, at chosen depth increments, at chosen time increments or upon command from a surface computer or terminal.

Plotting options allow up to five parameters to be displayed on the same graph with the range and colour of each being user selectable.

The unit can support a maximum of eight sensor channels. Sample rates can vary from a maximum of 15 scans per second to a minimum of one scan per nine hours.

In standard configuration the STD-12 Plus is powered by alkaline D-cell (nine) batteries. Alternatively, the instrument can be powered by rechargeable (or lithium) batteries.

Contractor
Applied Microsystems Ltd, Sidney, British Columbia.

VERIFIED

SV Plus

Type
Miniature sound velocity, pressure and temperature probe.

Description
The SV Plus is an intelligent, miniaturised sound velocity probe which records high-quality sound velocity profiles through the water column to depths of 5,000 m. Sound velocity is measured directly using a time of flight velocity sensor with an accuracy of ±0.06 m/s, it is not calculated from other measured parameters.

The probe has both military and scientific applications since it can be used to calibrate active and passive listening systems, bottom sounders, side scan sonar systems and associated surveying equipment. A built-in microprocessor and program enable samples to be logged at user-chosen intervals of depth or for every change of sound velocity greater than one m/s.

The instrument has two modes of operation: a logging mode in which all recorded data are initially stored in the unit's internal RAM memory for subsequent recovery; and a real-time mode in which data is transmitted to the surface, displayed in real time and saved to disk using any IBM-compatible computer, while simultaneously being saved in the probe's internal memory.

The probe is supported by an MS-DOS compatible software package Total System Software, which allows the user to program the instrument for deployment, monitor the data in real time, download the memory, process and then column print or plot the data.

Contractor
Applied Microsystems Ltd, Sidney, British Columbia.

VERIFIED

SC-12

Type
Solid-state salinity chain.

Description
The SC-12 is a solid-state salinity chain designed to record time series of the vertical distribution of seawater temperature and conductivity from which salinity is calculated. A typical configuration consists of 12 temperature/conductivity sensor pods spaced along a 100 m pressure balanced cable, terminating in a solid-state data logger.

The data logger has a standard battery-protected solid-state RAM of 128 kbytes. Optional memory upgrades to 1.1 Mbytes are available which allow the user to recover up to 40,000 scans of all 12 temperature/conductivity sensor pods.

The chain is powered with standard alkaline batteries and features a 'sleep mode' that enables the electronics to be switched off between samples. This combination of 'sleep mode', power and memory allows the instrument to be deployed continuously for periods of up to 15 months. Communication with the SC-12 is via any IBM-compatible computer and any simple terminal program.

Contractor
Applied Microsystems Ltd, Sidney, British Columbia.

VERIFIED

Accurate Surface Tracker

Type
Expendable Langrangian drifting tracker.

Description
The Accurate Surface Tracker is specifically designed to track within the top metre of a water column. The lightweight tracker weighs approximately 18 kg dry and is configurable with a variety of sensors.

Contractor
Seimac Ltd, Bedford, Nova Scotia.

VERIFIED

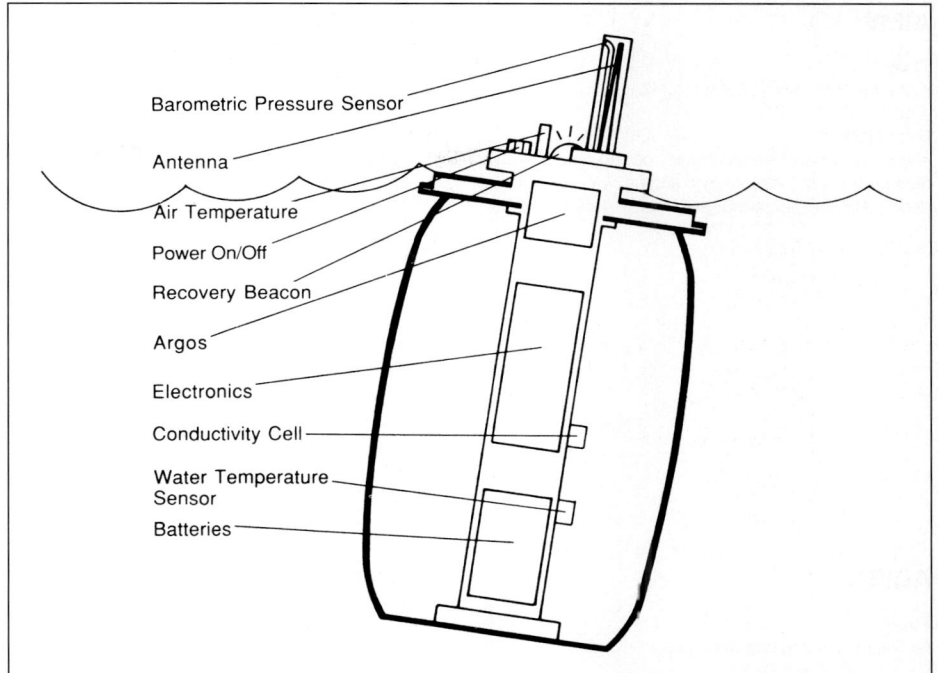

Accurate Surface Tracker

Nedal

Type
Navigation data collector.

Description
Nedal provides near real-time data collection of most shipborne navigation and environmental data at one central site, usually the ship's wheelhouse. Information typically needed for activities throughout the ship is thus available to all users either via a multiple RS-232 connection scheme or via a true network such as the scientific data highway. Data collected include time (GMT), depth, speed, heading, satellite navigation data, air temperature, humidity, sea temperature, windspeed, wind direction, air pressure and light density.

Contractor
Seimac Ltd, Bedford, Nova Scotia.

VERIFIED

RTM

Type
Remote Telemetry Module (RTM).

Description
The RTM is a totally self-contained telemetry module designed so that up to 32 RTM transmitters can be used with the same receiver module. A typical application for the RTM is the monitoring of ocean data, including buoys, where physical links are impractical.

Contractor
Seimac Ltd, Bedford, Nova Scotia.

VERIFIED

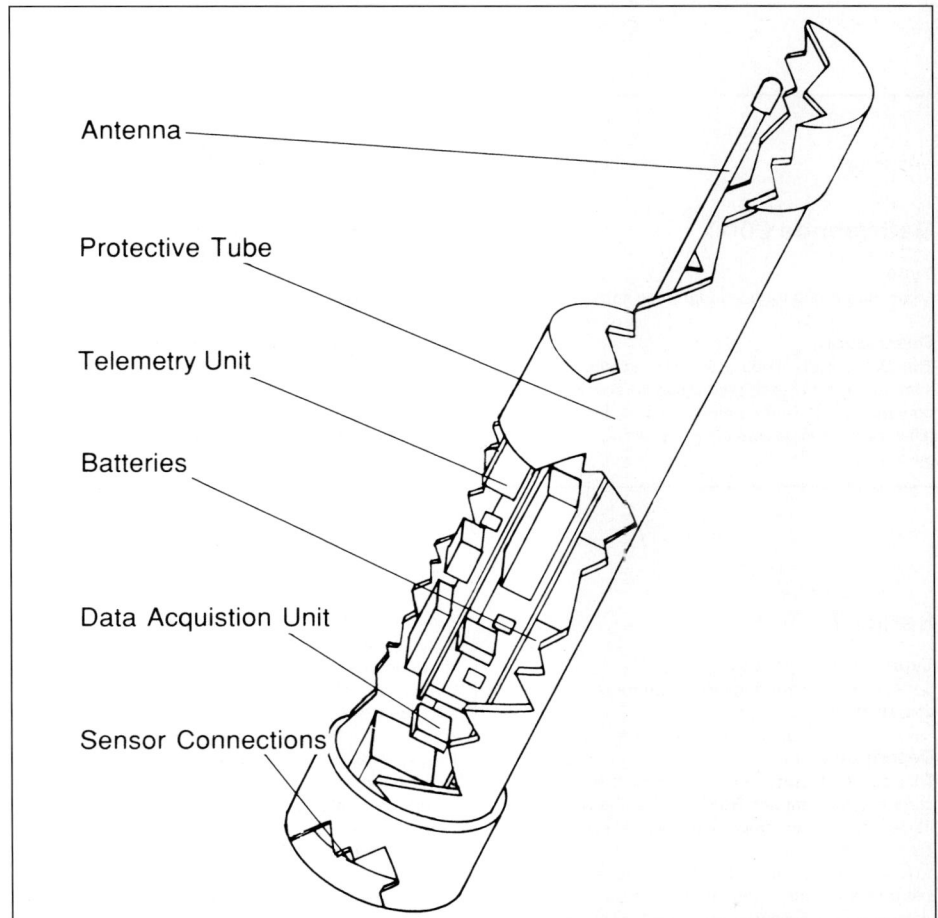

Remote Telemetry Module

AIMS

Type
Arctic Ice Monitoring System (AIMS).

Description
AIMS is capable of measuring air, ice and water column parameters and supporting strings for sensors at depth. The ice penetrator senses ice growth and ice temperature profiles, and can be supplied as a stand-alone sensor element. AIMS units are available with a large range of sensor suites for oceanographic and military applications.

Contractor
Seimac Ltd, Bedford, Nova Scotia.

VERIFIED

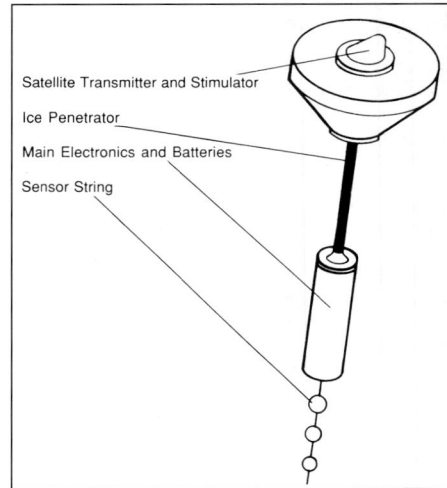

Satellite Transmitter and Stimulator
Ice Penetrator
Main Electronics and Batteries
Sensor String

Arctic Ice Monitoring System

ADIB

Type
Air Deployable Ice Beacon (ADIB).

Description
ADIB is one of a family of ice beacons. The ice-deployed version is designed to measure wind and other climate parameters, in addition to Argos satellite position, in Arctic areas. The ice beacon is floatable.

Contractor
Seimac Ltd, Bedford, Nova Scotia.

VERIFIED

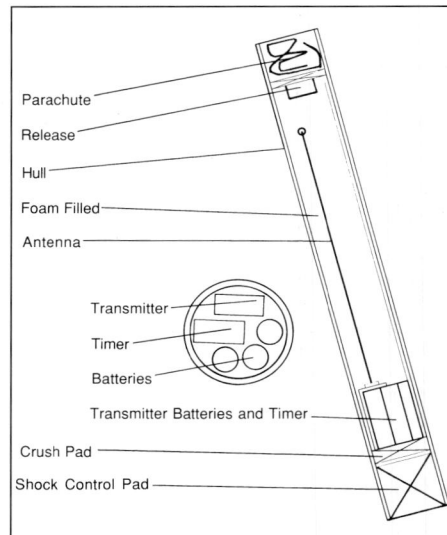

Parachute
Release
Hull
Foam Filled
Antenna
Transmitter
Timer
Batteries
Transmitter Batteries and Timer
Crush Pad
Shock Control Pad

Air Deployable Ice Beacon

GERMANY

Bathysonde 2000

Type
Measuring probe for oceanographic applications.

Description
The Bathysonde 2000 HS is an *in situ* measuring instrument for rapid recording of specific electric conductivity (C), temperature (T) and depth/pressure (D) in water. It is designed for use with an internal data recording system. A particular feature is the high measuring rate of 250 CTD values per second which, at a sinking rate of 4 m/s, reduces to a minimum the time a ship has to remain in position. The sensors measuring conductivity and temperature are designed to require only 5 ms to respond, with a spatial resolution capacity of 1 cm. The equipment can be operated down to a maximum depth of 6,000 m.

The Bathysonde 2000 LS is another high-precision *in situ* measuring instrument for recording of the same values. If fitted with fast CTD sensors, as in the 2000 HS, a data rate of 32/64 data sets or more can be achieved.

Operational status
In production.

Contractor
Salzgitter Elektronik GmbH, Flintbek.

VERIFIED

NORWAY

Series 7

Type
Series of oceanographic instruments for data collection.

Description
This range of instruments consists of the recording current meter model RCM 7, water level recorder model WLR 7 and temperature profile recorder model TR 7.

The RCM 7 and RCM 8 are self-contained instruments for recording speed, direction, temperature, pressure and conductivity of ocean currents. The RCM 7 can be moored to the seabed and record ocean data. It comprises a recording unit and vane assembly which is equipped with a rod that can be shackled to the mooring line. This arrangement allows the instrument to swing freely and align itself in the direction of the current. The recording unit contains all sensors, monitoring systems, battery and a detachable, reusable solid-state data storing unit. A built-in clock triggers the instrument at preset intervals and a total of six channels is sampled in sequence. The first channel is a fixed reference reading for control purposes. Channels 2, 3 and 4 represent measurements of temperature (three selectable ranges between u2.46 and 36.04°C plus an Arctic range (u2.46 to 5.62°C to special order), conductivity (0 to 74 mmho/cm standard with two other ranges available on request) and depth (down to 2,000 m) respectively; channels 5 and 6 represent measurement of vector averaged current speed and direction since the previous triggering of the instrument. The data are sequentially fed to the memory unit.

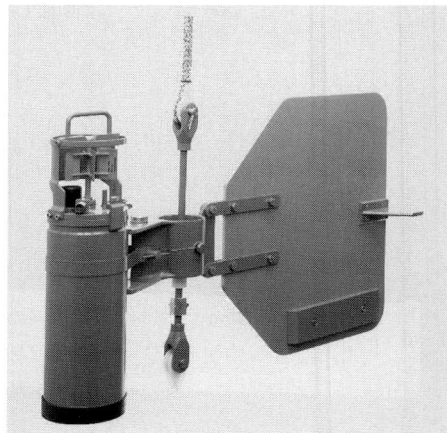

The RCM 7 vector averaging current meter

Simultaneously with the taking of the reading, the output pulse keys on and off an acoustic carrier transmitted by a transducer. This allows monitoring of the performance of moored instruments from the surface by a hydrophone and can be used for real-time telemetry of data.

The RCM 8 is identical to the RCM 7 except that it is a high-pressure model for measurement to depths of 6,000 m.

The RCM unit can be moored in one of two ways: a U-mooring which is best suited for use in relatively shallow water and the I-mooring involving the use of an acoustic release device and which can be used at any depth. It is essential that a subsurface float is used to avoid wave-induced motion on the mooring line.

The WLR 7 water level recorder is a self-contained high-precision instrument which is placed on the seabed to record water level by precise measurement of hydrostatic pressure, as well as measuring temperature and conductivity. The WLR averages the pressure over a period of 40 seconds in order to eliminate fluctuation in water level due to waves. When this integration time is completed the data words are recorded. The first data word is a fixed reference reading followed by the temperature of the ambient water. Pressure is recorded as two 10-bit words and finally a 10-bit word for the conductivity (optional sensor). Data are transmitted via an acoustic transducer to a hydrophone receiver hung from a relay buoy at the surface. The buoy then transmits the received data via VHF link to a shore base or parent ship.

The temperature profile recorder TR 7 is a fully self-contained instrument for recording temperature profiles in the sea, lakes or fjords. The vertical temperature profiles are used by oceanographers to study internal waves and other physical phenomena. In some applications *in situ* recording of data is required, while for other applications telemetry of data to shore or ship is required, and both functions can be provided by the TR 7. The profile recorder consists of a 12 channel recording unit and a thermistor string. The thermistor string employs 11 temperature sensing thermistors embedded in a polyurethane cable. The thermistors are spaced throughout the cable which can be from 5 up to 400 m maximum in length. Four different temperature ranges are available.

Contractor

Aanderaa Instruments, Bergen.

VERIFIED

RCM 9

Type

Recording current meter.

Description

Designed for use in both salt and fresh water, the RCM 9 recording current meter is based on Aanderaa's newly developed Doppler Current Sensor (DCS) which measures horizontal current speed and direction. The unit incorporates a new Hall-effect compass and a two-axis tilt sensor which permits accurate measurements at inclinations up to 35° from the vertical.

The unit is free from electrodes which are easily fouled, and features very low power consumption and measures current outwards from the sensor, reducing influences created by turbulence shed from flow around the instrument. The unit also features sensors for the profiling of temperature, conductivity, instrument depth and turbidity along with user selections for fresh water conductivity and Arctic temperature ranges. An optional sensor to measure, for example, dissolved oxygen may also be installed.

The RCM 9 can be deployed at depths of 2,000 m and will operate unattended for more than two years at 60 minute recording intervals.

Data is stored internally in a standard data storage unit (DSU 2990) or transmits data in real time via cable or acoustic telemetry.

The most common way to use the RCM 9 is in an in-line mooring configuration. As it operates under a tilt up to 35° from the vertical, it has a variety of in-line mooring applications using surface buoys or subsurface buoys. The instrument is installed in a mooring frame that allows easy installation and removal of the instrument without disassembly of the mooring line.

It is intended that the RCM 9 will, in time, replace mechanical rotor-type current meters.

Contractor

Aanderaa Instruments, Bergen.

VERIFIED

UNITED KINGDOM

Tidegauge

Type

Tidal measuring gauge.

Description

The Type 7751 telemetering offshore Tidegauge is a self-contained, precision pressure recording instrument intended for deployment on the seabed for short or long-term tidal measurements anywhere on the continental shelf to depths of 275 m or greater. The use of a 'DigiQuartz' transducer with temperature compensation provides accurate pressure measurements. The unit incorporates an acoustic release transponder facility. This allows the instrument to be moored on the seabed, supported by a flotation collar, without any surface marker buoys. The instrument can be relocated by acoustic ranging, and recovered by transmitting a secure command to drop the anchor weight. The tidal pressure data recorded in the instrument can be recovered while the instrument is still on the seabed, by acoustic telemetry.

Testing, deploying, relocating and extracting data and recovering the Tidegauge are achieved using a surface acoustic command system and control computer.

The unit uses a rechargeable battery which provides at least three months of recording life. An acoustic command is used to initiate logging, which is timed by the unit's internal clock/calendar. In the basic instrument, measurements are made according to a fixed programme, on the hour and at 10 minute intervals. Each measurement averages the 'DigiQuartz' pressure-dependant frequency output over a long period, to attenuate the effects of swell. The period may be preset up to a maximum of 8 minutes, with a default value of 4 minutes. The temperature-dependant frequency output is measured over a much shorter period and both the pressure and temperature frequency counts are used to calculate the absolute pressure in mbar which is stored in RAM with a date/time stamp. The basic instrument has a capacity of 50 days' logging with memory upgrade to 100, 200 or 400 days. During logging, battery usage is continuously calculated and logging stopped if capacity falls to a level where recovery might be endangered.

Operational status

In service with the Royal Navy.

Contractor

Sonardyne International Ltd, Yately.

VERIFIED

UNITED STATES OF AMERICA

XCTD

Type

Expendable conductivity, temperature and depth profiling system.

Description

The XCTD is designed to collect salinity profiles up to 1,000 m, while under way, from dedicated research vessels and craft of opportunity. The system comprises an expendable probe that measures conductivity and temperature, the Sippican Mk 12 data acquisition system, which is an expansion card that can be inserted into PCs, and a launcher.

The system can collect conductivity measurements to 0.03 milli-Siemens (mS/cm) and temperature to ±0.035°C. These measurements are sampled at a rate that provides a vertical resolution of 80 cm at a depth accuracy of ±5 m or 2 per cent of depth, whichever is greater.

The XCTD probe contains a conductivity cell, a thermistor, electronics and a BT wire link to the surface. The conductivity cell is a high-purity, alumina ceramic tube wired to form a four-electrode conductivity sensor. The cell is potted in a glass-filled, rigid epoxy compound.

The temperature sensor is a very stable glass

PERFORMANCE CHARACTERISATION
Nominal Accuracy to Standard*

Temperature		Conductivity		Depth		Vertical
Range	Accuracy**	Range	Accuracy**	Range	Accuracy***	resolution
u2.2-30°C	±0.035°C	20-75 mS/cm	±0.035 mS/cm	0-1,000 m	±5 m or 2% of depth	1 m

* Accuracy of temperature and conductivity standards used in XCTD calibration is ±0.005°C and mS/cm
** Nominal accuracy characterisation based on XCTD horizontal profiles against a calibrated transfer CTD (each comparison used 4 pt smoothing). 95% of tabulated data was within ±0.035°C and mS/cm of the transfer CTD
*** Drop rate characterisation has been conducted on vertical profiles referenced to a CTD. Maximum offset at 91% of full depth (910 m) for pairs of simultaneously launched probes was ±1.4%

encapsulated, fast-response temperature-cycled thermistor. The cell electronics package converts the measured resistance on the thermistor and conductivity cell into a frequency which is sent up the BT wire to the acquisition circuitry in the Mk 12.

Included in the probe electronics are two precision calibration resistors which are periodically sampled and their output transmitted up the BT wire in conjunction with the temperature and conductivity data to compensate for any induced changes in the electronics during the deployment.

Operational status
The XCTD has undergone extensive at-sea testing to verify its accuracy. The probe has been in full production since early 1994 and over 6,000 units have been deployed to date.

Specifications
XCTD-air launch version, AXCTD also available.

Contractor
Sippican Inc, Marion, Massachusetts.

VERIFIED

XCP profiler

Type
Horizontal ocean current profiler.

Description
The XCP is an expendable instrument capable of obtaining profiles of horizontal current direction and speed. The system obtains ocean current data down to depths of 1,500 m with an accuracy of ±1 cm/s RMS. The data acquisition and processing system samples eight parameters: north velocity component, east velocity component, temperature, rotational frequency, electric-field baseline, magnetic compass baseline, area ('tilt' correction) and velocity error. It may be deployed from either a surface or airborne platform since it uses an RF link to transmit data. Three channels are available for RF transmission: 170.5, 172 and 173.5 MHz (standard US Navy sonobuoy channels 12, 14 and 16). A complete XCP system includes expendable probes, a Sippican Mk 10 digital data interface, and any MS-DOS-compatible computer with IEEE-488 support.

The XCP has been used in studies of the mixed layer, ocean fronts, Gulf Stream rings, mid-ocean eddies, internal waves and acoustic propagation.

Operational status
Developed under the auspices of the US Naval Oceanic Research and Development Activity (now the office of Naval Research, Stennis Space Center) – over 5,000 XCPs have been successfully deployed. XCPs have been used in a number of oceanographic programmes and experiments, by agencies such as the National Oceanic and Atmospheric Administration (NOAA), the Office of Naval Research (ONR), and the National Science Foundation (NSF). An aircraft deployed XCP is also available.

Contractor
Sippican Inc, Marion, Massachusetts.

VERIFIED

RO-308/SSQ-36 data recorder

Type
Bathythermograph data recorder.

Description
The recorder is an integral part of the US Navy P-3C Orion aircraft ASW system. The equipment converts seawater temperature information provided by the AN/SSQ-36 bathythermograph buoy transmitter set and AN/AAR-72 radio receiving set to two output forms:

(a) a permanent record of the vertical temperature profile (temperature versus depth) on a paper strip chart
(b) a parallel mode, eight-bit binary-coded data word for delivery to the AN/AYA-8B data processing system.

The buoy is dropped from an aircraft in the target area. Seawater is utilised as the activating agent and after an initial, predictable delay, the buoy releases a temperature sensing probe (TS probe). The TS probe is the variable element in a frequency generation circuit. A radio frequency signal transmitted by the buoy is modulated at a frequency correlated to the temperature of the water. On board the aircraft, the ARR-72 radio is tuned to the buoy carrier frequency. Water temperature information is converted to an audio frequency signal and delivered to the recorder.

Contractor
Lockheed Martin, Archbald, Pennsylvania.

VERIFIED

XBT/XSV bathythermograph system

Type
Expendable temperature/sound velocity probes.

Description
The Sippican XBT/XSV system uses inexpensive, expendable probes to obtain vertical profiles of temperature and sound velocity. The temperature data obtained by XBT probes may be used to compute sound velocity, although the XSV probes measure sound velocity directly. Whether computed or measured directly, the sound velocity data obtained assists in the prediction of sound propagation and is crucial to the effective use of sonars. An XBT/XSV system includes XBT or XSV probes, a hand-held, deck-mounted or through-hull launcher, and a Mk 8 microprocessor-based data processor/recorder.

The temperature data obtained by the XBT may be used to compute sound velocity wherever salinity is constant. In areas where salinity may be highly variable, such as the Mediterranean and the Arctic, temperature data may not be sufficient to calculate velocity and to predict adequately acoustic propagation. In these areas sound velocity must be measured directly with the XSV probe. The XSV uses a unique sing-around transducer to obtain this direct measurement.

Both XBT and XSV probes obtain data in real time. Following launch, data collection begins at the surface and as the probe descends, wire unreels from a spool located within the probe, as well as unreeling from another spool in the launch canister. This enables the

Specifications

XBT	T-4	T-5	Fast Deep™	T-6	T-7	Deep Blue	T-10	T-11
Applications	standard probe used by the US Navy for ASW operations	deep ocean scientific and military applications	provides max depth capabilities at the highest possible ship speed of any XBT	oceanographic applications	increased depth for improved sonar prediction in ASW and other military applications	increased launch speed for oceanographic and naval applications	commercial fisheries applications	high resolution for US Navy mine counter-measures and physical oceanographics applications
Max depth	460 m 1,500 ft	1,830 m 6,000 ft	1,000 m 3,280 ft	460 m 1,500 ft	760 m 2,500 ft	760 m 2,500 ft	200 m 660 ft	460 m 1,500 ft
Rated ship speed*	30 kt	6 kt	20 kt	15 kt	15 kt	20 kt	10 kt	6 kt
Depth resolution	65 cm	65 cm	65 cm	65 cm	65 cm	65 cm	65 cm	18 cm

XSV	XSV-01	XSV-02	XSV-03
Applications	ASW application where salinity varies; naval and civilian oceanographic and acoustic applications	increased depth for improved ASW operation where salinity varies; naval and civilian oceanographic and acoustic applications	high-resolution data for improved mine counter-measures and ASW operations in shallow water, geophysical survey work, commercial oil industry support
Max depth	850 m 2,790 ft	2,000 m 6,560 ft	850 m 2,790 ft
Rated ship speed*	15 kt	8 kt	5 kt
Depth resolution	32 cm	32 cm	10 cm

System depth accuracy = 4.6 m or 2% of depth (whichever is greater)
*All probes may be used at speeds above rated maximum. However, there will be a proportional reduction in depth capability

ship to use the system without restrictions on speed or manoeuvrability.

The Mk 8 data processing system, US Navy nomenclature AN/BQH-7, is linked to the descending probe by the hard wire link. The processor provides profiles of temperature or sound velocity to three outputs: digital data stored on magnetic tape, analogue chart paper trace, and RS-232 digital interface to external data systems.

These probes may also be used with the IBIS – Integrated Bathythermal Information System (see later entry). This uses the Sippican Mk 12 data acquisition system in a ruggedised computer and is integrated with several prediction systems. This is in service with the Royal Navy as sonar 2090.

Operational status
In service.

Contractor
Sippican Inc, Marion, Massachusetts.

VERIFIED

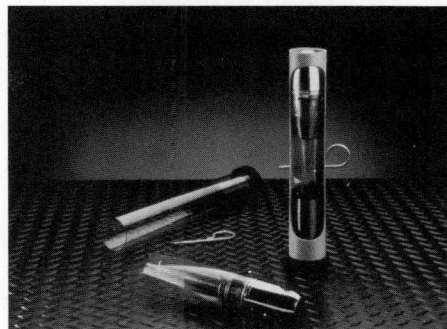

XBT/XSV expendable profiling system

SSXBT/SSXSV bathythermograph

Type
Submarine-launched expendable bathythermograph.

Description
Profiles of temperature and sound velocity versus depth may be obtained by a submerged submarine using the submarine-launched expendable bathythermograph/sound velocity (SSXBT/SSXSV) equipment. This is similar to the XBT/XSV system described previously except that the probes are launched through the signal ejector of the submarine and are carried to the surface by a buoy. The buoy releases the probe just before reaching the surface, and the probe then descends, measuring temperature or sound velocity on the way. The SSXBT expendable probe senses the water temperature profile from the surface to a depth of 760 m and transmits this to the moving submerged submarine. The SSXSV sound velocity probe provides a profile of measured sound velocity down to 850 m. Temperature or sound velocity data are transmitted to the submarine, via the fine wire tether, for recording and display.

The breech door/cable feed-through assembly provides the inner door for the signal ejector. It allows the tether wire to pass through the door and provides a seal around the wire. At the end of probe deployment a guillotine in the assembly cuts the tether wire allowing the SSXBT components to clear the submarine and scuttle.

The AN/BQH-7 and Mk 8 recorders/processors provide temperature and sound velocity profiles. The processor can convert measured temperature to compute sound velocity based on a standard salinity of 35 psu.

Operational status
In service.

Specifications
SSXBT
Range: u2.2 to +35.6°C
Accuracy: ±0.15°C
Depth range: 0-760 m
Depth accuracy: ±2%
Size: 97.6 × 7.6 cm

SSXSV
Range: 1,405-1,560 m/s
Accuracy: ±0.25 m/s
Depth range: 0-850 m
Depth accuracy: ±2%
Size: 97.6 × 7.6 cm

Contractor
Sippican Inc, Marion, Massachusetts.

VERIFIED

SSXCTD

Type
Submarine-launched conductivity, temperature and depth probe.

Description
An adaptation of the XCTD probe for launching from submarines. Salinity and density may be derived from the conductivity and temperature data. This is especially useful for submarine operations in areas where salinity is non-standard.

Operational status
In production.

Contractor
Sippican Inc, Marion, Massachusetts.

VERIFIED

AN/SSQ-36B

Type
Expendable bathythermograph.

Description
The 'A' size AN/SSQ-36B bathythermograph provides vertical temperature profiles of the ocean layer for research purposes, and for ASW evaluates local effects of seawater temperature on sonar propagation and acoustic range prediction.

The buoy provides 800 m temperature profiles and data can be transmitted on any one of 99 VHF channels. Thermistors located in the probe measure the changes in seawater temperature during the buoy's descent from the surface. An RF transmitter sends this data to the launch aircraft. On board the aircraft the data is processed and displayed as a temperature versus depth display.

The bathythermograph can be air launched at air speeds of 0 to 370 kt and at altitudes up to 9,144 m. Air descent is controlled and stabilised by a parachute that deploys when the buoy exits the aircraft.

Power is provided by five polycarbon monoflouride lithium cells which offer improved operational capability in sub-zero temperature seawater.

Operational status
In service with the US Navy.

Specifications
Length: 91.4 cm
Diameter: 12.4 cm
Weight: 7.3 kg
Descent rate: 1.5 m/s
Sensor temperature range: −2°C to +35°C
Temperature accuracy: ±0.55°C
Channels: 99 RF
RF Power: 0.25 W
Operating depth: 800 m

Contractor
Sparton Electronics, DeLeon Springs, Florida.

NEW ENTRY

HSCP-600

Type
Harbour surveillance current profiler.

Description
The HSCP-600 uses digital signal processing and wideband acoustic signals to measure water currents with high accuracy. The device is a real-time data gathering instrument, suitable for coastal, port or inland waterways applications where direct connection to the shore is required.

It is primarily designed for short-range, high-resolution applications, and with the addition of optional sensors, including a fifth vertical beam, it can also be configured to measure wave and tide variations, temperature and pressure.

The HSCP-600 uses a 600 kHz wideband signal with digital signal processing offering 50 m range, variable cell size and high resolution. It features a fast update rate (up to 4 pulses per second) and the transducer

Model HSCP-600 acoustic current profiler **1996**

uses four beams each 3° in width at an angle of u30° from the vertical in 90° azimuthal increments. HSCP-600 also incorporates a temperature sensor.

The data acquisition cycle is versatile and adaptive and provides a binary data output.

The pressure housing/transducer is constructed of hard anodised aluminium, reinforced plastic composite or stainless steel coated with anti-fouling paint, and is suitable for long-term deployment in harsh environments.

Optional features include a pressure sensor, right angle transducer kit, flux gate compass and dual-axis tilt sensors.

Specifications
Profiling range: 50 m (85% of distance of bottom or surface interface)
Current speed range: ±10 m/s
Depth cells: up to 128
Precision: 1 cm/s with 4 m cell
Wideband signal: 600 kHz ±40 kHz bandwidth
Sound power level: 213 dB re 1µPa at 1 m, programmable

Contractor
EDO Acoustic Products, Salt Lake City, Utah.

VERIFIED

Neptune RSVP/CTD-1020

Type
Sound velocimeter, self recording.

Description
The Neptune RSVP/CDT is a precision instrument for the direct measurement of sound velocity as well as conductivity, temperature and depth. The CTD measurements can be used for quality control of sound velocity measurements as well as for depth profiles of all parameters.

The RSVP/CDT uses the Ocean Data Equipment Corporation Model 1020 sing-around sound velocity sensor coupled with computer control for data logging and improved calibration, along with calibrated conductivity, temperature and pressure sensing.

The RSVP/CTD is packaged in a high pressure housing, with a standard 10,000 PSI/6,900 m capability. The system is entirely self-contained so that profiles with no electrical connections are possible. It includes 1/2 Mbyte of data storage and rechargeable NiCd batteries capable of 18 hours of profiling with 2 hours recharge. The Neptune RSVP/CTD is ideally suited for high accuracy sound velocity and CTD profiles for swath sonar mapping and for oceanographic surveys. It allows comparison of directly measured sound velocity with sound velocity calculated from conductivity, temperature and depth measurements from its own sensors. It can be used with or without a winch, and with or without electromechanical cable for deep or shallow water profiling. The unit is self-powered and self-recording with menu-driven sampling and setup options. Data is easily downloaded via a serial port and displayed using standard spreadsheet software.

Operational status
The product was introduced in 1997.

Contractor
Ocean Data Equipment Corporation, E Walpole, Massachusetts.

VERIFIED

ENVIRONMENT MEASURING BUOYS

BRAZIL

Oceanographic and meteorological buoy

Type

Metocean environment monitoring buoy.

Description

This buoy is designed to carry out metocean environment monitoring using satellite real-time

Oceanographic and meteorological buoy
1999/0024840

transmission. The buoy is constructed of naval aluminium and is designed for installation in positions where depths may extend down to 2,000 m. An extension of measuring depth to 5,000 m is available on request.

The solid-state electronic buoy incorporates a detachable 2.5 Mbyte memory while the receiver station can be equipped with non-directional antenna which is compatible with PC, XT and AT microcomputers.

Parameters measured by the buoy include PH value, salinity, temperature, conductivity, dissolved oxygen, current direction and current intensity. Meteorological measurements include atmospheric pressure, temperature, wind direction and wind intensity. Water measurements include wave directional spectra and wave statistics.

Telemetry is by satellite through the INMARSAT STD-C system, and there is no range limitation for the buoy/receiver distance.

Specifications
Diameter: 3 m
Height: 6 m
Weight: 1,500 kg
Measuring depth: 2,000 m (5,000 m on request)

Contractor
Consub SA, Rio de Janeiro.

UPDATED

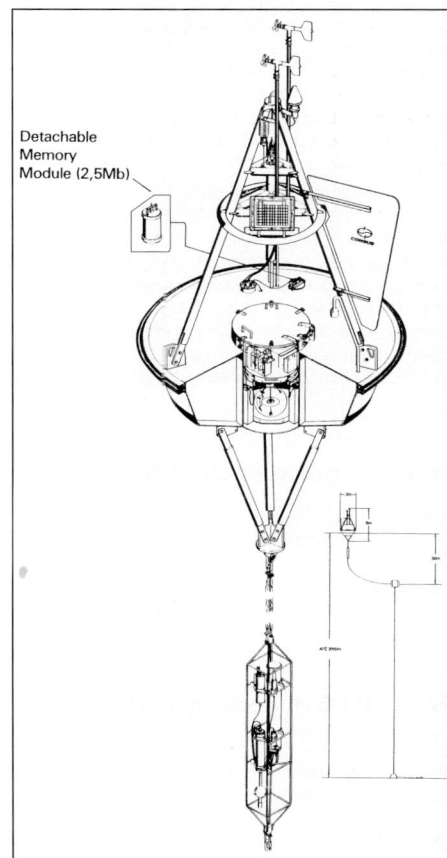

Detachable Memory Module (2,5Mb)

Oceanographic buoy
1996

CANADA

CMOD

Type

Compact Meteorological and Oceanographic Drifter (CMOD).

Description

The CMOD is a low-cost, easily transported and deployed 'A' size expendable Argos drifting buoy. Constructed of heavy gauge marine grade aluminium,

the buoy is compatible with the Argos satellite system. Standard onboard sensors provide for barometric pressure measurement, air temperature measurement, sea surface temperature measurement, battery voltage and data time history. The CMOD is a standard 'A' size sonobuoy and can be deployed by hand from virtually any size of aircraft or ship. The buoy is packaged for gravity tube launch from patrol aircraft at an indicated airspeed of 300 kt.

Specifications
Length: 91.4 cm (packaged)
Diameter: 12.2 cm
Weight: 12.73 kg

Contractor
Metocean Data Systems Ltd, Dartmouth, Nova Scotia.

VERIFIED

CMOD/Waves buoy

Type

Drifting buoy.

Description

The CMOD/Waves drifting buoy is a low-cost, expendable, standard 'A' size sonobuoy device designed to measure directional sea surface wave characteristics, process this data and telemeter results via Service Argos. The device can be transported and deployed from either aircraft or ship. In addition to the standard CMOD sensors, the buoy uses accelerometers, tilt sensors and a compass to determine directional wave information. Wave data is

sampled at 5.12 Hz for 800 seconds every hour and mean directional wave data is then computed.

Data is telemetered to the user via the integral MAT 906 Argos PTT using the NOAA Tiros satellite system. Meteorological data is sampled every 90 seconds and averaged for 10 minutes. New meteorological data is updated every 10 minutes, while wave data is updated once every hour.

The buoy housing is constructed of heavy gauge marine aluminium and flotation is provided by a gas-inflated collar attached to the top of the buoy housing.

Specifications
Air temperature measurements: u30 to +46.5°C

Sea surface temperature measurements: u5 to +35.8°C
Barometric pressure: 850-1,054.6
Tilt 2 axis: u70 to +70°
Accelerations 3 axis: u1.5 to +1.5
Operational lifetime: approx 20 days
Length: 22 in
Diameter: 4 in
Weight: 16 lb

Contractor
Metocean Data Systems Ltd, Dartmouth, Nova Scotia.

VERIFIED

Ice Drifter

Type

Ice drift measurement buoy.

Description

The Ice Drifter is designed and constructed to be a

through-ice position measuring platform, capable of supporting a wide variety of optional sensors. Upon ice break up/melt, it functions as a normal drifter. Sensor options include measurement of ice surface temperature, air temperature, subsurface temperature, barometric pressure, windspeed and direction, and current speed and direction.

Contractor
Metocean Data Systems Ltd, Dartmouth, Nova Scotia.

VERIFIED

CALIB

Type
Compact Air-Launched Ice Beacon.

Description
The CALIB is designed to provide a compact, lightweight and low-cost vehicle, capable of providing position information via the Argos satellite system. The

CALIB has a sonobuoy 'A' size configuration, which enables air launch from any type of aircraft, both fixed- and rotary-wing. Additional sensors to the standard position package are available. Options include barometric pressure sensor and surface temperature sensor and battery voltage.

Specifications
Length: 91.4 cm

Diameter: 12.4 cm
Weight: 8 kg

Contractor
Metocean Data Systems Ltd, Dartmouth, Nova Scotia.

VERIFIED

ANS

Type
Acoustic ambient noise sensor.

Description
The ANS is a flexible unit for use in systems requiring collection of underwater acoustic noise using one third octave filters. The unit comprises an ultra low-noise remote hydrophone with integral preamplifier, up to 300 m of user-supplied twin conductor cable and a Data Acquisition and Processing Unit (DAPU). The

DAPU has programmable filters, an integrating analogue to digital converter and a digital processor. It uses an MC146805 microcontroller, with calculations being performed with 16-bit floating point arithmetic.

The sensor is used to monitor surface wind conditions, rainfall studies, noise characterisation and ice cracking.

The device operates in the 10 Hz to 25 kHz frequency range.

Various options are available including: integral Argos PTT; high-resolution output; input of additional data; frequency counter inputs; voltmeter inputs; real-

time clock; internal calculations for windspeed, rain rate and data quality available.

Specifications
Diameter: 4.5 in (case); 1.5 in (hydrophone)
Height: 5 in (case); 4 in (hydrophone)

Contractor
Metocean Data Systems Ltd, Dartmouth, Nova Scotia.

VERIFIED

Ambient noise drifting buoy

Type
Acoustic measuring device.

Description
The ambient noise drifting buoy has been developed to

provide long-term statistical data on ocean ambient noise. Background noise is monitored in a number of frequency bands, spectral density levels are calculated and the data are telemetered via the Argos satellite system. A sophisticated suspension system ensures that self-generated noise from the buoy is minimised.

Contractor
Seimac Ltd, Bedford, Nova Scotia.

VERIFIED

AN/WSQ-6 mini-buoy series

Type
Multiparameter, satellite reporting, mini-buoy series for tactical oceanographic warfare support.

Description
All of these buoys are 'A' size and are air deployable.

They form the next-generation CMOD. The XAN-1 is similar to the standard CMOD; the XAN-2 has an acoustic sensor at 100 m; the XAN-3 has a thermistor string with a thermistor chain down to 100 m; the XAN-4 is a combination of XAN-2 and XAN-3; the XAN-5 has a windspeed and wind direction sensor; the XAN-6 is a wave buoy; the XAN6-M is a moored wave buoy.

Contractor
Metocean Data Systems Ltd, Dartmouth, Nova Scotia.

VERIFIED

FRANCE

Wadibuoy

Type
Wave directional buoy.

Description
The Wadibuoy is a comprehensive meteo/oceanographic real-time instrumentation system. It provides the user with directional spectra of the wave energy, the surface current speed and meteorological parameters such as windspeed, temperature, pressure, solar radiation and so on.

Wadibuoy is a flat-bottom toroidal buoy providing great stability and ability to remain on the surface of the wave. The buoy carries a protected radio transmission system for coastal application at ranges up to 30 km, and a satellite transmission system (Argos/Goes) for oceanic application.

Operational status
The buoy is in series production following five years of field testing in various oceans around the world. In excess of 30 buoys are in use in over 10 countries. It is used by the French Navy for determining the environmental conditions during firing trials, calibration of mathematical models which take into account wave frequency and oceanographic research.

Specifications
Diameter: 2.45 m
Height: 2.5 m
Weight: 550 kg (unequipped)
Buoyancy: 3,000 kg

Contractor
Société Nereides, Les Ulis.

VERIFIED

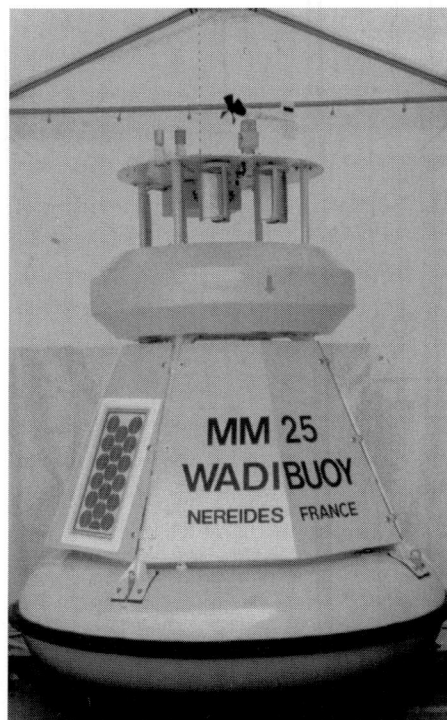

Wadibuoy wave directional buoy

NORWAY

DCM 12

Type
Doppler current meter.

Description
The DCM 12 current meter is designed specifically to measure currents in shallow harbour and coastal waters. The high-technology device is easy to use and handle and can be deployed without the need for divers, underwater foundations and heavy installation equipment.

The meter measures current speed and direction at the surface and five other depths, as well as measuring the depth of water and significant wave height.

The unit consists of an acoustic transducer head fitted on top of a recording unit for measuring the horizontal speed and direction of current. The transducer head has four acoustic transducers angled 30° in 90° azimuthal increments for measuring the horizontal speed and direction of current. A central transducer transmits data to a hydrophone on the surface or shore. The recording unit, housed in a pressure case, contains the main electronics, data storage unit, battery and a compass which enables current direction measurements to be referred to magnetic north.

The unit is installed in a self-levelling mooring frame that automatically ensures correct installation on seabeds with up to 20° inclination.

The unit samples over a 10 minute period before data is presented. Oppositely located transducers work together, one transducer pinging two seconds after the other. Three seconds later the same cycle is repeated by the other pair of transducers. As the ultrasonic pulses travel towards the surface, echoes from particles and organisms in the water are scattered back to the transducers. Echoes received from the scatterers change frequency due to their movement in correlation with the sea current. By measuring this frequency change (the Doppler shift) the unit can calculate the speed of the particles, and hence the speed of the current.

When 240 samples of the current speed have been taken data is calculated, transmitted or stored. The calculation takes place internally and the DCM presents the vector average speed and direction as two 10-bit binary words for the surface and each of five uniform segments called depth cells. These cells overlap each other by 50 per cent. The cell size is one third of the DCM's deployment depth minus illegible zone of 4 m or 2 m for depths above or under 15 m. The pulse energy is set by a three-position depth setting switch relative to the deployment depth.

Simultaneously with the current measurements, an internal pressure sensor measures once each second the hydrostatic pressure in the water column above the instrument. Based on these measurements, the instrument calculates the water level and the significant wave height on the surface. Data is transmitted in real time via underwater cable or acoustically to a receiving station on shore or on board a ship for immediate display. Power can be supplied via underwater cable or by batteries. Data may also be stored by the meter in a data storage unit.

Total weight of the instrument with mooring frame is 38.9 kg in water.

Operational status
Over 80 units have been supplied worldwide since its launch in May 1994.

Contractor
Aanderaa Instruments, Bergen.

UPDATED

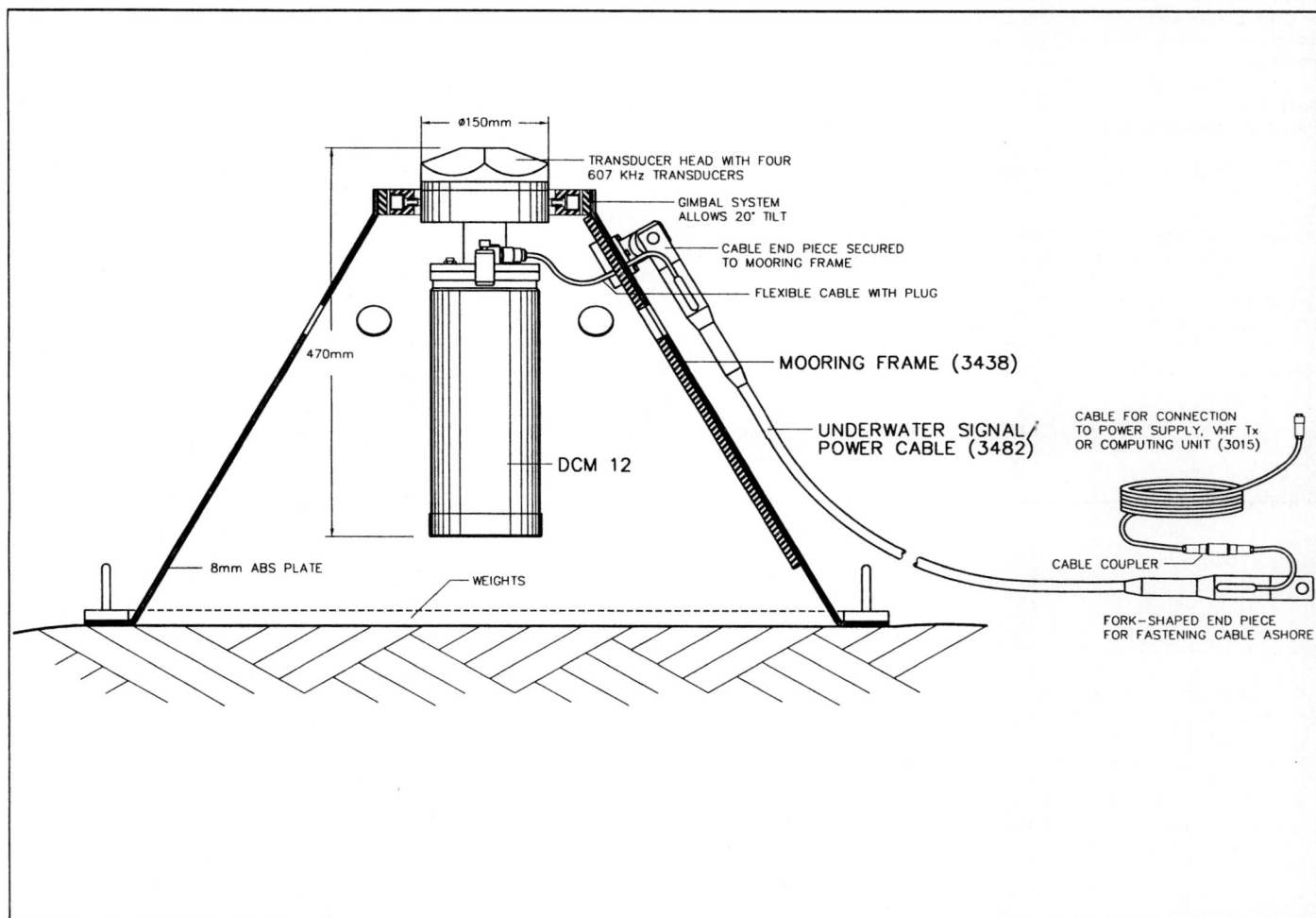

Ø150mm

TRANSDUCER HEAD WITH FOUR
607 KHz TRANSDUCERS

GIMBAL SYSTEM
ALLOWS 20° TILT

CABLE END PIECE SECURED
TO MOORING FRAME

FLEXIBLE CABLE WITH PLUG

470mm

MOORING FRAME (3438)

UNDERWATER SIGNAL/
POWER CABLE (3482)

DCM 12

CABLE FOR CONNECTION
TO POWER SUPPLY, VHF Tx
OR COMPUTING UNIT (3015)

8mm ABS PLATE

WEIGHTS

CABLE COUPLER

FORK-SHAPED END PIECE
FOR FASTENING CABLE ASHORE

DCM 12 current meter

1997

CMB 3280

Type
Coastal monitoring buoy.

Description
The rugged, lightweight and compact moored data buoy is intended for use along the coast and near off-shore platforms. It is designed to measure wave height, wave period, current speed, current direction and sea temperature, as well as the most important meteorological parameters and to transmit the information to the user in real time via VHF radio.

The device consists of a foam filled polyform buoy with a payload carrying the entire measuring system as well as all the sensors and the VHF radio transmitter. The payload can be removed from the buoy for service and maintenance without lifting the buoy out of the sea.

The buoy, which is moored in a fixed position, operates on solar cells. Its operation is controlled by the sensor scanning unit. An internal clock starts the measuring cycle every 10 minutes, reads the sensors and transmits the readings by radio in real time to the shore station or to a platform.

At the receiving end, the computing unit 3015 converts the raw data signals to engineering units which are subsequently displayed on an LCD or screen. On the screen, data are presented as a diurnal picture, which contains the last readings, the 3 hour average, and 24 hour maximum, minimum and average readings for all the parameters. Further relaying of data can be carried out by telephone both digitally and as a voice message.

Contractor
Aanderaa Instruments, Bergen.

UPDATED

CMB 3280 buoy
1997

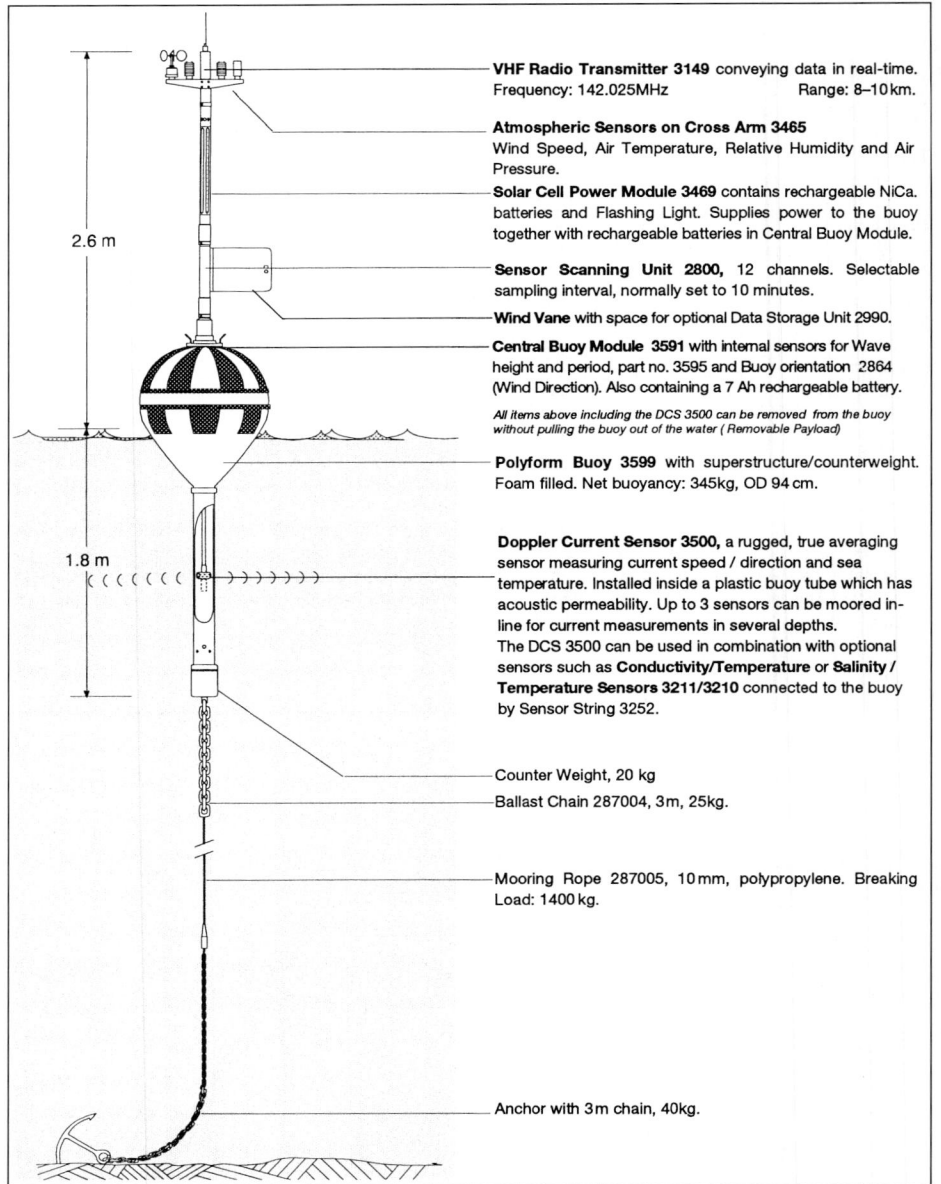

VHF Radio Transmitter 3149 conveying data in real-time.
Frequency: 142.025MHz   Range: 8–10km.

Atmospheric Sensors on Cross Arm 3465
Wind Speed, Air Temperature, Relative Humidity and Air Pressure.

Solar Cell Power Module 3469 contains rechargeable NiCa. batteries and Flashing Light. Supplies power to the buoy together with rechargeable batteries in Central Buoy Module.

Sensor Scanning Unit 2800, 12 channels. Selectable sampling interval, normally set to 10 minutes.

Wind Vane with space for optional Data Storage Unit 2990.

Central Buoy Module 3591 with internal sensors for Wave height and period, part no. 3595 and Buoy orientation 2864 (Wind Direction). Also containing a 7 Ah rechargeable battery.

All items above including the DCS 3500 can be removed from the buoy without pulling the buoy out of the water (Removable Payload)

Polyform Buoy 3599 with superstructure/counterweight. Foam filled. Net buoyancy: 345kg, OD 94 cm.

Doppler Current Sensor 3500, a rugged, true averaging sensor measuring current speed / direction and sea temperature. Installed inside a plastic buoy tube which has acoustic permeability. Up to 3 sensors can be moored in-line for current measurements in several depths.
The DCS 3500 can be used in combination with optional sensors such as **Conductivity/Temperature** or **Salinity / Temperature Sensors 3211/3210** connected to the buoy by Sensor String 3252.

Counter Weight, 20 kg
Ballast Chain 287004, 3m, 25kg.

Mooring Rope 287005, 10 mm, polypropylene. Breaking Load: 1400 kg.

Anchor with 3m chain, 40kg.

2.6 m

1.8 m

PREDICTION SYSTEMS

CANADA

Acoustic Range Prediction System (ARPS)

Type
Software acoustic prediction system.

Description
The ARPS is a software system designed to supply on-scene environmental support to ASW operations. Acoustic models have been developed which, when provided with environmental data such as temperature profiles, wave height and bottom reflectivity, compute acoustic transmission loss and thus predict sensor performance, expected ranges and detection probabilities.

Working from real-time or recorded data, the operator inputs source depth, receiver depth, parameters describing environmental conditions and a selection of acoustic frequencies. ARPS then computes propagation loss as a function of range for each frequency. Propagation loss curves are converted to probability of detection curves to obtain range information. Sound levels of source and ambient noise, detection threshold and standard deviation of signal excess are among the inputs.

These results can then be used to interpret where a sensor can detect a sound and where it is unable to do so because of oceanographic factors. The models have an extension beyond that of ASW applications and could be used to predict performance of any acoustic navigation, control or communication link or MCM operational conditions.

The database, which has been integrated into ARPS, includes ocean depth on a 5' (minutes of a degree) arc grid, and monthly temperature and salinity profiles on a 1° grid. Shipping density is also available on a 10° grid. The coverage of these is global. Monthly climatology features such as average windspeed are also included on a 10° grid for the North Atlantic and North Pacific.

Operational status
ARPS was developed for the Department of National Defence, Canada.

Contractor
Oceanroutes Canada Inc (a member of Swire Defence Systems), Halifax, Nova Scotia.

VERIFIED

Deployable Acoustic Calibration System (DACS)

Type
ASW research and calibration device.

Description
DACS consists of an onboard controller and power amplifier module with a deployable calibrated sound source. The system accurately generates low-frequency tones at operator-specified strengths and frequencies in actual operating conditions. These tones are received by the remote passive sensors being calibrated. From acoustic propagation loss and source strength, the signal level at the receiver can be accurately determined and thus provides a means of calibrating the passive sensor's sensitivity. The standard DACS comprises, at the wet end, a Model 18SA0325 low-frequency, wide bandwidth, depth compensated, ringshell projector with a matching network and a projector drive level feedback system.

The shipboard module comprises a winch and a PC-based controller and display terminal. The feedback subsystem monitors the projector's output and continuously transmits the formatted data to the control console which then makes the necessary adjustments.

The compact arrangement of DACS makes it suitable for installation on COOP vessels. The system can also be modified for moored applications.

Operational status
Systems delivered to Canadian Forces Maritime Command for field evaluation.

Specifications
Projector: RSP Model 18SA0325
Display: display terminal
Input device: keyboard

Number of tones: 4
Frequency accuracy: ±0.1 Hz
Frequency range [nominal]: 60-600 Hz
Frequency separation: 1 Hz
SPL* (nominal): 100-150 dB
SPL accuracy: ±1 dB
Operational depth: 3-300 m (10-1,000 ft)
Tow speed: up to 3 kt
Max sea state: 4
Power requirements: 220 V AC, 15 A 1 phase
*Not all frequency and SPL combinations available at all depths.

Contractor
Sparton of Canada Ltd, London, Ontario.

VERIFIED

FRANCE

Neptune

Type
Environmental tactical aid software.

Description
The modular Neptune system introduces a new concept of environmental tactical aid. The system contains different components, the complexity of which can be varied depending on a variety of customer requirements such as: type of vessel (submarine, surface ship and so on); integration into a combat management system; stand-alone configuration.

Through an intuitive GUI-based interface, Neptune can enable the user to assess his acoustic superiority or risk in the face of one or more hostile threats or targets. Operating modes range from fully automatic to operator specification of all variables.

The components of Neptune include: an environmental model (including measured, statistical and forecast data); ship and sensor parameters; acoustic confrontations (where the ship faces one or more threats in the current scenario); different ray tracing either in normal mode or using parabolic equation propagation loss models; and synthetic analysis of the acoustic situation from different graphical outputs – ray tracing, propagation loss (images and plots with panoramic views which are operator-selectable across various bearings) and signal to noise ratio and finally sophisticated tactical aid images. The propagation loss is displayed in global vision around the 'own ship' from which null points at all depths can be seen.

Among the various capabilities offered by Neptune are: operator selectable depth contours covering the whole world; an international database and the ability to read messages from shore bases using datalinks.

The displays can show surface temperatures for any geographical location and the final version will be able to display temperatures at different depths. It will also incorporate sonar equations and ship characteristics to which changes can be input by the operator. Other features include three-dimensional (3-D) pictures of the seabed, which can be oriented in three directions, and the ability to zoom in and out of displays.

Neptune can be colour coded according to customer requirements. Neptune is available in several versions based on 486 PCs or multiprocessor workstations with 3-D graphic accelerators. The system supports UNIX (including Motif) and Windows (NT or 95), as well as leading graphic libraries (Open-GL, Xgl, PEX).

Operational status
The first prototype has been built and is in exploratory development by DCN at Toulon.

Contractor
DCN International, Paris.

VERIFIED

NETHERLANDS

PRS-02/pathfinder

Type
Processor recorder for expendable bathythermograph (XBT) and sound velocity (XSV) probes.

Description
The PRS-02 system is used to process and record: seawater temperature in relation to depth (by a bathythermograph probe), and sound velocity as a function of the depth (by a velocimeter probe).

The system receives inputs from expendable probes. The probes transmit their data via signal wire to the combined processor recorder. The unit then translates the data into temperature/depth or sound velocity/ depth profiles and makes it available in both hard-copy or via an RS-232 or -422 interface.

Operational status
The system is now in production and delivered to several navies and is operational on board submarines and surface ships. In the Royal Netherlands Navy it is

aboard the 'M' class frigates and 'Walrus' class submarines; the system is on order for the German Type 124 frigates and the MEKO 200 frigates of the Turkish Navy.

Contractor
Signaal Special Products, Zoetermeer.

UPDATED

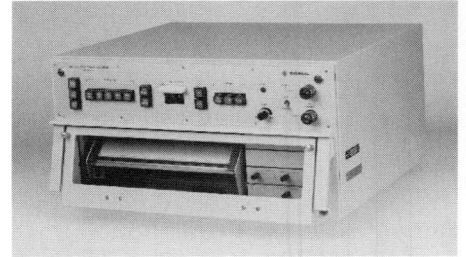

The PRS-02/Pathfinder processor/recorder

SWEDEN

SonTak

Type
Sonar tactical decision aid.

Description
SonTak is an onboard PC-based sonar tactical decision aid developed by SSPA Maritime Consulting AB, in conjunction with the Swedish Defence Materiel Administration (FMV). The system is designed to aid the command in deciding the best sonar to use, the best depth at which to deploy the sonar and how the various onboard sonars can best be combined to achieve the greatest probability of detection for possible targets in the vessel's operating environment. It is also valuable in submarines for calculating the possibility of detection by hostile ASW systems.

The sound velocity profile is measured *in situ*. SonTak was developed in Borland Pascal for Windows. The user interface is based on Microsoft Windows used together with ray tracing and normal mode algorithms in the SonTak system, determines the sound propagation transmission loss as a function of range and depth. Other data relating to both active and passive modes of sonar operations are estimated from environmental parameters (for example sea state, bottom depth and bottom type), a databank of possible targets and own sonar and platform data.

By combining information on anticipated targets, the environment in which the vessel is operating, and own sonar and own ship parameters, detection probability is calculated and displayed as a function of target range and depth.

SonTak consists of four main subsystems:
(1) sound velocity measurement and environmental database
(2) modem link to HAIS
(3) sonar and target database
(4) detection probability and transmission loss calculation.

SonTak both acquires and provides data for the shore-based HAIS (Hydro Acoustic Information System – purchased from BAeSEMA in the UK – see also separate entry on NECTA). Also forming part of the new ASW capability is a ship-based system for tactical use which is now being installed on all ASW units of the Swedish Navy. This latter system comprises small, rugged MIU 2000 PCs. These PCs are connected to one or two probes (XBTs on coastal corvettes), which measure various parameters such as conductivity, pressure and temperature, necessary for calculating sound velocity. The PCs are able to send and receive data from HAIS via modems.

To provide the necessary data for decision making, two programmes are run on board the ships – SonTak and HYDAB (Hydroacoustic Database). HYDAB allows the operator to run operational classification of passive sonar contacts (both radiated ship noise and pings from active sonars).

Two models are used for sound propagation calculations. Above 1 kHz a ray tracing algorithm including sound absorption is used. Surface bounces are taken into account but bottom bounces are ignored. Below 1 kHz a normal mode model with bottom sediment propagation is used. The programme can be set for various types of bottom to take into account reverberation.

SonTak is designed for use with hull, VDS and towed array sonars and detection probability is calculated and displayed as a function of target range and depth.

Operational status
Being installed on all ASW units of the Swedish Navy.

Contractor
SSPA Maritime Consulting AB, Gothenburg.

VERIFIED

UNITED KINGDOM

Minehunting performance prediction system

Type
Computer-based prediction system.

Development
The Thomson Marconi Sonar minehunting performance and prediction system has been developed to supplement modern minehunting sonars. Gathering data on underwater environmental and operating conditions, the system enables sonar operators to select optimum operating parameters and derive a probability factor for detection and classification of perceived mines using those parameters.

Description
The self-contained, computer-based system consists of a variable depth sound velocity measuring probe which is lowered and raised by an outboard winch and jib assembly sited on the upper deck, which is controlled either remotely from the operations room or from a local control panel on the winch. The computer, plotter and electronic panel and rack are located below deck in the operations room.

The probe measures the velocity of sound, pressure (indicating the depth) and temperature at intervals of time as the probe is raised from the operator-selected depth. An optional lightweight probe is available which can replace the winch and probe assembly.

Using this data together with keyed in data and data received from minehunting sonar, the computer applies preprogrammed algorithms to produce recommendations for the sonar parameters to be used, together with predictions of the classification and detection performance which can be obtained.

The sound measuring probe and handling system of the minehunting performance prediction system

The operator interacts with the menu-driven program software in the computer using a keyboard and screen displays to insert parameters, select options, deploy the probe and request recommendations and graphical and tabulated information.

The four-colour X-Y plotter is used to obtain hard-copy printout of the computer screen displays.

In computing the recommended sonar parameters and performance predictions, the system takes account of the ship's sonar characteristics (frequency, pulse length and so on); underwater environment (reverberation, ambient noise, sound profile and so on); ship characteristics (flow/propulsion noise levels and so on); detection, classification and operator performance models; and specific mission inputs (that is, ship speed, sonar operating mode, target characteristics and deployment). The system computes and displays detection range and depth characteristics (detection contour); probability of detection against range; probability of classification against range for moored and ground mines; real-time monitoring of sonar conditions; performance summary; velocity of sound versus depth; ray trace display; optimum towed body height for the variable depth sonar; and data for mission planning, allowing selection of optimum operating modes.

Operational status
The system is operational aboard 'Hunt' and 'Sandown' class MCMVs of the Royal Navy. It also forms part of Sonar 2093 systems sold to Saudi Arabia, Japan, Korea and Australia.

Contractor
Thomson Marconi Sonar Ltd, Templecombe.

VERIFIED

Type 2068 (SEPADS)

Type
Environmental tactical aid.

Description
The Type 2068 Sonar Environmental Prediction and Display System (SEPADS) has been designed to provide real-time advice to the command on the most effective deployment of units and sensors. This is a particular requirement for towed array sonars which need accurate information so that they can be used to the best effect.

SEPADS interfaces with onboard sensors for the direct input of satellite, radiosonde, navigational, bathythermograph, sonar and meteorological information which is stored on a computer disk. This information together with other tactical data provides the basis of input to environmental and acoustic models from which tactical decisions are evaluated and displayed. In conjunction with sound-propagation-loss models, it is used to forecast parameters such as the sonar ranges and the optimum depth at which to deploy the towed array. Processing is carried out in a PDP 11/44 computer.

The system also features advanced TMA packages which complement the tactical environmental decision facilities.

Operational status
In service with the Royal Navy.

Contractor
BAeSEMA Ltd, Dorchester.

VERIFIED

Typical display from the SEPADS system

NECTA

Type
Environmental tactical aid.

Description
In ASW sonar detection is a crucial consideration and the ability to predict sonar performance for realistic operational scenarios can be of great tactical significance.

The Naval Environmental Command Tactical Aid (NECTA) is a scenario-based sonar performance prediction system suitable for both shallow and deep water environments. It is available in variants addressing the requirements of research users, data gathering and analysis activities and operational use.

NECTA meets the following requirements for a sonar performance prediction system:
(a) provides a simple process for obtaining sonar range predictions quickly, whilst allowing detailed control of the process for more advanced users
(b) performs across the complete range of operational conditions (shallow or deep, LF or HF, passive or active, noise-limited or reverberation-limited)
(c) provides interactive and intuitive graphical visualisation of complex datasets, using both two-dimensional and three-dimensional displays
(d) handles the integration of range-dependent environmental data from variable-resolution grids and scattered observations
(e) allows the integration of different PL models and databases, according to the requirements of individual customers
(f) is open systems compliant enabling NECTA to run stand-alone or integrated within a full command support system.
NECTA provides the following key features:
(1) a chart display with worldwide coverage which provides the geographical context for scenarios. Chart overlays can be produced to show environmental and tactical data
(2) a fully integrated suite of acoustic propagation models which can be configured to suit individual customer requirements
(3) environmental data management and analysis facilities. These provide for the storage and display of both historical and observed oceanographic data with a global capability
(4) a tactical database for sonar and platform acoustic characteristics
(5) performance calculations based on evaluation of the sonar equation for the defined scenario.

Additional overlays available can display navigational data including shipping lanes, height of features, which fishing, traffic and topography data can also be inserted.

NECTA can be interfaced to onboard sensors (through the command system, the sonar system or directly as appropriate) to provide automatic input of navigational, oceanographic, sonar and noise data.

Operational status
NECTA systems have been delivered to the Swedish Navy and the Finnish Navy. A variant of NECTA will equip the new 'Visby' class corvettes of the Swedish Navy. A specially tailored variant is being installed on all Royal Navy submarines.

Contractor
BAeSEMA, Dorchester.

UPDATED

INSIGHT

Type
Propagation and sonar performance model.

Description
INSIGHT is a fast propagation and sonar performance model hosted on a PC. It provides a complete sonar performance modelling capability, which puts all the active and passive sonar equation terms at the disposal of the operator. At the heart of INSIGHT is a fast propagation and reverberation model which enables operators to understand acoustics in complex environments, displaying colour-coded acoustic components.

Sensitivity analysis is simple to use and special facilities are available for reconciling trials data with predications. A large number of plot options (against any parameter) allow the operator to investigate interference and finite frequency band affects, boundary losses, SSP details and beam patterns. These include various standard and user-defined volume and boundary loss laws.

Arbitrary speed profiles are easily handled and broken down into phase speed (or angle/wave number) bands which are shown explicitly and interactively on the SSP plot. The corresponding acoustic contributions are shown, using the same colour coding, on the transmission loss plot. Effects on convergence zones, surface ducts, bottom ducts and so on are seen instantaneously.

As well as propagation calculations, INSIGHT includes predictions of signals versus a background, which could be simply ambient noise for a passive sonar or reverberation plus noise for an active sonar.

An enhanced beam pattern facility provides a number of options for calculating beam responses. These options allow for input by selected beam pattern, user-specified beam shapes or array type.

Output files and displays can be sent to printer, plotter or disk.

INSIGHT runs on a 486 PC or above with colour display under Microsoft Windows.

Contractor
BAeSEMA, New Malden.

VERIFIED

Type 2090/IBIS (Integrated Bathy Information System)

Type
Recorder, processor and sonar prediction system for use with XBT, XSV and XCTD expendable probes.

Description
Designed jointly by Ultra Electronics Sonar and Communications and Sippican Inc (USA) to meet the Royal Navy's requirement for a powerful oceanographic data acquisition and sonar range prediction system (MoD designation Sonar 2090).

IBIS interfaces directly with standard Sippican launchers and probes (XBT, XSV and XCTD and submarine equivalents). It may easily be installed in place of all Sippican processors.

Its key features include: data acquisition, display and recording with zoom and edit facilities, and reduced breakpoint algorithms; Wader – worldwide oceanographic database (with zoom) of bottom profile, and monthly temperature and salinity profiles; Hodgson – range dependent high-speed propagation loss model; mimic – very low frequency, wave summation model, complementary to Hodgson.

Operational status
Production units available and in service with the Royal Navy as sonar Type 2090. The survey vessel *Scott* of the Royal Navy is equipped with the Type 2090 system. A total of 88 systems have been delivered.

Contractors
Ultra Electronics Limited, Sonar & Communication Systems, Greenford, Middlesex.
Sippican Inc, Marion, Massachusetts.

UPDATED

Mermaid

Type
Software and database package for predicting in-water environmental conditions in shallow water areas.

Development
Mermaid (Marine Environmental Model and Integrated Database) concept has been developed by the Defence Research Agency and the Unit for Coastal and Estuarine Studies at the University of Wales, Bangor.

Description
A forecast model enables Mermaid to make predictions over large or small shallow water regions for such parameters as temperature, sound speed, tides and currents. It also incorporates a search and rescue package for predicting the track of drifting objects or the movement of oil slicks, for example. The integral database contains an extensive set of temperature profiles, tidal components and bathymetry.

In naval applications, Mermaid can be used in support of shallow water ASW where it provides predictions of thermal structure and sound speed profiles for input to sonar performance models. It also predicts the position and strength of oceanographic features (for example fronts) from database and other considerations. Data on the depth of the surface mixed layer and estimates of diurnal heating effects can be provided together with estimates of environmental factors affecting ASW tactics.

In support of mine countermeasures Mermaid predicts the strength and direction of tidal currents in order to determine safe standoff distances from hostile objects. It also predicts tidal currents in support of diver and ROV operations. Waves and tides are predicted in order to determine the MCM platform's dynamics and response. It also provides estimates of near-seabed tidal and wave-induced currents for mine burial prediction and forecasts temperatures and sound speed profiles for input to MCM sonar performance prediction models. The search and rescue package can also be used to predict mine drift.

Operational status
The forecast model component of Mermaid is incorporated in the HAIS (Hydro Acoustic Information System) developed by BAeSEMA Ltd for the Swedish Navy.

Contractor
Defence Research Agency, Winfrith.

VERIFIED

SEATDA (Sonar Environmental and Acoustic Tactical Decision Aid)

Type
Prototype system for making optimum use of the ocean environment.

Development
Developed by the Defence Research Agency to enable the user to make optimum use of the ocean environment for tactical purposes in terms of maximising the detective or evasive capability of the unit.

Description
SEATDA has been developed as a result of a comprehensive programme of work designed to identify, investigate and demonstrate effective tactical exploitation. It uses extensive oceanographic databases, some of which incorporate actual trials measurements, from a variety of different sensors such as expendable bathythermographs, environmental sensors (gathering temperature conductivity versus depth data), undulating towed vehicles and high-resolution thermistor chain. The system is hosted on a SUN Sparc 2 system, incorporating high-resolution graphics with flexible multiwindowing facilities. SEATDA offers both plan and vertical sections of probability of detection on any chosen target. The plan sections allow own and target locations to be superimposed. Sophisticated acoustic prediction models are used which can generate range dependent propagation loss values. The split-screen facility allows simultaneous viewing of an acoustic ray trace with the corresponding estimates of acoustic propagation loss. This enables the comparison of qualitative and quantitative views of the sound field respectively.

Contractor
Defence Research Agency, Winfrith.

VERIFIED

OCIDAS-1 (Ocean Intervention and Data Assimilation System, Phase 1)

Type
Software and database package for forecasting in-water environmental conditions using numerical models and observations.

Development
Developed by the Defence Research Agency as a research prototype to enable a forecaster to compare historical data with numerical model forecasts and observations, and to generate predictions of in-water conditions in standard format signals.

Description
OCIDAS-1 is part of a continuing development by the Defence Research Agency of prototype software tools for oceanographic and acoustic forecasters. Incorporating a mixed layer model, the software enables a forecaster to make the best possible use of scarce in-water observations which can predict changes in the upper ocean in response to atmospheric effects, and historical databases of typical conditions for different ocean regions and different times of the year. Four acoustic propagation loss models are also incorporated to cover deep and shallow water regions, and high- and low-frequency regimes.

Driven by forecast fluxes from a numerical weather prediction model, the mixed layer model can generate forecasts on a grid of locations, or for irregularly defined regions. Contoured synoptic charts can be displayed, and maps can also be selected showing observation locations and the positions of ocean fronts and the ice edge. A 'quick-look' assessment of the acoustically important features of temperature profiles can be used to give an overall impression of the location of good, medium and poor acoustic propagation areas.

OCIDAS-1 is hosted on a SUN Sparc 2 system, incorporating high-resolution graphics and multiwindowing facilities. The software has been written in a modular form so that various combinations of components can be transported to other systems.

Operational status
OCIDAS-1 has been evaluated by the Defence Research Agency in collaboration with the Royal Navy, and is being integrated into RN operational systems.

Contractor
Defence Research Agency, Winfrith.

VERIFIED

OCIDAS-2

Type
Ocean forecast assessment and assimilation (3-D).

Description
With ever increasing computing power and improving numerical model codes, the ability to predict the time evolving three-dimensional (3-D) environment around a naval task force, quickly and accurately enough to provide a tactical advantage, is becoming a reality. Using the latest advances in ocean modelling, data visualisation and database architecture (from BAeSEMA), the DRA is developing a new ocean/acoustic forecaster's support tool. The Ocean Intervention and Data Assimilation System Phase 2 (OCIDAS-2) is a prototype for the forecaster's 'intervention desks' which will be required later this decade to enable the Royal Navy to exploit the oceanographic and sonar environment fully in any part

of the world, and to disseminate operational products to support command at sea.
OCIDAS-2 provides:
(1) forecaster's intervention desk
(2) 3-D ocean forecasting
(3) sophisticated data assimilation with automated quality control
(4) feature modelling of fronts and eddies
(5) products tailored to user's operational role
(6) communications management.

In a joint programme with the UK Meteorological Office, the DRA is developing the Forecast Ocean Atmosphere Model (FOAM). Interacting dynamically with the Meteorological Office's atmosphere models, FOAM will provide high-resolution 3-D predictions of the ocean structure for several days ahead.

OCIDAS-2 will be used by the Royal Navy forecasters to display, manipulate and interrogate FOAM's predictions.

The OCIDAS-2 system has been developed by

BAeSEMA using modular software and 'open systems' concepts. It will provide an integrated workstation-based forecaster's intervention desk with many of its important functions in 'black box' algorithms which use standard input/output formats to maximise interoperability and can be replaced or updated as required. While being designed specifically for FOAM, OCIDAS-2 could also take predictions from other numerical ocean models.

Operational status
OCIDAS-2 is being evaluated by the Defence Research Agency in collaboration with the Royal Navy.

Contractors
Defence Research Agency (OCIDAS-2), Winfrith.
BAeSEMA (database), Dorchester.

VERIFIED

Acoustic Support Function (ASF)

Type
Environmental tactical aid.

Description
The ASF provides environmental data processing and performance prediction for ASW sonar applications. Initially developed to be embedded in the Royal Netherlands Navy's 'M' class frigate command system, the ASF is flexible and portable, designed and implemented to operational standards to provide:
(1) environmental data access and manipulation
(2) environmental and sonar sensor data interfaces
(3) signal data interfaces
(4) propagation loss models and ray tracing

(5) sonar performance and probability of detection
(6) sonar deployment assessment.

At the heart of ASF is a suite of complementary propagation loss models from a number of leading NATO naval and research organisations incorporated as 'black boxes'. This is supported by comprehensive environmental and contact databases, interfaces to on-line data sources and oceanographic and acoustic processing functions to ensure that the models are supplied with the best available input data.

The ASF contains its own local database of data for the current operational area and sonars/platforms. A software presentation layer interfaces to the host environment, providing the means to populate the local database from an external master database. The interface also supports operator inputs and responses

to ASF, integrated sensor inputs, signal data and datalogger outputs from ASF, and the result from ASF processing functions.

Sensor inputs include observed profiles, directional ambient noise and reverberation.

Predictions can be made for any known combinations of platforms and sonars and are not limited to 'own sh p' calculations.

Operational status
In service with the Royal Netherlands Navy.

Contractor
BAeSEMA, Dorchester.

VERIFIED

UNITED STATES OF AMERICA

ATMS Track Management System

Type
ASW database management system.

Description
The ASW Track Management System (ATMS) receives undersea ocean surveillance information from supporting sensors and sources, and correlates and tracks this information to maintain a database of user-selected surface, subsurface and air reports.

ATMS provides the user with colour graphics, maps, decision aids and dynamic displays of all platforms and

ambiguous contacts within a specified region of the globe using NTDS symbology. Real-time tracking is supported by sophisticated correlation and tracker algorithms. These algorithms produce a database available to the operator. Through this database, the user can access track/contact, library, communications, decision tools, correlator information and automated message generation. The user can edit the information as required.

The baseline ATMS is hosted on a network of DTC II (or Sun 4) workstations over a standard Ethernet local area network. The maximum number of workstations is currently limited to 50. The preferred workstation

configuration is a 19 in colour monitor, 32 Mbytes RAM and 250 Mbytes of available disk storage. User interaction is via both keyboard and trackerball or mouse. ATMS is hosted under SunOS version 4.0.3. ATMS is written in C and FORTRAN.

Contractor
Lockheed Martin Missiles & Space Company, Sunnyvale, California.

VERIFIED

OCEANOGRAPHIC/HYDROGRAPHIC SURVEY SYSTEMS

SINGLE-BEAM ECHO-SOUNDERS

GERMANY

DESO 11/14/15/25 echo-sounders

Type
Survey echo-sounders.

Description
DESO 11
The DESO 11 is a single-channel echo-sounder with a built-in LCD display. Via a serial interface, it provides depth data to and is fully remote-controlled from an external computer. The frequency is user-adjustable, and the depth range extends to 650 m.

DESO 14
The DESO 14 is a portable, single-channel, dual-frequency, hydrographic echo-sounder. The compact, rugged but lightweight housing makes it convenient to use in portable applications. It can operate from 24 V DC batteries. The depth is recorded by means of a high-resolution thermo-electric printer with over 800 pixels in three grey levels across 10 cm recording width. The operation is menu-controlled; parameters and depths are displayed on dedicated LCD displays. Survey computers can be connected via a serial data interface. The operating range extends to 650 m.

DESO 15
The DESO 15 is offered as the standard echo-sounder for hydrographic surveys on the continental shelf. It is available in a single- or dual-channel version. Frequencies can be adjusted by the user within a low-frequency and a high-frequency band. Recording is performed by a high-resolution thermo-electric printer

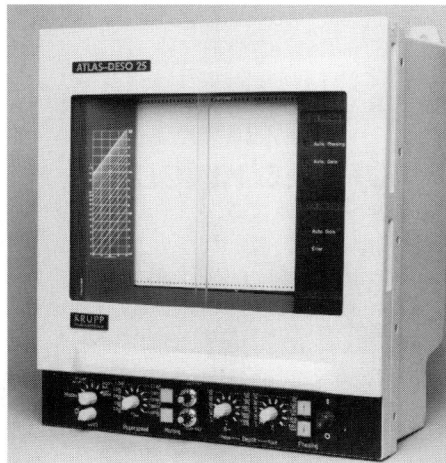

DESO 25 dual-channel precision sounder   **1994**

by means of 1,400 pixels with four grey levels or 700 pixels with eight grey levels across a recording width of 17.5 cm. The recording is annotated with the main measurement parameters. The system is operated via rotary switches for the main functions and menu-led control using a rotary knob and an LCD display. The measurements of both depth channels are indicated on separate LCD displays. The system offers flexible computer interfacing via RS-232 C, RS-422 and GP-IB bus. The operating range extends to 650 m. DESO 15 DS is a new extended model for surveying over depths

down to a maximum of 1,500 m with a resolution of 1 cm. It operates on standard frequencies of 33 and 210 kHz. Other frequencies are available as an option.

DESO 25
The DESO 25 is a dual-channel hydrographic echo-sounder covering the full application range from inshore to deep ocean surveying. The particularly refined sediment recording functions are supported by a high-resolution thermo-electric printer and a net recording width of 20 cm.

The strength of the echo can be recorded as an additional line providing information on ocean floor properties. The chart is fully annotated with all measurement parameters including date and time from an internal clock. The measurement range extends to 15,000 m. Optionally, one of the two channels can be fitted with a side scan electronics board and transducer. This provides an imaging function in addition to the depth recording.

Operational status
These echo-sounders are in widespread use with the navies of Australia, Brazil, CIS, France, Greece, India, South Korea, Netherlands, New Zealand, Portugal, Sweden, Thailand Turkey, UK and USA. The DESO 25 is used by various navies, including the Royal Navy, French Navy and the Portuguese Navy.

Contractor
STN ATLAS Marine Electronics GmbH, Bremen.

UPDATED

VE 59 series

Type
Echo-sounder for submarine navigation.

Description
The VE 59 family of naval echo-sounders is based on an improved design of the NATO DSQN-11 echo-sounder. Of compact, non-magnetic construction, the VE family uses many of the proven features of the DSQN-11, but with enhanced functions through the integration of modern microprocessor technology. The echo-sounder is designed primarily to display depth information according to requirements and mode selected: water depth and, in the case of submarines, diving depth, displayed either separately or simultaneously, with the height of the submarine shown to scale on the recording between water surface and seabed. To achieve this the submarine uses one transducer sounding down and one sounding up. Measurements are indicated both on recording paper and digitally. The selected mode of operation and measurement range are marked on the recording paper. Important recordings can be marked accurately with a real-time graduation.

For each application, various options allow adaptation to specific requirements. Among these are

digital slaves with alarm devices and special sensors. The family features: a high-frequency mode for measuring small distances minimising the risk of detection; a sound velocity measurement with display and recording of sound velocity versus diving depth or versus time; measurement of Conductivity, Temperature and Depth (CTD) for calculation of density, sound velocity and salinity; integrated and remote digital displays; digital data output.

To ensure a stable and even display, there is an Automatic Gain Control (AGC and TVG – Time Varied Gain) and an automatic switching of the pulse length and bandwidth coupled to measurement ranges. The paper advance is either coupled to the measurement range or can be controlled by ship's log and matched to the actual speed of the ship.

By use of a central digital switch the VE 59 can be set for the required sound velocity in water, thus ensuring measurement accuracy under all conditions.

In addition, data provided by special sensors can be called off for display of different bathythermal layers as well as sound velocity, salinity and water density.

Contractor
L-3 Communications ELAC Nautik GmbH, Kiel.

UPDATED

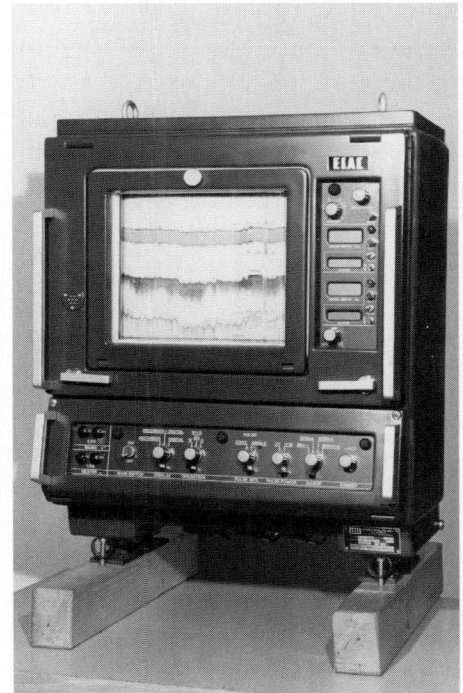

VE 59 echo-sounder
1999/0024838

NORWAY

EA 502

Type
Single beam echo-sounder.

Description
The EA 502 is a modular, dual channel (38 or 200 kHz), single beam, hydrographic echo-sounder based on microprocessor technology with an instantaneous dynamic range of 160 dB.

The echo-sounder incorporates a separate digitiser for each channel and uses sophisticated software algorithms for bottom tracking based on multicriterion decision theory. Features include an adjustable ping rate (up to 10 pings per second), digital data output for post processing, remote computer command control, echogram presentation in 12 colours and interface to heave, pitch and roll sensors. Sound velocity compensation is either through manual input or via profile input.

As an option, additional analysis capabilities can be incorporated for silt measurements and sub-bottom profiling.

Operational status
Acquired for installation in the new survey launches of the US Navy.

Contractor
Kongsberg Simrad AS, Horten.

UPDATED

UNITED KINGDOM

Series 700 echo-sounder

Type
Echo-sounding depth recorders.

Description
AB Precision produces a number of echo-sounders to meet modern naval requirements. These include the Type 778, Type 780, Type 786 and Type 787 systems.

Type 778
This equipment has been designed to meet the requirements of all vessels from coastal minesweepers upwards. It consists of a general service depth recorder operating with a single 10 kHz transducer; a bridge unit with numerical depth display as an optional fitting, which operates with its own two-element 48 kHz transducer; and a precision depth recorder for oceanographic survey purposes which operates with the transmitter of the depth recorder and its associated transducer.

Type 780
This is a high definition shallow sounder fitted in surface ships and submarines. Its maximum depths capability is 1,000 m. The set is available in the following versions: 30 kHz (surface vessels), 48 kHz

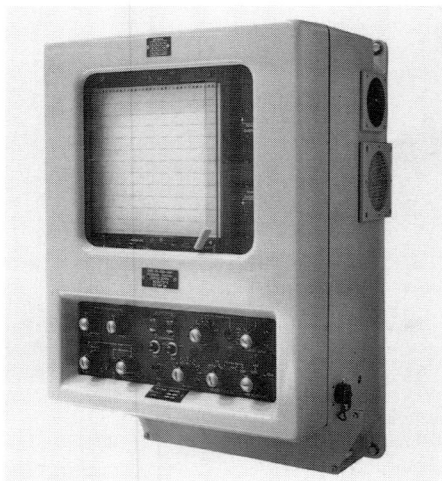

Type 778 echo-sounding equipment

(reduced magnetic signature for surface ships) and 48 kHz (for submarines). The sounder consists of a chart recorder and digitised and remote display.

Type 786
The Type 786 has been developed for use with flat-bottomed landing craft. Main features are that the transducer is extendable below the layer of aerated water, which forms beneath a flat-bottomed craft, to give good performance, and rapid automatic retraction of the transducer as the craft beaches. The system consists of a main recorder unit with chart facility and digital display which operates with a single 200 kHz transducer, a remote digital display, and an automatic transducer raise and lower mechanism with its control panel.

Type 787
The Type 787 is intended to meet the requirements for all vessels in the Royal Fleet Auxiliary Service and the Maritime Auxiliary Service. It consists of a main recorder unit with chart facility and digital readout which operates with a single 50 kHz transducer and remote digital readout(s).

Operational status
In operational service.

Contractor
AB Precision (Poole) Limited, Poole.

UPDATED

UNITED STATES OF AMERICA

Model 9057 (AN/UQN-4) sonar sounding set

Type
Depth sounder.

Description
The Model 9057 sonar sounding set (US Navy designation AN/UQN-4) is an echo-sounder which incorporates proven design and increased accuracy together with field change upgrades to improve shallow water capability. The Model 9057 has been developed as a replacement for the Model 185 (AN/UQN-1) set which has been the standard equipment on virtually every vessel and submarine in the US Navy, US Coast Guard and Coastal Geodetic Service, as well as on domestic and foreign research and offshore vessels, for over 20 years.

The Model 9057 system uses a 12 kHz transducer and visually presents digital and graphic displays of water depths by means of a digital numeric display and permanent strip chart recorder. It may also be used as a passive listening device.

Operational status
In service with Australia, France, Korea, Taiwan and the US Navy.

Contractor
EDO Acoustic Products, Salt Lake City, Utah.

VERIFIED

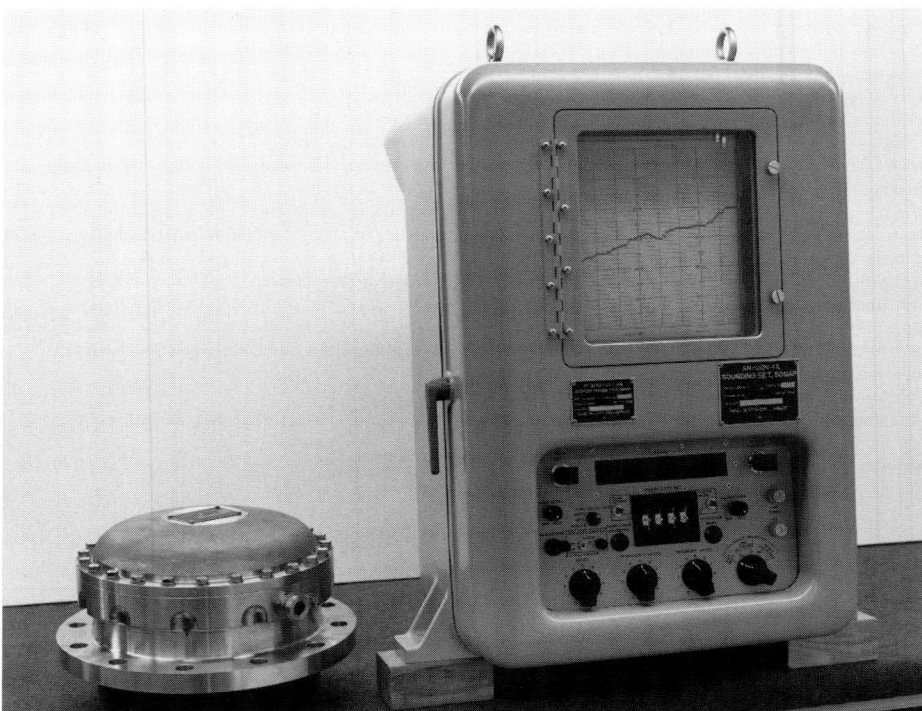

Model 9057 sonar sounding set

DSF-6000

Type
Digital survey echo-sounder.

Description
The DSF-6000 digital survey echo-sounder is a dual-frequency, portable depth sounder. Water depth is shown on a four-digit display and on an analogue chart recording and is sent out of an RS-232 port for digital recording. The analogue record accurately depicts the bottom profile displayed for the operator. This record can be used to verify the stored digital data or, in an emergency, can be manually digitised to reconstruct the whole digital record. Grid lines are written onto the initially blank chart paper during the bottom recording cycle.

The digitiser is programmed to acquire the bottom automatically under control of the system's computer. Once locked onto the bottom, an automatic tracking gate confines the search to the known vicinity of the bottom. Gate marks are printed on the chart. Digital depth data can be communicated to a tape recorder, data logger or computer through either the serial RS-232C or parallel port. The sounder operates at two frequencies; a high frequency of either 100 or 200 kHz and a low frequency of 12, 24, 33 or 40 kHz for greater depths, or to measure to firm bottom beneath the mud or fluff layers.

The unit can be operated at either frequency separately or at both frequencies simultaneously. The LF transmitter delivers a full 2 kW of power. Full depth capability, selectable in seven steps, is from 58 to 5,800 feet, fathoms or meters. Full depth on the chart display is covered in seven overlapping phases. A specific phase can be chosen manually for display on the chart, otherwise the automatic mode displays that phase which includes the ocean bottom. Factors affecting calculated and display depth are written directly on the chart and include: selected sound speed, tide and draft correction, chart scale, chart speed and a two-digit unit identification selected by the owner. Time ticks at one minute intervals are printed along the top and bottom margins of the chart.

Operational status
The DSF-6000 model has been replaced in production by the BATHY 1000 Dual-Frequency Echo-Sounder. The DSF-6000 echo-sounder is still available as a system and is supported for spares, repairs and consumable items. The BATHY-1000 unit uses more reliable and modern digital signal processing technology. The BATHY-1000 includes a thermal printer, LCD display and keypad data entry and is fully compatible with existing DSF-6000 installations.

Contractor
Ocean Data Equipment Corporation, E Walpole, Massachusetts.

VERIFIED

SIDE SCAN SONARS

UNITED KINGDOM

Type 2034 sonar

Type

Dual side scan sonar for hydrographic work.

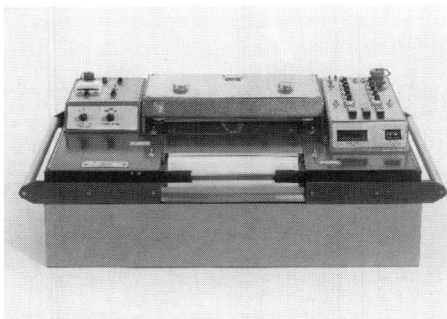

Type 2034 sonar

Description

The sonar Type 2034 is a short-range, high-definition, dual side scan sonar, suitable for: establishing the position of ships, aircraft or other objects lost at sea; charting obstructions in shipping channels or at sea floor engineering sites and evaluating underwater topography. Designed for the utmost reliability and to meet exacting military standards, Type 2034 sonars have been in use with the Royal Navy since 1976. The system is standard equipment on ocean-going and coastal vessels of the survey fleet.

The sideways-looking left and right transducers are housed in a towed body, called the towfish, the current version of which is the Mk 3. Some of the electronics are contained in the towfish but the majority of this part of the system is housed in the recorder unit which provides for operation and control of the equipment, and display by hard-copy, printout chart recordings.

The sonar operating frequency is 110 kHz with a nominal transmitted pulse length of 100 μs. A choice of three range scales, 75, 150 or 300 m, and one of three recording paper speeds, 30, 60 or 150 lines/cm, can be selected to suit the prevailing operating conditions. The transmission repetition rate depends on the selected range scale and is 10 pps for 75 m range, 5 pps for 150 m range or 2.5 pps for 300 m range. For all range settings, scale lines are recorded at 15 m range intervals to aid interpretation of the recordings. Facilities are provided for putting event markers on the recording.

Operational status

This system is now obsolescent, but units remain in service with the Royal Navy and other users worldwide.

Contractor

Ultra Electronics, Ocean Systems, Weymouth.

UPDATED

Type 2053/Sonar 3000

Type

Hydrographic survey sonar system.

Description

The Sonar 3000 family of side scan sonars has been developed from the Type 2034 high-definition sonar. The system uses modular construction which allows the assembly of variants to suit a range of applications. The system, originally aimed at the commercial market, is also in use for military applications. One variant known as the Type 2053 is in use with the Royal Navy for inshore hydrographic survey work.

The Sonar 3000 family includes a variety of options and accessories including lightweight and heavyweight towfish and a range of recorders and winch systems.

Operational status

Sonar 2053 is in service with the Royal Navy Hydrographic Service (30 systems).

About 50 Sonar 3000 systems are in use with various other hydrographic survey organisations. The Sonar 3000 family has now been replaced by the Widescan family described in the following entry.

Manufacturer

Ultra Electronics, Ocean Systems, Weymouth.

VERIFIED

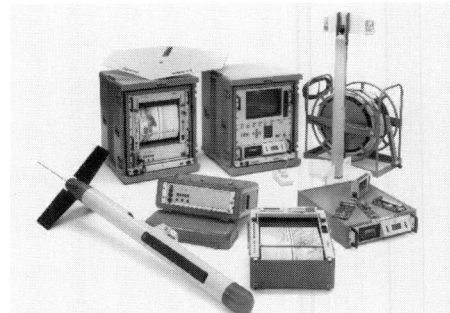

RN Type 2053 Sonar 3000

Sonar 3050 (Widescan)

Type

Dual-frequency image corrected side scan sonar for hydrographic survey, pipeline survey, and explosive ordnance disposal applications.

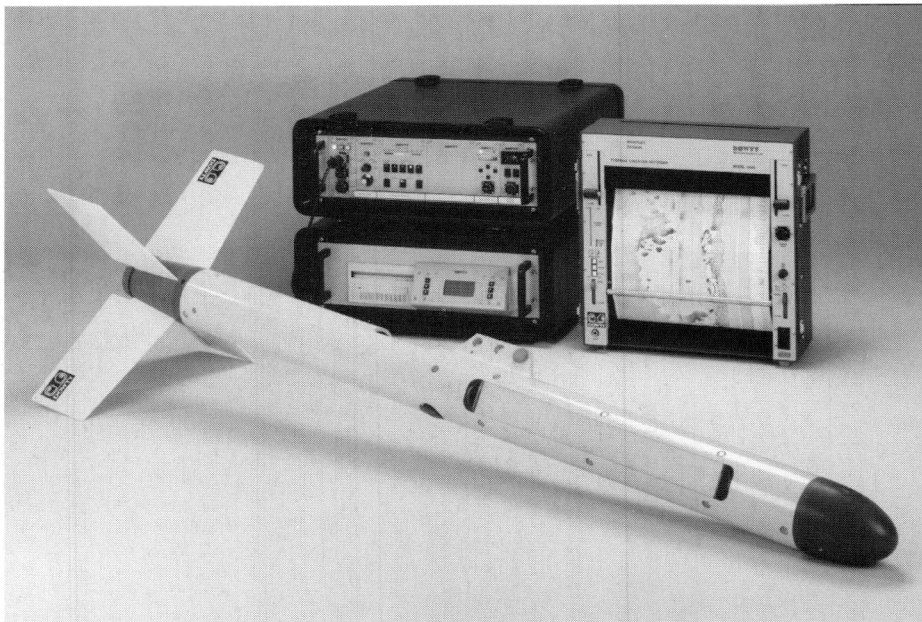

Description

The Widescan sonar operates at frequencies of 100 kHz and 325 kHz to provide a combination of long-range and high-resolution coverage. Ranges of up to 400 m per side are achievable at 100 kHz, and up to 200 m per side at 325 kHz. The 0.6° horizontal beamwidth at 325 kHz provides very high resolution seabed and bottom object imagery.

The Widescan stainless steel towfish has a built-in acoustic altimeter and features a third channel towfish height display on the thermal printing recorder unit. Image correction for slant range and speed-over-ground can be selected from the front panel.

The system features an advanced self-adaptive gain system which guarantees optimum records without the need for operator intervention, making Widescan ideal for non-expert users as well as for experienced users such as hydrographic surveyors.

The modular nature of the system facilitates field maintenance and the 3050 can be supplied with either a 12 in or a 20 in thermal recorder.

Options include a lightweight Kevlar towcable for small boat/shallow water operations and a wide range of sonar processing workstations for digital display recording and processing.

Operational status

The Widescan sonar is in use with the Royal Netherlands Navy, the Sultan of Oman's Navy, the UK Naval Hydrographer and more than 30 civil survey organisations.

Contractor

Ultra Electronics, Ocean Systems, Weymouth.

VERIFIED

Elements of the Widescan II Model 3050 sonar

Widescan 3

Type

Digital side scan sonar.

Description

Widescan 3 is a high-performance digital side scan sonar. Key features include: digital two-wire telemetry; dual simultaneous frequencies; composite transducer technology; built-in motion sensing; built-in altimeter; auxiliary channels for acoustic and serial data; and compatibility with all workstations.

Developed in response to market demands, Widescan 3 is ideal for all hydrographic and offshore survey requirements.

Designed to operate in conjunction with the many popular sonar workstations in use today, Widescan 3 comprises a high-performance dual-frequency towfish and a simple surface interface unit.

The all new digital two-wire towcable telemetry system allows operation on commonly used armoured coaxial cables at lengths of 3,000 m or more. All acoustic data (16-bit resolution) together with bidirectional serial data and power is multiplexed on to the single coaxial cable. With one spare acoustic data channel and three serial data channels, additional instrumentation is easily integrated. Requiring only one coaxial, Widescan 3 is also easily integrated onto RQVs and so on.

Advanced composite transducer technology offers many advantages over standard transducers including improvements in beam patterns and power output, resulting in better image quality. Longer transducers can be fitted as an option, providing improved resolution and selectable beam patterns. With longer transducers fitted, the user can select either standard or long aperture online.

Control of the system is provided by the host sonar workstation via the sonar interface unit. All towfish parameters including frequency, pulse length, range, repetition rate and TVG are controlled from the surface and can be changed at any time during operation. Other facilities such as digital recording and playback, target identification and logging and real-time mosaicing are all available as part of the workstation software package. Auxiliary towfish information including altitude, heading, pitch, roll, temperature and status is telemetered to the surface on the serial uplink. A basic Windows control programme is provided allowing user control of the system independent of the workstation.

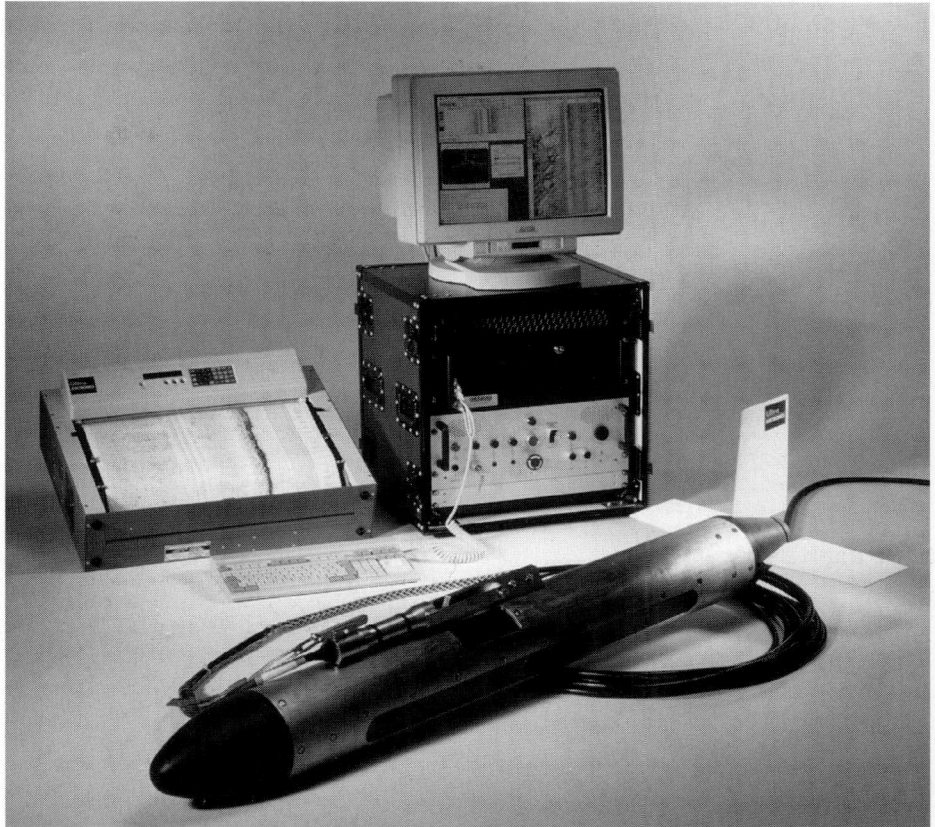

Widescan 3 towfish and shipboard elements   ***1998**/0009852*

Specifications

Towfish
Long array
Length: 1.61 m
Diameter: 11.2 cm
Weight: 35 kg
Short array
Length: 1.56 m
Diameter: 16.8 cm
Weight: 45 kg
Operating depth: 1,000 m
Towing speed: 6 kt
Frequencies: 100 and 325 kHz

Pulse length: 30-150 µs (100 kHz), 100-500 µs (325 kHz)
Beamwidth: horizontal long array: 0.76° (100 kHz) and 0.21° (325 kHz)
  horizontal half length array: 1.3° (100 kHz) and 0.42° (325 kHz)
  vertical: 35°

Contractor

Ultra Electronics, Ocean Systems, Weymouth.

VERIFIED

Deepscan 60 sonar

Type

Survey sonar for continental shelf and deep ocean areas.

Description

The Deepscan 60 sonar has been developed for surveying continental shelf and deep ocean areas down to a maximum depth of around 6,000 m. The system combines a high-resolution 60 kHz side scan sonar, swath bathymetry technology and a sub-bottom profiler in a towfish designed for deployment from ships or craft of opportunity. The towfish can be operated at speeds of up to 8 kt at shallow depths, or at slower speeds for deep deployment. The system uses the latest digital signal processing techniques combined with fibre optic towcable telemetry for display on the general purpose Tessa sonar workstation. Sufficient spare processing and telemetry capacity is included to provide flexibility and upgrade capability. The towfish is negatively buoyant and does not need a separate depressor. The tail and drogue arrangement improve stability and simplify deployment and recovery. The fish houses the two side scan/bathymetry arrays, the sub-bottom profiler array, an attitude heading reference unit, and a front end processing and telemetry unit.

Each side scan transducer array consists of seven identical sections, each section being separately connected to facilitate electronic beam-forming in transmit and receive. Transmit modulation can be either a short CW pulse or a long swept frequency (Chirp) pulse. CW operation is suitable for slant ranges of up to 800 m while the Chirp pulse can be used to extend this to around 1,600 m per side. A dual vernier interferometer operating in conjunction with the main side scan array is used to provide swath bathymetry.

The Deepscan 60 has a dual-role sub-bottom profiling system aimed at providing very high-resolution shallow penetration for cable route survey and similar applications, combined with a deeper penetration performance suitable for site surveys for underwater structures and so on. All beam-forming, filtering, gain processing, bathymetry and image processing functions are performed on the surface under software control in the Tessa sonar workstation. This compact unit contains a powerful 32-bit Unix computer and an array of high-speed digital processors which carry out all the real-time signal processing.

The workstation can be networked, interfaced to a variety of other sensors and can be expanded through additional plug-in modules.

Specifications

(Side scan sonar)
Frequency: 60 kHz
Beamwidth: horizontal 1°; vertical 35°
Source level: +231 dB rel to 1 µPa at 1 m
Modulation: CW or optional FM sweep (chirp)
Range: 100/200/400/800/1,600 m per side
Swath width: up to 800 m each side
(Sub-bottom profiler)
Operating frequency: 7.5-12.5 kHz
Beamwidth: 30° conical from 7 element array
Source level: +212 dB rel to 1 µPa at 1 m
Modulation: CW or FM sweep (chirp)
Resolution: better than 10 cm
Penetration: up to 75 m
(Towfish)
Length: 3.5 m
Diameter: 0.5 m
Weight: 500 kg (approx)
Tow depth: 50-6,000 m

Contractor

Ultra Electronics, Ocean Systems, Weymouth.

VERIFIED

Ultrascan 5000

Type
Side scan seabed survey sonar.

Description
Ultrascan 5000 is a seabed survey sonar developed to satisfy the requirements of commercial surveys and military applications alike where high-resolution long range images of seabed artifacts are of primary importance. The system combines a high-resolution, focused, 120 kHz side scan, and swath bathymetry system generating data out to 400 m per side at resolutions normally associated with 300 to 500 kHz sonars.

Ultrascan 5000 is derived from technology developed for Deepscan 60 and uses the latest digital signal processing techniques combined with fibre optic towcable telemetry and the general purpose Tessa sonar workstation. Sufficient spare processing and telemetry capacity has been included to facilitate system flexibility and future growth.

Each side scan transducer array is made up of separate electrical sections, each separately connected to facilitate electronic beam-forming and focusing in real time to achieve extremely high resolutions. A dual vernier interferometer operating in conjunction with the main side scan array is used to provide swath bathymetry.

All Ultrascan 5000 beam-forming, filtering, gain processing, bathymetry and image processing functions are performed on the surface, under software control, in the Tessa sonar workstation. This compact unit contains a powerful 32-bit computer and an array of high-speed digital signal processors which carry out all the real-time signal processing.

Operational status
Two systems based on Ultrascan 5000 have been supplied to Johns Hopkins University, Applied Physics Lab, USA, and are in service with the US Navy.

Contractor
Ultra Electronics, Ocean Systems, Weymouth.

VERIFIED

UNITED STATES OF AMERICA

Side scan sonar

Type
Side scan sonar for oceanographic research and survey.

Description
The side scan sonar system provides the operator with the capability to create high-resolution images of the sea bottom and sub-bottom for applications such as route survey for the determination of bottom conditions and the selection of Q routes, the detection of cables and pipelines on and beneath the sea bottom, oceanographic survey for the determination of geological or biological conditions, as well as other applications requiring high-resolution images of the seabed.

The design of the Klein sonar equipment permits the operation and real-time display of up to six channels of sonar data, including simultaneous 100 and 500 kHz side scan sonar for multispectral insonification of targets and 3.5 kHz for penetration into the near sub-bottom area of the seabed. These data are displayed in real time correlated on the data medium for ease of operator interpretation. The simultaneous display of the sonar data provides information on both surface and subsurface targets which could be missed if a single frequency were utilised.

The sonar system utilises a Variable Depth Sonar (VDS) transducer which can be selectively deployed to avoid interference from thermoclines. Various frequencies are available for the VDS transducer including 50 kHz for long-range, medium-resolution applications, 100 kHz for high-resolution applications, 100/500 kHz for simultaneous insonification of target areas, and 3.5 kHz for penetration of the sea bottom. Transducers are available for operations as deep as 12,000 m.

The sonar images are transmitted up a VDS towcable to a combination sonar transceiver and graphic recorder for processing and display. Accessories are available which permit advanced image processing of the sonar data, as well as data reduction. Fully integrated sonar systems which interface with navigation and shipboard equipment are available.

The sonar system is lightweight, weighing less than 100 kg (225 lb) and is easily transported and installed aboard vessels of various sizes down to 5 m in length. The VDS transducer weighs less than 25 kg (55 lb) and is easily deployed by a single crewman without specialised handling equipment.

500 kHz very high-resolution side scan sonar image of a magnetic sensing device on the ocean bottom made with Klein high-resolution side scan sonar and displayed on a Klein graphic recorder

Full accessories are available from Klein Associates including towcables, towing winches, and various related systems including acoustic positioning, surface navigation, and data management equipment.

The sonar system to be supplied under the terms of the contract noted below is a second-generation system which utilises the latest sonar technology to improve performance and reduce the size of the towfish. The sonar system creates five beams per side, port and starboard, with each beam being focused and steered to produce a 20 cm footprint of the bottom. At a sonar operating range of 150 m per side (300 m swath), the sonar system can be operated at speeds up to 10 kt with 100 per cent bottom coverage.

The sonar system consists of a 10-beam side scan sonar towfish (variable depth transducer), a coaxial towcable, a topside sonar processor, and a Triton Technology Isis sonar processor. The Triton Isis processor will be utilised for display of the sonar data.

Operational status
In production and in service with hydrographic and oceanographic agencies worldwide, both government and navy.

Klein Associates Inc has won a competitive contract to supply up to 11 high-speed, high-resolution, multibeam, side scan sonar systems to the US National Oceanic and Atmospheric Administration (NOAA).

Contractor
Klein Associates Inc, Salem, New Hampshire.

VERIFIED

Model 260-TH

Type
Dual-frequency, image correcting side scan sonar.

Description
The Model 260-TH microprocessor-based image correcting side scan sonar generates plan view images of the seabed depicting the size, shape and location of various sea floor materials and manmade objects. In addition to generating corrected hard-copy maps in real time, the system can store minimally processed data on analogue or digital magnetic tape for reprocessing by direct playback through the system.

Sonar images are fully corrected for slant range, ship speed and amplitude giving a plan view of the topographical features of the seabed. Each pixel is individually corrected which results in a high-quality, high-resolution image. The acoustics, signal processing, data recording and graphic recording are consistent for a display with a pixel size of 1/800th of the selected range to each side, and an amplitude dynamic range of 64 dB (of acoustic backscattering strength variations).

The system is small, portable and easy to operate and its components include a surface processing and graphics unit, a subsea towfish and a towcable.

Operational status
In production.

Contractor
EdgeTech, Milford.

VERIFIED

MULTIBEAM MAPPING SYSTEMS

FRANCE

TSM 5260/5265 echo-sounders

Type

Shallow and deep water multibeam echo-sounders.

Description

The TSM 5260 and 5265 are multibeam echo-sounders designed for shallow and deep water operations respectively. This multibeam equipment is essential for effective sea mapping and allow for simultaneous depth measurements on a line perpendicular to the ship's axis. Other features include real-time layout of bathymetric maps in depths over 6,000 m and logging of raw data for instant replay or offline processing.

The TSM 5260 and 5265 incorporate a full range of digital technologies and allow for onboard replay with modification of parameters where required, real-time testing and maintenance of all channels and optional modification of interfaces and peripherals. Optional equipment includes digital optical disk, side scan sonar and towed arrays.

The TSM 5260 Lennermor echo-sounder uses 20

channels for shallow water (down to below 500 m) sounding on an operating frequency of 100 kHz. The TSM 5265 Nadzomor is a deep water sounder operating at 12 kHz at depths greater than 10,000 m.

The TSM 5265 incorporates a new 6 m array with cylindrical ceramics, specially designed for deep water operations. It produces a very flat beam perpendicular to the ship with a wide beam to port and starboard. Two flank receiving arrays are mounted on the hull of the ship, these being of the same type as in the submarine *Le Triomphant*. The TSM 5265 is a high-precision system with a resolution of 0.1° and produces a 20,000 m swath at a depth of 5,000 m. The system also incorporates a high-resolution, integrated deep water sub-bottom profiler.

The system underwent sea trials in 1995 and the equipment is undergoing acceptance tests.

Operational status

In service on board *Marine II,* for the 1,000 Territories Administration.

Specifications

	TSM 5260	TSM 5265
Operating frequency	100 kHz	12 kHz
Pulse length	± 10 ms	1 ± 10 ms
Source level (μPa/m)	210 dB	235 dB
Number of beams	20	60
Max ranges		
Vertical beams	>500 m	>10,000 m
Lateral beams	At 50°	at 45°
	Approx 400 m	approx 9,500 m
Angular resolution	3°	2° with 1.5° intervals
Distance resolution	30 cm	1 m

Contractor

Thomson Marconi Sonar SAS, Sophia Antipolis.

UPDATED

TSM 6265

Type

Deep ocean seabed mapping and imaging system.

Description

The TSM 6265 is a seabed mapping and imaging system which is designed to carry out deep ocean surveys using the multibeam technique. An optional sub-bottom profiling capability is also available.

In transmission mode, up to five beams covering up to 140° transversely can be simultaneously transmitted with a longitudinal beamwidth of 1.4°. In receive mode the beamwidth covers 1.7 to 4° measured in 0.06° steps. The number of soundings is up to 3,000 evenly distributed.

The maximum speed at which the system can be operated is as operationally required, although nominally it is between 12 and 14 kt. The system can be operated in up to Sea State 5.

Specifications

Range: 100 – 11,000 m
Frequency: 12 kHz
Pulsewidth: 2.5 – 100 ms
Accuracy: better than 0.5%

Contractor

Thomson Marconi Sonar SAS, Sophia Antipolis.

VERIFIED

GERMANY

Dynabase CRU

Type

Motion sensor

Description

Dynabase CRU is an attitude measuring platform designed for real-time motion/correction or active beam-steering of multibeam echo-sounder systems, as well as for real-time heave correction of single-beam

survey echo-sounders. It uses state-of-the-art solid-state gyro principles and consists of three quartz rate sensors, three accelerometers and one temperature sensor. Data output includes processed raw data and derived data for roll angle, pitch angle and heave. Roll and pitch accuracy is better than 0.05°.

The Dynabase CRU, is designed for installation directly above transducers up to 10 m below water surfaces.

Operational status

Systems are operational with various multibeam echo-sounders of differing frequency, depth range and coverage.

Contractor

STN ATLAS Marine Electronics GmbH, Bremen.

UPDATED

Fansweep 15

Type

Multibeam swath surveying system.

Description

Fansweep 15 is a new portable multibeam swath surveying system which combines bathymetric measurement and side scan imaging for shallow water applications at depths down to 300 m operating at a frequency of 200 kHz.

The unit incorporates advanced hydroacoustic and electronic signal processing facilities and meets the

latest accuracy requirements of the IHO, relating to IHO guideline SP44.

The portable unit features a single transducer assembly which is designed for flat bottom installation on small vehicles or survey launches.

The bathymetric coverage is freely selectable up to 3.5 times the vertical depth, which corresponds to an angle of 120°. At the same time, and fully independent of the bathymetric measurements, the system is able to collect high-resolution side scan values with an angle of up to 150°.

Fansweep 15 can generate up to 600 beams per sweep and up to 12 sweeps per second. A total of 4,096

sidescan measurements are available for each sweep. Accuracy is typically ±0.1 m plus 0.2 per cent of water depth.

It can be interfaced to a wide range of processing facilities and sensors, including a proprietary high precision motion sensor.

Contractor

STN ATLAS Marine Electronics GmbH, Bremen.

NEW ENTRY

Fansweep 20

Type

Multibeam swath surveying system.

Description

A new portable multibeam swath sounding system which combines bathymetric measurement and side

scan imaging for shallow water applications at depths down to 600 m.

It is available in two versions for 100 kHz and 200 kHz operation with equivalent key technical parameters for both configurations.

Incorporating advanced hydroacoustic and electronic signal processing facilities, the system

provides high accuracy. The system meets the latest accuracy requirements of the IHO, relating to IHO guideline SP44.

The bathymetric coverage is freely selectable up to × 12 of the vertical depth, which corresponds to an angle of 161°. At the same time and fully independent of the bathymetric measurement, the system is able to collect high-resolution side scan values within an angle of up to 180°.

It can be interfaced to a wide range of processing facilities and sensors, including a proprietary high-precision motion sensor.

Operational status
In service with the Royal Navy and on order for the Australian Navy.

Contractor
STN ATLAS Marine Electronics GmbH, Bremen.

UPDATED

Hydrosweep DS-2

Type
Multibeam surveying system.

Description
The full ocean depth DS-2 sonar system with incorporated bottom imaging and classification functions has been developed for large-scale surveying over depths down to 11,000 m, the multibeam Hydrosweep DS-2 system uses advanced electronic beam-forming techniques to provide a swath width equivalent to almost four times water depth. The fully automatic system is self-calibrating, with mean sound

velocity applied for depth and slant angle correction in real time.

For system control and operation as well as data acquisition and quality control, a powerful workstation serves as operator interface. Geometrically corrected side scan imaging and seafloor backscatter analysis for feature display and bottom properties are also available. A fully IMO/IHO compliant electronic chart database can be overlaid to view real-time swath data in their true geographic reference. Interfaces to various sensors and navigation systems as well as powerful data processing systems are provided for developing fully integrated hydrographic survey systems.

Operational status
Various systems are in use by oceanographic institutions. A system is installed in the US Navy's Naval Sea Systems Command *Thomas G Thompson*, 263 ft oceanographic research vessel.

Contractor
STN ATLAS Marine Electronics GmbH, Bremen.

VERIFIED

Hydrosweep MD-2

Type
Hydrographic multibeam surveying sonar with incorporated side scan imaging function for medium water application.

Description
The Hydrosweep MD-2 system is available in two frequency versions (30 kHz and 50 kHz) and covers a maximum swath width of eight times water depth in a depth range from 3 to 5,000 m. A novel function is the slant range corrected side scan image which can be

produced in real time with an overlay of isobathic depth contour lines and backscatter analysis of the seabed. A powerful workstation serves as operation interface and provides online data quality control. A fully functional electronic chart can be overlaid to view real-time depth data with their true geographic reference. Interfaces to navigation systems and hydrographic survey systems provide the possibility for total integrated system structure.

Besides the hydrographic application, the system is also suitable for object search, for example wrecks, pipelines and the like.

Operational status
The first system has been operational since the beginning of 1991 on the German hydrographic vessel *Wega*. Systems have also been supplied to other hydrographic services and the NATO SACLANT centre.

A system (MD-2/30) has been supplied to the New Zealand Navy and is installed on the survey vessel *Resolution*.

Contractor
STN ATLAS Marine Electronics GmbH, Bremen.

UPDATED

NORWAY

EM 100

Type
Multibeam echo-sounder.

Description
This multibeam swath echo-sounder uses computerised post-processing to produce seabed maps in coastal areas and on the continental shelf down to 600 m. It produces records of the total area of the seabed rather than the single lines obtained from conventional echo-sounders. It uses three selectable swath widths to give optimum combination of swath coverage and horizontal resolution in water depths to 600 m.

The system uses a simple and easy to understand menu technique based on a joystick control. All other subsystems are controlled from the operator unit, the software carrying out calculations of relative position of

depths from each beam, calculation of data for presentation on the VDU, and preparation of data for storage on tape and/or for input to the real-time contour map processor. The graphic section of the display shows the athwartships depth profile of the swath as well as the along-track profile for any one of the 32 beams for a selected time interval. In addition, the signal strength of each receiving beam is shown below the profiles. Maps can be produced in real time while the echo-sounder is in operation, or offline, using recorded data as input.

The compact design and low weight enable the EM 100 to be mounted in small vessels down to launch size. The transducer may be retracted into the hull for protection when not in use.

The main features of the system are the beam-forming and interferometric processing within each beam, the unique bottom tracking algorithms, the utilisation of the sound velocity profile for real-time

correction for beam deflection and the compact, curved transducer which is lowered below the aerated water during the survey. Post-processing of the bathymetric data is carried out on a DEC MicroVax 3400. The computer, together with peripherals and Simrad's SP100 post-processing software go to make up the rest of the system. A full ocean depth version, the EM 12, has been developed.

Operational status
No longer in production. Two shallow water systems are installed in the US Navy's T-AGS 51 and 52 coastal survey vessels.

Contractor
Kongsberg Simrad AS, Horten.

UPDATED

EM 1000

Type
Multibeam echo-sounder.

Description
Developed from experience with the EM 100 multibeam sounder, the EM 1000 uses the same operating frequency, 95 kHz, and is intended for high-precision surveys in water depths between 5 and 800 m.

The system produces bathymetric data and, as an option, a geometrically correct sonar image of the acoustical reflectivity of the seabed. The sounder incorporates full corrections for vessel movements and acoustic raybending in real time.

Like other Kongsberg Simrad echo-sounders, the EM 1000 uses a common computer and software system, Neptune, to take care of post-processing of data. The system can communicate with other equipments via Ethernet and/or RS 232 connections. A synchronising unit can be added to eliminate problems

of interference between different acoustic devices on the ship.

The transducer is based on the EM 100 transducer design, using a single transducer for both transmission and reception. The transducer is cylinder-shaped with a radius of 45 cm. It is configured from 128 ceramic staves covering an arc of 160°. Each stave consists of five elements with a fixed weighting in the fore and aft plane, and spaced by 1.25° in the athwartships direction. In the fore and aft direction the fixed beam-forming is a 3° opening angle, and the beam centre direction is perpendicular to the transducer surface.

Received beams are formed in real time during the reception phase according to the beam-forming mode which has been selected. Each beam is formed as a symmetrically weighted sum of the signal received over a selected group of staves. This eliminates any influence on the beam's direction from the sound velocity at the transducer, and makes it possible to factory calibrate each beam with an accuracy of 0.1°. All receiver beams are electronically roll-stabilised.

Split aperture is used so that measurement within each beam of the instantaneous direction to the point of backscatter is possible. Depths are measured in the centre line of footprint of each beam.

The receive beam-forming modes are:
(a) Narrow – intended for deep water surveys typically 600 to 800 m, 48 roll-stabilised beams with a spacing of 1.25° are formed covering a 60° sector and mapping a swath of × 1.15 the depth of water
(b) Wide – intended for medium water depth surveys typically 200 to 600 m, 48 beams with a spacing of 2.5° are formed during each ping. All beams are shifted 1.25° every second ping, so that over a two-ping period the system obtains 96 soundings with a spacing of 1.25°. The measurement sector is 120°, corresponding to a swath width of × 3.4 the depth of water
(c) Ultra Wide – intended for shallow water surveys between 5 and 200 m, 60 beams with a spacing of 2.5° are formed during each ping. All beams are shifted 1.25° every second ping, so that over a

two-ping period the system obtains 120 soundings with a spacing of 1.25°. The measurement sector is 150°, corresponding to a swath width of × 7.4 the depth of water.

Split aperture phase detection allows more than one detection per beam and is to be introduced in future versions of the sounder.

Operational status
A trial survey was carried out aboard the 65 ft swath hydrographic survey ship *Frederick G Creed* of the Canadian Hydrographic Service during the spring of 1990. At speeds between 14 and 18 kt the sounder produced very successful results. Now in service on several vessels worldwide. Two EM 1000 systems were ordered in 1998 by the US Navy for installation on 'T-AGS 60' class survey vessels.

Contractor
Kongsberg Simrad AS, Horten.

UPDATED

UNITED KINGDOM

Bathyscan

Type
Precision ultra-wide swath depth sounder.

Development
Bathyscan is a seafloor mapping system which operates at frequencies of 100 kHz and 300 kHz. In the 100 kHz mode, it can map in full continental shelf water depths with a swath width of up to 500 m, whilst at 300 kHz it is more suitable for rivers, harbours and estuaries. The system has been successfully used in a variety of environments worldwide for hydrographic surveys, pipeline route surveys, rig site inspection, harbour surveys and dredging operations. It provides an accuracy better than 30 cm at 300 kHz. The towing speed of the survey vessel is normally 6 kt.

Description
The Bathyscan swath sounding system provides high-density depth information on both sides of a ship's track. Side scan signals are transmitted and received on multiple interferometers providing about 1,500 soundings per second. These are combined with attitude data of the towed 'fish' to provide real-time display of seabed profiles. With the navigation data, a 'footprint' plot is produced in real time so that survey coverage can be monitored.

Other attitude data, tide corrections and swath merging are applied during post-survey processing.

Sophisticated fast post-processing software merges the depth swaths, taking account of the tidal correction, and provides detailed quality control data about the resulting depth matrix. The latter can be presented on a scale suitable for detailed site investigation or ready for chart compilation.

Side scan sonar signals are available and can be displayed on a suitable recorder.

Real-time seabed profiles, corrected for roll and heave, are shown on a colour graphics display. About 50 profiles on both sides are shown and scrolled downwards with each transmission. The control VDU displays towed fish attitude, roll, pitch, heave, depth,

Bathyscan precision ultra-wide swath depth sounder

heading, system condition, mode of operation and system and error messages.

Operational status
The hydrographic service of the Royal Netherlands Navy has been operating Bathyscan on the survey vessel Hr Ms *Blommendal* and her sister ship, Hr Ms *Buyskes*. Another Bathyscan system is in service with the Royal Thai Navy hydrographic department. GEC-Marconi also offers Bathyscan for lease. A wide range of surveys has been undertaken in the North Sea for the major oil companies and operations have also extended to overseas surveys including the Caspian Sea.

Contractor
GEC-Marconi Underwater Systems Group, Waterlooville.

UPDATED

UNITED STATES OF AMERICA

EM 121A

Type
Multibeam bathymetric mapping sonar system.

Description
The EM 121A multibeam echo-sounder is the first commercially available 1° bathymetric mapping sonar system. It operates at a frequency of 12 kHz with a variable transmit pulse length from 1 to 20 ms, providing precise mapping and calibrated backscatter measurement capability from 10 m to full ocean depth (11,000 m). When operated with maximum resolution, the sonar transmits a beam measuring 1° fore and aft by 140° athwartships and forms 121 × 1° receive beams covering a swath across the vessel track. The transmit beam is stabilised for pitch and the receive beams are stabilised for roll which is important in minimising the position, depth and signal amplitude offset corrections. This feature extends the sonar's full system performance to Sea State 4 and also contributes to the system's typical depth sounding accuracies of 0.1 per cent of water depth or 30 cm (whichever is greater) and the precise bottom reflectivity measurements. Bottom detection is based on combined amplitude envelope detection and phase

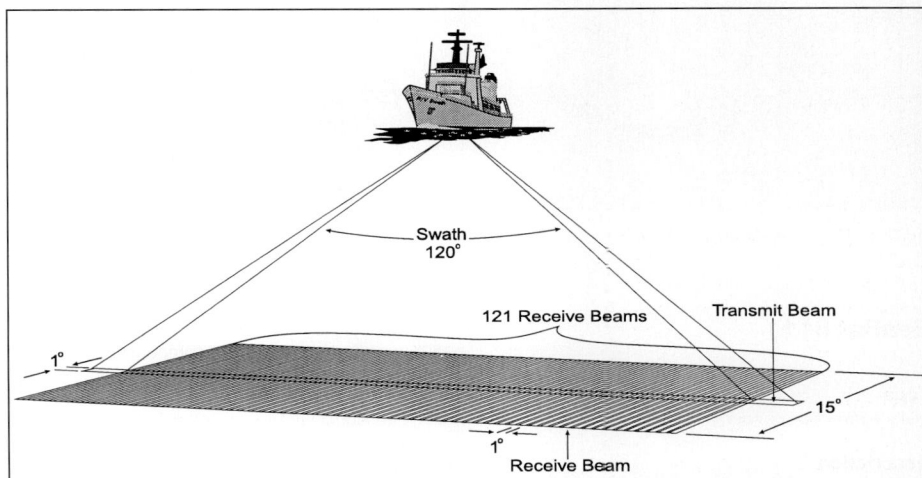

Swath coverage of the EM 121A multibeam echo-sounder

1995

detection (interferometric principle), a technique which has proved to be highly reliable and productive on previous Simrad designs. Also included is full and automatic compensation for ray bending caused by varying sound velocities in the water column.

The EM 121A functions without operator intervention

under almost any condition. The system automatically produces, in real time, bathymetric contour maps and sonar images of the seabed (equivalent to those produced by towed side scan sonars). The system features extensive built-in automatic test and calibration equipment to ensure that it functions within specification continuously and which also facilitates maintenance. For example, the performance of each of the 58 power amplifiers is measured during each ping and each of the 128 preamplifiers is monitored for overloads at three different stages. This quality assurance information is displayed for the operator and also recorded with the other datagrams. The volume and consistency of the recorded data provides a far

more accurate basis for construction of the desired digital terrain models than is possible with any earlier generation of equipment. From these models contour maps and three-dimensional views of any size and scale can be produced. These products can also be combined with the very high-resolution bottom reflectivity data. As an option, the raw data, complex samples from all 128 hydrophone channels, can be output from the receiver to an external system or logged on high-density tape (along with vessel attitude data) for other research and development applications. All output data formats include every vital system and ship parameter thus ensuring exact correlation and calibration of the data during post-processing.

Operational status
The first EM 121 (a variant of the EM 121A which includes Simrad's complete post-processing package Neptune) was commissioned into the US Navy *Zeus* in January 1994. Simrad has been contracted to build and install five more EM 121A systems for NAVOCEANO's new class of multimission oceanographic research vessels: USNS *Pathfinder* (T-AGOS 60) USNS *Sumner* (T-AGS 61), USNS *Bowditch* (T-AGS 62), USNS *Henson* (T-AGS 63) and T-AGS 64. The first three of these systems have been commissioned.

Contractor
Kongsberg Simrad Inc, Lynnwood, Washington.

Specifications

	Shallow	Intermediate	Deep	Very deep
Typical depths	10-300 m	200-1,500 m	1,200-5,000 m	4,500-11,000 m
Transmission sector coverage	120°	120°	120°	120° (90° at 11,000 m)
Transmission beamwidth	4°	2°	1 or 2°	1 or 2°
Source level (u30 dB sidelobe shading)	226 dB/μPa/m	232 dB/μPa/m	238 dB/μPa/m	238 dB/μPa/m
Default pulse length	1 ms	2 ms	7 ms	20 ms
Number of receive beams	61	121	121	121
Receive beamwidth (u30 dB sidelobe shading)	4°	2°	1 or 2°	1 or 2°
Range sampling interval	30 cm	60 cm	240 cm	480 cm

VERIFIED

SeaBat 8101 multibeam echo-sounder

Type
Hydrographic sonar.

Description
The new generation SeaBat 8101 multibeam echo-sounder is designed for continental shelf survey operations, measuring a wider swath at higher survey speeds and complying with International Hydrographic Organization (IHO) and US Army Corps of Engineers (USACE) Class 1 standards.

Operating at 240 kHz, the SeaBat 8101 measures a

150° swath consisting of 1.5 × 1.5° beams. Each beam is both phase and amplitude analysed, with one or both methods used to digitise the seafloor. Dense coverage is achieved utilising up to 3,000 soundings per second for 100 per cent coverage of a swath that can be over 500 m wide, even as the survey vessel travels at speeds of 18 kt.

Being lightweight and portable, the SeaBat 8101 transducer head can be mounted temporarily on vessels of opportunity, mounted permanently through the hull of a survey vessel, or mounted on an underwater platform such as an ROV, AUV, or ROTVs. The head can be ordered pressure-rated for operating depths of 300, 1,500, 3,000 or 6,000 m.

The system utilises RESON's advanced 8100 Series processor. This unit can download new firmware through a communications port or Ethernet – enabling changes to the processor firmware to be made in just 30 seconds. Thus, the processor can accommodate a 455 kHz, 240, 120, 60, 30 or 12 kHz SeaBat swath bathymetry system. It will sense automatically whether the right transducer is mounted, set itself up and calibrate the total system by introducing a test signal into all transducer elements. This arrangement gives maximum operational flexibility, minimises requirements for operator training and spare parts inventory, and means that keeping pace with the latest bottom detection methods and beam geometries requires just a firmware upgrade of the processor unit.

The processor is just one example of the modular design philosophy underlying the SeaBat 8101. A wide range of upgrades and options is available to tailor the SeaBat to the user's current and future needs.

Operational status
The SeaBat 8101 is in operation with hydrographic surveyors worldwide, including the Danish Coastal Authority, the Chinese Academy of Sciences, and the UK North Sea oil and gas industry.

Specifications
Operating frequency: 240 kHz
Resolution: 5 cm
No of beams: 101
Horizontal beamwidth: 170° (transmit); 1.5° (receive)
Vertical beamwidth: 1.5/3° (transmit); 20° (receive)
Range settings: 2.5, 5, 10, 25, 50, 100, 200, 300 m

Contractor
RESON Inc, Goleta, California.
RESON A/S, Slangerup, Denmark
RESON Offshore Ltd, UK.

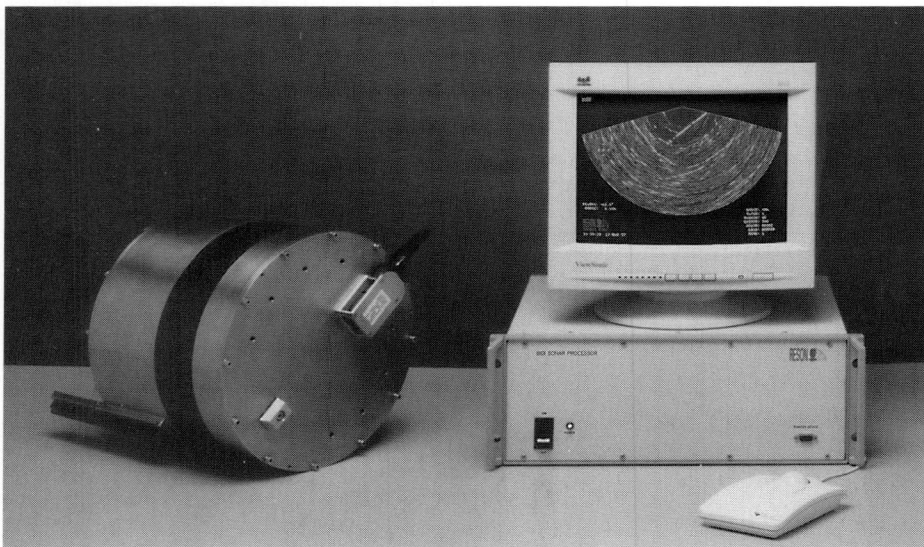

SeaBat 8101 display and wet end

1998/0009853

UPDATED

SeaBat 8111

Type
Multibeam echo sounder.

Description
The new SeaBat 8111 is a modular, multibeam echo sounder system operating at 100 kHz producing high-density, high-accuracy soundings of the seafloor over a

150° swath. Major system components include a transducer array, a transceiver unit and a processor unit.

The transducer array consists of a cylindrical receive array and a linear transmitter array, mounted together on a support cradle which provides mounting points to the vessel. Lightweight and portable, the array can be installed temporarily over the side of a COOP.

The SeaBat 8111 transceiver features plug-in cards

for easy maintenance and is controlled from the sonar processor. The 81-P processor is compatible with other SeaBat sonar heads, can be updated in minutes to accommodate future requirements and features a user-friendly point-and-click interface.

Operational status
Surveyors in Europe, Asia and North America use the SeaBat 8111.

Specifications
Number of beams: 101
Beam size: 1.5° × 1.5°
Resolution: 5 cm
Swath width: 150°

Coverage: up to 7.4 times water depth
Update rate: up to 30 times per second (full swath)
Stabilisation: pitch within ±10°
Measurement range: 3 to 600 m

Contractor
RESON Inc, Goleta.
RESON A/S, Slangerup, Denmark.
RESON Offshore Ltd, UK.

NEW ENTRY

SeaBat 9001

Type
Hydrographic sonar.

Description
The SeaBat 9001 is a lightweight, portable, multibeam echo-sounder designed to measure accurately a swath of the seabed from small vessels or ROVs.

The system measures a 90° swath consisting of 60 soundings. The measured profile is output to an onboard computer system for display and storage. Additionally, a real-time video profile is provided for quality control. If desired, it can be recorded to a standard VHS recorder or data collection system for archiving and replay.

The SeaBat 9001's 60 1.5 × 1.5° narrow 455 kHz beams update up to 15 times per second. With this high-frequency, narrow footprint, and fast update rate, the SeaBat can operate at vessel speeds in excess of 12 kt while maintaining 100 per cent coverage of the seabed.

The SeaBat 9001 transducer head is very small and weighs less than 15 lb wet. Accordingly, the system can quickly be mobilised on virtually any vessel of opportunity to measure water depths to 100 m. Or, it can be mounted on ROVs or towed bodies and submerged to depths of 500 m.

Numerous factory options and upgrades are available for the SeaBat 9001. Two of the most popular are the 'S' option and Option 019. When equipped with the 'S' option, the flick of a switch on the processor changes the SeaBat's beam geometry and presentation display from bottom-mapping echo sounder function (SeaBat 9001 mode) to forward-looking sonar function (SeaBat 6012 mode). When equipped with Option 019, the SeaBat 9001 measures a side scan swath from 4 to 400 m. The operator may select two of three side scan beams, each at an angle of 40°. The analogue side scan data may be viewed and correlated with the bathymetry data in real time.

Operational status
Independent tests of the system's accuracy and performance have been carried out by the National Ocean Service, a division of the National Oceanic and Atmospheric Administration (NOAA). The SeaBat 9001 is in service with a number of customers including the US Army Corps of Engineers, the Port of Los Angeles, the Japan Maritime Safety Administration and the Brazilian Navy Hydrographic Department.

Specifications
Operating frequency: 455 kHz
Number of beams: 60
Range settings: 2.5, 5, 10, 20, 25, 50, 100, 200 m
Range resolution: 5 cm
Horizontal beamwidth: receive 1.5° (−3 dB points for each beam); transmit 100°
Vertical beamwidth: receive 15°; transmit 1.5 °

Contractor
RESON Inc, Goleta, California.
RESON A/S, Slangerup, Denmark.
RESON Offshore Ltd, UK.

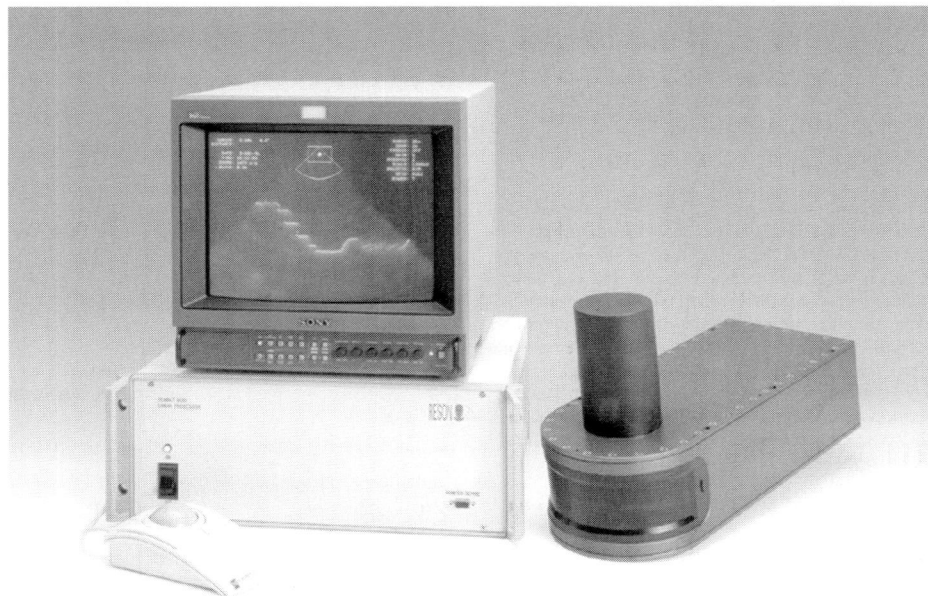

SeaBat 9001 display and wet end

1998/0007420

UPDATED

SeaBat 9002

Type
Hydrographic sonar.

Description
The SeaBat 9002 multibeam echo-sounder is a dual-head system that provides synchronised returns from two separate sonar heads. The system, like the single-head SeaBat 9001, is designed to measure high accuracy swaths of the seabed; however, the SeaBat 9002 can be configured to measure a total swath of 180°. Even with its two sonar heads, it is still compact, lightweight and portable enough to be used on small surface vessels or ROVs.

The system comprises two complete SeaBat 9001 systems, each sonar head measuring a complete swath of 60 soundings at a rate of up to 15 times per second. The two sonar heads operate independently of each other: each one provides its own reception, transmission and bottom detection; however, they are interconnected and time synchronised to alternate measurements between one another. The measured profiles are fed to an onboard computer, where the images can be viewed and simultaneously transmitted via RS-232 to a data acquisition system. Real-time quality control is provided by the updating video profiles, compiled by the measured swaths and raw sonar data.

Operating at a frequency of 455 kHz, each sonar head measures a total of 60 soundings (120 combined) each 1.5° × 1.5° across the seabed. One hundred per

A SeaBat 9002 system mounted on a MacArtney Focus 400 ROTV.

1998/0009854

cent coverage of the seabed, both along track and across track, can be obtained by the fast updating SeaBat, even at speeds greater than 12 kt.

When installed on a surface vessel or ROV, the sonar heads can be mounted in various positions, from separated locations angled towards each other and at varying angles for full coverage up to a 180° field of view. They are also capable of being mounted from a single mounting pole and from towed vessels/vehicles.

Operational status
Used daily by the National Ocean Service, a division of the NOAA, as its bathymetric system for least depth surveys.

Specifications
Operating frequency: 455 kHz
Number of beams: 120
Range settings: 2.5, 5, 10, 20, 25, 50, 100, 200 m

Range resolution: 5 cm
Horizontal beamwidth: receive 1.5° (–3 dB points for each beam); transmit up to 200°
Vertical beamwidth: receive 15°; transmit 1.5 °

Contractor
RESON Inc, Goleta, California.
RESON A/S, Slangerup, Denmark.
RESON Offshore, UK.

UPDATED

SeaBat 9003

Type
Hydrographic sonar.

Description
The SeaBat 9003 multi-beam echo sounder is similar to the SeaBat 9001 in function and form, except that it features a 120° swath made up of 40 × 3° beams rather than a 90° swath made up of 60 × 1.5° beams. With swath width approaching that of the dual-head SeaBat 9002 at the price of the SeaBat 9001, the SeaBat 9003 offers hydrographic surveyors an intriguing new option.

The SeaBat 9003 is available with all the same options and upgrades as the SeaBat 9001.

Operational status
The SeaBat 9003 is in operation in Australia, Brazil and the USA.

Contractor
RESON Inc, Goleta, California.
RESON A/S, Slangerup, Denmark.
RESON Offshore, UK.

UPDATED

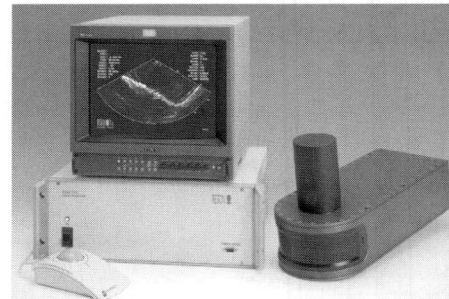

SeaBat 9003 system display.   **1998**/0009855

Sea Beam

Type
Ocean survey system.

Description
Sea Beam is a high-resolution bathymetric survey system which combines the Model 853-E narrow beam echo-sounder with the Model 875 echo processor for processing sounding data to generate bottom contour charts in real time. The chart is generated by means of soundings from 16 preformed beams positioned perpendicular to the ship's axis. Processed or unprocessed sonar echo signals are displayed on a VDU to provide a cross-track profile for each beam. The stabilised vertical beam depth data are displayed in digital form at the echo processor and on multiple depth display repeaters as required. The analogue signal from the vertical beam is displayed on a graphic recorder to provide a survey line directly under the ship's track. A contour plotter provides a continuous 11 in wide plot of the ocean bottom contours in real time. Time, heading and 16 pairs of depth and cross-track sounding co-ordinates are provided in digital form for recording.

Operational status
There are currently 16 Sea Beam systems in operation worldwide.

Contractor
SeaBeam Instruments Inc, East Walpole, Massachusetts.

VERIFIED

Sea Beam series 2100

Type
Multibeam high-speed seabed mapping system.

Description
The Sea Beam 2100 is a multibeam, high-speed mapping system which can be installed with multiple frequencies for complete deep and shallow water bathymetric capabilities. In addition to wide swath bathymetry, the system is configured for collocated and geometrically corrected side scan in both deep and shallow waters. The series can easily and economically be upgraded for sub-bottom and narrow beam profiling operations.

All systems electronics, including transmitter, receiver, sonar processor and optional mapping processor are capable of multiple frequency operations. Provisions can be made, as an option, to operate the system remotely using INMARSAT.

The Series 2100 systems are based on a SeaBeam Instruments' patented cross-fan beam technique relying on Mills Cross-type T-shaped arrays, with systems electronically steering beams in port and starboard directions. In a cross-fan system a projector array forms a single transmit beam which is broad in the athwartship direction and narrow in the along-track direction. A separate hydrophone array, perpendicular to the projector array, forms multiple receive beams simultaneously.

These beams are narrow in the athwartship direction and are steered electronically to intersect the transmit beam across its entire swath. Thus, each received beam detects only the echo from its portion of the transmitted swath. Taking all receive beams together, a complete cross-section of the bottom, perpendicular to the direction of the vessel, is measured on each ping cycle.

The Model 2112 of the Series 2100 is digitally controlled and features beamwidths of 2° × 2°.

The Model 2112 operates on a 12 kHz frequency and surveys from 50 m to full ocean depths. It provides swath coverage of at least 150° at depths of 50 to 300 m, and at least 120° to depths of 4,500 m.

The Model 2136 uses a 36 kHz frequency to survey in shallow to intermediate depths. It generates swath coverage of at least 150° at depths of 5 to 150 m, and at least 120° coverage to depths of 400 m. All Series 2100 systems are available with dual frequencies for expanded capabilities. The Model 2112.360, for example, features both a 12 kHz and a 36 kHz frequency.

The Series 2100 systems are powered up via a single switch. System initialisation and self-test are automatic. After initialisation, the operator can enter values for parameters such as chart scales, or default to values used the last time the system was run. The operator enters an approximate depth and starts the system pinging. Once the bottom is acquired, the system runs automatically, adjusting all sonar parameters (pulse width, receiver gain, and so on) for optimum operation as depths and bottom backscatter change.

In the dual-frequency system mode, the frequency switches automatically as the depth changes. The operator can set the depth at which the frequency

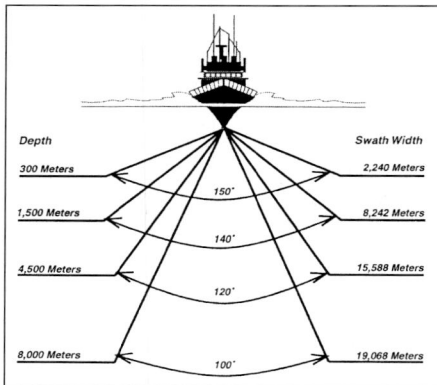

Sea Beam 2112 – Swath versus depth   **1996**

changes or manually control the system to remain at either frequency using menu selections.

RMS depth errors for the central beams are less than 0.2 per cent (±1 m) and the outermost beams are less than 0.5 per cent (±1 m) of the depth reported by the corresponding beams, for depths greater than 100 m. Swaths up to 120° still permit a 2:1 redundancy on the near vertical beams along-track at a speed in excess of 11 kt, a common survey speed.

However, it is necessary to gradually decrease the speed of the ship to 6 kt if the swath width increases from 120 to 150°, in order to maintain a 50 per cent overlap.

The echo processor comprises a receiver and processor controlled by a Motorola 68040 processor which receives its operator programme from the sonar processor at system start-up via Ethernet. This allows for receiver software upgrades via the modem port on the sonar processor. Control commands from the sonar processor to the receiver, as well as the transmission of hydrophone data from the receiver to the sonar processor, is via high-speed serial communication link.

The receiver portion amplifies and detects echoes received by the hydrophone arrays and sends sampled digital quadrature data to the sonar processor to produce bathymetric and side scan records. It provides operator interface, displays and plots, and control system timing.

The operator station comprises a menu control station and display unit, each housing a 20 in high resolution X-Windows display monitor. Menu and display functions can be performed at either station. The redundant stations run the Lynx operating system on 486-based PCs. Displays available include: profile, waterfall, side scan, selected beam, contour, and the sub-bottom option.

Operational status
Systems are in service with the National Science Foundation, the Korean Office of Hydrographic Affairs, the Ministry of Geology, People's Republic of China,

Scipps Institution of Oceanography, Woods Hole Oceanographic Institute, University of Hawaii and NOAA. A Sea Beam 2112 12 kHz version has been delivered to the India Naval Hydrographic Office for ocean research and survey duties and is installed on the survey vessel *Investigator*.

Contractor
SeaBeam Instruments Inc, East Walpole, Massachusetts.

Specifications

	Sea Beam 2112 **(12 kHz)**	**Sea Beam 2136** **(36 kHz)**
Resolution		
Bathymetry	2° × 2° at all depths	2° × 2° at all depths
Side scan	(Optional 1° × 1°)	(Optional 1° × 1°)
Frequency	12 kHz nominal	36 kHz nominal
Coverage	150° Swath – 50-300 m	150° Swath – 5-150 m
Swath vs depth	140° Swath – to 1,500 m	120° Swath – to 400 m
	120° Swath – to 4,500 m	100° Swath – to 800 m
	100° Swath – 8,000 m	Max depth – 1,500 m
Projectors	14 moulded modules, each containing ceramic resonators. Active array length is approximately 4.3 m	Single potted module. Active array length is approximately 1.4 m
Hydrophones	10 moulded modules, each containing 8 ceramic line hydrophones. Active array length is approximately 5.0 m	2 potted modules. Active array length is approximately 1.5 m
Number of beams	151	151
Power output	20 kW PEP (electrical)	1 kW
Pulsewidth	3 ms (shallow water) to 20 ms (deep water), automatically selected; operator override can be used to favour bathymetry or side scan	1 ms (shallow water) to 20 ms (deep water), automatically selected; operator override can be used to favour bathymetry or side scan
Transmit beamwidth	2° fore and aft at the −3 dB points	2° fore and aft at the −3 dB points
Receive beamwidth	2° athwartship at the −3 dB points	2° athwartship at the −3 dB points
Roll and pitch	Accommodates ± 10° of Roll and ± 7.5° of Pitch	Accommodates ±20° of Roll and ± 10° of Pitch
Data display	X – Y display of cross-track distance vs depth; graphic presentation of survey area, showing ship's track and near real-time ocean bottom contours of the surveyed area. Side scan imaging, 3-D, and waterfall displays	X – Y display of cross-track distance vs depth; graphic presentation of survey area, showing ship's track and near real-time ocean bottom contours of the surveyed area. Side scan imaging, 3-D, and waterfall displays
Data recording	Data is recorded on exabyte tape Sonar data includes depth, cross-track, and amplitude for each beam. Other data includes vertical sensor data, heading, time and navigation. Side scan data includes 1,000 points each side, equally spaced. Data for each point covers a 60 dB range in 0.25 dB increments	Data is recorded on exabyte tape Sonar data includes depth, cross-track, and amplitude for each beam. Other data includes vertical sensor data, heading, time and navigation. Side scan data includes 1,000 points each side, equally spaced. Data for each point covers a 60 dB range in 0.25 dB increments

VERIFIED

DEEP TOWED SURVEY SYSTEMS

UNITED STATES OF AMERICA

AMS sonar systems

Type
Towed wide swath imaging and bathymetric mapping sonar systems rated for operation to full ocean depth.

Development
The development of the AMS series of systems started in 1988 with the AMS 120SI in which Acoustic Marine Systems Inc (Redmond, Washington) introduced the 'passive interferometric' technique for obtaining simultaneous co-registered images and high-resolution bathymetric information. Later models included the AMS 120SP, AMS 36/120SI and AMS 60SI. Acoustic Marine Systems' technology was acquired by Simrad Inc in 1994.

Description
All AMS systems are designed for a two-body towing arrangement; the neutrally buoyant sonar vehicle towed behind a heavy depressor by means of a neutrally buoyant umbilical cable. This towing arrangement provides the sonar vehicle with the stability required for obtaining quality data even in high sea states.

AMS sonars are custom-configured to meet the user's specific requirements, with installed options that

AMS 60SI deep tow sonar for COMRA undergoing tests in Puget Sound   **1996**

include sub-bottom profiler, altimeter, three-axis magnetometer, navigation transceiver, various navigation interfaces and acoustic release, in addition to the standard pitch, roll, heading and depth sensors. Other configurations and options are available.

The SI models, using the 'passive interferometric' technique will operate over a 10,000 m, or longer, standard single coaxial oceanographic towcable, or optionally over a fibre optic cable. The AMS 120SP

uses a quadrature phase down-sampling technique and operates over a fibre optic cable in simultaneous imaging/topographic mapping mode, but also over a 10,000 m coaxial cable in swath imaging only mode. The AMS 36/120SI is a dual-frequency system, allowing simultaneous medium- and high-frequency swath imaging and swath bathymetric mapping at the high frequency. All systems are rated for operation to 6,000 m.

In addition to Simrad's AMS processor, the systems are also compatible with Triton Technology's Q-MIPS, Williamson & Associates' GeoMAP and others for processing of both image and bathymetric data.

Operational status
Including high-resolution topographic mapping, AMS systems are used for a variety of commercial surveys, deep sea search operations, geophysical and archaeological exploration, as well as limited military applications. AMS systems are operated by Williamson and Associates (International Deep Sea Survey), Oceaneering, Woods Hole Oceanographic Institution and China Ocean Mineral Resources R & D Association (COMRA).

Contractor
Kongsberg Simrad Inc, Lynnwood, Washington.

Specifications

	AMS 120SI AMS 120SP	AMS 60SI	AMS 36/120SI
Transmit frequency	120 kHz Port 120 kHz Stbd	57.6 kHz Port 57.6 kHz Stbd	33.3 kHz & 120 kHz Port 36.0 kHz & 120 kHz Stbd
Swath width, nominal (m) (km)	60, 125, 250, 500, 1,000	125, 250, 500, 1,000 1.5, 2.0, 2.5	60, 125, 500, 1,000 1.5 to 5.0 in 0.5 km steps
Transmit interval (ms) (s)	50, 100, 200, 400, 800	100, 200, 400, 800 1.2 to 2.0 in 0.4 s steps	100, 200, 400, 800 1.2 to 4.0 in 0.4 s steps
Transmit power, RMS	1 kW per side	2 kW per side	1 kW per side (120 kHz) 4 kW per side (33/36 kHz)
Beamwidth, horizontal vertical	1.5° 60°	1.5° 60°	1.5° 60° (120 kHz) 40° (33/36 kHz)
Transmit source level	224 dB/μPa/m	227 dB/μPa/m	224 dB/μPa/m (120 kHz) 228 dB/μPa/m (33/36 kHz)
OCV response	−180 dBV/μPa	−179 dBV/μPa	−180 dBV/μPa (120 kHz) −174 dBV/μPa (33/36 kHz)
Receiver sensitivity		−172 dBV/root Hz typical	
Time varied gain	Automatic correction for spreading and attenuation over 100 dB in 0.2 dB steps Operator selectable 'Search' or 'Survey' TVG modes		
Gain adjustment	Operator selectable in 3 dB steps over 42 dB range		
Transmit pulse length	Programmable/operator selectable from 1 to 80 cycles Automatically adjusted for swath selected, with manual override		
Across-track resolution	Typically 1/2,000 of swath width selected		
Bathymetry resolution	Typically 1/100 of tow vehicle altitude or 1/1,000 of swath width independent of water depth		

SeaMARC®

Type
Towed wide swath bathymetric and backscatter imaging sonar.

Development
The original SeaMARC® (Sea Mapping and Remote Characterisation system) was developed from a high-speed exploration system produced for INCO's sea-floor mining activity in 1978. The design was improved by International Submarine Technology throughout the 1980s. SeaMARC® was the first operational sea-going system to combine swath bathymetry and backscatter imagery in a single sensor. Successive models have incorporated current generation digital signal processing, all digital data output, deep tow hardening to 11,000 m, and operating frequencies from 11 to 150 kHz and higher.

Description
The system consists of: a neutrally buoyant towbody; power, tow and control cable (a single, contrahelically armoured coaxial cable with multiplexed power and data) with depressor and handling system; power supplies; system control console; and data displays and processing hardware. Light guide cables are used for very high-resolution applications. Heavy towbody versions are also available. All range data are compensated for towbody unsteadiness.

Dual hydrophones mounted on each side of the streamlined towfish produce a highly directional azimuth side-looking sonar pulse. Precise correlation of return-signal phase angle with time allows determination of bathymetric surface within a 160° zone directly beneath the towfish and perpendicular to the track. Quantified absolute signal strength values allow simultaneous backscatter intensity logging. The result is a single-ping, fan-shaped return divided into

some 2,000 pixels, each showing a bathymetric surface value and a backscatter image value. Three-dimensional imaging is induced with water volume targets shown suspended over a bathymetric surface.

Operating speeds vary from 1 to 10 kt and overall swath widths vary with application from less than 100 m to more than 20 km. The system can be operated in Sea State 6, with retrieval in Sea State 5.

SeaMARC® is capable of modification for parametric low-frequency operation, as well as multibeam configurations. It is already proven in such applications as seafloor route surveys and geophysical exploration. Units may be configured for deep towing or high-tow-altitude missions, and for resolutions varying from 5 cm to 10 m.

All towbodies provide co-registered bathymetry and backscatter imagery and are controlled from a common shipboard control system and data processing suite.

Operational status

A total of 14 SeaMARCs® has been built between 1981 and 1992. Present users include university oceanography departments and geophysical exploration firms. Mine countermeasures applications are being pursued. Development and sales continue.

Specifications

Performance: across track resolution 0.05% of swath width; bathymetric accuracy with respect to towfish is from 0.5% to 1% of swath width

Contractor

Raytheon Systems Company, Mukilteo, Washington.

VERIFIED

Deploying a SeaMARC® towbody

Sea Beam 2000

Type

Ocean survey system.

Description

Sea Beam 2000 is an advanced, wide swath bathymetric sonar whose design approach makes it possible to customise the system to meet the user's mission requirements and minimise the individual ship's limitations.

Sea Beam 2000 employs a modular approach to both transducer and electronic design. This creates a sonar which may be modified so as to have a different frequency, beamwidths, number of beams and array sizes.

Sea Beam 2000 utilises software-controlled beam-forming. For a given frequency, the sonar beam dimensions are determined by the array lengths which, in turn, determine the number of transducers and associated electronics required for the hydrophone and projector arrays. This modular system design provides the flexibility required to optimise frequency and array dimensions for a variety of system requirements and ship sizes. Multiple formed beams of 1 to 3° are accommodated. In all cases, the swath width is 92°.

Special features of the system include: full 92° swath coverage from 10 to 11,000 m; three operational depth modes – shallow, intermediate and deep; a hydrophone array which conforms to hull angles up to 10°; field sheet output showing ship's track against latitude and longitude with depth contour displayed; graphic recorder output of along-track profile; on-track strip chart recording of depth contours; beam intensity displayed with cross-track profile and graphics terminal with three-dimensional waterfall display of most recent cross-track profiles. During 1993 SeaBeam introduced the latest system in this series, the series 2100. In addition to the capabilities of the Sea Beam 2000, the 2100 offers: multifrequency operating, sub-bottom profiling (collocated), diver replaceable acoustic sensors, compact design and low cost.

Operational status

Contracts for Sea Beam 2000 have been awarded to SeaBeam Instruments Inc by the Japan Maritime Safety Agency (JMSA) for the Research Vessel (R/V) *Meiyo*, the Scripps Institution of Oceanography for the R/V *Melville* and the Korean Ocean Research and Development Institute.

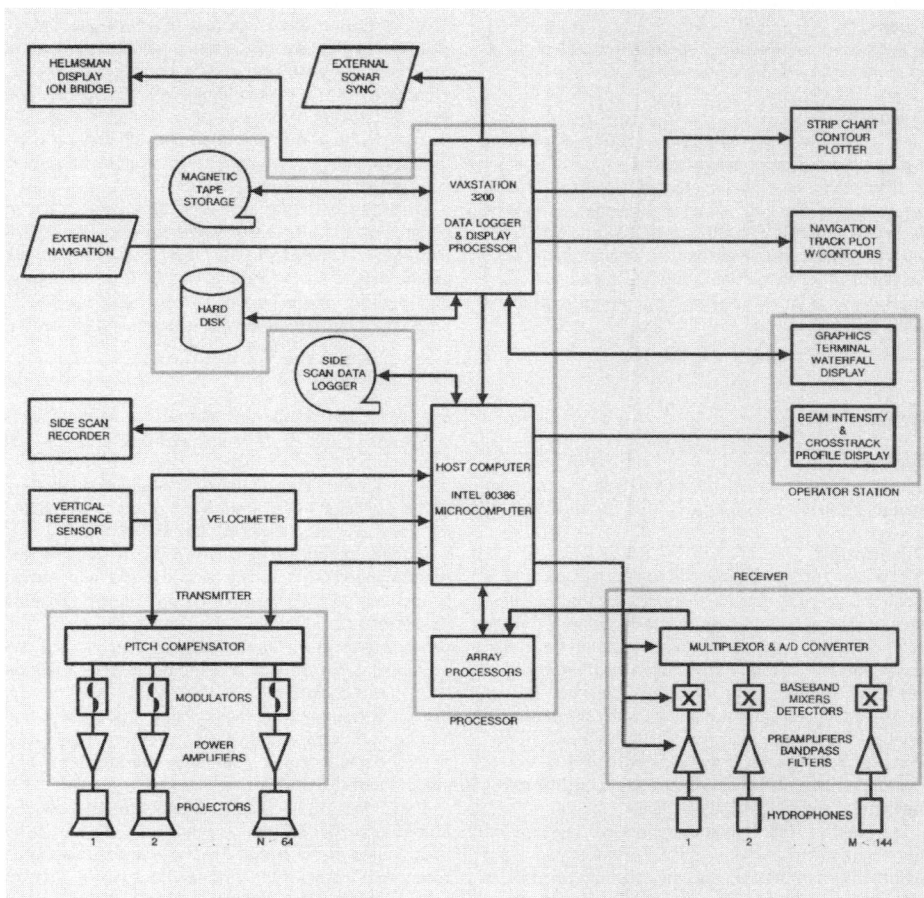

Block diagram of Sea Beam 2000

Depth mode	Shallow	Intermediate	Deep
Depth	10-400 m	200-5,000 m	300-11,000 m
Transmit beamwidth (fore and aft)	4°	2°	2°
Receive beams (per ping)	23	46	46
Receive beamwidth (athwartship)	4°	2°	2°
Receive beamwidth (fore and aft)	15°	15°	15°
Pulse length	2 ms	7 ms	20 ms
Ping period	0.5-2 s	1-11 s	4-22 s
Projectors/power amplifiers	14	28	28
Hydrophones/preamplifiers	42	84	84

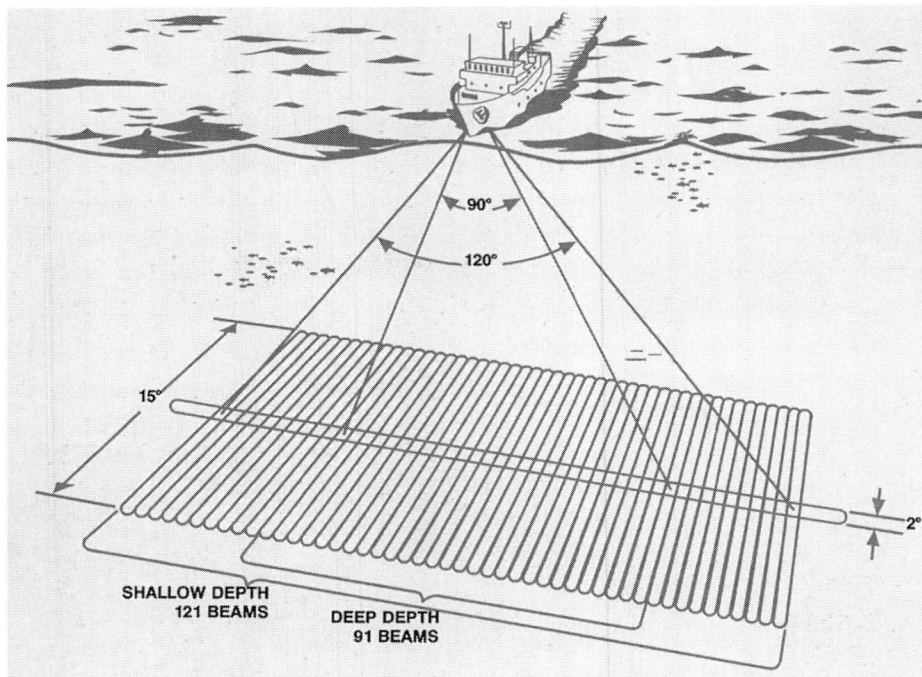

Specifications
General
Depth (3 modes): 10-11,000 m
Swath width: 92°
Transmit beams: 1 (for 92° swath)
Transmit beamwidth: 100° (athwartship)
Receive beamwidth: 2° (athwartship)
Receive beamwidth: 15° (fore and aft)
Pulse length: 2-20 ms
Frequency: 12 kHz
Ping period: 0.5-22 s
Projectors/power amplifiers: 28
Source level at 30°: 234 dB/μPa/m
Hydrophones/preamplifiers: 84

Contractor
SeaBeam   Instruments   Inc,   East   Walpole,
Massachusetts.

VERIFIED

Sea Beam 2000 beam patterns

Bathy 500

Type

Single- or dual-frequency, thermal recording echo-sounder for hydrographic surveying.

Description

The lightweight, low-power Bathy 500 is a low-cost, high-functionality echo-sounder that is easy to operate and maintain. Its ease of integration with shipboard systems and compact construction offer portability, accuracy and reliability in generating precision chart recordings and digital data output. Designed for depth ranges to 1,200 m, the Bathy 500 offers a wide range of applications from hydrographic surveying to medium-depth bathymetry.

The echo-sounder provides advanced digital signal processing for reliable bottom detection and automatic chart annotation. Automatic and manual controls allow the user to control time and date, position, and alarms. Equipped with a built-in digitiser with RS-232 and RS-422 data output interface ports and an NMEA 0183 navigation input interface, the echo-sounder provides the user with a complete integrated hydrographic survey system.

The Bathy 500 uses modern microprocessor-based electronics and a thermal chart recorder mechanism. It features non-volatile internal clock and parameter set up memory, a sealed keypad and LCD for data entry and remote mark input. It is housed within a splashproof aluminium enclosure weighing 10 kg. The

unit consumes less than 40 W and offers user selectable DC or AC input power.

Operational status

The Bathy 500 was introduced in 1996. Over 200 units have been delivered worldwide.

Contractor

Ocean Data Equipment Corporation, E Walpole, Massachusetts.

VERIFIED

Bathy 1000

Type

Dual-frequency, thermal recording echo-sounder for hydrographic surveying.

Description

The Bathy 1000 hydrographic survey system offers modern computer-based technology with resulting ease of operation and high reliability. In addition, the open architecture of the system allows interface with many peripherals and other shipboard networks. It provides accurate data through advanced signal processing and is easy to maintain with built-in tests. It is a full featured, research quality survey system offering maximum flexibility to meet a wide variety of mission requirements including sub-bottom profiling, bathymetry and bottom classification.

The Bathy 1000 includes controller, processor, thermal recorder, LCD display, sealed keypad and integral 2 kW amplifiers. It is designed for applications in rivers, harbours, and coastal waters and performs to depths of 6,000 m. The flat-panel LCD display provides a time continuous visual record of most recent depth, eliminating the need for paper records, except where a permanent hard-copy archive is required. The 2 kW power amplifiers provide 25 ms pulse widths for operation to 6,000 m at 12 kHz. Maintainability and reliability are improved due to fewer failures of high-reliability, low-power digital electronics, modular power supplies and a thermal printer without stylus or moving belt.

Various options for transducers and mounting include towfish, sea chest and pole mount allowing

installation on COOP in support of hydrographic surveying, pipeline routeing surveys, engineering evaluations for construction sites, pre- and post-dredging volumetric/material classification surveys and scientific studies. Standard frequencies available include 3.5, 7, 12, 24, 33, 40, 100 and 200 kHz.

Processed data are displayed on an integral thermal recorder, the LCD graphics display or are recorded on an optional 275 Mbyte internal data logger. Flexible data formats including SEG-Y are compatible with most data processing systems. Sound velocity profiles can be downloaded through a standard RS-232 interface from profiling units such as the SeaBird SBE 19 unit. Event marks, time markers and external annotation text can be sent to the Bathy 1000 via one of the RS-232 ports from an external computer through the data logger interface.

The Bathy 1000 uses the same highly flexible, embedded PC/DSP open architecture as the Bathy 2000 series of sounders. This allows advanced real-time digital processing and expansion. A 30-point sound velocity profile permits accurate corrections for water column variability. Reliable data are obtained using Ocean Data Equipment Corporation's proven detection algorithms generating robust bottom tracking in sloping terrain with excellent performance in high-noise environments and in the presence of surface fluff layers.

Bottom detection uses selectable leading edge or peak detection for the best estimate of depth. Bottom tracking uses ship's speed, and past depth history for a predictive estimate of the next expected depth. Automatic operation includes depth dependent power output level and pulse width along with adaptive

receiver automatic and time varying gain to reduce extraneous acoustic signals to maximise signal recognition.

Operator control is through a series of simple and logical dialogue boxes, data entry forms and menus. Data entry and control is via an integral sealed keypad or standard keyboard. Automatic, built-in tests simplify troubleshooting. This has resulted in nearly automatic start up and bottom acquisition with little operator intervention, although manual operation is still available for those who want to make their own adjustments.

The system is ideally suited to the needs of multimission research vessels. It has eight external interfaces for integrating with GPS navigation, heave sensors, sound velocimeters, IRIG-B or other devices along with remote VGA monitors, SCSI devices or ethernet systems.

The Bathy 1000 can be incorporated into systems as a replacement for the older DSF-6000 and will drive existing transducers. It will interface with existing computer networks using the DSF-6000 interface protocols.

Operational status

The Bathy 1000 was introduced in 1994 and has been adopted worldwide.

Contractor

Ocean Data Equipment Corporation, E Walpole, Massachusetts.

VERIFIED

Bathy 2000/2000P

Type

CHIRP dual-frequency echo-sounding for bathymetry and sub-bottom profiling.

Description

The Bathy 2000/2000P series are full featured, research quality survey systems offering maximum flexibility to meet a wide variety of mission requirements including CHIRP sub-bottom profiling, bathymetry, bottom type classification, and sub-bottom density calculations. They provide simultaneous dual-frequency operation and 8 cm resolution of sediment strata with bottom penetrations to 100 m.

The Bathy 2000 is configured with a controller, processor, 5 kW amplifier, video monitor and keyboard mounted in a standard rack for shipboard mounting. It is designed for full ocean depth operation to 12,000 m. The Bathy 2000P includes the same controller and processor, an LCD Display, sealed keypad, and 2 kW amplifiers contained in a single chassis for portable operation. It is designed for applications in rivers, harbours and coastal waters and performs to depths of 6,000 m.

Various options for transducers and mounting include towfish, sea chest and pole mount allowing installation on COOP in support of pipeline routeing surveys, engineering evaluations for construction sites, pre- and post-dredging volumetric/material classification survey and scientific studies.

Standard configurations include an array of six 7 kHz transducers mounted in a small, easy to handle towfish. This array provides 8 cm resolution sub-bottom profiles using an 8 kHz bandwidth FM CHIRP waveform, centred on 7 kHz for optimum penetration and resolution. Also available are arrays of 3.5 kHz CHIRP arrays for maximum penetration. CHIRP signal processing provides a processed SNR gain of greater than 23 dB for greater depth and enhanced performance in noisy environments. Other frequencies available include 12, 24, 33, 40, 100 and 200 kHz.

The 2,000 W power amplifier in the Bathy 2000P provides 25 ms CHIRP waveforms for operation to 6,000 m with typical sub-bottom penetration of 30 to 50 m. The 5,000 W amplifier in the Bathy 2000 extends operation to 12,000 m or gives deeper penetration in the sub-bottom.

Processed data is displayed on an integral colour LCD graphics display and recorded on an internal data logger. Optional recording on a colour plotter, thermal paper recorders or optical disk is available. Flexible data formats including SEG-Y are compatible with most post-survey data processing systems.

The Bathy 2000P uses a highly flexible, embedded PC/DSP open architecture which allows advanced real-time digital processing and expansion. Time domain digital correlation, coupled with variable length FM CHIRP waveforms, enables the system to perform extended depth operations while preserving sub-bottom layer resolution. A 30-point sound velocity profile permits accurate corrections for water column variability. Reliable data are obtained using Ocean Data Equipment Corporation's proven detection algorithms generating robust bottom tracking in sloping terrain and excellent performance in high-noise environments. Bottom detection uses selectable leading edge or peak detection for the best estimate of depth. Bottom tracking uses ship's speed and past depth history for a predictive estimate of the next expected depth. Pinger track and parametric modes are available.

Operator control is through a series of simple and logical dialogue boxes, data entry forms, pull down menus and buttons. Data entry and control is via an integral sealed keypad or standard keyboard. Automatic, built-in tests simplify troubleshooting.

The system is ideally suited to the needs of multi-mission research vessels. It has eight external interfaces for integrating with GPS navigation, heave sensors, sound velocimeters, IRIG-B or other devices along with remote VGA monitors, SCSI devices or ethernet systems.

Operational status

Thirty nine B2000/2000P systems have been sold worldwide and the system has become the standard deep water system in the US on AGOR and T-AGS class vessels and is being adopted worldwide.

Contractor

Ocean Data Equipment Corporation, E Walpole, Massachusetts.

VERIFIED

MISCELLANEOUS SURVEY SYSTEMS

GERMANY

Parasound

Type
Shipborne high-resolution sub-bottom profiler.

Description
Parasound is a hull-mounted combined narrow beam deep sea sounder and sub-bottom profiler with an overall depth sounding capability extending from 3 to 10,000 m. Designed essentially for research and geological survey applications, it comprises a Narrow Beam Sounder (NBS) and a low-frequency sediment sounder or Sub-Bottom Profiler (SBP) for high-resolution presentation of sediment layers; both generate highly directional sound beams electronically stabilised against ship's motion.

Incorporating an ice-tested, flush-mounted transducer assembly, the system ensures particularly high vertical and horizontal resolution, enabling detection of very fine layers of sediment while maintaining high ground penetration without necessitating traditional costly towed-fish methods of measurement. It has a deep sea sediment penetration capability in excess of 100 m and can be operated at vessel speeds of around 10 kt, even in gale force conditions.

Contractor
STN ATLAS Marine Electronics GmbH, Bremen.

VERIFIED

Hydromap

Type
Hydrographic software system.

Description
Hydromap is a comprehensive and modular hydrographic data processing system. It consists of the parts Hydromap Online and Offline, which are able to record, process, display and manage hydrographic and oceanographic survey data. The various functions of the ECDIS-based Hydromap Online cover all aspects of a hydrographic survey featuring sensor interfacing, time synchronisation, data validation, data storage, real-time visualisation, quality control and navigation. Hydromap Offline provides automatic and interactive tools to clean and edit all kinds of survey data like position, vessel motion data, depth, tide, sound velocity, side scan and backscatter data. The post-processing functions of Hydromap Offline allow digital terrain modelling, contouring, profiling and final bathymetry plots.

The software is designed for workstations with a UNIX operating system. The graphical user interface is based on the international software standard X-Windows and OSF Motif. It provides application-oriented and user-friendly menu operation and guarantees widely uniform and self-explanatory handling. Integrated systems with extended workstations solutions are possible.

Operational status
In service with the Royal Navy and other navies.

Contractor
STN ATLAS Marine Electronics GmbH, Bremen.

VERIFIED

UNITED KINGDOM

Acoustic tide monitor

Type
Tide and sea state monitor system.

Description
This acoustic above the water, temperature-corrected tide monitoring system is now installed in many ports around the world to provide accurate tide readings. Such measuring equipment is of value to hydrographers. The system is an 'in air' echo-sounder measuring the distance from the transducer to the water surface, taking account of the temperature dependent variation of the speed of sound by using a reference target. The system features an in-built clock and calendar, comprehensive data storage and communication facilities.

The system measurement resolution is 0.2 mm and displayed resolution is 10 mm. System accuracy is 0.1 per cent or within 1 cm, whichever is the smaller. The system also offers a peak to trough wave height measurement which reflects the sea state.

Software is available to store data from many tide monitors on desktop computers, including facilities to display predictive data and to provide automatic remote station data collection.

Operational status
In production for 10 years and in widespread service. Units are also available for hire. A fully portable, battery-operated unit is now available.

Contractor
Sonar Research and Development Ltd, Beverley.

UPDATED

UNITED STATES OF AMERICA

LS-4096

Type
Active laser underwater imaging system.

Description
The LS-4096 is a synchronous optical scanning system for two-dimensional reflectance maps of seabeds and submerged objects. A visual image of scanned objects, which can have a resolution as high as 4096 points across a 70° field of view, is obtained by displaying reflectance data on a high-resolution CRT monitor.

The system uses a diode-pumped solid-state frequency doubled Nd:YAG laser as a light source. The laser produces approximately 200 mW of light at a wavelength of 532 nanometers (green). Optical reflectance data are digitised in the sensor with 14-bit dynamic range and transmitted to the surface control console via a digital fibre optic link.

The surface console comprises two modules: a real-time processor and a display processor. The real-time processor compresses the dynamic range of the raw video signal while performing other signal processing functions and then transmits data to the display processor. The display processor presents processed reflectance data in a visual format and provides the operator interface as a graphic overlay. Unprocessed digital data can be stored on a conventional magnetic hard disk, a removable magneto-optical disk, or a high-capacity digital tape.

Operational status
Under development for the US Navy's Coastal Systems Station, Panama City, Florida, for subsea mine reconnaissance applications. Commercial versions in service with offshore survey companies in association with SubSea Survey of Aberdeen, UK.

Contractor
Applied Remote Technology Inca Raytheon Company, San Diego, California.

VERIFIED

OCEAN SURVEY SYSTEMS

CANADA

SHOALS-Hawkeye

Type
Airborne lidar bathymeter.

Description
SHOALS (Scanning Hydrographic Operational Airborne Lidar Survey) is a laser-based scanning lidar bathymeter that offers hydrographers a cost-effective and rapid means of surveying shallow coastal or inland waters with depth and position accuracy to International Hydrographic Organisation standards. The equipment is designed to operate in a relatively small fixed-wing aircraft or helicopter and provides an improvement over acoustic systems both in area coverage and in uniformity of sounding density.

Effective over most coastal areas that are of critical concern to the hydrographer, the system's speed and flexibility in deployment, and convenience in sounding dangerously shallow or highly confined and irregular areas, offer unique advantages in survey planning and deployment strategy. The system measures water depths 200 times per second from an aircraft flying at an altitude of 200 to 300 m.

The system uses two short duration laser pulses, one green and one IR in wavelength, transmitted simultaneously and coaxially from the aircraft down onto the surface of the water. Scattering of the IR pulse from the water surface is detected by a receiver located at the laser source. The green pulse is partially transmitted through the water, and its scattering from the bottom is also detected by the receiver at a later time. The water depth is determined by measuring the elapsed time between receiving the IR and green return signals. This depth value is subsequently corrected for geometric and environmental effects.

To provide wide ground coverage, the consecutive laser pulses are directed by a scanner sequentially across the flight path into preselected orientations that optimise the sounding pattern.

The location of each sounding is derived from data provided by aircraft positioning, attitude reference and scan angle control systems. Depth penetration is strongly dependent on the water quality, varying from several metres in turbid water to greater than 50 m in very clear water. Typically the system has a penetration capability of 20 to 35 m in most coastal areas.

The system is compatible with the Global Positioning System (GPS), which is valuable when surveys have to be conducted in remote areas.

Similar equipment may be used in ASW and MCM applications.

Operational status
A SHOALS system was delivered to the US Army Corps of Engineers. This was immediately followed by the delivery of a major part of two additional systems called Hawkeye to Saab Instruments of Sweden for use by the Swedish Navy and the Swedish Hydrographic Department.

Contractor
Optech Inc, North York, Ontario.

VERIFIED

UNITED KINGDOM

Aquashuttle Mk III

Type
Towed undulating data acquisition vehicle.

Description
The Aquashuttle is a towed undulating data acquisition vehicle developed by Plymouth Marine Laboratory and product-engineered by Chelsea Instruments to gather a wide range of oceanographic data.

The vehicle features a strong Glass Reinforced Plastic (GRP) body with a stainless steel chassis which increases overall vehicle strength and offers better distribution of towing loads, enabling the carriage of a multisensor package with 16 channels of data logged. The vehicle's flight profile is maintained by a servo-controlled elevator, the servo being powered by the vehicle's own impeller-driven alternator which is housed in a polypropylene package. The flight profiles are controlled in real time via an interface unit and PC housed in the laboratory. The vehicle is shipped to the port of departure, deployed upon leaving port by the ship's crew and can automatically commence sampling on entering the water. The speed of tow and period/amplitude of operation will provide full two-dimensional (2-D) coverage of the mixed layer to >100 m over long-range ocean transects.

On recent data gathering missions the Aquashuttle collected information on chlorophyll-a and thermoclines in the North Atlantic and in the South West approaches to Britain. Comparison of this data has shown the development of the seasonal thermocline from south to north, paralleled by the progression of the phytoplankton production northwards and the shallowing in depth of the euphotic zone. The relative importance and interaction of these influences on production are subjects of continuing study. Among studies currently being undertaken are: nutrient supply and availability; light limitation and cloud cover (incident solar radiation); and mixing and stratification of the surface waters. Apart from chlorophyll-a, conductivity and temperature measurement, the Aquashuttle data package can measure *in situ* particle concentration; nitrate/nitrite determination; upwelling and downwelling; light; and bioluminescence, pH, redox, dissolved oxygen and plankton sampling.

The data so gathered will be of considerable benefit in attempting to predict ocean fronts, and climatic conditions which affect sonar propagation.

In addition to the commercial fleet, the Royal Navy intends to deploy Aquashuttle from its survey ships. On these platforms, the system will integrate navigation data and includes a tow cable strain monitor. Real-time control and data display enables the operator to observe the 2-D oceanographic conditions and adjust the flight parameters accordingly. The vehicle is capable of being towed at speeds from 8 to 25 kt and can undulate from the surface to >100 m depending on cable configuration.

Operational status
In service with the Royal Navy aboard HMS *Scott*.

Specifications
Length: 1.39 m
Width: 0.72 m
Height: 0.50 m
Weight: 66 kg (in air)
Depth range: 0->100 m
Undulation length: 800 m – 40 km
Towing speed: 8-25 kt
Dive/climb speed: 1.0 m/s (max)

Contractor
Chelsea Instruments, West Molesey.

HMS Herald with the Aquashuttle Mk III

1996

UPDATED

Subpack/SSMS

Type
Submarine-mounted and ship-fitted multiparameter oceanographic sensor suite.

Description
Subpack/SSMS (Surface Ship Monitoring System) is based on the oceanographic sensor system developed by Chelsea Instruments in conjunction with the MoD since 1979. The device carries sensors to monitor several different oceanographic parameters for the detection of changes in water properties associated with ocean fronts, eddies, stratification, upwelling, coastal currents and the marginal ice zone. Changes in the speed of sound and seawater density are also monitored. All data is recorded and displayed in real time.

The sensors, which include conductivity, temperature, depth and several fluorimeters operating at different wavelengths are all housed in a single (small) pressure housing.

Operational status
Subpack is in the process of being fitted fleet-wide in the Royal Navy's submarine flotilla. SSMS is being used for research purposes, primarily by the UK MoD.

Contractor
Chelsea Instruments, West Molesey.

UPDATED

Bubble Photometer

Type
Self-contained towable bubble photometer.

Description
The bubble photometer system was designed in 1990 for the UK Royal Navy after a detailed feasibility study. It is capable of imaging entrained air bubbles down to 15 µm radius. The system is designed to determine the ambient entrained bubble size distribution in open seas and coastal areas with the ability to determine the bubble size distribution in large volumes of seawater. It uses photometric (rather than metrological) techniques whereby integrated image intensity measurements are subsequently analysed using an astronomical star-late measuring facility. The system is designed to be towed at speeds from 2 to 4 kt and at depths from 1 to 20 m using a 100 m tow cable.

The bubble measurement system comprises three major subsystems: the deck towing gear, deck control gear, and the bubble photometer. The photometer is a self-contained towable underwater body housing the flow chamber, a Hasselblad camera (with optics designed by Chelsea instruments), flashlamp and electronics. The instrument is connected to the surface vessel by a tow cable down which command signals are sent initiating the taking of photographs. A deck cradle is provided for the stowage of the bubble photometer while film magazine and batteries are changed.

A simple camera deck unit is provided to initiate the taking of photographs. In addition, a battery charger is supplied to recharge the photometer batteries between deployments.

Operational status
In service with the Royal Navy.

Contractor
Chelsea Instruments, West Molesey.

NEW ENTRY

SIGNATURE MANAGEMENT

Managing and controlling unwanted signatures on warships is now a major consideration for ship designers and operators. The advent of new generations of acoustic and wake homing torpedoes and modern acoustic/magnetic mines has led to an intense effort being devoted to reducing or modifying signatures created by ships and submarines in order to increase their stealth against the underwater threat.

Signature control slots naturally into a layered defence system. It can be used to counter a threat in all phases of the engagement from the detection down to the terminal attack phase. In the detection phase careful signature control can force an opponent to deploy additional resources in order to achieve detection. In certain circumstances it may inhibit detection altogether. In the localisation phase it can inhibit the performance of the hostile platform's sensors, causing it to delay or preventing it from launching any hostile weapons. In this instance the hostile platform will be forced to launch its weapons at a much shorter range than desired, thus enhancing the hard kill defences of the target. Finally, in the terminal homing phase, careful signature control can increase the effectiveness of decoys and jammers enabling the target to evade the attacking weapon.

Some signatures, such as the magnetic signature, have been the subject of considerable attention for many years. Even during the Second World War technology was being applied either to reduce to a minimum the magnetic signature, or to install compensation methods which would cancel out the effect of such signatures. Development was spurred on by the increasing losses suffered by merchant shipping from magnetic mines, and the need to build minesweepers which would be virtually immune to the magnetic mine.

With the post-war development of acoustic mines, and acoustic homing torpedoes, attention turned towards reducing the effect of a ship's acoustic as well as its magnetic signature. This was achieved principally through the damping of systems which created noise such as propulsion machinery and so on, by the use of rubber shock-absorbers. Using these protective measures had the added effect that if the ship were to suffer an explosion then equipment would be given some measure of protection from effects of the blast and whip.

However, ships also betray their presence in other ways. Such indications include unintentional electromagnetic emissions, chemical and bubble traces remaining in the wake, bioluminescence caused by the disturbance of minute organisms, and hydrodynamic pressure and surface wave patterns. Compared to acoustic, magnetic, IR and radar signatures, these are relatively insignificant. However as signature management of acoustic and magnetic emissions improves as well as reduction in IR and radar, these lesser signatures will gain in importance as secondary means of confirming the presence of a target.

Stealth technology
The new science of controlling and modifying these signatures is now referred to as stealth technology. Currently, effort is not only being devoted towards reducing these signatures, but also to adapting them so that a ship is not necessarily made invisible to hostile sensors but appears as something totally different. Using stealth technology is referred to as signature management.

The relative importance of different types of signatures can be expected to change with time, as new and more sensitive sensors and more sophisticated signal processing techniques become available. The art of signature management lies in the ability to obtain the right balance between the differing signatures. The object of signature management is to achieve the optimum combination of external shape, coatings and internal treatments that will give the desired overall result at an acceptable cost.

Acoustic signatures
Surface ships and even more importantly submarines emit high levels of underwater radiated noise which can be detected and tracked by passive sonars, possibly hundreds of miles away in the case of towed array sonars. Vessels which strongly reflect incident sound waves are easily detected by active sonar systems. Not only does a high level of radiated noise make a vessel far more susceptible to detection, it also inhibits the vessel's ability to operate its own acoustic sensors.

Underwater acoustic radiation is still the primary means by which submarines betray their presence when they are submerged. Hostile forces also use the underwater noise signature radiated by surface ships for detection and classification of targets. Cavitation and broadband noise are created by propellers and hulls, while various internal pumps, generators and diesel engines produce distinctive sounds at specific frequencies and amplitudes that can both aid and hamper detection. So sophisticated are modern detection methods that, in many instances, not only can the hunter classify a target as a specific class of ship, but even the individual ship within a class can be positively identified from its very own distinctive sound signature. For example, it may have a shaft alignment problem which has been identified by intelligence or a pump with an uneven bearing and the specific sound signature so produced can be recorded for storage in a sound library for future comparison and so on.

Minimum acceptable noise levels are constantly being driven downwards by developments in sonar sensors and signal processing technology. Hence the techniques used to control radiated noise must also become more subtle and refined. One of the most important elements in achieving a 'quiet' design is the ability to predict the noise levels while a ship is still at the design stage. Minimum noise levels are further achieved by the design of high-performance machinery isolation systems and methods for reducing cavitation around propellers and so on. The hydrodynamic shape of the hull in water is also subjected to close scrutiny in order to minimise any sound signature which may be created by its passage through the water.

Noise isolation systems for surface ships and submarines employ a wide range of techniques, including double-elastic mounting systems where, in the case of diesel engines, high intrinsic source levels have to be reconciled with stringent target levels for underwater radiated noise. Increasingly, serious attention is having to be paid to the secondary transmission paths in the design of such advanced isolation systems.

Active control techniques are currently under development to enhance the high degree of isolation which can already be achieved through well balanced, passive isolation treatments. These include systems designed for use with propellers and propulsors and their associated shafting systems, where dynamic interaction with the hull can play an extremely important part in determining the overall level of underwater noise radiation. In rotating machinery, for example, moving parts can be dynamically balanced to reduce the noise from shafts and connections to other machinery. In addition, the equipment can be mounted in special acoustically insulated housings which are, in turn, carried on flexible mounts to isolate them from the hull. In some instances, large items of machinery are additionally flexibly mounted on rafts which are themselves flexibly mounted to isolate the whole machinery installation from the hull. As well as these detailed arrangements, items such as water inlets/outlets and so on, must also be isolated from the hull in flexible mountings to prevent any vibration from the pipe being transmitted to the hull and thence to the surrounding water in the form of noise.

One of the most important areas for noise reduction concerns the propeller. Propeller noise is generally divided into two main types:
(a) blade noise which is generated by the thickness of the blade and the loading on the blade as it rotates, which gives rise to low-frequency noise
(b) cavitation noise generated by the implosion of cavitation bubbles which form around the leading edge of the blade, across the low-pressure region of the blade and in the vortex behind the hub of the blade. Cavitation is a broadband high-frequency noise which can only be avoided by delaying the onset of cavitation.

Coatings applied to hulls in the form of decoupling or damping layers designed to reduce the level of machinery-induced noise radiated into the water, or anechoic coatings designed to absorb as much as possible of the incident energy generated by active sonar equipment, are becoming increasingly important in the science of controlling unwanted noise signatures. For some years the hulls of nuclear submarines have been treated with sound-absorbing anechoic tiles and this is now being extended to diesel-electric powered submarines. Anechoic tiles are not the same as decoupling tiles, being designed to absorb or scatter incident sound energy from hostile active sonars. They operate on the principle that the material absorbs the incident sound, which is then turned into heat; hence no sound, or only minimal sound, is reflected off the hull. In order not to inhibit own-boat sonars, sonar 'windows' are incorporated into the anechoic tiling system.

Other treatments include continuous coatings in which acoustic and anechoic material is applied to large flat or curved surfaces. RhoC (PC) materials are designed to match the impedance and speed of sound in seawater. Sonar materials can replace GRP for sonar domes and windows, and encapsulation in which large and small case encapsulation of arrays is achieved with controlled exotherm.

The purpose of using acoustic control measures such as decoupling tiles (whose design is frequency and noise level dependent) and anechoic coatings and so on, is to reduce the detection range and classification capabilities of hostile sensors, at the same time improving the performance of 'own ship' sensors. However, the law of diminishing returns does apply and there comes a time when no amount of treatment will either (a) improve own detection range or (b) significantly reduce enemy detection range.

Air bubble screening has proved to be particularly valuable at higher frequencies and speeds.

Radar cross-section
Radar cross-section signature control can be approached in a variety of ways ranging from the use of appropriate design techniques (avoiding re-entrant corners, shaping, sloping and reducing the overall size of superstructures and so on), to the addition of temporary or permanent Radar Absorbent Materials (RAMs) and coatings of various types.

The design of structures to avoid specular reflections is an obvious means, but this may not be fully effective on mobile structures and cannot be applied easily to existing vessels. Absorbent coatings represent a versatile solution which can be applied in conjunction with shaping to new designs or retrospectively to existing vessels.

Coatings can be designed to absorb radar signals over a wide or narrow frequency bandwidth, as appropriate to the application. Radar signature can also be controlled by use of structural Radar Absorbent Material (RAM) and materials of this kind are emerging and can be considered at the design stage of new vessels. By appropriate formulation RAM coatings can be endowed with various special properties such as high friction. RAM structures operate on the thickness of the material used according to the angle of incidence of the beam.

Magnetic and electric signature
The need to control and manage a ship's magnetic signature has long been the subject of study by naval architects, scientists and operators and so on. Passive countermeasures involve measuring the ferromagnetic mass of the ship itself (including all the individual items of equipment carried by the ship) and reducing the magnetic signature. This is achieved either by magnetically treating the ship in special magnetic treatment facilities (deperming); or by fitting the ship with special active degaussing coils through which electric current is passed, the strength of which can be controlled from within the ship itself, thus ensuring that at any time, in any geographic position the ship's magnetic signature (that is, its own inherent magnetic signature plus that caused by its movement through the earth's magnetic field) is cancelled out; or by a combination of the deperming and active degaussing processes. Most modern warships are fitted with active degaussing systems, and are also regularly depermed. Merchant ships, on the other hand, rely primarily on deperming.

Deperming involves putting the ship inside an arrangement of coils or placing an arrangement of coils around the ship, and then passing a powerful electric current through the coils to create a magnetic field in opposition to the magnetic field of the ship, thus cancelling out the ship's own magnetic signature. Alternatively, deperming can be used to create a permanent magnetic field on the ship which is matched to the area in which it will operate. However, the disadvantage of deperming is that it is not permanent, and the vessel will need to be checked periodically in order to maintain the desired effect .

Active degaussing coils, on the other hand, are built into the ship during construction to provide magnetic field correction facilities. The coils are continually fed with electric current provided from special computer-controlled generators to create an opposing magnetic field which is continually matched to the ship's changing magnetic field as it crosses the ocean. Even with this system, however, periodic checks are still necessary and are carried out on special magnetic ranges.

For vessels such as MCMVs it is necessary to deal with the basic problem of magnetic content right from the onset of design and wherever possible to substitute non-magnetic for magnetic material. In such work it is essential to ensure that the substitution of magnetic materials by those of low permeability does not compromise the performance of the equipment.

Galvanic currents flowing in the hull and in the water around the hull of a ship build up underwater potentials which cause a ship to act as an electric dipole. When currents flow through a propeller shaft they are modulated in the bearings by the revolving shaft, and an Extremely Low-Frequency (ELF) electrical field is created and radiated into the water. The frequencies generated are dependent on the fundamental and on the harmonics of the number of revolutions.

Improvements in sensor technology and data analysis in recent years enable these very small signals to be detected and analysed. ELF signatures can be suppressed by short-circuiting the current loop, and through Passive Ground Shafting (PGS). Active Shaft Grounding (ASG) systems have also been developed to counter ELF signatures.

These produce a current counteracting the galvanic current, resulting in a zero current in the bearings.

Ranges

The sophistication of modern sensors, fire-control systems and weapons demands that the various signatures of warships be reduced to the lowest level possible, in order to reduce the risk of detection and lower their vulnerability to modern weapons. While considerable attention is paid to reducing a ship's various signatures to a minimum, both at the design stage and during construction, it is an undisputed fact that a ship still retains a definite signature, which, like the human fingerprint, enables it to be detected and identified. In order to overcome this limitation, it is essential that the user knows the precise nature of the signatures of his ships, and has available the means to reduce or nullify them or, as is so in some cases, to actually turn them into an advantage.

It is not only ships' signatures which are a vital factor in modern naval operations. Weapon performance and behaviour is continuously under close scrutiny, especially as the ranges and speeds of weapons increase. The discriminating user demands to be reassured as to precisely how a weapon behaves under various operating procedures. This is particularly so in the case of underwater-launched weapons. However, the user doesn't only want to know how the weapon

behaves, but also how effective it is in various situations against differing types of target, and how well it can overcome countermeasures which might be directed against it.

The question of knowing precisely what a ship's signature is and how an underwater weapon behaves is achieved using the underwater range. These ranges are so specialised, using such advanced technology to make the delicate measuring equipment, that only a few companies are capable of manufacturing them.

The two major signatures about which every user should have precise knowledge concerning his fleet are the acoustic and magnetic signatures. In addition to ranges for measuring these signatures, there are also tracking ranges for weapon system development and proving, mine warfare ranges and a very specialised type of calibration range known as FORACS (Fleet Operational Readiness Accuracy and Check System), of which two main ranges for NATO are sited in Norway and Crete.

Noise ranges

Noise ranges for static and underway measurements of surface vessels and submarines are among the largest and most sophisticated systems worldwide. These ranges are primarily designed for the measurement of radiated noise emissions in order to allow reliable and quantitative assessments of signature reduction measures.

There are three main types of noise range:
(a) the underway fixed range – used to measure the acoustic signature of ships at various speeds
(b) the portable range – a somewhat less sophisticated version of the fixed range, which can be set up at different sites to measure acoustic signatures.
The portable range comprises a number of hydrophones, suitably deployed from a surface ship or temporarily attached to fixed moorings on the seabed, to measure the noise characteristics of a submarine or surface ship at various ranges, speeds and depths. It is useful for measuring acoustic signatures under different environmental conditions, or when it may not be convenient to send a ship back to a fixed range, or where budgetary restrictions or environmental conditions may preclude the setting up of a fixed range. Portable systems typically cover a circular area of a 2 km radius and can operate suspended from the surface, in depths down to 300 or 400 m. The hydrophones in a portable range must be specially customised to overcome the problem of pressure changes acting on the ceramic head resulting from wave motion which, if severe, can saturate the front end of the preamp in the hydrophone.

Tracking ranges

Modern tracking ranges are capable of tracking up to six vehicles at speeds up to 75 kt with a maximum turn of about 20°/s. Positional accuracy is in the region of 5 m. The hydrophones are laid in a carefully planned arrangement and are calibrated from a surface vessel whose precise position can be fixed with an accuracy of about 2 m.

Magnetic ranges

Equally as important as a ship's acoustic signature is its magnetic signature. In the case of MCMVs it is vital that this signature be known in detail and checked at frequent and regular intervals. It is also essential that such a signature be measured whenever a vessel is assigned to another theatre of operations hundreds of miles away from its current area of operations. A ship's magnetic signature is not only governed by where it was built and with what materials, but also by the

earth's magnetic field at that point. Even moving a relatively short distance can considerably alter a ship's magnetic signature. It is essential, therefore, to protect ships from the threat of magnetic influence mines. This can be achieved in three ways by:
(a) designing a ship for low magnetic content (of paramount importance in MCMV design)
(b) providing the ship with degaussing coils to compensate for its magnetic field
(c) treating the ship electromagnetically by the application of a powerful external magnetic field to cancel out its magnetic field.
Like acoustic ranges, magnetic ranges are configured to fulfil different functions. There are five main types of magnetic range currently in service around the world:
(a) the open range – used to check the magnetic level of all types of ship and to adjust degaussing coil settings
(b) the fixed range – used to measure the magnetic effects due to eddy currents, stray fields and induced magnetic fields from which the magnetic field components can be separated using a Z-loop
(c) the land-based range – used to carry out magnetic measurements on ship equipments, or for modelling purposes on new designs or modifications to determine the coil system necessary to reduce the magnetic signature
(d) the transportable range – used to process MCMV signatures in the area of operations
(e) the treatment range – used to deperm, or wipe, a ship using an externally created magnetic field. With the increasing sophistication of modern influence mines, not only MCMVs but drone sweepers and hunters and ROVs must now have their signatures carefully measured.
The tasks of the degaussing range are to measure the ship's magnetic fields and analyse them to provide resolved components on three axes, synthesise the degaussing coil settings required, and record the measured values for archive purposes and reporting.

Early ranges were of the large fixed type and relied on manual tracking and analysis, which required the vessel to make numerous runs over the range in order for the shore-based operator to manually interpret the plotted signatures and to overcome accuracy limitations. These ranges relied on the comparison of signatures in opposing headings to differentiate between induced and permanent magnetisation. They suffered from the limitation that the horizontal field could not be determined. The next generation of range used automated computerised tracking to achieve faster and more accurate tracking of the ship, determining its position within 1 m. Using triaxial sensors the range provided data on the horizontal field. The computer analysis of the results provided, in addition to a much more accurate signature plot than the manual system, contour and three-dimensional plots. A major step forward with these systems was the ability to carry out coil current synthesis, from which the coil current on the ship could be set to counter the ship's magnetism.

Fixed ranges of any description, however, are not easy to put down, requiring a considerable civil engineering effort and extensive surveying. Nor are they cheap, although they are most cost-effective.

Description of specific types of range is almost impossible as ranges are tailor-designed, built and laid out to suit each specific requirement. Any description of a system must therefore be in general terms.

UPDATED

ACOUSTIC RANGES

AUSTRALIA

Acoustic Tracking System (ATS)

Type

Underwater acoustic tracking and positioning system.

Description

This series of Acoustic Tracking Systems (ATS) is particularly relevant to minehunting and mine clearance operations. It is also used for underwater vehicle tracking, equipment handling and for tracking towed objects such as variable depth sonars. By using a sophisticated eight-chirp signal in the 15 to 18 kHz frequency range, ATS offers high accuracy, for example better than 0.25 per cent of slant range in X and Y co-ordinates.

ATS is designed to provide solutions in harsh acoustic environments where other systems will not function, such as in shallow water, highly reflective conditions around underwater structures, and for operations near the water surface. It minimises problems such as background noise, reverberation and multipathing effects particularly in shallow waters.

The ATS series is rack-mounted and easy to operate, and needs only a minimum set up time with maximum flexibility. It is available in three very short baseline models: S04 and S08 are selected according to the number of beacons being tracked; D08 also provides riser angle information for the offshore drilling market. All models use the same hydrophone and beacons.

The hydrophone incorporates all receiving and transmitting acoustic elements in a compact self-contained unit. The ATS operates in a full hemisphere below the hydrophone and tracks two, four or eight fixed or moving beacons simultaneously, depending on the model.

Operational status

In operational service. ATS has accurately and reliably tracked Scarab work vehicles to depths of 2.6 km. ATS II will be used to track the VDS and mine disposal vehicles in the Australian 'Huon' class and Spanish 'Segura' class minehunters.

Contractor

Nautronix Limited, Fremantle, Western Australia.

UPDATED

RS5D

Type

Short baseline acoustic positioning system.

Description

The RS5D short baseline acoustic positioning system is designed for deep-water tracking and can be supplied to operate in one of two frequency ranges (22 to 30 kHz or 48 to 56 kHz), the lower frequency band providing a greater operating range.

High accuracy is achieved using multiple hydrophone arrays in a short baseline configuration, offering an accuracy of 0.5 per cent of water depth under normal operating conditions. The basic system uses four hydrophones as a minimum and is capable of operating with up to eight hydrophones.

Signal processing is performed by multiple DSP processors with the main processing power provided by an industrial standard, PC-based processor. The user interface is Windows-based.

The MMI features a keypad on the control display unit, with numeric keys for entering parameter values with software menu options being selected by use of a cursor moved by arrow keys.

The unit retains data relating to the hydrophone and beacon offsets, which obviates the need to re-enter this data when moving to a new location. All the operator has to do is enter the water depth.

The RS5D is ideally suited for dual redundant configurations in which two systems operate independently of each other but in a side-by-side configuration. The RS5D can be configured with up to eight hydrophones, but needs signals from only three of these (selected automatically) to operate; thus, the system has multiple redundant solutions.

Data from one tilt telemetry and three position beacons can be received simultaneously. Position information is compensated for vessel pitch and roll from the data provided by the system's VRU.

The system is available with beacons having source levels which range from 186 dB to 192 dB. In normal circumstances where there are no adverse conditions affecting acoustics, the unit has an operating depth of 2,000 m.

Specifications

Range: 2,000 m slant range (22-30 kHz); 1,000 m slant range (48-56 kHz) with 192 dB beacon
Frequency band: 22-30 kHz (standard); 48-56 kHz (optional)

Contractor

Nautronix Ltd, Fremantle, Western Australia.
Nautronix Inc, San Diego, California.

VERIFIED

SAUR

Type

Submarine underway radiated noise range.

Description

The innovative underway radiated noise range SAUR (Shallow Acoustic Underway Range) is based on synthetic aperture sonar processing (SYNAPS) techniques. Whereas conventional ranges rely on combining signals from a large number of underwater hydrophones, the Nautronix approach exploits the motion of the submarine which allows the use of a smaller number of hydrophones combined with sophisticated new processing algorithms. This vastly reduces the complexity of the in-water system, with the added bonus of being able to accurately locate discrete noise sources on the vessel being ranged.

A prototype system is installed off the South Australian coast. It was tested with both surface vessels and a 'Collins' class submarine, giving excellent results for both the levels of radiated noise and the locations of acoustic sources.

In May 1996, the company signed a contract with the office of the New Submarine Project in Australia to build and install a fully operational SAUR. The range will consist of a conventional acoustic hydrophone array, deployable either horizontally or vertically, connected via an electrical/fibre optic underwater cable to a data processing system housed in a transportable building on land. The on-shore processing system will include both conventional beam-forming and synthetic aperture processing on all elements in the array to provide both temporal and spatial processing gain. An added advantage of providing both systems is that the kinematic correction factors derived for the synthetic aperature processing can be used to take account of environmental factors and focus the spatial array, thereby providing close-to-theoretical gain from the spatial processing.

The SAUR system will be operated entirely from the desktop of a Sun Workstation, incorporating a data processing system and a sophisticated user interface. All software is being developed in C++ using commercial class libraries where available and appropriate.

The data processing platform utilises a multi-processor concept based on industry standard VME boards. The system architecture has been carefully designed for high data throughput and reliability. Acoustic data from individual hydrophones are stored to a disk array, and to tape for backup and to allow the data to be moved to other processing platforms.

The data processing software is divided between the real-time and post-processing functions. Real-time processing includes beam-forming, shading and steering of the beam to produce various spectral outputs. Post-processing incorporates synthetic aperture processing, transient and spectral analysis, together with generalised database manipulation. The real-time processing will provide the user with sufficient information to judge the value of the data taken from the current run. In addition to the output products available from conventional ranges, for example Lofargams, FRAZ and so on, the SYNAPS technology provides visualisation of the intensity and location of acoustic sources superimposed on a plan view of the submarine being ranged.

Operational status

A prototype is in operation and a fully operational model is being installed for the Royal Australian Navy.

Contractor

Nautronix Ltd, Fremantle, Western Australia.

VERIFIED

RS 910 series

Type

Ultrashort baseline/long baseline acoustic positioning system.

Description

The RS 910 series of acoustic tracking systems have been the industry standard in ultrashort/long baseline acoustic tracking since first introduced in 1978. The RS 910 models are an upgrade from the RS 900 resulting from replacement of the control console and central processor electronics.

The RS 910 series uses very robust, reliable, free-running pingers, transponders or responders to ensure that the system is tolerant to high ambient noise levels. Applications include ROV and towfish tracking, manned submersible tracking, subsea object positioning, and drilling rig positioning with riser angle differential capability.

The RS 910 electronics upgrade is based on the latest developments in digital signal processing and computer technology, featuring industry standard, PC-based processors, Windows format with high-resolution graphic displays, user-friendly trackerball keypad or mouse interface, a built-in spectrum analyser for online trouble shooting and continuous automatic self-test.

Three modules are available in the RS 910 series.

The RS 912 is a pinger-based USBL positioning

system which determines position by phase comparison. It can track up to six beacons with tilt and depth telemetry and differential display of two beacons showing their positions relative to one another. It operates in the 22 to 30 kHz band (17 channels at 0.5 kHz intervals). A low-band frequency option is available. The system is accurate to within 0.5 per cent of slant range for horizontal ranges up to 100 per cent of water depth and to within 1 per cent of slant range for horizontal ranges up to 200 per cent of water depth.

RS 914 is a pinger/transponder based USBL positioning system which determines position using a combination of ultrashort baseline techniques. It offers all the capabilities of the RS 912 plus transponder mode operation for determining position by phase comparison and slant range. It can interrogate five channels at 7, 9, 11, 13, 15 and 17 kHz and respond in 17 channels in the 22 to 30 kHz band at 0.5 kHz intervals. Again, a low-frequency band is an option. The RS 914 is accurate to within 0.5 per cent of slant range for horizontal offsets up to 200 per cent of water depth.

The RS 916 is a pinger/transponder, combined USB/LBL system which can handle the entire spectrum of offshore applications. It offers all the capabilities of the RS 912 and RS 914 plus long baseline operations to determine position relative to a three or four transponder grid. It allows simultaneous operation with both long and ultrashort baselines, and submersible-mode operation using a relay transponder on the submersible. The system can operate in both high (22 to 30 kHz) and low (6 to 15.25 kHz) bands. For high-band frequencies accuracy is 1 m or less with a three or four transponder grid. For low-band frequencies accuracy is 2 to 5 m with a three or four transponder grid.

Operational status
In service with the USN Submarine Development Group for ROV, DSV and DSRV submarine tracking. The latest oceanographic research ships of the US Navy 'Thomas G Thompson' class (AGOR 24, and 25) and NOAA oceanographic ships have combined RS 916 USBL/ LBL tracking systems.

Contractor
Nautronix Ltd, Fremantle, Western Australia.
Nautronix Inc, San Diego, California.

VERIFIED

FRANCE

Acoustic ranges

Acoustic Ranges
Using high-dynamic signals input, these ranges are used to locate, identify and measure noise sources on ships. The immersed measurement arrays are made up of hydrophones and/or LF and HF passive antennas.

Contractor
Thomson Marconi Sonar SAS, Sophia Antipolis.

VERIFIED

ITALY

PTM

Type
Tracking range.

Description
The PTM is a mobile three-dimensional (3-D) underwater tracking range used for testing modern torpedoes in different environmental conditions. It is able to track simultaneously (in real time) the torpedo, the launching unit and the target both in shallow and deep waters and it is capable of offline analysis of the trajectories.

The PTM consists of a surface vessel (dedicated or fitted for) a multihydrophone array, acquisition/ recording and analysis equipment, acoustic transmitters and ancillary equipment.

The tracking range acoustic technology is based on a long baseline system with SFSK (Space Frequency Shift Keying) acoustic signalling suitable for high-speed vehicle tracking. The mobile 3-D underwater tracking range can be arranged in different configurations to meet specific requirements and can be laid down in different water depths.

Operational status
In service.

Contractor
Whitehead Alenia Sistemi Subacquei, Livorno.

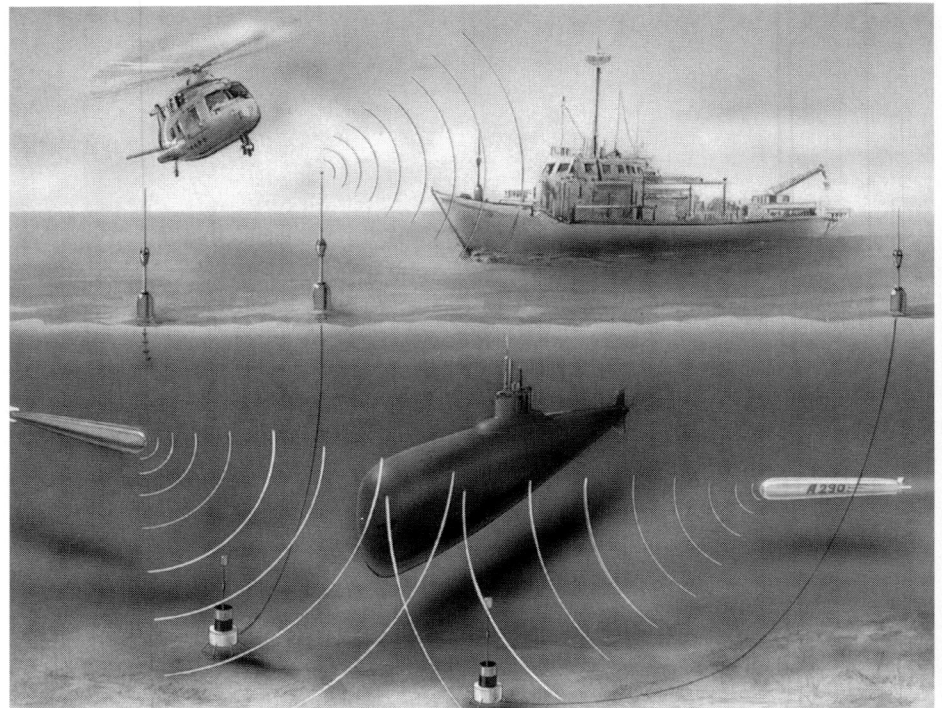

UPDATED   An artist's impression of the PTM

C304

Type
Retrievable three-dimensional underwater tracking range.

Description
The C304 is a retrievable acoustic range designed to carry out real-time tracking of mobile underwater vehicles. It is primarily aimed at the testing of modern underwater weapons in different environmental scenarios and for the training of personnel. It allows the simultaneous tracking of both surface and underwater units, including artifical targets. The range is currently in service with the Italian Navy for the evaluation of the new MU90 torpedo.

The range consists of: a minimum of three and up to 20 retrievable units composed of seabed operating acoustic sensors; sea surface floating devices and UHF transceivers; a surface control platform housing a 60 Mflops signal processor; UHF links for the retrievable units; the hull-mounted acoustic calibration device; and a synchronised acoustic transmitter located on each unit to be tracked.

The sequence of operation comprises: selection of the trial range (area, bottom depth); deployment of the retrievable units (positioning automatically guided by the computerised system of the surface control platform); acoustic calibration of the range by the self-tracking of the surface control platform; synchronisation of the underwater devices to be tracked; firing of the underwater devices; real-time tracking of the underwater devices; localisation and recovery of the underwater devices at the end of the test; recovery of the retrievable units.

Operational status
In service with the Italian Navy and being used in trials of the MU90 torpedo.

Specifications
Operational depth: 10-300 m
Frequency: 5 operating frequencies
Targets: up to 3 targets tracked simultaneously
Sea state: operates in up to Sea State 4

Contractor
Whitehead Alenia Sistemi Subacquei, Livorno.

VERIFIED

PUAPS

Type
Precise underwater acoustic positioning system.

Description
The PUAPS is a baseline underwater acoustic positioning system, conceived for oceanographic and offshore applications where high operational accuracy is required. It can position and navigate sea surface and/or underwater vehicles at depths from a few metres down to 1,000 m.

Computed positions are sent in real time by serial link to the external world, allowing operations integrated with other systems.

Two standard configurations are available – long baseline and short baseline configuration provides from 3 up to 8 ITR 100M transponder/releasers, an optional ITR 100M transponder releaser, an acoustic module mounted on the vessel, an acoustic module fitted on the underwater vehicle, a DMR 100M digital range meter mounted on the vessel and connected to a PC system, and a system management based on the PC. This system provides an area coverage of 1,500 m² with a positioning accuracy of ±5 cm (with sound velocity compensation).

The short baseline configuration provides four hull-mounted acoustic modules, a transponder fitted on board each target, a DRM 100M digital range meter mounted on the vessel and connected to a PC, and a

system management based on the PC. The system provides range coverage up to 2,000 m with accuracy less than 0.2 per cent of slant range with ray bending compensation and short baseline dimensions of 14 m.

The systems can track and position up to six units.

Contractor
Whitehead Alenia Sistemi Subacquei, Livorno.

VERIFIED

Mobile range

Type
Transportable acoustic range.

Development
The mobile acoustic range has been developed and built for the Italian Navy to record and analyse radiated noise from ships and submarines.

Description
The range consists of four main subsystems:
(a) the underwater subsystem comprises two hydrophone arrays connected to a surface buoy which is linked to a support ship by radio or by floating cable
(b) a data acquisition subsystem used to collect and record data on board the support ship. The signal recording and control equipment is contained in an easily transportable shelter which can be installed on the support ship or on the surface vessel being measured. It includes an emergency acoustic navigation system. To ensure complete acoustic silencing of the support ship the equipment is completely powered by batteries
(c) a computer-operated navigation subsystem for navigation control of the ship or submarine under test. This is located on the ship being measured and includes compensation for the sound speed profile and deformation of the moorings due to tidal streams. To measure the deformation the moorings are fitted with transducers. If the ship under test is equipped with its own underwater telephone these are not required
(d) a data processing subsystem for signal analysis. This is based on a modern computer system with appropriate peripherals and specially developed software. One third octave and line spectral data are compensated for frequency dependent propagation loss and bottom and surface interference, prior to time and space averaging.

The compact, flexible system can be launched and recovered from small support vessels the size of a minesweeper.

Operational status
The range has been in operation with the Italian Navy since 1987. The flexibility of the design has allowed the system to be used, with minor modifications, for near-field and far-field acoustic measurements of the NATO ASW research ship *Alliance*.

Specifications
Max operating depth: 400 m
Frequency range: 10-40,000 Hz
Acoustic navigation accuracy: better than 1 m

UPDATED

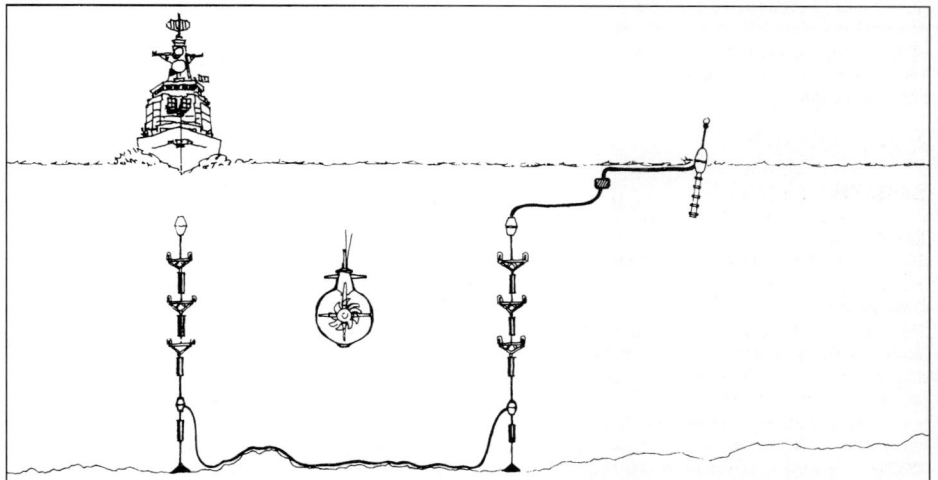

Configuration utilising hydrophone stations moored on the seabed and a radio link for data transmission to a support ship

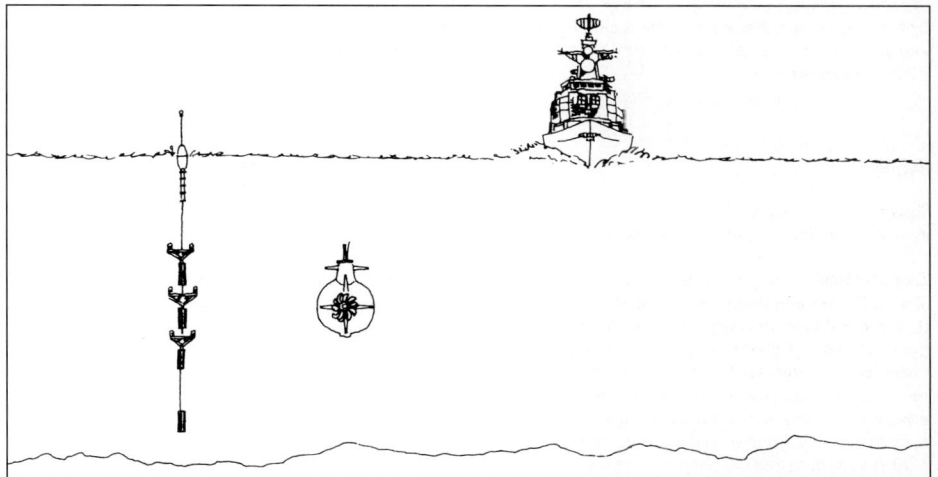

Drifting configuration utilising one or two hydrophone arrays suspended from the sea surface and a radio link for data transmission to a support ship

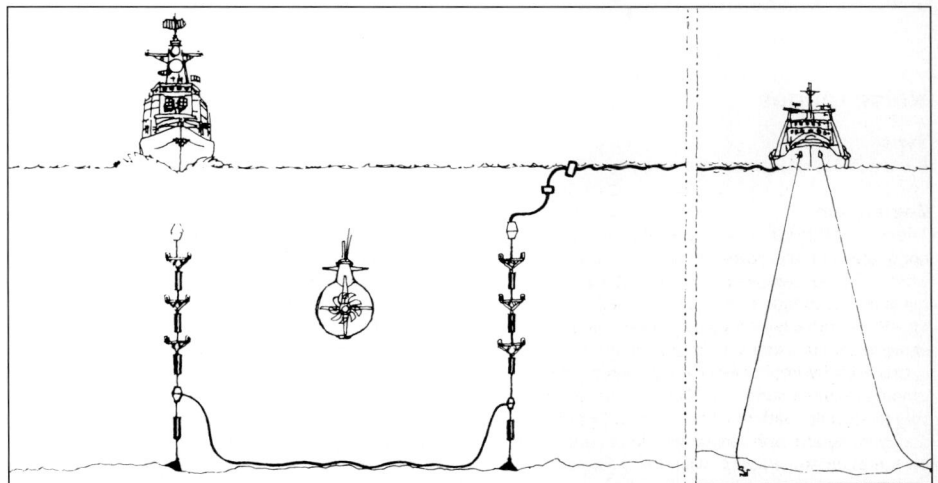

Floating cable configuration utilising hydrophone stations moored on the seabed and a floating cable about 1 km long for data transmission to a support ship

UNITED KINGDOM

SMARTTRAK

Type
Open-ocean acoustic tracking system.

Description
SmartTrak is a new, portable underwater acoustic tracking system that combines well-proven sonar and acoustic technology with specific tracking techniques.

The system comprises four buoyed receiver units and a tracking and display station. The receivers are battery powered and free floating and feature integral GPS. They remain operational in up to Sea State 4. SmartTrak provides accurate positioning to within ±5 m and tracking of up to five separate frequency pingers within a 25 km² area. This can be expanded if required. The system covers all locations from shallow water to the open ocean.

SmartTrak requires no cabling, no seabed deployment and no special lifting equipment or transport. There is also no need for shore-based infrastructure. All items are compact, man-portable and easily deployed from a small vessel.

Operational status
In service.

Specifications
Frequency range: 7-25 kHz

Contractor
PMES Ltd, Rugeley.

NEW ENTRY

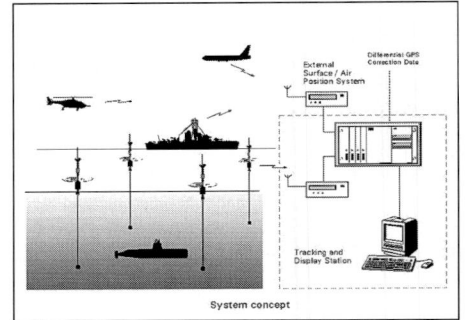

Schematic of operation of SmartTrak (Courtesy PMES Ltd)   *1999*/0043176

SMUTS

Type
Submarine-Mounted Underwater Tracking System.

Description
SMUTS is an in-service Royal Navy system developed for the evaluation of underwater weapon systems. It is a dual-mode acoustic tracking system capable of short-range (300 m) high resolution at 125 kHz, high accuracy, and three-dimensional (3-D) tracking with a long-range (4.8 km) distance measuring mode at 20 kHz. The system may also be used to measure the instant of fuze activation of a suitably instrumented weapon.

SMUTS may be used during the development and proving stages of a new weapon system and for routine in-service weapon effectiveness firings. The 'inboard' system is fully portable and may be rapidly installed on any vessel or installation fitted with an appropriate hydrophone array.

The prime uses are:
(a) to measure and display the approach and 3-D trajectory of the terminal phases of an underwater attack on static or moving target submarines, unmanned mobile targets, static targets or fixed installations
(b) to measure and display the discharge trajectory of any underwater weapon (such as a sub-launched missile, standoff weapon, torpedo, mine or decoy) from a submarine or fixed underwater tubes or dischargers. This may include measurement of the deployment/fall back of torpedo cable dispensers or missile shrouds after water exit separation and so on
(c) to measure and display the deployment and trajectory performance of bottom-launched weapon systems.

SMUTS may be supplied as a 'stand-alone' system, incorporating onboard computing and display together with digital data storage and recording for subsequent offline processing. Alternatively, it may be supplied as

part of a total attack/discharge system, interfacing with the ship's weapon and navigation systems either via discrete interfaces or via a highway. The submarine, which can operate either as the weapon launch platform or target, is fitted with up to eight hydrophones and associated onboard electronics.

Operational status
It has been used extensively by the Royal Navy, both for torpedo attack and discharge measurement, and for sub-launched (Sub-Harpoon) discharge, both in the research and development and fleet exercise role. The system has been used in an open ocean environment and on instrumented weapon ranges where compatibility with the weapon range system is required. It is in operation with, or ordered for, the Royal Navy and other NATO navies.

Contractor
PMES Ltd, Rugeley.

VERIFIED

ASR

Type
Acoustic Self-Ranging system (ASR).

Description
The ASR has been developed to meet the need for the acoustic self-ranging of warships to check and evaluate their acoustic signatures during periods between calibration at fixed noise range facilities or locations dictated by operational necessity. ASR is a cost-effective, portable acoustic measurement system. The system is highly effective in war zone conditions.

ASR uses data transmitted from an acoustic buoy, or from a bottom-mounted hydrophone, to provide a complete noise range facility. The underwater radiated noise signature is received, recorded and analysed in one-third octaves, Lofargrams or narrow frequency bands to produce results which are directly

comparable with those obtained on a fixed-calibration noise range, or obtained from other noise measuring systems.

The development of the self-contained acoustic measurement system gives rise to a family of applications which are possible with ASR:
(a) self-ranging – by installing ASR and deploying an acoustic buoy, a vessel is able to measure its own acoustic signature under varying operational conditions
(b) transportable ranging – ASR can form the basis of a relocateable land-based noise range facility to provide signature measurement and analysis for both surface ships and submarines
(c) remote ranging – a ship fitted with ASR can provide a floating noise range facility which can be deployed wherever ships or submarines are operational
(d) covert monitoring – being fully self-contained,

easily portable and if required, battery-operated, ASR can rapidly be deployed to monitor sensitive waterways and channels.

Operational status
The ASR has been successfully demonstrated and delivered to the UK Royal Navy and an Asian navy.

Specifications
Frequency: 10 Hz-40 kHz
Analysis: Lofargram, narrowband spectra, one third octave analysis
Power consumption: 110 V-240 V single phase AC or +24 V battery supply

Contractor
PMES Ltd, Rugeley.

VERIFIED

Noise ranges

Type
Fixed and transportable noise range.

Description
Thomson Marconi Sonar Ltd has manufactured a complete portable noise range for a European navy, which can be deployed in deep water to accurately measure submarine signatures. The data is recorded on IRIG-compatible magnetic tape for subsequent analysis ashore using a mainframe computer.

Standard hydrophones specially designed for noise ranging, manufactured by Sonar Systems, produce a very good polar pattern up to 150 kHz. Different gains, acoustic heads and other variations are available on the standard product, which is suitable for either

fixed or portable ranges. These hydrophones are used on a number of UK ranges, including the underway noise range at Rona and the static noise range at Loch Goil, as well as portable ranges for two European navies.

The transportable noise range is highly modular and in its standard form provides three noise measuring hydrophones at selectable depths suspended from a surface buoy some 400 m from the deployment vessel. An active transducer is suspended from the deployment vessel which emits pings for the tracking system. The vessel being ranged carries a self-contained transponder which replies to the active transducer ping. The noise measuring hydrophones receive both the ping and the reply, enabling dynamic position fixing of all elements of the system. The data processing and recording equipment is fitted on board

the deployment boat, enabling on and offline position line plotting and noise analysis. Software is also included for offline range normalisation of recorded noise. To assist submariners, the system includes a sonar beacon which may be switched on and off as required.

The fixed underwater acoustic noise range incorporates five variable depth underwater sensors, each connected by its own five-core cable to processing equipment in an onshore, prefabricated, air conditioned building.

The position of vessels being ranged relative to the underwater sensor array is determined by a laser tracking system for surface ships and an acoustic tracking system for submerged submarines, both independent of the normal equipment carried by either platform.

The three-dimensional tracking system has been designed to cover an area of up to approximately 70 km^2. Capable of operating in up to Sea State 4 the range comprises 20 easily deployed nodes each consisting of a surface buoy and seabed unit which form a distributed system. The nodes accept synchronised acoustic transmissions from vehicles operating over the range. These signals are compressed and sent via radio link to a central receiving and processing vessel. The nodes can be deployed in depths to 300 m and are designed for operations for up to three days before recovery. The system can be quickly deployed and recovered by a small vessel.

Operational status
In service.

Contractor
Thomson Marconi Sonar Ltd, Templecombe.

UPDATED

MAGNETIC RANGES

FINLAND

MGS-900-PFO

Type
Portable signature measurement range.

Description
The MGS-900-PFO is an advanced underwater measurement range for the determination of ship's magnetic, pressure and optionally acoustic signatures. The MGS-900-P was introduced in 1991 and the latest fibre optic enhancement, MGS-900-PFO, was introduced in 1995.

The MGS-900-PFO comprises a shore-based or onboard Range Processing and Display Unit (RPDU) and several MG-930 sensor units arranged in an array on the seabed. The sensor units are connected to the RPDU by a fibre optic hybrid underwater cable. The sensor unit array can be configured with a maximum of 8 (optionally 10) sensor units at a maximum cable distance up to 2,800 m between the seabed array and the RPDU.

Each MG-930 sensor unit has a 16-bit microcontroller which controls the operation according to the messages received from the RPDU. All measured signatures are digitised and buffered in the sensor unit's memory before sending to the RPDU. The basic configuration of the sensor contains the following sensors: three-axis fluxgate magnetometer for magnetic signature measurement; two-axis inclinometer for the tilt correction of magnetic sensors; pressure sensor for depth and pressure signature measurement; and optional hydrophone with required circuits for acoustic signature measurement (sensor units Type MG-930A).

The optional reference sensor (Type MG-930) is for the measurement of atmospheric pressure for accurate depth determination and the Earth's magnetic field during overrun in order to compensate for the natural changes in the magnetic field.

Each component of the magnetic signature can be measured in four selectable ranges between 0 and ±1,000,000 nT with DC to 10 Hz bandwidth. Pressure signature data has a dynamic range of 0 to ±15 kPa (static range is 0 to 500 kPa) with DC to 1 Hz bandwidth. Acoustic signature data has three measurement ranges

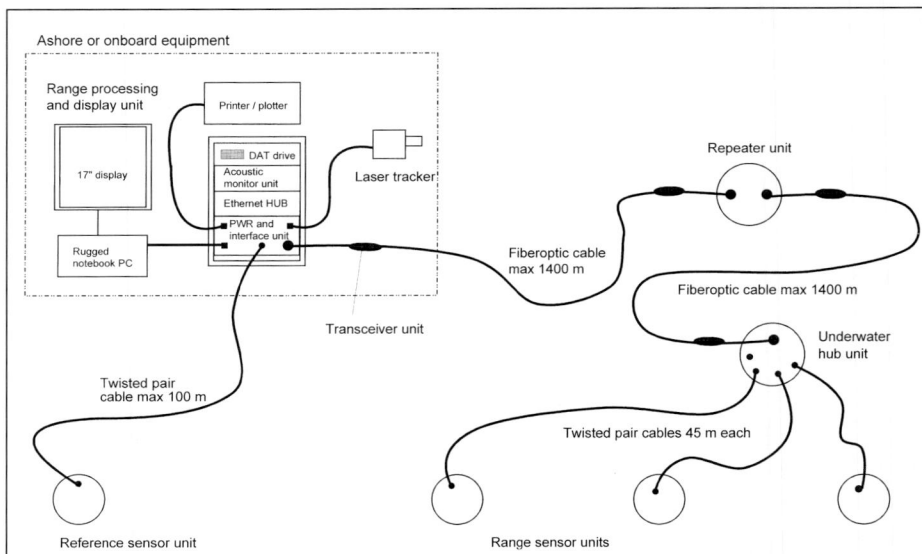

Schematic overview of portable signature measurement range MGS-900-PFO **1997**

between 60 and 180 dBµPa with 1 Hz to 10 kHz bandwidth.

The RPDU is based on a ruggedised laptop Pentium PC running an SCO Open Desktop operating system. The display is a high-resolution 17 in colour monitor mounted in a shockproof transport case. The online display shows the measured signatures and tracking data and processed displays have contour plots, profile plots and three-dimensional perspective plots. For hard copies the operating system and software supports different kinds of standard commercial printers and plotters.

The interface unit MI-901 supplies DC power to all sensor units and isolates them electrically from other equipment and contains protection circuits in case of voltage surges. The MI-901 is mounted in a case with power distribution panel, network hub, optional DAT tape drive and acoustic monitor.

The optional software modules include magnetic modelling of the ship, magnetic modelling of the degaussing coils and optimisation of the coil currents, acoustic monitoring and playback and acoustic signature processing and analyses. The standard system includes diagnostics, calibration and simulation.

Operational status
Three systems have been delivered since 1994 and more are on order.

Contractor
Elesco Oy, Espoo.

UPDATED

FRANCE

Magnetic ranges

Type
Overrun and component ranges.

Sea degaussing range
The two types of degaussing range manufactured are the open range, in which the vessel follows a track across a line of sensors, and the fixed range, in which the vessel is moored in a stationary position above the sensors. The ranges feature a limited number of sensors and highly sophisticated modelling software. These are used to predict a ship's magnetic signature under varying conditions and to prepare the design of degaussing systems. Ranges are operational in the UK and Norway.

Magnetic land range systems
These systems are used to carry out a full analysis of the magnetic characteristics of all equipments to be installed on a ship. The systems comprise a fixed magnetic range to measure magnetic signatures from which the necessary degaussing corrections can be calculated, deperming equipment studied, and the effects of roll and pitch on magnetic signature together with stray field effects measured.

Contractor
Thomson Marconi Sonar SAS, Sophia Antipolis.

UPDATED

MIR 2000

Type
Portable multi-influence ranging system for surface ships and submarines.

Description
The MIR 2000 is a multi-influence portable ranging system for any kind of vessel. It can be used either in a fixed range mode or in a portable mode.

As a result of new environmental conditions or the installation of new equipment configurations, the signature of a MCMV may be degraded. Before commencing mine sweeping or minehunting operations, the officer-in-command must evaluate the vessel's signature to determine safety limits under which it can carry out the operation.

The MIR 2000 is an easily transported system which can be fitted on board any vessel including MCMVs or smaller vessels. It is a fully autonomous system which is deployable in less than 4 hours. The system carries out simultaneous measurement of: magnetic AC and DC fields; electric AC and DC fields; acoustic influence; pressure/water depth; seismic influence; and the gradients of magnetic fields.

The range can be configured for use either as a portable range, a fixed range, or for use with exercise mines.

The control operation is divided into two phases: measurement from a small craft involving three runs with a duration of 30 minutes for collection of data followed by a 10 minute phase for the checking of the measurements and data transfer on to magnetic medium for analysis on board the parent ship. The analysis phase on board the parent vessel involves analysis of the data on a work station and comparison with previously recorded measurements.

The outboard equipment comprises a Thomson Marconi Sonar SAS TSM 3207 multi-influence sensor and a set of cables with drums. Inboard equipment consists of a power generator, an interface electronic unit, DGPS (Differential GPS), and a workstation.

Various options are available including: additional cable; mechanical protection against shocks (titanium construction); a non-magnetic releasing system;

additional sensors and software for modelling; and data transmission via radio link.

Specifications

Sensors

Type	Technology	Bandwidth
DC magnetic field	3-axis fluxgate	DC to 5 Hz
AC magnetic field	3-axis fluxmeter	5 – 3,000 Hz

LF electric field	3-axis silver/silver chloride electrodes	0.01 – 5 Hz
AC electric field	3-axis silver/silver chloride electrodes	5 – 3,000 Hz
Acoustic influence	Piezoelectric hydrophone	10 – 10 kHz
Depth/pressure	Absolute sensor influence	DC to 2 Hz

Physical characteristics
Operating depth: <50 m
Multicoaxial cable: Length 100 m
Inboard equipment: Weight 100 kg

Contractor
Thomson Marconi Sonar SAS, Sophia Antipolis.

VERIFIED

GERMANY

Land range

Type
Magnetic measurement system for ships' components.

Description
The magnetic measurement of items of equipment for installation in a ship constitutes part of its proof of performance to appropriate specifications.

The measurement, deperming and degaussing are consecutive operating processes which may need to be repeated until the minimum or acceptable interference field is achieved.

The equipment to be measured is traversed over a specially defined arrangement of sensors on a measuring track arranged in a north-south direction. The non-ferromagnetic test carriage is provided with a platform which can be rotated through 360°. The object-specific permanent fields, the fields induced by the earth's field and, where necessary, the electromagnetic stray fields produced by the object under test itself are measured, analysed and recorded. The sensors and the measuring electronic equipment are all components of the Forster Magnetomat magnetic measuring system.

Contractor
Institute Dr Forster, Reutlingen.

VERIFIED

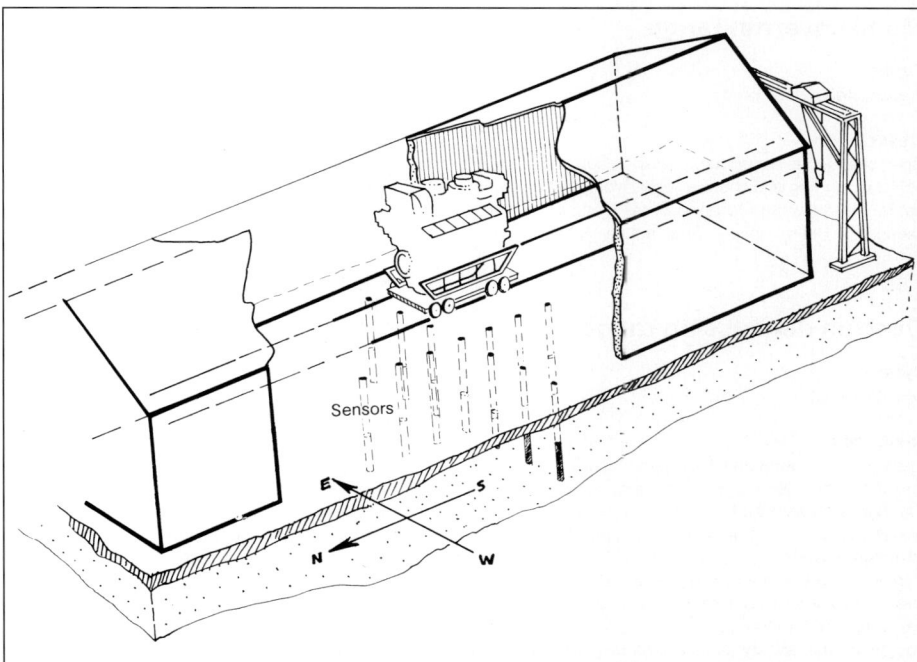

Magnetic measurement system for ships' components

Stationary range

Type
Stationary magnetic measuring system.

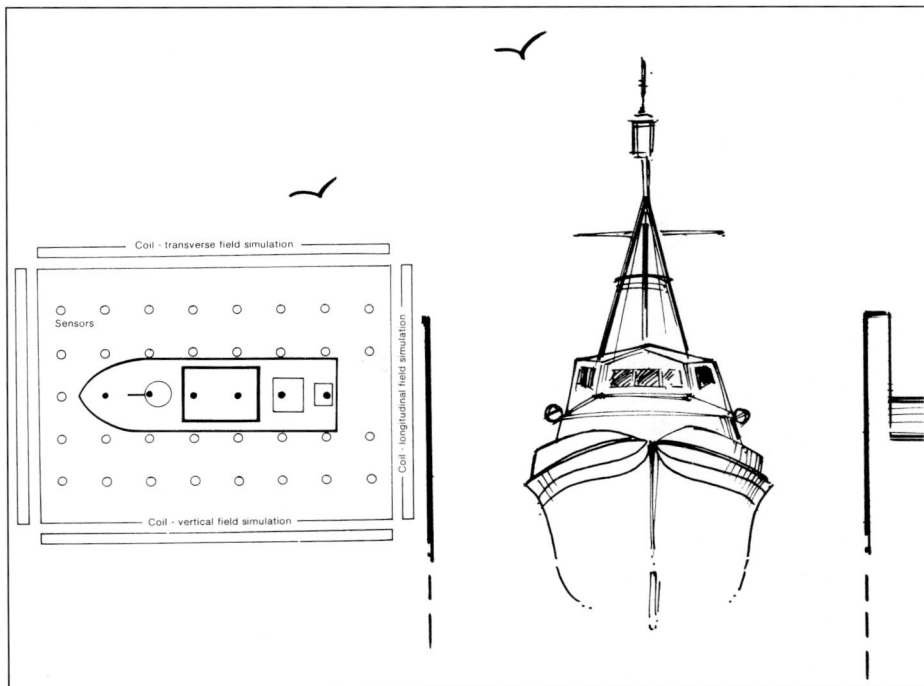

Description
A stationary measuring system with simulation equipment enables all magnetic fields occurring during operation of the ship to be measured and analytically separated in rest conditions. These comprise the ship's own permanent fields, course-dependent induced fields, electromagnetic stray fields, and course and position-dependent eddy current fields resulting from pitching or rolling.

The ship rests above a carpet comprising multi-axial sensors whereby the integral interference field is measured with one, two or three axes in a horizontal plane beneath the ship.

Because of the array of sensors distributed over a large area comparable with the size of the ship and owing to measurement in several axial directions, the interference field analyses and the results of monitoring are more precise, provide more information and are thus more reliable. The advantage of a stationary system as compared with the dynamic overrun measuring system lies in the fact that any particular and temporary change in the earth's fields can be simulated with the aid of large area current coils arranged horizontally and vertically. This also enables the stray field of the entire ship to be monitored in the rest condition for pitching and rolling. It is also possible to simulate known earth field data of intended zones of operation and thus obtain and program the degaussing variables required for self-protection.

Contractor
Institute Dr Forster, Reutlingen.

VERIFIED

Stationary magnetic measuring system

Open range

Type
Dynamic overrun measuring system.

Description
Dynamic overrun measuring systems are permanently installed monitoring stations for strategically important ships endangered by mines.

If the interference field exceeds known critical limit values, degaussing measures must be taken. Further overrun checks qualify the ship for the intended operation.

Following a north-south course and then an east-west course, the ship passes over a perpendicularly arranged chain of magnetic field sensors. This enables measurement to be conducted in two extreme courses with respect to the earth's field. The vertical Z component of the interference field is usually measured with uni-axial sensors. The depth and spacing between the sensors are matched to the type of ship in question.

Measured value acquisition, data processing and representation are carried out using several channels, analogous to the number of sensors or their measuring axes.

This produces, for example, graphical weighted pictorial relief representations in the longitudinal and cross-direction of the ship and these can be assigned geometrically by tracking the ship visually.

Contractor
Institute Dr Forster, Reutlingen.

VERIFIED

Mobile overrun range

Type
Transportable open range.

Description
Open sea ranges are primarily check ranges. The initial object of ranging is to check a vessel's degaussed state before operation and to perform magnetic calibration exercises. Being simple and rapidly installed the transportable degaussing (overrun) range which requires only one range array provides a cost-effective solution to providing facilities to meet logistic and operational requirements in areas where fixed facilities are not available. The transportable range can be set up in a very short time using a limited number of sensors and structures. Such a system can provide measurement, recording, analysis and reporting of actual magnetic states of ships.

The main components of the range are a transportable framework, an ISO container used as the range house, the Forster magnetic measuring system Magnetomat and the tracking system (no separate system is necessary as specialised software and range sensors are used).

Contractor
Institute Dr Forster, Reutlingen.

VERIFIED

Overrun degaussing range

Type
Overrun range.

Description
The basic configuration for ranging of modern MCMVs consists of an intercardinal probe array which enables the measurement of the magnetic signature while the vessel is sailing on an intercardinal course (45°) related to magnetic north.

The ranging procedure is conducted on opposite overruns in order to determine the different origins of the ship's magnetisation. During measurement the course of the MCMV is permanently tracked by a Differential Global Positioning System (DGPS) and course plots are fed to the evaluation system for further consideration. At the same time course data are transmitted to the MCMV and converted into course information which serves to support the ship's command.

The entire probe array consists of a number of active triple probes (X-, Y-, Z-sensors) with 'built-in' electronics. Three probes are installed at the same depth and form the measuring level.

Data from the probes are processed by software packages which enable the vessel's induced and permanent magnetisation to be separated. A prerequisite for separation is the ranging of the vessel on opposite intercardinal overruns. The software also enables the required coil settings of the ship's degaussing system to be calculated and the determination of the 'depth law' and conversion of the measured interference field into various depths.

Ranges with reduced numbers of probes on intercardinal alignments offer easy installation and avoid the complex underwater structures associated with stationary ranges. Such systems are suitable for both mobile and permanent employment and containerised versions are available.

Contractor
STN ATLAS Marine Electronics GmbH, Bremen.

UPDATED

Stationary degaussing range

Type
Stationary range.

Description
With this type of range, measurements are carried out while the vessel is moored over an array of probes. The probes are arranged along the longitudinal and transverse axes of the vessel and enable the interference field to be sampled at various points beneath the ship.

The essential feature of the stationary range is its ability to measure the ship's interference fields caused by eddy currents and to predict the vessel's field at various geographic locations. For this the stationary range is equipped with a coil arrangement which allows the vertical and horizontal magnetic fields to be generated to simulate the roll motions of the ship and its geographic location, including heading. Another important feature of a stationary range is its ability to directly measure and assess stray fields.

Contractor
STN ATLAS Elektronik GmbH, Bremen.

VERIFIED

Component range

Type
Magnetic ranging facilities for components.

Description
This type of range is designed to provide measurement of magnetic conditions and magnetic treatment – if considered necessary – of components such as diesel engines, generators and so on, prior to installation on board the MCMV.

As an additional feature, the range provides facilities for the evaluation of magnetic models concerning the ship's magnetic behaviour.

Measurements and treatment activities are computer-controlled and supported by suitable software measures.

Contractor
STN ATLAS Marine Electronics GmbH, Bremen.

VERIFIED

Magnetic ranging facility for components

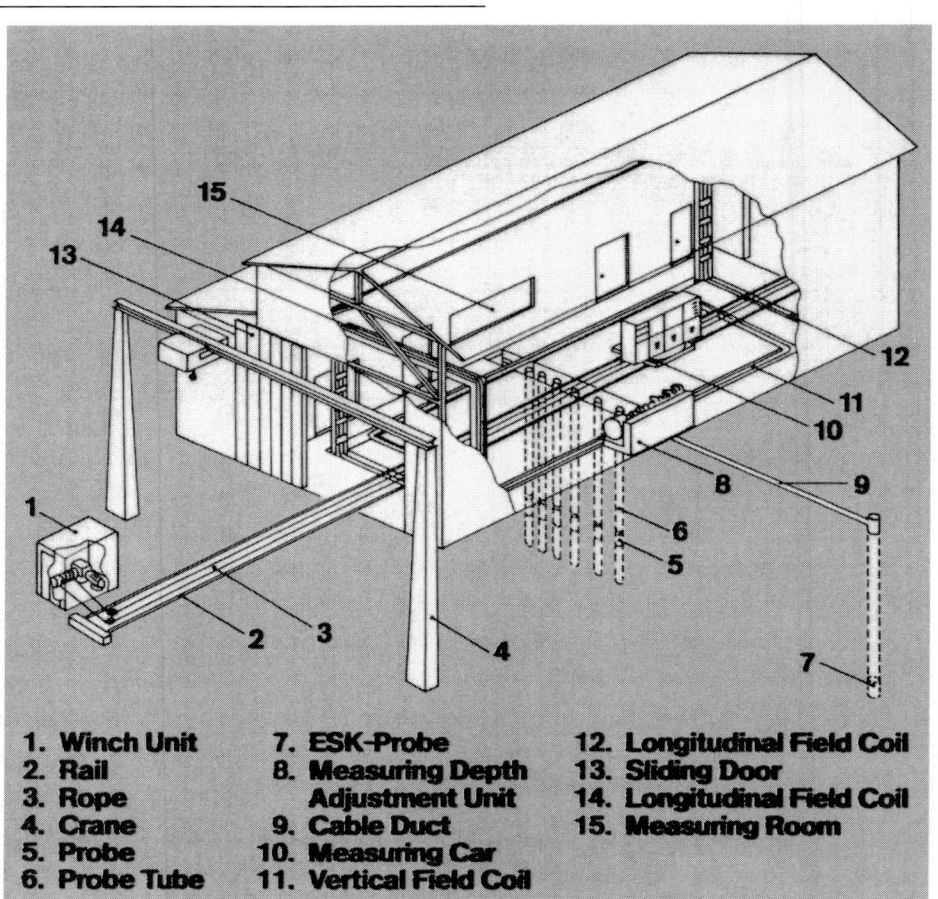

1. Winch Unit	7. ESK-Probe	12. Longitudinal Field Coil
2. Rail	8. Measuring Depth Adjustment Unit	13. Sliding Door
3. Rope		14. Longitudinal Field Coil
4. Crane	9. Cable Duct	15. Measuring Room
5. Probe	10. Measuring Car	
6. Probe Tube	11. Vertical Field Coil	

ITALY

MWR

Type
Mine Warfare Range.

Description
These systems comprise a group of mobile and fixed equipments that can produce several configurations of range for mine warfare. The ranges are basically used to study mine-target countermeasure interaction and to assess operational data necessary to planners in order to use available resources to achieve the best results.

The MWR system has been designed to operate with mines manufactured by Whitehead, and to obtain the maximum information relevant to the desired parameters. This includes: planning the precise location of mines within minefields and their general location; selecting the most effective route for MCMVs as well as targets; recording the responses of the mine; and analytically estimating the required parameters related to mine detonation as well as to units transiting minefields. All this is carried out automatically.

The system comprises the exercise mine linked by cable to a floating buoy, which receives signals from the mine either by the cable or by acoustic detection, and transmits this data by radio to a shore station within a range of 5 km.

Contractor
Whitehead Alenia Sistemi Subacquei, Livorno.

VERIFIED

Mine warfare testing range

Type
Mine warfare range.

Description
The range is designed to evaluate the effectiveness of mines against naval targets, the effectiveness of MCM operations against various types of mines, and the level of risk for MCM vessels and military and civil traffic.

The range consists of various equipments deployed at sea together with a shore base comprising a master and two remote stations.

At sea a series of exercise mines is acoustically linked with control buoys, and with a mine positioning system installed on board an auxiliary ship and able to compute, before each exercise, the exact position of all the mines in the range.

A communication system links the buoys to the computing centre in the master station. A ship positioning system tracks all the surface ships (targets) with high accuracy using interrogators in the master and remote stations and transponders installed on the target ships.

The computing centre acquires the information from the mines (activation, self-protection) and correlates the data with the time and with the ship positioning information. Data is displayed in real time on a graphic workstation and recorded on magnetic tape for subsequent analysis. Plottings and colour printouts are available to the operators.

The computer system comprises a DEC high-resolution graphic workstation equipped with disks and tape units. Peripherals include a printer, colour hard-copy, plotter and an alphanumeric terminal. The application software is written in Ada and it has been designed with the HOOD methodology with extensive support of CASE tools (Teamwork).

Operational status
The range has been installed near the naval base at La Spezia and has been in use with the Italian and other NATO navies since 1997.

Contractor
Datamat Ingegneria dei Sistemi, Rome.

UPDATED

SWEDEN

BMP 200

Type
Underwater measuring platform.

Description
The measuring platform BMP 200 is made for underwater surveillance and recording of different signals from vessels and/or minesweep and/or environment influence. It is designed for easy operation, versatile use and can be equipped with a variety of sensors depending on use and customer demand.

BMP 200 is autonomous and can be placed on the seabed for long periods. The storage of data from the sensors is controlled remotely by means of an acoustic link. Recovery is also remotely initiated by the acoustic link. After recovery, all collected data can be transferred to other systems for evaluation.

Operational status
In production.

Specifications
Length: 2,500 mm
Width: 2,100 mm
Height: 1,300 mm
Weight (in air): 700 kg
Operating depth: 10-200 m

Contractor
Bofors SA Marine, Landskrona.

VERIFIED

BMP 200 system

1998/0009856

BMP 201

Type
Underwater measuring platform.

Description
The measuring platform BMP 201 is made for underwater measurement of magnetic, acoustic and pressure signatures from a ship and/or a minesweep. It also registers the static water pressure at the sensor's position. The system consists of a sea module and a land module connected by a cable. To every land module, it is possible to connect four sea modules. The measured data is transferred from the sea module via the land module to a standard PC where it will be recorded. The evaluation takes place in the PC which is equipped with specially designed evaluation software.

Operational status
In production.

Specifications
Length: 700 mm
Width: 700 mm
Height: 700 mm
Operating depth: 5-200 m
Weight (in air): 85 kg
Cable length: >2,000 m

Contractor
Bofors SA Marine, Landskrona.

UPDATED

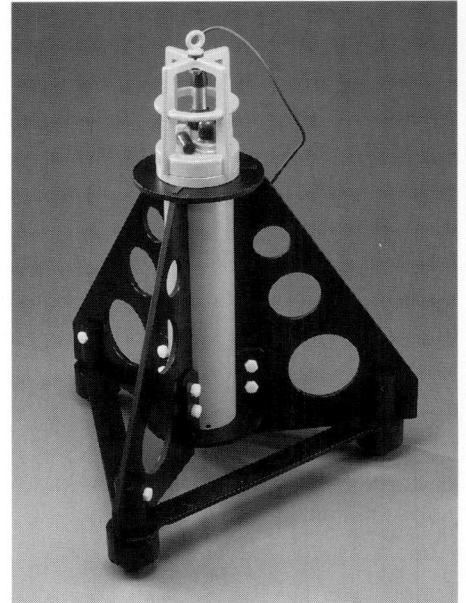

BMP 201 platform
1999/0038678

UNITED KINGDOM

MMR

Type
Magnetic Measurement Range.

Description
The Magnetic Measurement Range (MMR) uses two underwater sensors (minimum) to categorise the state of a vessel's magnetisation.

The sensors are easy to deploy and require minimal diver support. The system's design lends itself to high portability.

The determination of the sensors' positions and orientations do not require underwater surveying activity, surveying from the sea surface only being used.

The sensors and other underwater units are designed to withstand permanent immersion in over 50 m of seawater. Sensor tilt of up to 30° of arc is automatically compensated.

The system comprises (typically): the underwater sensors, data transmission to the shore (either radio telemetry or cable), a shore interface and the system's computer with its peripherals.

Vessel track, as it passes over the underwater array, is by the vessel's own magnetisation or by an explicit tracker (GPS or other).

The system's software allows in-depth examination and analysis of a vessel's magnetisation. The results are presented in comprehensive colour graphical and tabular formats.

Access to software operation is by mouse point-and click procedure in a 'windows-style' environment. Help is always available from the system by operator demand.

Prediction of fields at any distance and 'self-danger width' are some of the features of the system's capability.

Australian MMR/MTF Ranges   *1996*

Deployment of the system can be achieved in two days. Results, following ship transit, are presented in under 2 minutes.

The MMR can be configured to meet all requirements including combined multi-influence measurement.

Operational status
In service with the Royal Australian Navy. This system comprises two sensors with inclinometer and depth sensing laid in 15 m depth of water approximately 1 km offshore. Another land-based magnetic measurement and treatment range has been installed for the Royal Norwegian Navy to support the 'Oksoy/Alta' class. The facility makes extensive use of magnetic modelling techniques. It features magnetic sensors mounted on a 3-axis field coil assembly to measure both magnetostatic and alternating magnetic fields.

Contractor
PMES Ltd, Rugeley.

UPDATED

PMAR

Type
Portable magnetic and acoustic range.

Description
The PMAR (portable magnetic and acoustic range) has been designed as a transportable system that is easy to deploy and recover. Performance has been enhanced to be compatible with, or superior to, the large fixed installations previously required to carry out this type of measurement.

The equipment comprises underwater equipment and a shore analysis facility. The underwater equipment consists of two multi-influence sensors and a data-gathering unit (referred to as the underwater control unit), plus interconnecting cable. Inside the underwater control unit, acoustic and magnetic signatures are digitised before being sent via the underwater cable or a radio datalink to the shore facility.

PMAR underwater sensor unit
(Courtesy PMES Ltd)
1999/0043175

On shore this data is displayed using colour graphics on a large screen, high performance PC housed in an air-conditioned office space. Comprehensive analysis software is provided to process and display the data in real time with acoustic monitoring of the audible signature always available.

Operational status
In service with a Southeast Asian navy.

Contractor
PMES Ltd, Rugeley.

NEW ENTRY

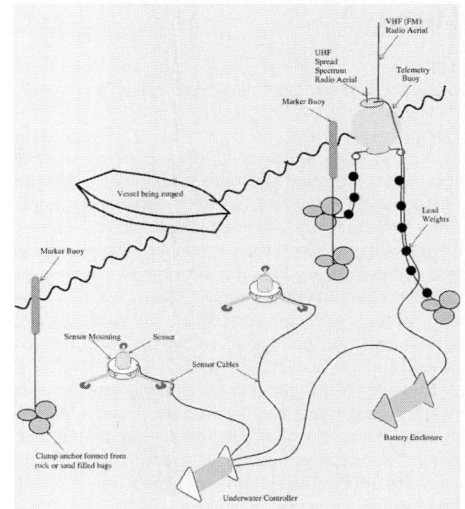

Block schematic of PMAR
(Courtesy PMES Ltd)
1999/0043174

Magnetic signature measurement

Type
Magnetic sensor assemblies.

Description
The fixed range system includes a linear intercardinal array of equispaced three-axis magnetic sensor assemblies individually mounted either on concrete limpets or on pile-type structures in the seabed.

At each end of the line array additional seabed structures incorporate an acoustic transponder for unit position measurement and a depth measuring sensor. Optionally, this unit can also be modified to monitor noise signals of the vessel. The data processing approach used does not demand a high alignment accuracy but does require an accurate knowledge of the actual position of each sensor. Digitised data from the sensors are transmitted by cable to a shore instrumentation facility.

The control instrumentation both supplies the array sensors with power and translates the communication link for use by the range computer. The cable has been designed to ensure very reliable operation after long-term immersion on the seabed. It is of small diameter with single cables being used for each sensor. This minimises the problems in cable laying, recovering and relaying, and enables proven wave zone protection techniques to be used. Avoidance of the use of junction boxes and use of single cables significantly increases range availability, particularly when cable damage occurs.

The computer system is based on the MicroVAX II, as this meets the processing requirements, providing the power necessary to implement the data processing approach and to make available the required signature and coil prediction data to the operator on demand.

Associated with, and digitally linked to the shore instrumentation facility is a laser-based tracking system which also provides data for the helmsman. The data, in the form of distance to go to array and distance off track, are communicated to the vessel by RF link. The laser tracker also requires the temporary addition of two omnidirectional prism reflectors to the vessel so that heading information can be obtained. For the transportable system, which would use single-axis magnetic field sensors assembled into independent but coupled sub-arrays, and as an option on the permanent range facility, magnetic tracking can be provided.

Contractor
Thomson Marconi Sonar Ltd, Templecombe.

VERIFIED

Transmag 2000

Type
Transportable degaussing range.

Development
Transmag 2000 has been developed as a totally transportable degaussing range system which can be moved readily between locations and deployed with the minimum of effort.

Description
Transmag 2000 can be used with all classes of ship, but is especially adaptable to MCMVs, its two modes enabling it to be used as a ship's magnetic signature check facility or as a ship's calibration facility.

Small and of modular design, Transmag 2000 can be packed into two 3 m standard ISO containers for transport by land, sea or air. Alternatively, the equipment can be mounted in a rugged cross-country vehicle or trailer of the customer's choice. Thus, a task such as helping to keep the Persian Gulf clear of mines can be made safer by carrying a transportable degaussing system to the actual area of operations.

The new system consists of an arrangement of magnetic field sensors located in a line on the seabed. These are connected by a secure cable link to a data acquisition and computer system in a control module, which can be located on shore or on a support vessel. Close by the control module, there is a unit which tracks the ship being ranged.

The computer system acquires magnetic and water depth data along with the ship's track data, and this is processed to advise the ship of the optimum course to steer over the range to resolve the ship's magnetic components and predict coil current settings. It can also predict fields remote from the ship to give safe distance analysis and Magnetic Anomaly Detection effects.

Transmag 2000 does not require the use of heavy mechanical handling equipment, the deployment of sensors being accomplished by manpower alone operating from an inflatable dinghy which is supplied as part of the plan-packed equipment. Three sensors and a junction box are deployed on the seabed, with one lightweight cable returning to the shore. Three sensors form a good compromise between the ease of deployment, adequate ship's navigation width, availability of magnetic information and a degree of built-in redundancy. Transmag 2000 is a robust system which enables ship's position and heading to be derived throughout its run.

This information is used to provide an onboard navaid, giving assistance to the helmsman to get the ship over the range as quickly and accurately as possible. In the magnetic signature check mode, the ship is sailed over the sensor array on one pair of reciprocal headings. From the data thus acquired, the permanent and induced components of the ship's magnetic signature can be isolated and resolved into longitudinal, athwartship and vertical components.

Should the permanent components be found greater than those which can be comfortably controlled by the onboard degaussing coils, or they are outside predefined limits, then the ship may be instructed to undergo treatment.

In the ship's calibration mode, a coiled ship can carry out a number of passes with its coils either on or off. Providing this is performed in a controlled sequence, the effect on the signature of the ship by energising each coil can be determined. In this way, a ranging can be implemented with onboard degaussing coils currents being predicted so as to reduce the magnetic signature down to the lowest possible value.

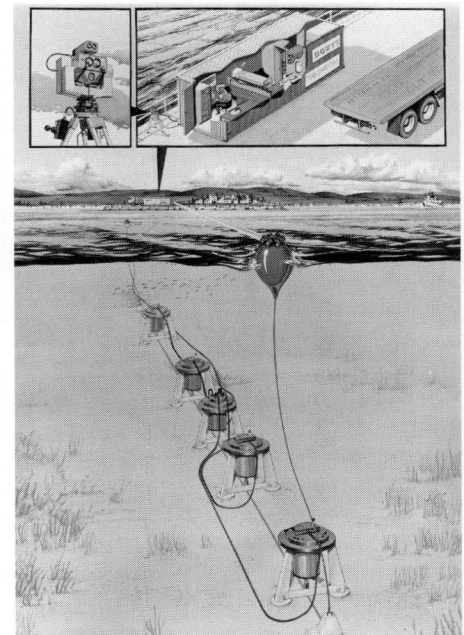

Transmag 2000 transportable degaussing range

Contractor
Ultra Electronics, Magnetics Division, Hednesford.

VERIFIED

MERS

Type
Signature measurement range and evaluation mines.

Description
MERS (MCM Evaluation and Range System) is a dual role system providing a cableless underwater signature measurement range and an MCM evaluation range. The measurement range allows multi-influence signature measurement and modelling of co-operative and non-co-operative vessels. The range sensor units are themselves Versatile Exercise Mines, which provide the system with an evaluation mine facility for assessment and training of MCM against simulated threat mines.

The system is transportable, allowing forward area use, and can be operated from a small vessel or MCMV. It can be rapidly deployed, recovered and turned around on deck without diver support and with no cables to be laid. The system allows measurement of all potential threat mine influences – static/alternating magnetic, acoustic, seismic, pressure, static/alternating electric (UEP/ELFE). These influences are all recorded on a single transit of the ship being measured, and with no equipment fitted to the ship.

The system also provides all of the functions of the Versatile Exercise Mine System (VEMS), with the addition of alternating magnetic and static/alternating electric field sensors.

Operational status
In production for the Belgian Navy and is entering service.

Contractor
BAeSEMA, Bristol.

VERIFIED

MERS undergoing trials   *1998*/0009858

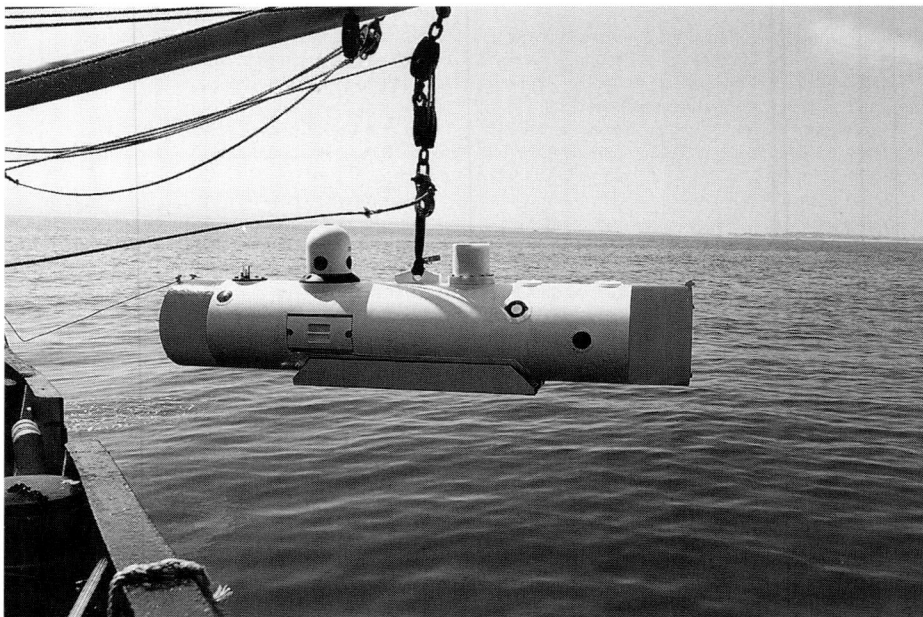

MERS undergoing trials
1998/0009859

UNITED STATES OF AMERICA

MSMS

Type
Magnetic Signature Measurement System (MSMS).

Description
This advanced system measures with precision the magnetic signature of ships and gathers, records, processes, stores and displays the data during magnetic ranging and treatment operations.

The system accurately and rapidly ranges both submarines and surface ships to determine the need for magnetic treatment. It also provides critical magnetic measurement readings throughout the deperming process.

With this system, magnetic signals emitted by ships are sensed by underwater magnetometers and recorded simultaneously. Shore-based data acquisition and analysis equipment receives and processes the sensor data to provide, in minimum time, an accurate profile of a ship's magnetic state.

Field measurements are achieved through the use of submerged magnetometers placed in tubes located in the ocean floor. These sensors can also be inserted in a bottom-mounted tube structure, at a customer's option, to give transportability to the system.

Each magnetometer provides a DC voltage that is proportional to the magnitude of the magnetic field component along its magnetic axis.

These sensors are capable of measurement of dynamic range of ±200,000 nT, noise of less than 0.1 nT in a DC to 10 Hz bandwidth, resolution equal to or less than 0.1 nT, and linearity of 0.005 per cent.

Magnetometer outputs can be digital or analogue. These outputs are routed through junction boxes to the interface electronics in the data acquisition console and to a digital computer.

Signatures for keel and athwartship are processed, analysed and displayed. A determination is then made as to whether to record the information.

Raytheon measurement facilities are interactive systems. Software controls the intra-system interface between magnetometers, the interface electronics and standard computer peripherals.

The operator directs the system computer to collect and analyse magnetic field data using appropriate software programs. These programs direct system activity and query the operator for data inputs and desired measurement parameters for ships undergoing treatment.

In response to data and computer commands entered by the operator, the system collects and stores magnetometer data, performs calculations and displays results relative to the magnetic treatment process.

The system is highly automated and includes data processing and display equipment and specially designed underwater magnetic sensors. It consists of several distinct subsystems.

The Magnetic Range System relies on two arrays of magnetometers, shallow and deep, to range ships. User vessels pass over the sensor arrays at least once in each direction, with signatures continuously recorded via shore-based equipment. This subsystem can also be used to calibrate ship degaussing systems.

The MSMS operates in conjunction with a fixed deperming facility. It uses a keel line of magnetometers to measure signature levels before, during and after deperming operations. User ships are moored over the sensor array and signatures recorded on command and at the appropriate time via shore-based equipment.

These subsystems, although independent, share common data acquisition and analysis equipment. Off-the-shelf hardware is used extensively throughout system designs. A modular approach to design, with key components separate from one another, will enable new features to be incorporated in the future with minimal system upgrade.

Operational status

MSMS is fully operational and meets US Navy requirements for performance, logistics support and life cycle costs. Early in 1991 Raytheon was awarded a US$21.9 million contract to design, build and install an advanced magnetic treatment at the US Navy's Trident Magnetic Silencing Facility in Kings Bay, Georgia. The automated system rapidly scans the submarines to determine the required magnetic treatment and then deperms them by removing or altering their permanent magnetism. In May 1991, Raytheon won a US$16 million contract to build two Type IV magnetic silencing ranges at San Diego, California and Mayport, Florida. The two systems ordered are now operational.

Contractor

Raytheon Company, Electronic Systems, Portsmouth, Rhode Island.

VERIFIED

Raytheon MSMS system

FACDAR

Type

Forward Area Combined Degaussing and Acoustic Range.

Description

The FACDAR is a transportable magnetic and acoustic signature measurement range designed to provide ships with protection against influence mines. The range is based on the highly successful transportable degaussing range used in the 1990-91 Gulf War by all the allied navies.

The range comprises Alliant Techsystems' Marine Systems (now part of Hughes Aircraft, Naval and Maritime Systems) acoustic instrumentation and power distribution system combined with Ultra Electronic's Transmag degaussing system.

The system, which is housed in three standard ISO containers, is designed to measure accurately a ship's magnetic and acoustic signatures using proven, reliable, COTS and non-development equipment. It features a single, digital, dual-fibre optic range-to-shore cable and a Differential Global Positioning System (DGPS)-based, real-time vessel tracking capability.

There is capability for additional sensors, including Underwater Electric Potential (UEP) and Alternating Magnetic (AM).

FACDAR measures a ship's magnetic, acoustic and pressure signatures as it passes over an undersea sensor array. The signature data is analysed and displayed on IBM-compatible computers in the system's portable range house. Recorded in digital format, the data can be archived and retrieved for future analysis.

The signature analysis allows a ship to:

(a) set the onboard degaussing system to minimise the magnetic signature
(b) check for satisfactory operation of the degaussing coil system
(c) set up the vessel for operation at any global location
(d) set up the onboard system to accommodate failed or faulty individual coils
(e) measure the noise-radiated acoustic signature of the vessel.
(f) compare the vessel's acoustic database to monitor degradation
(g) complement any onboard vibration monitoring and analysis system
(h) identify any item of equipment that causes the vessel to be out of its acoustic radiated noise specification.

Operational status

A US$7.4 million contract has been awarded to Hughes Aircraft Techsystems by the US Naval Surface Warfare Center, Caderock, for the manufacture of FACDAR. The contract provides for one prototype FACDAR and five production units, including spares and tracking equipment. The contract also contains provisions for up to eight additional units for possible sale to allied navies.

The first two systems have been delivered to the US Navy.

Specifications
Magnetic sensor
Dynamic range: 0 to ±100,000 nT
Accuracy: 0.5 nT or 0.5% (whichever is greater)
Bandwidth: DC to 15 Hz
Resolution: ≤3.05 nT at ±100,000 nT; ≤0.76 nT at ±25,000 nT; ≤0.095 nT at ±3,125 nT
Acoustic sensor
Max signal level: 160 dB re 1μPa
Accuracy: 0.5 dB
Bandwidth: 2 Hz to 30 kHz
Resolution: 0.1 dB max from full scale to 51 dB below full scale
Pressure sensor
Dynamic range: 10 bar
Resolution: <300 Pa 0.1 dB maximum

Contractors
Raytheon Systems Company, Mukilteo, Washington.
Ultra Electronics, Hednesford, UK.

UPDATED

DEGAUSSING SYSTEMS

FINLAND

FDS

Type
Onboard degaussing system.

Description
The FDS is an advanced modular degaussing system that can be used aboard a wide range of ships. The solid state power units are PLC controlled switching mode current amplifiers with high accuracy and efficiency. The system has several different control and back-up modes.

The FDS 700 Series onboard degaussing equipment is based on industrial and marine grade commercial off-the-shelf (COTS) products. The power amplifiers are controlled by the programmable logic controllers (PLC) that are connected to each other and to the operator's console by a Profibus field bus. The PLCs and the power amplifiers are located in the power cabinets (DCU) in the equipment room. The operator's console is typically on the bridge or in the machinery supervision room.

The console is a ruggedised workstation which operates under standard MS Windows NT. The MMI interface allows system configuration and start/stop, the entry of calibration parameters, manual control over the system, and gives system alarms with clear descriptions. There is also a special approximation software which calculates an estimate of the local magnetic field based on the position of the ship. In the normal mode the operator's console supervises the power controllers and gives indicative figures of the system status without any input from the operator.

The coil current control is based on a modern multiparameter DG (MPDG) control algorithm which uses the information from several different sources such as magnetometers, gyrocompass, pitch/roll sensors and the ship's navigation system. The current control mode is either automatically or manually selected by the operator. The manual heading control with the local earth field approximation is always available for the operator. Several calibration parameter sets can be stored into the system's memory and be recalled when the ship's load state is changed.

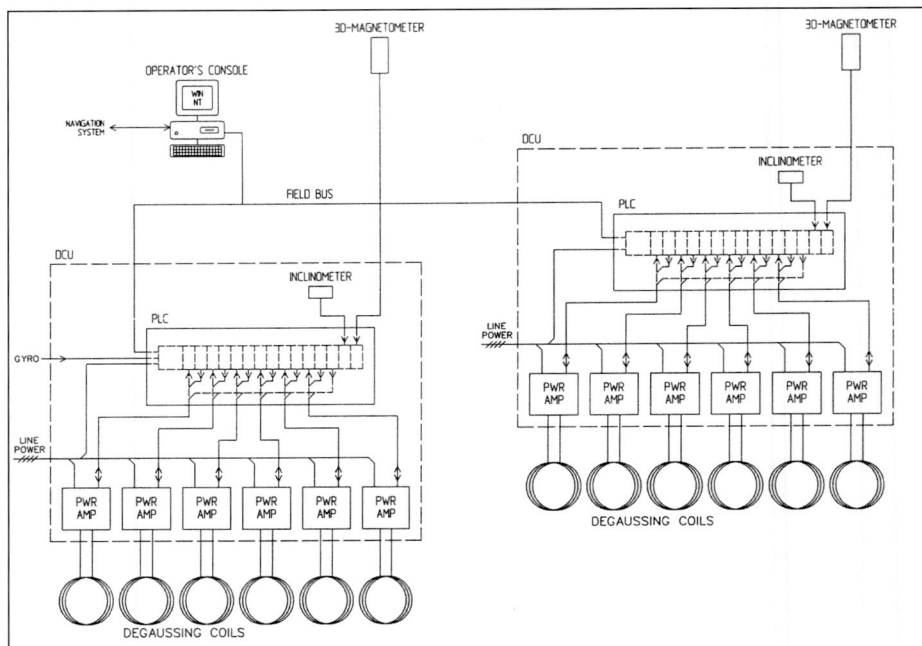

Block schematic of FDS 700

1999/0017881

The system features built-in diagnostics with clear alarm functions and fault-tolerant distributed architecture. All calibration parameters are stored in non-volatile memories both in the PLCs and the operator's console. In the event of major damage the system can change the coil current parameters into reduced magnetic silencing with only part of the power cabinets working. The field bus is not vital to the system; the PLCs will maintain full functionality if the connection to other DCUs or the operator's console is lost. Multiple back-up modes assure high system availability. The equipment may also contain uninterruptible power supply for signature reduction in a blackout situation.

The FDS 700 Series onboard degaussing systems are based extensively on COTS hardware, which ensures cost-efficient maintenance over the system lifetime.

Operational status
Since 1980 over 20 FDS systems have been delivered. Five of the above described FDS-700 systems were delivered in 1998.

Contractor
Elesco Oy, Espoo.

UPDATED

FRANCE

Magnetic treatment systems

Description
Thomson Marconi Sonar SAS manufactures magnetic treatment facilities which are used to reduce and/or normalise permanent magnetic signatures using a deperming system. These systems are of particular use for vessels lacking onboard degaussing equipment. Stations can be of the static type, or dynamic, in which the vessel to be depermed makes a run over the range.

Operations status
Systems have been sold to the Danish Navy.

Contractor
Thomson Marconi Sonar SAS, Sophia Antipolis.

UPDATED

GERMANY

Degaussing system

Description
In its smallest configuration the degaussing system consists of a triple probe, a miniaturised control cabinet and nine coil systems to control the permanent, induced and eddy current components of the ship's magnetic field in the ship's three axes: vertical, longitudinal and athwartship. The control cabinet includes a three-channel flux gate type magnetometer and nine-channel amplifiers that generate the output power required for non-magnetic vessels.

The structure of the standardised and miniaturised degaussing system is hierarchical. The system comprises self-contained non-interlinked components which are provided with their respective input values by the hierarchically superior components. This ensures a high degree of reliability and system availability, or a high degree of probability that the system will function effectively and can be easily maintained even under adverse conditions.

The complete system is composed of independently functioning and testable modules, such as:
(a) degaussing probe with magnetometer
(b) channel amplifiers
(c) operating panel
(d) manual adjustment device
(e) power supply module.

Every module forms a potential island by the separation and screening of power units and by the employment of a new method for floating measurement of DC currents, which has proved itself in this system. Thus there is no mutual influence of the system's individual channels, so that in the event of failures in the coil network the reduction in protection is kept to a minimum.

Provision is made for standby operation of the system with the aid of manual control equipment, which maintains a fixed value for the permanent channels and regulates the induced channels as a function by means of a manual adjustment device.

The system output is sufficient for non-magnetic submarines, small fast patrol boats and mine warfare vessels. Certain components, such as magnetometer and channel amplifier, can also be used for subordinate degaussing systems, such as for separate protection of guns.

However, for larger ships such as frigates and for ferromagnetic submarines, the system output must be increased. The following system components are supplied for this purpose:
(a) power supply units (inverters, rectifiers)
(b) additional amplifiers
(c) generator sets
(d) feedback compensation devices
(e) electric compass compensation switchboxes
(f) manual course adjustment devices.

All these components have interfaces compatible with the degaussing control cabinet, so that economical interconnection of the system is ensured.

These components are connected into the control loop according to their function, so that high accuracy of the system and a correspondingly high degree of ship protection is provided.

Contractor
STN ATLAS Marine Electronics GmbH, Bremen.

VERIFIED

DEG-COMP

Type
Decentralised degaussing system.

Development
This is a future orientated system developed for use on MCMVs, FACs and submarines. It is designed to meet increased demands for optimum magnetic protection against sea mines.

Description
The system uses common coils for compensating induced magnetism, permanent magnetism and eddy current components. Individual amplifiers are used to feed the decentralised coils. Geomagnetic field data in the ship operating area are synthetically generated. The system carries out digital control and monitoring of data transfer which uses a star configured data transfer structure between the central unit and amplifiers.

The configuration optimises energy requirements by summarising components of compensation and results in reduced cable and installation requirements for the degaussing coils.

The system comprises a central unit and a number of decentralised amplifiers which feed the degaussing coil system with controlled direct currents. The computer-supported central unit is equipped with an internal data memory for archiving the geomagnetic field data occurring in the area of operations.

The central unit is provided with suitable interfaces to the ship's navigation system to receive all relevant data regarding the current operational conditions (for example ship's position, heading, roll and pitch angle and so on) via the ship's internal databus or via direct interfacing.

The geomagnetic field components affecting the ship are constantly assessed with the aid of the stored geomagnetic field data and taking into consideration the existing operational conditions. These components are converted into set values for the induced current portion of the degaussing amplifiers.

By summarising the set values for the induced, permanent and eddy current portions of magnetic field components, the central unit generates individual set values for each amplifier to effect proportional compensation currents for each coil.

The decentralised amplifiers are directly assigned to the individual coils.

Use of an earthfield triple probe is possible as redundancy to the geomagnetic data memory further increases the system reliability. The man/machine interface relies on a clearly structured user interface incorporating a monitor with associated function keys and hierarchically organised menus for interactive system inputs and outputs within the various operating modes.

Contractor
STN ATLAS Marine Electronics GmbH, Bremen.

VERIFIED

UNITED KINGDOM

OBDG

Type
Shipborne degaussing system.

Description
The OBDG (On Board DeGaussing) is based on a distributed system architecture that uses a dedicated coil-driven power supply for each degaussing coil. This allows compact power supplies to be located in the most convenient position near the degaussing coils. Communication between the degaussing controller and each of the coil drivers is via serial digital databus.

The coils, installed in the ship's hull during construction, are energised by a DC current from the coil drivers to produce a magnetic field to oppose the existing ship influence caused by permanent or positional effects.

The onboard degaussing system uses information from an internal geomagnetic database to control ships' magnetic signatures. The fully automatic system correlates position and heading with geomagnetic data and computes the optimum degaussing coil currents for a minimum magnetic signature. A masthead magnetometer (earth field measurement) is available in addition to the geomagnetic map. Geomagnetic data may be used as one of the fallback modes of operation if a masthead magnetometer is installed.

The control unit is configured for a specific system by easily reconfigurable software. The unit uses microprocessor hardware. It accepts heading information in synchro data or serial data formats, plus inputs of latitude and longitude as well as the (optional) masthead magnetometer. The controller uses this information to compute the coil currents for optimum degaussing at the given location. The information is updated in real time, and the degaussing coil drivers continuously change coil current levels to provide minimum magnetic signature.

The system is suitable for steel ship, MCMV or submarine installation.

System and coil design services are available.

Operational status
In service. Other types of degaussing systems have been supplied to the navies of Australia, Canada, South Korea and the United Kingdom.

Contractor
PMES Ltd, Rugeley.

UPDATED

ACOUSTIC CONTROL

FRANCE

QSUA-4A

Type
Acoustic noise monitoring system.

Description
This equipment provides automatic detection and indication of propeller cavitation, and automatic monitoring and localisation of vibration and superstructure noise along the ship with preset alarm levels. It is an independent system fitted with its own display, but able to be integrated into a sonar system.

The system consists of a processor and display console, one or two remote cavitation indicators and up to 20 hydrophones and accelerometer sensors. The system performs continuously and without manual intervention: the detection and indication of propeller cavitation; monitoring and localisation of superstructure noise and vibration with synthesised

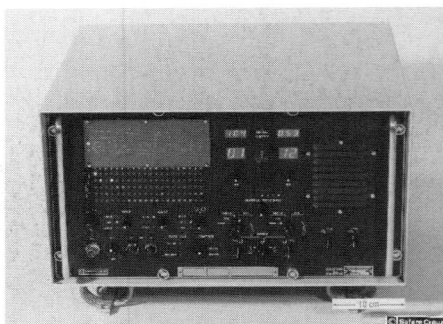

Control unit of the QSUA-4A noise monitoring system

display of broadband noise levels throughout the length of the vessel; the triggering of an alarm on detection of propeller cavitation or when a noise

level, preset for each sensor, is exceeded; and the optional automatic recording, on any external recorder, of the signals present on the two noisiest channels when one of the preset levels is exceeded. Manual operations available include the precise measurement, analysis and monitoring of noise or vibrations received by two sensors. This is used for determining any correlation between two noises. Other options available include the selection of two channels for external processing (spectral analysis), and the simultaneous external recording of signals present in two or more channels.

Operational status
In service with the French Navy.

Contractor
Safare Crouzet, Nice.

VERIFIED

ISRAEL

Adaptive noise cancellation

Description
The adaptive noise cancellation system is designed to improve the performance of submarine flank array sonars by providing real-time reduction in self-noise received by the sonars. This in turn results in a

substantial increase in target detection ranges together with improved target classification and identification.

The system uses advanced algorithms and state-of-the-art technology.

The equipment is available for new build sonars or to upgrade existing submarine sonar systems.

Contractor
Rafael-Armament Development Authority, Ordnance Systems Division, Haifa.

VERIFIED

UNITED KINGDOM

VIMOSPlus monitoring system

Type
Vibration and noise monitoring system.

Description
VIMOSPlus is a computer-controlled vibration and noise monitoring system designed to enable the continuous measurement of radiated noise levels at selected positions throughout the ship. In warships, for which the system was originally designed, this allows the increase in these levels to be detected before they adversely affect the ship's sonar capability or increase the risk of detection. Applications include use by the

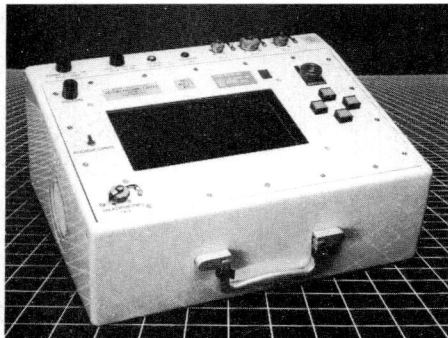

Operator's desk unit of VIMOS

ship's sonar operators and monitoring of machinery wear and of detection.

By measuring hull and equipment vibrations via sets of distributed sensors, VIMOSPlus performs a unique combination of spectrum analysis and data processing, providing an ability to quantify self-generated radiated noise, continuously monitor structure-borne noise and monitor machinery health.

A combination of accelerometers and hydrophones, placed at requisite internal and external points, is employed, from which signals are transmitted through amplifiers and switch equipment to a central processing unit based on a central processing unit 486 PC, modified to interface to a Thomson Marconi Sonar DSP board featuring the Lucent Technologies DSP32C hosted by the INMOS T802 transputer. These give data acquisition and signal switch control, Fast Fourier Transform (FFT) signal processing providing both narrow and broadband analysis and back-up facilities.

Results are displayed using Windows technology and pull-down menus.

A typical surface vessel system would deploy around 216 individual sensors grouped into nine arrays. In addition, a portable hand-held sensor is available for specific tracking of noise sources.

Operational status
VIMOS systems have been ordered by the UK MoD for the Royal Navy Type 23 frigates, 'Sandown' class minehunters and 'Fort Victoria' class Fleet Auxiliary vessels. Three earlier versions of this system, Hull

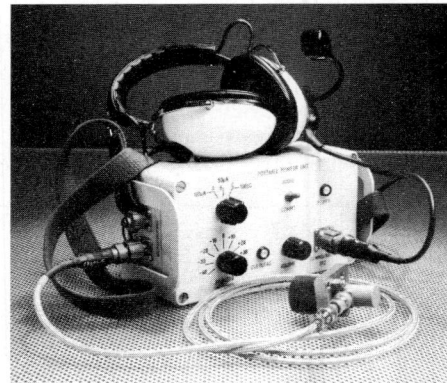

VIMOS portable monitoring equipment

Vibration Monitoring Equipment (HVME) Mk 1, were delivered to the Royal Navy in 1985.

The system is now fully operational in the Canadian Forces Maritime Command, Royal Navy and Royal Saudi Navy. During 1995 the company won a contract to supply VIMOSPlus to HDW in Germany for equipping submarines.

Contractor
Thomson Marconi Sonar Ltd, Templecombe.

UPDATED

TONES

Type
Transportable acoustic noise range.

Description
TONES (Transportable Overside Noise Evaluation System) provides the capability for a vessel to measure its own radiated noise within a matter of 2 to 3 hours. This may be particularly useful if the vessel is operating away from fixed ranging facilities.

TONES comprises two self-contained units, a wet end unit and a shipborne unit.

The wet end unit is deployed in the sea when the system is in use. It comprises a free-floating buoy (either a recoverable, reusable buoy or an SSQ 906 type sonobuoy) with a suspended hydrophone. The radiated noise received by the hydrophone is transmitted by radio to the shipborne unit using one of 40 channels in the 136 to 174 MHz frequency band. After use the wet end unit is brought back on board.

The shipborne unit is typically carried as a temporary installation on the vessel to be ranged, although it can alternatively be a permanent installation or can be operated from an escort. The unit incorporates a radio receiver, signal analyser and a VGA flat screen monitor. The monitor displays the analysed noise in the form of a narrowband spectrum, or a one-third octave analysis.

The system is straightforward in use with the minimum of operator input being required. The operator is required to initiate an event mark when a specific point on the ship passes the wet end unit buoy a few tens of metres abeam.

The data is analysed in real time, and the resulting plots displayed and stored. The stored plots can be printed out when required.

TONES will measure radiated noise in the acoustic band 20 Hz to 20 kHz, over a dynamic range of 60dB/1 µPa with a 1 Hz resolution. It will perform narrowband (FFT) and one-third octave analysis in real time on the incoming data.

Self-contained shipboard unit of TONES   1997

The hydrophone may be set at any depth between 10 m and 60 m, and the analysis software will automatically reference the data to the separation set by the operator. The standard deployment is at 10 m.

Operational status
In service with a Middle East customer. In a collaborative venture Ultra and the DRA are jointly to develop, market and sell TONES. TONES has been supplied to a number of navies and has been proven during acceptance testing on the DRA/DTEO Loch Goil facility.

Contractor
Ultra Electronics Ltd, Ocean Systems, Weymouth.

UPDATED

ATARS

Type
Transportable acoustic noise range.

Description
The ATARS System is designed as a transportable acoustic ranging facility, used to measure and analyse underwater radiated noise from a wide variety of platforms including warships, auxiliaries, ROVs and weapons. Its primary purpose is to obtain the underwater acoustic characteristics of the vessel to determine the probability of detection by mines and passive sonars.

The system comprises two underwater acoustic sensors, one mounted along the track of the vessel and one which is separated from the track by about 25 m to allow measurement of the beam signature. Sensors are easily deployed from a small craft.

The beam sensor is connected to the acoustic track sensor by a multiway copper cable. The track sensor has a fibre optic cable to the shore.

A signal processing system provides frequency domain analysis of the vessel noise characteristics and allows the processed data to be stored on floppy disk for subsequent retrieval and comparison.

A high-resolution colour display is provided for display of the processed data, and control of the system is via a computer workstation. The operator

Units of the ATARS system   1997

interacts with the system via a standard keyboard and trackball. All system commands are selected and executed using the trackerball via the displayed menus.

Ship tracking is achieved by means of a laser tracking system and a reflecting prism. As each ranging run is executed, all data is both displayed in waterfall format, and retained in high-capacity hard disk for temporary storage. Before the vessel enters the data collection area, the tracking system is 'locked on' to the vessel by the operator.

On successful completion of ranging, the processed data is archived to floppy disk. Processed data is displayed in several formats designed to assist the operator in identifying the nature and cause of possible extraneous acoustic noise sources. Display formats include both narrowband spectra, derived using FFT techniques, and true one-third octave broadband responses. The acoustic frequency range covered is 10 Hz to 56 kHz.

The whole system can be supplied in a 6 m ISO container which may be fitted out as a ranging control office and also used to store the cable and sensors when not in use.

Operational status
ATARS systems have been delivered to the Middle East and are in service in Asia.

Contractor
Ultra Electronics Ltd, Ocean Systems, Weymouth.

VERIFIED

Acoustic tiles

Type
Noise suppression tiles.

Description
Decoupling tiles are applied to surface ship and submarine hulls to reduce their radiated noise levels and provide a similar level of performance to the Prairie Masker system. In a submarine the most vulnerable areas to detection by sonar are the sail and other flat surfaces such as the hydroplanes and rudder as well as the region around the machinery compartment.

Decoupling tiles are made out of polymers (in some cases urethane-based polymer technology) and rubber compounds that absorb acoustic energy and create an impedance mismatch. This mismatch brakes the sound path into the ocean in the same way that the air layer along the hull does in the Prairie Masker system. In addition to providing this mismatch, decoupling tiles also scatter active sonar transmissions resulting in a poor return to the sonar receiver in the hostile vehicle.

Anechoic tiles operate in the same way and use the same techniques as radar absorbent material, absorbing the sonar transmission at discrete frequencies. In the same way as RAM, anechoic tiles are available in both dual band and triple band versions. Anechoic tiles measure approximately 305 mm^2 in area and are around 30 to 75 mm thick. They are attached to the outer hull by use of an adhesive. Alternatively some anechoic materials can be cast directly onto the hull or sprayed on.

Both types of tile are applied to those areas of the hull which are most vulnerable from a target strength point of view.

In service, several different types of tile are likely to be applied to different areas of the hull, depending on the type of protection deemed necessary. Generally materials are tuned to counter the anticipated threat frequencies and optimised for the depth and temperature of the operating environment. As temperature rises, anechoic material softens and its optimum performance shifts to the lower frequencies. Likewise the effect of depth changes the performance of the tiles Anechoic tiles are of cellular construction to maximise energy loss. As depth increases so the tiles are compressed and, to overcome this effect, tiles must be of stiffer material in order to maintain their operating frequency, or of softer material to reduce the frequency at which they operate.

Operational status
In service on many classes of submarine and surface ship worldwide.

NEW ENTRY

FSM signature management

Type
Acoustic, anechoic and radar signature management.

Description
Design, development and manufacture of signature management control of acoustic and radar signatures.

Acoustic management
This includes the fitting of properly designed acoustic material to submarines and surface ships to achieve a significant reduction in radiated and self-noise. This will decrease the range at which the platforms can be detected and classified, whilst increasing the performance of their own sonar systems.

Anechoic management
In order to reduce the probability of detection by active sonar operated from surface ships, submarines, sonobuoys or dipping sonars, it is possible to scatter or absorb the sonar energy by the application of special coatings. These coatings reduce target echo strength returns and considerably reduce the range at which detection is achieved.

Radar signature management
As the sophistication of acquisition sensors increases, the reduction in Radar Cross-Section (RCS) from periscopes, conning towers and induction masts becomes more important. RCS is the complex summation of the discrete radar reflections arising from various features of the structure, and Radar Absorbing Material (RAM) can be designed and fitted to minimise these radar reflections.

Contractor
Forsheda Signature Management, Ross-on-Wye.

UPDATED

UNITED STATES OF AMERICA

Multiplexed Extended Sensor Array (MESA)

Type
Fibre optic undersea acoustic tracking range.

Description
MESA is the first US Navy fibre optics undersea acoustic tracking range and was installed in AUTEC in August 1996. MESA makes use of Common Undersea Platform (CUP) components to provide a fully digital, time division multiplexed, wide bandwidth sensor system. The telemetry architecture is based on the SONET standard that supports data rates that allow the hundreds of sensor channels to be placed on a single optical fibre. MESA is a 20 year life system with fully undersea qualified redundant lasers. Also for redundancy, the design includes optical and power bypass features which allow for up to two consecutive multiplex/repeaters nodes to be bypassed without failure of the system. The system can be deployed to depths of 5,500 meters. MESA also includes a power and demultiplexer shore- or ship-based subsystem that provides power to the undersea cable and sensor subsystem and provides both digital and analogue data output for all sensors. A display panel continuously monitors the state of the undersea system with regard to laser operation, bypass status, and system current and voltage. The panel also allows for the operator to communicate to each sensor node through a supervisor channel allowing for greater flexibility in node status and control.

Contractor
Lucent Technologies, Advanced Technology Systems, Greensboro, North Carolina.

VERIFIED

Prairie Masker

Type
Radiated noise suppression system.

Description
The Prairie Masker radiated noise suppression system is designed to create an impedance mismatch around the parts of the hull which are inherently noisy. Through this mismatch the sound path from the hull or propeller to the surrounding water is blocked or distorted resulting in a very much reduced level of radiated noise or a noise which is very difficult to detect and identify because of its distorted nature.

In sensitive areas such as the region surrounding the machinery compartment, the Prairie Masker system blows air through a series of small nozzles mounted in the hull. The air supply is bled off from the ship's high-pressure gas turbine (if fitted) or supplied by dedicated compressors. The bubbles so created remain trapped along these sensitive regions of the hull and mask the noise that would otherwise be directly radiated into the ocean from the vibrating hull plating.

The other area where noise is a major problem is the propeller. In this case air is fed down the propeller shaft and through grooves in the propeller's leading edge to delay the onset of cavitation.

The disadvantage of this noise suppression system is that at lower speeds the layer of bubbles tends to rise to the surface rather than remain entrained along the hull. A similar effect occurs when the ship experiences a heavy seaway and the bubbles are sheared off by the violent pitching of the hull. In addition the bubbles contribute more noise to the ship's wake, which can increase its vulnerability if it is likely to be attacked by wake homing torpedoes.

Operational status
Mounted on many surface warships in many navies.

NEW ENTRY

EMTC

Type
Acoustic signature reduction technology.

Description
The electromagnetic turbulence control (EMTC) technology being developed by the US Defense Advanced Research Projects Agency (DARPA) and Naval Undersea Warfare Center (NUWC) is designed to reduce the acoustic signatures of submarine and possibly surface ships.

The technology uses an electromagnetic field that controls the boundary layer around the hull of the vessel that reduces drag-induced noise,

Tests have shown that it is possible to counteract turbulence using an array of magnets and electrodes embedded in the hull. It has been suggested that the viscous drag of the hull could be reduced by as much as 45 per cent using this technology.

It is also possible that EMTC could be applied to propellers to reduce cavitation.

Operational status
Under development.

NEW ENTRY

TRAINING AND SIMULATION SYSTEMS

In recent years a great deal of time and effort has gone into the development and production of simulation and training systems for the underwater scene. This has resulted in a much better appreciation of the operational requirements and has led to a greater efficiency on the part of the operators and designers.

Simulators and trainers in the underwater warfare scenario cover four main areas:

(a) submarine handling and manoeuvring
(b) trainers for machinery control
(c) systems trainers for combat team training
(d) equipment trainers for operator training.

This section deals primarily with system and equipment trainers, in particular those devoted to submarine use. It should be remembered, however, that many sonar systems incorporate a training facility within the main equipment itself. Details of these will be found in the appropriate section of this book.

Full details on trainers and simulators can be found in *Jane's Simulation and Training Systems*.

UPDATED

COMMAND TEAM TRAINERS

GERMANY

AGUS ASW trainer

Type
Command team trainer for Anti-Submarine Warfare (ASW) operations.

Description
AGUS is designed to train and assess command information teams in ASW. It also trains and assesses operators of different grades in the use of sensors and weapon systems and provides training in tactics. It uses distributed computer architecture techniques employing STN ATLAS MPR/EPR processors which allow modular construction to meet customers' requirements.

The system consists of a number of command information cubicles, an instructor's station, a debriefing auditorium with a large screen display and an electronics room. It can be equipped with either real operating equipment or with general purpose consoles which have intelligent graphic displays.

The simulator provides dynamic operating behaviour of own ship's course, speed and mission duration; realistic visual and acoustic presentation for active, passive and intercept sonar; presentation and operation on radar and ASW consoles; and weapon data together with combat situation displays. Also provided is realistic contact association, target motion analysis and data display, and weapon firing tactics, together with simulation of weapons and decoys.

For the German Navy four operations rooms are provided in the trainer: 'Bremen' class, 'Hamburg' class, 'Thetis' class and Mk 88 Sea Lynx helicopter. Equipments simulated include: sonars – Sonar 80, 1 BV2, AN/AQS-18; fire-control systems – SATIR, M 5/4, 9 AV with M 9/3; navigation systems – TANS, radar; weapons – DM 3 torpedo, Mk 46 torpedo, DM 4 torpedo, DM 21 missile, DM 11 depth charge.

Operational status
Operational with the German Navy.

Contractor
STN ATLAS Elektronik GmbH, Bremen.

VERIFIED

AWU 206/206A SCTT attack trainer

Type
Submarine Command Team Trainer (SCTT).

Description
The SCTT is a shore-based computerised simulation system for training all grades of submarine teams in order to achieve maximum combat power at sea. The trainer simulates tactical situations which can be presented to the trainees who are working in the submarine's Combat Information Centre (CIC) installed at the naval base. The weapon officers and operators are trained in operations and response to specific tactical problems under the supervision of exercise controllers. The trainer also provides for development and analysis of naval procedures for torpedo firing and guidance methods, and optimum operation of onboard equipment.

Through co-operation between the simulator and the CIC, optimum training can be achieved for:
(a) attack of surface vessels and submarines
(b) defence against attacking surface vessels, submarines, ASW aircraft and helicopters
(c) operations in minefields.

The system consists of an auditorium with large screen displays, instructor consoles and communications equipment, the CIC with real equipment and a computer room. The latter generates the sonar, tactical data handling, weapon handling, periscope, navigation, ship's controls, radar and ESM inputs for the system.

Operational status
In service with the German Navy.

Contractor
STN ATLAS Elektronik GmbH, Bremen.

VERIFIED

AWU attack trainer instructor consoles

SCOTT

Type
Submarine COMbat Team Trainer (SCOTT).

Description
SCOTT belongs to the family of shore-based simulation systems for training all levels of submarine personnel.

The system provides simulation of a submarine's combat information centre with all working functions using multipurpose consoles which are independent of any specific type of console. SCOTT is thus virtually unlimited in its simulation capabilities.

SCOTT offers the following functions: panoramic passive sonar, active sonar, intercept sonar, flank array sonar, passive ranging sonar, underwater telephony, attack periscope with visual system (CGI), fire-control systems, tactical data handling and radar with ESM and steering unit.

SCOTT can also be interfaced with a submarine diving trainer. Officers and operators are trained to operate and combine all available sensors using the right tactics and deployment of weapons, optimising, improving and developing tactical doctrines without having to use a real submarine and possibly putting it and its crew in a dangerous situation. SCOTT is supported by a comprehensive playback system and detailed briefing/debriefing can be carried out.

Using SCOTT enables command teams to be trained to carry out the full range of submarine operations against enemy forces including: surveillance and strikes against surface ships; convoys and/or submarines involving detection; classification; threat analysis; tactical navigation; weapon control and weapon deployment.

Detection and countermeasures against hostile attacks by surface vessels, submarines and ASW aircraft and helicopters can be practised and in addition submariners can be trained in minelaying operations, operations in mine areas, intercommunications during all combat situations and tactical manoeuvres.

The system uses the latest MMI and standard software tools. Hardware and software is of modular design, allowing a wide range of different types of boat to be simulated. The flexible design approach allows subsequent additions of new equipments to be easily integrated and potential for growth is available for future extensions.

Operational status
In service.

Contractor
STN ATLAS Elektronik, Bremen.

UPDATED

SCTT

Type
Submarine Command Team Trainer (SCTT).

Description
The SCTT is a shore-based, computerised simulator system for training all grades of the submarine team in order to achieve maximum combat efficiency at sea.

SCTT simulates real tactical situations which can be presented to trainees working in the submarine's combat information centre. The weapons officers and operators are trained in operation and response to specific tactical problems under the supervision of exercise controllers. SCTT also provides capabilities for the development and analysis of submarine tactics, torpedo firing and guidance methods and optimum operation of onboard equipment.

The trainer comprises an auditorium with a large screen display, instructor's console and communication equipment, combat information centre with real equipment, periscope with visual system (CGI) and computer room.

Operational status
In service with the navies of Germany (three systems) and South Korea.

Contractor
STN ATLAS Elektronik, Bremen.

UPDATED

CIC simulator

Type
Combat Information Centre (CIC) simulator for surface ships.

Description
The CIC simulator is a shore-based training system designed for individual operator training, subteam training, team training, maintainer training and research and development applications.

It simulates the original equipment of a surface ship's CIC with subsystems which include active and passive sonar, ECM/ECCM, radar and weapon systems (gun, missile, torpedo, fire-control, visual track capability). An entire system comprises an auditorium with large screen display, CIC with real equipment, computer room and offices.

Operational status
In service with the Republic of Singapore Navy.

Contractor
STN ATLAS Elektronik GmbH, Bremen.

VERIFIED

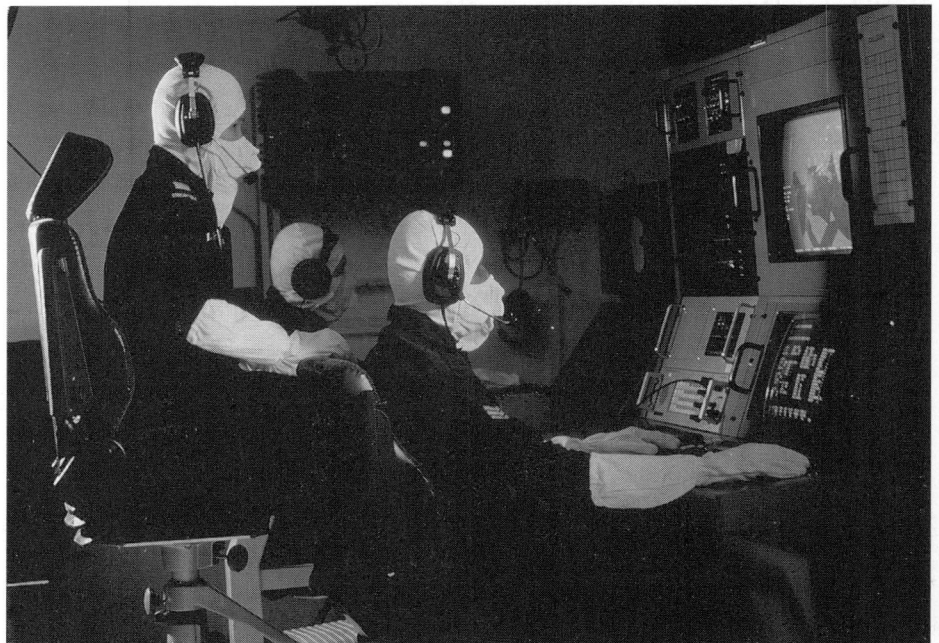

Combat Information Centre (CIC) simulator   *1995*

NATAT

Type
Naval tactical and command team trainer.

Description
A shore-based training system for command teams, post teams, midshipmen, system R & D teams and planning departments. It provides a full range of simulation facilities from communication, exchange of information, distribution of missions, co-operation between forces, to use of air support. The system can simulate any type of craft such as submarines, frigates and merchant ships over a typical exercise area of 2,048 × 2,048 n miles inclusive of shore lines, navigation marks, coastlines and bottom depths, wind, seastate, currents and bathymetrics.

Environmental type simulation includes wind, sea state, cloud, current, rain and bathymetric conditions.

Of modular design and expandable according to user requirements, a complete system comprises an instructor's console, trainee cubicles (typically 10 cubicles), computer facilities, auditorium, maintenance room, lobby and offices.

Contractor
STN ATLAS Elektronik GmbH, Bremen.

UPDATED

ITALY

Multi-Application Sonar Trainer (MAST)

Type
Training system for sonar operators.

Development
MAST is a multi-application sonar trainer developed in co-operation with Raytheon, USA and has been delivered to the Italian Navy training centre.

Description
MAST consists of the following units:
(a) trainee equipment consisting of a set of units identical to those used operationally on board
(b) an instructor control station from which the instructor provides for the insertion of data related to own ship, and the target's course, speed, position and so on
(c) a medium-size mini-computer
(d) simulation hardware producing signal and noise to be injected into the front-end transducer array interface. These signals simulate those which would have been generated by the acoustic environment when converted from pressure to voltage by the receiving transducer array.

These include all possible variations due to motion, mutual interference, propagation characteristics and the environmental conditions simulated.

The MAST system is used for the training of operators and maintenance personnel on the DE 1160 and DE 1164 sonar systems (see entry in the USA section of Sonar systems – Surface ship sonar systems). Further improvements will allow training facilities for other passive sonar systems produced and also for the DE 1167 being manufactured under licence from Raytheon.

Operational status
In service with the Italian Navy.

Specifications
Number of targets: up to 8
Target types: own ship, surface ship, submarine, torpedo decoys, biologics, wakes
Own ship simulation: self-noise, spoke noise, VDS simulation

Ocean effects: propagation loss, sea state, reverberation
Passive simulation: broadband and narrowband noise
Active simulation: highlights, target strength

Contractor
Alenia Sistemi Navali, Rome.

VERIFIED

Submarine trainer

Type
Attack team trainer.

Description
The trainer is used to train attack teams of the 'Sauro' class submarines.

The system comprises real onboard equipment (the command and control system MM/BSN-716(V)2 and the fire-control system for A184 wire-guided torpedoes) and the simulation and control facilities.

The simulation system comprises a computing subsystem for tactical scenario simulation based on a DEC mainframe with magnetic storage units and peripherals (terminals, printer, plotter digitiser). The ship simulation subsystem is based on the Motorola 68020 microprocessor and simulates sensors and

equipment interfaces (NTDS, TTL digital I/O, synchro, analogue and so on). A torpedo simulator for A184 torpedoes is also included. Hardware comprises a number of consoles for controlling sensors and navigation equipment and includes a video terminal, multifunction keyboard and a control panel for internal and external communication networks. Exercise control is conducted from two instructor consoles based on a DEC VAX station. Each includes three video terminals (one high-resolution colour), three multifunction keyboards, trackerball and communications control panel. A debriefing system includes a graphics workstation and a large screen projector.

System functions include: scenario preparation with full support provided for storage and retrieval of data including full exercise description, sensors and target characteristics, geographic information and so on;

tactical simulation, own ship and target's sensors simulation; data exchange and real onboard equipment; torpedo simulation (logical and electrical); and data recording and playback for debriefing.

A visual periscope simulator is integrated into the system, incorporating an SGI Onyx Reality Engine 2 workstation and a periscope mockup.

Operational status
The system was installed at the Italian Navy Training Centre at Taranto in June 1991. The periscope simulator entered service in early 1996.

Contractor
Datamat, Ingegneria dei Sistemi, Rome.

VERIFIED

NORWAY

MASWT (MPS Anti-Submarine Warfare Trainer)

Type
Anti-submarine warfare trainer for commanding officers and sonar operators.

Description
The MPS Anti-Submarine Warfare Trainer (MASWT) provides realistic and efficient training of commanding officers and sensor and weapon operators on board ASW destroyers, frigates, corvettes, fast patrol boats

and helicopters for the use of active sonars and other sensors. The MASWT also provides training in compilation of information, analysis and decision making in a realistic operational environment.

The MASWT concept gives an advanced simulation of the platforms, weapons, active sonars and other sensors. Consoles and weapon control panels are designed to look like the real consoles. The scenario definition is done via menu-driven programs. The concept has an easy to use Man/Machine Interface (MMI) and includes full recording and playback/debriefing facilities and a voice communication system.

The sonar equipment consists of the MPS Multibeam Sonar Simulator (MMSS).

Operational status
In service with the Norwegian and Swedish navies.

Contractor
Kongsberg Defence & Aerospace AS, Simulation & Training, Kongsberg.

VERIFIED

A student's console with platforms, weapons and sensors operated via dedicated easy to use panels

MASWT students under instruction

MASTT (MPS Action Speed Tactical Trainer)

Type
Action speed tactical trainer.

Description
The MPS Action Speed Tactical Trainer (MASTT) is a flexible system for building naval tactical trainers and simulators extendable to almost any level of complexity, sophistication and standard. It is designed to fulfil the training requirements of seamen (reporting procedures, sensors and weapon operating procedures, classification training and operator training on real equipment), officers under training

(decision making, team work, reporting procedures and weapon employment), and high-ranking officers (tactical disposition training, tactical analysis and exercise planning, analysis and debriefing). The system is stated to be very easy to use with students able to operate the cubicles by themselves after one or two 1-hour introductory sessions.

MASTT was developed in close co-operation with the Royal Norwegian Navy and installed at the Royal Norwegian Training Establishment in Bergen.

The installation in Bergen consists of 10 cubicle consoles, an extended control section, a standard control section, a system console, two recorder/playback computers, two weapon computers, a communication control, three sonar simulators and two

anti-submarine warfare consoles. Each trainer console is software configurable and the user has full control of model and scenario generation. The distributed concept, using an Ethernet network, makes it easy to expand the trainer by adding more cubicles or enhance it by interfacing with the real equipment simulators (such as radar, sonar and weapon control).

The model preparation and storage for the complete game is performed using a commercial computer. The user interacts with the system console via a menu-driven operator dialogue. All game settings and scenarios can be prepared and tested offline prior to the actual game, so it is possible to build up a library of preplanned games settings on disk.

Scenarios are based on digitised map contours

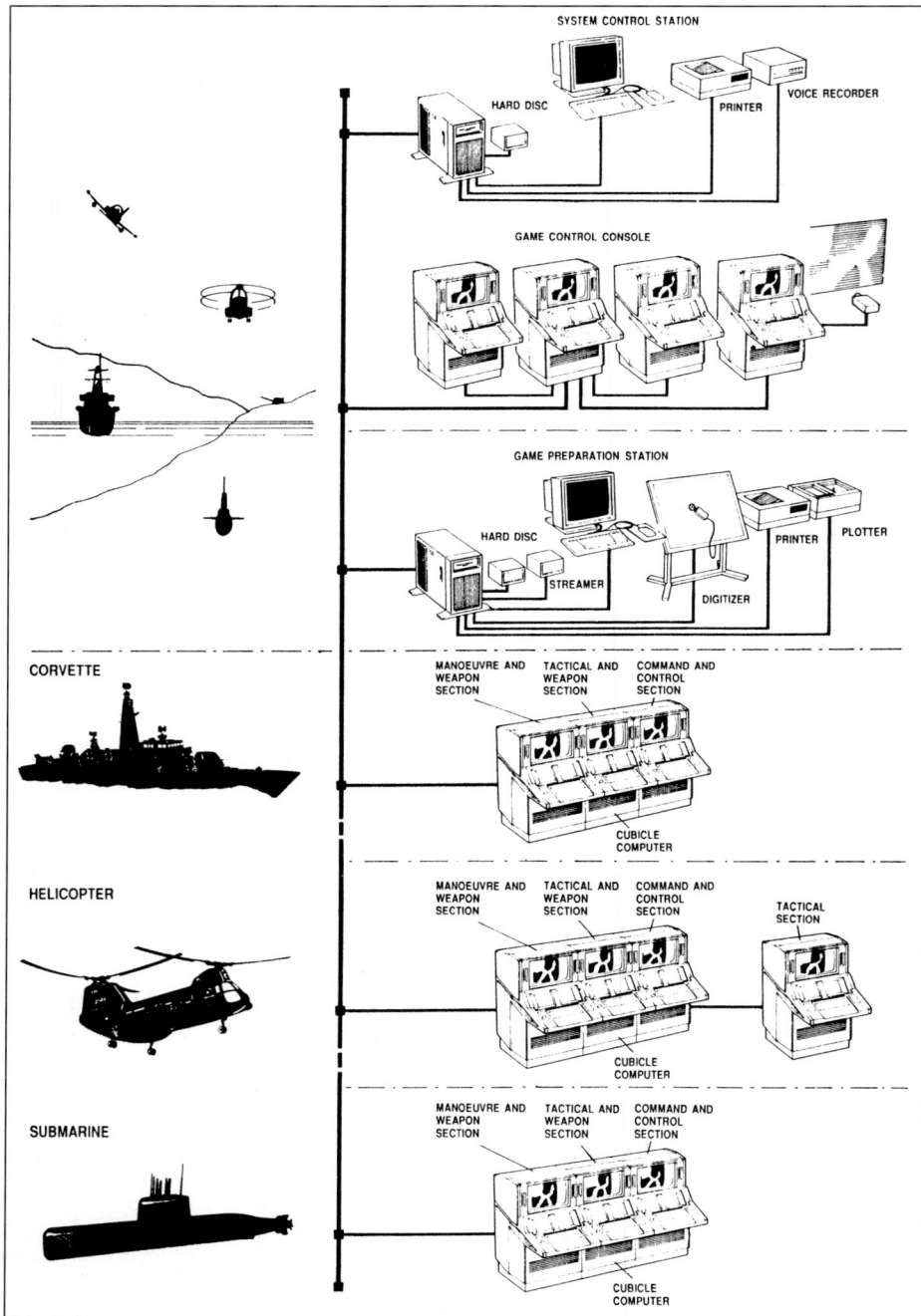

Schematic layout of an MASTT

overlaid with the various optional synthetic information. The actual exercise area is generated from a map library. Two files are made: one for the three-dimensional terrain model and the other for a two-dimensional model for presentation. These files can be stored for later use.

Each cubicle console can be configured as a surface ship, a submarine, an aircraft, a land-based coastal defence installation, or a coastal radar station. A unit configuration consists of a set of parameters describing the unit type and its dynamic behaviour. In addition, the user specifies the sensors and weapons which the unit will use.

Wind, sea current, precipitation and sea temperature layers are among the data provided.

The exercises are initiated from the same console used for set up and preparation. The system software is loaded into the different computers in each cubicle via the Local Area Network (LAN) data highway and the complete trainer is started. The user then selects the game set up to be used, downloads it and the exercise commences.

Following an exercise, a playback programme provides synchronised audio and video playback with freeze and fast wind/rewind functions. During replay the screen in the cubicles and control will contain the same information as during the exercise. Recorded information can be used during a game. The playback programme can also use recordings from real exercises.

Operational status
The MASTT system has been operational at the Royal Norwegian Training Establishment since 1986.

A MASTT system with 12 cubicles was delivered to the Swedish Navy in 1991.

Contractor
Kongsberg Defence & Aerospace AS, Simulation & Training, Kongsberg.

VERIFIED

SPAIN

STTS (Submarine Tactical Training Simulator)

Type
Submarine trainer for commanding officers and combat team.

Description
STTS is a shore-based computerised simulation system for training all grades of submarine combat team in tactics and operational procedures.

The primary training functions that the STTS achieved are:
(a) the training of submarine combat teams in various situations such as surveillance or attack through the gathering and analysis of data on hostile forces, the correct operational use of own fire-control system and weapons and defence against attacking surface vessels, submarines, helicopters and ASW patrol aircraft.

STTS system   *1999*/0043235

(b) the development and assessment of new naval tactics and procedures, primarily for torpedo firing and guidance

(c) training to respond to specific tactical problems.

The system consists of an auditorium with large screen displays for briefing and post-analysis, instructor's console, CIC with simulated equipment that is an exact replica of those on board, attack/surveillance periscope, torpedo fire-control system, plotting table, active sonar, passive sonar, ranging sonar, ESM, time-bearing recorder and so on.

On board environmental noise is also simulated to create a realistic atmosphere.

The system is of modular design that allows for further incorporation of additional equipments.

Operational status
In service with the Spanish Navy.

Contractor
SAES, Cartagena.

UPDATED

UNITED KINGDOM

Submarine Command Team Trainer (SCTT)

Type
Command team trainer.

Description
This submarine command team trainer is designed to provide effective training in command, target detection and surveillance, weapons control, and the use of combat data systems. The heart of the SCTT is a replica of a submarine control room with microprocessor-driven facsimiles of the command, weapon control and sensors consoles, together with a periscope. Visual simulation for the periscope uses computer-generated imagery. In a total training package for the Royal Norwegian Navy's 'Ula' class, Thomson Training & Simulation has supplied SCTT, SCS, MCT (Machinery Control Trainer, a comprehensive Technical and Operational Trainer) plus computer-based training to make up 90 per cent of the total requirements.

Operational status
In service with the Royal Norwegian Navy.

Contractor
Thomson Training & Simulation, Crawley.

UPDATED

Instructor's operating station for 'Ula' class

SCTT

Type
Submarine command team trainer.

Description
This trainer provides the control element of the 'Swiftsure' and 'Trafalgar' class submarine command team trainer modernisation programme.
The trainer accurately represents the control room that

is easily supported and containerised. It will be used to deliver command and weapon system training for the 'Swiftsure' and 'Trafalgar' classes. The control room will link to the periscope simulator (also supplied by GEC-Marconi) and the generic sound room simulator – which models the new Type 2076 sonar suite.

Operational status
Contract for development and production awarded in 1997.

Contractor
GEC-Marconi, Simulation & Training Division, Dunfermline.

NEW ENTRY

Submarine command trainer

Type
Operator trainer.

Description
The operator trainer is designed to train operators in the use of the new submarine command system SMCS. The trainer is based upon extensive reuse of sea-going software from the SMCS system. This has been achieved through work pioneered by BAeSEMA, the designer of the trainer, during further development of

the SMCS system for the nuclear submarine update programme of the Royal Navy. In this project, elements of SMCS software are being ported to run under the Unix operating system on COTS hardware. The SMCS operator trainer will continue this development to enable all SMCS applications to run under Unix in a totally COTS environment.
Compatibility with the developments of the SMCS operational system for the S & T Update Programme of the Royal Navy will allow the operator trainer to be upgraded in very short time scales.

Operational status
On order for the Royal Navy for installation at the Royal Naval Submarine School at HMS *Dolphin*.

Contractor
BAeSEMA, New Malden.

VERIFIED

UNITED STATES OF AMERICA

ASW team tactical trainer (Device 14A12)

Type
Tactical team trainer.

Description
The ASW team tactical trainer (Device 14A12) provides procedural, team tactical training and evaluation for conventional ASW warships and those equipped with tactical data systems. Simulated sonars, command information centres, bridges and platforms (aircraft and ships) are used to provide trainees with the capability of exercising essential anti-submarine warfare engagement procedures in a simulated multithreat environment.
These simulated systems contain general purpose programmable consoles which can be reconfigured under software control to represent the anti-submarine capabilities inherent in any of the US Navy's 16 major

classes of anti-submarine warfare ships. The simulated Underwater Battery Plot feature reprogrammable sonar general purpose consoles. The system's design covers six major categories of simulation: platform simulation (friendly and hostile), sensor detection, weapons employment, tactical environment, probability of interception calculations and damage assessment. A modular, database-driven design allows for ease of future modifications and additions as new platforms, weapons and sensors are added to the fleet.

Operational status
Five 14A12 Devices are in service at the US Navy Fleet Training Centres located at Norfolk Virginia, San Diego California, Mayport Florida, Pearl Harbor Hawaii and Charleston South Carolina.
Device 14A12J for the Japanese Maritime Self-Defence Force was delivered to Yokosuka Naval Base, headquarters of the Escort Fleet, in the spring of 1994.

A student console in a Device 14A12 system 1994

Manufacturer
Hughes Training Inc, Arlington, Texas.

VERIFIED

Submarine trainer

Type
Trident command team trainer.

Description
Hughes Training Inc is a major supplier of electronic equipment to the US Navy and has supplied its own submarine trainers. There are three types of Trident training device and four different simulators for various classes of nuclear attack submarines. They are designed to maintain proficiency and to increase crew effectiveness in a variety of tactical situations through highly advanced simulation.

The Trident command and control trainer (Device 21A42/A) simulates the Trident submarine tactical environment which consists of ocean models, contacts, weapons, sonar, the periscope visual targets and countermeasures. A second device, the defensive weapons subsystem operator trainer (Device 21B67/A), is used to train personnel in countermeasures procedures and tactics. The third device known as the Trident Unique Software Support Facility (TUSSF) supports future hardware and software development for the two operator-team trainers. Currently all trainers have undergone retrofits to reflect submarine modernisation programmes. In 1995 Hughes Training Inc upgraded the instruction stations of the above three devices to use Hughes Training Inc True Guide™ software on a Silicon Graphics workstation.

Submarine combat system trainers provide fire-control team training. All four devices simulate the tactical environment which includes oceans, targets, 'own ship' dynamics, sonar sensors, weapons and countermeasures. Device 21A37/A uses multiple attack centres to train crews, while Device 21A41, has a single attack centre to train one fire-control team. Devices 21A38/A and 21A39/A were produced as more capable upgrades to older attack and ballistic missile combat system simulators. Personnel can be trained on the Mark 113 and the Mark 117 fire-control systems using realistic simulations for the Mark 48 and the Mark 37 torpedoes, the Sub-Harpoon anti-ship missile and the Subroc anti-submarine system. Target data simulations for up to 57 targets with surface, submarine and air contacts are possible.

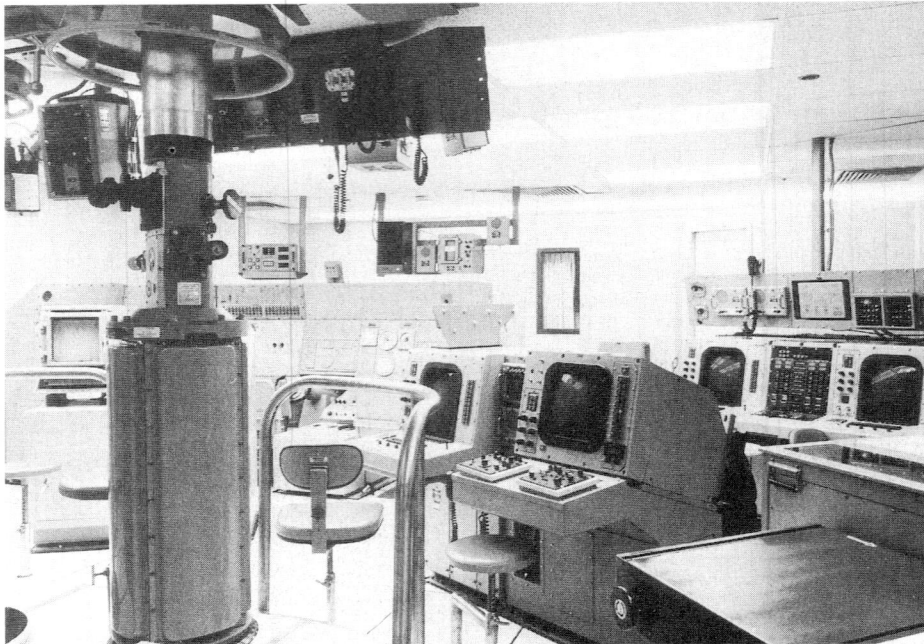

The simulated control room in a Hughes Training Inc Trident submarine command team trainer (Device 21A42)
1994

Operational status
Hughes Training Inc has supplied systems to act as the Trident Command and Control Team Trainer (CCTT) and the Defensive Weapons Subsystem Operator Trainer (DWSOT) to the US Navy's Trident Training Facility at Bangor, Washington and King's Bay, Georgia (one each per site). The TUSSF is located at the Naval Undersea Warfare Center, Newport, Rhode Island. Attack and ballistic missile submarine combat system trainers have been supplied to the Fleet Ballistic Missile Training Center at Charleston, South Carolina, the Pacific Submarine Training Center at Pearl Harbor, Hawaii, the Fleet Anti-Submarine Warfare Training Centers at Pearl Harbor, Hawaii and Norfolk, Virginia and the Naval Submarine base at New London, Connecticut.

In 1995 Hughes Training Inc completed upgrading the Trident trainers, which involved converting the Mark 118 fire-control system to a CCS Mark II system, conversion from AN/BQQ-6 to AN/BQQ-5E sonar, conversion from the WLR-17 to the WLR-9, ESM upgrades on the 214A2/A and DWSOT upgrades (including adding periscopes and ESM upgrades). The simulators involved are the CCTT, the DWSOT and the TUSSF.

Manufacturer
Hughes Training Inc, Arlington, Texas.

VERIFIED

Tactics trainer

Type
ASW tactical trainer.

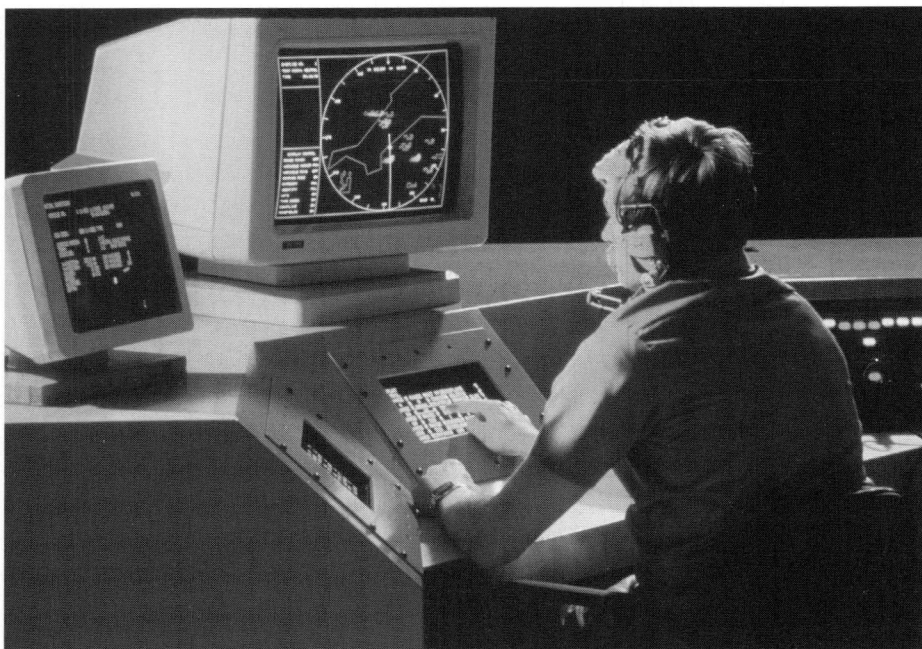

Description
The Hughes Training Inc tactics trainers are team trainers and consist of a central auditorium and debriefing room with large screen displays and up to seven instructor consoles together with one or two communications controller consoles, a computer room and a number of individual instruction areas.

The trainers can produce up to 24 surface, submarine and air vehicles and can engage up to 280 targets of which 80 can 'fight back'. A full range of sensors and weapon systems can be simulated with voice, teletype and datalink communications in a game area of 2,000 × 2,000 n miles (3,700 × 3,700 km), 30,480 m in height and 1,000 m in depth. The speed of the action can be accelerated 20 times.

One trainer provides simulation of 10 combat information centres together with a land-based headquarters.

Operational status
Tactics trainers have been supplied to the Italian Navy at Taranto and to both the Portuguese and Turkish navies, while the Royal Danish Navy has a Tactics trainer at Frederikshavn. In addition, several trainers have been delivered to Asian nations. The US Navy operates a number of the company's surface, subsurface and airborne tactics trainers.

Manufacturer
Hughes Training Inc, Arlington, Texas.

UPDATED

The instructor's station in the Link ASTT
1994

AN/BSY-2 SCSTT

Type
Submarine combat system team trainer.

Description
This shore-based electronics-based training system will provide individual and team proficiency training and skill reinforcement to sonar and fire-control subteams, as well as overall combat team training to all personnel manning the US Navy's AN/BSY-2 submarine combat system.

The design is based on distributed modular architecture and unique, hybrid simulation/stimulation and signal processing emulation techniques. The trainer will closely replicate large numbers of complex contacts, the submarine operating environment, the weapons system, periscope, radar, sonar suite and other AN/BSY-2 sensors to provide controlled, tactically realistic training exercises.

New displays, contacts that automatically react to the environment and student performance monitoring are some of the features which will aid the instructor during training exercises.

Operational status
A competitive contract worth US$80.9 million was awarded to Raytheon early in 1991. The project has now been cancelled due to changes in the scope of the AN/BSY-2 Programme.

Contractor
Raytheon Company, Submarine Signal Division, Portsmouth, Rhode Island.

VERIFIED

ASTT

Type
Action Speed Tactical Trainer.

Description
The ASTT is a strategic and tactical combat command trainer used to enhance the decision making and teamwork skills of junior and senior officers. The system provides simulated war games between two opposing forces. It has a game control suite and a number of student and command control stations representing platforms such as ships, submarines, fixed- and rotary-wing aircraft. Actual and simulated radio voice channels allow for interplatform communications under instructor control. The exercise is played in a 2,048 × 2,048 n miles (3,795 × 3,795 km) gaming area that can include landmasses and minefields.

Sensor data is displayed graphically and in alphanumeric form on student stations whose Man/Machine Interface (MMI) for platform, sensor and weapon control is based on keyboards and a rollerball. In addition to receiving processed electronic and visual data, each platform also has available unprocessed radar and sonar data using the Ship Analytics radar simulator. The simulated weapons systems include air, surface and subsurface launched missiles, depth charges, torpedoes and guns in quantities appropriate to the platform.

Platform movement through the exercise area is controlled by the students and the platform continues to participate in the game until it is destroyed or the exercise is terminated. The instructor station creates and monitors all aspects of the exercises and the instructor monitors the data from individual platform sensors and weapon systems. The ASTT allows the instructor to simulate a variety of malfunctions, including craft performance, weapons, sensors and communications failures. The instructor can also introduce and control simulated platforms that may be assigned to either force or be neutral.

Operational status
A system has been produced for a major Middle Eastern navy.

Contractor
Ship Analytics, North Stonnington, Connecticut.

VERIFIED

Typical ASTT cubicle and instructor equipment   **1995**

Naval ship Combat Information Centre

Type
Co-ordinated bridge/CIC simulator.

Description
The naval ship Combat Information Centre (CIC) has been developed as a companion option to its full mission ship-handling simulator. The CIC extends the scope of training to include co-ordinated bridge/CIC operations in a realistic tactical environment, as required to 'fight the ship'. Modular in design and engineered for installation in a separate space adjacent to the bridge, the CIC may be procured as part of the original FMSS system or added as an upgrade at a later date. The CIC includes six subsets of equipment:

(1) radar/sonar
(2) plotting
(3) ship tactical status display
(4) ship operational status display
(5) interior ship communications, and
(6) exterior ship communications.

The standard CIC is designed to provide the equipment functionality necessary to implement an interactive bridge/CIC capability, suitable for the conduct of effective operational procedure and tactics training for modern warship bridge and CIC personnel.

Operational status
The simulator has been delivered to the US Navy and the Turkish Navy.

Contractor
Ship Analytics, North Stonnington, Connecticut.

VERIFIED

The 'Spruance' class CIC in US Navy Device 20B6D with air plots and simulated AN/SPA-25F navigation radar display
1995

MINE WARFARE TRAINERS

GERMANY

MCOT

Type
Minehunting Command and Operator Trainer (MCOT).

Description
The MCOT comprises a minehunter's combat information centre, a mine disposal vehicle room, computer room and auditorium.

The Combat Information Centre (CIC) is an exact replica with identical equipment found on board the minehunter and consists of: a commanding officer's position; tactical display; command and documentation equipment; minehunting sonar equipment (DSQS-11M) operator's console; operator's console for the remotely controlled underwater vehicle;

a ship control console; plotter and printer and instructor's console.

The maintenance room contains a real underwater vehicle and associated electronics cabinets.

The computer room houses the computer system consisting of the host computer (which carries out central management of targets and exercise areas), a digital sonar simulator and the sonar image generator for the active identification sonar. This is supplemented by a TV simulator for the underwater vehicle and a database generation station.

The auditorium features the usual equipment associated with a lecture theatre.

The MCOT carries out operator training, command team training, procedural and tactical training for minehunting and maintainer training.

The simulator incorporates ship handling from the CIC, detection and classification of underwater objects, the identification, combat and documentation of mines and control of the underwater vehicle.

Operational status
In service with the German navy simulating the 'Frankenthal' class MCMV with DSQS-11M sonar and Pinguin ROV.

Contractor
STN ATLAS Elektronik, Bremen.

UPDATED

ROV simulator

Type
Minehunting ROV simulator.

Description
Designed for training operators of minehunting ROVs, the system consists of an instructor station, a replica of

the main control console, computerised sonar data calculation, sonar data display, TV image generation, ROV dynamic calculation and offline database generation facilities for three-dimensional modelling. Sonar data display and TV generation functions use Silicon Graphics' Indigo and Crimson workstations respectively.

Operational status
In service with an Asian navy.

Contractor
STN ATLAS Elektronik GmbH, Bremen.

VERIFIED

UNITED KINGDOM

FEMS

Type
Fleet Exercise Minelaying System (FEMS).

Description
FEMS is designed to be rapidly installed and removed from a host ship converting it to an exercise minelayer within 24 hours. The system is self-contained and requires only simple deck fixing and a standard ship's electrical supply. A variety of exercise mines can be laid in a realistic and accurate pattern. The ship can also assist in the recovery and interrogation of the mines, providing real-time exercise analysis. On completion of the exercise, mines are stowed onboard the host ship for return to the mine depot. FEMS can be removed from the ship in approximately eight hours.

The system carries the exercise mines on rails (one to port and one to starboard) with a midships stowage rail for either defective or extra mines. The mines are hauled aft to the minetraps by electrohydraulic winches operated from a remote-control unit. The minetrap enables the mines to be released on command (manual or automatic) and a sponson carries the mine clear of the ship. Up to 30 exercise or live mines (depending on type) can be deployed from the system.

Operational status
In service with the Royal Navy since 1992.

Contractor
MSI-Defence Systems Ltd, Norwich.

VERIFIED

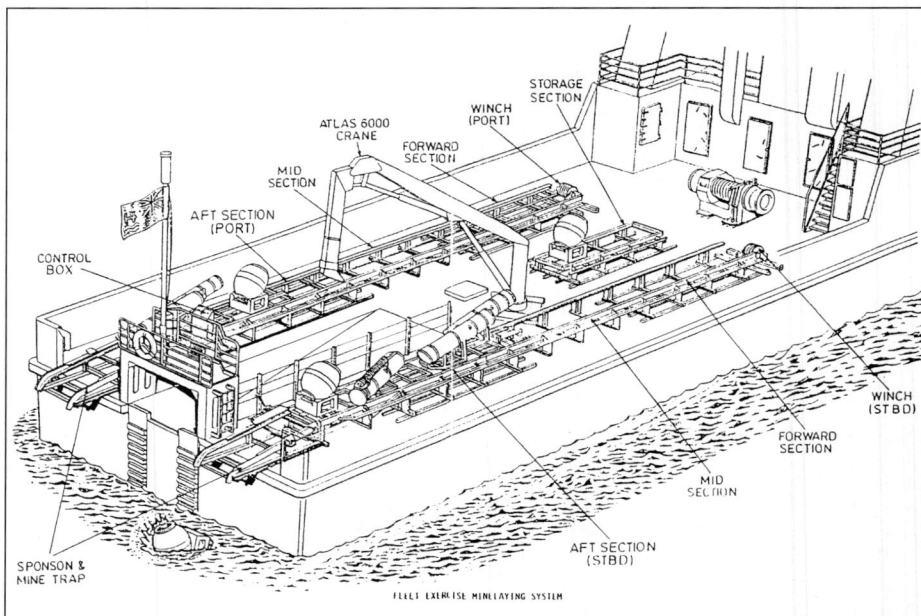

Fleet exercise minelaying system

1998/0009862

UNITED STATES OF AMERICA

MWSIM

Type
Mine warfare training simulator.

Development
In November 1976, the Supervisory Council of Eguermin, the Mine Warfare School at Ostend, Belgium, set up a working group to investigate the training needs with regard to the operational aspects of minehunting. In May 1978, the joint working group comprising members of the Belgian and Dutch navies, formulated an operational request for a minehunting simulator. An official request to include the project in SACLANT's budgetary planning was sent to CINCHAN in June 1986, and the NATO Infrastructure Committee approved the budget in September 1987. The US$12 million contract was awarded to Hughes Training Inc in December 1990, and the purpose-designed mine warfare training simulator was inaugurated on 3 February 1994 at Eguermin.

The simulator will be used for basic and intermediate training, to be followed shortly by advanced training at command team and staff level. It can simulate all current types of MCMV, minehunting combat systems, sonars and sweep gear, the use of precise navigation equipment, powerful data processing, high-resolution sonars and ROVs used by NATO navies.

Description
To allow for objective-oriented training, three coherently interacting models are available; a natural environment model, a ship model and mine model. The interaction of these three models offers the possibility of simultaneously simulating four different mine warfare worlds: one for each of the three minehunting cubicles and one for the six minesweeping cubicles. The natural environment model comprises a sea area including a coastline, varying water depth, currents and sea bottom (the latter represented with a composition of the bottom relief, rocks, and manmade objects). The current seabed model can create 300 subzones in a 20×20 m operational area generated from the database. The ship model creates the type of ships with all their characteristics, while the mine model creates interactions between the mine and the ship.

The simulator comprises an instructor's cubicle, nine training cubicles (three for minehunting, six for minesweeping) and a debriefing room. The nine cubicles comprise equipment identical to that found on board most types of MCMV and fitted with generic Commercial-Off-The-Shelf (COTS) workstations showing indicators and displays according to their appearance in the type of equipment simulated.

Input devices such as control buttons have been replaced by programmable touchscreen panels.

The core of the system is the Ethernet TCP-ICP network with 47 Silicon Graphics computers using COTS hardware linked with the custom-designed software via Ethernet.

The main control computer (a modular 32 Mbyte IRIS SS-V35M16 with graphic cards, a 780 Mbyte and two 1.2 Gbyte hard disks, a 150 Mbyte tape drive and CD-ROM) provides global management of the system and shares the general data.

Exercises are controlled, commanded and created from eight instructor consoles, from which either standard or specially designed exercises, aimed at rehearsing standard procedures and drills for an MCM warfare team, can be composed. The instructor controls the initial moves of the exercise such as the positioning of ships, availability of assets, presence of targets and the simulation of system failures. He is also able to record the presentation on any cubicle screen during the exercise. Ambient sounds can be recorded at the same time. With each recording of a particular screen, data from the ship's computer is also recorded.

Each of the three minehunting cubicles features six computers: a server for general computation duties; one to simulate data presentation and processing and to control the automatic pilot console; one for radar simulation; a computer to simulate sonar detection; one to simulate sonar classification; and a computer to simulate ROVs (including camera images and near field sonars).

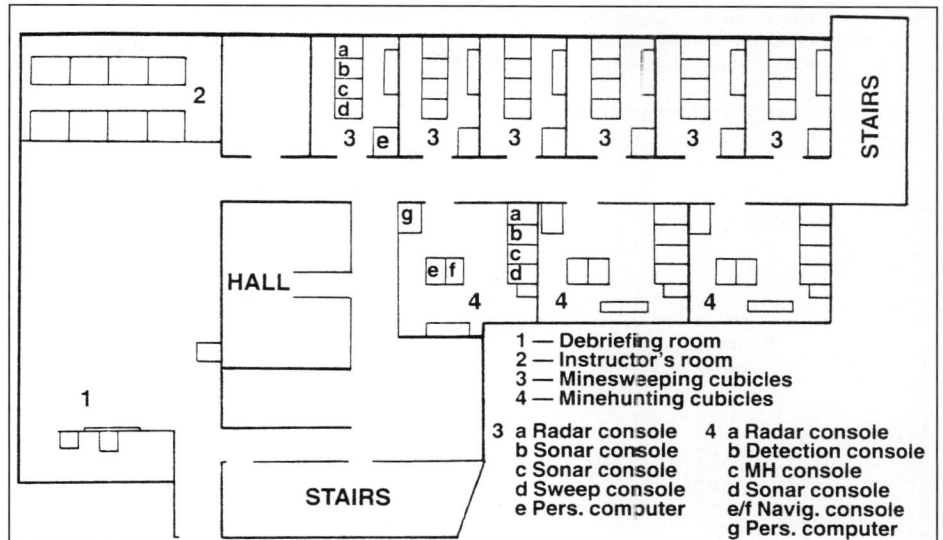

General layout of the MWSIM   *1996*

1 — Debriefing room
2 — Instructor's room
3 — Minesweeping cubicles
4 — Minehunting cubicles

3 a Radar console   4 a Radar console
   b Sonar console   b Detection console
   c Sonar console   c MH console
   d Sweep console   d Sonar console
   e Pers. computer   e/f Navig. console
                           g Pers. computer

Belgian/Netherlands mine countermeasures simulator   *1994*

The six minesweeping cubicles each feature three computers: one to handle data processing and general computation tasks; one to simulate data presentation and processing; and one for radar simulation.

Additional items comprise: communication equipment (to simulate both internal and external communications); closed-circuit telephone; printers and plotters (one of which is a colour plotter for the hard copying of the sea bottom); three PCs for simulation of real-time performance monitors; a BARCO video projector for debriefing; trackballs; and joysticks for the consoles.

The main driver is the IRIX System V release 4 (a version of the Silicon Graphics UNIX system). The simulator uses COTS software programmes such as the MultiGen Visual Database and Visionworks. The simulation software is mainly written in Ada, with the exception of some applications using COTS software. The simulated MCMVs can be designed using menu-driven programmes, selecting the onboard systems (typically detection and classification sonar, ROV, automatic data processing, autopilot and manoeuvring) and by inputting the corresponding ship characteristics (radiated noise, propulsion system

radar reflectivity, hull parameters for shock resistance model and so on). Data such as wind, current, rain, sound velocity, underwater visibility and so on, can be modified by the instructor. In principle the simulation runs as a fully interactive programme, reacting to operations made by the trainees. Specific equipment or events are simulated by inputting the corresponding parameters.

For minesweeping, the sonar and ROV consoles are replaced by a minesweeping console offering simulation of a variety of magnetic, acoustic and mechanical sweep gears.

The simulator is equipped with an audio-visual recording facility which allows exercises to be recorded for replay in the auditorium when it is finished. All exercises are recorded and at the end of each one the instructor can generate post-game reports on navigation, detection identification, threat evaluation, sweep evaluation and classification.

Currently MWSIM is capable of simulating the following MCMVs: 'Tripartite' type (Belgium, France, Indonesia, Netherlands, Pakistan); 'Sandown' class (UK and Saudi Arabia); 'Hunt' class (UK); 'Avenger' class (USA); 'Lerici' class (Italy, Malaysia and now

Australia); 'Frankenthal', 'Hameln' and 'Lindau' classes (Germany); 'Oksoy/Alta' class (Norway); ex-US 'Agressive' class (Belgium); and the 'Ton' class (UK).

An upgrade to MWSIM is already being planned and will include a mine avoidance/moored mine detection sonar for the minesweeper cubicles. By the end of the decade it will also be possible to use the MWSIM as a minefield performance analysis tool, allowing the planning and preplay MC contingency operation in any area in the world.

Operational status

Inaugurated on 3 February 1994 at Eguermin, Ostend, Belgium.

Contractor

Hughes Training Inc, Arlington, Texas.

UPDATED

SUBMARINE EQUIPMENT TRAINERS

GERMANY

Submarine control simulator

Type

Trainer for submarine diving and control room operations.

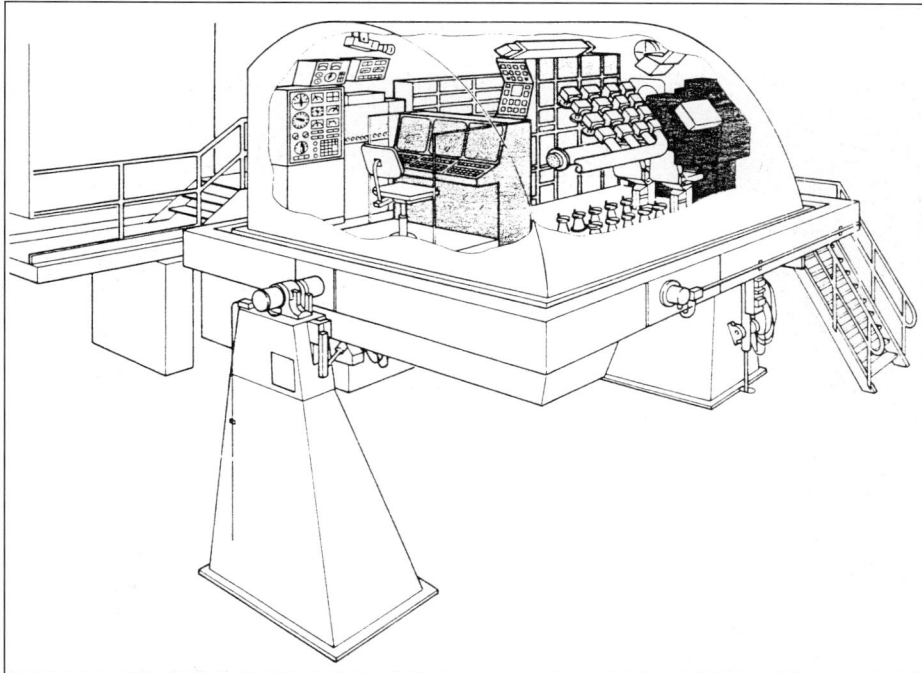

Description

The trainer consists of a replica of the control room of a submarine, an instructor station and a two-axis motion (±30° roll and ±45° pitch) platform. The control room consists of all the original submarine control and monitoring equipment necessary to train personnel in diving procedures and control equipment handling. A mathematical model performs the simulation of the submarine dynamics in 6 Degrees of Freedom (DoF) in real time, within the entire mission envelope including sea states, salinity and water temperature. All the effects of the crew's activities at the control consoles and the submarine's environment are realistically simulated and verified with original sea trial data.

The modular concept allows for a customer-tailored design of the submarine control room equipment and training requirements. The instructor station provides the tools necessary for the briefing, exercise control and debriefing activities, which are equivalent to the onboard procedures. The instructor can set parameters for environmental conditions and operating modes and can programme faults and system failures. The system can be installed in a building with minimised infrastructural requirements.

Operational status

Fully developed and in operation in various navies including those of Germany, Greece, Italy, South Korea and Turkey.

Contractor

STN ATLAS Elektronik GmbH, Bremen.

UPDATED

Schematic of submarine control simulator
1994

Submarine diving simulator

Type

Shore-based computerised trainer for diving crews.

Description

The system consists of the fore and aft hydroplane operator and the diving officer in order to secure maximum submarine handling capability. It comprises an air conditioned simulator cabin mounted on a dual-axis motion (±45° pitch and ±30° roll) platform, an instructor console and a monitoring station containing a computer and associated peripherals.

Main operational features include real-time simulation of a submarine model verification conforming to original shipyard construction and sea trial data.

Operational status

In service.

Contractor

STN ATLAS Elektronik GmbH, Bremen.

UPDATED

UNITED KINGDOM

Submarine Control Simulator (SCS)

Description

The Thomson Training & Simulation Submarine Control Simulator (SCS) incorporates a full-size replica of the forward and port side sections of a control room which is mounted, with the instructor's facilities and computer, on an electrically driven, two-axis (±45° pitch and roll) motion platform and includes an onboard training management system. The SCS is designed to produce as authentic an environment as possible including sea motion, a simulated intercom system and authentic submarine lighting. A distributed architecture with microprocessors is used in the system.

The instructor's console and management station is separated from the students' compartments, but students can be observed by the instructor through a one-way mirror. Each instructor's position includes consoles for the instructor and the unit operator. The consoles can display detailed schematics which permit the instructor to adapt a wide variety of lesson plans. These facilities permit student-paced training as well as precise monitoring of performance.

Submarine Control Simulator for Royal Navy's 'Upholder' class showing two-axis electrical motion system

Operational status

In service with the Royal Navy ('Upholder' and 'Vanguard' classes), Royal Netherlands Navy ('Walrus' class), Royal Norwegian Navy ('Ula' class), Royal Australian Navy ('Collins' class), Indian Navy ('Shishumar' class) and Brazilian Navy ('Tupi' class).

Contractor

Thomson Training & Simulation, Crawley.

UPDATED

Submarine Control Trainer (SCT)

Type

Nuclear submarine handling trainer.

Description

The SCT features two compartments, each configured to represent a different submarine, but both sharing the same motion platform giving ±40° pitch and roll motion. Only one student at a time undergoes training on the simulator.

Operational status

Two systems in service with the UK Royal Navy with one cabin replicating a 'Vanguard' class submarine and the other a 'Swiftsure/Trafalgar' submarine.

Contractor

Thomson Training & Simulation, Crawley.

NEW ENTRY

Marconi periscope simulator

Type

Periscope training system.

Description

Training operators in the use of periscopes at sea ties up operational resources and is limited in scope as it can place the submarine in potentially hazardous situations. The periscope trainer enables either initial or continuation training in both recognition and operational procedures related to any type or design of periscope to be simulated in a safe shore environment.

The simulator provides realistic training in search routines and attack procedures, with inshore or open water navigation in a multiship scenario, under normal or emergency conditions. To the operator the periscope feels and operates just as it would in a submarine, even rotational inertia being simulated for added realism.

The fidelity of ship-modelling provides sufficient detail of superstructure and major features to enable vessel identification and accurate assessment of range, bearing and angle on the bow. The trainer uses advanced visual systems based on commercially available graphics engines of high performance to provide the best means of day, dusk, night scene representation in a variety of weather and sea state conditions. All aspects of the periscope picture are produced, such as graticule and data displays, providing total control from a single system.

As well as being offered as a stand-alone trainer, the periscope trainer can be integrated with existing command team trainers. In team training, information on 'own boat', targets and the environment is passed from the command team trainer to the periscope simulator to co-ordinate the periscope visuals with the data received from other sensors.

Among the features provided for the simulator for the Royal Navy are: underwater views of surface ships;

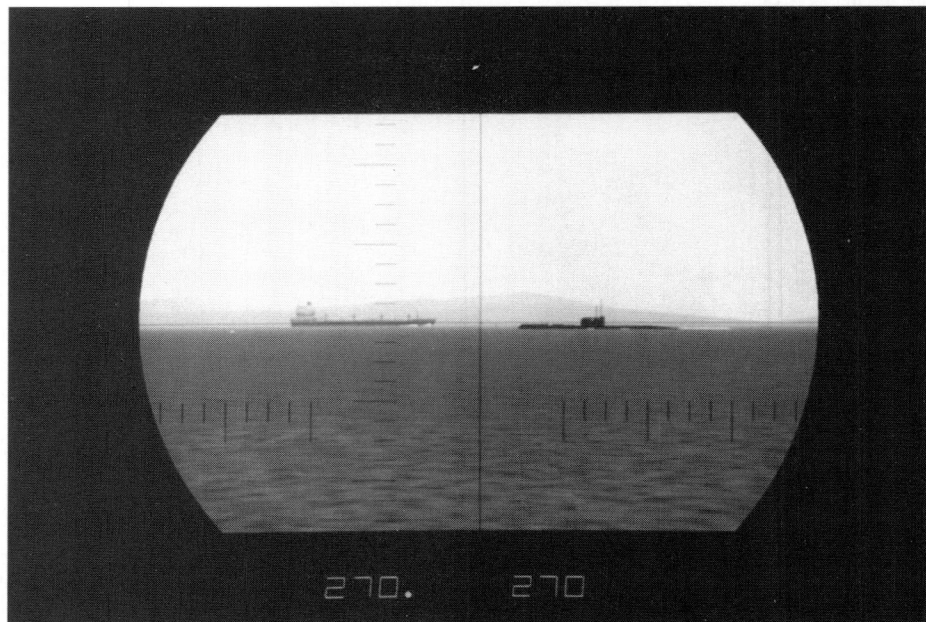

Periscope simulated image

simulation of drain-down effects occurring after periscope surfacing; infra-red and LLTV simulation. These effects enhance the realism offered by the visual system which includes a full three-dimensional textured wave motion to simulate various sea states and washover.

Operational status

Periscope simulators have been supplied to the Royal Navy's submarine command team trainers at HMS *Drake* and HMS *Neptune* and a generic simulator to the attack teacher at the submarine school at HMS *Dolphin*.

Further systems have also been supplied to the Royal Netherlands Navy and the Canadian Forces Maritime Command.

Contractor

GEC-Marconi Simulation and Training Division, Dunfermline.

VERIFIED

UNITED STATES OF AMERICA

STVTS periscope training system

Type

Training device for periscope operation.

Description

Kollmorgen has been designing periscopes and periscope training systems for some years. The most recent example is a Submarine Tactical Visual Training System (STVTS) for the US Naval Training Systems Center at Orlando, Florida. Using a Type 18 periscope, STVTS provides a training device for individual operator, submarine attack centre subteam or submarine combat system team training in an at-sea environment.

Operational status

STVTS is in production and service for the US Navy. Kollmorgen has supplied other periscope trainers for the US Navy during the 1980s.

Contractor

Kollmorgen Corporation, Electro-Optical Division, Northampton, Massachusetts.

VERIFIED

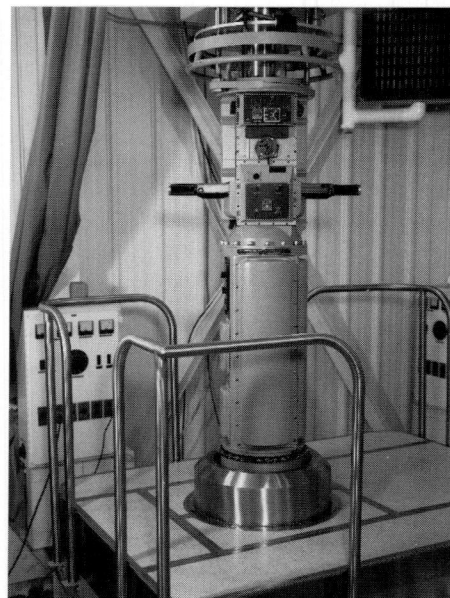

Submarine Tactical Visual Training System

Simulation and training system

Type
Navigation subsystem operational and maintenance training.

Description
This laboratory complex supports the Strategic Weapons System (SWS) of the Trident II submarine tactical navigation centre. It consists of an inertial laboratory, navigation aids laboratory, computer/ electrostatic gyro navigator laboratory, computer-based training laboratory and navigation operational trainer laboratory.

The Inertial Laboratory is a functional replica of the navigation centre in which tactical Electrostatically Supported Gyro Navigators (ESGNs), computers and support equipment and training-unique equipment simulators and stimulators are driven by tactical and training-unique simulation software. In this static laboratory, use of tactical technical documentation is emphasised during familiarisation and basic subsystem operational training conducted in real time. Specific subsystem and ancillary equipment maintenance training is also provided.

The Navaids Laboratory is a tear-down maintenance laboratory and contains tactical equipment and stimulators. Maintenance training on the Navigation Sonar System (NSS) and Global Position System (GPS) equipment emphasises tactical documentation and software diagnostic procedures. The Computer/ESGN (C/ESGN) Laboratory is also a maintenance laboratory

containing tactical Memory Processor (MP) computers, an ESGN system and ancillary support equipment. Maintenance training on this equipment uses tactical documentation and software diagnostic procedures.

The Computer-Based Trainer (CBT) Laboratory augments and supports certain training requirements that could not be conducted in the Inertial, Navaids, or C/ESGN laboratories because of their static nature, the physical constraints imposed by the hardware configuration and design, or operational and training time limitations. The CBT Laboratory is a training-unique Personal Computer (PC)-based complex of 10 student workstations and an instructor workstation, each composed of commercial hardware, with trainer-unique software and courseware. The system integrates two-dimensional dual-screen simulation, Interactive Video Disk (IVD) images and a tactical Auxiliary Display Terminal (ADT) keyboard for man/machine interface. Forty one hours of courseware training modules exist for the CBT Laboratory. One student or an entire class can exercise any module. Courseware modules provide theory, operational and maintenance instruction on various topics, such as basic inertial principles and ESGN calibration procedures. Although the latter requires 9 to 10 days in a tactical environment to perform, it can be accomplished in the CBT Laboratory in only 5 hours. IVD simulation of tactical equipment parallels the tactical operational and maintenance documentation procedures utilised in each module.

The Navigational Operational Trainer (NOT) Laboratory was developed to support dynamic

subsystem operational team training requirements. The NOT Laboratory is an array of training-unique simulators emulating and physically arranged to duplicate the Trident II submarine's navigation centre. Driven by simulation software, the trainer can functionally replicate all navigation subsystem normal operation as well as many system casualty operations. Submarine navigation crew members operate and man this trainer during each training period in the same team manner as they do on the ship. The laboratory interactively responds to the team's performance of tasks just as the real ship would, whether at sea or in port. Unlike the Inertial Laboratory, which supports only limited, real-time static operation training, the NOT Laboratory can provide a dynamic training capability at any location in which the submarine may operate. Using the available time compression capability, up to 24 hours of real-time performance can be achieved in 1 hour of laboratory time, thus shortening lengthy, real-time procedures. Training scenarios can also be saved for replay and evaluation at a later date. Instructors can interject malfunctions for the simulated tactical equipment support subsystem casualty operation. Adherence to documented operational and casualty procedures is emphasised.

Operational status
Five laboratories, consisting of tactical and training-unique hardware and software necessary to support the Trident II navigation community training requirements were installed at the Naval Submarine Center in Kings Bay, Georgia in 1991. Formal curricula were developed and, from 1987 to 1991, instruction was provided to the initial group of naval personnel assigned to the newly commissioned Trident II submarines and to the Trident Training Facility (TRITRAFAC) in Kings Bay, Georgia, using these laboratories.

These laboratories have been updated to support the Trident Navigation Commonality Program (TNCP) in the early part of 1995 and continue to maintain SWS navigation training.

Four laboratories, consisting of tactical and training-unique hardware and software necessary to support the TNCP training requirements were installed at the Naval Submarine Center in Bangor, Washington in the 1994-97 time frame. These laboratories were required when three Trident I submarines were converted to TNCP submarines. The Bangor Facility consolidated the tactical NSS and GPS equipment into the inertial laboratory to eliminate the need for a separate Navaids laboratory. This reduction in laboratories was achievable based on the fewer number of TNCP SSBNs on the West Coast.

Contractor
Lockheed Martin, Federal Systems, Great Neck, New York.

Kings Bay TRIDENT II navigation laboratories

1994

VERIFIED

AIRCRAFT EQUIPMENT TRAINERS

UNITED KINGDOM

ASW Crew Trainers (ACTs)

Type
Anti-submarine training systems.

Description
GEC-Marconi Avionics' ASW Crew Trainers (ACTs) provide real-time ASW training for maritime patrol aircraft and helicopter crews, ACT systems offer comprehensive acoustic, tactical and crew co-operational and co-ordination training by simulating sonobuoy, dipping sonar, target and ocean environmental data. Data can be controlled, processed and displayed on the aircraft's acoustic and tactical systems in exactly the same way as real data. Training is therefore achieved without the need to deploy sonobuoys or dipping sonar, or use co-operating submarines and surface vessels, thus substantially reducing costs and increasing training opportunities.

There are two basic variants of ACT.

ACT 1 is a software-based system developed specifically for the GEC-Marconi Avionics' AQS-901 acoustic processing systems now in service on Royal Air Force Nimrod MR Mk 2 and Royal Australian Air Force P-3C Orion maritime patrol aircraft. It comprises an Exercise Control Unit (ECU) and the computer program which is downloaded into the AQS-901 from a magnetic tape unit. The crew member acting as the exercise controller uses the ECU to set the target's initial start position, course, speed and depth and, if desired, control of subsequent target manoeuvres and ocean conditions. Otherwise, the facilities and functions necessary to simulate fully dynamic ASW scenarios are automatic.

ACT 2 is a hardware-based system which is far more flexible than ACT 1 and can be interfaced with any acoustic processor resident in the platform or, by means of a VHF radio link, with an ACT ground station. ACT 2 can simulate the characteristics of any

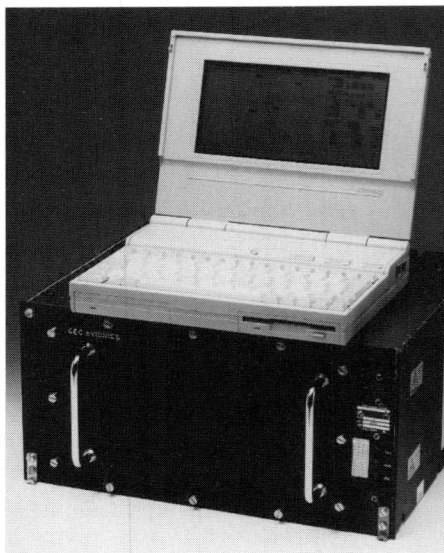

ACT simulation unit and exercise control unit

sonobuoy, dipping sonar, target and ocean environment specified by the user.

ACT 2 comprises an Acoustic Simulation Unit (ASU) and ECU. The ASU, which executes the ASW scenarios and generates synthetic passive and active acoustic data, can be configured to simulate any number of independent sonobuoy and dipping sonar channels. The ECU is an intelligent control terminal with an integral keyboard and display. In addition to performing all exercise control functions, the battery-supported ECU is used for the temporary storage of exercise scenario data.

ACT 2 systems currently available can simulate up to 32 independent data channels from any combination of LOFAR, Difar, Barra, Ranger, Dicass and bathythermal sonobuoys, and generate simultaneously the acoustic signature characteristics of up to three independent targets. Simulation of specific dipping sonars and other types of sonobuoy can be accommodated easily with the existing hardware.

Unlike ACT 1, there are no limitations with ACT 2 on the potential number and variety of exercise scenarios available for training. These are created and maintained by each customer, using an Exercise Support System (ESS). The ECU and ESS functions can be performed using a single/common IBM PC-compatible computer, which can also be used as a classroom trainer and for conducting post-flight debriefs. Scenarios created from target and ocean characteristics defined and stored in the ESS library files by the customer are downloaded into the ECU before the flight. On board the aircraft the scenario data are transferred to the ASU. The ECU is then used to control and record the results of each exercise for subsequent in-flight or post-flight debriefs.

Operational status
ACT 1 is in service with the RAF and RAAF. ACT 2 has completed development and is in full production and initial deliveries have been made to several customers.

A contract has been awarded for the development of the sonics simulation/stimulation equipment which is based on ACT 2. This system is designed to prove the functionality and evaluate the performance of the Royal Navy Merlin helicopter acoustic processor and can be used to assess any acoustic processing system.

Contractor
GEC-Marconi Avionics Ltd, Maritime Systems Division, Rochester.

VERIFIED

UNITED STATES OF AMERICA

International air training systems

Type
Combat trainer, crew trainer and so on.

Description
Training systems have been developed by Lockheed Martin Federal Systems for the LAMPS Mk III SH-60B Seahawk helicopter for the US Navy. The LAMPS training suite includes: avionics maintenance trainer, mobile team training unit, operational flight trainer, weapons systems trainer and weapons tactics trainer.

The company is developing full-mission simulators for training the Royal Navy Merlin helicopter operators. The training elements include: cockpit dynamics simulator, cockpit procedures trainer, computer-based trainer, common control unit/tactical situation display trainer, engine change unit trainer, mechanical systems trainer, rear crew trainer and weapon system trainer.

For the Spanish Navy Lockheed Martin Federal Systems, in co-operation with CESELSA, has provided hardware and technical information to assist in the development of a tactical simulator for the Spanish LAMPS helicopter system. The simulator suite will train pilots and aircraft tactical officers to operate the LAMPS helicopters.

Lockheed Martin Federal Systems has provided the acoustic signal generator, electronic flight control system, system databus adaptor, converter multiplexer amplifier and motion system and seat shakers. The

suite includes a version of the LAMPS trainer software which comprises the operational flight trainer and the weapon tactics trainer.

Operational status
Systems in operation for the Spanish Navy for LAMPS helicopters, and under development for the Royal Navy for the Merlin helicopter. All LAMPS systems for the US Navy are in operation.

Contractor
Lockheed Martin Federal Systems, Manassas, Virginia.

VERIFIED

Airborne ASW trainers

Hughes Training Inc has provided most of the airborne anti-submarine warfare training systems used by the US Navy and also for use by the Canadian Forces. The systems include acoustic and non-acoustic sensor operator trainers for individual training together with team trainers. These programmes include the following.

P-3C Tactical Operational Readiness Trainers (TORT). These systems, designated Device 2F140(T) by the US Navy, are designed to provide integrated team training for the tactical crews of the Lockheed P-3C Orion. Link has provided four of these systems to the US Navy. The simulator provides training in a high-density acoustic/RF environment and has true operational simulation capabilities.

The trainer is a facsimile of the anti-submarine warfare compartment which uses actual hardware. There is also an instructor's station divided into non-acoustic, acoustic, tactical/navigation and pseudo-pilot areas. The simulated operational area is 1,326 km² and up to 22,860 m deep. Up to 35 targets, including eight acoustic, are available at any time, while up to 31 possible radar contacts are available together with up to 12 radar jammers.

S-3 position trainer
Designated Device 14B49, this is a three-station, nine-position trainer currently being upgraded from Lockheed S-3A Viking to S-3B (Device 14B49B) status. The stations simulated are those of the co-pilot, the tactical co-ordinator and the sensor operator each of

which is triplicated for multiple training. The system is designed to provide each crew member with individual skill levels. The enhancements include increased acoustic and Electronic Support Measures (ESM) capabilities, and the addition of a Harpoon anti-ship missile system capability.

P-3C ASW tactics trainer
This system, designated Device 2F87(T) by the US Navy, is designed to provide crew members of the Lockheed P-3C Orion with complete tactical training in the Anti-Submarine Warfare (ASW) mission. The system features high-fidelity simulation of the search radar, acoustics, ESM, magnetic anomaly detection, navigation and communications systems.

CP-140 operational mission simulator

This system is similar to the tactical trainers used to train US Navy P-3 and S-3 aircrew but has been adapted to incorporate specific Canadian systems as used in the CP-140.

P-3 acoustic target generator

Hughes Training Inc provided an acoustic target generator to the Mitsubishi Precision Company Limited which was integrated into the third P-3 Operational Flight Tactical Trainer built by Mitsubishi for the Japanese Maritime Self-Defence Agency. The generator provides training in a high-density acoustic/RF environment for the P-3 tactical crew members. It simulates the real-world acoustic signals from sonobuoys used to detect, identify and track submarines. The Japanese trainer is based upon the P-3C Update III TORT.

Operational status

Device 2F140(T) is operational at NAS (Naval Air Station) Barbers Point, Hawaii; Brunswick, Maine; Jacksonville, Florida; and Moffet Field, California.

Device 14B49 is operational at NAS North Island, California and Cecil Field, Florida, with two Device 14B49B systems also deployed in Florida.

Contractor

Hughes Training Inc, Arlington, Texas.

VERIFIED

A sensor station in a Device 2F140(T) TORT
1994

SONAR EQUIPMENT TRAINERS

CANADA

CMS (Canadian towed array sonar system Mission Simulator)

Type
Towed array mission simulator.

Description
The primary objective of the CMS is to provide an advanced training platform for the training of CANTASS

operators. CMS provides the operator with a comprehensive advanced training tool using high-fidelity simulation techniques.

The simulator consists of an instructor's workstation and a number of student stations. The instructor station simulates the student stations by generating and transmitting raw acoustic data which reflects a realistic tactical problem. Students then analyse the simulated acoustic data.

The instructor's workstation allows the instructor to create, execute and control a simulated mission and simultaneously monitor the students' peformance. Using the station the instructor can debrief the students on their performance by rapidly replaying the mission simulation on a debrief monitor at the student station.

The instructor's workstation displays tactical and environmental data as well as displaying the student's decisions on the debrief monitor. Any of the students' acoustic screen monitors can be viewed from a local monitor at the instructor's workstation. The instructor can also communicate verbally with the student while a mission simulation is in progress.

Each student station is similar in function and performance to a passive tactical towed array sonar system. It comprises a sonar signal processor and data arrangement/display units and a debrief recorder. The sonar signal processor and data management/display unit is functionally equivalent to an actual shipboard system.

The debrief recorder captures and stores on the disk array, digital images of the video being displayed on each of the student station's four acoustic screen monitors. These images can later be replayed, under control of the instructor, and compared with a simultaneous recreation of the mission simulation. The audio commands from the student are also recorded.

Operational status
In service with the Canadian Navy.

Contractor
Array Systems Computing Inc, North York, Ontario.

UPDATED

*Consoles for the CMS
(Array Systems Computing)*
1998/0009872

OH-5001/SQS-505 (V) 6 simulator stimulator

Type
Onboard sonar simulator stimulator.

Description
The simulator stimulator injects noise and signal into the front end of a sonar receiver, thereby acting as a substitute for the sonar's transducer outputs. The noise and target signals feed the processing circuits in the sonar receiver to provide video and audio outputs such as operators would experience during a live action.

The system can generate four targets as either ships/submarines/wrecks or torpedoes together with a full range of associated operational noises. If the target simulated is a submarine, an air bubble may also be

deployed to simulate an attempt to mask the submarine's echoes. The operator can change or cancel any of the simulated parameters or targets, unless the selected target is a torpedo which is homed in on the ship by the simulation scenario.

The system has four operational modes:
(1) at sea, transmitter on – active and passive targets are simulated and signals superimposed on the ship's noise, flow noise and reverberation received through the transducer from normal operation
(2) at sea, transmitter off – no power is transmitted through the transducer and signals and reverberation are superimposed on own ship's noise and flow noise from the staves
(3) harbour operation – the simulator stimulator provides all signals, noise, reverberation and target

data as well as supplying the receiver with artificial data to represent ship's course and speed
(4) test set – performs tests in the sonar receiver on:
(a) preformed beam patterns
(b) fine video calibration
(c) Doppler filter
(d) passive and active signal processing gain.

Operational status
In production. A total of 13 sets has been delivered for the Canadian Patrol Frigate programme.

Contractor
Northrop Grumman-Canada Inc, Burlington, Ontario.

VERIFIED

AQQ-T501 tactical acoustic trainer

Type
Acoustic simulation and training system.

Description
The AQQ-T501 tactical acoustic trainer is an advanced modular acoustic stimulator that provides realistic, dynamic and effective ASW acoustic training in a manner that is indistinguishable from actual live target in-contact time. Realistic signatures and operating scenarios are achieved through a multiple target environment that includes oceanographic and shipping interference and fully responsive and manoeuvring targets. Targets can be defined with each target following independent manoeuvring patterns for

position, course, speed and depth. Also, each target is reactive to programmable oceanographic conditions and sonobuoy type and location. The result is type specific submarine signatures that are identical to, and indistinguishable from, actual live target signatures. Operators can easily build and store a library of realistic target signatures.

Both speed-dependent narrowband and non-speed related narrowband are accurately portrayed using over 1,500 simultaneously produced lines, with harmonic and gear ratio relationships. Broadband and transients are produced in an equally realistic manner. Detection opportunities are aspect dependent as individual lobing patterns are produced for each line family. Realism is further enhanced by variations in the passive acoustic environment such as sea state, interfering shipping, convergence zones, multipath and

layering which can all be controlled by preset and/or online manipulation and in the active environment by ambient noise, reverberation, propagation loss, vertical attenuation, Doppler shift and by the use of multiple target reflective points.

These features make the AQQ-T501 a thoroughly flexible training system that can be used airborne or land-based in a classroom/laboratory environment. A hands-on instructor control can produce instantaneous target manoeuvres within the actual submarine operating envelopes, or preprogrammed scenarios can initiate random manoeuvres with predetermined submarine responses to external stimuli. Meaningful training can now be achieved without the significant costs associated with major submarine deployments, and these training opportunities can be available on a daily (rather than yearly) basis. Further cost savings exist in the area of consumable stores as the AQQ-T501 simulates all existing buoy types. Due to the unique architecture, any number of buoys can be simulated.

The AQQ-T501 provides dynamic and accurate responses throughout all phases of the ASW mission and all features are available using existing acoustic processors.

Specifications
MIL-E-5400
MIL-STD-1553B compatible
8 Difar configuration
Size: 13.2 × 26.2 × 54.6 cm (W × H × D)
Weight: 16 kg

Contractor
Computing Devices Canada Ltd, Ottawa, Ontario, Canada.

UPDATED

HI-TASS

Type
Acoustic simulation and training system.

Description
The HI-TASS ASW training system is a portable, high-fidelity, COTS simulator specifically designed to generate the full ASW array mission, by including acoustic signatures, sensor performance, environmental conditions and tactical scenarios. Replicating the most demanding array operations (including towed and fixed arrays), HI-TASS generates acoustic and non-acoustic data in an operational environment, thereby providing operators and command teams alike with the opportunity to attain the highest proficiency levels. Complementing the training capability, HI-TASS provides a wide variety of user-selectable configurations which can be utilised in a support role for system maintenance and own platform noise assessment.

Operational status
HI-TASS systems have been delivered to the Canadian Navy and Republic of South Korea Navy.

Contractor
Computing Devices Canada Ltd, Ottawa, Ontario.

UPDATED

Elements of the COTS simulator HI-TASS
1997

FRANCE

Sygame

Type
Audio training system.

Description
The Sygame (TSM 9001) acoustic training system comprises a sound generator and a workstation with a set of software tools. Menu-driven software enables progressive training from elementary sound recognition to target classification in a complex noise environment.

The system can be used either as a stand-alone system on land or onboard ship at sea, or as a network consisting of two to eight workstations.

Sygame synthesises sounds from active or passive sonars that are then fed to the students to train them in recognising different sonar pulses and noises generated by propellers, engines, auxiliary machinery, hulls and transients. Up to 12 items of machinery can be mixed together in the system to provide realistic acoustic signatures for submarines (both diesel and nuclear) and surface ships (both military and civilian). Users can create acoustic signatures to define platform characteristics (dimensions, number of shafts and blades per shaft, asynchrony between shafts) and to mix various sound effects such as normal cavitation (strong, medium, hard), compressed cavitation (bubble, whip), shaft rub and singing propellers.

Using Sygame students are taught to develop effective search, localisation and classification skills. To further improve training the instructor can insert environmental conditions including noise created be traffic, weather conditions and the sea itself, as well as biological and reverberation noise.

The Sygame VI N is an improved version that can be integrated with a network and has an Internet protocol.

Operational status
In service with the navies of Australia, France and Italy.

Contractor
Thomson Marconi SAS, Sophia Antipolis.

NEW ENTRY

NORWAY

MMSS (MPS Multibeam Sonar Simulator)

Type
Training system for commanding officers and sonar operators in the use of active sonar variable depth or dipping.

Description
The MPS Multibeam Sonar Simulator (MMSS) is a training system for the operators of hull-mounted and variable depth multibeam sonars. The system generates both sonar-image and audio signals in real time so that the student can use the controls of the sonar console, learn to understand the working

principles of the equipment and assess the results when the control settings are altered. The system is based on the 32-bit Multibus II real-time computer using the Intel 80386 as the main processor.

Echoes from underwater objects are produced by a digitised representation of reflected sound at different aspect angles. This produces a very accurate generation of the acoustic characteristics of the sound return, which takes into account the size, geometry, location, aspect angle and Doppler effects. The operational characteristics of the sonar, including the

source level, transmission modes, scanning sectors and pulse lengths, together with the actual operator settings such as range scale and tilt, are used as the simulator model.

The system comprises an instructor's position, a student steering console position, a student sonar console (using a specific system), a game preparation console/system control station, laser printer, digitiser for map and terrain-model preparation.

The operator can use either an actual sonar console or an MPS tactical console simulating the man/ machine interface, presentation modes and signal processing capabilities of a specific, or a set of real sonar equipments. In order to steer and control the simulated sonar-carrying vessel, a manoeuvre console is connected for setting speed, course and manoeuvre models. The sensor platform and all the targets are controlled by the instructor.

Optional extras include a network interface with an action speed tactical trainer such as MASTT (see earlier entry), a plotter for analysis purposes, additional tactical displays and an interface with a video-audio recorder.

Operational status
In service with the Royal Norwegian Navy (integrated with the MASWT) since 1989 and with the Swedish Navy since 1990. Upgrades with additional emulated sonar operator consoles for both the Royal Norwegian Navy and Swedish Navy were supplied in 1993.

Contractor
Kongsberg Defence & Aerospace AS, Simulation & Training, Kongsberg.

MPS multibeam sonar simulator at the Swedish Navy Training School

VERIFIED

UNITED KINGDOM

AS 2105 sonar stimulators

Type
Sonar test and training for ships and submarines.

Description
The AS 2105 sonar stimulators have been designed to generate the accurate signals required to test comprehensively modern, complex, passive sonar systems. They also provide basic training for sonar operators, target motion analysis teams, command teams and maintenance personnel. More comprehensive training facilities are available with the AS 2107 sonar trainer (see next entry).

The AS 2105 equipments are compact, robust and lightweight, which makes them suitable for use ashore or on board surface ships and submarines, either at sea or in harbour. The AS 2105 can emulate a wide range of hull-mounted and towed hydrophone array geometrics (including towed, flank and conformal arrays), and by generating appropriate electrical signals, stimulate the inboard processing and display equipments.

The AS 2105 consists of two parts; a small lap computer which provides the man/machine interface allowing the user to construct scenarios for training and testing and the electronics unit which generates the acoustic signals.

The Mk II variant is designed for integrated sonar systems and uses a central computer controlling several simulators, each linked to a separate sonar. PASSIM (Passive Sonar Array Simulator) is a variant which simulates signals received by towed arrays.

AS2105 sonar stimulator

Operational status
A number of systems is in service with the Royal Navy and a NATO navy.

Contractor
Thomson Marconi Sonar Ltd, Templecombe.

UPDATED

AS 2107 sonar trainer

Type
Sonar training and system test for surface ships and submarines.

Description
The AS 2107 sonar trainer has been developed from the AS 2105 sonar stimulator (see previous entry), and will generate set-piece or free-play exercises with either real or simulated own ship parameters.

By synthesising the narrowband, broadband, Demon and aural emissions of platforms, the AS 2107 can describe complex signatures, together with vehicle dynamics, weapon configuration, environmental characteristics and other parameters, which would normally be recalled from the user's data library. Alternatively, the AS 2107 allows offline generation to permit the user to develop or modify data. The scenario author/command can perform a fast time review of the scenario using the built-in tactical display, this facility

being particularly useful for the initial construction of scenarios and for debriefing.

The AS 2107 consists of two parts; a laptop computer which provides the man/machine interface allowing the user to construct scenarios for training and testing, and the electronics unit which generates the acoustic signals.

The AS 2107 is intended primarily for onboard continuation training of sonar operators, target motion analysis teams and command teams, in conjunction with towed and hull-mounted sonar processing systems. It may also be used for extensive testing of sonar systems, particularly within a shore development establishment.

Operational status
In service with the Royal Navy, Canadian Forces Maritime Command and the Royal Netherlands Navy.

Contractor
Thomson Marconi Sonar Ltd, Templecombe.

VERIFIED

AS2107 sonar trainer

Mandarin/Solo sonar trainers

Type
Computer-based sonar training systems.

Description
The Mandarin/Solo for Windows integrated sonar training suite offers courseware products which represent a comprehensive answer to basic training requirements. These products provide training in sonar theory, lofargram analysis, target classification, aural recognition and operator procedures. Delivery of courseware is compatible with IBM Standard PCs.

Mandarin for Windows
This is a comprehensive toolset necessary to create, deliver and manage specific computer-based training courseware. It is an authority system designed by Marconi Simulation and Training and sold throughout the world. Access to facilities and movement between them is readily available by familiar menu-pointer interfaces and courseware production is via an interconnected series of on-screen editors for graphics, video, course planning and course management.

Solo for Windows
This is a unique sonar lofargram trainer providing complex acoustic environment and signature generation tools which can display detailed sonar lofargrams in real or accelerated time. This system enables the generation of scenarios involving a number of simulated targets and entities necessary for the training of lofar analysis and classification. The system also includes integrated aural and high-resolution graphics.

Sonar Principles Trainer (SPT)
This is an interactive training package covering sonar fundamentals and principles. It consists of five modular lessons covering all aspects of basic sonar theory. Students are fully interactive through graphic and animated sequences. (Mandarin for Windows).

Basic Operator Trainer (BOT)
This is a flexible sonar procedures trainer using facsimile or emulated keyboard representations and high-resolution displays for the production of sonar

screens. It enables the development of operator procedural skills in a classroom environment. (Mandarin for Windows).

Basic Acoustic Training System (BATS)
This is a comprehensive sonar acoustic skill trainer incorporating full facsimile or graphic representation of the sonar consoles in a classroom environment. The sonar type can be selected from a library to include hull, towed arrays or sonobuoys as required and utilises enhanced facilities of Mandarin Solo for Windows.

Operational status
Supplied to the Royal Navy (for ASW Sea King crews) and Royal Air Force (for Nimrod crews), the US Navy and the Royal Netherlands Navy.

Contractor
GEC-Marconi Simulation & Training Division, Dunfermline.

UPDATED

Sonar principles trainer

Type
Comprehensive sonar operator training system.

Description
The Sonar principles trainer has been designed to provide comprehensive facilities for the training of sonar operators in understanding the various elements of complex sonar signals and the skills associated with detection, evaluation and tracking of targets in a variety of scenarios and environmental conditions.

Consisting of an instructor's console and up to 10 identical trainee operator consoles, the system offers a standardised hardware concept which can be customised by software to meet the specific training needs of the user. Each trainee console consists of a

multipurpose display unit, a sonar control panel, sonar aural facilities and communications facilities. Each trainee has control over his own sonar with respect to mode of operation, frequency, pulsewidth and beam training, and the sonar pictures and aural effects are responsive to individual trainee actions.

The trainer provides the sonar instructor with a flexible training tool which can be used to prepare trainee operators for sonar operations across the range of equipments existing in his particular fleet. Through the means of a user-friendly console, the instructor can prepare, execute, monitor and debrief exercises. The instructor has full control over the ocean environment, movement of 'own ship', selection of sonar types for training, selection and movement of target vessels and weapon deployment.

A repeater monitor and audio system at the console allows the instructor to select any trainee operator's sonar picture and associated sonar aural for online monitoring of individual sonar performance. Each trainee operator's use of the relevant functional controls is also monitored as part of the performance analysis.

Operational status
In service.

Contractor
Thomson Marconi Sonar Ltd, Stockport.

VERIFIED

SAINT training system

Type
Initial training system for sonar operators.

Description
The Sonar Analysis INitial Trainer (SAINT) is intended to train operators in the principles of underwater noise generation, signal display presentation and narrowband analysis from basic principles to advanced levels. The system consists of a dual-position instructor's console linked to as many as 12 students' positions, each for two students. As a result of this

organisation, each instructor can carry out independent exercises with six positions.

Each instructor position has two video monitors, one for monitoring the trainer control menu and the other to repeat any picture seen by the students. There is also a joystick-controlled pointer to identify important features

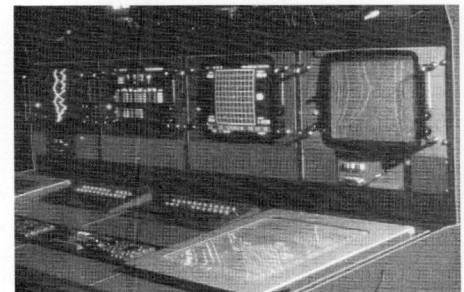

The SAINT (Sonar Analysis INitial Trainer) in operational mode

on any video seen by students, a digitiser pad and a stylus for repositioning of frequency lines in a Lofargram, or hand-free graphics 'sketching' on to video pictures. Video record facilities and communications with the students are also provided.

Each student position has a paper chart recorder with numeric display to identify the frequency window covered by the Lofargram being recorded. The positions also have a video monitor and cursor generation electronics. Cursors can be superimposed on a picture relayed by the instructor to allow the student to identify both frequencies and harmonic relationships. Numeric display can also be incorporated.

Operational status
Fully developed.

Contractor
Thomson Marconi Sonar Ltd, Stockport.

VERIFIED

Onboard sonar trainer

Type
Onboard trainer for the Type 2050 sonar.

Description
This simulator has been designed to provide training facilities for the Type 2050 sonar system in service aboard surface ships of the Royal Navy. The trainer provides high-fidelity acoustic contacts to stimulate the operational sonar processor. Controlled by an intelligent graphics terminal, it incorporates a comprehensive ocean environment model to ensure realistic detection scenarios under different thermal and acoustic conditions.

Several contacts can be injected simultaneously into the sonar processor. Exercises can be constructed to include either a real-world, real-time environment, or alternatively contacts can be generated together with a large variety of synthetic operational scenarios for continuation training either at sea or in harbour. A data recorder is included in the system.

Operational status
In service with the Royal Navy.

Contractor
Thomson Marconi Sonar Ltd, Stockport.

VERIFIED

AS 1092 sonar/radar trainer

Type
Shipborne sonar and radar training system.

Development
The AS 1092 is a development of the AS 1077.

Description
The AS 1092 is a multisensor onboard trainer designed to train command team sensor operators within their operational environment using their normal equipment. The systems are modular and compact to fit in the limited space normally available. They can be used to provide procedures training, or as part of command team training with simulated weapons firing.

The AS 1092 consists of a main control console, a sonar simulation cabinet and Initial Detection and Classification Trainer (IDCT), a second control unit, weapon system and navigation radar interface units, an IDCT remote-control unit, a second control position and a hand-held helmsman's control unit.

Synthetic targets can be inserted onto screens receiving real data or an exercise may be carried out using simulated targets and simulated background, with the latter covering an area of 2,048 × 2,048 n miles (3,788 × 3,788 km) to heights of 24,384 m (80,000 ft) and depths of 6,096 m (20,000 ft). A ship movement simulator is included with high-integrity changeover devices for switching between real and simulated data. Up to 70 platforms (surface ships, submarines, aircraft, torpedoes) can be simulated simultaneously from a library with a capacity for 254 classes.

Operational status
In service with the Royal Navy, but no longer in production.

VERIFIED

The AS 1092 shipborne radar and sonar trainer

Sonar procedures trainer

Type
Basic and advanced trainer for sonar operators.

Description
These are intelligent procedure trainers which provide students with a full range of training facilities from basic to complete tactical and weapons operations. Each student is guided automatically by verbal and visual commands in accordance with a predefined training programme. The system features high-resolution animated graphic presentations, full touchscreen control operations, complete emulation of main system displays and operating control panels, together with emulation of targets and the environment.

Two dedicated products have been produced; the IPT-32 and the IPT-2022. The first of these is an anti-submarine warfare sonar simulator for training operators of the PHS-32 hull-mounted sonar, while the IPT-2022 is a minehunting sonar and mine disposal system trainer for operators of the TSM 2022 sonars. Both systems have been supplied to Dornier for use with its training systems.

Operational status
In service with the Nigerian Navy.

Contractor
Digital Systems and Design Ltd, Eastleigh.

VERIFIED

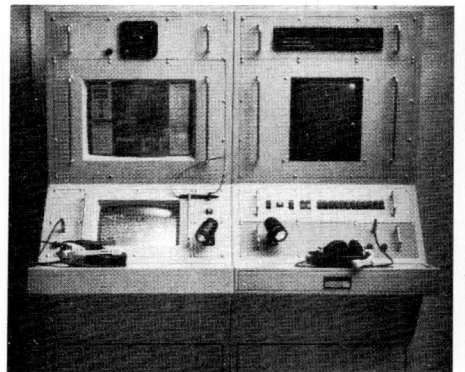

The IPT-32 anti-submarine sonar trainer

AST

Type
Acoustic Systems Trainer (AST) for teaching sonar principles and techniques.

Description
The AST is a commercial system designed to teach real-time sonar techniques in a laboratory. Previously this subject has been confined to verbal instruction and simulation due to the necessity to use large water tanks to achieve measurable results. By using a modest increase in transducer frequency, Instrutek have enabled the demonstration of a wide range of sonar techniques in a water tank measuring only 1 m in length.

The tank includes a heater for showing temperature effects, a transport mechanism for moving targets and a number of resin encapsulated reversible electrostictive transducers for transmitting and receiving signals. Various active and passive targets can be mounted in the tank to show and allow experimentation on the following topics:

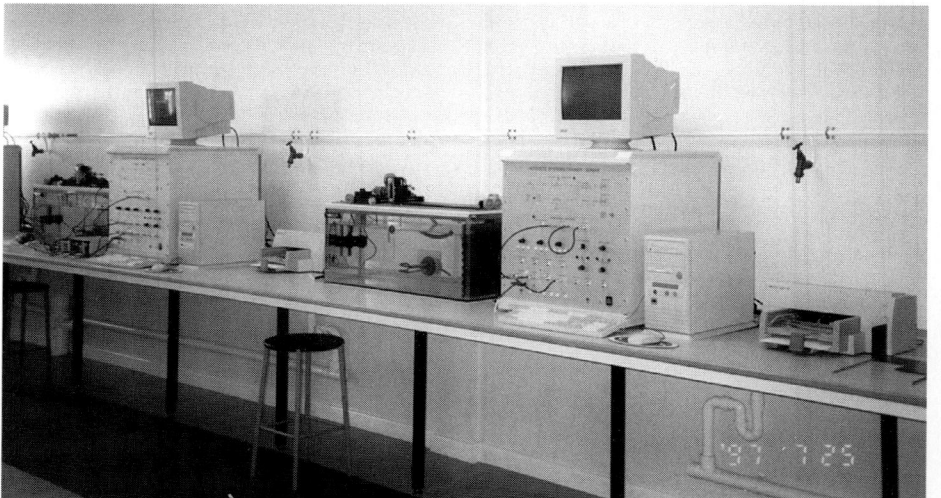

The AST workstation and water tank (Instrutek)

- Speed of sound in water
- Temperature dependence of the speed of sound
- Range measurement
- Interference effects
- Beam formation and steering
- Scattering
- Target signatures
- Jamming
- Cavitation
- Noise
- Reverberation

- Refraction
- Salinity effects.

The system comprises a robust 19 in console containing power supplies, signal sources and the sonar signal analyser which provides a range of computer based virtual instruments including an oscilloscope, multimeter and spectrum analyser. These enable the student to investigate and condition all the signals associated with acoustics and sonar. Connections to the transducers are made via standard BNC connectors and the front panel allows access to all

test points and signals.

Operational status

The AST is in service with the Royal Navy.

Contractor

Instrutek (UK), St. Ives.

VERIFIED

UNITED STATES OF AMERICA

AN/SQQ-T1 sonar training set

Type

Training equipment for ASW sonar operators.

Description

Designed to provide passive and active acoustic training for teams and subteams, the AN/SQQ-T1 stimulates sonar systems to generate high-fidelity acoustic signals resembling those of actual targets under various sea and environmental conditions. It provides sonar operators with the ability to improve their skills in target identification, analysis and classification during long periods at sea or at ground-based stations.

The system can be operated on an individual basis using preprogrammed scenarios, or on an interactive team basis with an instructor. Up to 64 preprogrammed scenarios may be stored within the system and displayed on the instructor control unit. Two modes are available to the instructor; one with all simulated data

and the other with live sensor data overlaid with simulated data. In addition, the instructor can select from 16 preprogrammed ocean areas and sea state and shipping density levels. All databases are stored on a removable hard disk unit to facilitate update. The gaming area covered is 512 n miles^2 with depths to 3,000 m and altitudes to 6,000 m.

The AN/SQQ-T1 simulates up to four high-fidelity acoustics simultaneously, selected from the US government-based Common Acoustic Database, or from other acoustic databases. This extensive library is expandable and can accommodate up to 40 acoustic target types. Each acoustic contact has both passive and active signatures, which can be modified by the instructor for added realism.

The set supports onboard training by providing high-fidelity simulated data separately or simultaneously to the AN/SQR-17A and AN/SQR-18A. The inputs to the AN/SQR-17A are two selectable sonobuoy channels, eight beams for the AN/SQS-26CX PEC and a DIMUS channel. In addition, the AN/SQQ-T1 provides

hydrophone group data for injection into the beamformer of the AN/SQR-18A.

The instructor interfaces with the AN/SQQ-T1 via a graphics terminal, keyboard and colour display. The set can simulate own ship data under preprogrammed scenario or instructor control, or use live own ship data. Information available to the instructor for training scenarios includes own ship course and speed and AN/SQR-18A array depth.

Operational status

The system is fitted in 'Knox' class frigates now in service in various navies around the world.

Contractor

DRS Electronic Systems Inc, a subsidiary of DRS Technologies Inc, Gaithersburg, Maryland.

UPDATED

AN/SQQ-89(V) trainer

Type

Onboard sonar trainer.

Description

The OnBoard Trainer (OBT) is designed to provide individual and team proficiency training and skill reinforment to embarked combat system personnel. The trainer uses simulated targets to stimulate the hull, towed array, sonobuoys and electronic sensors of the AN/SQQ-89 surface ship ASW combat system. The OBT allows personnel to carry out realistic training exercises both in port and while at sea.

The system uses electronic synthesis to generate realistic high-fidelity active and passive, three-dimensional moving acoustic contacts. To achieve this the OBT features front end injection and sonar system stimulation.

The trainer interfaces with the AN/SQS-53B/C active and passive hull-mounted sonar, the AN/SQR-19 tactical towed array system, the AN/SQQ-28 and LAMPS Mk III systems, and the Mk 116 fire-control system.

Realistic passive and active targets stimulate these while underway and without interfering with the ability of the sensors to detect, classify and track naturally appearing targets. The OBT provides stimulation of the LAMPS Mk III sensors with the helicopter's navigation information for heading, altitude and speed. The trainer is provided in four configurations to support the different AN/SQQ-89 configurations aboard the various ship platforms. Together with ASW training capability, the device supports training activities for AN/SLQ-32 operators, AN/ALQ-142 operators, EW supervisors and other ESM personnel.

Using an Acoustic Signal Generator (ASG), the 14E35 device trainers for the AN/SQQ-89 provide up to 20 dynamic surface and subsurface targets simultaneously in a tactical environment for sea level to 15,000 ft and an ocean depth of 50,000 ft.

The ASG simulates signatures of diesel, diesel/electric, steam, turbine, nuclear and auxiliary machinery.

Contractor

Northrop Grumman Corporation, Electrical Sensors and Systems Division, Sykesville, Maryland

Scientific Atlantic, Signal Processing Center, San Diego, California.

VERIFIED

Raytheon sonar trainers

Type

Training systems for both surface ships and submarines.

Development

The prototype DS1200 system was first used by the US Navy's submarine forces to successfully demonstrate an onboard sonar training capability for a passive subsystem. The production version, designated the DS1210, now provides ASW target training capability on 31 fleet ballistic missile submarines.

Two production DS1210 trainers, modified and designated DS1213, were later applied by US Navy submarine forces to totally integrated sonar suites. Although installed for test purposes and on a temporary basis, overall performance proved so effective that both trainers remain in service to this date aboard US Navy ships.

The Submarine Active Detection System – Transmit Group (SADS-TG) was developed and produced by Raytheon for the US Navy's advanced submarine

combat system. SADS-TG contains a fully integrated onboard trainer which supports training for all shipboard sonar arrays and provides a moving contact for fire-control team training.

In addition to the SADS-TG OBT capability, the company has developed a shore-based Multiple Array Test Set (MATS). It uses the same technology applied to the onboard trainers and will interface with the acoustic sensor group for these new submarine combat systems. MATS provides acoustic sensor stimulation and simulation of the high-frequency array, top and bottom sounders, sound velocity and noise monitoring systems and acoustic communications during system level performance and operability testing.

The Generalised Simulation/Stimulation (GSS) system was developed to facilitate the US Navy's programme of preplanned product improvements to the fast attack submarine Combat Control System (CCS). GSS necessitated development of facilities and tools to support the evolution, certification, life cycle support and training for major CCS components.

Description

Raytheon Company is a leader in advanced 'stimulation' technologies, with many trainer systems incorporating distinct simulation techniques to provide controlled, tactically 'real' training that is dynamic and highly effective.

The trainers provide operational and maintenance support to shipboard sonar and combat systems. They function at sea in port, on shore or in the air and address the training needs of surface ship, submarine and airborne platforms.

These trainers are engineered for versatility. They have successfully interfaced with the acoustic sensors of widely varying systems, both aboard multiple ship platform classes and in shore-based training facilities. They rely on advanced techniques for sonar stimulation and target simulation to generate, for both individuals and teams, the most realistic ASW training exercises available anywhere.

The entire ASW trainer series is based on established concepts of proven value. Expandable system architecture; the use of common hardware and software modules; programmable, functional partitioning of system processing – these techniques enhance realism and ensure design flexibility and manageable system upgrades in the future.

Raytheon trainers have high-fidelity, electronically synthesised active and passive contacts that are manoeuvrable, three-dimensional and ultra-realistic. The multiple targets found in systems are sensor co-ordinated. Onboard trainers feature 'front end' target injection and sonar system stimulation.

Operational status
In service. Over 40 trainers from the company's DS1200 Trainer Series, representing four system generations, have collectively logged in excess of 80,000 operating hours.

Raytheon has delivered nine GSS laboratories for use with CCS components and two trainers which allow a shore-based CCS to function as it would aboard an operating submarine.

Contractor
Raytheon Company, Electronic Systems, Portsmouth, Rhode Island.

VERIFIED

TSMT FES

Type
Shore-based trainer.

Description
TSMT FES (Trident Sonar Maintenance Trainer Front End Simulator) will offer personnel basic, intermediate and advanced instruction in sonar system troubleshooting, repair and calibration. Realism will be achieved through the injection of TSMT-generated signals into the front end of AN/BQQ-6 and AN/WLR-17 sonar systems.

The system is totally integrated and will initiate and control the exercises, providing the required signal generation, propagation processing and sonar interfacing.

The trainer has four electronically synthesised manoeuvrable, active and passive three-dimensional, realistic targets.

The system can be modified to provide personnel with operational training in addition to maintenance training. It can be further developed into a compact, onboard trainer for all submarine classes or can be used to establish a realistic fully capable standard evaluation facility.

Operational status
In service with the US Navy.

Contractor
Raytheon Company, Electronic Systems, Portsmouth, Rhode Island.

UPDATED

Device 14E35

Type
Acoustic Operator Trainer.

Description
Hughes Training Inc has produced many acoustic operator trainers in support of the sensor systems in the AN/SQQ-89 underwater combat system. These training devices have four components: the AN/SQQ-28 LAMPS Mark III sonobuoy signal processor, the AN/SQR-19 Tactical Towed Array System (TACTAS), the AN/SQS-53B hull-mounted sonar and the AN/UYQ-25 Sonar In-situ Mode Assessment System (SIMAS). The training device allows for training in the crew configuration, or the four trainee stations can be configured to one of the three sonars for individual operator training. The trainer also provides reduced capability for training in the casualty modes of the operational AN/SQQ-89 Underwater Sensor System. Later production trainers have incorporated an improved AN/SQS-53B sonar together with an interface with the Mark 116 Underwater Fire-Control System.

The trainer simulates the performance of the operational systems and can also provide a variety of exercises from simple target identification and location

Device 14E35 acoustic operator trainer

to complex tactical operations. Each device comprises an instructor station, four trainee stations, the computer system (based upon Harris Series 800 and 1200 mainframe computers with Fortran language software) and an acoustic signal generator. The acoustic signal generator is a special purpose digital signal generator utilising data pipelining and parallel processing techniques to produce the target composite digital time waveform which can generate up to 20 dynamic air, surface and subsurface targets to depths of 4,572 m

(15,000 ft) and to altitudes of 15,240 m (50,000 ft). It can also generate the signatures of diesel, diesel-electric, steam, turbine, nuclear and auxiliary machinery. Each device is configured to a class of ship, such as the 'Ticonderoga' (CG47) class cruiser, 'Spruance' (DD963) and 'Kidd' (DDG993) class destroyers, or 'Oliver Hazard Perry' (FFG7) class frigates.

Operational status
Hughes Training Inc received the first contract for these systems in 1982 configured to the 'Spruance' and 'Oliver Hazard Perry' classes. The first two units entered service with the Fleet Anti-Submarine Training Center at San Diego, California, late in 1987 and two more followed by 1990. The fifth unit was shipped to San Diego in February 1993. Two of the San Diego devices are designated 14E35C and are used to train operators of the AN/SQS-53C. Five systems have been supplied to San Diego in a programme which cost approximately US$107 million.

Contractor
Hughes Training Inc, Arlington, Texas.

VERIFIED

Device 14G1 sonar trainer

Type
Sonar operator and diagnostic basic trainer.

Description
The sonar and diagnostic trainer, Device 14G1, is designed to provide student sonar operators with familiarisation training on the AN/SQQ-89(V) sonar.

It consists of four student stations, each of which simulates the appearance and operations of the AN/UYK-21 console associated with the AN/SQQ-89(V) sonar. The students receive 6 hours of fully automated, self-paced, computer-controlled instruction. The instructor selects either individual exercises or a common exercise for all the students. Each includes recorded narrations and visual presentations of the selected exercise.

Operational status
In service with the US Navy.

Contractor
EMS Development Corporation, Farmingdale, New York.

VERIFIED

Hughes Training Inc – sonar trainers

Type
Range of sonar operator trainers.

Description
The former Honeywell (now Hughes Training Inc) designed and manufactured a number of training systems intended for use by sonar operators and teams. Each system is dedicated to a particular sonar equipment; 14E36 for the AN/SQR-18A, 14E19 for the AN/SQS-26CX, 14E23 for the AN/SQS-35, 14E24 for the AN/SQQ-23 and 14E25 for the AN/SQS-53.

Depending on the operational system, the student stations are either actual sonar consoles or facsimiles. In either case, video and acoustic signals are fed to the student stations. The underwater acoustic environment is modelled in all operator trainers. Simulated sonar propagation is generated based on ocean variables such as regional seasonal velocity profiles, layer depth and sea state as well as bottom type, depth and slope. These simulations use complex algorithms to account for the effects of propagation loss, reverberation, surface duct, time delays, convergence zones, multipath returns and shallow water operation.

The instructor consoles are used to set up problem

exercises, to monitor student performance and to modify the tactical situation in real time. The instructor is also able to insert malfunctions and replay the exercise.

Operational status
In service with the US Navy.

Contractor
Hughes Training Inc, Arlington, Texas.

VERIFIED

Passive acoustic analysis trainers

Type

Analysis and classification trainers for sonar operators.

Description

The US Navy Passive Acoustic Analysis trainers (Devices 14E40 and 21H14) are designed for use by sonar operators for both surface vessels and submarines. They are used for training in analysis and classification of various surface ship, submarine and commercial ship contacts, both aurally and visually. The displays include Lofargram, Demon and Vernier frequency data, together with a student worksheet. The displays are time synchronised with raw and processed audio signal presentation. The worksheet data entered by each student provide a record of performance for training feedback and course grading.

The trainer generates realistic sonar audio and visual displays by preprocessing user-supplied acoustic analogue tape recordings from a variety of user-deployed tactical sonar systems. The preprocessing facility produces VHS tape cassettes for audio signals and computer disk files of synchronised generic visual displays for replay on the trainer. The trainer includes hardware for up to 12 student stations, a single instructor station and a preprocessing facility.

Operational status

In service with the US Navy.

Contractor

Ship Analytics, North Stonington, Connecticut.

VERIFIED

Two Passive Acoustic Analysis (PAA) trainer student displays

Sonar simulator

Type

Sonar and radar simulator.

Description

The sonar simulator is a generic sonar and radar simulator that may be configured to represent specific displays such as the AN/SPA-25F or the Canadian Marconi LN-66. The unit duplicates all the functions of the real display and can also simulate an Automatic Radar Plotting Aid (ARPA) including the touchscreen technology found in many modern shipboard ARPA radars. The simulator can be easily configured to simulate other specific radar units.

The system consists of a simulation unit based on a personal computer, a display unit and a date acquisition subsystem with controls. The simulator may be configured as either an air search or a sea search unit and can display tactical contacts, weather, chaff, sea return, jamming and landmass features. The simulator sonar is active, operating in either a steerable sector or an omni mode. The radar simulator can be interfaced with any host simulator.

Operational status

The simulator is used in Ship Analytics' full mission ship handling simulator systems and action speed tactical trainers. Over 70 units are currently in operation throughout the world for both military and commercial applications.

Contractor

Ship Analytics, North Stonington, Connecticut.

UPDATED

A radar simulator in an ARPA configuration
1995

Rockwell onboard trainer

Type

Onboard sonar training system.

Description

Rockwell has designed an onboard trainer, based on a trainer developed for the US Navy, which simulates sonar hydrophones and array geometry before beam-forming. It is therefore not affected by changes to hardware or software in the processing or display part of the system. The trainer can operate either under way with the contacts generated by the trainer overlaid on the environmental background noise from the array, or in harbour with the trainer generating independent background noise to simulate at-sea conditions.

The trainer design incorporates commercial VME-based processors programmed with software to provide maximum flexibility for adaptation to various sonar and array conditions, growth features which allow for multiple contacts to satisfy the most demanding training scenarios and programmable contact signature generation to allow the user to create a library of specific contacts.

The Rockwell trainer can be tailored to meet the most basic or the most complex training needs of the customer.

Operational status

In service.

Contractor

Rockwell International Corporation, Autonetics Marine Systems Division, Anaheim, California.

VERIFIED

TARGETS

FRANCE

STAR

Type
Submarine training target for torpedoes.

Description
STAR (an acronym for Submarine TARget) is an artificial towed target designed to replace the use of submarines as torpedo targets. It simulates the acoustic behaviour of a real submarine when under attack from an active torpedo. It has been developed for both the training of naval forces and for the operational evaluation of torpedoes. It can also be used for the technical qualification of new torpedoes, when fitted with specific components, as well as towed decoy under certain conditions.

STAR consists of two main parts, the submerged part and the onboard installation on a surface ship. The submerged part comprises five parts:
(a) the antenna target made up of a linear chain of between 5 and 15 acoustic repeater modules and a tail module trailed behind a towfish
(b) the underwater body (fish) containing the electronics, power supplies, navigation sensors and external trajectography sensors
(c) a passive hydrodynamic depressor, weighing about 350 kg in water, which is used to obtain the desired immersion since the antenna, tether cable and fish have a slightly positive buoyancy
(d) a tether cable of approximately 400 m connecting the fish to the depressor
(e) a towing cable of approximately 600 m linking the depressor to the ship.

The onboard installation consists of a 1 tonne hydraulic winch, a 15 kW hydraulic power unit and an electronics rack containing the control, checking, visualisation and recording instruments.

STAR is a rugged system which can be launched in up to Sea States 3 or 4 and will remain operational in up to Sea State 5. The target can be towed at speeds from 3 to 20 kt and at depths varying from 20 to 400 m, the depth being set up by the paid out length of the cable. In the event of cable breakage, the tail module ensures hydrodynamic stability of the antenna, as well as its recovery.

Sensors in the fish transmit the heading and depth to the surface ship. The fish can also be equipped with specific sensors to match any external trajectography requirements. When the target antenna is partially or totally illuminated by the torpedo sonar emission, each of the repeater modules retransmits the signal received at the appropriate level, to simulate the acoustic return from a submarine. The simulation operates within a wide frequency band to protect the confidentiality of the torpedo's exact frequency. The torpedo signals received and transmitted by each module are sent via a multiplex cable and are then recorded on board.

Operational status
In service.

Contractor
Société ECA, Boulogne-Billancourt, (design and manufacture).
Safare Crouzet (acoustic components), Nice.

UPDATED

Calas

Type
ASW training target vehicle.

Description
Calas is designed to provide realistic submarine simulation for the training of active and passive sonar operators. It is a reusable torpedo shaped vehicle which tows a small hydrophone and which is easy to use and requires no maintenance. The vehicle is easy to program using a laptop computer interface. Signature transmission is one to three lines within the 300 to 800 Hz frequency band. It offers the simultaneous processing of two active sonars within the frequency band 4 to 33 kHz.

Launch from the surface platform is automatic and the vehicle, towed by an 18 m cable, has an endurance of up to 2 hours. Pinger functions are incorporated for location by a tracking range.

Calas is designed to simulate a submarine's behaviour both in manoeuvring and acoustic signature.

Operational status
In service with the French navy.

Specifications
Length: 1.85 m
Diameter: 150 mm
Weight: 30 kg (in air)
Speed: 7, 10, 12 kt
Operating depth: 20-300 m

Contractor
Thomson Marconi Sonar SAS, Sophia Antipolis.
Constructions Industrielles de la Mediterranée (CNIM), La Seyne sur Mer.

UPDATED

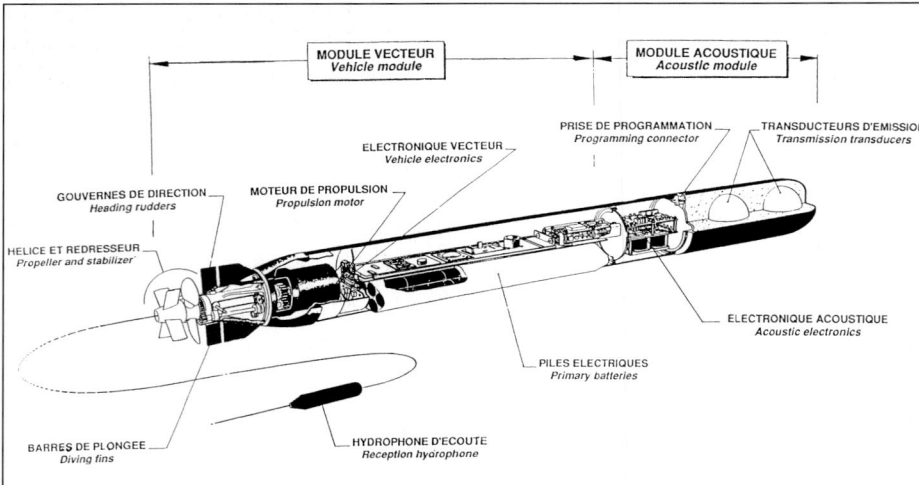

MODULE VECTEUR *Vehicle module* — MODULE ACOUSTIQUE *Acoustic module*
GOUVERNES DE DIRECTION *Heading rudders*; MOTEUR DE PROPULSION *Propulsion motor*; PRISE DE PROGRAMMATION *Programming connector*; TRANSDUCTEURS D'EMISSION *Transmission transducers*; ELECTRONIQUE VECTEUR *Vehicle electronics*; HELICE ET REDRESSEUR *Propeller and stabilizer*; ELECTRONIQUE ACOUSTIQUE *Acoustic electronics*; PILES ELECTRIQUES *Primary batteries*; BARRES DE PLONGEE *Diving fins*; HYDROPHONE D'ECOUTE *Reception hydrophone*

Calas reusable submarine target
1998/0009863

GERMANY

Subtas

Type
Submarine target simulator.

Description
Subtas is designed for fleet Anti-Submarine Warfare (ASW) training in the open ocean and can replace the need for an instrumented range.

The vehicle operates in preselected run patterns designed according to customer requirements, simulating the dynamic behaviour of a submarine and its acoustic characteristics. A variety of run patterns is available for selection prior to launching the vehicle. The predicted run can be monitored on the control unit aboard the launch vessel. Preset operational characteristics include prerun parameters, run pattern, speed, depth limits, run time, acoustic mode/levels, carrier course/speed and latitude/longitude.

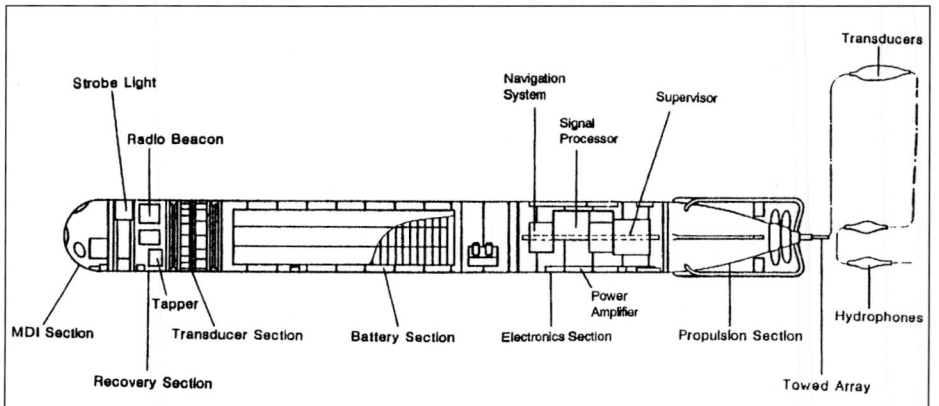

Strobe Light; Radio Beacon; Navigation System; Signal Processor; Supervisor; Transducers; Tapper; MDI Section; Transducer Section; Battery Section; Electronics Section; Propulsion Section; Power Amplifier; Hydrophones; Recovery Section; Towed Array

Block schematic of Subtas
1996

Recovery of Subtas unit (L-3 Communications ELAC)   *1999*/0038855

Three microprocessor-controlled main units supervise the entire system: a navigation computer; an acoustic computer; and a supervisor unit. The navigation computer is a guidance and control system which uses the latest solid-state digital technology. The computer-controlled acoustic system is especially designed to operate in both passive as well as active mode. It provides sophisticated true simulation characteristics which are detectable by all types of sonar systems, including homing heads of torpedoes, dipping sonars and sonobuoys. The acoustic performance is achieved by sensors and underwater sound projectors (formed from two transducer rings) situated both on the vehicle itself and within a 40 m towed array (in which five sound projectors are mounted). The sound projectors simulate a submarine's low-frequency propulsion noise and sonar echoes (the LF section provides up to six tonals between 200 to 400 Hz and 5 to 60 kHz). Four hydrophones are mounted in the head of Subtas with two more mounted at the end of the array. These latter hydrophones provide a miss-distance indication system. The main computer controls the overall performance of the system and its functionality. All guidance and control functions are software supported allowing subsequent alteration as required.

The hull and propulsion system of Subtas are taken from the Mk 37 torpedo. To reduce acquisition costs and enable multiple recharging, a nickel cadmium battery has been installed instead of the silver-zinc

Subtas   *1998*/0009864

Subtas launch   *1998*/0009865

batteries of the Mk 37, which can only be recharged a limited number of times. Recharging is carried out on deck without the need to remove the batteries.

Several comprehensive test functions are incorporated in all systems, providing prelaunch tests as well as continuous failure monitoring of all systems during the run which is carried out by several BITE functions.

Operational status
The system is in production and 5 units were delivered to the German Navy at the end of 1997. The system is also available for export.

Specifications
Length: 5.2 m
Diameter: 483 mm
Preset speeds: 6-8 or 10-13 kt
Operating depth: 300 m
Operation time: 2.5 h

Contractor
L-3 Communications ELAC Nautik GmbH, Kiel.

UPDATED

ITALY

BSS

Type
Underwater mobile target.

Description
The BSS has been conceived to provide a high-fidelity emulation of the real dynamics and acoustic signatures of all types of submarines. It is a self-propelled, multispeed, deep running active and/or passive acoustic target, suitable for any type of acoustic torpedo. The BSS is derived from the A184 exercise torpedo and tows an advanced acoustic array tail providing physical dimensions equivalent to the emulated echo. The system comprises the towed array cannister and deployment system and a 49 m towed array consisting of 16 independent transducers. The propulsion system consists of two contrarotating

propellers, a three-speed DC motor, speed commutator and a servo system to contol the fins. Power is derived from rechargeable silver oxide/zinc batteries rated at 100 kW. The acoustic control and target guidance electronics use microprocessor-based technology and the synchronised acoustic transmitter is used for three-dimensional underwater tracking range. The target can perform a wide selection of preprogrammed patterns and automatically initiates evasive manoeuvres when under torpedo attack. Remote-control is exercised by means of an acoustic underwater telephone link.

The system is designed for use in ASW training scenarios, for advanced operational evaluation, development proving trials of weapons, sea acceptance trials and the warstock surveillance of torpedoes.

Operational status
In service with the Italian Navy. Being used for the operational evaluation of the MU90 torpedo.

Specifications
Length: 6,085 mm
Diameter: 534 mm
Weight: 1,122 kg
Speed: 14, 21 or 28 kt
Endurance: 7, 15 or 30 min
Operating depth: 10-400 m

Contractor
Whitehead Alenia Sistemi Subacquei SpA, Livorno.

UPDATED

CTS 105B

Type
Floating artificial target for torpedoes and sonar.

Description
The CTS 105B is an unmanned floating artificial acoustic target designed for fleet exercise trials and ASW training. It was conceived to provide a high-fidelity emulation of the acoustic signatures of real submarines and is suitable for any type of acoustic torpedo and the majority of sonars.

It consists of a submerged acoustic unit housing the sonar projector, the torpedo projector, hydrophone and power supply; a floating cylindrical unit with underwater connection cable, winch, remote VHF radio control link; and the remote-control console housing the power supply, finer-touch MMI, recording system and VHF radio-control system.

The target is able to emulate target strength, submarine length, submarine high lights, target Doppler, coherent range rate and target radiated noise.

Contractor
Whitehead Alenia Sistemi Subacquei SpA, Livorno.

UPDATED

CTS 106

Type
Dual-purpose artificial target for torpedoes.

Description
The CTS 106 is a stationary/towed manually controlled artificial acoustic target suitable for any type of acoustic torpedo and designed to provide a high-fidelity emulation of the real echo of all types of submarine. Applications include ASW training, operational evaluation, sea acceptance trials and war stock surveillance of torpedoes.

The target consists of a submerged body of fusiform shape with stabilising ring, incorporating the projector, the towed hydrophone and power supply, an armoured connecting cable for towing and for transmitting electrical communications, a handling winch and the control console housing the electronics rack, processing units, man/machine interface and graphic recorder.

The vehicle operates on a frequency between 15 to 60 kHz, with Doppler, -5 dB to + 15 dB and uses a time/frequency code which can be received by a 3-D tracking range. Maximum towing speed is 16 kt with an operating depth of 20 to 50 m. In stationary mode the vehicle can go down to 200 m.

Operational status
In service with the Italian navy.

Contractor
Whitehead Alenia Sistemi Subacquei SpA, Livorno.

UPDATED

UNITED KINGDOM

Type 2058/SoundTrak series

Type
Sonar target simulators.

Description
The Type 2058/SoundTrak family of acoustic target simulators is designed to be used with surface ships, submarines and aircraft in ASW training, system performance assessment and other operational applications, which require a versatile, acoustic source.

The system consists of a towed body, containing the electro-acoustic transducers, which is towed behind a surface ship and is connected to an inboard electronics rack containing the signal generation and drive circuits.

The system is able to produce realistic acoustic target signatures, transmitting simultaneously a mixture of narrowband tones and broadband noise.

The nature of the output, in terms of narrowband frequencies, noise bands, signal amplitudes, modulation, can be programmed manually and the system can store up to eight preprogrammed pages of data. Additionally, the system can transmit acoustic signatures which have been previously recorded or which have been generated using large-scale computer generated simulations.

SoundTrak
This system features exceptional low-frequency performance in relation to its very small physical size and has been designed to operate with ships' and submarines' passive sonar systems.

SoundTrak II
In addition to low-frequency performance, this system includes both narrowband and broadband high-frequency signals and is able to echo-repeat signals received from ships' or submarines' active sonars.

Systems can be supplied to specific requirements either as an enclosed body capable of being towed at high speed, or as an open frame structure.

The SoundTrak target simulators reduce the need to deploy ships and submarines as targets in exercise roles. They may also be used in applications which include acoustic countermeasures, own signature confusion, force countermeasures/decoys and so on.

Type 2058 towed sonar source

The reliability of the acoustic source, provided by the SoundTrak target simulators, is also well suited to assessment of the performance of sonar systems, sonobuoys and noise ranges.

The SoundTrak target simulators can be deployed from a variety of vessels using standard winch systems.

Operational status
SoundTrak is currently in service with the Royal Navy (as Sonar 2058) and other European (French) and Asian navies. SoundTrak II is also currently in service in Asia.

Contractor
PMES Ltd, Rugeley.

UPDATED

RASAT sonar target

Type
Underwater sonar target.

Description
RASAT (Radar And Sonar Alignment Target) is a programmable, self-contained transponder sonar target which is compatible with a wide range of modern, high-technology sonars, including hull-mounted and dipping types. It can be used for the alignment of surface ship or helicopter systems, can act both as a transponder for active sonar and as a CW noise source for passive sonar and it can be deployed from a ship or helicopter. Operating depth is selectable from 15 to 50 m and bandwidth is 3 to 12.5 kHz for the basic model; other operating depths and bandwidths are available.

Operational status
Many systems in service with the Royal Navy and other Asian navies.

Contractor
Ultra Electronics, Ocean Systems, Weymouth.

VERIFIED

RASAT sonar target

SLUTT type G 733

Type
Programmable Ship-Launched Underwater Transponder Target (SLUTT).

Description
SLUTT is designed to simulate underwater targets enabling both active and passive functional testing of a wide range of surface ship, submarine and helicopter sonar systems checking sonar range and bearing accuracies. The transponder can also be used for sonar/radar alignment checks and operator training and as a sea acceptance trials target.

It features user-programmable sonar parameters, multiple output signals in FM or CW for realistic sonar simulation, provides a noise source for checking passive sonars and may be used as a transponder or a free-running pinger.

SLUTT is a self-contained unit suspended at depths of between 10 and 120 m beneath a free-floating buoy fitted with a radar reflector. This method of deployment allows interrogation of the transponder by both surface duct and variable depth sonars. A towed version can be supplied where accurate Doppler measurements are required.

All incoming acoustic signals are monitored and SLUTT responds to signals which have the correct frequency and continuity. A suitable time-simulated output echo pulse is then transmitted. SLUTT is immune to the reverberations set up by the transponder pulse and to incorrect inputs. SLUTT simulates a typical beam aspect submarine target or an incoming torpedo, by providing suitable FM, CW and pulsed noise outputs to enable both active and passive sonars to be checked for range and bearing accuracy against relevant radar readings.

The system is provided with a battery charger, hand-held programming unit and lead, handbook and stowage case.

Operational status
In service use with the Royal Navy and many other navies.

Contractor
Graseby Dynamics Ltd, Watford.

VERIFIED

Topat

Type
Towed passive torpedo target.

Description
The ultimate capability of a torpedo can only be assessed by conducting in-water trials and exercises against realistic targets. The availability of suitable ships and submarines for these trials is severely limited. Free-swimming target vehicles, on the other hand, have limited endurance, little real-time command and control and may not be cost-effective.

The Topat system is designed to overcome these disadvantages. It provides a simple, adaptable set of equipment which will meet torpedo trials requirements.

The system emulates the acoustic returns expected from real targets. This is achieved by placing two acoustic reflectors in the same relative positions and of the same strength, as the expected acoustic highlights from a real target. Two towed bodies carry the acoustic reflectors to obtain the required towing configuration and provide a 'moving target' capability.

The bodies can be deployed and recovered from a range of towing vessels, including stern trawlers, without any need for specialised handling equipment.

Operational status
In service with the Royal Navy.

Contractor
BAeSEMA, Bristol.

VERIFIED

Topat passive torpedo target

Subtrack

Type
Sonar target system.

Description
Subtrack is designed for training sonar operators and complete command teams using passive and active sonars onboard ships, submarines, helicopters and maritime patrol aircraft. It can also be used to prove sonar systems and aid new developments, prove new installations and post refit installations for all types of sonars – including hull mounted arrays, towed arrays, flank arrays, intercept sonars and active sonars. In addition, it has applications in mine warfare – as part of an influence minesweeping system, including a very low-frequency sound source producing outputs from as low as 4 Hz at high power. Other applications include acoustic and oceanographic research, range calibration and pop-up targets on tactical ranges.

Subtrack essentially comprises the following system components: a low-frequency acoustic projector (hydrosounder) to generate frequencies from <30 Hz to as high as 3 kHz; additional transducers to operate from 3 up to 100 kHz, or higher; a computer controlled signal generator to produce multiple narrow band spectra, broadband noise, modulation and transient effects for use with passive sonars; software and processing to produce an echo repeat facility for use with active sonars; power amplifiers to drive each transducer; and acoustic monitoring and measurement facilities. Optional equipment such as a hydrodynamic towfish, towing cable, winch and so on may also be added depending on the method of deployment.

Subtrack may be supplied in the following configurations: towed system for deployment from COOP; surface ship hull-mounted system; submarine hull-mounted system to augment the submarine signature, thus providing a submarine target with variable acoustic components; minesweeping variant for towing configuration below an otter, or hull-mounted in a drone; statically deployed system for research applications; seabed mounted systems for use on underway and tactical ranges; free swimming systems, battery operated, deployed below a buoy to provide a single or multistatic sound source.

The low-frequency capability and output source level have recently been enhanced by the introduction of a new transducer. Hydrosounder model UW600 is able to generate output signals at frequencies as low as 4 Hz, with source levels as high as 190 dB re 1µP @ 1 m.

Typical Subtrack towfish

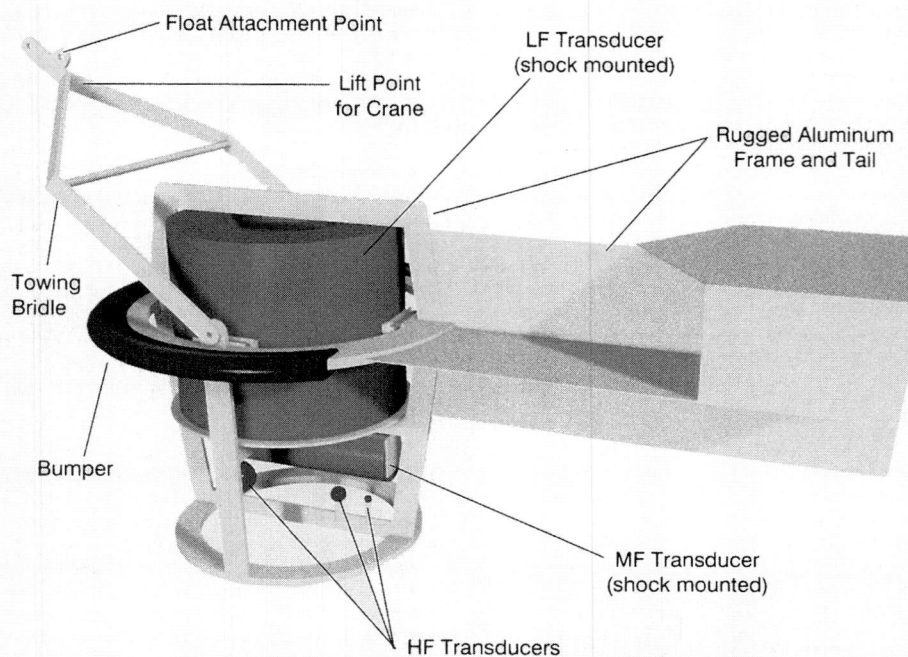

Typical minesweeping towfish   *1998*/0009868

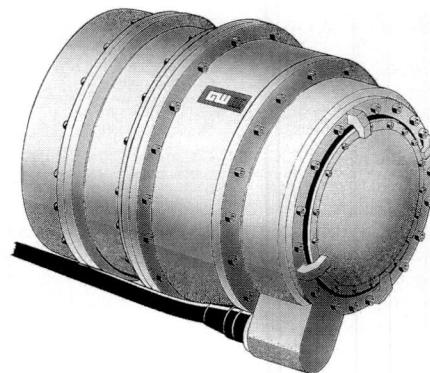

Model UW600 hydrosounder   *1998*/0009867

Variants of the UW600 include a shock rated device for minesweeping activities, as part of an influence sweep system.

Operational status
Variants of Subtrack systems are in use with the navies of 10 countries, throughout the world. Subtrack systems are also specified by many defence associated research agencies.

Contractor
Gearing & Watson Electronics Ltd, Hailsham.

VERIFIED

THOR torpedo simulation

Type
Torpedo weapon system assessment system.

Description
THOR is a generic, multi-application Monte Carlo simulation model for the statistical evaluation of torpedo weapon system performance and the assessment of anti-torpedo and anti-submarine countermeasure effectiveness.

THOR may simulate an engagement involving attacking platforms which fire torpedoes and target platforms which may become alerted to the threat weapon and attempt to evade by manoeuvres and/or by the employment of countermeasures. The attacking platforms may be submarines, surface vessels or aircraft in any number or combination. Target detection, closure, TMA refinement, weapon launch and guidance are simulated. Torpedo firing on the target vessels may be in single shot or salvo, involving wire-guided or unguided acoustic homing torpedoes, other homing torpedoes, or simple straight/pattern runners. There is a capability to simulate the deployment of missiles with a lightweight torpedo payload, the direct deployment of lightweight torpedoes from aircraft and tube launch from submarines and surface ships.

THOR incorporates the following representations:
(a) platform sensors – active and passive hull-mounted sonar, towed array, sonobuoy and dipping sonar systems for target or torpedo detection
(b) weapon sensors – active and passive sonar and other sensors
(c) combat information systems – data fusion, target or threat torpedo tracking, weapon and countermeasure targeting and guidance and tactical decision making
(d) acoustic countermeasures – towed, thrown and directly launched (static or mobile) jammers and decoys operating at weapon or platform frequencies
(e) other anti-torpedo 'soft-kill' countermeasures
(f) 'hard-kill' countermeasures
(g) physical characteristics – broadband and narrowband radiated noise, target strength, manoeuvrability and device deployment error and limitations
(h) environment and interference effects, including masking.

THOR has been developed as a combined continuous and discrete event-based simulation model and has been designed in accordance with rigorous programming standards and development methodology to form a fully self-consistent and robust product. The facilities are currently hosted on the VAX range of mini- and super mini-computers and operate under VMS. The software is programmed largely in ANSI-standard Fortran 77, but with some VAX specific management and graphics functions to enhance the user interface.

THOR has an established and growing user base currently comprising government research establishments, defence contractors and navies in the UK and in other NATO countries. Its applications are wide-ranging from current system tactical development to the assessment of desirable options for future systems.

Operational status
THOR was originally developed under contract to the UK Admiralty Research Establishment, and early versions of it have been supplied to UK industry and to US defence establishments. Also supplied to the French government.

Contractor
EDS Defence Ltd, Hook.

UPDATED

UNITED STATES OF AMERICA

Mk 38

Type
Miniature mobile target.

Description
The Mk 38 Miniature Mobile Target (MMT) is a self-propelled submersible acoustic transponder. It is normally launched over the side of a surface ship and glides to its preset operating depth, where the propulsion motor turns on. The MMT maintains its operating depth within ±60 ft and travels at a speed of 4 kt on a random course. The acoustic receiver and signal processor is programmed to recognise different acoustic interrogation frequencies and will respond with either an acoustic CW pulse or an FM pulse at the interrogation frequencies desired by the user. During periods when the target is not interrogated, it transmits two CW tones simultaneously. The MMT can be used as a training device with any underwater acoustic detection system which has a transmission frequency within the receiving frequency band of the target. The MMT can also be used as a target for the Mk 46 torpedo.

Specifications
Length: 129.5 cm
Diameter: 8.9 cm
Weight: 7.7 kg
Speed: 4 kt
Endurance: 2-3 h
Operating depth: 30 and 106 m

Contractor
GEC-Marconi Hazeltine Corporation, Greenlawn, New York.

UPDATED

Mk 39 EMATT

Type
Expendable Mobile ASW Training Target (EMATT).

Description
The Mk 39 EMATT is easily deployed from surface ships or ASW aircraft. EMATT offers ASW forces the opportunity to practise the complete ASW problem from detection to attack in the open ocean in any area of operations.

The capability for launch from ASW aircraft gives EMATT an added dimension previously lacking in ASW training. The Mk 39 also has the capability to simultaneously interact with all sonars, towed arrays, dipping sonars and sonobuoys, both active and passive. Operational commanders have the opportunity to conduct co-ordinated exercises with several different air and surface ASW platforms.

EMATT contributes to fleet readiness by:

(a) increasing ASW tactical proficiency in anticipation of actual submarine operations

(b) providing a means to complete readiness exercises when submarines or trainers are not available, especially on deployment

(c) being compatible with most major acoustic underwater ranges

(d) giving operational commanders the option to conduct ASW training anytime, anywhere, under demanding, at sea conditions.

EMATT is a versatile all-weather, day and night target. The new upgraded Mk 39 is the same size as a standard 36 in sonobuoy and may be loaded in any sonobuoy chute for deployment. For surface ships, it may be slipped over the side by hand. Additional features include field programmability, variable tonal level, variable speed, integrated pinger and autonomous manoeuvre.

These features enhance ASW training at all levels for both surface ship and air crews.

Prior to launch, one of three preprogrammed run geometries may be selected. Each run may be set for operation with or without the MAD capability. After deployment, the vehicle runs for up to 3 hours or more depending on the speed selected. Acoustically, it generates four discrete frequencies. It has a unique echo repeat system which receives, stores, then retransmits active sonar signals enhanced to simulate the echo from an actual submarine. It is also compatible with the Mk 46 torpedo to provide acquisition and homing response.

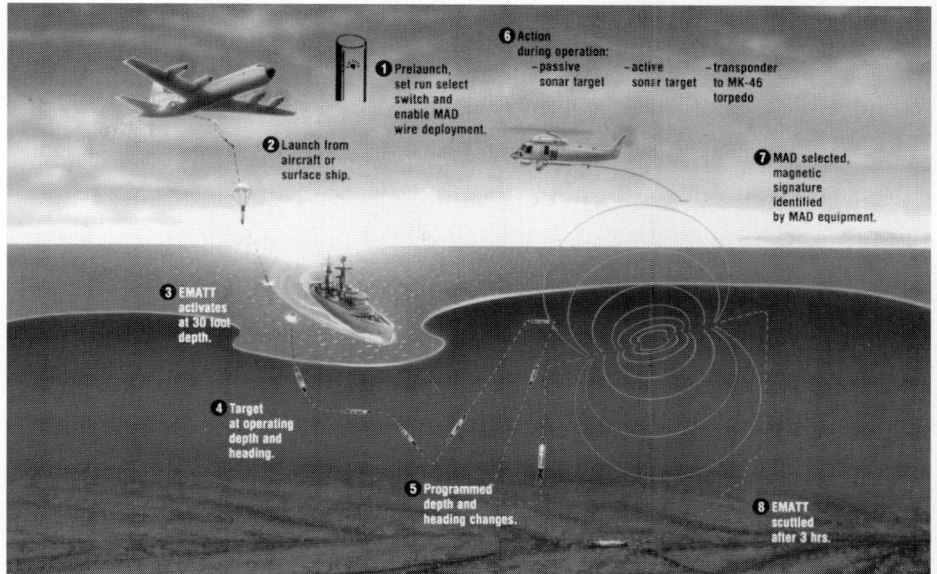

Sequence of operation of Mk 39 EMATT

For use with MAD-equipped aircraft, EMATT deploys a 100 ft wire through which DC power is pulsed to create a magnetic field around the vehicle. The signature is unique and provides excellent tactical training for aircrews.

Operational status

Over 7,000 Mk 39s have been produced for the US and foreign navies. The target has been purchased by 17 nations and navies in service worldwide.

Specifications
Length: 900 mm
Diameter: 122 mm
Weight: 10 kg
Operating depth: 23 – 183 m
Operating time: 3 hours

Contractor
Sippican Inc, Marion, Massachusetts.

UPDATED

SPAT

Type
Self-Propelled Acoustic Target (SPAT).

Description
The SPAT underwater vehicle simulates the acoustic and dynamic characteristics of a diesel electric submarine responding to torpedo and sonar activity to provide a realistic training target for surface ships, submarines, helicopters, fixed-wing aircraft and acoustic homing torpedoes.

Among the features incorporated in SPAT are programmable acoustic signature (broad band and narrow band, tonals) and vehicle dynamics (flow noise and propeller noise), together with training levels, all of which can be prepared in advance on any PC. The transmission response features various types of signal including FM, swept FM and pulse coded. The outputs vary according to the speed of the vehicle to provide a realistic response. A true echo repeater is incorporated with programmable characteristics. The vehicle is able to multiple incoming sonar signals. Course plots can be generated for run verification. An optional MAD capability is available.

SPAT contains an eight-hour solid state data recorder with 14 data channels, active acoustic analysis, three programmable band pass filters and programmable thresholds. Acoustic activity and vehicle data versus time is recorded together with automatic run identification showing date, time and vehicle serial number and vehicle status shutdown. It has a graphics output for easy post run analysis.

The vehicle can be deployed from the deck (gravity slide, automatic or manual), submarine (torpedo tube launch) or helicopter deployment and incorporates a portable, lightweight fire control unit. It is compatible with most ranges and incorporates retrieval aids for use in open water.

Operational status
Three units have been ordered by the UK Royal Navy, and over 60 other units are in service around the world..

Specifications
Length: 3,937 mm
Diameter: 254 mm
Weight: 181.5 kg
Speed: 4 to 12 kt
Operating depth: 3-300 m

Operating time: 2-6 h
Target strength: -12 to 21 dB
Output (max): +185 dB/1μPa/m
Dynamic range: 60 dB
Broadband
Frequency: two bands (user defined) range in each band 1 to 85 kHz
Levels: band 1 – 90 to 130 dB/1μPa/m at lowest frequemcy; band 2 – 8 0to 125 dB/1μPa/m lowest frequency
Narrowband
Frequency: 100 Hz to 2 kHz
Number of tones: 6
Tonal level: 100 to 150 dB/1μPa/m

Contractor
Northrop Grumman Corporation, Electronic Sensors and Systems Division, Cleveland, Ohio.

NEW ENTRY

GENERAL NAVIGATION EQUIPMENT

FRANCE

NUBS-8A/NUUS-8A

Type
Navigation echo-sounder.

Description
This navigation echo-sounder is designed to measure the distance of the seabed, on a vertical line beneath the ship for normal navigation purposes.

The sounder provides a graphic recording of the seabed and is equipped with a minimum depth alarm device.

The difference between the two versions lies in the type of transducer used.

Maximum range is 1,400 m.

Operational status
In service with the French Navy.

Contractor
Safare Crouzet, Nice.

VERIFIED

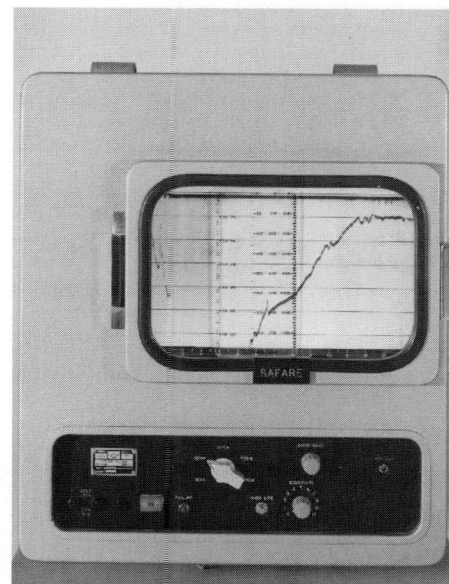

Control panel and display of the NUBS-8A/NUUS-8A

Minicin

Type
Submarine inertial navigation system.

Description
Minicin has been developed for submarines requiring high-accuracy position and attitude information. It was therefore designed to provide continuous heading and position (latitude and longitude) data for navigation purposes as well as attitude (roll and pitch) and velocities (north, east and vertical) readings required as inputs by modern combat and weapon systems. Each system consists of: an inertial platform with its adjustable support; an electronic unit; and a remote-control unit.

The inertial platform comprises three gimbals. It is fitted with three dry gyros (each having two degrees of freedom) and two high-accuracy accelerometers. The essential feature of the system is the 'gyro monitor', an effective continuous gyro drift supervisor.

The inertial system's electronic package is composed of a set of replaceable cards and modules. It uses standardised components designed to meet military standards and specifications and relies heavily on efficient microprocessors. The remote-control unit provides the interface between the operator and the system. This element, which can be installed at the most appropriate point in the vessel, provides the operator with a simple yet effective means of dialogue (use of menus and function keys) and provides all the resources required for system initialisation, recalibration, monitoring and hardware status, display functions and so on.

Operational status
Minicin is currently in service on numerous vessels in the French Navy, including nuclear-powered attack submarines, the *Charles de Gaulle* nuclear-powered carrier (three systems), anti-aircraft and anti-submarine frigates and destroyers, aircraft carriers and so on. Units are also operational on board different foreign navy submarines including the new 'Agosta 90 B' of the Pakistan Navy.

Contractor
SAGEM SA, Defence and Security Division, Paris.

UPDATED

SLINS

Type
Submarine ring Laser gyro Inertial Navigation System.

Description
SLINS has been developed for submarines requiring accurate position and attitude information together with a high level of data availability. The system is based on a dual SIGMA 40 INS configuration, mounted in a single cabinet and operating in parallel to measure the same heading, roll, pitch, acceleration, velocity and position. Two independent navigation algorithms (KALMAN filters) continuously cross-check the data measured by each SIGMA 40 INS, so as to identify and calibrate most of the inertial errors of the Ring Laser Gyroscopes (RLG) and accelerometers while on the move. Position accuracy is better than 1 n mile when the system position is updated every 24 hours by means of an external position reference such as GPS.

The strapdown architecture of the RLG-based SIGMA INS allows the system to deliver, from a very compact equipment, the following redundant data:
(a) heading, roll and pitch, as well as their angular derivatives
(b) position, speed and heave
(c) accelerations
(d) status and accuracy estimates.

The equipment is easy to install and maintain on board any type of submarine. It is designed to be fully integrated into the combat system via an ethernet bus and is remotely controlled from a multifunction common console.

Operational status
As of April 1997, the SIGMA 40 INS has been in service in the French Navy and three foreign navies. SLINS is now available.

Contractor
SAGEM SA, Defence and Security Division, Paris.

VERIFIED

GPS 40-18 receiver

Type
Global Positioning System (GPS) receiver.

Description
The GPS 40-18 receiver is an all-in-view, parallel tracking, 18 channel receiver, with 12 channels for the P(Y) mode. It is a fast acquisition receiver which provides navigation data through the combination of GPS pseudo-ranges, attitude angles and log speed, achieved by Kalman filter. It has a high resistance to jamming and a RAIM function.

All navigation parameters appear clearly to the user who can easily select a large number of data screens. 99 waypoints can be stored together with 10 routes consisting of up to 20 waypoints each. The system is capable of operating either in absolute SPS (Standard Positioning Service) or PPS (Precise Positioning Service) mode or in differential SPS mode.

Operational status
Operational on board surface ships of foreign navies.

Contractor
SAGEM SA, Defence and Security Division, Paris.

UPDATED

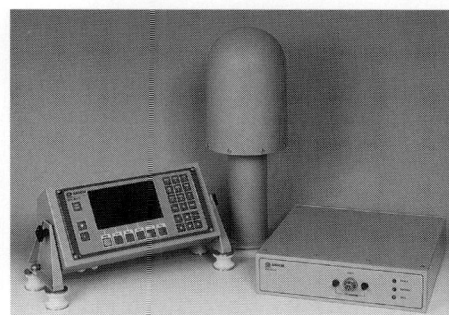

GPS 40-18 receiver, naval P(Y) code GPS navigator
1995

SIGMA 40 INS

Type
Inertial Navigation System (INS).

Description
The SIGMA 40 is an INS based on Ring Laser Gyro (RLG) technology, designed for naval applications. Its high level of performance meets both navigation and combat system requirements on board submarines and unmanned underwater vehicles.

The strapdown platform architecture used, allows the SIGMA 40 to provide: heading, roll, pitch and their angular rates; position and heave; horizontal/vertical velocity and acceleration.

The compact SIGMA 40 is rugged, sea proven, and uses software and hardware which allow the inertial navigation unit to interface with the Doppler log and GPS receivers and to deliver even more accurate navigation data. It is easy to install, operate and maintain.

Operational status
In service with French and foreign navies including the Norwegian Navy.

Contractor
SAGEM SA, Defence and Security Division, Paris.

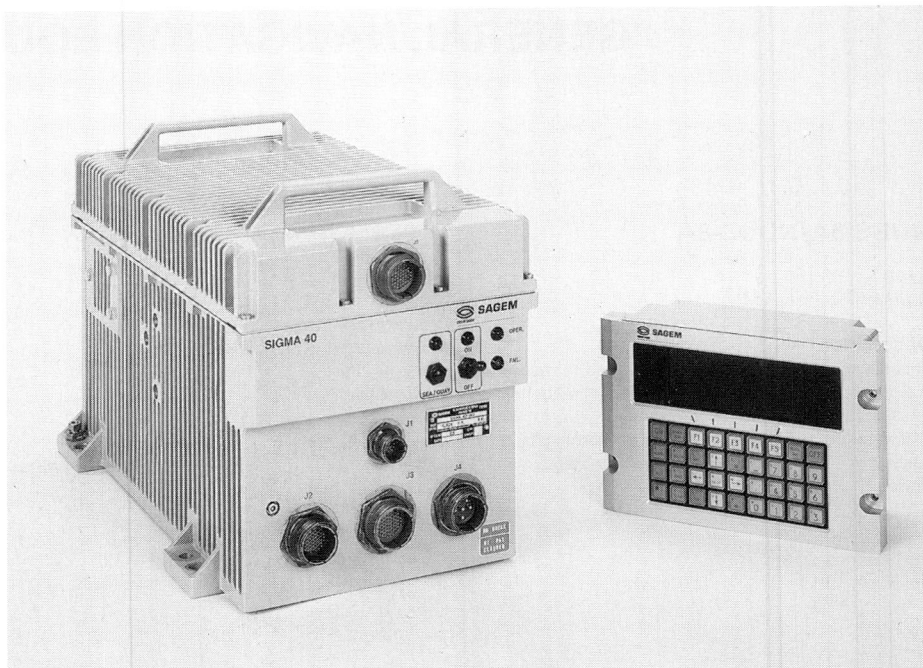

UPDATED   *SIGMA 40 Ring Laser Gyro (RLG) navigation system*   *1997*

TSM 5750/TSM 5740

Type
Doppler log for surface ships/submarines.

Description
These two logs, respectively for installation on surface ships and submarines, provide Doppler measurement of speed relative to the seabed. The systems can measure up to 40 kt in longitudinal direction and up to 20 kt in transverse. Operational depth for the two systems is: 2 to 200 m for the TSM 5750 and 1.5 to 150 m or 0.5 to 80 m for the TSM 5740 unit. Speed accuracy as measured is 0.01 kt at speeds up to 9.99 kt and 0.1 kt between 9.99 and 40 kt. Depth accuracy is 0.1 m down to 99.9 m and between 99.9 m and 200 m accuracy is 1 m.

Contractor
Thomson Marconi SAS, Sophia Antipolis.

VERIFIED

ITALY

RS-100

Type
Underwater transponder.

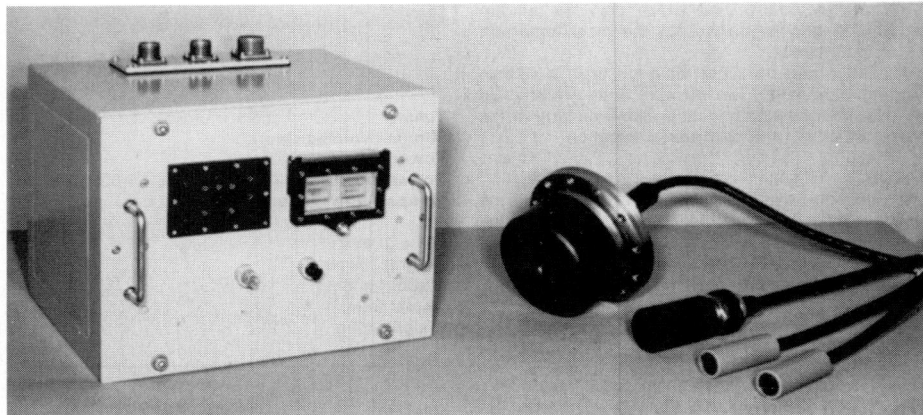

Description
The RS-100 underwater transponder has been designed for installation on submarines as an emergency localisation device. Should a damaged vessel be lying on the seabed, the RS-100 will automatically respond to suitable acoustic interrogation pulses received.

Reception and transmission of acoustic pulses is accomplished by means of a cylindrical piezoelectric transducer which is installed on the outside of the hull.

The RS-100 is sensitive to acoustic pulses at three frequencies in the 6 to 10 kHz band. When a pulse at one of these frequencies and of the expected duration is received, a corresponding pulse at the same frequency is sent back by the transponder.

Operational status
Currently installed on board 'Sauro' class submarines.

Contractor
USEA, La Spezia.

VERIFIED

RS-100 underwater transponder

UNITED KINGDOM

Master Yeoman

Type
Electronic chart table.

Description
Master Yeoman integrates electronic and conventional navigation on the nautical chart, combining the labour saving of electronics with the safety of manual plotting and pilotage.

Having fixed and referenced his chart to the chart table, the navigator has all the vessel's electronic aids output data at his fingertips on the chart.

The user interface is the Puck which acts as the system control and display unit. Around the window of the Puck are four cardinal point lights which indicate the direction in which the Puck needs to be moved on the chart to locate the navigation position or other locations, depending in which mode the Puck is activated. To assist speed of plotting, two colours are used – red when larger distances are involved and green when the Puck is close to the target radius. When the lights are extinguished the target position is located and plotted.

Master Yeoman operates on latitude and longitude and in its basic form takes output data from a selectable single navigation aid, GPS, Loran, Decca and so on, which is displayed in the Puck window. Whenever the Puck is moved across the chart the position is displayed and with the press of a key, range and bearing to and from navigation position, a mark position or waypoint is displayed.

Master Yeoman provides the navigator with all

information required for position plotting, route planning, waypoint acquisition, range and bearing calculations and multiple chart navigation.

Master Yeoman can be expanded to take in speed, log and gyro outputs, range and bearing from an ARPA radar and is able to download all information to video plotter or printer.

An alarm card can be inserted into the system which includes both plot and polygon alarm features allowing up to 350 'no go' areas to be geographically defined and alarmed.

The intelligent Puck of Master Yeoman

Operational status

In service with the Royal Navy, Commonwealth navies, NATO, Middle East and Asian navies, as well as many commercial users.

Contractor

Kelvin Hughes Ltd, Hainault.

VERIFIED

INTPS

Type

Automatic plotting table with integral navigation computer.

Description

The INTPS (Integrated Navigation and Tactical Plotting System) is an electromechanical X-Y plotter which projects a light spot cursor through a transparent top surface to illuminate ships or target positions onto a standard navigation chart or plotsheet.

It incorporates a computer system which carries out a range of navigational calculations and presents the results on a hand-held keyboard display unit.

It is capable of accepting navigation data from a wide range of sensors and processing the information to provide accurate position information. It is also capable of combining these with visual fixes from the navigator to allow calibration of the navigation aids.

Comprehensive navigation facilities include passage planning, waypoint calculation, great circle or rhumb line sailing, passage of intended movement, closest point of approach, pool of errors, station-keeping and sea area control.

Positions are maintained internally in World Geodetic 84 Datum but compatible data may be accommodated. Positions may be plotted on chart projections such as mercator, transverse mercator, polyconic, polar stereographic or gnomonic.

A database of tracks is maintained which can be updated either automatically from sensors such as radar, periscope, sonar or manually via the keyboard. Track positions can be held to an accuracy of better than 6 m worldwide and used for preparing the tactical plot. Positions may be entered and displayed in either latitude/longitude, range and bearing or grid co-ordinates. A relative velocity calculation is provided to assist in tactical manoeuvres.

Tactical functions include track projection (forward in time), moving havens, combat system navigation support, missile waypoint planning, naval gunfire support, target localisation (target motion analysis) and maintenance of Economic Exclusion Zones.

INTPS automatic plotting table on order for the US Navy

An optional data logger may be connected to record position and track information for the purposes of the operational report, mission debrief or training. Passage plans or navigation marks may be stored on magnetic tape for reuse on regular passages such as port egress.

Equipment maintenance is organised through self-contained test facilities which include background (non-interruptive) tests and interface tests. Comprehensive maintainer tests are also provided to allow fault isolation down to individual replacement modules.

Facilities are built-in to allow navigation training to be carried out in port or at fixed shore establishments.

Operational status

Currently in production for the US Navy, and already in service with a number of NATO navies.

Contractor

Kelvin Hughes Ltd, Hainault.

VERIFIED

Navpac

Type

Integrated navigation system for precise positioning.

Description

The Navpac system is an electronic system fitted in a ruggedised workstation which provides a graphical presentation of position derived from a variety of navigation sensors on a 21 in high-definition colour raster scan CRT display. Intended particularly for applications where accurate navigation is required, the system is optimised for MCM, surveying and the management of EEZs.

A simple graphical user interface is provided on a high-resolution screen to allow the operator to view and select individual navigation sensors and determine their inclusion in the navigation calculation.

Interfacing to radio navigation aids such as Syledis, Hyperfix, Microfix, GPS and Trisponder allows the Navpac to achieve better than 9 m absolute repeatable accuracy. Corrections are applied to the navigation aid positions to correct for datum offsets, antenna offsets and sonar or towfish relative positions. A true error ellipse is continuously calculated, superimposed over all the available position lines to give an indication of accuracy and the performance of the navigation sensors. The accuracy of the fix is also calculated and provided.

The system can transform the position into any selected datum or spheroid in use worldwide so that the position output can be matched to local chart data.

The system provides the capability to plan and maintain a comprehensive route database and a track-keeping facility which can be routed to a separate helmsman's display. Up to 999 waypoints may be stored in the system. Built-in mapping and drawing facilities are provided to accurately display the surveyed or swept area and a hard-copy printer and plotter can be provided to allow the preparation of chart information, swept area management, or evidence of position.

Contractor

Kelvin Hughes, Hainault.

VERIFIED

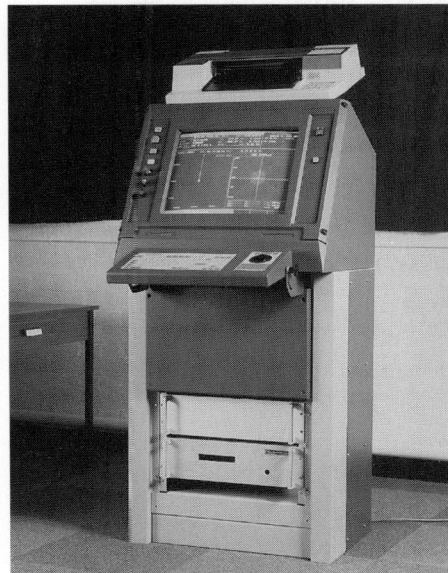

Console and display of the Navpac system

1995

Chart display unit

Type
Electronic chart display system.

Description
The Kelvin Hughes Multifunction Feature Display (MFD) is a high-resolution display for bridge navigation and operations room planning purposes. It is capable of displaying official chart data from hydrographic offices including the international S57 format and the UK Hydrographers Admiralty Raster Chart Service.

User facilities allow navigation tasks such as position plotting, passage planning, visual fixing and route monitoring to be carried out using graphical symbology overlaid on the chart display. The system can be self-contained with direct interfaces to the navigation sensors and will meet the International Maritime Organisation (IMO) requirements for Electronic Chart Display and Information System (ECDIS) which provides it with the legal status equivalent to a paper chart.

In addition, the radar interface can be enhanced to allow both ARPA or combat system tracks to be overlaid on the chart and to have the complete radar picture superimposed.

The database of navigation chart information can be added to at any time and updated easily to incorporate the latest notices to mariners. Data and corrections can be supplied either on a CD-ROM or electronically via communications satellites direct to the ship.

Operational status
The unit is currently in production and running the ARCS application. Trials are continuing of the ECDIS upgrade awaiting IMO regulations regarding functionality and test methods.

Contractor
Kelvin Hughes, Hainault.

VERIFIED

Electronic chart display unit
1997

Distress submarine location equipment

Type
Submarine indicator buoys.

Description
The role of the Type 639 Submarine Indicator Buoy (SIB) is to alert sea rescue authorities to the plight of a submarine in the event of its inability to surface.

The distress submarine indicator system, developed by GEC-Marconi in collaboration with Bradley Electronics and Patrick Engineering, is designed to provide an integrated distress submarine indicator for new build and in-service submarines.

One or two units are installed on each submarine, stored in a nest outside the pressure hull. When two are installed, each can be released independently from within the vessel, one tethered forward, one aft. On release, the unit is automatically activated and transmits for a period of at least 72 hours.

The distress message is preprogrammed to identify the buoy and the submarine, and broadcast at 406 MHz. The distress message is processed by the Global Maritime Distress System for Ships (GMDSS) utilising the COSPAS-SARSAT satellite system.

In addition, the SIB radiates the international SARBE tone at 243 MHz for radio direction-finding, and as a high-intensity beacon to assist final location. The SARBE tone is a swept frequency dropping from

Type 639 indicator buoy

1,320 Hz to 440 Hz (nominal) in 40 ms, repeated twice followed by a 0.8 second quiet period. The cycle is then repeated.

The modular system comprises a GRP moulded nest, buoyant covers (to protect and streamline the Type 639 installation), retaining lockbolts, 'buoy gone' indication, mooring line dispensing system and associated internal control unit. The GRP moulded nest and streamlining covers can be customised to suit specific requirements. A single, customer selectable, pressure hull penetration is required.

A GMDSS modification kit is available for existing in-service Type 639 SIBs and the GRP moulded nest and associated features are available for retrofitting to existing Type 639 installations.

Operational status
The buoy, originally developed as the Type 639 for the Royal Navy is also in service with the Brazilian, Canadian, Indian, Royal Netherlands and many other navies. It is now available with full GMDSS (COSPAS-SARSAT) capability.

Specifications
COSPAS-SARSAT transmitter
Frequency: 406.025 MHz
Transmission mode: BI phase modulation
COSPAS SARSAT message
Format: digital message with submarine and SIB identity
SARBE tone transmitter
Frequency: 243 MHz + 7.5 kHz
Effective radiated power: 250 mW minimum
Transmission mode: amplitude shift keying

Contractor
GEC-Marconi Radar and Defence Systems, Command and Information Systems Division, Camberley.
Patrick Engineering Ltd, Freshwater.
Bradley Electronics Ltd, London.

VERIFIED

Honeypot beacon

Type
VHF location beacon.

Description
Honeypot is a VHF location beacon operating at standard sonobuoy VHF frequencies which provides the ASW aircraft with a direction-finding location capability to its home base or ship.

The beacon is indistinguishable from normal LOFAR sonobuoy transmission to provide a tactical navigation aid in support of anti-submarine warfare operations. It is a low-cost system designed for high reliability and minimum maintenance, and gives the ship's ASW commander flexibility in the choice of navigation aid to recover the ASW helicopter.

Honeypot operates over the VHF frequency band 136 to 173.5 MHz, to transmit 1 W as an omnidirectional beacon. A channel switch on the

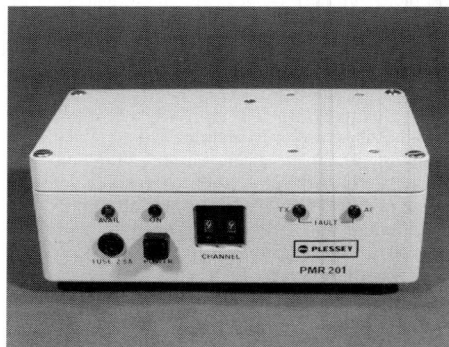

Honeypot VHF location beacon

transmitter unit allows transmission on any of the 99 sonobuoy channels. The unit is powered from the standard 24/28 V DC supplies.

Honeypot is a portable system, the transmitter unit is designed for installation in a shipborne, naval enclosed environment. The unit is a sealed splashproof housing which may be fitted in any suitable compartment. The system can be supplied complete with a coaxial feed cable and antenna, which may be clamped to any suitable point on the ship's structure.

Operational status
In service with the Royal Navy.

Contractor
Thomson Marconi Sonar Ltd, Templecombe.
VERIFIED

Model 2433A

Type
Submarine warning beacon.

Description
The Model 2433A submarine warning beacon is a heavy-duty underwater acoustic beacon designed to meet MoD specifications to protect oil/gas rigs operating in NATO authorised submarine transit lanes.

The pinger is activated by water immersion or manually using a shorting link and emits sufficient power to alert dived submarines of the rig's position, allowing safe submersible navigation past the rig without risk of collision.

The beacon operates for up to 60 days.

Specifications
Operating depth: 1,200 m
Frequency: 9.5 kHz

Coverage: omnidirectional
Pulse length: 115 ms
Pulse repetition rate: 0.5 pps
Dimensions: 89.5 cm (long) × 13.03 cm (diameter)
Weight: 16 kg (in air)

Contractor
Nautronix Ltd, Aberdeen.
VERIFIED

Sonar locator beacon

Type
Locator beacon for downed aircraft.

Description
The sonar locator beacon will withstand the impact of any aircraft crashing into the sea, the mechanical design being such that it will remain undamaged when subjected to impact shocks of up to 600 g and crush forces of more than 2,200 kg. On submersion in water, the beacon is switched on automatically when a depth of 5 to 10 m is reached. It then transmits acoustic signals which can be received by ship or airborne sonar equipment. The beacon is designed to withstand pressure corresponding to a water depth of 600 fathoms (1,100 m).

Operational status
In service.

Specifications
Operating frequency: 9.5 kHz
Operating life: 10 days minimum
Size: 75 × 75 × 150 mm
Weight: 1.2 kg

Contractor
Ultra Electronics Sonar and Communication Systems, Greenford.
VERIFIED

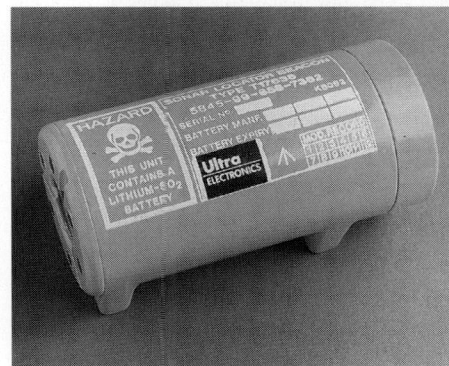
Sonar locator beacon 1995

Pingers

Type
Underwater location devices.

Description
The Nautronix range of acoustic pingers allows valuable equipment, test apparatus, divers, torpedoes and ROVs to be marked and easily relocated using Nautronix surface diver, or ROV-mounted locators. The range of frequencies from 12 to 50 kHz allows easy differentiation between equipment in local areas.

A variety of models are available, with battery lives between four days and one year, as well as ROV power loss units which start pinging when power to the vehicle is interrupted.

All units are switched on via through-water contacts which guarantees operation in all circumstances. An option is also available for manual switches if required.

All units have a minimum depth rating of 1,860 m with options beyond 4,000 m.

Operational status
In use with various navies' MCMVs, acoustic ranges, ROVs and subsea equipment marking.

Contractor
Nautronix Ltd, Aberdeen.
VERIFIED

Pingers for ROV mounting 1997

UNITED STATES OF AMERICA

3951 under-ice sonar system

Type
Dual-beam, high-resolution side scan for under-ice applications.

Description
This is a high-frequency, dual-beam, side scan sonar system for use in imaging the underside of the polar ice pack. The transducers utilise an inverted horizontal beam pattern which images the underside of the ice pack providing the sonar operator with a high-resolution image.

The transducers, being less than 50 cm (20 in) in length and 13 cm² (2.2 in²) in cross-sectional area, are sufficiently small to permit installation on board the submarine in any available location. Preprocessing electronics are located exterior to the pressure hull and required penetrations into the pressure hull are minimal. The processing electronics can be located at locations up to 500 m (1,500 ft) from the sonar transceiver for ease in installation.

The sonar transceiver is 19 in rack-mountable and requires only 10 in of vertical rack space. The processed sonar data are displayed on a Video Display Unit (VDU) which can be located remote from the sonar transceiver.

Three frequencies are available for operation with the sonar system to optimise the installation for the specific mission requirements: 50 kHz provides long range with medium resolution; 100 kHz provides high resolution and simultaneous dual frequency; 100/500 kHz provides simultaneous insonification of the ice pack for maximum resolution of the ice/water interface.

Operational status
The under-ice sonar system is currently in production and in operation aboard various units of the US Navy.

Contractor
Klein Associates Inc, Salem, New Hampshire.
VERIFIED

Video Display Unit (VDU) for under-ice sonar

Model 3040 Doppler velocity log

Type
Doppler velocity log.

Description
The Model 3040 Doppler velocity log is EDO Electro-Acoustic Division's fifth-generation Doppler system.

Digital signal processing results in the least variance estimate and eliminates the need for periodic alignment and calibration. The system is microprocessor-controlled, and automatically selects its operating parameters for optimum bottom tracking performance.

The Model 3040's main features are:
(a) full MIL parts selection, environmental testing (EMI/RFI, shock, vibration) and electrical interface
(b) simultaneous, two axes, water track and bottom track, providing measurement of platform set and drift
(c) BIT functions to circuit board level
(d) phased-array beam-forming provides automatic compensation for sound velocity variations. The transducer face forms a flat surface which installs flush with the hull.

The system is crystal-controlled and completely coherent. All system frequencies are derived from a common source. Minimum velocity variance is achieved through the use of an optimal (maximum likelihood) digital spectral estimator (called the Rummler estimator). Bias errors are suppressed through the use of a phased-array flat-faced transducer, an unbiased pulsed Doppler frequency discriminator and platform trim compensation. High reliability is achieved through a substantial reduction in card count as compared to analogue designs.

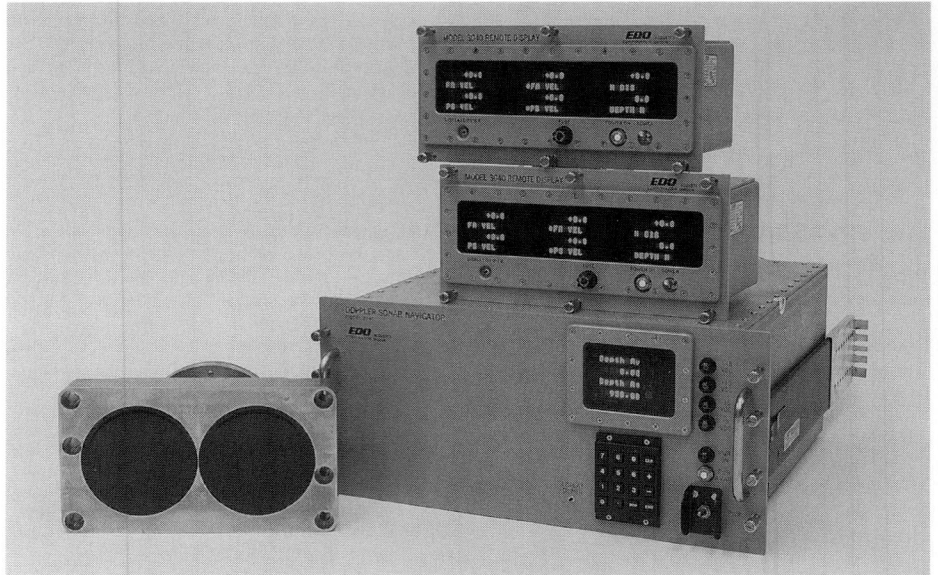

Model 3040 Doppler velocity log   **1995**

Operational status
In service with Australian and Turkish navy submarines. Also used in US Navy projects.

Contractor
EDO Acoustic Products, Salt Lake City, Utah.

VERIFIED

Model 3050

Type
Doppler sonar velocity system.

Description
The Model 3050 Doppler sonar velocity system provides accurate navigational data that eliminates velocity of sound measurement using new technology phased-array design with pulsed Doppler techniques. It is a self-contained system that uses the Doppler principle to measure velocities of a submersible or surface platform in three axes plus water depth, current profiling and intensity information.

Operational status
The Model 3050 is used in a number of new unmanned underwater vehicle programmes.

Contractor
EDO Acoustic Products, Salt Lake City, Utah.

VERIFIED

Model 3050 Doppler system
1997

Model 4208 Correlation Velocity Log (CVL)

Type
Full ocean depth velocity log.

Description
The Model 4208 provides a method of measuring absolute velocity relative to the ocean floor. Its ability to make these measurements in deep water, with practical acoustic power levels, high accuracy and with total autonomy, make it an ideal sensor for commercial, military, surface ship, submarine and UUV applications. The CVL is capable of establishing its velocity in motion or stationary at system power-up. The transmitted waveform, power level and duty cycle are dynamically adjusted with depth and platform velocity. Everything needed for operation is continued in EPROM embedded software.

Unlike a Doppler velocity measuring system which operates in the frequency domain, the correlation velocity log operates in the time domain. Relative velocity between the vehicular-mounted sonar and the bottom is determined by correlating sonar returns in time and position. The CVL's measurement is not dependent on the speed of sound.

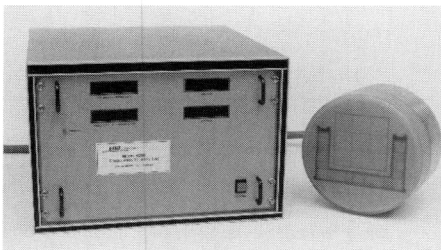

Model 4208 correlation velocity log

The acoustic arrays are housed in a single unit, requiring a gate valve installation of only 12 in. The transducer generates one broad acoustic beam directed downward, which returns to the hydrophone after striking the bottom. The CVL is most sensitive to positional changes that are parallel to the transducer array plane. In this respect, the CVL provides a better match to the propagation geometry and platform constraints than other sonar velocity measuring techniques.

Operational status
In production.

Contractor
EDO Acoustics Products, Salt Lake City, Utah.

VERIFIED

Mk 6 Dead Reckoning Tracer

Type
Plotting table.

Description
Two types of Dead Reckoning Tracers (DRT) are available to match equipment requirements. The digital Mk 6 Mod 4D DRT (DDRT) accepts distance inputs in digital NTDS slow format while the standard Mk 6 Mod 4C DRT operates from distant north and east step inputs. Both DRTs trace 'own ship' track and compute and display position (latitude and longitude) on counters. A compass rose is projected upward on to a glass plotting surface so that 'own ship' and target ship locations may be plotted.

Operational status
In service.

Contractor
Lockheed Martin, Syracuse, New York.

VERIFIED

DSN-450/DSL-250

Type
Doppler sonar speed log.

Description
The DSN-450 is a dual-axis Doppler speed log for application on board ships to measure speed along course and distance travelled, as well as measuring thwart ship speed when docking. It provides true speed over the ground from 0 to 40 kt in depths to 300 m. It automatically switches from bottom track mode to watermass mode when poor bottom returns are received using the water column 15 to 50 m below the transducer for its calculation.

The DSL-250 is a single-axis model for measuring speed along track.

The bridge control unit has large, easy to read digital displays, with depth readout selectable in feet, fathoms or metres. An adjustable depth warning provides visual and aural alarms whenever the actual depth goes to less than the selected minimum limit. Speed values and depth alarm output signal are available to provide speed and depth to other ships' systems. Digital speed and depth repeater displays can be installed in any shipboard location.

Contractor
Ocean Data Equipment Corporation, E Walpole, Massachussetts.

VERIFIED

Speed log system

Type
Electromagnetic (EM) speed log sensor and associated indicator/transmitter.

Description
EM speed log sensors include retractable 3 in and 6 in widths with lengths up to 10 ft and fixed 6 in and 10 in sensors up to 46 in long. In operation, the sensor coil is energised with AC current and establishes an AC magnetic field in the water. Because the water is a conductor, the water flow past the sensor cuts the magnetic field and generates a voltage that is directly proportional to speed. This AC voltage is picked up by two electrodes in the sensor and sent back to the indicator/transmitter where it is converted for display and transmission to remote indicators, navigation, sonar and fire-control systems.

Operational status
EM logs, together with the associated Mk 4 series indicator/transmitters have been in service with the US Navy for many years and are installed on most major ships. A commercial version of the indicator/transmitter, designated the UL-100-4, which employs solid-state, no moving parts construction, is also available.

Contractor
Lockheed Martin, Syracuse, New York.

VERIFIED

C-100 sensor

Type
Miniature digital compass sensor.

Description
The C-100 miniature heading sensor is a very small, extremely accurate heading sensor, ideal for use in systems where space is at a premium and where direction information is essential. Typical applications include oceanographic equipment, underwater warfare systems, vehicle navigation and mapping systems, night vision devices, and weapon aiming systems. The C-100 provides heading information with ±0.5° accuracy.

The digital flux gate compass sensor, which uses no moving parts, measures the earth's magnetic field electronically, thus eliminating card movement which makes conventional compasses inaccurate and unstable. It comprises a remote sensor coil and electronics board featuring a microprocessor-controlled calibration system. The sensor is compatible with multiple floating ring coil sensors. Two coil arrangements are offered – the SE25 for ±16° tilt and the SE10 for ±45° tilt. The modular design enables sensor coils and electronics to be separately mounted in different locations. The sensor communicates with other electronics with a variety of analogue and digital interfaces (RS-232) in any of six user-selectable formats.

Operational status
In service on board ROVs.

Contractor
KVH Industries Inc, Middletown, Rhode Island.

UPDATED

C-100 miniature digital compass sensor **1995**

Azimuth Digital Gyro Compass (ADGC)

Type
Digital gyro compass.

Description
KVH's ADGC outputs stable heading during even the most demanding sea conditions. The 3-axis gyro stabilised unit uses patented electronic gyros backed by a super-accurate, digital compass. Its rate gyro corrects for short-term acceleration errors while the digital magnetic compass and inclinometer provide long-term, drift-free pitch, roll and yaw. A microprocessor accesses navigational data from the compass and inclinometer sensors and the gyroscope. Through a software process, it is calculated into stabilised azimuth, pitch and roll information. By stabilising or eliminating the negative effects of pitch, roll and yaw, the ADGC outputs heading within 1° accuracy. This stable data is then used to drive and direct other equipment like autopilots, radars and satellite communication products.

Contractor
KVH Industries Inc, Middletown, Rhode Island.

UPDATED

ADGC **1998**/0009861

SUBMARINE RADAR

CANADA

CMR-85 radar system

Type
Lightweight air surveillance system.

Description
Although designed for operation in a submarine environment, the CMR-85 can be used in any role requiring a lightweight/portable system, either as the primary radar in a small craft or as a secondary radar in larger vessels. It is another variant of the LN66 series.

The CMR-85 is a 6 kW radar comprising only two units: a scanner unit incorporating the transmitter/receiver and the antenna (either a 4 ft slotted waveguide or a 3 ft antenna enclosed in a radome), and an 8 in display unit. With the 4 ft unit, the antenna can be removed from the transmitter/receiver unit by means of a quick-release mechanism for easy transportation through hatches.

Operational status
Various versions are in widespread use with the US Navy and other customers.

Specifications
Frequency: 9,410 ± 30 MHz
Peak power: 6 kW
Rotation speed: 30 rpm
Pulsewidth: 0.15 and 0.12 µs
PRF: 1,200 and 3,000 Hz
Topside weight: 13.6 kg (including antenna)

Contractor
Canadian Marconi Company, Kanata, Ontario.

VERIFIED

FRANCE

TRS 3100 (Calypso III) submarine radar

Type
Submarine air and surface surveillance radar.

Description
Calypso III is a higher-performance radar than Calypso II, with the same general characteristics and same functions (navigation, surveillance, target designation), but with a greater detection range and the additional capability of accurate range measurement. It is also of compact design and well fitted to the severe environment characteristics of a submarine.

Calypso III comprises four main parts: an antenna (the Calypso II antenna); a non-rotating periscopic mast

(hoisting is achieved using the ship's hydraulic system pressure); a transmitter/receiver cabinet (secured at the foot of the mast) and an operational console.

The transmitter is a conventional magnetron transmitter with a frequency adjustable magnetron and the same I/J-band (18 to 20 GHz) transmitter employed by Calypso II. The receiver is fitted with RF elements of long service life, a modern mixer and a logarithmic anti-clutter chain. The operational console comprises the control panel for radar and antenna, the display (16 in (406 mm) CRT) with an accurate digital range-finding device and an operational panel. The optical-periscope direction ESM and sonar data can be displayed on the PPI. The accurate range measurement of a detected target can be made during short time transmission.

The detection range capability of Calypso III for a typical ASW aircraft at 2,500 m altitude is 18 n miles but depends on the radar horizon for a surface vessel.

Operational status
In service with three navies (including the Brazilian 'Tupi' class submarine). No longer in production.

Contractor
Thomson-CSF, Division Systèmes, Défense et Contrôle, Bagneux.

UPDATED

TRS 3110 (Calypso IV) submarine radar

Type
Submarine air and surface surveillance radar.

Description
The TRS 3110 Calypso IV radar is an I/J-band (8 to 20 GHz) equipment designed to provide surveillance and navigation facilities for submarines. It comprises three units: an antenna, a transmitter/receiver cabinet and an operating console. The antenna and operating console are similar to those of Calypso III. To fulfil its functions, the Calypso IV is built around a simple navigation radar equipment, and modular design enables it to carry out additional functions merely by the addition of appropriate functional circuit cards, without changing the basic equipment.

In using a 1 ms pulse, it has a good detection range, limited only by the radar horizon (at least 10 n miles in free space), and using the narrow (50 ns) pulse, it has very high resolution over shorter detection ranges (down to about 15 m). The receiver design has been developed to give good capabilities and performance in the presence of clutter.

The transmitter is a 25 kW (peak) klystron-driven unit providing several transmission options (continuous, sectoral, burst, or receive only – radar silence) giving maximum operational flexibility to meet specific requirements. The overall radar transmission parameters have been chosen to enhance the submarine's security by arranging for them to be similar to those of merchant ship radars, frequency and pulse length in particular.

The operating console and its data processor permit the presentation, in either raw radar or synthetic form, of information such as target echoes, range and bearing labels, transmission sectors and so on, as well as providing for operational control of the radar equipment itself.

Operational status
In service but no longer in production.

Contractor
Thomson-CSF, Division Systèmes, Défense et Contrôle, Bagneux.

UPDATED

INTERNATIONAL

CSR

Type
Compact Submarine Radar (CSR).

Description
Kelvin Hughes in the UK and SAGEM SA of France are producing a new concept for submarine navigation radars. In this concept, the transceiver is housed in a watertight pod, directly beneath the antenna to which it is rigidly fixed. Both transmitter and antenna rotate together at the top of the radar mast which functions in the normal manner. As a result of this design, the

conventional peak power of 25 kW can be reduced to 5 kW, ensuring the same performance but with an important gain in reliability. This arrangement eliminates the need for a complex waveguide and its associated disadvantages (waveguide hull penetration, and losses due to waveguide length, condensation and the need for a rotary joint), allowing the entire system to be mounted outside the pressure hull. Furthermore, the antenna height above sea level is very easy to command and adjust. Finally, only one cable penetrates the pressure hull, the rest of the equipment (display monitor and processor unit) is installed within the submarine control room.

The CSR is suitable for installation on board small, as well as larger submarines.

Operational status
The CSR is currently being installed on board French nuclear attack submarines.

Contractor
SAGEM SA, Defence and Security Division, Paris, France.
Kelvin Hughes, Hainault, UK.

VERIFIED

ISRAEL

EA-20/SRD radar antenna

Type
Submarine navigation radar antenna.

Description
The EA-20/SRD radar antenna has been developed for integration into a submarine navigation and surveillance radar system. The dimension restraints led to the design of a hog horn feeding a cylindrical parabolic dish. The complete antenna will withstand a pressure of 40 atmospheres and is constructed of anti-corrosive stainless steel.

Operational status
Fully developed.

Specifications
Frequency range: 9.44 ± 0.02 GHz
Polarisation: horizontal
Peak power: 200 kW
Average power: 200 W
Gain: >25 dBi
Azimuth beamwidth: <3° (3 dB)
Elevation beamwidth: >15°
VSWR: 1.2:1
Dimensions: 175 mm (high) × 800 mm (wide)
Weight: 32 kg

Contractor
MTI Technologies and Engineering Ltd, Ashdod.

VERIFIED

EA-20/SRD submarine radar antenna

ITALY

MM/BPS-704 submarine radar

Type
Submarine search and navigation radar.

Description
The MM/BPS-704 I-band (8 to 10 GHz) is a naval search and navigation radar for use aboard submarines. The general characteristics are similar to the MM/SPN-703 surface ship version, using the same 20 kW transmitter and pulse length/PRF combinations. A specially designed high-pressure-resistant antenna (up to 60 kg/cm²) mounted on a telescopic mast, is part of the equipment. The principal differences are indicated in the specifications.

MM/BPS-704 antenna

A new improved version has been introduced which features a more powerful transmitter (90 kW peak power) with pulse-to-pulse frequency agility. Performance has been enhanced which allows special modes of operation for the submarine.

Operational status
In service with Italian Navy 'Sauro' class submarines.

Specifications
Antenna span: 1 m
Beamwidth: 2.2° (horizontal); 11° (vertical)
Gain: at least 27 dB
Noise figure: better than 11 dB

Contractor
Alenia Difesa Avionic Systems and Equipment Division, Florence.

UPDATED

NETHERLANDS

ZW07 submarine radar

Type
Surface search and navigation radar.

Description
The ZW07 is one of the family of the ZW series radars produced by Signaal. This particular model is designed for use by submarines for navigation, distance measuring, surface search and limited air warning. A single-shot mode has been adapted for a range of surface targets. Other operational features include: surface coverage up to the radar horizon, limited air warning and high resolution for navigation. Anti-clutter measures and ECCM provisions include sector-scan facilities, logarithmic receiver with pulse length discriminator, suppression of non-correlated pulses, and a tunable transmitter.

Operational status
In operational service on board Netherlands Navy 'Walrus' class submarines.

Specifications
Frequency: I-band
Polarisation: horizontal
Transmitter: tunable magnetron
Power output: 60 kW (peak); 100 W (average)
Maximum range: 22 km (against small surface craft)

Contractor
Hollandse Signaalapparaten BV, Hengelo.

UPDATED

RUSSIAN FEDERATION AND ASSOCIATED STATES (CIS)

Snoop type

Type
Submarine surface search radars.

Description
A number of surveillance radars have been fitted to Russian submarines since the Second World War, some of which have been given a NATO designation starting with Snoop. The early submarines were equipped with a small antenna designated Snoop Plate. This was apparently fitted to all boats up to the 'Golf'/'Foxtrot'/'Romeo' classes and was reputed to have a range of 25 n miles against aircraft and 12 n miles against surface vessels. Later, 'Foxtrot' class vessels and the 'Hotel' generation were fitted with a larger and more powerful I-band radar known as Snoop Tray which entered service in the early 1960s.

Since the early 1960s, nearly all new submarines have been fitted with Snoop Tray, with the exception of the 'Juliett' and 'Echo' classes of cruise missile submarines which were equipped with a much larger antenna known as Snoop Slab. This antenna is twice the size of the Snoop Tray unit and must be capable of greater definition at longer ranges.

Two other Snoop class radars are the Snoop Pair fitted to 'Typhoon' and 'Akula' class submarines, and Snoop Head fitted to the 'Oscar' and 'Alfa' classes.

Snoop-Tray radars operate in the I-band (8 to 10 GHz).

Operational status
In operational service. Virtually all Russian built submarines are fitted with a Snoop class surveillance radar. Also operational on Chinese 'Xia' 'Han' and 'Ming' classes and Yugoslav 'Sava' and 'Heroj' classes of submarine.

UPDATED

SWEDEN

Subfar 100 submarine radar antenna

Type
Submarine air and surface search radar antenna.

Subfar 100 submarine antenna

Description
The Subfar 100 is an air and surface search radar antenna for submarines.

The 9GA 300 antenna is designed for mounting with its hydraulic turntable on top of the submarine radar mast. This way of mounting eliminates bearing error sources from torsion effects in the radar mast. The hydraulic turntable gives a very low noise level as well as flexible control of antenna rotation speed. The antenna direction can be manually set or slaved to external sources such as the periscope.

The other main elements of the system are the I-band frequency agile transceiver and display console. Sector transmission as well as short time transmission down to a single pulse can be selected. Automatic target detection can be included.

Operational status
In service.

Specifications
(9GA 300 antenna)
Frequency range: 8.5-9.6 GHz
Rotation speed: 0.5-24 rpm

Beamwidths
Horizontal: 2.4°
Vertical: 16°
Gain: 26 dB
Sidelobes: 6-18 dB
Antenna aperture: 1,000 × 140 mm
Height above mast tube: 600 mm
Rotating circle diameter: 1,040 mm
Antenna system weight: 95 kg

Contractor
CelsiusTech Electronics AB, Järfälla.

UPDATED

UNITED KINGDOM

Type 1006 radar

Type
Navigation and surface search radar.

Description
The Type 1006 was the standard I-band navigational radar of the Royal Navy until the arrival of the Type 1007.

Antenna outfit AZJ
The surface role antenna outfit consists of a 2.4 m slotted waveguide linear array rotated at 24 rpm by a turning mechanism. The array has a horizontal beamwidth of 1° and low sidelobe levels to give good bearing discrimination. The vertical beamwidth of 18° gives good performance in conditions of roll and pitch.

Antenna outfit AZK
This antenna outfit consists of a 3.1 m slotted waveguide linear array rotated by the same type of turning mechanism as outfit AZJ. It is used when the improved bearing discrimination given by the 0.75° beamwidth is required.

Transmitter/receiver (surface)
The equipment is non-thermionic with the exception of the magnetron, and operates at a frequency of approximately 9,445 MHz.

Transmitter/receiver (submarine)
In order to be compatible with the existing submarine antenna system, a variant of the transmitter/receiver operating at 9,650 MHz is available.

Display unit JUD
This unit is non-thermionic with the exception of the cathode ray tube. The display unit uses a rotating scanning coil system to provide range scales from 0.5 to 96 data miles.

Operational status
The Type 1006 radar has been in series production for the Royal Navy and other navies since 1971. Production ceased in 1986. The radar equips 'Oberon' class submarines It is now being superseded by the Type 1007.

Contractor
Kelvin Hughes Ltd, Hainault.

UPDATED

Type 1007 radar

Type
F- (3-4 GHz) and I-band (8-10 GHz) navigation and surface/air search radar.

Description
The Type 1007 is the standard I-band navigation radar of the Royal Navy. It consists of a range of navigation, surface and air search equipments for naval use. It includes a choice of antennas, I-band and F-band transmitter/receivers and a range of displays. Reduced magnetic signature variants are available for use on MCMVs.

Antennas
Three versions of the antenna outfit are available for surface vessels: a 2.4 m single array; a 3.1 m single array; and a 2.4 m dual array for use with helicopter transponders and outfit RRB. All are horizontally polarised, end-fed slotted line arrays, incorporating vertical polarisation filters to give low sidelobe and back radiation levels. The 3.1 m array has a horizontal beamwidth of 0.75°. Surface ship antenna outfits can operate in winds up to 185 km/h and withstand funnel gas temperatures up to 120°C, as well as gun and shock effects of blast. De-icing systems are available. A 4.1 m antenna is used with the F-band radar and an Identification Friend-or-Foe (IFF) antenna can be surmounted as an option. A fully pressure-tested submarine antenna is also available for fitting to a variety of submarine masts.

Transmitter/receivers
The transmitter/receivers are solid-state, with the exception of the magnetron, and operate at frequencies of 9,410 MHz and 3,050 MHz with a

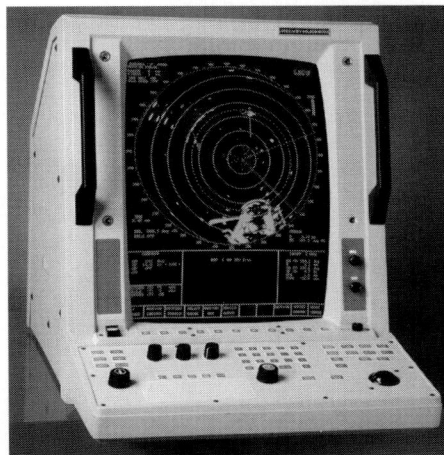

The Type 1007 colour tactical display

transmitter power output of 25 kW. A wide dynamic range logarithmic receiver is provided. A built-in monitoring system is included to check that the equipment is operating at peak performance. A low-leakage dummy load allows for system testing during periods of radar silence. Centralised emission control circuitry enables the command to inhibit transmission immediately. Sector transmission is also incorporated with direct control from the main display (optional for the F-band radar). Blanking pulses are incorporated to safeguard sensitive ESM equipment.

Display
The colour tactical display (CTD) is a highly capable navigation display with a wide selection of operational/

tactical facilities. It has a built-in tracking capability of up to 50 automatically tracked targets and 20 manually tracked targets. The CTD gives the operator a clear, sharp, colour tactical picture out to over 300 km with the ability to label tracks with ships' names/numbers. Symbology and a choice of colours are used to indicate hostile/friendly/neutral/unknown, and whether air/ surface or subsurface.

Standard interfacing includes most log, gyro and GPS equipment. The CTD can receive four radar video inputs and three auxiliary inputs. Optional facilities include the control and display of frequency modulated continuous wave radars, ability to receive plot-extracted data, VESTA, carry out IFF active gate interrogation and interfacing to most combat systems, fire-control systems, infra-red and ESM equipments. A high-speed air tracking facility is available as an option.

The CTD displays can be networked to share track data and form a highly cost-effective command system.

A weatherproof auxiliary raster scan display is available as an option. This unit is portable and is designed for use on an open bridge or submarine fin.

A built-in simulator package allows onboard training to be carried out in the minimum of time.

Operational status
The Type 1007 radar associated with the colour tactical display is in service throughout the world with over 30 navies on all sizes of vessel, including submarines and MCMVs. Submarine fitments include Norwegian 'Kobben' and 'Ula' classes, Portuguese 'Daphné' class, UK 'Trafalgar' and 'Vanguard' classes.

Contractor
Kelvin Hughes Ltd, Hainault.

UPDATED

AZL mast

Type
Radar submarine mast.

Description
This mast outfit is spigot-mounted in a cylindrical pressure pod, built into the free-flooding area of a submarine fin. A hydraulically retractable mast supports the antenna head which is rotated electrically and connected to the ship's radar system via a waveguide.

The outfit does not penetrate the pressure hull of the submarine. Waveguide and electric connections are made via a service pipe at the bottom of the pod. Hydraulic connections are made via a casting casing forming the upper end of the pod. Remote indications of mast, drive and antenna status are provided. If water enters the pod, a detector signals the ship's alarm system.

If the mast shares the fin with other masts where fouling could occur, interlocking is included in the control circuits. Interlocking also ensures correct operation of the indexing mechanism, mast and antenna operating gear, including antenna head alignment.

Operational status
The AZL radar submarine mast is in service with all submarines in the Royal Navy.

Contractor
MSI Defence Systems Ltd, Norwich.

VERIFIED

AZL mast as fitted on Royal Navy submarines
1997

UNITED STATES OF AMERICA

AN/BPS-15 submarine radar

Type
Submarine surface search and navigation radar.

Description
The AN/BPS-15 is a submarine radar designed for surface search, navigation and limited air warning facilities. It operates in I-band using a horn array antenna.

Operational status
In widespread use on US Navy submarines. No longer in production.

Specifications
Frequency: I-band
Peak power: 35 kW
Pulsewidth: 0.1; 0.5 μs
PRF: 1,500; 750 pps
Scan rate: up to 9.5 rpm

Antenna dimensions: 101 cm (aperture)
Antenna weight: 76 kg
Range resolution: 10 m (in short pulse mode); 30 m (in long pulse mode)

Contractor
Litton Sperry Marine, Charlottesville, Virginia.

UPDATED

AN/BPS-16 submarine radar

Type
Submarine navigation and surface search radar.

Description
The AN/BPS-16 (previously known as the AN/BPS-XX) is a submarine I-band (8 to 10 GHz) radar designed to provide nuclear-powered fast attack and ballistic missile submarines with a navigation and search radar capability when operating on the surface. The AN/BPS-16 features a new 50 kW frequency-agile transmitter in I-band and the latest in signal processing techniques to enhance operational performance in heavy weather. Unlike the AN/BPS-15, the new system is supplied with a new unique radar mast assembly to 'raise and retract' more effectively and reliably a state-of-the-art antenna.

Operational status
The AN/BPS-16 has completed First Article Testing and Navy Operational Evaluation. The equipment to be furnished under Production Options to the present contract will be installed in SSN 21 and 22, SSBN 741, 742 and 743 and shore installations at TRF Kings Bay and NSWC, Damneck. At the end of 1991, Sperry Marine was awarded a contract for 10 AN/BPS-16 radars for 'Ohio' and 'Seawolf' class submarines and training sites.

Contractor
Litton Sperry Marine, Charlottesville, Virginia.

UPDATED

CONSOLES AND DISPLAYS

BELGIUM

Q-70

Type
Advanced display subsystem.

Description
The subsystem for the US AN/UYQ-70 is the advanced high-performance IVS4100 graphics display controller with RGN639 and RGN651 rugged CRT displays which is being supplied to Lockheed Martin for integration

into the US Navy's UYQ-70 advanced display system. The advanced workstation is the first to use commercial-off-the-shelf (COTS) components which eliminate development and cost factors from those of previous proprietary architectures.

Operational status
Being supplied to Lockheed Martin for integration into the AN/UYQ-70 advanced display system which will be installed in the NSSN. The subsystem will also be

installed in new UYQ-70 equipment to be fitted into other US Navy platforms, including the Hawkeye E2-C aircraft and Landing Craft Air Cushion.

Contractor
BARCO nv, Display Systems, Kortrijk.

VERIFIED

CANADA

AN/UYQ-501(V) Shinpads display

Type
Multisensor display system.

Description
The MSD-7001, developed by Computing Devices for the Canadian Department of National Defence as part of a distributed processing system, is a true multifunction display that accepts inputs from all shipboard sensor systems and computers. In its naval configuration, it is designated AN/UYQ-501(V) and provides both sensor information and complex graphical overlays on high-resolution full-colour television monitors. It interfaces with any general

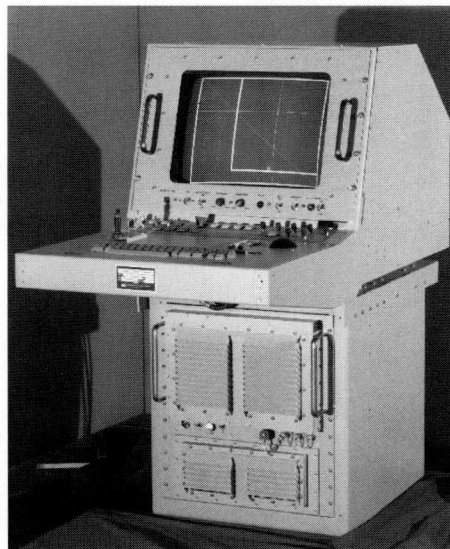

AN/UYQ-501(V) Shinpads display

purpose NTDS-capable computer functioning as a display processor and hence communicates over any standard databus.

Standardisation of hardware, software and interfacing has been achieved to the point where this display satisfies all the requirements for operator interface with any sensor, weapon or machinery control function. It is a powerful tactical and command situation display providing the command and control team with instant access to all data available on board.

The AN/UYQ-501(V)'s unique video features optimise the match between sensor and display for all sensors on board ship. Careful human engineering, a high-resolution display and judicious use of colour combine to give the operator all the information required to carry out his tasks effectively on a single-display CRT.

Video processing
(a) 4-megabit RAM video memory
(b) 1,024 × 1,024 pixels each having 4-bit intensity
(c) video area controlled in size and position anywhere within complete raster area.

Radar characteristics
(a) accepts normal radar PPI range and bearing and converts these to X-Y co-ordinate display
(b) presents section of PPI radar in B-scan format
(c) presents PPI and fire-control (A + R-scan) radar simultaneously
(d) offset centre of PPI outside display field of view
(e) any scan rate 0 to 60 rpm
(f) video pulsewidth, 50 ns minimum.

Line scan characteristics
High-speed serial interface using STANAG 4153 format accepts preformatted data from sensors such as sonar and infra-red at 10 Mbits/s rate directly into memory.

Television characteristics
(a) accepts TV composite video via RS-343A or RS-170

interface at standard 525, 625, 875 and 1,075-line rates
(b) TV video stored digitally in video memory permitting annotation, magnification, freeze-frame and image processing.

Graphic processing
(a) 8-megabit RAM graphics memory
(b) covers full extent of 1,225-line raster scan format. 1,152 active lines at 1,536 pixels/line with 4-bits/pixel for colour and blink
(c) eight colours selectable from palette of 4,096
(d) graphics repertoire includes ASCII characters, NTDS characters, vectors, arcs and circles, in both positive and inverted video format
(e) graphics are overlaid without loss of information
(f) graphics scrolling is independent of video presentation
(g) graphics addressing is user-friendly and simple; built-in generator does the housekeeping.

Application processing
(a) interfaces with any TDS-capable computer of the user's choice by means of a 16 unit NATO Standard NTDS parallel interface (STANAG 4146)
(b) specifically designed for universal application in the combat system environment with distributed processing interconnected by a standard high-speed serial databus.

Operational status
In service with the Canadian Forces Maritime Command (Canadian Patrol Frigate and TRUMP). Special variants have been developed for the AN/SQR-501 (CANTASS) and AN/SQS-510 (see separate entries).

Contractor
Computing Devices Canada Ltd, Ottawa, Ontario.

VERIFIED

Magic 2000

Type
Multimode, advanced graphics and imaging for command and control display system.

Description
The Magic 2000 display system, developed from the Shinpads standard display (see previous entry), is a multisensor display that accepts video, radar and high-speed digital input data from a variety of sensor systems. The display combines digital scan conversion techniques and a dedicated, powerful graphics processor to provide sensor information and a series of complex graphical overlays on a high-resolution 1,280 × 1,024 colour monitor. The single display screen provides simultaneous display of radar/TV, video and tactical graphics in a modern Windows environment. The system is available as a complete display workstation or as a set of modular blocks.

The display is based on VME-standard architecture with minimum six-slot expansion capability. The use of the VME architecture permits the inclusion of complementary third-party hardware to meet specific customer requirements.

The system comprises seven major functional blocks – display controller; radar scan converter; radar interface; IFF/Emulator; graphics engine; TV driver; and TV monitor.

The display controller subsystem comprises a VME 68030-based single-board computer which acts as VME bus master and interfaces with all the system slaves – graphic engine, radar scan converter, TV driver, and IFF driver. The computer incorporates a Motorola 68030 and 68882 co-processor operating at 16, 25 or 32 MHz. Memory capability comprises up to 4 Mbytes of static RAM with optional battery back-up and up to 2 Mbytes of EPROM. Two counter/timers, a real-time clock and eight serial ports are also provided. Up to six MMI devices can be connected to the display

controller. These currently include – trackerball, touchscreen, electroluminescent panel and QWERTY keyboard.

The advanced radar scan converter (1,024 × 1,024 by 16 intensities) provides real-time radar video with selectable variable persistence and processing. It comprises three VME CCAs based on a Motorola 68020 central processor, fast static RAM and several custom gate arrays. The converter is designed to provide a high-resolution radar video PPI display in near-realtime, that is, in synchronism with current radar trigger position. A unique true motion feature permits small radar offsets to be implemented without loss of screen data while complex range-scaling circuits permit a continuous quasi-analogue range scale capability. The range scale is continuously variable from 2 to 512 miles with continuously variable offset up to 512 miles.

The high-resolution graphics engine (with 256 displayed colours) provides a comprehensive

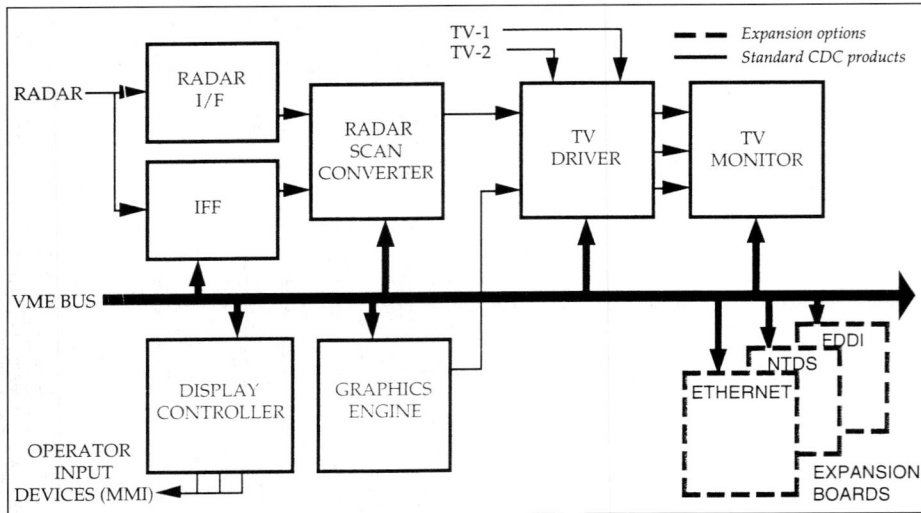

multilayered tactical situation display using two tightly coupled VME CCAs based on AMD 29000 RISC technology, display VRAM graphics frame buffer and a number of proprietary VLSI ASICs.

TV sensors may include other spectral bands such as IR as well as standard and low-light cameras. The TV input is 525/625-line with 8-bit monochrome or pseudo-colour.

Operational status
In production.

Contractor
Computing Devices Canada Ltd, Ottawa, Ontario.

VERIFIED

Magic 2000 functional block diagram showing typical configuration

ISRAEL

Terminal

Type
Ruggedised PC-compatible terminal.

Description
These ruggedised PC-compatible terminals provide a hardware solution for C³ systems.

They enable field use of the latest commercial-off-the-shelf (COTS) laptop, notebook and hand-held computers. The devices incorporate a 9 in colour display and a full range of I/O devices (touchpanel, pen, trackball, mouse, joystick and standard keyboard with function keys). The integrated communications controller supports up to four communication networks with programmable protocols, digital and analogue interfaces to various radios (VHF, UHF, HF and so on) and wire.

The systems incorporate special purpose software for a variety of applications, including communication controllers (such as the Sacu).

Optional kits can be provided and include GPS, user special card (ISA bus), encryption and external display.

Contractor
Elbit Systems Ltd, Haifa.

VERIFIED

PC-compatible terminal
1996

UNITED KINGDOM

Ultra Compact Console

Type
Flat-screen workstation.

Description
The Ultra Compact Console (UCC) is a multifunctional twin-screen naval workstation which incorporates the latest flat-screen developments using plasma and LCD technology to achieve the most compact dimensions. The concept has been developed from studies into the Royal Navy's future requirements for submarine and surface ship operator consoles and is based on Ultra's experience of the Royal Navy's current command and control systems (SMCS and SSCS). The UCC provides improved operator facilities, saves space, is simple to install and significantly reduces heat dissipated in operational compartments, whilst leading to improvements in operational effectiveness.

The flat screens allow a considerable reduction in overall size of the workstation. It can now be mounted on any suitable bulkhead or pillar and free up to 75 per cent of the volume required by conventional consoles. This represents a significant improvement in operational space usage and allows innovative control room layouts to be realised. The UCC concept offers significant operational space and technical benefits over consoles using conventional CRT displays.

The folding desk ensures that when out of use or when information has only to be monitored, the UCC does not intrude unnecessarily into operations room space.

In existing systems the role of the workstation or console is defined by the communication and control equipment which surrounds it. The multifunction console, being totally modular, allows the incorporation of these facilities within the displays themselves. This allows the operator to configure the console for his own task. For example, tactical picture compiler air, when logging on and the workstation will automatically configure to provide the facilities for this role. This enables the workstations available in the combat system to be deployed in the best way to monitor the control of the operations being carried out. This flexibility will improve operational effectiveness.

The system uses open system architecture and hence the updating of processors, with new ones as they are developed, is straightforward. If required, the customer can define the applications processor hardware to meet purchasing commitments. The design allows the flexibility to call up any picture using Windows and Motif techniques. Digitised maps are easily projected and the display can incorporate and superimpose video images from various sensors. The size of the individual window displays on the screen can be varied and conventional windows overlaying or tiling can be offered.

The Man/Machine Interface (MMI) can be tailored towards the applications, for example, the correct choice of screens can allow viewing by a number of users from a distance. The desk area incorporates a touchpanel for tote pages, together with a joystick or trackerball. The keyboard, which is provided primarily for data entry, can be revolved to provide a clear writing/working area. The console processors, which may be sited remotely from the console, are linked by a LAN to other parts of the combat system. The reduced power consumption of the flat screens and the facility to put the processors elsewhere means that the need for chilled water in the operational area can be eliminated and the installation considerably simplified.

Operational status
In production.

Contractor
Ultra Electronics Command and Control Systems, Loudwater.

UPDATED

UNITED STATES OF AMERICA

MRCS

Type
Militarised Reconfigurable Console System.

Description
The MRCS is a tactical console capable of being reconfigured for a wide range of applications. The hardware has been designed and tested in accordance with stringent US Navy requirements for shock, vibration, EMI, humidity, sand, dust, salty fog, fungus and acoustic noise.

Up to four MC68020/68881 processors are built around the VME bus architecture, providing a large, commercial support base which is expandable in both processing power and memory. Processing speeds up to 6 Mips are possible and the system has a 4 to 24 Mbyte RAM memory. Up to 12 interfaces are available in the single display, depending on the configuration selected.

It provides a flexible I/O configuration which can be expanded or modified to fit the application. A powerful embedded graphics generator can provide multiple, independent graphics output and input video overlays and the system supports two independent monitor drivers with four graphics planes per driver (expandable to eight).

Operational status
The consoles are being supplied under contract to the US Navy for the MHC-51 programme where they are used in the navigation/command and control system, the AN/SYQ-13.

Contractor
Lockheed Martin, Tactical Defense Systems, Great Neck, New York.

VERIFIED

AN/UYQ-70 ADS

Type
Advanced display system.

Description
The hardware and software architecture of the AN/UYQ-70 is supported by high-performance commercial-off-the-shelf (COTS) and modified COTS modules and is housed in militarised racks and cabinets to provide the foundation for a complete system. Using state-of-the-art open system architecture, the modular high-performance system offers a flexible, rapid deployment, cost-effective system.

The AN/UYQ-70 uses a 100 MHz HP 743 single-board 6U VME processor with 64 Mbytes of error-correcting RAM which can be expanded to 256 Mbytes. ROM is user-definable and interfaces include integrated SCS-II, Ethernet, RS-232 and Centronics. Onboard graphics is available as an option.

The graphics system features a 1.5 million vectors/s capability with eight or 12 memory planes and eight or 12 overlay planes. The radar scan converter unit interfaces to a variety of naval radars.

The displays include 14 and 19 in units and the MMI includes keyboard and trackerball, a communications panel emulator, and Electroluminescent Category Select and CCAEP touchpanels.

The system is available in three configurations – a basic combat system console, rack-mounted console and an unattended equipment rack. The equipment is fully compatible with tactical surface and subsurface ship environments.

Operational status
Lockheed Martin awarded DRS Technologies further contracts worth approximately US$9.3 million in the spring of 1997 to manufacture additional quantities of AN/UYQ-70 tactical workstations. Initial deployment of the AN/UYQ-70 display consoles include the US Navy 'Arleigh Burke' class DDGs.

Under a separate contract, DRS has developed a Horizontal Large Screen Display (HLSD) workstation to replace the current Mk 19 manual plotting table for the NSSN Advanced Technology Demonstration programme.

Contractor
Lockheed Martin, Tactical Defense Systems, Great Neck, New York
DRS Electronic Systems Inc, a Division of DRS Technologies Inc, Gaithersburg, Maryland.

VERIFIED

MMC

Type
Modular Multifunction Console (MMC).

Description
This is a real-time, tactical, general purpose computer and high-resolution graphics system using the design philosophy of open systems architecture. The MMC design is based on the VME-VSB bus architecture, which allows the system to be easily configured to any application by integration of extra printed circuit cards in the VME Euro-card form.

The MMC's modular construction allows it to meet almost any customer requirement. It is configured in three basic units: the processor unit, the Operator Entry Panel (OEP) and the monitors. These units are capable of operation as a single integrated unit or of being split into its major sub-units, allowing the monitors and OEP to be operated remotely from the processor unit. This flexibility allows for a variety of configurations to be formed. These include installation and operation as a personal workstation, as a unit interface and node in a distributed system, or as an interface to a central mainframe.

The Monitor Unit is capable of displaying multicolour text, radar video, CCTV data, high-speed graphics such as Windows, overlays, cartographics, and tactical symbology on a militarised, high-resolution, 19 in raster scan, colour Cathode Ray Tube (CRT). The monitor unit also provides Finger On Glass (FOG) touchscreen capability, interfacing with the graphics/display processors, which are components of the processor unit.

The processor unit is a self-contained, self-cooled unit that is capable of operating independently of the other MMC units. The processor unit houses the graphics processor, display generator, applications processor, and external I/O interfaces. The processor unit uses a RISC processor and/or processors from the 680X0 family, combined with an optional floating point co-processor to perform its application and graphics

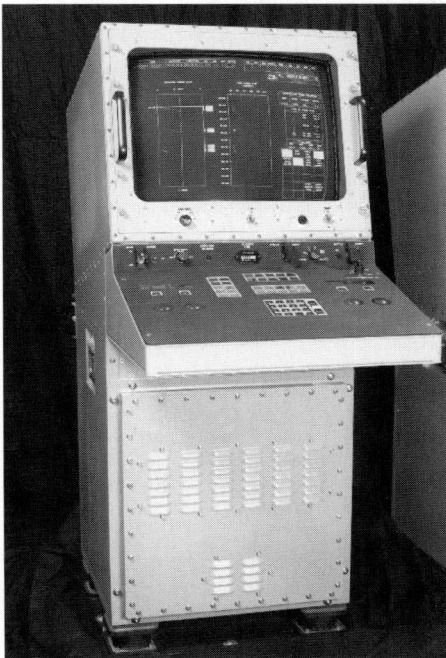
The MMC which can be used in a variety of configurations

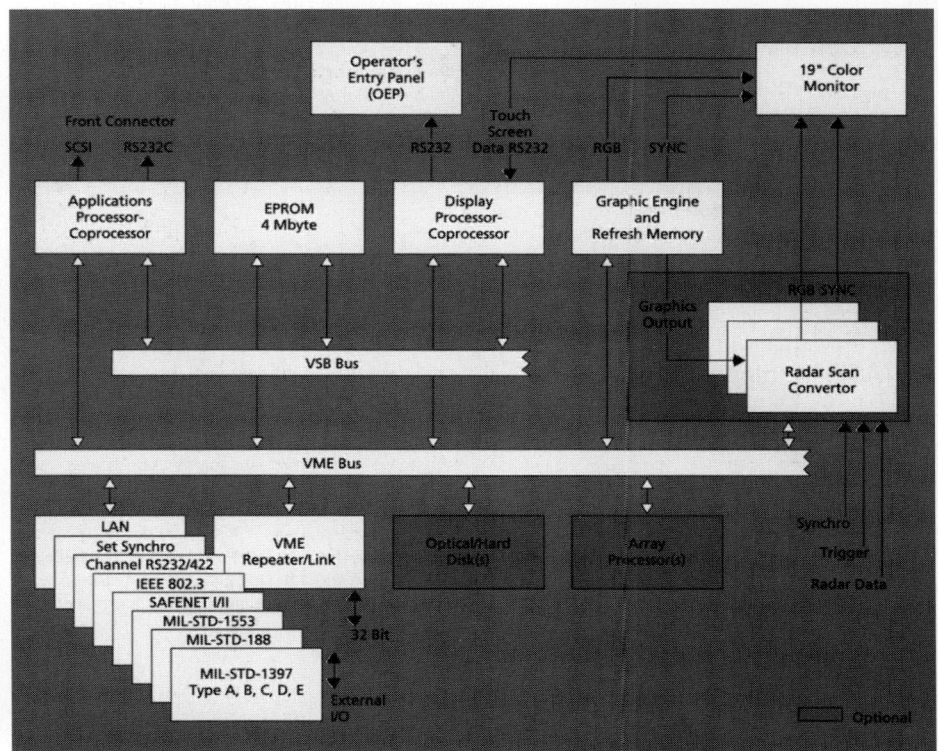
MMC architecture

1994

processing. Software packages presently in use include: X-Windows, GKS, CGI, CGM, C, Unix, and Ada. External interfaces supported include: Synchro, RS-232C, RS-422, MIL-STD-1397 A to E, MIL-STD-188, MIL-STD-1553, SAFENET, IEEE 802.3, SCSI and Centronics.

The OEP is the primary device for operator control of the MMC. The OEP is tailored to meet customer requirements and MMI interfaces, and includes: FOG; trackerball/marble; electroluminescent panel with infra-red touch panel; QWERTY keyboard; fixed and programmable keys; audio and communication controls; and numeric keypad.

Operational status
Produced for the PISCES combat information system and in service aboard Egyptian Navy 'Romeo' class submarines.

Contractor
Lockheed Martin, Tactical Defense Systems, Great Neck, New York.

UPDATED

AN/UYQ-65 data processing and display set

Type
Data processing and display workstation.

Description
The AN/UYQ-65 workstation which is compatible with the US Navy's AN/UYQ-70 Advanced Display System is the first totally commercial-off-the-shelf (COTS) based workstation packaged to comply with the stringent environmental requirements of the AEGIS 'Arleigh Burke' class guided missile destroyer shipbuilding programme. The workstation has been designed to replace the tactical and acoustic sensor displays in the AN/SQQ-89 ASW combat system. Using modern open systems architecture, with two VME backplanes and dual processors, the AN/UYQ-65 workstation provides a UNIX/POSIX-compliant operating system with X-Windows for graphics and a Motif Windows manager for ease in operator control.

The system fully supports the US Navy's Joint Maritime Command Information System/Unified Build programme for TAC-3/4 software applications and is available in HP 743 processor configurations.

The system hosts the US Navy's Sonar In-Situ Mode Assessment System (SIMAS), developed by the Navy's Advanced Systems Technology Office, which uses in-situ and real-time data, models, and predicts sonar sensor performance. SIMAS II systems will be installed on all 'Ticonderoga' class cruisers and 'Arleigh Burke' and 'Spruance' class destroyers. In its SIMAS II configuration, the AN/UYQ-65 has over 20 spare VME slots available for other applications.

The AN/UYQ-65 is being further developed to host the US Navy's Tactical Data Support System (TDSS) offering passive sonar management, MSTRAP (torpedo alert), active intercept system, bistatics, onboard trainer, EC-16 and the SQS-56 upgrade.

The workstation incorporates a high-performance, high-resolution, dual monitor with integrated touchscreens, colour, raster, and multiprocessor capable of supporting both tactical graphics and acoustic sensor display formats. It can accept all AN/SQQ-89 sonar sensor (SQS-53C, SQR-19 and SQQ-28) display, tactical and alert data over its interfaces, performing all display updates in less than 100 ms. It is available to replace existing OJ-452 display stations when the AN/SQQ-89 system architecture evolves sufficiently to allow insertion of modern computers and UNIX-based workstations.

Operational status
Being manufactured under a US$65 million contract for the US Navy. Being fitted to 'Tarawa' and 'Wasp' class amphibious vessels and 'Kidd' and 'Spruance' class destroyers.

Contractor
DRS Electronic Systems Inc, A Division of DRS Technologies Inc, Gaithersburg, Maryland.

UPDATED

MDE

Type
Military Display Equipment (MDE).

Description
The MDE product line provides plug-compatible emulation of the US Navy standard AN/UYA-4 and AN/UYQ-21 display equipment at a very much reduced cost compared to the military equipment. The line includes commercial-grade equivalents of the fully militarised tactical and acoustic display equipment, video signal simulator and communication systems.

The display consoles are based on the modular design of the military OJ-535 terminal, digital TV graphics generator and radar scan converter. The MDE products provide command interpreter functions to support emulation of the military OJ-194, OJ-197, OJ-451 and OJ-452 consoles in a physical replication of the military OJ-535 terminal and base unit. The products feature reconfigurable operator panels that support multiple platform baselines in test and training applications.

The radar video simulator supports plug-compatible emulation of the SM-441 video signal simulator and features programmable radar parameters to simulate any surveillance or navigation radar in the Navy's inventory. Additional features include complex scenario generation and control of target, clutter and jamming environments which are anticipated beyond the year 2000.

The communication system offers connectivity to the military sound-powered interphone and radio-telephone switchboards. Both the AN/UYA-4 and AN/UYQ-21 communications panels are supported and may be interfaced to the patch panel to provide a stand-alone communication system.

Operational status
Under contract for the US Navy.

Contractor
DRS Electronic Systems Inc, A Division of DRS Technologies Inc, Gaithersburg, Maryland.

VERIFIED

AN/ASA-66 display

Type
Tactical data display for the P-3C Orion.

Description
The AN/ASA-66 is a multipurpose cockpit display for the P-3C Orion maritime patrol aircraft. A 9 in (23 cm) diameter CRT provides pilot and co-pilot with real-time presentation of tactical situations. Graphic and alphanumeric data are presented on the high-brightness and contrast screen for viewing in either high ambient or controlled lighting conditions. The display is designed to be driven from a remote display generator to provide analogue deflection and video drive from computer data or direct from aircraft sensors.

Operational status
In operational service aboard P-3C Orion aircraft.

Contractor
Lockheed Martin, Yonkers, New York.

UPDATED

AN/ASA-66 tactical data display system

AN/ASA-82 display system

Type
Tactical data display system for the S-3A.

Description
The AN/ASA-82 tactical data display system (TDS) is the primary data display for the US Navy S-3A Viking maritime aircraft. It serves as a real-time link between the four-man crew and the various electronic sensors which the aircraft carries for its task of maritime reconnaissance and anti-submarine warfare. High-speed, high-density data in the form of alphanumeric symbols, vectors, conic projections and other appropriate display formats from both acoustic and non-acoustic sensors are presented to the crew. Information is stored, updated and refreshed by the onboard general purpose digital computer and selectively displayed. Tactical and tabular data, controlled by the computer, are routed to the display via the Display Generator Unit (DGU), which also provides the computer with display fault status information on a priority basis, controls the routeing of display information and generates the system built-in test functions.

The equipment consists of five CRT displays in addition to the DGU. The TACCO and SENSO (tactical co-ordinator and sensor operator) are each provided with identical multipurpose display units in their respective consoles. In addition, the SENSO has an Auxiliary Readout Unit (ARU), the co-pilot has a multipurpose display, and the pilot has a display which presents a summary tactical plot. The DGU provides all of the displays with digital computer data except the ARU which receives acoustic data direct from the acoustic data processor.

Operational status
In operational service.

Contractor
Lockheed Martin, Yonkers, New York.

VERIFIED

US Navy S-3A Viking ASW maritime patrol aircraft, showing location of various items of equipment – comprising AN/ASA-82 tactical data display system

TRANSDUCERS

BRAZIL

ASM-1/50

Type
Multifrequency sonar transponder.

Development
The multifrequency sonar target Model ASM-1/50 is a microprocessor-based acoustic transponder which can identify the sound frequency sourced by the ship's sonar in the 3 to 15 kHz continuous frequency range. It transmits back to the ship an identical frequency response signal with a determined interval, from a waterproofed ultrasonic transducer.

Powered by 16 alkaline size D batteries, the transponder can operate down to a maximum depth of

50 m in Sea State 4. It has an operating autonomy of up to 100 hours for one transmission in each 10 seconds.

Specifications
Length: 1.4 m
Width: 260 mm
Weight: 29.2 kg
Max operating depth: 50 m
Sea State: 4
Autonomy: Up to 100 h for 1 transmission each 10 s

Contractor
CONSUB SA, Rio de Janeiro.

VERIFIED

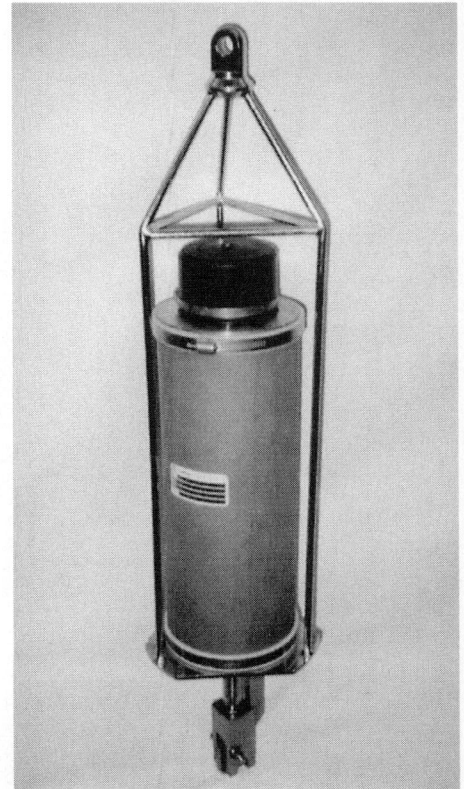

The ASM-1/50 sonar transponder
1997

CANADA

Depth-compensated RingShell Projector (RSP)

Type
ASW research and training device.

Development
The ringshell underwater sound projector was originally developed by the Defence Research Establishment Atlantic in Dartmouth, Nova Scotia, Canada. Transfer of the technology to Sparton of Canada Ltd was initiated in 1981. Sparton is the sole licensee for production.

Models of various sizes, weights and performance characteristics have been produced incorporating industrial design improvements to increase the overall efficiency. Design for models resistant to underwater explosive shock are in development.

Description
The depth compensated RSP is a Class V flextensional transducer that can efficiently produce a high acoustic output over a broad operating band. With its passive pressure compensation system, a ringshell projector has an operating depth exceeding 400 m and a safe immersion depth exceeding 500 m.

RSPs are being used in a variety of applications where high power and efficiency are required at relatively low frequencies. As a research tool, the projectors have been employed as single units or as part of large arrays. Further, passive ASW training is available without the use of a target submarine by generating signals of desired frequencies and levels. The projector can also be adapted for use as a separate active adjunct element to current passive arrays, or as part of a low-frequency, high-power, deep submergence sonar. Applications also involve using RSP-based systems to calibrate and evaluate the performance of towed arrays.

Because the Class V RSP has only two fundamental modes of vibration (ring and shell modes), its element-to-element interactions are minimised when compared to other technologies. Reduction of these interactions is extremely significant when operating projectors in large arrays, and allows for a much tighter projector spacing. In addition, the design results in low mechanical stresses in the shell and ring.

The RSP consists of a glass fibre-wrapped, piezoelectric ceramic ring fastened between two convex-domed spherical metals (see diagram). Stimulation of the ceramic ring by an electrical signal causes vibration of the ring in the radial plane. The

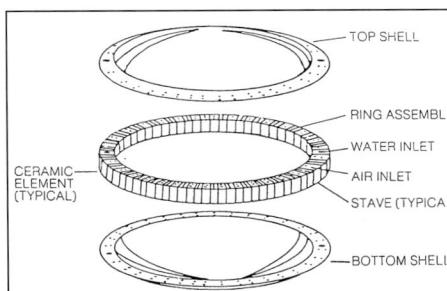

Ringshell component diagram

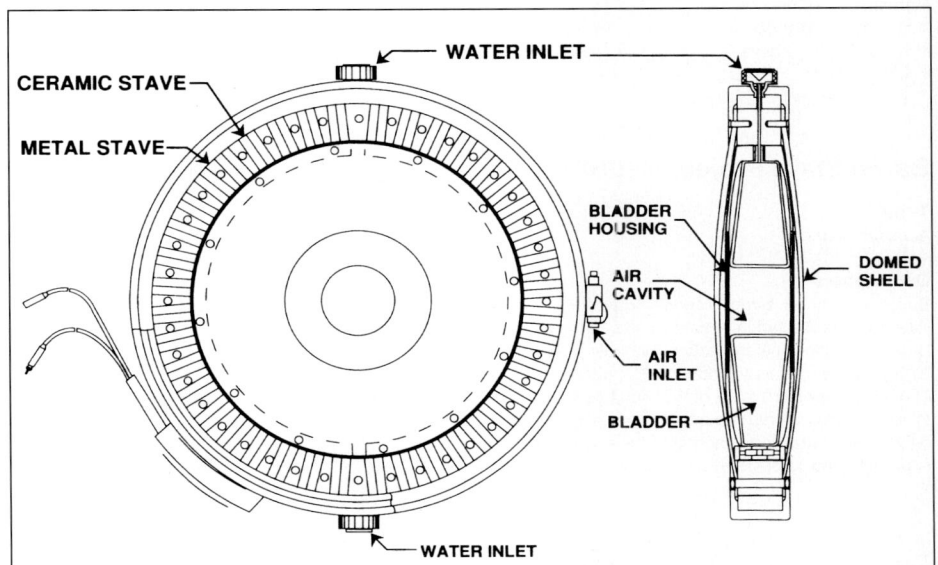

Depth compensated ringshell projector

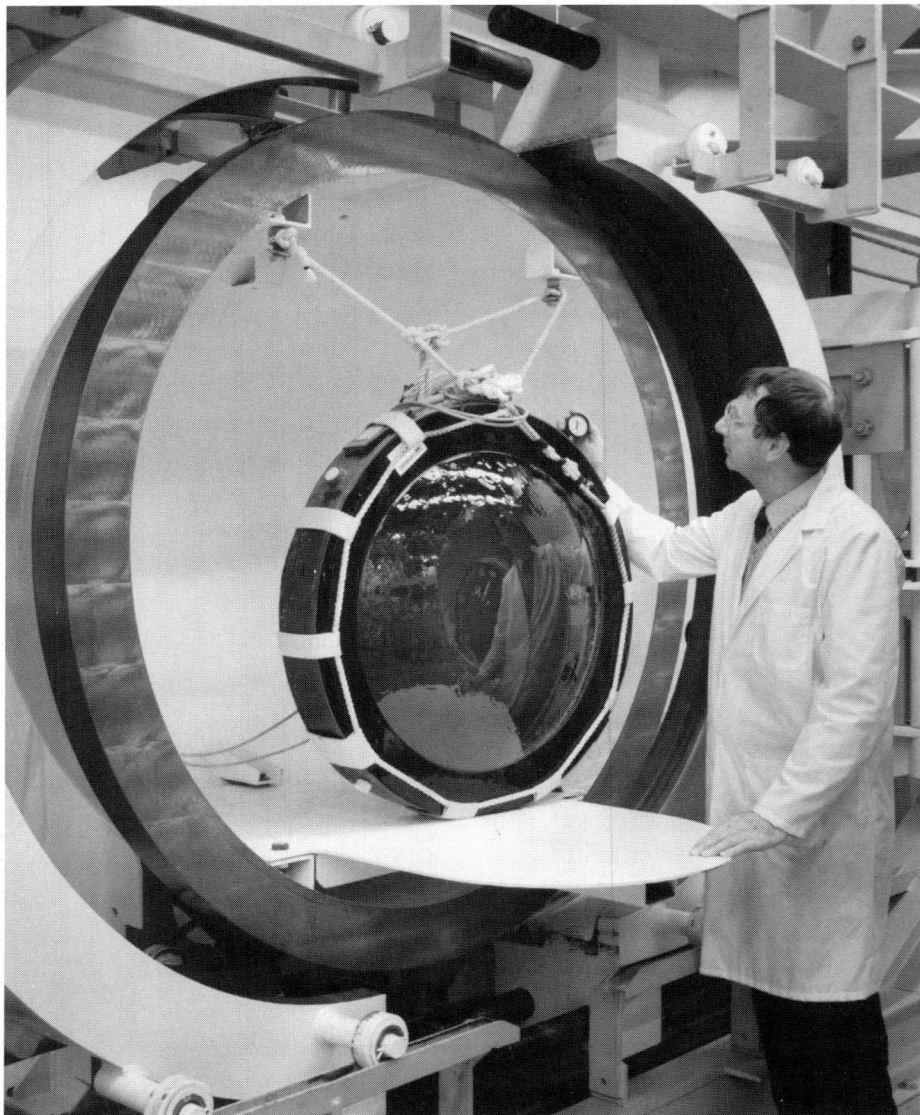

radial motion of the ceramic ring drives the shells in a flexure mode with a significant enhancement of the volume velocity at the shell surfaces.

The projector design is flexible and can be tailored to specific requirements for frequency, bandwidth and power. The resonant frequency can be adjusted for specific ring diameters over a three octave band, by altering the shell curvature and thickness. Acoustic power for a given ring size is controlled by the ceramic ring cross-sectional area. Bandwidth is adjusted by modifying the acoustic radiating surface.

Passive depth compensation is effected through the use of a bladder, housed within the inner shell of the RSP. The bladder is open to the sea and is allowed to expand or contract within the pressurised chamber of the RSP. Because of the depth compensating system and the pressurised air cavity, the depth capability of the RSP exceeds three times the range of one without these features.

The design of ringshell projectors by computer is the result of extensive finite element modelling, of which the accuracy and correlation with test results have been confirmed over several years.

Good thermal contact between seawater and the ceramic ring, both outside and inside the projector, allows the use of projectors for extended durations. The standard operating drive voltage of 3,000 V provides a duty cycle of greater than 10 per cent.

The table lists some of the characteristics of units manufactured and delivered to various customers. These units tracked computer data with extremely good fidelity (model number codes list the diameter in inches as the first two digits and resonant frequency as the last four digits. Resonant frequencies are listed at 30 m depth and source levels are based on a 3,000 V RMS drive).

Operational status
In production. RSPs have been deployed from surface ships and submarines as hull-mounted or VDS transmitters, in single or multi-element arrays.

Contractor
Sparton of Canada Ltd, London, Ontario.

Cutaway of depth compensated RSP

Model number	Resonance Frequency*	Resonance Source level†	Bandwidth at resonance‡	A	B	Ring size diameters C	D	E Min	Comments
18A0325	325 Hz	205 dB	32 Hz	18.6 in (47.2 cm)	21.6 in (54.9 cm)	3.8 in (9.6 cm)	2.9 in (7.4 cm)	24.0 in (61.0 cm)	219 dB possible at 3.6 kHz
18A1000	1000 Hz	207 dB	235 Hz	18.6 in (47.2 cm)	21.6 in (54.9 cm)	6.5 in (16.5 cm)	3.1 in (7.9 cm)	24.0 in (61.0 cm)	217 dB possible at 3.6 kHz
34A0400	400 Hz	211 dB	85 Hz	34.6 in (87.9 cm)	37.6 in (95.5 cm)	9.3 in (23.6 cm)	4.5 in (11.4 cm)	24.0 in (61.0 cm)	220 dB possible at 2.5 kHz
34A0610	610 Hz	213 dB	160 Hz	34.6 in (87.9 cm)	37.6 in (95.5 cm)	12.2 in (30.1 cm)	4.7 in (11.9 cm)	24.0 in (61.0 cm)	220 dB possible at 2.5 kHz

Nominal technical data and design parameters provided for guidance only. Contact Sparton of Canada Ltd for specific requirements.
Notes:
* at nominal 30 m depth
† at 3,000 V RMS drive
‡ at 3dB points

VERIFIED

Barrel Stave Projector (BSP)

Type
Acoustic projector.

Description
Built by Sparton under licence from the Defence Research Establishment, Atlantic (DREA), the BSP is a Class III flextensional transducer capable of providing high power and broad bandwidth in a very small unit. The unit comprises a stack of PZT ceramic rings bolted to end plates. Staves mounted in an inverted barrel configuration are also attached to the end plates. The expansion and contraction of the stack is transmitted to

the staves and results in an enhanced volume velocity. With diameters of 10 cm to 30 cm, BSPs are able to radiate efficiently at frequencies below 1 kHz. The transducer is ideal for air-deployable applications and for use in multi-element towed or volumetric arrays.

Operational status
In production.

Contractor
Sparton of Canada Ltd, London, Ontario.

VERIFIED

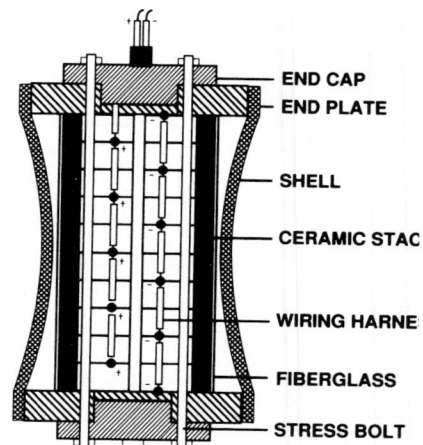

- END CAP
- END PLATE
- SHELL
- CERAMIC STAC
- WIRING HARNE
- FIBERGLASS
- STRESS BOLT

Cutaway of BSP

Free-Flooding Ring (FFR) projector

Type
Acoustic projector.

Description
The FFR projector consists of a ring of piezoelectric ceramics contained within a neoprene boot which produces a toroidal beam pattern around the horizontal plane. The unit has exceptional properties for low-frequency active sonar applications, very wide bandwidth ($Q^7 = {\sim}^1$), very high efficiency (greater than 90 per cent), and unlimited depth capability. The FFR projector operates in the 500 Hz to 8 kHz frequency band and can produce source levels in excess of 220 dB per element at 1 kHz.

Operational status
In production.

Contractor
Sparton of Canada Ltd, London, Ontario.

VERIFIED

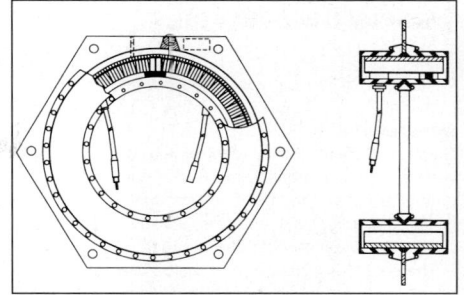

Cutaway of FFR projector

Resonant Pipe Projector (RPP)

Type
Acoustic projector.

Description
The RPP consists of a ring of piezoelectric ceramics attached to the centre of a large pipe. The ring excites the pipe, which acts as an acoustic waveguide. Because of its free-flooding design, the RPP has unlimited depth capability and can operate at or below 250 Hz. The unit was developed specifically for the oceanographic community and is being used as a source for SOFAR (Sound Ocean Fixing And Ranging) experiments.

Operational status
In production.

Contractor
Sparton of Canada Ltd, London, Ontario.

Specifications

	Frequency range*	Source level†	Dry weight	Efficiency	Maximum operating depth	Cavitation depth	Bandwidth at resonance‡	Outside diameter	Thickness
Free flooded ring									
13FA2000	2,000-4,000 Hz	216 dB		90%	Unlimited		2,000 Hz	31 cm	15 cm
27FA200	1,000-2,000 Hz	222 dB		90%	Unlimited		1,000 Hz	68 cm	
Resonant pipe projector									
28PA0260	260 Hz	205 dB	242 kg	75%	Unlimited	111 m	4 Hz	71 cm	178 cm
Barrel stave projector									
03BA1,100	1,100 Hz	191 dB	2 kg	70%	200 m		260 Hz	8 cm	18 cm
06BA0550	550 Hz	200 dB	10 kg	70%	250 m		110 Hz	8 cm	26 cm

Nominal technical data and design parameters provided for guidance only. Contact Sparton of Canada Ltd for specific requirements.
Notes:
* at nominal 30 m depth
† at 3,500 V RMS drive
‡ at 3 dB points

VERIFIED

SUBEX

Type
Acoustic noise augmenter.

Description
The SUBEX system has been designed to provide training for passive sonar operators in submarine detection, localisation and tracking techniques at actual submarine depths and speed, by imitating and augmenting the sound generated by submarines. The signal generated is used during naval training exercises to enhance or disguise the boat's radiated noise. The level and type of augmentation can be adjusted as the scenario requires. The standard SUBEX system comprises a Model 18SA0350 RSP housed within the sail of the submarine and a small, remote, lightweight display console which interfaces with an RSP controller. The RSP is a depth compensated, low-frequency, wide bandwidth acoustic source.

The flexible mounting requirements of the standard SUBEX system enable it to be easily modified to fit most existing submarines with minimum intrusion in the main command centre area. An event recorder allows future recall of settings for comparison with other sensor information. Analogue inputs to the control unit allow the system to generate signals other than discrete tones. Lockouts on the main control unit ensure only authorised use of the system.

Specifications
Number of tones: 4
Frequency range: 60-600 Hz (nominal)
Frequency separation: 1 Hz
Operational depth: 2-300 m

Contractor
Sparton of Canada Ltd, London, Ontario.

VERIFIED

UNITED KINGDOM

Thomson Marconi transducers

Description
Thomson Marconi Sonar Ltd manufactures a variety of transducers, arrays and acoustic devices. Many of the latest elements are modular devices designed to be configured into a variety of array shapes to meet individual needs. Emphasis has centred on the development of specialist polymer and ceramic techniques for sonar devices.

The design and manufacture of sonar systems and acoustic devices include large hull-mounted ship and submarine active/passive transducer arrays, as well as flank and towed arrays for the Royal Navy. The company manufactures a wide range of underwater transducers covering the frequency range from 1 Hz to 6 MHz, including barrel staves, piston stacks, tubular and spherical ceramic transceivers and omnidirectional wideband hydrophones.

The latest transducer development concerns the use of fibre optic hydrophones for sonars. This technology, although still in its infancy, is considered to offer a number of benefits over traditional ceramic hydrophones, depending on the particular application. Among benefits noted are a reduction in overall power requirements and the ability to feed power over long distances to wet end components (as in towed arrays) with reduction in strength of power.

Contractor
Thomson Marconi Sonar Ltd, Templecombe.

UPDATED

Thomson Marconi Sonar Ltd towed arrays being assembled

Graseby transducers

Type
Wide range of various types of transducer.

Description
Graseby Dynamics manufactures a wide range of various types of transducer. The standard range of omnidirectional spherical hydrophones is suited to a variety of applications, and specialised devices can be developed and manufactured to order.

The spherical hydrophones can be used individually or mounted in arrays to achieve specific directional characteristics. The hydrophones are manufactured from two silver-plated electrostrictive hemispheres which are bonded together and encapsulated in polythene, polyurethane or rubber, using specialised techniques. The connecting cable is moulded to the hydrophones during the process. The hydrophones range in size from 12.5 to 40 mm with resonant frequencies from 140 kHz to 42 kHz respectively.

A wide range of piston-type transducers are currently in service with surface and underwater vessels of the Royal Navy and the navies of other countries worldwide. They are used in a variety of modes, from single elements for underwater telephones through to complete arrays for sonar systems. These transducers are normally associated with high power and low frequency (that is, 2 to 50 kHz).

Tubular transducers are suitable for transponders and pingers. These rugged, lightweight transducers have a low Q and can be used as a single element or in an array configuration, depending on the bandwidth required. The transducers consist of a tube of lead zirconate titanate as the active material, encapsulated in polyurethane for acoustic coupling. The transducers can be supplied with different mounting arrangements and cable lengths to customer requirements.

Bimorph transducers are low-frequency devices suitable for communication and calibration applications. They comprise a trilaminar bimorph of lead zirconate titanate encapsulated in epoxy resin. Units can be used as a single element or in an array configuration to obtain directional characteristics.

Operational status
Graseby transducers are used in the following sonars: Royal Navy Type 162, 184, 195, 778, 2001, 2007, 2008 and 2016; Graseby Type G750; and Signaal Type PHS32.

Contractor
Graseby Dynamics Ltd, Watford.

UPDATED

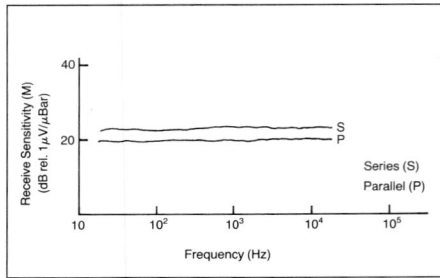

Typical projector sensitivity curve for a tubular transducer (Courtesy Graseby Dynamics)
1999/0043266

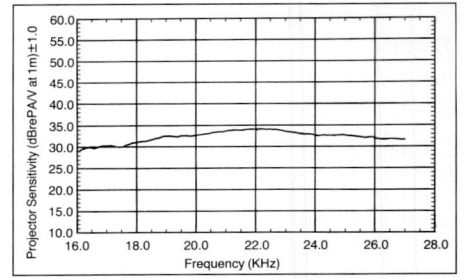

Typical spherical hydrophone sensitivity curve (Courtesy Graseby Dynamics)   *1999*/0043265

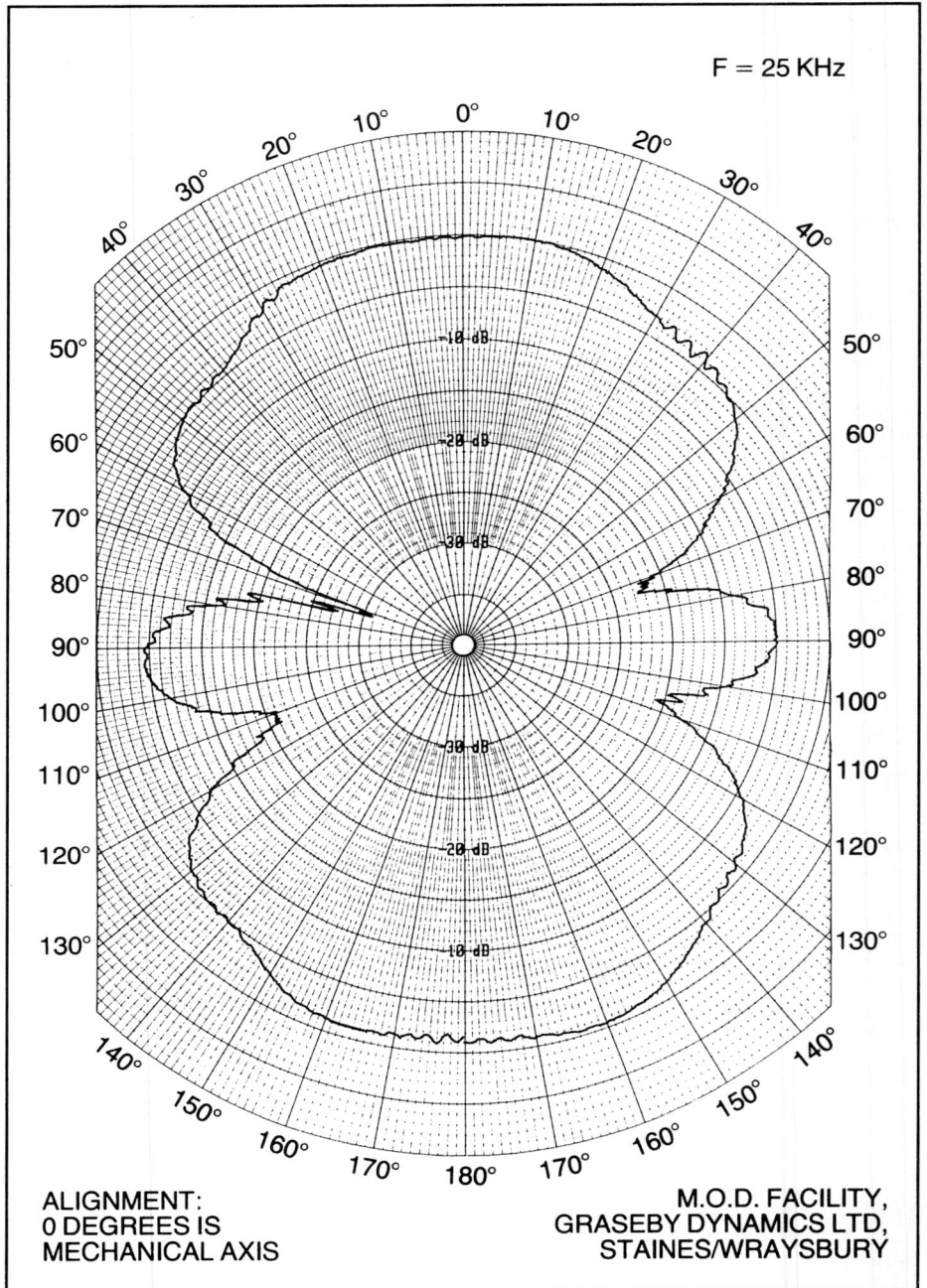

F = 25 KHz

ALIGNMENT:
0 DEGREES IS
MECHANICAL AXIS

M.O.D. FACILITY,
GRASEBY DYNAMICS LTD,
STAINES/WRAYSBURY

Directivity beam pattern for a tubular transducer (Courtesy Graseby Dynamics)
1999/0043247

Flextensional transducers

Type
Underwater acoustic transducer devices.

Description
BAeSEMA has developed a range of flextensional transducers for underwater applications. These comprise a piezoelectric ceramic stack operating in piston mode and enclosed in an elliptical shell. These transducers can be built for operation at frequencies from 300 Hz to 3,500 Hz, either from a range of standard designs, or to suit individual requirements. They can be assembled into staves or arrays to achieve specific directivity requirements. Research is continuing, both on new types of transducers, particularly those using magnetostrictive rare earth metals, and on increasing the depth performance of the present range of transducers.

Operational status
BAeSEMA has supplied low-frequency flextensional transducers to the US Naval Underwater Systems Center. Transducers have also been supplied to France, Japan, Korea, Norway and to the UK MoD. Flextensional transducers are a key component in the low-frequency ATAS (Active Towed Array Sonar).

Contractor
BAeSEMA, Bristol.

VERIFIED

UNITED STATES OF AMERICA

Low-frequency Transmit Subsystem (LTS)

Type
Flextensional transducer acoustic source array.

Description
The LTS is a large, hydrodynamically stabilised array of low-frequency, active, high-powered projectors with associated power amplifiers, handling equipment, software, data processing and a unique transmit control beamformer.

The LTS is claimed to be the largest flextensional system ever built. LTS forms part of the US Navy's SURTASS system, the total system including a towed hydrophone array, processing and display equipment, communications and navigation segments and a shore facility (qv).

Operational status
Two LTS sets have been built under a contract issued in October 1990. One set is for technical evaluation and the other for operational evaluation. Hardware endurance testing commenced in May 1991 and was completed in the summer of 1991. The first production module was delivered early in July 1991 and the system is in full-scale production.

Contractor
Sanders, a Lockheed Martin Company, Manchester, New Hampshire.

VERIFIED

Model 30

Type
High-powered projector.

Description
The Model 30 is a high-powered, low-frequency, broadband, flextensional transducer with an uncompensated operating pressure of 300 psi. The transducer weighs 96 lb in air with the coil, and has a minimum efficiency rating of 80 per cent.

Operational status
The Model 30 has been used by the US Navy in a variety of exercises and operational tests.

Contractor
Sanders, a Lockheed Martin Company, Manchester, New Hampshire.

VERIFIED

Model 40

Type
Flextensional transducer.

Description
The Model 40 flextensional transducer is a high-power, low-frequency, broadband flextensional transducer.

It has an uncompensated operating pressure of 300 psi, weighs 185 lb in air with its coil and has a minimum efficiency rating of 80 per cent.

Operational status
The Model 40 has been used by the US Navy in a variety of exercises and operational tests.

Contractor
Sanders, a Lockheed Martin Company, Manchester, New Hampshire.

VERIFIED

RESON transducers and hydrophones

Type
Acoustic transducers and reference hydrophones.

Description
RESON manufactures transducers and hydrophones for a wide range of underwater measuring systems and industrial products. The equipments are available for systems and products such as echo-sounders, flow-, level- and velocity meters, side scan systems and sonar arrays.

A continuing programme of research and development ensures that equipments are available to meet the most demanding customer requirements as well as for simple transmit/receive operations. The product line encompasses both omnidirectional and directional transducers in a wide frequency range covering 3 kHz to 2 MHz.

Contractor
RESON Inc, Goleta, California.
RESON A/S, Slangerup, Denmark.
RESON Offshore Ltd, UK.

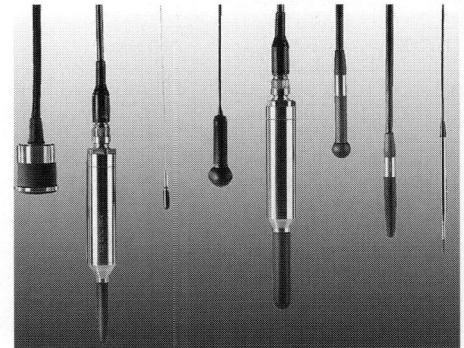

RESON standard hydrophones
1998/0009871

Model	Description	Resonant Frequency (Usable Bandwidth)	Vertical Directivity Pattern	Horizontal Directivity Pattern	Transmitting Sensitivity (+/- 3dB)	Receiving Sensitivity (+/-3dB)	Maximum Power	Depth (Survival)	Weight in air with cable
TC 1010	Telephone/ Distress Pinger	12 kHz (7-16)	200°	Omni	+ 129 dB	− 185 dB	300 W	800 m	3.2 kg
TC 1032	Telephone Transducer	8 kHz (6-15)	80°	Conical	+ 139 dB	− 162 dB	400 W	800 m	5.8 kg
TC 1035	Communication Transducer	9.5 kHz (8-11)	70°	Omni	+ 148 dB	− 184 dB	1,000 W	3,000 m	9.6 kg
TC 1037	Telephone Transducer	8 kHz (6-16)	80°	Conical	+ 145 dB	− 171 dB	400 W	800 m	
TC 2016	Survey Transducer Echo-sounder	200 kHz +/−10	9°	Conical	+ 173 dB	− 187 dB	300 W	30 m	1.8 kg
TC 2024	Survey Transducer Echo-sounder	200 kHz +/−10	9°	Conical	+ 173 dB	− 187 dB	300 W	30 m	2.3 kg
TC 2032	Side-scan Transducer	60 kHz +/−6	53°	1.8°	+ 175 dB	− 179 dB	1,000 W	600 m	9.6 kg
TC 2046	Echo-sounder Transducer	200 kHz	18°	Conical	+ 160 dB	− 192 dB	50 W	30 m	0.5 kg
TC 2053	Triple Beam Width Transducer for Survey Purposes w/Adaptable Footprint	100 kHz +/−5 kHz		Conical				50 m	10 kg
A			30°		+ 162 dB	− 192 dB	200 W		
B1			10°		+ 171 dB	− 183 dB	500 W		
B2			5.5°		+ 176 dB	− 178 dB	1,500 W		
TC 2078	Directional Transducer for Echo-sounder	300 kHz +/−15	2.6°	Conical	+ 179 dB	− 184 dB	1,000 W	400 m	7.8 kg
TC 2080	Echo-sounder Transducer	30 kHz (27-35)	45°	Conical	+ 161 dB	− 182 dB	300 W	250 m	3.6 kg
TC 2084	Echo-sounder Transducer	33 kHz (28-34)	21°	Conical	+ 166 dB	− 178 dB	1,000 W	30 m	5 kg

Model	Description	Resonant Frequency (Usable Bandwidth)	Vertical Directivity Pattern	Horizontal Directivity Pattern	Transmitting Sensitivity (+/- 3dB)	Receiving Sensitivity (+/-3dB)	Maximum Power	Depth (Survival)	Weight in air with cable
TC 2088	Side scan Transducer	100 kHz +/−20	30-35°	1.9°				600 m	
TC 2105	Positioning/ Navigation Echo-sounding	430 kHz +/−20 (430-500 kHz)	50°	Conical	+ 166 dB	− 178 dB	10 W	50 m	
TC 2111	Compact Echo-sounder Transducer	200 kHz +/−10	18.3°	Conical	+ 163 dB	− 189 dB	50 W	30 m	0.4 kg
TC 2115	Broad-band Transducer for Echo-sound	30 kHz (28-34)	23°	Conical	+ 168 dB	− 178 dB	1,000 W	50 m	5 kg
TC 2116	Broad-band Transducer for Echo-sound	50 kHz (40-110)	13.5°	Conical	+ 172 dB	− 177 dB	1,000 W	40 m	5 kg
TC 2122	Dual Frequency Survey Transducer	33 kHz 200 kHz	20° ,9°	Conical, Conical	+ 167 dB + 170 dB	− 178 dB − 187 dB	1,000 W, 450 W	50 m	5 kg
TC 2127	Narrow Beam Survey Transducer	600 kHz +/−10	3.2°	Conical	+ 173 dB	− 187 dB	450 W	50 m	2.3 kg
TC 2130	'BASS' Broad-band Transducer	100 kHz, 150 kHz, 200 kHz	16.4°, 12.4°, 8.1°	Conical, Conical, Conical	+ 160 dB + 158.4 dB + 155.5 dB	− 177 dB − 180 dB − 185 dB			2.3 kg
TC 3012	Transducer for Sound Velocity Equip	1 MHz +/−50 kHz	5.8°	Conical	+ 173 dB	− 202 dB	10 W	500 m	0.075 kg
TC 3017	2 MHz Narrow Beam-width Transducer	2 MHz	2.2°	Conical					0.075 kg
TC 3019	Air Transducer for Pulse-Echo	2 MHz +/−100 k	5°	Conical	upon request	request	5 W	N/A	0.010 kg
TC 3021	Transducer for Sound Velocity Equip	2 MHz +/−150 k	2°	Conical	+ 186 dB	− 205 dB	5 W	700 m	0.035 kg
TC 3022	Universal Transducer in CuNi10-Alloy	1 MHz +/−50 kH	5.8°	Conical	+ 174 dB	− 202 dB	10 W	1,000 m	0.1 kg
TC 3027	Transducer for Sound Velocity Equip	1 MHz +/−100 k	5.8°	Conical	+ 170 dB	− 200 dB	10 W	500 m	0.04 kg
TC 3029	Universal Transducer	500 kHz	11°	Conical	+ 165 dB	− 192 dB	5 W	500 m	0.04 kg
TC 4003	Pinger Location Hydrophone w/Line Driver w/Preamp, box	37 kHz, 45 kHz, 70 kHz +/−10 k	84.1°, 86.3° 48.3°	34.7° 28.4° 16.1°	N/A, N/A, N/A	−157 dB, −158 dB, −144 dB		900 m	2 kg
TC 4004	Explosion tested Submarine Hydrophone	1 - 11 kHz (17)	Omni	Omni	+ 107 dB	−197 dB	–	800 m	0.35 kg
TC 4013	Miniature Reference Hydrophone	120 kHz (1 Hz-170 kHz)	270° (Omni)	Omni	+ 133 dB (80- 130 dB)	− 210 dB	–	900 m	0.075 kg
TC 4014	Broad-band Spherical HY w/26dB Preamp	15 Hz - 480 kHz	270° (Omni)	Omni	N/A	− 187 dB (Linear)	<50 mA (12 - 24 V)	1,000 m	0.65 kg
TC 4023	Robust Reference Hydrophone	100 kHz (10- 200 kHz)	270° (Omni)	Omni	149.5 dB (100- 30 dB)	− 200 dB (−207 dB)	–	800 m	1.5 kg
TC 4027	General Purpose HY w/20dB Preamp	10 Hz - 10 kHz	Omni	Omni	N/A	− 180 dB	60 mic amp (steady st.)	300 m	0.133 kg
TC 4032	Low Noise / High Sensitivity HY w/ 10dB Preamplifier	5 Hz - 120 kHz (flat 1 - 80 kHz)	270° (Omni)	Omni	N/A	− 170 dB	<40 mA (12 - 24 V)	900 m	0.825 kg
TC 4033	Robust, Spherical Reference Hydrophone	100 kHz (1 Hz - 160 kHz)	270° (Omni)	Omni	+ 149 dB (96- 136 dB)	−203 dB	–	900 m	1.5 kg
TC 4034	Ultra Broad-band Spherical Reference HY	300 kHz (1 Hz - 430 kHz)	270° (Omni)	Omni	+ 145 dB (75- 135 dB)	− 218 dB (Linear)	–	800 m	1.6 kg
TC 4035	Miniature Probe HY w/10dB preamp, 4mm tip, High Freq	50 - 900kHz	270° (Omni)	Omni	N/A	− 214 dB (Linear)	<15 mA (10 - 18 V)	200 m	0.7 kg
TC 4037	Spherical Ref, HY, Low Frequencies, High Pressure	1 Hz - 100 kHz	270° (Omni)	Omni	–	− 195 dB		2,000 m	
TC 4038	Miniature Reference Hydrophone - 4mm (TC 4035 w/o preamp)	10 - 800 kHz (linear 100-500 kHz)	270° (Omni)	Omni	–	− 227 dB @100 kHz		20 m	0.023 kg
TC 4050	Mini Flush-mounted HY 10mm tip	1 Hz-120 kHz	>180 (Omni)	Omni	N/A	− 216 dB		400 m	0.7 kg

UPDATED

Transducers

Type

Bathymetric and hydrographic survey echo-sounder transducers.

Description

Ocean Data Equipment Corporation offers a complete line of acoustic transducers for shallow and deep water bathymetric and hydrographic surveying and for sub-bottom profiling. Transducers are available in the following frequencies in narrow and wide beamwidths: 3.5, 7, 12, 24, 33, 40, 100 and 200 kHz. Single transducers and arrays of transducers for sea chest or towfish installation can be provided. Wide frequency bandwidth transducers for CHIRP sub-bottom profiling are standard.

Contractor

Ocean Data Equipment Corporation, E Walpole, Massachusetts.

VERIFIED

Model number	Frequency	Beamwidth	Depth*	Comment
TR-109	3.5 kHz	**	10,000 m	4 kHz CHIRP; arrays replaces TR 75A
TC-4NB	4 kHz	60°	5,000 m	CHIRP
TC-2040	7 kHz	65°***	10,000 m	8 kHz CHIRP; arrays
TC-12NB	12 kHz	18°	10,000 m	CHIRP
TC-12/34	12/34 kHz	30/10°	6,000 m	CHIRP; replaces AN/UQN-1
TC-24	24 kHz	22°	2,000 m	
7426C	24 kHz	22°	2,000 m	Through hull mount
TC-33	33 kHz	21°	1,400 m	
7510	40 kHz	20°	1,000 m	
2572	40 kHz	43°	1,000 m	
7511	100 kHz	9°	500 m	
7245A	208 kHz	3°	350 m	
200T5H	200 kHz	10°	250 m	

Note: * System depth performance is a function of frequency, power level transducer beamwidth, efficiency, transducer mounting depth and ship specific acoustic environment

** TR-109 transducers are designed to be mounted in arrays having the following beamwidths; array of 4/53°; array of 7/43°; array of 9/35°; array of 12/30°; array of 16/26°

*** An array of 6 TC 2040 transducers has a beamwidth of 28°.

MISCELLANEOUS EQUIPMENT

CANADA

CTASP

Type
Commercial-off-the-shelf (COTS) Towed Array Signal Processor.

Description
CTASP is an advanced signal processing and display system used to detect, localise and track surface and subsurface targets. CTASP interfaces with the command and control system in order to provide fire-

control solutions. The system uses commercial high-speed DSPs within a VME chassis. CTASP uses an open system architecture and can be made to integrate with most passive towed arrays. It can be supplied either as a complete end-to-end solution, including handling systems and towed arrays, or be integrated into an existing towed array as a system upgrade. CTASP can also be offered with a Low-Frequency Active Sonar (LFAS) solution, either as original fit or as a future enhancement.

Operational status
Commercially available.

Contractor
Array Systems Computing Inc, North York, Ontario.

VERIFIED

Reconfigurable, Scalable Multiprocessor (RSM)

Type
Embedded signal processor.

Description
The RSM is a modular, fault tolerant, reconfigurable integrated data acquisition and processing system, intended for use as an embedded signal processor in a variety of sonar applications including partial digitisation and signal processing on existing sonar systems or as the basis for the new sonar systems. The RSM consists of a reconfigurable parallel array of Processing Elements (PEs), all of which are connected

across several databusses. Interprocessor data communication is achieved through time-division multiplexing on a proprietary bus. System control communications are performed over a VMEbus. The scalable PE array can be arbitrarily configured to any parallel and/or series topology suitable for a given application. Both online system health monitoring and detailed offline diagnostics are available. In the event of a recoverable error, the RSM will restore its original configuration through the dynamic reallocation of system resources.

Specifications
Input: analogue channels up to 80
total Dynamic Range >100 dB/channel

Output: analogue channels up to 80
bandwidth 500 Hz to 10.8 kHz
Processing: computational capacity 3.6 Gflops (peak) Integer and/or IEEE 32 bit floating point
Host Interface: Ethernet (802.3) and TCP/IP >500 kbytes/s

Contractor
Northrop Grumman - Canada, Burlington, Ontario.

VERIFIED

FRANCE

PIC PL13

Type
Single helmsman submarine control console.

Steering station for the 'Le Triomphant' class (Courtesy of SAGEM) **1999**/0024693

Description
The PL13 submarine control console provides all the necessary functions to control safely a submarine in the three dimensions. Depth, trim, heading and speed are

controlled in a centralised manner by the single helmsman. Platform control is achieved through the management of the submarine's diving and rudder planes, compensation and trim tanks, and propulsion system, which are under full control of the PL13 console.

A very high level of integration and safety (reliability and availability) is achieved through the PL13 architecture. The main features of the system are:
(1) three modes of operation (automatic, manual, emergency with components and modules separately dedicated to each mode both for display and control). Reversion to a lower category mode is carried out either manually or automatically in case of a fault, while changing back to a higher mode, can only be carried out automatically
(2) constant system monitoring and automatic reversal to a safer mode
(3) three-channel fault-tolerant control computer
(4) three-channel sensors (depth and trim)
(5) two-channel plane management.

The PL13 incorporates a built-in autopilot which controls accurately and silently the depth, trim and heading either in deep submerged conditions or at periscope depth under swell. This is achieved through control of the submarine's upper and lower rudders, two aft diving planes with vertical stabiliser surfaces and two forward diving planes carried on the sail.

Maintainability is a key feature of the system. It offers multilayered, built-in tests, leading straight to the identification of the defective PCB in most cases. Its modular arrangement enables a fast access to the internal racks.

Operational status
Operational on board the French Navy 'Le Triomphant' class SSBNs since 1993.

Contractor
SAGEM SA, Defence and Security Division, Paris.

UPDATED

SS Mk 1

Type
Single helmsman submarine control console.

Description
The SS Mk 1 console provides all the necessary features to control safely a submarine in the three dimensions. Depth, trim, heading and speed are controlled in a centralised manner by the single helmsman. The centralised control and monitoring station has been selected for the new 'Agosta 90B' programme. Due to its total functional and structural flexibility, it meets the present and future requirements in submarine control for boats in the 200 to 2,000 tonne displacement range.

The main functions of the system are:
(1) integrated course and depth control in three modes – automatic, manual servo-controlled, and manual non-servo-controlled

(2) calculation of settings and controls to accomplish manoeuvres
(3) remote control of buoyancy and trim tanks
(4) remote control of propulsion
(5) monitoring of all parameters relating to submarine control and safety (heading, depth, trim, speed and steering).

A high level of operational safety is ensured by integrated architecture of the system, which includes differentiated operating modes, multiple redundancies and automatic reconfiguration capability. The MMI is based on modern technology using colour active matrix LCDs and the modular management of the presentation screens.

The system integrates all hardware functions needed for submarine control including the multimode autopilot, redundant servo-control systems, monitoring devices and a large number of sensors. A high level of functional integration together with the very compact nature of the design ensure optimal installation

conditions and maintainability. The system can be handled through a 700 mm diameter hatchway and allows separate installation of the upper and lower lockers inside the submarine. The design makes the system ideally suited to upgrade and modernisation programmes.

Operational status
Prototype developed and tested in 1993. 'Agosta 90B' version currently in production to equip the new submarines for Pakistan.

Contractor
SAGEM SA, Defence and Security Division, Paris.

UPDATED

Primary batteries

Type
Batteries for heavyweight torpedoes.

Description
Developed by DCN and SAFT, these new batteries have been designed to replace batteries on current heavyweight torpedoes. SAFT is responsible for the electrochemical cells and DCN for the activation

system which is similar to the PB32 in service on the F 17 Mod 2 torpedo.

The battery features high performance and a very safe activation system. To meet the UK requirement, the same components will be used as in the F 17, but configured to meet the needs of Tigerfish.

Operational status
On offer for the upgrading of the Royal Navy Tigerfish torpedo, and has been proposed to a foreign navy to

upgrade its German SST4 torpedoes. It is also available for German-manufactured SUT torpedoes.

Contractor
DCN International, Paris.
SAFT, Romainville.

VERIFIED

GERMANY

Shore-Based Information Centre

Type
Analysis system for hydroacoustic noise and signals, and for the creation and maintenance of databases.

Description
The Shore-Based Information Centre (SOBIC) has been developed and manufactured by STN ATLAS to enable naval forces to get maximum use from the passive sonar sensors installed in surface ships, submarines, ASW aircraft, helicopters, shore-based underwater surveillance and harbour protection facilities.

Intercepted hydroacoustic noise and signals of target vessels not only reveal the bearing but also carry numerous items of information regarding the source of the sound. This information can be extracted and

understood only when it has been processed, evaluated and classified. To perform this, SOBIC acts in a central role:
(a) raw acoustic recordings from all available hydroacoustic sensors are transferred to the SOBIC, accompanied by detailed descriptions of the classified source of noise if possible
(b) the analysis personnel of SOBIC evaluate the recordings with all different kinds of signals processing available, such as Lofar (with different algorithms of pre- and post-FFT processing), Demon, spectrum analysis, audio analysis, transient analysis and sonar signal analysis. The optimum output is the identification after correlation of the classified signals with information from other sources
(c) SOBIC establishes and updates an acoustic

database of all sources possible in the operational area
(d) SOBIC distributes user-adapted databases to all vessels, shore stations and training centres equipped with passive acoustic sensors
(e) feedback reports and recordings are used to provide continuous updating of the SOBIC-controlled database.

Operational status
In service.

Contractor
STN ATLAS Elektronik GmbH, Bremen.

VERIFIED

MSC (Mine warfare Support Centre)

Type
Land-based data evaluation centre for mine warfare data.

Description
The MSC is a land-based data evaluation facility based on commercial-off-the-shelf (COTS) workstation hardware and peripherals with the following main characteristics: a central database for underwater objects; the merging and evaluation of underwater data from different sources (minehunting sonars, side scan sonars, manual inputs, and so on); and the preparation of mine warfare missions.

The MSC is fitted with the electronic sea chart capability (ECDIS). This feature allows display overlay of geographical and mine warfare data and thus easy correlation for evaluation, planning and storage. Search area and track planning for minehunting and minesweeping missions can be centralised with the aid of the MSC. Underwater information and results from previous missions can be merged with mission planning data and stored on a portable data media for use on board MCMVs to improve their MCM capability.

The flexible database organisation gives a high-storage capacity and enables the adaptation to various input sources and sensors. Likewise the output format and media for use on board MCMVs or other evaluation centres is characterised by a high degree of flexibility.

The STN ATLAS MSC allows for bidirectional data exchange with the integrated minehunting weapons system MWS 80/MWS 90 and other data sources and users.

Operational status
In operational service.

Contractor
STN ATLAS Elektronik GmbH, Bremen.

UPDATED

INDIA

MDB

Type
Modular databus.

Description
The overall mission-effectiveness of the modern

warship is critically dependent on the interconnection and intercommunication between sophisticated sensors, weapons, guidance and control, communication and navigation systems. The current trends in naval system integration worldwide is evolving towards intelligent, computer-based, semi-

autonomous subsystems integrated by loosely coupled, distributed computer architectures.

The MDB is one such general purpose multiplexed data network designed for the data communication needs of a wide variety of naval platforms. It features modular design, distributed control, redundant data paths and a centralised health monitoring system. The major advantage of this system over point-to-point cabling is in the enormous reduction in cable length, reduction in cost, weight and volume and ease of system reconfiguration for future expansion. MDB receives data inputs in analogue, synchro, resolver, discrete or digital forms and switches or distributes them to one or multiple datalinks after optional format conversion. MDB can sample data at preprogrammed rates.

Operational status
Being introduced into service in the Indian Navy.

Contractor
Bharat Electronics Ltd, Bangalore.

VERIFIED

ITALY

ORACOM

Type
Operational research tool for anti-torpedo countermeasures.

Description
The modelling of heavyweight torpedo firings incorporating wire guidance, the characteristics of surface ships and submarines, their features and torpedo alert capabilities, together with soft/hard kill anti-torpedo countermeasures systems demand advanced computer simulation techniques. The multirole, stand-alone, computer-based ORACOM simulator system is designed for underwater warfare research, development and training in this field. It offers a comprehensive tool for statistical and deterministic evaluation of various possible solutions to the torpedo threat. It incorporates, as default packages, detailed models of heavyweight torpedoes, anti-torpedo countermeasures systems, surface ships and submarines, and allows simultaneous running of all models in a wide variety of environments and scenarios. Two working consoles, assigned to the attacking submarine, and to the threatened ship/submarine, allow continuous deterministic optimisation of the tactics of both platforms together with the assessment of torpedo engagement trajectories and an assessment of their hit/kill probability.

The simulator is based on a Digital VMS workstation with DEC Windows and VAX GKS graphic library with presentation on a 21 in colour graphic display. The high-level programming language is Fortran 77.

Contractor
Whitehead Alenia Sistemi Subacquei SpA, Genoa.

VERIFIED

TETA

Type
Torpedo Engagement Training Aid.

Description
The TETA (Torpedo Engagement Training Aid) system is a stand-alone computer-based simulator designed for research, development and training in ASW and provides a comprehensive tool set for statistical and deterministic evaluation. It incorporates, as default packages, the full-scale models of an existing lightweight torpedo and anti-torpedo countermeasures system, as well as conventional diesel-electric submarines, including their torpedo alert capability. The system allows the simultaneous running of all the models in a wide variety of environments and scenarios. Using TETA, the operator can identify the optimum tactical deployment of lightweight torpedoes against a submarine, and determine the launch profile for airborne and torpedo tube-fired lightweight torpedoes. TETA will carry out an assessment of torpedo engagement trajectories and assess the success probability of lightweight torpedoes. It can also be used to determine and identify the optimum tactical use of anti-torpedo countermeasures against lightweight torpedoes. Other features include the determination of the torpedo alert range requirements of submarines and the manoeuvres the submarine must make in order to optimise the effect of the anti-torpedo countermeasures and ensure its own safety, together with an assessment of the escape probability of the submarine following these manoeuvres.

The simulator is based on a Digital VMS workstation with DEC Windows and VAX GKS graphic library with presentations displayed on a 21 in colour graphic display. The high-level language is Fortran 77.

Contractor
Whitehead Alenia Sistemi Subacquei SpA, Genoa.

VERIFIED

NETHERLANDS

Reserve lithium batteries

Type
Batteries for mines, depth charges and sonobuoys.

Description
Reserve lithium batteries for mines and depth charges are hermetically sealed and provide a guaranteed storage period of 10 years and an expected storage period of 30 years. During storage the batteries can be left unattended. Several methods can be used for activation, which occurs in less than 1 second. They are fully operable in extreme temperatures ranging from −40 to +63°C.

Operational status
In production and in use for mines, fuzes and underwater applications.

Specifications
Nominal current: from mA to several A
Nominal voltage: 2.5-40 V
Activation time: <1 s
Shelf life: >10 years
Operating temperature: -40 to +63°C

Contractor
Signaal-USFA, Eindhoven.

VERIFIED

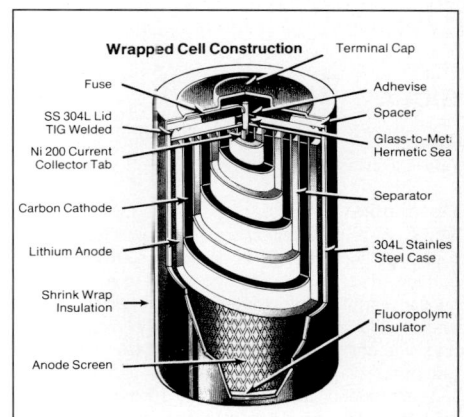

Wrapped Cell Construction

Active lithium batteries

Type
Batteries for mines, depth charges and sonobuoys.

Description
Active lithium batteries use various lithium chemistries to suit a wide range of requirements and applications. They have very high voltage per cell and can operate in extreme temperatures. Special designs are available for very high sustained current drains, including pulsed loads. The batteries are available in standard sizes and as battery packs of series and parallel combinations.

Operational status
In production.

Specifications
Capacity: ranging from 0.8 to 30 Ah per cell*
Rated current: ranging from 1 mA to 4 A per cell*

Storage life: up to 10 years
Operating temperature: -40 to +63°C per cell*
*depending on chemical system used

Contractor
Signaal-USFA, Eindhoven.

VERIFIED

NORWAY

MFPS

Type
Minefield planning, evaluation and realisation description tool.

Description
The MineField Planning System (MFPS) is a tool designed to assist the process of planning a controllable minefield. The system can be used in the various stages of planning, to create the most cost-effective minefield. When planning has been completed, the system produces complete lists of all the items needed to realise the field.

MFPS is a commercial-off-the-shelf (COTS) based system running on a standard workstation.

The system uses a very high-resolution three-dimensional bottom terrain model and a set of automatic and manual functions to assist the minefield designer in the task of designing an effective minefield. Additional visual aids are provided to assess the actual placement of equipment on the seabed.

A set of system-defined parameters, affecting minefield content and behaviour, may be altered by the minefield designer when specific details need to be studied.

MFPS allows a set of different threat scenarios to be created for simulation purposes. Using these scenarios, the system may be used to perform simulated target runs through a designed minefield. A simulation run includes target engagement, mine detonation and damage evaluation. The simulation produces data on inflicted target losses and overall mine usage and this is used by the minefield designer to evaluate the behaviour and performance of the minefield.

When a minefield layout has been completed, the system computes the cost of the complete minefield. All equipment used in the field will be included in this analysis, and any additional operator-defined overhead costs (rates for ships, man-hours and so on) not directly related to the equipment can be taken into account if required.

MFPS system equipment   *1996*

A complete description of the minefield may then be produced on paper. This includes a complete inventory of all the equipment with sub-parts in the minefield, and a colour map of the minefield area. An additional equipment list featuring the planned co-ordinates of all the equipment in the field can be produced on paper and/or on a data cassette tape for data transfer to the minelaying vessel (KNM *Tyr* of the Royal Norwegian Navy).

Operational status
In service at KNM *Tordenskjold*.

Contractor
Kongsberg Defence & Aerospace AS, Kongsberg.

VERIFIED

SPAIN

SICLA

Type
Acoustic classification system.

Description
SICLA is a powerful computer-aided classification system of acoustic targets on board submarines and surface ships. It is directly connected to any acoustic sensor output (sonar, sonobuoy, recorder and so on) and provides a powerful aid by means of simultaneous and interactive analysis, graphic tools and target database.

The frequency band covers 0 to 16 kHz, processing simultaneously in real time more than 10 Lofar, Demon, Vernier, Ali, transients analysis as well as a high-fidelity audio signal.

SICLA allows the digital recording of acquired signal and onboard or onshore quick post-analysis at up to 20 times the real time.

The modular design of SICLA allows configurations of up to 4 input channels in a reduced volume and weight.

The operational use is greatly simplified by the extensive use of the latest technologies in man/machine interface and database management, allowing powerful computer-aided hypothesis generation and verification.

Operational status
In production and in service.

Sicla display (Courtesy SAES)   *1999*/0043251

Specifications
Weight: 54 kg
Dimensions:
  (rack) 60 × 60 × 50 cm
  (monitor) 50 × 45 × 60 cm
Power: 0.4 kVA (115 V/60 Hz)

Contractor
SAES, Cartagena.

UPDATED

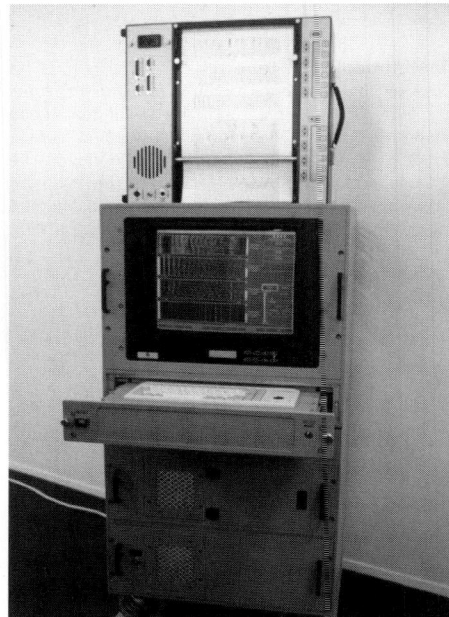

Sicla console (Courtesy SAES)   *1999*/0043252

SWEDEN

Safe Barrier System

Type
Intruder security system.

Description
Developed in co-operation with the Swedish Navy, the Safe Barrier System is a strong physical shield in combination with sophisticated alarm and surveillance techniques. It is designed to protect vital and vulnerable coastal installations against surface and subsurface intrusion. The system can be adapted to suit the anticipated type of intruder whether divers, torpedoes, mini-submarines, ROVs or surface vessels.

The system consists of: the alarm net system; the alarm; the gate; buoyancy units; anchoring devices; surface- and sub-surface surveillance; and the command centre.

Each net section is manufactured using automated construction techniques. The net mesh consists of a double insulated galvanised wire coated with polyethylene for electrical integrity and with a durable polyurethane jacket for additional protection. The knots are made of SS-reinforced solid polyurethane.

The Safe Barrier System allows authorised vessels to pass through vertically controlled gates. The gate is an alarm net section connected to the net system at the sides and the anchoring devices. At the surface the net is connected to inflatable twin hoses, stretching over the entire gate's width. When closed, the hoses are filled with compressed air. When opening the gate, remotely controlled discharge valves evacuate the air, which causes the net to sink.

The monitoring and alarm system covers all components of the alarm net system. These include the net panels, anchors, interconnections, buoyancy units and access gates. The alarm system consists of both a sabotage/intrusion alarm and a service guard.

Cutting, short circuiting or bypassing the net system(s) activates the sabotage alarm system in the command centre. The service guard ensures that the complete system can be maintained in a timely manner.

The surface surveillance and sub-surface surveillance includes acoustic, IR, radar, camera and magnetic, state-of-the-art technology.

The command centre collects and registers data for analysis from all the different sensor systems and alarm systems in the protected area. Sophisticated software integrates the different systems and presents relevant information in a logical and user-friendly format as a support for decision making and for active countermeasures.

Operational status
More than 40 installations operational worldwide.

Contractor
Safe Barrier System AB, Solna.

UPDATED

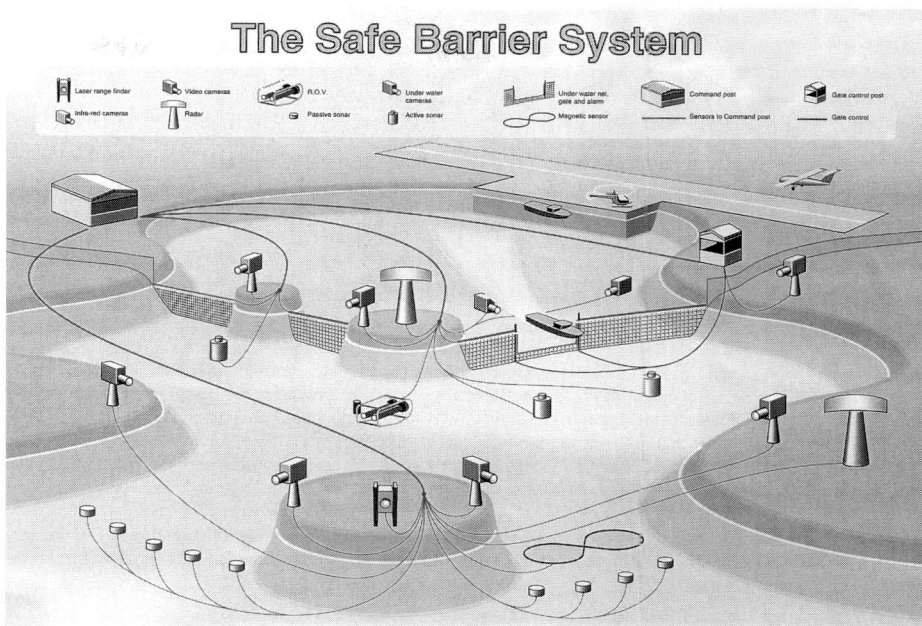

Safe Barrier System *1999*/0043248

The alarm net system lowered to allow a vessel to pass through *1999*/0043250

UNITED KINGDOM

MTSS

Type
Mine warfare Tactical Support System.

Description
MTSS is an autonomous stand-alone (or integrated) series of mine warfare tasking, planning and evaluation systems. It has flexibility to be deployed in the following variants:

(a) in a standard ISO container fitted as a command HQ for the MCM Commander

(b) as an office system version for training or shore command HQ usage

(c) as a portable version in a series of ruggedised boxes for deployment on any COOP or integrated into standard racking in any MCMV or warship.

The equipment comprises commercial-off-the-shelf (COTS) hardware and software where appropriate and is the result of many years of successful MCM operations by the Royal Navy and Mod UK/Computing Devices Company expertise in aviation electronics, software development and integration. The MTSS is a multiseat system with the information presented to the operator on high-resolution colour computer screens using the industry standard X-Windows/Motif Graphical User Interface. MTSS enables the operator to perform the required functions through the various stages of MCM route generation, planning, tasking and evaluation. MTSS communicates with the MCMVs (and other units/authorities), through its comprehensive signals package. This facility enables the transfer of formatted, structured and free text signals and utilises specialised MCMV combat system data.

Additionally, bulk route survey data can be downloaded or passed to the appropriate tactical display to ensure the integrity and completeness of the underwater picture. MTSS also displays and updates numerous totes, including mine, equipment, operation defects and ship planning. The system is UNIX-based, currently using SPARC servers and can be operated using any ship or shore supply.

Operational status
The MTSS has been in service with the Royal Navy since October 1994. Since that date it has been operated successfully on joint United States Navy exercises off the USA coast and with NATO and own national forces. The MTSS will shortly be purchased by Middle East and Asian navies with a similar requirement and other NATO countries have registered their interest. Selected modules from the MTSS are planned to be incorporated into other command system support programmes.

Contractor
Computing Devices Company Ltd, St Leonards-on-Sea, East Sussex.

VERIFIED

SPUD

Type
Spectral multichannel analytical system.

Description
The SPUD analysis equipment has been developed, using the Thomson Marconi Sonar Series 5 technology, for detailed spectral analysis of data in the field or laboratory. The system features 24 inputs with 40 independent processing channnels with spectrum or modulation processing selectable on each channel with up to 15,360 Hz bandwidth per channel. The man/machine interface is an easy-to-use menu system presented on a high-definition colour VDU which controls the processing carried out by an array of digital signal processors. The system architecture allows the efficient incorporation of the user's specialist algorithms and requirements. A built-in 105 Mbyte hard-disk stores output data for rapid redisplay of sections of interest. A removable hard-disk recorder can be included for archive and reference purposes.

Two linescan printers can be controlled by the system for continuous hard-copy output.

Operational status
In service with the Royal Navy.

Contractor
Thomson Marconi Sonar Ltd, Templecombe.

VERIFIED

CRISP

Type
Portable mission support acoustic analyser.

Description
The CRISP (Compact, Reconfigurable Interactive Signal Processor) equipment has been developed using the Thomson Marconi Sonar Series 5 technology to provide a compact portable mission support system. The current version of CRISP is now available to support the airborne processing community.

A high-resolution colour graphics display with pop-up menus and mouse roll-and-click selection allows interactive and user-friendly operation. The system uses a Windows-style environment.

Using modular boards already in operational naval service CRISP offers the following capabilities:
(1) simultaneous broadband and narrowband analysis
(2) digital Difar processing (16 channels - real time in omni and steered cardiod with bearing extraction)
(3) frequency range from infrasonic to intercept
(4) very fast time replay (×8 of standard)
(5) short-term event capture and analysis
(6) rapid software reconfigurability.

Spectral processing analysis bands cover 2,400, 1,200, 800, 400, 200, 100, 80, 40, 20 and 10 Hz. Demon processing covers input bands of 300, 600 and 1,200 Hz with analysis bands of 40, 80 and 200 Hz. In Lofar the CRISP system processes eight channels.

The CRISP architecture allows it to be an onboard data gatherer or a mission support, faster than real time analyser. This flexibility has allowed t to be successfully operated by NATO navies and air forces.

Operational status
In service with Royal Navy, RAF and Canadian Forces.

Contractor
Thomson Marconi Sonar Ltd, Templecombe.

VERIFIED

OSDI

Type
Own Ship's Data Interface system.

Description
The OSDI equipment is designed to gather information via a number of differing interfacing standards. As part of its function the equipment provides an intelligent interface to the submarine's navigational systems and the speed and depth sensors. Data is collected and distributed to the remainder of the platform weapon and sensor systems over a fibre optic data highway so that they are able to use this information in the format which they require.

The equipment comprises two identical systems which provide dual redundancy so that the resilience and availability of the navigational information distribution system is greatly enhanced. The equipment also provides a remote display of submarine depth, speed and heading for use by the command.

Operational status
Under contract for development and initial production of equipment for installation as part of the tactical weapon system in the Royal Navy's 'Trafalgar' class submarines.

Contractor
BAeSEMA Ltd, Dorchester, Dorset.

VERIFIED

Wideline 200 series

Type
Three-channel analogue and digital recorder.

Description
The Wideline 200 Series three-channel recorder provides 20 in, near photographic quality images using a direct thermal print process with 256 pixels/in with up to 256 shades of grey and 64 lines per second print speed. The machine is controlled via a 48-character alphanumeric LCD for all status and error messages, and push-button selection of range, sweep speed, trig level, delay, print direction and grey scale thresholds. The unit provides a display of any one, two or all three analogue channels with independent selection of operating functions.

Operational status
Over 500 systems have been sold worldwide and are in service with a number of navies including the Royal Navy (Type 2031 system), Royal Air Force, French Navy (Fast Time Analysis Programme – FTAS), Italian Navy (FTAS), Royal Netherlands Navy (FTAS), US Navy (FTAS), and US Navy Surface Ship Analysis Centers (SSAC).

Contractor
Ultra Electronics Ocean Systems, Weymouth, Dorset.

VERIFIED

Wideline 200 digital recorder
***1998**/0009869*

SATAID

Type
Sonar Auto-detection and Tracking, AI and Data fusion (SATAID).

Description
A sonar detection and tracking system that uses transputer technology and Artificial Intelligence (AI) to assist in the analysis of data from towed arrays has been developed. This new technology is built into a product called SATAID (Sonar Auto-detection and Tracking, AI and Data fusion). Further developments planned for this programme include the actual identification of vessel types through the use of AI.

Operational status
In service.

Contractor
EDS Defence Ltd, Hook.

VERIFIED

MOAT

Type
Mine warfare analysis tool.

Description
A Mine warfare Operational Analysis Tool (MOAT), programmed in Fortran 77 and hosted on the DEC VAX range of mini and super mini-computers, has been developed to improve the effectiveness of MCM systems. The system provides facilities for the display and annotation of MCM vessels' operational plots. The operator may interactively display selected vessel positions, distances and bearings for time or event-based replay. Minefield data may be created, edited and plotted, and MOAT can perform MCM vessels' performance analysis using nominated mine targets. Performance data are displayed in either graphical form or as a tabular summary. A feature of MOAT is a package that permits the operator to recall a complete MCM operation on either a time or event basis.

Operational status
In service.

Contractor
EDS Defence Ltd, Hook.

VERIFIED

Magnetic sensors

PMES has a complete capability in magnetic detection and measurement with over 20 years' involvement in the design and manufacture of magnetic sensors and systems.

Magnetic sensors for anomaly detection, heading reference or instrumentation feature small size, low power and robust construction. Applications are predominantly fuzing, surveillance and heading reference (for example, sonobuoys, ROVs). Either one-, two- or three-sensing axis types are available.

PMES's magnetic sensors use very little electrical power (for instance only 0.5 mW is required for the type LPM2).

There are various sensors in the range for differing applications: for example, LNS4-3, LNR3-3 and LNR1 for fuzing/surveillance, STAGS, TAG and HR3MA for heading reference.

Contractor
PMES Ltd, Rugeley.

VERIFIED

TAHS magnetic sensor

Type
Towed array heading reference sensor.

Description
The heading sensor has been specifically designed to meet the requirements of current and future towed sonar arrays. Housed in a titanium pressure vessel, a serial datalink provides outputs of heading and three-axis acceleration (that is, attitude, with respect to the gravity vector). The package is a tube of 25 mm diameter by 130 mm long. The design is solid state and based around a three-axis magnetic sensor and a three-axis accelerometer. All processing is performed within the device using high-performance micro controllers to convert the sensor data into heading with respect to magnetic north.

The device is designed to withstand the stresses of the towed array environment, and is suitable for indefinite immersion at pressures up to 350 bar.

Operational status
In production, selected for Royal Navy Type 2076 sonar.

Contractor
PMES Ltd, Rugeley

UPDATED

Type 618 Link 11 data terminal set

Type
Modem and network controller for Link 11 data systems for ASW helicopters and other platforms.

Description
The T 618 data terminal set provides all modem and network control functions defined by MIL-STD-188-203-1A, STANAG 5511 and ADat P-11 in picket and network control modes.

As a modem, the data terminal set converts the Link 11 data into audio tones suitable for transmission over radio circuits. It applies Doppler correction and synchronisation to received audio tones, and converts them into Link 11 data with error detection and correction. As a network controller it provides all addresses and roll-call management functions in the net.

The T 618 includes features which are of great importance to Link 11 system operators. The unit provides an outlet to display the performance of every station in the network in real time. The link monitor displays the received signal quality on both upper and lower sidebands (SSB for UHF). The percentage of time each picket responds to interrogation and the analysis of the data being received are also displayed for each picket unit.

For the system maintainer, the T 618 contains BITE which isolates malfunctions to the card level. In addition, various loop-back modes provide signal paths which help to isolate difficulties in system configuration, including multiple stations. The combination of these controls and the link quality assessment provides a high visibility into the system operation.

The single-tone Link 11 system enhancement available with the T 618, provided as a switchable option, will enable the data terminal set to be backwards compatible with current conventional Link 11 modes. The data terminal set is also programmable to operate in some combination of TDMA network protocol and frequency hopping. The T 618 can be easily upgraded to provide NATO Improve Link 11 (NILE) performance and Interim Link Eleven improvements.

Operational status
In production for Merlin EH 101 helicopter and selected for Nimrod 2000 and Australian helicopter for ANZAC class.

Contractor
Ultra Electronics Sonar and Communication Systems, Greenford.

UPDATED

Type 619 Link 11 datalink processor

Type
Data processor for ASW helicopters and other platforms.

Description
The T 619 datalink processor is a companion unit to the T 618 and uses similar up-to-date technology to provide a small lightweight unit suitable for helicopter, fixed-wing and other applications.

The datalink processor is compliant with STANAG 5511 and provides track database management, track correlation, gridlock, message assembly/disassembly, transmit and receive filtering, operator control and monitoring.

The T 619 consists of a single processor card and an interface card, both of which are VME-based. A spare slot is available to double the processor throughput and memory capability or provide additional functions such as integration of a single card crypto. For this purpose the processor is equipped with all the required front panel controls.

A comprehensive BITE facility is provided.

Operational status
In production for Merlin EH 101 ASW helicopter. Selected for DLP/DTS for Nimrod 2000 (RMPA) and for Australian 'ANZAC'/OPC helicopter.

Contractor
Ultra Electronics Sonar and Communication Systems, Greenford.

UPDATED

DS25/30 naval gun

Type
Small calibre gun for close-range air and surface defence.

Description
The DS25/30 series of naval guns, which is used as a primary armament for MCMVs, is stabilised and completely self-contained and requires no deck penetration for mounting. They feature low-magnetic, radar and IR signatures utilising advanced materials and control techniques. Three methods of control are available to the command – remote via external E/O or radar-based fire-control system; local via joystick control by an on-mount seated operator (in this mode tracking is enhanced by stabilisation provided by two rate integrating gyros); and emergency where, in the event of ship's power failure, the mounting can still perform fully in local control mode using on-mount battery back-up.

A DS25/30 30 mm gun on an MCMV
(Courtesy MSI-Defence Systems Ltd)
***1999**/0043249*

There is a choice of 25 mm or 30 mm cannon, for example Oerlikon KCB for the Type DS30B, the M242 (25 mm) or Bushmaster II (30 mm) for the Type DS25/30M, and the Mauser 173 for the Type DS30F.

With various sight options there is a day/night capability.

Operational status

To date, approximately 70 type DS30B naval guns have been supplied to the Royal Navy where it is the standard fit on the 'Hunt' and 'Sandown' classes, and more than 20 are in service or on order for several other navies, including the Australian 'Huon' class MCMVs.

Specifications
Operational elevation arc: -25 to +70°
Operational training arc: -175 to +175°
Velocity (training and elevation): 55°/s
Acceleration (training and elevation): 80°/s²
Swept radius: 2.365 m

Ready use ammunition: 16 rds
Oerlikon KCB 30 mm
Rate of fire: 650 rpm (max)
Muzzle velocity: 1,080 m/s
Recoil force: 1,400 kg (max)

Contractor
MSI-Defence Systems, Norwich, Norfolk.

UPDATED

UNITED STATES OF AMERICA

Isis sonar data acquisition and processing system

Type
Modular sonar image processing system.

Description
Isis sonar digitises, stores and processes side scan sonar signals and combines the sonar imagery with navigation inputs to geocode the data in real time. The system interfaces with most conventional analogue and digital side scan sonars and runs under the Windows operating system. Isis sonar includes many image processing capabilities such as beam and grazing angle corrections, edge detection through spatial filters and target mensuration.

Post processing modules for Isis sonar include Triton Elics International Delph Map, a geoprocessing and mapping package that builds a georeferenced visual database of images, vectorised sonar feature extractions and other spatial data. An operator can process side scan sonar images acquired by Isis sonar into geocoded sonar mosaics and combine them with swath bathymetry data in a single software environment.

Isis sonar applications include MCM and route surveillance, research, harbour and ship security and hydrographic survey.

Operational status
Isis has been selected by the US Navy (Deep Submergence Unit, NAVOCEANO, SUPSALV, EOD) and by other navies for route surveillance and MCM operations. For hydrographic applications, Isis is in use with the Australian, Brazilian and US navies and with the Hydrographic departments of Denmark and Germany.

Contractor
Triton Elics International, Watsonville, California.

UPDATED

Isis terminal and display
1996

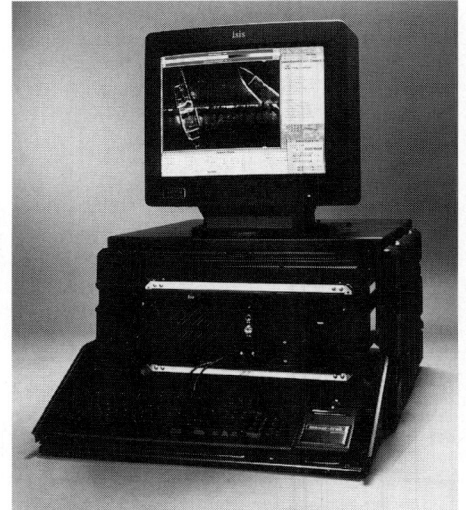

MUDSS

Type
Mobile Underwater Debris Survey System.

Description
Designed by the Jet Propulsion Laboratory of NASA, MUDSS has been developed using space programming technology and computer applications to assist in the identification of underwater mines and ordnance using data from existing US Navy sonar, laser and magnetic instruments. The laboratory also provides a chemical detector that can sniff out small traces of explosives in the water.

Five instruments are towed beneath the surface of the water on cables strung from a catamaran. The chemical sensor is towed behind the vessel on another cable and samples the water to detect the presence of explosives.

Once the navy finds debris, the results are turned over to the US Army which has the responsibility for disposing of all unexploded military waste.

The aim of the MUDSS programme is to demonstrate various technologies that can be used to survey former defence sites for unexploded waste. Some of the explosive debris in various bays and harbours under US jurisdiction has been there since long before the Second World War. MUDSS will help the US Navy clean up existing sites and clear other areas being returned to civilian authorities.

Operational status
The first phase of the data analysis programme has been handed over to the US Navy following its first year's feasibility demonstration at St Andrews Bay, Panama City, Florida, the location of the US Navy's Coastal Systems Station's test site. Much of the hardware tested in the demonstration was developed at the Naval Surface Warfare Center. MUDSS is now completing a three-year programme under contract to the US DoD, with funding provided by the department's Strategic Environmental Research and Development Program. The work is being carried out in partnership with the Naval Coastal System Station, Naval Surface Warfare Center in Florida.

Contractor
Jet Propulsion Laboratory, Pasadena, California.

UPDATED

ELP-362D

Type
Emergency locator beacon.

Description
The ELP-362-D is designed for applications where size and reliability in severe environmental conditions are critical. It can be used to mark aircraft or aircraft voice and data recorders for recovery in case of loss over water and for marking mines, torpedoes and 'black boxes'.

Contractor
RJE International Inc, Irvine, California.

VERIFIED

ALP-365 Flexi-Pinger™

Type
Acoustic pinger.

Description
The Datasonics Flexi-Pinger™ uses state-of-the-art electronics which enables the user to quickly and easily customise the pinger for their own specific applications.

The electronics are protected by a tough, rugged, aluminium housing which ensures long operating life in a hostile marine environment. The unit is powered by two alkaline or lithium 9 V batteries (six with optional long-life housing for four additional batteries.

Specifications
Frequency: 25-40 Khz (user selectable)
Output: 0.125, 0.5, 2 or 5 W (user selectable)
Pulse length: 4 ms

Pulse repetition: 2 pulse/s, 1 pulse/s or 1 pulse/2 s (user selectable)

Contractor
Datasonics (manufacturer).
RJE International Inc, Ine, California (distributor).

VERIFIED

UAT-370

Type

Acoustic multifunction transponder.

Description

The transponder is three units in one. It can be user-programmed to be a pinger, responder or transponder. It is fully compatible with Trackpoint System, as well as Kongsberg Simrad's and Raytheon Systems Company's positioning systems. The unit features user-programmable frequency, pulse length and pulse repetition rates.

The unit is suitable for use as a shallow water transponder or in any situation where the positioning transponder can be used with an ultra-short baseline positioning system. Most commonly this will involve tracking an ROV whereby the transponder would be mounted on the ROV itself.

Contractor

Datasonics (manufacturer).
RJE International Inc, Irvine, California (distributor).

VERIFIED

UAT-376

Type

Underwater acoustic transponder.

Description

The general purpose unit is designed to be used with a variety of diver and ship installed acoustic interrogators and is compatible with the Trackpoint II and LTX navigation systems. The unit operates in the mid-range frequency band of 20 to 35 Khz and is powered by two 9 V alkaline or lithium transistor-type batteries and can be deployed in depths up to 1,000 m.

The unit in its aluminium housing is suitable for marking, relocating and tracking objects, equipment or sites, for measuring distance to an object or divers underwater and for marking underwater sites or objects where stealth and secrecy are necessary.

Specifications

Frequency
  receive: 26 Khz
  transmit: 25, 27, 28, 29, 30, 31, 32 Khz
Output: 180 dB ref 1μPa @ 1 m

Contractor

Datasonics (manufacturer).
RJE International Inc, Irvine, California (distributor).

VERIFIED

DRI-267

Type

Diver operated transponder interrogator.

Description

This is a seven-channel acoustic transponder interrogator which features a backlit LCD display showing range to the transponder, as well as bearing (derived using phase-comparison technology) information to the transponder. An optional surface conversion kit (ACU-266) allows the user to turn the unit into a surface unit. It is designed for use with the UAT-376, UAT-387 and UAT-387E acoustic transponders. The system can interrogate and track seven different transponders, each operating on a different frequency. It can be factory-programmed to be compatible with customers' existing acoustic transponders.

The unit is suitable for use where an underwater acoustic transponder needs to be located, tracked or homed in on by a diver and is useful for underwater tracking and navigation where stealth is a key concern.

Contractor

Datasonics (manufacturer).
RJE International Inc, Irvine, California (distributor).

VERIFIED

ADDENDA

ADDENDA

SUBMARINE DESIGN

NORTH KOREA

'Yugo' class

Type
Midget submarine.

Description
The 'Yugo' midget submarine is designed for covert operations transporting infiltration teams into hostile territory.

The design is based on midget submarines imported from Yugoslavia during the 1960s and 1970s incorporating updated technology developed by North Korea. From the sparse details available it appears that a number of distinct variants of this type are in existence in which displacement, dimensions, power plant and weapons fit vary.

The description given here is based on information gleaned from the capture of the 'Yugo' off the coast of South Korea on 22 June 1998. In the case of this vessel

the length is in the region of 22 m, and displacement around 70 tonnes. The power plant drives an advanced single shaft twin propeller. The main propeller consists of five specially designed highly skewed blades measuring approximately 130 cm in diameter, while the second auxiliary propeller measures some 30 cm in diameter. The use of highly skewed blades in this midget submarine would considerably reduce the radiated noise signature while cruising and the auxiliary propeller is probably used during the silent, slow run in phase to the hostile beach area as it exhibits an even lower noise signature. In addition stealth is further aided by the dark camouflage colouring (green/dark green mottled pattern) and fibre resin plastic (FRP) materials which are used to coat the conning tower.

The small conning tower houses the periscope and radio antennas, and also serves as a diver lockout chamber with a hand operated outer hatch and an inner hatch that is controlled by a hydraulic arm that can only

be operated from inside the midget. Between the two hatches there is sufficient room to allow the infiltration team to put on or remove their scuba diving equipment.

A pivoted snorkel device is mounted on the stern deck which, when not in use, lays flush with the deck. Two small torpedo tubes are also said to be fitted, but this has not yet been verified.

The interior of the boat is divided into 10 watertight compartments. The vessel was manned by a crew of six and carried an infiltration team of two men and their escort.

Operational status
Between 30 to 50 midgets are believed to be operational in North Korea and two are in service with the Vietnamese People's Navy. It is also believed that one of these craft has been exported to Iran.

NEW ENTRY

UNITED KINGDOM

'Astute' class

Type
Nuclear powered attack submarine.

Description
The Astute class for the Royal Navy is being developed to replace the 'Swiftsure' class SSNs which, in spite of a major update programme, will reach the end of their life around 2005-2010. The basic design, with similar dimensions will be an evolution of the existing 'Trafalgar' class, but will offer an enhanced performance with increased weapon payload and reduced radiated noise signature. The fin will be slightly longer, housing non-hull penetrating masts while the hull diameter will be increased to provide space for the larger PWR 2 pressurised water-cooled reactor powering two GEC turbines driving a single shaft with a pump jet propulsor. The reactor was originally developed for the 'Vanguard' class SSBNs and also

equips the Batch II boats of the 'Trafalgar' class. In addition the boats will be fitted with two diesel alternators, an emergency drive motor and an auxiliary, retractable propeller.

The combat information and weapon control system will be the BAeSEMA SMCS tactical data handling system that is currently being back fitted to the 'Trafalgar' class boats.

Weapons will include the Tomahawk Block III missile and Sub Harpoon in addition to a mix of Spearfish and Tigerfish torpedoes. Weapons will be launched from five or six 533 mm torpedo tubes and a total of 38 weapons will be carried. Countermeasures will include decoys (both active and passive) as well as ESM.

The sensor suite will include an optronic non-hull penetrating mast (systems currently being evaluated include Kollmorgen, Pilkington and SAGEM) rather than conventional periscope systems, ESM and I-band navigation radar. The sonar suite will be the Type 2076 developed for the 'Trafalgar' class Update programme.

Operational status
A contract for three boats worth £2 billion was awarded to GEC-Marconi VSEL shipyard on 17 March 1997. First steel is due to be cut during 1999 and in service date is scheduled for 2006. The first three boats will be named *Astute*, *Ambush* and *Artful* and there is an option on two more boats.

Specifications
Displacement: 6,300/6,800 t
Length: 91.7 m
Beam: 10.8 m
Draught: 10 m

Contractor
VSEL (a division of GEC-Marconi), Barrow-in-Furness.

NEW ENTRY

SUBMARINE AND SURFACE SHIP ASW COMBAT INFORMATION SYSTEMS

NORWAY

MSI-90U

The MSI-90U Mk 2 command system is a modified derivative of the system developed for, and fitted in, the Norwegian 'Ula' class submarines and on order for the German Navy Type 212 submarines.

The MSI-90U Mk 2 is based on COTS technology using the new MFC-2000M multifunction console with UltraSPARC processors and NEC active matrix thin film transistors and high-resolution (1,280 x 1,024 pixels)

colour LCDs with backlight. The software is also based on COTS including POSIX, Open GL, X-Windows, OSF/Motif, digital maps, audio-video modules and asynchronous transfer mode (ATM) network protocol.

The Mk 2 is capable of interfacing with the DM2A4 and Type 184 Mod 3 torpedoes and Sub Harpoon and SM 39 Exocet submarine launched missiles. In addition it is planned to interface the Sub-Polyphem fibre-optic guided anti-ship/anti-helicopter/land attack missile with the fire-control system. The system can also

optionally interface with the SST4, SUT, Mk 37, F17, L5 or other torpedoes.

The MSI-90U Mk 2 is said to be capable of simultaneous automatic and operator interactive TMA computation on up to 25 targets, with firing preparation and guidance of up to eight torpedoes, and preparation and control of up to four missiles.

NEW ENTRY

SURFACE SHIP SONAR SYSTEMS

CANADA

SM 900

Type
Long-range obstacle avoidance surveillance and target designation sonar.

Description
The SM 900 multibeam sonar was developed for the detection of sunken submarines, and for long-range target detection. Its relatively small size (600mm maximum width) and lightweight configuration (10.7 kg in water) make the sonar ideal for deployment from a surface vessel, submersible or ROV.

The sonar head is operated using a standard SM 900 or SM 2000 surface processor.

The SM 900 has a range of 800 m, 90° coverage (receive beam angle of 1.8°), and an operating depth rating of 1,000 m.

Options for the SM 900 include audio output, data recording and an integrated pinger receiver system.

Operational status
In full-scale production and in service.

Contractor
Kongsberg Simrad Mesotech Ltd, Port Coquitlam, British Columbia.

NEW ENTRY

SM 2000

Type
Route survey forward-looking obstacle avoidance sonar.

Description
SM 2000 multibeam obstacle avoidance sonar was developed for the detection of bottom and mid-water depth mine-like objects. Mounted on a deployment staff, the sonar is equipped with a pan-drive unit that allows the operator to change the azimuthal angle of coverage and follow a target.

The sonar head and pan drives are controlled by the SM 2000 surface processor. The sonar display is pitch-stabilised to minimise the effect of vessel movement.

The SM 2000 systems supplied to naval customers have a frequency of 200 kHz, a 60° field of view (with receive beam angles less than 1°), and a range up to 400 m.

Subsurface electronics are mounted on a pan mechanism capable of 4°/s rotation at speeds over 4 kt.

Operational status
In full-scale production and in service.

Contractor
Kongsberg Simrad Mesotech Ltd, Port Coquitlam, British Columbia.

NEW ENTRY

UNITED STATES OF AMERICA

HAS-1254

Type
Hull mounted LF sonar.

Description
Developed from the USN AN/SQS-53C hull-mounted sonar, the HAS-1254 is designed to meet multimission requirements in both open ocean and shallow water environments.

Using a 5 kHz transducer design, the HAS-1254 provides long range underwater search and surveillance, detection, tracking, localisation and classification, torpedo alert and mine detection and avoidance capabilities.

The system uses a number of menu-controlled operating modes and sub-modes to meet the demands of various tactical situations. Included among these are simultaneous active/passive operation, multiple frequency band allocations, and waveform signal processing and display enhancements. The waveforms can be matched to cover the type of threat anticipated and the sea state likely to be encountered. The transmit and receive beams can be steered vertically to optimise performance and are fully stabilised to compensate for roll, pitch and yaw.

Other features include computer-aided detection, automatic target tracking, and automatic contact management.

Operational status
Development completed.

Contractor
Lockheed Martin, Ocean Radar and Sensor Systems, Syracuse, New York.

NEW ENTRY

SUBMARINE SONAR SYSTEMS

SPAIN

Solarsub

Type
Submarine towed array sonar.

Description
The Solarsub towed array sonar is designed to complement submarine hull-mounted sonars and offers greater range of detection and easy platform integration as stand-alone equipment.

The system offers wide area surveillance; long range detection; multiple track capability; and improved classification aids.

The modularity of the system with its small footprint components and low power consumption enables the system to be easily installed on board any type of submarine.

Solarsub comprises an antenna and junction box, input signal conditioner, operator console, and optionally an outboard winch for deployment and recovery of the antenna, depending on operational requirements.

The system uses COTS hardware components and a new generation streamer that, due to its reduced diameter, allows a wide range of towing speeds.

The sonar is fitted for standard interface with combat systems.

Operational status
Entering service with the Spanish Navy.

Contractor
SAES, Cartagena.

NEW ENTRY

ADDENDA 555

MINESWEEPING SYSTEMS

UNITED STATES OF AMERICA

AMNS

Type
Airborne mine neutralising system (AMNS).

Description
Lockheed Martin in partnership with STN ATLAS Elektronik is adapting the German Seafox expendable mine neutralisation vehicle to meet the USN requirement for an MH-53 helicopter-towed remotely operated device for the rapid destruction of mines at day or night.

In this configuration the Seafox self-propelled, unmanned, wire-guided one-shot vehicle will be integrated with the AQS-14A sonar and a laser line scan system providing interim mine identification capability.

The operator-controlled vehicle exhibits a high degree of manoeuvrability in hover, with backing and precise control in both pitch and yaw. This is achieved using four conventional, independent, battery-powered, variable speed, fully reversible electrical motor propellers arranged in four quadrants at the rear of the vehicle. In addition a vertical tunnel thruster mounted amidships with a fully reversible impeller provides movement in the vertical plane.

Vehicle tracking is carried out using the AQS-14A sonar to activate a dorsal-mounted transponder on the vehicle. The vehicle sonar provides acquisition, homing and classification of targets and is mechanically scanned in the horizontal plane providing resolution down to 0.9°. A TV camera and spotlight are used for visual identification of targets.

The operator in the helicopter carries out control and guidance of the vehicle using attitude, heading, and depth and altitude sensors.

Data transfer and vehicle control signals are passed through a free spooling fibre optic cable.

The warhead comprises a shaped charge and method of detonation is operator selectable.

Operational status
The engineering, manufacturing and development contract was awarded in the latter part of 1998. The research and development phase is scheduled for the year 2000.

Contractor
Lockheed Martin, Ocean Radar & Sensor Systems, Syracuse, New York.
STN ATLAS Elektronik, Bremen, Germany.

NEW ENTRY

DIVING EQUIPMENT

UNITED STATES OF AMERICA

LIMIS

Type
Diver sonar.

Description
LIMIS has been developed under sponsorship from the US Office of Special Technology and supervised by the US Naval EOD Technology Division.

The hand-held sonar provides underwater images down to zero visibility. LIMIS (Limpet Mine imaging Sonar) operates at a frequency of 2 MHz with 0.3° resolution at a maximum range of 15 m. Range resolution is 1.8 mm at a distance of 3 m. A mask-mounted video monitor displays images from the sonar, which are updated 10 times a second for ranges up to 9 m. The imagery can be transmitted in real time to a display unit aboard a ship or quayside and can store up to 50 images for subsequent analysis.

LIMIS measures 18 cm (wide) by 20 cm (high) and 30 cm (long) including a 10 cm handle. The unit is neutrally buoyant in water which allows the diver to manage the instrument with one hand leaving the other free for other purposes.

The sonar can operate for up to 4 hours on a single battery charge, or for longer periods using auxiliary power.

Operational status
Development completed.

Contractor
Applied Physics Laboratory, University of Washington.

NEW ENTRY

MULTIBEAM MAPPING SYSTEMS

UNITED STATES OF AMERICA

8111 Multibeam Echo-sounder

Type
Hydrographic sonar.

Description
Operating at 100 kHz, the portable wide-swath multibeam SeaBat 8111 features an extended depth range to support full continental shelf mapping projects, as well as general hydrographic surveying and dredging support.

The modular design offers permanent through-hull installations and also temporary over-the-side mountings. The transducer assembly is fully rubberised for corrosion protection.

Like all 8100 Series models, the SeaBat 8111 uses the RESON advanced 81-P processor, which allows the operator to fine-tune settings for local conditions, provides advanced self-diagnostics and can quickly download new firmware through a communications port.

The sonar complies to IHO accuracy which is maximised by the system's fast update rate, its use of narrow beams and ability to automatically select the most accurate bottom detection method for individual beam – that is phase, amplitude, or a combination of both. The system also features a pitch stabilisation capability operating within ±10°.

The sonar surveys a swath that is 7.4 times the water depth in width down to 150 m water depth. Narrower swaths can be measured to reach depths down to 600 m.

Operational status
The SeaBat 8111 is used by numerous hydrographic survey authorities and private survey contractors in Asia, Europe and the United States.

Specifications
Operating frequency: 100 kHz
Resolution: 5 cm
Number of beams: 101
Beam size: 1.5° × 1.5°
Operating depth: 3 to 600 m

Contractor
RESON Inc, Goleta, California.
RESON A/S, Slangerup, Denmark.
RESON Offshore Ltd, UK.

NEW ENTRY

COMMAND TEAM TRAINERS

UNITED KINGDOM

Submarine team trainers

Type
Command team trainers.

Description
These trainers comprise one or two submarine control rooms with simulated sonar, radar, ESM, fire-control systems, attack periscope, plotting facilities and ship's instruments such as log and compass. Within the periscope a variety of aircraft and ship models can be displayed in daylight, LLTV or thermal imaging modes.

The trainer for the Royal Navy 'Vanguard' class installed at the Clyde submarine base features two training rooms, an instructor's station, viewing area and debriefing facilities. One room houses a 10-position command team trainer with simulated fire control, weapons and tactical data handling system consoles. Computer-generated images are displayed on a screen at a two-position command desk.

The seven-position acoustic sensor area is housed in a separate room and incorporates a simulated Type 2054 sonar suite including ambient, biological and 'own ship' noise as affected by sea state, weather and bathythermal conditions. An additional CBT sonar trainer is incorporated.

The four-position instructor station provides monitoring, control and record/playback facilities.

Operational status
Command team trainers are in service with the navies of Australia, Brazil, Norway and the UK.

Contractor
GEC-Marconi Simulation and Training Division, Dunfermline.

NEW ENTRY

MINE WARFARE TRAINERS

AUSTRALIA

Visem

Type
Mine exercise training aid.

Description
Visem has been developed to provide enhanced realism and situational awareness for ships' crews during MCM exercises. The system is based on an underwater flare launcher that provides visual and audio indication to the crews of surface vessels that they have triggered an exercise mine.

Currently, as ships navigate a path through an exercise minefield, the only indication that a mine has been triggered is a signal detected by the ship's own exercise monitors.

Visem releases a streamlined payload from the seabed as a ship triggers a mine. As the payload broaches the surface a flare is released giving off three loud reports alerting the crew to the fact that the ship has detonated an exercise mine.

Operational Status
On order for the Australian Navy to equip the 'Adelaide' and 'Anzac' class frigates. Development is continuing.

Contractor
ADI Ltd, Carrington, New South Wales.

NEW ENTRY

UNITED KINGDOM

'Hunt' class trainer

Type
Operations room simulator.

Description
This is a modular system for individual operators and command teams. The system is contained in two semi-mobile trailers, one of which houses the computer and simulation hardware and the other the Computer Aided Action Information System (CAAIS) computer, instructor and student cubicle.

The student station features two CAAIS displays with an overhead visual control system, remote-control mine destruction system (RCMDS – PAP ROV), a Type 193M minesweeping sonar console, simulated acoustic and magnetic sweep equipment, chart table and echo-sounder.

The sonar display features sea bottom pictures, with models of mines and mine-like objects that are inserted by the instructor. The instructors console is a 2-position station with an adjacent helmsman's console separated from the instructor by a curtain.

The trainer also incorporates a combat data system, communications facilities and ship control. Orders from the student commander are injected into the simulation computer at the helmsman's console.

Operational Status
In service with the Royal Navy.

Contractor
GEC-Marconi Simulation and Training Division, Dunfermline.

NEW ENTRY

'Sandown' class trainer

Type
Operations room simulator.

Description
This trainer features an operations room outfitted with the Nautis command system, a Type 2093 sonar and the Remote Control Mine Disposal system (RCMDS – PAP ROV). The trainer also incorporates part of the bridge area of the 'Sandown' with a replicated Nautis position and ship's position control station. This allows bridge and operations room crew to be trained as a team.

Operational status
In service with the Royal Navy and a variant with Nautis and Type 2093 sonar supplied to the Australian Navy.

Contractor
GEC-Marconi Simulation and Training Division, Dunfermline.

NEW ENTRY

SUBMARINE EQUIPMENT TRAINERS

UNITED KINGDOM

Machinery simulators

Type
Nuclear submarine propulsion simulators.

Description
Various simulators have been developed to provide engineering watch-keepers with training in start up, shut down and emergency procedures covering nuclear power plants and diesel generators in nuclear submarines of the Royal Navy. The simulators include sound effects that provide background noise for diesel generator run-up, rod drive transmitters, circuit breaker actions and instructor-induced steam leaks.

The instructor's console allows the input of initial conditions for the reactor including core age, power level of pervious operations, time since shutdown and desired pressuriser and heat exchange levels. The console overlooks the simulator control room through a one-way glass and incorporates facilities for record and post exercise playback.

Operational status
Seven units in service with the Royal Navy covering 'Swiftsure', 'Trafalgar' and 'Vanguard' class submarines.

Contractor
GEC-Marconi Simulation and Training Division, Dunfermline.

NEW ENTRY

APPENDIX

TABLE 1

Inventory of submarine systems in service (not listed by class)

Class	Sonars	Command	Missiles	Torpedoes	Radars	Countermeasures
EUROPE & NORTH AMERICA (including Canada, RFAS (CIS) & USA)						
BULGARIA						
Romeo				SAET-60	Snoop Plate	Stop Light
CANADA						
Oberon	BQG 501 Type 2007 Type 2051	TFCS		Mk 48	Type 1006	Guardian Star
DENMARK						
Kobben (Type 207) Narhvalen	PSU 83	M8		Type 41 Type 61		Sea Lion
FRANCE						
Agosta Daphne L'Inflexible Rubis Le Triomphant	DMUX 80 DMUX 20 DSUV 2 DSUV 22 DSUV 61 DSUV 62A DSUV 62C DSUX 21 DUUA 1D DUUA 2D DUUX 2 DUUX 5	DLA 2A DLA 4A DLT D3 SAD/SAT TIT	M45 M4 SM 39 Exocet	F17 Mod 2 L5 Mod 3	DRUA 33 Calypso Type 1007	DR 3000U/Arur 13 Arud
GERMANY						
Type 205 Type 206 Type 206A Type 212	DBQS-21D DBQS-40 DUUX 2 FAS-3 SRS M1H TAS-3	CSU 83 Mk 8 MSI-90U		DM2A3 DM2A4	Type 1007 Calypso II	FL 1800U DR 2000U Sarie 2 TAU 2000
GREECE						
Glavkos (209/1100)	CSU 3-4 CSU 83-90 (DBQS-21) PRS-3-4	Kanaris Sinbads	Sub Harpoon	SUT Mod 0	Calypso II	AR-700-S5
ITALY						
Improved Sauro Sauro (Type 1081) Type 212A	DBQS-40 IPD 70/S MD 100S FAS-3 TAS-3	BSN 716 MSI-90U		A184 DM2A4	BPS 704	BLD-727 FL1800U
NETHERLANDS						
Walrus	DUUX 5 TSM 2272 Type 2026	SEWACO VIII Gipsy	Sub Harpoon	Mk 48 Mod 4 NT 37D	ZW 07	ARGOS 700
NORWAY						
Modernised Kobben Ula (Type P 6071)	CSU83	MSI-90(U)		DM 2A3 NT37C Type 61	Type 1007	Sealion
POLAND						
Foxtrot (Type 641) Kilo (Type 877E)	Shark Teeth Mouse Roar Herkules			53-65 SAET 60 SET-65E Test 71	Snoop Tray	Brick Group Quad Loop Stop Light
PORTUGAL						
Albacora (Daphne)	DSUV 2 DUUA 2	DLT D3		E14/15 L3	Type 1007	Arur

Class	Sonars	Command	Missiles	Torpedoes	Radars	Countermeasures
ROMANIA						
Kilo (Type 877E)	Mouse Roar	Punch Bowl		53-65	Snoop Tray	Brick Group
	Shark Teeth			Test-71		Quad Loop
RUSSIAN FEDERATION AND ASSOCIATED STATES (CIS)						
Akula I-II (Type 971)	Mouse Roar	3R65	SA-N-5/8	SAET 40	Snoop Pair	Bald Head
Delta I-IV (Type 667)	Pike Jaw	Punch Bowl	SS-N-8	SAET 60	Snoop Tray	Brick Pulp
Foxtrot (Type 641)	Shark Gill	MVU-110EM	SS-N-15	SET 92K	Snoop Half	Brick Spit
Kilo (Type 877/636)	Shark Teeth		SS-N-16	TEST 71		Park Lamp
Oscar I-II (Type 949)	Shark Hide		SS-N-18	USET/TE2		Rim Hat
Sierra I-II (Type 945A)	Shark Rib		SS-N-19	53-65		Squid Head
Tango (Type 641)	Pelamida		SS-N-20	65		Stop Light
Typhoon (Type 941)	Skat 3		SS-N-21	Shkval		
Victor III (Type 671)			SS-N-23			
Yankee Notch (Type 667)						
SPAIN						
Delfin (Daphne)	DSUV 22	DLA 2A		F17 Mod 2	DRUA 31	Manta E
Galerna (Agosta)	DSUV-62	DLT 3A		L5 Mod 3/4	DRUA 33A	
	DUUA 2A/2B				DRUA 33C	
	DUUX 2A					
SWEDEN						
Gotland (A 19)	CSU-83	IPS-17		Type 431		AR-700-S5
Nacken (A 14)	CSU-90-2	IPS-19		Type 432/451		Manta S
Vastergotland (A 17)				Type 613/62		
TURKEY						
Atilay (209/1200)	BQG 3	ISUS 83-2	Sub Harpoon	Mk 37	BPS 12	DR 2000
Guppy IIA	BQG 4	M8		SST 4	SS 2A	Porpoise
Guppy III	BQR 2B	Mk 106		Tigerfish	S 63B	WLR-1
Preveze (209/1400)	BQS 4	Sinbads				
Tang	CSU 3					
	CSU 83					
	TAS 3					
UNITED KINGDOM						
Swiftsure	Type 2007	DCB	Sub Harpoon	Spearfish	Type 1006	UAP 3
Trafalgar	Type 2019	DCG	Tomahawk	Tigerfish	Type 1007	Type 2066
Vanguard	Type 2020	SAFS 3	Trident 2 (D5)			Type 2071
	Type 2026	SMCS				
	Type 2046					
	Type 2054					
	Type 2072					
	Type 2074					
	Type 2076					
	Type 2077					
	Type 2082					
UNITED STATES OF AMERICA						
Benjamin Franklin	BQG 5D	BSY-1	Sub Harpoon	Mk 48	BPS 14	BRD-7
Los Angeles	BQQ 5D	BSY-2	Tomahawk	Mk 48 ADCAP	BPS 15A	WLR-1H
Narwhal	BQQ 6	CCS Mk 2	Trident I (C4)		BPS 16	WLQ-4(V)1
Ohio	BQQ 7	Mk 2	Trident II (D5)			WLR-8
Seawolf	BQR 15	Mk 113				WLR-10
Sturgeon	BQR 19	Mk 117				BLD-1
	BQR 21	Mk 118				WLQ-4
	BQS 4	Mk 98				
	BQS 8					
	BQS 13					
	BQS 14A					
	BQS 15					
	BQS 24					
	MIDAS					
	SADS-TG					
	TB-16					
	TB-23					
	TB-29					
YUGOSLAVIA						
Heroj	PRS-3			Test-71ME	Snoop Group	Stop Light
Sava	Eledone			Set-65E		
AFRICA						
SOUTH AFRICA						
Daphne	DSUV 2	DCSC-2		E15	Calypso II	Arud
	DUUA 2			L4/5		
	DUUX 2					

Class	Sonars	Command	Missiles	Torpedoes	Radars	Countermeasures
MIDDLE EAST & NORTH AFRICA						
ALGERIA						
Kilo (Type 877E)	Mouse Roar Shark Teeth Shark Fin			53-65 Test-71ME	Snoop Tray	Brick Group
EGYPT						
Romeo	CSU 83	Mk 2	Sub Harpoon	Mk 37F Mod 2	Snoop Plate	AR 700-55
IRAN						
Kilo (Type 877 EKM)	Mouse Roar Shark Teeth			53-65 Test-71/96	Snoop Tray	Quad Loop Squid Head
ISRAEL						
Dolphin Gal (Vickers Type 540)	PRS-3 CSU-90 FAS-3 Type 1110	ISUS 90-1	Sub Harpoon	DM A23 NT 37E		Timnex 4CH(V)1
LIBYA						
Foxtrot (Type 641)	Feniks Herkules			SAET-60 SET-65E	Snoop Tray	Stop Light
FAR EAST & AUSTRALASIA						
AUSTRALIA						
Collins Oberon	BQQ 4 CSU3-41 Type 2007 Kariwara Narama Scylla	SFCS Mk 1	Sub Harpoon	Mk 48 Mod 4	Type 1006 Type 1007	AR 740 Mavis
CHINA, PEOPLE'S REPUBLIC						
Golf Han (Type 091) Kilo (Type 877EKM) Ming (Type 035) Romeo (Type 033) Modified Romeo Xia (Type 092) Song (Type 039)	DUUX-5 Herkules Pike Jaw Mouse Roar Shark Teeth Tamir 5 Trout Cheek		C-801 Ying Ji JL-1 JL-2	SAET-60 SET-65E Test 71/96 Type 53-51	Snoop Plate Snoop Tray	Brick Pulp Squid Head Type 921-A
INDIA						
Foxtrot (Type 641) Kilo (Type 877EM) Shishumar (209/1500)	CSU 83 DUUX-5 Feniks Herkules Mouse Roar Shark Fin	Mk 1	SA-N-8	SAET-60 SET-65E SUT Mod 1 Test 71/96 Type 53-65	Calypso Snoop Tray	Phoenix II AR 700 Sea Sentry Stop Light Squid Head C 303
INDONESIA						
Cakra (209/1300) Type 206	DUUX 2 CSU 3-2 PRS 3/4 410 A4	Sinbad Mk 8		SUT Mod 0 DM 2A1	Calypso II	DR 2000 U
JAPAN						
Harushio Yuushio Oyashio	ZQQ 5 ZQQ 5B ZQR 1		Sub Harpoon	Type 89 Type 80	ZPS 6	ZLR 3-6 ZLR 7
KOREA, NORTH						
Romeo (Type 033) Sang-O	Feniks Pike Jaw			SAET-60? Type 53-56?	Snoop Plate	921A
KOREA, SOUTH						
Chang Bogo (209/1200)	CSU 83	ISUS 83		SUT Mod 2		

Class	Sonars	Command	Missiles	Torpedoes	Radars	Countermeasures
PAKISTAN Hangor (Daphne) Hashmat (Agosta) Agosta 90B	DSUV 1 DSUV 2H DUUA 1D DUUA 2A/2B DUUX 2A TSM 2233	SUBICS Mk 2	Exocet SM 39 Sub Harpoon	F17P L5 Mod 3	DRUA 31 DRUA 33	DR-3000U Arud
SINGAPORE Sjoormen	Hydra	IPS-12		Type 613 Type 431		
TAIWAN Guppy II Hai Lung	BQR 2B BQS 4C DUUG 1B SIASS-Z	Sinbads M		SUT	SS 2 ZW 06	WLR-1/3 AR 700 SF Timnex 4CH(V)2
LATIN AMERICA & CARIBBEAN						
ARGENTINA Salta (209/1200) Santa Cruz (TR 1700)	CSU 3 CSU 3/4 DUUG 1D DUUX 2C DUUX 5	M8 Sinbads		Mk 37 SST 4	Calypso II Calypso IV	DR 2000 Sea Sentry III
BRAZIL Humaita (Oberon) Tupi (209/1400) Improved Tupi	CSU-83/1 Type 2007 CSU 90-61	DCH KAFS ISUS 83-13		Mk 8 Mk 37 Tigerfish	Type 1006 Calypso III	DR 4000 UA 4
CHILE Oberon Thomson (209/1300) Scorpene	CSU 3 CSU 90 Type 2007	Subtics		SUT Mod 1	Type 1006 Calypso II	DR 2000 U
COLOMBIA Pijao (209/1200)	CSU 3-2 PRS 3-4	M8/24		SUT	Calypso II	DR 2000
CUBA Foxtrot (Type 641)	Feniks Herkules			53-65	Snoop Tray	Stop Light
ECUADOR Type (209/1300)	CSU 3 DUUX 2	M8/24		SUT	Calypso	DR 2000 U
PERU Abtao Casma (209/1200)	CSU 3 DUUX 2C PRS 3 Eledone 1102/5	M8/24 Mk 3		A184 Mk 37	SS-2A Calypso	
VENEZUELA Cabalo (209/1300)	CSU 3-32 DUUX 2	ISUS		Mk 37 SST 4	Scanter	DR 2000

UPDATED

TABLE 2

NATO designations of Russian underwater warfare systems and equipment

Designation	Description
Alfa	Submarine-launched cruise missile
Bald Head	Submarine ESM system
Ball End	Minehunter surveillance radar
Boat Sail	Submarine radar
Brick Group	Submarine ECM system incorporating Brick Pulp and Brick Spit
Brick Pulp	Submarine EW antenna
Brick Spit	Submarine EW antenna
Buck Toe	Searchlight sonar system
Bull Horn	Active/passive surface ship sonar
Bull Nose	Medium-frequency surface ship bow sonar
Cluster Bay	'Rising' sea mine
Clusterguard	Anti-sonar hull coating for submarines
Cluster Gulf	'Rising' sea mine for deep water
Cod Eye A/B	Submarine sensors
Drum Tilt	Mine warfare vessel fire-control radar
Dustbin	Submarine mast sensor
Elk Tail	Variable depth sonar
Feniks	Submarine sonar
Fez	Underwater telephone
Fin Curve	Mine warfare vessel surveillance radar
Fin Trough	Mine warfare vessel navigation radar
Foal Tail	Variable depth sonar
Golf Ball	Submarine mast sensor
Herkules	Submarine and surface ship HF scanning sonar
Horse Jaw	Low-frequency surface ship sonar
Horse Tail	Variable depth sonar
Lamb Tail	Variable depth sonar
Low Trough	Mine warfare vessel navigation radar
Mare Tail	Variable depth sonar
Moose Jaw	Low-frequency hull sonar
Mouse Roar	Submarine active HF sonar
Mouse Tail	Variable depth sonar
Neptune	Mine warfare vessel navigation radar
Ox Tail	Variable depth sonar
Ox Yoke	Active MF surface ship sonar
Park Lamp	Submarine mast direction-finding loop antenna
Perch Gill	Submarine HF attack sonar
Pike Jaw	Searchlight attack sonar
Port Spring	Submarine mast conical spiral antenna
Punch Bowl	Submarine mast radome
Quad Loop	Submarine direction-finding system
Rat Tail	Variable depth/helicopter dipping sonar
Rim Hat	Submarine ESM system
Sampson	SS-N-21 intermediate-range SLBM
Sandbox	SS-N-12 Cruise missile
Sark	SS-N-5 submarine ballistic missile
Sawfly	SS-N-8 SLBM
Scorpion	SS-NX-24 Cruise missile
Seal Skin	Submarine sonar
Serb	Submarine-launched ballistic missile (SS-N-6)
Shark Fin	Submarine medium-frequency sonar
Shark Teeth	Submarine LF bow sonar
Shipwreck	SS-N-19 Cruise missile
Silex	SS-N-14 ASW missile
Siren	SS-N-9 submarine-launched anti-ship missile
Skiff	SS-N-23 SLBM
Skin Head	Mine warfare vessel navigation radar
Snipe	SS-N-17 SLBM
Snoop Plate	Submarine surveillance radar
Snoop Slab	Submarine surface search radar
Snoop Tray	Submarine surveillance radar
Spin Trough	Mine warfare vessel navigation radar
Square Head	Mine warfare vessel IFF
Squid Head	Submarine ESM system
Stag Ear	Minehunting HF searchlight sonar
Stag Hoof	HF searchlight sonar
Stallion	SS-N-16 ASW missile
Starfish	SS-N-15 ASW missile
Steer Hide	Active LF/MF surface ship sonar
Stingray	SS-N-18 SLBM
Stop Light	Submarine ESM system
Sturgeon	SS-N-20 SLBM
Tamir	Submarine sonar
Trout Cheek	Submarine passive array bow sonar
Whale Series	Active MF surface ship sonar

CONTRACTORS

Australia

ADI Ltd
Major Projects Group
Level 6-Westfield Towers
100 William Street, East Sydney
New South Wales 2011
Australia
Tel: (+61 2) 99 16 49 00
Fax: (+61 2) 93 68 09 75

Australian Submarine Corporation Pty Ltd
Mersey Road
Osborne
Adelaide
South Australia 5001
Australia
Tel: (+61 8) 348 70 00
Fax: (+61 8) 348 70 01

British Aerospace Australia
Head Office
PO Box 180
Salisbury
South Australia 5108
Australia
Tel: (+61 8) 82 90 88 88
Fax: (+61 8) 82 90 88 00

Nautronix Ltd
108 Marine Terrace
PO Box 1352
Fremantle
Western Australia 6160
Australia
Tel: (+61 9) 430 59 00
Fax: (+61 9) 430 59 01

Thomson Marconi Sonar Pty Ltd
274 Victoria Road
Rydalmere
New South Wales 2116
Australia
Tel: (+61 2) 98 48 35 00
Fax: (+61 2) 98 48 38 88

Belgium

BARCO nv
Display Systems
The President Kennedy park 35
B-8500 Kortrijk
Belgium
Tel: (+32 56) 26 26 11
Fax: (+32 56) 26 22 62

Brazil

CONSUB
Rua Pesqueira
108 à 118 Bonsucesso
21041 150 Rio de Janeiro
Brazil
Tel: (+55 21) 280 80 96
Fax: (+55 21) 290 50 05
email: consub@ax.apc.org

Canada

Amphibico Inc
9563 Côte de Liesse
Dorval
Quebec H9P 1A3
Canada
Tel: (+1 514) 636 99 10
Fax: (+1 514) 636 87 04

Array Systems Computing Inc
1120 Finch Avenue West
8th Floor
North York
Ontario M3J 3H7
Canada
Tel: (+1 416) 736 09 00
Fax: (+1 416) 736 47 15

Barringer Research Ltd
304 Carlingview Drive
Metropolitan Toronto
Rexdale
Ontario M9W 5GZ
Canada
Tel: (+1 416) 675 38 70
Fax: (+1 416) 675 38 76

C-Tech Ltd
PO Box 1960
525 Boundary Road
Cornwall
Ontario K6H 6N7
Canada
Tel: (+1 613) 933 79 70
Fax: (+1 613) 933 79 77

Canadian Marconi Company
415 Leggett Drive
PO Box 13330
Kanata
Ontario K2K 2B2
Canada

CAE Electronics Ltd
PO Box 1800
St Laurent
Montreal
Quebec H4L 4X4
Canada
Tel: (+1 514) 341 67 80
Fax: (+1 514) 341 76 99

Computing Devices Canada Ltd
PO Box 8508
3758 Richmond Road
Nepean
Ontario K1G 3M9
Canada
Tel: (+1 613) 596 70 00; 596 70 59
Fax: (+1 613) 596 73 96

Fullerton Sherwood Engineering Ltd
6450 van Deemter Court
Mississauga
Ontario L5T 1S1
Canada
Tel: (+1 905) 670 06 56
Fax: (+1 905) 670 83 18

Halifax-Dartmouth Industries Ltd
PO Box 9110
Halifax
Nova Scotia B3K 5M7
Canada
Tel: (+1 506) 632 59 39
Fax: (+1 506) 632 59 12

Hermes Electronics Inc
40 Atlantic Street
PO Box 1005
Dartmouth
Nova Scotia B2Y 4A1
Canada
Tel: (+1 902) 466 74 91
Fax: (+1 902) 463 60 98

Indal Technologies Inc
3570 Hawkestone Road
Mississauga
Ontario L5C 2V8
Canada
Tel: (+1 905) 275 53 00
Fax: (+1 905) 273 70 04

International Submarine Engineering Research Ltd
1734 Broadway Street
Port Coquitlam
British Columbia
Canada V3C 2M8
Tel: (+1 604) 942 52 23
Fax: (+1 604) 942 75 77

Kongsberg Simrad Mesotech Ltd
1598 Kebet Way
Port Coquitlam
British Columbia V3C 5M5
Canada
Tel: (+1 604) 464 81 44
Fax: (+1 604) 941 54 23

Macdonald Dettwiler
13800 Commerce Parkway
Richmond
British Columbia V6V 2J3
Canada
Tel: (+1 604) 278 34 11
Fax: (+1 604) 278 12 85

Northrop Grumman - Canada Ltd
777 Walkers Line
Burlington
Ontario L7N 2G1
Canada
Tel: (+1 905) 333 60 07
Fax: (+1 905) 333 60 14

Oceanroutes Canada Inc
Halifax
Nova Scotia
Canada
Tel: (+1 902) 468 30 08

Optech Inc
100 Wildcat Road
North York (Toronto)
Ontario M3J 2Z9
Canada
Tel: (+1 416) 661 59 04
Fax: (+1 416) 661 41 68

Orcatron Communications Ltd
1595 Kebet Way
Port Coquitlam
British Columbia V3C 5W9
Canada
Tel: (+1 604) 941 79 09
Fax: (+1 604) 941 75 17

Saint John Shipbuilding Ltd
PO Box 5111
Saint John
New Brunswick E2L 4L4
Canada
Tel: (+1 506) 632 59 39
Fax: (+1 506) 632 59 12

Seimac Ltd
271 Brownlow Avenue
Dartmouth
Nova Scotia B3B 1W6
Canada
Tel: (+1 902) 468 30 07
Fax: (+1 902) 468 30 09

Sparton of Canada Ltd
PO Box 5125
99 Ash Street
London
Ontario N6A 4N2
Canada
Tel: (+1 519) 455 53 20
Fax: (+1 519) 452 39 67

Chile

Industrias Cardoen Ltda
Avda Providencia 2237
660 Piso
Santiago
Chile
Tel: (+56) 232 10 81/232 10 82/251 58 84
Tx: 340997 INCAR CK

China, People's Republic

China National Electronics Import & Export Corporation
49 Fuxing Road
Beijing
People's Republic of China
Tel: (+86) 81 09 10
Tx: 22475

China Precision Machinery Import & Export Corporation
22 Fu Cheng Lu
PO Box 129
Beijing 100036
People's Republic of China
Tel: (+86 106) 837 16 50
Tx: 22484 CPMC CN

Dalian Shipbuilding Industry Corporation
16 Zhuqingje
Dalian
People's Republic of China
Tel: (+86) 262 77
Tx: 86171

Denmark

Danyard A/S
Kradegholmen 4
PO Box 719
DK-9900
Frederikshavn
Denmark
Tel: (+45) 98 42 22 99
Fax: (+45) 98 43 29 30

Eiva A/S
Teglbaekvej 8
DK-8361 Hasselager
Denmark
Tel: (+45) 86 28 20 11
Fax: (+45) 86 28 21 11

Nea-Lindberg A/S
Industriparken 39-43
Post Box 226
DK-2750 Ballerup
Denmark
Tel: (+45) 44 97 22 00
Fax: (+45) 44 66 09 10

RESON A/S
Fabriksvangen 13
DK-3550 Slangerup
Denmark
Tel: (+45) 47 38 00 22
Fax: (+45) 47 38 00 66

Terma Elektronik A/S
Hovmarken 4
DK-8520 Lystrup
Denmark
Tel: (+45) 86 22 20 00
Fax: (+45) 86 22 27 99

Finland

Elesco Oy
Luomannotko 4
PO Box 128
FIN-02201 Espoo
Finland
Tel: (+358) 420 86 00
Fax: (+358) 420 86 10

Finnyards Ltd Electronics
Naulakatu 3
FIN-33100 Tampere
Finland
Tel: (+358 32) 45 01 11
Fax: (+358 32) 13 01 88

France

Aerospatiale
Division Engins Tactiques
2 rue Béranger BP 84
F-92322 Chatillon Cedex
France
Tel: (+33 2) 47 46 33 97
Fax: (+33 2) 47 46 33 19

Aerospatiale
Espace & Défense
BP 3002
F-78133 Les Mureaux
France
Tel: (+33 1) 34 92 12 34
Fax: (+33 1) 34 75 09 15

DCN International
19/21 rue du Colonel Pierre-Avia
BP 532
F-75725 Paris Cedex 15
France
Tel: (+33 1) 41 08 71 71
Fax: (+33 1) 41 08 00 27

Engins Matra
Matra SA
37 Avenue Louis Bréguet, BPI
F-78140 Velizy Villacoublay Cedex
France
Tel: (+33 1) 34 88 30 00
Tx: 698130

Eurotorp
525 route des Dolines
Les Bouillides
BP 113
F-06902 Sophia Antipolis Cedex
France
Tel: (+33 4) 92 96 38 50
Fax: (+33 4) 92 96 38 55

Mécanique Creusot-Loire
Immeuble Ile de France
Paris La Défense
France
Tel: (+33 1) 49 00 60 50
Fax: (+33 1) 49 00 57 30

R Alkan & Cie
rue de 8 Mai 1945
F-94460 Valenton
France
Tel: (+33 2) 43 89 39 90
Tx: 203876

Safare Crouzet SA
98 avenue Saint Lambert
F-06105 Nice Cedex 2
France
Tel: (+33 4) 92 09 76 76
Fax: (+33 4) 93 51 56 94

SAGEM SA
Defence and Security Division
Paris La Défense
61 rue Salvador Allende
F-92751 Nanterre Cedex
France
Tel: (+33 1) 40 70 63 63
Fax: (+33 1) 40 70 64 54

Serrico Ltd
La Hache
F-27450 St Martin St Firmin
France
Tel: (+33 2) 32 42 75 23
Fax: (+33 2) 32 57 39 60

Sextant Avionique
BP 200
F-78141 Vélizy Villacoublay Cedex
France
Tel: (+33 1) 46 29 70 00
Fax: (+33 1) 46 32 85 96

SFIM industries
13 avenue Marcel Ramolfo Garnier
F-91344
Massy Cedex
France
Tel: (+33 1) 69 19 66 00
Fax: (+33 1) 69 19 69 19

Sillinger SA
BP 23
F-75560 Paris Cedex 12
France
Tel: (+33 1) 43 07 21 55
Fax: (+33 1) 43 40 70 49

Société ECA
76 blvd de la République
F-92772 Boulogne-Billancourt Cedex
France
Tel: (+33 1) 46 10 90 60
Fax: (+33 1) 46 10 90 70

Société Industrielle d'Aviation Latecoere
(SILAT)
79 avenue Marceau
F-75116 Paris
France
Tel: (+33 1) 47 20 01 05
Tx: 631712

Société Nereides
4 avenue des Indes
ZA de Courtaboeuf
F-91969 Les Ulis Cedex 13
France
Tel: (+33 4) 69 07 20 48
Fax: (+33 4) 69 07 19 14

Thomson-CSF Radars et Contre-Mesures
La Clef de Saint Pierre
1 blvd Jean Moulin
F-78852 Elancourt Cedex
France
Tel: (+33 1) 34 59 60 00
Fax: (+33 1) 34 59 63 42

Thomson-CSF
Aerospace Group
173 boulevard Haussmann
F-75415 Paris Cedex 08
France
Tel: (+33 1) 53 77 80 00
Fax: (+33 1) 53 77 81 18

Thomson-CSF
Division Systèmes Défense et Contrôle
7-9 rue des Mathurins
F-92221 Bagneux Cedex
France
Tel: (+33 1) 40 84 17 62
Fax: (+33 1) 40 84 16 72

Thomson Marconi Sonar SAS
525 route des Dolines
BP 157
F-06903 Sophia Antipolis Cedex
France
Tel: (+33 4) 92 96 30 00
Fax: (+33 4) 92 96 41 24

Underwater Defence Systems (UDS) International
525 route des Dolines
BP 157
F-06903 Sophia Antipolis Cedex
France
Tel: (+33 4) 92 96 34 86
Fax: (+33 4) 92 96 37 69

Germany

Abeking & Rasmussen
PO Box 1160
Lemwerder
D-27805
Germany
Tel: (+49 421) 673 35 32
Fax: (+49 421) 673 31 15

DaimlerChrysler Aerospace AG
Wörthstr 85
D-89077 Ulm
Germany
Tel: (+49 731) 392 53 92
Fax: (+49 731) 392 37 55

Diehl GmbH
Ammunition
Strephanstrasse 49
D-90478 Nuremburg
Germany
Tel: (+49 911) 94 70
Fax: (+49 911) 947 34 29

Faun-Werke
PO Box 8
D-8560 Lauf ad Pegnitz
Germany
Tel: (+49 912) 318 50
Tx: 626093
Fax: (+49 912) 37 53 20

Fr Lürssen Werft GmbH & Co
Friedrich-klippert-Str 1
28759 Bremen
Germany
Tel: (+49 421) 660 40
Fax: (+49 421) 660 44 43

Haux-Life Support GmbH
Descostrasse 19
D-76307 Karlsbad-Iltersbach
Germany
Tel: (+49 72) 489 16 00
Fax: (+49 72) 48 53 57

Howaldtswerke-Deutsche Werft AG
PO Box 63 09
D-24124 Kiel
Germany
Tel: (+49 431) 70 00
Fax: (+49 431) 700 23 12

IBAK Helmut Hunger GmbH
Wehdenweg 1224
PO Box 6260
D-2300 Kiel 1
Germany
Tel: (+49 431) 727 00
Fax: (+49 431) 727 02 70

Institut Dr Förster GmbH
Postfach 1564
D-72705 Reutlingen
Germany
Tel: (+49 712) 114 04 87
Fax: (+49 712) 114 04 88

Kröger Werft GmbH & Co KG
PO Box 460
24 754 Rendsburg
Germany
Tel: (+49 4331) 95 12 07
Fax: (+49 4331) 95 12 05

L-3 Communications ELAC Nautik GmbH
Neufeldstrasse
D-24118 Kiel
Germany
Tel: (+49 431) 88 34 14
Fax: (+49 431) 88 32 24

Rheinmetall W & M
Pempelfurtstrasse 1
PO Box 1663
D-40880 Ratingen
Germany
Tel: (+49 21) 029 00
Fax: (+49 21) 02 47 35 53

Salzgitter Elektronik GmbH
PO Box 160
D-2302 Flintbek
Germany
Tel: (+49) 43 47/90 80
Tx: 292976

STN ATLAS Elektronik GmbH
Sebaldsbrücker Heerstrasse 235
28305 Bremen
Germany
Tel: (+49 421) 45 70
Fax: (+49 421) 457 29 00

Thyssen Werften GmbH
PO Box 2351
D-26703 Emden
Germany

Vallon GmbH
Im Grund 3
D-72800 Eningen
Germany
Tel: (+49 71) 219 85 50
Fax: (+49 71) 218 36 43

Zeiss-Eltro Optronic GmbH
Carl-Zeiss-Strasse 22
D-73442 Oberkochen
Germany
Tel: (+49 736) 42 01
Fax: (+49 736) 420 45 88

India

Bharat Electronics Ltd
Shankaranarayan Bldg, 11 Floor
25 M.G. Road
Bangalore 560 001
India
Tel: (+91 80) 558 38 51
Fax: (+91 80) 558 49 11

Israel

Elbit Systems Ltd
Advanced Technology Center
PO Box 539
IL-31053 Haifa
Israel
Tel: (+972 4) 31 53 15
Fax: (+972 4) 55 00 02

Elisra Electronic Systems Ltd
(a subsidiary of Tadiran Ltd)
48 Mivtza Kadesh Street
IL-51203 Bene Beraq
Israel
Tel: (+972 3) 754 51 11
Tx: 33553

Rafael
Israel Armament Development Authority
PO Box 2082
IL-31021 Haifa
Israel
Tel: (+972 4) 77 69 65
Fax: (+972 4) 79 46 57

Italy

Alenia Difesa Naval System Divsion
Via di S. Alessandro 28/30
I-00131 Rome
Italy
Tel: (+39 6) 41 88 31
Fax: (+39 6) 41 88 32 97

CONSORZIO SMIN
Via Hermada 6/B
I-16154 Genova
Italy
Tel: (+39 10) 654 81
Fax: (+39 10) 654 85 90

COSMOS SpA
Via Della Padula 83
I-57124 Livorno
Italy
Tel: (+39 586) 86 80 11
Fax: (+39 586) 85 93 73

Datamat Ingegneria Dei Sistemi SpA
Via Laurentina 760
I-00143 Rome
Italy
Tel: (+39 6) 502 71
Fax: (+39 6) 50 51 14 07

Elettronica SpA
Via Tiburtina Km 13.7
I-00131 Rome
Italy
Tel: (+39 6) 415 41
Fax: (+39 6) 419 13 20

Elmer SpA
Vialle dell'Industria 4
I-00040 Pomezia
Italy
Tel: (+39 6) 91 29 71
Tx: 610112

Fincantieri
Naval Shipbuilding Division
Via Cipro II
I-16129 Genoa
Italy
Tel: (+39 10) 599 51
Fax: (+39 10) 599 53 79

FIAR
Via Montefeltro 8
I-20156 Milano
Italy
Tel: (+39 2) 35 79 01
Tx: 331140 FIARMO I

GF Galileo SMA Srl
(a Finmeccanica Company)
Via del Ferrone 5
I-50124 Florence
Italy
Tel: (+39 55) 275 01
Fax: (+39 55) 71 49 34

Intermarine
I-19038 Sargana
PO Box 185
La Spezia
Italy
Tel: (+39 187) 61 71 11
Fax: (+39 187) 67 42 49

Riva Calzoni SpA
Via Emilia Ponente 72
I-40133 Bologna
Italy
Tel: (+39 51) 52 75 11
Fax: (+39 51) 43 72 33

Servomeccanismi
Via Mediana Km 29.3
I-00040 Pomezia
Italy

SEI SpA
Via Gavardo, 3
I-25016 Ghedi (Brescia)
Italy
Tel: (+39 30) 904 11
Fax: (+39 30) 903 14 61

SEPA
Societa di Elettronica per L'Automazione SpA
Corso Giulio Cesare 294
I-10154 Turin
Italy
Tel: (+39 11) 205 33 71
Tx: 221527 SEPA 1

Sistemi Subacquei WELSE SpA
Consortile
Via L Manara 2
I-16154 Genova-Sestri
Italy
Tel: (+39 10) 651 13 21
Fax: (+39 10) 651 21 47

USEA SpA
Via delle Pianazze 74
I-19027 Termo Della
La Spezia
Italy
Tel: (+39 187) 98 27 50
Tx: 281216
Fax: (+39 187) 98 07 48

Whitehead Alenia Sistemi Subacquei
Via Hermada 6/B
I-16154 Genova
Italy
Tel: (+39 10) 654 81
Fax: (+39 10) 654 85 90

Japan

NEC Corporation
Legal and Administration Division
7-1 Shiba 5-chome
Minato-ku
Tokyo 108-01
Japan
Tel: (+81 3) 37 98 65 31
Fax: (+81 3) 37 98 65 34

Netherlands

Hollandse Signaalapparaten BV
PO Box 42
NL-7550 GD Hengelo
Netherlands
Tel: (+31 74) 248 81 11
Fax: (+31 74) 242 59 36

Signaal Special Products
PO Box 241
NL-2700 AE Zoetermeer
Netherlands
Tel: (+31 79) 344 59 99
Fax: (+31 79) 344 59 38

Signaal USFA
PO Box 6034
NL-5600 HA Eindhoven
Netherlands
Tel: (+31 40) 250 36 03
Fax: (+31 40) 250 37 77
email: info@usfa.nl

Van der Giessen-de-Noord Marinbouw BV
PO Box 13
NL-2950 AA Alblasserdam
Netherlands

Norway

Aanderaa Instruments
Fanaveien 13 B
N-5050 Nesttun
Norway
Tel: (+47) 55 13 25 00
Fax: (+47) 55 13 79 50

Kongsberg Gruppen AS
Defence Systems
PO Box 1003
N-3601 Kongsberg
Norway
Tel: (+47) 32 73 82 00
Fax: (+47) 32 73 86 20

Kongsberg Simrad AS
Hydrographic Department
PO Box 111
N-3191 Horten
Norway
Tel: (+47) 33 03 40 00
Fax: (+47) 33 04 44 24

Kvaerner Mandal AS
PO Box 283
N-4501 Mandal
Norway
Tel: (+47) 38 27 92 00
Fax: (+47) 38 26 03 88

PAG - Automasjon A/S
PO Box 150
N-8601 Moi Rana
Norway
Tel: (+47) 75 15 88 88
Fax: (+47) 75 15 88 80
email: pag@okstind.monet.no

Russian Federation and Associated States (CIS)

Lomo plc
20 Chugunnaya 194044
St Petersburg
Russia
Tel: (+7 812) 248 52 42
Fax: (+7 812) 542 18 39

South Africa

Southern Oceanics (Pty) Ltd
PO Box 36541
Chempet 7441
Cape Town
South Africa
Tel: (+27 21) 551 22 33
Fax: (+27 21) 551 22 75

Spain

SAES
Carretera de la Algameca, s/n
E-30205 Cartegena
Murcia
Spain
Tel: (+34 68) 50 82 14
Fax: (+34 68) 50 77 13

Sweden

Bofors AB
SE-691 80 Karlskoga
Sweden
Tel: (+46 586) 810 00
Fax: (+46 586) 857 00

Bofors SA Marine AB
PO Box 627
SE-261 26 Landskrona
Sweden
Tel: (+46 418) 516 00
Fax: (+46 418) 229 52

CelsiusTech Systems AB
SE-175 88 Nettovägen
Sweden
Tel: (+46 8) 58 08 53 70
Fax: (+46 8) 59 08 72 16

Karlskronavarvet
SE-37182 Karlskrona
Sweden
Tel: (+46 455) 33 41 00
Fax: (+46 455) 31 79 34

Kockums Submarine Systems
SE-205 55 Malmö
Sweden
Tel: (+46 40) 34 80 00
Fax: (+46 40) 97 32 81

Saab Dynamics
SE-581 88 Linköping
Sweden
Tel: (+46 13) 28 60 00
Tx: 50001

Safe Bridge AB
Vallgatan 3
SE-170 80 Solna
Sweden
Tel: (+46 8) 655 32 50
Fax: (+46 8) 624 00 70

Switzerland

Gayrobot
Via Magazzini Generali 13/A
CH-6828 Balerna
Switzerland
Tel: (+41 91) 43 92 84
Fax: (+49 91) 43 92 85

Taiwan

Keng Chieh Enterprises Co Inc
PO Box 11-006
Pei-Tou
Taipei 112
Taiwan
Tel: (+886 8) 21 68 05
Tx: 17326

United Kingdom

AB Precision (Poole) Ltd
Stanley Green Road
Poole
Dorset BH15 3AL
UK
Tel: (+44 1202) 67 31 85
Tx: 417102

BAeSEMA
Apex Tower
7 High Street
New Malden
Surrey KT3 4LH
UK
Tel: (+44 181) 942 96 61
Fax: (+44 181) 942 97 71

Bradley Electronics Ltd
Neasdon Lane
London NW10 1RR
UK
Tel: (+44 181) 450 78 11
Fax: (+44 181) 450 83 57

Bridport Aviation Products Ltd
The Court
West Street
Bridport
Dorset DT6 3QU
UK
Tel: (+44 1308) 45 66 66
Fax: (+44 1308) 45 66 05

Cable & Wireless (Marine) Ltd
Berth 203 Western Docks
Southampton
Hampshire SO15 0HH
UK
Tel: (+44 1703) 35 86 17
Fax: (+44 1703) 35 87 43

Chelsea Instruments Ltd
55 Central Avenue
West Molesey
Surrey KT8 2QZ
UK
Tel: (+44 181) 941 00 44
Fax: (+44 181) 941 93 19
email: 100731.2345@compuserve.com

Clarke-Chapman Ltd
Saltmeadows Road
PO Box 9
Gateshead
Tyne & Wear
NE8 1SW
UK
Tel: (+44 191) 477 22 71
Tx: 53239

Computing Devices Company Ltd
Castleham Road
St Leonards-on-Sea
E Sussex TN38 9NJ
UK
Tel: (+44 1424) 85 34 81
Fax: (+44 1424) 85 15 20

COGENT Defence Systems
Wednesbury Street
Newport
Gwent NP9 0WS
UK
Tel: (+44 1633) 24 42 44
Tx: 498368

DERA
Reports Centre, Bldg A22
DRA Winfrith
Dorchester
Dorset DT2 8XJ
UK

Digital Systems and Design Ltd
18 Shakespeare Business Centre
Hathaway Close
Eastleigh
Hampshire SO5 4SR
UK
Tel: (+44 1703) 62 04 99
Tx: 477575

Divex
Pressure Products House
Westhill Industrial Estate
Westhill
Aberdeen AB32 6TQ
UK
Tel: (+44 1224) 74 01 45
Fax: (+44 1224) 74 01 72

EDS
1-3 Bartley Way
Bartley Wood
Hook
Hampshire RG27 9XA
UK
Tel: (+44 1256) 74 24 61
Fax: (+44 1256) 74 23 49

Fairey Hydraulics Ltd
Claverham
Bristol BS19 4NF
UK
Tel: (+44 1934) 83 52 24
Fax: (+44 1934) 83 53 37

Fugro - UDI Ltd
Denmore Road
Bridge of Don
Aberdeen AB23 8JW
UK
Tel: (+44 1224) 25 75 00
Fax: (+44 1224) 25 75 01

Gearing & Watson Electronics Ltd
South Road
Hailsham
East Sussex BN27 3JJ
UK
Tel: (+44 1323) 84 64 64
Fax: (+44 1323) 84 75 50
email: steve@gearing-watson.com

Graseby Dynamics Ltd
459 Park Avenue
Bushey
Watford
Hertfordshire WD2 2BW
UK
Tel: (+44 1923) 22 85 66
Fax: (+44 1923) 24 02 85

Hydrovision Ltd
Howe Moss Avenue
Kirkhill Industrial Estate
Dyce
Aberdeen AB21 0GP
UK
Tel: (+44 1224) 77 21 50
Fax: (+44 1224) 77 21 66
email: rovs@hydrovision.co.uk

Hyspec Systems Ltd
Peter Green Way
Furness Business Park
Barrow-in-Furness
Cumbria LA14 2PE
UK
Tel: (+44 1229) 43 11 43
Fax: (+44 1229) 87 16 00

Intech Corporation Ltd
Leandes House
Hurst Road
West Molesey
Surrey KT8 1RF
UK
Tel: (+44 181) 974 53 64
Fax: (+44 181) 979 16 75
email: intechcorp@msn.com

Interspiro Ltd
Halesfield 19
Telford
Shropshire TF7 4QT
UK
Tel: (+44 1952) 58 28 22
Fax: (+44 1952) 58 18 89

John Crane Signature Management
Alton Road
Ross-on-Wye
Herefordshire HR9 5NF
UK
Tel: (+44 1989) 76 55 55
Fax: (+44 1989) 76 35 02

Kelvin Hughes Ltd
New North Road
Hainault
Ilford
Essex IG6 2UR
UK
Tel: (+44 181) 500 10 20
Fax: (+44 181) 500 08 37

Logica Defence & Civil Government Ltd
Cobham Park
Downside Road
Cobham
Surrey KT11 3LG
UK
Tel: (+44 171) 637 91 11
Tx: 27200

Marconi Avionics Ltd
Mission Avionics Division
1 South Gyle Crescent
Edinburgh EH12 9HQ
UK
Tel: (+44 131) 332 24 11
Fax: (+44 131) 314 82 37

Marconi Avionics Ltd
Airport Works
Rochester
Kent ME1 2XX
UK
Tel: (+44 1634) 84 44 00
Fax: (+44 1634) 82 73 32

Marconi Communications
Military Communications Division
Marconi House
New Street
Chelmsford
Essex CM1 1PL
UK
Tel: (+44 1245) 27 56 22
Fax: (+44 1245) 27 59 20

Marconi Electro-Optics Ltd
Sensors Division
Christopher Martin Rd
Basildon
Essex SS14 3EL
UK
Tel: (+44 1268) 52 28 22
Fax: (+44 1268) 88 31 40

Marconi Radar and Defence Systems Ltd
New Parks
Leicester LE3 1UF
UK
Tel: (+44 116) 256 15 37
Fax: (+44 116) 256 15 10

Marconi Radar and Defence Systems
Command and Information Systems Division
PO Box 133
Chobham Road
Frimley
Camberley
Surrey GU16 5PE
UK
Tel: (+44 1276) 633 11
Fax: (+44 1276) 69 54 98

Marconi Electronic Systems Limited
Underwater Weapons Division
Elettra Avenue
Waterlooville
Hampshire PO7 7XS
UK
Tel: (+44 1705) 26 44 66
Fax: (+44 1705) 26 02 46

MSI - Defence Systems Ltd
Salhouse Road
Norwich
Norfolk NR7 9AY
UK
Tel: (+44 1603) 48 43 65
Fax: (+44 1603) 41 56 49

Nautronix Ltd
Nautronix House
Howe Moss Ave
Kirkhill
Dyce
Aberdeen AB21 0GF
UK
Tel: (+44 1224) 77 57 00
Fax: (+44 1224) 77 58 00

Patrick Engineering Ltd
Golden Hill Park
Freshwater
Isle of Wight PO40 9UJ
Tel: (+44 1933) 75 43 11
Fax: (+44 1933) 75 41 86

Pilkington Optronics
1 Linthouse Road
Glasgow G51 4BZ
UK
Tel: (+44 141) 440 40 00
Fax: (+44 141) 440 43 01

PMES Ltd
Armitage Road
Rugeley
Staffordshire WS15 1DR
UK
Tel: (+44 1889) 58 51 51
Fax: (+44 1889) 57 82 09

Polaris International Ltd
North Star Business Park
Woodrow Way
Gloucester GL2 6DX
UK
Tel: (+44 1452) 38 00 00
Fax: (+44 1452) 38 44 38

Racal Group Services Ltd
Burlington House
118 Burlington Road
New Malden
Surrey KT3 4NR
UK
Tel: (+44 181) 942 24 64
Fax: (+44 181) 942 08 35

Racal Radar Defence Systems Ltd
Manor Royal
Crawley
W Sussex RH10 2PZ
UK
Tel: (+44 1293) 52 87 87
Fax: (+44 1293) 54 28 18

Redifon MEL Limited
Newton Road
Crawley
West Sussex RH10 2TU
UK
Tel: (+44 1293) 51 88 55
Fax: (+44 1293) 54 00 45

RESON Offshore Ltd
Howeness Crescent
Kirkhill Industrial Estate
Dyce
Aberdeen AB21 0GN
UK
Tel: (+44 1224) 72 74 27
Fax: (+44 1224) 72 74 28

Siemens Plessey Systems
Grange Road
Christchurch
Dorset BH23 4JE
UK
Tel: (+44 1202) 48 63 44
Fax: (+44 1202) 40 42 21

Simrad Ltd
Campus 1
Aberdeen Science and Technology Park
Balgownie Road
Bridge of Don
Aberdeen AB22 8GT
UK
Tel: (+44 1224) 22 65 00
Fax: (+44 1224) 22 65 01

Slingsby Engineering Ltd
Kirkbymoorside
York YO6 6EZ
UK
Tel: (+44 1751) 43 17 57
Fax: (+44 1751) 43 13 88

Sonar Research & Development Ltd
Unit 1B
Grovehill Industrial Estate
Beverley
North Humberside HU17 0JW
UK
Tel: (+44 1482) 86 95 59
Fax: (+44 1482) 87 21 84

Sonardyne International Ltd
Ocean House
Blackbushe Business Park
Yately
Hampshire GU46 6GD
UK
Tel: (+44 1252) 87 22 88
Fax: (+44 1252) 87 61 00

Strachan & Henshaw
Ashton House
Ashton Vale Rd
PO Box 103
Bristol BS99 7TJ
UK
Tel: (+44 117) 966 46 77
Fax: (+44 117) 963 42 59

Systems Engineering (now owned by Kongsberg Simrad)
4 Waterview
White Cross
Lancaster LA1 4XQ
UK
Tel: (+44 1524) 84 48 64
Fax: (+44 1524) 84 92 73
email: sales@ se-ltd.demon.co.uk

The Gates Rubber Company Ltd
Edinburgh Road
Dumfries DG1 1QA
UK
Tel: (+44 1387) 26 95 91
Fax: (+44 1387) 25 09 95

Thomson Marconi Sonar Ltd
Bird Hall Lane
Cheadle Heath
Stockport SK3 0XQ
UK
Tel: (+44 161) 491 40 01
Fax: (+44 161) 491 17 96

Thomson Marconi Sonar Ltd
Wilkinthroop House
Templecombe
Somerset BA8 0DH
UK
Tel: (+44 1963) 37 05 51
Fax: (+44 1935) 44 22 00

Thomson Training & Simulation
Gatwick Road
Crawley
W Sussex RH10 2RL
UK
Tel: (+44 1293) 56 28 22
Fax: (+44 1293) 56 33 66

Tilt Measurement Ltd
Horizon House
London Road
Baldock
Hertfordshire SG7 6NG
UK
Tel: (+44 1462) 89 45 66
Fax: (+44 1462) 89 59 90

Typhoon International Ltd
Mortlake Court
28 Sheen Lane
East Sheen
London SW14 8LW
UK
Tel: (+44 181) 876 98 18
Fax: (+44 181) 876 92 23

Ultra Electronics
Command and Control Systems
Knaves Beech Centre
Loudwater
High Wycombe
Buckinghamshire HP10 9UT
UK
Tel: (+44 1628) 53 00 00
Fax: (+44 1628) 52 45 57

Ultra Electronics
Ocean Systems
Waverley House
Hampshire Road
Granby Estate
Weymouth
Dorset DT4 9XD
UK
Tel: (+44 1305) 78 47 38
Fax: (+44 1305) 77 79 04

Ultra Electronics
Sonar and Communications Systems
419 Bridport Road
Greenford Industrial Estate
Greenford
Middlesex UB6 8UA
UK
Tel: (+44 181) 813 45 67
Fax: (+44 181) 813 45 68

Vosper Thornycroft (UK) Ltd
Victoria Road
Woolston
Southampton
Hampshire SO19 9RR
UK
Tel: (+44 1703) 44 51 44
Fax: (+44 1703) 42 15 39

United States of America

Allen Osborne Associates
756 Lakefield Road, Bldg J
Westlake Village
California 91361-2624
USA
Tel: (+1 805) 495 84 20
TWX: 910 494 1710
Fax: (+1 805) 373 60 67

AlliedSignal Inc
Ocean Systems
15825 Roxford Street
Sylmar
California 91342-3597
USA
Tel: (+1 818) 367 01 11
Fax: (+1 818) 367 04 03

ARGOSystems Inc
PO Box 3452
Sunnyvale
California 94088-3452
USA
Tel: (+1 408) 737 20 00
Tx: 6711100
Fax: (+1 415) 737 92 36

BBN Systems and Technology Corporation
10 Moulton Street
Cambridge
Massachusetts 02238
USA
Tel: (+1 617) 873 30 00
Tx: 921470
Fax: (+1 617) 873 37 76

Benthos Inc
49 Edgerton Drive
North Falmouth
Massachusetts 02556-2826
USA
Tel: (+1 508) 563 10 00
Fax: (+1 508) 563 64 44

Boeing Aerospace
PO Box 3999, M/S 85-19
Seattle
Washington 98124
USA
Tel: (+1 206) 773 28 16

Concurrent Computer Corporation
One Technology Way
Westford
Massachusetts 01886
USA
Tel: (+1 508) 692 62 00

Deep Ocean Engineering Co
1431 Doolittle Drive
San Leandro
California 94577
USA
Tel: (+1 510) 562 93 00
Fax: (+1 510) 430 82 49

Deep Sea Power & Light
4819 Ronson Ct
San Diego
California 92111-1803
USA
Tel: (+1 619) 576 12 61
Fax: (+1 619) 576 02 19

Dowty Avionics
300E Live Oak Avenue
Arcadia
California 91006-5617
USA
Tel: (+1 818) 445 59 55
Fax: (+1 818) 447 08 80

DRS Inc
5 Sylvan Way
Parsippany
New Jersey 07054
USA
Tel: (+1 201) 898 15 00
Fax: (+1 201) 898 47 30

DRS Inc
Military Systems
138 Bauer Drive
Oakland
New Jersey 07436
USA
Tel: (+1 201) 337 38 00
Fax: (+1 201) 337 47 75

Dukane Corporation
Seacom Division
2900 Dukane Drive
St Charles
Illinois 60174
Tel: (+1 630) 584 23 00
Fax: (+1 630) 584 57 54

EdgeTech
455 Fortune Boulevard
Milford
Massachusetts 01757
USA
Tel: (+1 508) 478 95 00
Fax: (+1 508) 478 14 56

EDO Acoustic Products
2645 South 300 West
Salt Lake City
Utah 84115
USA
Tel: (+1 801) 461 94 35
Fax: (+1 801) 484 33 01

EDO Corporation
Combat Systems
1801-E Sara Drive
Chesapeake
Virginia 23320
USA
Tel: (+1 757) 424 10 04
Fax: (+1 747) 424 16 02

EDO Corporation
Marine and Aircraft Systems
14-04 111th Street
College Point
New York 11356-1434
USA
Tel: (+1 718) 321 40 00
Fax: (+1 718) 939 01 19

EM Systems Inc
Textron
45757 W Northport Loop
Fremont
California 94538
USA
Tel: (+1 415) 657 99 60
Tx: 4970260

Flightline Electronics Inc
7500 Main St
PO Box 750
Fishers
New York 14453-0750
USA
Tel: (+1 716) 924 40 00
Fax: (+1 716) 924 57 32

General Dynamics Defense Systems
100 Plastics Avenue
Pittsfield
Massachusetts 01201-3698
USA
Tel: (+1 413) 494 11 10
Fax: (+1 413) 494 64 13

General Dynamics
Electric Boat Division
Groton
Connecticut
USA

GTE Electronic Defense Systems Division
100 Ferguson Drive
PO Box 7188
Mountain View
California 94039
USA
Tel: (+1 415) 966 34 61
Fax: (+1 415) 966 23 03

GTE Government Systems Corp
77 'A' Street
Needham Heights
Massachusetts 02194-2892
USA
Tel: (+1 617) 449 20 00
Tx: 922497

GEC Marconi - Hazeltine Corporation
50 Pulaski Road
Greenlawn
New York 11740-1606
USA
Tel: (+1 516) 351 40 00
Tx: 510221 2113

Hughes Aircraft Co
Ground Systems Group
PO Box 3310
Fullerton
California 92634
USA
Tel: (+1 714) 871 32 32
Tx: 685504

Hughes Aircraft
Naval and Maritime Systems
6500 Harbour Heights Parkway
Mukilteo
Washington 98275-4862
USA
Tel: (+1 206) 356 30 00
Fax: (+1 206) 356 31 85

Hughes Defense Communications
1313 Production Road
Fort Wayne
Indiana 46808
USA
Tel: (+1 219) 429 60 00
Fax: (+1 219) 429 52 45

Hughes Defense Communications
2829 Maricopa Street
Torrance
California 90503
Tel: (+1 213) 618 12 00
Tx: 674373; 696101
TWX: 910349 6657
Fax: (+1 213) 618 70 01

Hughes Training Inc
PO Box 6171
Arlington
Texas 76005-6171
USA
Tel: (+1 817) 619 22 00
Fax: (+1 817) 619 35 55

International Transducer Corp
869 Ward Drive
Santa Barbara
California 93111
USA
Tel: (+1 805) 683 25 75
Fax: (+1 805) 967 81 99

Kaman Corporation
1332 Blue Hills Avenue
Bloomfield
Connecticut 06002
USA
Tel: (+1 860) 243 71 00

Klein Associates Inc
11 Klein Drive
Salem
New Hampshire 03079-1249
USA
Tel: (+1 603) 893 61 31
Fax: (+1 603) 893 88 07

Kollmorgen Corp
Electro-Optical Division
347 King Street
Northampton
Massachusetts 01060
USA
Tel: (+1 413) 586 23 30
Fax: (+1 413) 586 13 24

Kongsberg Simrad Inc
19210 33rd Avenue West
Lynnwood
Washington 98036-4707
USA
Tel: (+1 206) 778 88 21
Fax: (+1 206) 771 72 11

Kongsberg Simrad Inc
PO Box 1749
1225 Stone Drive
San Marcos
California 92069
USA
Tel: (+1 619) 471 22 23
Tx: 695409
Fax: (+1 619) 471 11 21

KVH Industries Inc
110 Enterprise Center
Middletown
Rhode Island 02842
USA
Tel: (+1 401) 847 33 27
Fax: (+1 401) 849 00 45
email: ejoiner@kvh.com

Litton Systems Inc
Amecon Division
5115 Calvert Road
College Park
Maryland 20740
USA
Tel: (+1 301) 864 56 00
Tx: 710826 9650

Lockheed Martin Feceral Systems Inc
Building 400/Mail Drop 043
9500 Godwin Drive
Manassas
Virginia 22110-4157
USA
Tel: (+1 703) 367 23 23
Fax: (+1 703) 367 32 36

Lockheed Martin Librascope
833 Sonora Avenue
Glendale
California 91201-2433
Tel: (+1 818) 244 65 41
Tx: 715620
TWX: 910497 2266

Lockheed Martin SAR Imaging Systems
Defense Systems Division
PO Box 85
Litchfield Park
Arizona 85340-0085
USA
Tel: (+1 602) 925 32 32

Lockheed Martin
6801 Rockledge Drive
Bethesda
Maryland 20817
USA
Tel: (+1 301) 897 60 00
Tx: 898437; 248934
TWX: 7108249 050
Fax: (+1 301) 897 62 52

Lockheed Martin
Aero and Naval Systems
103 Chesapeake Plaza
Baltimore
Maryland 21220
USA
Tel: (+1 301) 338 50 00
TWX: 710239 9049

Lockheed Martin
Ocean Radar and Sensor Systems
6711 Baymeadow Drive
Glen Burnie
Maryland 21060-6493
USA
Tel: (+1 410) 760 31 00
Tx: 908310

Lockheed Martin
Ocean Radar & Sensor Systems
PO Box 4840
EP7-MD31
Syracuse
New York 13221-4840
USA
Tel: (+1 315) 456 74 72
Fax: (+1 315) 456 35 15

Lockheed Martin
Perry Technologies
100 East 17th Street
Riviera Beach
Florida 33404
USA
Tel: (+1 561) 840 97 98
Fax: (+1 561) 842 53 03

Lockheed Martin
Tactical Defense Systems
1210 Massillon Road
Akron
Ohio 44315-0001
USA
Tel: (+1 330) 796 28 00

Lockheed Martin
Tactical Defense Systems
365 Lakeville Road
Great Neck
New York 11020-1696
USA
Tel: (+1 516) 574 29 75
Fax: (+1 516) 574 19 67

Lockheed Martin
Tactical Defense Systems
Archbald
Pennsylvania 18403-1598
USA
Tel: (+1 717) 876 15 00
Fax: (+1 717) 876 57 42

Lucent Technologies
Advanced Technology Systems
1919 S Eads Street
Suite 300
Arlington
Virginia 22202-3028
USA
Tel: (+1 800) 767 72 30

Metocean Data Systems Inc
Building 1103
Suite 149
Stennis Space Center
Mississippi 39529
USA
Tel: (+1 601) 688 18 19
Fax: (+1 601) 688 28 39

Morton Thiokol Inc
PO Box 241
Elkton
Maryland 21921
USA
Tel: (+1 301) 398 30 00

Navtek Inc
PO Box 10998
West Palm Beach
Florida 33419
USA
Tel: (+1 407) 881 96 02
Fax: (+1 407) 881 91 93

Newport News Shipbuilding
4101 Washington Avenue
Newport News
Virginia 23607
USA
Tel: (+1 757) 380 20 00

Northrop Grumman Corporation
Electronic Sensors and Systems Division
Oceanic Systems
PO Box 1488
Oceanic Drive
Annapolis
Maryland 21404-1488
USA
Tel: (+1 410) 260 56 56
Fax: (+1 410) 260 53 93

Northrop Grumman Corporation
Electronic Sensors & Systems Division
PO Box 17319-MSA255
Baltimore
Maryland 21203-7319
USA
Tel: (+1 410) 765 44 41
Fax: (+1 410) 993 87 71

Ocean Data Equipment Corporation
141 Washington Street
E Walpole
Massachusetts 02032-1155
USA
Tel: (+1 508) 660 60 10
Fax: (+1 508) 660 60 61

Perry Tritech Inc
821 Jupiter Park Drive
Jupiter
Florida 33458-8946
USA
Tel: (+1 407) 743 70 00
Fax: (+1 407) 743 13 13

Raytheon Electronic Systems
Portsmouth Facility
1847 West Main Road
PO Box 360
Portsmouth
Rhode Island 02871-1087
USA
Tel: (+1 401) 842 80 00
Fax: (+1 401) 842 52 00

Raytheon TI Systems
2501 South Highway 121
PO Box 405
MS 3500
Lewisville
Texas 75067-8122
USA
Tel: (+1 972) 462 65 01
Fax: (+1 972) 462 65 08

RESON Inc
300 Lopez Road
Goleta
California 93117
USA
Tel: (+1 805) 964 62 60
Fax: (+1 805) 964 75 37
email: reson@reson.com

RJE International
2192 Dupont Drive
Suite 110
Irvine
California 92612
USA
Tel: (+1 714) 833 84 23
Fax: (+1 714) 833 85 57
email: sales@rjeint.com

Rockwell Corporation
Autonetics Maritime Systems Division
PO Box 4921
3370 Miraloma Avenue
Anaheim
California
USA
Tel: (+1 714) 762 33 27
Fax: (+1 714) 762 01 46

Sanders
a Lockheed Martin Company
65 Spit Brook Road
PO Box 868
Nashua
New Hampshire 03061-0868
USA
Tel: (+1 603) 885 43 21

Science Applications International Corporation
3990 Old Town Avenue
Suite 303C
San Diego
California 92110
USA
Tel: (+1 619) 686 56 35
Fax: (+1 619) 296 57 44

Sea Hydro Products Inc
11803 Sorrento Valley Road
San Diego
California 92121-1006
USA
Tel: (+1 619) 453 23 45
Fax: (+1 619) 793 26 35

SeaBeam Instruments Inc
141 Washington Street
E Walpole
Massachusetts 02032-1155
USA
Tel: (+1 508) 660 60 00
Fax: (+1 508) 660 60 61

Ship Analytics
183 Providence-New London Turnpike
North Stonington
Connecticut 06359-0250
USA
Tel: (+1 203) 535 30 92
Fax: (+1 203) 535 05 60

Sippican Inc
Seven Barnabas Road
Marion
Massachusetts 02738-1499
USA
Tel: (+1 508) 748 11 60
Fax: (+1 508) 748 36 26

Sparton Corporation
Electronics Division
5612 Johnson Lake Road
PO Box 788
DeLeon Springs
Florida 32130
USA
Tel: (+1 904) 985 46 31
Fax: (+1 904) 985 50 36

Spears Communications Group
Sippican Inc
Seven Barnabas Road
Marion
Massachusetts 02738-1499
USA
Tel: (+1 508) 748 11 60
Fax: (+1 508) 748 95 69

Sperry Marine
1070 Seminole Trail
Charlottesville
Virginia 22901-2891
USA
Tel: (+1 804) 974 20 00
Fax: (+1 804) 974 22 59

Texas Instruments
Defense Systems & Electronics Group
8505 Forest Lane
PO Box 660246
Mail Station 3134
Dallas
Texas 75266
USA
Tel: (+1 214) 480 68 67
Fax: (+1 214) 480 32 81

The Boeing Company
PO Box 516
St Louis
Missouri 63166-0516
USA
Tel: (+1 314) 232 02 32
Fax: (+1 314) 233 54 45

Triton Elics Inc
125 Westridge Drive
Watsonville
California 95076
USA
Tel: (+1 408) 722 73 73
Fax: (+1 408) 722 14 05
email: triton@tritontech.com

United Technologies Corporation
Norden Systems
Norden Place
Box 5300
Norwalk
Connecticut 06856
USA
Tel: (+1 203) 852 50 00
Tx: 240005

Western Electric
Guildford Centre
PO Box 20046
Greenboro
North Carolina 27420
USA
Tel: (+1 91) 96 97 67 70

Yugoslavia, Federal Republic

Yugoimport SDPR
Bulevar umetnosti 2
11070 Novi Beograd
Yugoslavia
Tel: (+381 11) 311 27 43
Fax: (+381 11) 324 87 38
email: jugoimport@sezampro.yu

Manufacturers Index

Alphabetical Index

Alphabetical list of advertisers